Student's Solution

to accompany

Precalculus:
Functions and Graphs

Seventh Edition

Raymond A. Barnett
Merritt College

Michael R. Ziegler
Marquette University

Karl E. Byleen
Marquette University

Fred Safier
City College of San Francisco

Special thanks to Jeanne Wallace for the preparation of this manuscript.

Boston   Burr Ridge, IL   Dubuque, IA   Madison, WI   New York   San Francisco   St. Louis
Bangkok   Bogotá   Caracas   Lisbon   London   Madrid
Mexico City   Milan   New Delhi   Seoul   Singapore   Sydney   Taipei   Toronto

McGraw-Hill Higher Education

*A Division of The **McGraw-Hill** Companies*

Student's Solutions Manual to accompany
PRECALCULUS: FUNCTIONS AND GRAPHS, FIFTH EDITION
RAYMOND A. BARNETT/MICHAEL R. ZIEGLER/KARL E. BYLEEN

Published by McGraw-Hill Higher Education, an imprint of The McGraw-Hill Companies, Inc.,
1221 Avenue of the Americas, New York, NY 10020. Copyright © The McGraw-Hill Companies,
Inc., 2001, 1999, 1993, 1989, 1984, 1979, 1974. All rights reserved.

This book is printed on acid-free paper.

3 4 5 6 7 8 9 0 BKM BKM 0 3 2 1 0

ISBN 0-07-242737-X

www.mhhe.com

Suggestions for Success in a Precalculus Course

1. Go to class, participate in the class discussions, and don't hesitate to ask questions.

2. Do all assigned homework, as soon as possible after class. Try each assigned problem on your own. If you need help, first check the text for similar examples. Once you have found an answer, check it with the answer in the back of the text. If you have difficulty, study the solution given here, then try to work the problem again without looking at the solution.

3. Use the answers in the texts and the solutions in this manual as guides. However, do not always try to work toward the answer in the back of the text. You should be able to do problems without looking at the answers. On a test (and in real life) you are seldom provided with the answers, so you cannot work toward them.

4. At the end of each chapter, work the Chapter Review exercises. These exercises provide an excellent review for a test. In areas of weakness, return to an appropriate section in the chapter for a refresher.

5. If you are still having difficulty, be sure to talk to your instructor and to use any tutorial help that is available to you. However, do not let a tutor do the work for you.

6. Take all tests and quizzes that the instructor gives.

This manual contains solutions for all odd-numbered problems in the text exercises plus all problems in the Chapter Review exercises. The Chapter Review exercises are keyed to corresponding text sections.

TABLE OF CONTENTS

CHAPTER 1

Exercise 1-1

Key Ideas and Formulas

The **solution set** for an equation is defined to be the set of elements in the domain of the variable that makes the equation true. Each element of the solution set is called a **solution**, or **root**, of the equation. To **solve an equation** is to find the solution set for the equation.

An equation is called an **identity** if the equation is true for all elements from the domain of the variable. An equation is called a **conditional equation** if it is true for certain domain values and false for others.

Any equation that can be written in the form $ax + b = 0$ $a \neq 0$ is called a linear, or first degree, equation. To solve a linear equation, use properties of equality:

If $a = b$, then $a + c = b + c$

If $a = b$, then $a - c = b - c$

If $a = b$, then $ca = cb$ $c \neq 0$

If $a = b$, then $\dfrac{a}{c} = \dfrac{b}{c}$ $c \neq 0$

If $a = b$, then $b = a$

If $a = b$, then either may replace the other in any statement without changing the truth or falsity of the statement.

To solve equations involving variables in the denominator, we must exclude any value of the variable that will make a denominator 0. With these values excluded, we may multiply through by the LCD.

---

1. $4(x + 5) = 6(x - 2)$
$4x + 20 = 6x - 12$
$-2x = -32$
$x = 16$
Solution: 16

3. $5 + 4(w - 1) = 2w + 2(w + 4)$
$5 + 4w - 4 = 2w + 2w + 4$
$4w + 1 = 4w + 4$
$1 = 4$
No solution

5.
$$5 - \frac{3a - 4}{5} = \frac{7 - 2a}{2} \qquad \text{LCD} = 10$$
$$10 \cdot 5 - 10 \cdot \frac{(3a - 4)}{5} = 10 \cdot \frac{(7 - 2a)}{2}$$
$$50 - 2(3a - 4) = 5(7 - 2a)$$
$$50 - 6a + 8 = 35 - 10a$$
$$-6a + 58 = 35 - 10a$$
$$4a = -23$$
$$a = -\frac{23}{4} \text{ or } -5.75$$

> **Common Error:**
> After line 2, students often write
> $50 - \overset{2}{\cancel{10}}\dfrac{3a - 4}{\cancel{5}} = \ldots$
> $50 - 6a - 4 = \ldots$
> forgetting to distribute the -2. Put compound numerators in parentheses to avoid this.

7.
$$\frac{x}{2} + \frac{2x - 1}{3} = \frac{3x + 4}{4} \qquad \text{LCD} = 12$$
$$12 \cdot \frac{x}{2} + 12 \cdot \frac{(2x - 1)}{3} = 12 \cdot \frac{(3x + 4)}{4}$$
$$6x + 4(2x - 1) = 3(3x + 4)$$
$$6x + 8x - 4 = 9x + 12$$
$$14x - 4 = 9x + 12$$
$$5x = 16$$
$$x = \frac{16}{5} \text{ or } 3.2$$

9. $0.1(t + 0.5) + 0.2t = 0.3(t - 0.4)$
$0.1t + 0.05 + 0.2t = 0.3t - 0.12$
$0.3t + 0.05 = 0.3t - 0.12$
$0.05 = -0.12$
No solution

11. $0.35(s + 0.34) + 0.15s = 0.2s - 1.66$
$0.35s + 0.119 + 0.15s = 0.2s - 1.66$
$0.5s + 0.119 = 0.2s - 1.66$
$0.3s = -1.779$
$s = -5.93$

Solution: -5.93

13. $\dfrac{2}{y} + \dfrac{5}{2} = 4 - \dfrac{2}{3y}$

Excluded value: $y \neq 0$ LCD $= 6y$

$6y \cdot \dfrac{2}{y} + 6y \cdot \dfrac{5}{2} = 6y(4) - 6y \cdot \dfrac{2}{3y}$

$12 + 15y = 24y - 4$
$-9y = -16$
$y = \dfrac{16}{9}$ or $1.\overline{7}$

15. $\dfrac{z}{z - 1} = \dfrac{1}{z - 1} + 2$

Excluded value: $z \neq 1$ LCD $= z - 1$

$(z - 1)\dfrac{z}{z - 1} = (z - 1)\dfrac{1}{z - 1} + 2(z - 1)$

$z = 1 + 2z - 2$
$-z = -1$
$z = 1$

No solution: 1 is excluded

17. $\dfrac{2m}{5} + \dfrac{m - 4}{6} = \dfrac{4m + 1}{4} - 2$ LCD $= 60$

$60 \cdot \dfrac{2m}{5} + 60 \cdot \dfrac{(m - 4)}{6} = 60 \cdot \dfrac{(4m + 1)}{4} - 60(2)$

$24m + 10(m - 4) = 15(4m + 1) - 120$
$24m + 10m - 40 = 60m + 15 - 120$
$34m - 40 = 60m - 105$
$-26m = -65$
$m = \dfrac{65}{26}$
$m = \dfrac{5}{2}$ or 2.5

19. $1 - \dfrac{x - 3}{x - 2} = \dfrac{2x - 3}{x - 2}$

Excluded value: $x \neq 2$ LCD $= x - 2$

$(x - 2)1 - (x - 2)\dfrac{(x - 3)}{x - 2} = (x - 2)\dfrac{(2x - 3)}{x - 2}$

$x - 2 - (x - 3) = 2x - 3$
$x - 2 - x + 3 = 2x - 3$
$1 = 2x - 3$
$4 = 2x$
$2 = x$

No solution: 2 is excluded

21. $\dfrac{6}{y + 4} + 1 = \dfrac{5}{2y + 8}$

Excluded value: $y \neq -4$ LCD $2(y + 4)$

$2(y + 4)\dfrac{6}{y + 4} + 2(y + 4)1 = 2(y + 4)\dfrac{5}{2(y + 4)}$

$12 + 2y + 8 = 5$
$2y + 20 = 5$
$2y = -15$
$y = -\dfrac{15}{2}$ or -7.5

23. $\dfrac{3a - 1}{a^2 + 4a + 4} - \dfrac{3}{a^2 + 2a} = \dfrac{3}{a}$

$\dfrac{3a - 1}{(a + 2)^2} - \dfrac{3}{a(a + 2)} = \dfrac{3}{a}$ Excluded values: $a \neq 0, -2$

LCD $= a(a + 2)^2$

$a(a + 2)^2 \dfrac{(3a - 1)}{(a + 2)^2} - a(a + 2)^2 \dfrac{3}{a(a + 2)} = a(a + 2)^2 \dfrac{3}{a}$

$a(3a - 1) - (a + 2)3 = (a + 2)^2 3$

$3a^2 - a - 3a - 6 = 3(a^2 + 4a + 4)$

$3a^2 - 4a - 6 = 3a^2 + 12a + 12$

$-16a = 18$

$a = -\dfrac{9}{8}$ or -1.125

25.

$3.142x - 0.4835(x - 4) = 6.795$

$3.142x - 0.4835x + 1.934 = 6.795$

$2.6585x + 1.934 = 6.795$

$2.6585x = 4.861$

$x = \dfrac{4.861}{2.6585}$

$x = 1.83$ to 3 significant digits

Solution: 1.83

27.

$\dfrac{2.32x}{x - 2} - \dfrac{3.76}{x} = 2.32$

Excluded values : $x \neq 0, 2$ LCD $= x(x - 2)$

$x(x - 2)\dfrac{2.32x}{x - 2} - x(x - 2)\dfrac{3.76}{x} = 2.32x(x - 2)$

$2.32x^2 - 3.76(x - 2) = 2.32x(x - 2)$

$2.32x^2 - 3.76x + 7.52 = 2.32x^2 - 4.64x$

$-3.76x + 7.52 = -4.64x$

$7.52 = -0.88x$

$x = -8.55$

Solution: -8.55

29.

$a_n = a_1 + (n - 1)d$

$a_1 + (n - 1)d = a_n$

$(n - 1)d = a_n - a_1$

$d = \dfrac{a_n - a_1}{n - 1}$

31.

$\dfrac{1}{f} = \dfrac{1}{d_1} + \dfrac{1}{d_2}$ LCD $= d_1 d_2 f$

$d_1 d_2 f \dfrac{1}{f} = d_1 d_2 f \dfrac{1}{d_1} + d_1 d_2 f \dfrac{1}{d_2}$

$d_1 d_2 = d_2 f + d_1 f$

$d_2 f + d_1 f = d_1 d_2$

$(d_2 + d_1)f = d_1 d_2$

$f = \dfrac{d_1 d_2}{d_2 + d_1}$

33.

$A = 2ab + 2ac + 2bc$

$2ab + 2ac + 2bc = A$

$2ab + 2ac = A - 2bc$

$a(2b + 2c) = A - 2bc$

$a = \dfrac{A - 2bc}{2b + 2c}$

35.

$y = \dfrac{2x - 3}{3x + 5}$

$(3x + 5)y = 2x - 3$

$3xy + 5y = 2x - 3$

$5y + 3 = 2x - 3xy$

$5y + 3 = x(2 - 3y)$

$\dfrac{5y + 3}{2 - 3y} = x$

$x = \dfrac{5y + 3}{2 - 3y}$

37. The "solution" is incorrect. Although 3 is a solution of the two last equations, they are not equivalent to the first equation because both sides have been multiplied by $x - 3$, which is zero when $x = 3$. It is not permitted to multiply both sides of an equation by zero. When $x = 3$, the first equation involves division by zero. Since 3, the only possible solution, is not a solution, the given (first) equation has no solution.

39.
$$\frac{x - \frac{1}{x}}{1 + \frac{1}{x}} = 3 \quad \text{Excl. val.: } x \neq 0$$

$$\frac{x(x - \frac{1}{x})}{x(1 + \frac{1}{x})} = 3$$

$$\frac{x^2 - 1}{x + 1} = 3 \quad \text{Excl. val.: } x \neq -1$$

$$\frac{\overset{1}{(x - 1)}\,\cancel{(x + 1)}}{\underset{1}{\cancel{x + 1}}} = 3$$

$$x - 1 = 3$$
$$x = 4$$

Solution: 4

41.
$$\frac{x + 1 - \frac{2}{x}}{1 - \frac{1}{x}} = x + 2 \quad \text{Excl. val.: } x \neq 0$$

$$\frac{x(x + 1 - \frac{2}{x})}{x(1 - \frac{1}{x})} = x + 2$$

$$\frac{x^2 + x - 2}{x - 1} = x + 2 \quad \text{Excl. val.: } x \neq 1$$

$$\frac{\cancel{(x - 1)}\,(x + 2)}{\underset{1}{\cancel{x - 1}}} = x + 2$$

$$x + 2 = x + 2$$

Solution: All real numbers except the excluded numbers 0 and 1.

43.
$$y = \frac{a}{1 + \frac{b}{x + c}}$$

$$y = \frac{a(x + c)}{(x + c)(1 + \frac{b}{x + c})}$$

$$y = \frac{a(x + c)}{x + c + b}$$

$$y = \frac{ax + ac}{x + c + b}$$

$$y(x + c + b) = ax + ac$$
$$xy + cy + by = ax + ac$$
$$cy + by - ac = ax - xy$$
$$cy + by - ac = x(a - y)$$

$$\frac{cy + by - ac}{a - y} = x$$

$$x = \frac{cy + by - ac}{a - y}$$

45. Let x = the number,
Then 10 less than two thirds the number is one fourth the number

$$\frac{2}{3}x - 10 \qquad = \qquad \frac{1}{4}x$$

$$\frac{2}{3}x - 10 = \frac{1}{4}x$$

$$12\left(\frac{2}{3}x\right) - 12(10) = 12\left(\frac{1}{4}x\right)$$

$$8x - 120 = 3x$$
$$-120 = -5x$$
$$x = 24$$

The number is 24.

47.
Let x = first of the consecutive even numbers
$x + 2$ = second of the numbers
$x + 4$ = third of the numbers
$x + 6$ = fourth of the numbers
first + second + third = 2 more than twice fourth
$$x + x + 2 + x + 4 = 2 + 2(x + 6)$$
$$3x + 6 = 2 + 2x + 12$$
$$3x + 6 = 2x + 14$$
$$x = 8$$
The four consecutive numbers are 8, 10, 12, 14.

49. Let w = width of rectangle
$2w - 3$ = length of rectangle
We use the perimeter formula
$$P = 2a + 2b.$$
$$54 = 2w + 2(2w - 3)$$
$$54 = 2w + 4w - 6 \qquad 10 = w$$
$$54 = 6w - 6 \qquad\quad 17 = 2w - 3$$
$$60 = 6w \qquad\qquad \text{dimensions: 17 meters} \times \text{10 meters}$$

51. Let P = perimeter of triangle
$\quad$ 16 = length of one side
$\quad \dfrac{2}{7}P$ = length of second side
$\quad \dfrac{1}{3}P$ = length of third side

We use the perimeter formula
$$P = a + b + c$$
$$P = 16 + \frac{2}{7}P + \frac{1}{3}P$$
$$21P = 21(16) + 21\left(\frac{2}{7}P\right) + 21\left(\frac{1}{3}P\right)$$
$$21P = 336 + 6P + 7P$$
$$21P = 336 + 13P$$
$$8P = 336$$
$$P = 42 \text{ feet}$$

53.
$$\text{Let } P = \text{price before discount}$$
$$0.20P = 20 \text{ percent discount on } P$$
Then price before discount − discount = price after discount
$$P - 0.20P = 72$$
$$0.8P = 72$$
$$P = \frac{72}{0.8}$$
$$P = \$90$$

55. Let x = sales of employee
Then $x - 7{,}000$ = sales on which 8% commission is paid
$0.08(x - 7{,}000)$ = (rate of commission) × (sales) = (amount of commission)
$2{,}150 + 0.08(x - 7{,}000)$ = (base salary) + (amount of commission) = earnings
$$\text{Earnings} = 3{,}170$$
$$2{,}150 + 0.08(x - 7{,}000) = 3{,}170$$
$$2{,}150 + 0.08x - 560 = 3{,}170$$
$$0.08x + 1{,}590 = 3{,}170$$
$$0.08x = 1{,}580$$
$$x = \frac{1{,}580}{0.08}$$
$$x = \$19{,}750$$

57. (A) We note: The temperature increased 2.5°C for each additional 100 meters of depth. Hence, the temperature increased 25 degrees for each additional kilometer of depth.

$\quad$ Let x = the depth (in kilometers), then $x - 3$ = the depth beyond 3 kilometers.
$\quad 25(x - 3)$ = the temperature increase for $x - 3$ kilometers of depth.
$\quad\quad T$ = temperature at 3 kilometers + temperature increase.
$\quad\quad T = 30 + 25(x - 3)$

(B) We are to find T when $x = 15$. We use the above relationship as a formula.
$$T = 30 + 25(15 - 3)$$
$$= 330°C$$

(C) We are to find x when $T = 280$. We use the above relationship as an equation.
$$280 = 30 + 25(x - 3)$$
$$280 = 30 + 25x - 75$$
$$280 = -45 + 25x$$
$$325 = 25x$$
$$x = 13 \text{ kilometers}$$

59. Let D = distance from earthquake to station

Then $\dfrac{D}{5}$ = time of primary wave.

$\dfrac{D}{3}$ = time of secondary wave.

Time difference = time of *slower* secondary wave - time of *faster* primary wave

$$12 = \dfrac{D}{3} - \dfrac{D}{5}$$

$$15(12) = 15\left(\dfrac{D}{3}\right) - 15\left(\dfrac{D}{5}\right)$$

$$180 = 5D - 3D$$

$$180 = 2D$$

$$D = 90 \text{ miles}$$

> **Common Error:**
> Time difference is *not* time of fast wave (short) - time of slow wave (long)

61. We set up the proportion

$$\dfrac{\text{marked trout in second sample}}{\text{total number in second sample}} = \dfrac{\text{marked trout in first sample}}{\text{total trout population}}$$

Let x = total population

$$\dfrac{8}{200} = \dfrac{200}{x}$$

$$200x\left(\dfrac{8}{200}\right) = 200x\left(\dfrac{200}{x}\right)$$

$$8x = 40,000$$

$$x = 5,000 \text{ trout}$$

63.

Let x = amount of distilled water

50 = amount of 30% solution

Then $50 + x$ = amount of 25% solution

acid in 30% solution + acid in distilled water = acid in 25% solution

$$0.3(50) + 0 = 0.25(50 + x)$$

$$0.3(50) = 0.25(50 + x)$$

$$15 = 12.5 + 0.25x$$

$$2.5 = 0.25x$$

$$x = 10 \text{ gallons}$$

65.

Let x = amount of 50% solution

5 = amount of distilled water

Then $x - 5$ = amount of 90% solution

acid in 90% solution + acid in distilled water = acid in 50% solution

$$0.9(x - 5) + 0 = 0.5x$$

$$0.9x - 4.5 = 0.5x$$

$$-4.5 = -0.4x$$

$$x = 11.25 \text{ liters}$$

67.   Let t = time for both computers to finish the job

Then $t + 1$ = time worked by old computer

t = time worked by new computer

Since the old computer can do 1 job in 5 hours, it works at a rate

(1 job) ÷ (5 hours) = $\dfrac{1}{5}$ job per hour

Similarly the new computer works at a rate of $\dfrac{1}{3}$ job per hour.

Part of job completed by old computer in $t + 1$ hours	+	Part of job completed by new computer in t hours	= 1 whole job.

$$\text{(Rate of old)(time of old)} + \text{(Rate of new)(Time of new)} = 1$$
$$\frac{1}{5}(t + 1) + \frac{1}{3}(t) = 1$$
$$15\left(\frac{1}{5}\right)(t + 1) + 15\left(\frac{1}{3}t\right) = 15$$
$$3(t + 1) + 5t = 15$$
$$3t + 3 + 5t = 15$$
$$8t + 3 = 15$$
$$8t = 12$$
$$t = 1.5 \text{ hours}$$

69. Let d = distance flown north

(A) Using $t = \dfrac{d}{r}$, we note:

$$\text{rate flying north} = 150 - 30 = 120 \text{ miles per hour}$$
$$\text{rate flying south} = 150 + 30 = 180 \text{ miles per hour}$$
$$\text{time flying north} + \text{time flying south} = 3 \text{ hours}$$
$$\frac{d}{120} + \frac{d}{180} = 3$$
$$360\frac{d}{120} + 360\frac{d}{180} = 3(360)$$
$$3d + 2d = 1080$$
$$5d = 1080$$
$$d = 216 \text{ miles}$$

(B) We still use the above ideas, except that rate flying north = rate flying south = 150 miles per hour.

$$\frac{d}{150} + \frac{d}{150} = 3$$
$$\frac{2d}{150} = 3$$
$$\frac{d}{75} = 3$$
$$d = 225 \text{ miles}$$

71. Let x = frequency of second note
y = frequency of third note

$$\frac{264}{4} = \frac{x}{5} = \frac{y}{6}$$
$$\frac{264}{4} = \frac{x}{5} \qquad \frac{264}{4} = \frac{y}{6}$$
$$66 = \frac{x}{5} \qquad 66 = \frac{y}{6}$$
$$x = 330 \text{ hertz} \quad y = 396 \text{ hertz}$$

73. We are to find d when $p = 40$.

$$40 = -\frac{1}{5}d + 70$$
$$5(40) = 5\left(-\frac{1}{5}d\right) + 5(70)$$
$$200 = -d + 350$$
$$-150 = -d$$
$$d = 150 \text{ centimeters}$$

75. total height = height in sand + height in water + height in air
Let h = total height

$$h = \frac{1}{5}h + 20 + \frac{2}{3}h$$
$$15h = 3h + 300 + 10h$$
$$15h = 13h + 300$$
$$2h = 300$$
$$h = 150 \text{ feet}$$

77. The hands will meet when the minute hand has made exactly 1 more revolution than the hour hand.

$$\text{Let } t = \text{time after 12 o'clock in hours}$$
$$\text{rate of hour hand} = \tfrac{1}{12} \text{ revolution per hour}$$
$$\text{rate of minute hand} = 1 \text{ revolution per hour}$$
$$\text{distance of minute hand} = 1 + \text{distance of hour hand}$$
$$1(t) = 1 + \tfrac{1}{12}(t)$$
$$t = 1 + \tfrac{1}{12}t$$
$$12t = 12 + t$$
$$11t = 12$$
$$t = \tfrac{12}{11} \text{ hours}$$

That clock will read $12 + \tfrac{12}{11}$ o'clock, that is $1\tfrac{1}{11}$ hours after 12, or 1 hour $\tfrac{60}{11}$ minutes after 12. This will be $\tfrac{60}{11}$ minutes, or $5\tfrac{5}{11}$ minutes after 1 PM.

Exercise 1-2

Key Ideas and Formulas

A **system** of two linear equations in two variables can be written in the form:

$$ax + by = h$$
$$cx + dy = k$$

where x and y are variables and a, b, c, d, h, and k are real constants. A pair of numbers $x = x_0$ and $y = y_0$ is a **solution** of this system if each equation is satisfied by the pair. The set of all such pairs of numbers is called the **solution set** for the system. To **solve** a system is to find its solution set.

Such systems can be solved by the method of substitution (see text).

---

1. $y = 3x + 5$
$y = 4x + 7$

Substitute y from the first equation into the second equation to eliminate y.

$3x + 5 = 4x + 7$
$-x + 5 = 7$
$\quad -x = 2$
$\quad\; x = -2$

Now replace x with -2 in the first equation to find y.

$\quad y = 3(-2) + 5$
$\quad y = -1$

Solution: $x = -2$, $y = -1$

3. $4x + 5y = -8$
$3x + y = 5$

Solve the second equation for y in terms of x.
$\quad y = 5 - 3x$

Substitute into the first equation to eliminate y.

$4x + 5(5 - 3x) = -8$
$\quad 4x + 25 - 15x = -8$
$\qquad\qquad -11x = -33$
$\qquad\qquad\quad x = 3$

Now replace x with 3 in the second equation to find y.
$\qquad 3(3) + y = 5$
$\qquad\quad 9 + y = 5$
$\qquad\qquad\quad y = -4$
Solution: $x = 3$, $y = -4$

5. $6x + 11y = 16$
$2x - 3y = -8$
Solve the second equation for x in terms of y.
$2x = 3y - 8$
$x = \dfrac{3y - 8}{2}$
Substitute into the first equation to eliminate x.
$6\left(\dfrac{3y - 8}{2}\right) + 11y = 16$
$3(3y - 8) + 11y = 16$
$9y - 24 + 11y = 16$
$20y = 40$
$y = 2$
Now replace y with 2 in the second equation to find x.
$2x - 3(2) = -8$
$2x - 6 = -8$
$2x = -2$
$x = -1$
Solution: $x = -1$, $y = 2$

7. $3s - 5t = -30$
$7s + 11t = 32$
Solve the first equation for s in terms of t.
$3s = 5t - 30$
$s = \dfrac{5t - 30}{3}$
Substitute into the second equation to eliminate s.
$7\left(\dfrac{5t - 30}{3}\right) + 11t = 32$
$\dfrac{35t - 210}{3} + 11t = 32$
$35t - 210 + 33t = 96$
$68t = 306$
$t = \dfrac{9}{2}$ or 4.5
Now replace t with 4.5 in the first equation to find s.
$3s - 5(4.5) = -30$
$3s - 22.5 = -30$
$3s = -7.5$
$s = -2.5$ or $-\dfrac{5}{2}$
Solution: $s = -2.5$ or $-\dfrac{5}{2}$,
$t = 4.5$ or $\dfrac{9}{2}$

9. $13m + 3n = 10$
$7m - 3n = -10$
Solve the first equation for n in terms of m.
$3n = 10 - 13m$
$n = \dfrac{10 - 13m}{3}$
Substitute into the second equation to eliminate n.
$7m - 3\left(\dfrac{10 - 13m}{3}\right) = -10$
$7m - (10 - 13m) = -10$
$7m - 10 + 13m = -10$
$20m = 0$
$m = 0$
Now replace m by 0 in the first equation to find n.
$13(0) + 3n = 10$
$3n = 10$
$n = \dfrac{10}{3}$
Solution: $m = 0$, $n = \dfrac{10}{3}$ or $3.\overline{3}$

11. $y = 5.46x$
$y = 6{,}300 + 4.86x$
Substitute y from the first equation into the second equation to eliminate y.
$5.46x = 6{,}300 + 4.86x$
$0.6x = 6{,}300$
$x = 10{,}500$
Now replace x by 10,500 in the first equation to find y.
$y = 5.46(10{,}500)$
$y = 57{,}330$
Solution: $x = 10{,}500$, $y = 57{,}330$

13. $0.7u - 0.2v = 0.5$
$0.3u - 0.5v = 0.09$

For convenience, eliminate decimals by multiplying both sides of each equation by 100.

$70u - 20v = 50$ | **Common Error**: $3u - 5v \neq 9$ |
$30u - 50v = 9$

Solve the second equation for u in terms of v and substitute into the first equation to eliminate u.

$$30u = 9 + 50v$$
$$u = \frac{9 + 50v}{30}$$

$$70\left(\frac{9 + 50v}{30}\right) - 20v = 50$$

$$7\left(\frac{9 + 50v}{3}\right) - 20v = 50$$

$$\frac{63 + 350v}{3} - 20v = 50$$

$$63 + 350v - 60v = 150$$
$$290v = 87$$
$$v = 0.3$$
$$u = \frac{9 + 50(0.3)}{30}$$
$$u = 0.8$$

Solution: $u = 0.8$, $v = 0.3$

15. $\frac{6}{5}a + \frac{3}{2}b = -4$

$\frac{1}{3}a - \frac{5}{4}b = 1$

Eliminate fractions by multiplying both sides of the first equation by 10 and both sides of the second equation by 12.

$$10\left(\frac{6}{5}a + \frac{3}{2}b\right) = -40$$
$$12a + 15b = -40$$

$$12\left(\frac{1}{3}a - \frac{5}{4}b\right) = 12$$
$$4a - 15b = 12$$

Solve the second equation for a in terms of b and substitute into the first equation to eliminate a.

$$4a = 15b + 12$$
$$a = \frac{15b + 12}{4}$$

$$12\left(\frac{15b + 12}{4}\right) + 15b = -40$$
$$3(15b + 12) + 15b = -40$$
$$45b + 36 + 15b = -40$$
$$60b = -76$$
$$b = -\frac{19}{15} \text{ or } -1.2\overline{6}$$

$$a = \frac{15\left(-\frac{19}{15}\right) + 12}{4}$$
$$a = \frac{-19 + 12}{4}$$
$$a = -\frac{7}{4} \text{ or } -1.75$$

Solution: $a = -\frac{7}{4}$ or -1.75, $b = -\frac{19}{15}$ or $-1.2\overline{6}$

17. If a contradiction such as $0 = 1$ is encountered, this indicates that the system has no solutions. In the example, if we solve the first equation for x in terms of y we obtain $x = 2y - 3$. Substituting this into the second equation yields

$$-2(2y - 3) + 4y = 7$$
$$-4y + 6 + 4y = 7$$
$$6 = 7$$

Thus, this system has no solutions.

19. $x = 2 + p - 2q$
$y = 3 - p + 3q$

Solve the first equation for p in terms of q, x, and y and substitute into the second equation to eliminate p, then solve for q in terms of x and y.

$$p = x - 2 + 2q$$
$$y = 3 - (x - 2 + 2q) + 3q$$
$$y = 3 - x + 2 - 2q + 3q$$
$$y = 5 - x + q$$
$$q = x + y - 5$$

Now substitute this expression for q into $p = x - 2 + 2q$ to find p in terms of x and y.

$$p = x - 2 + 2(x + y - 5)$$
$$p = x - 2 + 2x + 2y - 10$$
$$p = 3x + 2y - 12$$

Solution: $p = 3x + 2y - 12$, $q = x + y - 5$

To check this solution substitute into the original equations to see if true statements result:

$x = 2 + p - 2q$ $\qquad\qquad$ $y = 3 - p + 3q$
$x \stackrel{?}{=} 2 + (3x + 2y - 12) - 2(x + y - 5)$ $\quad$ $y \stackrel{?}{=} 3 - (3x + 2y - 12) + 3(x + y - 5)$
$x \stackrel{?}{=} 2 + 3x + 2y - 12 - 2x - 2y + 10$ $\quad$ $y \stackrel{?}{=} 3 - 3x - 2y + 12 + 3x + 3y - 15$
$x \stackrel{\surd}{=} x$ $\qquad\qquad\qquad\qquad\qquad\qquad$ $y \stackrel{\surd}{=} y$

21. $ax + by = h$
$cx + dy = k$

Solve the first equation for x in terms of y and the constants.

$$ax = h - by$$
$$x = \frac{h - by}{a} \qquad (a \neq 0)$$

Substitute this expression into the second equation to eliminate x.

$$c\left(\frac{h - by}{a}\right) + dy = k$$

$$ac\left(\frac{h - by}{a}\right) + ady = ak$$

$$c(h - by) + ady = ak$$
$$ch - bcy + ady = ak$$
$$(ad - bc)y = ak - ch$$
$$y = \frac{ak - ch}{ad - bc} \qquad ad - bc \neq 0$$

Similarly, solve the first equation for y in terms of x and the constants.

$$by = h - ax$$
$$y = \frac{h - ax}{b} \qquad (b \neq 0)$$

Substitute this expression into the second equation to eliminate y.

$$cx + d\left(\frac{h - ax}{b}\right) = k$$

$$bcx + bd\left(\frac{h - ax}{b}\right) = bk$$

$$bcx + d(h - ax) = bk$$

$$bcx + dh - adx = bk$$

$$(bc - ad)x = bk - dh$$

$$x = \frac{bk - dh}{bc - ad} \qquad bc - ad \neq 0$$

or, for consistency with the expression for y,

$$x = \frac{dh - bk}{ad - bc}$$

Solution: $x = \dfrac{dh - bk}{ad - bc}$, $y = \dfrac{ak - ch}{ad - bc}$ $ad - bc \neq 0$

23. Let x = airspeed of the plane
 y = rate at which wind is blowing
Then
 $x - y$ = ground speed flying from Atlanta to Los Angeles (head wind)
 $x + y$ = ground speed flying from Los Angeles to Atlanta (tail wind)
Then, applying Distance = Rate × Time, we have
 $2{,}100 = 8.75(x - y)$
 $2{,}100 = 5(x + y)$

After simplification, we have
 $x - y = 240$
 $x + y = 420$
Solve the first equation for x in terms of y and substitute into the second equation.

$$x = 240 + y$$
$$240 + y + y = 420$$
$$2y = 180$$
$$y = 90 \text{ mph} = \text{wind rate}$$
$$x = 240 + y$$
$$x = 240 + 90$$
$$x = 330 \text{ mph} = \text{airspeed}$$

25. Let x = time rowed upstream
 y = time rowed downstream
Then $x + y = \dfrac{1}{4}$ (15 min = $\dfrac{1}{4}$ hr.)
Since   rate upstream = $20 - 2 = 18$ mph and
    rate downstream = $20 + 2 = 22$ mph,
applying Distance = Rate × Time to the equal distances upstream and downstream, we have
 $18x = 22y$
Solve the first equation for y in terms of x and substitute into the second equation.

$$y = \frac{1}{4} - x$$

$$18x = 22\left(\frac{1}{4} - x\right)$$

$$18x = 5.5 - 22x$$

$$40x = 5.5$$

$$x = 0.1375 \text{ hr.}$$

Then the distance rowed upstream = $18x = 18(0.1375) = 2.475$ km.

27. Let x = amount of 50% solution
  y = amount of 80% solution
100 milliliters are required, hence
  $x + y = 100$
68% of the 100 milliliters must be acid, hence
  $0.50x + 0.80y = 0.68(100)$

Solve the first equation for y in terms of x and substitute into the second equation.
  $y = 100 - x$
$0.50x + 0.80(100 - x) = 0.68(100)$
$\qquad 0.5x + 80 - 0.8x = 68$
$\qquad\qquad\qquad -0.3x = -12$
$\qquad\qquad\qquad\quad x = 40$ milliliters of 50% solution
$\qquad\qquad\qquad\quad y = 100 - x$
$\qquad\qquad\qquad\quad y = 100 - 40 = 60$ milliliters of 80% solution

29. "Break even" means Cost = Revenue.
  Let y = Cost = Revenue.
  Let x = number of CDs sold
  y = Revenue = number of CDs sold $\times$ price per CD
  $y = x(8.00)$
  y = Cost = Fixed Cost + Variable Cost
  $\qquad$ = 17,680 + number of CDs $\times$ cost per CD
  $\qquad y = 17,680 + x(4.60)$

Substitute y from the first equation into the second equation to eliminate y.
  $8.00x = 17,680 + 4.60x$
  $3.40x = 17,680$
  $\quad x = \dfrac{17,680}{3.40}$
  $\quad x = 5,200$ CDs

31. Let x = amount invested at 10% $\quad 0.1x$ = yield on amount invested at 10%
  y = amount invested at 15% $\quad 0.15y$ = yield on amount invested at 15%
Then
  $\qquad\qquad x + y = 12,000$ (amount invested)
  $0.1x + 0.15y = 0.12(12,000)$ (total yield)

Solve the first equation for y in terms of x and substitute into the second equation.
  $0.1x + 0.15(12,000 - x) \qquad = 0.12(12,000)$
  $0.1x + 0.15(12,000) - 0.15x = 0.12(12,000)$
  $0.1x + 1,800 - 0.15x \qquad = 1,440$
  $\qquad\qquad -0.05x + 1,800 = 1,440$
  $\qquad\qquad -0.05x \qquad\qquad = -360$
  $\qquad\qquad\qquad\qquad\quad x = \$7,200$ invested at 10%
  $\qquad\quad y = 12,000 - x = \$4,800$ invested at 15%

33. Let x = number of hours Mexico plant is operated
  y = number of hours Taiwan plant is operated
Then (Production at Mexico plant) + (Production at Taiwan plant) = (Total Production)
  $\qquad\qquad 40x \qquad\qquad + \qquad\qquad 20y \qquad\qquad = 4000$ (keyboards)
  $\qquad\qquad 32x \qquad\qquad + \qquad\qquad 32y \qquad\qquad = 4000$ (screens)

Solve the first equation for y in terms of x and substitute into the second equation.

$$20y = 4,000 - 40x$$
$$y = 200 - 2x$$
$$32x + 32(200 - 2x) = 4,000$$
$$32x + 6,400 - 64x = 4,000$$
$$-32x = -2,400$$
$$x = 75 \text{ hours Mexico plant}$$
$$y = 200 - 2x = 200 - 2(75)$$
$$= 50 \text{ hours Taiwan plant}$$

35. Let x = number of grams of Mix A
    y = number of grams of Mix B
Then (Nutrition from Mix A) +   (Nutrition for Mix B) = (Total Nutrition)

| $0.10x$ | + | $0.20y$ | = | 20 (Total protein) |
| $0.06x$ | + | $0.02y$ | = | 6 (Total fat) |

For convenience, eliminate decimals by multiplying both sides of the first equation by 10 and the second equation by 100.

$$x + 2y = 200$$
$$6x + 2y = 600$$

Solve the first equation for x in terms of y and substitute into the second equation.

$$x = 200 - 2y$$
$$6(200 - 2y) + 2y = 600$$
$$1200 - 12y + 2y = 600$$
$$-10y = -600$$
$$y = 60 \text{ grams Mix } A$$
$$x = 200 - 2y = 200 - 2(60)$$
$$= 80 \text{ grams Mix } B$$

37. (A) Following the hint, write $p = aq + b$.
    Since $p = 0.60$ corresponds to supply $q = 450$,
        $0.60 = 450a + b$
    Since $p = 0.90$ corresponds to supply $q = 750$,
        $0.90 = 750a + b$

Solve the first equation for b in terms of a and substitute into the second equation.

$$b = 0.60 - 450a$$
$$0.90 = 750a + 0.60 - 450a$$
$$0.30 = 300a$$
$$a = 0.001$$
$$b = 0.60 - 450a = 0.60 - 450(0.001)$$
$$= 0.15$$

Thus, the supply equation is $p = 0.001q + 0.15$.

(B) Use the hint for part (A) to write $p = cq + d$.
    Since $p = 0.60$ corresponds to demand $q = 645$,
        $0.60 = 645c + d$
    Since $p = 0.90$ corresponds to demand $q = 495$,
        $0.90 = 495c + d$
    Solve the first equation for d in terms of c and substitute into the second equation.

$$d = 0.60 - 645c$$
$$0.90 = 495c + 0.60 - 645c$$
$$0.30 = -150c$$
$$c = -0.002$$
$$d = 0.60 - 645c = 0.60 - 645(-0.002)$$
$$= 1.89$$

Thus, the demand equation is $p = -0.002q + 1.89$

(C) Solve the system of equations
$$p = 0.001q + 0.15$$
$$p = -0.002q + 1.89$$
Substitute p from the first equation into the second equation to eliminate p.
$$0.001q + 0.15 = -0.002q + 1.89$$
$$0.003q = 1.74$$
$$q = 580 \text{ bushels = equilibrium quantity}$$
$$p = 0.001q + 0.15 = 0.001(580) + 0.15$$
$$= \$0.73 \text{ equilibrium price}$$

39. $s = a + bt^2$

(A) We are given: When $t = 1$, $s = 180$
When $t = 2$, $s = 132$
Substituting these values in the given equation, we have
$$180 = a + b(1)^2$$
$$132 = a + b(2)^2 \quad \text{or}$$
$$180 = a + b$$
$$132 = a + 4b$$
Solve the first equation for a in terms of b and substitute into the second equation to eliminate a.
$$a = 180 - b$$
$$132 = 180 - b + 4b$$
$$132 = 180 + 3b$$
$$-48 = 3b$$
$$b = -16$$
$$180 = a - 16$$
$$a = 196$$

(B) The height of the building is represented by s, the distance of the object above the ground, when $t = 0$. Since we now know
$$s = 196 - 16t^2$$
from part (A), when $t = 0$, $s = 196$ feet is the height of the building.

(C) The object falls until s, its distance above the ground, is zero. Since
$$s = 196 - 16t^2$$
we substitute $s = 0$ and solve for t.
$$0 = 196 - 16t^2$$
$$16t^2 = 196$$
$$t^2 = \frac{196}{16}$$
$$t = \frac{14}{4} \quad \text{(discarding the negative solution)}$$
$$t = 3.5 \text{ seconds}$$

41. Let p = time of primary wave
s = time for secondary wave
We know
$s - p = 16$ (time difference)
To find a second equation, we have to use Distance = Rate × Time
$5p$ = distance for primary wave
$3s$ = distance for secondary wave
These distances are equal, hence
$5p = 3s$
Solve the first equation for s in terms of p and substitute into the second equation to eliminate s.
$$s = p + 16$$
$$5p = 3(p + 16)$$
$$5p = 3p + 48$$
$$2p = 48$$
$$p = 24 \text{ seconds}$$
$$s = 24 + 16$$
$$s = 40 \text{ seconds}$$
The distance traveled = $5p = 3s = 120$ miles

Exercise 1-3

Key Ideas and Formulas

a is less than b ($a < b$) or b is greater than a ($b > a$) if there exists a positive real number p such that $a + p = b$.

For any two real numbers a and b, $a < b$, $a > b$, or $a = b$. (Trichotomy principle)

For any real numbers a, b, and c with $a < b$, then:

   1. $a + c < b + c$
   2. $a - c < b - c$
   3. If c is positive, then $ac < bc$ and $\dfrac{a}{c} < \dfrac{b}{c}$
   4. If c is negative, then $ac > bc$ and $\dfrac{a}{c} > \dfrac{b}{c}$

Similar properties hold if $a > b$, then $a + c > b + c$ and so on. The order of an inequality reverses if we multiply or divide both sides of an inequality statement by a negative number.

---

1. $-8 \leq x \leq 7$

$-10 \quad -5 \quad 0 \quad 5 \quad 10$

3. $-6 \leq x < 6$

$-10 \quad -5 \quad 0 \quad 5 \quad 10$

5. $x \geq -6$

$-10 \quad -5 \quad 0 \quad 5 \quad 10$

7. $(-2, 6]$

$-10 \quad -5 \quad 0 \quad 5 \quad 10$

9. $(-7, 8)$

$-10 \quad -5 \quad 0 \quad 5 \quad 10$

11. $(-\infty, -2]$

$-10 \quad -5 \quad 0 \quad 5 \quad 10$

13. $[-7, 2)$; $-7 \leq x < 2$

15. $(-\infty, 0]$; $x \leq 0$

17. $7x - 8 < 4x + 7$
$3x < 15$
$x < 5$ or $(-\infty, 5)$

5

19. $3 - x \geq 5(3 - x)$
$3 - x \geq 15 - 5x$
$4x \geq 12$
$x \geq 3$ or $[3, \infty)$

3

21. $\dfrac{N}{-2} > 4$
$N < -8$ or $(-\infty, -8)$

-8

23. $-5t < -10$
$t > 2$ or $(2, \infty)$

2

25. $3 - m < 4(m - 3)$
$3 - m < 4m - 12$
$-5m < -15$
$m > 3$ or $(3, \infty)$

3

> **Common Error:**
> Neglecting to reverse the order after division by -5.

27. $-2 - \dfrac{B}{4} \leq \dfrac{1 + B}{3}$

$12\left(-2 - \dfrac{B}{4}\right) \leq 12\dfrac{(1 + B)}{3}$

$-24 - 3B \leq 4(1 + B)$
$-24 - 3B \leq 4 + 4B$
$-7B \leq 28$
$B \geq -4$ or $[-4, \infty)$

-4

29. $-4 < 5t + 6 \leq 21$
$-10 < 5t \leq 15$
$-2 < t \leq 3$ or $(-2, 3]$

$-2 \qquad 3$

31.

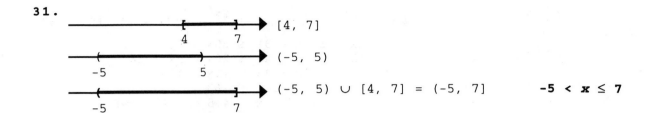

[4, 7]

(-5, 5)

(-5, 5) ∪ [4, 7] = (-5, 7]   **-5 < x ≤ 7**

33.

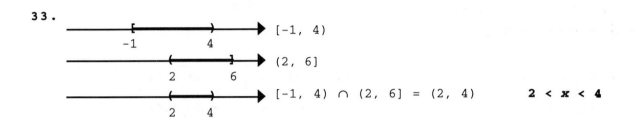

[-1, 4)

(2, 6]

[-1, 4) ∩ (2, 6] = (2, 4)   **2 < x < 4**

35.

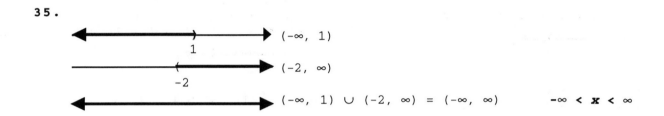

(-∞, 1)

(-2, ∞)

(-∞, 1) ∪ (-2, ∞) = (-∞, ∞)   **-∞ < x < ∞**

37.

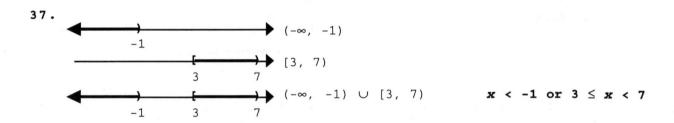

(-∞, -1)

[3, 7)

(-∞, -1) ∪ [3, 7)   **x < -1 or 3 ≤ x < 7**

39.

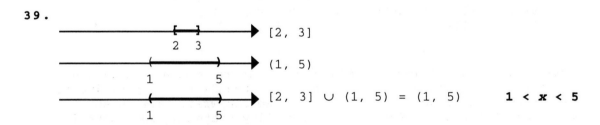

[2, 3]

(1, 5)

[2, 3] ∪ (1, 5) = (1, 5)   **1 < x < 5**

41.

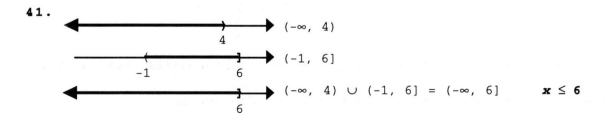

(-∞, 4)

(-1, 6]

(-∞, 4) ∪ (-1, 6] = (-∞, 6]   **x ≤ 6**

43. $\dfrac{q}{7} - 3 > \dfrac{q - 4}{3} + 1$

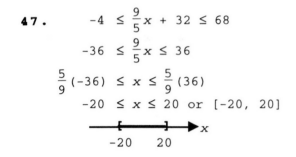

$21\left(\dfrac{q}{7} - 3\right) > 21\left(\dfrac{(q - 4)}{3} + 1\right)$

$3q - 63 > 7(q - 4) + 21$

$3q - 63 > 7q - 28 + 21$

$3q - 63 > 7q - 7$

$-4q > 56$

$q < -14$ or $(-\infty, -14)$

-14

45. $\dfrac{2x}{5} - \dfrac{1}{2}(x - 3) \leq \dfrac{2x}{3} - \dfrac{3}{10}(x + 2)$ LCD = 30

$12x - 15(x - 3) \leq 20x - 9(x + 2)$

$12x - 15x + 45 \leq 20x - 9x - 18$

$-3x + 45 \leq 11x - 18$

$-14x \leq -63$

$x \geq 4.5$ or $[4.5, \infty)$

4.5

47. $-4 \leq \dfrac{9}{5}x + 32 \leq 68$

$-36 \leq \dfrac{9}{5}x \leq 36$

$\dfrac{5}{9}(-36) \leq x \leq \dfrac{5}{9}(36)$

$-20 \leq x \leq 20$ or $[-20, 20]$

$-20 \quad 20$

49. $-12 < \dfrac{3}{4}(2 - x) \leq 24$

$\dfrac{4}{3}(-12) < 2 - x \leq \dfrac{4}{3}(24)$

$-16 < 2 - x \leq 32$

$-18 < -x \leq 30$

$18 > x \geq -30$

$-30 \leq x < 18$ or $[-30, 18)$

$-30 \quad 18$

51. $16 < 7 - 3x \leq 31$

$9 < -3x \leq 24$

$-3 > x \geq -8$

$-8 \leq x < -3$ or $[-8, -3)$

$-8 \quad -3$

53. $-6 < -\dfrac{2}{5}(1 - x) \leq 4$

$-30 < -2(1 - x) \leq 20$

$-30 < -2 + 2x \leq 20$

$-28 < 2x \leq 22$

$-14 < x \leq 11$ or $(-14, 11]$

$-14 \quad 11$

55. $5.23(x - 0.172) \leq 6.02x - 0.427$

$5.23x - 0.89956 \leq 6.02x - 0.427$

$5.23x - 6.02x - 0.89956 \leq -0.427$

$-0.79x - 0.89956 \leq -0.427$

$-0.79x \leq -0.427 + 0.89956$

$-0.79x \leq 0.47256$

$x \geq -0.60$

57. $-0.703 < 0.112 - 2.28x < 0.703$

$-0.703 - 0.112 < -2.28x < 0.703 - 0.112$

$-0.815 < -2.28x < 0.591$

$0.357 > x > -0.259$

$-0.259 < x < 0.357$

59. $\sqrt{1 - x}$ represents a real number exactly when $1 - x$ is positive or zero. We can write this as an inequality statement and solve for x.

$1 - x \geq 0$

$-x \geq -1$

$x \leq 1$

61. $\sqrt{3x + 5}$ represents a real number exactly when $3x + 5$ is positive or zero. We can write this as an inequality statement and solve for x.

$3x + 5 \geq 0$

$3x \geq -5$

$x \geq -\dfrac{5}{3}$

63. $\dfrac{1}{\sqrt[4]{2x + 3}}$ represents a real number exactly when $2x + 3$ is positive. (*not* zero).

We can write this as an inequality statement and solve for x.

$2x + 3 > 0$

$2x > -3$

$x > -\dfrac{3}{2}$

65. (A) For $ab > 0$, ab must be positive, hence a and b must have the same sign. Either
1. $a > 0$ and $b > 0$ or
2. $a < 0$ and $b < 0$

(B) For $ab < 0$, ab must be negative, hence a and b must have opposite signs. Either
1. $a > 0$ and $b < 0$ or
2. $a < 0$ and $b > 0$

(C) For $\frac{a}{b} > 0$, $\frac{a}{b}$ must be positive, hence a and b must have the same sign. Answer as in (A).

(D) For $\frac{a}{b} < 0$, $\frac{a}{b}$ must be negative, hence a and b must have opposite signs. Answer as in (B).

67. (A) If $a - b = 1$, then $a = b + 1$. Therefore, a is greater than b. >

(B) If $u - v = -2$, then $v = u + 2$. Therefore, u is less than v. <

69. If $\frac{b}{a}$ is greater than 1

$\frac{b}{a} > 1$

$a \cdot \frac{b}{a} < a \cdot 1$. (since a is negative.)

$b < a$

$0 < a - b$

$a - b$ is positive

71. (A) F   (B) T   (C) T

73. If $a < b$, then by definition of $<$, there exists a positive number p such that $a + p = b$. Then, adding c to both sides, we obtain $(a + c) + p = b + c$, where p is positive. Hence, by definition of $<$, we have
$a + c < b + c$

75. (A) If $a < b$, then by definition of $<$, there exists a positive number p such that $a + p = b$. If we multiply both sides of this by the positive number c, we obtain $(a + p)c = bc$, or $ac + pc = bc$, where pc is positive. Hence, by definition of $<$, we have $ac < bc$.

(B) If $a < b$, then by definition of $<$, there exists a positive number p such that $a + p = b$. If we multiply both sides of this by the negative number c, we obtain $(a + p)c = bc$, or $ac + pc = bc$, where pc is negative. Hence, by definition of $<$, we have $ac > bc$.

77. We want $200 \leq T \leq 300$. We are given $T = 30 + 25(x - 3)$. Substituting, we must solve
$200 \leq 30 + 25(x - 3) \leq 300$
$200 \leq 30 + 25x - 75 \leq 300$
$200 \leq -45 + 25x \leq 300$
$245 \leq 25x \leq 345$
$9.8 \leq x \leq 13.8$
(from 9.8 km to 13.8 km)

79. Let x = number of calculators sold
Then Revenue = (price per calculator) × (number of calculators sold) = $63x$
Cost = fixed cost + variable cost
= 650,000 + (Cost per calculator) × (number sold)
= $650,000 + 47x$

(A) We want Revenue > Cost
$$63x > 650,000 + 47x$$
$$16x > 650,000$$
$$x > 40,625$$
More than 40,625 calculators must be sold for the company to make a profit.

(B) We want Revenue = Cost
$$63x = 650,000 + 47x$$
$$16x = 650,000$$
$$x = 40,625$$

(C) 40,625 calculators sold represents the break-even point, the boundary between profit and loss.

81. (A) The company might try to increase sales and keep the price the same (see part B). It might try to increase the price and keep the sales the same (see part C). Either of these strategies would need further analysis and implementation that are out of place in a discussion here.

(B) Here the cost has been changed to $650,000 + 50.5x$, but the revenue is still $63x$.
    Revenue > Cost
$$63x > 650,000 + 50.5x$$
$$12.5x > 650,000$$
$$x > 52,000 \text{ calculators}$$

(C) Let p = the new price. Here the cost is still $650,000 + 50.5x$ as in part (B) where x is now known to be 40,625. Thus, cost = $650,000 + 50.5(40,625)$. The revenue is (price per calculator) × (number of calculators) = $p(40,625)$.
    Revenue > Cost
$$p(40,625) > 650,000 + 50.5(40,625)$$
$$p > \frac{650,000 + 50.5(40,625)}{40,625}$$
$$p > 66.50$$
The price could be raised by \$3.50 to \$66.50.

83. We want $220 \leq W \leq 2,750$. We are given $W = 110I$. Substituting, we must solve
$$220 \leq 110I \leq 2,750$$
$$2 \leq I \leq 25 \text{ or } [2, 25].$$

85. We are given that the benefit reduction (B) is \$1 for each \$2 that earnings (E) exceed \$8,880. Thus, $B = \frac{1}{2}(E - 8,880)$.

In this situation, we are dealing with the range of values for E of between \$13,000 and \$16,000, that is:
$$13,000 \leq E \leq 16,000$$
Then
$$13,000 - 8,880 \leq E - 8,880 \leq 16,000 - 8,880$$
$$4,120 \leq E - 8,880 \leq 7,120$$
$$\frac{1}{2}(4,120) \leq \frac{1}{2}(E - 8,880) \leq \frac{1}{2}(7,120)$$
$$2,060 \leq B \leq 3,560$$
Thus, the benefit reduction is in the range between \$2,060 and \$3,560.

Exercise 1-4

Key Ideas and Formulas

$$|x| = \begin{cases} x \text{ if } x \geq 0 \\ -x \text{ if } x < 0 \end{cases}$$ [Note: $-x$ is positive if x is negative]

The distance between two points A and B on a number line with coordinates a and b respectively is given by

$$d(A, B) = |b - a| = |a - b|$$

Geometric Interpretation of
Absolute Value Equations and Inequalities

Form ($d > 0$)	Geometric Interpretation	Solution	Graph
$|x-c|=d$	Distance between x and c is equal to d.	$\{c-d, c+d\}$	
$|x-c|<d$	Distance between x and c is less than d.	$(c-d, c+d)$	
$0<|x-c|<d$	Distance between x and c is less than d but $x \neq c$.	$(c-d, c) \cup (c, c+d)$	
$|x-c|>d$	Distance between x and c is greater than d.	$(-\infty, c-d) \cup (c+d, \infty)$	

For $p > 0$

$|x| = p$ is equivalent to $x = p$ or $x = -p$

$|x| < p$ is equivalent to $-p < x < p$

$|x| > p$ is equivalent to $x < -p$ or $x > p$

> **Common Error:**
> Writing: $p < x < -p$
> This is true for no real value of x,
> since $p < -p$ is false.

To solve: (p assumed positive)

$|ax + b| = p$, solve $ax + b = p$ or $ax + b = -p$

$|ax + b| < p$, solve $-p < ax + b < p$

$|ax + b| > p$, solve $ax + b < -p$ or $ax + b > p$

1. $\sqrt{5}$

3. $|(-6) - (-2)| = |-4| = 4$

5. $|5 - \sqrt{5}| = 5 - \sqrt{5}$ since $5 - \sqrt{5}$ is positive.

7. $|\sqrt{5} - 5| = -(\sqrt{5} - 5) = 5 - \sqrt{5}$ since $\sqrt{5} - 5$ is negative.

9. $d(A,B) = |b - a|$
$= |5 - (-7)|$
$= |12|$
$= 12$

11. $d(A,B) = |b - a|$
$= |-7 - 5|$
$= |-12|$
$= 12$

13. $d(B,0) = |0 - (-4)|$
$= |4|$
$= 4$

15. $d(0,B) = |-4 - 0|$
$= |-4|$
$= 4$

17. $d(B,C) = |5 - (-4)|$
$= |9|$
$= 9$

19. The distance between x and 3 is equal to 4.
$|x - 3| = 4$

21. The distance between m and -2 is equal to 5.
$|m - (-2)| = 5$
$|m + 2| = 5$

23. The distance between x and 3 is less than 5.
$|x - 3| < 5$

25. The distance between p and -2 is more than 6.
$|p - (-2)| > 6$
$|p + 2| > 6$

27. The distance between q and 1 is not less than 2.
$$|q - 1| \geq 2$$

29. x is no more than 7 units from the origin.
$$-7 \leq x \leq 7 \quad [-7, 7]$$

31. x is at least 7 units from the origin.
$$x \leq -7 \text{ or } x \geq 7$$
$$(-\infty, -7] \cup [7, \infty)$$

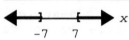

Common Error: $7 \leq x \leq -7$
This is meaningless

33. y is 3 units from 5.
$$|y - 5| = 3$$
$$y - 5 = \pm 3$$
$$y = 5 \pm 3$$
$$y = 2, 8$$

35. y is less than 3 units from 5.
$$|y - 5| < 3$$
$$-3 < y - 5 < 3$$
$$2 < y < 8$$
$$(2, 8)$$

37. y is more than 3 units from 5.
$$|y - 5| > 3$$
$$y - 5 < -3 \text{ or } y - 5 > 3$$
$$y < 2 \text{ or } y > 8$$
$$(-\infty, 2) \cup (8, \infty)$$

39. $|u - (-8)| = 3$
u is 3 units from -8.
$$|u + 8| = 3$$
$$u + 8 = \pm 3$$
$$u = -8 \pm 3$$
$$u = -11 \text{ or } -5$$

41. $|u - (-8)| \leq 3$
u is no more than 3 units from -8.
$$|u + 8| \leq 3$$
$$-3 \leq u + 8 \leq 3$$
$$-11 \leq u \leq -5$$
$$[-11, -5]$$

43. $|u - (-8)| \geq 3$
u is at least 3 units from -8.
$$|u + 8| \geq 3$$
$$u + 8 \leq -3 \text{ or } u + 8 \geq 3$$
$$u \leq -11 \text{ or } u \geq -5$$
$$(-\infty, -11] \cup [-5, \infty)$$

45. $|3x - 7| \leq 4$
$$-4 \leq 3x - 7 \leq 4$$
$$3 \leq 3x \leq 11$$
$$1 \leq x \leq \frac{11}{3}$$
$$\left[1, \frac{11}{3}\right]$$

47. $|4 - 2t| > 6$
$$4 - 2t > 6 \text{ or } 4 - 2t < -6$$
$$-2t > 2 \text{ or } -2t < -10$$
$$t < -1 \text{ or } t > 5$$
$$(-\infty, -1) \cup (5, \infty)$$

49. $|7m + 11| = 3$
$$7m + 11 = 3 \text{ or } 7m + 11 = -3$$
$$7m = -8 \text{ or } 7m = -14$$
$$m = -\frac{8}{7} \text{ or } m = -2$$

51. $\left|\frac{1}{2}w - \frac{3}{4}\right| < 2$
$$-2 < \frac{1}{2}w - \frac{3}{4} < 2$$
$$-8 < 2w - 3 < 8$$
$$-5 < 2w < 11$$
$$-\frac{5}{2} < w < \frac{11}{2}$$
$$-2.5 < w < 5.5$$
$$(-2.5, 5.5)$$

53. $|0.2u + 1.7| \geq 0.5$
$$0.2u + 1.7 \geq 0.5 \text{ or } 0.2u + 1.7 \leq -0.5$$
$$0.2u \geq -1.2 \text{ or } 0.2u \leq -2.2$$
$$u \geq -6 \text{ or } u \leq -11$$
$$[-6, \infty) \cup (-\infty, -11]$$

55. $\left|\dfrac{9}{5}C + 32\right| < 31$

$-31 < \dfrac{9}{5}C + 32 < 31$

$-63 < \dfrac{9}{5}C < -1$

$-35 < C < -\dfrac{5}{9}$

$\left(-35, \ -\dfrac{5}{9}\right)$

57. $\sqrt{x^2} < 2$
$|x| < 2$
$-2 < x < 2$
$(-2, \ 2)$

59. $\sqrt{(1 - 3t)^2} \le 2$
$|1 - 3t| \le 2$
$-2 \le 1 - 3t \le 2$
$-3 \le -3t \le 1$
$1 \ge t \ge -\dfrac{1}{3}$
$-\dfrac{1}{3} \le t \le 1$

$\left[-\dfrac{1}{3}, \ 1\right]$

61. $\sqrt{(2t - 3)^2} > 3$
$|2t - 3| > 3$
$2t - 3 < -3 \ \text{or} \ 2t - 3 > 3$
$\quad 2t < 0 \qquad\qquad 2t > 6$
$\quad\ \ t < 0 \qquad\qquad\ \ t > 3$
$(-\infty, \ 0) \ \cup \ (3, \ \infty)$

63. $0 < |x - 3| < 0.1$
The distance between x and 3 is less than 0.1 but $x \ne 3$.
$-0.1 < x - 3 < 0.1$ except $x \ne 3$
$2.9 < x < 3.1$ but $x \ne 3$
$2.9 < x < 3$ or $3 < x < 3.1$
$(2.9, \ 3) \ \cup \ (3, \ 3.1)$

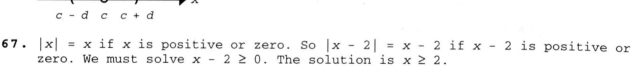

65. $0 < |x - c| < d$
The distance between x and c is less than d but $x \ne c$.
$\quad -d < x - c < d$ except $x \ne c$
$c - d < x < c + d$ but $x \ne c$
$c - d < x < c$ or $c < x < c + d$
$(c - d, \ c) \ \cup \ (c, \ c + d)$

67. $|x| = x$ if x is positive or zero. So $|x - 2| = x - 2$ if $x - 2$ is positive or zero. We must solve $x - 2 \ge 0$. The solution is $x \ge 2$.

69. $|x| = -x$ if x is negative or zero. So $|2x - 3| = -(2x - 3)$ or $3 - 2x$ if $2x - 3$ is negative or zero. We must solve $2x - 3 \le 0$.

The solution is $x \le \dfrac{3}{2}$ or $x \le 1.5$.

71. We consider two cases for $|3x + 5| = 2x + 6$

Case 1: $3x + 5 \ge 0$

For this case, the possible values of x are in the set $x \ge -\dfrac{5}{3}$.

$\quad$ Then $|3x + 5| = 3x + 5$.
$\quad$ We solve $\quad 3x + 5 = 2x + 6$
$\qquad\qquad\qquad\quad x = 1 \qquad$ A solution, since 1 is among the possible
$\qquad\qquad\qquad\qquad\qquad\qquad$ values of x.

Case 2: $3x + 5 < 0$

For this case, the possible values of x are in the set $x < -\dfrac{5}{3}$.

$\quad$ Then $|3x + 5| = -(3x + 5)$
$\quad$ We solve $\quad -(3x + 5) = 2x + 6$
$\qquad\qquad\qquad\ -3x - 5 = 2x + 6$
$\qquad\qquad\qquad\qquad\ -5x = 11$
$\qquad\qquad\qquad\qquad\qquad x = -2.2 \ $ A solution, since -2.2 is among the possible
$\qquad\qquad\qquad\qquad\qquad\qquad\qquad$ values of x.

Solution $x = 1, \ -2.2$

73. We consider four cases for $|x| + |x + 3| = 3$.

Case 1: $x \geq 0$ and $x + 3 \geq 0$, that is, $x \geq 0$ and $x \geq -3$ (or simply $x \geq 0$)
$|x| = x$ and $|x + 3| = x + 3$
Hence: $x + x + 3 = 3$
                    $2x + 3 = 3$
                          $x = 0$
which is a possible value for x in this case.

Case 2: $x < 0$ and $x + 3 \geq 0$, that is, $x < 0$ and $x \geq -3$ (or simply $-3 \leq x < 0$)
$|x| = -x$ and $|x + 3| = x + 3$
Hence: $-x + x + 3 = 3$
                          $3 = 3$
This is satisfied by all x, but the condition $-3 \leq x < 0$ must be imposed.

Case 3: $x \geq 0$ and $x + 3 < 0$, that is, $x \geq 0$ and $x < -3$. These are mutually contradictory, so no solution is possible in this case.

Case 4: $x < 0$ and $x + 3 \leq 0$, that is, $x < 0$ and $x \leq -3$ (or simply $x \leq -3$)
$|x| = -x$ and $|x + 3| = -(x + 3)$
Hence: $-x + -(x + 3) = 3$
                  $-x - x - 3 = 3$
                          $-2x = 6$
                            $x = -3$
Combining the results of the four cases, $-3 \leq x \leq 0$ is the solution.

75. We consider four cases for $|2x + 7| - |6 - 3x| = 8$.

Case 1: $2x + 7 \geq 0$ and $6 - 3x \geq 0$, that is, $x \geq -\dfrac{7}{2}$ and $x \leq 2$
(or simply $-\dfrac{7}{2} \leq x \leq 2$)
$|2x + 7| = 2x + 7$ and $|6 - 3x| = 6 - 3x$
Hence: $2x + 7 - (6 - 3x) = 8$
                $2x + 7 - 6 + 3x = 8$
                              $5x + 1 = 8$
                                   $5x = 7$
                                     $x = 1.4$
which is a possible value for x in this case.

Case 2: $2x + 7 < 0$ and $6 - 3x \geq 0$, that is, $x < -\dfrac{7}{2}$ and $x \leq 2$
$\left(\text{or simply } x < -\dfrac{7}{2}\right)$
$|2x + 7| = -(2x + 7)$ and $|6 - 3x| = 6 - 3x$
Hence: $-(2x + 7) - (6 - 3x) = 8$
                  $-2x - 7 - 6 + 3x = 8$
                                $x - 13 = 8$
                                      $x = 21$
which is not a possible value for x in this case.

Case 3: $2x + 7 \geq 0$ and $6 - 3x < 0$, that is, $x \geq -\dfrac{7}{2}$ and $x > 2$ (or simply $x > 2$)
$|2x + 7| = 2x + 7$ $|6 - 3x| = -(6 - 3x)$
Hence: $2x + 7 - [-(6 - 3x)] = 8$
                  $2x + 7 + 6 - 3x = 8$
                            $-x + 13 = 8$
                                  $-x = -5$
                                    $x = 5$
which is a possible value for x in this case.

Case 4: $2x + 7 < 0$ and $6 - 3x < 0$, that is, $x < -\dfrac{7}{2}$ and $x > 2$. These are mutually contradictory, so no solution is possible in this case.
Combining the results of the four cases, $x = 1.4$ or $x = 5$ are the solutions.

77. *Case 1:* $x > 0$. Then $|x| = x$. Hence $\dfrac{x}{|x|} = \dfrac{x}{x} = 1$.

Case 2: $x = 0$. Then $|x| = 0$. Hence $\dfrac{x}{|x|}$ is not defined.

Case 3: $x < 0$. Then $|x| = -x$. Hence $\dfrac{x}{|x|} = \dfrac{x}{-x} = -1$.

Thus, the possible values of $\dfrac{x}{|x|}$ are 1 and -1.

79. There are three possible relations between real numbers a and b; either $a = b$, $a > b$, or $a < b$. We examine each case separately.

Case 1: $a = b$ *Case 2:* $a > b$

$|b - a| = |0| = 0;$ $|b - a| = -(b - a) = a - b$

$|a - b| = |0| = 0$ $|a - b| = a - b$

Case 3: $b > a$

$|b - a| = b - a$

$|a - b| = -(a - b) = b - a$

Thus in all three cases $|b - a| = |a - b|$.

81. If $m < n$, then $m + m < m + n$ (adding m to both sides)

Also, $m + n < n + n$ (adding n to both sides). Hence,

$m + m < m + n < n + n$

$\qquad 2m < m + n < 2n$

$\qquad\quad m < \dfrac{m + n}{2} < n$

83. *Case 1.* $m > 0$. Then $|m| = m$ $|-m| = -(-m) = m$. Hence $|m| = |-m|$

Case 2. $m < 0$. Then $|m| = -m$ $|-m| = -m$. Hence $|m| = |-m|$

Case 3. $m = 0$. Then $0 = m = -m$, hence $|m| = |-m| = 0$.

85. If $n \neq 0$, $n > 0$ or $n < 0$.

Case 1. $n > 0$. If $m \geq 0$ $|m| = m$, $\dfrac{m}{n} \geq 0$ $\left|\dfrac{m}{n}\right| = \dfrac{m}{n}$ $|n| = n$.

Hence: $\left|\dfrac{m}{n}\right| = \dfrac{m}{n} = \dfrac{|m|}{|n|}$

If $m < 0$ $|m| = -m$, $\dfrac{m}{n} < 0$ $\left|\dfrac{m}{n}\right| = -\dfrac{m}{n}$ $|n| = n$.

Hence: $\left|\dfrac{m}{n}\right| = -\dfrac{m}{n} = \dfrac{-m}{n} = \dfrac{|m|}{|n|}$

Case 2. $n < 0$. If $m > 0$ $|m| = m$, $\dfrac{m}{n} < 0$ $\left|\dfrac{m}{n}\right| = -\dfrac{m}{n}$ $|n| = -n$

Hence: $\left|\dfrac{m}{n}\right| = -\dfrac{m}{n} = \dfrac{m}{-n} = \dfrac{|m|}{|n|}$

If $m \leq 0$ $|m| = -m$ $\dfrac{m}{n} > 0$ $\left|\dfrac{m}{n}\right| = \dfrac{m}{n}$ $|n| = -n$

Hence: $\left|\dfrac{m}{n}\right| = \dfrac{m}{n} = \dfrac{-m}{-n} = \dfrac{|m|}{|n|}$

87. First note that $a \leq b$ is true if $a < b$ or if $a = b$. Hence $a < b$ implies $a \leq b$.
Also $a = b$ implies $a \leq b$. Now consider three cases ($m > 0$, $m = 0$, $m < 0$).

Case 1. $m > 0$. Then $|m| = m$. Also $-|m| < 0$

Hence $-|m| < 0 < m = |m|$

$\qquad\quad -|m| < m = |m|$

$\qquad\quad -|m| \leq m \leq |m|$

Case 2. $m = 0$. Then $-|m| = m = |m| = 0$.

Hence $-|m| \leq m \leq |m|$

Case 3. $m < 0$. Then $|m| = -m$, hence $-|m| = m$. Also $|m| > 0$.

Hence $-|m| = m < 0 < |m|$

$\qquad\quad -|m| = m < |m|$

$\qquad\quad -|m| \leq m \leq |m|$

89. Consider three cases ($a < b$, $a = b$, $a > b$)

Case 1. $a < b$. Then $\max(a,b) = b$. Also $a - b < 0$, hence $|a - b| = b - a$.

Hence $\frac{1}{2}[a + b + |a - b|] = \frac{1}{2}[a + b + b - a] = \frac{1}{2}(2b) = b$.

Thus $\max(a,b) = b = \frac{1}{2}[a + b + |a - b|]$

Case 2. $a = b$. Then $\max(a,b) = a = b$. Also $a - b = 0$, hence $|a - b| = 0$.

Hence $\frac{1}{2}[a + b + |a - b|] = \frac{1}{2}[a + b] = \frac{1}{2}[2a] = a$ or $\frac{1}{2}[2b] = b$.

Thus $\max(a,b) = a = b = \frac{1}{2}[a + b + |a - b|]$

Case 3. $a > b$. Then $\max(a,b) = a$. Also $a - b > 0$, hence $|a - b| = a - b$

Hence $\frac{1}{2}[a + b + |a - b|] = \frac{1}{2}[a + b + a - b] = \frac{1}{2}(2a) = a$.

Thus $\max(a,b) = a = \frac{1}{2}[a + b + |a - b|]$

91. $\left|\dfrac{x - 45.4}{3.2}\right| < 1$

$-1 < \dfrac{x - 45.4}{3.2} < 1$

$-3.2 < x - 45.4 < 3.2$

$42.2 < x < 48.6$

93. The difference between P and 500 has an absolute value of no more than 20.
$|P - 500| \leq 20$

95. The difference between A and 12.436 has an absolute value of less than the error of 0.001.
$|A - 12.436| < 0.001$
$-0.001 < A - 12.436 < 0.001$
$12.435 < A < 12.437$ or, in interval notation, (12.435, 12.437)

97. The difference between N and 2.37 has an absolute value of no more than 0.005.
$|N - 2.37| \leq 0.005$.

Exercise 1-5
Key Ideas and Formulas

For a, b real numbers.

i = the imaginary unit = $\sqrt{-1}$

$a + bi$ = complex number (or imaginary number) in standard form

$0 + bi$ = pure imaginary number

$a = a + 0i$ = real number

$0 + 0i$ = zero

$a - bi$ = conjugate of $a + bi$

$a + bi = c + di$ if and only if $a = c$ and $b = d$

$(a + bi) + (c + di) = (a + c) + (b + d)i$

$(a + bi)(c + di) = (ac - bd) + (ad + bc)i$

$(a + bi)(a - bi) = a^2 + b^2$

$\dfrac{a + bi}{c} = \dfrac{a}{c} + \dfrac{b}{c}i$

$\dfrac{a + bi}{c + di} = \dfrac{(a + bi)}{(c + di)}\dfrac{(c - di)}{(c - di)}$ which is simplified to

$\dfrac{ac + bd}{c^2 + d^2} + \dfrac{bc - ad}{c^2 + d^2}i$

The principal square root of a negative real number, denoted by $\sqrt{-a}$ where a is positive, is defined by $\sqrt{-a} = i\sqrt{a}$

$a + bi$, where a and b are real numbers and i is the imaginary unit, is called a complex number in standard form.

1. $(3 + 5i) + (2 + 4i) = 3 + 5i + 2 + 4i = 5 + 9i$

3. $(8 - 3i) + (-5 + 6i) = 8 - 3i - 5 + 6i = 3 + 3i$

5. $(9 + 5i) - (6 + 2i) = 9 + 5i - 6 - 2i = 3 + 3i$

7. $(3 - 4i) - (-5 + 6i) = 3 - 4i + 5 - 6i = 8 - 10i$

9. $2 + (3i + 5) = 2 + 3i + 5 = 7 + 3i$

11. $(2i)(4i) = 8i^2 = 8(-1) = -8$

13. $-2i(4 - 6i) = -8i + 12i^2 = -8i + 12(-1) = -12 - 8i$

15. $(1 + 2i)(3 - 4i) = 3 - 4i + 6i - 8i^2 = 3 + 2i - 8(-1) = 3 + 2i + 8 = 11 + 2i$

17. $(3 - i)(4 + i) = 12 + 3i - 4i - i^2 = 12 - i - (-1) = 12 - i + 1 = 13 - i$

19. $(2 + 9i)(2 - 9i) = 4 - 81i^2 = 4 + 81 = 85$ or $85 + 0i$

21. $\dfrac{1}{2 + 4i} = \dfrac{1}{(2 + 4i)} \dfrac{(2 - 4i)}{(2 - 4i)} = \dfrac{2 - 4i}{4 - 16i^2} = \dfrac{2 - 4i}{4 + 16} = \dfrac{2 - 4i}{20} = 0.1 - 0.2i$

23. $\dfrac{4 + 3i}{1 + 2i} = \dfrac{(4 + 3i)}{(1 + 2i)} \dfrac{(1 - 2i)}{(1 - 2i)} = \dfrac{4 - 5i - 6i^2}{1 - 4i^2} = \dfrac{4 - 5i + 6}{1 + 4} = \dfrac{10 - 5i}{5} = 2 - i$

25. $\dfrac{7 + i}{2 + i} = \dfrac{(7 + i)}{(2 + i)} \dfrac{(2 - i)}{(2 - i)} = \dfrac{14 - 5i - i^2}{4 - i^2} = \dfrac{14 - 5i + 1}{4 + 1} = \dfrac{15 - 5i}{5} = 3 - i$

27. $(2 - \sqrt{-4}) + (5 - \sqrt{-9}) = (2 - i\sqrt{4}) + (5 - i\sqrt{9}) = 2 - 2i + 5 - 3i = 7 - 5i$

29. $(9 - \sqrt{-9}) - (12 - \sqrt{-25}) = (9 - i\sqrt{9}) - (12 - i\sqrt{25})$
$$= (9 - 3i) - (12 - 5i) = 9 - 3i - 12 + 5i = -3 + 2i$$

31. $(3 - \sqrt{-4})(-2 + \sqrt{-49}) = (3 - i\sqrt{4})(-2 + i\sqrt{49})$
$$= (3 - 2i)(-2 + 7i) = -6 + 25i - 14i^2 = -6 + 25i + 14 = 8 + 25i$$

33. $\dfrac{5 - \sqrt{-4}}{7} = \dfrac{5 - i\sqrt{4}}{7} = \dfrac{5 - 2i}{7} = \dfrac{5}{7} - \dfrac{2}{7}i$

35. $\dfrac{1}{2 - \sqrt{-9}} = \dfrac{1}{2 - i\sqrt{9}} = \dfrac{1}{2 - 3i} = \dfrac{1}{(2 - 3i)} \dfrac{(2 + 3i)}{(2 + 3i)} = \dfrac{2 + 3i}{4 - 9i^2}$
$$= \dfrac{2 + 3i}{4 + 9} = \dfrac{2 + 3i}{13} = \dfrac{2}{13} + \dfrac{3}{13}i$$

37. $\dfrac{2}{5i} = \dfrac{2}{5i} \cdot \dfrac{i}{i} = \dfrac{2i}{5i^2} = \dfrac{2i}{-5} = -\dfrac{2}{5}i$ or $0 - \dfrac{2}{5}i$

39. $\dfrac{1 + 3i}{2i} = \dfrac{(1 + 3i)}{(2i)} \cdot \dfrac{i}{i} = \dfrac{i + 3i^2}{2i^2} = \dfrac{i - 3}{-2} = \dfrac{3}{2} - \dfrac{1}{2}i$

41. $(2 - 3i)^2 - 2(2 - 3i) + 9 = 4 - 12i + 9i^2 - 4 + 6i + 9$
$$= 4 - 12i - 9 - 4 + 6i + 9 = -6i \text{ or } 0 - 6i$$

43. $x^2 - 2x + 2 = (1 - i)^2 - 2(1 - i) + 2 = 1 - 2i + i^2 - 2 + 2i + 2$
$$= 1 - 2i - 1 - 2 + 2i + 2 = 0 \text{ or } 0 + 0i$$

45. $i^{18} = \underbrace{i^{16}} \cdot i^2 = (i^4)^4 \cdot i^2 = 1^4(-1) = -1$

　　　largest integer in 18 exactly divisible by 4
$i^{32} = (i^4)^8 = 1^8 = 1$
$i^{67} = \underbrace{i^{64}} \cdot i^3 = (i^4)^{16} \cdot i^2 \cdot i = 1^{16}(-1)i = -i$

　　　largest integer in 67 exactly divisible by 4

47. According to the definition of equality for complex numbers

$(2x - 1) + (3y + 2)i = 5 - 4i = 5 + (-4)i$

if and only if

$2x - 1 = 5$ and $3y + 2 = -4$

So $2x = 6$ and $3y = -6$

$x = 3$ $y = -2$

49. $\sqrt{3 - x}$ represents an imaginary number when $3 - x$ is negative. We solve:

$3 - x < 0$

$\quad 3 < x$ or $x > 3$

51. $\sqrt{2 - 3x}$ represents an imaginary number when $2 - 3x$ is negative. We solve:

$2 - 3x < 0$

$\quad -3x < -2$

$\quad\quad x > \dfrac{2}{3}$

53. $(3.17 - 4.08i)(7.14 + 2.76i)$

$\quad = (3.17)(7.14) + (3.17)(2.76i) - (4.08i)(7.14) - (4.08i)(2.76i)$

$\quad = 22.6338 + 8.7492i - 29.1312i - 11.2608i^2$

$\quad = 22.6338 - 20.382i + 11.2608$

$\quad = 33.89 - 20.38i$

55. $\dfrac{8.14 + 2.63i}{3.04 + 6.27i} = \dfrac{(8.14 + 2.63i)}{(3.04 + 6.27i)} \cdot \dfrac{(3.04 - 6.27i)}{(3.04 - 6.27i)}$

$\quad = \dfrac{(8.14)(3.04) - (8.14)(6.27i) + (2.63i)(3.04) - (2.63i)(6.27i)}{(3.04)^2 + (6.27)^2}$

$\quad = \dfrac{24.7456 - 51.0378i + 7.9952i - 16.4901i^2}{9.2416 + 39.3129}$

$\quad = \dfrac{24.7456 - 43.0426i + 16.4901}{48.55455} = \dfrac{41.2357 - 43.0426i}{48.5545} = 0.85 - 0.89i$

57. $(a + bi) + (c + di) = a + bi + c + di = a + c + bi + di = (a + c) + (b + d)i$

59. $(a + bi)(a - bi) = a^2 - b^2i^2 = a^2 + b^2$ or $(a^2 + b^2) + 0i$

61. $(a + bi)(c + di) = ac + adi + bci + bdi^2 = ac + (ad + bc)i - bd$

$\quad\quad\quad\quad\quad = (ac - bd) + (ad + bc)i$

63. $i^{4k} = (i^4)^k = (i^2 \cdot i^2)^k = [(-1)(-1)]^k = 1^k = 1$

65. 1. Definition of addition
   2. Commutative property for addition of real numbers.
   3. Definition of addition (read from right to left).

67. The product of a complex number and its conjugate is a real number.

$z\bar{z} = (x + yi)(x - yi)$

$\quad = x^2 - (yi)^2$

$\quad = x^2 - y^2i^2$

$\quad = x^2 + y^2$ or

$\quad\quad (x^2 + y^2) + 0i$.

This is a real number.

69. The conjugate of a complex number is equal to the complex number if and only if the number is real. To prove a theorem containing the phrase "if and only if", it is often helpful to prove two parts separately. Thus: $\bar{z} = z$ if z is real; $\bar{z} = z$ only if z is real

Hypothesis: z is real

Conclusion: $\bar{z} = z$

Proof: Assume z is real, then
$z = x + 0i = x$
$\bar{z} = x - 0i = x$

Hence $z = \bar{z}$.

Hypothesis: $\bar{z} = z$

Conclusion: z is real

Proof: Assume $\bar{z} = z$,
that is, $x - yi = x + yi$
Then by the definition of equality
$x = x \quad -y = y$
$\quad\quad -2y = 0$
$\quad\quad y = 0$
Hence $z = x + 0i$, that is, z is real.

71. The conjugate of the sum of two complex numbers is equal to the sum of their conjugates.

$$
\begin{aligned}
\overline{z + w} &= \overline{(x + yi) + (u + vi)} \\
&= \overline{x + yi + u + vi} \\
&= \overline{x + u + (y + v)i} \\
&= (x + u) - (y + v)i \\
&= x + u - yi - vi \\
&= (x - yi) + (u - vi) \\
&= \bar{z} + \bar{w}
\end{aligned}
$$

73. The conjugate of the product of two complex numbers is equal to the product of their conjugates.

$$
\begin{aligned}
\overline{zw} &= \overline{(x + yi)(u + vi)} \\
&= \overline{xu + xvi + yui + yvi^2} \\
&= \overline{xu + (xv + yu)i - yv} \\
&= xu - yv - (xv + yu)i \\
&= xu - xvi - yv - yui \\
&= x(u - vi) - yui + yv(-1) \\
&= x(u - vi) - yui + yvi^2 \\
&= x(u - vi) - yi(u - vi) \\
&= (x - yi)(u - vi) \\
&= \bar{z}\,\bar{w}
\end{aligned}
$$

Exercise 1-6

Key Ideas and Formulas

A quadratic equation in one variable is any equation that can be written in the form

$$ax^2 + bx + c = 0 \quad\quad a \neq 0 \quad\quad \text{(standard form)}$$

where x is a variable and a, b, c are constants.

To solve a quadratic equation in standard form, check if it can be solved using factoring and the zero property:

$$m \cdot n = 0 \text{ if and only if } m = 0 \text{ or } n = 0 \text{ or both}$$

If not, generally the quadratic formula is applied:

$$x = \frac{-b \pm \sqrt{b^2 - 4ac}}{2a} \quad\quad (a \neq 0)$$

Occasionally the square root property

$$\text{if } A^2 = C, \text{ then } A = \pm\sqrt{C}$$

can be useful, especially if the equation is in the form

$$ax^2 = b \text{ or } (ax + b)^2 = d$$

To complete the square of a quadratic of the form $x^2 + bx$, add the square of one-half the coefficient of x; that is, add $(b/2)^2$. Thus,

$$x^2 + bx$$

$$x^2 + bx + \left(\frac{b}{2}\right)^2 = \left(x + \frac{b}{2}\right)^2$$

To solve a quadratic equation by completing the square, then, put the equation in the form $x^2 + bx = m$, add $(b/2)^2$ to both sides and use the square root property to complete the solution.

The quantity $b^2 - 4ac$, appearing in the quadratic formula, is called the discriminant. If the discriminant is positive, the equation has two distinct real roots. If the discriminant is 0, the equation has one real root (a double root). If the discriminant is negative, the equation has two imaginary roots, one the conjugate of the other.

1. $2x^2 = 8x$
$$2x^2 - 8x = 0$$
$$2x(x - 4) = 0$$
$$2x = 0 \text{ or } x - 4 = 0$$
$$x = 0 \qquad\qquad x = 4$$

3. $4t^2 + 9 = 12t$
$$4t^2 - 12t + 9 = 0$$
$$(2t - 3)^2 = 0$$
$$2t - 3 = 0$$
$$2t = 3$$
$$t = \frac{3}{2} \text{ (double root)}$$

5. $3w^2 + 13w = 10$
$$3w^2 + 13w - 10 = 0$$
$$(3w - 2)(w + 5) = 0$$
$$3w - 2 = 0 \text{ or } w + 5 = 0$$
$$3w = 2 \qquad\qquad w = -5$$
$$w = \frac{2}{3}$$

7. $m^2 - 25 = 0$
$$m^2 = 25$$
$$m = \pm\sqrt{25}$$
$$m = \pm 5$$

9. $c^2 + 9 = 0$
$$c^2 = -9$$
$$c = \pm\sqrt{-9}$$
$$c = \pm 3i$$

11. $4y^2 + 9 = 0$
$$4y^2 = -9$$
$$y^2 = -\frac{9}{4}$$
$$y = \pm\sqrt{-\frac{9}{4}}$$
$$y = \pm\frac{i\sqrt{9}}{\sqrt{4}}$$
$$y = \pm\frac{3i}{2}$$

13. $25z^2 - 32 = 0$
$$25z^2 = 32$$
$$z^2 = \frac{32}{25}$$
$$z = \pm\sqrt{\frac{32}{25}}$$
$$z = \pm\frac{\sqrt{32}}{\sqrt{25}}$$
$$z = \pm\frac{4\sqrt{2}}{5}$$

15. $(s + 1)^2 = 5$
$$s + 1 = \pm\sqrt{5}$$
$$s = -1 \pm \sqrt{5}$$

17. $(n - 3)^2 = -4$
$$n - 3 = \pm\sqrt{-4}$$
$$n - 3 = \pm 2i$$
$$n = 3 \pm 2i$$

19. $x^2 - 2x - 1 = 0$
$$x = \frac{-b \pm \sqrt{b^2 - 4ac}}{2a} \qquad a = 1, \; b = -2, \; c = -1$$
$$x = \frac{-(-2) \pm \sqrt{(-2)^2 - 4(1)(-1)}}{2(1)}$$
$$x = \frac{2 \pm \sqrt{8}}{2}$$
$$x = \frac{2 \pm 2\sqrt{2}}{2}$$
$$x = 1 \pm \sqrt{2}$$

> **Common Errors:**
> It is incorrect to cancel this way:
> $$\frac{2 \pm 2\sqrt{2}}{2} \neq \pm 2\sqrt{2}$$
> or this way $\dfrac{2 \pm 2\sqrt{2}}{2} \neq 1 \pm 2\sqrt{2}$

21. $x^2 - 2x + 3 = 0$

$x = \dfrac{-b \pm \sqrt{b^2 - 4ac}}{2a}$ $a = 1, \ b = -2,$
$\quad\quad\quad\quad\quad\quad\quad\quad\quad\quad c = 3$

$x = \dfrac{-(-2) \pm \sqrt{(-2)^2 - 4(1)(3)}}{2(1)}$

$x = \dfrac{2 \pm \sqrt{-8}}{2}$

$x = \dfrac{2 \pm 2i\sqrt{2}}{2}$

$x = 1 \pm i\sqrt{2}$

23. $\quad\quad 2t^2 + 8 = 6t$
$\quad 2t^2 - 6t + 8 = 0$
$\quad\ \ t^2 - 3t + 4 = 0$

$t = \dfrac{-b \pm \sqrt{b^2 - 4ac}}{2a}$ $a = 1, \ b = -3,$
$\quad\quad\quad\quad\quad\quad\quad\quad\quad\quad c = 4$

$t = \dfrac{-(-3) \pm \sqrt{(-3)^2 - 4(1)(4)}}{2(1)}$

$t = \dfrac{3 \pm \sqrt{-7}}{2}$

$t = \dfrac{3 \pm i\sqrt{7}}{2}$

25. $2t^2 + 1 = 6t$
$\ 2t^2 - 6t + 1 = 0$

$t = \dfrac{-b \pm \sqrt{b^2 - 4ac}}{2a}$ $a = 2, \ b = -6, \ c = 1$

$t = \dfrac{-(-6) \pm \sqrt{(-6)^2 - 4(2)(1)}}{2(2)}$

$t = \dfrac{6 \pm \sqrt{28}}{4}$

$t = \dfrac{6 \pm 2\sqrt{7}}{4}$

$t = \dfrac{3 \pm \sqrt{7}}{2}$

27. $x^2 - 4x - 1 = 0$
$\quad x^2 - 4x = 1$
$\quad x^2 - 4x + 4 = 5$
$\quad (x - 2)^2 = 5$
$\quad\quad x - 2 = \pm\sqrt{5}$
$\quad\quad\quad\quad x = 2 \pm \sqrt{5}$

29. $2r^2 + 10r + 11 = 0$

$r^2 + 5r + \dfrac{11}{2} = 0$

$r^2 + 5r = -\dfrac{11}{2}$

$r^2 + 5r + \dfrac{25}{4} = -\dfrac{11}{2} + \dfrac{25}{4}$

$\left(r + \dfrac{5}{2}\right)^2 = \dfrac{3}{4}$

$r + \dfrac{5}{2} = \pm\sqrt{\dfrac{3}{4}}$

$r = -\dfrac{5}{2} \pm \dfrac{\sqrt{3}}{\sqrt{4}}$

$r = -\dfrac{5}{2} \pm \dfrac{\sqrt{3}}{2}$

$r = \dfrac{-5 \pm \sqrt{3}}{2}$

31. $4u^2 + 8u + 15 = 0$

$u^2 + 2u + \dfrac{15}{4} = 0$

$u^2 + 2u = -\dfrac{15}{4}$

$u^2 + 2u + 1 = -\dfrac{15}{4} + 1$

$(u + 1)^2 = -\dfrac{11}{4}$

$u + 1 = \pm\sqrt{-\dfrac{11}{4}}$

$u = -1 \pm \dfrac{i\sqrt{11}}{\sqrt{4}}$

$u = -\dfrac{2}{2} \pm \dfrac{i\sqrt{11}}{2}$

$u = \dfrac{-2 \pm i\sqrt{11}}{2}$

33.
$$3w^2 + 4w + 3 = 0$$
$$w^2 + \frac{4}{3}w + 1 = 0$$
$$w^2 + \frac{4}{3}w \qquad = -1$$
$$w^2 + \frac{4}{3}w + \frac{4}{9} = -1 + \frac{4}{9}$$
$$\left(w + \frac{2}{3}\right)^2 = -\frac{5}{9}$$
$$w + \frac{2}{3} = \pm\sqrt{-\frac{5}{9}}$$
$$w = -\frac{2}{3} \pm \frac{i\sqrt{5}}{\sqrt{9}}$$
$$w = -\frac{2}{3} \pm \frac{i\sqrt{5}}{3}$$
$$w = \frac{-2 \pm i\sqrt{5}}{3}$$

35.
$$12x^2 + 7x = 10$$
$$12x^2 + 7x - 10 = 0$$
$$(4x + 5)(3x - 2) = 0 \quad \text{Polynomial is factorable.}$$
$$4x + 5 = 0 \text{ or } 3x - 2 = 0$$
$$4x = -5 \qquad 3x = 2$$
$$x = -\frac{5}{4} \qquad x = \frac{2}{3}$$

37. $(2y - 3)^2 = 5$ Format for the square root method.
$$2y - 3 = \pm\sqrt{5}$$
$$2y = 3 \pm \sqrt{5}$$
$$y = \frac{3 \pm \sqrt{5}}{2}$$

39.
$$x^2 = 3x + 1$$
$$x^2 - 3x - 1 = 0 \quad \text{Polynomial is not factorable, use quadratic formula.}$$
$$x = \frac{-b \pm \sqrt{b^2 - 4ac}}{2a} \quad a = 1, \ b = -3, \ c = -1$$
$$x = \frac{-(-3) \pm \sqrt{(-3)^2 - 4(1)(-1)}}{2(1)}$$
$$x = \frac{3 \pm \sqrt{13}}{2}$$

41.
$$7n^2 = -4n$$
$$7n^2 + 4n = 0$$
$$n(7n + 4) = 0 \quad \text{Polynomial is factorable.}$$
$$n = 0 \text{ or } 7n + 4 = 0$$
$$7n = -4$$
$$n = -\frac{4}{7}$$

43.
$$1 + \frac{8}{x^2} = \frac{4}{x} \quad \text{Excluded value: } x \neq 0$$
$$x^2 + 8 = 4x$$
$$x^2 - 4x + 8 = 0 \quad \text{Polynomial is not factorable, use quadratic formula, or complete the square.}$$
$$x^2 - 4x = -8$$
$$x^2 - 4x + 4 = -4$$
$$(x - 2)^2 = -4$$
$$x - 2 = \pm\sqrt{-4}$$
$$x - 2 = \pm i\sqrt{4}$$
$$x - 2 = \pm 2i$$
$$x = 2 \pm 2i$$

45.
$$\frac{24}{10 + m} + 1 = \frac{24}{10 - m} \quad \text{Excluded value: } m \neq -10, \ 10: \text{ LCD is } (10 + m)(10 - m)$$
$$(10 + m)(10 - m)\frac{24}{10 + m} + (10 + m)(10 - m) = (10 + m)(10 - m)\frac{24}{10 - m}$$
$$24(10 - m) + 100 - m^2 = 24(10 + m)$$
$$240 - 24m + 100 - m^2 = 240 + 24m$$
$$340 - 24m - m^2 = 240 + 24m$$
$$0 = m^2 + 48m - 100$$
$$m^2 + 48m - 100 = 0 \quad \text{Polynomial is factorable.}$$
$$(m + 50)(m - 2) = 0$$
$$m + 50 = 0 \quad \text{or} \quad m - 2 = 0$$
$$m = -50 \qquad m = 2$$

47. $\dfrac{2}{x - 2} = \dfrac{4}{x - 3} - \dfrac{1}{x + 1}$ Excluded values: $x \ne 2,\ 3,\ -1$

$$(x - 2)(x - 3)(x + 1)\dfrac{2}{x - 2} = (x - 2)(x - 3)(x + 1)\dfrac{4}{x - 3} - (x - 2)(x - 3)(x + 1)\dfrac{1}{x + 1}$$

$$2(x - 3)(x + 1) = 4(x - 2)(x + 1) - (x - 2)(x - 3)$$
$$2(x^2 - 2x - 3) = 4(x^2 - x - 2) - (x^2 - 5x + 6)$$
$$2x^2 - 4x - 6 = 4x^2 - 4x - 8 - x^2 + 5x - 6$$
$$2x^2 - 4x - 6 = 3x^2 + x - 14$$
$$0 = x^2 + 5x - 8$$

$x^2 + 5x - 8 = 0$ Polynomial is not factorable, use quadratic formula.

$$x = \dfrac{-b \pm \sqrt{b^2 - 4ac}}{2a} \quad a = 1,\ b = 5,\ c = -8$$

$$x = \dfrac{-5 \pm \sqrt{(5)^2 - 4(1)(-8)}}{2(1)} = \dfrac{-5 \pm \sqrt{57}}{2}$$

49. $\dfrac{x + 2}{x + 3} - \dfrac{x^2}{x^2 - 9} = 1 - \dfrac{x - 1}{3 - x}$ Excluded values: $x \ne 3,\ -3$

$$(x - 3)(x + 3)\dfrac{(x + 2)}{x + 3} - (x - 3)(x + 3)\dfrac{x^2}{x^2 - 9} = (x - 3)(x + 3) - (x - 3)(x + 3)\dfrac{x - 1}{3 - x}$$

$$(x - 3)(x + 2) - x^2 = x^2 - 9 + (x - 1)(x + 3)$$
$$x^2 - x - 6 - x^2 = x^2 - 9 + x^2 + 2x - 3$$
$$-x - 6 = 2x^2 + 2x - 12$$
$$0 = 2x^2 + 3x - 6$$

$2x^2 + 3x - 6 = 0$ Polynomial is not factorable, use quadratic formula.

$$x = \dfrac{-b \pm \sqrt{b^2 - 4ac}}{2a} \quad a = 2,\ b = 3,\ c = -6$$

$$x = \dfrac{-3 \pm \sqrt{(3)^2 - 4(2)(-6)}}{2(2)}$$

$$x = \dfrac{-3 \pm \sqrt{57}}{4}$$

51. According to Theorem 3, Section 1-4, $|3u - 2| = u^2$ is equivalent to $3u - 2 = u^2$ or $3u - 2 = -u^2$. Hence we must solve:

$3u - 2 = u^2$ or $3u - 2 = -u^2$

$0 = u^2 - 3u + 2$ $u^2 + 3u - 2 = 0$

$0 = (u - 1)(u - 2)$ $u = \dfrac{-b \pm \sqrt{b^2 - 4ac}}{2a} \quad a = 1,\ b = 3,\ c = -2$

$u - 1 = 0$ or $u - 2 = 0$

$u = 1$ $u = 2$

$$u = \dfrac{-3 \pm \sqrt{(3)^2 - 4(1)(-2)}}{2(1)}$$

$$u = \dfrac{-3 \pm \sqrt{17}}{2}$$

53. $s = \dfrac{1}{2}gt^2$

$\dfrac{1}{2}gt^2 = s$

$gt^2 = 2s$

$t^2 = \dfrac{2s}{g}$

$t = \sqrt{\dfrac{2s}{g}}$

55. $P = EI - RI^2$

$RI^2 - EI + P = 0$

$$I = \dfrac{-b \pm \sqrt{b^2 - 4ac}}{2a} \quad a = R,\ b = -E,\ c = P$$

$$I = \dfrac{-(-E) \pm \sqrt{(-E)^2 - 4(R)(P)}}{2(R)}$$

$$I = \dfrac{E + \sqrt{E^2 - 4RP}}{2R} \quad \text{(positive square root)}$$

57. $2.07x^2 - 3.79x + 1.34 = 0$

$$x = \frac{-b \pm \sqrt{b^2 - 4ac}}{2a} \quad a = 2.07,\ b = -3.79,\ c = 1.34$$

$$x = \frac{-(-3.79) \pm \sqrt{(-3.79)^2 - 4(2.07)(1.34)}}{2(2.07)} = \frac{3.79 \pm 1.81}{4.14}$$

$$x = 1.35,\ 0.48$$

59. $4.83x^2 + 2.04x - 3.18 = 0$

$$x = \frac{-b \pm \sqrt{b^2 - 4ac}}{2a} \quad a = 4.83,\ b = 2.04,\ c = -3.18$$

$$x = \frac{-2.04 \pm \sqrt{(2.04)^2 - 4(4.83)(-3.18)}}{2(4.83)} = \frac{-2.04 \pm 8.10}{9.66}$$

$$x = -1.05,\ 0.63$$

61. In this problem, $a = 1$, $b = 4$, $c = c$. Thus, the discriminant $b^2 - 4ac = (4)^2 - 4(1)(c) = 16 - 4c$. Hence,
if $16 - 4c > 0$, thus $16 > 4c$ or $c < 4$, there are two distinct real roots.
if $16 - 4c = 0$, thus $c = 4$, there is one real double root,
and if $16 - 4c < 0$, thus $16 < 4c$ or $c > 4$, there are two distinct imaginary roots.

63. $0.0134x^2 + 0.0414x + 0.0304 = 0$
The discriminant is $b^2 - 4ac$, where $a = 0.0134$, $b = 0.0414$, $c = 0.0304$.
$b^2 - 4ac = (0.0414)^2 - 4(0.0134)(0.0304) = 8.45 \times 10^{-5} > 0$
Since the discriminant is positive, the equation has real solutions.

65. $0.0134x^2 + 0.0214x + 0.0304 = 0$
Here $a = 0.0134$, $b = 0.0214$, $c = 0.0304$.
$b^2 - 4ac = (0.0214)^2 - 4(0.0134)(0.0304) = -1.17 \times 10^{-3} < 0$
Since $b^2 - 4ac$, the discriminant, is negative, the equation has no real solutions.

67. $\sqrt{3}x^2 = 8\sqrt{2}x - 4\sqrt{3}$

$\sqrt{3}x^2 - 8\sqrt{2}x + 4\sqrt{3} = 0$

$$x = \frac{-b \pm \sqrt{b^2 - 4ac}}{2a} \quad a = \sqrt{3},$$
$$b = -8\sqrt{2}$$
$$c = 4\sqrt{3}$$

$$x = \frac{-(-8\sqrt{2}) \pm \sqrt{(-8\sqrt{2})^2 - 4(\sqrt{3})(4\sqrt{3})}}{2(\sqrt{3})}$$

$$x = \frac{8\sqrt{2} \pm \sqrt{(64)(2) - (16)(3)}}{2\sqrt{3}}$$

$$x = \frac{8\sqrt{2} \pm \sqrt{80}}{2\sqrt{3}}$$

$$x = \frac{8\sqrt{2} \pm 4\sqrt{5}}{2\sqrt{3}}$$

$$x = \frac{2(4\sqrt{2} \pm 2\sqrt{5})}{2\sqrt{3}} \cdot \frac{\sqrt{3}}{\sqrt{3}}$$

$$x = \frac{4\sqrt{6} \pm 2\sqrt{15}}{3} \text{ or } \frac{4}{3}\sqrt{6} \pm \frac{2}{3}\sqrt{15}$$

69. $x^2 + 2ix = 3$

$x^2 + 2ix - 3 = 0$

$$x = \frac{-b \pm \sqrt{b^2 - 4ac}}{2a} \quad a = 1$$
$$b = 2i$$
$$c = -3$$

$$x = \frac{-2i \pm \sqrt{(2i)^2 - 4(1)(-3)}}{2(1)}$$

$$x = \frac{-2i \pm \sqrt{-4 + 12}}{2}$$

$$x = \frac{-2i \pm \sqrt{8}}{2}$$

$$x = \frac{-2i \pm 2\sqrt{2}}{2}$$

$$x = \frac{2(-i \pm \sqrt{2})}{2}$$

$$x = -i \pm \sqrt{2}$$

$$x = \sqrt{2} - i,\ -\sqrt{2} - i$$

71. $x^3 - 1 = 0$

$(x - 1)(x^2 + x + 1) = 0$

$x - 1 = 0$ or $x^2 + x + 1 = 0$

$x = 1$ $x = \dfrac{-b \pm \sqrt{b^2 - 4ac}}{2a}$ $a = 1, \ b = 1, \ c = 1$

$\qquad\qquad x = \dfrac{-1 \pm \sqrt{(1)^2 - 4(1)(1)}}{2(1)}$

$\qquad\qquad x = \dfrac{-1 \pm \sqrt{1 - 4}}{2}$

$\qquad\qquad x = \dfrac{-1 \pm \sqrt{-3}}{2}$

$\qquad\qquad x = \dfrac{-1 \pm i\sqrt{3}}{2}$ or $-\dfrac{1}{2} \pm \dfrac{1}{2}i\sqrt{3}$

73. If a quadratic equation has two roots, they are $\dfrac{-b + \sqrt{b^2 - 4ac}}{2a}$ and

$\dfrac{-b - \sqrt{b^2 - 4ac}}{2a}$. If a, b, c are rational, then so are $-b$, $2a$, and $b^2 - 4ac$.

Then, *either* $\sqrt{b^2 - 4ac}$ is rational, hence $\dfrac{-b + \sqrt{b^2 - 4ac}}{2a}$ and $\dfrac{-b - \sqrt{b^2 - 4ac}}{2a}$ are

both rational, *or*, $\sqrt{b^2 - 4ac}$ is irrational, hence $\dfrac{-b + \sqrt{b^2 - 4ac}}{2a}$ and

$\dfrac{-b - \sqrt{b^2 - 4ac}}{2a}$ are both irrational , *or*, $\sqrt{b^2 - 4ac}$ is imaginary, hence

$\dfrac{-b + \sqrt{b^2 - 4ac}}{2a}$ and $\dfrac{-b + \sqrt{b^2 - 4ac}}{2a}$ are both imaginary. There is no other

possibility; hence, one root cannot be rational while the other is irrational.

75. $r_1 = \dfrac{-b + \sqrt{b^2 - 4ac}}{2a}$ $r_2 = \dfrac{-b - \sqrt{b^2 - 4ac}}{2a}$

$r_1 r_2 = \dfrac{(-b + \sqrt{b^2 - 4ac})}{2a} \dfrac{(-b - \sqrt{b^2 - 4ac})}{2a}$

$= \dfrac{(-b)^2 - (\sqrt{b^2 - 4ac})^2}{4a^2} = \dfrac{b^2 - (b^2 - 4ac)}{4a^2} = \dfrac{b^2 - b^2 + 4ac}{4a^2} = \dfrac{4ac}{4a^2} = \dfrac{c}{a}$

77. The $\pm$ in front still yields the same two numbers even if a is negative.

79. Let x = one number.
Since their sum is 21,
$21 - x$ = other number
Then, since their product is 104,
$\qquad x(21 - x) = 104$
$\qquad\qquad 21x - x^2 = 104$
$\qquad\qquad\quad 0 = x^2 - 21x + 104$
$\quad x^2 - 21x + 104 = 0$
$\quad (x - 13)(x - 8) = 0$
$\qquad\quad x - 13 = 0$ or $x - 8 = 0$
$\qquad\qquad\quad x = 13 \qquad\qquad x = 8$
The numbers are 8 and 13.

81. Let x = first of the two
consecutive even integers.
Then $x + 2$ = second of these integers
Since their product is 168,
$\qquad\quad x(x + 2) = 168$
$\qquad\quad x^2 + 2x = 168$
$\quad x^2 + 2x - 168 = 0$
$(x - 12)(x + 14) = 0$
$\qquad x - 12 = 0$ or $x + 14 = 0$
$\qquad\qquad x = 12 \qquad\qquad x = -14$
If $x = 12$, the two consecutive positive
even integers must be 12 and 14. We
discard the other solution, since the
numbers must be positive.

83. Let x = amount of increase
Then $x + 4$ = length of new rectangle
    $x + 2$ = width of new rectangle

Using $A = ab$, we have

Area of new rectangle = 2 × Area of old rectangle.

$(x + 4)(x + 2) = 2(4)(2)$
$x^2 + 6x + 8 = 16$
$x^2 + 6x - 8 = 0$
  $x^2 + 6x = 8$
$x^2 + 6x + 9 = 17$
  $(x + 3)^2 = 17$
    $x + 3 = \pm\sqrt{17}$
      $x = -3 \pm \sqrt{17}$

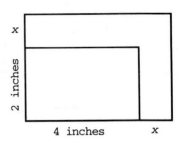

4 inches

We discard the negative solution and take $x = -3 + \sqrt{17} = 1.12$. Then dimensions of new rectangle are $x + 4$ by $x + 2$ or 5.12 by 3.12 inches.

85. Let r = interest rate. Applying the given formula, we have
    $1440 = 1000(1 + r)^2$
    $1.44 = (1 + r)^2$
$(1 + r)^2 = 1.44$
  $1 + r = \pm\sqrt{1.44}$
  $1 + r = \pm 1.2$
    $r = -1 \pm 1.2$
Discarding the negative root, we have
$r = .2$ or 20%.

87. Let r = rate of slow plane.
Then $r + 140$ = rate of fast plane.
After 1 hour $r(1) = r$ = distance traveled by slow plane.
$(r + 140)(1) = r + 140$ = distance traveled by fast plane.

Applying the Pythagorean theorem, we have
    $r^2 + (r + 140)^2 = 260^2$
$r^2 + r^2 + 280r + 19,600 = 67,600$
  $2r^2 + 280r - 48,000 = 0$
    $r^2 + 140r - 24,000 = 0$
    $(r + 240)(r - 100) = 0$
        $r + 240 = 0$ or $r - 100 = 0$
            $r = -240$ $r = 100$

Slow Plane after 1 hour

260 miles

Airport   Fast Plane after 1 hour

Discarding the negative solution, we have

    $r = 100$ miles per hour = rate of slow plane
$r + 140 = 240$ miles per hour = rate of fast plane.

89. Let t = time for smaller pipe to fill tank alone
    $t - 5$ = time for larger pipe to fill tank alone
    5 = time for both pipes to fill tank together

Then $\dfrac{1}{t}$ = rate for smaller pipe

$\dfrac{1}{t - 5}$ = rate for larger pipe

$\begin{pmatrix}\text{Part of job}\\\text{completed by}\\\text{smaller pipe}\end{pmatrix} + \begin{pmatrix}\text{Part of job}\\\text{completed by}\\\text{larger pipe}\end{pmatrix} = 1$ whole job

    $\dfrac{1}{t}(5)$ + $\dfrac{1}{t - 5}(5)$ $= 1$

    $\dfrac{5}{t}$ + $\dfrac{5}{t - 5}$ $= 1$ Excluded values: $t \neq 0, 5$

$$t(t - 5)\frac{5}{t} \quad + t(t - 5)\frac{5}{t - 5} = t(t - 5)$$
$$5(t - 5) + 5t = t(t - 5)$$
$$5t - 25 + 5t = t^2 - 5t$$
$$10t - 25 = t^2 - 5t$$
$$0 = t^2 - 15t + 25$$
$$t^2 - 15t + 25 = 0$$
$$t = \frac{-b \pm \sqrt{b^2 - 4ac}}{2a} \quad a = 1, \ b = -15, \ c = 25$$
$$t = \frac{-(-15) \pm \sqrt{(-15)^2 - 4(1)(25)}}{2(1)} = \frac{15 \pm \sqrt{125}}{2}$$
$$t = 13.09, \ 1.91$$
$$t - 5 = 8.09, \ -3.09$$

Discarding the answer for t which results in a negative answer for $t - 5$, we have 13.09 hours for smaller pipe alone, 8.09 hours for larger pipe alone.

91. Let v = speed of car. Applying the given formula, we have
$$165 = 0.044v^2 + 1.1v$$
$$0 = 0.044v^2 + 1.1v - 165$$
$$0.044v^2 + 1.1v - 165 = 0$$
$$v = \frac{-b \pm \sqrt{b^2 - 4ac}}{2a} \quad a = 0.044, \ b = 1.1, \ c = -165$$
$$v = \frac{-1.1 \pm \sqrt{(1.1)^2 - 4(0.044)(-165)}}{2(0.044)} = \frac{-1.1 \pm 5.5}{0.088}$$
$$v = -75 \text{ or } 50$$

Discarding the negative answer, we have $v = 50$ miles per hour.

93. Let ℓ = length of building
w = width of building.
Then, using the hint, in the similar triangles
ABC and AFE, we have

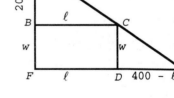

$$\frac{\ell}{200 - w} = \frac{400}{200}$$
$$\frac{\ell}{200 - w} = 2$$
$$\ell = 2(200 - w)$$
$$\ell = 400 - 2w$$
Since the cross-sectional area of the building is given as 15,000 ft^2, we have
$$\ell w = 15,000$$
Since the cross-sectional area of the building is given as 15,000 ft^2, we have
$$\ell w = 15,000$$
Substituting, we get,
$$(400 - 2w)w = 15,000$$
$$400w - 2w^2 = 15,000$$
$$0 = 2w^2 - 400w + 15,000$$
$$0 = w^2 - 200w + 7,500$$
$$0 = (w - 50)(w - 150)$$
$$w - 50 = 0 \quad \text{or } w - 150 = 0$$
$$w = 50 \qquad\qquad w = 150$$
$$\ell = 400 - 2w \qquad \ell = 400 - 2w$$
$$= 400 - 2(50) \qquad = 400 - 2(150)$$
$$= 300 \qquad\qquad = 100$$

Thus, there are two solutions: the building is 50 ft wide and 300 ft long or 150 ft wide and 100 ft long.

95. Let x = distance from the warehouse to Factory A. Since the distance from the warehouse to Factory B via Factory A is known (it is the difference in odometer readings: 52937—52846) to be 91 miles, then

$91 - x$ = distance from Factory A to Factory B.

The distance from Factory B to the warehouse is known (it is the difference in odometer readings: 53002—52937) to be 65 miles. Applying the Pythagorean theorem, we have

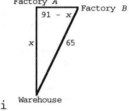

$$x^2 + (91 - x)^2 = 65^2$$
$$x^2 + 8281 - 182x + x^2 = 4225$$
$$2x^2 - 182x + 4056 = 0$$
$$x^2 - 91x + 2028 = 0$$
$$(x - 52)(x - 39) = 0$$
$$x - 52 = 0 \quad \text{or} \quad x - 39 = 0$$
$$x = 52 \text{ mi} \qquad x = 39 \text{ mi}$$

Since we are told that the distance from the warehouse to Factory A was greater than the distance from Factory A to Factory B, we discard the solution $x = 39$, which would lead to $91 - x = 52$ miles, a contradiction.

52 miles.

Exercise 1-7

Key Ideas and Formulas

1. If both sides of an equation are squared, then the solution set of the original equation is a subset of the solution set of the new equation.

   Any new equation obtained by raising both members of an equation to the same power may have solutions (called *extraneous solutions*) that are not solutions of the original equation.

   Thus, every solution of the new equation must be checked in the original equation to eliminate extraneous solutions.

2. To solve an equation that is not quadratic, it may be possible to transform it to the form

$$au^2 + bu + c = 0$$

   where u is an expression in some other variable.

---

1. This statement is true, since $\pm\sqrt{5}$ are the only two solutions of $x^2 = 5$. T

3. This statement is false.

$(\sqrt{x - 1} + 1)^2$ is not equal to $x - 1 + 1$ or x, in general.
In fact
$(\sqrt{x - 1} + 1)^2 = x - 1 + 2\sqrt{x - 1} + 1 = x + 2\sqrt{x - 1}$.
This is only equal to x in case $x = 1$.   F

5. This statement is false. If $x^3 = 2$ then $x = \sqrt[3]{2}$ or x is equal to one of two non-real complex numbers whose cube is 2. If $x = 8$, x^3 must equal 512.   F

7. $\sqrt{x + 2} = 4$
$x + 2 = 16$
$x = 14$
Check:
$\sqrt{14 + 2} \overset{?}{=} 4$
$\sqrt{16} \overset{?}{=} 4$
$4 \overset{\checkmark}{=} 4$
Solution: 14

9. $\sqrt{3y - 2} = y - 2$
$3y - 2 = y^2 - 4y + 4$
$0 = y^2 - 7y + 6$
$y^2 - 7y + 6 = 0$
$(y - 1)(y - 6) = 0$
$y = 1, 6$

Check: 1: $\sqrt{3(1) - 2} \overset{?}{=} 1 - 2$ 6: $\sqrt{3(6) - 2} \overset{?}{=} 6 - 2$
$\sqrt{1} \overset{?}{=} -1$ $\sqrt{16} \overset{?}{=} 4$
$1 \neq -1$ $4 \overset{\checkmark}{=} 4$
Not a solution   A solution
Solution: 6

11.
$$\sqrt{5w + 6} - w = 2$$
$$\sqrt{5w + 6} = w + 2$$
$$5w + 6 = w^2 + 4w + 4$$
$$0 = w^2 - w - 2$$
$$w^2 - w - 2 = 0$$
$$(w - 2)(w + 1) = 0$$
$$w = 2, -1$$

Common Error: $5w + 6 - w^2 = 4$
or $5w + 6 + w^2 = 4$
are not equivalent to the original equation.

Check: 2: $\sqrt{5(2) + 6} - 2 \overset{?}{=} 2$ -1: $\sqrt{5(-1) + 6} - (-1) \overset{?}{=} 2$
$$4 - 2 = 2 \quad \checkmark \qquad\qquad 1 + 1 = 2 \quad \checkmark$$
A solution   A solution
Solution: 2, -1

13.
$$\sqrt{3t - 2} = 1 - 2\sqrt{t}$$
$$3t - 2 = 1 - 4\sqrt{t} + 4t$$
$$-t - 3 = -4\sqrt{t}$$
$$t^2 + 6t + 9 = 16t$$
$$t^2 - 10t + 9 = 0$$
$$(t - 9)(t - 1) = 0$$
$$t = 9, 1$$

Check: 9: $\sqrt{3 \cdot 9 - 2} \overset{?}{=} 1 - 2\sqrt{9}$ 1: $\sqrt{3 \cdot 1 - 2} \overset{?}{=} 1 - 2\sqrt{1}$
$$\sqrt{25} \overset{\checkmark}{=} 1 - 6 \qquad\qquad \sqrt{1} \overset{\checkmark}{=} 1 - 2$$
$$5 \neq -5 \qquad\qquad\qquad 1 \neq -1$$
Not a solution   Not a solution

No solution

15. $m^4 + 2m^2 - 15 = 0$
Let $u = m^2$, then
$$u^2 + 2u - 15 = 0$$
$$(u + 5)(u - 3) = 0$$
$$u = -5, 3$$
$$m^2 = -5 \qquad m^2 = 3$$
$$m = \pm i\sqrt{5} \qquad m = \pm\sqrt{3}$$

17. $3x = \sqrt{x^2 - 2}$
$$9x^2 = x^2 - 2$$
$$8x^2 = -2$$
$$x^2 = -\frac{1}{4}$$
$$x = \pm\sqrt{-\frac{1}{4}}$$
$$x = \pm\frac{1}{2}i$$

Check: $\frac{1}{2}i$: $3\left(\frac{1}{2}i\right) \overset{?}{=} \sqrt{\left(\frac{1}{2}i\right)^2 - 2}$ $-\frac{1}{2}i$: $3\left(-\frac{1}{2}i\right) \overset{?}{=} \sqrt{\left(-\frac{1}{2}i\right)^2 - 2}$
$$\frac{3}{2}i \overset{?}{=} \sqrt{-\frac{1}{4} - 2} \qquad\qquad -\frac{3}{2}i \overset{?}{=} \sqrt{-\frac{1}{4} - 2}$$
$$\frac{3}{2}i \overset{?}{=} \sqrt{-\frac{9}{4}} \qquad\qquad -\frac{3}{2}i \overset{?}{=} \sqrt{-\frac{9}{4}}$$
$$\frac{3}{2}i \overset{\checkmark}{=} \frac{3}{2}i \qquad\qquad -\frac{3}{2}i \neq \frac{3}{2}i$$
A solution   Not a solution
Solution: $\frac{1}{2}i$

19. $2y^{2/3} + 5y^{1/3} - 12 = 0$
Let $u = y^{1/3}$, then
$$2u^2 + 5u - 12 = 0$$
$$(2u - 3)(u + 4) = 0$$
$$2u - 3 = 0 \qquad u + 4 = 0$$
$$u = \frac{3}{2} \qquad\qquad u = -4$$
$$y^{1/3} = \frac{3}{2} \qquad y^{1/3} = -4$$
$$y = \frac{27}{8} \qquad\qquad y = -64$$

21. $(m^2 - 2m)^2 + 2(m^2 - 2m) = 15$
Let $u = m^2 - 2m$, then
$$u^2 + 2u = 15$$
$$u^2 + 2u - 15 = 0$$
$$(u + 5)(u - 3) = 0$$
$$u = -5 \qquad\qquad\qquad u = 3$$
$$m^2 - 2m = -5 \qquad\qquad m^2 - 2m = 3$$
$$m^2 - 2m + 1 = -4 \qquad m^2 - 2m - 3 = 0$$
$$(m - 1)^2 = -4 \quad (m - 3)(m + 1) = 0$$
$$m - 1 = \pm 2i \qquad m - 3 = 0 \qquad m + 1 = 0$$
$$m = 1 \pm 2i \qquad m = 3 \qquad\qquad m = -1$$

23. $\sqrt{2t + 3} + 2 = \sqrt{t - 2}$

$2t + 3 + 4\sqrt{2t + 3} + 4 = t - 2$

$2t + 7 + 4\sqrt{2t + 3} = t - 2$

$4\sqrt{2t + 3} = -t - 9$

$16(2t + 3) = t^2 + 18t + 81$

$32t + 48 = t^2 + 18t + 81$

$0 = t^2 - 14t + 33$

$0 = (t - 11)(t - 3)$

$t = 11 \qquad t = 3$

Common Error: $2t + 3 + 4 = t - 2$ is not an equivalent equation to the given equation $(\sqrt{2t + 3} + 2)^2 \neq 2t + 3 + 4$

Check: 11: $\sqrt{2(11) + 3} + 2 \overset{?}{=} \sqrt{11 - 2}$

$\sqrt{25} + 2 \overset{?}{=} \sqrt{9}$

$5 + 2 \neq 3$

Not a solution

3: $\sqrt{2(3) + 3} + 2 \overset{?}{=} \sqrt{3 - 2}$

$\sqrt{9} + 2 \overset{?}{=} \sqrt{1}$

$3 + 2 \neq 1$

Not a solution

No solution

25. $\sqrt{w + 3} + \sqrt{2 - w} = 3$

$\sqrt{w + 3} = 3 - \sqrt{2 - w}$

$w + 3 = 9 - 6\sqrt{2 - w} + 2 - w$

$w + 3 = 11 - w - 6\sqrt{2 - w}$

$2w - 8 = -6\sqrt{2 - w}$

$w - 4 = -3\sqrt{2 - w}$

$w^2 - 8w + 16 = 9(2 - w)$

$w^2 - 8w + 16 = 18 - 9w$

$w^2 + w - 2 = 0$

$(w + 2)(w - 1) = 0$

$w = -2 \qquad w = 1$

Check: -2: $\sqrt{(-2) + 3} + \sqrt{2 - (-2)} \overset{?}{=} 3$

$\sqrt{1} + \sqrt{4} \overset{?}{=} 3$

$1 + 2 \overset{\checkmark}{=} 3$

A solution

1: $\sqrt{1 + 3} + \sqrt{2 - 1} \overset{?}{=} 3$

$\sqrt{4} + \sqrt{1} \overset{\checkmark}{=} 3$

A solution

Solution: -2, 1

27. $\sqrt{8 - z} = 1 + \sqrt{z + 5}$

$8 - z = 1 + 2\sqrt{z + 5} + z + 5$

$8 - z = z + 6 + 2\sqrt{z + 5}$

$2 - 2z = 2\sqrt{z + 5}$

$1 - z = \sqrt{z + 5}$

$z^2 - 2z + 1 = z + 5$

$z^2 - 3z - 4 = 0$

$(z - 4)(z + 1) = 0$

$z = 4 \qquad z = -1$

Check: 4: $\sqrt{8 - 4} \overset{?}{=} 1 + \sqrt{4 + 5}$

$\sqrt{4} \overset{?}{=} 1 + \sqrt{9}$

$2 \neq 1 + 3$

Not a solution

-1: $\sqrt{8 - (-1)} \overset{?}{=} 1 + \sqrt{(-1) + 5}$

$\sqrt{9} \overset{?}{=} 1 + \sqrt{4}$

$3 \overset{\checkmark}{=} 1 + 2$

A solution

Solution: -1

29. $\sqrt{4x^2 + 12x + 1} - 6x = 9$

$\sqrt{4x^2 + 12x + 1} = 6x + 9$

$4x^2 + 12x + 1 = 36x^2 + 108x + 81$

$0 = 32x^2 + 96x + 80$

$0 = 2x^2 + 6x + 5$

$x = \dfrac{-b \pm \sqrt{b^2 - 4ac}}{2a} \qquad a = 2 \quad b = 6 \quad c = 5$

$x = \dfrac{-6 \pm \sqrt{6^2 - 4(2)(5)}}{2(2)}$

$x = \dfrac{-6 \pm \sqrt{36 - 40}}{4}$

$x = \dfrac{-6 \pm \sqrt{-4}}{4}$

$x = \dfrac{-6 \pm 2i}{4}$

$x = -\dfrac{3}{2} \pm \dfrac{1}{2}i$

Check: $-\frac{3}{2} + \frac{1}{2}i$: $\sqrt{4\left(-\frac{3}{2} + \frac{1}{2}i\right)^2 + 12\left(-\frac{3}{2} + \frac{1}{2}i\right) + 1} - 6\left(-\frac{3}{2} + \frac{1}{2}i\right) \stackrel{?}{=} 9$

$$\sqrt{(-3 + i)^2 - 18 + 6i + 1} + 9 - 3i \stackrel{?}{=} 9$$

$$\sqrt{9 - 6i + i^2 - 18 + 6i + 1} + 9 - 3i \stackrel{?}{=} 9$$

$$\sqrt{9 - 6i - 1 - 18 + 6i + 1} + 9 - 3i \stackrel{?}{=} 9$$

$$\sqrt{-9} + 9 - 3i \stackrel{?}{=} 9$$

$$3i + 9 - 3i \stackrel{\sqrt{}}{=} 9$$

A solution

$-\frac{3}{2} - \frac{1}{2}i$: $\sqrt{4\left(-\frac{3}{2} - \frac{1}{2}i\right)^2 + 12\left(-\frac{3}{2} - \frac{1}{2}i\right) + 1} - 6\left(-\frac{3}{2} - \frac{1}{2}i\right) \stackrel{?}{=} 9$

$$\sqrt{(-3 - i)^2 - 18 - 6i + 1} + 9 + 3i \stackrel{?}{=} 9$$

$$\sqrt{9 + 6i + i^2 - 18 - 6i + 1} + 9 + 3i \stackrel{?}{=} 9$$

$$\sqrt{9 + 6i - 1 - 18 - 6i + 1} + 9 + 3i \stackrel{?}{=} 9$$

$$\sqrt{-9} + 9 + 3i \stackrel{?}{=} 9$$

$$3i + 9 + 3i \stackrel{?}{=} 9$$

$$9 + 6i \neq 9 \quad \text{Not a solution}$$

Solution: $-\frac{3}{2} + \frac{1}{2}i$

31. $y^{-2} - 2y^{-1} + 3 = 0$ or alternatively, write

Let $u = y^{-1}$, then

$u^2 - 2u + 3 = 0$ $\frac{1}{y^2} - \frac{2}{y} + 3 = 0$ $y \neq 0$ LCD $= y^2$

$u^2 - 2u = -3$ $1 - 2y + 3y^2 = 0$

$u^2 - 2u + 1 = -2$ $3y^2 - 2y + 1 = 0$

$(u - 1)^2 = -2$

$u - 1 = \pm i\sqrt{2}$ $y = \frac{-b \pm \sqrt{b^2 - 4ac}}{2a}$ $a = 3, \; b = -2, \; c = 1$

$u = 1 \pm i\sqrt{2}$

$y^{-1} = 1 \pm i\sqrt{2}$ $y = \frac{-(-2) \pm \sqrt{(-2)^2 - 4(1)(3)}}{2(3)}$

$\frac{1}{y} = 1 \pm i\sqrt{2}$

$y = \frac{1}{1 \pm i\sqrt{2}}$ $y = \frac{2 \pm \sqrt{-8}}{6}$

$y = \frac{1}{(1 \pm i\sqrt{2})} \frac{(1 \mp i\sqrt{2})}{(1 \mp i\sqrt{2})}$ $y = \frac{2 \pm i2\sqrt{2}}{6}$

$y = \frac{1 \mp i\sqrt{2}}{1 - (-2)}$ $y = \frac{1 \pm i\sqrt{2}}{3}$

$y = \frac{1 \pm i\sqrt{2}}{3}$

33. $2t^{-4} - 5t^{-2} + 2 = 0$
Let $u = t^{-2}$, then
$$2u^2 - 5u + 2 = 0$$
$$(2u - 1)(u - 2) = 0$$

$2u - 1 = 0$ or $u - 2 = 0$

$u = \dfrac{1}{2}$ $u = 2$

$t^{-2} = \dfrac{1}{2}$ $t^{-2} = 2$

$t^2 = 2$ $t^2 = \dfrac{1}{2}$

$t = \pm\sqrt{2}$ $t = \pm\sqrt{\dfrac{1}{2}}$

$t = \pm\dfrac{1}{\sqrt{2}}$

$t = \pm\dfrac{\sqrt{2}}{2}$

35. $3z^{-1} - 3z^{-1/2} + 1 = 0$
Let $u = z^{-1/2}$, then
$3u^2 - 3u + 1 = 0$

$u = \dfrac{-b \pm \sqrt{b^2 - 4ac}}{2a}$ $a = 3,\ b = -3,$
$c = 1$

$u = \dfrac{-(-3) \pm \sqrt{(-3)^2 - 4(3)(1)}}{2(3)}$

$u = \dfrac{3 \pm i\sqrt{3}}{6}$

$z^{-1/2} = \dfrac{3 \pm i\sqrt{3}}{6}$

$z^{-1} = \left(\dfrac{3 \pm i\sqrt{3}}{6}\right)^2$

$= \dfrac{9 \pm 2(3)i\sqrt{3} + (i\sqrt{3})^2}{36}$

$z^{-1} = \dfrac{9 \pm 6i\sqrt{3} - 3}{36}$

$z^{-1} = \dfrac{6 \pm 6i\sqrt{3}}{36}$

$z^{-1} = \dfrac{1 \pm i\sqrt{3}}{6}$

$z = \dfrac{6}{1 \pm i\sqrt{3}}$

$z = \dfrac{6}{(1 \pm i\sqrt{3})}\dfrac{(1 \mp i\sqrt{3})}{(1 \mp i\sqrt{3})}$

$z = \dfrac{6(1 \mp i\sqrt{3})}{1 - (-3)}$

$z = \dfrac{6(1 \pm i\sqrt{3})}{4}$ or $\dfrac{3 \pm 3i\sqrt{3}}{2}$

37. By squaring: $4m + 8\sqrt{m} - 5 = 0$
$4m - 5 = -8\sqrt{m}$
$16m^2 - 40m + 25 = 64m$
$16m^2 - 104m + 25 = 0$
$(4m - 1)(4m - 25) = 0$
$4m - 1 = 0$ or $4m - 25 = 0$
    $4m = 1$ $4m = 25$
     $m = 0.25$ $m = 6.25$

Check: $m = 0.25$

$4(0.25) + 8\sqrt{0.25} - 5 \overset{?}{=} 0$

$1 + 8(0.5) - 5 \overset{?}{=} 0$

$1 + 4 - 5 \overset{\checkmark}{=} 0$

A solution

$m = 6.25$

$4(6.25) + 8\sqrt{6.25} - 5 \overset{?}{=} 0$

$25 + 8(2.5) - 5 \overset{?}{=} 0$

$25 + 20 - 5 \overset{?}{=} 0$

$40 \neq 0$

Not a solution

By substitution:
$$4m + 8\sqrt{m} - 5 = 0$$
Let $u = \sqrt{m}$, then
$$4u^2 + 8u - 5 = 0$$
$$(2u + 5)(2u - 1) = 0$$
$$2u + 5 = 0 \quad \text{or} \quad 2u - 1 = 0$$
$$u = -2.5 \qquad\qquad u = 0.5$$
$$\sqrt{w} = -2.5 \qquad\qquad \sqrt{w} = 0.5$$
Impossible: $\qquad\qquad w = 0.25$
$\sqrt{w}$ must be positive   This solution has already been checked.

Solution: 0.25

39. $2w + 3\sqrt{w} = 14$

By squaring:
$$2w + 3\sqrt{w} - 14 = 0$$
$$2w - 14 = -3\sqrt{w}$$
$$4w^2 - 56w + 196 = 9w$$
$$4w^2 - 65w + 196 = 0$$
$$(4w - 49)(w - 4) = 0$$
$$4w - 49 = 0 \qquad w - 4 = 0$$
$$w = \frac{49}{4} \qquad\qquad w = 4$$

Check:

$w = \dfrac{49}{4}:\ 2\left(\dfrac{49}{4}\right) + 3\sqrt{\dfrac{49}{4}} \overset{?}{=} 14$

$\qquad\qquad \dfrac{49}{2} + 3\left(\dfrac{7}{2}\right) \overset{?}{=} 14$

$\qquad\qquad\qquad \dfrac{70}{2} \neq 14$

Not a solution

$w = 4:\ 2(4) + 3\sqrt{4} \overset{?}{=} 14$

$\qquad\qquad 8 + 6 \overset{\sqrt{}}{=} 14$

A solution

Solution: 4

By substitution:

Let $u = \sqrt{w}$, then
$$2u^2 + 3u = 14$$
$$2u^2 + 3u - 14 = 0$$
$$(2u + 7)(u - 2) = 0$$
$$2u + 7 = 0 \qquad u - 2 = 0$$
$$u = -\frac{7}{2} \qquad\qquad u = 2$$
$$\sqrt{w} = -\frac{7}{2} \qquad\qquad \sqrt{w} = 2$$
Impossible: $\qquad\qquad w = 4$
$\sqrt{w}$ must
be positive

Solution: 4 (This has already been checked.)

41. $\sqrt{7 - 2x} - \sqrt{x + 2} = \sqrt{x + 5}$
$$7 - 2x - 2\sqrt{7 - 2x}\sqrt{x + 2} + x + 2 = x + 5$$
$$-2\sqrt{7 - 2x}\sqrt{x + 2} - x + 9 = x + 5$$
$$-2\sqrt{7 - 2x}\sqrt{x + 2} = 2x - 4$$
$$-\sqrt{7 - 2x}\sqrt{x + 2} = x - 2$$
$$(7 - 2x)(x + 2) = x^2 - 4x + 4$$
$$7x + 14 - 2x^2 - 4x = x^2 - 4x + 4$$
$$0 = 3x^2 - 7x - 10$$
$$0 = (3x - 10)(x + 1)$$
$$3x - 10 = 0 \qquad\qquad x + 1 = 0$$
$$x = \frac{10}{3} \qquad\qquad\qquad x = -1$$

Check: $x = \frac{10}{3}$: $\sqrt{7 - 2\left(\frac{10}{3}\right)} - \sqrt{\frac{10}{3} + 2} \overset{?}{=} \sqrt{\frac{10}{3} + 5}$

$$\sqrt{\frac{21}{3} - \frac{20}{3}} - \sqrt{\frac{10}{3} + \frac{6}{3}} \overset{?}{=} \sqrt{\frac{10}{3} + \frac{15}{3}}$$

$$\sqrt{\frac{1}{3}} - \sqrt{\frac{16}{3}} \overset{?}{=} \sqrt{\frac{25}{3}}$$

$$\frac{1}{\sqrt{3}} - \frac{4}{\sqrt{3}} \neq \frac{5}{\sqrt{3}}$$

Not a solution

$$x = -1: \sqrt{7 - 2(-1)} - \sqrt{(-1) + 2} \overset{?}{=} \sqrt{(-1) + 5}$$

$$\sqrt{9} - \sqrt{1} \overset{?}{=} \sqrt{4}$$

$$3 - 1 \overset{\sqrt{}}{=} 2$$

A solution

Solution: -1

43.
$$3 + x^{-4} = 5x^{-2} \quad x \neq 0, \quad \text{LCD} = x^4$$
$$3x^4 + 1 = 5x^2$$
$$3x^4 - 5x^2 + 1 = 0$$
Let $u = x^2$, then
$$3u^2 - 5u + 1 = 0$$

$$u = \frac{-b \pm \sqrt{b^2 - 4ac}}{2a} \quad a = 3, \ b = -5, \ c = 1$$

$$u = \frac{-(-5) \pm \sqrt{(-5)^2 - 4(3)(1)}}{2(3)}$$

$$u = \frac{5 \pm \sqrt{13}}{6}$$

$$x^2 = \frac{5 \pm \sqrt{13}}{6}$$

$$x = \pm\sqrt{\frac{5 \pm \sqrt{13}}{6}} \quad \text{(four roots)}$$

45.
$$2\sqrt{x + 5} = 0.01x + 2.04$$
$$200\sqrt{x + 5} = x + 204$$
$$40,000(x + 5) = (x + 204)^2$$
$$40,000x + 200,000 = x^2 + 408x + 41,616$$
$$0 = x^2 - 39,592x - 158,384$$

Although this is factorable in the integers, one is unlikely to notice this or to detect the factors, so the quadratic formula is used

$$x = \frac{-b \pm \sqrt{b^2 - 4ac}}{2a} \quad a = 1$$
$$b = -39,592$$
$$c = -158,384$$

$$x = \frac{-(-39,592) \pm \sqrt{(-39,592)^2 - 4(1)(-158,384)}}{2(1)}$$

$$x = \frac{39,592 \pm 39,600}{2}$$

$$x = -4, \ 39,596$$

Check:

-4: $2\sqrt{-4 + 5} \overset{?}{=} 0.01(-4) + 2.04$ $39,596$: $2\sqrt{39,596 + 5} \overset{?}{=} 0.01(39,596) + 2.04$

$2 \overset{\sqrt{}}{=} 2$ $398 \overset{\sqrt{}}{=} 398$

Solution: -4, 39596

47. $2x^{-2/5} - 5x^{-1/5} + 1 = 0$

Let $u = x^{-1/5}$, then
$$2u^2 - 5u + 1 = 0$$

$$u = \frac{-b \pm \sqrt{b^2 - 4ac}}{2a} \quad a = 2, \ b = -5, \ c = 1$$

$$u = \frac{-(-5) \pm \sqrt{(-5)^2 - 4(2)(1)}}{2(2)}$$

$$u = \frac{5 \pm \sqrt{17}}{4}$$

$$x^{-1/5} = \frac{5 + \sqrt{17}}{4} \qquad\qquad x^{-1/5} = \frac{5 - \sqrt{17}}{4}$$

$$x^{1/5} = \frac{4}{5 + \sqrt{17}} \qquad\qquad x^{1/5} = \frac{4}{5 - \sqrt{17}}$$

$$x = \left(\frac{4}{5 + \sqrt{17}}\right)^5 \qquad x = \left(\frac{4}{5 - \sqrt{17}}\right)^5$$

$$x \approx 0.016203 \qquad\qquad x \approx 1974.98$$

49. Let x = width of cross-section of the beam
y = depth of cross-section of the beam
From the Pythagorean theorem
$$x^2 + y^2 = 16^2$$
Thus,
$$y = \sqrt{256 - x^2}$$

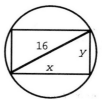

Since the area of the rectangle is given by xy, we have
$$xy = 120$$
$$x\sqrt{256 - x^2} = 120$$
$$x^2(256 - x^2) = 14,400$$
$$256x^2 - x^4 = 14,400$$
$$-x^4 + 256x^2 - 14,400 = 0$$
$$(x^2)^2 - 256x^2 + 14,400 = 0$$

$$x^2 = \frac{-b \pm \sqrt{b^2 - 4ac}}{2a} \quad a = 1, \ b = -256, \ c = 14,400$$

$$x^2 = \frac{-(-256) \pm \sqrt{(-256)^2 - 4(1)(14,400)}}{2(1)}$$

$$x^2 = \frac{256 \pm \sqrt{65,536 - 57,600}}{2}$$

$$x^2 = \frac{256 \pm \sqrt{7,936}}{2}$$

$$x^2 = 128 \pm \sqrt{1,984}$$

$$x = \sqrt{128 \pm \sqrt{1,984}}$$

If $x = \sqrt{128 + \sqrt{1,984}} \approx 13.1$ then
$$y = \sqrt{256 - x^2}$$
$$= \sqrt{256 - (128 + \sqrt{1,984})}$$
$$= \sqrt{128 - \sqrt{1,984}} \approx 9.1$$

Thus the dimensions of the rectangle are 13.1 inches by 9.1 inches. Notice that if $x = \sqrt{128 - \sqrt{1,984}}$, then $y = \sqrt{128 + \sqrt{1,984}}$ and the dimensions are still 13.1 inches by 9.1 inches.

51.

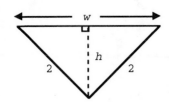

Let w = width of trough
h = altitude of triangular end

Examining the triangular end of the trough sketched above, we see that

$h^2 + \left(\dfrac{1}{2}w\right)^2 = 2^2$. The area of this end, $A = \dfrac{1}{2}wh$. Since the volume of the trough

V is given by $V = A \cdot 6$, we have

$$9 = 6A$$

$$9 = 6\left(\dfrac{1}{2}wh\right)$$

$$9 = 3wh$$

$$3 = wh$$

Since $h^2 = 2^2 - \left(\dfrac{1}{2}w\right)^2$

$$h^2 = 2^2 - \dfrac{1}{4}w^2$$

$$h = \sqrt{4 - \dfrac{1}{4}w^2}$$

Hence we solve

$$3 = w\sqrt{4 - \dfrac{1}{4}w^2}$$

$$9 = w^2\left(4 - \dfrac{1}{4}w^2\right)$$

$$9 = 4w^2 - \dfrac{1}{4}(w^2)^2$$

$$36 = 16w^2 - (w^2)^2$$

$$(w^2)^2 - 16w^2 + 36 = 0$$

$$w^2 = \dfrac{-b \pm \sqrt{b^2 - 4ac}}{2a} \quad a = 1, \; b = -16, \; c = 36$$

$$w^2 = \dfrac{-(-16) \pm \sqrt{(-16)^2 - 4(1)(36)}}{2(1)}$$

$$w^2 = \dfrac{16 \pm \sqrt{256 - 144}}{2}$$

$$w^2 = \dfrac{16 \pm \sqrt{112}}{2}$$

$$w^2 = 8 \pm 2\sqrt{7}$$

$$w = \sqrt{8 \pm 2\sqrt{7}}$$

$$w = 1.65 \text{ ft or } 3.65 \text{ ft}$$

53. $p = 14 + 0.01q$

$p = 50 - 0.5\sqrt{q}$

Substitute p from the first equation into the second to eliminate p.

$$14 + 0.01q = 50 - 0.5\sqrt{q}$$

$$-36 + 0.01q = -0.5\sqrt{q}$$

$$(-36 + 0.01q)^2 = (-0.5\sqrt{q})^2$$

$$1296 - 0.72q + 0.0001q^2 = 0.25q$$

$$0.0001q^2 - 0.97q + 1296 = 0$$

$$q = \frac{-b \pm \sqrt{b^2 - 4ac}}{2a} \qquad a = 0.0001, \ b = -0.97, \ c = 1296$$

$$q = \frac{-(-0.97) \pm \sqrt{(-0.97)^2 - 4(0.0001)(1296)}}{2(0.0001)}$$

$$q = \frac{0.97 \pm \sqrt{0.9409 - 0.5184}}{0.0002}$$

$$q = \frac{0.97 \pm \sqrt{0.4225}}{0.0002}$$

$$q = \frac{0.97 \pm 0.65}{0.0002}$$

$$q = 1600 \text{ or } 8100$$

Check: $14 + 0.01(1600) \overset{?}{=} 50 - 0.5\sqrt{1600}$ $14 + 0.01(8100) \overset{?}{=} 50 - 0.5\sqrt{8100}$

$\qquad\qquad\qquad\quad 30 \overset{\checkmark}{=} 30 \qquad\qquad\qquad\qquad\qquad\qquad\qquad 95 \neq 5$

$\qquad q = 1600$ telephones

$\qquad p = 14 + 0.01q = 14 + 0.01(1600)$

$\qquad\quad = \$30$

Exercise 1-8

Key Ideas and Formulas

A non-zero polynomial will have a constant sign (either always positive or always negative) within each interval determined by its real zeros plotted on a number line. If a polynomial has no real zeros, then the polynomial is either positive over the whole line or negative over the whole line.

The rational expression P/Q, where P and Q are non-zero polynomials, will have a constant sign (either always positive or always negative) within each interval determined by the real zeros of P and Q plotted on a number line. If neither P nor Q have real zeros, then the rational expression P/Q is either positive over the whole line or negative over the whole line.

To solve a polynomial inequality:

Step 1: Write the polynomial inequality in standard form.

Step 2: Find all real zeros of the polynomial.

Step 3: Plot the real zeros on a number line, dividing the number line into intervals.

Step 4: Choose a test number (that is easy to compute with) in each interval and evaluate the polynomial for each number.

Step 5: Using the results of Step 4 construct a sign chart, showing the sign of the polynomial in each interval.

Step 6: From the sign chart, write down the solution (and draw the graph, if required) of the original polynomial inequality.

A similar sequence of steps is required to solve a rational inequality.

1.
$$x^2 < 10 - 3x$$
$$x^2 + 3x - 10 < 0$$
$$(x + 5)(x - 2) < 0$$
Zeros: −5, 2

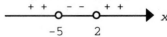

$x^2 + 3x - 10 = (x + 5)(x - 2)$			
Test Number	−6	0	3
Value of Polynomial for Test Number	8	−10	8
Sign of Polynomial in Interval	+	−	+
Interval	$(-\infty, -5)$	$(-5, 2)$	$(2, \infty)$

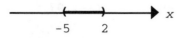

$x^2 + 3x - 10$ is negative within the interval $(-5, 2)$. $-5 < x < 2$.

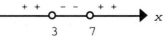

3.
$$x^2 + 21 > 10x$$
$$x^2 - 10x + 21 > 0$$
$$(x - 3)(x - 7) > 0$$
Zeros: 3, 7

$x^2 - 10x + 21 = (x - 3)(x - 7)$			
Test Number	2	5	8
Value of Polynomial for Test Number	5	−4	5
Sign of Polynomial in Interval	+	−	+
Interval	$(-\infty, 3)$	$(3, 7)$	$(7, \infty)$

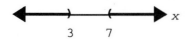

$x^2 - 10x + 21$ is positive within the intervals $(-\infty, 3)$ and $(7, \infty)$.
$x < 3$ or $x > 7$

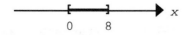

5.
$$x^2 \le 8x$$
$$x^2 - 8x \le 0$$
$$x(x - 8) \le 0$$
Zeros: 0,8

0,8 are part of the solution set, so we use solid dots.

$x^2 - 8x = x(x - 8)$			
Test Number	−1	4	9
Value of Polynomial for Test Number	9	−16	9
Sign of Polynomial in Interval	+	−	+
Interval	$(-\infty, 0)$	$(0, 8)$	$(8, \infty)$

$x^2 - 8x$ is non-positive within the interval $[0,8]$. $0 \le x \le 8$

7.
$$x^2 + 5x \le 0$$
$$x(x + 5) \le 0$$
Zeros: −5, 0
−5, 0 are part of the solution set, so we use solid dots.

$x^2 + 5x = x(x + 5)$			
Test Number	−6	−1	1
Value of Polynomial for Test Number	6	−4	6
Sign of Polynomial in Interval	+	−	+
Interval	$(-\infty, -5)$	$(-5, 0)$	$(0, \infty)$

$x^2 + 5x$ is non-positive within the interval $[-5,0]$. $-5 \le x \le 0$

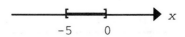

9.
$$x^2 > 4$$
$$x^2 - 4 > 0$$
$$(x + 2)(x - 2) > 0$$
Zeros: -2, 2

$(-\infty, -2)$ $(-2, 2)$ $(2, \infty)$

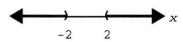

$x^2 - 4 = (x + 2)(x - 2)$			
Test Number	-3	0	3
Value of Polynomial for Test Number	5	-4	5
Sign of Polynomial in Interval	+	–	+
Interval	$(-\infty, -2)$	$(-2, 2)$	$(2, \infty)$

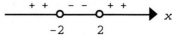

$x^2 - 4$ is positive within the intervals $(-\infty, -2)$ and $(2, \infty)$. $x < -2$ or $x > 2$

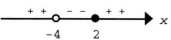

11. $\dfrac{x - 2}{x + 4} \le 0$

Zeros of P, Q: -4, 2
2 is a part of the solution set, so we use a solid dot there. -4 is not part of the solution set, ($\frac{P}{Q}$ is not defined there) so we use an open dot there.

$(-\infty, -4)$ $(-4, 2)$ $(2, \infty)$

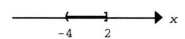

$\frac{P}{Q} = \frac{x - 2}{x + 4}$			
Test Number	-5	0	3
Value of $\frac{P}{Q}$	7	$-\frac{1}{2}$	$\frac{1}{7}$
Sign of $\frac{P}{Q}$	+	–	+
Interval	$(-\infty, -4)$	$(-4, 2)$	$(2, \infty)$

$\dfrac{x - 2}{x + 4}$ is non-positive within the interval $(-4, 2]$. $-4 < x \le 2$

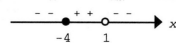

13. $\dfrac{x + 4}{1 - x} \le 0$

Zeros of P, Q: -4, 1
-4 is a part of the solution set, so we use a solid dot there. 1 is not part of the solution set, ($\frac{P}{Q}$ is not defined there) so we use an open dot there.

$(-\infty, -4)$ $(-4, 1)$ $(1, \infty)$

$\frac{P}{Q} = \frac{x + 4}{1 - x}$			
Test Number	-5	0	2
Value of $\frac{P}{Q}$	$-\frac{1}{6}$	4	-6
Sign of $\frac{P}{Q}$	–	+	–
Interval	$(-\infty, -4)$	$(-4, 1)$	$(1, \infty)$

$\dfrac{x + 4}{1 - x}$ is non-positive within the intervals $(-\infty, -4]$ and $(1, \infty)$. $x \le -4$ or $x > 1$

15. $\dfrac{x^2 + 5x}{x - 3} \geq 0$

$\dfrac{x(x + 5)}{x - 3} \geq 0$

Zeros of P, Q: -5, 0, 3
-5 and 0 are part of the solution set, so we use solid dots there. 3 is not part of the solution set, ($\frac{P}{Q}$ is not defined there) so we use an open dot there.

(−∞, −5) (−5, 0) (0, 3) (3, ∞)

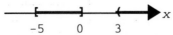

	$\dfrac{P}{Q} = \dfrac{x(x + 5)}{x - 3}$			
Test Number	-6	-1	1	4
Value of $\frac{P}{Q}$	$-\frac{1}{9}$	1	-3	36
Sign of $\frac{P}{Q}$	$-$	$+$	$-$	$+$
Interval	$(-\infty, -5)$	$(-5, 0)$	$(0, 3)$	$(3, \infty)$

$\dfrac{x^2 + 5x}{x - 3}$ is non-negative within the intervals $[-5, 0]$ and $(3, \infty)$.
$-5 \leq x \leq 0$ or $3 < x$.

17. $\dfrac{(x + 1)^2}{x^2 + 2x - 3} \leq 0$

$\dfrac{(x + 1)(x + 1)}{(x - 1)(x + 3)} \geq 0$

Zeros of P, Q: -1, 1, -3
-1 is part of the solution set, so we use a solid dot there.
-3 and 1 are not part of the solution set, ($\frac{P}{Q}$ is not defined there) so we use open dots there.

(−∞, −3) (−3, −1) (−1, 1) (1, ∞)

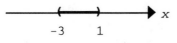

	$\dfrac{P}{Q} = \dfrac{(x + 1)(x + 1)}{(x - 1)(x + 3)}$			
Test Number	-4	-2	0	2
Value of $\frac{P}{Q}$	$\frac{9}{5}$	$-\frac{1}{3}$	$-\frac{1}{3}$	$\frac{9}{5}$
Sign of $\frac{P}{Q}$	$+$	$-$	$-$	$+$
Interval	$(-\infty, -3)$	$(-3, -1)$	$(-1, 1)$	$(1, \infty)$

$\dfrac{(x + 1)^2}{x^2 + 2x - 3}$ is non-positive within the intervals $(-3, -1)$ and $(-1, 1)$ as well as at $x = -1$. More simply, we can write $-3 < x < 1$ or $(-3, 1)$.

19. $\dfrac{1}{x} < 4$

$\dfrac{1}{x} - 4 < 0$

$\dfrac{1 - 4x}{x} < 0$

Zeros of P, Q: 0, $\dfrac{1}{4}$

(−∞, 0) (0, 1/4) (1/4, ∞)

	$\dfrac{P}{Q} = \dfrac{1 - 4x}{x}$		
Test Number	-1	0.1	1
Value of $\frac{P}{Q}$	-5	6	-3
Sign of $\frac{P}{Q}$	$-$	$+$	$-$
Interval	$(-\infty, 0)$	$(0, \frac{1}{4})$	$(\frac{1}{4}, \infty)$

$\dfrac{1 - 4x}{x} < 0$ and $\dfrac{1}{x} < 4$ within the intervals $(-\infty, 0)$ and $\left(\dfrac{1}{4}, \infty\right)$. $x < 0$ or $\dfrac{1}{4} < x$

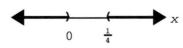

21.

$$\frac{3x + 1}{x + 4} \le 1$$

$$\frac{3x + 1}{x + 4} - 1 \le 0$$

$$\frac{3x + 1 - (x + 4)}{x + 4} \le 0$$

$$\frac{2x - 3}{x + 4} \le 0$$

> **Common Error:**
> $3x + 1 \le x + 4$
> This is not equivalent
> to the given inequality.

> **Common Error:**
> $3x + 1 - x + 4$ is an
> incorrect numerator.

$(-\infty, -4)$ $(-4, 3/2)$ $(3/2, \infty)$

(number line with open dot at -4, solid dot at $\frac{3}{2}$)

Zeros of P, Q: -4, $\frac{3}{2}$

$\frac{3}{2}$ is part of the solution set so we
use a solid dot there. -4 is not
part of the solution set ($\frac{P}{Q}$ is not
defined there) so we use an open dot
there.

	$\frac{P}{Q} = \frac{2x - 3}{x + 4}$		
Test Number	-5	0	2
Value of $\frac{P}{Q}$	13	$-\frac{3}{4}$	$\frac{1}{6}$
Sign of $\frac{P}{Q}$	$+$	$-$	$+$
Interval	$(-\infty, -4)$	$(-4, \frac{3}{2})$	$(\frac{3}{2}, \infty)$

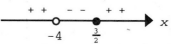

$\frac{2x - 3}{x + 4} \le 0$ and $\frac{3x + 1}{x + 4} \le 1$ within the interval $\left(-4, \frac{3}{2}\right]$. $-4 < x \le \frac{3}{2}$

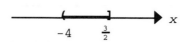

23.

$$\frac{2}{x + 1} \ge \frac{1}{x - 2}$$

$$\frac{2}{x + 1} - \frac{1}{x - 2} \ge 0$$

$$\frac{2(x - 2) - (x + 1)}{(x + 1)(x - 2)} \ge 0$$

$$\frac{2x - 4 - x - 1}{(x + 1)(x - 2)} \ge 0$$

$$\frac{x - 5}{(x + 1)(x - 2)} \ge 0$$

Zeros of P, Q: 5, -1, 2

5 is a part of the solution set, so we use a solid dot there. -1 and 2 are not
part of the solution set, ($\frac{P}{Q}$ is not defined there)
so we use an open dot there.

$(-\infty, -1)$ $(-1, 2)$ $(2, 5)$ $(5, \infty)$

(number line with open dots at -1 and 2, solid dot at 5)

$\frac{x - 5}{(x - 1)(x - 2)} \ge 0$ and

$\frac{2}{x + 1} \ge \frac{1}{x - 2}$ within the
intervals $(-1, 2)$ and $[5, \infty)$.

$-1 < x < 2$ or $5 \le x$

(number line)

	$\frac{P}{Q} = \frac{x - 5}{(x + 1)(x - 2)}$			
Test Number	-2	0	3	6
Value of $\frac{P}{Q}$	$-\frac{7}{4}$	$\frac{5}{2}$	$-\frac{1}{2}$	$\frac{1}{28}$
Sign of $\frac{P}{Q}$	$-$	$+$	$-$	$+$
Interval	$(-\infty, -1)$	$(-1, 2)$	$(2, 5)$	$(5, \infty)$

(number line with signs $- - + + - - + +$, open dots at -1 and 2, solid dot at 5)

25.
$$x^3 + 2x^2 \leq 8x$$
$$x^3 + 2x^2 - 8x \leq 0$$
$$x(x^2 + 2x - 8) \leq 0$$
$$x(x + 4)(x - 2) \leq 0$$
Zeros: 0, -4, 2
0, -4, 2 are part of the solution set, so we use solid dots there.

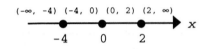

$x^3 + 2x^2 - 8x = x(x + 4)(x - 2)$				
Test Number	–5	–1	1	3
Value of Polynomial for Test Number	–35	9	–5	21
Sign of Polynomial in Interval	–	+	–	+
Interval	$(-\infty, -4)$	$(-4, 0)$	$(0, 2)$	$(2, \infty)$

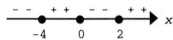

$x(x + 4)(x - 2) \leq 0$ and $x^3 + 2x^2 \leq 8x$ within the intervals $(-\infty, -4]$ and $[0, 2]$.
$x \leq -4$ or $0 \leq x \leq 2$

27. $\sqrt{x^2 - 9}$ will represent a real number when $x^2 - 9 \geq 0$.
$$x^2 - 9 \geq 0$$
$$(x + 3)(x - 3) \geq 0$$
Zeros of P, Q: -3, 3
-3 and 3 are part of the solution set, so we use solid dots.

$x^2 - 9 = (x + 3)(x - 3)$			
Test Number	–4	0	4
Value of Polynomial for Test Number	7	–9	7
Sign of Polynomial in Interval	+	–	+
Interval	$(-\infty, -3)$	$(-3, 3)$	$(3, \infty)$

$x^2 - 9 \geq 0$ and $\sqrt{x^2 - 9}$ will represent a real number within the intervals $(-\infty, -3]$ and $[3, \infty)$. $x \leq -3$ or $x \geq 3$

29. $\sqrt{2x^2 + x - 6}$ will represent a real number when $2x^2 + x - 6 \geq 0$
$$2x^2 + x - 6 \geq 0$$
$$(2x - 3)(x + 2) \geq 0$$
Zeros: -2, $\dfrac{3}{2}$

-2 and $\dfrac{3}{2}$ are part of the solution set, so we use solid dots.

$2x^2 + x - 6 = (2x - 3)(x + 2)$			
Test Number	–3	0	2
Value of Polynomial for Test Number	9	–6	4
Sign of Polynomial in Interval	+	–	+
Interval	$(-\infty, -2)$	$(-2, \frac{3}{2})$	$(\frac{3}{2}, \infty)$

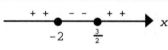

$2x^2 + x - 6 \geq 0$ and $\sqrt{2x^2 + x - 6}$ will represent a real number within the intervals $(-\infty, -2]$ and $\left[\dfrac{3}{2}, \infty\right)$. $x \leq -2$ or $x \geq \dfrac{3}{2}$

31. $\sqrt{\dfrac{x + 7}{3 - x}}$ will represent a real number when $\dfrac{x + 7}{3 - x} \geq 0$.

$\dfrac{x + 7}{3 - x} \geq 0$

Zeros of $\dfrac{P}{Q}$: -7, 3

-7 is a part of the solution set, so we use a solid dot there. 3 is not part of the solution set, ($\dfrac{P}{Q}$ is not defined there) so we use an open dot there.

	$\dfrac{P}{Q} = \dfrac{x + 7}{3 - x}$		
Test Number	-8	0	4
Value of $\dfrac{P}{Q}$	$-\dfrac{1}{11}$	$\dfrac{7}{3}$	-11
Sign of $\dfrac{P}{Q}$	$-$	$+$	$-$
Interval	$(-\infty, -7)$	$(-7, 3)$	$(3, \infty)$

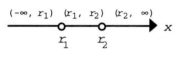

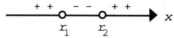

$\dfrac{x + 7}{3 - x} \geq 0$ and $\sqrt{\dfrac{x + 7}{3 - x}}$ will represent a real number within the interval $[-7, 3)$.

$-7 \leq x < 3$

33. Since r_1 and r_2 are distinct real roots of $ax^2 + bx + c = 0$, the inequality can be written:

$a(x - r_1)(x - r_2) > 0$

If $a > 0$, this is equivalent to $(x - r_1)(x - r_2) > 0$

If $a < 0$, this is equivalent to $(x - r_1)(x - r_2) < 0$

Zeros: r_1, r_2

	$(x - r_1)(x - r_2)$		
Test Number	$t_1 < r_1$	$r_1 < t_2 < r_2$	$r_2 < t_3$
Sign of Polynomial in Interval	$+$	$-$	$+$
Interval	$(-\infty, r_1)$	(r_1, r_2)	(r_2, ∞)

Thus if $a > 0$, $(x - r_1)(x - r_2) > 0$ and $ax^2 + bx + c > 0$ within the intervals $(-\infty, r_1)$ and (r_2, ∞).

Example: $x^2 - 5x - 6 > 0$ or $(x + 1)(x - 6) > 0$. Solution: $(-\infty, -1) \cup (6, \infty)$

If $a < 0$, $(x - r_1)(x - r_2) < 0$ and $ax^2 + bx + c > 0$ within the interval (r_1, r_2)

Example: $6 - x - x^2 > 0$ or $(x + 3)(x - 2) < 0$. Solution: $(-3, 2)$

35. Since r is a double root of $ax^2 + bx + c = 0$, the inequality can be written

$a(x - r)^2 \geq 0$

If $a > 0$, this is true for all real x, and the solution set is R, the set of all real numbers.

If $a < 0$, this is equivalent to $(x - r)^2 \leq 0$, which is true only if $x = r$. The solution set is $\{r\}$.

37. One example is $x^2 \geq 0$.

39. $x^2 + 1 < 2x$

$x^2 - 2x + 1 < 0$

$(x - 1)^2 < 0$

Since the square of no real number is negative, these statements are never true for any real number x. No solution (and no graph). $\varnothing$ is the solution set.

41. $x^2 < 3x - 3$
$x^2 - 3x + 3 < 0$
We attempt to find all real zeros of the polynomial.
$x^2 - 3x + 3 = 0$

$$x = \frac{-b \pm \sqrt{b^2 - 4ac}}{2a} \quad a = 1, \ b = -3, \ c = 3$$

$$x = \frac{-(-3) \pm \sqrt{(-3)^2 - 4(1)(3)}}{2(1)}$$

$$x = \frac{3 \pm \sqrt{-3}}{2}$$

The polynomial has no real zeros. Hence the statement is either true for all real x or for no real x. To determine which, we choose a test number, say 0.
$x^2 < 3x - 3$
$0^2 \overset{?}{<} 3(0) - 3$
$0 \overset{?}{<} -3$ False.
The statement is never true for any real number x. No solution (and no graph).
$\varnothing$ is the solution set.

43. $\quad x^2 - 1 \geq 4x$
$x^2 - 4x - 1 \geq 0$
Find all real zeros of the polynomial.
$x^2 - 4x - 1 = 0$

$$x = \frac{-b \pm \sqrt{b^2 - 4ac}}{2a} \quad a = 1, \ b = -4, \ c = -1$$

$$x = \frac{-(-4) \pm \sqrt{(-4)^2 - 4(1)(-1)}}{2(1)}$$

$$x = \frac{4 \pm \sqrt{16 + 4}}{2}$$

$$x = \frac{4 \pm \sqrt{20}}{2}$$

$$x = 2 \pm \sqrt{5}$$

Common Error:
$x \neq 2 \pm \sqrt{20}$

$\approx -0.236, \ 4.236$

Plot the real zeros on a number line.

Polynomial $x^2 - 4x - 1$			
Test Number	-1	0	5
Value of Polynomial for Test Number	4	-1	4
Sign of Polynomial in Interval	+	-	+
Interval	$(-\infty, 2 - \sqrt{5})$	$(2 - \sqrt{5}, 2 + \sqrt{5})$	$(2 + \sqrt{5}, \infty)$

$x^2 - 4x - 1 \geq 0$ and $x^2 - 1 \geq 4x$ within the intervals $(-\infty, \ 2 - \sqrt{5}]$ and $[2 + \sqrt{5}, \ \infty)$.
$x \leq 2 - \sqrt{5}$ or $x \geq 2 + \sqrt{5}$

45.
$$x^3 > 2x^2 + x$$
$$x^3 - 2x^2 - x > 0$$

Find all real zeros of the polynomial.
$$x^3 - 2x^2 - x = 0$$
$$x(x^2 - 2x - 1) = 0$$
$$x = 0 \text{ or } x^2 - 2x - 1 = 0$$
$$x = \frac{-b \pm \sqrt{b^2 - 4ac}}{2a} \quad a = 1, \ b = -2, \ c = -1$$
$$x = \frac{-(-2) \pm \sqrt{(-2)^2 - 4(1)(-1)}}{2(1)}$$
$$x = \frac{2 \pm \sqrt{8}}{2}$$
$$x = 1 \pm \sqrt{2}$$
$$\approx -0.414, \ 2.414$$

Plot the real zeros on a number line.

Polynomial $x^3 - 2x^2 - x$				
Test Number	-1	-0.1	1	3
Value of Polynomial for Test Number	-2	0.079	-2	6
Sign of Polynomial in Interval	$-$	$+$	$-$	$+$
Interval	$(-\infty, 1 - \sqrt{2})$	$(1 - \sqrt{2}, 0)$	$(0, 1 + \sqrt{2})$	$(1 + \sqrt{2}, \infty)$

$x^3 - 2x^2 - x > 0$ and $x^3 > 2x^2 + x$ within the intervals $(1 - \sqrt{2}, 0)$, and $(1 + \sqrt{2}, \infty)$. $1 - \sqrt{2} < x < 0$ or $x > 1 + \sqrt{2}$

47.
$$4x^4 + 4 \leq 17x^2$$
$$4x^4 - 17x^2 + 4 \leq 0$$
$$(4x^2 - 1)(x^2 - 4) \leq 0$$
$$(2x - 1)(2x + 1)(x - 2)(x + 2) \leq 0$$

Zeros: $-2, \ -\frac{1}{2}, \ \frac{1}{2}, \ 2$

$-2, \ -\frac{1}{2}, \ \frac{1}{2},$ and 2 are part of the solution set, so we use solid dots.

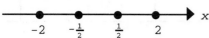

$4x^4 - 17x^2 + 4 = (2x - 1)(2x + 1)(x - 2)(x + 2)$					
Test Number	-3	-1	0	1	3
Value of Polynomial for Test Number	175	-9	4	-9	175
Sign of Polynomial in Interval	$+$	$-$	$+$	$-$	$+$
Interval	$(-\infty, -2)$	$(-2, -\frac{1}{2})$	$(-\frac{1}{2}, \frac{1}{2})$	$(\frac{1}{2}, 2)$	$(2, \infty)$

$4x^4 - 17x^2 + 4 \leq 0$ and $4x^4 + 4 \leq 17x^2$ within the intervals $\left[-2, \ -\frac{1}{2}\right]$ and $\left[\frac{1}{2}, \ 2\right]$.

$-2 \leq x \leq -\frac{1}{2}$ or $\frac{1}{2} \leq x \leq 2$.

49. $|x^2 - 1| \leq 3$ According to Theorem 2, Section 1-4, we must consider the equations $x^2 - 1 = 3$ and $x^2 - 1 = -3$ to solve the equality part of the inequality statement, and the double inequality $-3 < x^2 - 1 < 3$ to solve the inequality part.

Case 1:
$x^2 - 1 = 3$
   $x^2 = 4$
There are two solutions: -2 and 2.

Case 2:
$x^2 - 1 = -3$
   $x^2 = -2$
There are no real solutions.

Case 3:
$-3 < x^2 - 1 < 3$
To satisfy these relations, both statements $-3 < x^2 - 1$ and $x^2 - 1 < 3$ must hold. But $-3 < x^2 - 1$ or $-4 < x^2$ holds for all real x, so we need only examine $x^2 - 1 < 3$ or $x^2 - 4 < 0$. $(x - 2)(x + 2) < 0$

Zeros: -2, 2

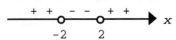

$x^2 - 4 = (x - 2)(x + 2)$			
Test Number	-3	0	3
Value of Polynomial for Test Number	5	-4	5
Sign of Polynomial in Interval	$+$	$-$	$+$
Interval	$(-\infty, -2)$	$(-2, 2)$	$(2, \infty)$

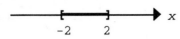

$(x - 2)(x + 2)$ is negative within the interval $(-2, 2)$. Combining the solutions from the separate cases, $|x^2 - 1| \leq 3$ when $-2 \leq x \leq 2$. $[-2, 2]$

51. (A) A profit will result if cost is less than revenue; that is, if
$$C < R$$
$$28 - 2p < 9p - p^2$$
$p^2 - 11p + 28 < 0$
$(p - 7)(p - 4) < 0$
Zeros: 4, 7

$p^2 - 11p + 28 = (p - 7)(p - 4)$			
Test Number	3	5	8
Value of Polynomial for Test Number	4	-2	4
Sign of Polynomial in Interval	$+$	$-$	$+$
Interval	$(-\infty, 4)$	$(4, 7)$	$(7, \infty)$

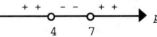

$p^2 - 11p + 28 < 0$ and a profit will occur $(C < R)$, for $\$4 < p < \$7$ or $(\$4, \$7)$

(B) A loss will result if cost is greater than revenue; that is, if
$$C > R$$
$$28 - 2p > 9p - p^2$$
$p^2 - 11p + 28 \qquad > 0$
Referring to the sign chart in part (A), we see that $p^2 - 11p + 28 > 0$, and a loss will occur $(C > R)$, for $p < \$4$ or $p > \$7$.
Since a negative price doesn't make sense we delete any number to the left of 0.
Thus, a loss will occur for $\$0 \leq p < \$4$ or $p > \$7$. $[\$0, \$4) \cup (\$7, \infty)$.

53. The object will be 160 feet or higher while $160 \leq d$.

$$160 \leq 112t - 16t^2$$
$$16t^2 - 112t + 160 \leq 0$$
$$t^2 - 7t + 10 \leq 0 \text{ (dividing both sides by 16)}$$
$$(t - 2)(t - 5) \leq 0$$

Zeros: 2, 5

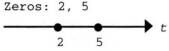

$t^2 - 7t + 10 = (t - 2)(t - 5)$			
Test Number	0	3	6
Value of Polynomial for Test Number	10	-2	4
Sign of Polynomial in Interval	+	-	+
Interval	$(-\infty, 2)$	$(2, 5)$	$(5, \infty)$

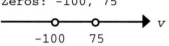

$t^2 - 7t + 10 \leq 0$, and the object will be 160 feet or higher, for $2 \leq t \leq 5$. [2, 5]

55. It will take the car more than 330 feet to stop when $d > 330$.

$$0.044v^2 + 1.1v > 330$$
$$0.044v^2 + 1.1v - 330 > 0$$
$$v^2 + 25v - 7,500 > 0 \text{ (dividing both sides by 0.044)}$$
$$(v - 75)(v + 100) > 0$$

Zeros: -100, 75

$v^2 + 25v - 7,500 = (v - 75)(v + 100)$			
Test Number	-200	0	100
Value of Polynomial for Test Number	27,500	-7,500	5,000
Sign of Polynomial in Interval	+	-	+
Interval	$(-\infty, -100)$	$(-100, 75)$	$(75, \infty)$

$v^2 + 25v - 7,500 > 0$, and it will take the car more than 330 feet to stop, for $v < -100$ or $v > 75$.

Since a negative speed doesn't make sense we delete any number to the left of 0. Thus, $v > 75$ miles per hour.

57. Sales will be 8,000 units or more when $8 \leq S$.

$$8 \leq \frac{200t}{t^2 + 100}$$
$$8 - \frac{200t}{t^2 + 100} \leq 0$$
$$\frac{8(t^2 + 100) - 200t}{t^2 + 100} \leq 0$$
$$\frac{8t^2 - 200t + 800}{t^2 + 100} \leq 0$$

Find all real zeros of P and Q,

$$\frac{P}{Q} = \frac{8t^2 - 200t + 800}{t^2 + 100}.$$

$$8t^2 - 200t + 800 = 0$$
$$t^2 - 25t + 100 = 0$$
$$(t - 5)(t - 20) = 0$$
$$t = 5, 20 \text{ (Zeros of } P)$$

$t^2 + 100$ has no real zero.

	$\frac{P}{Q} = \frac{8t^2 - 200t + 800}{t^2 + 100}$		
Test Number	0	10	30
Value of $\frac{P}{Q}$	8	-2	2
Sign of $\frac{P}{Q}$	+	-	+
Interval	$(-\infty, 5)$	$(5, 20)$	$(20, \infty)$

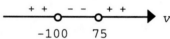

$\frac{8t^2 - 200t + 800}{t^2 + 100} \leq 0$ and sales will be 8 thousand units or more, for $5 \leq t \leq 20$.

CHAPTER 1 REVIEW

1. $0.05x + 0.25(30 - x) = 3.3$
$0.05x + 7.5 - 0.25x = 3.3$
$-0.2x + 7.5 = 3.3$
$-0.2x = -4.2$
$x = \dfrac{-4.2}{-0.2}$
$x = 21$ *(1-1)*

2. $\dfrac{5x}{3} - \dfrac{4 + x}{2} = \dfrac{x - 2}{4} + 1$

$12\dfrac{5x}{3} - 12\dfrac{(4 + x)}{2} = 12\dfrac{(x - 2)}{4} + 12$
$20x - 6(4 + x) = 3(x - 2) + 12$
$20x - 24 - 6x = 3x - 6 + 12$
$14x - 24 = 3x + 6$
$11x = 30$
$x = \dfrac{30}{11}$ *(1-1)*

3. $y = 4x - 9$
$y = -x + 6$
Substitute y from the first equation
into the second equation to
eliminate y.
$4x - 9 = -x + 6$
$5x - 9 = 6$
$5x = 15$
$x = 3$
$y = -x + 6 = -3 + 6 = 3$
$x = 3, \; y = 3$ *(1-2)*

4. $3(2 - x) - 2 \le 2x - 1$
$6 - 3x - 2 \le 2x - 1$
$-3x + 4 \le 2x - 1$
$-5x \le -5$
$x \ge 1$
$[1, \infty)$

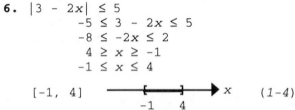

    (1-3)

5. $|y + 9| < 5$
$-5 < y + 9 < 5$
$-14 < y < -4$
$(-14, -4)$

    (1-4)

6. $|3 - 2x| \le 5$
$-5 \le 3 - 2x \le 5$
$-8 \le -2x \le 2$
$4 \ge x \ge -1$
$-1 \le x \le 4$
$[-1, 4]$

    (1-4)

7. $x^2 + x < 20$
$x^2 + x - 20 < 0$
$(x + 5)(x - 4) < 0$
Zeros: $-5, 4$
$(-\infty, -5) \; (-5, 4) \; (4, \infty)$

$x^2 + x - 20 = (x + 5)(x - 4)$			
Test Number	-6	0	5
Value of Polynomial for Test Number	10	-20	10
Sign of Polynomial in Interval	$+$	$-$	$+$
Interval	$(-\infty, -5)$	$(-5, 4)$	$(4, \infty)$

$x^2 + x - 20$ is negative within the interval $(-5, 4)$. $-5 < x < 4$.

    (1-8)

8. $x^2 \ge 4x + 21$
$x^2 - 4x - 21 \ge 0$
$(x + 3)(x - 7) \ge 0$
Zeros: $-3, 7$
$-3, 7$ are part of the
solution set, so we use
solid dots.
$(-\infty, -3) \; (-3, 7) \; (7, \infty)$

$x^2 - 4x - 21 = (x + 3)(x - 7)$			
Test Number	-4	0	8
Value of Polynomial for Test Number	11	-21	11
Sign of Polynomial in Interval	$+$	$-$	$+$
Interval	$(-\infty, -3)$	$(-3, 7)$	$(7, \infty)$

$x^2 - 4x - 21 \ge 0$ is non-negative within the intervals $(-\infty, -3]$ and $[7, \infty)$.
$x \le -3$ or $x \ge 7$ *(1-8)*

9. (A) $(-3 + 2i) + (6 - 8i) = -3 + 2i + 6 - 8i = 3 - 6i$

(B) $(3 - 3i)(2 + 3i) = 6 + 3i - 9i^2 = 6 + 3i + 9 = 15 + 3i$

(C) $\dfrac{13 - i}{5 - 3i} = \dfrac{(13 - i)}{(5 - 3i)} \dfrac{(5 + 3i)}{(5 + 3i)}$

$$= \frac{65 + 34i - 3i^2}{25 - 9i^2} = \frac{65 + 34i + 3}{25 + 9} = \frac{68 + 34i}{34} = 2 + i \qquad (1\text{-}5)$$

10. $2x^2 - 7 = 0$

$2x^2 = 7$

$x^2 = \dfrac{7}{2}$

$x = \pm\sqrt{\dfrac{7}{2}}$

$x = \pm\dfrac{\sqrt{14}}{2}$ $(1\text{-}6)$

11. $\qquad 2x^2 = 4x$

$2x^2 - 4x = 0$

$2x(x - 2) = 0$

$2x = 0 \quad x - 2 = 0$

$x = 0 \qquad x = 2$

$\qquad\qquad (1\text{-}6)$

12. $\qquad 2x^2 = 7x - 3$

$2x^2 - 7x + 3 = 0$

$(2x - 1)(x - 3) = 0$

$2x - 1 = 0 \quad x - 3 = 0$

$x = \dfrac{1}{2} \qquad x = 3$

$\qquad\qquad\qquad (1\text{-}6)$

13. $m^2 + m + 1 = 0$

$m = \dfrac{-b \pm \sqrt{b^2 - 4ac}}{2a} \quad \begin{aligned} a &= 1 \\ b &= 1 \\ c &= 1 \end{aligned}$

$m = \dfrac{-1 \pm \sqrt{(1)^2 - 4(1)(1)}}{2(1)}$

$m = \dfrac{-1 \pm \sqrt{-3}}{2}$

$m = \dfrac{-1 \pm i\sqrt{3}}{2}$

$m = -\dfrac{1}{2} \pm \dfrac{\sqrt{3}}{2}i \qquad (1\text{-}6)$

14. $\qquad y^2 = \dfrac{3}{2}(y + 1)$

$2y^2 = 3(y + 1)$

$2y^2 = 3y + 3$

$2y^2 - 3y - 3 = 0$

$y = \dfrac{-b \pm \sqrt{b^2 - 4ac}}{2a} \quad \begin{aligned} a &= 2 \\ b &= -3 \\ c &= -3 \end{aligned}$

$y = \dfrac{-(-3) \pm \sqrt{(-3)^2 - 4(2)(-3)}}{2(2)}$

$y = \dfrac{3 \pm \sqrt{33}}{4}$

$\qquad\qquad\qquad (1\text{-}6)$

15. $\quad \sqrt{5x - 6} - x = 0$

$\sqrt{5x - 6} = x$

$5x - 6 = x^2$

$0 = x^2 - 5x + 6$

$x^2 - 5x + 6 = 0$

$(x - 3)(x - 2) = 0$

$x = 2, 3$

Check: $\sqrt{5(2) - 6} - 2 \overset{?}{=} 0$

$0 \overset{\checkmark}{=} 0$

$\sqrt{5(3) - 6} - 3 \overset{?}{=} 0$

$0 \overset{\checkmark}{=} 0$

Solution: 2, 3 $\qquad (1\text{-}7)$

16. $3x + 2y = 5$

$4x - y = 14$

Solve the first equation for y in terms of x and substitute into the second equation

$2y = 5 - 3x$

$y = \dfrac{5 - 3x}{2}$

$4x - \left(\dfrac{5 - 3x}{2}\right) = 14$

$8x - \dfrac{2}{1}\left(\dfrac{5 - 3x}{2}\right) = 28$

$8x - (5 - 3x) = 28$

$8x - 5 + 3x = 28$

$11x = 33$

$x = 3$

$y = \dfrac{5 - 3x}{2} = \dfrac{5 - 3 \cdot 3}{2}$

$y = -2$

$x = 3, \ y = -2 \qquad (1\text{-}2)$

17. $\sqrt{3 - 5x}$ represents a real number exactly when $3 - 5x$ is positive or zero. We can write this as an inequality statement and solve for x.

$$3 - 5x \geq 0$$
$$-5x \geq -3$$
$$x \leq \frac{3}{5} \qquad (1\text{-}3)$$

18.

$$\frac{7}{2 - x} = \frac{10 - 4x}{x^2 + 3x - 10}$$

$$\frac{7}{2 - x} = \frac{10 - 4x}{(x - 2)(x + 5)} \qquad \text{Excluded values: } x \neq 2,\ -5$$

$$\overset{-1}{\cancel{(x - 2)}}(x + 5)\frac{7}{\underset{1}{\cancel{2 - x}}} = (x - 2)(x + 5)\frac{10 - 4x}{(x - 2)(x + 5)}$$

$$-7(x + 5) = 10 - 4x$$
$$-7x - 35 = 10 - 4x$$
$$-3x = 45$$
$$x = -15 \qquad (1\text{-}1)$$

19.

$$\frac{u - 3}{2u - 2} = \frac{1}{6} - \frac{1 - u}{3u - 3}$$

$$\frac{u - 3}{2(u - 1)} = \frac{1}{6} - \frac{1 - u}{3(u - 1)} \qquad \text{Excluded value: } u \neq 1$$

$$6(u - 1)\frac{(u - 3)}{2(u - 1)} = 6(u - 1)\frac{1}{6} - 6(u - 1)\frac{(1 - u)}{3(u - 1)}$$

$$3(u - 3) = u - 1 - 2(1 - u)$$
$$3u - 9 = u - 1 - 2 + 2u$$
$$3u - 9 = 3u - 3$$
$$-9 = -3$$

No solution $\qquad (1\text{-}1)$

20. $5m + 6n = 2$
$4m - 9n = 20$

Solve the first equation for m in terms of n and substitute into the second equation.

$$5m = 2 - 6n$$
$$m = \frac{2 - 6n}{5}$$

$$4\left(\frac{2 - 6n}{5}\right) - 9n = 20$$

$$\frac{8 - 24n}{5} - 9n = 20$$

$$\frac{5}{1}\left(\frac{8 - 24n}{5}\right) - 45n = 100$$

$$8 - 24n - 45n = 100$$
$$-69n = 92$$
$$n = -\frac{4}{3}$$

$$m = \frac{2 - 6\left(-\frac{4}{3}\right)}{5} = \frac{2 + 8}{5} = 2$$

$$m = 2,\ n = -\frac{4}{3} \qquad (1\text{-}2)$$

21.

$$\frac{x + 3}{8} \leq 5 - \frac{2 - x}{3}$$

$$24\frac{(x + 3)}{8} \leq 120 - 24\frac{(2 - x)}{3}$$

$$3(x + 3) \leq 120 - 8(2 - x)$$
$$3x + 9 \leq 120 - 16 + 8x$$
$$3x + 9 \leq 8x + 104$$
$$-5x \leq 95$$
$$x \geq -19$$

$$[-19,\ \infty)$$

$$\qquad (1\text{-}3)$$

22. $|3x - 8| > 2$

$$3x - 8 < -2 \quad \text{or} \quad 3x - 8 > 2$$
$$3x < 6 \qquad\qquad 3x > 10$$
$$x < 2 \quad \text{or} \quad x > \frac{10}{3}$$

$$(-\infty,\ 2) \cup \left(\frac{10}{3},\ \infty\right)$$

$$\qquad (1\text{-}4)$$

23.
$$\frac{1}{x} < 2$$

$$\frac{1}{x} - 2 < 0$$

$$\frac{1 - 2x}{x} < 0$$

Zeros of P, Q: 0, $\frac{1}{2}$

	$\frac{P}{Q} = \frac{1 - 2x}{x}$		
Test Number	-1	0.1	1
Value of $\frac{P}{Q}$	-3	8	-1
Sign of $\frac{P}{Q}$	$-$	$+$	$-$
Interval	$(-\infty, 0)$	$(0, \frac{1}{2})$	$(\frac{1}{2}, \infty)$

$\dfrac{1 - 2x}{x} < 0$ and $\dfrac{1}{x} < 2$ within the intervals $(-\infty, 0)$ and $\left(\dfrac{1}{2}, \infty\right)$. $x < 0$ or $x > \dfrac{1}{2}$

$(1-8)$

24.
$$\frac{3}{x - 4} \le \frac{2}{x - 3}$$

$$\frac{3}{x - 4} - \frac{2}{x - 3} \le 0$$

$$\frac{3(x - 3) - 2(x - 4)}{(x - 4)(x - 3)} \le 0$$

$$\frac{3x - 9 - 2x + 8}{(x - 4)(x - 3)} \le 0$$

$$\frac{x - 1}{(x - 4)(x - 3)} \le 0$$

Zeros of P, Q: 1, 3, 4
1 is part of the solution set so
we use a solid dot there. 3 and 4
are not part of the solution set
($\frac{P}{Q}$ is not defined there) so we
use open dots there.

	$\frac{P}{Q} = \frac{x - 1}{(x - 4)(x - 3)}$			
Test Number	0	2	3.5	5
Value of $\frac{P}{Q}$	$-\frac{1}{12}$	$\frac{1}{2}$	-10	2
Sign of $\frac{P}{Q}$	$-$	$+$	$-$	$+$
Interval	$(-\infty, 1)$	$(1, 3)$	$(3, 4)$	$(4, \infty)$

$\dfrac{x - 1}{(x - 4)(x - 3)} \le 0$ and $\dfrac{3}{x - 4} \le \dfrac{2}{x - 3}$ within the intervals $(-\infty, 1]$ and $(3, 4)$.
$x \le 1$ or $3 < x < 4$

$(1-8)$

25.
$$\sqrt{(1 - 2m)^2} \le 3$$

$$|1 - 2m| \le 3$$

$$-3 \le 1 - 2m \le 3$$

$$-4 \le -2m \le 2$$

$$2 \ge m \ge -1$$

$$-1 \le m \le 2$$

$[-1, 2]$

$(1-4)$

26. $\sqrt{\dfrac{x + 4}{2 - x}}$ will represent a real number when $\dfrac{x + 4}{2 - x} \geq 0$.

$\dfrac{x + 4}{2 - x} \geq 0$

Zeros of P, Q: -4, 2

-4 is part of the solution set so we use a solid dot there. 2 is not part of the solution set ($\frac{P}{Q}$ is not defined there) so we use an open dot there.

	$\frac{P}{Q} = \frac{x+4}{2-x}$		
Test Number	-5	0	3
Value of $\frac{P}{Q}$	$-\frac{1}{7}$	2	-7
Sign of $\frac{P}{Q}$	$-$	$+$	$-$
Interval	$(-\infty, -4)$	$(-4, 2)$	$(2, \infty)$

$\dfrac{x + 4}{2 - x} \geq 0$ and $\sqrt{\dfrac{x + 4}{2 - x}}$ will represent a real number within the interval $[-4, 2)$.
$-4 \leq x < 2$ $\hfill (1\text{-}8)$

27. (A) $d(A,B) = |-2 - (-8)| = |6| = 6$ (B) $d(B,A) = |-8 - (-2)| = |-6| = 6$ $\quad (1\text{-}4)$

28. (A) $(3 + i)^2 - 2(3 + i) + 3 = 9 + 6i + i^2 - 6 - 2i + 3$
$= 9 + 6i - 1 - 6 - 2i + 3$
$= 5 + 4i$

(B) $i^{27} = i^{26}i = (i^2)^{13}i = (-1)^{13}i = (-1)i = -i$ $\hfill (1\text{-}5)$

29. (A) $(2 - \sqrt{-4}) - (3 - \sqrt{-9}) = (2 - i\sqrt{4}) - (3 - i\sqrt{9}) = (2 - 2i) - (3 - 3i)$
$= 2 - 2i - 3 + 3i = -1 + i$

(B) $\dfrac{2 - \sqrt{-1}}{3 + \sqrt{-4}} = \dfrac{2 - i\sqrt{1}}{3 + i\sqrt{4}} = \dfrac{2 - i}{3 + 2i} = \dfrac{(2 - i)}{(3 + 2i)}\dfrac{(3 - 2i)}{(3 - 2i)} = \dfrac{6 - 7i + 2i^2}{9 - 4i^2} = \dfrac{6 - 7i - 2}{9 + 4}$

$= \dfrac{4 - 7i}{13} = \dfrac{4}{13} - \dfrac{7}{13}i$

(C) $\dfrac{4 + \sqrt{-25}}{\sqrt{-4}} = \dfrac{4 + i\sqrt{25}}{i\sqrt{4}} = \dfrac{4 + 5i}{2i} = \dfrac{4 + 5i}{2i}\dfrac{i}{i} = \dfrac{4i + 5i^2}{2i^2} = \dfrac{4i - 5}{-2} = \dfrac{5}{2} - 2i$ $\quad (1\text{-}5)$

30. $\left(u + \dfrac{5}{2}\right)^2 = \dfrac{5}{4}$

$u + \dfrac{5}{2} = \pm\sqrt{\dfrac{5}{4}}$

$u + \dfrac{5}{2} = \pm\dfrac{\sqrt{5}}{2}$

$u = -\dfrac{5}{2} \pm \dfrac{\sqrt{5}}{2}$

$u = \dfrac{-5 \pm \sqrt{5}}{2}$ $\hfill (1\text{-}6)$

31. $1 + \dfrac{3}{u^2} = \dfrac{2}{u}$ Excluded value: $u \neq 0$

$u^2 + 3 = 2u$

$u^2 - 2u = -3$

$u^2 - 2u + 1 = -2$

$(u - 1)^2 = -2$

$u - 1 = \pm\sqrt{-2}$

$u = 1 \pm \sqrt{-2}$

$u = 1 \pm i\sqrt{2}$ $\hfill (1\text{-}6)$

32.

$$\frac{x}{x^2 - x - 6} - \frac{2}{x - 3} = 3$$

$$\frac{x}{(x - 3)(x + 2)} - \frac{2}{x - 3} = 3 \quad \text{Excluded values: } x \neq 3, -2$$

$$(x - 3)(x + 2)\frac{x}{(x - 3)(x + 2)} - (x - 3)(x + 2)\frac{2}{x - 3} = 3(x - 3)(x + 2)$$

$$x - 2(x + 2) = 3(x - 3)(x + 2)$$
$$x - 2x - 4 = 3(x^2 - x - 6)$$
$$-x - 4 = 3x^2 - 3x - 18$$
$$0 = 3x^2 - 2x - 14$$
$$3x^2 - 2x - 14 = 0$$

$$x = \frac{-b \pm \sqrt{b^2 - 4ac}}{2a} \qquad a = 3$$
$$b = -2$$
$$c = -14$$

$$x = \frac{-(-2) \pm \sqrt{(-2)^2 - 4(3)(-14)}}{2(3)}$$

$$x = \frac{2 \pm \sqrt{172}}{6}$$

$$x = \frac{2 \pm 2\sqrt{43}}{6}$$

$$x = \frac{1 \pm \sqrt{43}}{3} \qquad \qquad (1\text{-}6)$$

33. $2x^{2/3} - 5x^{1/3} - 12 = 0$
Let $u = x^{1/3}$, then
$$2u^2 - 5u - 12 = 0$$
$$(2u + 3)(u - 4) = 0$$
$$u = -\frac{3}{2},\ 4$$

$$x^{1/3} = -\frac{3}{2} \qquad x^{1/3} = 4$$

$$x = -\frac{27}{8} \qquad x = 64 \qquad (1\text{-}7)$$

34. $m^4 + 5m^2 - 36 = 0$
Let $u = m^2$, then
$$u^2 + 5u - 36 = 0$$
$$(u + 9)(u - 4) = 0$$
$$u = -9,\ 4$$
$$m^2 = -9 \qquad m^2 = 4$$
$$m = \pm 3i \qquad m = \pm 2 \qquad (1\text{-}7)$$

35. $\sqrt{y - 2} - \sqrt{5y + 1} = -3$
$$-\sqrt{5y + 1} = -3 - \sqrt{y - 2}$$
$$5y + 1 = 9 + 6\sqrt{y - 2} + y - 2$$
$$5y + 1 = y + 7 + 6\sqrt{y - 2}$$
$$4y - 6 = 6\sqrt{y - 2}$$
$$2y - 3 = 3\sqrt{y - 2}$$
$$4y^2 - 12y + 9 = 9(y - 2)$$
$$4y^2 - 12y + 9 = 9y - 18$$
$$4y^2 - 21y + 27 = 0$$
$$(4y - 9)(y - 3) = 0$$
$$y = \frac{9}{4},\ 3$$

> **Common Error:**
> $y - 2 - 5y + 1 = 9$
> is not equivalent to the
> equation formed by
> squaring both members
> of the given equation.

Check: $\sqrt{\frac{9}{4} - 2} - \sqrt{5(\frac{9}{4}) + 1} \overset{?}{=} -3$

$$\sqrt{\frac{1}{4}} - \sqrt{\frac{49}{4}} \overset{?}{=} -3$$

$$-3 \overset{\sqrt{}}{=} -3$$

$$\sqrt{3 - 2} - \sqrt{5(3) + 1} \overset{?}{=} -3$$

$$-3 \overset{\sqrt{}}{=} -3$$

Solution: $\frac{9}{4},\ 3$ $\qquad (1\text{-}7)$

36. $2.15x - 3.73(x - 0.93) = 6.11x$
$$2.15x - 3.73x + 3.4689 = 6.11x$$
$$-1.58x + 3.4689 = 6.11x$$
$$3.4689 = 7.69x$$
$$x = 0.45 \qquad (1\text{-}1)$$

37. $-1.52 \leq 0.77 - 2.04x \leq 5.33$
$$-2.29 \leq -2.04x \leq 4.56$$
$$1.12 \geq x \geq -2.24$$
$$-2.24 \leq x \leq 1.12 \text{ or } [-2.24, 1.12]$$
$$(1\text{-}3)$$

38. $\dfrac{3.77 - 8.47i}{6.82 - 7.06i} = \dfrac{(3.77 - 8.47i)\ (6.82 + 7.06i)}{(6.82 - 7.06i)\ (6.82 + 7.06i)}$

$\qquad\qquad\quad = \dfrac{(3.77)(6.82) + (3.77)(7.06i) - (6.82)(8.47i) - (8.47)(7.06)i^2}{(6.82)^2 + (7.06)^2}$

$\qquad\qquad\quad = \dfrac{25.7114 + 26.6162i - 57.7654i + 59.7982}{46.5124 + 49.8436}$

$\qquad\qquad\quad = \dfrac{85.5096 - 31.1492i}{96.356}$

$\qquad\qquad\quad = 0.89 - 0.32i \qquad\qquad\qquad\qquad\qquad\qquad\qquad (1\text{-}5)$

39. $6.09x^2 + 4.57x - 8.86 = 0$

$\qquad\qquad x = \dfrac{-b \pm \sqrt{b^2 - 4ac}}{2a} \quad a = 6.09,\ b = 4.57,\ c = -8.86$

$\qquad\qquad x = \dfrac{-4.57 \pm \sqrt{(4.57)^2 - 4(6.09)(-8.86)}}{2(6.09)}$

$\qquad\qquad x = \dfrac{-4.57 \pm \sqrt{236.7145}}{12.18}$

$\qquad\qquad x = \dfrac{-4.57 \pm 15.3855}{12.18}$

$\qquad\qquad x = -1.64,\ 0.89 \qquad\qquad\qquad\qquad\qquad\qquad\qquad (1\text{-}6)$

40. $15.2x + 5.6y = 20$
$\quad\ 2.5x + 7.5y = 10$

Solve the first equation for y in terms of x and substitute into the second equation.

$\qquad\quad 5.6y = 20 - 15.2x$

$\qquad\qquad y = \dfrac{20 - 15.2x}{5.6}$

$\qquad 2.5x + 7.5\left(\dfrac{20 - 15.2x}{5.6}\right) = 10$

$\qquad\qquad 2.5x + \dfrac{150 - 114x}{5.6} = 10$

$\qquad 5.6(2.5x) + \dfrac{5.6}{1}\left(\dfrac{150 - 114x}{5.6}\right) = 56$

$\qquad\qquad 14x + 150 - 114x = 56$

$\qquad\qquad\qquad\qquad -100x = -94$

$\qquad\qquad\qquad\qquad\quad\ x = 0.94$

$\qquad\qquad\qquad\qquad\quad\ y = \dfrac{20 - 15.2(0.94)}{5.6}$

$\qquad\qquad\qquad\qquad\quad\ y = 1.02$

$\quad x = 0.94,\ y = 1.02 \qquad\qquad\qquad\qquad\qquad\qquad\qquad (1\text{-}2)$

41. $\qquad\quad P = M - Mdt$
$\qquad M - Mdt = P$
$\qquad M(1 - dt) = P$
$\qquad\qquad\quad M = \dfrac{P}{1 - dt} \qquad (1\text{-}1)$

42. $\qquad\qquad\quad P = EI - RI^2$
$\qquad RI^2 - EI + P = 0$

$\qquad\qquad I = \dfrac{-b \pm \sqrt{b^2 - 4ac}}{2a} \qquad a = R$
$\qquad\qquad\qquad\qquad\qquad\qquad\qquad\qquad b = -E$
$\qquad\qquad\qquad\qquad\qquad\qquad\qquad\qquad c = P$

$\qquad\qquad I = \dfrac{-(-E) \pm \sqrt{(-E)^2 - 4(R)(P)}}{2(R)}$

$\qquad\qquad I = \dfrac{E \pm \sqrt{E^2 - 4PR}}{2R} \qquad\qquad (1\text{-}6)$

43.
$$x = \frac{4y + 5}{2y + 1}$$
$$x(2y + 1) = (2y + 1)\frac{4y + 5}{2y + 1}$$
$$2xy + x = 4y + 5$$
$$2xy + x - 4y = 5$$
$$2xy - 4y = 5 - x$$
$$y(2x - 4) = 5 - x$$
$$y = \frac{5 - x}{2x - 4} \qquad (1\text{-}1)$$

44. The original equation can be rewritten as
$$\frac{4}{(x - 1)(x - 3)} = \frac{3}{(x - 1)(x - 2)}$$
Thus, $x = 1$ cannot be a solution of this equation. This extraneous solution was introduced when both sides were multiplied by $x - 1$ in the second line. $x = 1$ must be discarded and the only correct solution is $x = -1$. $\qquad (1\text{-}1)$

45. In this problem, $a = 1$, $b = -6$, $c = c$. Thus, the discriminant $b^2 - 4ac = (-6)^2 - 4(1)(c) = 36 - 4c$. Hence,
if $36 - 4c > 0$, thus $36 > 4c$ or $c < 9$, there are two distinct real roots.
if $36 - 4c = 0$, thus $c = 9$, there is one real double root.
if $36 - 4c < 0$, thus $36 < 4c$ or $c > 9$, there are two distinct imaginary roots. $(1\text{-}6)$

46. The given inequality $a + b < b - a$ is equivalent to, successively,
$\quad a < -a$
$2a < 0$
$\quad a < 0$
Thus its truth is independent of the value of b, and dependent on a being negative. True for all real b and all negative a. $\qquad (1\text{-}3)$

47. If $a > b$ and b is negative, then $\frac{a}{b} < \frac{b}{b}$, that is, $\frac{a}{b} < 1$, since dividing both sides by b reverses the order of the inequality. $\frac{a}{b}$ is less than 1. $\qquad (1\text{-}3)$

48.
$$y = \frac{1}{1 - \frac{1}{1 - x}}$$
$$y = \frac{1(1 - x)}{(1 - x)1 - (1 - x)\frac{1}{1 - x}}$$
$$y = \frac{1 - x}{1 - x - 1}$$
$$y = \frac{1 - x}{-x}$$
$$-xy = 1 - x$$
$$x - xy = 1$$
$$x(1 - y) = 1$$
$$x = \frac{1}{1 - y} \qquad (1\text{-}1)$$

49. $0 < |x - 6| < d$ means: the distance between x and 6 is less than d but $x \neq 6$.
$\quad -d < x - 6 < d$ except $x \neq 6$
$6 - d < x < 6 + d$ but $x \neq 6$
$6 - d < x < 6$ or $6 < x < 6 + d$
$(6 - d, 6) \cup (6, 6 + d)$

$$6 - d \quad 6 \quad 6 + d$$
$\qquad (1\text{-}4)$

50.
$$2x^2 = \sqrt{3}x - \frac{1}{2}$$
$$4x^2 = 2\sqrt{3}x - 1$$
$$4x^2 - 2\sqrt{3}x + 1 = 0$$
$$x = \frac{-b \pm \sqrt{b^2 - 4ac}}{2a} \quad a = 4, \ b = -2\sqrt{3}, \ c = 1$$
$$x = \frac{-(-2\sqrt{3}) \pm \sqrt{(-2\sqrt{3})^2 - 4(4)(1)}}{2(4)}$$
$$x = \frac{2\sqrt{3} \pm \sqrt{-4}}{8}$$
$$x = \frac{2\sqrt{3} \pm 2i}{8}$$
$$x = \frac{\sqrt{3} \pm i}{4} \quad \text{or} \quad \frac{\sqrt{3}}{4} \pm \frac{1}{4}i$$

$(1-6)$

51.
$$4 = 8x^{-2} - x^{-4}$$
$$4x^4 = 8x^2 - 1 \qquad x \neq 0 \quad \text{LCD} = x^4$$
$$4x^4 - 8x^2 + 1 = 0$$
Let $u = x^2$, then
$$4u^2 - 8u + 1 = 0$$
$$u = \frac{-b \pm \sqrt{b^2 - 4ac}}{2a} \quad a = 4, \ b = -8, \ c = 1$$
$$u = \frac{-(-8) \pm \sqrt{(-8)^2 - 4(4)(1)}}{2(4)}$$
$$u = \frac{8 \pm \sqrt{48}}{8}$$

> **Common Error**: It is incorrect to "cancel" the 8's at this point.

$$u = \frac{8 \pm 4\sqrt{3}}{8}$$
$$u = \frac{4(2 \pm \sqrt{3})}{8}$$
$$u = \frac{2 \pm \sqrt{3}}{2}$$
$$x^2 = \frac{2 \pm \sqrt{3}}{2}$$
$$x = \pm\sqrt{\frac{2 \pm \sqrt{3}}{2}} \quad \text{(four real roots)}$$

$(1-7)$

52. $(a + bi)\left(\dfrac{a}{a^2 + b^2} - \dfrac{b}{a^2 + b^2}i\right)$

$= \dfrac{(a + bi)}{1}\left(\dfrac{a}{a^2 + b^2} - \dfrac{bi}{a^2 + b^2}\right)$

$= \dfrac{a(a + bi)}{a^2 + b^2} - \dfrac{bi(a + bi)}{a^2 + b^2}$

$= \dfrac{a^2 + abi - abi - b^2i^2}{a^2 + b^2}$

$= \dfrac{a^2 + b^2}{a^2 + b^2} = 1$

$(1-5)$

53. $2x > \dfrac{x^2}{5} + 5$

$10x > x^2 + 25$

$0 > x^2 - 10x + 25$

$0 > (x - 5)^2$

Since the right side is never negative for x a real number, this statement is satisfied by no real number.

$(1-8)$

54.
$$\frac{x^2}{4} + 4 \geq 2x$$
$$x^2 + 16 \geq 8x$$
$$x^2 - 8x + 16 \geq 0$$
$$(x - 4)^2 \geq 0$$

Since the left side is always positive or zero for x a real number, this statement is satisfied by all real numbers x. *(1-8)*

55. $\left| x - \dfrac{8}{x} \right| \geq 2$

There are two ways for this statement to hold. The solution set is the union of the solution sets for:

$x - \dfrac{8}{x} \leq -2$ and $x - \dfrac{8}{x} \geq 2$

Case 1: $x - \dfrac{8}{x} \leq -2$

$$x - \frac{8}{x} + 2 \leq 0$$
$$\frac{x^2 - 8 + 2x}{x} \leq 0$$
$$\frac{x^2 + 2x - 8}{x} \leq 0$$
$$\frac{(x - 2)(x + 4)}{x} \leq 0$$

Zeros of P, Q: 2, -4, 0

2 and -4 are part of the solution set, so we use solid dots there. 0 is not part of the solution set, so we use an open dot there.

	$\frac{P}{Q} = \frac{(x-2)(x+4)}{x}$			
Test Number	-5	-1	1	3
Value of $\frac{P}{Q}$	$-\frac{7}{5}$	9	-5	$\frac{7}{3}$
Sign of $\frac{P}{Q}$	–	+	–	+
Interval	$(-\infty, -4)$	$(-4, 0)$	$(0, 2)$	$(2, \infty)$

$\dfrac{(x - 2)(x + 4)}{x} \leq 0$ and $x - \dfrac{8}{x} \leq -2$ within the intervals $(-\infty, -4]$ and $(0, 2]$.

Case 2: $x - \dfrac{8}{x} \geq 2$

$$x - \frac{8}{x} - 2 \geq 0$$
$$\frac{x^2 - 8 - 2x}{x} \geq 0$$
$$\frac{x^2 - 2x - 8}{x} \geq 0$$
$$\frac{(x - 4)(x + 2)}{x} \geq 0$$

Zeros of P, Q: -2, 4, 0. -2 and 4 are part of the solution set, so we use solid dots there. 0 is not part of the solution set, so we use an open dot there.

	$\frac{P}{Q} = \frac{(x-4)(x+2)}{x}$			
Test Number	-3	-1	1	5
Value of $\frac{P}{Q}$	$-\frac{7}{3}$	5	-9	$\frac{7}{5}$
Sign of $\frac{P}{Q}$	–	+	–	+
Interval	$(-\infty, -2)$	$(-2, 0)$	$(0, 4)$	$(4, \infty)$

$\dfrac{(x - 4)(x + 2)}{x} \geq 0$ and $x - \dfrac{8}{x} \geq 2$ within the intervals $[-2, 0)$ and $[4, \infty)$.

Combining Case 1 and Case 2, we have: $x \leq -4$ or $-2 \leq x < 0$ or $0 < x \leq 2$ or $x \geq 4$

$(-\infty, -4] \cup [-2, 0) \cup (0, 2] \cup [4, \infty)$ *(1-8)*

56. $x = 2 + 3u + 7v$
$y = -3 + 2u + 5v$
Solve the first equation for u in terms of the other variables and substitute into the second equation.

$x - 2 - 7v = 3u$

$$u = \frac{x - 2 - 7v}{3}$$

$$y = -3 + 2\left(\frac{x - 2 - 7v}{3}\right) + 5v$$

$$y = -3 + \frac{2x - 4 - 14v}{3} + 5v$$

$$3y = -9 + \frac{3}{1}\left(\frac{2x - 4 - 14v}{3}\right) + 15v$$

$$3y = -9 + 2x - 4 - 14v + 15v$$

$$13 - 2x + 3y = v$$

$$u = \frac{x - 2 - 7v}{3} = \frac{x - 2 - 7(13 - 2x + 3y)}{3} = \frac{x - 2 - 91 + 14x - 21y}{3}$$

$$= \frac{15x - 21y - 93}{3} = 5x - 7y - 31$$

The checking steps are omitted for lack of space.   *(1-2)*

57. (A) $2x - y = -5$
    $-6x + 3y = 15$
Solve the first equation for y in terms of x and substitute into the second equation.

$-y = -2x - 5$
$y = 2x + 5$
$-6x + 3(2x + 5) = 15$
            $15 = 15$

Thus, the system has an infinite number of solutions.

(B) $2x - y = -5$
    $-6x + 3y = 10$
Solve the first equation for y in terms of x and substitute into the second equation.

$-y = -2x - 5$
$y = 2x + 5$
$-6x + 3(2x + 5) = 10$
            $15 = 10$

Thus, the system has no solution.

(1-2)

58. Let x = the number
$\dfrac{1}{x}$ = its reciprocal

Then $x - \dfrac{1}{x} = \dfrac{16}{15}$ Excluded value: $x \neq 0$

$15x^2 - 15 = 16x$
$15x^2 - 16x - 15 = 0$
$(5x + 3)(3x - 5) = 0$
$5x + 3 = 0$ or $3x - 5 = 0$
    $x = -\dfrac{3}{5}$ $x = \dfrac{5}{3}$ *(1-6)*

59. (A) $H = 0.7(220 - A)$

(B) We are to find H when $A = 20$.
$H = 0.7(220 - 20)$
$H = 140$ beats per minute.
(C) We are to find A when $H = 126$.
$126 = 0.7(220 - A)$
$126 = 154 - 0.7A$
$-28 = -0.7A$
$A = 40$ years old.   *(1-2)*

60. Let x = amount of 80% solution
Then $50 - x$ = amount of 30% solution
since 50 = amount of 60% solution

acid in 80% solution	+	acid in 30% solution	=	acid in 60% solution

$0.8(x) + 0.3(50 - x) = 0.6(50)$
$0.8x + 15 - 0.3x = 30$
$15 + 0.5x = 30$
$0.5x = 15$
$x = \dfrac{15}{0.5}$
$x = 30$ milliliters of 80% solution
$50 - x = 20$ milliliters of 30% solution   *(1-2)*

61. Let $\quad x$ = the rate of the current
Then $12 - x$ = the rate of the boat upstream
$\quad\quad 12 + x$ = the rate of the boat downstream

Solving $d = rt$ for t, we have $t = \dfrac{d}{r}$. We use this formula, together with

time upstream = 2 + time downstream

time upstream = $\dfrac{\text{distance upstream}}{\text{rate upstream}}$ = $\dfrac{45}{12 - x}$

time downstream = $\dfrac{\text{distance downstream}}{\text{rate downstream}}$ = $\dfrac{45}{12 + x}$

So, $\quad\quad \dfrac{45}{12 - x} = 2 + \dfrac{45}{12 + x}$ Excluded values: $x \neq -12,\ 12$

$\quad\quad 45(12 + x) = (12 + x)(12 - x)2 + 45(12 - x)$
$\quad\quad\quad 540 + 45x = 288 - 2x^2 + 540 - 45x$
$\quad 2x^2 + 90x - 288 = 0$
$\quad 2(x + 48)(x - 3) = 0$
$\quad\quad\quad\quad\quad\quad x = -48,\ 3$

Discarding the negative answer, we have rate = 3 miles per hour. $\quad$ *(1-6)*

62. $\quad$ (A) Let x = distance rowed
$\quad\quad$ then $15 - 3 = 12$ km/hr = the rate rowed upstream
$\quad\quad\quad\quad 15 + 3 = 18$ km/hr = the rate rowed downstream

Using $t = \dfrac{d}{r}$ as in the previous problem yields

$\quad$ time upstream = $\dfrac{x}{12}$

time downstream = $\dfrac{x}{18}$

So $\quad \dfrac{x}{12} + \dfrac{x}{18} = \dfrac{25}{60}$ $\quad$ LCD = 180

$\dfrac{180}{1} \cdot \dfrac{x}{12} + \dfrac{180}{1} \cdot \dfrac{x}{18} = \dfrac{180}{1} \cdot \dfrac{25}{60}$

$\quad\quad\quad 15x + 10x = 75$
$\quad\quad\quad\quad\quad 25x = 75$
$\quad\quad\quad\quad\quad\quad x = 3$ km

(B) Now let x = still-water speed
$\quad\quad x - 3$ = the rate rowed upstream
$\quad\quad x + 3$ = the rate rowed downstream

$\quad$ time upstream = $\dfrac{3}{x - 3}$

time downstream = $\dfrac{3}{x + 3}$

So $\quad \dfrac{3}{x - 3} + \dfrac{3}{x + 3} = \dfrac{23}{60}$ $\quad$ Excluded values: $x = 3,\ -3$

$60(x + 3)3 + 60(x - 3)3 = 23(x + 3)(x - 3)$
$180x + 540 + 180x - 540 = 23x^2 - 207$
$\quad\quad\quad\quad\quad\quad 0 = 23x^2 - 360x - 207$

$\quad\quad\quad\quad x = \dfrac{-b \pm \sqrt{b^2 - 4ac}}{2a}$ $\quad a = 23,\ b = -360,\ c = -207$

$\quad\quad\quad\quad x = \dfrac{-(-360) \pm \sqrt{(-360)^2 - 4(23)(-207)}}{2(23)}$

$\quad\quad\quad\quad x = 16.2$ or -0.6

Discarding the negative answer, we have $x = 16.2$ km/hr.

(C) Now $18 - 3 = 15$ km/hr = the rate rowed upstream

$18 + 3 = 21$ km/hr = the rate rowed downstream

So $\dfrac{3}{15} + \dfrac{3}{21}$ = round trip time

$= 0.343$ hr

$= 0.343 \times 60$ min

$= 20.6$ min   (1-1, 1-6)

63. Let x = number of bags of brand A

y = number of bags of brand B

Then (Nutrition from brand A) + (Nutrition from brand B) = (Total nutrition)

$8x$	+	$4y$	=	860 (Total nitrogen)
$9x$	+	$7y$	=	1,080 (Total phosphoric acid)

Solve the first equation for y in terms of x and substitute into the second equation.

$4y = 860 - 8x$

$y = 215 - 2x$

$9x + 7(215 - 2x) = 1,080$

$9x + 1,505 - 14x = 1,080$

$-5x = -425$

$x = 85$ bags of brand A

$y = 215 - 2(85)$

$y = 45$ bags of brand B (1-2)

64. (A) Apply the given formula with $C = 15$.

$15 = x^2 - 10x + 31$

$0 = x^2 - 10x + 16$

$x^2 - 10x + 16 = 0$

$(x - 8)(x - 2) = 0$

$x = 2$ or 8

Thus the output could be either 2000 or 8000 units.

(B) Apply the given formula with $C = 6$.

$6 = x^2 - 10x + 31$

$0 = x^2 - 10x + 25$

$x^2 - 10x + 25 = 0$

$(x - 5)^2 = 0$

$x = 5$

Thus the output must be 5000 units.   (1-6)

65. The break-even points are defined by $C = R$ (cost = revenue). Applying the formulas in this problem and the previous one, we have

$x^2 - 10x + 31 = 3x$

$x^2 - 13x + 31 = 0$

$x = \dfrac{-b \pm \sqrt{b^2 - 4ac}}{2a}$ $a = 1,\ b = -13,\ c = 31$

$x = \dfrac{-(-13) \pm \sqrt{(-13)^2 - 4(1)(31)}}{2(1)}$

$x = \dfrac{13 \pm \sqrt{45}}{2}$ thousand or approximately 3,146 and 9,854 units (1-6)

66. A profit will result if $R > C$, that is, if

$3x > x^2 - 10x + 31$

$x^2 - 10x + 31 < 3x$

$x^2 - 13x + 31 < 0$

The zeros of this polynomial were determined in problem 65 to be $\dfrac{13 \pm \sqrt{45}}{2}$ (in thousands)

$x^2 - 13x + 31$			
Test Number	0	5	10
Value of Polynomial for Test Number	31	-9	1
Sign of Polynomial in Interval	+	-	+
Interval	$(-\infty, \frac{13 - \sqrt{45}}{2})$	$(\frac{13 - \sqrt{45}}{2}, \frac{13 + \sqrt{45}}{2})$	$(\frac{13 + \sqrt{45}}{2}, \infty)$
approximately	$(-\infty, 3.146)$	$(3.146, 9.854)$	$(9.854, \infty)$

$x^2 - 13x + 31 < 0$, and a profit will result if $\frac{13 - \sqrt{45}}{2} < x < \frac{13 + \sqrt{45}}{2}$, or, approximately, $3.146 < x < 9.854$, in thousands. $(1-8)$

67. The distance of T from 110 must be no greater than 5. $|T - 110| \le 5$ $(1-4)$

68. Let x = width of page
$\qquad y$ = height of page

Then $xy = 480$, thus $y = \dfrac{480}{x}$.

Since the printed portion is surrounded by margins of 2 cm on each side, we have
$x - 4$ = width of printed portion
$y - 4$ = height of printed portion
Hence
$\quad (x - 4)(y - 4) = 320$, that is

$(x - 4)\left(\dfrac{480}{x} - 4\right) = 320$

Solving this, we obtain:

$x\left(\dfrac{480}{x}\right) - 4x - 4\left(\dfrac{480}{x}\right) + 16 = 320$

$\quad 480 - 4x - \dfrac{1,920}{x} + 16 = 320$

$\qquad\qquad -4x - \dfrac{1,920}{x} = -176 \quad \text{LCD: } x \quad x \ne 0$

$\qquad\quad -4x^2 - 1,920 = -176x$

$\qquad\qquad\qquad 0 = 4x^2 - 176x + 1,920$

$\qquad\qquad\qquad 0 = x^2 - 44x + 480$

$\qquad\qquad\qquad 0 = (x - 20)(x - 24)$

$\qquad\qquad x - 20 = 0 \quad \text{or} \quad x - 24 = 0$

$\qquad\qquad\qquad x = 20 \qquad\qquad x = 24$

$\qquad\qquad \dfrac{480}{x} = 24 \qquad\qquad \dfrac{480}{x} = 20$

Thus, the dimensions of the page are 20 cm by 24 cm. $(1-6)$

69.

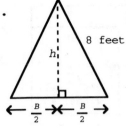

8 feet

h

$\frac{B}{2}$ $\frac{B}{2}$

In the isosceles triangle we note:

$$\frac{1}{2}Bh = A = 24$$

Hence

$$h = \frac{48}{B}$$

Applying the Pythagorean theorem, we have

$$h^2 + \left(\frac{B}{2}\right)^2 = 8^2$$

$$\left(\frac{48}{B}\right)^2 + \left(\frac{B}{2}\right)^2 = 8^2$$

$$\frac{2,304}{B^2} + \frac{B^2}{4} = 64$$

$$4B^2\left(\frac{2,304}{B^2}\right) + 4B^2\left(\frac{B^2}{4}\right) = 4B^2(64)$$

$$9,216 + B^4 = 256B^2$$

$$(B^2)^2 - 256B^2 + 9,216 = 0$$

$$B^2 = \frac{-b \pm \sqrt{b^2 - 4ac}}{2a} \quad a = 1, \ b = -256, \ c = 9,216$$

$$B^2 = \frac{-(-256) \pm \sqrt{(-256)^2 - 4(1)(9,216)}}{2(1)}$$

$$B^2 = \frac{256 \pm \sqrt{65,536 - 36,864}}{2}$$

$$B^2 = \frac{256 \pm \sqrt{28,672}}{2}$$

$$B^2 = 128 \pm 32\sqrt{7}$$

$$B = \sqrt{128 \pm 32\sqrt{7}}$$

$$B = 14.58 \text{ ft or } 6.58 \text{ ft}$$

(1-7)

CHAPTER 2
Exercise 2-1
Key Ideas and Formulas

There is a one-to-one correspondence between the points in a plane and the elements in the set of all ordered pairs of real numbers.

The graph of an equation in two variables is the graph of its solution set, the set of all ordered pairs of real numbers (a, b) that satisfy the equation.

To sketch the graph of an equation, we include enough points from its solution set so that the total graph is apparent. We may use point-to-point plotting and aids to graphing, such as symmetry.

A graph is symmetric with respect to the y axis if $(-a, b)$ is on the graph whenever (a, b) is on the graph.

A graph is symmetric with respect to the x axis if $(a, -b)$ is on the graph whenever (a, b) is on the graph.

A graph is symmetric with respect to the origin if $(-a, -b)$ is on the graph whenever (a, b) is on the graph.

Test for symmetry with respect to the	if equation is equivalent when
y axis	x is replaced with $-x$
x axis	y is replaced with $-y$
origin	x and y are replaced with $-x$ and $-y$.

Distance Formula:

Distance between $P_1(x_1, y_1)$ and $P_2(x_2, y_2)$: $d(P_1, P_2) = \sqrt{(x_2 - x_1)^2 + (y_2 - y_1)^2}$

Standard Equations of a Circle

1. Circle with radius r and center at (h, k): $(x - h)^2 + (y - k)^2 = r^2$ $r > 0$
2. Circle with radius r and center at $(0, 0)$: $x^2 + y^2 = r^2$ $r > 0$

---

1. The set of all points for which the x coordinate is 0 is the y axis.

3. The set of all points for which the x and y coordinates are negative is quadrant III.

5. The set of all points for which the x coordinate is positive and the y coordinate is negative is quadrant IV.

7. The set of all points for which $xy < 0$ includes those points for which the x coordinate is positive and the y coordinate is negative (quadrant IV) and also those points for which the x coordinate is negative and the y coordinate is positive (quadrant II).

9. The set of all points for which x is positive, excluding those points for which $y = 0$ (positive x axis), includes quadrants I and IV.

11. $y = 2x - 4$ y axis symmetry? $y = 2(-x) - 4$ $y = -2x - 4$ No, not equivalent
x axis symmetry? $-y = 2x - 4$ $y = 4 - 2x$ No, not equivalent
origin symmetry? $-y = 2(-x) - 4$ $-y = -2x - 4$
$y = 2x + 4$ No, not equivalent

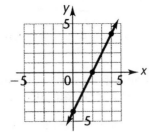

x	y
0	-4
2	0
4	4

The graph has none of the three symmetry properties.

13. $y = \frac{1}{2}x$ y axis symmetry? $y = \frac{1}{2}(-x)$ $y = -\frac{1}{2}x$ No, not equivalent

x axis symmetry? $-y = \frac{1}{2}x$ $y = -\frac{1}{2}x$ No, not equivalent

origin symmetry? $-y = \frac{1}{2}(-x)$ $y = \frac{1}{2}x$ Yes, equivalent

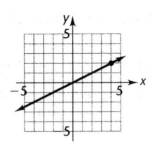

x	y
0	0
4	2

The graph has symmetry with respect to the origin.

We reflect the portion of the graph in quadrant I through the origin, using the origin symmetry.

15. $|y| = x$ y axis symmetry? $|y| = -x$ No, not equivalent
x axis symmetry? $|-y| = x$ $|y| = x$,   Yes, equivalent
origin symmetry? $|-y| = -x$ $|y| = -x$ No, not equivalent

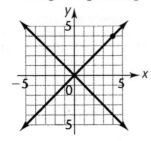

x	y
0	0
4	4

The graph has symmetry with respect to the x axis.

We reflect the portion of the graph where $y \geq 0$ through the x axis, using the x axis symmetry.

17. $|x| = |y|$ y axis symmetry? $|-x| = |y|$ $|x| = |y|$ Yes, equivalent
x axis symmetry? $|x| = |-y|$ $|x| = |y|$ Yes, equivalent
origin symmetry follows automatically.

x	y
0	0
4	4

The graph has all three symmetries.

We reflect the portion of the graph in quadrant I through the y axis, and the x axis, and the origin, using all three symmetries.

19. $d = \sqrt{[4 - (-5)]^2 + [2 - (-3)]^2}$
$= \sqrt{9^2 + 5^2} = \sqrt{106}$

Common Error: *not* $9 + 5$.

21. $d = \sqrt{(2 - 3)^2 + [(-4) - 5]^2}$
$= \sqrt{(-1)^2 + (-9)^2} = \sqrt{82}$

23. $(x - 0)^2 + (y - 0)^2 = 4^2$
$x^2 + y^2 = 16$

25. $(x - 3)^2 + [y - (-2)]^2 = 1^2$
$(x - 3)^2 + (y + 2)^2 = 1$

Common Error: *not* $(y - 2)^2$.

27. $(x - 2)^2 + (y - 6)^2 = \sqrt{3}^2$
$(x - 2)^2 + (y - 6)^2 = 3$

29. (A) When $x = -3$, the corresponding y value on the graph is 3, to the nearest integer.

(B) When $x = 2$, the corresponding y value on the graph is -2, to the nearest integer.

(C) Three values of x correspond to $y = 3$ on the graph. To the nearest integer they are -3, -1, and 4.

(D) Three values of x correspond to $y = -1$ on the graph. To the nearest integer, they are -4, 1, and 3.

31. (A) Reflect the given graph across the x axis.

(B) Reflect the given graph across the y axis.

(C) Reflect the given graph through the origin.

(D) Reflect the given graph across the y axis, then reflect the resulting curve across the x axis.

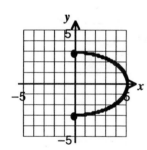

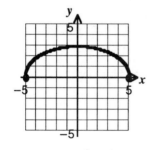

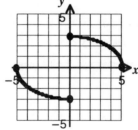

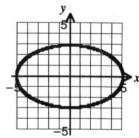

33. $y^2 = x + 2$

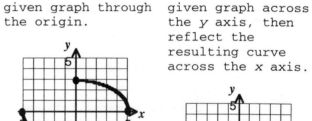

y axis symmetry?	$y^2 = (-x) + 2$	$y^2 = 2 - x$	No, not equivalent
x axis symmetry?	$(-y)^2 = x + 2$	$y^2 = x + 2$	Yes, equivalent
origin symmetry?	$(-y)^2 = (-x) + 2$	$y^2 = 2 - x$	No, not equivalent

x	y
-2	0
-1	1
2	2

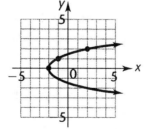

The graph has x axis symmetry.
To obtain the portion of the graph for $y \geq 0$, we sketch $y = \sqrt{x + 2}$, $x \geq -2$.
We reflect the portion of the graph for $y \geq 0$ across the x axis, using the x axis symmetry.

35. $y = x^2 + 1$ 　　y axis symmetry? $y = (-x)^2 + 1$ 　$y = x^2 + 1$ 　　Yes, equivalent

　　　　　　　　　　x axis symmetry? $-y = x^2 + 1$ 　　$y = -x^2 - 1$ 　　No, not equivalent

　　　　　　　　　　origin symmetry? $-y = (-x)^2 + 1$ 　$y = -x^2 - 1$ 　　No, not equivalent

x	y
0	1
1	2
2	5

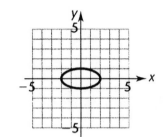

The graph has y axis symmetry.

We reflect the portion of the graph for $x \geq 0$ across the y axis, using the y axis symmetry.

37. $x^2 + 4y^2 = 4$ 　y axis symmetry? $(-x)^2 + 4y^2 = 4$ 　$x^2 + 4y^2 = 4$ Yes, equivalent

$4y^2 = 4 - x^2$ 　　x axis symmetry? $x^2 + 4(-y)^2 = 4$ 　$x^2 + 4y^2 = 4$ Yes, equivalent

$y^2 = \dfrac{4 - x^2}{4}$ 　origin symmetry follows automatically.

$y = \pm \dfrac{1}{2}\sqrt{4 - x^2}$

x	y
0	1
1	$\dfrac{\sqrt{3}}{2} \approx .9$
2	0

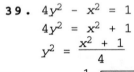

The graph has all three symmetries.

To obtain the quadrant I portion of the graph, we sketch $y = \dfrac{1}{2}\sqrt{4 - x^2}$,

$0 \leq x \leq 2$. We reflect this graph across the y axis, then reflect everything across the x axis.

39. $4y^2 - x^2 = 1$ 　y axis symmetry? $4y^2 - (-x)^2 = 1$ $4y^2 - x^2 = 1$ 　Yes, equivalent

$4y^2 = x^2 + 1$ 　　x axis symmetry? $4(-y)^2 - x^2 = 1$ $4y^2 - x^2 = 1$ 　Yes, equivalent

$y^2 = \dfrac{x^2 + 1}{4}$ 　origin symmetry follows automatically.

$y = \pm\dfrac{1}{2}\sqrt{x^2 + 1}$

x	y
0	$\dfrac{1}{2}$
2	$\dfrac{1}{2}\sqrt{5} \approx 1.1$
4	$\dfrac{1}{2}\sqrt{17} \approx 2.0$

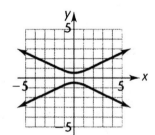

The graph has all three symmetries.

To obtain the quadrant I portion of the graph, we sketch $y = \dfrac{1}{2}\sqrt{x^2 + 1}$, $x \geq 0$.

We reflect this graph across the y axis, then reflect everything across the x axis.

41. $y^3 = x$ 　y axis symmetry? $y^3 = -x$ 　　　　　　No, not equivalent

　　　　　x axis symmetry? $(-y)^3 = x$ 　$y^3 = -x$ 　No, not equivalent

　　　　　origin symmetry? $(-y)^3 = -x$ 　$y^3 = x$ 　Yes, equivalent

x	y
0	0
1	1
8	2

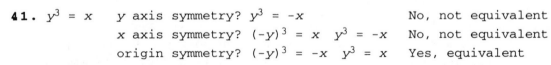

The graph has symmetry with respect to the origin.

We reflect the portion of the graph in quadrant I through the origin, using the origin symmetry.

43. $y = 0.6x^2 - 4.5$ y axis symmetry? $y = 0.6(-x)^2 - 4.5$ $y = 0.6x^2 - 4.5$ Yes, equivalent

x axis symmetry? $-y = 0.6x^2 - 4.5$ $y = -0.6x^2 + 4.5$ No, not equivalent

origin symmetry? $-y = 0.6(-x^2) - 4.5$ $y = -0.6x^2 + 4.5$ No, not equivalent

x	y
0	-4.5
1	-3.9
2	-2.1
3	0.9
4	5.1

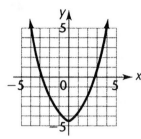

The graph has y axis symmetry.

We reflect the portion of the graph for $x \geq 0$ across the y axis, using the y axis symmetry.

45. $y = \sqrt{17 - x^2}$ y axis symmetry? $y = \sqrt{17 - (-x)^2}$ $y = \sqrt{17 - x^2}$ Yes, equivalent

x axis symmetry? $-y = \sqrt{17 - x^2}$ $y = -\sqrt{17 - x^2}$ No, not equivalent

origin symmetry? $-y = \sqrt{17 - (-x)^2}$ $y = -\sqrt{17 - x^2}$ No, not equivalent

x	y
0	$\sqrt{17} \approx 4.1$
1	$\sqrt{16} = 4$
2	$\sqrt{13} \approx 3.6$
3	$\sqrt{8} \approx 2.8$
4	$\sqrt{1} = 1$
$\sqrt{17}$	0

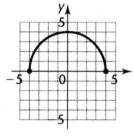

The graph has y axis symmetry.

We reflect the portion of the graph for $x \geq 0$ across the y axis, using the y axis symmetry.

47. $y = x^{2/3}$ y axis symmetry? $y = (-x)^{2/3}$ $y = x^{2/3}$ Yes, equivalent

x axis symmetry? $-y = x^{2/3}$ $y = -(x^{2/3})$ No, not equivalent

origin symmetry? $-y = (-x)^{2/3}$ $y = -(x^{2/3})$ No, not equivalent

x	y
0	0
1	1
2	$2^{2/3} \approx 1.6$
3	$3^{2/3} \approx 2.1$

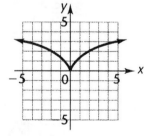

The graph has y axis symmetry.

We reflect the portion of the graph for $x \geq 0$ across the y axis, using the y axis symmetry.

49. The lengths of the three sides are found from the distance formula to be:

$$\sqrt{[1 - (-3)]^2 + [(-2) - (2)]^2} = \sqrt{4^2 + (-4)^2} = \sqrt{32}$$

$$\sqrt{[8 - 1]^2 + [5 - (-2)]^2} = \sqrt{7^2 + 7^2} = \sqrt{98}$$

$$\sqrt{[8 - (-3)]^2 + [5 - 2]^2} = \sqrt{11^2 + 3^2} = \sqrt{130}$$

Since $(\sqrt{130})^2 = (\sqrt{32})^2 + (\sqrt{98})^2$, the triangle is a right triangle.

Area $= \frac{1}{2} ab = \frac{1}{2}\sqrt{32}\sqrt{98} = 28$

Perimeter $= a + b + c = \sqrt{32} + \sqrt{98} + \sqrt{130} \approx 26.96$

Common Error:
Confusing these true statements with the *false* statement
$\sqrt{130} = \sqrt{32} + \sqrt{98}$.

51. The distance formula requires that $\sqrt{[x - (-4)]^2 + (7 - 1)^2} = 10$
Solving, we have $\sqrt{(x + 4)^2 + (6)^2} = 10$
$$(x + 4)^2 + (6)^2 = 10^2$$
$$(x + 4)^2 + 36 = 100$$
$$(x + 4)^2 = 64$$
$$x + 4 = \pm 8$$
$$x = -4 \pm 8$$
$$x = -12, \ 4$$

53. The distance formula requires that $\sqrt{[2 - (-1)]^2 + (y - 4)^2} = 3$
Solving, we have $\sqrt{(3)^2 + (y - 4)^2} = 3$
$$(3)^2 + (y - 4)^2 = (3)^2$$
$$(y - 4)^2 = 0$$
$$y - 4 = 0$$
$$y = 4$$

55. $(x + 4)^2 + (y - 2)^2 = 7$
$[x - (-4)]^2 + (y - 2)^2 = (\sqrt{7})^2$
Center $(-4, 2)$; Radius $= \sqrt{7}$

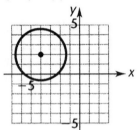

57. $x^2 + y^2 - 6x - 4y = 36$
$x^2 - 6x + y^2 - 4y = 36$
$x^2 - 6x + 9 + y^2 - 4y + 4 = 36 + 9 + 4$
$(x - 3)^2 + (y - 2)^2 = 49$
Center $(3, 2)$; Radius $= 7$

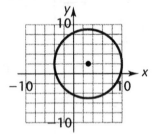

59. $x^2 + y^2 + 8x - 6y + 8 = 0$
$x^2 + 8x + y^2 - 6y = -8$
$x^2 + 8x + 16 + y^2 - 6y + 9 = -8 + 16 + 9$
$(x + 4)^2 + (y - 3)^2 = 17$
Center $(-4, 3)$; Radius $= \sqrt{17}$

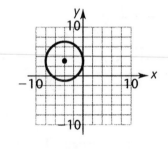

61.

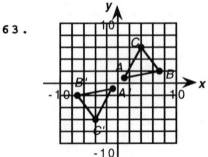

Each point, and therefore the whole graph, is transformed into its reflection across the x axis.

63.

Each point, and therefore the whole graph, is transformed into its reflection across the origin.

65. $x^2 + y^2 = 3$
$\qquad\quad y^2 = 3 - x^2$
$\qquad\quad y = \pm\sqrt{3 - x^2}$

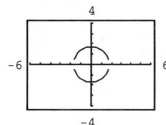

67. $(x + 3)^2 + (y + 1)^2 = 2$
$\qquad\qquad\quad (y + 1)^2 = 2 - (x + 3)^2$
$\qquad\qquad\qquad\; y + 1 = \pm\sqrt{2 - (x + 3)^2}$
$\qquad\qquad\qquad\quad\; y = -1 \pm \sqrt{2 - (x + 3)^2}$

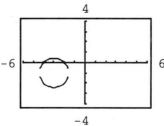

69.

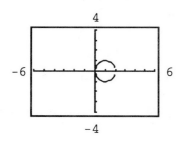

From the graph, the center of the circle is at $(1, 0)$ and its radius is 1. The equation of the circle is therefore $(x - 1)^2 + (y - 0)^2 = 1^2$ or
$\qquad (x - 1)^2 + y^2 = 1$
To check this observation, solve this equation for y in terms of x and note that this gives the original equations for $0 \le x \le 2$.

Solving for y, we obtain
$\qquad y^2 = 1 - (x - 1)^2$
$\qquad y^2 = 1 - (x^2 - 2x + 1)$
$\qquad y^2 = 1 - x^2 + 2x - 1$
$\qquad y^2 = 2x - x^2$
$\qquad\; y = \pm\sqrt{2x - x^2}$

71.

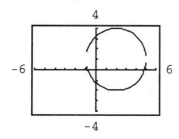

From the graph, the center of the circle is at $(2, 1)$ and its radius is 3. The equation of the circle is therefore $(x - 2)^2 + (y - 1)^2 = 3^2$ or
$\qquad (x - 2)^2 + (y - 1)^2 = 9$
To check this observation, solve this equation for y in terms of x and note that this gives the original equations for $-1 \le x \le 5$.

Solving for y, we obtain
$\qquad (y - 1)^2 = 9 - (x - 2)^2$
$\qquad (y - 1)^2 = 9 - (x^2 - 4x + 4)$
$\qquad (y - 1)^2 = 9 - x^2 + 4x - 4$
$\qquad (y - 1)^2 = 5 + 4x - x^2$
$\qquad\quad\; y - 1 = \pm\sqrt{5 + 4x - x^2}$
$\qquad\qquad\; y = 1 \pm \sqrt{5 + 4x - x^2}$

73. $y^3 = |x|$ $\quad$ y axis symmetry? $y^3 = |-x|$ $\qquad$ $y^3 = |x|$ $\qquad$ Yes, equivalent
$\qquad\qquad\qquad\;\; x$ axis symmetry? $(-y)^3 = |x|$ $\qquad$ $-y^3 = |x|$ $\qquad$ No, not equivalent
$\qquad\qquad\qquad\;\;$ origin symmetry? $(-y)^3 = |-x|$ $\qquad$ $-y^3 = |x|$ $\qquad$ No, not equivalent

x	y
0	0
1	1
8	2

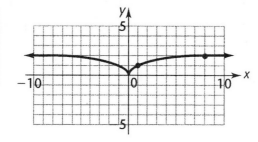

The graph has y axis symmetry.

We reflect the portion of the graph for $x \ge 0$ across the y axis, using the y axis symmetry.

75. $xy = 1$ y axis symmetry? $-xy = 1$ No, not equivalent
              x axis symmetry? $x(-y) = 1$ $-xy = 1$ No, not equivalent
              origin symmetry? $(-x)(-y) = 1$ $xy = 1$ Yes, equivalent

x	y
1	1
2	$\frac{1}{2}$
3	$\frac{1}{3}$
$\frac{1}{2}$	2
$\frac{1}{3}$	3

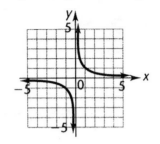

The graph has origin symmetry.

We reflect the portion of the graph in quadrant I through the origin, using the origin symmetry.

77. $y = 6x - x^2$ y axis symmetry? $y = 6(-x) - (-x)^2$ $y = -6x - x^2$ No, not equivalent
                   x axis symmetry? $-y = 6x - x^2$ $y = x^2 - 6x$ No, not equivalent
                   origin symmetry? $-y = 6(-x) - (-x)^2$ $y = x^2 + 6x$ No, not equivalent

x	y
-1	-7
0	0
1	5
2	8
3	9
4	8
5	5
6	0
7	-7

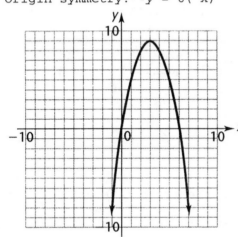

The graph has none of the three symmetries.

A larger table of values is needed since we have no symmetry information.

79. From geometry we know that the perpendicular bisector of a line segment is the set of points whose distances to the end points of the segment are equal. Using the distance formula, let (x, y) be a typical point on the perpendicular bisector. Then the distance from (x, y) to $(-6, -2)$ equals the distance from (x, y) to $(4, 4)$.

$$\sqrt{[x - (-6)]^2 + [y - (-2)]^2} = \sqrt{(x - 4)^2 + (y - 4)^2}$$
$$(x + 6)^2 + (y + 2)^2 = (x - 4)^2 + (y - 4)^2$$
$$x^2 + 12x + 36 + y^2 + 4y + 4 = x^2 - 8x + 16 + y^2 - 8y + 16$$
$$20x + 12y = -8$$
$$5x + 3y = -2$$

81. The center of the circle must be the midpoint of the diameter segment. Using the midpoint formula from problem 80, we have: center(= midpoint) has

coordinates $\left(\dfrac{7 + 1}{2}, \dfrac{-3 + 7}{2} \right) = (4, 2)$. The radius of the circle must be the

distance from the center to one of the endpoints of the diameter. It does not matter which endpoint we choose. For example,
$r = \sqrt{(7 - 4)^2 + [-3 - 2]^2} = \sqrt{3^2 + (-5)^2} = \sqrt{34}$.
The equation of the circle is
$(x - 4)^2 + (y - 2)^2 = (\sqrt{34})^2$ or $(x - 4)^2 + (y - 2)^2 = 34$.

83. The equation of a circle with center (h, k) and radius r is $(x - h)^2 + (y - k)^2$ $= r^2$. We are given a circle with center $(2, 2)$; therefore the equation must be of form $(x - 2)^2 + (y - 2)^2 = r^2$. Since the graph passes through the point $(3, -5)$, its coordinates must satisfy the equation, hence:

$$(3 - 2)^2 + [(-5) - 2]^2 = r^2$$
$$1^2 + (-7)^2 = r^2$$
$$50 = r^2$$

Thus the equation of the circle is

$$(x - 2)^2 + (y - 2)^2 = 50$$

85. Reflecting a point (x, y) through the x axis yields the point $(x, -y)$. Reflecting this point through the origin yields the point $(-x, y)$. This point is the same point that would result from reflecting the original point through the y axis. Therefore, if the graph is unchanged by reflecting through the x axis and through the origin, it will be unchanged by reflecting through the y axis and will necessarily have y axis symmetry.

87. (A) $6.00 on the price scale corresponds to 3,000 cases on the demand scale.

(B) The demand decreases from 3,000 to 2,600 cases, that is, by 400 cases.

(C) The demand increases from 3,000 to 3,600 cases, that is, by 600 cases.

(D) Demand decreases with increasing price and increases with decreasing price. To increase demand from 2,000 to 4,000 cases, a price decrease from $6.90 to $5.60 is necessary.

89. (A) 9:00 is halfway from 6 AM to noon, and corresponds to a temperature of 53°.

(B) The highest temperature occurs halfway from noon to 6 PM, at 3 PM. This temperature is 68°.

(C) This temperature occurs at 1 AM, 7 AM, and 11 PM.

91. (A) There is no obvious symmetry. A table of values yields the following approximate values:

x	0	0.5	1	1.5	2
v	0.7	0.6	0.5	0.35	0

(C) For a displacement of 2 cm, the ball is stationary ($v = 0$). As the vertical displacement approaches 0, the ball gathers speed until $v = 0.5\sqrt{2} \approx 0.7$ m/sec. The velocity carries the ball through the equilibrium position ($x = 0$) to rise again to displacement 2 cm and velocity 0.

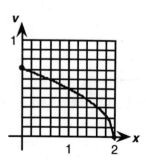

93. Using the hint, we note that $(2, r - 1)$ must satisfy $x^2 + y^2 = r^2$, that is

$$2^2 + (r - 1)^2 = r^2$$
$$4 + r^2 - 2r + 1 = r^2$$
$$-2r + 5 = 0$$
$$r = \frac{5}{2} \text{ or } 2.5 \text{ ft.}$$

95. (A) From the drawing, we can write:

$$\begin{pmatrix}\text{Distance from tower} \\ \text{to town } B\end{pmatrix} = 2 \times \begin{pmatrix}\text{Distance from tower} \\ \text{to town } A\end{pmatrix}$$

$$\begin{pmatrix}\text{Distance from } (x,\ y) \\ \text{to } (36,\ 15)\end{pmatrix} = 2 \times \begin{pmatrix}\text{Distance from } (x,\ y) \\ \text{to } (0,\ 0)\end{pmatrix}$$

$$\sqrt{(36 - x)^2 + (15 - y)^2} = 2\sqrt{(0 - x)^2 + (0 - y)^2}$$
$$\sqrt{(36 - x)^2 + (15 - y)^2} = 2\sqrt{x^2 + y^2}$$
$$(36 - x)^2 + (15 - y)^2 = 4(x^2 + y^2)$$
$$1{,}296 - 72x + x^2 + 225 - 30y + y^2 = 4x^2 + 4y^2$$
$$1{,}521 = 3x^2 + 3y^2 + 72x + 30y$$
$$507 = x^2 + y^2 + 24x + 10y$$
$$144 + 25 + 507 = x^2 + 24x + 144 + y^2 + 10y + 25$$
$$676 = (x + 12)^2 + (y + 5)^2$$

The circle has center (-12, -5) and radius 26.

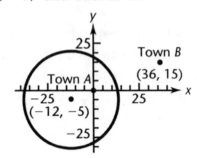

(B) All points due east of Town A have y coordinate 0 in this coordinate system. The points on the circle for which $y = 0$ are found by substituting $y = 0$ into the equation of the circle and solving for x.

$$(x + 12)^2 + (y + 5)^2 = 676$$
$$(x + 12)^2 + 25 = 676$$
$$(x + 12)^2 = 651$$
$$x + 12 = \pm\sqrt{651}$$
$$x = -12 \pm \sqrt{651}$$

Choosing the positive square root so that x is greater than -12 (east rather than west) we have $x = -12 + \sqrt{651} \approx 13.5$ miles.

Exercise 2-2

Key Ideas and Formulas

The graph of the equation

$$Ax + By = C$$

where A, B, and C are constants (A and B not both 0), is a straight line. Any straight line in a rectangular coordinate system has an equation of this form. If a line passes through two distinct points $P_1(x_1,\ y_1)$ and $P_2(x_2,\ y_2)$, its slope is given by

$$m = \frac{y_2 - y_1}{x_2 - x_1} \qquad x_2 \neq x_1$$
$$= \frac{\text{vertical change}}{\text{horizontal change}} = \frac{\text{rise}}{\text{run}}.$$

For a vertical line, $x_2 = x_1$, hence slope is not defined. Equation: $x = a$.
For a horizontal line, $y_2 = y_1$, slope = 0. Equation: $y = b$.

Slope-intercept form of an equation of a non-vertical line:

$$y = mx + b \qquad m = \frac{\text{Rise}}{\text{Run}} = \text{Slope} \qquad b = y \text{ intercept}$$

Point-slope form of an equation of a non-vertical line through (x_1, y_1) with slope m:

$$y - y_1 = m(x - x_1)$$

Given two nonvertical lines L_1 and L_2 with slopes m_1 and m_2, respectively, then

L_1 is parallel to L_2 ($L_1 \parallel L_2$) if and only if $m_1 = m_2$

L_1 is perpendicular to L_2 ($L_1 \perp L_2$) if and only if $m_1 m_2 = -1$. $(m_1 = -\frac{1}{m_2})$

Summary, Equations of a Line

Standard form	$Ax + By = C$	A and B not both 0
Slope-intercept form	$y = mx + b$	Slope: m; y intercept: b
Point-slope form	$y - y_1 = m(x - x_1)$	Slope: m; Point (x_1, y_1)
Horizontal line	$y = b$	Slope 0
Vertical line	$x = a$	Slope: Undefined

---

1. The x intercept is -2. The y intercept is 2. From the point $(-2, 0)$ to the point $(0, 2)$, the value of y increases by 2 units as the value of x increases by 2 units. Thus slope $= \dfrac{\text{change in } y}{\text{change in } x} = \dfrac{2}{2} = 1$.

Equation: $y = mx + b$
$\qquad\quad y = 1x + 2$ or $y = x + 2$

3. The x intercept is -2. The y intercept is -4. From the point $(-2, 0)$ to the point $(0, -4)$ the value of y decreases by 4 units as the value of x increases by 2 units. Thus, the slope $= \dfrac{\text{change in } y}{\text{change in } x} = \dfrac{-4}{2} = -2$.

Equation: $y = mx + b$
$\qquad\quad y = -2x + (-4)$ or $y = -2x - 4$

5. The x intercept is 3. The y intercept is -1. From the point $(0, -1)$ to the point $(3, 0)$ the value of y increases by 1 unit as the value of x increases by 3 units. Thus, the slope $= \dfrac{\text{change in } y}{\text{change in } x} = \dfrac{1}{3}$.

Equation: $y = mx + b$
$\qquad\quad y = \dfrac{1}{3}x + (-1)$ or $y = \dfrac{1}{3}x - 1$

7. $y = -\dfrac{3}{5}x + 4$ slope $-\dfrac{3}{5}$

x	y
0	4
5	1
-5	7

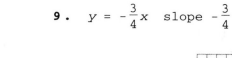

9. $y = -\dfrac{3}{4}x$ slope $-\dfrac{3}{4}$

x	y
0	0
4	-3
-4	3

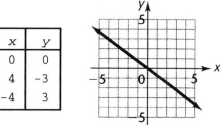

11. $2x - 3y = 15$

$\qquad -3y = -2x + 15$

$\qquad y = \frac{2}{3}x - 5 \quad$ slope $\frac{2}{3}$

x	y
0	-5
3	-3
-3	-7

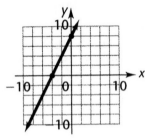

13. $4x - 5y = -24$

$\qquad -5y = -4x - 24$

$\qquad y = \frac{4}{5}x + \frac{24}{5} \quad$ slope $\frac{4}{5}$

x	y
-1	4
4	8
9	12

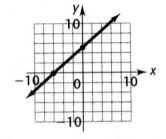

15. $\frac{y}{8} - \frac{x}{4} = 1$

$\qquad \frac{y}{8} = \frac{x}{4} + 1$

$\qquad y = 2x + 8 \quad$ slope 2

x	y
0	8
-4	0
-8	-8

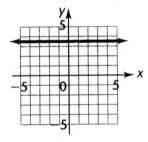

17. $x = -3$ slope not defined
vertical line

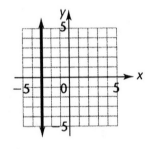

19. $y = 3.5$ slope 0
horizontal line

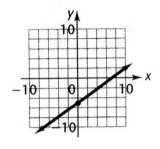

21. Slope and y intercept are given; we use slope-intercept form.

$y = 1x + 0$

$y = x$

$0 = x - y$

$x - y = 0$

23. Slope and y intercept are given; we use slope-intercept form.

$y = -\frac{2}{3}x + (-4)$

$y = -\frac{2}{3}x - 4$

$3y = -2x - 12$

$2x + 3y = -12$

25. A point and the slope are given; we use point-slope form.

$y - 3 = -2(x - 0)$

$y - 3 = -2x$

$\qquad y = -2x + 3$

27. A point and the slope are given; we use point-slope form.

$$y - 4 = \frac{3}{2}[x - (-5)]$$

$$y - 4 = \frac{3}{2}[x + 5]$$

$$y - 4 = \frac{3}{2}x + \frac{15}{2}$$

$$y = \frac{3}{2}x + \frac{23}{2}$$

29. Two points are given; we first find the slope, then we use point-slope form.

$$m = \frac{-3 - 5}{4 - 2} = \frac{-8}{2} = -4$$

$$y - 5 = -4(x - 2)$$
$$y - 5 = -4x + 8$$
$$y = -4x + 13$$

or

$$y - (-3) = -4(x - 4)$$
$$y + 3 = -4x + 16$$
$$y = -4x + 13$$

Thus, it does not matter which point is chosen in substituting into the point-slope form; both points must give rise to the same equation.

31. We proceed as in problem 29.

$$m = \frac{5 - 2}{(-3) - (-3)} = \frac{3}{0}$$

Slope is undefined.
A vertical line through (-3, 2) has equation $x = -3$.

33. We proceed as in problem 29.

$$m = \frac{2 - 2}{0 - 4} = \frac{0}{-4} = 0$$

$$y - 2 = 0(x - 0)$$
$$y - 2 = 0$$
$$y = 2$$

35. A line parallel to $y = -3x + 7$ will have the same slope, namely -3. We now use the point-slope form.

$$y - (-1) = -3(x - 2)$$
$$y + 1 = -3x + 6$$
$$y = -3x + 5$$
$$3x + y = 5$$

37. A line parallel to $2x + 3y = 9$ will have the same slope; to determine the slope we write $2x + 3y = 9$ in slope-intercept form:

$$2x + 3y = 9$$
$$3y = -2x + 9$$
$$y = \frac{-2x + 9}{3}$$
$$y = -\frac{2}{3}x + 3$$

The slope is $-\frac{2}{3}$. We now use the point-slope form.

$$y - (-4) = -\frac{2}{3}(x - 0)$$

$$y + 4 = -\frac{2}{3}x$$

$$y = -\frac{2}{3}x - 4$$

$$3y = -2x - 12$$
$$2x + 3y = -12$$

39. A line parallel to the x axis is horizontal. The equation of a horizontal line through (3, 3) is $y = 3$.

41. A line perpendicular to $y = \frac{3}{2}x - 4$ will have slope satisfying $\frac{3}{2}m = -1$, or $m = -\frac{2}{3}$. We use the point-slope form.

$$y - 5 = -\frac{2}{3}(x - 4)$$

$$y - 5 = -\frac{2}{3}x + \frac{8}{3}$$

$$y = -\frac{2}{3}x + \frac{23}{3}$$

$$3y = -2x + 23$$
$$2x + 3y = 23$$

43. First we determine the slope of $-5x + 2y = 1$ by writing it in slope-intercept form:

$$-5x + 2y = 1$$
$$2y = 5x + 1$$
$$y = \frac{5x + 1}{2}$$
$$y = \frac{5}{2}x + \frac{1}{2}$$

A line perpendicular to this will have slope satisfying $\frac{5}{2}m = -1$, or $m = -\frac{2}{5}$. We use the point-slope form.

$$y - 0 = -\frac{2}{5}(x - 5)$$
$$y = -\frac{2}{5}x + 2$$
$$5y = -2x + 10$$
$$2x + 5y = 10$$

45. A line perpendicular to the x axis is vertical. The equation of a vertical line through $(-2, -3)$ is $x = -2$.

47. We calculate the slopes of the four sides:

$$m_{AB} = \frac{7 - 2}{8 - (-2)} = \frac{5}{10} = \frac{1}{2} \qquad m_{BC} = \frac{1 - 7}{10 - 8} = \frac{-6}{2} = -3$$

$$m_{CD} = \frac{(-6) - 1}{(-4) - 10} = \frac{-7}{-14} = \frac{1}{2} \qquad m_{DA} = \frac{2 - (-6)}{(-2) - (-4)} = \frac{8}{2} = 4$$

Since the quadrilateral has one pair of parallel sides, but not two pairs (AB and CD), it is a trapezoid.

49. We calculate the slopes of the four sides.

$$m_{AB} = \frac{(-1) - 2}{4 - 0} = \frac{-3}{4} \qquad m_{BC} = \frac{(-5) - (-1)}{1 - 4} = \frac{4}{3}$$

$$m_{CD} = \frac{(-2) - (-5)}{(-3) - 1} = \frac{3}{-4} = \frac{-3}{4} \qquad m_{DA} = \frac{2 - (-2)}{0 - (-3)} = \frac{4}{3}$$

Since the quadrilateral has two pairs of opposite sides parallel, it is a parallelogram. Since moreover adjacent sides are perpendicular $\left[\left(-\frac{3}{4}\right)\left(\frac{4}{3}\right) = -1\right]$ it is a rectangle.

51. midpoint of the line segment $= \left(\frac{(-4) + 2}{2}, \frac{(-3) + 4}{2}\right) = \left(-1, \frac{1}{2}\right)$

slope of the line segment $= \frac{4 - (-3)}{2 - (-4)} = \frac{7}{6}$

We require the equation of a line, through the midpoint of the line segment, which is perpendicular to the line segment. Its slope will satisfy $\frac{7}{6}m = -1$, or $m = -\frac{6}{7}$. We use the point-slope form.

$$y - \frac{1}{2} = -\frac{6}{7}[x - (-1)]$$
$$y - \frac{1}{2} = -\frac{6}{7}(x + 1)$$
$$y - \frac{1}{2} = -\frac{6}{7}x - \frac{6}{7}$$
$$y = -\frac{6}{7}x - \frac{6}{7} + \frac{1}{2}$$
$$y = -\frac{6}{7}x - \frac{5}{14}$$

53. The circle has center (0, 0). The radius drawn from (0, 0) to the given point (3, 4) has slope given by

$$M_R = \frac{4 - 0}{3 - 0} = \frac{4}{3}$$

Therefore, the slope of the tangent line is given by

$$\left(\frac{4}{3}\right)m = -1 \quad \text{or} \quad m = -\frac{3}{4}$$

We require the equation of a line through (3, 4) with slope $-\frac{3}{4}$. We use the point-slope form.

$$y - 4 = -\frac{3}{4}(x - 3)$$

$$y - 4 = -\frac{3}{4}x + \frac{9}{4}$$

$$4y - 16 = -3x + 9$$

$$3x + 4y = 25$$

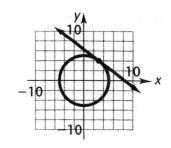

55. The circle has center (0, 0). The radius drawn from (0, 0) to the given point (5, -5) has slope given by

$$M_R = \frac{-5 - 0}{5 - 0} = -1$$

Therefore, the slope of the tangent line is given by

$$(-1)m = -1 \quad \text{or} \quad m = 1$$

We require the equation of a line through (5, -5) with slope 1. We use the point-slope form.

$$y - (-5) = 1(x - 5)$$

$$y + 5 = x - 5$$

$$x - y = 10$$

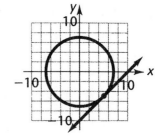

57. The circle has center (3, -4). The radius drawn from (3, -4) to the given point (8, -16) has slope given by

$$M_R = \frac{-16 - (-4)}{8 - 3} = -\frac{12}{5}$$

Therefore, the slope of the tangent line is given by

$$\left(-\frac{12}{5}\right)m = -1 \quad \text{or} \quad m = \frac{5}{12}$$

We require the equation of a line through (8, -16) with slope $\frac{5}{12}$. We use the point-slope form.

$$y - (-16) = \frac{5}{12}(x - 8)$$

$$y + 16 = \frac{5}{12}x - \frac{40}{12}$$

$$12y + 192 = 5x - 40$$

$$232 = 5x - 12y$$

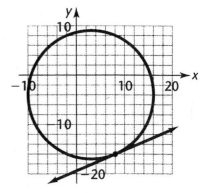

59. (A) Solve each equation for y in terms of x, then graph.

$$3x + 2y = 6 \qquad 3x + 2y = -6 \qquad 3x + 2y = 3 \qquad 3x + 2y = -3$$
$$2y = 6 - 3x \qquad 2y = -6 - 3x \qquad 2y = 3 - 3x \qquad 2y = -3 - 3x$$
$$y = \frac{6 - 3x}{2} \qquad y = \frac{-6 - 3x}{2} \qquad y = \frac{3 - 3x}{2} \qquad y = \frac{-3 - 3x}{2}$$

x	$y = \dfrac{6 - 3x}{2}$	$y = \dfrac{-6 - 3x}{2}$	$y = \dfrac{3 - 3x}{2}$	$y = \dfrac{-3 - 3x}{2}$
0	3	-3	$\frac{3}{2}$	$-\frac{3}{2}$
2	0	-6	$-\frac{3}{2}$	$-\frac{9}{2}$
-2	6	0	$\frac{9}{2}$	$\frac{3}{2}$

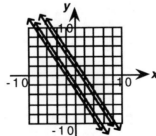

(B) The family of lines all have slope $y = -\dfrac{A}{B}$ and y intercept $\dfrac{C}{B}$. The slopes are therefore equal and the lines are parallel.

(C) Solve $Ax + By = C$ for y in terms of x.
$$By = -Ax + C$$
$$y = \frac{-Ax + C}{B}$$
$$y = -\frac{A}{B}x + \frac{C}{B}$$

For fixed A and B and varying C this is a family of lines as described in part (B).

61. $y = \left|\dfrac{1}{2}x\right|$ y axis symmetry? $y = \left|\dfrac{1}{2}(-x)\right|$ $y = \left|\dfrac{1}{2}x\right|$ Yes, equivalent

x axis symmetry? $-y = \left|\dfrac{1}{2}x\right|$ $y = -\left|\dfrac{1}{2}x\right|$ No, not equivalent

origin symmetry? $-y = \left|\dfrac{1}{2}(-x)\right|$ $y = -\left|\dfrac{1}{2}x\right|$ No, not equivalent

x	y
0	0
2	1
4	2

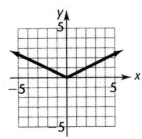

The graph has y axis symmetry.

We reflect the portion of the graph in quadrant I through the y axis, using the y axis symmetry.

63. $y = 2|x| - 4$ y axis symmetry $y = 2|-x| - 4$ $y = 2|x| - 4$ Yes, equivalent
          x axis symmetry $-y = 2|x| - 4$ $y = -2|x| + 4$ No, not equivalent
          origin symmetry $-y = 2|-x| - 4$ $y = -2|x| + 4$ No, not equivalent

x	y
0	-4
1	-2
2	0
3	2

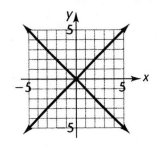

The graph has y axis symmetry. We reflect the portion of the graph to the right of the y axis through the y axis, using the y axis symmetry.

65. $x^2 - y^2 = 0$ y axis symmetry? $(-x)^2 - y^2 = 0$ $x^2 - y^2 = 0$ Yes, equivalent
          x axis symmetry? $x^2 - (-y)^2 = 0$ $x^2 - y^2 = 0$ Yes, equivalent
          origin symmetry follows automatically.

x	y
0	0
4	4

The graph has all three symmetries.

We reflect the portion of the graph in quadrant I through the y axis, and the x axis, and the origin, using all three symmetries.

67. The graph of $y = |mx + b|$ is the same as the graph of $y = mx + b$ for $mx + b$ positive, that is, for $x > -\dfrac{b}{m}$ (m positive) or for $x < -\dfrac{b}{m}$ (m negative). It is however, the reflection of the graph of $y = mx + b$ across the x axis for the remaining portion of the graph. Thus the graph has the shape of a v with vertex at $\left(-\dfrac{b}{m},\ 0\right)$.

69. Two points given; we first find slope, then use point-slope form.

$$m = \frac{0 - b}{a - 0} = -\frac{b}{a} \quad a \neq 0$$

$$y - b = -\frac{b}{a}(x - 0)$$

$$y - b = -\frac{bx}{a}$$

Divide both sides by b, then ($b \neq 0$)

$$\frac{y - b}{b} = -\frac{x}{a}$$

$$\frac{y}{b} - 1 = -\frac{x}{a}$$

$$\frac{y}{b} = 1 - \frac{x}{a}$$

$$\frac{x}{a} + \frac{y}{b} = 1$$

71. Using the result of problem 69, we have:
 a = -2, b = 7
The equation is

$$\frac{x}{-2} + \frac{y}{7} = 1$$

$$7x - 2y = -14$$

73. (A)

x	0	5,000	10,000	15,000	20,000	25,000	30,000
212 - 0.0018x = B	212	203	194	185	176	167	158

(B) The boiling point drops 9°F for each 5,000 foot increase in altitude.

75. The rental charges are $25 per day plus $0.25 per mile driven.

77. (A)

x	0	1	2	3	4
Sales	5.9	6.5	7.7	8.6	9.7
5.74 + 0.97x = y	5.7	6.7	7.7	8.6	9.6

(B)

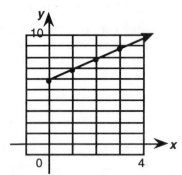

(C) Since 1988 corresponds to $x = 0$, 1993 corresponds to $x = 5$ and 2000 corresponds to $x = 12$.

When $x = 5$, $y = 5.74 + 0.97(5)$
$y \approx 10.6$ (billion dollars)

When $x = 12$, $y = 5.74 + 0.97(12)$
$y \approx 17.4$ (billion dollars)

(D) The sales rose approximately linearly from $5.9 billion in 1988 to $9.7 billion in 1992, a rate of 0.97 billion per year.

79. (A) If F is linearly related to C, then we are looking for an equation whose graph passes through $(C_1, F_1) = (0, 32)$ and $(C_2, F_2) = (100, 212)$. We find the slope, and then use the point-slope form to find the equation.

$$m = \frac{F_2 - F_1}{C_2 - C_1} = \frac{212 - 32}{100 - 0} = \frac{180}{100} = \frac{9}{5}$$

$$F - F_1 = m(C - C_1)$$

$$F - 32 = \frac{9}{5}(C - 0)$$

$$F - 32 = \frac{9}{5}C$$

$$F = \frac{9}{5}C + 32$$

(B) We are asked for F when $C = 20$.

$$F = \frac{9}{5}(20) + 32 = 36 + 32 = 68°$$

We are then asked for C when $F = 86°$

$$86 = \frac{9}{5}C + 32$$

$$430 = 9C + 160$$
$$270 = 9C$$
$$C = 30°$$

(C) The slope is $m = \frac{9}{5}$.

81. (A) If V is linearly related to t, then we are looking for an equation whose graph passes through $(t_1, V_1) = (0, 8,000)$ and $(t_2, V_2) = (5, 0)$. We find the slope, and then we use the point-slope form to find the equation.

$$m = \frac{V_2 - V_1}{t_2 - t_1} = \frac{0 - 8,000}{5 - 0} = \frac{-8,000}{5} = -1,600$$

$$V - V_1 = m(t - t_1)$$
$$V - 8,000 = -1,600(t - 0)$$
$$V - 8,000 = -1,600t$$
$$V = -1,600t + 8,000$$

(B) We are asked for V when $t = 3$
$$V = -1,600(3) + 8,000$$
$$V = -4,800 + 8,000$$
$$V = 3,200$$

(C) The slope is $m = -1,600$

83. (A) If T is linearly related to A, then we are looking for an equation whose graph passes through $(A_1, T_1) = (0, 70)$ and $(A_2, T_2) = (18, -20)$. We find the slope, and then we use the point-slope form to find the equation.

$$m = \frac{T_2 - T_1}{A_2 - A_1} = \frac{(-20) - 70}{18 - 0} = \frac{-90}{18} = -5$$

$$T - T_1 = m(A - A_1)$$
$$T - 70 = -5(A - 0)$$
$$T - 70 = -5A$$
$$T = -5A + 70 \quad A \geq 0$$

(B) We are asked for A when $T = 0$.
$$0 = -5A + 70$$
$$5A = 70$$
$$A = 14 \text{ thousand feet}$$

(C) The slope is $m = -5$; the temperature changes $-5°F$ for each $1,000$ foot rise in altitude.

85. (A) If h is linearly related to t, then we are looking for an equation whose graph passes through $(t_1, h_1) = (9, 23)$ and $(t_2, h_2) = (24, 40)$. We find the slope, and then we use the point-slope form to find the equation.

$$m = \frac{h_2 - h_1}{t_2 - t_1} = \frac{40 - 23}{24 - 9} = \frac{17}{15} \approx 1.13$$

$$h - h_1 = m(t - t_1)$$

$$h - 23 = \frac{17}{15}(t - 9)$$

$$h - 23 = \frac{17}{15}t - \frac{51}{5}$$

$$h = \frac{17}{15}t - \frac{51}{5} + 23$$

$$h = 1.13t + 12.8$$

(B) We are asked for t when $h = 50$.
$$50 = 1.13t + 12.8$$
$$37.2 = 1.13t$$
$$t = 32.9 \text{ hours}$$

87. (A) If R is linearly related to C, then we are looking for an equation whose graph passes through $(C_1, R_1) = (210, 0.160)$ and $(C_2, R_2) = (231, 0.192)$. We find the slope, and then we use the point-slope form to find the equation.

$$m = \frac{R_2 - R_1}{C_2 - C_1} = \frac{0.192 - 0.160}{231 - 210} = \frac{0.032}{21} = 0.00152$$

$$R - R_1 = m(C - C_1)$$
$$R - 0.160 = 0.00152(C - 210)$$
$$R - 0.160 = 0.00152C - 0.319$$
$$R = 0.00152C - 0.159, \quad C \geq 210$$

(B) We are asked for R when $C = 260$.
$$R = 0.00152(260) - 0.159$$
$$R = 0.236$$

(C) The slope is $m = 0.00152$; coronary risk increases 0.00152 per unit increase in cholesterol above the 210 cholesterol level.

Exercise 2-3

Key Ideas and Formulas

A **function** is a rule that produces a correspondence between two sets of elements such that to each element in the first set there corresponds one and only one element in the second set. The first set is called the **domain** and the second set is called the **range**.

In an equation in two variables, if to each value of the independent variable (input) there corresponds exactly one value of the dependent variable (output), then the equation defines a function. If there is any value of the independent variable to which there corresponds more than one value of the dependent variable, then the equation does not define a function.

An equation defines a function if each vertical line in the coordinate system passes through at most one point on the graph of the equation. If any vertical line passes through two or more points on the graph of an equation, then the equation does not define a function. (Vertical Line Test)

If a function is defined by an equation and the domain is not indicated, we assume that the domain is the set of all real number replacements of the independent variable (inputs) that produce real values for the dependent variable (outputs).

The symbol $f(x)$ represents the real number in the range of the function f corresponding to the domain value x. Symbolically, $f:x \rightarrow f(x)$. The ordered pair $(x, f(x))$ belongs to the function f. If x is a real number that is not in the domain of f, then f is not defined at x and $f(x)$ does not exist.

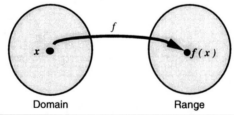

Domain   Range

---

1. A function   **3.** Not a function (two range values correspond to some domain values)

5. A function   **7.** A function; domain = {2, 3, 4, 5}; range = {4, 6, 8, 10}

9. Not a function (two range values correspond to some domain values)

11. A function; domain = {0, 1, 2, 3, 4, 5}; range = {1, 2}

13. A function   **15.** Not a function (fails vertical line test)

17. Not a function (fails vertical line test)   **19.** $f(-1) = 2(-1) + 6 = 4$

21. $G(-2) = (-2)^2 + (-2) - 2 = 0$ **23.** $F(-1) + f(3) = 2(-1)^2 + 3(-1) - 1 + 2(3) + 6$
$$= 2 - 3 - 1 + 6 + 6 = 10$$

25. $2F(-2) - G(-1) = 2[2(-2)^2 + 3(-2) - 1] - [(-1)^2 + (-1) - 2]$
$$= 2[8 - 6 - 1] - [1 - 1 - 2] = 4$$

27. $\dfrac{f(0) \cdot g(-2)}{F(-3)} = \dfrac{[2(0) + 6] \cdot [3 - 2(-2)]}{2(-3)^2 + 3(-3) - 1} = \dfrac{(6)(7)}{8} = 5.25$

29. When $x = -4$, $y = 7$, so $f(-4) = 7$

31. When $y = 4$, x can equal -5, -1, or 6.

33. Solving for the dependent variable y, we have
$$y - x^2 = 1$$
$$y = x^2 + 1$$
Since $x^2 + 1$ is a real number for each real number x, the equation defines a function with domain all real numbers.

35. Solving for the dependent variable y, we have
$$2x^3 + y^2 = 4$$
$$y^2 = 4 - 2x^3$$
$$y = \pm\sqrt{4 - 2x^3}$$
Since each positive number has two real square roots, the equation does not define a function. For example, when $x = 0$, $y = \pm\sqrt{4 - 2(0)^3} = \pm\sqrt{4} = \pm 2$.

37. Solving for the dependent variable y, we have
$$x^3 - y = 2$$
$$-y = 2 - x^3$$
$$y = x^3 - 2$$
Since $x^3 - 2$ is a real number for each real number x, the equation defines a function with domain all real numbers.

39. If $xy = 0$, then either $x = 0$ or $y = 0$. This equation does not define a function since if $x = 0$, y can be any real number.

41. Solving for the dependent variable y, we have
$$x^2 + xy = 1$$
$$xy = 1 - x^2$$
$$y = \frac{1 - x^2}{x}$$
Since $\dfrac{1 - x^2}{x}$ is a real number for each real number x, except when $x = 0$, the equation defines a function with domain all real numbers except 0.

43. The domain is the set of all real numbers x such that $\sqrt{x - 2}$ is a real number, that is, such that $x - 2 \geq 0$ or $x \geq 2$.

45. The domain is the set of all real numbers u, since $\sqrt[3]{u + 9}$ represents a real number for all replacements of u by real numbers.

47. $\dfrac{2w + 3}{w - 4}$ represents a real number for all replacements of w by real numbers except for $w = 4$. All real numbers except 4.

> **Common Error:** $2w + 3 = 0$, $w = -\dfrac{3}{2}$ is *not* excluded from the domain. Zero *can* be divided by a non-zero real number.

49. $\dfrac{3n + 7}{n^2 + n - 2} = \dfrac{3n + 7}{(n + 2)(n - 1)}$. This represents a real number for all replacements of n by real numbers except -2 and 1 (division by zero is not defined). All real numbers except -2 and 1.

51. The domain is the set of all real numbers such that $\sqrt{y^2 - 2y - 3}$ is a real number, that is, such that $y^2 - 2y - 3 \geq 0$.
Solving by the methods of Section 2-8, we obtain the sign chart:
Thus, $y^2 - 2y - 3$ is non-negative

and $\sqrt{y^2 - 2y - 3}$ is real for $y \leq -1$ or $y \geq 3$.

53. The domain is the set of all real numbers such that $\sqrt{\dfrac{x^2 - 4x + 5}{x^2 + 2x + 3}}$ is a real number, that is, such that $\dfrac{x^2 - 4x + 5}{x^2 + 2x + 3}$ is defined and non-negative.
We note: $\dfrac{x^2 - 4x + 5}{x^2 + 2x + 3} = \dfrac{x^2 - 4x + 4 + 1}{x^2 + 2x + 1 + 2} = \dfrac{(x - 2)^2 + 1}{(x + 1)^2 + 2}$. Both numerator and denominator are always defined and always positive for any real replacement of x; in fact $(x - 2)^2 + 1$ is never less than 1 and $(x + 1)^2 + 2$ is never less than 2. Thus, the domain of the function is all real numbers.

55. The domain is the set of all real numbers such that $\sqrt{\dfrac{y - 3}{y + 2}}$ is a real number, that is, such that $\dfrac{y - 3}{y + 2} \geq 0$.

Solving by the methods of Section 2-8, we obtain the sign chart:

Thus, $\dfrac{y - 3}{y + 2}$ is non-negative and $\sqrt{\dfrac{y - 3}{y + 2}}$ is real for $y < -2$ or $y \geq 3$.

57. $g(x) = 2x^3 - 5$ **59.** $G(x) = 2\sqrt{x} - x^2$

61. Function f multiplies the domain element by 2 and subtracts 3 from the result.

63. Function F multiplies the cube of the domain element by 3 and subtracts twice the square root of the domain element from the result.

65.
$$F(s) = 3s + 15$$
$$F(2 + h) = 3(2 + h) + 15$$
$$F(2) = 3(2) + 15$$
$$\dfrac{F(2 + h) - F(2)}{h} = \dfrac{[3(2 + h) + 15] - [3(2) + 15]}{h} = \dfrac{[6 + 3h + 15] - [21]}{h}$$
$$= \dfrac{3h + 21 - 21}{h} = \dfrac{3h}{h} = 3$$

67.

$$g(x) = 2 - x^2$$
$$g(3 + h) = 2 - (3 + h)^2$$
$$g(3) = 2 - (3)^2$$
$$\frac{g(3 + h) - g(3)}{h} = \frac{[2 - (3 + h)]^2 - [2 - (3)^2]}{h} = \frac{[2 - 9 - 6h - h^2] - [-7]}{h}$$
$$= \frac{-6h - h^2}{h} = \frac{h(-6 - h)}{h} = -6 - h$$

69.

$$L(w) = -2w^2 + 3w - 1$$
$$L(-2 + h) = -2(-2 + h)^2 + 3(-2 + h) - 1$$
$$L(-2) = -2(-2)^2 + 3(-2) - 1$$
$$\frac{L(-2 + h) - L(-2)}{h} = \frac{[-2(-2 + h)^2 + 3(-2 + h) - 1] - [-2(-2)^2 + 3(-2) - 1]}{h}$$
$$= \frac{[-2(4 - 4h + h^2) - 6 + 3h - 1] - [-15]}{h}$$
$$= \frac{-8 + 8h - 2h^2 - 6 + 3h - 1 + 15}{h}$$
$$= \frac{11h - 2h^2}{h} = \frac{h(11 - 2h)}{h} = 11 - 2h$$

71. To find $f(x)$, replace $x + h$ by x wherever $x + h$ occurs. $f(x) = 3x^2 - 5x + 9$.

73. To find $m(t)$, replace $t + h$ by t wherever $t + h$ occurs. $m(t) = -2t - 5\sqrt{t} - 2$.

75. (A)

$$f(x) = 4x - 7$$
$$f(x + h) = 4(x + h) - 7$$
$$\frac{f(x + h) - f(x)}{h} = \frac{[4(x + h) - 7] - [4x - 7]}{h}$$
$$= \frac{4x + 4h - 7 - 4x + 7}{h}$$
$$= \frac{4h}{h}$$
$$= 4$$

(B)

$$f(x) = 4x - 7$$
$$f(a) = 4a - 7$$
$$\frac{f(x) - f(a)}{x - a} = \frac{(4x - 7) - (4a - 7)}{x - a}$$
$$= \frac{4x - 7 - 4a + 7}{x - a}$$
$$= \frac{4x - 4a}{x - a}$$
$$= \frac{4(x - a)}{x - a}$$
$$= 4$$

Common Errors: $f(x + h) \neq f(x) + f(h)$
or $4x - 7 + 4h - 7$
$f(x + h) \neq f(x) + h$
or $4x - 7 + h$
Also note: $-f(x) \neq -4x - 7$
Parentheses must be supplied.

77. (A)

$$f(x) = 2x^2 - 4$$
$$f(x + h) = 2(x + h)^2 - 4$$
$$\frac{f(x + h) - f(x)}{h} = \frac{[2(x + h)^2 - 4] - [2x^2 - 4]}{h}$$
$$= \frac{2(x^2 + 2xh + h^2) - 4 - 2x^2 + 4}{h}$$
$$= \frac{2x^2 + 4xh + 2h^2 - 4 - 2x^2 + 4}{h}$$
$$= \frac{4xh + 2h^2}{h}$$
$$= \frac{h(4x + 2h)}{h}$$
$$= 4x + 2h$$

(B) $f(x) = 2x^2 - 4$

$f(a) = 2a^2 - 4$

$\dfrac{f(x) - f(a)}{x - a} = \dfrac{(2x^2 - 4) - (2a^2 - 4)}{x - a}$

$= \dfrac{2x^2 - 4 - 2a^2 + 4}{x - a}$

$= \dfrac{2x^2 - 2a^2}{x - a}$

$= \dfrac{2(x - a)(x + a)}{x - a}$

$= 2(x + a)$

$= 2x + 2a$

79. (A) $f(x) = -4x^2 + 3x - 2$

$f(x + h) = -4(x + h)^2 + 3(x + h) - 2$

$\dfrac{f(x + h) - f(x)}{h} = \dfrac{[-4(x + h)^2 + 3(x + h) - 2] - [-4x^2 + 3x - 2]}{h}$

$= \dfrac{-4(x^2 + 2xh + h^2) + 3(x + h) - 2 + 4x^2 - 3x + 2}{h}$

$= \dfrac{-4x^2 - 8xh - 4h^2 + 3x + 3h - 2 + 4x^2 - 3x + 2}{h}$

$= \dfrac{-8xh - 4h^2 + 3h}{h}$

$= \dfrac{h(-8x - 4h + 3)}{h}$

$= -8x - 4h + 3$

(B) $f(x) = -4x^2 + 3x - 2$

$f(a) = -4a^2 + 3a - 2$

$\dfrac{f(x) - f(a)}{x - a} = \dfrac{(-4x^2 + 3x - 2) - (-4a^2 + 3a - 2)}{x - a}$

$= \dfrac{-4x^2 + 3x - 2 + 4a^2 - 3a + 2}{x - a}$

$= \dfrac{-4x^2 + 4a^2 + 3x - 3a}{x - a}$

$= \dfrac{-4(x - a)(x + a) + 3(x - a)}{x - a}$

$= \dfrac{(x - a)[-4(x + a) + 3]}{x - a}$

$= -4(x + a) + 3$

$= -4x - 4a + 3$

81. (A) $f(x) = x^3 - 2x$

$f(x + h) = (x + h)^3 - 2(x + h)$

$\dfrac{f(x + h) - f(x)}{h} = \dfrac{[(x + h)^3 - 2(x + h)] - [x^3 - 2x]}{h}$

$= \dfrac{x^3 + 3x^2h + 3xh^2 + h^3 - 2x - 2h - x^3 + 2x}{h}$

$= \dfrac{3x^2h + 3xh^2 + h^3 - 2h}{h}$

$= \dfrac{h(3x^2 + 3xh + h^2 - 2)}{h}$

$= 3x^2 + 3xh + h^2 - 2$

(B) $f(x) = x^3 - 2x$
        $f(a) = a^3 - 2a$

$$\frac{f(x) - f(a)}{x - a} = \frac{(x^3 - 2x) - (a^3 - 2a)}{x - a}$$

$$= \frac{x^3 - 2x - a^3 + 2a}{x - a}$$

$$= \frac{x^3 - a^3 - 2x + 2a}{x - a}$$

$$= \frac{(x - a)(x^2 + ax + a^2) - 2(x - a)}{x - a}$$

$$= \frac{(x - a)[x^2 + ax + a^2 - 2]}{x - a}$$

$$= x^2 + ax + a^2 - 2$$

83. Given w = width and Area = 64, we use $A = \ell w$ to write $\ell = \dfrac{A}{w} = \dfrac{64}{w}$.

Then $P = 2w + 2\ell = 2w + 2\left(\dfrac{64}{w}\right) = 2w + \dfrac{128}{w}$. Since w must be positive, the domain of $P(w)$ is $w > 0$.

85.

Using the given letters, the Pythagorean theorem gives
$h^2 = b^2 + 5^2$
$h^2 = b^2 + 25$
  $h = \sqrt{b^2 + 25}$ (since h is positive)

Since b must be positive, the domain of $h(b)$ is $b > 0$.

87. Daily cost = fixed cost + variable cost
        $C(x)$ = \$300 + (\$1.75 per dozen doughnuts) $\times$ (number of dozen doughnuts)
        $C(x) = 300 + 1.75x$

89. (A) $s(t) = 16t^2$
        $s(0) = 16(0)^2 = 0$
        $s(1) = 16(1)^2 = 16$
        $s(2) = 16(2)^2 = 64$
        $s(3) = 16(3)^2 = 144$

(B) $s(2 + h) = 16(2 + h)^2$
        $s(2)\ \ \ \ \ = 64$ from (A)

$$\frac{s(2 + h) - s(2)}{h} = \frac{16(2 + h)^2 - 64}{h} = \frac{16(4 + 4h + h^2) - 64}{h}$$

$$= \frac{64 + 64h + 16h^2 - 64}{h} = \frac{64h + 16h^2}{h} = \frac{h(64 + 16h)}{h} = 64 + 16h$$

(C) As h tends to 0, $16h$ tends to 0, so $64 + 16h$ tends to 64. If we interpret $s(2 + h)$ as the distance fallen after $2 + h$ seconds, and $s(2)$ as the distance fallen after 2 seconds, then we can think of $s(2 + h) - s(2)$ as the distance fallen in h seconds, starting from $t = 2$. Then

$$\frac{s(2 + h) - s(2)}{h} = \frac{\text{distance fallen in } h \text{ seconds}}{h \text{ seconds}}$$

$$= \text{speed during } h \text{ seconds after } t = 2$$

As h tends to zero, this appears to tend to the speed of the object at the end of 2 seconds.

91.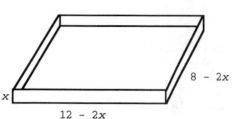

From the above figures it should be clear that
V = length × width × height = $(12 - 2x)(8 - 2x)x$.
Since all distances must be positive, $x > 0$, $8 - 2x > 0$, $12 - 2x > 0$.
Thus, $0 < x$, $4 > x$, $6 > x$, or $0 < x < 4$ (the last condition, $6 > x$,
will be automatically satisfied if $x < 4$.) Domain: $0 < x < 4$.

93. From the text diagram, since each pen must have area 50 square feet, we see
Area = (length)(width) or 50 = (length)x. Thus, the length of each pen is $\dfrac{50}{x}$ feet.
The total amount of fencing = 4(width) + 5(length) + 4(width − gate width)

$$F(x) = 4x + 5\left(\frac{50}{x}\right) + 4(x - 3)$$

$$F(x) = 4x + \frac{250}{x} + 4x - 12$$

$$F(x) = 8x + \frac{250}{x} - 12$$

Then $F(4) = 8(4) + \dfrac{250}{4} - 12 = 82.5$ $F(5) = 8(5) + \dfrac{250}{5} - 12 = 78$

$F(6) = 8(6) + \dfrac{250}{6} - 12 = 77.7$ $F(7) = 8(7) + \dfrac{250}{7} - 12 = 79.7$

95.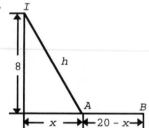

We note that the pipeline consists of the lake section IA,
and shore section AB. The shore section has length
$20 - x$. The lake section has length h, where $h^2 = 8^2 + x^2$,
thus $h = \sqrt{64 + x^2}$.
The cost of all the pipeline =

$$\left(\begin{array}{c}\text{cost of shore}\\\text{section per mile}\end{array}\right)\left(\begin{array}{c}\text{number of}\\\text{shore miles}\end{array}\right) + \left(\begin{array}{c}\text{cost of lake}\\\text{section per mile}\end{array}\right)\left(\begin{array}{c}\text{number of}\\\text{lake miles}\end{array}\right)$$

$C(x) = 10,000(20 - x) + 15,000\sqrt{64 + x^2}$
From the diagram we see that x must be non-negative, but no more than 20.
Domain: $0 \le x \le 20$

97. Cost = (cost per hour) × number of hours
     = (Fuel cost per hour + Fixed cost per hour) × number of hours.
For a 500 mile trip, since $D = rt$, $t = D/r = 500/v$ hours.

Hence cost = $\left(\dfrac{v^2}{5} + 400\right)\left(\dfrac{500}{v}\right)$

$$C(v) = \frac{v^2}{5} \cdot \frac{500}{v} + 400 \cdot \frac{500}{v}$$

$$= 100v + \frac{200,000}{v}$$

Exercise 2-4

Key Ideas and Formulas

The graph of a function f is the same as the graph of the equation $y = f(x)$. The abscissa of a point where the graph of a function intersects the x axis is called an **x intercept** or **zero** of the function. The x intercept is also a real solution or **root** of the equation $f(x) = 0$. The ordinate of a point where the graph of a function crosses the y axis is called the **y intercept** of the function. The y intercept is given by $f(0)$, provided 0 is in the domain of f. The domain of a function is the set of all the x coordinates of points on the graph of the function and the range is the set of all the y coordinates.

Let I be an interval in the domain of a function f. Then:

1. f is increasing on I if $f(b) > f(a)$ whenever $b > a$ in I.

2. f is decreasing on I if $f(b) < f(a)$ whenever $b > a$ in I.

3. f is constant on I if $f(a) = f(b)$ for all a and b in I.

A function f is a linear function if $f(x) = mx + b$ $m \neq 0$ where m and b are real numbers.

The graph of a linear function f is a nonvertical and nonhorizontal straight line with slope m and y intercept b.

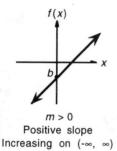

$m > 0$
Positive slope
Increasing on $(-\infty, \infty)$

Domain: All real numbers

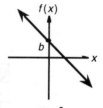

$m < 0$
Negative slope
Decreasing on $(-\infty, \infty)$

Range: All real numbers

A function f is a quadratic function if $f(x) = ax^2 + bx + c$ $a \neq 0$ where a, b, and c are real numbers.

The graph of a quadratic function is a parabola. For $f(x) = ax^2 + bx + c = a(x - h)^2 + k$ $a \neq 0$. The parabola is as shown, with the properties given.

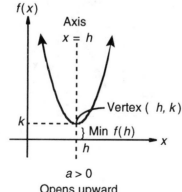

$a > 0$
Opens upward

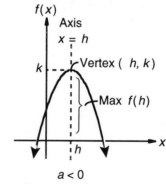

$a < 0$
Opens downward

Vertex: (h, k) (Parabola increases on one side of the vertex and decreases on the other.)

Axis (of symmetry): $x = h$ (Parallel to y axis)

$f(h) = k$ is the minimum if $a > 0$ and the maximum if $a < 0$.

Domain: All real numbers; Range: $(-\infty, k]$ if $a < 0$ or $[k, \infty)$ if $a > 0$

Functions whose definitions involve more than one formula are called piecewise defined functions. Examples include:

1. the absolute value function

$$f(x) = |x| = \begin{cases} -x \text{ if } x < 0 \\ x \text{ if } x \geq 0 \end{cases}$$

2. the greatest integer function

$f(x) = [\![x]\!] = $ the integer n such that $n \leq x < n + 1$

---

1. (A) $[-4, 4)$ (B) $[-3, 3)$ (C) 0 (D) 0 (E) $[-4, 4)$ (F) None
    (G) None   (H) None

3. (A) $(-\infty, \infty)$ (B) $[-4, \infty)$ (C) $-3, 1$ (D) -3 (E) $[-1, \infty)$ (F) $(-\infty, -1]$
    (G) None   (H) None

5. (A) $(-\infty, 2) \cup (2, \infty)$ (The function is not defined at $x = 2$.)
    (B) $(-\infty, -1) \cup [1, \infty)$ (C) None   (D) 1 (E) None   (F) $(-\infty, -2] \cup (2, \infty)$
    (G) $[-2, 2)$ (H) $x = 2$

7. One possible answer: **9.** One possible answer: **11.** One possible answer:

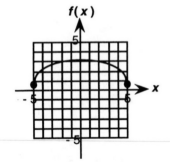 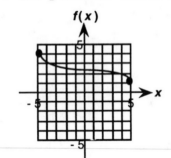 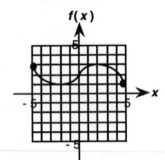

13. $f(x) = \underset{\text{slope}}{2}x + 4$

The y intercept is $f(0) = 4$, and the slope is 2. To find the x intercept, we solve the equation $f(x) = 0$ for x.

$$\begin{aligned} f(x) &= 0 \\ 2x + 4 &= 0 \\ 2x &= -4 \\ x &= -2 \end{aligned}$$

The x intercept is -2.

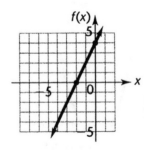

15. $f(x) = -\underbrace{\dfrac{1}{2}}_{\text{slope}} x - \dfrac{5}{3}$

The y intercept is $f(0) = -\dfrac{5}{3}$, and the slope is $-\dfrac{1}{2}$. To find the x intercept, we solve the equation $f(x) = 0$ for x.

$$f(x) = 0$$
$$-\dfrac{1}{2}x - \dfrac{5}{3} = 0$$
$$-\dfrac{1}{2}x = \dfrac{5}{3}$$
$$x = (-2)\dfrac{5}{3}$$
$$x = \dfrac{-10}{3}$$

The x intercept is $\dfrac{-10}{3}$.

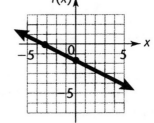

17. A linear function must have the form $f(x) = mx + b$. We are given $f(-2) = 7$, hence $7 = f(-2) = m(-2) + b$. Thus $b = 7 + 2m$. Since $f(4) = -2$, we also have $-2 = f(4) = 4m + b$. Substituting, we have

$$b = 7 + 2m$$
$$-2 = 4m + b$$
$$-2 = 4m + 7 + 2m$$
$$-2 = 6m + 7$$
$$-9 = 6m$$
$$m = -\dfrac{3}{2}$$

$$b = 7 + 2m = 7 + 2\left(-\dfrac{3}{2}\right) = 4$$

Hence $f(x) = mx + b$ becomes $f(x) = -\dfrac{3}{2}x + 4$.

19. $f(x) = (x - 3)^2 + 2$
Comparing with $f(x) = a(x - h)^2 + k$, $h = 3$ and $k = 2$. Thus, the vertex is $(3, 2)$, the axis of symmetry is $x = 3$, and the minimum value is 2. $y = f(x)$ can be any number greater than or equal to 2, so the range is $[2, \infty)$.
Graph: Locate axis and vertex, then plot several points on either side of the axis.

x	$f(x)$
0	11
1	6
2	3
3	2
4	3
5	6

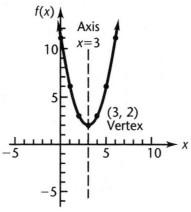

21. $f(x) = -(x + 3)^2 - 2$
Comparing with $f(x) = a(x - h)^2 + k$, $h = -3$ and $k = -2$. Thus, the vertex is $(-3, -2)$, the axis of symmetry is $x = -3$, and the maximum value is -2. $y = f(x)$ can be any number less than or equal to -2, so the range is $(-\infty, -2]$.
Graph: Locate axis and vertex, then plot several points on either side of the axis.

x	$f(x)$
-5	-6
-4	-3
-3	-2
-2	-3
-1	-6
0	-11

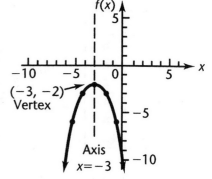

23. $f(x) = x^2 - 4x - 5$. Complete the square.

$f(x) = (x^2 - 4x + 4) - 4 - 5$
  $= (x - 2)^2 - 9$

Comparing with $f(x) = a(x - h)^2 + k$, $h = 2$ and $k = 9$. Thus, the vertex is $(2, 9)$ and the axis of symmetry is $x = 2$.

y intercept: Set $x = 0$, then $f(0) = -5$ is the y intercept.

x intercepts: Set $f(x) = 0$, then
$x^2 - 4x - 5 = 0$
$(x - 5)(x + 1) = 0$
$x = 5$ or $x = -1$ are the x intercepts

Graph: Locate axis and vertex, then plot several points on either side of the axis.

x	$f(x)$
-1	0
0	-5
1	-8
2	-9
3	-8
4	-5
5	0

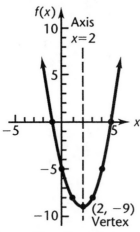

25. $f(x) = -x^2 + 6x$. Complete the square.

$f(x) = -(x^2 - 6x)$
  $= -(x^2 - 6x + 9) + 9$
  $= -(x - 3)^2 + 9$

Comparing with $f(x) = a(x - h)^2 + k$, $h = 3$ and $k = 9$. Thus, the vertex is $(3, 9)$ and the axis of symmetry is $x = 3$.

y intercept: Set $x = 0$, then $f(0) = 0$ is the y intercept.

x intercepts: Set $f(x) = 0$, then
$-x^2 + 6x = 0$
$x(-x + 6) = 0$
$x = 0$ or $x = 6$ are the x intercepts.

Graph: Locate axis and vertex, then plot several points on either side of the axis.

x	$f(x)$
0	0
1	5
2	8
3	9
4	8
5	5
6	0

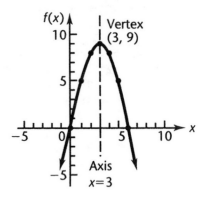

27. $f(x) = x^2 + 6x + 11$. Complete the square.

$f(x) = (x^2 + 6x + 9) - 9 + 11$
  $= (x + 3)^2 + 2$

Comparing with $f(x) = a(x - h)^2 + k$, $h = -3$ and $k = 2$. Thus, the vertex is $(-3, 2)$ and the axis of symmetry is $x = -3$.

Graph: Locate axis and vertex, then plot several points on either side of the axis.

x	$f(x)$
-5	6
-4	3
-3	2
-2	3
-1	6
0	11

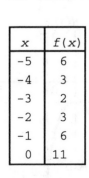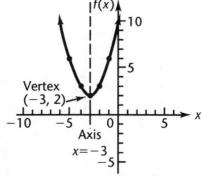

From the graph we see that f is decreasing on $(-\infty, -3]$ and increasing on $[-3, \infty)$.

29. $f(x) = -x^2 + 6x - 6$. Complete the square.

$f(x) = -(x^2 - 6x + 9) + 9 - 6$
  $= -(x - 3)^2 + 3$

Comparing with $f(x) = a(x - h)^2 + k$, $h = 3$ and $k = 3$. Thus, the vertex is $(3, 3)$ and the axis of symmetry is $x = 3$.

Graph: Locate axis and vertex, then plot several points on either side of the axis.

x	$f(x)$
0	-6
1	-1
2	2
3	3
4	2
5	-1

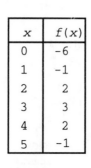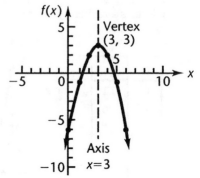

From the graph we see that f is increasing on $(-\infty, 3]$ and decreasing on $[3, \infty)$.

31.

x	y = x + 1		x	y = -x + 1
-1	0		0	1
$-\frac{1}{2}$	$\frac{1}{2}$		$\frac{1}{2}$	$\frac{1}{2}$
			1	0

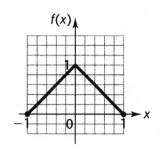

Domain: [-1, 1]
Range: [0, 1]

33.

x	y = -2	x	y = 4
-3	-2	0	4
-2	-2	1	4
		2	4

Domain: [-3, -1) ∪ (-1, 2]
We graph using open dots at (-1, -2) and (-1, 4) to indicate that these points do not belong to the graph of f. In fact f is not defined at -1.

Range: The only possible values of y are -2 and 4. The range consists of these two values, in set notation: {-2, 4}. (This does not mean the interval (-2, 4).)

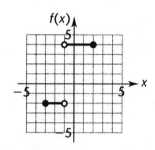

Common Error: Not including the portions of line segments between (-2, -2) and (-2, -1), or (-1, 4) and (0, 4). The domain extends up to -1 on both sides.

Discontinuous: at x = -1

35.

x	y = x + 2	x	y = x - 2
-5	-3	-1	-3
-2	0	5	3

Domain: All real numbers
We graph using an open dot at (-1, 1) and a solid dot at (-1, -3) to show that only the latter belongs to the graph of f.

Range: All real numbers   Discontinuous: at x = -1

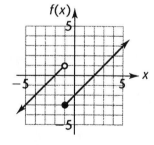

37.

x	y = x² + 1	x	y = -x² - 1
-2	5	1	-2
-1	2	2	-5

Domain: (-∞, 0) ∪ (0, ∞)
We graph using open dots at (0, 1), and (0, -1) to indicate that these points do not belong to the graph of g. g is discontinuous at x = 0.

Range: (-∞, -1) ∪ (1, ∞)

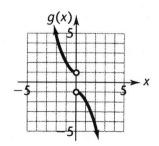

39. $f(x) = \frac{1}{2}x^2 + 2x + 3$. Complete the square.

$f(x) = \frac{1}{2}(x^2 + 4x + 4) - \frac{1}{2}(4) + 3$

$\quad\ = \frac{1}{2}(x + 2)^2 + 1$

Comparing with $f(x) = a(x - h)^2 + k$, $h = -2$ and $k = 1$. Thus, the vertex is $(-2, 1)$, the axis of symmetry is $x = -2$, and the minimum value is 1. $y = f(x)$ can be any number greater then or equal to 1, so the range is $[1, \infty)$.

y intercept: Set $x = 0$, then $f(0) = 3$ is the y intercept.

x intercept: Solutions of

$\quad 0 = \frac{1}{2}x^2 + 2x + 3$

$\quad 0 = x^2 + 4x + 6$

$\quad x = \dfrac{-4 \pm \sqrt{-8}}{2}$

Graph:

x	$f(x)$
-5	5.5
-4	3
-3	1.5
-2	1
-1	1.5
0	3
1	5.5

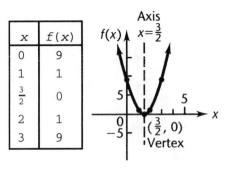

no real solutions, hence no x intercepts
From the graph, we see that f is decreasing on $(-\infty, -2]$ and increasing on $[-2, \infty)$.

41. $f(x) = 4x^2 - 12x + 9$. Complete the square.

$f(x) = 4(x^2 - 3x + \frac{9}{4}) - 4 \cdot \frac{9}{4} + 9$

$\quad\ = 4(x - \frac{3}{2})^2$

Comparing with $f(x) = a(x - h)^2 + k$, $h = \frac{3}{2}$ and $k = 0$. Thus, the vertex is $(\frac{3}{2}, 0)$, the axis of symmetry is $x = \frac{3}{2}$, and the minimum value is 0. $y = f(x)$ can be any number greater then or equal to 0, so the range is $[0, \infty)$.

y intercept: Set $x = 0$, then $f(0) = 9$ is the y intercept.

x intercept: Solutions of $\quad 0 = 4x^2 - 12x + 9$

$\quad\quad\quad\quad\quad\quad (2x - 3)^2 = 0$

$\quad\quad\quad\quad\quad\quad\ \ 2x - 3 = 0$

$\quad\quad\quad\quad\quad\quad\quad\quad\ x = \frac{3}{2}$

Graph:

x	$f(x)$
0	9
1	1
$\frac{3}{2}$	0
2	1
3	9

From the graph, we see that f is decreasing on $(-\infty, \frac{3}{2}]$ and increasing on $[\frac{3}{2}, \infty)$.

43. $f(x) = -2x^2 - 8x - 2$. Complete the square.

$f(x) = -2(x^2 + 4x + 4) + 2 \cdot 4 - 2$

$\quad\ = -2(x + 2)^2 + 6$

Comparing with $f(x) = a(x - h)^2 + k$, $h = -2$ and $k = 6$. Thus, the vertex is $(-2, 6)$, the axis of symmetry is $x = -2$, and the maximum value is 6. $y = f(x)$ can be any number less then or equal to 6, so the range is $(-\infty, 6]$.

y intercept: Set $x = 0$, then $f(0) = -2$ is the y intercept.

x intercept: Solutions of

$\quad 0 = -2x^2 - 8x - 2$

$\quad 0 = x^2 + 4x + 1$

$\quad x = \dfrac{-4 \pm \sqrt{12}}{2} = -2 \pm \sqrt{3}$

Graph:

x	$f(x)$
-4	-2
-3	4
-2	6
-1	4
0	-2

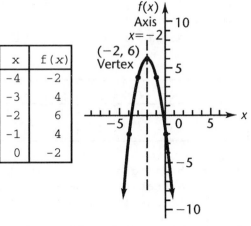

From the graph, we see that f is increasing on $(-\infty, -2]$ and decreasing on $[-2, \infty)$.

45. $f(x) = \dfrac{|x|}{x}$

If $x < 0$ then $|x| = -x$ and $f(x)$ $\dfrac{|x|}{x} = \dfrac{-x}{x} = -1$

If $x = 0$, then f is not defined. f is discontinuous at $x = 0$.

If $x > 0$, then $|x| = x$ and $f(x) = \dfrac{|x|}{x} = \dfrac{x}{x} = 1$

Domain: $(-\infty, 0) \cup (0, \infty)$, that is, $x \neq 0$
Range: $\{-1, 1\}$, from the graph we see that only two
values of f are possible.
Piecewise definition for f:

$$f(x) = \begin{cases} -1 \text{ if } x < 0 \\ 1 \text{ if } x > 0 \end{cases}$$

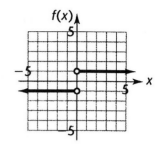

47. $f(x) = x + \dfrac{|x - 1|}{x - 1}$

If $x < 1$, then $|x - 1| = -(x - 1)$ and $f(x) = x + \dfrac{-(x - 1)}{x - 1} = x - 1$

If $x = 1$, then $f(x)$ is not defined. f is discontinuous at $x = 1$.

If $x > 1$, then $|x - 1| = x - 1$ and $f(x) = x + \dfrac{x - 1}{x - 1} = x + 1$

x	$y = x - 1$	x	$y = x + 1$
-1	-2	2	3
0	-1	3	4

Domain: $(-\infty, 1) \cup (1, \infty)$, that is, $x \neq 1$
Piecewise definition for f:

$$f(x) = \begin{cases} x - 1 \text{ if } x < 1 \\ x + 1 \text{ if } x > 1 \end{cases}$$

Range: From the graph, we see that y is less than 0
or y is greater than 2:
$(-\infty, 0) \cup (2, \infty)$

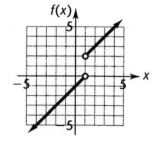

49. $f(x) = |x| + |x - 2|$
If $x < 0$, then $|x| = -x$ and $|x - 2| = -(x - 2) = 2 - x$ and
$f(x) = |x| + |x - 2| = -x + (2 - x) = 2 - 2x$
If $0 \leq x < 2$, then $|x| = x$ but $|x - 2| = -(x - 2) = 2 - x$ and
$f(x) = |x| + |x - 2| = x + (2 - x) = 2$
If $2 \leq x$, then $|x| = x$ and $|x - 2| = x - 2$ and
$f(x) = |x| + |x - 2| = x + (x - 2) = 2x - 2$
Domain: All real numbers
Piecewise definition for f:

$$f(x) = \begin{cases} 2 - 2x & x < 0 \\ 2 & 0 \leq x < 2 \\ 2x - 2 & x \geq 2 \end{cases}$$

There are no discontinuities for f.

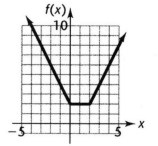

x	$y = 2 - 2x$	x	$y = 2$	x	$y = 2x - 2$
-2	6	0	2	2	2
-1	4	1	2	3	4

Range: From the graph we see that y is never less than 2. $[2, \infty)$

51. We note: $f(0) = [\![0/2]\!] = [\![0]\!] = 0$

$f(1) = [\![1/2]\!] = 0$

$f(2) = [\![2/2]\!] = [\![1]\!] = 1$

$f(3) = [\![3/2]\!] = 1$ $f(x) = [\![x/2]\!]$ appears to jump at intervals of 2 units.

Generalizing, we can write: $f(x) = \begin{cases} \vdots & & \vdots \\ -2 & \text{if} & -4 \le x < -2 \\ -1 & \text{if} & -2 \le x < 0 \\ 0 & \text{if} & 0 \le x < 2 \\ 1 & \text{if} & 2 \le x < 4 \\ 2 & \text{if} & 4 \le x < 6 \\ \vdots & & \vdots \end{cases}$

Domain: All real numbers
Range: All integers
Discontinuous at the even integers

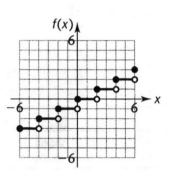

53. Having noted in problem 51 that $[\![\frac{x}{2}]\!]$ jumps at intervals of 2 units, we can investigate $[\![3x]\!] = [\![x \div \frac{1}{3}]\!]$ to see if it jumps at intervals of $\frac{1}{3}$ unit.

$f(0) = [\![3 \cdot 0]\!] = [\![0]\!] = 0$

$f(\frac{1}{6}) = [\![3 \cdot \frac{1}{6}]\!] = [\![\frac{1}{2}]\!] = 0$

$f(\frac{1}{3}) = [\![3 \cdot \frac{1}{3}]\!] = [\![1]\!] = 1$

$f(\frac{1}{2}) = [\![3 \cdot \frac{1}{2}]\!] = [\![\frac{3}{2}]\!] = 1$

$f(\frac{2}{3}) = [\![3 \cdot \frac{2}{3}]\!] = [\![2]\!] = 2$

$f(\frac{5}{6}) = [\![3 \cdot \frac{5}{6}]\!] = [\![\frac{5}{2}]\!] = 2$

Generalizing, we can write: $f(x) = \begin{cases} \vdots & & \vdots \\ -2 & \text{if} & -\frac{2}{3} \le x < -\frac{1}{3} \\ -1 & \text{if} & -\frac{1}{3} \le x < 0 \\ 0 & \text{if} & 0 \le x < \frac{1}{3} \\ 1 & \text{if} & \frac{1}{3} \le x < \frac{2}{3} \\ 2 & \text{if} & \frac{2}{3} \le x < 1 \\ \vdots & & \vdots \end{cases}$

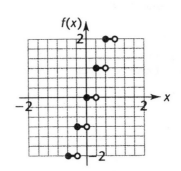

Domain: All real numbers
Range: All integers
Discontinuous at rational numbers
of the form $\frac{k}{3}$ where k is an integer

55. This is discontinuous at every integer value of x. If x is an integer, $x - [\![x]\!]$
$= x - x = 0$. If $x = n + \varepsilon$, where $0 < \varepsilon < 1$, then $[\![x]\!] = n$, $-[\![x]\!] = -n$, and
$x - [\![x]\!] = \varepsilon = x - n$. So the graph between $x = n$ and $x = n + 1$ is a line segment
of form $y = x - n$, connecting $(n, 0)$ and $(n + 1, 1)$. However, $f(n + 1) \neq 1$
since $f(n + 1) = 0$. We can write:

$$f(x) = \begin{cases} \vdots & \vdots \\ x + 2 & \text{if} & -2 \leq x < -1 \\ x + 1 & \text{if} & -1 \leq x < 0 \\ x & \text{if} & 0 \leq x < 1 \\ x - 1 & \text{if} & 1 \leq x < 2 \\ x - 2 & \text{if} & 2 \leq x < 3 \\ \vdots & \vdots \end{cases}$$

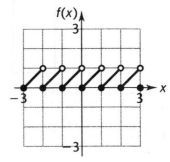

Domain: all real numbers
Range: [0, 1)

57. $f(x) = a(x - h)^2 + k$. Since Min $f(x) = f(2) = 4$, $h = 2$, $k = 4$, and the vertex
is $(2, 4)$. The axis is $x = h$ or $x = 2$. Since y is never less than 4, the range
is $[4, \infty)$ and there are no x intercepts since y cannot be equal to 0.

59. (A) One possible answer:

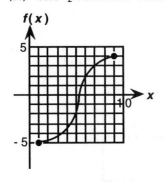

(B) This graph crosses the x axis once.
To meet the conditions specified a graph
must cross the x axis exactly once. If
it crossed more times the function would
have to be decreasing somewhere; if it
did not cross at all the function would
have to be discontinuous somewhere.

61. (A) One possible answer:

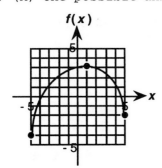

(B) This graph crosses the x axis twice.
To meet the conditions specified a graph
must cross the x axis at least twice. If
it crossed fewer times the function
would have to be discontinuous
somewhere. However, the graph could
cross more times; in fact there is no
upper limit on the number of times it
can cross the x axis.

63. The secant line is the line through (-1, -3) and (3, 5). To find its equation we find its slope, then use the point-slope form of the equation of a line.

$$m = \frac{f(x_2) - f(x_1)}{x_2 - x_1} = \frac{5 - (-3)}{3 - (-1)} = \frac{8}{4} = 2$$

$y - f(x_1) = m(x - x_1)$
$y - (-3) = 2[x - (-1)]$
$y + 3 = 2[x + 1]$
$y + 3 = 2x + 2$
$y = 2x - 1$

Graph:

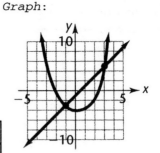

x	$y = 2x - 1$	$y = x^2 - 4$
-2	-5	0
-1	-3	-3
0	-1	-4
1	1	-3
2	3	0
3	5	5

65. (A) The slope of the secant line is given by

$$m = \frac{f(2 + h) - f(2)}{(2 + h) - 2} = \frac{f(2 + h) - f(2)}{h}$$

Since $f(2 + h) = (2 + h)^2 - 3(2 + h) + 5 = 4 + 4h + h^2 - 6 - 3h + 5 = 3 + h + h^2$
and $f(2) = (2)^2 - 3(2) + 5 = 4 - 6 + 5 = 3$
we have

$$m = \frac{f(2 + h) - f(2)}{h} = \frac{(3 + h + h^2) - (3)}{h} = \frac{h + h^2}{h} = \frac{h(1 + h)}{h} = 1 + h$$

(B)

h	$slope = 1 + h$
1	2
0.1	1.1
0.01	1.01
0.001	1.001

The slope seems to be approaching 1.

67. Graphs of f and g

Graph of $m(x) = 0.5[-2x + 0.5x + |-2x - 0.5x|]$
$= 0.5[-1.5x + |-2.5x|]$

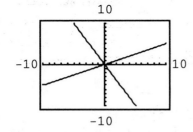

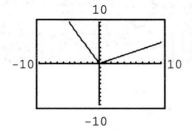

Graph of $n(x) = 0.5[-2x + 0.5x - |-2x - 0.5x|] = 0.5[-1.5x - |-2.5x|]$

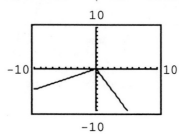

69. Graphs of f and g

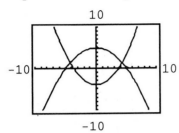

Graph of $m(x) =$
$0.5[5 - 0.2x^2 + 0.3x^2 - 4 +$
$\qquad |5 - 0.2x^2 - (0.3x^2 - 4)|]$
$= 0.5[1 + 0.1x^2 + |9 - 0.5x^2|]$

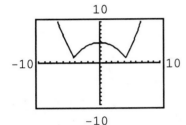

Graph of $n(x) =$
$0.5[5 - 0.2x^2 + 0.3x^2 - 4 -$
$\qquad |5 - 0.2x^2 - (0.3x^2 - 4)|]$
$= 0.5[1 + 0.1x^2 - |9 - 0.5x^2|]$

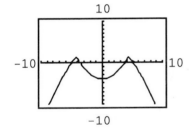

71. Graphs of f and g

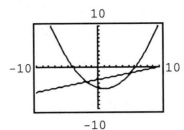

Graph of $m(x) =$
$0.5[0.2x^2 - 0.4x - 5 + 0.3x - 3 +$
$\qquad |0.2x^2 - 0.4x - 5 - (0.3x - 3)|]$
$\quad = 0.5[0.2x^2 - 0.1x - 8 + |0.2x^2 - 0.7x - 2|]$

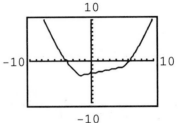

Graph of $n(x) =$
$0.5[0.2x^2 - 0.4x - 5 + 0.3x - 3 -$
$\qquad |0.2x^2 - 0.4x - 5 - (0.3x - 3)|]$
$\quad = 0.5[0.2x^2 - 0.1x - 8 - |0.2x^2 - 0.7x - 2|]$

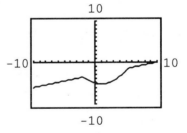

73. The graphs of $m(x)$ show that the value of $m(x)$ is always the larger of the two values for $f(x)$ and $g(x)$. In other words, $m(x) = \max[f(x), g(x)]$. This is confirmed in Exercise 1-4, Problem 89 where it was shown that
$\max(a, b) = 0.5[a + b + |a - b|]$,
hence $\max[f(x), g(x)] = 0.5[f(x) + g(x) + |f(x) - g(x)|] = m(x)$.

75. (A)

x	28	30	32	34	36
Mileage	45	52	55	51	47
$-0.518x^2 + 33.3x - 481 = f(x)$	45.3	51.8	54.2	52.4	46.5

(B)

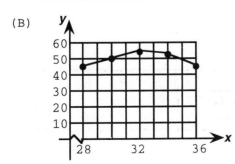

(C) When $x = 31$, $y = -0.518(31)^2 + 33.3(31) - 481$
$\qquad\qquad\qquad y \approx 53.50$ thousand miles
When $x = 35$, $y = -0.518(35)^2 + 33.3(35) - 481$
$\qquad\qquad\qquad y \approx 49.95$ thousand miles

(D) The mileage increases with tire pressure to a maximum of 55 thousand miles at tire pressure 32 pounds per square inch and decreases with greater tire pressure after that.

77. (A) Using the hint, we have $f(10) = 1$ and $f(0) = 0$. Since $f(w) = mw + b$, we must have $1 = f(10) = m(10) + b$ and $0 = f(0) = m(0) + b$.

Therefore $b = 0$ and $m = \dfrac{1}{10}$. So $s = f(w) = \dfrac{w}{10}$.

(B) $f(w) = \dfrac{w}{10}$. Therefore $f(15) = \dfrac{15}{10} = 1.5$ inches and $f(30) = \dfrac{30}{10} = 3$ inches.

(C) The slope is the coefficient of w, or $\dfrac{1}{10}$.

(D)

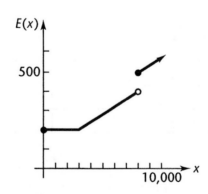

79. If $0 \le x \le 3{,}000$, $E(x) = 200$

	Base Salary	+	Commission on Sales over $3,000
If $3{,}000 < x < 8{,}000$, $E(x)$	$= 200$	$+$	$0.04(x - 3{,}000)$
	$= 200$	$+$	$0.04x - 120$
	$= 80$	$+$	$0.04x$

Common Error:
Commission is not $0.04x$ (4% of sales) nor is it $0.04x + 200$ (base salary plus 4% of sales).

There is a point of discontinuity at $x = 8{,}000$.

	Salary	+	Bonus
If $x \ge 8{,}000$, $E(x)$	$= 80 + 0.04x$	$+$	100
	$= 180$		$+ 0.04x$

Summarizing, $E(x) = \begin{cases} 200 & \text{if } 0 \le x \le 3{,}000 \\ 80 + 0.04x & \text{if } 3{,}000 < x < 8{,}000 \\ 180 + 0.04x & \text{if } 8{,}000 \le x \end{cases}$

$E(5{,}750) = 80 + 0.04(5{,}750) = \$310$

$E(9{,}200) = 180 + 0.04(9{,}200) = \$548$

x	$y = 200$	x	$y = 80 + 0.04x$	x	$y = 180 + 0.04x$
0	200	3,000	200	8,000	500
2,000	200	7,000	360	10,000	580

81. (A) Let x = width of pen
Then $2x + 2\ell = 100$
        so $2\ell = 100 - 2x$
Therefore $\ell = 50 - x$ = length of pen
        $A(x) = x(50 - x) = 50x - x^2$

(B) Since all distances must be positive, $x > 0$ and $50 - x > 0$.
Therefore, $0 < x < 50$ or $(0, 50)$ is the domain of $A(x)$.

(C) Note: 0, 50 were excluded from the domain for geometrical reasons; but can be used to help draw the graph since the *polynomial* $50x - x^2$ has domain including these values.

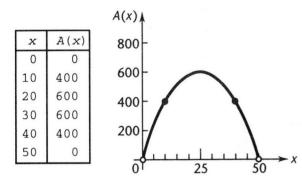

x	$A(x)$
0	0
10	400
20	600
30	600
40	400
50	0

(D) From the graph, the maximum value of $A(x)$ seems to occur when $x = 25$. Checking, we see that $50x - x^2 = -x^2 + 50x$ is a quadratic function that can be written as $-(x^2 - 50x) = -(x^2 - 50x + 625) + 625 = -(x - 25)^2 + 625$. Thus it has its maximum value at $x = 25$. Then $50 - x = 25$, and the dimensions are 25 feet by 25 feet.

83.

$$f(4) = 10[\![0.5 + 0.4]\!] = 10(0) = 0$$
$$f(-4) = 10[\![0.5 - 0.4]\!] = 10(0) = 0$$
$$f(6) = 10[\![0.5 + 0.6]\!] = 10(1) = 10$$
$$f(-6) = 10[\![0.5 - 0.6]\!] = 10(-1) = -10$$
$$f(24) = 10[\![0.5 + 2.4]\!] = 10(2) = 20$$
$$f(25) = 10[\![0.5 + 2.5]\!] = 10(3) = 30$$
$$f(247) = 10[\![0.5 + 24.7]\!] = 10(25) = 250$$
$$f(-243) = 10[\![0.5 - 24.3]\!] = 10(-24) = -240$$
$$f(-245) = 10[\![0.5 - 24.5]\!] = 10(-24) = -240$$
$$f(-246) = 10[\![0.5 - 24.6]\!] = 10(-25) = -250$$

f rounds numbers to the tens place

85. Since $f(x) = [\![10x + 0.5]\!]/10$ rounds numbers to the nearest tenth, (see text example 6) we try $[\![100x + 0.5]\!]/100 = f(x)$ to round to the nearest hundredth.
$$f(3.274) = [\![327.9]\!]/100 = 3.27$$
$$f(7.846) = [\![785.1]\!]/100 = 7.85$$
$$f(-2.8783) = [\![-287.33]\!]/100 = -2.88$$
A few examples suffice to convince us that this is probably correct.
(A proof would be out of place in this book.)
$$f(x) = [\![100x + 0.5]\!]/100.$$

87. (A)

$$C(x) = \begin{cases} 15 & 0 < x \le 1 \\ 18 & 1 < x \le 2 \\ 21 & 2 < x \le 3 \\ 24 & 3 < x \le 4 \\ 27 & 4 < x \le 5 \\ 30 & 5 < x \le 6 \end{cases}$$

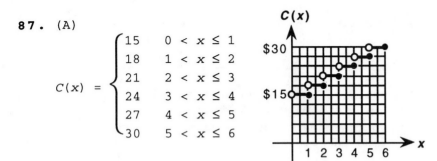

(B) The two functions appear to coincide, for example

$$C(3.5) = 24 \qquad f(3.5) = 15 + 3[\![3.5]\!] = 15 + 3 \cdot 3 = 24$$

However,

$$C(1) = 15 \qquad f(1) = 15 + 3[\![1]\!] = 15 \cdot 3 \cdot 1 = 18$$

The functions are not the same, therefore. In fact, $f(x) \ne C(x)$ at $x = 1, 2, 3, 4, 5, 6$.

89. Let x = rate per car per day.
Then (number of cars rented) = 300 - 5(increase of rate over \$40)

$$\begin{aligned} &= 300 - 5(x - 40) \\ &= 300 - 5x + 200 \\ &= 500 - 5x \end{aligned}$$

$$\text{Income} = I(x) = \begin{pmatrix} \text{Numbers of} \\ \text{Cars Rented} \end{pmatrix}\begin{pmatrix} \text{Rate Per} \\ \text{Car Per Day} \end{pmatrix}$$

$$I(x) = (500 - 5x)x$$
$$I(x) = 500x - 5x^2$$

This is a quadratic function. Completing the square yields

$$\begin{aligned} I(x) &= -5x^2 + 500x \\ &= -5(x^2 - 100x) \\ &= -5(x^2 - 100x + 2500) + 5(2500) \\ &= -5(x - 50)^2 + 12{,}500 \end{aligned}$$

Thus the maximum value of this function occurs when $x = 50$. Hence the rate should be \$50 per day and the resulting maximum income is $I(50) = \$12{,}500$.

91. (A) The given function $f(x) = \dfrac{1}{4}x - \left(\dfrac{16}{v^2}\right)x^2$ is a quadratic function for every positive value of v. In the coordinate system given, this function has x intercepts 0 and 80, and maximum value when $x = 40$. Completing the square yields

$$\begin{aligned} f(x) &= -\frac{16}{v^2}x^2 + \frac{1}{4}x \\ &= -\frac{16}{v^2}\left(x^2 - \frac{v^2}{64}x\right) \\ &= -\frac{16}{v^2}\left(x^2 - \frac{v^2}{64}x + \left(\frac{v^2}{128}\right)^2\right) + \frac{16}{v^2}\left(\frac{v^2}{128}\right)^2 \\ &= -\frac{16}{v^2}\left(x - \frac{v^2}{128}\right)^2 + \frac{v^2}{1024} \end{aligned}$$

Comparing with $a(x - h)^2 + k$, the maximum value occurs when $x = h = 40$, thus

$$\frac{v^2}{128} = 40$$
$$v^2 = 5120$$
$$v = \sqrt{5120} \quad \text{(discarding the negative solution)}$$
$$v = 32\sqrt{5} \approx 71.55 \text{ ft/sec}$$

(B) The maximum height of the cycle above the ground is $10 + k = 10 + \dfrac{v^2}{1024}$.

Substituting $v = 32\sqrt{5}$ yields $10 + k = 10 + \dfrac{v^2}{1024} = 10 + \dfrac{(32\sqrt{5})^2}{1024} = 10 + \dfrac{5120}{1024} = 10 + 5 = 15$ feet.

Exercise 2-5

Key Ideas and Formulas

The intersection of sets A and B is the set of all elements in A that are also in B. The sum, difference, product and quotient of the function f and g are the functions defined by:

$$(f + g)(x) = f(x) + g(x) \quad \text{Sum function}$$
$$(f - g)(x) = f(x) - g(x) \quad \text{Difference function}$$
$$(fg)(x) = f(x)g(x) \quad \text{Product function}$$
$$\left(\frac{f}{g}\right)(x) = \frac{f(x)}{g(x)} \quad \text{Quotient function}$$

Each function is defined on the intersection of the domains of f and g, with the exception that the values of x where $g(x) = 0$ must be excluded from the domain of the quotient function.

Given functions f and g, then $f \cdot g$ is called their composite and is defined by the equation

$$(f \cdot g)(x) = f[g(x)]$$

The domain of $f \cdot g$ is the set of all real numbers x in the domain of g where $g(x)$ is in the domain of f.

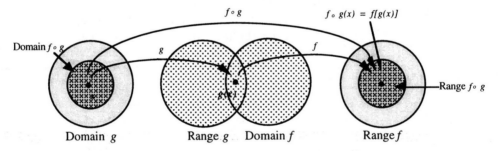

Graphs of Basic Functions:

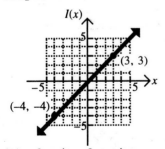

(a) Identity function
$I(x) = x$
Domain: R
Range: R

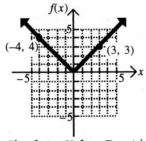

(b) Absolute Value Function
$f(x) = |x|$
Domain: R
Range: $[0, \infty)$

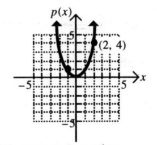

(c) Square Function
$p(x) = x^2$
Domain: R
Range: $[0, \infty)$

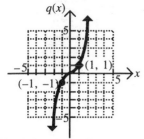

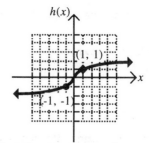

(d) Cube Function

$q(x) = x^3$
Domain: R
Range: R

(e) Square Root Function

$g(x) = \sqrt{x}$
Domain: $[0, \infty)$
Range: $[0, \infty)$

(f) Cube Root Function

$h(x) = \sqrt[3]{x}$
Domain: R
Range: R

Graph Transformations:

Vertical Translation:

$$y = f(x) + k \quad \begin{cases} k > 0 & \text{Shift graph of } y = f(x) \text{ up } k \text{ units} \\ k < 0 & \text{Shift graph of } y = f(x) \text{ down } |k| \text{ units} \end{cases}$$

Horizontal Translation:

$$y = f(x + h) \quad \begin{cases} h > 0 & \text{Shift graph of } y = f(x) \text{ left } h \text{ units} \\ h < 0 & \text{Shift graph of } y = f(x) \text{ right } |h| \text{ units} \end{cases}$$

Reflection:

$$y = -f(x) \qquad \text{Reflect the graph of } y = f(x) \text{ in the } x \text{ axis}$$

Vertical Expansion and Contraction:

$$y = Af(x) \quad \begin{cases} A > 1 & \text{Vertically expand graph of } y = f(x) \\ & \text{by multiplying each ordinate value by } A \\ 0 < A < 1 & \text{Vertically contract graph of } y = f(x) \\ & \text{by multiplying each ordinate value by } A \end{cases}$$

1. Domain: Since $x \geq 0$, the domain is $[0, \infty)$

  Range: Since the range of $f(x) = \sqrt{x}$ is $y \geq 0$, for $h(x) = -\sqrt{x}$, $y \leq 0$.
  Thus, the range of h is $(-\infty, 0]$.

3. Domain: R

  Range: Since the range of $f(x) = x^2$ is $y \geq 0$, for $g(x) = -2x^2$, $y \leq 0$.
  Thus, the range of g is $(-\infty, 0]$.

5. Domain: R; Range: R

7.
$(f + g)(x) = f(x) + g(x) = 4x + x + 1 = 5x + 1$ Domain: $(-\infty, \infty)$
$(f - g)(x) = f(x) - g(x) = 4x - (x + 1) = 3x - 1$ Domain: $(-\infty, \infty)$
$(fg)(x) = f(x)g(x) = 4x(x + 1) = 4x^2 + 4x$ Domain: $(-\infty, \infty)$

$$\left(\frac{f}{g}\right)(x) = \frac{f(x)}{g(x)} = \frac{4x}{x + 1}$$

  Domain: $\{x \mid x \neq -1\}$, or

  $(-\infty, -1) \cup (-1, \infty)$

> **Common Error:**
> $f(x) - g(x) \neq 4x - x + 1$. The parentheses are necessary.

9.
$(f + g)(x) = f(x) + g(x) = 2x^2 + x^2 + 1 = 3x^2 + 1$ Domain: $(-\infty, \infty)$
$(f - g)(x) = f(x) - g(x) = 2x^2 - (x^2 + 1) = x^2 - 1$ Domain: $(-\infty, \infty)$
$(fg)(x) = f(x)g(x) = 2x^2(x^2 + 1) = 2x^4 + 2x^2$ Domain: $(-\infty, \infty)$

$$\left(\frac{f}{g}\right)(x) = \frac{f(x)}{g(x)} = \frac{2x^2}{x^2 + 1}$$

  Domain: $(-\infty, \infty)$

  (since $g(x)$ is never 0.)

11. $(f \circ g)(x) = f[g(x)] = f(x^2 - 4x) = (x^2 - 4x)^2 + 3 = x^4 - 8x^3 + 16x^2 + 3$

Domain: $(-\infty, \infty)$

$(g \circ f)(x) = g[f(x)] = g(x^2 + 3) = (x^2 + 3)^2 - 4(x^2 + 3)$

$= x^4 + 6x^2 + 9 - 4x^2 - 12 = x^4 + 2x^2 - 3$

Domain: $(-\infty, \infty)$

13. $(f \circ g)(x) = f[g(x)] = f(x^3 - 1) = 2(x^3 - 1)^{2/3}$ Domain: $(-\infty, \infty)$

$(g \circ f)(x) = g[f(x)] = g(2x^{2/3}) = (2x^{2/3})^3 - 1 = 8x^2 - 1$ Domain: $(-\infty, \infty)$

15. The graph of $f(x)$ is shifted up 2 units

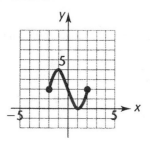

17. The graph of $g(x)$ is shifted left 2 units.

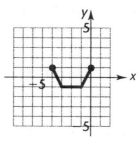

Common Error:
Confusing the positive 2 in $g(x + 2)$ with a shift right.

19. The graph of $f(x)$ is reflected in the x axis.

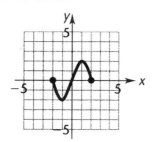

21. The graph of $g(x)$ is vertically expanded by multiplying each ordinate by 2.

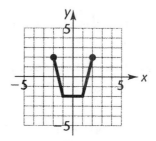

23. The graph of $y = |x|$ is shifted two units to the left and reflected in the x axis.

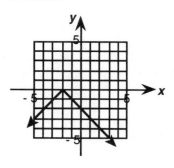

25. The graph of $y = x^2$ is shifted 2 units to the right and 4 units down.

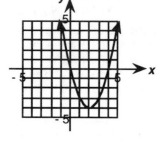

27. The graph of $y = \sqrt{x}$ is vertically expanded by a factor of 2, reflected in the x axis, and shifted 4 units up.

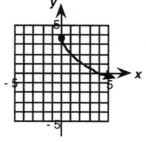

29. $(f + g)(x) = f(x) + g(x) = \sqrt{x + 2} + \sqrt{4 - x}$

$(f - g)(x) = f(x) - g(x) = \sqrt{x + 2} - \sqrt{4 - x}$

$(fg)(x) = f(x)g(x) = \sqrt{x + 2}\sqrt{4 - x} = \sqrt{(x + 2)(4 - x)} = \sqrt{8 + 2x - x^2}$

$\left(\dfrac{f}{g}\right)(x) = \dfrac{f(x)}{g(x)} = \dfrac{\sqrt{x + 2}}{\sqrt{4 - x}} = \sqrt{\dfrac{x + 2}{4 - x}}$

The domains of f and g are:

Domain of $f = \{x \mid x + 2 \geq 0\} = [-2, \infty)$

Domain of $g = \{x \mid 4 - x \geq 0\} = (-\infty, 4]$

The intersection of these domains is $[-2, 4]$.

This is the domain of $f + g$, $f - g$, and fg.

Since $g(4) = 0$, $x = 4$ must be excluded from the domain of $\dfrac{f}{g}$, so its domain is $[-2, 4)$.

31. $(f + g)(x) = f(x) + g(x) = 5 - 2\sqrt{x} + 3 - \sqrt{x} = 8 - 3\sqrt{x}$

$(f - g)(x) = f(x) - g(x) = 5 - 2\sqrt{x} - (3 - \sqrt{x}) = 5 - 2\sqrt{x} - 3 + \sqrt{x} = 2 - \sqrt{x}$

$(fg)(x) = f(x)g(x) = (5 - 2\sqrt{x})(3 - \sqrt{x}) = 15 - 11\sqrt{x} + 2x$

$\left(\dfrac{f}{g}\right)(x) = \dfrac{f(x)}{g(x)} = \dfrac{5 - 2\sqrt{x}}{3 - \sqrt{x}}$

The domains of f and g are both $\{x \mid x \geq 0\} = [0, \infty)$.

This is the domain of $f + g$, $f - g$, and fg. We note that in the domain of $\dfrac{f}{g}$, $g(x) \neq 0$. Thus $3 - \sqrt{x} \neq 0$. To solve this we solve

$3 - \sqrt{x} = 0$

$3 = \sqrt{x}$ $\boxed{\text{Common Error: } \sqrt{3} \neq x}$

$9 = x$

Hence, 9 must be excluded from $\{x \mid x \geq 0\}$ to find the domain of $\dfrac{f}{g}$.

Domain of $\dfrac{f}{g} = \{x \mid x \geq 0, x \neq 9\} = [0, 9) \cup (9, \infty)$.

33. $(f + g)(x) = f(x) + g(x) = \sqrt{x^2 + x - 2} + \sqrt{24 + 2x - x^2}$

$(f - g)(x) = f(x) - g(x) = \sqrt{x^2 + x - 2} - \sqrt{24 + 2x - x^2}$

$(fg)(x) = f(x)g(x) = \sqrt{x^2 + x - 2}\sqrt{24 + 2x - x^2} = \sqrt{-x^4 + x^3 + 28x^2 + 20x - 48}$

$\left(\dfrac{f}{g}\right)(x) = \dfrac{f(x)}{g(x)} = \dfrac{\sqrt{x^2 + x - 2}}{\sqrt{24 + 2x - x^2}}$

The domains of f and g are:

Domain of $f = \{x \mid x^2 + x - 2 \geq 0\} = \{x \mid (x + 2)(x - 1) \geq 0\} = (-\infty, -2] \cup [1, \infty)$

Domain of $g = \{x \mid 24 + 2x - x^2 \geq 0\} = \{x \mid (6 - x)(4 + x) \geq 0\} = [-4, 6]$

The intersection of these domains is $[-4, -2] \cup [1, 6]$.

This is the domain of $f + g$, $f - g$, and fg.

Since $g(x) = \sqrt{24 + 2x - x^2} = \sqrt{(6 - x)(4 + x)}$, $g(6) = 0$ and $g(-4) = 0$,

hence 6 and -4 must be excluded from the domain of $\dfrac{f}{g}$,

so its domain is $(-4, -2] \cup [1, 6)$.

35. $(f \circ g)(x) = f[g(x)] = f(\sqrt{4 - x}) = \sqrt{4 - x} + 2$ \qquad Domain: $\{x \mid x \leq 4\}$ or $(-\infty, 4]$

$(g \circ f)(x) = g[f(x)] = g(x + 2) = \sqrt{4 - (x + 2)} = \sqrt{2 - x}$

\qquad\qquad Domain: $\{x \mid x \leq 2\}$ or $[-\infty, 2]$

37. $(f \circ g)(x) = f[g(x)] = f\left(\dfrac{1}{x-2}\right) = \dfrac{1}{x-2} + 3 = \dfrac{1 + 3(x-2)}{x-2} = \dfrac{3x-5}{x-2}$

Domain: $\{x \mid x \neq 2\}$ or $(-\infty, 2) \cup (2, \infty)$

$(g \circ f)(x) = g[f(x)] = g(x+3) = \dfrac{1}{x+3-2} = \dfrac{1}{x+1}$

Domain: $\{x \mid x \neq -1\}$ or $(-\infty, -1) \cup (-1, \infty)$

39. $(f \circ g)(x) = f[g(x)] = f\left(\dfrac{x}{x-3}\right) = \left|\dfrac{x}{x-3} + 2\right| = \left|\dfrac{x}{x-3} + \dfrac{2(x-3)}{x-3}\right|$

$= \left|\dfrac{3x-6}{x-3}\right|$ or $3\left|\dfrac{x-2}{x-3}\right|$

Domain: $\{x \mid x \neq 3\}$ or $(-\infty, 3) \cup (3, \infty)$

$(g \circ f)(x) = g[f(x)] = g(|x+2|) = \dfrac{|x+2|}{|x+2|-3}$

To find the domain of g, we must find where $|x+2| - 3 \neq 0$.
To solve this, we must solve

$\begin{aligned}|x+2| - 3 &= 0\\ |x+2| &= 3\\ x+2 = 3 \quad &\text{or} \quad x+2 = -3\\ x = 1 \quad\quad & \quad\quad x = -5\end{aligned}$

Thus, the domain of g is $\{x \mid x \neq 1 \text{ and } x \neq -5\}$ or $(-\infty, -5) \cup (-5, 1) \cup (1, \infty)$.

41. The graph of the function $f(x) = |x|$ has been shifted one unit to the left and two units down. Equation: $y = |x+1| - 2$.

43. The graph of the function $f(x) = \sqrt[3]{x}$ has been reflected in the x axis and shifted 3 units up. Equation: $y = -\sqrt[3]{x} + 3$ or $y = 3 - \sqrt[3]{x}$.

45. The graph of the function $y = x^3$ has been reflected in the x axis and shifted two units to the left and one unit up. Equation: $y = -(x+2)^3 + 1$ or $y = 1 - (x+2)^3$.

47. $y = \sqrt{x+2} + 3$ **49.** $y = -|x-3|$ **51.** $y = -(x+2)^3 + 1$

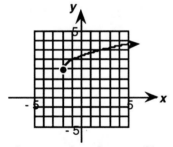 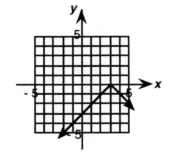 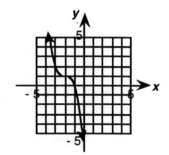

53. The graph of $y = f(x - h)$ represents a horizontal shift from the graph of $y = f(x)$. The graph of $y = f(x) + k$ represents a vertical shift from the graph of $y = f(x)$. The graph of $y = f(x - h) + k$ represents both a horizontal and a vertical shift but the order does not matter:
Vertical first then horizontal: $y = f(x) \rightarrow y = f(x) + k \rightarrow y = f(x - h) + k$
Horizontal first then vertical: $y = f(x) \rightarrow y = f(x - h) \rightarrow y = f(x - h) + k$
The same result is achieved; reversing the order does not change the result.

55. Consider the graph of $y = x^2$.
If a vertical shift is performed the equation becomes $y = x^2 + k$.
If a reflection is now performed the equation becomes
$y = -(x^2 + k)$ or $y = -x^2 - k$.
If the reflection is performed first the equation becomes $y = -x^2$.
If the vertical shift is now performed the equation becomes $y = -x^2 + k$.
Since $y = -x^2 - k$ and $y = -x^2 + k$ differ (unless $k = 0$), reversing the order changes the result.

57. The graph of $y = f(x - h)$ represents a horizontal shift from the graph of $y = f(x)$. The graph of $y = -f(x)$ represents a reflection of the graph of $y = f(x)$ in the x axis. The graph of $y = -f(x - h)$ represents both a horizontal shift and a reflection but the order does not matter:
Shift first then reflection: $y = f(x) \rightarrow y = f(x - h) \rightarrow y = -f(x - h)$
Reflection first then shift: $y = f(x) \rightarrow y = -f(x) \rightarrow y = -f(x - h)$
The same result is achieved; reversing the order does not change the result.

59. If we let $g(x) = 2x - 7$, then
$h(x) = [g(x)]^4$
Now if we let $f(x) = x^4$, we have
$h(x) = [g(x)]^4 = f[g(x)] = (f \circ g)(x)$

61. If we let $g(x) = 4 + 2x$, then
$h(x) = \sqrt{g(x)}$
Now if we let $f(x) = x^{1/2}$, we have
$h(x) = \sqrt{g(x)} = [g(x)]^{1/2}$
$\qquad = f[g(x)] = (f \circ g)(x).$

63. If we let $f(x) = x^7$, then
$h(x) = 3f(x) - 5$
Now if we let $g(x) = 3x - 5$, we have
$h(x) = 3f(x) - 5 = g[f(x)] = (g \circ f)(x)$

65. If we let $f(x) = x^{-1/2}$, then
$h(x) = 4f(x) + 3$
Now if we let $g(x) = 4x + 3$, we have
$h(x) = 4f(x) + 3 = g[f(x)] = (g \circ f)(x)$

67. The graph of $y = |x|$ is reflected in the x axis and vertically expanded by a factor of 3. Equation: $y = -3|x|$.

69. The graph of $y = x^3$ is reflected in the x axis and vertically contracted by a factor of 0.5. Equation: $y = -0.5x^3$.

71. fg and gf are identical, since
$(fg)(x) = f(x)g(x) = g(x)f(x)$ by the commutative law for multiplication
$\qquad\qquad\qquad\qquad\qquad$ of real numbers
$\qquad = (gf)(x)$

73. Yes, the function $g(x) = x$ satisfies these conditions.
$(f \circ g)(x) = f(g(x)) = f(x)$, so $f \circ g = f$
$(g \circ f)(x) = g(f(x)) = f(x)$, so $g \circ f = f$

75. $(f + g)(x) = f(x) + g(x) = x + \dfrac{1}{x} + x - \dfrac{1}{x} = 2x$

> **Common Error:**
> Domain is not $(-\infty, \infty)$. See below.

$(f - g)(x) = f(x) - g(x) = x + \dfrac{1}{x} - \left(x - \dfrac{1}{x}\right) = \dfrac{2}{x}$

$(fg)(x) = f(x)g(x) = \left(x + \dfrac{1}{x}\right)\left(x - \dfrac{1}{x}\right) = x^2 - \dfrac{1}{x^2}$

$\left(\dfrac{f}{g}\right)(x) = \dfrac{f(x)}{g(x)} = \dfrac{x + \dfrac{1}{x}}{x - \dfrac{1}{x}} = \dfrac{x^2 + 1}{x^2 - 1}$

The domains of f and g are both $\{x \mid x \neq 0\} = (-\infty, 0) \cup (0, \infty)$
This is therefore the domain of $f + g$, $f - g$, and fg. To find the domain of $\dfrac{f}{g}$, we must exclude from this domain the set of values of x for which $g(x) = 0$.

$x - \dfrac{1}{x} = 0$
$x^2 - 1 = 0$
$x^2 = 1$
$x = -1, 1$

Hence, the domain of $\dfrac{f}{g}$ is $\{x \mid x \neq 0, -1, \text{ or } 1\}$ or
$(-\infty, -1) \cup (-1, 0) \cup (0, 1) \cup (1, \infty)$.

77. $(f + g)(x) = f(x) + g(x) = 1 - \dfrac{x}{|x|} + 1 + \dfrac{x}{|x|} = 2$

$(f - g)(x) = f(x) - g(x) = 1 - \dfrac{x}{|x|} - \left(1 + \dfrac{x}{|x|}\right) = 1 - \dfrac{x}{|x|} - 1 - \dfrac{x}{|x|} = \dfrac{-2x}{|x|}$

$(fg)(x) = f(x)g(x) = \left(1 - \dfrac{x}{|x|}\right)\left(1 + \dfrac{x}{|x|}\right) = (1)^2 - \left(\dfrac{x}{|x|}\right)^2 = 1 - \dfrac{x^2}{|x|^2}$

$= 1 - \dfrac{x^2}{x^2} = 1 - 1 = 0$

$\left(\dfrac{f}{g}\right)(x) = \dfrac{f(x)}{g(x)} = \dfrac{1 - \frac{x}{|x|}}{1 + \frac{x}{|x|}} = \dfrac{|x| - x}{|x| + x}$. This can be further simplified

however, when we examine the domain of $\dfrac{f}{g}$ below.
The domains of f and g are both
$\{x \mid x \neq 0\} = (-\infty, 0) \cup (0, \infty)$
This is therefore the domain of $f + g$, $f - g$, and fg. To find the domain of $\dfrac{f}{g}$,
we must exclude from this domain the set of values of x for which $g(x) = 0$.

$1 + \dfrac{x}{|x|} = 0$

$|x| + x = 0$

$|x| = -x$

$\quad x$ is negative

Thus the domain of $\dfrac{f}{g}$ is the positive numbers, $(0, \infty)$. On this domain,

$|x| = x$, hence $\left(\dfrac{f}{g}\right)(x) = \dfrac{|x| - x}{|x| + x} = \dfrac{x - x}{x + x} = \dfrac{0}{2x} = 0$

79. $(f \circ g)(x) = f[g(x)] = f(\sqrt{9 - x}) = (\sqrt{9 - x})^2 = 9 - x$
The domain of f is all real numbers. The domain of g is $(-\infty, 9]$. Hence the
domain of $f \circ g$ is the set of those numbers x in $(-\infty, 9]$ for which $g(x)$ is
in $(-\infty, \infty)$, that is, $(-\infty, 9]$.

$(g \circ f)(x) = g[f(x)] = g(x^2) = \sqrt{9 - x^2}$.
The domain of $g \circ f$ is the set of those real numbers x for which $f(x)$ is
in $(-\infty, 9]$, that is, for which $x^2 \leq 9$, or $-3 \leq x \leq 3$.
Domain of $g \circ f = \{x \mid -3 \leq x \leq 3\}$ or $[-3, 3]$.

81. $(f \circ g)(x) = f[g(x)] = f\left(\dfrac{x + 2}{x - 3}\right) = \dfrac{2\left(\frac{x + 2}{x - 3}\right) + 1}{\frac{x + 2}{x - 3} - 2} = \dfrac{2(x + 2) + x - 3}{x + 2 - 2(x - 3)}$

$= \dfrac{2x + 4 + x - 3}{x + 2 - 2x + 6} = \dfrac{3x + 1}{8 - x}$

The domain of f is $\{x \mid x \neq 2\}$. The domain of g is $\{x \mid x \neq 3\}$. Hence the domain
of $f \circ g$ is the set of those numbers in $\{x \mid x \neq 3\}$ for which $g(x)$ is in
$\{x \mid x \neq 2\}$. Thus we must exclude from $\{x \mid x \neq 3\}$ those numbers x for which
$\dfrac{x + 2}{x - 3} = 2$, or $x + 2 = 2(x - 3)$, or $x = 8$.
Hence the domain of $f \circ g$ is $\{x \mid x \neq 3, x \neq 8\}$ or $(-\infty, 3) \cup (3, 8) \cup (8, \infty)$.

$(g \circ f)(x) = g[f(x)] = g\left(\dfrac{2x + 1}{x - 2}\right) = \dfrac{\frac{2x + 1}{x - 2} + 2}{\frac{2x + 1}{x - 2} - 3} = \dfrac{2x + 1 + 2(x - 2)}{2x + 2 - 3(x - 2)}$

$= \dfrac{2x + 1 + 2x - 4}{2x + 1 - 3x + 6} = \dfrac{4x - 3}{7 - x}$

The domain of $g \circ f$ is the set of those numbers in $\{x \mid x \neq 2\}$ for which $g(x)$ is
in $\{x \mid x \neq 3\}$. Hence we must exclude from $\{x \mid x \neq 2\}$ those numbers x for which
$\dfrac{2x + 1}{x - 2} = 3$, or $2x + 1 = 3(x - 2)$, or $x = 7$.
Hence the domain of $g \circ f$ is $\{x \mid x \neq 2, x \neq 7\}$ or $(-\infty, 2) \cup (2, 7) \cup (7, \infty)$.

83. $(f \circ g)(x) = f[g(x)] = f(\sqrt{x^2 - 4}) = \sqrt{(\sqrt{x^2 - 4})^2 + 5} = \sqrt{x^2 - 4 + 5} = \sqrt{x^2 + 1}$.
The domain of f is all real numbers. The domain of g is $(-\infty, -2] \cup [2, \infty)$.
Hence the domain of $f \circ g$ is the set of those numbers in $(-\infty, -2] \cup [2, \infty)$
for which $f(x)$ is real. Since $f(x)$ is real for all x, the domain of
$f \circ g$ is $(-\infty, -2] \cup [2, \infty)$.

> Common Error: The domain of $f \circ g$ is not evident from the final form $\sqrt{x^2 + 1}$. It is not $(-\infty, \infty)$.

$(g \circ f)(x) = g[f(x)] = g[\sqrt{x^2 + 5}] = \sqrt{(\sqrt{x^2 + 5})^2 - 4}$
$$= \sqrt{x^2 + 5 - 4} = \sqrt{x^2 + 1}$$

The domain of $g \circ f$ is the set of those real numbers for which $f(x)$ is in
$(-\infty, -2] \cup [2, \infty)$. Since we can write:
$$x^2 \geq 0$$
$$x^2 + 5 \geq 5$$
$$\sqrt{x^2 + 5} \geq \sqrt{5} \geq 2,$$
$\sqrt{x^2 + 5}$ is in $(-\infty, -2] \cup [2, \infty)$ for all real x. The domain of $g \circ f$ is $(-\infty, \infty)$.

85. The profit function is the difference of the revenue and the cost functions.
$P = R - C$
Hence $P(x) = R(x) - C(x)$
$$= \left(20x - \frac{1}{200}x^2\right) - (10x + 30,000)$$
$$= 20x - \frac{1}{200}x^2 - 10x - 30,000$$
$$= 10x - \frac{1}{200}x^2 - 30,000$$

Next we use composition to express P as a function of the price p.
$(P \circ f)(p) = P[f(p)]$
$$= P(4,000 - 200p)$$
$$= 10(4,000 - 200p) - \frac{1}{200}(4,000 - 200p)^2 - 30,000$$
$$= 40,000 - 2,000p - \frac{1}{200}(16,000,000 - 1,600,000p + 40,000p^2) - 30,000$$
$$= 40,000 - 2,000p - 80,000 + 8,000p - 200p^2 - 30,000$$
$$= -70,000 + 6,000p - 200p^2$$

87. $y = \frac{1}{C}x^2 - C$

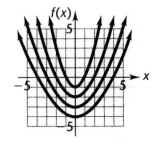

$C = 1$: $y = x^2 - 1$ We sketch the graph of $y = x^2$
shifted down 1 unit

$C = 2$: $y = \frac{1}{2}x^2 - 2$ We sketch the graph of $y = x^2$
contracted by a factor of $\frac{1}{2}$,
and shifted down 2 units.

$C = 3$: $y = \frac{1}{3}x^2 - 3$ We sketch the graph of $y = x^2$
contracted by a factor of $\frac{1}{3}$,
and shifted down 3 units.

$C = 4$: $y = \frac{1}{4}x^2 - 4$ We sketch the graph of $y = x^2$ contracted by a factor of $\frac{1}{4}$,
and shifted down 4 units.

89. $V(t) = \dfrac{64}{c^2}(C - t)^2 \qquad 0 \le t \le C$

t	$C = 1$ $V = 64(1 - t)^2$	$C = 2$ $V = 16(2 - t)^2$	$C = 4$ $V = 4(4 - t)^2$	$C = 8$ $V = (8 - t)^2$
0	64	64	64	64
1	0	16	36	49
2	not defined for $t > 1$	0	16	36
4	arc of a parabola	not defined for $t > 2$	0	16
6	with vertex at (1,0)	arc of a parabola with	not defined for $t > 4$	4
8		vertex at (2,0)	arc of a parabola	0
			with vertex at (4,0)	arc of a parabola
				with vertex at (8,0)

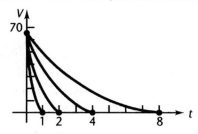

91.

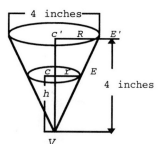

(A) We note: In the figure, triangles VCE and $VC'E'$ are similar. Moreover

R = radius of cup = $\dfrac{1}{2}$ diameter of cup = $\dfrac{1}{2}(4)$ = 2 inches.

Hence $\dfrac{r}{2} = \dfrac{h}{4}$ or $r = \dfrac{1}{2}h$. We write $r(h) = \dfrac{1}{2}h$.

(B) Since $V = \dfrac{1}{3}\pi r^2 h$ and $r = \dfrac{1}{2}h$, $V = \dfrac{1}{3}\pi\left(\dfrac{1}{2}h\right)^2 h = \dfrac{1}{3}\pi\dfrac{1}{4}h^2 h = \dfrac{1}{12}\pi h^3$.

We write $V(h) = \dfrac{1}{12}\pi h^3$.

(C) Since $V(h) = \dfrac{1}{12}\pi h^3$ and $h(t) = 0.5\sqrt{t}$, we use composition to express V as a function of t.

$(V \circ h)(t) = V[h(t)]$

$\qquad = \dfrac{1}{12}\pi [h(t)]^3$

$\qquad = \dfrac{1}{12}\pi [0.5\sqrt{t}]^3$

$\qquad = \dfrac{1}{12}\pi(0.125)(t^{1/2})^3$

$\qquad = \dfrac{0.125}{12}\pi t^{3/2}$

We write $V(t) = \dfrac{0.125}{12}\pi t^{3/2}$.

Exercise 2-6

Key Ideas and Formulas

A function is one-to-one if no two ordered pairs in the function have the same second component and different first components.

If $f(a) = f(b)$ for at least one pair of domain values a and b, $a \neq b$, then f is not one-to-one.

If the assumption of $f(a) = f(b)$ always implies that the domain values a and b are equal, then f is one-to-one.

A function is one-to-one if and only if each horizontal line intersects the graph of the function in at most one point. (Horizontal line test.)

If a function f is increasing throughout its domain or decreasing throughout its domain, then f is a one-to-one function.

If f is a one-to-one function, then the inverse of f, denoted f^{-1}, is the function formed by reversing all the ordered pairs in f. $f^{-1} = \{(y, x) \mid (x, y)$ is in $f\}$.

If f is not one-to-one, then f does not have an inverse and f^{-1} does not exist.

If f^{-1} exists, then

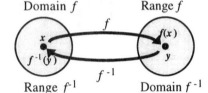

1. f^{-1} is a one-to-one function.
2. Domain of f^{-1} = Range of f.
3. Range of f^{-1} = Domain of f

If f^{-1} exists, then

1. $x = f^{-1}(y)$ if and only if $y = f(x)$
2. $f^{-1}[f(x)] = x$ for all x in the domain of f
3. $f[f^{-1}(y)] = y$ for all y in the domain of f^{-1}

To find the inverse of a function f

1. Find the domain of f and verify that f is one-to-one. If f is not one-to-one, then stop—f^{-1} does not exist.

2. Solve the equation $y = f(x)$ for x. The result is an equation of the form $x = f^{-1}(y)$.

3. Interchange x and y in the equation found in Step 2. The result is an equation of the form $y = f^{-1}(x)$.

4. Find the domain of f^{-1}. (This is the same as the range of f.)

5. Check by showing that
$f^{-1}[f(x)] = x$ for all x in domain of f
$f[f^{-1}(x)] = x$ for all x in the domain of f^{-1}

The graphs of $y = f(x)$ and $y = f^{-1}(x)$ are symmetric with respect to the line $y = x$.

---

1. One-to-one

3. The range element 3 corresponds to more than one domain element. Not one-to-one.

5. One-to-one

7. The range element 7 corresponds to more than one domain element. Not one-to-one.

9. One-to-one

11. Some range elements (0, for example) correspond to more than one domain element. Not one-to-one.

13. One-to-one   **15.** One-to-one   **17.** Assume $F(a) = F(b)$
$$\frac{1}{2}a + 1 = \frac{1}{2}b + 1$$
Then $\frac{1}{2}a = \frac{1}{2}b$
$$a = b$$
Therefore F is one-to-one.

19. $H(x) = 4 - x^2$
Since $H(1) = 4 - 1^2 = 3$ and $H(-1) = 4 - (-1)^2$
$= 3$, both $(1, 3)$ and $(-1, 3)$ belong to H.
H is not one-to-one.

21. Assume $M(a) = M(b)$
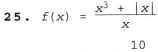
$$\sqrt{a + 1} = \sqrt{b + 1}$$
Then $a + 1 = b + 1$
$$a = b$$
M is one-to-one.

23. $f(x) = \dfrac{x^2 + |x|}{x}$

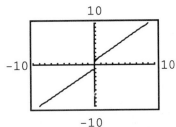

This graph passes the horizontal line test. f is one-to-one.

25. $f(x) = \dfrac{x^3 + |x|}{x}$

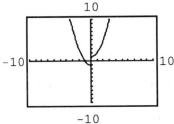

This graph does not pass the horizontal line test. f is not one-to-one.

27. $f(x) = \dfrac{x^2 - 4}{|x - 2|}$

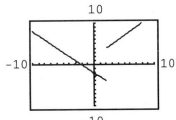

This graph does not pass the horizontal line test. f is not one-to-one.

29. $f(x) = \dfrac{x^3 - 9x}{|x^2 - 9|}$

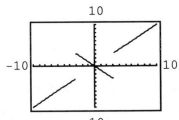

This graph passes the horizontal line test. f is one-to-one.

31. From the given graph of f, we see
Domain of $f = [-4, 4]$ Range of $f = [1, 5]$
Hence
Range of $f^{-1} = [-4, 4]$ Domain of $f^{-1} = [1, 5]$
The graph of f^{-1} is drawn by reflecting the
given graph of f in the line $y = x$.

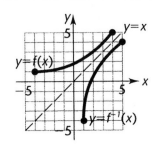

33. From the given graph of f, we see
Domain of $f = [-5, 3]$ Range of $f = [-3, 5]$
Hence
Range of $f^{-1} = [-5, 3]$ Domain of $f^{-1} = [-3, 5]$
The graph of f^{-1} is drawn by reflecting the given graph
of f in the line $y = x$.

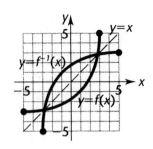

35. $f(x) = 3x + 6$ $g(x) = \frac{1}{3}x - 2$

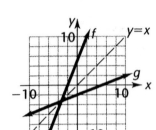

$$g[f(x)] = g(3x + 6) = \frac{1}{3}(3x + 6) - 2 = x + 2 - 2 = x$$

$$f[g(x)] = f\left(\frac{1}{3}x - 2\right) = 3\left(\frac{1}{3}x - 2\right) + 6 = x - 6 + 6 = x$$

37. $f(x) = 4 + x^2$ $x \geq 0$ $g(x) = \sqrt{x - 4}$

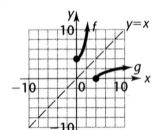

$$g[f(x)] = g(4 + x^2) = \sqrt{4 + x^2 - 4} = \sqrt{x^2} = x \text{ since } x \geq 0$$

$$f[g(x)] = f(\sqrt{x - 4}) = 4 + (\sqrt{x - 4})^2 = 4 + x - 4 = x$$

39. $f(x) = -\sqrt{x - 2}$ $g(x) = x^2 + 2$, $x \leq 0$

$$f[g(x)] = f(x^2 + 2) = -\sqrt{x^2 + 2 - 2} = -\sqrt{x^2} = -(-x) = x$$

(Note: $\sqrt{x^2} = -x$ since $x \leq 0$)

$$g[f(x)] = g(-\sqrt{x - 2}) = (-\sqrt{x - 2})^2 + 2 = x - 2 + 2 = x$$

Common Error: $(-\sqrt{x - 2})^2 \neq -(x - 2) \text{ nor } -x - 2$

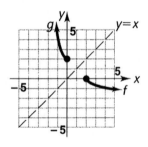

41. $f(x) = \dfrac{1}{5}x$

Solve $y = f(x)$ for x:

$$y = \frac{1}{5}x$$

$$x = 5y = f^{-1}(y)$$

Interchange x and y:

$$y = f^{-1}(x) = 5x$$

Domain of f^{-1} = Range of $f = (-\infty, \infty)$

Check: $f^{-1}[f(x)] = f^{-1}\left(\dfrac{1}{5}x\right) = 5\left(\dfrac{1}{5}x\right) = x$

$$f[f^{-1}(x)] = f(5x) = \frac{1}{5}(5x) = x$$

$$f^{-1}(x) = 5x$$

43. $f(x) = 2x + 7$

Solve $y = f(x)$ for x:

$$y = 2x + 7$$

$$y - 7 = 2x$$

$$x = \frac{y-7}{2} = \frac{1}{2}y - \frac{7}{2} = f^{-1}(y)$$

Interchange x and y:

$$y = f^{-1}(x) = \frac{1}{2}x - \frac{7}{2}$$

Domain of f^{-1} = Range of $f = (-\infty, \infty)$

Check: $f^{-1}[f(x)] = f^{-1}(2x + 7)$

$$= \frac{1}{2}(2x + 7) - \frac{7}{2}$$

$$= x + \frac{7}{2} - \frac{7}{2}$$

$$= x$$

$$f[f^{-1}(x)] = f\left(\frac{1}{2}x - \frac{7}{2}\right)$$

$$= 2\left(\frac{1}{2}x - \frac{7}{2}\right) + 7$$

$$= x - 7 + 7$$

$$= x$$

$$f^{-1}(x) = \frac{1}{2}x - \frac{7}{2}$$

45. $f(x) = 0.2x + 0.4$

Solve $y = f(x)$ for x:

$$y = 0.2x + 0.4$$

$$y - 0.4 = 0.2x$$

$$x = \frac{y - 0.4}{0.2}$$

$$x = \frac{y}{0.2} - 2$$

$$x = 5y - 2 = f^{-1}(y)$$

Interchange x and y:

$$y = f^{-1}(x) = 5x - 2$$

Domain of f^{-1} = Range of $f = (-\infty, \infty)$

Check: $f^{-1}[f(x)] = f^{-1}(0.2x + 0.4)$

$$= 5(0.2x + 0.4) - 2$$

$$= x + 2 - 2$$

$$= x$$

$$f[f^{-1}(x)] = f(5x - 2)$$

$$= 0.2(5x - 2) + 0.4$$

$$= x - 0.4 + 0.4$$

$$= x$$

$$f^{-1}(x) = 5x - 2$$

47. $f(x) = 3 - \dfrac{2}{x}$

Solve $y = f(x)$ for x:

$$y = 3 - \frac{2}{x} \quad x \neq 0$$

$$xy = 3x - 2$$

$$xy - 3x = -2$$

$$x(y - 3) = -2$$

$$x = \frac{2}{3 - y} = f^{-1}(y)$$

Interchange x and y:

$$y = f^{-1}(x) = \frac{2}{3 - x}$$

Domain of f^{-1} = Range of f

$$= (-\infty, 3) \cup (3, \infty)$$

Check: $f^{-1}[f(x)] = f^{-1}\left(3 - \dfrac{2}{x}\right)$

$$= \dfrac{2}{3 - \left(3 - \frac{2}{x}\right)}$$

$$= \dfrac{2}{3 - 3 + \frac{2}{x}}$$

$$= \dfrac{2}{\frac{2}{x}}$$

$$= x$$

$f[f^{-1}(x)] = f\left(\dfrac{2}{3 - x}\right)$

$$= 3 - \dfrac{2}{\frac{2}{(3 - x)}}$$

$$= 3 - \dfrac{2(3 - x)}{2}$$

$$= 3 - (3 - x)$$

$$= 3 - 3 + x$$

$$= x$$

$$f^{-1}(x) = \dfrac{2}{3 - x}$$

49. $f(x) = \dfrac{2x}{x + 1}$

Solve $y = f(x)$ for x:

$$y = \dfrac{2x}{x + 1} \quad x \neq -1$$

$$y(x + 1) = 2x$$

$$xy + y = 2x$$

$$y = 2x - xy$$

$$y = x(2 - y)$$

$$x = \dfrac{y}{2 - y} = f^{-1}(y)$$

Interchange x and y:

$$y = \dfrac{x}{2 - x} = f^{-1}(x)$$

Domain of f^{-1} = Range of f = $(-\infty, 2) \cup (2, \infty)$

Check: $f^{-1}[f(x)] = f^{-1}\left(\dfrac{2x}{x + 1}\right)$

$$= \dfrac{\frac{2x}{x + 1}}{2 - \frac{2x}{x + 1}}$$

$$= \dfrac{2x}{2(x + 1) - 2x}$$

$$= \dfrac{2x}{2x + 2 - 2x}$$

$$= \dfrac{2x}{2} = x$$

$$f^{-1}(x) = \dfrac{x}{2 - x}$$

$f[f^{-1}(x)] = f\left(\dfrac{x}{2 - x}\right)$

$$= \dfrac{2 \cdot \frac{x}{2 - x}}{\frac{x}{2 - x} + 1}$$

$$= \dfrac{\frac{2x}{2 - x}}{\frac{x}{2 - x} + 1}$$

$$= \dfrac{2x}{x + 2 - x}$$

$$= \dfrac{2x}{2}$$

$$= x$$

51. $f(x) = \dfrac{0.2x - 0.4}{0.1x + 0.5}$

Solve $y = f(x)$ for x:

$$y = \dfrac{0.2x - 0.4}{0.1x + 0.5}$$

$$y = \dfrac{2x - 4}{x + 5} \quad x \neq -5$$

$$y(x + 5) = 2x - 4$$

$$xy + 5y = 2x - 4$$

$$5y + 4 = 2x - xy$$

$$5y + 4 = x(2 - y)$$

$$x = \dfrac{5y + 4}{2 - y} = f^{-1}(y)$$

Interchange x and y:
$$y = \frac{5x + 4}{2 - x} = f^{-1}(x)$$
Domain of f^{-1} = Range of $f = (-\infty, 2) \cup (2, \infty)$

Check:

$$f^{-1}[f(x)] = f^{-1}\left[\frac{0.2x - 0.4}{0.1x + 0.5}\right]$$

$$= \frac{5\left[\frac{0.2x - 0.4}{0.1x + 0.5}\right] + 4}{2 - \left[\frac{0.2x - 0.4}{0.1x + 0.5}\right]}$$

$$= \frac{\frac{x - 2}{0.1x + 0.5} + 4}{2 - \left[\frac{0.2x - 0.4}{0.1x + 0.5}\right]}$$

$$= \frac{x - 2 + 4(0.1x + 0.5)}{2(0.1x + 0.5) - (0.2x - 0.4)}$$

$$= \frac{x - 2 + 0.4x + 2}{0.2x + 1 - 0.2x + 0.4}$$

$$= \frac{1.4x}{1.4}$$

$$= x$$

$$f^{-1}(x) = \frac{5x + 4}{2 - x}$$

$$f[f^{-1}(x)] = f\left(\frac{5x + 4}{2 - x}\right)$$

$$= \frac{0.2\left(\frac{5x + 4}{2 - x}\right) - 0.4}{0.1\left(\frac{5x + 4}{2 - x}\right) + 0.5}$$

$$= \frac{0.2(5x + 4) - 0.4(2 - x)}{0.1(5x + 4) + 0.5(2 - x)}$$

$$= \frac{x + 0.8 - 0.8 + 0.4x}{0.5x + 0.4 + 1 - 0.5x}$$

$$= \frac{1.4x}{1.4}$$

$$= x$$

53. $f(x) = 8x^3 - 5$
Solve $y = f(x)$ for x:
$$y = 8x^3 - 5$$
$$y + 5 = 8x^3$$
$$\frac{y + 5}{8} = x^3$$

$$x = \sqrt[3]{\frac{y + 5}{8}}$$

$$x = \frac{\sqrt[3]{y + 5}}{\sqrt[3]{8}} = \frac{\sqrt[3]{y + 5}}{2}$$

$$x = 0.5\sqrt[3]{y + 5} = f^{-1}(y)$$

Interchange x and y:
$$y = 0.5\sqrt[3]{x + 5} = f^{-1}(x)$$
Domain of f^{-1} = Range of $f = (-\infty, \infty)$

Check: $f^{-1}[f(x)] = f^{-1}(8x^3 - 5)$

$$= 0.5\sqrt[3]{8x^3 - 5 + 5}$$

$$= 0.5\sqrt[3]{8x^3}$$

$$= 0.5(2x)$$

$$= x$$

$$f[f^{-1}(x)] = f[0.5\sqrt[3]{x + 5}]$$

$$= 8[0.5\sqrt[3]{x + 5}]^3 - 5$$

$$= 8(0.125)(x + 5) - 5$$

$$= x + 5 - 5$$

$$= x$$

$$f^{-1}(x) = 0.5\sqrt[3]{x + 5}$$

55. $f(x) = 2 + \sqrt[5]{3x - 7}$

Solve $y = f(x)$ for x:

$$y = 2 + \sqrt[5]{3x - 7}$$

$$y - 2 = \sqrt[5]{3x - 7}$$

$$(y - 2)^5 = 3x - 7$$

$$(y - 2)^5 + 7 = 3x$$

$$x = \frac{1}{3}(y - 2)^5 + \frac{7}{3} = f^{-1}(y)$$

Interchange x and y:

$$y = \frac{1}{3}(x - 2)^5 + \frac{7}{3} = f^{-1}(x)$$

Domain of f^{-1} = Range of f = $(-\infty, \infty)$

Check:

$$f^{-1}[f(x)] = f^{-1}[2 + \sqrt[5]{3x - 7}]$$

$$= \frac{1}{3}[2 + \sqrt[5]{3x - 7} - 2]^5 + \frac{7}{3}$$

$$= \frac{1}{3}(\sqrt[5]{3x - 7})^5 + \frac{7}{3}$$

$$= \frac{1}{3}(3x - 7) + \frac{7}{3}$$

$$= x - \frac{7}{3} + \frac{7}{3}$$

$$= x$$

$$f[f^{-1}(x)] = f\left[\frac{1}{3}(x - 2)^5 + \frac{7}{3}\right]$$

$$= 2 + \sqrt[5]{3[\frac{1}{3}(x - 2)^5 + \frac{7}{3}] - 7}$$

$$= 2 + \sqrt[5]{(x - 2)^5 + 7 - 7}$$

$$= 2 + \sqrt[5]{(x - 2)^5}$$

$$= 2 + x - 2$$

$$= x$$

$$f^{-1}(x) = \frac{1}{3}(x - 2)^5 + \frac{7}{3}$$

57. $f(x) = 2\sqrt{9 - x}$

Solve $y = f(x)$ for x:

$\left.\begin{array}{l} y = 2\sqrt{9 - x} \\ y^2 = 4(9 - x) \end{array}\right\}$ these are equivalent only if $y \geq 0$

$$\frac{y^2}{4} = 9 - x$$

$$\frac{1}{4}y^2 - 9 = -x$$

$$x = 9 - \frac{1}{4}y^2 \quad y \geq 0 \quad f^{-1}(y) = 9 - \frac{1}{4}y^2$$

Interchange x and y:

$$y = f^{-1}(x) = 9 - \frac{1}{4}x^2 \quad x \geq 0$$

Domain: $[0, \infty)$

Check: $f^{-1}[f(x)] = 9 - \frac{1}{4}[2\sqrt{9 - x}]^2$

$$= 9 - \frac{1}{4}[4(9 - x)]$$

$$= 9 - (9 - x)$$

$$= 9 - 9 + x$$

$$= x$$

$$f[f^{-1}(x)] = 2\sqrt{9 - (9 - \frac{1}{4}x^2)}$$

$$= 2\sqrt{9 - 9 + \frac{1}{4}x^2}$$

$$= 2\sqrt{\frac{1}{4}x^2}; \quad \sqrt{\frac{1}{4}x^2} = \frac{1}{2}x$$

$$\text{since } x \geq 0$$

$$= 2(\frac{1}{2}x)$$

$$= x$$

$$f^{-1}(x) = 9 - \frac{1}{4}x^2, \quad x \geq 0$$

59. $f(x) = 2 + \sqrt{3 - x}$

Solve $y = f(x)$ for x:

$$y = 2 + \sqrt{3 - x}$$

$\left.\begin{array}{l} y - 2 = \sqrt{3 - x} \\ (y - 2)^2 = 3 - x \end{array}\right\}$ these are equivalent only if $y \geq 2$

$$x = 3 - (y - 2)^2$$

$$x = 3 - (y^2 - 4y + 4)$$

$$x = -y^2 + 4y - 1 \qquad y \geq 2 \quad f^{-1}(y) = -y^2 + 4y - 1$$

Interchange x and y:

$$y = f^{-1}(x) = -x^2 + 4x - 1 \quad x \geq 2 \quad \text{Domain: } [2, \infty)$$

Check: $f^{-1}[f(x)] = f^{-1}[2 + \sqrt{3 - x}\,]$

$\qquad\qquad\quad = -[2 + \sqrt{3 - x}\,]^2 + 4[2 + \sqrt{3 - x}\,] - 1$

$\qquad\qquad\quad = -(4 + 4\sqrt{3 - x} + 3 - x) + 8 + 4\sqrt{3 - x} - 1$

$\qquad\qquad\quad = -4 - 4\sqrt{3 - x} - 3 + x + 8 + 4\sqrt{3 - x} - 1$

$\qquad\qquad\quad = -8 + x + 8$

$\qquad\qquad\quad = x$

$\qquad f[f^{-1}(x)] = f(-x^2 + 4x - 1)$

$\qquad\qquad\quad = 2 + \sqrt{3 - (-x^2 + 4x - 1)}$

$\qquad\qquad\quad = 2 + \sqrt{3 + x^2 - 4x + 1}$

$\qquad\qquad\quad = 2 + \sqrt{x^2 - 4x + 4}$

$\qquad\qquad\quad = 2 + \sqrt{(x - 2)^2}; \quad \sqrt{(x - 2)^2} = x - 2$ since $x \geq 2$

$\qquad\qquad\quad = 2 + x - 2$

$\qquad\qquad\quad = x$

$f^{1-}(x) = -x^2 + 4x - 1 \quad x \geq 2$

61. Since in passing from a function to its inverse, x and y are interchanged, the x intercept of f is the y intercept of f^{-1} and the y intercept of f is the x intercept of f^{-1}.

63. $f(x) = (x - 1)^2 + 2 \quad x \geq 1$

Solve $y = f(x)$ for x:

$\qquad y = (x - 1)^2 + 2 \quad x \geq 1$

$\qquad y - 2 = (x - 1)^2 \quad x \geq 1$

$\sqrt{y - 2} = \sqrt{(x - 1)^2}$ Since $x \geq 1$, $\sqrt{(x - 1)^2} = x - 1$

$\sqrt{y - 2} = x - 1$

$\qquad x = 1 + \sqrt{y - 2} = f^{-1}(y)$

Interchange x and y:

$y = f^{-1}(x) = 1 + \sqrt{x - 2}$ Domain: $x \geq 2$

Check:

$f^{-1}[f(x)] = 1 + \sqrt{(x - 1)^2 + 2 - 2}$ $\qquad\qquad f[f^{-1}(x)] = (1 + \sqrt{x - 2} - 1)^2 + 2$

$\qquad\qquad = 1 + \sqrt{(x - 1)^2}$ $\qquad\qquad\qquad\qquad\qquad = (\sqrt{x - 2})^2 + 2$

$\qquad\qquad\quad \sqrt{(x - 1)^2} = x - 1$ since $x \geq 1$ $\qquad\qquad = x - 2 + 2$

$\qquad\qquad = 1 + x - 1$ $\qquad\qquad\qquad\qquad\qquad\qquad = x$

$\qquad\qquad = x$

$f^{-1}(x) = 1 + \sqrt{x - 2}, \; x \geq 2$

65. $f(x) = x^2 + 2x - 2 \quad x \leq -1$

Solve $y = f(x)$ for x:

$\qquad y = x^2 + 2x - 2 \qquad x \leq -1$

$\qquad y + 3 = x^2 + 2x + 1 \qquad x \leq -1$

$\qquad y + 3 = (x + 1)^2 \qquad x \leq -1$

$\sqrt{y + 3} = \sqrt{(x + 1)^2}$ Since $x \leq -1$ $\sqrt{(x + 1)^2} = -(x + 1)$

$\sqrt{y + 3} = -(x + 1)$

$-\sqrt{y + 3} = x + 1$

$\qquad x = -1 - \sqrt{y + 3} = f^{-1}(y)$

Interchange x and y:

$y = f^{-1}(x) = -1 - \sqrt{x + 3} \quad x \geq -3$ Domain: $[-3, \infty)$

Check: $f^{-1}[f(x)] = -1 - \sqrt{x^2 + 2x - 2 + 3}$ $x \leq -1$

$= -1 - \sqrt{x^2 + 2x + 1}$ $x \leq -1$

$= -1 - \sqrt{(x + 1)^2}$ $x \leq -1$

$\sqrt{(x + 1)^2} = -(x + 1)$ since $x \leq -1$

$= -1 - [-(x + 1)]$

$= -1 + x + 1$

$= x$

$f[f^{-1}(x)] = (-1 - \sqrt{x + 3})^2 + 2(-1 - \sqrt{x + 3}) - 2$

$= 1 + 2\sqrt{x + 3} + x + 3 - 2 - 2\sqrt{x + 3} - 2$

$= 4 + x - 4$

$= x$

$f^{-1}(x) = -1 - \sqrt{x + 3}, \; x \geq -3$

67. $f(x) = -\sqrt{9 - x^2} \;\; 0 \leq x \leq 3$ f is one-to-one. See graph below.

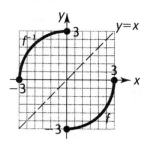

Solve $y = f(x)$ for x:

$y = -\sqrt{9 - x^2}$ $0 \leq x \leq 3$

$y^2 = 9 - x^2$ $y \leq 0$

$y^2 - 9 = -x^2$ $y \leq 0$

$9 - y^2 = x^2$ $y \leq 0$

$x = \sqrt{9 - y^2}$ $y \leq 0$ $f^{-1}(y) = \sqrt{9 - y^2}$

positive square root only because $0 \leq x$

Interchange x and y:

$y = f^{-1}(x) = \sqrt{9 - x^2}$ Domain: $-3 \leq x \leq 0$

Check: $f^{-1}[f(x)] = \sqrt{9 - (-\sqrt{9 - x^2})^2}$

$= \sqrt{9 - (9 - x^2)}$

$= \sqrt{9 - 9 + x^2}$

$= \sqrt{x^2}$

$\sqrt{x^2} = x$ since $x \geq 0$ *in the domain of* f

$= x$

$f[f^{-1}(x)] = -\sqrt{9 - (\sqrt{9 - x^2})^2}$

$= -\sqrt{9 - (9 - x^2)}$

$= -\sqrt{9 - 9 + x^2}$

$= -\sqrt{x^2}$

$\sqrt{x^2} = -x$ since $x \leq 0$ *in the domain of* f^{-1}

$= -(-x)$

$= x$

$f^{-1}(x) = \sqrt{9 - x^2}$ Domain of $f^{-1} = [-3, 0]$

Range of f^{-1} = Domain of $f = [0, 3]$

69. $f(x) = \sqrt{9 - x^2}$ $-3 \le x \le 0$ f is one-to-one. See graph below.

Solve $y = f(x)$ for x

$$y = \sqrt{9 - x^2} \quad -3 \le x \le 0$$
$$y^2 = 9 - x^2 \quad y \ge 0$$
$$y^2 - 9 = -x^2 \quad y \ge 0$$
$$9 - y^2 = x^2 \quad y \ge 0$$
$$x = -\underbrace{\sqrt{9 - y^2}} \quad y \ge 0 \quad f^{-1}(y) = -\sqrt{9 - y^2}$$

negative square root only because $x \le 0$

Interchange x and y:

$y = f^{-1}(x) = -\sqrt{9 - x^2}$ Domain: $0 \le x \le 3$

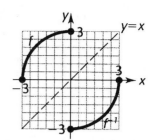

Check: $f^{-1}[f(x)] = -\sqrt{9 - (\sqrt{9 - x^2})^2}$

$$= -\sqrt{9 - (9 - x^2)}$$
$$= -\sqrt{9 - 9 + x^2}$$
$$= -\sqrt{x^2}$$
$$\sqrt{x^2} = -x \text{ since } x \le 0 \text{ in the domain of } f$$
$$= -(-x)$$
$$= x$$

$f[f^{-1}(x)] = \sqrt{9 - (-\sqrt{9 - x^2})^2}$

$$= \sqrt{9 - (9 - x^2)}$$
$$= \sqrt{9 - 9 + x^2}$$
$$= \sqrt{x^2}$$
$$\sqrt{x^2} = x \text{ since } x \ge 0 \text{ in the domain of } f^{-1}$$
$$= x$$

$f^{-1}(x) = -\sqrt{9 - x^2}$ Domain: $f^{-1} = [0, 3]$
Range of f^{-1} = Domain of f = $[-3, 0]$

71. $f(x) = 1 + \sqrt{1 - x^2}$ $0 \le x \le 1$ f is one-to-one. See graph below.

Solve $y = f(x)$ for x:

$$y = 1 + \sqrt{1 - x^2} \quad 0 \le x \le 1$$
$$y - 1 = \sqrt{1 - x^2}$$
$$(y - 1)^2 = 1 - x^2 \quad y \ge 1$$
$$y^2 - 2y + 1 = 1 - x^2$$
$$y^2 - 2y = -x^2$$
$$2y - y^2 = x^2$$
$$x = \underbrace{\sqrt{2y - y^2}} \quad y \ge 1 \quad f^{-1}(y) = \sqrt{2y - y^2}$$

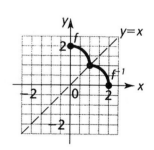

positive square root only because $0 \le x$

Interchange x and y:

$y = f^{-1}(x) = \sqrt{2x - x^2}$ Domain: $1 \le x \le 2$

Check: $f[f^{-1}(x)] = 1 + \sqrt{1 - (\sqrt{2x - x^2})^2}$

$$= 1 + \sqrt{1 - (2x - x^2)}$$
$$= 1 + \sqrt{1 - 2x + x^2}$$
$$= 1 + \sqrt{(1 - x)^2}$$
$$= 1 + [-(1 - x)] \text{ because } 1 \le x \text{ in the domain of } f^{-1}$$
$$= 1 - 1 + x$$
$$= x$$

$$f^{-1}[f(x)] = \sqrt{2(1 + \sqrt{1 - x^2}) - (1 + \sqrt{1 - x^2})^2}$$
$$= \sqrt{2 + 2\sqrt{1 - x^2} - (1 + 2\sqrt{1 - x^2} + 1 - x^2)}$$
$$= \sqrt{2 + 2\sqrt{1 - x^2} - 1 - 2\sqrt{1 - x^2} - 1 + x^2}$$
$$= \sqrt{x^2}$$
$$= x$$

because $x \geq 0$ *in the domain of f*

$f^{-1}(x) = \sqrt{2x - x^2}$ Domain of $f^{-1} = [1, 2]$
Range of f^{-1} = Domain of $f = [0, 1]$

73. $f(x) = 1 - \sqrt{1 - x^2}$ $-1 \leq x \leq 0$ f is one-to-one. See graph below.

Solve $y = f(x)$ for x:

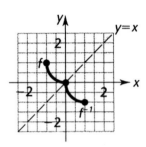

$$y = 1 - \sqrt{1 - x^2} \quad -1 \leq x \leq 0$$
$$y - 1 = -\sqrt{1 - x^2}$$
$$1 - y = \sqrt{1 - x^2}$$
$$(1 - y)^2 = 1 - x^2 \qquad y \leq 1$$
$$1 - 2y + y^2 = 1 - x^2$$
$$-2y + y^2 = -x^2$$
$$2y - y^2 = x^2$$
$$x = -\underbrace{\sqrt{2y - y^2}} \qquad y \leq 1 \quad f^{-1}(y) = -\sqrt{2y - y^2}$$

negative square root only because $x \leq 0$

Interchange x and y:

$$y = f^{-1}(x) = -\sqrt{2x - x^2} \quad \text{Domain: } 0 \leq x \leq 1$$

Check: $f[f^{-1}(x)] = 1 - \sqrt{1 - (-\sqrt{2x - x^2})^2}$
$$= 1 - \sqrt{1 - (2x - x^2)}$$
$$= 1 - \sqrt{1 - 2x + x^2}$$
$$= 1 - \sqrt{(1 - x)^2}$$
$$= 1 - (1 - x) \text{ because } x \leq 1 \text{ in the domain of } f^{-1}$$
$$= 1 - 1 + x$$
$$= x$$

$$f^{-1}[f(x)] = -\sqrt{2(1 - \sqrt{1 - x^2}) - (1 - \sqrt{1 - x^2})^2}$$
$$= -\sqrt{2 - 2\sqrt{1 - x^2} - (1 - 2\sqrt{1 - x^2} + 1 - x^2)}$$
$$= -\sqrt{2 - 2\sqrt{1 - x^2} - 1 + 2\sqrt{1 - x^2} - 1 + x^2}$$
$$= -\sqrt{x^2}$$
$$= -(-x) \text{ because } x \leq 0 \text{ in the domain of } f$$
$$= x$$

$f^{-1}(x) = -\sqrt{2x - x^2}$ Domain of $f^{-1} = [0, 1]$
Range of f^{-1} = Domain of $f = [-1, 0]$

75. $f(x) = ax + b$ $a \neq 0$ f is one-to-one.   Proof: Assume $f(u) = f(v)$
Solve $y = f(x)$ for x: $au + b = av + b$
$\quad y = ax + b$ Then $au = av$
$y - b = ax$ $u = v$
$\quad\quad x = \dfrac{y - b}{a} = f^{-1}(y)$

Interchange x and y:

$y = f^{-1}(x) = \dfrac{x - b}{a}$ Domain: $(-\infty, \infty)$

Check: $f^{-1}[f(x)] = \dfrac{ax + b - b}{a} = \dfrac{ax}{a} = x$

$\quad\quad f[f^{-1}(x)] = a\dfrac{x - b}{a} + b = x - b + b = x$

$\quad\quad\quad f^{-1}(x) = \dfrac{x - b}{a}$

77. For f to be its own inverse, $f(x)$ must equal $f^{-1}(x)$ for all x.

Thus $ax + b = \dfrac{x - b}{a}$ or

$\quad\quad a^2x + ab = x - b$ or
$\quad\quad a^2x - x = -ab - b$ or
$\quad (a^2 - 1)x = -ab - b$

This can be true for all x only if $a^2 - 1 = 0$ and $-ab - b = 0$.
If $a^2 - 1 = 0$, then $a = -1$ or $a = 1$.
If $a = -1$, then $-ab - b = -(-1)b - b = 0$ for any b.
If $a = 1$, then $-ab - b = -b - b = -2b = 0$ only if $b = 0$.
Summarizing, f will be its own inverse if either $a = -1$ (b arbitrary), or $a = 1$ and $b = 0$.

79. The slope of the line through (a, b) and (b, a) is
$\quad m_1 = \dfrac{a - b}{b - a} = \dfrac{-(b - a)}{b - a} = -1$
The slope of the line $y = x$ is $1 = m_2$.
Thus, $m_1 m_2 = (-1)(1) = -1$. Hence the lines are perpendicular.

81. $f(x) = (2 - x)^2$

(A) $x \leq 2$
Solve $y = f(x)$ for x:
$\quad\quad y = (2 - x)^2$ $x \leq 2$
$\quad\sqrt{y} = 2 - x$ Since $2 - x \geq 0$
$\sqrt{y} - 2 = -x$
$\quad\quad x = 2 - \sqrt{y} = f^{-1}(y)$

Interchange x and y:

$y = f^{-1}(x) = 2 - \sqrt{x}$ Domain: $x \geq 0$
Check: $f^{-1}[f(x)] = 2 - \sqrt{(2 - x)^2}$
$\quad\quad\quad\quad\quad = 2 - (2 - x)$ since $2 - x \geq 0$ *in the domain of f*
$\quad\quad\quad\quad\quad = 2 - 2 + x$
$\quad\quad\quad\quad\quad = x$
$\quad\quad f[f^{-1}(x)] = [2 - (2 - \sqrt{x})]^2$
$\quad\quad\quad\quad\quad = [2 - 2 + \sqrt{x}]^2$
$\quad\quad\quad\quad\quad = [\sqrt{x}]^2$
$\quad\quad\quad\quad\quad = x$
$\quad f^{-1}(x) = 2 - \sqrt{x}$

(B) $x \geq 2$

Solve $y = f(x)$ for x:

$\begin{aligned} y &= (2 - x)^2 && x \geq 2 \\ \sqrt{y} &= -(2 - x) && \text{Since } 2 - x \leq 0 \\ \sqrt{y} &= -2 + x \\ x &= 2 + \sqrt{y} = f^{-1}(y) \end{aligned}$

Interchange x and y:

$y = f^{-1}(x) = 2 + \sqrt{x}$ ⠀Domain: $x \geq 0$

Check: $\begin{aligned} f^{-1}[f(x)] &= 2 + \sqrt{(2 - x)^2} \\ &= 2 + [-(2 - x)] \text{ since } 2 - x \leq 0 \text{ in the domain of } f \\ &= 2 - 2 + x \\ &= x \end{aligned}$

$\begin{aligned} f[f^{-1}(x)] &= [2 - (2 + \sqrt{x})]^2 \\ &= [2 - 2 - \sqrt{x}]^2 \\ &= [-\sqrt{x}]^2 \\ &= x \end{aligned}$

$f^{-1}(x) = 2 + \sqrt{x}$

83. $f(x) = \sqrt{4x - x^2}$

(A) $0 \leq x \leq 2$

Solve $y = f(x)$ for x:

$\begin{aligned} y &= \sqrt{4x - x^2} \\ y^2 &= 4x - x^2 && y \geq 0 \\ -y^2 &= x^2 - 4x \\ 4 - y^2 &= x^2 - 4x + 4 \\ 4 - y^2 &= (x - 2)^2 \\ -\underbrace{\sqrt{4 - y^2}}_{} &= x - 2 \end{aligned}$

negative square root only because $x \leq 2$

$x = 2 - \sqrt{4 - y^2}$ ⠀$y \geq 0$ ⠀$f^{-1}(y) = 2 - \sqrt{4 - y^2}$

Interchange x and y:

$y = f^{-1}(x) = 2 - \sqrt{4 - x^2}$ ⠀Domain: $0 \leq x \leq 2$

Check: $\begin{aligned} f^{-1}[f(x)] &= 2 - \sqrt{4 - (\sqrt{4x - x^2})^2} \\ &= 2 - \sqrt{4 - (4x - x^2)} \\ &= 2 - \sqrt{4 - 4x + x^2} \\ &= 2 - \sqrt{(2 - x)^2} \\ &= 2 - (2 - x) \text{ since } 2 - x \geq 0 \\ &= x \end{aligned}$

$\begin{aligned} f[f^{-1}(x)] &= \sqrt{4(2 - \sqrt{4 - x^2}) - (2 - \sqrt{4 - x^2})^2} \\ &= \sqrt{8 - 4\sqrt{4 - x^2} - (4 - 4\sqrt{4 - x^2} + 4 - x^2)} \\ &= \sqrt{8 - 4\sqrt{4 - x^2} - 4 + 4\sqrt{4 - x^2} - 4 + x^2} \\ &= \sqrt{x^2} \\ &= x \text{ since } 0 \leq x \end{aligned}$

$f^{-1}(x) = 2 - \sqrt{4 - x^2}, \ 0 \leq x \leq 2$

(B) $2 \leq x \leq 4$

Solve $y = f(x)$ for x:

$$y = \sqrt{4x - x^2}$$
$$y^2 = 4x - x^2 \quad y \geq 0$$
$$-y^2 = x^2 - 4x$$
$$4 - y^2 = x^2 - 4x + 4$$
$$4 - y^2 = (x - 2)^2$$
$$\underbrace{\sqrt{4 - y^2}} = x - 2$$

positive square root only because $x \geq 2$

$x = 2 + \sqrt{4 - y^2}$ $\quad y \geq 0$ $\quad f^{-1}(y) = 2 + \sqrt{4 - y^2}$

Interchange x and y:

$y = f^{-1}(x) = 2 + \sqrt{4 - x^2}$ $\quad$ Domain: $0 \leq x \leq 2$

Check: $f^{-1}[f(x)] = 2 + \sqrt{4 - (\sqrt{4x - x^2})^2}$

$$= 2 + \sqrt{4 - (4x - x^2)}$$
$$= 2 + \sqrt{4 - 4x + x^2}$$
$$= 2 + \sqrt{(2 - x)^2}$$
$$= 2 + [-(2 - x)] \text{ since } 2 \leq x \text{ in the domain of } f$$
$$= 2 - 2 + x$$
$$= x$$

$$f[f^{-1}(x)] = \sqrt{4(2 + \sqrt{4 - x^2}) - (2 + \sqrt{4 - x^2})^2}$$
$$= \sqrt{8 + 4\sqrt{4 - x^2} - (4 + 4\sqrt{4 - x^2} + 4 - x^2)}$$
$$= \sqrt{8 + 4\sqrt{4 - x^2} - 4 - 4\sqrt{4 - x^2} - 4 + x^2}$$
$$= \sqrt{x^2}$$
$$= x \text{ since } 0 \leq x \text{ in the domain of } f^{-1}$$

$f^{-1}(x) = 2 + \sqrt{4 - x^2}, \ 0 \leq x \leq 2$

CHAPTER 2 REVIEW

1. (A) $d(A, B) = \sqrt{[4 - (-2)]^2 + (0 - 3)^2}$
$$= \sqrt{36 + 9} = \sqrt{45}$$

(B) $m = \dfrac{0 - 3}{4 - (-2)} = \dfrac{-3}{6} = -\dfrac{1}{2}$

(C) The slope m_1 of a line perpendicular to AB must satisfy

$$m_1\left(-\frac{1}{2}\right) = -1.$$

Therefore, $m_1 = 2.$ $\quad$ (2-1, 2-2)

2. (A) Center at $(0, 0)$ and radius $\sqrt{7}$
$$x^2 + y^2 = r^2$$
$$x^2 + y^2 = (\sqrt{7})^2$$
$$x^2 + y^2 = 7$$

(B) Center at $(3, -2)$ and radius $\sqrt{7}$
$$(h, k) = (3, -2)$$
$$r = \sqrt{7}$$
$$(x - h)^2 + (y - k)^2 = r^2$$
$$(x - 3)^2 + [y - (-2)]^2 = (\sqrt{7})^2$$
$$(x - 3)^2 + (y + 2)^2 = 7 \quad (2-1)$$

3. $(x + 3)^2 + (y - 2)^2 = 5$
$$[x - (-3)]^2 + (y - 2)^2 = (\sqrt{5})^2$$
Center: $C(h, k) = (-3, 2)$
Radius: $r = \sqrt{5}$ $\quad$ (2-1)

4.
$$3x + 2y = 9$$
$$2y = -3x + 9$$
$$y = -\frac{3}{2}x + \frac{9}{2}$$

slope: $-\frac{3}{2}$

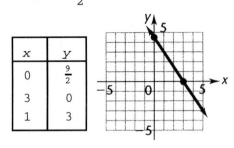

x	y
0	$\frac{9}{2}$
3	0
1	3

(2-2)

5. The line passes through the two given points, (6, 0) and (0, 4). Thus, its slope is given by

$$m = \frac{0 - 4}{6 - 0} = \frac{-4}{6} = -\frac{2}{3}$$

The equation of the line is, therefore, using the point-slope form,

$$y - 0 = -\frac{2}{3}(x - 6)$$

or $\quad 3y = -2(x - 6)$

or $\quad 3y = -2x + 12$.

$2x + 3y = 12$ (2-2)

6. $y = mx + b \quad m = -\frac{2}{3} \quad b = 2$

$$y = -\frac{2}{3}x + 2$$ (2-2)

7. *vertical:* $x = -3$, slope not defined; *horizontal:* $y = 4$, slope = 0 (2-2)

8. (A) A function; domain = {1, 2, 3}; range = {1, 4, 9}
(B) Not a function (two range values correspond to some domain values)
(C) A function; domain = {-2, -1, 0, 1, 2}; range = {2} (2-3)

9. (A) Not a function (fails vertical line test)
(B) A function
(C) A function
(D) Not a function (fails vertical line test) (2-3)

10. (A) Function

(B) Not a function—two range elements correspond to some domain elements; for example 2 and -2 correspond to 4.

(C) Function

(D) Not a function—two range elements correspond to some domain elements; for example 2 and -2 correspond to 2. (2-3)

11. $f(2) = 3(2) + 5 = 11$
$g(-2) = 4 - (-2)^2 = 0$
$k(0) = 5$
Therefore
$f(2) + g(-2) + k(0)$
$\quad = 11 + 0 + 5 = 16$ (2-3)

12. $m(-2) = 2|-2| - 1 = 3$
$\quad g(2) = 4 - (2)^2 = 0$

Therefore $\dfrac{m(-2) + 1}{g(2) + 4} = \dfrac{3 + 1}{0 + 4} = 1$ (2-3)

13.
$$\frac{f(2 + h) - f(2)}{h} = \frac{[3(2 + h) + 5] - [3(2) + 5]}{h}$$
$$= \frac{6 + 3h + 5 - 11}{h}$$
$$= \frac{3h}{h}$$
$$= 3$$ (2-3)

14. $\dfrac{g(a + h) - g(a)}{h}$ $= \dfrac{[4 - (a + h)^2] - [4 - a^2]}{h}$

$= \dfrac{4 - a^2 - 2ah - h^2 - 4 + a^2}{h}$

$= \dfrac{-2ah - h^2}{h}$

$= \dfrac{h(-2a - h)}{h}$

$= -2a - h$ *(2-3)*

15. $(f + g)(x) = f(x) + g(x)$
$= 3x + 5 + 4 - x^2 = 9 + 3x - x^2$ *(2-5)*

16. $(f - g)(x) = f(x) - g(x) = 3x + 5 - (4 - x^2) = 3x + 5 - 4 + x^2 = x^2 + 3x + 1$
(2-5)

17. $(fg)(x) = f(x)g(x) = (3x + 5)(4 - x^2)$
$= 12x - 3x^3 + 20 - 5x^2 = 20 + 12x - 5x^2 - 3x^3$ *(2-5)*

18. $\left(\dfrac{f}{g}\right)(x) = \dfrac{f(x)}{g(x)} = \dfrac{3x + 5}{4 - x^2}$
Domain: $\{x \mid 4 - x^2 \neq 0\}$ or $\{x \mid x \neq \pm 2\}$ *(2-5)*

19. $(f \circ g)(x) = f[g(x)] = f(4 - x^2) = 3(4 - x^2) + 5 = 12 - 3x^2 + 5 = 17 - 3x^2$ *(2-5)*

20. $(g \circ f)(x) = g[f(x)] = g(3x + 5) = 4 - (3x + 5)^2 = 4 - (9x^2 + 30x + 25)$
$= 4 - 9x^2 - 30x - 25 = -21 - 30x - 9x^2$ *(2-5)*

21. (A) The graph of $f(x)$ is reflected across the x axis.

(B) The graph of $f(x)$ is shifted up 4 units.

(C) The graph of $f(x)$ is shifted right 2 units.

(D) The graph of $f(x)$ is shifted left 3 units, reflected across the x axis and shifted down 3 units.

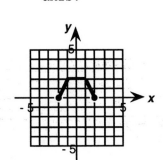
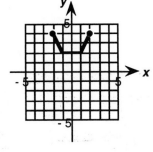
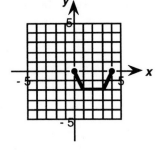
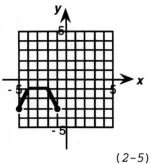

(2-5)

22. (A) The graph that opens up and has a vertex at $(2, -4)$ is g.
(B) The graph that opens down and has vertex at $(-2, 4)$ is m.
(C) The graph that opens down and has vertex at $(2, 4)$ is n.
(D) The graph that opens up and has vertex at $(-2, -4)$ is f. *(2-4, 2-5)*

23. The equation corresponding to graph f is $y = (x + 2)^2 - 4$.

(A) y intercept: Set $x = 0$, then $y = (0 + 2)^2 - 4 = 0$
  x intercepts: Set $y = 0$, then $0 = (x + 2)^2 - 4$
$$(x + 2)^2 = 4$$
$$x + 2 = \pm 2$$
$$x = 0, -4$$

(B) $(-2, -4)$

(C) The minimum of -4 occurs at the vertex.

(D) Since y is never less than -4, the range is $[-4, \infty)$.

(E) y is increasing on $[-2, \infty)$.

(F) y is decreasing on $(-\infty, -2]$. *(2-4)*

24. $f(x) = x^2 - 6x + 11$. Complete the square:
$$f(x) = (x^2 - 6x + 9) - 9 + 11$$
$$= (x - 3)^2 + 2$$

Comparing with $f(x) = a(x - h)^2 + k$; $h = 3$ and $k = 2$. Thus, the minimum value is 2 and the vertex is $(3, 2)$. *(2-4)*

25. (A) Reflected across x axis   (B) Shifted down 3 units

(C) Shifted left 3 units *(2-5)*

26. (A) 0   (B) 1   (C) 2   (D) 0 *(2-4)*

27. (A) $f(x) = 0$ when $x = -2$ or $x = 0$

(B) $f(x) = 1$ when $x = -1$ or $x = 1$

(C) There is no value of x for which $f(x) = -3$. No solution.

(D) $f(x) = 3$ when $x = 3$ and also for any value of $x < -2$. *(2-4)*

28. Domain: $(-\infty, \infty)$ **29.** $[-2, -1]$, $[1, \infty)$ **30.** $[-1, 1)$ **31.** $(-\infty, -2)$

Range $= (-3, \infty)$ *(2-4)* *(2-4)* *(2-4)*

(2-4)

32. $x = -2$, $x = 1$ *(2-4)*   **33.** $f(x) = 4x^3 - \sqrt{x}$ *(2-3)*

34. The function f multiplies the square of the domain element by 3, adds 4 times the domain element, and then subtracts 6. *(2-3)*

35. (A) Since two points are given, we find the slope, then apply the point-slope form.
$$m = \frac{-3 - 3}{0 - (-4)} = \frac{-6}{4} = -\frac{3}{2}$$
$$y - 3 = -\frac{3}{2}[x - (-4)]$$
$$2(y - 3) = -3(x + 4)$$
$$2y - 6 = -3x - 12$$
$$3x + 2y = -6$$

(B) $d(P, Q) = \sqrt{(-3 - 3)^2 + [0 - (-4)]^2} = \sqrt{36 + 16} = \sqrt{52} = 2\sqrt{13}$ *(2-1, 2-2)*

36. The line $6x + 3y = 5$, or
$3y = -6x + 5$,

or $y = -2x + \frac{5}{3}$, has slope -2.

(A) We require a line through $(-2, 1)$, with slope -2. Applying the point-slope form, we have
$$y - 1 = -2[x - (-2)]$$
$$y - 1 = -2x - 4$$
$$y = -2x - 3$$

(B) We require a line with slope m satisfying $-2m = -1$, or $m = \frac{1}{2}$. Again applying the point-slope form, we have

$$y - 1 = \frac{1}{2}[x - (-2)]$$
$$y - 1 = \frac{1}{2}x + 1$$
$$y = \frac{1}{2}x + 2 \qquad (2-2)$$

37. Since the equation is unchanged by any substitution of $-x$ for x or $-y$ for y, or both, the graph must be symmetric with respect to all three. $(2-1)$

38. The domain is the set of all real numbers x such that $\dfrac{1}{\sqrt{3 - x}}$ is a real number—that is, such that $3 - x > 0$, or $x < 3$. Domain: $(-\infty, 3)$ $(2-3)$

39. $f(x) = x^2 - 6x + 5$. Complete the square.
$$f(x) = (x^2 - 6x + 9) - 9 + 5$$
$$= (x - 3)^2 - 4$$

Comparing with $f(x) = a(x - h)^2 + k$, $h = 3$ and $k = -4$. Thus, the vertex is $(3, -4)$, the axis of symmetry is $x = 3$, and the minimum value is -4. $y = f(x)$ can be any number greater then or equal to -4, so the range is $[-4, \infty)$.
y intercept: Set $x = 0$, then $f(0) = 5$ is the y intercept.
x intercepts: Set $f(x) = 0$, then
$$x^2 - 6x + 5 = 0$$
$$(x - 5)(x - 1) = 0$$
$$x = 5 \text{ or } x = 1 \text{ are the } x \text{ intercepts.}$$

x	$f(x)$
0	5
1	0
2	-3
3	-4
4	-3
5	0

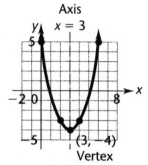

Graph: Locate axis and vertex, then plot several points on either side of the axis. $(2-4)$

40. The domain is the set of all real numbers x such that $\dfrac{1}{4 - \sqrt{x}}$ is a real number, that is, such that $x \geq 0$ and $4 - \sqrt{x} \neq 0$. The latter condition is equivalent to $\sqrt{x} \neq 4$ or $x \neq 16$. Thus, the domain is all x such that $x \geq 0$ except $x \neq 16$. $[0, 16) \cup (16, \infty)$. $(2-3)$

41. $f(x) = \sqrt{x} - 8 \qquad g(x) = |x|$
(A) $(f \circ g)(x) = f[g(x)] = f(|x|) = \sqrt{|x|} - 8$
$(g \circ f)(x) = g[f(x)] = g(\sqrt{x} - 8) = |\sqrt{x} - 8|$

(B) The domain of f is $\{x \mid x \geq 0\}$. The domain of g is all real numbers. Hence the domain of $f \circ g$ is the set of those real numbers x for which $g(x)$ is non-negative, that is, all real numbers.
The domain of $(g \circ f)$ is the set of all those non-negative numbers x for which $f(x)$ is real, that is all $\{x \mid x \geq 0\}$ or $[0, \infty)$ $(2-5)$

42. (A) $f(x) = x^3$. The graph passes the horizontal line test, so f is one-to-one.
Also, assume
$$f(a) = f(b)$$
$$a^3 = b^3$$
$$a^3 - b^3 = 0$$
$$(a - b)(a^2 + ab + b^2) = 0$$
The only real solutions of this equation are those for which $a - b = 0$, hence $a = b$. Thus $f(x)$ is one-to-one.

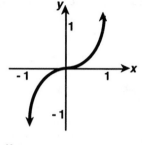

(B) $g(x) = (x - 2)^2$. Since $g(3) = g(1) = 1$, g is not one-to-one.

(C) $h(x) = 2x - 3$
Assume $h(a) = h(b)$
$$2a - 3 = 2b - 3$$
Then $\quad 2a = 2b$
$$a = b$$
Thus h is one-to-one.

(D) $F(x) = (x + 3)^2 \quad x \geq -3$
The graph passes the horizontal line test, so F is one-to-one.

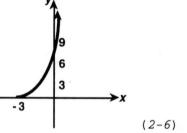

$(2-6)$

43. (A) $f(x) = 3x - 7$
Assume $f(a) = f(b)$
$$3a - 7 = 3b - 7$$
$$3a = 3b$$
$$a = b$$
Hence f is one-to-one.
Solve $y = f(x)$ for x:
$$y = 3x - 7$$
$$y + 7 = 3x$$
$$x = \frac{1}{3}y + \frac{7}{3} = f^{-1}(y)$$

Interchange x and y:
$$y = f^{-1}(x) = \frac{1}{3}x + \frac{7}{3}$$
Domain: $(-\infty, \infty)$

Check: $\quad f^{-1}[f(x)] = \frac{1}{3}(3x - 7) + \frac{7}{3}$
$$= x - \frac{7}{3} + \frac{7}{3} = x$$
$$f[f^{-1}(x)] = 3\left(\frac{1}{3}x + \frac{7}{3}\right) - 7$$
$$= x + 7 - 7 = x$$
$$f^{-1}(x) = \frac{1}{3}x + \frac{7}{3} = \frac{x + 7}{3}$$

(B) $f^{-1}(5) = \dfrac{5 + 7}{3} = 4$

(C) $f^{-1}[f(x)] = x$ (See part A).

(D) Since $a < b$ implies $3a < 3b$, which implies $3a - 7 < 3b - 7$, or $f(a) < f(b)$, f is increasing.

$(2-6)$

44.

x	$y = 2 - x$	x	$y = x^2$
-1	3	0	0
		$\frac{1}{2}$	$\frac{1}{4}$
$-\frac{1}{2}$	$2\frac{1}{2}$	1	1

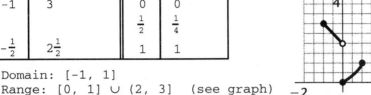

Domain: $[-1, 1]$
Range: $[0, 1] \cup (2, 3]$ (see graph)
Discontinuous at $x = 0$. $(2\text{-}4)$

45. The graph of $y = x^2$ is vertically expanded by a factor of 2, reflected in the x axis and shifted to the left 3 units. Equation: $y = -2(x + 3)^2$. $(2\text{-}5)$

46. $g(x) = 5 - 3|x - 2|$

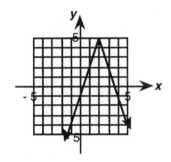

$(2\text{-}5)$

47. The graph of $y = x^2$ has been reflected across the x axis, shifted right 4 units and up 3 units so that the parabola has vertex $(4, 3)$.
Equation: $y = -(x - 4)^2 + 3$. $(2\text{-}4, 2\text{-}5)$

48. (A) This is the same as the graph of $y = |x|$ shifted down 2 units.

(B) This is the same as the graph of $y = |x|$ shifted left 1 unit.

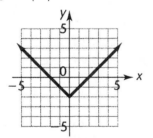

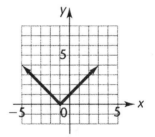

(C) This is the same as the graph of $y = |x|$ contracted by a factor of $\frac{1}{2}$.

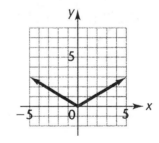

$(2\text{-}5)$

49. $f(x) = \sqrt{x - 1}$.
Assume $f(a) = f(b)$
$$\sqrt{a - 1} = \sqrt{b - 1}$$
$$a - 1 = b - 1$$
$$a = b$$
Thus f is one-to-one.

(A) Solve $y = f(x)$ for x

$y = \sqrt{x - 1}$

$y^2 = x - 1$ $y \geq 0$

$x = 1 + y^2$ $y \geq 0$ $f^{-1}(y) = 1 + y^2$

Interchange x and y:

$y = f^{-1}(x) = 1 + x^2$ Domain: $x \geq 0$

Check:

$f^{-1}[f(x)] = 1 + (\sqrt{x - 1})^2 = 1 + x - 1 = x$

$f[f^{-1}(x)] = \sqrt{1 + x^2 - 1} = \sqrt{x^2} = x$

since $x \geq 0$ in the domain of f^{-1}

(B) Domain of $f = [1, \infty) =$ Range of f^{-1}
Range of $f = [0, \infty) =$ Domain of f^{-1}

(C)

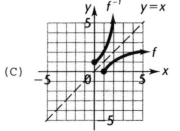

$(2-6)$

---

50. We are given $C(h, k) = (3, 0)$. To find r we use the distance formula. $r =$ distance from the center to $(-1, 4)$

$= \sqrt{[(-1) - 3]^2 + (0 - 4)^2}$

$= \sqrt{16 + 16}$

$= \sqrt{32}$

Then the equation of the circle is

$(x - h)^2 + (y - k)^2 = r^2$

$(x - 3)^2 + (y - 0)^2 = (\sqrt{32})^2$

$(x - 3)^2 + y^2 = 32$ $(2-1)$

51.

$x^2 + y^2 + 4x - 6y = 3$

$(x^2 + 4x + ?) + (y^2 - 6y + ?) = 3$

$(x^2 + 4x + 4) + (y^2 - 6y + 9) = 3 + 4 + 9$

$(x + 2)^2 + (y - 3)^2 = 16$

$[x - (-2)]^2 + (y - 3)^2 = 4^2$

Center: $C(h, k) = C(-2, 3)$

Radius $r = \sqrt{16} = 4$ $(2-1)$

---

52. $xy = 4$ y axis symmetry? $(-x)y = 4$ $-xy = 4$ No, not equivalent

x axis symmetry? $x(-y) = 4$ $-xy = 4$ No, not equivalent

origin symmetry? $(-x)(-y) = 4$ $xy = 4$ Yes, equivalent

The graph has symmetry with respect to the origin.

We reflect the portion of the graph in the first quadrant through the origin, using origin symmetry.

x	y
1	4
2	2
3	$\frac{4}{3}$
4	1

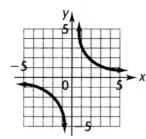

$(2-1)$

53. decreasing $(2-2, 2-3)$

54. (A) Domain of $f = [0, \infty)$ = Range of f^{-1}
Since $x^2 \geq 0$, $x^2 - 1 \geq -1$, so Range of $f = [-1, \infty)$ = Domain of f^{-1}

(B) $f(x) = x^2 - 1$ $x \geq 0$
f is one-to-one on its domain (steps omitted)

Solve $y = f(x)$ for x:
$y = x^2 - 1$ $x \geq 0$
$x^2 = y + 1$ $x \geq 0$
$\underbrace{x = \sqrt{y + 1}}_{} = f^{-1}(y)$

positive square root since $x \geq 0$

Interchange x and y:
$y = f^{-1}(x) = \sqrt{x + 1}$ Domain: $[-1, \infty)$

Check:
$f^{-1}[f(x)] = \sqrt{x^2 - 1 + 1}$ $x \geq 0$
$= \sqrt{x^2}$ $x \geq 0$
$= x$ since $x \geq 0$
$f[f^{-1}(x)] = (\sqrt{x + 1})^2 - 1$
$= x + 1 - 1$
$= x$

(C) $f^{-1}(3) = \sqrt{3 + 1} = 2$ (D) $f^{-1}[f(4)] = 4$ (E) $f^{-1}[f(x)] = x$ *(2-6)*

55. The graph of $y = \sqrt[3]{x}$ is vertically expanded by a factor of 2, reflected in the x axis, shifted 1 unit left and 1 unit down. Equation: $y = -2\sqrt[3]{x + 1} - 1$.   *(2-4)*

56. It is the same as the graph of g shifted to the right 2 units and down 1 unit, then reflected in the x axis.   *(2-5)*

57. This is the same as the graph of $y = |x|$, reflected across the x axis, and shifted to the left 1 unit and down 1 unit.

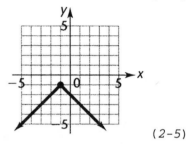

(2-5)

58. The domain of the set of all real numbers x such that $\sqrt{25 - x^2}$ is a real number—that is, such that $25 - x^2 \geq 0$. Solving by the methods of Section 1-8, we obtain $-5 \leq x \leq 5$ or $[-5, 5]$ *(1-8, 2-3)*

59. (A) $(fg)(x) = f(x)g(x) = x^2\sqrt{1 - x}$
The domain of f is $(-\infty, \infty)$. The domain of g is $(-\infty, 1]$. Hence the domain of fg is the intersection of these sets, that is, $(-\infty, 1]$.

(B) $\left(\dfrac{f}{g}\right)(x) = \dfrac{f(x)}{g(x)} = \dfrac{x^2}{\sqrt{1 - x}}$

To find the domain of $\dfrac{f}{g}$, we exclude from $(-\infty, 1]$ the set of values of x for which $g(x) = 0$
$\sqrt{1 - x} = 0$
$1 - x = 0$
$x = 1$

Thus the domain of $\dfrac{f}{g}$ is $(-\infty, 1)$

(C) $(f \circ g)(x) = f[g(x)] = f(\sqrt{1 - x}) = [\sqrt{1 - x}]^2 = 1 - x$.
The doman of $f \circ g$ is the set of those numbers in $(-\infty, 1]$ for which $g(x)$ is real, that is $(-\infty, 1]$.

(D) $(g \circ f)(x) = g[f(x)] = g(x^2) = \sqrt{1 - x^2}$
The domain of $g \circ f$ is the set of those real numbers for which $f(x)$ is in $(-\infty, 1]$, that is, $x^2 \leq 1$, or $-1 \leq x \leq 1$. $[-1, 1]$.   *(2-5)*

60. (A) $f(x) = \dfrac{x + 2}{x - 3}$

Solve $y = f(x)$ for x:

$$y = \dfrac{x + 2}{x - 3}$$

$$y(x - 3) = x + 2$$

$$xy - 3y = x + 2$$

$$xy - x = 3y + 2$$

$$x(y - 1) = 3y + 2$$

$$x = \dfrac{3y + 2}{y - 1} = f^{-1}(y)$$

Interchange x and y:

$y = f^{-1}(x) = \dfrac{3x + 2}{x - 1}$ Domain: $x \neq 1$

Check: $f^{-1}[f(x)] = \dfrac{3\frac{x + 2}{x - 3} + 2}{\frac{x + 2}{x - 3} - 1}$ $f[f^{-1}(x)] = \dfrac{3\frac{3x + 2}{x - 1} + 2}{\frac{3x + 2}{x - 1} - 3}$

$\qquad\qquad = \dfrac{3(x + 2) + 2(x - 3)}{(x + 2) - (x - 3)}$ $= \dfrac{3x + 2 + 2(x - 1)}{3x + 2 - 3(x - 1)}$

$\qquad\qquad = \dfrac{3x + 6 + 2x - 6}{x + 2 - x + 3}$ $= \dfrac{3x + 2 + 2x - 2}{3x + 2 - 3x + 3}$

$\qquad\qquad = \dfrac{5x}{5}$ $= \dfrac{5x}{5}$

$\qquad\qquad = x$ $= x$

(B) $f^{-1}(3) = \dfrac{3(3) + 2}{3 - 1} = \dfrac{11}{2}$ (C) $f^{-1}[f(x)] = x$ (See part A) $(2\text{-}6)$

61. $f(x) = |x + 1| - |x - 1|$

If $x < -1$, then $|x + 1| = -(x + 1)$ and $|x - 1| = -(x - 1)$, hence

$f(x) = -(x + 1) - [-(x - 1)]$

$\qquad = -x - 1 + x - 1$

$\qquad = -2$

If $-1 \leq x < 1$, then $|x + 1| = x + 1$ but $|x - 1| = -(x - 1)$, hence

$f(x) = x + 1 - [-(x - 1)]$

$\qquad = x + 1 + x - 1$

$\qquad = 2x$

If $x \geq 1$, then $|x + 1| = x + 1$ and $|x - 1| = x - 1$, hence

$f(x) = x + 1 - (x - 1)$

$\qquad = x + 1 - x + 1$

$\qquad = 2$

Domain: $(-\infty, \infty)$

Piecewise definition for f: $f(x) = \begin{cases} -2 & \text{if } x < -1 \\ 2x & \text{if } -1 \leq x < 1 \\ 2 & \text{if } x \geq 1 \end{cases}$

Range: Since if $-1 \leq x < 1$, then $-2 \leq 2x < 2$, $f(x)$ is always between -2 and 2.
The range is $[-2, 2]$. $(2\text{-}4)$

62. Let (x, y) be a point equidistant from $(3, 3)$ and $(6, 0)$. Then

$$\sqrt{(x - 3)^2 + (y - 3)^2} = \sqrt{(x - 6)^2 + (y - 0)^2}$$

$$(x - 3)^2 + (y - 3)^2 = (x - 6)^2 + y^2$$

$$x^2 - 6x + 9 + y^2 - 6y + 9 = x^2 - 12x + 36 + y^2$$

$$-6x - 6y + 18 = -12x + 36$$

$$6x - 6y = 18$$

$$x - y = 3$$

This is the equation of a line. $(2\text{-}1, \ 2\text{-}2)$

63. We will show separately:
(A) If two nonvertical lines are parallel, then they have the same slope.
(B) If two lines have the same slope they are nonvertical and parallel.

(A) Let $y = m_1x + b_1$ and $y = m_2x + b_2$ be parallel. Then there is no point with coordinates that satisfy both equations. But for any y, $m_1x + b_1 = m_2x + b_2$. Then (x, y) will satisfy both equations unless this equation has no solution. But this equation will have a solution
$$m_1x - m_2x = b_2 - b_1$$

$(m_1 - m_2)x = b_2 - b_1,$ that is, $x = \dfrac{b_2 - b_1}{m_1 - m_2}$

unless $m_1 - m_2 = 0$. So $m_1 = m_2$. So the lines have the same slope.

(B) Assume the two lines have equations $y = mx + b_1$, and $y = mx + b_2$. Then both have slopes, hence are nonvertical. For any y a point that lies on both lines must have coordinates that satisfy $mx + b_1 = mx + b_2$ or $b_1 = b_2$. So unless the lines are the same line there is no point which lies on both of them. Hence the lines do not intersect, that is, they are parallel.   (2-2)

64. If $m = 0$, the two lines have equations $-y = b$ and $x = c$. The first line is horizontal and the second is vertical, hence they are perpendicular. Otherwise, $m \neq 0$, and neither line is horizontal or vertical. The two equations can be written as

$\ell_1: y = mx - b$ (slope m)

$\ell_2: y = -\dfrac{1}{m}x + \dfrac{c}{m}$ (slope $= -\dfrac{1}{m}$)

Note that $m = -\dfrac{1}{m}$ has no real solutions, hence the lines have unequal slopes, are not parallel, and must intersect.

Sketch a figure.

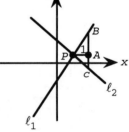

The case $m > 0$ is illustrated; if $m < 0$ the argument is similar (and left to the student).

In the figure, PA is constructed horizontal with length 1 unit. BC is constructed vertical, perpendicular to PA at A. We will show that BPC is a right triangle.

1. Since ℓ_1 has slope $m = \dfrac{\text{rise}}{\text{run}} = \dfrac{AB}{AP} = \dfrac{AB}{1}$, AB has length m.
2. Since ℓ_2 has slope $-\dfrac{1}{m} = \dfrac{\text{rise}}{\text{run}} = -\dfrac{AC}{AP} = -\dfrac{AC}{1}$, AC has length $\dfrac{1}{m}$.
3. PAB is constructed as a right triangle, hence by the Pythagorean theorem
$$PB = \sqrt{PA^2 + AB^2} = \sqrt{1 + m^2}.$$
4. PAC is constructed as a right triangle, hence by the Pythagorean theorem,
$$PC = \sqrt{PA^2 + AC^2} = \sqrt{1 + \tfrac{1}{m^2}}$$
5. $BC = BA + AC = m + \dfrac{1}{m}$.
6. Therefore $PB^2 + PC^2 = BC^2$ since
$$(\sqrt{1 + m^2})^2 + \left(\sqrt{1 + \tfrac{1}{m^2}}\right)^2 = (m + \tfrac{1}{m})^2 \qquad \text{(check!)}$$

7. Hence, by the converse of the Pythagorean theorem, BPC is a right triangle and the two lines are perpendicular.   (2-2)

65. (A) The portion of the graph in the first quadrant is the same as the graph of $[\![x]\!]$, since $|x| = x$ for $x \geq 0$. We draw this portion and reflect it in the y axis.

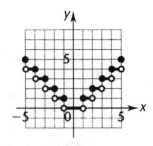

(B) If $x \geq 0$, $[\![x]\!] \geq 0$, $\big|[\![x]\!]\big| = [\![x]\!]$.
If $x < 0$, $[\![x]\!] \leq 0$, $\big|[\![x]\!]\big| = -[\![x]\!]$.
Hence for non-negative x we draw the graph of $[\![x]\!]$, and for negative x we draw the graph of $-[\![x]\!]$.

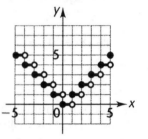

(2-5)

66. Domain: All real numbers except $x = 2$; Range: $y > -3$ or $(-3, \infty)$

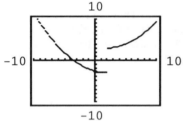

The function is discontinuous at $x = 2$.   (2-4)

67. (A) Reflect the given graph across the y axis:

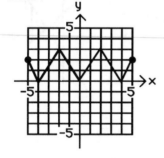

(B) Reflect the given graph across the origin:

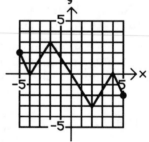

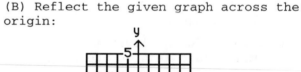
(2-1, 2-5)

68. (A) The graph must cross the x axis exactly once. Some possible graphs are shown:

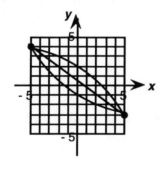

(B) The graph may cross the x axis once, but it may fail to cross the x axis at all. A possible graph of the latter type is shown:

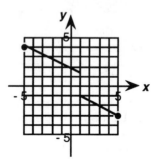

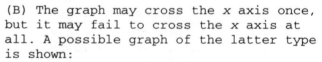
(2-4)

69. (A) If V is linearly related to t, then we are looking for an equation whose graph passes through $(t_1, V_1) = (0, 12,000)$ and $(t_2, V_2) = (8, 2,000)$. We find the slope, and then we use the point-slope form to find the equation.

$$m = \frac{V_2 - V_1}{t_2 - t_1} = \frac{2,000 - 12,000}{8 - 0} = \frac{-10,000}{8} = -1,250$$

$$V - V_1 = m(t - t_1)$$
$$V - 12,000 = -1,250(t - 0)$$
$$V - 12,000 = -1,250t$$
$$V = -1,250t + 12,000$$

(B) We are asked for V when $t = 5$
$$v = -1,250(5) + 12,000$$
$$V = -6,250 + 12,000$$
$$V = \$5,750$$

(2-2)

70. (A) If R is linearly related to C, then we are looking for an equation whose graph passes through $(C_1, R_1) = (30, 48)$ and $(C_2, R_2) = (20, 32)$. We find the slope, and then we use the point-slope form to find the equation.

$$m = \frac{R_2 - R_1}{C_2 - C_1} = \frac{32 - 48}{20 - 30} = \frac{-16}{-10} = 1.6$$

$$R - R_1 = m(C - C_1)$$
$$R - 48 = 1.6(C - 30)$$
$$R - 48 = 1.6C - 48$$
$$R = 1.6C$$

(B) We are asked for R when $C = 105$.
$$R = 1.6(105)$$
$$= \$168$$

(2-2)

71. If $0 \leq x \leq 3,000$, $E(x) = 200$

$$\begin{pmatrix} \text{Base} \\ \text{Salary} \end{pmatrix} \quad + \quad \begin{pmatrix} \text{Commission on} \\ \text{sales over } \$3,000 \end{pmatrix}$$

If $x > 3,000$, $E(x) = 200 \quad + \quad 0.1(x - 3,000)$
$$= 200 \quad + \quad 0.1x - 300$$
$$= 0.1x - 100$$

Summarizing,

$$E(x) = \begin{cases} 200 & \text{if } 0 \leq x \leq 3,000 \\ 0.1x - 100 & \text{if } x > 3,000 \end{cases}$$

$$E(2,000) = 200$$
$$E(5,000) = 0.1(5,000) - 100 = 500 - 100 = 400$$

(2-4)

72. (A)

x	0	5	10	15	20
Consumption	309	276	271	255	233
$303.4 - 3.46x = f(x)$	303	286	269	252	234

(B)

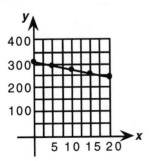

(C) In 1995, $x = 25$, hence $y = 303.4 - 3.46(25) \approx 217$
In 2000, $x = 30$, hence $y = 303.4 - 3.46(30) \approx 200$

(D) Per capita egg consumption is dropping about 17 eggs every five years. (2-4)

73. (A) If $0 \le x < 36$, $C(x) = 0.49x$
If $36 \le x < 72$, $C(x) = 0.44x$
If $72 \le x$, $C(x) = 0.39x$

Summarizing,
$$C(x) = \begin{cases} 0.49x & \text{for } 0 \le x < 36 \\ 0.44x & \text{for } 36 \le x < 72 \\ 0.39x & \text{for } 72 \le x \end{cases}$$

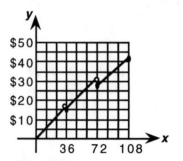

(B) There are points of discontinuity at $x = 36$ and $x = 72$.

(2-4)

x	$y = 0.49x$	x	$y = 0.44x$	x	$y = 0.39x$
0	0	36	15.84	72	28.08
18	8.82	54	23.76	108	42.12

74. (A) Let x = number of units sold. Then
$$C = \begin{pmatrix} \text{cost of shooting} \\ \text{video} \end{pmatrix} + \begin{pmatrix} \text{number of} \\ \text{units} \end{pmatrix} \times \begin{pmatrix} \text{cost per} \\ \text{unit} \end{pmatrix}$$
$$= \quad 84,000 \quad + \quad x \quad \cdot \quad 15$$
$$= \quad 84,000 + 15x$$

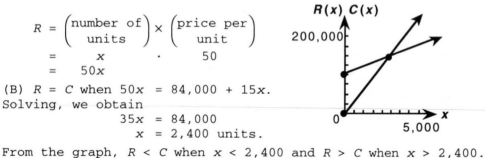

$$R = \begin{pmatrix} \text{number of} \\ \text{units} \end{pmatrix} \times \begin{pmatrix} \text{price per} \\ \text{unit} \end{pmatrix}$$
$$= \quad x \quad \cdot \quad 50$$
$$= \quad 50x$$

(B) $R = C$ when $50x = 84,000 + 15x$.
Solving, we obtain
$$35x = 84,000$$
$$x = 2,400 \text{ units.}$$

From the graph, $R < C$ when $x < 2,400$ and $R > C$ when $x > 2,400$. (2-2)

75. The profit function is the difference of the revenue and the cost functions
$P = R - C$
Hence
$P(x) = R(x) - C(x)$

$$= \left(50x - \frac{1}{10}x^2\right) - (20x + 4,000)$$

$$= 50x - \frac{1}{10}x^2 - 20x - 4,000$$

$$= 30x - \frac{1}{10}x^2 - 4,000$$

Next we use composition to express P as a function of the price p.
$(P \circ f)(p) = P[f(p)]$

$$= P(500 - 10p)$$

$$= 30(500 - 10p) - \frac{1}{10}(500 - 10p)^2 - 4,000$$

$$= 15,000 - 300p - \frac{1}{10}(250,000 - 10,000p + 100p^2) - 4,000$$

$$= 15,000 - 300p - 25,000 + 1,000p - 10p^2 - 4,000$$
$$= -14,000 + 700p - 10p^2$$

(2-5)

76. In the sketch, we note that the point $(4, r - 2)$ is on the circle with equation $x^2 + y^2 = r^2$, hence $(4, r - 2)$ must satisfy this equation.

$$4^2 + (r - 2)^2 = r^2$$
$$16 + r^2 - 4r + 4 = r^2$$
$$-4r + 20 = 0$$
$$r = 5 \text{ feet}$$

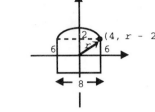

$(2-1)$

77. (A) From the figure, we see that $A = x(y + y) = 2xy$. Since the fence consists of four pieces of length y and three pieces of length x, we have $3x + 4y = 120$. Hence $4y = 120 - 3x$

$$y = 30 - \frac{3}{4}x$$

$$A = 2x\left(30 - \frac{3}{4}x\right)$$

$$A(x) = 60x - \frac{3}{2}x^2$$

(B) Since both x and y must be positive, we have $x > 0$

$30 - \frac{3}{4}x > 0$ or $-\frac{3}{4}x > -30$ or $x < 40$

Hence $0 < x < 40$ is the domain of A.

(C) The function A is a quadratic function. Completing the square yields:

$$A(x) = -\frac{3}{2}x^2 + 60x$$

$$= -\frac{3}{2}(x^2 - 40x)$$

$$= -\frac{3}{2}(x^2 - 40x + 400) + \frac{3}{2} \cdot 400$$

$$= -\frac{3}{2}(x - 20)^2 + 600$$

Comparing with $f(x) = a(x - h)^2 + k$, the total area will be maximum when $x = 20$. Then

$y = 30 - \frac{3}{4}x = 30 - \frac{3}{4}(20) = 15$

$(2-4)$

78. (A) $f(1) = 1 - (⟦\sqrt{1}⟧)^2 = 1 - 1 = 0$

(B) $f(2) = 2 - (⟦\sqrt{2}⟧)^2 = 2 - 1 = 1$

(C) $f(3) = 3 - (⟦\sqrt{3}⟧)^2 = 3 - 1 = 2$

(D) $f(4) = 4 - (⟦\sqrt{4}⟧)^2 = 4 - 4 = 0$

(E) $f(5) = 5 - (⟦\sqrt{5}⟧)^2 = 5 - 4 = 1$

(F) $f(n^2) = n^2 - (⟦\sqrt{n^2}⟧)^2$

$\qquad = n^2 - (⟦n⟧)^2$ since $\sqrt{n^2} = n$ if n is positive

$\qquad = n^2 - (n)^2$ since $⟦n⟧ = n$ if n is a (positive) integer.

$\qquad = 0$

$(2-4)$

CUMULATIVE REVIEW EXERCISE (Chapters 1 and 2)

1.
$$\frac{7x}{5} - \frac{3 + 2x}{2} = \frac{x - 10}{3} + 2$$

$$30\frac{7x}{5} - 30\frac{(3 + 2x)}{2} = 30\frac{(x - 10)}{3} + 2(30)$$

$$42x - 15(3 + 2x) = 10(x - 10) + 60$$

$$42x - 45 - 30x = 10x - 100 + 60$$

$$12x - 45 = 10x - 40$$

$$2x = 5$$

$$x = \frac{5}{2} \qquad (1\text{-}1)$$

2. $2x - 3y = 8$
$4x + y = 2$
Solve the second equation for y in terms of x and substitute into the first equation.

$$y = 2 - 4x$$
$$2x - 3(2 - 4x) = 8$$
$$2x - 6 + 12x = 8$$
$$14x = 14$$
$$x = 1$$
$$y = 2 - 4x = 2 - 4(1) = -2$$
$x = 1, \; y = -2$

$(1\text{-}2)$

3.
$$2(3 - y) + 4 \leq 5 - y$$
$$6 - 2y + 4 \leq 5 - y$$
$$-2y + 10 \leq 5 - y$$
$$-y \leq -5$$
$$y \geq 5$$
$[5, \infty)$

$(1\text{-}3)$

4. $|x - 2| < 7$
$$-7 < x - 2 < 7$$
$$-5 < x < 9$$
$(-5, 9)$

$(1\text{-}4)$

5.
$$x^2 + 3x \geq 10$$
$$x^2 + 3x - 10 \geq 0$$
$$(x + 5)(x - 2) \geq 0$$
Zeros: $-5, 2$
$-5, 2$ are part of the solution set, so we use solid dots.

$x^2 + 3x - 10 = (x + 5)(x - 2)$			
Test Number	-6	0	3
Value of Polynomial for Test Number	8	-10	8
Sign of Polynomial in Interval	$+$	$-$	$+$
Interval	$(-\infty, -5)$	$(-5, 2)$	$(2, \infty)$

$x^2 + 3x - 10$ is non-negative within the intervals $(-\infty, -5]$ and $[2, \infty)$.
$(-\infty, -5] \cup [2, \infty)$

$(1\text{-}8)$

6. (A) $(2 - 3i) - (-5 + 7i) = 2 - 3i + 5 - 7i = 7 - 10i$

(B) $(1 + 4i)(3 - 5i) = 3 + 7i - 20i^2 = 3 + 7i + 20 = 23 + 7i$

(C) $\dfrac{5 + i}{2 + 3i} = \dfrac{(5 + i)}{(2 + 3i)} \dfrac{(2 - 3i)}{(2 - 3i)} = \dfrac{10 - 13i - 3i^2}{4 - 9i^2}$

$$= \frac{10 - 13i + 3}{4 + 9} = \frac{13 - 13i}{13} = 1 - i$$

$(1\text{-}5)$

7.
$$3x^2 = -12x$$
$$3x^2 + 12x = 0$$
$$3x(x + 4) = 0$$
$$3x = 0 \quad \text{or} \quad x + 4 = 0$$
$$x = 0 \qquad\qquad x = -4$$
$$(1-6)$$

8. $4x^2 - 20 = 0$
$$4x^2 = 20$$
$$x^2 = 5$$
$$x = \pm\sqrt{5}$$
$$(1-6)$$

9. $x^2 - 6x + 2 = 0$
$$x^2 - 6x = -2$$
$$x^2 - 6x + 9 = 7$$
$$(x - 3)^2 = 7$$
$$x - 3 = \pm\sqrt{7}$$
$$x = 3 \pm \sqrt{7}$$
$$(1-6)$$

10.
$$x - \sqrt{12 - x} = 0$$
$$x = \sqrt{12 - x}$$
$$x^2 = 12 - x$$
$$x^2 + x - 12 = 0$$
$$(x + 4)(x - 3) = 0$$
$$x = -4, \; 3$$

Check: $-4 - \sqrt{12 - (-4)} \overset{?}{=} 0$
$$-4 - 4 \overset{?}{=} 0$$
$$-8 \neq 0$$
$$3 - \sqrt{12 - 3} \overset{?}{=} 0$$
$$3 - 3 \overset{?}{=} 0$$
$$0 \overset{\sqrt{}}{=} 0$$

Solution: 3 $\qquad\qquad (1-7)$

11. $\sqrt{2 + 3x}$ represents a real number exactly when $2 + 3x$ is positive or zero. We can write this as an inequality statement and solve for x.
$$2 + 3x \geq 0$$
$$3x \geq -2$$
$$x \geq -\frac{2}{3} \quad \text{or} \quad \left[-\frac{2}{3}, \; \infty\right)$$
$$(1-3)$$

12. (A) $d(A, \; B) = \sqrt{(5 - 3)^2 + (6 - 2)^2} = \sqrt{4 + 16} = \sqrt{20} = 2\sqrt{5}$

(B) $m = \dfrac{6 - 2}{5 - 3} = \dfrac{4}{2} = 2$

(C) The slope m_1 of a line perpendicular to AB must satisfy $m_1(2) = -1$.

Therefore $m_1 = -\dfrac{1}{2}$ $\qquad\qquad (2-1, \; 2-2)$

13. (A) Center at $(0, \; 0)$ and radius $\sqrt{2}$
$$x^2 + y^2 = r^2$$
$$x^2 + y^2 = (\sqrt{2})^2$$
$$x^2 + y^2 = 2$$

(B) Center at $(-3, \; 1)$ and radius $\sqrt{2}$
$$(h, \; k) = (-3, \; 1)$$
$$r = \sqrt{2}$$
$$(x - h)^2 + (y - k)^2 = r^2$$
$$[x - (-3)]^2 + (y - 1)^2 = (\sqrt{2})^2$$
$$(x + 3)^2 + (y - 1)^2 = 2 \qquad (2-1)$$

14. $2x - 3y = 6$
$$-3y = -2x + 6$$
$$y = \frac{2}{3}x - 2$$

slope: $\dfrac{2}{3}$ y intercept: -2 x intercept: 3 (if $y = 0$, $2x = 6$, hence $x = 3$)

x	y
-3	-4
0	-2
3	0

$$(2-2)$$

15. (A) A function; domain = {1, 2, 3}; range = {1}
(B) Not a function (three range values correspond to the only domain value)
(C) A function; domain = {-2, -1, 0, 1, 2}; range = {-1, 0, 2} (*2-3*)

16. (A) $f(-2) = (-2)^2 - 2(-2) + 5 = 13$
$\quad\quad g(3) = 3 \cdot 3 - 2 = 7$
Therefore
$f(-2) + g(3) = 13 + 7 = 20$

(B) $(f + g)(x) = f(x) + g(x)$
$\quad\quad\quad\quad\quad = x^2 - 2x + 5 + 3x - 2$
$\quad\quad\quad\quad\quad = x^2 + x + 3$

(C) $(f \circ g)(x) = f[g(x)] = f(3x - 2)$
$\quad\quad\quad\quad\quad\quad\quad\quad = (3x - 2)^2 - 2(3x - 2) + 5$
$\quad\quad\quad\quad\quad\quad\quad\quad = 9x^2 - 12x + 4 - 6x + 4 + 5$
$\quad\quad\quad\quad\quad\quad\quad\quad = 9x^2 - 18x + 13$

(D) $\dfrac{f(a + h) - f(a)}{h} = \dfrac{[(a + h)^2 - 2(a + h) + 5] - (a^2 - 2a + 5)}{h}$

$\quad\quad\quad\quad\quad\quad\quad\quad = \dfrac{a^2 + 2ah + h^2 - 2a - 2h + 5 - a^2 + 2a - 5}{h}$

$\quad\quad\quad\quad\quad\quad\quad\quad = \dfrac{2ah + h^2 - 2h}{h}$

$\quad\quad\quad\quad\quad\quad\quad\quad = \dfrac{h(2a + h - 2)}{h}$

$\quad\quad\quad\quad\quad\quad\quad\quad = 2a + h - 2$ (*2-2, 2-5*)

17. (A) Expanded by a factor of 2   (B) Shifted right 2 units
(C) Shifted down 2 units (*2-5*)

18. (A) The graph of $f(x)$ is shifted left one unit and reflected across the x axis.
(B) The graph of $f(x)$ is stretched by a factor of two with respect to the y axis and shifted down two units.

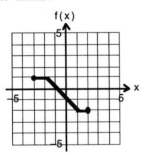

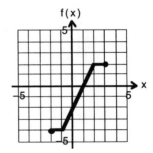

(*2-5*)

19.

$$\frac{x + 3}{2x + 2} + \frac{5x + 2}{3x + 3} = \frac{5}{6}$$

$$\frac{(x + 3)}{2(x + 1)} + \frac{(5x + 2)}{3(x + 1)} = \frac{5}{6} \quad \text{LCD: } 6(x + 1) \quad \text{Excluded value: } x \neq -1$$

$$6(x + 1)\frac{(x + 3)}{2(x + 1)} + 6(x + 1)\frac{(5x + 2)}{3(x + 1)} = 6(x + 1)\frac{5}{6}$$

$$3(x + 3) + 2(5x + 2) = 5(x + 1)$$
$$3x + 9 + 10x + 4 = 5x + 5$$
$$13x + 13 = 5x + 5$$
$$8x = -8$$
$$x = -1$$

No solution: -1 is excluded (*1-1*)

20.
$$\frac{3}{x} = \frac{6}{x + 1} - \frac{1}{x - 1} \qquad \text{Excluded values: } x \neq 0, \; 1, \; -1$$

$$x(x + 1)(x - 1)\frac{3}{x} = x(x + 1)(x - 1)\frac{6}{x + 1} - x(x + 1)(x - 1)\frac{1}{x - 1}$$

$$3(x + 1)(x - 1) = 6x(x - 1) - x(x + 1)$$
$$3(x^2 - 1) = 6x^2 - 6x - x^2 - x$$
$$3x^2 - 3 = 5x^2 - 7x$$
$$0 = 2x^2 - 7x + 3$$
$$0 = (2x - 1)(x - 3)$$
$$2x - 1 = 0 \quad \text{or} \quad x - 3 = 0$$
$$2x = 1 \qquad\qquad x = 3$$
$$x = \frac{1}{2} \qquad\qquad\qquad\qquad (1\text{-}6)$$

21.
$$2x + 1 = 3\sqrt{2x - 1}$$
$$(2x + 1)^2 = 9(2x - 1)$$
$$4x^2 + 4x + 1 = 18x - 9$$
$$4x^2 - 14x + 10 = 0$$
$$2x^2 - 7x + 5 = 0$$
$$(2x - 5)(x - 1) = 0$$
$$2x - 5 = 0 \quad \text{or} \quad x - 1 = 0$$
$$x = \frac{5}{2} \qquad\qquad x = 1$$

Check: $2\left(\frac{5}{2}\right) + 1 \overset{?}{=} 3\sqrt{2\left(\frac{5}{2}\right) - 1}$

$$5 + 1 \overset{?}{=} 3\sqrt{5 - 1}$$
$$6 \overset{\checkmark}{=} 6$$
$$2 \cdot 1 + 1 \overset{?}{=} 3\sqrt{2 \cdot 1 - 1}$$
$$3 \overset{\checkmark}{=} 3$$

Solution: $1, \; \frac{5}{2}$ $\qquad (1\text{-}7)$

22. $2x - 3y = 9$
$4x + 2y = 23$
Solve the first equation for x in terms of y and substitute into the second equation.

$$2x = 3y + 9$$
$$x = \frac{3y + 9}{2}$$

$$\frac{4}{1}\left(\frac{3y + 9}{2}\right) + 2y = 23$$
$$2(3y + 9) + 2y = 23$$
$$6y + 18 + 2y = 23$$
$$8y = 5$$
$$y = \frac{5}{8}$$

$$x = \frac{3y + 9}{2} = \frac{3\left(\frac{5}{8}\right) + 9}{2} = \frac{87}{16}$$

$$x = \frac{87}{16}, \; y = \frac{5}{8} \qquad\qquad (1\text{-}2)$$

23. $|4x - 9| > 3$
$$4x - 9 < -3 \quad \text{or} \quad 4x - 9 > 3$$
$$4x < 6 \qquad\qquad 4x > 12$$
$$x < \frac{3}{2} \quad \text{or} \qquad x > 3$$

$$\left(-\infty, \; \frac{3}{2}\right) \cup (3, \; \infty)$$

$\qquad (1\text{-}4)$

24. $\sqrt{(3m - 4)^2} \leq 2$
$$|3m - 4| \leq 2$$
$$-2 \leq 3m - 4 \leq 2$$
$$2 \leq 3m \leq 6$$
$$\frac{2}{3} \leq m \leq 2$$

$$\left[\frac{2}{3}, \; 2\right]$$

$\qquad (1\text{-}4)$

25.
$$\frac{2}{x + 1} \geq \frac{1}{x - 2}$$

$$\frac{2}{x + 1} - \frac{1}{x - 2} \geq 0$$

$$\frac{2(x - 2) - (x + 1)}{(x + 1)(x - 2)} \geq 0$$

$$\frac{2x - 4 - x - 1}{(x + 1)(x - 2)} \geq 0$$

$$\frac{x - 5}{(x + 1)(x - 2)} \geq 0$$

Common Error:
$2(x - 2) \geq x + 1$ is not equivalent to the given inequality.

Zeros of P, Q: -1, 2, 5
5 is part of the solution set
so we use a solid dot there.
-1 and 2 are not part of the
solution set ($\frac{P}{Q}$ is not defined
there) so we use open dots
there.

(−∞, −1) (−1, 2) (2, 5) (5, ∞)

	$\dfrac{P}{Q} = \dfrac{x - 5}{(x + 1)(x - 2)}$			
Test Number	-2	0	3	6
Value of $\frac{P}{Q}$	$-\frac{7}{4}$	$\frac{5}{2}$	$-\frac{1}{2}$	$\frac{1}{28}$
Sign of $\frac{P}{Q}$	$-$	$+$	$-$	$+$
Interval	$(-\infty,\ -1)$	$(-1,\ 2)$	$(2,\ 5)$	$(5,\ \infty)$

$$\frac{x - 5}{(x + 1)(x - 2)} \geq 0 \text{ and }$$

$$\frac{2}{x + 1} \geq \frac{1}{x - 2} \text{ within the intervals } (-1,\ 2) \text{ and } [5,\ \infty).$$

$-1 < x < 2$ or $5 \leq x$

(1–8)

26. $\dfrac{\sqrt{x - 2}}{x - 4}$ represents a real number if $x - 2$ is positive or zero, except if $x = 4$.
Thus, $x \geq 2$, $x \neq 4$, or $[2,\ 4) \cup (4,\ \infty)$.

(1–3)

27. (A) $(2 - 3i)^2 - (4 - 5i)(2 - 3i) - (2 + 10i)$
$\quad = (2)^2 - 2(2)(3i) + (3i)^2 - (8 - 12i - 10i + 15i^2) - 2 - 10i$
$\quad = 4 - 12i + 9i^2 - (8 - 22i + 15i^2) - 2 - 10i$
$\quad = 4 - 12i + 9i^2 - 8 + 22i - 15i^2 - 2 - 10i$
$\quad = 4 - 12i - 9 - 8 + 22i + 15 - 2 - 10i$
$\quad = 0 + 0i \quad \text{or} \quad 0$

(B) $\dfrac{3}{5} + \dfrac{4}{5}i + \dfrac{1}{\frac{3}{5} + \frac{4}{5}i} = \dfrac{3}{5} + \dfrac{4}{5}i + \dfrac{1}{(\frac{3}{5} + \frac{4}{5}i)}\dfrac{(\frac{3}{5} - \frac{4}{5}i)}{(\frac{3}{5} - \frac{4}{5}i)}$

$\qquad\qquad\qquad\quad = \dfrac{3}{5} + \dfrac{4}{5}i + \dfrac{\frac{3}{5} - \frac{4}{5}i}{\frac{9}{25} - \frac{16}{25}i^2}$

$\qquad\qquad\qquad\quad = \dfrac{3}{5} + \dfrac{4}{5}i + \dfrac{\frac{3}{5} - \frac{4}{5}i}{\frac{9}{25} + \frac{16}{25}}$

$\qquad\qquad\qquad\quad = \dfrac{3}{5} + \dfrac{4}{5}i + \dfrac{3}{5} - \dfrac{4}{5}i$

$\qquad\qquad\qquad\quad = \dfrac{6}{5}$

(C) $i^{35} = i^{32}i^3$
$\qquad = (i^4)^8(-i)$
$\qquad = 1^8(-i)$
$\qquad = -i$

(1–5)

28. (A) $(5 + 2\sqrt{-9}) - (2 - 3\sqrt{-16}) = (5 + 2i\sqrt{9}) - (2 - 3i\sqrt{16})$
$\qquad\qquad\qquad\qquad\qquad\quad = (5 + 6i) - (2 - 12i)$
$\qquad\qquad\qquad\qquad\qquad\quad = 5 + 6i - 2 + 12i$
$\qquad\qquad\qquad\qquad\qquad\quad = 3 + 18i$

(B) $\dfrac{2 + 7\sqrt{-25}}{3 - \sqrt{-1}} = \dfrac{2 + 7i\sqrt{25}}{3 - i}$

$\qquad = \dfrac{2 + 35i}{3 - i}$

$\qquad = \dfrac{(2 + 35i)}{(3 - i)} \dfrac{(3 + i)}{(3 + i)}$

$\qquad = \dfrac{6 + 107i + 35i^2}{9 - i^2}$

$\qquad = \dfrac{6 + 107i - 35}{9 + 1}$

$\qquad = \dfrac{-29 + 107i}{10}$

$\qquad = -2.9 + 10.7i$

(C) $\dfrac{12 - \sqrt{-64}}{\sqrt{-4}} = \dfrac{12 - i\sqrt{64}}{i\sqrt{4}}$

$\qquad = \dfrac{12 - 8i}{2i}$

$\qquad = \dfrac{12 - 8i}{2i} \dfrac{i}{i}$

$\qquad = \dfrac{12i - 8i^2}{2i^2}$

$\qquad = \dfrac{12i + 8}{-2}$

$\qquad = -4 - 6i$

$\hfill (1\text{-}5)$

29. (A) All real numbers $(-\infty, \infty)$

(B) From the graph, the possible function values include -2 (only) and all numbers greater than or equal to 1. In set and interval notation: $\{-2\} \cup [1, \infty)$.

(C) $f(-3) = 1 \quad f(-2) = 2$ (not -2) $\quad f(2) = -2$ (not 2).
Thus, $f(-3) + f(-2) + f(2) = 1 + 2 + (-2) = 1$

(D) $[-3, -2]$ and $[2, \infty)$

(E) f is discontinuous at $x = -2$ and at $x = 2$. $\hfill (2\text{-}3, \ 2\text{-}4)$

30. The line $3x + 2y = 12$, or
$\qquad 2y = -3x + 12$,
or $y = -\dfrac{3}{2}x + 6$, has slope $-\dfrac{3}{2}$.

(A) We require a line through $(-6, 1)$ with slope $-\dfrac{3}{2}$. Applying the point-slope form, we have

$y - 1 = -\dfrac{3}{2}(x + 6)$

$y - 1 = -\dfrac{3}{2}x - 9$

$\qquad y = -\dfrac{3}{2}x - 8$

(B) We require a line with slope m satisfying $-\dfrac{3}{2}m = -1$, or $m = \dfrac{2}{3}$. Again applying the point-slope form, we have

$y - 1 = \dfrac{2}{3}(x + 6)$

$y - 1 = \dfrac{2}{3}x + 4$

$\qquad y = \dfrac{2}{3}x + 5 \hfill (2\text{-}2)$

31. $x + 4 \geq 0$ is equivalent to $x \geq -4$. Domain of $g(x) = \sqrt{x + 4}$ is $[-4, \infty)$ $\hfill (2\text{-}3)$

32. $f(x) = x^2 - 2x - 8$. Complete the square.
$f(x) = (x^2 - 2x + 1) - 1 - 8$
$\qquad = (x - 1)^2 - 9$

Comparing with $f(x) = a(x - h)^2 + k$, $h = 1$ and $k = -9$. Thus, the vertex is $(1, -9)$, the axis of symmetry is $x = 1$, and the minimum value is -9. $y = f(x)$ can be any number greater than or equal to -9, so the range is $[-9, \infty)$.
y intercept: Set $x = 0$, then $f(0) = -8$ is the y intercept.
x intercepts: Set $f(x) = 0$, then
$\quad 0 = x^2 - 2x - 8$
$\quad 0 = (x - 4)(x + 2)$
$\quad x = 4 \quad$ or $\quad x = -2$ are the x intercepts.

Graph: Locate axis and vertex, then plot several points on either side of the axis.

x	$f(x)$
-3	7
-2	0
-1	-5
0	-8
1	-9
2	-8
3	-5
4	0
5	7

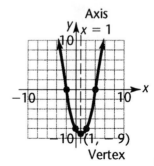

$\hfill (2\text{-}4)$

33. $(f \circ g)(x) = f[g(x)] = f\left(\dfrac{x + 3}{x}\right) = \dfrac{1}{\frac{x+3}{x} - 2} = \dfrac{x(1)}{x\left(\frac{x+3}{x} - 2\right)} = \dfrac{x}{x + 3 - 2x} = \dfrac{x}{3 - x}$

The domain of f is all real numbers except 2.
The domain of g is all non-zero real numbers.
The domain of $f \circ g$ is the set of all non-zero real numbers for which $g(x) \neq 2$.

$g(x) = 2$ only if $\dfrac{x + 3}{x} = 2$, that is $x + 3 = 2x$, or $x = 3$.

Thus the domain of $f \circ g$ is the set of all non-zero real numbers except 3, that is, $(-\infty, 0) \cup (0, 3) \cup (3, \infty)$.

> Common Error:
> The domain of $f \circ g$ cannot be found by looking at the final form $\dfrac{x}{3 - x}$. It is not $\{x \mid x \neq 3\}$.

$(2\text{-}5)$

34. $f(x) = 2x + 5$
Assume $f(a) = f(b)$
$\qquad 2a + 5 = 2b + 5$
$\qquad\quad 2a = 2b$
$\qquad\quad\ a = b$
Thus, f is one-to-one.

Solve $y = f(x)$ for x:
$\qquad y = 2x + 5$
$\quad y - 5 = 2x$
$\qquad\quad x = \dfrac{y - 5}{2} = f^{-1}(y)$

Interchange x and y

$y = f^{-1}(x) = \dfrac{x - 5}{2}$ or $\dfrac{1}{2}x - \dfrac{5}{2}$ Domain: $(-\infty, \infty)$ $\qquad$ $(2\text{-}6)$

35.

x	$y = x - 1$	x	$y = x^2 + 1$
-1	-2	0	1
-2	-3	1	2
		2	5

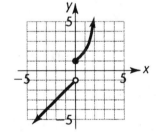

Domain: all real numbers
Range: $(-\infty, -1) \cup [1, \infty)$
Discontinuous at: $x = 0$ $\qquad\qquad\qquad\qquad\qquad\qquad\qquad\qquad\qquad\qquad$ $(2\text{-}4)$

36. (A) This is the same as the graph of $y = \sqrt{x}$ expanded by a factor of 2 and shifted up one unit.

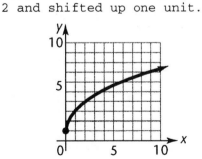

(B) This is the same as the graph of $y = \sqrt{x}$ shifted left 1 unit and reflected with respect to the x axis.

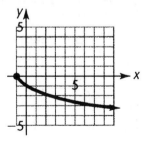

$(2\text{-}5)$

37. The graph of $y = |x|$ is contracted by $\dfrac{1}{2}$, reflected in the x axis, shifted two units to the right and three units up; $y = -\dfrac{1}{2}|x - 2| + 3$. $\qquad$ $(2\text{-}5)$

38. $f(x) = \sqrt{x + 4}$

Assume $f(a) = f(b)$
$$\sqrt{a + 4} = \sqrt{b + 4}$$
$$a + 4 = b + 4$$
$$a = b$$

Thus f is one-to-one.

(A) Solve $y = f(x)$ for x

$y = \sqrt{x + 4}$
$y^2 = x + 4 \quad y \geq 0$
$x = y^2 - 4 \quad y \geq 0 \quad f^{-1}(y) = y^2 - 4$

Interchange x and y:
$y = f^{-1}(x) = x^2 - 4 \quad$ Domain: $x \geq 0$

Check: $f^{-1}[f(x)] = (\sqrt{x + 4})^2 - 4 = x + 4 - 4 = x$
$\qquad f[f^{-1}(x)] = \sqrt{x^2 - 4 + 4} = \sqrt{x^2} = x \quad$ since $x \geq 0$ in the domain of f^{-1}.

(B) Domain of $f = [-4, \infty) =$ Range of f^{-1} (C)
  Range of $f = [0, \infty) =$ Domain of f^{-1}

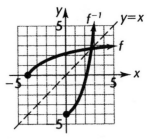

$(2-6)$

39.
$$x^2 - 6x + y^2 + 2y = 0$$
$$(x^2 - 6x + ?) + (y^2 + 2y + ?) = 0$$
$$(x^2 - 6x + 9) + (y^2 + 2y + 1) = 9 + 1$$
$$(x - 3)^2 + (y + 1)^2 = 10$$
$$(x - 3)^2 + [y - (-1)]^2 = (\sqrt{10})^2$$

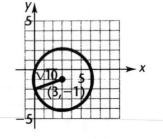

Center: $C(h, k) = (3, -1)$ Radius $r = \sqrt{10}$

$(2-1)$

40. $xy + |xy| = 5$ y axis symmetry? $(-x)y + |(-x)y| = 5$ $-xy + |xy| = 5$
                                         No, not equivalent

            x axis symmetry? $x(-y) + |x(-y)| = 5$ $-xy + |xy| = 5$
                                         No, not equivalent

            origin symmetry? $(-x)(-y) + |(-x)(-y)| = 5$ $xy + |xy| = 5$
                                         Yes, equivalent

The graph has symmetry with respect to the origin. $(2-1)$

41. Since the graph is a parabola opening up, $a = +1$. Since the vertex is at $(-2, -3)$ $= (h, k)$, $h = -2$ and $k = -3$. Thus, the equation is $y = (x + 2)^2 - 3$. $(2-4)$

42. $1 + \dfrac{14}{y^2} = \dfrac{6}{y}$ Excluded value: $y \neq 0$

$y^2(1) + y^2\dfrac{14}{y^2} = y^2\dfrac{6}{y}$

$\qquad y^2 + 14 = 6y$
$y^2 - 6y + 14 = 0$
$\qquad y^2 - 6y = -14$
$y^2 - 6y + 9 = -5$
$\qquad (y - 3)^2 = -5$

$\qquad\quad y - 3 = \pm\sqrt{-5}$
$\qquad\qquad y = 3 \pm \sqrt{-5}$
$\qquad\qquad y = 3 \pm i\sqrt{5}$ $(1-6)$

43. $4x^{2/3} - 4x^{1/3} - 3 = 0$
Let $u = x^{1/3}$, then
$$4u^2 - 4u - 3 = 0$$
$$(2u - 3)(2u + 1) = 0$$
$$u = \frac{3}{2}, -\frac{1}{2}$$

$x^{1/3} = \frac{3}{2}$ $x^{1/3} = -\frac{1}{2}$

$x = \frac{27}{8}$ $x = -\frac{1}{8}$ *(1-7)*

44. $u^4 + u^2 - 12 = 0$
Let $w = u^2$, then
$$w^2 + w - 12 = 0$$
$$(w + 4)(w - 3) = 0$$
$$w = -4, 3$$
$u^2 = -4$ $u^2 = 3$
$u = \pm 2i$ $u = \pm\sqrt{3}$ *(1-7)*

45. $\sqrt{8t - 2} - 2\sqrt{t} = 1$
$$\sqrt{8t - 2} = 2\sqrt{t} + 1$$
$$8t - 2 = 4t + 4\sqrt{t} + 1$$
$$4t - 3 = 4\sqrt{t}$$
$$16t^2 - 24t + 9 = 16t$$
$$16t^2 - 40t + 9 = 0$$
$$(4t - 1)(4t - 9) = 0$$
$$t = \tfrac{1}{4}, \tfrac{9}{4}$$

Check: $\sqrt{8(\frac{1}{4}) - 2} - 2\sqrt{\frac{1}{4}} \overset{?}{=} 1$

$0 - 2(\frac{1}{2}) \overset{?}{=} 1$

$-1 \neq 1$

$\sqrt{8(\frac{9}{4}) - 2} - 2\sqrt{\frac{9}{4}} \overset{?}{=} 1$

$\sqrt{16} - 2\sqrt{\frac{9}{4}} \overset{?}{=} 1$

$4 - 2(\frac{3}{2}) \overset{?}{=} 1$

$1 \overset{\checkmark}{=} 1$

Solution: $\frac{9}{4}$ *(1-7)*

46. $-3.45 < 1.86 - 0.33x \leq 7.92$
$-5.31 < -0.33x \leq 6.06$
$16.09 > x \geq -18.36$
$-18.36 \leq x < 16.09$ or $[-18.36, 16.09)$ *(1-3)*

47. $2.35x^2 + 10.44x - 16.47 = 0$

$x = \dfrac{-b \pm \sqrt{b^2 - 4ac}}{2a}$ $a = 2.35$, $b = 10.44$, $c = -16.47$

$x = \dfrac{-10.44 \pm \sqrt{(10.44)^2 - 4(2.35)(-16.47)}}{2(2.35)}$

$= \dfrac{-10.44 \pm \sqrt{263.8116}}{4.70}$

$= \dfrac{-10.44 \pm 16.2423}{4.70}$

$= -5.68, 1.23$ *(1-6)*

48. $12.5x + 2.5y = 20$
$3.5x + 8.7y = 10$
Solve the first equation for y in terms of x and substitute into the second equation.
$$2.5y = 20 - 12.5x$$
$$y = \frac{20 - 12.5x}{2.5}$$
$$y = 8 - 5x$$
$$3.5x + 8.7(8 - 5x) = 10$$
$$3.5x + 69.6 - 43.5x = 10$$
$$-40x = -59.6$$
$$x = 1.49$$
$$y = 8 - 5(1.49) = 0.55$$
$x = 1.49$, $y = 0.55$ *(1-2)*

49.

$$\frac{x - 2}{x + 1} = \frac{2y + 1}{y - 2}$$

$$(x + 1)(y - 2)\frac{x - 2}{x + 1} = (x + 1)(y - 2)\frac{2y + 1}{y - 2}$$

$$(y - 2)(x - 2) = (x + 1)(2y + 1)$$

$$xy - 2y - 2x + 4 = 2xy + x + 2y + 1$$

$$3 - 3x = xy + 4y$$

$$3 - 3x = y(x + 4)$$

$$y = \frac{3 - 3x}{x + 4} \qquad\qquad (1\text{-}1)$$

50. $x = -1 + 5s + 2t$
$y = 2 + 2s + t$
Solve the second equation for t in terms of the other variables and substitute into the first equation.

$$t = y - 2 - 2s$$
$$x = -1 + 5s + 2(y - 2 - 2s)$$
$$x = -1 + 5s + 2y - 4 - 4s$$
$$x = -5 + s + 2y$$
$$s = 5 + x - 2y$$
$$t = y - 2 - 2s = y - 2 - 2(5 + x - 2y) = y - 2 - 10 - 2x + 4y$$
$$t = -12 - 2x + 5y$$

The checking steps are omitted for lack of space. $\qquad (1\text{-}2)$

51. The graph is the same as the graph of $y = \sqrt{x}$ shifted 1 unit to the left, expanded by a factor of 2, reflected with respect to the x axis, and shifted up 3 units.

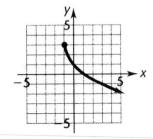

$(2\text{-}5)$

52. $x^2 - x + 2 = \left(\frac{1}{2} - \frac{i}{2}\sqrt{7}\right)^2 - \left(\frac{1}{2} - \frac{i}{2}\sqrt{7}\right) + 2$

$$= \left(\frac{1}{2}\right)^2 - 2\left(\frac{1}{2}\right)\left(\frac{i}{2}\sqrt{7}\right) + \left(\frac{i}{2}\sqrt{7}\right)^2 - \frac{1}{2} + \frac{i}{2}\sqrt{7} + 2$$

$$= \frac{1}{4} - \frac{i}{2}\sqrt{7} + \frac{7i^2}{4} - \frac{1}{2} + \frac{i}{2}\sqrt{7} + 2$$

$$= \frac{1}{4} - \frac{7}{4} - \frac{1}{2} + 2$$

$$= 0 \qquad\qquad (1\text{-}5)$$

53. The given inequality
$$a - b < b - a$$
is equivalent to, successively,
$$2a - b < b$$
$$2a < 2b$$
$$a < b$$
Thus it is true for all real a and b such that $a < b$. $\qquad (1\text{-}3)$

54.

$$\frac{x + y}{y - \frac{x + y}{x - y}} = 1$$

$$\frac{(x + y)(x - y)}{y(x - y) - (x + y)} = 1$$

$$\frac{x^2 - y^2}{xy - y^2 - x - y} = 1$$

$$x^2 - y^2 = xy - y^2 - x - y$$

$$x^2 + x = xy - y$$

$$x^2 + x = y(x - 1)$$

$$y = \frac{x^2 + x}{x - 1} \qquad (1\text{-}1)$$

55.

$$3x^2 = 2\sqrt{2}\,x - 1$$

$$3x^2 - 2\sqrt{2}\,x + 1 = 0$$

$$x = \frac{-b \pm \sqrt{b^2 - 4ac}}{2a}$$

$a = 3$

$b = -2\sqrt{2}$

$c = 1$

$$x = \frac{-(-2\sqrt{2}) \pm \sqrt{(-2\sqrt{2})^2 - 4(3)(1)}}{2(3)}$$

$$x = \frac{2\sqrt{2} \pm \sqrt{8 - 12}}{6}$$

$$x = \frac{2\sqrt{2} \pm \sqrt{-4}}{6}$$

$$x = \frac{2\sqrt{2} \pm 2i}{6}$$

$$x = \frac{2(\sqrt{2} \pm i)}{6}$$

$$x = \frac{\sqrt{2} \pm i}{3} \qquad (1\text{-}6)$$

56. In this problem, $a = 1$, $b = b$, $c = 1$. Thus, the discriminant $b^2 - 4ac = b^2 - 4 \cdot 1 \cdot 1 = b^2 - 4$. Hence, the number and types of roots depend on the sign of $b^2 - 4 = (b + 2)(b - 2)$. Charting the sign behavior of $b^2 - 4 = (b - 2)(b + 2)$, note:

Zeros: -2, 2

$(-\infty, -2)$ $(-2, 2)$ $(2, \infty)$

-2 $\quad$ 2 $\longrightarrow b$

	$b^2 - 4 = (b + 2)(b - 2)$		
Test Number	-3	0	3
Value of Polynomial for Test Number	5	-4	5
Sign of Polynomial in Interval	$+$	$-$	$+$
Interval	$(-\infty, -2)$	$(-2, 2)$	$(2, \infty)$

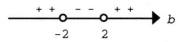

$+\ +$ $\quad$ $-\ -$ $\quad$ $+\ +$

-2 $\quad$ 2 $\longrightarrow b$

Hence,

if $b^2 - 4 > 0$, there are two distinct real roots. This occurs if $b < -2$ or $b > 2$.

if $b^2 - 4 = 0$, there is one real double root. This occurs if $b = -2$ or $b = 2$.

if $b^2 - 4 < 0$, there are two distinct imaginary roots. This occurs if $-2 < b < 2$.

$(1\text{-}6,\ 1\text{-}8)$

57.

$$1 = 6x^{-2} + 9x^{-4}$$

$$1 = \frac{6}{x^2} + \frac{9}{x^4} \quad \text{Excluded value: } x \neq 0$$

$$x^4 = 6x^2 + 9$$

$$x^4 - 6x^2 - 9 = 0$$

$$x^4 - 6x^2 = 9$$

$$x^4 - 6x^2 + 9 = 18$$

$$(x^2 - 3)^2 = 18$$

$$x^2 - 3 = \pm\sqrt{18}$$

$$x^2 = 3 \pm \sqrt{18}$$

$$x^2 = 3 \pm 3\sqrt{2}$$

$3 - 3\sqrt{2}$ is negative; there can be no real solutions of $x^2 = 3 - 3\sqrt{2}$. Since we are asked to find all real roots of the original equation, we discard $x^2 = 3 - 3\sqrt{2}$.

$$x^2 = 3 + 3\sqrt{2}$$
$$x = \pm\sqrt{3 + 3\sqrt{2}} \tag{1-7}$$

58. $\dfrac{a + bi}{a - bi} = \dfrac{(a + bi)}{(a - bi)}\dfrac{(a + bi)}{(a + bi)} = \dfrac{(a + bi)^2}{a^2 - (bi)^2} = \dfrac{a^2 + 2abi + b^2i^2}{a^2 - b^2i^2}$

$$= \dfrac{a^2 + 2abi - b^2}{a^2 + b^2}$$

$$= \dfrac{a^2 - b^2 + 2abi}{a^2 + b^2}$$

$$= \dfrac{a^2 - b^2}{a^2 + b^2} + \dfrac{2ab}{a^2 + b^2}i \tag{1-5}$$

59. $\left| \dfrac{x + 4}{x} \right| < 3$

$-3 < \dfrac{x + 4}{x} < 3$

It is difficult (and a rich source of possible errors) to solve this double inequality as a double inequality. We choose to solve the inequalities $-3 < \dfrac{x + 4}{x}$ and $\dfrac{x + 4}{x} < 3$ separately. Then the solution set of the given inequality is the intersection of the solution sets of the two inequalities.

Part 1: $-3 < \dfrac{x + 4}{x}$

$0 < \dfrac{x + 4}{x} + 3$

$0 < \dfrac{x + 4 + 3x}{x}$

$0 < \dfrac{4x + 4}{x}$

Part 2: $\dfrac{x + 4}{x} < 3$

$\dfrac{x + 4}{x} - 3 < 0$

$\dfrac{x + 4 - 3x}{x} < 0$

$\dfrac{4 - 2x}{x} < 0$

Zeros of $\dfrac{P}{Q}$: -1, 0

	$\dfrac{P}{Q} = \dfrac{4x + 4}{x}$		
Test Number	-2	$-\dfrac{1}{2}$	1
Value of $\dfrac{P}{Q}$	2	-4	8
Sign of $\dfrac{P}{Q}$	$+$	$-$	$+$
Interval	$(-\infty, -1)$	$(-1, 0)$	$(0, \infty)$

Zeros of $\dfrac{P}{Q}$: 0, 2

	$\dfrac{P}{Q} = \dfrac{4 - 2x}{x}$		
Test Number	-1	1	3
Value of $\dfrac{P}{Q}$	-6	2	$-\dfrac{2}{3}$
Sign of $\dfrac{P}{Q}$	$-$	$+$	$-$
Interval	$(-\infty, 0)$	$(0, 2)$	$(2, \infty)$

$0 < \dfrac{4x + 4}{4}$ and $-3 < \dfrac{x + 4}{x}$ within the intervals $(-\infty, -1)$ and $(0, \infty)$.

$\dfrac{4 - 2x}{x} < 0$ and $\dfrac{x + 4}{x} < 3$ within the intervals $(-\infty, 0)$ and $(2, \infty)$.

Combining the results of parts 1 and 2, the solution set of $\left| \dfrac{x + 4}{x} \right| < 3$ is the intersection of the two subsets of the real number line, all points which lie in *both* sets.

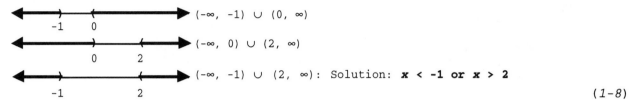

$(-\infty,\ -1) \cup (0,\ \infty)$

$(-\infty,\ 0) \cup (2,\ \infty)$

$(-\infty,\ -1) \cup (2,\ \infty)$: Solution: **x < -1 or x > 2**

$(1-8)$

60. $f(x) = |x + 2| + |x - 2|$

If $x < -2$, then $|x + 2| = -(x + 2)$ and $|x - 2| = -(x - 2)$, hence
$$f(x) = -(x + 2) + -(x - 2)$$
$$= -x - 2 - x + 2$$
$$= -2x$$

If $-2 \le x \le 2$ then $|x + 2| = x + 2$ but $|x - 2| = -(x - 2)$, hence
$$f(x) = x + 2 + -(x - 2)$$
$$= x + 2 - x + 2$$
$$= 4$$

If $x > 2$, then $|x + 2| = x + 2$ and $|x - 2| = x - 2$, hence
$$f(x) = x + 2 + x - 2$$
$$= 2x$$
Domain: $(-\infty,\ \infty)$

x	$f(x) = -2x$	x	$f(x) = 4$	x	$f(x) = 2x$
-5	10	-2	4	3	6
-3	6	0	4	5	10
		2	4		

Piecewise definition for f:
$$f(x) = \begin{cases} -2x & \text{if } x < -2 \\ 4 & \text{if } -2 \le x \le 2 \\ 2x & \text{if } x > 2 \end{cases}$$

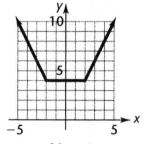

Range: $[4,\ \infty)$ $(2-4)$

61. (A) $g(x) = \sqrt{4 - x^2}$

The domain of g is those values of x for which $4 - x^2 \ge 0$. Solving by the methods of Section 1-8, we obtain:

Thus, the domain of g is $-2 \le x \le 2$, or $[-2,\ 2]$

(B) $\left(\dfrac{f}{g} \right)(x) = \dfrac{f(x)}{g(x)} = \dfrac{x^2}{\sqrt{4 - x^2}}$. The domain of $\dfrac{f}{g}$ is the intersection of the domains of f (all real numbers) and g ($[-2,\ 2]$) with the exclusion of those points (-2 and 2) where $g(x) = 0$. Thus, domain of $f/g = (-2,\ 2)$.

(C) $(f \circ g)(x) = f[g(x)] = f(\sqrt{4 - x^2}) = (\sqrt{4 - x^2})^2 = 4 - x^2$.
The domain of $f \circ g$ is the set of all real numbers in the domain of g ($[-2,\ 2]$) for which $f(x)$ is real, that is, all numbers in $[-2,\ 2]$. $(2-5)$

62. (A) $f(x) = x^2 - 2x - 3 \quad x \geq 1$

f passes the horizontal line test (see graph below, in part C), hence f is one-to-one.

Solve $y = f(x)$ for x:
$$y = x^2 - 2x - 3 \quad x \geq 1$$
$$x^2 - 2x = y + 3 \quad x \geq 1$$
$$x^2 - 2x + 1 = y + 4 \quad x \geq 1$$
$$(x - 1)^2 = y + 4 \quad x \geq 1$$
$$x - 1 = \underbrace{\sqrt{y + 4}} \quad x \geq 1$$

positive square root only because $x \geq 1$

$x = 1 + \sqrt{y + 4} = f^{-1}(y)$

Interchange x and y:

$y = f^{-1}(x) = 1 + \sqrt{x + 4}$ Domain: $x \geq -4$
Check omitted for lack of space.

(B) Domain of $f^{-1} = [-4, \infty)$ (C)
   Range of f^{-1} = Domain of $f = [1, \infty)$.

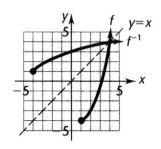

(2-6)

63. In problems 51 and 53, Section 2-4, we noted that $[\![x/2]\!]$ jumps at intervals of 2 units, and $[\![3x]\!] = [\![x \div \frac{1}{3}]\!]$ jumps at intervals of $\frac{1}{3}$ unit. We expect $[\![2x]\!]$ to jump at intervals of $\frac{1}{2}$ unit.

$f(x) = 2x - [\![2x]\!]$

$f(0) = 2 \cdot 0 - [\![2 \cdot 0]\!] = 0 - [\![0]\!] = 0 - 0 = 0$

$f(\frac{1}{4}) = 2 \cdot \frac{1}{4} - [\![2 \cdot \frac{1}{4}]\!] = \frac{1}{2} - [\![\frac{1}{2}]\!] = \frac{1}{2} - 0 = \frac{1}{2}$

$f(\frac{1}{2}) = 2 \cdot \frac{1}{2} - [\![2 \cdot \frac{1}{2}]\!] = 1 - [\![1]\!] = 1 - 1 = 0$

$f(\frac{3}{4}) = 2 \cdot \frac{3}{4} - [\![2 \cdot \frac{3}{4}]\!] = \frac{3}{2} - [\![\frac{3}{2}]\!] = \frac{3}{2} - 1 = \frac{1}{2}$

$f(1) = 2 \cdot 1 - [\![2 \cdot 1]\!] = 2 - [\![2]\!] = 2 - 2 = 0$

Thus, $f(x) = 2x$ if $0 \leq x \leq \frac{1}{2}$, $f(x) = 2x - 1$ if $\frac{1}{2} \leq x < 1$

Generalizing, we can write:

$$f(x) = \begin{cases} \vdots & \quad \vdots \\ 2x + 2 & \text{if} \quad -1 \leq x < -\frac{1}{2} \\ 2x + 1 & \text{if} \quad -\frac{1}{2} \leq x < 0 \\ 2x & \text{if} \quad 0 \leq x < \frac{1}{2} \\ 2x - 1 & \text{if} \quad \frac{1}{2} \leq x < 1 \\ 2x - 2 & \text{if} \quad 1 \leq x < \frac{3}{2} \\ 2x - 3 & \text{if} \quad \frac{3}{2} \leq x < 2 \\ \vdots & \quad \vdots \end{cases}$$

Domain: All real numbers
Range: $[0, 1)$
Discontinuous at $x = k/2$, k an integer

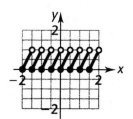

(2-4)

64. Let x = the number

$\dfrac{1}{x}$ = its reciprocal

Then

$$x - \dfrac{1}{x} = \dfrac{3}{2} \quad \text{Excl. value: } x \neq 0$$
$$2x^2 - 2 = 3x$$
$$2x^2 - 3x - 2 = 0$$
$$(2x + 1)(x - 2) = 0$$
$$2x + 1 = 0 \quad \text{or} \quad x - 2 = 0$$
$$x = -\dfrac{1}{2} \qquad\qquad x = 2 \qquad (1\text{-}6)$$

65. Let t = time for newer belt to fill car alone

$\quad$ 14 = time for older belt to fill car alone

$\quad$ 6 = time for both belts to fill car together

Then $\dfrac{1}{t}$ = rate for newer belt

$\dfrac{1}{14}$ = rate for older belt

$$\left(\begin{array}{c}\text{Part of job}\\\text{completed by}\\\text{newer belt}\end{array}\right) + \left(\begin{array}{c}\text{Part of job}\\\text{completed by}\\\text{older belt}\end{array}\right) = 1 \text{ whole job}$$

$$\dfrac{1}{t}(6) \quad + \quad \dfrac{1}{14}(6) \quad = 1$$

$$\dfrac{6}{t} + \dfrac{6}{14} = 1 \quad \text{Excluded value: } t \neq 0$$

$$14t\dfrac{6}{t} + 14t\dfrac{6}{14} = 14t$$

$$84 + 6t = 14t$$

$$84 = 8t$$

$$t = 10.5 \text{ minutes} \qquad (1\text{-}2)$$

66. Let $\quad x$ = the rate of the current

Then $15 - x$ = the rate of the boat upstream

$\quad\quad 15 + x$ = the rate of the boat downstream

Solving $d = rt$ for t, we have $t = \dfrac{d}{r}$. We use this formula, together with

time upstream + time downstream = 4.8.

$$\text{time upstream} = \dfrac{\text{distance upstream}}{\text{rate upstream}} = \dfrac{35}{15 - x}$$

$$\text{time downstream} = \dfrac{\text{distance downstream}}{\text{rate downstream}} = \dfrac{35}{15 + x}$$

So,

$$\dfrac{35}{15 - x} + \dfrac{35}{15 + x} = 4.8 \quad \text{Excluded values: } x \neq 15, -15$$
$$35(15 + x) + 35(15 - x) = 4.8(15 + x)(15 - x)$$
$$525 + 35x + 525 - 35x = 4.8(225 - x^2)$$
$$1050 = 1080 - 4.8x^2$$
$$-30 = -4.8x^2$$
$$6.25 = x^2$$
$$x = 2.5 \text{ miles per hour (discarding the negative answer)}$$
$$(1\text{-}6)$$

67. Let x = amount of distilled water (0% acid)
          24 = amount of 90% solution
   Then x + 24 = amount of 60% solution

$$\begin{array}{ccc} \text{acid in} & \text{acid in} & \text{acid in} \\ \text{distilled} & + & \text{90\% solution} & = & \text{60\% solution} \\ \text{water} \end{array}$$

$$0 \quad + \quad 0.9(24) \quad = 0.6(x + 24)$$
$$21.6 = 0.6x + 14.4$$
$$7.2 = 0.6x$$
$$x = 12 \text{ gallons} \qquad (1\text{-}2)$$

68. "Break even" means Cost = Revenue
   Let x = number of books sold
   Revenue = number of books sold × price per book
   $\qquad = x(9.65)$
   Cost = Fixed Cost + Variable Cost
   $\qquad = 41,800 + \text{number of books} \times \text{cost per book}$
   $\qquad = 41,800 + x(4.90)$
   $9.65x = 41,800 + 4.90x$
   $4.75x = 41,800$
   $x = \dfrac{41,800}{4.75}$
   $x = 8,800 \text{ books}$ $\qquad\qquad (1\text{-}2)$

69. The distance of p from 200 must be no greater than 10.
   $|p - 200| \leq 10$ $\qquad\qquad\qquad (1\text{-}4)$

70. Solve the system of equations
   $\qquad p = 5.5 + 0.002q$
   $\qquad p = 22 - 0.001q$
   Substitute p from the first equation into the second equation to eliminate p.
   $\qquad 5.5 + 0.002q = 22 - 0.001q$
   $\qquad\qquad 0.003q = 16.5$
   $\qquad\qquad\qquad q = 5,500 \text{ cheeseheads} = \text{equilibrium quantity}$
   $\qquad\qquad p = 5.5 + 0.002q = 5.5 + 0.002(5,500)$
   $\qquad\qquad\quad = \$16.50 \text{ equilibrium price}$ $\qquad\qquad (1\text{-}2)$

71. (A) A profit will result if revenue is greater than cost; that is, if
   $\qquad\qquad R > C$
   $\qquad 15p - 2p^2 > 88 - 12p$

   $-88 + 27p - 2p^2 > 0$
   $2p^2 - 27p + 88 < 0$
   $(2p - 11)(p - 8) < 0$
   Zeros: 5.5, 8

   $(-\infty, 5.5)$ $(5.5,8)$ $(8,\infty)$

$2p^2 - 27p + 88 = (2p - 11)(p - 8)$			
Test Number	0	6	9
Value of Polynomial for Test Number	88	-2	7
Sign of Polynomial in Interval	+	–	+
Interval	$(-\infty, 5.5)$	$(5.5, 8)$	$(8, \infty)$

$2p^2 - 27p + 88 < 0$ and a profit will occur ($R > C$) for $\$5.5 < p < \$8$ or ($\$5.5, \$8$).

(B) A loss will result if cost is greater than revenue, that is if

$$C > R$$
$$88 - 12p > 15p - 2p^2$$
$$2p^2 - 27p + 88 > 0$$

Referring to the sign chart in part (A), we see that $2p^2 - 27p + 88 > 0$, and a loss will occur ($C > R$), for $p < \$5.5$ or $p > \$8$. Since a negative price doesn't make sense, we delete any number to the left of 0. Thus, a loss will occur for $\$0 \le p < \$5.5$ or $p > \$8$. [$\$0, \$5.5$) $\cup$ ($\$8, \infty$)   (1-8)

72.   Let x = the distance from port A to port B

Then $115 - x$ = the distance from port B to port C

Applying the Pythagorean theorem, we have

$$x^2 + (115 - x)^2 = 85^2$$
$$x^2 + 13,225 - 230x + x^2 = 7,225$$
$$2x^2 - 230x + 6,000 = 0$$
$$x^2 - 115x + 3,000 = 0$$
$$(x - 40)(x - 75) = 0$$
$$x = 40, 75$$
$$115 - x = 75, 40$$

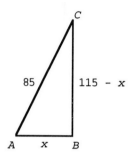

Based on the given information, there are two possible solutions:

40 miles from A to B and 75 miles from B to C or
75 miles from A to B and 40 miles from B to C (1-6)

73. If x is linearly related to p, then we are looking for an equation whose graph passes through $(p_1, x_1) = (3.79, 1,160)$ and $(p_2, x_2) = (3.59, 1,340)$. We find the slope, and then we use the point-slope form to find the equation.

$$m = \frac{x_2 - x_1}{p_2 - p_1} = \frac{1,340 - 1,160}{3.59 - 3.79} = \frac{180}{-0.2} = -900$$
$$x - x_1 = m(p - p_1)$$
$$x - 1,160 = -900(p - 3.79)$$
$$x - 1,160 = -900p + 3,411$$
$$x = -900p + 4,571$$

If the price is lowered to $3.29, we are asked for x when $p = 3.29$
$$x = -900(3.29) + 4,571$$
$$= 1,610 \text{ bottles}$$ (2-2)

74. If $0 \le x \le 60$, $C(x) = 0.06x$

If $60 < x \le 60 + 90$,

that is, $\begin{pmatrix} \text{Cost of first} \\ 60 \text{ calls} \end{pmatrix} + \begin{pmatrix} \text{Cost of next } x - 60 \\ \text{calls at } 0.05 \text{ per call} \end{pmatrix}$

$60 < x \le 150$, $C(x)$ $= 0.06(60)$ $+ 0.05(x - 60)$
                   $= 3.60 + 0.05x - 3$
                   $= 0.05x + 0.6$

If $60 + 90 < x \le 60 + 90 + 150$,

that is, $\begin{pmatrix} \text{Cost of first} \\ 150 \text{ calls} \end{pmatrix} + \begin{pmatrix} \text{Cost of next } x - 150 \\ \text{calls at } 0.04 \text{ per call} \end{pmatrix}$

$150 < x \le 300$, $C(x)$ $= 0.05(150) + 0.6 + 0.04(x - 150)$
                   $= 7.5 + 0.6 + 0.04x - 6$
                   $= 0.04x + 2.1$

Finally, if $x > 300$, $\left(\begin{array}{c}\text{Cost of first} \\ 300 \text{ calls}\end{array}\right) + \left(\begin{array}{c}\text{Cost of next } x - 300 \\ \text{calls at } 0.03 \text{ per call}\end{array}\right)$

$$C(x) = 0.04(300) + 2.1 + 0.03(x - 300)$$
$$= 12 + 2.1 + 0.03x - 9$$
$$= 0.03x + 5.1$$

Summarizing, $C(x) = \begin{cases} 0.06x & \text{if} & 0 \le x \le 60 \\ 0.05x + 0.6 & \text{if} & 60 < x \le 150 \\ 0.04x + 2.1 & \text{if} & 150 < x \le 300 \\ 0.03x + 5.1 & \text{if} & 300 < x \end{cases}$

x	$C(x) = 0.06x$	x	$C(x) = 0.05x + 0.6$	x	$C(x) = 0.04x + 2.1$	x	$C(x) = 0.03x + 5.1$
0	0	90	5.1	160	8.5	310	14.4
30	1.8	140	7.6	200	10.1	400	17.1
60	3.6						

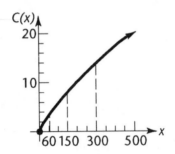

(2-4)

75. (A) Let x = width of pen
Then $2x + \ell = 80$
So $\ell = 80 - 2x$ = length of pen
$A(x) = x\ell = x(80 - 2x) = 80x - 2x^2$

(B) Since all distances must be positive, $x > 0$ and $80 - 2x > 0$, hence $80 > 2x$ or $x < 40$. Therefore $0 < x < 40$ or $(0, 40)$ is the domain of $A(x)$.

(C) Note: 0, 40 were excluded from the domain for geometrical reasons; but can be used to help draw the graph since the *polynomial* $80x - 2x^2$ has domain including these values.

x	$A(x)$
0	0
10	600
20	800
30	600
40	0

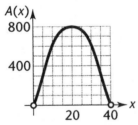

From the graph, the maximum value of A seems to occur when $x = 20$. Checking, we see that $80x - 2x^2 = -2x^2 + 80x = -2(x^2 - 40x + 400) + 800 = -2(x - 20)^2 + 800$. Thus the maximum value of A indeed occurs when $x = 20$.
Then $80 - 2x = 40$, and the dimensions are 20 feet by 40 feet.   (2-4)

76. (A) $f(1) = 1 - 2[\![1/2]\!] = 1 - 2(0) = 1$
$f(2) = 2 - 2[\![2/2]\!] = 2 - 2(1) = 0$
$f(3) = 3 - 2[\![3/2]\!] = 3 - 2(1) = 1$
$f(4) = 4 - 2[\![4/2]\!] = 4 - 2(2) = 0$

(B) If n is an integer, n is either odd or even.
If n is even, it can be written as $2k$, where k is an integer.
Then $f(n) = f(2k) = 2k - 2[\![2k/2]\!] = 2k - 2[\![k]\!] = 2k - 2k = 0$.
Otherwise, n is odd, and n can be written as $2k + 1$, where k is an integer.

Then $f(n) = f(2k + 1) = 2k + 1 - 2[\![(2k+1)/2]\!] = 2k + 1 - 2[\![k + \frac{1}{2}]\!]$
$$= 2k + 1 - 2k$$
$$= 1$$

Summarizing, if n is an integer, $f(n) = \begin{cases} 1 & \text{if } n \text{ is an odd integer} \\ 0 & \text{if } n \text{ is an even integer} \end{cases}$ *(2-4)*

77. $s = a + bt^2$

(A) We are given: When $t = 5$, $s = 2{,}100$
                 When $t = 10$, $s = 900$

Substituting these values into the given equation, we have

$2{,}100 = a + b(5)^2$
$\ \ \ 900 = a + b(10)^2$ or
$2{,}100 = a + 25b$
$\ \ \ 900 = a + 100b$

Solve the first equation for a in terms of b and substitute into the second
equation to eliminate a.

$\ \ \ \ \ a = 2{,}100 - 25b$
$\ \ \ 900 = 2{,}100 - 25b + 100b$
$-1{,}200 = 75b$
$\ \ \ \ \ \ b = -16$
$\ \ \ \ \ \ a = 2{,}100 - 25b = 2{,}100 - 25(-16) = 2{,}500$

(B) The height of the balloon is represented by s, the distance of the object
above the ground, when $t = 0$. Since we now know

$\ \ \ \ \ s = 2{,}500 - 16t^2$

from part(A), when $t = 0$, $s = 2{,}500$ feet is the height of the balloon.

(C) The object falls until s, its distance above the ground, is zero. Since

$\ \ \ \ \ s = 2{,}500 - 16t^2$

we substitute $s = 0$ and solve for t.

$\ \ \ \ \ \ \ 0 = 2{,}500 - 16t^2$
$\ \ 16t^2 = 2{,}500$
$\ \ \ \ \ t^2 = \dfrac{2{,}500}{16}$
$\ \ \ \ \ \ t = \dfrac{50}{4}$ (discarding the negative solution)
$\ \ \ \ \ \ t = 12.5$ seconds   *(1-1, 1-2, 1-6)*

CHAPTER 3

Exercise 3-1

Key Ideas and Formulas

For the nth degree polynomial function P given by

$$P(x) = a_n x^n + a_{n-1}x^{n-1} + \cdots + a_1 x + a_0 \qquad a_n \neq 0$$

The number r is said to be a zero of the function P, or a zero of the polynomial $P(x)$, or a solution or root of the equation $P(x) = 0$, if $P(r) = 0$.

If the coefficients of a polynomial $P(x)$ are real, then the x intercepts of the graph of $y = P(x)$ are real zeros of P and $P(x)$ and real solutions or roots for the equation $P(x) = 0$.

Synthetic Division: To divide $P(x)$ by $x - r$ arrange the coefficients of $P(x)$ in order of descending powers of x (write 0 as the coefficient for each missing power).

Use the bring down, multiply, add scheme below:

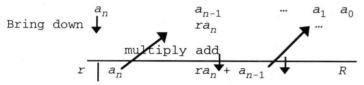

The last number to the right in the third row of numbers is the remainder R; the other numbers in the third row are the coefficients of the quotient $Q(x)$, which is of degree one less than $P(x)$.

$$\frac{P(x)}{x - r} = Q(x) + \frac{R}{x - r}$$

Division Algorithm:

For each polynomial $P(x)$ of degree greater than 0 and each number r, there exists a unique polynomial $Q(x)$ of degree 1 less than $P(x)$ and a unique number R such that

$$P(x) = (x - r)Q(x) + R$$

$Q(x)$ is called the quotient, $x - r$ the divisor, and R the remainder. Note: R may be zero.

Remainder Theorem:

If R is the remainder after dividing $P(x)$ by $x - r$, then $P(r) = R$.

Left and Right Behavior of a Polynomial:

$$P(x) = a_n x^n + a_{n-1}x^{n-1} + \cdots + a_1 x + a_0, \qquad a_n \neq 0$$

1. $a_n > 0$ and n even

Graph of $P(x)$ increases without bound as x decreases to the left and as x increases to the right.

$$P(x) \to \infty \text{ as } x \to -\infty$$
$$P(x) \to \infty \text{ as } x \to \infty$$

2. $a_n > 0$ and n odd

Graph of $P(x)$ decreases without bound as x decreases to the left and increases without bound as x increases to the right.

$$P(x) \to -\infty \text{ as } x \to -\infty$$
$$P(x) \to \infty \text{ as } x \to \infty$$

3. $a_n < 0$ and n even

Graph of $P(x)$ decreases without bound as x decreases to the left and as x increases to the right.

$$P(x) \to -\infty \text{ as } x \to -\infty$$
$$P(x) \to -\infty \text{ as } x \to \infty$$

4. $a_n < 0$ and n odd

Graph of $P(x)$ increases without bound as x decreases to the left and decreases without bound as x increases to the right.

$$P(x) \to \infty \text{ as } x \to -\infty$$
$$P(x) \to -\infty \text{ as } x \to \infty$$

A **turning point** on a continuous graph is a point that separates an increasing portion from a decreasing portion.

Graph Properties of Polynomial Functions:

Let P be an nth degree polynomial function with real coefficients.

1. P is continuous for all real numbers.
2. The graph of P is a smooth curve.
3. The graph of P has at most n x intercepts.
4. P has at most $n - 1$ turning points.

---

1. Since $P(x) \to \infty$ as $x \to \infty$ and $P(x) \to -\infty$ as $x \to -\infty$, this matches graph c.

3. Since $P(x) \to \infty$ as $x \to \infty$ and $P(x) \to \infty$ as $x \to -\infty$, this matches graph d.

5. Since the graph of a second-degree polynomial has at most 1 turning point, only graph h could be an answer. h.

7. Since the graph of a fourth-degree polynomial can have 3 turning points, graph k could be an answer. But graph k is not the only graph which decreases without bound as x decreases to the left and as x increases to the right; graph h could also be an answer. h, k.

9.
$$
\begin{array}{r}
a - 2 \\
a + 2 \overline{\smash{\big)}\, a^2 + 0a - 4} \\
\underline{a^2 + 2a} \\
-2a - 4 \\
\underline{-2a - 4} \\
0
\end{array}
$$
$a - 2$, $R = 0$

11.
$$
\begin{array}{r}
b - 3 \\
b - 3 \overline{\smash{\big)}\, b^2 - 6b - 9} \\
\underline{b^2 - 3b} \\
-3b - 9 \\
\underline{-3b + 9} \\
-18
\end{array}
$$
$b - 3$, $R = -18$

13.
$$
\begin{array}{r}
x^2 + 3 \\
x - 1 \overline{\smash{\big)}\, x^3 - x^2 + 3x + 2} \\
\underline{x^3 - x^2} \\
3x + 2 \\
\underline{3x - 3} \\
5
\end{array}
$$
$x^2 + 3$, $R = 5$

15.
$$
\begin{array}{r}
4y^2 - 4y + 1 \\
2y + 1 \overline{\smash{\big)}\, 8y^3 - 4y^2 - 2y + 1} \\
\underline{8y^3 + 4y^2} \\
-8y^2 - 2y \\
\underline{-8y^2 - 4y} \\
2y + 1 \\
\underline{2y + 1} \\
0
\end{array}
$$
$4y^2 - 4y + 1$; $R = 0$

17.
$$
\begin{array}{r}
1 \quad 4 \quad -15 \\
3 \quad 21 \\
\hline
3\, \rvert\, 1 \quad 7 \quad 6
\end{array}
$$
$$\frac{x^2 + 4x - 15}{x - 3} = x + 7 + \frac{6}{x - 3}$$

19.
$$
\begin{array}{r}
3 \quad -1 \quad -7 \\
-6 \quad 14 \\
\hline
-2\, \rvert\, 3 \quad -7 \quad 7
\end{array}
$$
$$\frac{3x^2 - x - 7}{x + 2} = 3x - 7 + \frac{7}{x + 2}$$

21.
$$
\begin{array}{r}
2 \quad 3 \quad -8 \quad 1 \\
-6 \quad 9 \quad -3 \\
\hline
-3\, \rvert\, 2 \quad -3 \quad 1 \quad -2
\end{array}
$$
$$\frac{2x^3 + 3x^2 - 8x + 1}{x + 3} = 2x^2 - 3x + 1 - \frac{2}{x + 3}$$

23.
$$
\begin{array}{r}
2 \quad 8 \quad -6 \\
-10 \quad 10 \\
\hline
-5\, \rvert\, 2 \quad -2 \quad 4
\end{array}
$$
$P(-5) = 4$

25.
$$
\begin{array}{r}
4 \quad 12 \quad -8 \quad 25 \\
-16 \quad 16 \quad -32 \\
\hline
-4\, \rvert\, 4 \quad -4 \quad 8 \quad -7
\end{array}
$$
$P(-4) = -7$

27.
$$
\begin{array}{r}
2 \quad -5 \quad 2 \quad -11 \quad -14 \\
6 \quad 3 \quad 15 \quad 12 \\
\hline
3\, \rvert\, 2 \quad 1 \quad 5 \quad 4 \quad -2
\end{array}
$$
$P(3) = -2$

29.
$$
\begin{array}{r}
2 \quad 0 \quad 0 \quad -5 \quad 0 \quad 3 \\
2 \quad 2 \quad 2 \quad -3 \quad -3 \\
\hline
1\, \rvert\, 2 \quad 2 \quad 2 \quad -3 \quad -3 \quad 0
\end{array}
$$
$2x^4 + 2x^3 + 2x^2 - 3x - 3$, $R = 0$

> **Common Error:**
> The first row is *not*
> 2   -5   3
> 0's must be inserted for the missing powers.

31.
$$
\begin{array}{r}
1 \quad 0 \quad 0 \quad 0 \quad -16 \\
-4 \quad 16 \quad -64 \quad 256 \\
\hline
-4\, \rvert\, 1 \quad -4 \quad 16 \quad -64 \quad 240
\end{array}
$$
$x^3 - 4x^2 + 16x - 64$, $R = 240$

33.
$$
\begin{array}{r}
4 \quad -9 \quad -8 \quad -2 \quad -7 \\
12 \quad 9 \quad 3 \quad 3 \\
\hline
3\, \rvert\, 4 \quad 3 \quad 1 \quad 1 \quad -4
\end{array}
$$
$4x^3 + 3x^2 + x + 1$, $R = -4$

35.

```
    1   7   10    0   -1   -5    0
       -5  -10    0    0    5    0
  -5│ 1   2    0    0   -1    0    0
```

$x^5 + 2x^4 - x, \ R = 0$

37.

```
         2    9    5   -4    3
             -3   -9    6   -3
  - 3/2 │ 2    6   -4    2    0
```

$2x^3 + 6x^2 - 4x + 2, \ R = 0$

39.

```
        3    5   -5   10   -1
             1    2   -1    3
  1/3 │ 3    6   -3    9    2
```

$3x^3 + 6x^2 - 3x + 9, \ R = 2$

41.

```
         5   -4    0      2      -5
             1   -0.6  -0.12   0.376
  0.2│ 5   -3  -0.6   1.88   -4.624
```

$5x^3 - 3x^2 - 0.6x + 1.88, \ R = -4.624$

43.

```
        5    2    4      6       0        -6
            -3   0.6  -2.76   -1.944    1.1664
  -0.6│ 5   -1   4.6   3.24   -1.944   -4.8336
```

$5x^4 - x^3 + 4.6x^2 + 3.24x - 1.944, \ R = -4.8336$

45. We form a synthetic division table:

	1	-5	2	8	
-2	1	-7	16	-24	= P(-2)
-1	1	-6	8	0	= P(-1)
0	1	-5	2	8	= P(0)
1	1	-4	-2	6	= P(1)
2	1	-3	-4	0	= P(2)
3	1	-2	-4	-4	= P(3)
4	1	-1	-2	0	= P(4)
5	1	0	2	18	= P(5)

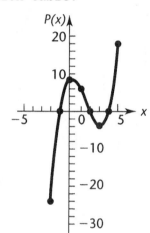

The graph has three x intercepts and two turning points.

$P(x) \to \infty$ as $x \to \infty$ and $P(x) \to -\infty$ as $x \to -\infty$.

47. We form a synthetic division table:

	1	4	-1	-4	
-5	1	-1	4	-24	= P(-5)
-4	1	0	-1	0	= P(-4)
-3	1	1	-4	8	= P(-3)
-2	1	2	-5	6	= P(-2)
-1	1	3	-4	0	= P(-1)
0	1	4	-1	-4	= P(0)
1	1	5	4	0	= P(1)
2	1	6	11	18	= P(2)

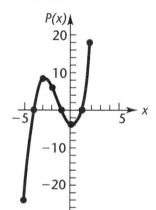

The graph has three x intercepts and two turning points.

$P(x) \to \infty$ as $x \to \infty$ and $P(x) \to -\infty$ as $x \to -\infty$.

49. We form a synthetic division table:

	−1	2	0	−3	
−2	−1	4	−8	13	= P(−2)
−1	−1	3	−3	0	= P(−1)
0	−1	2	0	−3	= P(0)
1	−1	1	−1	−2	= P(1)
2	−1	0	0	−3	= P(2)
3	−1	−1	−3	−12	= P(3)

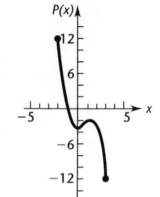

The graph has one x intercept and two turning points.

$P(x) \to -\infty$ as $x \to \infty$ and $P(x) \to \infty$ as $x \to -\infty$.

51. We form a synthetic division table:

	−1	3	−3	2	
−1	−1	4	−7	9	= P(−1)
−0.5	−1	3.5	−4.75	4.375	= P(−0.5)
0	−1	3	0	2	= P(0)
0.5	−1	2.5	−1.75	1.125	= P(0.5)
1	−1	2	−1	1	= P(1)
1.5	−1	1.5	−0.75	0.875	= P(1.5)
2	−1	1	−1	0	= P(2)
2.5	−1	0.5	−1.75	−2.375	= P(2.5)
3	−1	0	−3	−7	= P(3)

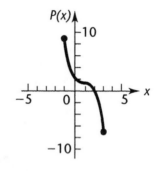

53. $P(x) = x^3$ is an example of a third-degree polynomial with one x intercept $(x = 0)$.

55. No such polynomial exists; the graph of a third-degree polynomial must cross the x axis at least once.

57.
$$
\begin{array}{r}
x^2 + x + 2 \\
x^2 - 1 \overline{\smash{)}\, x^4 + x^3 + x^2 - x - 2} \\
\underline{x^4 \qquad\quad - x^2} \\
x^3 + 2x^2 - x \\
\underline{x^3 \qquad\quad - x} \\
2x^2 \qquad - 2 \\
\underline{2x^2 \qquad - 2} \\
0
\end{array}
$$
$x^2 + x + 2$, $R = 0$

59.
$$
\begin{array}{r}
x^2 - 2x + 1 \\
x^2 + 3x - 2 \overline{\smash{)}\, x^4 + x^3 - 7x^2 + 8x + 1} \\
\underline{x^4 + 3x^3 - 2x^2} \\
-2x^3 - 5x^2 + 8x \\
\underline{-2x^3 - 6x^2 + 4x} \\
x^2 + 4x + 1 \\
\underline{x^2 + 3x - 2} \\
x + 3
\end{array}
$$
$x^2 - 2x + 1$, $R = x + 3$

61.
$$
\begin{array}{r|rrrrr}
 & 1 & 2 & -2 & 2 & -3 \\
 & & i & -1 + 2i & -2 - 3i & 3 \\
\hline
i & 1 & 2 + i & -3 + 2i & -3i & 0
\end{array}
$$
$x^3 + (2 + i)x^2 + (-3 + 2i)x - 3i$, $R = 0$

63. We use synthetic division.

(A)
$$
\begin{array}{r|rrr}
 & 1 & 2i & -10 \\
 & & 2 - i & 5 \\
\hline
2 - i & 1 & 2 + i & -5
\end{array}
$$
$P(2 - i) = -5$

(B)
$$
\begin{array}{r|rrr}
 & 1 & 2i & -10 \\
 & & 5 - 5i & 10 - 40i \\
\hline
5 - 5i & 1 & 5 - 3i & -40i
\end{array}
$$
$P(5 - 5i) = -40i$

(C)   1   2i -10
                3 - i 10
   ————————————————————
3 - i|1   3 + i 0

$P(3 - i) = 0$

(D)   1   2i -10
                -3 - i 10
   ————————————————————
-3 - i|1   -3 + i 0

$P(-3 - i) = 0$

65. We form a synthetic division table:

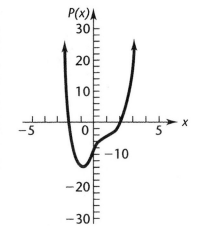

	1	-2	-2	8	-8	
-3	1	-5	13	-31	85	= $P(-3)$
-2	1	-4	6	-4	0	= $P(-2)$
-1	1	-3	1	7	-15	= $P(-1)$
0	1	-2	-2	8	-8	= $P(0)$
1	1	-1	-3	5	-3	= $P(1)$
2	1	0	-2	4	0	= $P(2)$
3	1	1	1	11	25	= $P(3)$

The graph has two x intercepts and one turning point.

$P(x) \to \infty$ as $x \to \infty$ and as $x \to -\infty$.

67. (A) We form a synthetic division table:

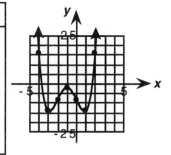

	1	4	-1	-10	-8	
-5	1	-1	4	-30	142	= $P(-5)$
-4	1	0	-1	-6	16	= $P(-4)$
-3	1	1	-4	2	-14	= $P(-3)$
-2	1	2	-5	0	-8	= $P(-2)$
-1	1	3	-4	-6	-2	= $P(-1)$
0	1	4	-1	-10	-8	= $P(0)$
1	1	5	4	-6	-14	= $P(1)$
2	1	6	11	12	16	= $P(2)$

The graph has two x intercepts and three turning points.
$P(x) \to \infty$ as $x \to \infty$ and as $x \to -\infty$.

69. (A) We form a synthetic division table:

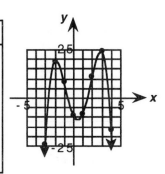

	-1	2	10	-10	-9	
-3	-1	5	-5	5	-24	= $P(-3)$
-2	-1	4	2	-14	19	= $P(-2)$
-1	-1	3	7	-17	8	= $P(-1)$
0	-1	2	10	-10	-9	= $P(0)$
1	-1	1	11	1	-8	= $P(1)$
2	-1	0	10	10	11	= $P(2)$
3	-1	-1	7	11	24	= $P(3)$
4	-1	-2	2	-2	-17	= $P(4)$

The graph has four x intercepts and three turning points.
$P(x) \to -\infty$ as $x \to \infty$ and as $x \to -\infty$.

71. (A) We form a synthetic division table:

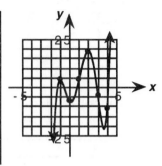

	1	-6	4	17	-5	-7	
-2	1	-8	20	-23	41	-89	= $P(-2)$
-1	1	-7	11	6	-11	4	= $P(-1)$
0	1	-6	4	17	-5	-7	= $P(0)$
1	1	-5	-1	16	11	4	= $P(1)$
2	1	-4	-4	9	13	19	= $P(2)$
3	1	-3	-5	2	1	-4	= $P(3)$
4	1	-2	-4	1	-1	-11	= $P(4)$
5	1	-1	-1	12	55	268	= $P(5)$

The graph has five x intercepts and four turning points.
$P(x) \to \infty$ as $x \to \infty$ and $P(x) \to -\infty$ as $x \to -\infty$.

73. (A)

$$
\begin{array}{r}
a_2(x) + (a_1 + a_2r) \\
x - r\overline{\smash{\big)}\ a_2x^2 + a_1x \qquad\qquad\qquad + a_0} \\
\underline{a_2x^2 - a_2rx} \\
(a_1 + a_2r)x \qquad\qquad + a_0 \\
\underline{(a_1 + a_2r)x - r(a_1 + a_2r)} \\
a_0 + r(a_1 + a_2r)
\end{array}
$$

$$
\begin{array}{r|cccc}
 & a_2 & a_1 & & a_0 \\
 & & a_2r & & (a_1 + a_2r)r \\
\hline
r & a_2 & a_1 + a_2r & & a_0 + (a_1 + a_2r)r
\end{array}
$$

In both cases the coefficient of x is a_2, the constant term $a_2r + a_1$, and the remainder is $(a_2r + a_1)r + a_0$.

(B) The remainder expanded is $a_2r^2 + a_1r + a_0 = P(r)$.

75.
$$
\begin{aligned}
P(x) &= \{[(2x - 3)x + 2]x - 5\}x + 7 \\
P(-2) &= \{[\{2(-2) - 3\}(-2) + 2](-2) - 5\}(-2) + 7 \\
&= \{[\{-7\}(-2) + 2](-2) - 5\}(-2) + 7 \\
&= \{[16](-2) - 5\}(-2) + 7 \\
&= \{-37\}(-2) + 7 \\
&= 81 \\
P(1.7) &= \{[\{2(1.7) - 3\}(1.7) + 2](1.7) - 5\}(1.7) + 7 \\
&= \{[\{0.4\}(1.7) + 2](1.7) - 5\}(1.7) + 7 \\
&= \{[2.68](1.7) - 5\}(1.7) + 7 \\
&= \{-0.444\}(1.7) + 7 \\
&= 6.2452 \text{ or } 6.2 \text{ to two significant digits.}
\end{aligned}
$$

Exercise 3-2

Key Ideas and Formulas

Factor Theorem:
If r is a zero of the polynomial $P(x)$, then $x - r$ is a factor of $P(x)$; conversely, if $x - r$ is a factor of $P(x)$, then r is a zero of $P(x)$.

Fundamental Theorem of Algebra:
Every polynomial $P(x)$ of degree $n > 0$ has at least one zero.

n Zeros Theorem:
Every polynomial $P(x)$ of degree $n > 0$ can be expressed as the product of n linear factors. Hence, $P(x)$ has exactly n zeros — not necessarily distinct.

If $P(x)$ is represented as the product of linear factors and $x - r$ occurs m times, then r is called a **zero of multiplicity *m*.**

Imaginary Zeros Theorem:

Imaginary zeros of polynomials with real coefficients, if they exist, occur in conjugate pairs.

A polynomial of odd degree with real coefficients always has at least one real zero.

Rational Zero Theorem:

If the rational number b/c, in lowest terms, is a zero of the polynomial

$$P(x) = a_n x^n + a_{n-1} x^{n-1} + \cdots + a_1 x + a_0 \qquad a_n \neq 0$$

with integer coefficients, then b must be an integer factor of a_0 and c must be an integer factor of a_n.

Strategy for Finding Rational Zeros:

Assume that $P(x)$ is a polynomial with integer coefficients and is of degree greater than 2.

Step 1. List the possible rational zeros of $P(x)$ using the rational zero theorem.

Step 2. Construct a synthetic division table. If a rational zero r is found, stop, write

$$P(x) = (x - r)Q(x)$$

and immediately proceed to find the rational zeros for $Q(x)$, the reduced polynomial relative to $P(x)$. If the degree of $Q(x)$ is greater than 2, return to step 1 using $Q(x)$ in place of $P(x)$. If $Q(x)$ is quadratic, find all its zeros using standard methods for solving quadratic equations.

1. -2 (multiplicity 5), 3 (multiplicity 4); degree of $P(x)$ is 9

3. -1, 1 (multiplicity 4), 7 (multiplicity 3); degree of $P(x)$ is 8

5. $P(x) = (x - 2)^3(x + 1)$; degree 4

7. $P(x) = x(x - 2)[x - (-1 + \sqrt{3})][x - (-1 - \sqrt{3})]$
$= x(x - 2)(x + 1 - \sqrt{3})(x + 1 + \sqrt{3})$; degree 4

9. $P(x) = (x + 2)^2(x - 2)^2[x - (3 - i)][x - (3 + i)]$
$= (x + 2)^2(x - 2)^2(x - 3 + i)(x - 3 - i)$; degree 6

11. Since -2, 1, and 3 are zeros, $P(x) = (x + 2)(x - 1)(x - 3)$ is the lowest degree polynomial that has this graph. The degree of $P(x)$ is 3.

13. Since -2 and 1 are zeros, each with multiplicity 2, $P(x) = (x + 2)^2(x - 1)^2$ is the lowest degree polynomial that has this graph. The degree of $P(x)$ is 4.

15. Since -3, -2, 0, 1, and 2 are zeros, $P(x) = (x + 3)(x + 2)x(x - 1)(x - 2)$ is the lowest degree polynomial that has this graph. The degree of $P(x)$ is 5.

17. $x - 1$ will be a factor of $P(x)$ if $P(1) = 0$. Since $P(x) = x^{23} - 1$, $P(1) = 1^{23} - 1 = 0$. Therefore $x - 1$ is a factor of $x^{23} - 1$.

19. $x + 1$ will be a factor of $P(x)$ if $P(-1) = 0$. Since $P(x) = 3x^3 - 5x^2 - 4x + 6$, $P(-1) = 3(-1)^3 - 5(-1)^2 - 4(-1) + 6 = -3 - 5 + 4 + 6 = 2$. Therefore $x + 1$ is not a factor of $3x^3 - 5x^2 - 4x + 6$.

21. Possible factors of -10 are ± 1, ± 2, ± 5, ± 10. Possible factors of 1 are ± 1. Therefore the possible rational zeros are ± 1, ± 2, ± 5, ± 10.

23. Possible factors of 5 are ± 1, ± 5. Possible factors of 2 are ± 2. Therefore the possible rational zeros are ± 1, ± 5, $\pm \dfrac{1}{2}$, $\pm \dfrac{5}{2}$.

25. Possible factors of -25 are ±1, ±5, ±25. Possible factors of 6 are ±1, ±2, ±3, ±6. Therefore the possible rational zeros are ±1, ±5, ±25, $\pm\frac{1}{2}$, $\pm\frac{5}{2}$, $\pm\frac{25}{2}$, $\pm\frac{1}{3}$, $\pm\frac{5}{3}$, $\pm\frac{25}{3}$, $\pm\frac{1}{6}$, $\pm\frac{5}{6}$, $\pm\frac{25}{6}$.

27. $P(x) = x^3 + 9x^2 + 24x + 16$. Since -1 is a zero of $P(x)$, we can write

```
        1     9     24    16
             -1    -8    -16
 -1 | 1      8     16     0
```

$P(x) = (x - 1)(x^2 + 8x + 16) = (x - 1)Q(x)$
Since $Q(x)$ is a quadratic with obvious factors we can write
$P(x) = (x - 1)(x + 4)(x + 4)$.

29. $P(x) = x^4 - 1$. Since 1 is a zero of $P(x)$, we can write

```
       1    0    0    0    -1
            1    1    1     1
  1 |  1    1    1    1     0
```

$P(x) = (x - 1)(x^3 + x^2 + x + 1)$. Since -1 is also a zero of $P(x)$, we can write

```
       1    1    1    1
           -1    0   -1
 -1 |  1    0    1    0
```

$P(x) = (x - 1)(x + 1)(x^2 + 1)$. Since $x^2 + 1 = x^2 - (-1) = x^2 - i^2$, we can write
$P(x) = (x - 1)(x + 1)(x - i)(x + i)$

31. $P(x) = 2x^3 - 17x^2 + 90x - 41$. Since $\frac{1}{2}$ is a zero of $P(x)$, we can write

```
              2    -17    90    -41
                     1    -8     41
 1/2 |        2    -16    82      0
```

$P(x) = \left(x - \frac{1}{2}\right)(2x^2 - 16x + 82) = \left(x - \frac{1}{2}\right)2(x^2 - 8x + 41) = (2x - 1)(x^2 - 8x + 41)$
$= (2x - 1)Q(x)$

Since $Q(x)$ is a quadratic, we can find its zeros by the quadratic formula
$x^2 - 8x + 41 = 0$ $a = 1$
 $b = -8$
 $c = 41$

$x = \dfrac{-b \pm \sqrt{b^2 - 4ac}}{2a}$

$= \dfrac{-(-8) \pm \sqrt{(-8)^2 - 4(1)(41)}}{2(1)}$

$= \dfrac{8 \pm \sqrt{-100}}{2}$

$= \dfrac{8 \pm 10i}{2}$

$= 4 \pm 5i$

Since the zeros of $Q(x)$ are $4 + 5i$ and $4 - 5i$, by the factor theorem
$Q(x) = [x - (4 + 5i)][x - (4 - 5i)]$
and $P(x) = (2x - 1)[x - (4 + 5i)][x - (4 - 5i)]$

33. Let $P(x) = 2x^3 - 7x^2 + 2x + 6$. The possible rational zeros are ± 1, ± 2, ± 3, ± 6, $\pm\frac{1}{2}$, $\pm\frac{3}{2}$.

We form a synthetic division table:

	2	-7	2	6	
1	2	-5	-3	3	
2	2	-3	-4	-2	
$\frac{3}{2}$	2	-4	-4	0	$\frac{3}{2}$ is a zero

So $P(x) = \left(x - \frac{3}{2}\right)(2x^2 - 4x - 4)$

To find the remaining zeros, we solve $2x^2 - 4x - 4 = 0$, or $x^2 - 2x - 2 = 0$, by completing the square.

$$x^2 - 2x = 2$$
$$x^2 - 2x + 1 = 3$$
$$(x - 1)^2 = 3$$
$$x - 1 = \pm\sqrt{3}$$
$$x = 1 \pm \sqrt{3}$$

The zeros are $\frac{3}{2}$, $1 \pm \sqrt{3}$. These are the roots of the equation.

35. Let $P(x) = x^4 + 2x^3 - 2x^2 - 6x - 3$. The possible rational zeros are ± 1, ± 3.
We form a synthetic division table:

	1	2	-2	-6	-3	
1	1	3	1	-5	-8	
3	1	5	13	33	96	
-1	1	1	-3	-3	0	-1 is a zero

We examine $Q(x) = x^3 + x^2 - 3x - 3$. This clearly factors by grouping into $(x^2 - 3)(x + 1)$, but if we don't notice this we can proceed as before. Then the possible remaining rational zeros are -1, -3.

We form a synthetic division table for $Q(x)$:

	1	1	-3	-3	
-1	1	0	-3	0	-1 is a zero

Thus -1 is a double zero of $P(x)$.

To find the remaining zeros, we solve $x^2 - 3 = 0$, or $x^2 = 3$, to obtain $x = \pm\sqrt{3}$.
The zeros are -1 (double) and $\pm\sqrt{3}$. These are the roots of the equation.

37. Let $P(x) = x^4 + 2x^3 - 10x^2 - 18x + 9$. The possible rational zeros are ± 1, ± 3, ± 9. We form a synthetic division table.

	1	2	-10	-18	9	
1	1	3	-7	-25	-16	
3	1	5	5	-3	0	3 is a zero

We examine $Q(x) = x^3 + 5x^2 + 5x - 3$. The possible remaining rational zeros are 3, -1, -3. We form a synthetic division table for $Q(x)$:

	1	5	5	-3	
3	1	8	29	84	
-1	1	4	1	-4	
-3	1	2	-1	0	-3 is a zero

So $P(x) = (x - 3)(x + 3)(x^2 + 2x - 1)$.

To find the remaining zeros, we solve $x^2 + 2x - 1 = 0$ by completing the square.

$$x^2 + 2x = 1$$
$$x^2 + 2x + 1 = 2$$
$$(x + 1)^2 = 2$$
$$x + 1 = \pm\sqrt{2}$$
$$x = -1 \pm \sqrt{2}$$

The zeros are 3, -3, $-1 \pm \sqrt{2}$. These are the roots of the equation.

39. Let $P(x) = 2x^5 - x^4 - 5x^3 + 10x^2 - 2x - 4$. The possible rational zeros are ± 1, ± 2, ± 4, $\pm\frac{1}{2}$. We form a synthetic division table.

	2	-1	-5	10	-2	-4	
1	2	1	-4	6	4	0	1 is a zero

We examine $Q(x) = 2x^4 + x^3 - 4x^2 + 6x + 4$. We form a synthetic division table:

	2	1	-4	6	4	
1	2	3	-1	5	9	
2	2	5	6	18	40	
4	2	9	32	134	540	
-1	2	-1	-3	9	-5	
-2	2	-3	2	2	0	-2 is a zero

We examine $R(x) = 2x^3 - 3x^2 + 2x + 2$. The possible remaining rational zeros are $\pm\frac{1}{2}$ and -2. We form a synthetic division table:

	2	-3	2	2	
-2	2	-7	16	-30	
$\frac{1}{2}$	2	-2	1	$\frac{5}{2}$	
$-\frac{1}{2}$	2	-4	4	0	$-\frac{1}{2}$ is a zero

So $P(x) = (x - 1)(x + 2)\left(x + \frac{1}{2}\right)(2x^2 - 4x + 4)$.

To find the remaining zeros, we solve $2x^2 - 4x + 4 = 0$, or $x^2 - 2x + 2 = 0$ by completing the square.

$$x^2 - 2x = -2$$
$$x^2 - 2x + 1 = -1$$
$$(x - 1)^2 = -1$$
$$x - 1 = \pm i$$
$$x = 1 \pm i$$

The zeros are 1, -2, $-\frac{1}{2}$, $1 \pm i$. These are the roots of the equation.

41. $P(x) = x^3 + 5x^2 - 2x - 24$. The possible rational zeros are ± 1, ± 2, ± 3, ± 4, ± 6, ± 8, ± 12, ± 24. We form a synthetic division table:

	1	5	-2	-24	
1	1	6	4	-20	
2	1	7	12	0	2 is a zero

So $P(x) = (x - 2)(x^2 + 7x + 12)$
$\qquad\quad = (x - 2)(x + 3)(x + 4)$

So the zeros of $P(x)$ are 2, -3, -4.

43. $P(x) = x^4 - 3.3x^3 + 2.3x^2 + 0.6x = \frac{1}{10}(10x^4 - 33x^3 + 23x^2 + 6x)$

$= \frac{1}{10}x(10x^3 - 33x^2 + 23x + 6)$. Clearly, 0 is a zero. We examine

$Q(x) = 10x^3 - 33x^2 + 23x + 6$. Possible factors of 6 are ±1, ±2, ±3, ±6. Possible factors of 10 are ±1, ±2, ±5, ±10. Hence the possible rational zeros of $P(x)$ are ±1, ±2, ±3, ±6, ±$\frac{1}{2}$, ±$\frac{3}{2}$, ±$\frac{1}{5}$, ±$\frac{2}{5}$, ±$\frac{3}{5}$, ±$\frac{6}{5}$, ±$\frac{1}{10}$, ±$\frac{3}{10}$.

We form a synthetic division table:

	10	-33	23	6	
1	10	-23	0	6	
2	10	-13	-3	0	2 is a zero

So $P(x) = \frac{1}{10}x(x - 2)(10x^2 - 13x - 3)$

$= \frac{1}{10}x(x - 2)(2x - 3)(5x + 1)$

So the zeros of $P(x)$ are 0, 2, 1.5, -0.2.

45. $P(x) = x^4 - 2x^3 - 14x^2 + 30x + 9$. The possible rational zeros are ±1, ±3, ±9.
We form a synthetic division table:

	1	-2	-14	30	9	
1	1	-1	-15	15	24	
3	1	1	-11	-3	0	3 is a zero

We examine $Q(x) = x^3 + x^2 - 11x - 3$. The only remaining possible rational zeros are ±3, -1. We form a synthetic division table for $Q(x)$.

	1	1	-11	-3	
3	1	4	1	0	3 is a zero

Thus 3 is a double zero of $P(x)$.

$P(x) = (x - 3)^2(x^2 + 4x + 1)$

To find the remaining zeros, we solve $x^2 + 4x + 1 = 0$ by completing the square.

$$x^2 + 4x = -1$$
$$x^2 + 4x + 4 = 3$$
$$(x + 2)^2 = 3$$
$$x + 2 = \pm\sqrt{3}$$
$$x = -2 \pm \sqrt{3}$$

So the zeros of $P(x)$ are 3 (double), $-2 \pm \sqrt{3}$.

47. $P(x) = 3x^5 - 2x^4 + 6x^3 + 20x^2 - x - 10$.
Possible factors of -10 are ±1, ±2, ±5, ±10. Possible factors of 3 are ±1, ±3.

Hence the possible rational zeros are ±1, ±2, ±5, ±10, ±$\frac{1}{3}$, ±$\frac{2}{3}$, ±$\frac{5}{3}$, ±$\frac{10}{3}$.

We form a synthetic division table:

	3	-2	6	20	-1	-10	
1	3	1	7	27	26	16	
-1	3	-5	11	9	-10	0	-1 is a zero

We examine $Q(x) = 3x^4 - 5x^3 + 11x^2 + 9x - 10$. We form a synthetic division table for $Q(x)$:

	3	-5	11	9	-10	
-1	3	-8	19	-10	0	-1 is a zero

Thus -1 is a double zero of $P(x)$.

We examine $R(x) = 3x^3 - 8x^2 + 19x - 10$. We form a synthetic division table for $R(x)$.

	3	-8	19	-10
-1	3	-11	30	-40
$\frac{1}{3}$	3	-7	$\frac{50}{3}$	$-\frac{40}{9}$
$\frac{2}{3}$	3	-6	15	0	$\frac{2}{3}$ is a zero

$$P(x) = (x + 1)^2\left(x - \frac{2}{3}\right)(3x^2 - 6x + 15)$$

To find the remaining zeros, we solve $3x^2 - 6x + 15 = 0$, or $x^2 - 2x + 5 = 0$, by completing the square.

$$x^2 - 2x = -5$$
$$x^2 - 2x + 1 = -4$$
$$(x - 1)^2 = -4$$
$$x - 1 = \pm 2i$$
$$x = 1 \pm 2i$$

So the zeros of $P(x)$ are -1 (double), $\frac{2}{3}$, $1 \pm 2i$.

49. The possible rational zeros are ± 1, ± 2, ± 3, ± 6, $\pm\frac{1}{2}$, $\pm\frac{3}{2}$, $\pm\frac{1}{3}$, $\pm\frac{2}{3}$, $\pm\frac{1}{6}$. We form a synthetic division table:

	6	19	11	-6
1	6	25	36	30
-1	6	13	-2	-4
-2	6	7	-3	0	-2 is a zero

$P(x) = (x + 2)(6x^2 + 7x - 3) = (x + 2)(3x - 1)(2x + 3)$.

51. The possible rational zeros are ± 1, ± 3. We form a synthetic division table:

	1	-1	-13	-3
1	1	0	-13	-16
3	1	2	-7	-24
-1	1	-2	-11	8
-3	1	-4	-1	0	-3 is a zero

So $P(x) = (x + 3)(x^2 - 4x - 1)$. We solve $x^2 - 4x - 1 = 0$ by completing the square.

$$x^2 - 4x = 1$$
$$x^2 - 4x + 4 = 5$$
$$(x - 2)^2 = 5$$
$$x - 2 = \pm\sqrt{5}$$
$$x = 2 \pm \sqrt{5}$$

So $P(x) = (x + 3)[x - (2 + \sqrt{5})][x - (2 - \sqrt{5})]$
$$= (x + 3)(x - 2 - \sqrt{5})(x - 2 + \sqrt{5})$$

53. The possible rational zeros are ± 1, ± 2, ± 3, ± 4, ± 6, ± 12, $\pm\frac{1}{2}$, $\pm\frac{3}{2}$, $\pm\frac{1}{4}$, $\pm\frac{3}{4}$.
We form a synthetic division table:

	4	-4	-19	16	12	
1	4	0	-19	-3	9	
2	4	4	-11	-6	0	2 is a zero

So $P(x) = (x - 2)(4x^3 + 4x^2 - 11x - 6)$. We examine $Q(x) = 4x^3 + 4x^2 - 11x - 6$.
We form a synthetic division table for $Q(x)$.

	4	4	-11	-6	
2	4	12	13	20	
-1	4	0	-11	5	
-2	4	-4	-3	0	-2 is a zero

So $P(x) = (x - 2)(x + 2)(4x^2 - 4x - 3) = (x - 2)(x + 2)(2x - 3)(2x + 1)$.

55. $x^2 \leq 4x - 1$
$x^2 - 4x + 1 \leq 0$
We factor $x^2 - 4x + 1$ by solving $x^2 - 4x + 1 = 0$ by completing the square.
$\quad x^2 - 4x = -1$
$x^2 - 4x + 4 = 3$
$\quad (x - 2)^2 = 3$
$\quad\quad x - 2 = \pm\sqrt{3}$
$\quad\quad\quad x = 2 \pm \sqrt{3}$
Hence $x^2 - 4x + 1 = [x - (2 + \sqrt{3})][x - (2 - \sqrt{3})]$.
To solve $[x - (2 + \sqrt{3})][x - (2 - \sqrt{3})] \leq 0$ we form a sign chart, noting
$2 - \sqrt{3} < 2 + \sqrt{3}$, hence $2 - \sqrt{3}$ is to the left of $2 + \sqrt{3}$.

$x^2 - 4x + 1 = [x - (2 + \sqrt{3})][x - (2 - \sqrt{3})]$			
Test Number	0	2	4
Value of Polynomial for Test Number	1	-3	1
Sign of Polynomial in Interval	+	-	+
Interval	$(-\infty, 2 - \sqrt{3})$	$(2 - \sqrt{3}, 2 + \sqrt{3})$	$(2 + \sqrt{3}, \infty)$

$x^2 - 4x + 1 \leq 0$ and $x^2 \leq 4x - 1$ within the interval $[2 - \sqrt{3}, 2 + \sqrt{3}]$, or
$2 - \sqrt{3} \leq x \leq 2 + \sqrt{3}$

57. $x^3 + 3 \leq 3x^2 + x$
$x^3 - 3x^2 - x + 3 \leq 0$
To factor $x^3 - 3x^2 - x + 3$, we can use factoring by grouping to obtain
$(x - 3)(x^2 - 1)$ or $(x - 3)(x - 1)(x + 1)$.
However, if we don't notice this, we can search for zeros by the methods of
this section. A synthetic division table immediately gives:

	1	-3	-1	3
1	1	-2	-3	0

Hence $x^3 - 3x^2 - x + 3 = (x - 1)(x^2 - 2x - 3) = (x - 1)(x - 3)(x + 1)$.
We form a sign chart.
Zeros: -1, 1, 3

$x^3 - 3x^2 - x + 3 = (x - 1)(x - 3)(x + 1)$				
Test Number	-2	0	2	4
Value of Polynomial for Test Number	-15	3	-3	15
Sign of Polynomial in Interval	$-$	$+$	$-$	$+$
Interval	$(-\infty, -1)$	$(-1, 1)$	$(1, 3)$	$(3, \infty)$

$x^3 - 3x^2 - x + 3 \leq 0$ and $x^3 + 3 \leq 3x^2 + x$ within the intervals $(-\infty, -1]$ and $[1, 3]$, or $x \leq -1$ or $1 \leq x \leq 3$.

59. $2x^3 + 6 \geq 13x - x^2$
$2x^3 + x^2 - 13x + 6 \geq 0$
To factor $2x^3 + x^2 - 13x + 6$, we search for zeros of the polynomial.

Possible rational zeros are ± 1, ± 2, ± 3, ± 6, $\pm \frac{1}{2}$, $\pm \frac{3}{2}$.

We form a synthetic division table:

	2	1	-13	6	
1	2	3	-10	-4	
2	2	5	-3	0	2 is a zero

$2x^3 + x^2 - 13x + 6 = (x - 2)(2x^2 + 5x - 3) = (x - 2)(2x - 1)(x + 3)$. Hence we must examine the sign behavior of $(x - 2)(2x - 1)(x + 3)$.

We form a sign chart.
Zeros: -3, $\frac{1}{2}$, 2

$2x^3 + x^2 - 13x + 6 = (x - 2)(2x - 1)(x + 3)$				
Test Number	-4	0	1	3
Value of Polynomial for Test Number	-54	6	-4	30
Sign of Polynomial in Interval	$-$	$+$	$-$	$+$
Interval	$(-\infty, -3)$	$(-3, \frac{1}{2})$	$(\frac{1}{2}, 2)$	$(2, \infty)$

$2x^3 + x^2 - 13x + 6 \geq 0$ and $2x^3 + 6 \geq 13x - x^2$ within the intervals $\left[-3, \frac{1}{2}\right]$ and $[2, \infty)$, or $-3 \leq x \leq \frac{1}{2}$ or $x \geq 2$.

61. $[x - (4 - 5i)][x - (4 + 5i)] = [x - 4 + 5i][x - 4 - 5i]$
$\qquad\qquad\qquad\qquad\qquad = [(x - 4) + 5i][(x - 4) - 5i]$
$\qquad\qquad\qquad\qquad\qquad = (x - 4)^2 - 25i^2$
$\qquad\qquad\qquad\qquad\qquad = x^2 - 8x + 16 + 25$
$\qquad\qquad\qquad\qquad\qquad = x^2 - 8x + 41$

63. $[x - (a + bi)][x - (a - bi)] = [x - a - bi][x - a + bi]$
$\qquad\qquad\qquad\qquad\qquad = [(x - a) - bi][(x - a) + bi]$
$\qquad\qquad\qquad\qquad\qquad = (x - a)^2 - b^2 i^2$
$\qquad\qquad\qquad\qquad\qquad = (x - a)^2 + b^2$
$\qquad\qquad\qquad\qquad\qquad = x^2 - 2ax + a^2 + b^2$

65. If $1 + 2i$ is a zero, then $1 - 2i$ is a zero.
So $Q(x) = [x - (1 + 2i)][x - (1 - 2i)]$ divides $P(x)$ evenly.
Applying problem 63, $Q(x) = x^2 - 2x + 1 + 4 = x^2 - 2x + 5$. Dividing, we see

$$
\begin{array}{r}
x + 2 \\
x^2 - 2x + 5 \overline{\smash{\big)}\ x^3 - 0x^2 + x + 10} \\
\underline{x^3 - 2x^2 + 5x} \\
2x^2 - 4x + 10 \\
\underline{2x^2 - 4x + 10} \\
0
\end{array}
$$

So $P(x) = (x + 2)Q(x)$ and the other zeros are $1 - 2i$ and -2.

67. If $-3i$ is a zero, then so is $3i$. So $Q(x) = (x + 3i)(x - 3i)$ divides $P(x)$
evenly. $Q(x) = x^2 - 9i^2 = x^2 + 9$. Dividing, we see

$$
\begin{array}{r}
x + 4 \\
x^2 + 9 \overline{\smash{\big)}\ x^3 + 4x^2 + 9x + 36} \\
\underline{x^3 \qquad\ + 9x} \\
4x^2 \qquad + 36 \\
\underline{4x^2 \qquad + 36} \\
0
\end{array}
$$

So $P(x) = (x + 4)Q(x)$ and the other zeros are $3i$ and -4.

69. If $3 - 2i$ is a zero, then so is $3 + 2i$. So $Q(x) = [x - (3 - 2i)][x - (3 + 2i)]$
divides $P(x)$ evenly. Applying problem 63, $Q(x) = x^2 - 6x + 9 + 4$
$= x^2 - 6x + 13$. Dividing, we see

$$
\begin{array}{r}
x^2 - 2x - 1 \\
x^2 - 6x + 13 \overline{\smash{\big)}\ x^4 - 8x^3 + 24x^2 - 20x - 13} \\
\underline{x^4 - 6x^3 + 13x^2} \\
- 2x^3 + 11x^2 - 20x \\
\underline{- 2x^3 + 12x^2 - 26x} \\
-x^2 + 6x - 13 \\
\underline{-x^2 + 6x - 13} \\
0
\end{array}
$$

So $P(x) = Q(x)(x^2 - 2x - 1)$. $x^2 - 2x - 1$ has two zeros; solving by completing
the square we obtain:
$$
\begin{aligned}
x^2 - 2x - 1 &= 0 \\
x^2 - 2x &= 1 \\
x^2 - 2x + 1 &= 2 \\
(x - 1)^2 &= 2 \\
x - 1 &= \pm\sqrt{2} \\
x &= 1 \pm \sqrt{2}
\end{aligned}
$$
So the remaining zeros of $P(x)$ are $3 + 2i$, $1 \pm \sqrt{2}$.

71. $\dfrac{4}{2x^3 + 5x^2 - 2x - 5} \geq 0$

We need to form a sign chart for $\dfrac{P}{Q} = \dfrac{4}{2x^3 + 5x^2 - 2x - 5}$. We first locate the
zeros of P and Q. $P = 4$ has no zeros. To find the zeros of $Q = 2x^3 + 5x^2 - 2x - 5$
we can factor by grouping into $(2x + 5)(x^2 - 1) = (2x + 5)(x - 1)(x + 1)$.
However, if we don't notice this, we can search for zeros by the methods of

this section. A synthetic division table immediately gives

	2	5	-2	-5
1	2	7	5	0

Hence $2x^3 + 5x^2 - 2x - 5 = (x - 1)(2x^2 + 7x + 5) = (x - 1)(x + 1)(2x + 5)$.

We form a sign chart.

Zeros of Q: $-\dfrac{5}{2}$, -1, 1

	$\dfrac{P}{Q} = \dfrac{4}{(x - 1)(x + 1)(2x + 5)}$			
Test Number	-3	-2	0	2
Value of $\dfrac{P}{Q}$	$-\dfrac{1}{2}$	$\dfrac{4}{3}$	$-\dfrac{4}{5}$	$\dfrac{4}{27}$
Sign of $\dfrac{P}{Q}$	$-$	$+$	$-$	$+$
Interval	$\left(-\infty, -\dfrac{5}{2}\right)$	$\left(-\dfrac{5}{2}, -1\right)$	$(-1, 1)$	$(1, \infty)$

$\dfrac{4}{2x^3 + 5x^2 - 2x - 5} \geq 0$ within the intervals $\left(-\dfrac{5}{2}, -1\right)$ and $(1, \infty)$, or $-\dfrac{5}{2} < x < -1$ or $x > 1$.

73. $\dfrac{x^2 - 3x - 10}{x^3 - 4x^2 + x + 6} \leq 0$

We need to form a sign chart for $\dfrac{P}{Q} = \dfrac{x^2 - 3x - 10}{x^3 - 4x^2 + x + 6}$. We first locate the zeros of P and Q. $x^2 - 3x - 10 = (x - 5)(x + 2)$, hence P has zeros at 5 and -2. To find the zeros of Q, we note that the possible rational zeros are ± 1, ± 2, ± 3 and ± 6. We form a synthetic division table:

	1	-4	1	6	
1	1	-3	-2	4	
2	1	-2	-3	0	2 is a zero

Hence $Q = (x - 2)(x^2 - 2x - 3) = (x - 2)(x - 3)(x + 1)$.

We form a sign chart.

Zeros of P, Q: -2, -1, 2, 3, 5

Open dots at zeros of Q

Solid dots at zeros of P

	$\dfrac{P}{Q} = \dfrac{(x - 5)(x + 2)}{(x - 2)(x - 3)(x + 1)}$					
Test Number	-3	$-\dfrac{3}{2}$	0	$\dfrac{5}{2}$	4	6
Value of $\dfrac{P}{Q}$	$-\dfrac{2}{15}$	$\dfrac{26}{63}$	$-\dfrac{5}{3}$	$\dfrac{90}{7}$	$-\dfrac{3}{5}$	$\dfrac{2}{21}$
Sign of $\dfrac{P}{Q}$	$-$	$+$	$-$	$+$	$-$	$+$
Interval	$(-\infty, -2)$	$(-2, -1)$	$(-1, 2)$	$(2, 3)$	$(3, 5)$	$(5, \infty)$

$\dfrac{x^2 - 3x - 10}{x^3 - 4x^2 + x + 6} \leq 0$ within the intervals $(-\infty, -2]$ or $(-1, 2)$ or $(3, 5]$. $x \leq -2$ or $-1 < x < 2$ or $3 < x \leq 5$.

75. Here is a computer-generated graph
of $P(x) = 3x^3 - 37x^2 + 84x - 24$.

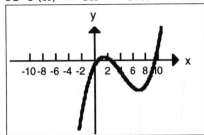

The rational zero theorem gives ± 1, ± 2,
± 3, ± 4, ± 6, ± 8, ± 12, ± 24, $\pm \frac{1}{3}$, $\pm \frac{2}{3}$, $\pm \frac{4}{3}$,
$\pm \frac{8}{3}$ as possible rational zeros. However,
the computer graph suggests that $P(x)$
has no integer zeros; there are no
negative zeros, and the positive zeros
would appear to lie between 0 and 1, 2 and 3, and 9 and 10. This suggests that
we consider only the possible zeros $\frac{1}{3}$, $\frac{2}{3}$, and $\frac{8}{3}$.

We form a synthetic division table:

	3	−37	84	−24	
$\frac{1}{3}$	3	−36	72	0	$\frac{1}{3}$ is a zero

So $P(x) = \left(x - \frac{1}{3}\right)(3x^2 - 36x + 72) = \left(x - \frac{1}{3}\right)3(x^2 - 12x + 24)$

To find the remaining zeros, we solve $x^2 - 12x + 24 = 0$ by completing the
square:
$$x^2 - 12x = -24$$
$$x^2 - 12x + 36 = 12$$
$$(x - 6)^2 = 12$$
$$x - 6 = \pm\sqrt{12}$$
$$x = 6 \pm \sqrt{12} \text{ or } 6 \pm 2\sqrt{3}$$

So the zeros of $P(x)$ are $\frac{1}{3}$, $6 \pm 2\sqrt{3}$.

77. Here is a computer-generated graph
of
$P(x) = 4x^4 + 4x^3 + 49x^2 + 64x - 240$.

The rational zero theorem gives ± 1, ± 2,
± 3, ± 4, ± 5, ± 6, ± 8, ± 10, ± 12, ± 15, ± 16,
± 20, ± 24, ± 30, ± 40, ± 48, ± 60, ± 80, ± 120,
± 240, $\pm \frac{1}{2}$, $\pm \frac{3}{2}$, $\pm \frac{5}{2}$, $\pm \frac{15}{2}$, $\pm \frac{1}{4}$, $\pm \frac{3}{4}$, $\pm \frac{5}{4}$, $\pm \frac{15}{4}$
as possible rational zeros. However, the
computer graph suggests that $P(x)$ has no
integer zeros, and the zeros appear to
lie between −3 and −2, and 1 and 2. This suggests that we consider only the
possible zeros $\frac{3}{2}$, $\frac{5}{4}$, and $-\frac{5}{2}$.

We form a synthetic division table:

	4	4	49	64	−240	
$\frac{3}{2}$	4	10	64	160	0	$\frac{3}{2}$ is a zero

So $P(x) = \left(x - \frac{3}{2}\right)(4x^3 + 10x^2 + 64x + 160) = \left(x - \frac{3}{2}\right)2(2x^3 + 5x^2 + 32x + 80)$

We consider the reduced polynomial $Q(x) = 2x^3 + 5x^2 + 32x + 80$. The computer
graph suggests that the other real zero is negative, so we try $-\frac{5}{2}$ next.

	2	5	32	80	
$-\frac{5}{2}$	2	0	32	0	$-\frac{5}{2}$ is a zero

So $P(x) = \left(x - \dfrac{3}{2}\right)2\left(x + \dfrac{5}{2}\right)(2x^2 + 32) = 4\left(x - \dfrac{3}{2}\right)\left(x + \dfrac{5}{2}\right)(x^2 + 16)$

The remaining zeros of $P(x)$ are the zeros of $x^2 + 16$, that is, $\pm 4i$. So the zeros of $P(x)$ are $\dfrac{3}{2}$, $-\dfrac{5}{2}$, $\pm 4i$.

79. Here is a computer-generated graph of $P(x) = 4x^4 - 44x^3 + 145x^2 - 192x + 90$.

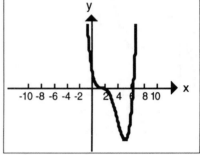

The rational zero theorem gives ± 1, ± 2, ± 3, ± 5, ± 6, ± 9, ± 10, ± 15, ± 18, ± 30, ± 45, ± 90, $\pm\dfrac{1}{2}$, $\pm\dfrac{3}{2}$, $\pm\dfrac{5}{2}$, $\pm\dfrac{9}{2}$, $\pm\dfrac{15}{2}$, $\pm\dfrac{45}{2}$, $\pm\dfrac{1}{4}$, $\pm\dfrac{3}{4}$, $\pm\dfrac{5}{4}$, $\pm\dfrac{9}{4}$, $\pm\dfrac{15}{4}$, $\pm\dfrac{45}{4}$ as possible rational zeros. However, the computer graph suggests that the only possible integer zeros are 1 and 2; there are no negative zeros, and the zeros appear to lie between 0 and 3, and between 6 and 7. This suggests that we consider only the possible zeros 1, 2, $\dfrac{1}{2}$, $\dfrac{3}{2}$, $\dfrac{5}{2}$, $\dfrac{1}{4}$, $\dfrac{3}{4}$, $\dfrac{5}{4}$, $\dfrac{9}{4}$.

We form a synthetic division table:

	4	-44	145	-192	90	
1	4	-40	105	-87	3	
2	4	-36	73	-46	-2	
$\dfrac{3}{2}$	4	-38	88	-60	0	$\dfrac{3}{2}$ is a zero

So $P(x) = \left(x - \dfrac{3}{2}\right)(4x^3 - 38x^2 + 88x - 60) = \left(x - \dfrac{3}{2}\right)2(2x^3 - 19x^2 + 44x - 30)$

We consider the reduced polynomial $Q(x) = 2x^3 - 19x^2 + 44x - 30$. The remaining possibilities from the reduced list are $\dfrac{1}{2}$, $\dfrac{3}{2}$, $\dfrac{5}{2}$.

We form a synthetic division table:

	2	-19	44	-30	
$\dfrac{1}{2}$	2	-18	35	$-\dfrac{25}{2}$	
$\dfrac{3}{2}$	2	-16	20	0	$\dfrac{3}{2}$ is a zero

So $P(x) = \left(x - \dfrac{3}{2}\right)^2 2(2x^2 - 16x + 20) = \left(x - \dfrac{3}{2}\right)^2 4(x^2 - 8x + 10)$

To find the remaining zeros, we solve $x^2 - 8x + 10 = 0$, by completing the square.

$$x^2 - 8x = -10$$
$$x^2 - 8x + 16 = 6$$
$$(x - 4)^2 = 6$$
$$x - 4 = \pm\sqrt{6}$$
$$x = 4 \pm \sqrt{6}$$

So the zeros of $P(x)$ are $\dfrac{3}{2}$ (double), $4 \pm \sqrt{6}$.

81. (A) Since there are 3 zeros of $x^3 - 1$, there are 3 cube roots of 1.

(B) $x^3 - 1 = (x - 1)(x^2 + x + 1)$. The other cube roots of 1 will be solutions to $x^2 + x + 1 = 0$. Applying the quadratic formula with $a = b = c = 1$, we have $x = \dfrac{-1 \pm \sqrt{(1)^2 - 4(1)(1)}}{2(1)} = \dfrac{-1 \pm \sqrt{-3}}{2} = \dfrac{-1 \pm i\sqrt{3}}{2}$. Thus $-\dfrac{1}{2} + \dfrac{\sqrt{3}}{2}i$ and $-\dfrac{1}{2} - \dfrac{\sqrt{3}}{2}i$ are the other cube roots of 1.

83. $P(x)$ can have at most n and must have at least one real zero. Each zero of $P(x)$ represents a point where $P(x) = y = 0$ so the graph of $P(x)$ will cross the x axis at and only at zeros of $P(x)$. Thus there can be a maximum of n axis crossings and there is a minimum of 1 axis crossing.

85. $P(2 + i) = (2 + i)^2 + 2i(2 + i) - 5$
$= 4 + 4i + i^2 + 4i + 2i^2 - 5$
$= 4 + 4i - 1 + 4i - 2 - 5$
$= -4 + 8i$

So $P(2 + i) \neq 0$ and $2 + i$ is not a zero of $P(x)$. This does not contradict the theorem, since $P(x)$ is not a polynomial with real coefficients (the coefficient of x is the imaginary number $2i$).

87. Let x = the amount of increase.
Then old volume = $1 \times 2 \times 3 = 6$
    new volume = $(x + 1)(x + 2)(x + 3) = x^3 + 6x^2 + 11x + 6$
Since (new volume) = 10 (old volume), we must solve
$x^3 + 6x^2 + 11x + 6 = 10(6)$
$x^3 + 6x^2 + 11x + 6 = 60$
$x^3 + 6x^2 + 11x - 54 = 0$
The possible rational zeros are ± 1, ± 2, ± 3, ± 6, ± 9, ± 18, ± 27, ± 54. We form a synthetic division table:

	1	6	11	-54	
1	1	7	18	-36	
2	1	8	27	0	2 is a zero

2 is the only positive zero. Hence the increase must equal 2 feet.

89.
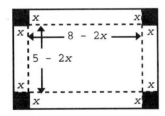 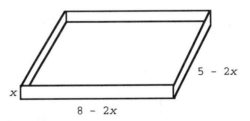

From the figure, it should be clear that
Volume = $x(5 - 2x)(8 - 2x) = 14$

Since x, $5 - 2x$, and $8 - 2x$ must all be positive, the domain of x is $0 < x < \dfrac{5}{2}$ or $(0, 2.5)$. We solve $x(5 - 2x)(8 - 2x) = 14$, or $4x^3 - 26x^2 + 40x = 14$, for x in this domain.
$4x^3 - 26x^2 + 40x = 14$
$4x^3 - 26x^2 + 40x - 14 = 0$
$2x^3 - 13x^2 + 20x - 7 = 0$

Possible rational zeros: ± 1, ± 7, $\pm \dfrac{1}{2}$, $\pm \dfrac{7}{2}$.

We form a synthetic division table

	2	-13	20	-7	
1	2	-11	9	2	
$\frac{1}{2}$	2	-12	14	0	$\frac{1}{2}$ is a zero

So $2x^3 - 13x^2 + 20x - 7 = \left(x - \dfrac{1}{2}\right)(2x^2 - 12x + 14) = (2x - 1)(x^2 - 6x + 7)$

To find the remaining zeros, we solve $x^2 - 6x + 7 = 0$, by completing the square:

$$x^2 - 6x = -7$$
$$x^2 - 6x + 9 = 2$$
$$(x - 3)^2 = 2$$
$$x - 3 = \pm\sqrt{2}$$
$$x = 3 \pm \sqrt{2}$$

Hence the zeros are $\dfrac{1}{2}$, $3 - \sqrt{2}$, $3 + \sqrt{2}$, or 0.5, 1.59, 4.44 to two significant digits. We discard $3 + \sqrt{2}$ or 4.44, since it is not in the interval (0, 2.5). The square should be 0.5 × 0.5 inches or 1.59 × 1.59 inches.

Exercise 3-3

Key Ideas and Formulas

Location Theorem

If f is continuous on an interval I, a and b are two numbers in I, and $f(a)$ and $f(b)$ are of opposite sign, then there is at least one x intercept for f between a and b.

Upper and Lower Bounds of Real Zeros:

Given an nth degree polynomial $P(x)$ with real coefficients, $n > 0$, $a_n > 0$ and $P(x)$ divided by $x - r$ using synthetic division:

1. **Upper bound.** If $r > 0$ and all numbers in the quotient row of the synthetic division including the remainder are nonnegative, then r is an upper bound of the real zeros of $P(x)$.

2. **Lower bound.** If $r < 0$ and all numbers in the quotient row of the synthetic division including the remainder alternate in sign, then r is a lower bound of the real zeros of $P(x)$.

Bisection Method used to approximate real zeros

Let $P(x)$ be a polynomial with real coefficients. If $P(x)$ has opposite signs at the endpoints of the interval (a, b), then a real zero r lies in this interval. We bisect this interval [find the midpoint $m = (a + b)/2$], check the sign of $P(m)$, and choose the interval (a, m) or (m, b) on which $P(x)$ has opposite signs at the endpoints. We repeat this bisecting process (producing a set of "nested" intervals each half the size of the preceding one, and each containing the real zero r) until we get the desired decimal accuracy for the zero approximation. At any point in the process if $P(m) = 0$, we stop, since m is a real zero.

---

1. Since $P(x)$ has opposite signs at -5 and -1, at -1 and -3, and at 5 and 8, there is at least one x intercept in each of the intervals (-5, -1), (-1, -3), and (5, 8).

3. Since $P(x)$ has opposite signs at -6 and -4, at -4 and 0, at 2 and 4, and at 4 and 7, there is at least one x intercept in each of the intervals (-6, -4), (-4, 0), (2, 4), and (4, 7).

5. We form a synthetic division table.

	1	-9	23	-14
0	1	-9	23	-14
1	1	-8	15	1
2	1	-7	9	4
3	1	-6	5	1
4	1	-5	3	-2
5	1	-4	3	1

Since $P(x)$ has opposite signs at 0 and 1, at 3 and 4, and at 4 and 5, it has at least one zero in each of the intervals $(0, 1)$, $(3, 4)$, $(4, 5)$. Since $P(x)$ is a third degree polynomial, there can be only 3 real zeros, and they must each lie in one of these intervals.

7. We form a synthetic division table.

	1	3	-1	-5
0	1	3	-1	-5
1	1	4	3	-2
2	1	5	9	13
-1	1	2	-3	-2
-2	1	1	-3	1
-3	1	0	-1	-2

Since $P(x)$ has opposite signs at 1 and 2, at -2 and -1, and at -3 and -2, it has at least one zero in each of the intervals $(1, 2)$, $(-2, -1)$, $(-3, -2)$. Since $P(x)$ is a third degree polynomial, there can be only 3 real zeros, and they must each lie in one of these intervals.

9. We form a synthetic division table.

	1	0	-3	1	
0	1	0	-3	1	
1	1	1	-2	-1	
2	1	2	1	3	an upper bound
-1	1	-1	-2	3	
-2	1	-2	1	-1	a lower bound

2 is an upper bound; -2 is a lower bound.

11. We form a synthetic division table.

	1	-3	4	2	-9	
0	1	-3	4	2	-9	
1	1	-2	2	4	-5	
2	1	-1	2	6	3	
3	1	0	4	14	33	an upper bound
-1	1	-4	8	-6	-3	
-2	1	-5	14	-26	43	a lower bound

3 is an upper bound; -2 is a lower bound.

13. We form a synthetic division table.

	1	0	-3	3	2	-2	
0	1	0	-3	3	2	-2	
1	1	1	-2	1	3	1	
2	1	2	1	5	12	22	an upper bound
-1	1	-1	-2	5	-3	1	
-2	1	-2	1	1	0	-2	
-3	1	-3	6	-15	47	-143	a lower bound

2 is an upper bound; -3 is a lower bound.

15. (A) We form a synthetic division table.

	1	-2	-5	4	
0	1	-2	-5	4	
1	1	-1	-6	-2	a zero in (0, 1)
2	1	0	-5	-6	
3	1	1	-2	-2	
4	1	2	3	16	an upper bound, also a zero in (3, 4)
-1	1	-3	-2	6	
-2	1	-4	3	-2	a lower bound, also a zero in (-2, -1)

(B) We search for the real zero in (3, 4). We organize our calculations in a table.

Sign Change Interval (a, b)	Midpoint m	P(a)	P(m)	P(b)
(3, 4)	3.5	−	+	+
(3, 3.5)	3.25	−	+	+
(3, 3.25)	3.125	−	−	+
(3.125, 3.25)	3.1875	−	+	+
(3.125, 3.1875)	3.15625	−	−	+
(3.156, 3.188)	We stop here	−		+

Since each endpoint rounds to 3.2, a real zero lies on this last interval and is given by 3.2 to one decimal place accuracy.

17. (A) We form a synthetic division table.

	1	-2	-1	5	
0	1	-2	-1	5	
1	1	-1	-2	3	
2	1	0	-1	3	
3	1	1	2	11	an upper bound
-1	1	-3	2	3	
-2	1	-4	7	-9	a lower bound, also a zero in (-2, -1)

(B) We search for the real zero in (-2, -1). We organize our calculations in a table.

Sign Change Interval (a, b)	Midpoint m	P(a)	P(m)	P(b)
(-2, -1)	-1.5	−	−	+
(-1.5, -1)	-1.25	−	+	+
(-1.5, -1.25)	-1.375	−	−	+
(-1.375, -1.25)	-1.3125	−	+	+
(-1.375, -1.3125)	-1.34375	−	+	+
(-1.375, -1.34375)	-1.359375	−	+	+
(-1.375, -1.359375)	We stop here	−		+

Since each endpoint rounds to -1.4, a real zero lies on this last interval and is given by -1.4 to one decimal place accuracy.

19. (A) We form a synthetic division table.

	1	-2	-7	9	7	
0	1	-2	-7	9	7	
1	1	-1	-8	1	8	
2	1	0	-7	-5	-3	a zero in (1, 2)
3	1	1	-4	-3	-2	
4	1	2	1	13	59	an upper bound, also a zero in (3, 4)
-1	1	-3	-4	13	-6	a zero in (-1, 0)
-2	1	-4	1	7	-7	
-3	1	-5	8	-15	52	a lower bound, also a zero in (-3, -2)

(B) We search for the real zero in (3, 4). We organize our calculations in a table.

Sign Change Interval (a, b)	Midpoint m	Sign of P		
		$P(a)$	$P(m)$	$P(b)$
(3, 4)	3.5	-	+	+
(3, 3.5)	3.25	-	+	+
(3, 3.25)	3.125	-	+	+
(3, 3.125)	3.0625	-	-	+
(3.0625, 3.125)	We stop here	-		+

Since each endpoint rounds to 3.1, a real zero lies on this last interval and is given by 3.1 to one decimal place accuracy.

21. (A) We form a synthetic division table.

	1	-1	-4	4	3	
0	1	-1	-4	4	3	
1	1	0	-4	0	3	
2	1	1	-2	0	3	
3	1	2	2	10	33	an upper bound
-1	1	-2	-2	6	-3	a zero in (-1, 0)
-2	1	-3	2	0	3	a lower bound, also a zero in (-2, -1)

(B) We search for the real zero in (-1, 0). We organize our calculations in a table.

Sign Change Interval (a, b)	Midpoint m	Sign of P		
		$P(a)$	$P(m)$	$P(b)$
(-1, 0)	-0.5	-	+	+
(-1, -0.5)	-0.75	-	-	+
(-0.75, -0.5)	-0.625	-	-	+
(-0.625, -0.5)	-0.5625	-	-	+
(-0.5625, -0.5)	-0.53125	-	-	+
(-0.53125, -0.5)	We stop here	-		+

Since each endpoint rounds to -0.5, a real zero lies on this last interval and is given by -0.5 to one decimal place accuracy.

23. (A) We form a synthetic division table.

	1	-2	3	-8
0	1	-2	3	-8
1	1	-1	2	-6
2	1	0	3	-2
3	1	1	6	10
-1	1	-3	6	-14

From the table, 3 is an upper bound and -1 is a lower bound.

(B) The only interval in which a real zero is indicated is (2, 3). We search for this real zero. We organize our calculations in a table.

Sign Change Interval (a, b)	Midpoint m	Sign of P P(a)	P(m)	P(b)
(2, 3)	2.5	−	+	+
(2, 2.5)	2.25	−	+	+
(2, 2.25)	2.125	−	−	+
(2.125, 2.25)	2.1875	−	−	+
(2.1875, 2.25)	2.21875	−	−	+
(2.21875, 2.25)	2.234375	−	−	+
(2.234375, 2.25)	2.2421875	−	−	+
(2.2421875, 2.25)	2.24609375	−	−	+
(2.24609375, 2.25)	We stop here	−		+

Since each endpoint rounds to 2.25, a real zero lies on this last interval and is given by 2.25 to two decimal place accuracy. A glance at a computer-generated graph of P(x) suggests that this is the only real zero.

25. (A) We form a synthetic division table.

	1	1	−5	7	−22
0	1	1	−5	7	−22
1	1	2	−3	4	−18
2	1	3	1	9	−4
3	1	4	7	28	62
−1	1	0	−5	12	−34
−2	1	−1	−3	13	−48
−3	1	−2	1	4	−34
−4	1	−3	7	−21	62

From the table, 3 is an upper bound and -4 is a lower bound.

(B) There are real zeros in the invervals (2, 3) and (-4, -3) indicated in the table. We search for the zero in (2, 3). We organize our calculations in a table.

Sign Change Interval (a, b)	Midpoint m	Sign of P P(a)	P(m)	P(b)
(2, 3)	2.5	−	+	+
(2, 2.5)	2.25	−	+	+
(2, 2.25)	2.125	−	+	+
(2, 2.125)	2.0625	−	−	+
(2.0625, 2.125)	2.09375	−	−	+
(2.09375, 2.125)	2.109375	−	−	+
(2.109375, 2.125)	2.1171875	−	−	+
(2.1171875, 2.125)	2.12109375	−	+	+
(2.1171875, 2.12109375)	We stop here	−		+

Since each endpoint rounds to 2.12, a real zero lies on this last interval and is given by 2.12 to two decimal place accuracy. A similar search (details omitted) leads to -3.51 as the other indicated real zero. A glance at a computer-generated graph of $P(x)$ suggests that these are the only real zeros.

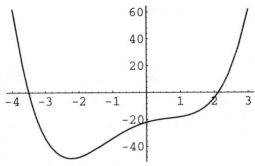

27. (A) We form a synthetic division table.

	1	0	-3	0	-4	4
0	1	0	-3	0	-4	4
1	1	1	-2	-2	-6	-2
2	1	2	1	2	0	4
-1	1	-1	-2	2	-6	10
-2	1	-2	1	-2	0	4
-3	1	-3	6	-18	50	-146

From the table, 2 is an upper bound and -3 is a lower bound.

(B) There are real zeros in the invervals (0, 1), (1, 2), and (-3, -2) indicated in the table. We search for the real zero in (0, 1). We organize our calculations in a table.

Sign Change Interval (a, b)	Midpoint m	Sign of P $P(a)$	$P(m)$	$P(b)$
(0, 1)	0.5	+	+	-
(0.5, 1)	0.75	+	-	-
(0.5, 0.75)	0.625	+	+	-
(0.625, 0.75)	0.6875	+	+	-
(0.6875, 0.75)	0.71875	+	+	-
(0.71875, 0.75)	0.734375	+	+	-
(0.734375, 0.75)	0.7421875	+	+	-
(0.7421875, 0.75)	0.74609875	+	+	-
(0.74609875, 0.75)	We stop here	+		-

Since each endpoint rounds to 0.75, a real zero lies on this last interval and is given by 0.75 to two decimal place accuracy. Similar searches (details omitted) lead to -2.09 and 1.88 as the other indicated real zeros. A glance at a computer-generated graph of $P(x)$ suggests that these are the only real zeros.

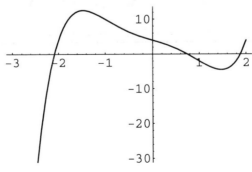

29. (A) We form a synthetic division table.

	1	1	3	1	2	-5
0	1	1	3	1	2	-5
1	1	2	5	6	8	3
-1	1	0	3	-2	4	-9

From the table, 1 is an upper bound and -1 is a lower bound.

(B) There is a real zero in the inverval (0, 1) indicated in the table. We search for this real zero. We organize our calculations in a table.

Sign Change Interval (a, b)	Midpoint m	Sign of P P(a)	P(m)	P(b)
(0, 1)	0.5	-	-	+
(0.5, 1)	0.75	-	-	+
(0.75, 1)	0.875	-	+	+
(0.75, 0.875)	0.8125	-	-	+
(0.8125, 0.875)	0.84375	-	+	+
(0.8125, 0.84375)	0.828125	-	-	+
(0.828125, 0.84375)	0.8359375	-	+	+
(0.828125, 0.8359375)	0.83203125	-	-	+
(0.83203125, 0.8359375)	0.833984375	-	-	+
(0.833984375, 0.8359375)	0.8349609375	-	+	+
(0.833984375, 0.8349609375)	We stop here	-		+

Since each endpoint rounds to 0.83, a real zero lies on this last interval and is given by 0.83 to two decimal place accuracy. A glance at a computer-generated graph of $P(x)$ suggests that this is the only real zero.

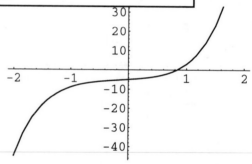

31. (A) We form a synthetic division table.

	1	-5	5	5	-10	5	
0	1	-5	5	5	-10	5	
1	1	-4	1	6	-4	1	
2	1	-3	-1	3	-4	-3	a zero in (1, 2)
3	1	-2	-1	2	-4	-7	
4	1	-1	1	9	26	109	a zero in (3, 4)
5	1	0	5	30	140	705	an upper bound
-1	1	-6	11	-6	-4	9	
-2	1	-7	19	-33	56	-107	a lower bound, also a zero in (-2, -1)

(B) We search for the real zero in (3, 4). We organize our calculations in a table.

Sign Change Interval (a, b)	Midpoint m	Sign of P		
		P(a)	P(m)	P(b)
(3, 4)	3.5	−	+	+
(3, 3.5)	3.25	−	+	+
(3, 3.25)	3.125	−	−	+
(3.125, 3.25)	3.1875	−	−	+
(3.1875, 3.25)	3.21875	−	+	+
(3.1875, 3.21875)	3.203125	−	−	+
(3.203125, 3.21875)	3.2109375	−	−	+
(3.2109375, 3.21875)	3.21484375	−	−	+
(3.21484375, 3.21875)	3.216796875	−	+	+
(3.21484375, 3.216796875)	3.2158203125	−	+	+
(3.2148375, 3.2158203125)	3.21532890625	−	−	+
(3.21532…, 3.2158…)	We stop here	−		+

Since each endpoint rounds to 3.22, a real zero lies on this last interval and is given by 3.22 to two decimal place accuracy.

33. (A) We form a synthetic division table.

	1	0	-10	0	9	10	
0	1	0	-10	0	9	10	
1	1	1	-9	-9	0	10	
2	1	2	-6	-12	-15	-20	a zero in (1, 2)
3	1	3	-1	-3	0	10	a zero in (2, 3)
4	1	4	6	24	105	430	an upper bound
-1	1	-1	-9	9	0	10	
-2	1	-2	-6	12	-15	40	
-3	1	-3	-1	3	0	10	
-4	1	-4	6	-24	105	-410	a lower bound, also a zero in (-4, -3)

(B) We search for the real zero in (2, 3). We organize our calculations in a table.

Sign Change Interval (a, b)	Midpoint m	Sign of P		
		P(a)	P(m)	P(b)
(2, 3)	2.5	−	−	+
(2.5, 3)	2.75	−	−	+
(2.75, 3)	2.875	−	−	+
(2.875, 3)	2.9375	−	+	+
(2.875, 2.9375)	2.90625	−	−	+
(2.90625, 2.9375)	2.921875	−	−	+
(2.921875, 2.9375)	2.9296875	−	+	+
(2.921875, 2.9296875)	2.92578125	−	+	+
(2.921875, 2.92578125)	2.923828125	−	+	+
(2.921875, 2.923828125)	We stop here	−		+

Since each endpoint rounds to 2.92, a real zero lies on this last interval and is given by 2.92 to two decimal place accuracy.

35. (A) We form a synthetic division table.

	1	-24	-25	10
0	1	-24	-25	10
10	1	-14	-165	-1640
20	1	-4	-105	-2090
30	1	6	155	4660
-10	1	-34	315	-3140

From the table, 30 is an upper bound and -10 is a lower bound.

(B) The real zeros are found by the bisection method to be -1.29, 0.31, and 24.98 to two decimal place accuracy (details omitted).

37. (A) We form a synthetic division table.

	1	12	-900	0	5,000
0	1	12	-900	0	5,000
10	1	22	-680	-680	-63,000
20	1	32	-260	-520	-99,000
30	1	42	360	10,800	329,000
-10	1	2	-920	9,200	-87,000
-20	1	-8	-740	14,800	-291,000
-30	1	-18	-360	10,800	-319,000
-40	1	-28	220	-8,800	357,000

From the table, 30 is an upper bound and -40 is a lower bound.

(B) The real zeros are found by the bisection method to be -36.53, -2.33, 2.40, and 24.46 to two decimal place accuracy (details omitted).

39. (A) We form a synthetic division table.

	1	0	-100	-1,000	-5,000
0	1	0	-100	-1,000	-5,000
10	1	10	0	-1,000	-15,000
20	1	20	300	5,000	95,000
-10	1	-10	0	-1,000	5,000

From the table, 20 is an upper bound and -10 is a lower bound.

(B) The real zeros are found by the bisection method to be -7.47 and 14.03 to two decimal place accuracy (details omitted).

41. (A) We form a synthetic division table.

	4	-40	-1,475	7,875	-10,000
0	4	-40	-1,475	7,875	-10,000
10	4	0	-1,475	-6,875	-78,750
20	4	40	-675	-5,625	-122,500
30	4	80	925	35,625	1,058,750
-10	4	-80	-675	14,625	-156,250
-20	4	-120	925	-10,625	202,500

From the table, 30 is an upper bound and -20 is a lower bound.

(B) Two real zeros can be found by the bisection method to be -17.66 and 22.66 to two decimal place accuracy. In finding the remaining zeros, we note:

	4	-40	-1,475	7,875	-10,000
2.5	4	-30	-1,550	4,000	0

2.5 is a zero. Moreover, $P(x) = (x - 2.5)(4x^3 - 30x^2 - 1,550x + 4,000)$. Testing 2.5 again,

	4	-30	-1,550	4,000
2.5	4	-20	-1,600	0

2.5 is in fact a double zero of $P(x)$.
Hence $P(x) = (x - 2.5)^2(4x^2 - 20x - 1600)$.
Solving $4x^2 - 20x - 1600 = 0$ yields exact solutions

$$x = \frac{20 \pm \sqrt{26,000}}{8}$$

or, again, $x = -17.66$ and 22.66 to two decimal place accuracy.

43. (A) We form a synthetic division table.

	0.01	-0.1	-12	0	0	9,000
0	0.01	-0.1	-12	0	0	9,000
10	0.01	0	-12	-120	-1,200	-3,000
20	0.01	0.1	-10	-200	-4,000	-71,000
30	0.01	0.2	-6	-180	-5,400	-153,000
40	0.01	0.3	0	0	0	9,000
-10	0.01	-0.2	-10	100	-1,000	19,000
-20	0.01	-0.3	-6	120	-2,400	57,000
-30	0.01	-0.4	0	0	0	9,000
-40	0.01	-0.5	8	-320	12,800	-503,000

From the table, 40 is an upper bound and -40 is a lower bound.

(B) The real zeros are found by the bisection method to be -30.45, 9.06, and 39.80 to two decimal place accuracy (details omitted).

45. Let (x, x^2) be a point on the graph of $y = x^2$. Then the distance from $(1, 2)$ to (x, x^2) must equal 1 unit. Applying the distance formula, we have,

$$\sqrt{(x - 1)^2 + (x^2 - 2)^2} = 1$$
$$(x - 1)^2 + (x^2 - 2)^2 = 1$$
$$x^2 - 2x + 1 + x^4 - 4x^2 + 4 = 1$$
$$x^4 - 3x^2 - 2x + 4 = 0$$

Let $P(x) = x^4 - 3x^2 - 2x + 4$
The only rational zero is 1.

	1	0	-3	-2	4
1	1	1	-2	-4	0

We examine $Q(x) = x^3 + x^2 - 2x - 4$.

We form a synthetic division table for $Q(x)$.

	1	1	-2	-4	
0	1	1	-2	-4	
1	1	2	0	-4	
2	1	3	4	4	a zero in (1, 2)
-1	1	0	-2	-2	
-2	1	-1	0	-4	
-3	1	-2	4	-16	a lower bound

The table is inconclusive as to the existence of negative zeros. However, the graph of $Q(x)$ strongly suggests that there are none. We can now apply the bisection method (details omitted) to locate the positive zero of $Q(x)$ to any desired accuracy. To one decimal place, it is 1.7. Therefore, the two real zeros of $P(x)$ are 1 and 1.7. Hence the two required points (x, x^2) are $(1, 1)$ and $(1.7, 2.9)$.

Graph of $y = Q(x) = x^3 + x^2 - 2x - 4$

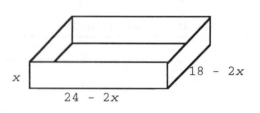

47.

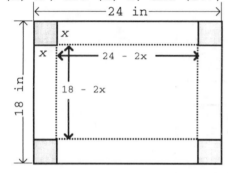

From the above figures it should be clear that
$V = $ length $\times$ width $\times$ height $= (24 - 2x)(18 - 2x)x \quad 0 < x < 9 \quad$ (why?)

We solve $(24 - 2x)(18 - 2x)x = 600$
$$432x - 84x^2 + 4x^3 = 600$$
$$4x^3 - 84x^2 + 432x - 600 = 0$$
$$x^3 - 21x^2 + 108x - 150 = 0$$
Let $P(x) = x^3 - 21x^2 + 108x - 150$. We search for positive rational zeros. We form a synthetic division table.

	1	-21	108	-150	
0	1	-21	108	-150	
1	1	-20	88	-62	
2	1	-19	70	-10	
3	1	-18	54	12	a zero in (2,3)
5	1	-16	28	-10	a zero in (3, 5)
6	1	-15	18	-42	
10	1	-11	-2	-170	
15	1	-6	18	120	a zero in (10, 15)
25	1	4	208	5050	an upper bound

There are no rational zeros. Applying the bisection method (details omitted), the positive zeros in (0, 9) are found to be 2.3 and 4.6 to one decimal place accuracy.
$x = 2.3$ inches or 4.6 inches.

49. We note:

$$\begin{pmatrix} \text{Volume} \\ \text{of} \\ \text{tank} \end{pmatrix} = \begin{pmatrix} \text{Volume of} \\ \text{two hemispheres} \\ \text{of radius } x \end{pmatrix} + \begin{pmatrix} \text{Volume of cylinder} \\ \text{with radius } x, \\ \text{height } 10 - 2x \end{pmatrix}$$

$$20\pi = \frac{4}{3}\pi x^3 + \pi x^2(10 - 2x)$$

$$20 = \frac{4}{3}x^3 + 10x^2 - 2x^3$$

$$60 = 4x^3 + 30x^2 - 6x^3$$

$$2x^3 - 30x^2 + 60 = 0$$

$$x^3 - 15x^2 + 30 = 0 \quad \text{From physical considerations, we are interested only in solutions in (0, 5).}$$

Let $P(x) = x^3 - 15x^2 + 30$
There are no rational zeros.
We form a synthetic division table. We are only interested in the positive real
zeros.

	1	-15	0	30	
0	1	-15	0	30	
1	1	-14	-14	16	
2	1	-13	-26	-22	There is a zero in (1, 2)
3	1	-12	-36	-78	
4	1	-11	-44	-146	
5	1	-10	-50	-220	There does not seem to be another zero in (0, 5)
15	1	0	0	30	There is a zero in (5, 15)

From the table, we see that the only zero of physical interest lies in (1, 2).
Applying the bisection method (details omitted) the zero in (1, 2) is found to
be 1.5 to one decimal place accuracy.
$x = 1.5$ feet.

Exercise 3-4

Key Ideas and Formulas

A function f is a rational function if $f(x) = \dfrac{n(x)}{d(x)}$, $d(x) \neq 0$, where $n(x)$ and $d(x)$
are polynomials. The domain of f is the set of all real numbers such that $d(x) \neq 0$.
If $f(x) = \dfrac{n(x)}{d(x)}$ and $d(a) = 0$, then f is discontinuous at $x = a$ and the graph of f has
a hole or break at $x = a$. If $n(c) = 0$ and $d(c) \neq 0$, then $x = c$ is an x intercept for
the graph of f.

Asymptotes: the line $x = a$ is a vertical asymptote for the graph of $y = f(x)$ if $f(x)$
either increases or decreases without bound as x approaches a from the right or from
the left, $f(x) \to \infty$ or $f(x) \to -\infty$ as $x \to a^+$ or $x \to a^-$.

If $f(x) = \dfrac{n(x)}{d(x)}$, $d(a) = 0$ and $n(a) \neq 0$, then the line $x = a$ is a vertical asymptote.

The line $y = b$ is a horizontal asymptote for the graph of $y = f(x)$ if $f(x)$
approaches b as x increases without bound or as x decreases without bound,
$f(x) \to b$ as $x \to \infty$ or $x \to -\infty$.

$$\text{If } f(x) = \frac{a_m x^m + \cdots + a_1 x + a_0}{b_n x^n + \cdots + b_1 x + b_0}, \quad a_m, \ b_n \neq 0, \text{ then}$$

1. for $m < n$ the x axis is a horizontal asymptote

2. for $m = n$ the line $y = \dfrac{a_m}{b_n}$ is a horizontal asymptote

3. for $m > n$ the graph will increase or decrease without bound, depending on m, n,
a_m, and b_n, and there are no horizontal asymptotes

If m, the degree of $n(x)$, is one more than n, the degree of $d(x)$, then $f(x)$ can be
written in the form $f(x) = mx + b + \dfrac{r(x)}{d(x)}$, where the degree of $f(x)$ is less than the
degree of $d(x)$. Then the line $y = mx + b$ is an oblique asymptote for the graph of f.
$$[f(x) - (mx + b)] \to 0 \text{ as } x \to -\infty \text{ or } x \to \infty$$

To graph a rational function, follow the steps
1. Intercepts
2. Vertical asymptotes
3. Sign chart
4. Horizontal asymptotes
5. Complete the sketch. For details, see text.

1. This graph has a vertical asymptote $x = 2$, and a horizontal asymptote $y = -2$. This corresponds to $g(x)$.

3. This graph has a vertical asymptote $x = 2$, and a horizontal asymptote $y = 2$. This corresponds to $h(x)$.

5. $\dfrac{2x - 4}{x + 1}$ *Domain:* $d(x) = x + 1$ zero: $x = -1$ domain: $(-\infty, -1) \cup (-1, \infty)$

x *intercepts:* $n(x) = 2x - 4$ zero: $x = 2$ x intercept: 2

7. $\dfrac{x^2 - 1}{x^2 - 16}$ *Domain:* $d(x) = x^2 - 16$ zeros: $x^2 - 16 = 0$

$$x^2 = 16$$
$$x = \pm 4$$

domain: $(-\infty, -4) \cup (-4, 4) \cup (4, \infty)$

x *intercepts:* $n(x) = x^2 - 1$ zeros: $x^2 - 1 = 0$
$$x^2 = 1$$
$$x = \pm 1$$

x intercepts: $-1, 1$

9. $\dfrac{x^2 - x - 6}{x^2 - x - 12}$ *Domain:* $d(x) = x^2 - x - 12$ zeros: $x^2 - x - 12 = 0$

$$(x + 3)(x - 4) = 0$$
$$x = -3, 4$$

domain: $(-\infty, -3) \cup (-3, 4) \cup (4, \infty)$

x *intercepts:* $n(x) = x^2 - x - 6$ zeros: $x^2 - x - 6 = 0$
$$(x + 2)(x - 3) = 0$$
$$x = -2, 3$$

x intercepts: $-2, 3$

11. $\dfrac{x}{x^2 + 4}$ *Domain:* $d(x) = x^2 + 4$ no real zeros

Domain: all real numbers

x *intercepts:* $n(x) = x$ zero: $x = 0$

x intercept: 0

13. $\dfrac{2x}{x - 4}$ *vertical asymptotes:* $d(x) = x - 4$ zero: $x = 4$

vertical asymptote $x = 4$

> **Common Error:**
> $x = 0$ is not a vertical asymptote.

horizontal asymptotes: Since $n(x)$ and $d(x)$ have the same degree, the line $y = 2$ is a horizontal asymptote.

15. $\dfrac{2x^2 + 3x}{3x^2 - 48}$ *vertical asymptotes:* $d(x) = 3x^2 - 48$ zeros: $3x^2 - 48 = 0$

$$3(x + 4)(x - 4) = 0$$
$$x = -4, 4$$

vertical asymptotes: $x = -4$, $x = 4$

horizontal asymptotes: Since $n(x)$ and $d(x)$ have the same degree, the line $y = \dfrac{2}{3}$ is a horizontal asymptote.

17. $\dfrac{2x}{x^4 + 1}$ *vertical asymptote:* $d(x) = x^4 + 1$ No real zeros:

No vertical asymptotes

horizontal asymptotes: Since the degree of $n(x)$ is less than the degree of $d(x)$, the x axis is a horizontal asymptote

horizontal asymptote: $y = 0$

19. $\dfrac{6x^4}{3x^2 - 2x - 5}$ *vertical asymptotes:* $d(x) = 3x^2 - 2x - 5$ zeros: $3x^2 - 2x - 5 = 0$

$$(3x - 5)(x + 1) = 0$$
$$x = \frac{5}{3}, -1$$

vertical asymptotes: $x = -1$, $x = \dfrac{5}{3}$

horizontal asymptotes: Since the degree of $n(x)$ is greater than the degree
of $d(x)$, there are no horizontal asymptotes.

21. $f(x) = \dfrac{1}{x - 4} = \dfrac{n(x)}{d(x)}$

Intercepts. There are no real zeros of $n(x) = 1$. No x intercept

$f(0) = -\dfrac{1}{4}$ $y = -\dfrac{1}{4}$ y intercept

Vertical asymptotes. $d(x) = x - 4$ zeros: 4 $x = 4$

Sign Chart.

Test numbers	3	5
Value of f	-1	1
Sign of f	–	+

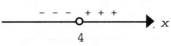

Since $x = 4$ is a vertical asymptote and $f(x) < 0$ for $x < 4$,
  $f(x) \rightarrow -\infty$ as $x \rightarrow 4^-$
Since $x = 4$ is a vertical asymptote and $f(x) > 0$ for $x > 4$,
  $f(x) \rightarrow \infty$ as $x \rightarrow 4^+$
Horizontal asymptotes. Since the degree of $n(x)$ is less that the degree of
$d(x)$, the x axis is a horizontal asymptote.
Complete the sketch.

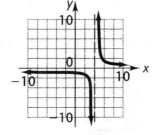

23. $f(x) = \dfrac{x}{x + 1} = \dfrac{n(x)}{d(x)}$

Intercepts. Real zeros of $n(x) = x$ $x = 0$ x intercept
                $f(0) = 0$ $y = 0$ y intercept
The graph crosses the coordinate axes only at the origin.
Vertical asymptotes. $d(x) = x + 1$ zeros: -1 $x = -1$
Sign chart.

Test numbers	-2	$-\dfrac{1}{2}$	1
Value of f	2	-1	$\dfrac{1}{2}$
Sign of f	+	–	+

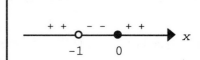

Since $x = -1$ is a vertical asymptote and $f(x) > 0$ for $x < -1$,
  $f(x) \rightarrow \infty$ as $x \rightarrow -1^-$
Since $x = -1$ is a vertical asymptote and $f(x) < 0$ for $-1 < x < 0$,
  $f(x) \rightarrow -\infty$ as $x \rightarrow -1^+$

Horizontal asymptotes. Since $n(x)$ and $d(x)$ have the same degree, the line $y = 1$ is a horizontal asymptote.
Complete the sketch.

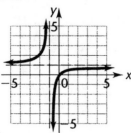

25. $h(x) = \dfrac{x}{2x - 2} = \dfrac{n(x)}{d(x)}$

Intercepts. Real zeros of $n(x) = x$ $x = 0$ x intercept
$\qquad\qquad\qquad\qquad\qquad\qquad h(0) = 0$ $y = 0$ y intercept
The graph crosses the coordinate axes only at the origin.
Vertical asymptotes. $d(x) = 2x - 2$ zeros: 1 $x = 1$
Sign Chart.

Test numbers	-1	$\frac{1}{2}$	2
Value of h	$\frac{1}{4}$	$-\frac{1}{2}$	1
Sign of h	$+$	$-$	$+$

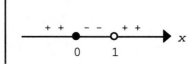

Since $x = 1$ is a vertical asymptote and $h(x) < 0$ for $0 < x < 1$,
$\qquad h(x) \to -\infty$ as $x \to 1^{-}$
Since $x = 1$ is a vertical asymptote and $h(x) > 0$ for $x > 1$
$\qquad h(x) \to \infty$ as $x \to 1^{+}$

Horizontal asymptotes. Since $n(x)$ and $d(x)$ have the same degree, the line $y = \dfrac{1}{2}$ is a horizontal asymptote.

Complete the sketch.

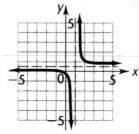

27. $f(x) = \dfrac{2x - 4}{x + 3} = \dfrac{n(x)}{d(x)}$

Intercepts. Real zeros of $n(x) = 2x - 4$ $x = 2$ x intercept
$\qquad\qquad\qquad\qquad\qquad f(0) = -\dfrac{4}{3}$ $\qquad\qquad\qquad$ y intercept
Vertical asymptotes. $d(x) = x + 3$ zeros: -3 $x = -3$
Sign Chart.

Test numbers	-4	0	3
Value of f	12	$-\frac{4}{3}$	$\frac{1}{3}$
Sign of f	$+$	$-$	$+$

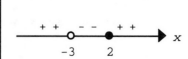

Since $x = -3$ is a vertical asymptote and $f(x) > 0$ for $x < -3$
$\qquad f(x) \to \infty$ as $x \to -3^{-}$
Since $x = -3$ is a vertical asymptote and $f(x) < 0$ for $-3 < x < 2$
$\qquad f(x) \to -\infty$ as $x \to -3^{+}$
Horizontal asymptotes. Since $n(x)$ and $d(x)$ have the same degree, the line $y = 2$ is a horizontal asymptote.

Complete the sketch.

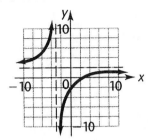

29. $g(x) = \dfrac{1 - x^2}{x^2} = \dfrac{n(x)}{d(x)}$

Intercepts. Real zeros of $n(x) = 1 - x^2$ $1 - x^2 = 0$

$$x^2 = 1$$
$$x = \pm 1 \quad x \text{ intercepts}$$

$g(0)$ is not defined   no y intercepts

Vertical asymptotes. $d(x) = x^2$ zeros: 0 $x = 0$

Sign Chart.

Test numbers	-2	$-\frac{1}{2}$	$\frac{1}{2}$	2
Value of g	$-\frac{3}{4}$	3	3	$-\frac{3}{4}$
Sign of g	$-$	$+$	$+$	$-$

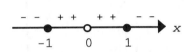

Since $x = 0$ is a vertical asymptote and $g(x) > 0$ for $-1 < x < 0$ and $0 < x < 1$,
   $g(x) \to \infty$ as $x \to 0^+$ and as $x \to 0^-$

Horizontal asymptotes. Since $n(x)$ and $d(x)$ have the same degree, the line $y = -1$ is a horizontal asymptote.

Complete the sketch.

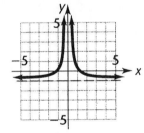

31. $f(x) = \dfrac{9}{x^2 - 9} = \dfrac{n(x)}{d(x)}$

Intercepts. There are no real zeros of $n(x) = 9$. No x intercept

$$f(0) = -1 \quad y \text{ intercept}$$

Vertical asymptotes. $d(x) = x^2 - 9$ zeros: $x^2 - 9 = 0$

$$x^2 = 9$$
$$x = \pm 3$$

Sign Chart.

Test numbers	-4	0	4
Value of f	$\frac{9}{7}$	-1	$\frac{9}{7}$
Sign of f	$+$	$-$	$+$

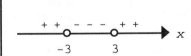

Since $x = -3$ is a vertical asymptote and $f(x) > 0$ for $x < -3$ and $f(x) < 0$ for $-3 < x < 3$,
   $f(x) \to \infty$ as $x \to -3^-$ and $f(x) \to -\infty$ as $x \to -3^+$
Since $x = 3$ is a vertical asymptote and $f(x) < 0$ for $-3 < x < 3$ and $f(x) > 0$ for $x > 3$,
   $f(x) \to -\infty$ as $x \to 3^-$ and $f(x) \to \infty$ as $x \to 3^+$

Horizontal asymptotes. Since the degree of $n(x)$ is less than the degree of $d(x)$, the x axis is a horizontal asymptote.

Complete the sketch.

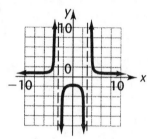

33. $f(x) = \dfrac{x}{x^2 - 1} = \dfrac{n(x)}{d(x)}$

Intercepts. Real zeros of $n(x) = x$ $x = 0$ x intercept

$f(0) = 0$ $y = 0$ y intercept

The graph crosses the coordinate axes only at the origin.

Vertical asymptotes. $d(x) = x^2 - 1$ zeros: $x^2 - 1 = 0$

$$x^2 = 1$$
$$x = \pm 1$$

Sign Chart.

Test numbers	-2	$-\frac{1}{2}$	$\frac{1}{2}$	2
Value of f	$-\frac{2}{3}$	$\frac{2}{3}$	$-\frac{2}{3}$	$\frac{2}{3}$
Sign of f	$-$	$+$	$-$	$+$

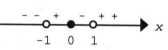

Since $x = -1$ is a vertical asymptote and $f(x) < 0$ for $x < -1$ and $f(x) > 0$ for $-1 < x < 0$,

$f(x) \to -\infty$ as $x \to -1^-$ and $f(x) \to \infty$ as $x \to -1^+$

Since $x = 1$ is a vertical asymptote and $f(x) < 0$ for $0 < x < 1$ and $f(x) > 0$ for $x > 1$,

$f(x) \to -\infty$ as $x \to 1^-$ and $f(x) \to \infty$ as $x \to 1^+$

Horizontal asymptotes. Since the degree of $n(x)$ is less than the degree of $d(x)$, the x axis is a horizontal asymptote.

Complete the sketch.

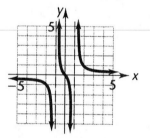

35. $g(x) = \dfrac{2}{x^2 + 1} = \dfrac{n(x)}{d(x)}$

Intercepts. There are no real zeros of $n(x) = 2$.   No x intercept

$g(0) = 2$ y intercept

Vertical asymptotes. There are no real zeros of $d(x) = x^2 + 1$

No vertical asymptotes

Sign behavior. $g(x)$ is always positive.

Horizontal asymptotes. Since the degree of $n(x)$ is less than the degree of $d(x)$, the x axis is a horizontal asymptote.

Complete the sketch.

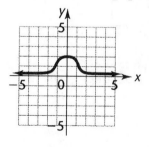

37. $f(x) = \dfrac{12x^2}{(3x + 5)^2} = \dfrac{n(x)}{d(x)}$

Intercepts. Real zeros of $n(x) = 12x^2$ $x = 0$ x intercept
$f(0) = 0$ $y = 0$ y intercept
The graph crosses the coordinate axes only at the origin.
Vertical asymptotes. $d(x) = (3x + 5)^2$ zeros: $(3x + 5)^2 = 0$
$3x + 5 = 0$
$x = -\dfrac{5}{3}$

Sign Chart.

Test numbers	-2	-1	1
Value of f	48	3	$\frac{3}{16}$
Sign of f	$+$	$+$	$+$

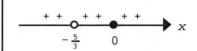

Since $x = -\dfrac{5}{3}$ is a vertical asymptote and $f(x) > 0$ for $x < -\dfrac{5}{3}$ and $-\dfrac{5}{3} < x < 0$,

$f(x) \to \infty$ as $x \to -\dfrac{5}{3}^-$ and as $x \to -\dfrac{5}{3}^+$

Horizontal asymptotes. Since $n(x)$ and $d(x)$ have the same degree, the line
$y = \dfrac{12}{3^2} = \dfrac{4}{3}$ is a horizontal asymptote.

Complete the sketch.

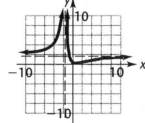

39. $f(x) = \dfrac{x^2 - 1}{x^2 + 7x + 10} = \dfrac{n(x)}{d(x)}$

Intercepts. Real zeros of $n(x) = x^2 - 1$ $x^2 - 1 = 0$
$x^2 = 1$
$x = \pm 1$ x intercepts

$f(0) = -\dfrac{1}{10}$ y intercept

Vertical asymptotes. Real zeros of $d(x) = x^2 + 7x + 10$ $x^2 + 7x + 10 = 0$
$(x + 2)(x + 5) = 0$
$x = -2, -5$

Sign Chart.

Test numbers	-6	-3	$-\frac{3}{2}$	0	2
Value of f	$\frac{35}{4}$	-4	$\frac{5}{7}$	$-\frac{1}{10}$	$\frac{3}{28}$
Sign of f	$+$	$-$	$+$	$-$	$+$

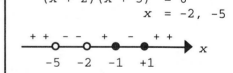

Since $x = -5$ is a vertical asymptote and $f(x) > 0$ for $x < -5$ and $f(x) < 0$ for
$-5 < x < -2$,
$f(x) \to \infty$ as $x \to -5^-$ and $f(x) \to -\infty$ as $x \to -5^+$
Since $x = -2$ is a vertical asymptote and $f(x) < 0$ for $-5 < x < -2$ and $f(x) > 0$
for $-2 < x < -1$,
$f(x) \to -\infty$ as $x \to -2^-$ and $f(x) \to \infty$ as $x \to -2^+$
Horizontal asymptotes. Since $n(x)$ and $d(x)$ have the same degree, the line $y = 1$
is a horizontal asymptote.

Complete the sketch.

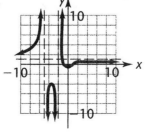

41. The x intercepts of $f(x) = \dfrac{n(x)}{d(x)}$ are the real zeros of $n(x)$. Since $n(x)$ is a quadratic function, the maximum number of real zeros is 2 and the minimum number is 0. Therefore, the maximum number of x intercepts is 2 and the minimum number is 0. For example, $\dfrac{x^2 - 1}{x^2}$ has two intercepts and $\dfrac{x^2 + 1}{x^2}$ has none.

43. $f(x) = \dfrac{2x^2}{x - 1} = \dfrac{n(x)}{d(x)}$

Vertical asymptotes. Real zeroes of $d(x) = x - 1$ $x = 1$
Horizontal asymptote. Since the degree of $n(x)$ is greater than the degree of $d(x)$, there is no horizontal asymptote.
Oblique asymptote.

$$
\begin{array}{r}
2x + 2 \\
x - 1 \overline{\smash{)}\ 2x^2} \\
\underline{2x^2 - 2x} \\
2x \\
\underline{2x - 2} \\
2
\end{array}
$$

Thus, $f(x) = 2x + 2 + \dfrac{2}{x - 1}$. Hence, the line $y = 2x + 2$ is an oblique asymptote.

45. $p(x) = \dfrac{x^3}{x^2 + 1} = \dfrac{n(x)}{d(x)}$

Vertical asymptotes. There are no real zeros of $d(x) = x^2 + 1$.
No vertical asymptotes.
Horizontal asymptotes. Since the degree of $n(x)$ is greater than the degree of $d(x)$, there is no horizontal asymptote.
Oblique asymptote:

$$
\begin{array}{r}
x \\
x^2 + 1 \overline{\smash{)}\ x^3} \\
\underline{x^3 + x} \\
- x
\end{array}
$$

Thus, $p(x) = x + \dfrac{-x}{x^2 + 1}$. Hence, the line $y = x$ is an oblique asymptote.

47. $r(x) = \dfrac{2x^2 - 3x + 5}{x} = \dfrac{n(x)}{d(x)}$

Vertical asymptotes. Real zeros of $d(x) = x$ $x = 0$
Horizontal asymptote. Since the degree of $n(x)$ is greater than the degree of $d(x)$, there is no horizontal asymptote.
Oblique asymptote. $\dfrac{2x^2 - 3x + 5}{x} = \dfrac{2x^2}{x} - \dfrac{3x}{x} + \dfrac{5}{x}$

$$= 2x - 3 + \dfrac{5}{x}$$

Thus $r(x) = 2x - 3 + \dfrac{5}{x}$. Hence the line $y = 2x - 3$ is an oblique asymptote.

49. Here is a computer-generated graph of $f(x)$.

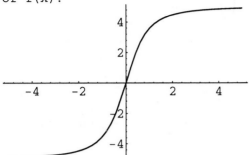

From the graph, we can see that $f(x) \to 5$ as $x \to \infty$ and $f(x) \to -5$ as $x \to -\infty$; the lines $y = 5$ and $y = -5$ are horizontal asymptotes.

51. Here is a computer-generated graph of $f(x)$.

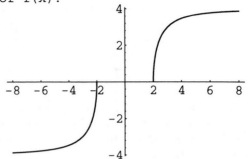

From the graph, we can see that $f(x) \to 4$ as $x \to \infty$ and $f(x) \to -4$ as $x \to -\infty$; the lines $y = 4$ and $y = -4$ are horizontal asymptotes.

53. $f(x) = \dfrac{x^2 + 1}{x} = \dfrac{n(x)}{d(x)}$

Intercepts. There are no real zeros of $n(x) = x^2 + 1$. No x intercept
$\qquad\qquad f(0)$ is not defined No y intercept

Vertical asymptotes. Real zeros of $d(x) = x$. $x = 0$

Sign Chart.

Test numbers	-1	1
Value of f	-2	2
Sign of f	–	+

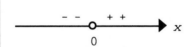

Since $x = 0$ is a vertical asymptote and $f(x) < 0$ for $x < 0$ and $f(x) > 0$ for $x > 0$,

$\quad f(x) \to -\infty$ as $x \to 0^-$ and $f(x) \to \infty$ as $x \to 0^+$

Horizontal asymptote. Since the degree of $n(x)$ is greater than the degree of $d(x)$, there is no horizontal asymptote.

Oblique asymptote. $f(x) = \dfrac{x^2 + 1}{x} = \dfrac{x^2}{x} + \dfrac{1}{x} = x + \dfrac{1}{x}$

Hence, the line $y = x$ is an oblique asymptote.

Complete the sketch.

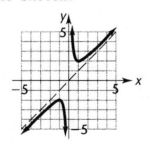

55. $k(x) = \dfrac{x^2 - 4x + 3}{2x - 4} = \dfrac{n(x)}{d(x)}$

Intercepts. Real zeros of $n(x) = x^2 - 4x + 3$ $x^2 - 4x + 3 = 0$
$\qquad\qquad\qquad\qquad\qquad\qquad\qquad\qquad\qquad (x - 1)(x - 3) = 0$
$\qquad\qquad\qquad\qquad\qquad\qquad\qquad\qquad\qquad\qquad x = 1, 3$ x intercepts

$$k(0) = -\frac{3}{4} \quad y \text{ intercept}$$

Vertical asymptotes. Real zeros of $d(x) = 2x - 4$ $2x - 4 = 0$
$\qquad\qquad\qquad\qquad\qquad\qquad\qquad\qquad\qquad\qquad\qquad\qquad 2x = 4$
$\qquad\qquad\qquad\qquad\qquad\qquad\qquad\qquad\qquad\qquad\qquad\qquad x = 2$

Sign Chart.

Test numbers	0	$1\frac{1}{2}$	$2\frac{1}{2}$	4
Value of k	$-\frac{3}{4}$	$\frac{3}{4}$	$-\frac{3}{4}$	$\frac{3}{4}$
Sign of k	$-$	$+$	$-$	$+$

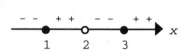

Since $x = 2$ is a vertical asymptote and $k(x) > 0$ for $1 < x < 2$ and $k(x) < 0$ for $2 < x < 3$,

$k(x) \to \infty$ as $x \to 2^-$ and $k(x) \to -\infty$ as $x \to 2^+$

Horizontal asymptote. Since the degree of $n(x)$ is greater than the degree of $d(x)$, there is no horizontal asymptote.

Oblique asymptote:

$$
\begin{array}{r}
\frac{1}{2}x - 1 \\
2x - 4 \overline{)\, x^2 - 4x + 3} \\
\underline{x^2 - 2x} \\
-2x + 3 \\
\underline{-2x + 4} \\
-1
\end{array}
$$

Thus, $k(x) = \frac{1}{2}x - 1 + \frac{-1}{2x - 4}$. Hence, the line $y = \frac{1}{2}x - 1$ is an oblique asymptote.

Complete the sketch.

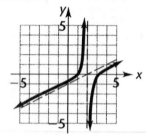

57. $F(x) = \dfrac{8 - x^3}{4x^2} = \dfrac{n(x)}{d(x)}$

Intercepts. Real zeros of $n(x) = 8 - x^3$ $\qquad 8 - x^3 = 0$

$\qquad\qquad\qquad\qquad\qquad\qquad (2 - x)(4 + 2x + x^2) = 0$

$\qquad\qquad\qquad\qquad\qquad\qquad\quad 2 - x = 0 \quad 4 + 2x + x^2 = 0$

$\qquad\qquad\qquad\qquad\qquad\qquad\qquad x = 2 \quad$ No real zeros

$\qquad\qquad x = 2 \quad x$ intercept

$\qquad\qquad\quad F(0)$ is not defined. No y intercept.

Vertical asymptotes. Real zeros of $d(x) = 4x^2$. $\quad x = 0$

Sign chart.

Test numbers	-1	1	3
Value of F	$\frac{9}{4}$	$\frac{7}{4}$	$-\frac{19}{36}$
Sign of F	$+$	$+$	$-$

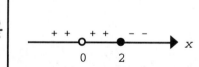

Since $x = 0$ is a vertical asymptote and $F(x) > 0$ for $x < 0$ and $0 < x < 2$,

$\quad F(x) \to \infty$ as $x \to 0^-$ and $F(x) \to \infty$ as $x \to 0^+$

Horizontal asymptote. Since the degree of $n(x)$ is greater than the degree of $d(x)$, there is no horizontal asymptote.

Oblique asymptote. $F(x) = \dfrac{8 - x^3}{4x^2} = \dfrac{8}{4x^2} - \dfrac{x^3}{4x^2} = -\dfrac{1}{4}x + \dfrac{2}{x^2}$. Hence, the line $y = -\dfrac{1}{4}x$ is an oblique asymptote.

Complete the sketch.

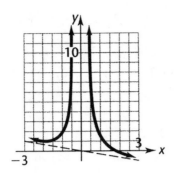

59. $x^2 + 1 \overline{\smash{\big)}\ \begin{array}{l} x^2 - 1 \\[2pt] x^4 \end{array}}$

$$\underline{x^4 + x^2}$$
$$-x^2$$
$$\underline{-x^2 - 1}$$
$$1$$

Thus, $f(x) = x^2 - 1 + \dfrac{1}{x^2 + 1}$.

Let $p(x) = x^2 - 1$; then $[f(x) - p(x)] = \dfrac{1}{x^2 + 1} \to 0$ as $x \to \infty$ and as $x \to -\infty$.

Thus, the graph of $f(x)$ approaches the graph of $p(x)$ (a parabola) asymptotically.

61. $x^2 - 1 \overline{\smash{\big)}\ \begin{array}{l} x^3 + x \\[2pt] x^5 \end{array}}$

$$\underline{x^5 - x^3}$$
$$x^3$$
$$\underline{x^3 - x}$$
$$x$$

Thus, $f(x) = x^3 + x + \dfrac{x}{x^2 - 1}$.

Let $p(x) = x^3 + x$; then $[f(x) - p(x)] = \dfrac{x}{x^2 - 1} \to 0$ as $x \to \infty$ and as $x \to -\infty$.

Thus, the graph of $f(x)$ approaches the graph of $p(x)$ asymptotically.

63. $f(x) = \dfrac{x^2 - 4}{x - 2}$. $f(x)$ is not defined if $x - 2 = 0$,
that is, $x = 2$

Domain: $(-\infty, 2) \cup (2, \infty)$

$f(x) = \dfrac{(x - 2)(x + 2)}{(x - 2)}$

$f(x) = x + 2$

The graph is a straight line with slope 1 and y intercept 2, except that the point $(2, 4)$ is not on the graph.

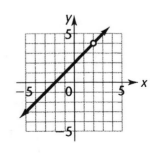

65. $r(x) = \dfrac{x + 2}{x^2 - 4}$. $f(x)$ is not defined if $x^2 - 4 = 0$, that is,

$$x^2 = 4$$
$$x = \pm 2$$

Domain: $(-\infty, -2) \cup (-2, 2) \cup (2, \infty)$

$r(x) = \dfrac{x + 2}{(x + 2)(x - 2)}$

$r(x) = \dfrac{1}{x - 2}$

The graph is the same as the graph of the function $\dfrac{1}{x - 2}$, except that the

point $\left(-2, -\dfrac{1}{4}\right)$ is not on the graph.

Intercepts: $y = -\dfrac{1}{2}$. No x intercept.

Vertical asymptote: $x = 2$

Sign chart.

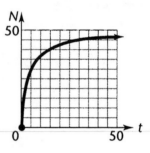

As $x \to 2^-$, $r(x) \to -\infty$. As $x \to 2^+$, $r(x) \to \infty$

Horizontal asymptote: $y = 0$

67. $N(t) = \dfrac{50t}{t + 4}$ $t \geq 0$

Intercepts: Real zeros of $50t$: $t = 0$ $N(0) = 0$

Vertical asymptotes: None, since -4, the only zero of $t + 4$, is not in the domain of N.

Sign behavior: $N(t)$ is always positive.

Horizontal asymptote: $N = 50$. As $t \to \infty$, $N \to 50$

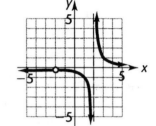

69. $N(t) = \dfrac{5t + 30}{t}$ $t \geq 1$

Intercepts: Real zeros of $5t + 30$, $t \geq 1$. None, since -6, the only zero of $5t + 30$, is not in the domain of N.

Vertical asymptotes: None, since 0, the only zero of t, is not in the domain of N.

Sign behavior: $N(t)$ is always positive.

Horizontal asymptote: $N = 5$. As $t \to \infty$, $N \to 5$

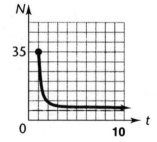

71. (A) $\overline{C}(n) = \dfrac{C(n)}{n} = \dfrac{2{,}500 + 175n + 25n^2}{n} = 25n + 175 + \dfrac{2{,}500}{n}$

(B) The minimum value of the function $\overline{C}(n)$ is $\overline{C}\left(\sqrt{\dfrac{c}{a}}\right)$, where $a = 25$ and $c = 2{,}500$

$\min \overline{C}(n) = \overline{C}\left(\sqrt{\dfrac{2{,}500}{25}}\right) = \overline{C}(\sqrt{100}) = \overline{C}(10)$

This minimum occurs when $n = 10$, after 10 years.

(C) *Intercepts:* Real zeros of $2,500 + 175n + 25n^2$.
None. No n intercepts.

0 is not in the domain of n, so there are no $\overline{C}$
intercepts.
Vertical asymptotes: Real zeros of n. The line $n = 0$
is a vertical asymptote.

Sign behavior: $\overline{C}$ is always positive since $n \geq 0$.
Horizontal asymptote: None, since the degree of $C(n)$
is greater than the degree of n.

Oblique asymptote: The line $\overline{C} = 25n + 175$ is an oblique asymptote.

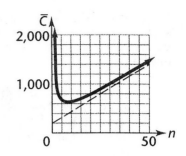

73. (A) Since Area = length × width, length = $\dfrac{\text{Area}}{\text{width}} = \dfrac{225}{x}$. Then total length of

fence = 2 × width + 2 × length
$$L(x) = 2x + \frac{450}{x} = \frac{2x^2 + 450}{x}$$

(B) x can be any positive number, thus, domain = $(0, \infty)$

(C) The minimum value of the function $L(x)$ is $L\left(\sqrt{\dfrac{c}{a}}\right)$ where $a = 2$ and $c = 450$.

$$\min L(x) = L\left(\sqrt{\frac{450}{2}}\right) = L(\sqrt{225}) = L(15)$$
This minimum occurs when $x = 15$.

Width = 15 feet. Length = $\dfrac{225}{15} = 15$ feet.

(D) *Intercepts:* Real zeros of $2x^2 + 450$. None, hence,
no x intercepts. 0 is not in the domain of L, so there
are no L intercepts.
Vertical asymptotes: Real zeros of x. The line $x = 0$
is a vertical asymptote.
Sign behavior: $L(x)$ is always positive since $x > 0$.
Horizontal asymptote: None, since the degree of
$2x^2 + 450$ is greater than the degree of x.
Oblique asymptote: The line $L = 2x$ is an oblique
asymptote.

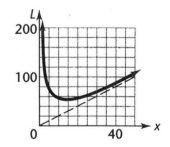

Exercise 3-5

Key Ideas and Formulas

Two polynomials are equal to each other if and only if the coefficients of like-degree terms are equal.

For a polynomial with real coefficients, there always exists a complete factoring involving only linear and/or quadratic factors with real coefficients where the linear and quadratic factors are prime relative to the real numbers.

Any proper fraction $P(x)/D(x)$ reduced to lowest terms can be decomposed into a sum of partial fractions as follows:

If $D(x)$ has:	then the decomposition of $P(x)/D(x)$ has a term of form:
1. a non-repeating linear factor of the form $ax + b$	$\dfrac{A}{ax + b}$ A a constant
2. a k-repeating linear factor of the form $(ax + b)^k$	$\dfrac{A_1}{ax + b} + \dfrac{A_2}{(ax + b)^2} + \cdots + \dfrac{A_k}{(ax + b)^k}$ $A_1, A_2, \cdots, A_k$ constant

3. a non-repeating quadratic factor of the form $ax^2 + bx + c$, prime relative to the real numbers

$$\frac{Ax + B}{ax^2 + bx + c}$$

A and B constants

4. a k-repeating quadratic factor of the form $(ax^2 + bx + c)^k$, where $ax^2 + bx + c$ is prime relative to the real numbers

$$\frac{A_1x + B_1}{ax^2 + bx + c} + \frac{A_2x + B_2}{(ax^2 + bx + c)^2} + \cdots + \frac{A_kx + B_k}{(ax^2 + bx + c)^k}$$

$A_1, \cdots, A_k$ and $B_1, \cdots, B_k$ constants

Common Errors:

1. Confusing $(ax + b)^2$ and $ax^2 + bx + c$. The first is a repeated linear factor, and leads to a partial fraction of form

$$\frac{A_1}{ax + b} + \frac{A_2}{(ax + b)^2}$$

The second is a quadratic factor and leads to a partial fraction of form

$$\frac{Ax + B}{ax^2 + bx + c}$$

2. Neglecting the first-degree terms in the numerator in cases 3 and 4.

Always consider $\dfrac{Ax + B}{ax^2 + bx + c}$. Never write $\dfrac{A}{ax^2 + bx + c}$.

---

1. If $x = -4$, then
$$-30 = 10B$$
$$B = 3$$
If $x = 6$, then
$$20 = 10A$$
$$A = 2$$
$$A = 2, \quad B = 3$$

3. If $x = -\dfrac{5}{2}$, then

$$-9\left(-\frac{5}{2}\right) + 5 = B\left[3\left(-\frac{5}{2}\right) + 2\right]$$

$$\frac{55}{2} = -\frac{11}{2}B$$

$$B = -5$$

If $x = -\dfrac{2}{3}$, then

$$-9\left(-\frac{2}{3}\right) + 5 = A\left[2\left(-\frac{2}{3}\right) + 5\right]$$

$$11 = \frac{11}{3}A$$

$$A = 3$$

$$A = 3, \quad B = -5$$

5. $\dfrac{5x^2 + 7x + 6}{(x - 1)(x + 2)^2} = \dfrac{A}{x - 1} + \dfrac{B}{x + 2} + \dfrac{C}{(x + 2)^2}$

$$= \frac{A(x + 2)^2 + B(x - 1)(x + 2) + C(x - 1)}{(x - 1)(x + 2)^2}$$

Thus, for all x

$$5x^2 + 7x + 6 = A(x + 2)^2 + B(x - 1)(x + 2) + C(x - 1)$$

If $x = -2$, then
$$12 = -3C$$
$$C = -4$$

If $x = 1$, then
$$18 = 9A$$
$$A = 2$$

If $x = 0$, then using $A = 2$ and $C = -4$, we have
$$6 = 8 - 2B + 4$$
$$B = 3$$

$$A = 2, \quad B = 3, \quad C = -4$$

7. $\dfrac{x^2 + 5x + 3}{x(x^2 + 1)} = \dfrac{A}{x} + \dfrac{Bx + C}{x^2 + 1}$

$= \dfrac{A(x^2 + 1) + (Bx + C)x}{x(x^2 + 1)}$

Thus, for all x

$x^2 + 5x + 3 = A(x^2 + 1) + (Bx + C)x$

Multiplying out the right side, we have

$x^2 + 5x + 3 = Ax^2 + A + Bx^2 + Cx$

$= (A + B)x^2 + Cx + A$

Equating coefficients of like terms, we have

$\quad A + B = 1$

$\qquad\quad C = 5$

$\qquad\quad A = 3$

Hence $A = 3$, $B = -2$, $C = 5$.

9. $\dfrac{2x^3 + x^2 - 2}{(x^2 + x + 2)^2} = \dfrac{Ax + B}{x^2 + x + 2} + \dfrac{Cx + D}{(x^2 + x + 2)^2}$

$= \dfrac{(Ax + B)(x^2 + x + 2) + Cx + D}{(x^2 + x + 2)^2}$

Thus, for all x

$2x^3 + x^2 - 2 = (Ax + B)(x^2 + x + 2) + Cx + D$

Multiplying out the right side, we have

$2x^3 + x^2 - 2 = Ax^3 + Ax^2 + 2Ax + Bx^2 + Bx + 2B + Cx + D$

$= Ax^3 + (A + B)x^2 + (2A + B + C)x + 2B + D$

Equating coefficients of like terms, we have

$\quad 2 = A$

$\quad 1 = A + B$

$\quad 0 = 2A + B + C$

$-2 = 2B + D$

Hence $A = 2$, $B = -1$, $C = -3$, $D = 0$.

11. Since $x^2 - x - 12 = (x - 4)(x + 3)$, we write

$\dfrac{3x - 40}{x^2 - x - 12} = \dfrac{A}{x - 4} + \dfrac{B}{x + 3}$

$= \dfrac{A(x + 3) + B(x - 4)}{(x - 4)(x + 3)}$

Thus, for all x

$3x - 40 = A(x + 3) + B(x - 4)$

If $x = 4$

$-28 = 7A$

$A = -4$

If $x = -3$

$-49 = -7B$

$B = 7$

So $\dfrac{3x - 40}{x^2 - x - 12} = \dfrac{-4}{x - 4} + \dfrac{7}{x + 3}$.

13. Since $6x^2 + 17x - 14 = (2x + 7)(3x - 2)$, we write

$\dfrac{8x - 22}{6x^2 + 17x - 14} = \dfrac{A}{2x + 7} + \dfrac{B}{3x - 2}$

$= \dfrac{A(3x - 2) + B(2x + 7)}{(2x + 7)(3x - 2)}$

Thus, for all x

$8x - 22 = A(3x - 2) + B(2x + 7)$

If $x = \dfrac{2}{3}$

$8\left(\dfrac{2}{3}\right) - 22 = B\left[2\left(\dfrac{2}{3}\right) + 7\right]$

$-\dfrac{50}{3} = \dfrac{25}{3}B$

$B = -2$

If $x = -\dfrac{7}{2}$

$8\left(-\dfrac{7}{2}\right) - 22 = A\left[3\left(-\dfrac{7}{2}\right) - 2\right]$

$-50 = -\dfrac{25}{2}A$

$A = 4$

So $\dfrac{8x - 22}{6x^2 + 17x - 14} = \dfrac{4}{2x + 7} - \dfrac{2}{3x - 2}$.

15. Since $4x^3 - 4x^2 + x = x(4x^2 - 4x + 1) = x(2x - 1)^2$, we write

$$\frac{6x^2 - 14x + 4}{4x^3 - 4x^2 + x} = \frac{A}{x} + \frac{B}{2x - 1} + \frac{C}{(2x - 1)^2}$$

$$= \frac{A(2x - 1)^2 + Bx(2x - 1) + Cx}{x(2x - 1)^2}$$

Thus, for all x

$$6x^2 - 14x + 4 = A(2x - 1)^2 + Bx(2x - 1) + Cx$$

If $x = \frac{1}{2}$

$$6\left(\frac{1}{2}\right)^2 - 14\left(\frac{1}{2}\right) + 4 = \frac{1}{2}C$$

$$-\frac{3}{2} = \frac{1}{2}C$$

$$C = -3$$

If $x = 0$

$$4 = A$$

If $x = 1$

$$-4 = A + B + C$$

$$-4 = 4 + B - 3 \quad \text{using } A = 4 \text{ and } C = -3$$

$$B = -5$$

So $\dfrac{6x^2 - 14x + 4}{4x^3 - 4x^2 + x} = \dfrac{4}{x} - \dfrac{5}{2x - 1} - \dfrac{3}{(2x - 1)^2}$.

17. Since $2x^3 + x^2 + x = x(2x^2 + x + 1)$, we write

$$\frac{10x^2 + 4x + 3}{2x^3 + x^2 + x} = \frac{A}{x} + \frac{Bx + C}{2x^2 + x + 1} = \frac{A(2x^2 + x + 1) + x(Bx + C)}{x(2x^2 + x + 1)}$$

> **Common Error:**
> Writing $\dfrac{B}{2x^2 + x + 1}$. Since $2x^2 + x + 1$ is quadratic, the numerator must be linear.

Thus, for all x

$$10x^2 + 4x + 3 = A(2x^2 + x + 1) + x(Bx + C)$$

$$= (2A + B)x^2 + (A + C)x + A$$

Equating coefficients of like terms, we have

$$10 = 2A + B$$

$$4 = A + C$$

$$3 = A$$

Hence $A = 3$, $B = 4$, $C = 1$.

So $\dfrac{10x^2 + 4x + 3}{2x^3 + x^2 + x} = \dfrac{3}{x} + \dfrac{4x + 1}{2x^2 + x + 1}$.

19. Since $x^4 + 2x^2 + 1 = (x^2 + 1)^2$, we write

$$\frac{4x^3 - 5x^2 + 6x - 5}{x^4 + 2x^2 + 1} = \frac{Ax + B}{x^2 + 1} + \frac{Cx + D}{(x^2 + 1)^2} = \frac{(Ax + B)(x^2 + 1) + (x + D)}{(x^2 + 1)^2}$$

Thus, for all x

$$4x^3 - 5x^2 + 6x - 5 = (Ax + B)(x^2 + 1) + Cx + D$$

$$= Ax^3 + Bx^2 + (A + C)x + B + D$$

Equating coefficients of like terms, we have

$$4 = A$$

$$-5 = B$$

$$6 = A + C$$

$$-5 = B + D$$

Hence $A = 4$, $B = -5$, $C = 2$, $D = 0$

So $\dfrac{4x^3 - 5x^2 + 6x - 5}{x^4 + 2x^2 + 1} = \dfrac{4x - 5}{x^2 + 1} + \dfrac{2x}{(x^2 + 1)^2}$.

21. First we must factor $x^3 + x + 2$. Possible rational zeros of this polynomial are ± 1, ± 2. Forming a synthetic division table, we see

	1	0	1	2
1	1	1	2	4
-1	1	-1	2	0

Thus, $x^3 + x + 2 = (x + 1)(x^2 - x + 2)$.

$x^2 - x + 2$ cannot be factored further in the real numbers.

So $\dfrac{4x^2 - 5x + 3}{x^3 + x + 2} = \dfrac{A}{x + 1} + \dfrac{Bx + C}{x^2 - x + 2} = \dfrac{A(x^2 - x + 2) + (Bx + C)(x + 1)}{(x + 1)(x^2 - x + 2)}$

Thus, for all x

$$
\begin{aligned}
4x^2 - 5x + 3 &= A(x^2 - x + 2) + (Bx + C)(x + 1) \\
&= Ax^2 - Ax + 2A + Bx^2 + Bx + Cx + C \\
&= (A + B)x^2 + (-A + B + C)x + 2A + C
\end{aligned}
$$

Before equating coefficients of like terms, we note that if $x = -1$

$$
\begin{aligned}
4(-1)^2 - 5(-1) + 3 &= A[(-1)^2 - (-1) + 2] \\
12 &= 4A \\
A &= 3
\end{aligned}
$$

Since $4 = A + B$

$\quad\quad -5 = -A + B + C$

$\quad\quad 3 = 2A + C$

we have $A = 3$, $B = 1$, $C = -3$.

So $\dfrac{4x^2 - 5x + 3}{x^3 + x + 2} = \dfrac{3}{x + 1} + \dfrac{x - 3}{x^2 - x + 2}$

23. First we divide to obtain a plynomial plus a proper fraction.

$$
\begin{array}{r}
x + 2 \\
x^4 - 2x^3 - 8x + 16\overline{\smash{\big)}\,x^5 + 0x^4 + 0x^3 - 4x^2 - 16x + 40} \\
\underline{x^5 - 2x^4 \qquad\qquad - 8x^2 + 16x} \\
2x^4 + 0x^3 + 4x^2 - 32x + 40 \\
\underline{2x^4 - 4x^3 \qquad\qquad - 16x + 32} \\
4x^3 + 4x^2 - 16x + 8
\end{array}
$$

Now we must decompose $\dfrac{4x^3 + 4x^2 - 16x + 8}{x^4 - 2x^3 - 8x + 16}$, starting by factoring

$x^4 - 2x^3 - 8x + 16$. We may use the methods of this chapter, or note that the polynomial factors by grouping:

$$
\begin{aligned}
x^4 - 2x^3 - 8x + 16 &= (x^4 - 2x^3) - (8x - 16) \\
&= x^3(x - 2) - 8(x - 2) \\
&= (x - 2)(x^3 - 8) \\
&= (x - 2)(x - 2)(x^2 + 2x + 4)
\end{aligned}
$$

$x^2 + 2x + 4$ cannot be factored further in the integers, so we can write:

$$
\begin{aligned}
\dfrac{4x^3 + 4x^2 - 16x + 8}{x^4 - 2x^3 - 8x + 16} &= \dfrac{A}{x - 2} + \dfrac{B}{(x - 2)^2} + \dfrac{Cx + D}{x^2 + 2x + 4} \\
&= \dfrac{A(x - 2)(x^2 + 2x + 4) + B(x^2 + 2x + 4) + (Cx + D)(x - 2)^2}{(x - 2)^2(x^2 + 2x + 4)}
\end{aligned}
$$

Thus, for all x

$4x^3 + 4x^2 - 16x + 8 = A(x - 2)(x^2 + 2x + 4) + B(x^2 + 2x + 4) + (Cx + D)(x - 2)^2$

If $x = 2$

$$4(2)^3 + 4(2)^2 - 16(2) + 8 = B(2^2 + 2 \cdot 2 + 4)$$
$$24 = 12B$$
$$B = 2$$

Now $4x^3 + 4x^2 - 16x + 8 = A(x^3 - 8) + 2(x^2 + 2x + 4) + (Cx + D)(x^2 - 4x + 4)$

$$= Ax^3 - 8A + 2x^2 + 4x + 8 + Cx^3 - 4Cx^2 + 4Cx + Dx^2 - 4Dx + 4D$$

$$= (A + C)x^3 + (2 - 4C + D)x^2 + (4 + 4C - 4D)x - 8A + 8 + 4D$$

Equating coefficients of like terms, we have

$$4 = A + C$$
$$4 = 2 - 4C + D$$
$$-16 = 4 + 4C - 4D$$
$$8 = -8A + 8 + 4D$$

Thus $-12 = 6 - 3D$

$$D = 6$$

Hence $C = 1$, $A = 3$.

So $\dfrac{x^5 - 4x^2 - 16x + 40}{x^4 - 2x^3 - 8x + 16} = x + 2 + \dfrac{3}{x - 2} + \dfrac{2}{(x - 2)^2} + \dfrac{x + 6}{x^2 + 2x + 4}$.

25. First we divide to obtain a polynomial plus a proper fraction.

$$
\begin{array}{r}
2x \\
x^4 - 4x^2 - 12x - 9\overline{\smash{\big)}\,2x^5 + 0x^4 + 0x^3 - 7x^2 + 20x + 21} \\
\underline{2x^5 - 8x^3 - 24x^2 - 18x } \\
8x^3 + 17x^2 + 38x + 21
\end{array}
$$

Now we must decompose $\dfrac{8x^3 + 17x^2 + 38x + 21}{x^4 - 4x^2 - 12x - 9}$, starting by factoring $x^4 - 4x^2 - 12x - 9$. Possible rational zeros are ± 1, ± 3, ± 9.
We form a synthetic division table:

	1	0	-4	-12	-9	
1	1	1	-3	-15	-24	
3	1	3	5	3	0	3 is a zero

We investigate $x^3 + 3x^2 + 5x + 3$; the only remaining possible rational zeros are 3, -1, -3.

	1	3	5	3	
3	1	6	23	72	
-1	1	2	3	0	-1 is a zero

So $x^4 - 4x^2 - 12x - 9 = (x - 3)(x + 1)(x^2 + 2x + 3)$.

$x^2 + 2x + 3$ cannot be factored further in the integers, so we write:

$$\frac{8x^3 + 17x^2 + 38x + 21}{x^4 - 4x^2 - 12x - 9} = \frac{A}{x - 3} + \frac{B}{x + 1} + \frac{Cx + D}{x^2 + 2x + 3}$$

$$= \frac{A(x + 1)(x^2 + 2x + 3) + B(x - 3)(x^2 + 2x + 3) + (Cx + D)(x - 3)(x + 1)}{(x - 3)(x + 1)(x^2 + 2x + 3)}$$

Thus, for all x

$8x^3 + 17x^2 + 38x + 21$

$$= A(x + 1)(x^2 + 2x + 3) + B(x - 3)(x^2 + 2x + 3) + (Cx + D)(x - 3)(x + 1)$$

If $x = 3$

$$8(3)^3 + 17(3)^2 + 38(3) + 21 = A(3 + 1)(3^2 + 2 \cdot 3 + 3)$$
$$504 = 72A$$
$$A = 7$$

If $x = -1$

$$8(-1)^3 + 17(-1)^2 + 38(-1) + 21 = B(-1 - 3)[(-1)^2 + 2(-1) + 3]$$
$$-8 = -8B$$
$$B = 1$$

If $x = 0$

$$21 = 3A - 9B - 3D$$
$$21 = 21 - 9 - 3D$$
$$D = -3$$

If $x = 1$

$$8(1)^3 + 17(1)^2 + 38(1) + 21$$
$$= A(1 + 1)(1^2 + 2 \cdot 1 + 3) + B(1 - 3)(1^2 + 2 \cdot 1 + 3) + (C + D)(1 - 3)(1 + 1)$$
$$84 = 12A - 12B - 4C - 4D$$
$$84 = 84 - 12 - 4C + 12$$
$$C = 0$$

So $\dfrac{2x^5 - 7x^2 + 20x + 21}{x^4 - 4x^2 - 12x - 9} = 2x + \dfrac{7}{x - 3} + \dfrac{1}{x + 1} - \dfrac{3}{x^2 + 2x + 3}$

27. $\dfrac{x}{(x + a)^2} = \dfrac{A}{x + a} + \dfrac{B}{(x + a)^2}$

$$= \dfrac{A(x + a) + B}{(x + a)^2}$$

Thus, for all x

$$x = A(x + a) + B$$

If $x = -a$

$$-a = B$$

If $x = 0$

$$0 = Aa + B$$
$$0 = Aa - a$$
$$A = 1$$

So $\dfrac{x}{(x + a)^2} = \dfrac{1}{x + a} - \dfrac{a}{(x + a)^2}$.

29. $\dfrac{1}{(x - a)(x - b)} = \dfrac{A}{x - a} + \dfrac{B}{x - b}$

$$= \dfrac{A(x - b) + B(x - a)}{(x - a)(x - b)}$$

Thus, for all x

$$1 = A(x - b) + B(x - a)$$

If $x = a$

$$1 = A(a - b)$$
$$A = \dfrac{1}{a - b}$$

If $x = b$

$$1 = B(b - a)$$
$$B = \dfrac{1}{b - a}$$

So $\dfrac{1}{(x - a)(x - b)}$

$$= \dfrac{1}{(a - b)(x - a)} + \dfrac{1}{(b - a)(x - b)}.$$

CHAPTER 3 REVIEW

1.

$$\begin{array}{r} 2 \quad\; 3 \quad\; 0 \quad -1 \\ -4 \quad\; 2 \quad -4 \\ \hline -2\,\overline{)\, 2 \quad -1 \quad\; 2 \quad -5} \end{array}$$

$$2x^3 + 3x^2 - 1$$
$$= (x + 2)(2x^2 - x + 2) - 5 \qquad (3-1)$$

2.

$$\begin{array}{r} 1 \quad -4 \quad\; 0 \quad\; 9 \quad\; 0 \quad -8 \\ 3 \quad -3 \quad -9 \quad\; 0 \quad\; 0 \\ \hline 3\,\overline{)\, 1 \quad -1 \quad -3 \quad\; 0 \quad\; 0 \quad -8} \end{array}$$

$$P(3) = -8 \qquad (3\text{-}1, 3\text{-}2)$$

3. $2, -4, -1$ \qquad $(3\text{-}2)$

4. Since complex zeros come in conjugate pairs, $1 - i$ is a zero. \qquad $(3\text{-}3)$

5. (A) Since the graph has x intercepts -2, 0, and 2, these are zeros of $P(x)$. Therefore, $P(x) = (x + 2)x(x - 2) = x^3 - 4x$.

(B) $P(x) \to \infty$ as $x \to \infty$ and $P(x) \to -\infty$ as $x \to -\infty$.   *(3-1)*

6. We form a synthetic division table:

	1	-4	0	2	
-2	1	-6	12	-22	both are lower bounds, since
-1	1	-5	5	-3	both rows alternate in sign
3	1	-1	-3	-7	
4	1	0	0	2	upper bound

(3-3)

7. We investigate $P(1)$ and $P(2)$ by forming a synthetic division table.

	2	-3	1	-5
1	2	-1	0	-5
2	2	1	3	1

Since $P(1)$ and $P(2)$ have opposite signs, there is at least one real zero between 1 and 2.   *(3-3)*

8. The factors of 6 are ± 1, ± 2, ± 3, ± 6.   *(3-2)*

9. Using the possibilities found in problem 8, we form a synthetic division table:

	1	-4	1	6	
1	1	-3	-2	4	
2	1	-2	-3	0	2 is a zero

Thus $x^3 - 4x^2 + x + 6 = (x - 2)(x^2 - 2x - 3) = (x - 2)(x - 3)(x + 1)$. The rational zeros are 2, 3, -1.   *(3-2)*

10. (A) $f(x) = \dfrac{2x - 3}{x + 4} = \dfrac{n(x)}{d(x)}$

The domain of f is the set of all real numbers x such that $d(x) = x + 4 \neq 0$, that is, $(-\infty, -4) \cup (-4, \infty)$. f has an x intercept where $n(x) = 2x - 3 = 0$, that is, $x = \dfrac{3}{2}$

(B) $g(x) = \dfrac{3x}{x^2 - x - 6} = \dfrac{n(x)}{d(x)}$

The domain of g is the set of all real numbers x such that $d(x) = x^2 - x - 6 \neq 0$, that is, $(x + 2)(x - 3) \neq 0$, that is, $x \neq -2$, 3, or $(-\infty, -2) \cup (-2, 3) \cup (3, \infty)$. g has an x intercept where $n(x) = 3x = 0$, that is, $x = 0$ *(3-4)*

11. (A) Horizontal asymptote: since $n(x)$ and $d(x)$ have the same degree, the line $y = 2$ is a horizontal asymptote. Vertical asymptotes: zeros of $d(x)$

$x + 4 = 0$
$\quad x = -4$

(B) Horizontal asymptote: since the degree of $n(x)$ is less than the degree of $d(x)$, the line $y = 0$ is a horizontal asymptote.
Vertical asymptotes: zeros of $d(x)$

$x^2 - x - 6 = 0$
$(x + 2)(x - 3) = 0$
$\quad x = -2, \quad x = 3$ *(3-4)*

12. We write
$$\frac{7x - 11}{(x - 3)(x + 2)} = \frac{A}{x - 3} + \frac{B}{x + 2}$$
$$= \frac{A(x + 2) + B(x - 3)}{(x - 3)(x + 2)}$$

Thus for all x
$$7x - 11 = A(x + 2) + B(x - 3)$$

If $x = -2$ If $x = 3$
$-25 = -5B$ $10 = 5A$
$B = 5$ $A = 2$

So
$$\frac{7x - 11}{(x - 3)(x + 2)} = \frac{2}{x - 3} + \frac{5}{x + 2} \qquad (3\text{-}5)$$

13. (A) We form a synthetic division table:

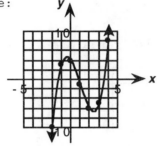

	1	-3	-3	4	
-2	1	-5	7	-10	= P(-2)
-1	1	-4	1	3	= P(-1)
0	1	-3	-3	4	= P(0)
1	1	-2	-5	-1	= P(1)
2	1	-1	-5	-6	= P(2)
3	1	0	-3	-5	= P(3)
4	1	1	1	8	= P(4)

The graph of $P(x)$ has three x intercepts and two turning points;
$P(x) \to \infty$ as $x \to \infty$ and $P(x) \to -\infty$ as $x \to -\infty$

(B) We search for the real zero in $(3, 4)$. We organize our calculations in a table.

Sign Change Interval (a, b)	Midpoint m	Sign of P		
		$P(a)$	$P(m)$	$P(b)$
(3, 4)	3.5	−	−	+
(3.5, 4)	3.75	−	+	+
(3.5, 3.75)	3.625	−	+	+
(3.5, 3.625)	3.5625	−	+	+
(3.5, 3.5625)	3.53125	−	+	+
(3.5, 3.53125)	We stop here	−		+

Since each endpoint rounds to 3.5, a real zero lies on this last interval and
is given by 3.5 to one decimal place accuracy. $(3\text{-}1,\ 3\text{-}3)$

14. We use synthetic divison:

```
     8  -14  -13   -4    7
          2   -3   -4   -2
  ──┬──────────────────────
  1/4│ 8  -12  -16   -8    5
```

Thus,
$$P(x) = \left(x - \frac{1}{4}\right)(8x^3 - 12x^2 - 16x - 8) + 5$$
$$P\left(\frac{1}{4}\right) = 5 \qquad\qquad (3\text{-}1)$$

15.

```
         4   -8   -3   -3
             -2    5   -1
   ──┬──────────────────────
 -1/2│ 4  -10    2   -4
```

$$P\left(-\frac{1}{2}\right) = -4 \qquad (3\text{-}1)$$

16. The quadratic formula tells us that $x^2 - 2x - 1 = 0$ if

$$x = \frac{-b \pm \sqrt{b^2 - 4ac}}{2a} \qquad a = 1, \ b = -2, \ c = -1$$

$$x = \frac{-(-2) \pm \sqrt{(-2)^2 - 4(1)(-1)}}{2(1)}$$

$$= \frac{2 \pm \sqrt{8}}{2}$$

$$= 1 \pm \sqrt{2}$$

Since $1 \pm \sqrt{2}$ are zeros of $x^2 - 2x - 1$, its factors are $x - (1 + \sqrt{2})$ and $x - (1 - \sqrt{2})$, that is, $x^2 - 2x - 1 = [x - (1 + \sqrt{2})][x - (1 - \sqrt{2})]$ *(3-2)*

17. $x + 1$ will be a factor of $P(x)$ if $P(-1) = 0$.
$P(-1) = 9(-1)^{26} - 11(-1)^{17} + 8(-1)^{11} - 5(-1)^4 - 7 = 9 + 11 - 8 - 5 - 7 = 0$, so the answer is yes, $x + 1$ is a factor.   *(3-2)*

18. The possible rational zeros are ± 1, ± 2, ± 4, ± 8, $\pm\frac{1}{2}$. We form a synthetic division table:

	2	-3	-18	-8
1	2	-1	-19	-27
2	2	1	-16	-40
4	2	5	2	0

So $2x^3 - 3x^2 - 18x - 8 = (x - 4)(2x^2 + 5x + 2)$
$$= (x - 4)(2x + 1)(x + 2)$$

Zeros: 4, $-\frac{1}{2}$, -2 *(3-2)*

19. $(x - 4)(2x + 1)(x + 2)$ *(3-2)*

20. The possible rational zeros are ± 1, ± 5. We form a synthetic division table:

	1	-3	0	5
1	1	-2	-2	3
5	1	2	10	55
-1	1	-4	4	1
-5	1	-8	40	-195

There are no rational zeros, since all possibilities fail.   *(3-2)*

21. $P(x) = 2x^4 - x^3 + 2x - 1$

We can factor $P(x)$ by grouping into $(2x - 1)(x^3 + 1) = (2x - 1)(x + 1)(x^2 - x + 1)$. However, if we don't notice this, we find the possible rational zeros to be ± 1, $\pm\frac{1}{2}$. We form a synthetic division table:

	2	-1	0	2	-1	
1	2	1	1	3	2	
-1	2	-3	3	-1	0	-1 is a zero

We now examine $2x^3 - 3x^2 + 3x - 1$. Only -1, $\frac{1}{2}$, and $-\frac{1}{2}$ remain as possible rational zeros; We form a synthetic division table:

	2	-3	3	-1	
-1	2	-5	8	-9	Not a double zero
$\frac{1}{2}$	2	-2	2	0	$\frac{1}{2}$ is a zero

So $P(x) = (x + 1)\left(x - \frac{1}{2}\right)(2x^2 - 2x + 2) = (x + 1)(2x - 1)(x^2 - x + 1)$

To find the remaining zeros, we solve $x^2 - x + 1 = 0$, by the quadratic formula.
$x^2 - x + 1 = 0$

$$x = \frac{-b \pm \sqrt{b^2 - 4ac}}{2a} \quad a = 1, \ b = -1, \ c = 1$$

$$x = \frac{-(-1) \pm \sqrt{(-1)^2 - 4(1)(1)}}{2(1)}$$

$$x = \frac{1 \pm \sqrt{-3}}{2}$$

$$x = \frac{1 \pm i\sqrt{3}}{2}$$

The four zeros are -1, $\frac{1}{2}$, and $\frac{1 \pm i\sqrt{3}}{2}$ (3-2)

22. $(x + 1)\left(x - \frac{1}{2}\right)2\left(x - \frac{1 + i\sqrt{3}}{2}\right)\left(x - \frac{1 - i\sqrt{3}}{2}\right) = (x + 1)(2x - 1)\left(x - \frac{1 + i\sqrt{3}}{2}\right)\left(x - \frac{1 - i\sqrt{3}}{2}\right)$
(3-2)

23. $2x^3 + 3x^2 \le 11x + 6$
$2x^3 + 3x^2 - 11x - 6 \le 0$
To factor $2x^3 + 3x^2 - 11x - 6$, we search for zeros of the polynomial. Possible rational zeros are ± 1, ± 2, ± 3, ± 6, $\pm\frac{1}{2}$, and $\pm\frac{3}{2}$. We form a synthetic division table:

	2	3	-11	-6	
1	2	5	-6	-12	
2	2	7	3	0	2 is a zero

$2x^3 + 3x^2 - 11x - 6 = (x - 2)(2x^2 + 7x + 3) = (x - 2)(x + 3)(2x + 1)$
Hence we must examine the sign behavior of $(x - 2)(x + 3)(2x + 1)$.
We form a sign chart.
Zeros: -3, $-\frac{1}{2}$, 2

$(-\infty, -3)\ (-3, -1/2)\ (-1/2, 2)\ (2, \infty)$

$2x^3 + 3x^2 - 11x - 6 = (x - 2)(x + 3)(2x + 1)$				
Test Number	-4	-1	0	3
Value of Polynomial for Test Number	-42	6	-6	42
Sign of Polynomial in Interval	−	+	−	+
Interval	$(-\infty, -3)$	$(-3, -\frac{1}{2})$	$(-\frac{1}{2}, 2)$	$(2, \infty)$

$2x^3 + 3x^2 - 11x - 6 \le 0$ and $2x^3 + 3x^2 \le 11x + 6$ within the intervals $(-\infty, -3]$, and $\left[-\frac{1}{2}, 2\right]$, or $x \le -3$ or $-\frac{1}{2} \le x \le 2$.   (3-2, 1-8)

24. (A) We form a synthetic division table.

	1	-2	-30	0	-25	
0	1	-2	-30	0	-25	
1	1	-1	-31	-31	-56	
2	1	0	-30	-60	-145	
3	1	1	-27	-81	-268	
4	1	2	-22	-88	-377	
5	1	3	-15	-75	-400	
6	1	4	-6	-36	-241	
7	1	5	5	35	220	7 is an upper bound
-1	1	-3	-27	27	-52	
-2	1	-4	-22	44	-113	
-3	1	-5	-15	45	-160	
-4	1	-6	-6	24	-121	
-5	1	-7	5	-25	100	-5 is a lower bound

(B) We search for the real zero in (6, 7) indicated in the table. We organize our calculations in a table.

Sign Change Interval (a, b)	Midpoint m	Sign of P $P(a)$	$P(m)$	$P(b)$
(6, 7)	6.5	−	−	+
(6.5, 7)	6.75	−	+	+
(6.5, 6.75)	6.625	−	+	+
(6.5, 6.625)	6.5625	−	−	+
(6.5625, 6.625)	6.59375	−	−	+
(6.59375, 6.625)	6.609375	−	−	+
(6.609375, 6.625)	6.6171875	−	−	+
(6.6171875, 6.625)	6.62109375	−	+	+
(6.6171875, 6.62109375)	We stop here	−		+

Since each endpoint rounds to 6.62, a real zero lies on this last interval and is given by 6.62 to two decimal place accuracy.

(C) Here is a computer-generated graph of P.

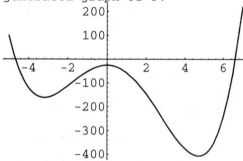

The x intercepts are on the intervals (−4.8, −4.4) and (6.4, 6.8). To estimate more accurately, we zoom in. Here are graphs of P on these intervals.

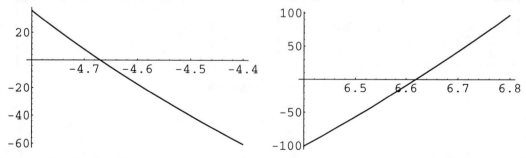

To two decimal place accuracy, the zeros are −4.67 and 6.62.

(3-3)

25. $f(x) = \dfrac{x - 1}{2x + 2} = \dfrac{n(x)}{d(x)}$

(A) The domain of f is the set of all real numbers x such that $d(x) = 2x + 2 \neq 0$, that is $(-\infty, -1) \cup (-1, \infty)$.

f has an x intercept where $n(x) = x - 1 = 0$, that is, $x = 1$. $f(0) = -\dfrac{1}{2}$, hence f has a y intercept at $y = -\dfrac{1}{2}$

(B) Vertical asymptote: $x = -1$. Horizontal asymptote: since $n(x)$ and $d(x)$ have the same degree, the line $y = \dfrac{1}{2}$ is a horizontal asymptote.

(C) *Sign Chart.*

Test numbers	-2	0	2
Value of f	$\frac{3}{2}$	$-\frac{1}{2}$	$\frac{1}{6}$
Sign of f	+	−	+

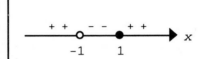

Since $x = -1$ is a vertical asymptote and $f(x) > 0$ for $x < -1$, and $f(x) < 0$ for $-1 < x < 1$

$$f(x) \rightarrow \infty \text{ as } x \rightarrow -1^-$$

and $f(x) \rightarrow -\infty$ as $x \rightarrow -1^+$

Complete the sketch.

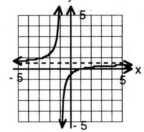

$(3\text{-}4)$

26. $\dfrac{-x^2 + 3x + 4}{x(x - 2)^2} = \dfrac{A}{x} + \dfrac{B}{x - 2} + \dfrac{C}{(x - 2)^2} = \dfrac{A(x - 2)^2 + Bx(x - 2) + Cx}{x(x - 2)^2}$

Thus for all x

$-x^2 + 3x + 4 = A(x - 2)^2 + Bx(x - 2) + Cx$

If $x = 0$

$4 = 4A$

$A = 1$

If $x = 2$

$6 = 2C$

$C = 3$

If $x = 1$, using $A = 1$ and $C = 3$

$6 = 1 - B + 3$

$B = -2$

So $\dfrac{-x^2 + 3x + 4}{x(x - 2)^2} = \dfrac{1}{x} - \dfrac{2}{x - 2} + \dfrac{3}{(x - 2)^2}$

$(3\text{-}5)$

27. First we factor $2x^3 - 3x^2 + 3x = x(2x^2 - 3x + 3)$. $2x^2 - 3x + 3$ cannot be factored further in the real numbers, so we write

$$\frac{8x^2 - 10x + 9}{2x^3 - 3x^2 + 3x} = \frac{A}{x} + \frac{Bx + C}{2x^2 - 3x + 3} = \frac{A(2x^2 - 3x + 3) + (Bx + C)x}{x(2x^2 - 3x + 3)}$$

Thus, for all x

$$8x^2 - 10x + 9 = A(2x^2 - 3x + 3) + (Bx + C)x$$

If $x = 0$

$$9 = 3A$$
$$A = 3$$

$$8x^2 - 10x + 9 = 3(2x^2 - 3x + 3) + Bx^2 + Cx = (6 + B)x^2 + (-9 + C)x + 9$$

Equating coefficients of like terms, we have

$$8 = 6 + B$$
$$-10 = -9 + C$$
$$B = 2$$
$$C = -1$$

So $\dfrac{8x^2 - 10x + 9}{2x^3 - 3x^2 + 3x} = \dfrac{3}{x} + \dfrac{2x - 1}{2x^2 - 3x + 3}$ $(3-5)$

28.

$$
\begin{array}{r}
\begin{array}{ccccc}
1 & 0 & 3 & 2 \\
 & 1 + i & 2i & 1 + 5i \\
\end{array} \\
\hline
1 + i\,|\,\begin{array}{cccc} 1 & 1 + i & 3 + 2i & 3 + 5i \end{array}
\end{array}
$$

$$(1 + i)^2 = (1 + i)(1 + i) = 1 + 2i + i^2 = 1 + 2i - 1 = 2i$$
$$(1 + i)(3 + 2i) = 3 + 5i + 2i^2 = 3 + 5i - 2 = 1 + 5i$$
$$P(x) = [x^2 + (1 + i)x + (3 + 2i)][x - (1 + i)] + 3 + 5i$$ $(3-1)$

29. $P(x) = \left(x + \dfrac{1}{2}\right)^2 (x + 3)(x - 1)^3$. The degree is 6. $(3-2)$

30. $P(x) = (x + 5)[x - (2 - 3i)][x - (2 + 3i)]$. The degree is 3. $(3-2)$

31. The possible rational zeros are ± 1, ± 2, ± 4, $\pm \dfrac{1}{2}$. We form a synthetic division table:

	2	-5	-8	21	0	-4	
1	2	-3	-11	10	10	6	
2	2	-1	-10	1	2	0	2 is a zero

We continue to examine $2x^4 - x^3 - 10x^2 + x + 2$. The possible rational zeros are -1, ± 2, and $\pm \dfrac{1}{2}$. We form a synthetic division table:

	2	-1	-10	1	2	
2	2	3	-4	-7	-12	Not a double zero
-1	2	-3	-7	8	-6	
-2	2	-5	0	1	0	-2 is a zero

Hence $P(x) = (x - 2)(x + 2)(2x^3 - 5x^2 + 1)$. $2x^3 - 5x^2 + 1$ has been shown (see Exercise 4-3, problem 27 for details) to have zeros $\dfrac{1}{2}$, $1 \pm \sqrt{2}$.

Hence $P(x)$ has zeros $\dfrac{1}{2}$, ± 2, $1 \pm \sqrt{2}$. $(3-2)$

32. $(x - 2)(x + 2)\left(x - \dfrac{1}{2}\right)2[x - (1 - \sqrt{2})][x - (1 + \sqrt{2})]$

$= (x - 2)(x + 2)(2x - 1)[x - (1 - \sqrt{2})][x - (1 + \sqrt{2})]$ $(3-2)$

33. $\dfrac{4x^2 + 4x - 3}{2x^3 + 3x^2 - 11x - 6} \geq 0$

We need to form a sign chart for $\dfrac{P}{Q} = \dfrac{4x^2 + 4x - 3}{2x^3 + 3x^2 - 11x - 6}$.

We first locate the zeros of P and Q.

$4x^2 + 4x - 3 = (2x - 1)(2x + 3)$, hence P has zeros at $\dfrac{1}{2}$ and $-\dfrac{3}{2}$. The zeros of Q were found in problem 23 to be -3, $-\dfrac{1}{2}$, and 2. We now form the sign chart, with zeros -3, $-\dfrac{3}{2}$, $-\dfrac{1}{2}$, $\dfrac{1}{2}$, and 2.

Open dots at zeros of Q.
Solid dots at zeros of P.

$\dfrac{P}{Q} = \dfrac{(2x - 1)(2x + 3)}{(x - 2)(2x + 1)(x + 3)}$						
Test Number	-4	-2	-1	0	1	3
Value of $\dfrac{P}{Q}$	$-\dfrac{15}{14}$	$\dfrac{5}{12}$	$-\dfrac{1}{2}$	$\dfrac{1}{2}$	$-\dfrac{5}{12}$	$\dfrac{15}{14}$
Sign of $\dfrac{P}{Q}$	$-$	$+$	$-$	$+$	$-$	$+$
Interval	$(-\infty, -3)$	$(-3, -\dfrac{3}{2})$	$(-\dfrac{3}{2}, -\dfrac{1}{2})$	$(-\dfrac{1}{2}, \dfrac{1}{2})$	$(\dfrac{1}{2}, 2)$	$(2, \infty)$

$\dfrac{4x^2 + 4x - 3}{2x^3 + 3x^2 - 11x - 6} \geq 0$ within the intervals $\left(-3, -\dfrac{3}{2}\right] \cup \left(-\dfrac{1}{2}, \dfrac{1}{2}\right] \cup (2, \infty)$, or

$-3 < x \leq -\dfrac{3}{2}$ or $-\dfrac{1}{2} < x \leq \dfrac{1}{2}$ or $x > 2$. *(3-2, 1-8)*

34. $P(x)$ changes sign three times. Therefore, it has three zeros and its minimal degree is 3. *(3-3)*

35. Since $1 + 2i$ is a zero, $1 - 2i$ is also a zero. Hence
$[x - (1 - 2i)][x - (1 + 2i)] = [(x - 1) + 2i][(x - 1) - 2i] = (x - 1)^2 - 4i^2$
$= x^2 - 2x + 5$ is a factor. Since $P(x)$ is a cubic polynomial, it must be of the form $a(x - r)(x^2 - 2x + 5)$. Since the constant term of this polynomial, $-5ar$, must be an integer, r must be a rational number. Thus there can be no irrational zeros. *(3-2)*

36. (A) Since $x^3 - 27$ is a cubic polynomial, it has 3 zeros and there are 3 cube roots of 27.

(B) $x^3 - 27 = (x - 3)(x^2 + 3x + 3^2) = (x - 3)(x^2 + 3x + 9)$. We solve $x^2 + 3x + 9 = 0$ by applying the quadratic formula with $a = 1$, $b = 3$, $c = 9$,

to obtain $x = \dfrac{-3 \pm \sqrt{3^2 - 4(1)(9)}}{2(1)} = \dfrac{-3 \pm \sqrt{-27}}{2} = \dfrac{-3 \pm 3i\sqrt{3}}{2}$ or $-\dfrac{3}{2} \pm \dfrac{3i}{2}\sqrt{3}$ *(3-2)*

37. (A) We form a synthetic division table.

	1	2	-500	0	-4,000
0	1	2	-500	0	-4,000
10	1	12	-380	-3,800	-42,000
20	1	22	-60	-1,200	-28,000
30	1	32	460	13,800	410,000
-10	1	-8	-420	4,200	-46,000
-20	1	-18	-140	2,800	-60,000
-30	1	-28	340	-10,200	302,000

From the table, 30 is an upper bound and -30 is a lower bound.

(B) Here is a computer-generated graph of P.

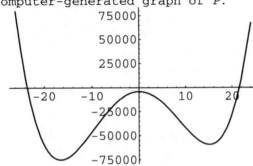

The x intercepts are on the intervals $(-24, -23)$ and $(21, 22)$. To estimate more accurately, we zoom in. Here are graphs of P on the intervals.

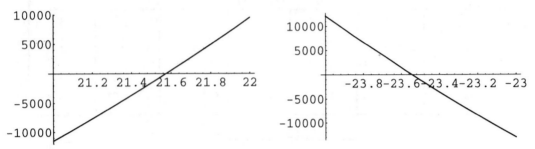

To two decimal place accuracy, the zeros are -23.54 and 21.57. $(3-3)$

38. $f(x) = \dfrac{x^2 + 2x + 3}{x + 1} = \dfrac{n(x)}{d(x)}$

Intercepts. There are no real zeros of $n(x) = x^2 + 2x + 3$. No x intercept
$f(0) = 3$ y intercept
Vertical asymptotes. Real zeros of $d(x) = x + 1$ $x = -1$
Sign Chart.

Test numbers	-2	0
Value of f	-3	3
Sign of f	–	+

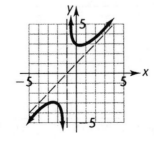

Since $x = -1$ is a vertical asymptote and $f(x) < 0$ for $x < -1$, and $f(x) > 0$ for $x > -1$
$$f(x) \to -\infty \text{ as } x \to -1^- \text{ and } f(x) \to \infty \text{ as } x \to -1^+$$
Horizontal asymptotes. Since the degree of $n(x)$ is greater than the degree of $d(x)$, there is no horizontal asymptote.
Oblique asymptote:

$$
\begin{array}{r}
x + 1 \\
x + 1 \overline{)\, x^2 + 2x + 3} \\
\underline{x^2 + x} \\
x + 3 \\
\underline{x + 1} \\
2
\end{array}
$$

Thus, $f(x) = x + 1 + \dfrac{2}{x + 1}$.
Hence, the line $y = x + 1$ is an oblique asymptote.

Complete the sketch

$(3-4)$

39. Here is a computer-generated graph of $f(x)$.

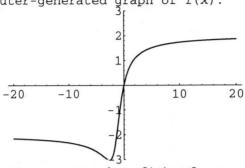

From the graph, we can see that $f(x) \to 2$ as $x \to \infty$ and $f(x) \to -2$ as $x \to -\infty$; the lines $y = 2$ and $y = -2$ are horizontal asymptotes. *(3-4)*

40. First we factor $x^4 - 3x^3 + x^2 - 3x = x(x^3 - 3x^2 + x - 3) = x[x^2(x - 3) + 1(x - 3)]$
$= x(x - 3)(x^2 + 1)$. We write

$$\frac{5x^2 + 2x + 9}{x^4 - 3x^3 + x^2 - 3x} = \frac{A}{x} + \frac{B}{x - 3} + \frac{Cx + D}{x^2 + 1}$$

$$= \frac{A(x - 3)(x^2 + 1) + Bx(x^2 + 1) + x(x - 3)(Cx + D)}{x(x - 3)(x^2 + 1)}$$

Thus, for all x
$$5x^2 + 2x + 9 = A(x - 3)(x^2 + 1) + Bx(x^2 + 1) + x(x - 3)(Cx + D)$$

If $x = 0$
$$9 = -3A$$
$$A = -3$$

If $x = 3$
$$60 = 30B$$
$$B = 2$$

$$5x^2 + 2x + 9 = -3(x - 3)(x^2 + 1) + 2x(x^2 + 1) + x(x - 3)(Cx + D)$$
$$= -3(x^3 - 3x^2 + x - 3) + 2x^3 + 2x + (x^2 - 3x)(Cx + D)$$
$$= -3x^3 + 9x^2 - 3x + 9 + 2x^3 + 2x + (x^2 - 3x)(Cx + D)$$
$$= -x^3 + 9x^2 - x + 9 + Cx^3 - 3Cx^2 + Dx^2 - 3Dx$$
$$= x^3(-1 + C) + x^2(9 - 3C + D) + x(-1 - 3D) + 9$$

Equating coefficients of like terms, we have
$$0 = -1 + C$$
$$5 = 9 - 3C + D$$
$$2 = -1 - 3D$$
So $C = 1$, $D = -1$.

So $\dfrac{5x^2 + 2x + 9}{x^4 - 3x^3 + x^2 - 3x} = \dfrac{-3}{x} + \dfrac{2}{x - 3} + \dfrac{x - 1}{x^2 + 1}$ *(3-5)*

41. In the given figure, let y = height of door
$$2x = \text{width of door}$$
Then Area of door = $48 = 2xy$

Since (x, y) is a point on the parabola $y = 16 - x^2$, its coordinates satisfy the equation of the parabola.

Hence
$$48 = 2x(16 - x^2)$$
$$48 = 32x - 2x^3$$
$$2x^3 - 32x + 48 = 0$$
$$x^3 - 16x + 24 = 0$$

The possible rational solutions of this equation are ±1, ±2, ±3, ±4, ±6, ±8, ±12, ±24. We form a synthetic division table.

	1	0	-16	24	
1	1	1	-15	9	
2	1	2	-12	0	2 is a zero

Thus the equation can be factored
$(x - 2)(x^2 + 2x - 12) = 0$.

To find the remaining zeros, we solve $x^2 + 2x - 12 = 0$, by completing the square.
$$x^2 + 2x = 12$$
$$x^2 + 2x + 1 = 13$$
$$(x + 1)^2 = 13$$
$$x + 1 = \pm\sqrt{13}$$
$$x = -1 + \sqrt{13} \text{ (discarding the negative solution)} \approx 2.61$$

Thus the positive zeros are $x = 2, 2.61$.

Thus the dimensions of the door are either $2x = 4$ feet by $16 - x^2 = 12$ feet, or $2x = 5.2$ feet by $16 - x^2 = 9.2$ feet.   *(3-2)*

42. We note:

$$\begin{pmatrix} \text{Volume} \\ \text{of} \\ \text{silo} \end{pmatrix} = \begin{pmatrix} \text{Volume of} \\ \text{hemisphere} \\ \text{of radius } x \end{pmatrix} + \begin{pmatrix} \text{Volume of} \\ \text{cylinder with} \\ \text{radius } x, \text{ height } 18 \end{pmatrix}$$

$$486\pi = \frac{2}{3}\pi x^3 + \pi x^2 \cdot 18$$

$$486 = \frac{2}{3}x^3 + 18x^2$$

$$0 = x^3 + 27x^2 - 729$$

There are no rational zeros of $P(x) = x^3 + 27x^2 - 729$. To search for irrational zeros, we form a synthetic division table.

	1	27	0	-729	
	1	27	0	-729	
0	1	27	0	-729	
1	1	28	28	-701	
2	1	29	58	-613	
3	1	30	90	-459	
4	1	31	124	-233	
5	1	32	160	71	There is a zero between 4 and 5

We have located a zero between successive integers. We now apply the bisection method to locate the zero to one decimal place.

We organize our calculations in a table.

Sign Change Interval (a, b)	Midpoint m	Sign of P $P(a)$	$P(m)$	$P(b)$
(4, 5)	4.5	-	-	+
(4.5, 5)	4.75	-	-	+
(4.75, 5)	4.875	-	+	+
(4.75, 4.875)	4.8125	-	+	+
(4.75, 4.8125)	4.78125	-	-	+
(4.78125, 4.8125)	We stop here	-		+

Since each endpoint rounds to 4.8, a real zero lies on this last interval and is given by 4.8 to one decimal place accuracy. To one decimal place accuracy, the radius is 4.8 feet.   *(3-3)*

43.

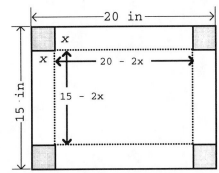

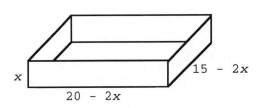

From the above figures it should be clear that

$V = $ length $\times$ width $\times$ height $= (20 - 2x)(15 - 2x)x \quad 0 < x < 7.5$

We solve $(20 - 2x)(15 - 2x)x = 300$

$$300x - 70x^2 + 4x^3 = 300$$

$$4x^3 - 70x^2 + 300x - 300 = 0$$

There are no rational zeros of $P(x) = 4x^3 - 70x^2 + 300x - 300$. To search for irrational zeros, we form a synthetic division table.

	4	-70	300	-300	
0	4	-70	300	-300	
1	4	-66	234	-66	
2	4	-62	176	52	There is a zero between 1 and 2
3	4	-58	126	78	
4	4	-54	84	36	
5	4	-50	50	-50	There is a zero between 4 and 5
6	4	-46	24	-156	
7	4	-42	6	-258	
8	4	-38	-4	-332	

Applying the bisection method (details omitted) the positive zeros in $(0, 7.5)$ are found to be 1.4 and 4.5 to one decimal place accuracy.

$x = 1.4$ inches or 4.5 inches

$(3-3)$

CHAPTER 4

Exercise 4-1

Key Ideas and Formulas

The equation $f(x) = b^x$ $b > 0$, $b \neq 1$ defines an exponential function for each different constant b, called the base. Domain: $(-\infty, \infty)$; Range: $(0, \infty)$

Basic Properties of the graph of $f(x) = b^x$ $b > 0$, $b \neq 1$
1. All graphs pass through $(0, 1)$
2. All graphs are continuous with no holes or jumps.
3. The x-axis is a horizontal asymptote.
4. If $b > 1$, then b^x increases as x increases.
5. If $0 < b < 1$, then b^x decreases as x increases.
6. The function f is one-to-one.

Additional Exponential Function Properties.

For a and b positive, $a \neq 1$, $b \neq 1$, and x and y real:
1. Exponent laws
   $a^x a^y = a^{x+y}$ $(a^x)^y = a^{xy}$ $(ab)^x = a^x b^x$ $(a/b)^x = a^x/b^x$ $a^x/a^y = a^{x-y}$
2. $a^x = a^y$ if and only if $x = y$
3. For $x \neq 0$, then $a^x = b^x$ if and only if $a = b$.

Doubling Time Growth Model:
$P = P_0 2^{t/d}$ where

P = Population at time t
P_0 = Population at time $t = 0$
d = doubling time

Half-Life Decay Model:

$A = A_0 \left(\dfrac{1}{2}\right)^{t/h}$

$\quad = A_0 2^{-t/h}$

where
A = amount at time t
A_0 = amount at time $t = 0$
h = half-life

Compound Interest:

If a principal P is invested at an annual rate r compounded n times a year, then the amount A in the account at the end of t years is given by

$A = P\left(1 + \dfrac{r}{n}\right)^{nt}$

1.

x	y
-3	0.04
-2	0.11
-1	0.33
0	1.00
1	3.00
2	9.00
3	27.00

3.

x	y
-3	27.00
-2	9.00
-1	3.00
0	1.00
1	0.33
2	0.11
3	0.04

5.

x	y = g(x)
-3	-27.00
-2	-9.00
-1	-3.00
0	-1.00
1	-0.33
2	-0.11
3	-0.04

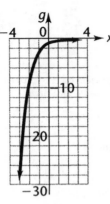

7.

x	y = h(x)
-3	0.19
-2	0.56
-1	1.67
0	5.00
1	15.00
2	45.00
3	135.00

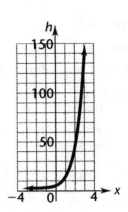

9.

x	y
-6	-4.96
-5	-4.89
-4	-4.67
-3	-4.00
-2	-2.00
-1	4.00
0	22.00

11. $2^{5x+1}2^{3-2x} = 2^{5x+1+3-2x} = 2^{3x + 4}$

13. $\dfrac{3^{2y-x}}{3^{3x-5y}} = 3^{(2y-x)-(3x-5y)} = 3^{2y-x-3x+5y} = 3^{-4x+7y}$

15. 4^{3x}

17. $(10^2)^x (10^x)^2 = 10^{2x}10^{2x} = 10^{2x+2x} = 10^{4x}$

19. $\left(\dfrac{5^x}{4^y}\right)^3 = \dfrac{5^{3x}}{4^{3y}}$

21. $\left(\dfrac{a^{-2}b^3c^{-1}}{a^{-3}b^2c^{-3}}\right)^2 = \dfrac{a^{-4}b^6c^{-2}}{a^{-6}b^4c^{-6}} = a^{(-4)-(-6)}b^{6-4}c^{(-2)-(-6)} = a^2b^2c^4$

23. $3^{2x-5} = 3^{4x+2}$ if and only if
$$2x - 5 = 4x + 2$$
$$-2x = 7$$
$$x = -3.5$$

25. $10^{x^2+2} = 10^{2x+2}$ if and only if
$$x^2 + 2 = 2x + 2$$
$$x^2 - 2x = 0$$
$$x(x - 2) = 0$$
$$x = 0, 2$$

27. $(2x + 1)^3 = 8$
$(2x + 1)^3 = 2^3$ if and only if
$$2x + 1 = 2$$
$$2x = 1$$
$$x = 0.5$$

29. $5^{3x} = 25^{x+3}$
$5^{3x} = (5^2)^{x+3}$
$5^{3x} = 5^{2(x+3)}$ if and only if
$$3x = 2(x + 3)$$
$$3x = 2x + 6$$
$$x = 6$$

31. $4^{2x+2} = 8^{x+2}$
$(2^2)^{2x+2} = (2^3)^{x+2}$
$2^{2(2x+2)} = 2^{3(x+2)}$ if and only if
$$2(2x + 2) = 3(x + 2)$$
$$4x + 4 = 3x + 6$$
$$x = 2$$

33. $100^{x^2} = 10^{5x-3}$
$(10^2)^{x^2} = 10^{5x-3}$
$10^{2x^2} = 10^{5x-3}$ if and only if
$$2x^2 = 5x - 3$$
$$2x^2 - 5x + 3 = 0$$
$$(2x - 3)(x - 1) = 0$$
$$x = \frac{3}{2} \text{ or } 1.5, 1$$

35.
$$a^2 = a^{-2}$$
$$a^2 = \frac{1}{a^2}$$
$$a^4 = 1 \quad (a \neq 0)$$
$$a^4 - 1 = 0$$
$$(a - 1)(a + 1)(a^2 + 1) = 0$$
$$a = 1 \text{ or } a = -1$$

This does not violate the exponential property mentioned because $a = 1$ and a negative are excluded from consideration in the statement of the property.

37.

t	$G(t)$
-200	0.11
-150	0.19
-100	0.33
-50	0.58
0	1.00
50	1.73
100	3.00
150	5.20
200	9.00

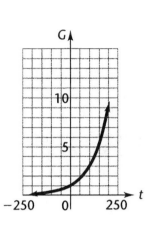

39.

x	y
-9	1543.3
-8	891.0
-7	514.4
-6	297.0
-5	171.5
-4	99
-3	57.2
-2	33
0	11
3	2.1
6	0.4
9	0.1

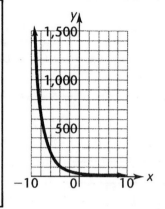

41.

x	$y = g(x)$
-3	0.13
-2	0.25
-1	0.5
0	1.0
1	0.5
2	0.25
3	0.13

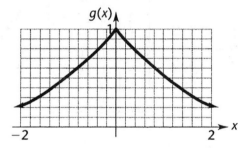

43.

x	y
-10	463
-8	540
-6	630
-4	735
-2	857
0	1,000
2	1,166
4	1,360
6	1,587
8	1,851
10	2,159

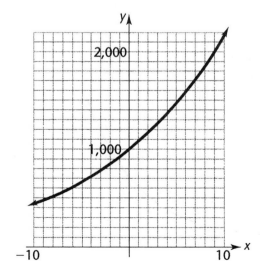

45.

x	y
-2	0.06
-1.5	0.21
-1	0.50
-0.5	0.84
0	1.00
0.5	0.84
1	0.50
1.5	0.21
2	0.06

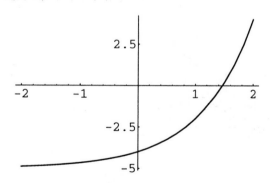

47. $(6^x + 6^{-x})(6^x - 6^{-x}) = (6^x)^2 - (6^{-x})^2$
$= 6^{2x} - 6^{-2x}$

(think: $(a + b)(a - b) = a^2 - b^2$)

> **Common Errors:**
> $(6^x)^2 \neq 6^{x^2}$
> $6^{2x} - 6^{-2x} \neq 6^{4x}$

49. $(6^x + 6^{-x})^2 - (6^x - 6^{-x})^2 = (6^x)^2 + 2(6^x)(6^{-x}) + (6^{-x})^2 - [(6^x)^2 - 2(6^x)(6^{-x}) + (6^{-x})^2]$
$= 6^{2x} + 2 + 6^{-2x} - [6^{2x} - 2 + 6^{-2x}]$
$= 6^{2x} + 2 + 6^{-2x} - 6^{2x} + 2 - 6^{-2x}$
$= 4$

51.

x	y = m(x)
-3	-81
-2.5	-39
-2	-18
-1.5	-7.8
-1	-3
-0.5	-0.87
0	0
0.5	0.29
1	0.33
1.5	0.29
2	0.22
2.5	0.16
3	0.11

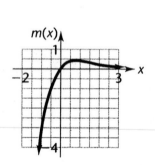

53.

x	y = f(x)
-3	4.06
-2	2.13
-1	1.25
0	1.00
1	1.25
2	2.13
3	4.06

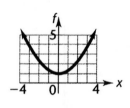

55. (A) Here is a computer-generated graph of $f(x) = 3^x - 5$.

The only zero apparent is on the interval (1.4, 1.5). To estimate more accurately, we zoom in. Here is a graph of f on this interval.

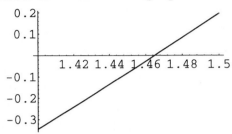

To two decimal place accuracy, the zero is 1.46.

(B) As $x \to \infty$, $f(x) \to \infty$. As $x \to -\infty$, it appears that $f(x) \to -5$. The line $y = -5$ is a horizontal asymptote.

57. (A) Here is a computer-generated graph of $f(x) = 1 + x + 10^x$.

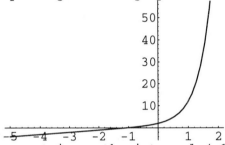

The only zero apparent is on the interval (-1.1, -1). To estimate more accurately, we zoom in. Here is a graph of f on this interval.

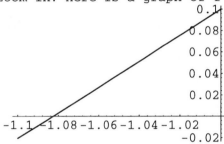

To two decimal place accuracy, the zero is -1.08.

(B) As $x \to \infty$, $f(x) \to \infty$. As $x \to -\infty$, $f(x) \to -\infty$. There is no horizontal asymptote. It appears that the line $y = x + 1$ may be an oblique asymptote as $x \to -\infty$.

59.

n	L
1	2
2	4
3	8
4	16
5	32
6	64
7	128
8	256
9	512
10	1,024

61. We use the Doubling Time Growth Model:

$P = P_0 2^{t/d}$

Substituting $P_0 = 10$ and $d = 2.4$, we have

$P = 10(2^{t/2.4})$

(A) $t = 7$, hence $P = 10(2^{7/2.4})$
$= 75.5$

76 flies

(B) $t = 14$, hence $P = 10(2^{14/2.4})$
$= 570.2$

570 flies

63. We use the Half-Life Decay Model

$A = A_0\left(\dfrac{1}{2}\right)^{t/h} = A_0 2^{-t/h}$

Substituting $A_0 = 25$ and $h = 12$, we have

$A = 25(2^{-t/12})$

(A) $t = 5$, hence $A = 25(2^{-5/12})$
$= 19$ pounds

(B) $t = 20$, hence $A = 25(2^{-20/12})$
$= 7.9$ pounds

65. We use the compound interest formula

$A = P\left(1 + \dfrac{r}{n}\right)^{nt} \qquad n = \dfrac{365}{7}$

$P = 4,000 \quad r = 0.11$

$A = 4,000\left(1 + \dfrac{0.11}{365/7}\right)^{365t/7}$

$= 4,000\left(1 + \dfrac{0.77}{365}\right)^{365t/7}$

(A) $t = 0.5$, hence

$A = 4,000\left(1 + \dfrac{0.77}{365}\right)^{365(0.5)/7}$
$= \$4,225.92$

(B) $t = 10$, hence

$A = 4,000\left(1 + \dfrac{0.77}{365}\right)^{365(10)/7}$
$= \$12,002.75$

67. We use the compound interest formula

$A = P\left(1 + \dfrac{r}{n}\right)^{nt}$ to find P: $P = \dfrac{A}{(1 + \frac{r}{n})^{nt}}$

$n = 365 \quad r = 0.0825 \quad A = 40,000 \quad t = 17$

$P\dfrac{40,000}{(1 + \frac{0.0825}{365})^{365\cdot17}} = \$9,841$

69. Using the compound interest formula with $P = 10,000$, $r = 0.09$, and $n = 4$, the value of the 9% investment is $A_1(t) = 10,000\left(1 + \dfrac{.09}{4}\right)^{4t}$.

Using the compound interest formula with $P = 10,000$, $r = 0.089$ and $n = 365$, the value of the 8.9% investment is $A_2(t) = 10,000\left(1 + \dfrac{.089}{365}\right)^{365t}$.

The two investments are equal at $t = 0$, but after $t = 0$, the values can be calculated to be as follows:

t (in years)	$\dfrac{1}{4}$	$\dfrac{2}{4}$	$\dfrac{3}{4}$	$\dfrac{4}{4} = 1$	2	5	10
$A_1(t)$	10225	10455.06	10690.30	10930.83	11948.31	15605.90	24351.89
$A_2(t)$	10224.97	10454.97	10690.19	10930.69	11947.39	15604.06	24348.65

Thus, the second investment is less than the first by an ever-increasing amount, and can never be worth more.

Exercise 4-2
Key Ideas and Formulas

As m increases without bound, the value of $\left(1 + \dfrac{1}{m}\right)^m$ approaches an irrational number called e. To twelve decimal places, $e = 2.718\ 281\ 828\ 459$.

For x a real number, the equation $f(x) = e^x$ defines the exponential function with base e.

Exponential Growth and Decay

Description	Equation	Graph	Uses
Unlimited Growth	$y = ce^{kt}$ $c,\ k > 0$		Short term population growth (people, bacteria, etc.) Growth of money at continuous compound interest. ($A = Pe^{rt}$)
Exponential Decay	$y = ce^{-kt}$ $c,\ k > 0$		Radioactive decay. Light absorption in water, glass, etc. Atmosphere pressure. Electric circuits.
Limited Growth	$y = c(1 - e^{-kt})$ $c,\ k > 0$		Learning skills. Sales fads. Company growth. Electric Circuits.
Logistic Growth	$y = \dfrac{M}{1 + ce^{-kt}}$ $c,\ k,\ M > 0$		Long-term population growth. Epidemics. Sales of new products. Company growth.

1.

x	y
-3	-0.05
-2	-0.14
-1	-0.37
0	-1.00
1	-2.72
2	-7.39
3	-20.09

3.

x	y
-5	3.68
-4	4.49
-3	5.49
-2	6.7
-1	8.19
0	10
1	12.21
2	14.92
3	18.22
4	22.26
5	27.18

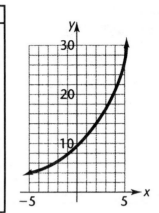

5.

t	f(t)
-5	164.87
-4	149.18
-3	134.99
-2	122.14
-1	110.52
0	100
1	90.48
2	81.87
3	74.08
4	67.03
5	60.65

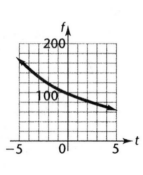

7. $e^{2x}e^{-3x} = e^{2x-3x} = e^{-x}$

9. e^{3x}

11. $\dfrac{e^{5x}}{e^{2x+1}} = e^{5x-(2x+1)} = e^{5x-2x-1} = e^{3x-1}$

13. (A) Although $1 + \dfrac{1}{m}$ approaches 1, $1 + \dfrac{1}{m}$ is not equal to 1 for any m, hence reasoning as if it were 1 is incorrect.

(B) As $m \to \infty$, $\left(1 + \dfrac{1}{m}\right)^m \to e$.

15.

x	y
-1	2.05
0	2.14
1	2.37
2	3.00
3	4.72
4	9.39
5	22.09

17.

x	y
-3	0.05
-2	0.14
-1	0.37
0	1.00
1	0.37
2	0.14
3	0.05

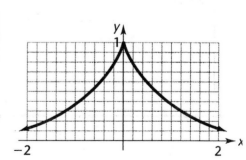

19.

x	y = M(x)
-5	12.3
-4	7.5
-3	4.7
-2	3.1
-1	2.3
0	2.0
1	2.3
2	3.1
3	4.7
4	7.5
5	12.3

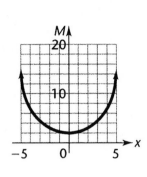

21.

t	N
-3	3.3
-2	8.6
-1	22
0	50
1	95
2	142
3	174
4	190
5	196

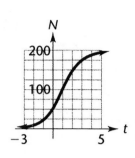

23. $\dfrac{-2x^3 e^{-2x} - 3x^2 e^{-2x}}{x^6} = \dfrac{x^2 e^{-2x}(-2x - 3)}{x^6} = \dfrac{e^{-2x}(-2x - 3)}{x^4}$

25. $(e^x + e^{-x})^2 + (e^x - e^{-x})^2 = (e^x)^2 + 2(e^x)(e^{-x}) + (e^{-x})^2 + (e^x)^2 - 2(e^x)(e^{-x}) + (e^{-x})^2$
$= e^{2x} + 2 + e^{-2x} + e^{2x} - 2 + e^{-2x}$
$= 2e^{2x} + 2e^{-2x}$

> **Common Errors:**
> $(e^x)^2 \neq e^{x^2}$
> $e^{2x} + e^{2x} \neq e^{4x}$

27. $\dfrac{e^{-x}(e^x - e^{-x}) + e^{-x}(e^x + e^{-x})}{e^{-2x}} = \dfrac{e^{-x}e^x - e^{-x}e^{-x} + e^{-x}e^x + e^{-x}e^{-x}}{e^{-2x}}$

$= \dfrac{1 - e^{-2x} + 1 + e^{-2x}}{e^{-2x}}$

$= \dfrac{2}{e^{-2x}}$

$= 2e^{2x}$

29. $2xe^{-x} = 0$ if $2x = 0$ or $e^{-x} = 0$. Since e^{-x} is never 0, the only solution is $x = 0$.

31. $x^2 e^x - 5xe^x = 0$
$xe^x(x - 5) = 0$
$\quad\quad x = 0$ or $e^x = 0$ or $x - 5 = 0$
$\quad\quad\quad\quad\quad\quad$ never
$\quad\quad x = 0, 5 \quad\quad\quad\quad\quad\quad x = 5$

33.

x	f(x)
-2	0.02
-1.5	0.11
-1	0.37
-0.5	0.78
0	1.00
0.5	0.78
1	0.37
1.5	0.11
2	0.02

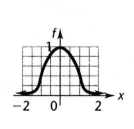

35. (A)

s	f(s)	s	f(s)
-0.5	4.0000	0.5	2.2500
-0.2	3.0518	0.2	2.4883
-0.1	2.8680	0.1	2.5937
-0.01	2.7320	0.01	2.7048
-0.001	2.7196	0.001	2.7169
-0.0001	2.7184	0.0001	2.7181

(B) Both tables are "closing in" on 2.7182… or e.

37.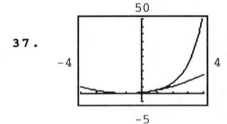

39.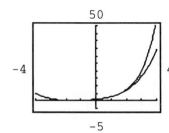

41. Here are computer-generated graphs of $f_1(x) = \dfrac{x}{e^x}$, $f_2(x) = \dfrac{x^2}{e^x}$, and $f_3(x) = \dfrac{x^3}{e^x}$.

In each case as $x \to \infty$, $f_n(x) \to 0$. The line $y = 0$ is a horizontal asymptote. As $x \to -\infty$, $f_1(x) \to -\infty$ and $f_3(x) \to -\infty$, while $f_2(x) \to \infty$. It appears that as $x \to -\infty$, $f_n(x) \to \infty$ if n is even and $f_n(x) \to -\infty$ if n is odd.

f_1: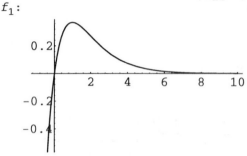

f_2: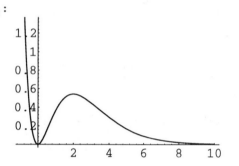

f_3: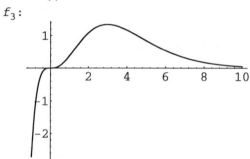

43. We use the continuous compounding formula.

$A = Pe^{rt}$
$P = 6$ billion $r = 0.017$ $t = 10$
$A = (6 \text{ billion}) e^{(0.017)(10)}$
$A = 7.1$ billion

45. We use the Continuous Compounding Formula

$A = Pe^{rt}$

For Russia, $P = 148$ million, $r = -0.0062$, $A_1 = (148 \text{ million}) e^{-0.0062t}$

For Nigeria, $P = 104$ million, $r = 0.03$, $A_2 = (104 \text{ million}) e^{0.03t}$

Here is a computer-generated graph of A_1 and A_2 (vertical axis understood in millions).

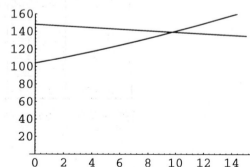

From the graph, assuming $t = 0$ in 1996, it appears that the two populations will be equal when t is approximately 10, in 2006. After that the population of Nigeria will be greater than the population of Russia.

47.

t	P
0	75
10	72
20	70
30	68
40	65
50	63
60	61
70	59
80	57
90	55
100	53

49. $I = I_0 e^{-0.00942d}$

    (A) $d = 50$ $I = I_0 e^{-0.00942(50)} = 0.62 I_0$ 62%

    (B) $d = 100$ $I = I_0 e^{-0.00942(100)} = 0.39 I_0$ 39%

51. We use the Continuous Compound Interest Formula

    $A = P e^{rt}$

    $P = 5,250$ $r = 0.1138$ $A = 5,250 e^{0.1138t}$

    (A) $t = 6.25$ $A = 5,250 e^{0.1138(6.25)} = \$10,691.81$

    (B) $t = 17$ $A = 5,250 e^{0.1138(17)} = \$36,336.69$

53. Gill Savings: Use the Continuous Compound Interest Formula

    $A = P e^{rt}$ $P = 1,000$ $r = 0.083$ $t = 2.5$

    $A = 1,000 e^{(0.083)(2.5)}$

    $A = \$1,230.60$

    Richardson S & L: Use the Compound Interest Formula

    $A = P\left(1 + \dfrac{r}{n}\right)^{nt}$ $P = 1,000$ $r = 0.084$ $n = 4$ $t = 2.5$

    $A = 1,000\left(1 + \dfrac{0.084}{4}\right)^{(4)(2.5)}$

    $A = \$1,231.00$

    U.S.A. Savings: Use the Compound Interest Formula

    $A = P\left(1 + \dfrac{r}{n}\right)^{nt}$ $P = 1,000$ $r = 0.0825$ $n = 365$ $t = 2.5$

    $A = 1,000\left(1 + \dfrac{0.0825}{365}\right)^{(365)(2.5)}$

    $A = \$1,229.03$

55. We use the Continuous Compound Interest Formula

    $A = P e^{rt}$

    $P = \dfrac{A}{e^{rt}}$ or $P = A e^{-rt}$

    $A = 30,000$ $r = 0.09$ $t = 10$

    $P = 30,000 e^{(-0.09)(10)}$

    $P = \$12,197.09$

57. We use the Continuous Compounding Formula

    $A = P e^{rt}$

    (A) $P = 7.7$ million $r = 0.17$ $t = 4$ (B) $B = 7.7$ million $r = 0.17$ $t = 8$

        $A = (7.7 \text{ million}) e^{0.17(4)}$ $A = (7.7 \text{ million}) e^{0.17(8)}$

        $A = 15$ million $A = 30$ million

59.

t	N
0	0
5	18
10	28
15	33
20	36
25	38
30	39

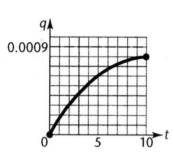

As t increases without bound, $e^{-0.12t}$ approaches 0, hence $N = 40(1 - e^{-0.12t})$ approaches 40. Hence 40 boards is the maximum number of boards an average person could be expected to produce in one day.

61. $T = T_m + (T_0 - T_m)e^{-kt}$
$T_m = 40°$ $T_0 = 72°$ $k = 0.4$ $t = 3$
$T = 40 + (72 - 40)e^{-0.4(3)}$
$T = 50°$

63.

t	q
0	0
1	0.00016
2	0.00030
3	0.00041
4	0.00050
5	0.00057
6	0.00063
7	0.00068
8	0.00072
9	0.00075
10	0.00078

As t increases without bound, $e^{-0.2t}$ approaches 0, hence $q = 0.0009(1 - e^{-0.2t})$ approaches 0.0009. Hence 0.0009 coulomb is the maximum charge on the capacitor.

65.

t	N
0	20
5	33
10	50
15	67
20	80
25	89
30	94

As t increases without bound, $e^{-0.14t}$ approaches 0, hence $N = \dfrac{100}{1 + 4e^{-0.14t}}$ approaches 100. Hence 100 is the number of deer the island can support.

67. $y = \dfrac{e^{0.25x} + e^{-0.25x}}{2(0.25)} = \dfrac{e^{0.25x} + e^{-0.25x}}{0.5}$

x	y
-5	7.6
-4	6.2
-3	5.2
-2	4.5
-1	4.1
0	4.0
1	4.1
2	4.5
3	5.2
4	6.2
5	7.6

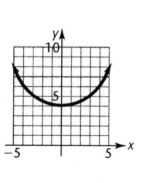

Exercise 4-3

Key Ideas and Formulas

Definition of Logarithmic Function:

For $b > 0$ and $b \neq 1$

   Logarithmic form   Exponential form

   $y = \log_b x$ is equivalent to $x = b^y$

   The log to the base b of x is the exponent to which b must be raised to obtain x. A logarithm is, therefore, an exponent.

Typical Logarithmic Graphs

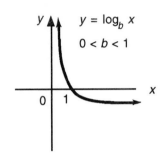

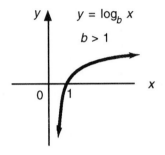

Domain = $(0, \infty)$
Range = $(-\infty, \infty)$

Domain = $(0, \infty)$
Range = $(-\infty, \infty)$

Properties of Logarithmic Functions:

If b, M, and N are positive real numbers, $b \neq 1$, and p and x are real numbers, then:

1. $\log_b 1 = 0$

2. $\log_b b = 1$

3. $\log_b b^x = x$

4. $b^{\log_b x} = x$

5. $\log_b MN = \log_b M + \log_b N$

6. $\log_b \dfrac{M}{N} = \log_b M - \log_b N$

7. $\log_b M^p = p \log_b M$

8. $\log_b M = \log_b N$ if and only if $M = N$

---

1. $64 = 2^6$ **3.** $100,000 = 10^5$ **5.** $9^{1/2} = 3$ **7.** $\left(\dfrac{1}{4}\right)^{-3} = 64$

9. $\log_{10}(0.001) = -3$ **11.** $\log_8 4 = \dfrac{2}{3}$ **13.** $\log_{81} \dfrac{1}{3} = -\dfrac{1}{4}$

15. $5 = \sqrt[3]{125}$ is rewritten $5 = 125^{1/3}$. In equivalent logarithmic form this becomes $\log_{125} 5 = \dfrac{1}{3}$.

17. 1 **19.** 0 **21.** $\log_{10} 1,000 = \log_{10} 10^3 = 3$

23. $\log_e \sqrt{e} = \log_e e^{1/2} = \dfrac{1}{2}$

25. $\log_{10} 0.001 = \log_{10} 10^{-3} = -3$ **27.** x^2

29. $e^{3 \log_e (x-1)} = e^{\log_e (x-1)^3} = (x - 1)^3$ $\boxed{\text{\textbf{Common Error:} } e^{3 \log_e (x-1)} \neq 3(x - 1)}$

31. Write $\log_2 x = 3$ in equivalent exponential form: $x = 2^3 = 8$.

33. Write $\log_5 x = -2$ in equivalent exponential form: $x = 5^{-2} = \dfrac{1}{25}$.

35. $\log_3 81 = \log_3 3^4 = 4$

37. Write $\log_{16} x = \dfrac{3}{2}$ in equivalent exponential form: $x = 16^{3/2} = 64$.

39. $\log_{1/2} 8 = \log_{1/2} 2^3 = \log_{1/2}\left(\dfrac{1}{2}\right)^{-3} = -3$

41. Write $\log_b 32 = 2.5$ in equivalent exponential form:
$$b^{2.5} = 32$$
$$b^{5/2} = 32$$
$$b^5 = 32^2$$
$$b = \sqrt[5]{32^2} = (\sqrt[5]{32})^2 = 2^2 = 4$$

43. Write $\log_b 0.0001 = -2$ in equivalent exponential form:
$$b^{-2} = 0.0001$$
$$b^2 = \dfrac{1}{0.0001}$$
$$b^2 = 10{,}000$$
$$b = 100 \text{ since bases are required to be positive}$$

45. $\log_b x^6 y^9 = \log_b x^6 + \log_b y^9 = 6 \log_b x + 9 \log_b y$

47. $\log_b \dfrac{u^{3/2}}{v^{5/3}} = \log_b u^{3/2} - \log_b v^{5/3} = \dfrac{3}{2} \log_b u - \dfrac{5}{3} \log_b v$

49. $\log_b \dfrac{mn}{pq} = \log_b mn - \log_b pq = \log_b m + \log_b n - (\log_b p + \log_b q)$
$$= \log_b m + \log_b n - \log_b p - \log_b q$$

> **Common Error:** forgetting the parentheses.
> $-\log_b p + \log_b q$ is incorrect

51. $\log_b \dfrac{1}{a^4} = \log_b a^{-4} = -4 \log_b a$

53. $\log_b \sqrt{c^2 + d^2} = \log_b (c^2 + d^2)^{1/2} = \dfrac{1}{2} \log_b (c^2 + d^2)$

> **Common Error:** $\log_b(c^2 + d^2)$ is not $\log_b c^2 + \log_b d^2$

55. $\log_b \dfrac{\sqrt{u}}{vw^2} = \log_b \sqrt{u} - \log_b (vw^2)$
$$= \log_b u^{1/2} - (\log_b v + \log_b w^2)$$
$$= \dfrac{1}{2} \log_b u - (\log_b v + 2 \log_b w)$$
$$= \dfrac{1}{2} \log_b u - \log_b v - 2 \log_b w$$

57. $\log_b \sqrt[3]{\dfrac{x^2 \sqrt{y}}{z^3}} = \log_b \left(\dfrac{x^2 y^{1/2}}{z^3}\right)^{1/3}$
$$= \log_b \left(\dfrac{x^{2/3} y^{1/6}}{z}\right)$$
$$= \log_b x^{2/3} + \log_b y^{1/6} - \log_b z$$
$$= \dfrac{2}{3} \log_b x + \dfrac{1}{6} \log_b y - \log_b z$$

59. $2 \log_b x - \log_b y = \log_b x^2 - \log_b y = \log_b \dfrac{x^2}{y}$

61. $\log_b w - \log_b x - \log_b y = \log_b \dfrac{w}{x} - \log_b y$

$$= \log_b \left(\dfrac{w}{x} \div y \right)$$

$$= \log_b \left(\dfrac{w}{x} \cdot \dfrac{1}{y} \right)$$

$$= \log_b \dfrac{w}{xy}$$

63. $3 \log_b x + 2 \log_b y - \dfrac{1}{4} \log_b z = \log_b x^3 + \log_b y^2 - \log_b z^{1/4}$

$$= \log_b x^3 y^2 - \log_b z^{1/4} = \log_b \dfrac{x^3 y^2}{z^{1/4}}$$

65. $5 \left(\dfrac{1}{2} \log_b u - 2 \log_b v \right) = 5 (\log_b u^{1/2} - \log_b v^2) = 5 \log_b \dfrac{u^{1/2}}{v^2} = \log_b \left(\dfrac{u^{1/2}}{v^2} \right)^5$

67. $\dfrac{1}{5} (2 \log_b x + 3 \log_b y) = \dfrac{1}{5} (\log_b x^2 + \log_b y^3) = \dfrac{1}{5} \log_b x^2 y^3 = \log_b (x^2 y^3)^{1/5}$

$$= \log_b \sqrt[5]{x^2 y^3}$$

69. $\log_b [(x + 3)^5 (2x - 7)^2] = \log_b (x + 3)^5 + \log_b (2x - 7)^2$

$$= 5 \log_b (x + 3) + 2 \log_b (2x - 7)$$

71. $\log_b \dfrac{(x + 10)^7}{(1 + 10x)^2} = \log_b (x + 10)^7 - \log_b (1 + 10x)^2 = 7 \log_b (x + 10) - 2 \log_b (1 + 10x)$

73. $\log_b \dfrac{x^2}{\sqrt{x + 1}} = \log_b x^2 - \log_b \sqrt{x + 1}$

$$= \log_b x^2 - \log_b (x + 1)^{1/2}$$

$$= 2 \log_b x - \dfrac{1}{2} \log_b (x + 1)$$

75. $\log_b (x^4 + x^3 - 20x^2) = \log_b [x^2 (x^2 + x - 20)]$

$$= \log_b [x^2 (x + 5)(x - 4)]$$

$$= \log_b x^2 + \log_b (x + 5) + \log_b (x - 4)$$

$$= 2 \log_b x + \log_b (x + 5) + \log_b (x - 4)$$

77. $\log_2 (x + 5) = 2 \log_2 3$

$\log_2 (x + 5) = \log_2 3^2$

$\quad x + 5 = 3^2$

$\quad x + 5 = 9$

$\qquad x = 4$

Check: $\log_2 (x + 5) = 2 \log_2 3$

$\log_2 (4 + 5) \overset{?}{=} 2 \log_2 3$

$\log_2 9 \overset{\checkmark}{=} \log_2 9$

79. $2 \log_5 x = \log_5 (x^2 - 6x + 2)$

$\log_5 x^2 = \log_5 (x^2 - 6x + 2)$

$\quad x^2 = x^2 - 6x + 2$

$\quad 6x = 2$

$\qquad x = \dfrac{1}{3}$

Check: $2 \log_5 x = \log_5 (x^2 - 6x + 2)$

$2 \log_5 \dfrac{1}{3} \overset{?}{=} \log_5 \left[\left(\dfrac{1}{3} \right)^2 - 6 \left(\dfrac{1}{3} \right) + 2 \right]$

$\log_5 \dfrac{1}{9} \overset{\checkmark}{=} \log_5 \left(\dfrac{1}{9} \right)$

81. $\log_e(x + 8) - \log_e x = 3 \log_e 2$

$$\log_e \frac{x + 8}{x} = \log_e 2^3$$

$$\frac{x + 8}{x} = 2^3$$

$$x + 8 = 8x$$

$$8 = 7x$$

$$x = \frac{8}{7}$$

Check: $\log_e(x + 8) - \log_e x = 3 \log_e 2$

$$\log_e\left(\frac{8}{7} + 8\right) - \log_e \frac{8}{7} \overset{?}{=} 3 \log_e 2$$

$$\log_e\left(\frac{64}{7}\right) - \log_e \frac{8}{7} \overset{?}{=} \log_e 8$$

$$\log_e\left(\frac{64}{7} \div \frac{8}{7}\right) \overset{?}{=} \log_e 8$$

$$\log_e 8 \overset{\sqrt{}}{=} \log_e 8$$

83. $2 \log_3 x = \log_3 2 + \log_3(4 - x)$

$$\log_3 x^2 = \log_3 2(4 - x)$$

$$x^2 = 2(4 - x)$$

$$x^2 = 8 - 2x$$

$$x^2 + 2x - 8 = 0$$

$$(x + 4)(x - 2) = 0$$

$$x = -4 \text{ or } x = 2$$

Check: $2 \log_3 x = \log_3 2 + \log_3(4 - x)$

$$x = -4$$

$$2 \log_3(-4) \overset{?}{=} \log_3 2 + \log_3[4 - (-4)]$$

False. -4 is not in the domain of $\log_3 x$.
-4 is not a solution.

$$x = 2$$

$$2 \log_3 2 \overset{?}{=} \log_3 2 + \log_3(4 - 2)$$

$$2 \log_3 2 \overset{\sqrt{}}{=} 2 \log_3 2$$

Solution: 2

85. $3 \log_b 2 + \frac{1}{2} \log_b 25 - \log_b 20 = \log_b x$

$$\log_b 2^3 + \log_b 25^{1/2} - \log_b 20 = \log_b x$$

$$\log_b \frac{2^3 \cdot 25^{1/2}}{20} = \log_b x$$

$$\frac{2^3 \cdot 25^{1/2}}{20} = x$$

$$x = \frac{8 \cdot 5}{20}$$

$$x = 2$$

Check:

$$3 \log_b 2 + \frac{1}{2} \log_b 25 - \log_b 20 = \log_b x$$

$$3 \log_b 2 + \frac{1}{2} \log_b 25 - \log_b 20 \overset{?}{=} \log_b 2$$

$$\log_b 8 + \log_b 5 - \log_b 20 \overset{?}{=} \log_b 2$$

$$\log_b \frac{40}{20} \overset{?}{=} \log_b 2$$

$$\log_b 2 \overset{\sqrt{}}{=} \log_b 2$$

87. $\log_b(30) = \log_b(2 \cdot 3 \cdot 5) = \log_b 2 + \log_b 3 + \log_b 5 = 0.69 + 1.10 + 1.61 = 3.40$

89. $\log_b \frac{2}{5} = \log_b 2 - \log_b 5 = 0.69 - 1.61 = -0.92$

91. $\log_b 27 = \log_b 3^3 = 3 \log_b 3 = 3(1.10) = 3.30$

93. $\log_b \sqrt[3]{2} = \log_b 2^{1/3} = \frac{1}{3} \log_b 2 = \frac{1}{3}(0.69) = 0.23$

95. $\log_b \sqrt{0.9} = \log_b\left(\frac{9}{10}\right)^{1/2} = \frac{1}{2} \log_b \frac{3^2}{2 \cdot 5} = \frac{1}{2}[\log_b 3^2 - \log_b 2 - \log_b 5]$

$$= \frac{1}{2}[2 \log_b 3 - \log_b 2 - \log_b 5] = \frac{1}{2}[2.20 - 0.69 - 1.61] = \frac{1}{2}(-0.10)$$

$$= -0.05$$

97.

x	y
10	3
6	2
4	1
3	0
$2\frac{1}{2}$	-1
$2\frac{1}{4}$	-2

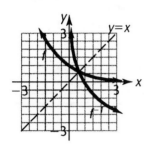

The line $x = 2$ is a vertical asymptote. Note: The graph is the same as the graph of $y = \log_2 x$ shifted 2 units to the right.

99.

x	y
8	1
4	0
2	-1
1	-2
$\frac{1}{2}$	-3
$\frac{1}{4}$	-4

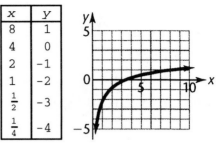

The y axis is a vertical asymptote. Note: The graph is the same as the graph of $y = \log_2 x$ shifted 2 units down.

101.

(B) Domain $f = (-\infty, \infty) = $ Range f^{-1}
Range $f = (0, \infty) = $ Domain f^{-1}

(C) $f^{-1}(x) = \log_{1/2} x = -\log_2 x$

103. $f(x) = 5^{3x-1} + 4$
Solve $y = f(x)$ for x
$$y = 5^{3x-1} + 4$$
$$y - 4 = 5^{3x-1}$$
$$\log_5(y - 4) = 3x - 1$$
$$3x - 1 = \log_5(y - 4)$$
$$3x = 1 + \log_5(y - 4)$$
$$x = \frac{1}{3}[1 + \log_5(y - 4)] = f^{-1}(y)$$

Interchange x and y:

$y = f^{-1}(x) = \frac{1}{3}[1 + \log_5(x - 4)]$
Domain: $x > 4$
(Check omitted for lack of space)

105. $g(x) = 3 \log_e(5x - 2)$
Solve $y = g(x)$ for x
$$y = 3 \log_e(5x - 2)$$
$$\frac{y}{3} = \log_e(5x - 2)$$
$$5x - 2 = e^{y/3}$$
$$5x = e^{y/3} + 2$$
$$x = \frac{1}{5}(e^{y/3} + 2) = g^{-1}(y)$$

Interchange x and y:

$y = g^{-1}(x) = \frac{1}{5}(e^{x/3} + 2)$
Domain: $(-\infty, \infty)$
(Check omitted for lack of space)

107. $y = 3^{x^2}$ is not a one-to-one function, since for $x = 1$ and $x = -1$, y has the same value ($y = 3^1 = 3$). Therefore the inverse function does not exist and the reflection is not the graph of a function.

109. $\log_e x - \log_e 100 = -0.08t$

$\log_e \dfrac{x}{100} = -0.08t$

Write this in equivalent exponential form

$\dfrac{x}{100} = e^{-0.08t}$

$x = 100e^{-0.08t}$

111. Let $u = \log_b M$ and $v = \log_b N$; then
$M = b^u$ and $N = b^v$. Thus,
$\log_b M/N = \log_b b^u/b^v = \log_b b^{u-v}$
$\qquad = u - v = \log_b M - \log_b N.$

Exercise 4-4

Key Ideas and Formulas

Logarithmic Notation

$\log x = \log_{10} x$ Common logarithm

$\ln x = \log_e x$ Natural logarithm

$\log x = y$ is equivalent to $x = 10^y$

$\ln x = y$ is equivalent to $x = e^y$

Sound Intensity	Earthquake Intensity	Rocket Flight Velocity
$D = 10 \log \dfrac{I}{I_0}$ D = decibel level I = intensity of sound I_0 = threshold of hearing	$M = \dfrac{2}{3} \log \dfrac{E}{E_0}$ M = magnitude on Richter Scale E = Energy released by earthquake E_0 = Energy released by small reference earthquake	$v = c \ln \dfrac{W_t}{W_b}$ v = Velocity at fuel burnout level c = exhaust velocity W_t = take off weight W_b = burnout weight

1. 4.4408   **3.** −2.3644   **5.** −7.3324   **7.** 6.1242

9. $\ln x = 3.4797$ **11.** $\ln x = -2.2643$ **13.** $\log x = -1.2543$ **15.** $\log x = 3.5324$
   $x = e^{3.4797}$ $x = e^{-2.2643}$ $x = 10^{-1.2543}$ $x = 10^{3.5324}$
   $x = 32.45$ $x = 0.1039$ $x = 0.05568$ $x = 3,407$

17. 1.238   **19.** 2.320   **21.** −51.083   **23.** 5.192

25. $x = \ln(3.4562 \times 10^{15}) = \ln 3.4562 + \ln 10^{15} = \ln 3.4562 + 15 \ln 10 = 35.779$
(Round off error is significantly reduced if we enter 3.4562 instead of 3.4562×10^{15}.)

27. $x = \log(6.7744 \times 10^{-13}) = \log 6.7744 + \log 10^{-13} = \log 6.7744 - 13 = -12.169$

29. $\log x = 21.667503 = .667503 + 21 = \log 4.6505 + \log 10^{21} = \log(4.6505 \times 10^{21})$
Hence $x = 4.6505 \times 10^{21}$

31. $\ln x = -11.112\ 445$
   $x = e^{-11.112\ 445}$
   $x = 1.4925 \times 10^{-5}$

33.

x	y
0.5	−0.7
1	0
2	0.7
3	1.1
4	1.4
5	1.6

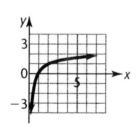

The y axis is a vertical asymptote.

35.

x	y
0.25	1.4
0.5	0.7
0.75	0.3
1	0
2	0.7
3	1.1
4	1.4
5	1.6

The y axis is a vertical asymptote.

37.

x	y
-1.5	-1.4
-1	0
-0.5	0.8
0	1.4
1	2.2
2	2.8
3	3.2

The line $x = -2$ is a vertical asymptote. Note: The graph is the same of $y = \ln x$ shifted 2 units to the left and expanded by a factor of 2.

39.

x	y
0.5	-5.8
1	-3.0
2	-0.2
3	1.4
4	2.5
5	3.4
6	4.2
7	4.8
8	5.3
9	5.8
10	6.2

The y axis is a vertical asymptote. Note: The graph is the same as the graph of $y = \ln x$ expanded by a factor of 4 and shifted 3 units down.

41. The inequality sign in the last step reverses because $\log \frac{1}{3}$ is negative.

43. (A)

x	log x	log(log x)
10	1	0
100	2	$\log 2 \approx 0.30$
1000	3	$\log 3 \approx 0.48$
10^6	6	$\log 6 \approx 0.78$
10^{100}	100	2
$10^{10^{10}}$	10^{10}	10

(B) Since the range of $\log x$ is all real numbers, the domain of $\log(\log x)$ is those real numbers for which $\log x$ is in the domain of $\log x$, that is, $\log x > 0$, or $x > 1$. The domain is thus $(1, \infty)$. Since if $y = \log(\log x)$, $10^y = \log x$ or $x = 10^{10^y}$, y can take on any real value and the range of g is $(-\infty, \infty)$.

(C) The graph of f has a vertical asymptote at $x = 0$ and no horizontal asymptote. The graph of g has a vertical asymptote at $x = 1$ and no horizontal asymptote. The graph of f has an x intercept at $x = 1$. The graph of g has an x intercept at $x = 10$. Here are computer-generated graphs of f and g in the same coordinate system.

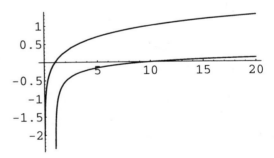

45. Here is a computer-generated graph showing $f(x) = \ln x$ and $g(x) = 0.1x - 0.2$.

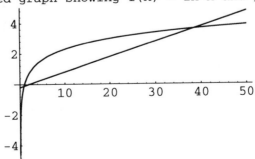

The graphs appear to intersect on the intervals (0, 2) and (38, 40). To estimate the coordinates of the points of intersection, we zoom in. Here are computer-generated graphs of f and g on (0.85, 0.95) and (38.4, 38.6).

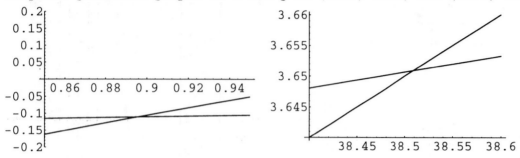

To two decimal places, the coordinates are (0.90, -0.11) and (38.51, 3.65).

47. Here is a computer-generated graph showing $f(x) = \ln x$ and $g(x) = x^{1/3}$.

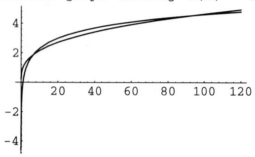

The graphs appear to intersect on the intervals (4, 8) and (92, 96). To estimate the coordinates of the points of intersection, we zoom in. Here are computer-generated graphs of f and g on (6.3, 6.5) and (93.3, 93.5).

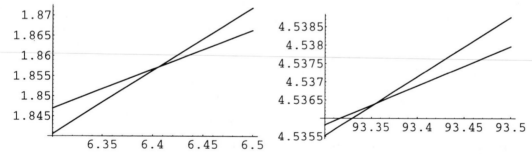

To two decimal places, the coordinates are (6.41, 1.86) and (93.35, 4.54).

49.

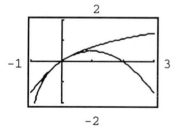

51.

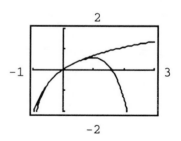

53. We use the decibel formula

$$D = 10 \log \frac{I}{I_0}$$

(A) $I = I_0$

$$D = 10 \log \frac{I_0}{I_0}$$

$$D = 10 \log 1$$

$$D = 0 \text{ decibels}$$

(B) $I_0 = 1.0 \times 10^{-12}$ $I = 1.0$

$$D = 10 \log \frac{1.0}{1.0 \times 10^{-12}}$$

$$D = 120 \text{ decibels}$$

55. We use the decibel formula

$$D = 10 \log \frac{I}{I_0}$$

$$I_2 = 1000 I_1$$

$$D_1 = 10 \log \frac{I_1}{I_0} \qquad D_2 = 10 \log \frac{I_2}{I_0}$$

$$D_2 - D_1 = 10 \log \frac{I_2}{I_0} - 10 \log \frac{I_1}{I_0}$$

$$= 10 \log \left(\frac{I_2}{I_0} \div \frac{I_1}{I_0} \right)$$

$$= 10 \log \frac{I_2}{I_1}$$

$$= 10 \log \frac{1000 I_1}{I_1}$$

$$= 10 \log 1000$$

$$= 30 \text{ decibels}$$

57. We use the magnitude formula

$$M = \frac{2}{3} \log \frac{E}{E_0}$$

$$E_0 = 10^{4.40} \qquad E = 1.99 \times 10^{17}$$

$$M = \frac{2}{3} \log \frac{1.99 \times 10^{17}}{10^{4.40}}$$

$$= \frac{2}{3} \log (1.99 \times 10^{12.6})$$

$$= \frac{2}{3} (\log 1.99 + \log 10^{12.6})$$

$$= \frac{2}{3} (0.299 + 12.6)$$

$$= 8.6$$

59. We use the magnitude formula

$$M = \frac{2}{3} \log \frac{E}{E_0}$$

For the Long Beach earthquake,

$$6.3 = \frac{2}{3} \log \frac{E_1}{E_0}$$

$$9.45 = \log \frac{E_1}{E_0}$$

$$\frac{E_1}{E_0} = 10^{9.45}$$

$$E_1 = E_0 \cdot 10^{9.45}$$

For the Anchorage earthquake,

$$8.3 = \frac{2}{3} \log \frac{E_2}{E_0}$$

$$12.45 = \log \frac{E_2}{E_0}$$

$$\frac{E_2}{E_0} = 10^{12.45}$$

$$E_2 = E_0 \cdot 10^{12.45}$$

Hence, the Anchorage earthquake compares to the Long Beach earthquake as follows:

$$\frac{E_2}{E_1} = \frac{E_0 \cdot 10^{12.45}}{E_0 \cdot 10^{9.45}} = 10^3$$

$$E_2 = 10^3 E_1, \text{ or } 1000 \text{ times as powerful}$$

61. We use the rocket equation.

$$v = c \ln \frac{W_t}{W_b}$$

$$v = 2.57 \ln (19.8)$$
$$v = 7.67 \text{ km/s}$$

63. (A) $pH = -\log[H^+] = -\log(4.63 \times 10^{-9}) = 8.3$. Since this is greater than 7, the substance is basic.

(B) $pH = -\log[H^+] = -\log(9.32 \times 10^{-4}) = 3.0$ Since this is less than 7, the substance is acidic.

65. Since $pH = -\log[H^+]$, we have
$$5.2 = -\log[H^+], \text{ or}$$
$$[H^+] = 10^{-5.2} = 6.3 \times 10^{-6} \text{ moles per liter}$$

Exercise 4-5

Key Ideas and Formulas

To solve equations in which the variable appears only in the exponent, use logarithms. If $a^x = b$, then $x \log a = \log b$.

To solve equations in which the variable appears only in the argument of a logarithm, change to equivalent exponential form.

If $\log_a x = b$, then $x = a^b$

Change-of-Base Formula: $\log_b N = \dfrac{\log_a N}{\log_a b}$

---

1. $10^x = 27.5$
$$x = \log_{10} 27.5$$
$$x = 1.44$$

3. $e^{-x} = 0.0028$
$$-x = \log_e 0.0028$$
$$x = -\log_e 0.0028$$
$$x = 5.88$$

5. $10^{2x+5} = 43.7$
$$2x + 5 = \log_{10} 43.7$$
$$2x = \log_{10} 43.7 - 5$$
$$x = \frac{\log_{10} 43.7 - 5}{2}$$
$$x = -1.68$$

7. $e^{4-2x} = 45$
$$4 - 2x = \log_e 45$$
$$-2x = \log_e 45 - 4$$
$$x = \frac{\log_e 45 - 4}{-2}$$
$$x = 0.967$$

9. $3^x = 35$
$$x \ln 3 = \ln 35$$
$$x = \frac{\ln 35}{\ln 3}$$
$$x = 3.24$$

11. $5^{-x} = 250$
$$-x \ln 5 = \ln 250$$
$$x = \frac{\ln 250}{-\ln 5}$$
$$x = -3.43$$

13. $\log x + \log 4 = 1$
$$\log(4x) = 1$$
$$4x = 10$$
$$x = \frac{5}{2}$$

15. $\ln 8 - \ln x = 2$
$$\ln \frac{8}{x} = 2$$
$$\frac{8}{x} = e^2$$
$$8 = xe^2$$
$$x = \frac{8}{e^2}$$

17. $\log(x + 10) + \log(x - 5) = 2$
$$\log[(x + 10)(x - 5)] = 2$$
$$(x + 10)(x - 5) = 10^2$$
$$x^2 + 5x - 50 = 100$$
$$x^2 + 5x - 150 = 0$$
$$(x - 10)(x + 15) = 0$$
$$x = 10, -15$$

Check:

$\log(10 + 10) + \log(10 - 5) = \log 20 + \log 5$
$$= \log 100 = 2 \quad \checkmark$$

$\log(-15 + 10)$ and $\log (-15 - 5)$ are not defined

Solution: $x = 10$

> **Common Errors:** $\log(x + 10) \neq \log x + \log 10$
> $(x + 10)(x - 5) \neq 2$

19. $2 = 1.002^{4x}$
$\ln 2 = \ln 1.002^{4x}$
$\ln 2 = 4x \ln 1.002$
$x = \dfrac{\ln 2}{4 \ln 1.002}$
$x = 86.7$

21. $e^{-0.005x} = 100$
$-0.005x = \ln 100$
$x = \dfrac{\ln 100}{-0.005}$
$x = -921$

23. $1{,}000 = 75e^{0.5x}$
$\dfrac{1{,}000}{75} = e^{0.5x}$
$0.5x = \ln \dfrac{1{,}000}{75}$
$x = \dfrac{1}{0.5} \ln \dfrac{1{,}000}{75}$
$x = 5.18$

25. $e^{-0.1x^2} = 0.2$
$-0.1x^2 = \ln 0.2$
$x^2 = -10 \ln 0.2$
$x = \pm\sqrt{-10 \ln 0.2}$
$x = \pm 4.01$

27. $\log x - \log 5 = \log 2 - \log(x - 3)$
$\log \dfrac{x}{5} = \log \dfrac{2}{x - 3}$
$\dfrac{x}{5} = \dfrac{2}{x - 3}$
Excluded value: $x \neq 3$
$5(x - 3)\dfrac{x}{5} = 5(x - 3)\dfrac{2}{x - 3}$
$(x - 3)x = 10$
$x^2 - 3x = 10$
$x^2 - 3x - 10 = 0$
$(x - 5)(x + 2) = 0$
$x = 5, -2$

Check:
$\log 5 - \log 5 \overset{\surd}{=} \log 2 - \log 2$
$\log(-2)$ is not defined
Solution: 5

29. $\ln x = \ln(2x - 1) - \ln(x - 2)$
$\ln x = \ln \dfrac{2x - 1}{x - 2}$
$x = \dfrac{2x - 1}{x - 2}$
Excluded value: $x \neq 2$
$x(x - 2) = (x - 2)\dfrac{2x - 1}{x - 2}$
$x(x - 2) = 2x - 1$
$x^2 - 2x = 2x - 1$
$x^2 - 4x + 1 = 0$
$x = \dfrac{-b \pm \sqrt{b^2 - 4ac}}{2a}$
$a = 1, \ b = -4, \ c = 1$
$x = \dfrac{-(-4) \pm \sqrt{(-4)^2 - 4(1)(1)}}{2(1)}$
$x = \dfrac{4 \pm \sqrt{12}}{2}$
$x = 2 \pm \sqrt{3}$

Check:
$\ln(2 + \sqrt{3}) \overset{?}{=} \ln[2(2 + \sqrt{3}) - 1]$
$\qquad - \ln[(2 + \sqrt{3}) - 2]$
$\ln(2 + \sqrt{3}) \overset{?}{=} \ln(3 + 2\sqrt{3}) - \ln\sqrt{3}$
$\ln(2 + \sqrt{3}) \overset{?}{=} \ln\left(\dfrac{3 + 2\sqrt{3}}{\sqrt{3}}\right)$
$\ln(2 + \sqrt{3}) \overset{\surd}{=} \ln(\sqrt{3} + 2)$
$\ln(x - 2)$ is not defined if $x = 2 - \sqrt{3}$
Solution: $2 + \sqrt{3}$

31.

$$\log(2x + 1) = 1 - \log(x - 1)$$
$$\log(2x + 1) + \log(x - 1) = 1$$
$$\log[(2x + 1)(x - 1)] = 1$$
$$(2x + 1)(x - 1) = 10$$
$$2x^2 - x - 1 = 10$$
$$2x^2 - x - 11 = 0$$
$$x = \frac{-b \pm \sqrt{b^2 - 4ac}}{2a}$$
$$a = 2, \ b = -1, \ c = -11$$
$$x = \frac{-(-1) \pm \sqrt{(-1)^2 - 4(2)(-11)}}{2(2)}$$
$$x = \frac{1 \pm \sqrt{89}}{4}$$

Check: $\log\left(2\dfrac{1 + \sqrt{89}}{4} + 1\right) \overset{?}{=} 1 - \log\left(\dfrac{1 + \sqrt{89}}{4} - 1\right)$

$$\log\left(\frac{1 + \sqrt{89} + 2}{2}\right) \overset{?}{=} 1 - \log\left(\frac{1 + \sqrt{89} - 4}{4}\right)$$
$$\log\left(\frac{3 + \sqrt{89}}{2}\right) \overset{?}{=} 1 - \log\left(\frac{\sqrt{89} - 3}{4}\right)$$
$$\log\left(\frac{3 + \sqrt{89}}{2}\right) \overset{?}{=} \log 10 - \log\left(\frac{\sqrt{89} - 3}{4}\right)$$
$$\overset{?}{=} \log\left(\frac{40}{\sqrt{89} - 3}\right)$$
$$\overset{?}{=} \log\left[\frac{40(\sqrt{89} + 3)}{89 - 9}\right]$$
$$\overset{\checkmark}{=} \log\left(\frac{\sqrt{89} + 3}{2}\right)$$

$\log(x - 1)$ is not defined if $x = \dfrac{1 - \sqrt{89}}{4}$

Solution: $\dfrac{1 + \sqrt{89}}{4}$

33.

$$(\ln x)^3 = \ln x^4$$
$$(\ln x)^3 = 4 \ln x$$
$$(\ln x)^3 - 4 \ln x = 0$$
$$\ln x[(\ln x)^2 - 4] = 0$$
$$\ln x(\ln x - 2)(\ln x + 2) = 0$$

$\ln x = 0$	$\ln x - 2 = 0$	$\ln x + 2 = 0$
$x = 1$	$\ln x = 2$	$\ln x = -2$
	$x = e^2$	$x = e^{-2}$

Check:

$(\ln 1)^3 \overset{?}{=} \ln 1^4 \qquad (\ln e^2)^3 \overset{?}{=} \ln(e^2)^4 \qquad (\ln e^{-2})^3 \overset{?}{=} \ln(e^{-2})^4$

$\quad 0 \overset{\checkmark}{=} 0 \qquad\qquad\qquad 8 \overset{\checkmark}{=} 8 \qquad\qquad\qquad -8 \overset{\checkmark}{=} -8$

Solution: $1, \ e^2, \ e^{-2}$

35. $\ln(\ln x) = 1$
$\quad\quad \ln x = e^1$
$\quad\quad \ln x = e$
$\quad\quad\quad x = e^e$

37. $x^{\log x} = 100x$
We start by taking logarithms of both sides.
$$\log(x^{\log x}) = \log 100x$$
$$\log x \log x = \log 100 + \log x$$
$$(\log x)^2 = 2 + \log x$$
$$(\log x)^2 - \log x - 2 = 0$$
$$(\log x - 2)(\log x + 1) = 0$$

$\log x - 2 = 0 \quad\quad\quad \log x + 1 = 0$
$\quad \log x = 2 \quad\quad\quad\quad \log x = -1$
$\quad\quad\quad x = 10^2 \quad\quad\quad\quad\quad x = 10^{-1}$
$\quad\quad\quad x = 100 \quad\quad\quad\quad\quad x = 0.1$

Check: $100^{\log 100} \overset{?}{=} 100\cdot 100 \quad\quad (0.1)^{\log(0.1)} \overset{?}{=} 100(0.1)$
$\quad\quad\quad\quad\quad 10^4 \overset{\sqrt{}}{=} 10^4 \quad\quad\quad\quad\quad\quad\quad 0.1^{-1} \overset{?}{=} 10$
$\quad\quad\quad\quad\quad\quad\quad\quad\quad\quad\quad\quad\quad\quad\quad\quad\quad 10 \overset{\sqrt{}}{=} 10$

39. (A) In solving an exponential equation, we often can get the variable out of the exponent by using logarithms. If we do that in this case, however, we have made no progress:
$$e^{x/2} = 5 \ln x$$
$$\ln e^{x/2} = \ln(5 \ln x)$$
$$\frac{x}{2} = \ln 5 + \ln(\ln x)$$

This equation is no easier than the original to solve. Other algebraic methods seem to be equally useless.

(B) Here is a computer-generated graph of $y = e^{x/2}$ and $y = 5 \ln x$.

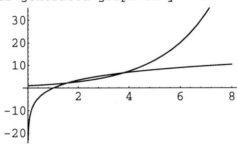

From the graph, it appears that there are two solutions of $e^{x/2} = 5 \ln x$.

41. (A) In solving an exponential equation, we often can get the variable out of the exponent by using logarithms. If we do that in this case, however, we have made no progress:
$$3^x + 2 = 7 + x - e^{-x}$$
$$3^x = -e^{-x} + x + 5$$
$$\log_3 3^x = \log_3(-e^{-x} + x + 5)$$
$$x = \log_3(-e^{-x} + x + 5)$$

This equation is no easier than the original to solve.

(B) Here is a computer-generated graph of $y = 3^x + 2$ and $y = 7 + x - e^{-x}$, $-2 \leq x \leq 2$.

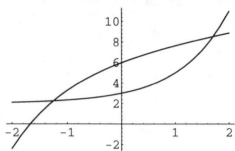

There are zeros between -1.3 and -1.2 and between 1.6 and 1.8. To estimate the zeros to three decimal places, we zoom in. Here are computer-generated graphs of the two functions on (-1.3, -1.2) and (1.7, 1.72).

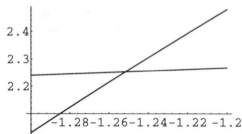
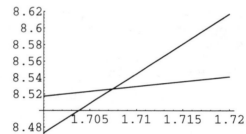

We can now estimate the solutions as -1.252 and 1.707.

43. $\log_5 372 = \dfrac{\ln 372}{\ln 5} = 3.6776$ or $\log_5 372 = \dfrac{\log_{10} 372}{\log_{10} 5} = 3.6776$

45. $\log_8 0.0352 = \dfrac{\ln 0.0352}{\ln 8} = -1.6094$ **47.** $\log_3 0.1483 = \dfrac{\ln 0.1483}{\ln 3} = -1.7372$

49.
$$A = Pe^{rt}$$
$$\frac{A}{P} = e^{rt}$$
$$\ln \frac{A}{P} = rt$$
$$\frac{1}{t} \ln \frac{A}{P} = r$$
$$r = \frac{1}{t} \ln \frac{A}{P}$$

51.
$$D = 10 \log \frac{I}{I_0}$$
$$\frac{D}{10} = \log \frac{I}{I_0}$$
$$\frac{I}{I_0} = 10^{D/10}$$
$$I = I_0(10^{D/10})$$

53.
$$M = 6 - 2.5 \log \frac{I}{I_0}$$
$$6 - M = 2.5 \log \frac{I}{I_0}$$
$$\frac{6 - M}{2.5} = \log \frac{I}{I_0}$$
$$\frac{I}{I_0} = 10^{(6-M)/2.5}$$
$$I = I_0[10^{(6-M)/2.5}]$$

55.

$$I = \frac{E}{R}(1 - e^{-Rt/L})$$

$$RI = E(1 - e^{-Rt/L})$$

$$\frac{RI}{E} = 1 - e^{-Rt/L}$$

$$\frac{RI}{E} - 1 = -e^{-Rt/L}$$

$$-\left(\frac{RI}{E} - 1\right) = e^{-Rt/L}$$

$$-\frac{RI}{E} + 1 = e^{-Rt/L}$$

$$1 - \frac{RI}{E} = e^{-Rt/L}$$

$$\ln\left(1 - \frac{RI}{E}\right) = -\frac{Rt}{L}$$

$$-\frac{L}{R}\ln\left(1 - \frac{RI}{E}\right) = t$$

$$t = -\frac{L}{R}\ln\left(1 - \frac{RI}{E}\right)$$

57.

$$y = \frac{e^x + e^{-x}}{2}$$

$$2y = e^x + e^{-x}$$

$$2y = e^x + \frac{1}{e^x}$$

$$2ye^x = (e^x)^2 + 1$$

$$0 = (e^x)^2 - 2ye^x + 1$$

This equation is quadratic in e^x

$$e^x = \frac{-b \pm \sqrt{b^2 - 4ac}}{2a}$$

$$a = 1, \quad b = -2y, \quad c = 1$$

$$e^x = \frac{-(-2y) \pm \sqrt{(-2y)^2 - 4(1)(1)}}{2(1)}$$

$$e^x = \frac{2y \pm \sqrt{4y^2 - 4}}{2}$$

$$e^x = \frac{2(y \pm \sqrt{y^2 - 1})}{2}$$

$$e^x = y \pm \sqrt{y^2 - 1}$$

$$x = \ln(y \pm \sqrt{y^2 - 1})$$

59.

$$y = \frac{e^x - e^{-x}}{e^x + e^{-x}}$$

$$y = \frac{e^x - \frac{1}{e^x}}{e^x + \frac{1}{e^x}}$$

$$y = \frac{e^x e^x - \frac{1}{e^x}e^x}{e^x e^x + \frac{1}{e^x}e^x}$$

$$y = \frac{e^{2x} - 1}{e^{2x} + 1}$$

$$y(e^{2x} + 1) = e^{2x} - 1$$

$$ye^{2x} + y = e^{2x} - 1$$

$$1 + y = e^{2x} - ye^{2x}$$

$$1 + y = (1 - y)e^{2x}$$

$$e^{2x} = \frac{1 + y}{1 - y}$$

$$2x = \ln\frac{1 + y}{1 - y}$$

$$x = \frac{1}{2}\ln\frac{1 + y}{1 - y}$$

61. $y = 3 + \log_2(2 - x)$
To enter this into the calculator, we use the change-of-base formula to write this as:
$$y = 3 + \frac{\log(2 - x)}{\log 2}$$

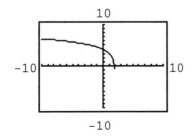

63. $y = \log_3 x - \log_2 x$
To enter this into the calculator, we use the change-of-base formula to write this as:
$$y = \frac{\log x}{\log 3} - \frac{\log x}{\log 2}$$

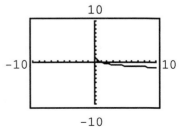

65. Here is a computer-generated graph of $2^{-x} - 2x$, $0 \leq x \leq 1$.

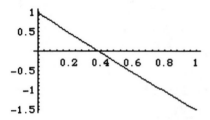

There is a zero between 0.36 and 0.4. To estimate the zero to two decimal places, we "zoom in" on it. Here is the computer-generated graph of $2^{-x} - 2x$, $0.36 \leq x \leq 0.4$

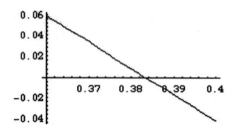

We can now estimate the zero as 0.38.

69. Here is a computer-generated graph of $e^{-x} - x$, $0 \leq x \leq 1$.

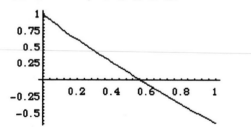

67. Here is a computer-generated graph of $x3^x - 1$, $0 \leq x \leq 1$.

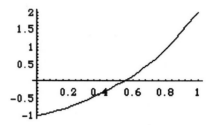

There is a zero between 0.52 and 0.56. To estimate the zero to two decimal places, we zoom in on it. Here is the computer-generated graph of $x3^x - 1$, $0.52 \leq x \leq 0.56$.

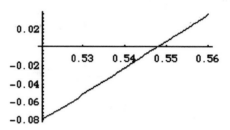

We can now estimate the zero as 0.55.

There is a zero between 0.56 and 0.6. To estimate the zero to two decimal places, we zoom in on it. Here is the computer-generated graph of $e^{-x} - x$, $0.56 \leq x \leq 0.6$.

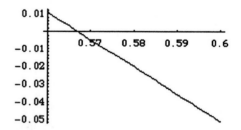

We can now estimate the zero as 0.57.

71. Here is a computer-generated graph of $xe^x - 2$, $0 \leq x \leq 1$.

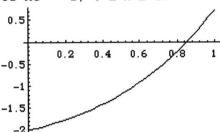

There is a zero between 0.84 and 0.88. To estimate the zero to two decimal places, we zoom in on it. Here is the computer-generated graph of $xe^x - 2$, $0.84 \leq x \leq 0.88$.

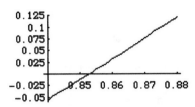

We can now estimate the zero as 0.85.

73. Here is a computer-generated graph of $\ln x + 2x$, $0 < x \leq 1$.

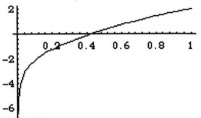

There is a zero between 0.4 and 0.44. To estimate the zero to two decimal places, we zoom in on it. Here is the computer-generated graph of $\ln x + 2x$, $0.4 \leq x \leq 0.44$.

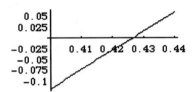

We can now estimate the zero as 0.43.

75. Here is a computer-generated graph of $\ln x + e^x$, $0 < x \leq 1$.

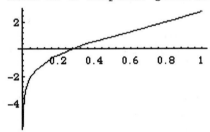

There is a zero between 0.24 and 0.28. To estimate the zero to two decimal places, we zoom in on it. Here is the computer-generated graph of $\ln x + e^x$, $0.24 \leq x \leq 0.28$.

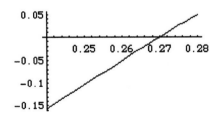

We can now estimate the zero as 0.27.

77. To find the doubling time we replace A in $A = P(1 + 0.15)^n$ with $2P$ and solve for n.

$$2P = P(1.15)^n$$
$$2 = (1.15)^n$$
$$\ln 2 = n \ln 1.15$$
$$n = \frac{\ln 2}{\ln 1.15}$$

$n = 5$ years to the nearest year

79. We solve $A = Pe^{rt}$ for r, with $A = 2{,}500$, $P = 1{,}000$, $t = 10$

$$2{,}500 = 1{,}000e^{r(10)}$$
$$2.5 = e^{10r}$$
$$10r = \ln (2.5)$$
$$r = \frac{1}{10} \ln 2.5$$
$$r = 0.0916 \text{ or } 9.16\%$$

81. $m = 6 - 2.5 \log \dfrac{L}{L_0}$

(A) We find m when $L = L_0$

$m = 6 - 2.5 \log \dfrac{L_0}{L_0}$

$m = 6 - 2.5 \log 1$

$m = 6$

(B) We compare L_1 for $m = 1$ with L_2 for $m = 6$

$1 = 6 - 2.5 \log \dfrac{L_1}{L_0}$ $\qquad$ $6 = 6 - 2.5 \log \dfrac{L_2}{L_0}$

$-5 = -2.5 \log \dfrac{L_1}{L_0}$ $\qquad$ $0 = -2.5 \log \dfrac{L_2}{L_0}$

$2 = \log \dfrac{L_1}{L_0}$ $\qquad\qquad$ $0 = \log \dfrac{L_2}{L_0}$

$\dfrac{L_1}{L_0} = 10^2$ $\qquad\qquad\qquad$ $\dfrac{L_2}{L_0} = 1$

$L_1 = 100L_0$ $\qquad\qquad$ $L_2 = L_0$

Hence $\dfrac{L_1}{L_2} = \dfrac{100L_0}{L_0} = 100$. The star of magnitude 1 is 100 times brighter.

83. We solve $P = P_0 e^{rt}$ for t with
$P = 2P_0$, $r = 0.02$.

$2P_0 = P_0 e^{0.02t}$

$2 = e^{0.02t}$

$\ln 2 = 0.02t$

$\dfrac{\ln 2}{0.02} = t$

$t = 35$ years to the nearest year

85. We solve $A = A_0 e^{-0.000124t}$ for t with
$A = 0.1A_0$.

$0.1A_0 = A_0 e^{-0.000124t}$

$0.1 = e^{-0.000124t}$

$\ln 0.1 = -0.000124t$

$t = \dfrac{\ln 0.1}{-0.000124}$

$t = 18{,}600$ years old

87. We solve $q = 0.0009(1 - e^{-0.2t})$ for t with $q = 0.0007$

$0.0007 = 0.0009(1 - e^{-0.2t})$

$\dfrac{0.0007}{0.0009} = 1 - e^{-0.2t}$

$\dfrac{7}{9} = 1 - e^{-0.2t}$

$-\dfrac{2}{9} = -e^{-0.2t}$

$\dfrac{2}{9} = e^{-0.2t}$

$\ln \dfrac{2}{9} = -0.2t$

$t = \dfrac{\ln \frac{2}{9}}{-0.2}$

$t = 7.52$ seconds

89. First, we solve $T = T_m + (T_0 - T_m)e^{-kt}$
for k, with $T = 61.5°$, $T_m = 40°$,
$T_0 = 72°$, $t = 1$

$61.5 = 40 + (72 - 40)e^{-k(1)}$

$21.5 = 32e^{-k}$

$\dfrac{21.5}{32} = e^{-k}$

$\ln \dfrac{21.5}{32} = -k$

$k = -\ln \dfrac{21.5}{32}$

$k = 0.40$

Now we solve $T = T_m + (T_0 - T_m)e^{-0.40t}$
for t, with $T = 50°$, $T_m = 40°$, $T_0 = 72°$

$50 = 40 + (72 - 40)e^{-0.40t}$

$10 = 32e^{-0.40t}$

$\dfrac{10}{32} = e^{-0.40t}$

$\ln \dfrac{10}{32} = -0.40t$

$t = \dfrac{\ln^{10/32}}{-0.40}$

$t = 2.9$ hours

91. We solve $N = \dfrac{100}{1 + 4e^{-0.14t}}$ for t, with $N = 50$

$$50 = \frac{100}{1 + 4e^{-0.14t}}$$

$$\frac{1}{50} = \frac{1 + 4e^{-0.14t}}{100}$$

$$2 = 1 + 4e^{-0.14t}$$

$$1 = 4e^{-0.14t}$$

$$0.25 = e^{-0.14t}$$

$$\ln 0.25 = -0.14t$$

$$t = \frac{\ln 0.25}{-0.14}$$

$$t = 10 \text{ years}$$

CHAPTER 4 REVIEW

1. $\log m = n$ *(4-3)*

2. $\ln x = y$ *(4-3)*

3. $x = 10^y$ *(4-3)*

4. $y = e^x$ *(4-3)*

5. $\dfrac{7^{x+2}}{7^{2-x}} = 7^{(x+2)-(2-x)}$

$$= 7^{x+2-2+x}$$

$$= 7^{2x}\quad (4-1)$$

6. $\left(\dfrac{e^x}{e^{-x}}\right)^x = [e^{x-(-x)}]^x$

$$= (e^{2x})^x = e^{2x \cdot x}$$

$$= e^{2x^2}\quad (4-1)$$

7. $\log_2 x = 3$

$$x = 2^3$$

$$x = 8\quad (4-3)$$

8. $\log_x 25 = 2$

$$25 = x^2$$

$$x = 5$$

since bases are
restricted positive
  (4-3)

9. $\log_3 27 = x$

$$\log_3 3^3 = x$$

$$3 = x\quad (4-3)$$

10. $10^x = 17.5$

$$x = \log_{10} 17.5$$

$$x = 1.24\quad (4-3)$$

11. $e^x = 143,000$

$$x = \ln 143,000$$

$$x = 11.9\quad (4-3)$$

12. $\ln x = -0.01573$

$$x = e^{-0.01573}$$

$$x = 0.984\quad (4-3)$$

13. $\log x = 2.013$

$$x = 10^{2.013}$$

$$x = 103\quad (4-3)$$

14. $\ln(2x - 1) = \ln(x + 3)$

$$2x - 1 = x + 3$$

$$x = 4$$

Check:

$$\ln(2 \cdot 4 - 1) \stackrel{?}{=} \ln(4 + 3)$$

$$\ln 7 \stackrel{\checkmark}{=} \ln 7\quad (4-3)$$

15. $\log(x^2 - 3) = 2\log(x - 1)$

$$\log(x^2 - 3) = \log(x - 1)^2$$

$$x^2 - 3 = (x - 1)^2$$

$$x^2 - 3 = x^2 - 2x + 1$$

$$-3 = -2x + 1$$

$$-4 = -2x$$

$$x = 2$$

Check:

$$\log(2^2 - 3) \stackrel{?}{=} \log(2 - 1)$$

$$\log 1 \stackrel{?}{=} 2\log 1$$

$$0 \stackrel{\checkmark}{=} 0\quad (4-3)$$

16.

$$e^{x^2-3} = e^{2x}$$

$$x^2 - 3 = 2x$$

$$x^2 - 2x - 3 = 0$$

$$(x - 3)(x + 1) = 0$$

$$x = 3, -1$$

  (4-2)

17.

$$4^{x-1} = 2^{1-x}$$

$$(2^2)^{x-1} = 2^{1-x}$$

$$2^{2(x-1)} = 2^{1-x}$$

$$2(x - 1) = 1 - x$$

$$2x - 2 = 1 - x$$

$$3x = 3$$

$$x = 1\quad (4-1)$$

18.
$$2x^2 e^{-x} = 18 e^{-x}$$
$$2x^2 e^{-x} - 18 e^{-x} = 0$$
$$2e^{-x}(x^2 - 9) = 0$$
$$2e^{-x}(x - 3)(x + 3) = 0$$
$$2e^{-x} = 0 \quad x - 3 = 0 \quad x + 3 = 0$$
never $\qquad x = 3 \qquad x = -3$
Solution: 3, -3 $\qquad$ (4-2)

19.
$$\log_{1/4} 16 = x$$
$$\log_{1/4} 4^2 = x$$
$$\log_{1/4} \left(\frac{1}{4}\right)^{-2} = x$$
$$-2 = x \qquad (4-3)$$

20. $\log_x 9 = -2$
$$x^{-2} = 9$$
$$\frac{1}{x^2} = 9$$
$$1 = 9x^2$$
$$\frac{1}{9} = x^2$$
$$x = \pm\sqrt{\frac{1}{9}}$$
$$x = \frac{1}{3}$$
since bases are
restricted positive
$\qquad$ (4-3)

21. $\log_{16} x = \frac{3}{2}$
$$16^{3/2} = x$$
$$64 = x$$
$$x = 64 \qquad (4-3)$$

22. $\log_x e^5 = 5$
$$e^5 = x^5$$
$$x = e \quad (4-3)$$

23. $10^{\log_{10} x} = 33$
$$\log_{10} x = \log_{10} 33$$
$$x = 33 \qquad (4-3)$$

24. $\ln x = 0$
$$e^0 = x$$
$$x = 1 \qquad (4-3)$$

25. 1.145 (4-3)

26. Not defined. ($-e$ is not in the domain of the logarithm function.) $\qquad$ (4-3)

27. 2.211 $\qquad$ (4-3)

28. 11.59 (4-3)

29. $x = 2(10^{1.32})$
$$x = 41.8 \qquad (4-1)$$

30. $x = \log_5 23$
$$x = \frac{\log 23}{\log 5} \text{ or } \frac{\ln 23}{\ln 5}$$
$$x = 1.95 \qquad (4-3)$$

31. $\ln x = -3.218$
$$x = e^{-3.218}$$
$$x = 0.0400 \quad (4-3)$$

32. $x = \log(2.156 \times 10^{-7})$
$$x = \log 2.156 + \log 10^{-7}$$
$$x = \log 2.156 - 7$$
$$x = -6.67 \qquad (4-3)$$

33. $x = \dfrac{\ln 4}{\ln 2.31}$
$$x = 1.66 \qquad (4-3)$$

34.
$$25 = 5(2)^x$$
$$\frac{25}{5} = 2^x$$
$$5 = 2^x$$
$$\ln 5 = x \ln 2$$
$$\frac{\ln 5}{\ln 2} = x$$
$$x = 2.32 \qquad (4-5)$$

35. $4,000 = 2,500 e^{0.12x}$
$$\frac{4,000}{2,500} = e^{0.12x}$$
$$0.12x = \ln \frac{4,000}{2,500}$$
$$x = \frac{1}{0.12} \ln \frac{4,000}{2,500}$$
$$x = 3.92 \qquad (4-5)$$

36.
$$0.01 = e^{-0.05x}$$
$$-0.05x = \ln 0.01$$
$$x = \frac{\ln 0.01}{-0.05}$$
$$x = 92.1 \qquad (4-5)$$

37.
$$5^{2x-3} = 7.08$$
$$(2x - 3)\log 5 = \log 7.08$$
$$2x - 3 = \frac{\log 7.08}{\log 5}$$
$$x = \frac{1}{2}\left[3 + \frac{\log 7.08}{\log 5}\right]$$
$$x = 2.11 \qquad (4-5)$$

38.
$$\frac{e^x - e^{-x}}{2} = 1$$
$$e^x - e^{-x} = 2$$
$$e^x - \frac{1}{e^x} = 2$$
$$e^x e^x - e^x\left(\frac{1}{e^x}\right) = 2e^x$$
$$(e^x)^2 - 1 = 2e^x$$
$$(e^x)^2 - 2e^x - 1 = 0$$

This equation is quadratic in e^x

$$e^x = \frac{-b \pm \sqrt{b^2 - 4ac}}{2a}$$
$$a = 1, \quad b = -2, \quad c = -1$$

$$e^x = \frac{-(-2) \pm \sqrt{(-2)^2 - 4(1)(-1)}}{2}$$

$$e^x = \frac{2 \pm \sqrt{8}}{2}$$

$$e^x = 1 \pm \sqrt{2}$$

$$x = \ln(1 \pm \sqrt{2})$$

$1 - \sqrt{2}$ is negative, hence not in the domain
of the logarithm function.

$$x = \ln(1 + \sqrt{2})$$
$$x = 0.881$$

(4-5)

39.
$$\log 3x^2 - \log 9x = 2$$
$$\log \frac{3x^2}{9x} = 2$$
$$\frac{3x^2}{9x} = 10^2$$
$$\frac{x}{3} = 100$$
$$x = 300$$

Check:

$$\log(3 \cdot 300^2) - \log(9 \cdot 300) \overset{?}{=} 2$$
$$\log(270{,}000) - \log(2{,}700) \overset{?}{=} 2$$
$$\log \frac{270{,}000}{2{,}700} \overset{?}{=} 2$$
$$\log 100 \overset{\sqrt{}}{=} 2 \qquad (4\text{-}5)$$

40.
$$\log x - \log 3 = \log 4 - \log(x + 4)$$
$$\log \frac{x}{3} = \log \frac{4}{x + 4}$$
$$\frac{x}{3} = \frac{4}{x + 4} \quad \text{excluded value:} \atop x \neq -4$$
$$3(x + 4)\frac{x}{3} = 3(x + 4)\frac{4}{x + 4}$$
$$(x + 4)x = 12$$
$$x^2 + 4x = 12$$
$$x^2 + 4x - 12 = 0$$
$$(x + 6)(x - 2) = 0$$
$$x = -6 \quad x = 2$$

Check: $\log(-6)$ is not defined

$$\log 2 - \log 3 \overset{?}{=} \log 4 - \log(2 + 4)$$
$$\log \frac{2}{3} \overset{?}{=} \log \frac{4}{6}$$
$$\log \frac{2}{3} \overset{\sqrt{}}{=} \log \frac{2}{3}$$

Solution: 2 (4-5)

41.
$$\ln(x + 3) - \ln x = 2 \ln 2$$
$$\ln \frac{x + 3}{x} = \ln 2^2$$
$$\frac{x + 3}{x} = 2^2$$
$$\frac{x + 3}{x} = 4$$
$$x + 3 = 4x$$
$$3 = 3x$$
$$x = 1$$

Check:

$$\ln(1 + 3) - \ln 1 \overset{?}{=} 2 \ln 2$$
$$\ln 4 - 0 \overset{?}{=} 2 \ln 2$$
$$\ln 4 \overset{\sqrt{}}{=} \ln 4 \qquad (4\text{-}5)$$

42.
$$\ln(2x + 1) - \ln(x - 1) = \ln x$$
$$\ln \frac{2x + 1}{x - 1} = \ln x$$
$$\frac{2x + 1}{x - 1} = x \quad \text{Excluded value: } x \neq 1$$
$$(x - 1)\frac{2x + 1}{x - 1} = x(x - 1)$$
$$2x + 1 = x^2 - x$$
$$0 = x^2 - 3x - 1$$

$$x = \frac{-b \pm \sqrt{b^2 - 4ac}}{2a} \quad a = 1, \ b = -3, \ c = -1$$

$$x = \frac{-(-3) \pm \sqrt{(-3)^2 - 4(1)(-1)}}{2(1)}$$

$$x = \frac{3 \pm \sqrt{13}}{2}$$

Check: $\ln\left(\dfrac{3 - \sqrt{13}}{2}\right)$ is not defined

$$\ln\left(2 \cdot \frac{3 + \sqrt{13}}{2} + 1\right) - \ln\left(\frac{3 + \sqrt{13}}{2} - 1\right) \stackrel{?}{=} \ln\left(\frac{3 + \sqrt{13}}{2}\right)$$

$$\ln(3 + \sqrt{13} + 1) - \ln\left(\frac{3 + \sqrt{13} - 2}{2}\right) \stackrel{?}{=} \ln\left(\frac{3 + \sqrt{13}}{2}\right)$$

$$\ln(4 + \sqrt{13}) - \ln\left(\frac{1 + \sqrt{13}}{2}\right) \stackrel{?}{=} \ln\left(\frac{3 + \sqrt{13}}{2}\right)$$

$$\ln\left(\frac{4 + \sqrt{13}}{1} \cdot \frac{2}{1 + \sqrt{13}}\right) \stackrel{?}{=} \ln\left(\frac{3 + \sqrt{13}}{2}\right)$$

$$\ln\left(\frac{(4 + \sqrt{13})2}{1 + \sqrt{13}}\right) \stackrel{?}{=} \ln\left(\frac{3 + \sqrt{13}}{2}\right)$$

$$\ln\left(\frac{(4 + \sqrt{13})2(1 - \sqrt{13})}{(1 + \sqrt{13})(1 - \sqrt{13})}\right) \stackrel{?}{=} \ln\left(\frac{3 + \sqrt{13}}{2}\right)$$

$$\ln\left(\frac{2(4 - 3\sqrt{13} - 13)}{1 - 13}\right) \stackrel{?}{=} \ln\left(\frac{3 + \sqrt{13}}{2}\right)$$

$$\ln\left(\frac{-18 - 6\sqrt{13}}{-12}\right) \stackrel{?}{=} \ln\left(\frac{3 + \sqrt{13}}{2}\right)$$

$$\ln\left(\frac{3 + \sqrt{13}}{2}\right) \stackrel{\surd}{=} \ln\left(\frac{3 + \sqrt{13}}{2}\right)$$

Solution: $\dfrac{3 + \sqrt{13}}{2}$ $\qquad\qquad\qquad\qquad\qquad\qquad\qquad\qquad$ (4-5)

43.

$$\begin{aligned}
(\log x)^3 &= \log x^9 \\
(\log x)^3 &= 9 \log x \\
(\log x)^3 - 9 \log x &= 0 \\
\log x[(\log x)^2 - 9] &= 0 \\
\log x(\log x - 3)(\log x + 3) &= 0
\end{aligned}$$

$$\log x = 0 \quad \log x - 3 = 0 \quad \log x + 3 = 0$$
$$x = 1 \qquad \log x = 3 \qquad \log x = -3$$
$$\qquad\qquad x = 10^3 \qquad\quad x = 10^{-3}$$

Check:

$$(\log 1)^3 \stackrel{?}{=} \log 1^9$$
$$0 \stackrel{\surd}{=} 0$$
$$(\log 10^3)^3 \stackrel{?}{=} \log(10^3)^9$$
$$27 \stackrel{\surd}{=} 27$$
$$(\log 10^{-3})^3 \stackrel{?}{=} \log(10^{-3})^9$$
$$-27 \stackrel{\surd}{=} -27$$

Solution: $1, \ 10^3, \ 10^{-3}$ $\quad$ (4-5)

44.
$$\begin{aligned}
\ln(\log x) &= 1 \\
\log x &= e \\
x &= 10^e
\end{aligned}$$
$\qquad$ (4-5)

45. $(e^x + 1)(e^{-x} - 1) - e^x(e^{-x} - 1) = e^x e^{-x} - e^x + e^{-x} - 1 - e^x e^{-x} + e^x$
$$= 1 - e^x + e^{-x} - 1 - 1 + e^x = e^{-x} - 1 \qquad (4-2)$$

46. $(e^x + e^{-x})(e^x - e^{-x}) - (e^x - e^{-x})^2 = (e^x)^2 - (e^{-x})^2 - [(e^x)^2 - 2e^x e^{-x} + (e^{-x})^2]$
$$= e^{2x} - e^{-2x} - [e^{2x} - 2 + e^{-2x}]$$
$$= e^{2x} - e^{-2x} - e^{2x} + 2 - e^{-2x}$$
$$= 2 - 2e^{-2x} \qquad (4-2)$$

47.

x	y
-2	0.13
-1	0.25
0	0.5
1	1
2	2
3	4
4	8

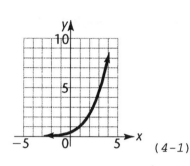

$(4-1)$

48.

t	f(t)
-25	74
-20	50
-15	33
-10	22
-5	15
0	10
5	6.7
10	4.5
15	3.0

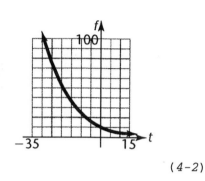

$(4-2)$

49.

x	y
-0.5	-0.7
0	0
1	0.7
2	1.1
4	1.6
6	1.9
8	2.2
10	2.4

The line $x = -1$ is a vertical asymptote. Note: The graph is the same as the graph of $y = \ln x$ shifted 1 unit to the left $(4-3)$

50.

t	N
-3	1.6
-2	4.3
-1	11
0	25
1	48
2	71
3	87
4	95
5	98

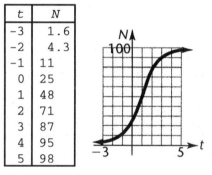

$(4-2)$

51. If the graph of $y = e^x$ is reflected in the x axis, y is replaced by $-y$ and the graph becomes the graph of $-y = e^x$ or $y = -e^x$.
If the graph of $y = e^x$ is reflected in the y axis, x is replaced by $-x$ and the graph becomes the graph of $y = e^{-x}$ or $y = \dfrac{1}{e^x}$ or $y = \left(\dfrac{1}{e}\right)^x$. $\qquad (4-3)$

52. (A) For $x > -1$, $y = e^{-x/3}$ decreases from $e^{1/3}$ to 0 while $\ln(x + 1)$ increases from $-\infty$ to ∞. Consequently, the graphs can intersect at exactly one point.

(B) Here is a computer-generated graph of $e^{-x/3}$ and $\ln(x + 1)$, $-1 < x \le 2$.

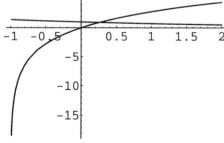

There is a point of intersection between 0.2 and 0.3. To estimate the solution to three decimal places, we zoom in on it. Here is the computer-generated graph of $e^{-x/3}$ and $\ln(x + 1)$, $0.25 \le x \le 0.27$.

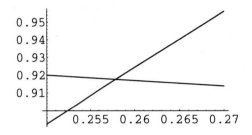

We can now estimate the solution as 0.258. (4-5)

53. Here is a computer-generated graph of $4 - x^2 + \ln x$, $0 < x \le 3$.

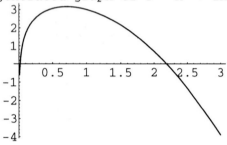

There are zeros between 0 and 0.1 and also between 2.1 and 2.2. To estimate the zeros to three decimal places, we zoom in. Here are computer-generated graphs of $4 - x^2 + \ln x$, $0.01 \le x \le 0.02$ and $2.18 \le x \le 2.19$.

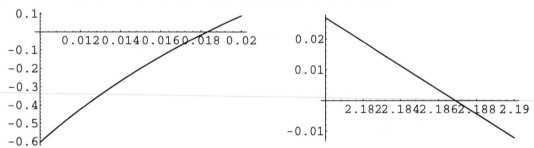

We can now estimate the zeros as 0.018 and 2.187. (4-3)

54. Here is a computer-generated graph of 10^{x-3} and $8 \log x$, $0 < x \le 4$.

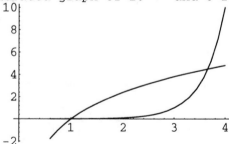

There are points of intersection just to the right of 1, and also between 3.6 and 3.8. To estimate the coordinates to three decimal places, we zoom in. Here are computer-generated graphs of 10^{x-3} and $8 \log x$, $1.00 \le x \le 1.01$ and $3.65 \le x \le 3.66$.

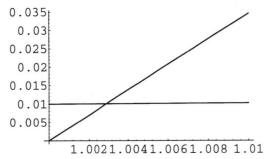
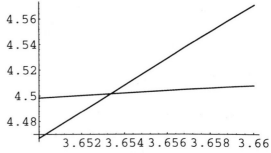

We can now estimate the coordinates of the points of intersection as (1.003, 0.010) and (3.653, 4.502).

(4-4)

55.

$$D = 10 \log \frac{I}{I_0}$$

$$\frac{D}{10} = \log \frac{I}{I_0}$$

$$10^{D/10} = \frac{I}{I_0}$$

$$I_0 10^{D/10} = I$$

$$I = I_0(10^{D/10})$$

(4-5)

56.

$$y = \frac{1}{\sqrt{2\pi}} e^{-x^2/2}$$

$$\sqrt{2\pi}y = e^{-x^2/2}$$

$$-\frac{x^2}{2} = \ln(\sqrt{2\pi}y)$$

$$x^2 = -2 \ln(\sqrt{2\pi}y)$$

$$x = \pm\sqrt{-2 \ln(\sqrt{2\pi}y)}$$

(4-5)

57.

$$x = -\frac{1}{k} \ln \frac{I}{I_0}$$

$$-kx = \ln \frac{I}{I_0}$$

$$\frac{I}{I_0} = e^{-kx}$$

$$I = I_0(e^{-kx})$$

(4-5)

58.

$$r = P \frac{i}{1 - (1 + i)^{-n}}$$

$$\frac{r}{P} = \frac{i}{1 - (1 + i)^{-n}}$$

$$\frac{P}{r} = \frac{1 - (1 + i)^{-n}}{i}$$

$$\frac{Pi}{r} = 1 - (1 + i)^{-n}$$

$$\frac{Pi}{r} - 1 = -(1 + i)^{-n}$$

$$1 - \frac{Pi}{r} = (1 + i)^{-n}$$

$$\ln\left(1 - \frac{Pi}{r}\right) = -n \ln (1 + i)$$

$$\frac{\ln (1 - \frac{Pi}{r})}{-\ln (1 + i)} = n$$

$$n = -\frac{\ln (1 - \frac{Pi}{r})}{\ln (1 + i)}$$

(4-5)

59. $f(x) = 2 \ln(x - 1)$

Since the graph of f is the same as the graph of $y = \ln x$ shifted right one unit and expanded by a factor of 2, it passes the horizontal line test, and f is one-to-one.

Solve $y = f(x)$ for x

$$y = 2 \ln(x - 1)$$

$$\frac{y}{2} = \ln (x - 1)$$

$$x - 1 = e^{y/2}$$

$$x = e^{y/2} + 1 = f^{-1}(y)$$

Interchange x and y:

$y = f^{-1}(x) = e^{x/2} + 1$ Domain: $(-\infty, \infty)$

Check: $f^{-1}[f(x)] = e^{2\ln(x-1)/2} + 1$

$$= e^{\ln(x-1)} + 1$$

$$= x - 1 + 1$$

$$= x$$

$$f[f^{-1}(x)] = 2 \ln(e^{x/2} + 1 - 1)$$

$$= 2 \ln e^{x/2}$$

$$= 2\left(\frac{x}{2}\right)$$

$$= x$$

(4-5, 2-6)

60. $f(x) = \dfrac{e^x - e^{-x}}{2}$ f is one-to-one (proof omitted)

Solve $y = f(x)$ for x

$$y = \frac{e^x - e^{-x}}{2}$$

$$2y = e^x - e^{-x}$$

$$2y = e^x - \frac{1}{e^x}$$

$$2ye^x = e^x e^x - e^x\left(\frac{1}{e^x}\right)$$

$$2ye^x = (e^x)^2 - 1$$

$$0 = (e^x)^2 - 2ye^x - 1$$

This equation is quadratic in e^x

$$e^x = \frac{-b \pm \sqrt{b^2 - 4ac}}{2a} \quad a = 1, \ b = -2y, \ c = -1$$

$$e^x = \frac{-(-2y) \pm \sqrt{(-2y)^2 - 4(1)(-1)}}{2(1)}$$

$$e^x = \frac{2y \pm \sqrt{4y^2 + 4}}{2}$$

$$e^x = y \pm \sqrt{y^2 + 1}$$

Note: Since $0 < 1$, $y^2 < y^2 + 1$, $\sqrt{y^2} < \sqrt{y^2 + 1}$ and $y < \sqrt{y^2 + 1}$ for all real y.
Hence $y - \sqrt{y^2 + 1}$ is always negative. Also, $y + \sqrt{y^2 + 1}$ is always positive.
$x = \ln(y + \sqrt{y^2 + 1}) = f^{-1}(y)$
Interchange x and y:
$y = f^{-1}(x) = \ln(x + \sqrt{x^2 + 1})$ Domain: $(-\infty, \infty)$ (since $x + \sqrt{x^2 + 1}$ is always positive)
(Check omitted for lack of space)   (*4-5, 2-6*)

61. $\ln y = -5t + \ln c$
$\ln y - \ln c = -5t$

$$\ln\left(\frac{y}{c}\right) = -5t$$

$$\frac{y}{c} = e^{-5t}$$

$$y = ce^{-5t} \qquad (4\text{-}3, \ 4\text{-}5)$$

62.

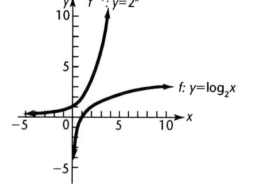

x	$y = \log_2 x$	$x = \log_2 y$	y
1	0	0	1
2	1	1	2
4	2	2	4
8	3	3	8

Domain $f = (0, \infty) = $ Range f^{-1}
Range $f = (-\infty, \infty) = $ Domain f^{-1} (*4-3*)

63. If $\log_1 x = y$, then we would have to have $1^y = x$; that is, $1 = x$ for arbitrary positive x, which is impossible.   (*4-3*)

64. Let $u = \log_b M$ and $v = \log_b N$; then $M = b^u$ and $N = b^v$. Thus,
$\log(M/N) = \log_b(b^u/b^v) = \log_b b^{u-v} = u - v = \log_b M - \log_b N.$ $\qquad$ *(4-3)*

65. We solve $P = P_0(1.03)^t$ for t,
using $P = 2P_0$.
$$2P_0 = P_0(1.03)^t$$
$$2 = (1.03)^t$$
$$\ln 2 = t \ln 1.03$$
$$\frac{\ln 2}{\ln 1.03} = t$$
$$t = 23.4 \text{ years} \qquad (4\text{-}5)$$

66. We solve $P = P_0 e^{0.03t}$ for t using
$P = 2P_0$.
$$2P_0 = P_0 e^{0.03t}$$
$$2 = e^{0.03t}$$
$$\ln 2 = 0.03t$$
$$\frac{\ln 2}{0.03} = t$$
$$t = 23.1 \text{ years} \qquad (4\text{-}5)$$

67. $\quad A_0 = $ original amount
$0.01A_0 = 1$ percent of original
$\qquad\qquad$ amount
We solve $A = A_0 e^{-0.000124t}$ for t,
using $A = 0.01A_0$.
$$0.01A_0 = A_0 e^{-0.000124t}$$
$$0.01 = e^{-0.000124t}$$
$$\ln 0.01 = -0.000124t$$
$$\frac{\ln 0.01}{-0.000124} = t$$
$$t = 37{,}100 \text{ years} \qquad (4\text{-}5)$$

68. (A) When $t = 0$, $N = 1$. As t increases
by $1/2$, N doubles.
Hence $N = 1 \cdot (2)^{t \div 1/2}$
$\qquad\qquad N = 2^{2t}$ (or $N = 4^t$)
(B) We solve $N = 4^t$ for t, using
$N = 10^9$
$$10^9 = 4^t$$
$$9 = t \log 4$$
$$t = \frac{9}{\log 4}$$
$$t = 15 \text{ days} \qquad (4\text{-}5)$$

69. We use $A = Pe^{rt}$ with $P = 1$, $r = 0.03$, and $t = 2000$.
$$A = 1e^{0.03(2000)}$$
$$A = 1.1 \times 10^{26} \text{ dollars} \qquad (4\text{-}2)$$

70. (A)

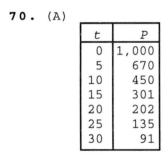

t	P
0	1,000
5	670
10	450
15	301
20	202
25	135
30	91

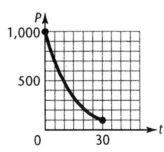

(B) As t tends to infinity, P appears to tend to 0. $\qquad$ *(4-2)*

71. $M = \frac{2}{3} \log \frac{E}{E_0}$ $\quad E_0 = 10^{4.40}$
We use $E = 1.99 \times 10^{14}$
$$M = \frac{2}{3} \log \frac{1.99 \times 10^{14}}{10^{4.40}}$$
$$M = \frac{2}{3} \log(1.99 \times 10^{9.6})$$
$$M = \frac{2}{3}(\log 1.99 + 9.6)$$
$$M = \frac{2}{3}(0.299 + 9.6)$$
$$M = 6.6 \qquad (4\text{-}4)$$

72. We solve $M = \frac{2}{3} \log \frac{I}{I_0}$ for I, using
$I_0 = 10^{4.40}$, $M = 8.3$
$$8.3 = \frac{2}{3} \log \frac{I}{10^{4.40}}$$
$$\frac{3}{2}(8.3) = \log \frac{I}{10^{4.40}}$$
$$12.45 = \log \frac{I}{10^{4.40}}$$
$$\frac{I}{10^{4.40}} = 10^{12.45}$$
$$I = 10^{4.40} \cdot 10^{12.45}$$
$$I = 10^{16.85} \text{ or } 7.08 \times 10^{16} \text{ joules}$$
$$(4\text{-}4)$$

73. We use the given formula twice, with

$$I_2 = 100,000 I_1$$

$$D_1 = 10 \log \frac{I_1}{I_0} \qquad D_2 = 10 \log \frac{I_2}{I_0}$$

$$D_2 - D_1 = 10 \log \frac{I_2}{I_0} - 10 \log \frac{I_1}{I_0}$$

$$= 10 \log \left(\frac{I_2}{I_0} \div \frac{I_1}{I_0} \right)$$

$$= 10 \log \frac{I_2}{I_1}$$

$$= 10 \log \frac{100,000 I_1}{I_1}$$

$$= 10 \log 100,000$$

$$= 50 \text{ decibels}$$

The level of the louder sound is 50 decibels more. (4-4)

74. $I = I_0 e^{-kd}$

To find k, we solve for k using

$I = \frac{1}{2} I_0$ and $d = 73.6$

$$\frac{1}{2} I_0 = I_0 e^{-k(73.6)}$$

$$\frac{1}{2} = e^{-73.6k}$$

$$-73.6k = \ln \frac{1}{2}$$

$$k = \frac{\ln \frac{1}{2}}{-73.6}$$

$$k = 0.00942$$

We now find the depth at which 1% of the surface light remains. We solve $I = I_0 e^{-0.00942d}$ for d with $I = 0.01 I_0$

$$0.01 I_0 = I_0 e^{-0.00942d}$$

$$0.01 = e^{-0.00942d}$$

$$-0.00942d = \ln 0.01$$

$$d = \frac{\ln 0.01}{-0.00942}$$

$$d = 489 \text{ feet}$$ (4-2)

75. We solve $N = \dfrac{30}{1 + 29e^{-1.35t}}$ for t with $N = 20$.

$$20 = \frac{30}{1 + 29e^{-1.35t}}$$

$$\frac{1}{20} = \frac{1 + 29e^{-1.35t}}{30}$$

$$1.5 = 1 + 29e^{-1.35t}$$

$$0.5 = 29e^{-1.35t}$$

$$\frac{0.5}{29} = e^{-1.35t}$$

$$-1.35t = \ln \frac{0.5}{29}$$

$$t = \frac{\ln \frac{0.5}{29}}{-1.35}$$

$$t = 3 \text{ years}$$ (4-5)

CUMULATIVE REVIEW EXERCISE (Chapters 3 and 4)

1. (A) Since the graph has x intercepts -1, 1, and 2, and -1 is at least a double zero (the graph is tangent to the x axis at $x = -1$), the lowest degree equation would be $P(x) = (x + 1)^2 (x - 1)(x - 2)$.

   (B) $P(x) \to \infty$ as $x \to \infty$ and as $x \to -\infty$. $\hspace{2cm}$ (3-1)

2.
$$
\begin{array}{r|rrrr}
 & 3 & 5 & -18 & -3 \\
 & & -9 & 12 & 18 \\
\hline
-3 & 3 & -4 & -6 & 15
\end{array}
$$

   $3x^3 + 5x^2 - 18x - 3 = (x + 3)(3x^2 - 4x - 6) + 15$ $\quad$ (3-1)

3. $-2, 3, 5$ $\hspace{2cm}$ (3-2)

4. We investigate $P(1)$ and $P(2)$ by forming a synthetic division table.

$$
\begin{array}{c|rrrr}
 & 4 & -5 & -3 & -1 \\
\hline
1 & 4 & -1 & -4 & -5 \\
2 & 4 & 3 & 3 & 5
\end{array}
$$

   Since $P(1)$ and $P(2)$ have opposite signs, there is at least one real zero between 1 and 2. $\hspace{2cm}$ (3-3)

5. The possible rational zeros are ± 1, ± 2, ± 4, ± 8. We form a synthetic division table.

$$
\begin{array}{c|rrrrl}
 & 1 & 1 & -10 & 8 & \\
\hline
1 & 1 & 2 & -8 & 0 & 1 \text{ is a zero}
\end{array}
$$

   Thus $x^3 + x^2 - 10x + 8 = (x - 1)(x^2 + 2x - 8) = (x - 1)(x - 2)(x + 4)$.
   The rational zeros are 1, 2, -4. $\hspace{2cm}$ (3-2)

6. We write
$$
\frac{5x - 4}{(x - 2)(x + 1)} = \frac{A}{x - 2} + \frac{B}{(x + 1)} = \frac{A(x + 1) + B(x - 2)}{(x - 2)(x + 1)}
$$

   Thus for all x
   $5x - 4 = A(x + 1) + B(x - 2)$
   If $x = -1$
   $\quad -9 = -3B$
   $\quad \; B = 3$
   If $x = 2$
   $\quad 6 = 3A$
   $\quad A = 2$

   So $\dfrac{5x - 4}{(x - 2)(x + 1)} = \dfrac{2}{x - 2} + \dfrac{3}{x + 1}$ $\hspace{2cm}$ (3-5)

7. (A) $x = \log_{10} y$ or $x = \log y$ $\hspace{1.5cm}$ (B) $x = e^y$ $\hspace{2cm}$ (4-4)

8. (A) $(2e^x)^3 = 2^3(e^x)^3 = 8e^{3x}$ $\hspace{1.5cm}$ (B) $\dfrac{e^{3x}}{e^{-2x}} = e^{3x-(-2x)} = e^{5x}$ $\hspace{1cm}$ (4-2)

9. (A) $\log_3 x = 2$ (B) $\log_3 81 = x$ (C) $\log_x 4 = -2$

             $x = 3^2$ $\log_3 3^4 = x$ $x^{-2} = 4$

             $x = 9$ $x = 4$ $\dfrac{1}{x^2} = 4$

$$1 = 4x^2$$

$$x^2 = \dfrac{1}{4}$$

$$x = \dfrac{1}{2}$$

since bases are restricted positive *(4-3)*

10. (A) $10^x = 2.35$ (B) $e^x = 87,500$

         $x = \log 2.35$ $x = \ln 87,500$

         $x = 0.371$ $x = 11.4$

  (C) $\log x = -1.25$ (D) $\ln x = 2.75$

         $x = 10^{-1.25}$ $x = e^{2.75}$

         $x = 0.0562$ $x = 15.6$ *(4-4)*

11. $f(x) = 3 \ln x - \sqrt{x}$ *(2-3, 4-3)*

12. The function f multiplies the base e raised to the power of one-half of the domain element by 100 and then subtracts 50.   *(2-3, 4-2)*

13.

$$
\begin{array}{r|rrrr}
 & 2 & -5 & 3 & 2 \\
 & & 1 & -2 & \frac{1}{2} \\
\hline
\frac{1}{2} & 2 & -4 & 1 & \frac{5}{2}
\end{array}
$$

$P\left(\dfrac{1}{2}\right) = \dfrac{5}{2}$ *(3-2)*

14. $x - 1$ will be a factor of $P(x)$ if $P(1) = 0$

$P(1) = 1^{25} - 1^{20} + 1^{15} + 1^{10} - 1^5 + 1$

     $= 1 - 1 + 1 + 1 - 1 + 1 = 2 \neq 0$,

so $x - 1$ is not a factor. $x + 1$ will be a factor of $P(x)$ if $P(-1) = 0$

$P(-1) = (-1)^{25} - (-1)^{20} + (-1)^{15} + (-1)^{10} - (-1)^5 + 1$

     $= -1 - 1 - 1 + 1 + 1 + 1 = 0$,

so $x + 1$ is a factor of $P(x)$.   *(3-2)*

15. (A) We form a synthetic division table:

	1	0	-8	0	3	
-3	1	-3	1	-3	12	$= P(-3)$
-2	1	-2	-4	8	-13	$= P(-2)$
-1	1	-1	-7	7	-4	$= P(-1)$
0	1	0	-8	0	3	$= P(0)$
1	1	1	-7	-7	-4	$= P(1)$
2	1	2	-4	-8	-13	$= P(2)$
3	1	3	1	3	12	$= P(3)$

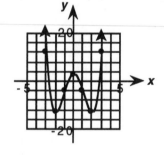

The graph of $P(x)$ has four x intercepts and three turning points; $P(x) \to \infty$ as $x \to \infty$ and as $x \to -\infty$.

(B) There are real zeros in the intervals $(-3, -2)$, $(-1, 0)$, $(0, 1)$, and $(2, 3)$ indicated in the table. We search for the real zero in $(2, 3)$. We organize our calculations in a table.

Sign Change Interval (a, b)	Midpoint m	Sign of P		
		$P(a)$	$P(m)$	$P(b)$
(2, 3)	2.5	−	−	+
(2.5, 3)	2.75	−	−	+
(2.75, 3)	2.875	−	+	+
(2.75, 2.875)	2.8125	−	+	+
(2.75, 2.8125)	We stop here	−		+

Since each endpoint rounds to 2.8, a real zero lies on this last interval and is given by 2.8 to one decimal place accuracy.   *(3-1, 3-3)*

16. (A) We form a synthetic division table:

	1	2	-20	0	-30
0	1	2	-20	0	-30
1	1	3	-17	-17	-47
2	1	4	-12	-24	-78
3	1	5	-5	-15	-75
4	1	6	4	16	34
-1	1	1	-21	21	-51
-2	1	0	-20	40	-110
-3	1	-1	-17	51	-183
-4	1	-2	-12	48	-222
-5	1	-3	-5	25	-155
-6	1	-4	4	-24	114

From the table, 4 is an upper bound and -6 is a lower bound.

(B) There are real zeros in the intervals $(-6, -5)$ and $(3, 4)$ indicated in the table. We search for the real zero in $(3, 4)$. We organize our calculations in a table.

Sign Change Interval (a, b)	Midpoint m	Sign of P		
		$P(a)$	$P(m)$	$P(b)$
(3, 4)	3.5	−	−	+
(3.5, 4)	3.75	−	−	+
(3.75, 4)	3.875	−	+	+
(3.75, 3.875)	3.8125	−	+	+
(3.75, 3.8125)	3.78125	−	−	+
(3.78125, 3.8125)	3.796875	−	−	+
(3.796875, 3.8125)	3.8046875	−	+	+
(3.796875, 3.8046875)	We stop here	−		+

Since each endpoint rounds to 3.80, a real zero lies on this last interval and is given by 3.80 to two decimal place accuracy.

(C) Here is a computer-generated graph of P.

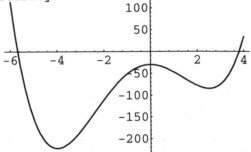

The x intercepts are on the invervals $(-6, -5.6)$ and $(3.6, 4)$. To estimate more accurately we zoom in. Here are graphs of P on the intervals $(-5.7, -5.6)$ and $(3.7, 3.9)$.

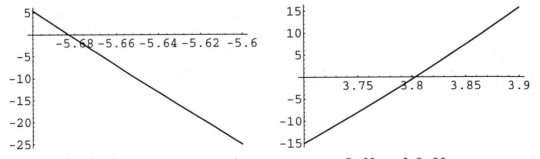

To two decimal place accuracy, the zeros are -5.68 and 3.80.   *(3-3)*

17. Use synthetic division

$$
\begin{array}{r|rrrrr}
 & 2 & -9 & 10 & 1 & -4 \\
 &   & 4 & -10 & 0 & 2 \\
\hline
2 & 2 & -5 & 0 & 1 & -2 \\
\end{array}
$$

$P(x) = (x - 2)(2x^3 - 5x^2 + 1) + (-2)$

By the Remainder Theorem, the remainder, when $P(x)$ is divided by $x - 2$, is $P(2)$. Hence, $P(2) = -2$.   (3-2)

18. The possible rational zeros are ±1, ±3, ±5, ±15, $\pm\frac{1}{2}$, $\pm\frac{3}{2}$, $\pm\frac{5}{2}$, $\pm\frac{15}{2}$, $\pm\frac{1}{4}$, $\pm\frac{3}{4}$, $\pm\frac{5}{4}$, $\pm\frac{15}{4}$. We form a synthetic division table.

	4	-20	29	-15	
1	4	-16	13	-2	
3	4	-8	5	0	3 is a zero

So $P(x) = (x - 3)(4x^2 - 8x + 5)$

To find the remaining zeros, we solve $4x^2 - 8x + 5 = 0$ by the quadratic formula.
$4x^2 - 8x + 5 = 0$

$$x = \frac{-b \pm \sqrt{b^2 - 4ac}}{2a} \quad a = 4,\ b = -8,\ c = 5$$

$$x = \frac{-(-8) \pm \sqrt{(-8)^2 - 4(4)(5)}}{2(4)}$$

$$x = \frac{8 \pm \sqrt{64 - 80}}{8}$$

$$x = \frac{8 \pm \sqrt{-16}}{8}$$

$$x = \frac{8 \pm 4i}{8}$$

$$x = 1 \pm \frac{1}{2}i$$

The zeros are 3, $1 \pm \frac{1}{2}i$ (3-2)

19. The possible rational zeros are ±1, ±2, ±3, ±4, ±6, ±12. We form a synthetic division table.

	1	5	1	-15	-12	
1	1	6	7	-8	-20	
2	1	7	15	15	18	a zero between 1 and 2; 2 is an upperbound
-1	1	4	-3	-12	0	-1 is a zero

We now examine $x^3 + 4x^2 - 3x - 12$. This factors by grouping into $(x + 4)(x^2 - 3)$, however, if we don't notice this, we find the remaining possible rational zeros to be -1, -2, -3, -4, -6, -12. We form a synthetic division table.

	1	4	-3	-12	
-1	1	3	-6	-6	not a double zero
-2	1	2	-7	2	a zero between -1 and -2
-3	1	1	-6	6	
-4	1	0	-3	0	-4 is a zero

So $P(x) = (x + 1)(x + 4)(x^2 - 3) = (x + 1)(x + 4)(x - \sqrt{3})(x + \sqrt{3})$.
The four zeros are -1, -4, $\pm\sqrt{3}$.   (3-2)

20. $x^3 + 36 \leq 7x^2$
$x^3 - 7x^2 + 36 \leq 0$
To factor $x^3 - 7x^2 + 36$, we search for zeros of the polynomial.
Possible rational zeros are ±1, ±2, ±3, ±4, ±6, ±9, ±12, ±18, ±36. We form a synthetic division table.

	1	-7	0	36	
1	1	-6	-6	30	
2	1	-5	-10	16	
3	1	-4	-12	0	3 is a zero

$x^3 - 7x^2 + 36 = (x - 3)(x^2 - 4x - 12) = (x - 3)(x - 6)(x + 2)$.

Hence we must examine the sign behavior of $(x - 3)(x - 6)(x + 2)$.

We form a sign chart.
Zeros: 3, 6, -2

$x^3 - 7x^2 + 36 = (x - 3)(x - 6)(x + 2)$				
Test Number	-3	0	4	7
Value of Polynomial for Test Number	-54	36	-12	36
Sign of Polynomial in Interval	-	+	-	+
Interval	$(-\infty, -2)$	$(-2, 3)$	$(3, 6)$	$(6, \infty)$

$x^3 - 7x^2 + 36 \leq 0$ and $x^3 + 36 \leq 7x^2$ within the intervals $(-\infty, -2]$ and $[3, 6]$ or $x \leq -2$ or $3 \leq x \leq 6$.   *(3-2, 1-8)*

21. The possible rational zeros are ±1, ±2, ±4, ±5, ±10, and ±20. We form a synthetic division table.

	1	0	4	-20	
0	1	0	4	-20	
1	1	1	5	-15	
2	1	2	8	-4	
4	1	4	20	60	an upper bound
-1	1	-1	5	-25	a lower bound

There are no rational zeros. Since $P(2)$ and $P(4)$ have opposite signs, the only real zero is between 2 and 4. We search for this real zero. We organize our calculations in a table.

Sign Change Interval (a, b)	Midpoint m	Sign of P $P(a)$	$P(m)$	$P(b)$
(2, 4)	3	-	+	+
(2, 3)	2.5	-	+	+
(2, 2.5)	2.25	-	+	+
(2, 2.25)	2.125	-	-	+
(2.125, 2.25)	2.1875	-	-	+
(2.1875, 2.25)	2.21875	-	-	+
(2.21875, 2.25)	2.234375	-	+	+
(2.21875, 2.234375)	2.2265625	-	-	+
(2.2265625, 2.234375)	We stop here	-		+

Since each endpoint rounds to 2.23, a real zero lies on this last interval and is given by 2.23 to two decimal places. A glance at a computer-generated graph of $P(x)$ suggests that this is the only real zero.

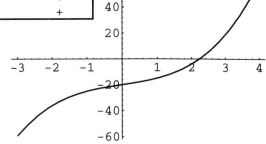

(3-3)

22. $\dfrac{3x^2 - x + 1}{x(x + 1)^2} = \dfrac{A}{x} + \dfrac{B}{x + 1} + \dfrac{C}{(x + 1)^2} = \dfrac{A(x + 1)^2 + Bx(x + 1) + Cx}{x(x + 1)^2}$

Thus for all x

$3x^2 - x + 1 = A(x + 1)^2 + Bx(x + 1) + Cx$

If $x = 0$	If $x = -1$	If $x = 1$, using $A = 1$ and $C = -5$
$1 = A$	$5 = C(-1)$	$3 = 4 + 2B - 5$
	$C = -5$	$B = 2$

So $\dfrac{3x^2 - x + 1}{x(x + 1)^2} = \dfrac{1}{x} + \dfrac{2}{x + 1} - \dfrac{5}{(x + 1)^2}$ (3-5)

23. First we factor $x^3 - x^2 + x = x(x^2 - x + 1)$. $x^2 - x + 1$ cannot be factored further in the real numbers, so we write

$\dfrac{x^2 + x - 2}{x^3 - x^2 + x} = \dfrac{A}{x} + \dfrac{Bx + C}{x^2 - x + 1} = \dfrac{A(x^2 - x + 1) + (Bx + C)x}{x(x^2 - x + 1)}$

Thus for all x

$x^2 + x - 2 = A(x^2 - x + 1) + (Bx + C)x$

If $x = 0$

$\qquad -2 = A$

$x^2 + x - 2 = -2(x^2 - x + 1) + Bx^2 + Cx$

$\qquad\qquad\quad = (B - 2)x^2 + (C + 2)x - 2$

Equating coefficients of like terms, we have

$\quad 1 = B - 2$

$\quad 1 = C + 2$

$\quad B = 3$

$\quad C = -1$

So $\dfrac{x^2 + x - 2}{x^3 - x^2 + x} = \dfrac{-2}{x} + \dfrac{3x - 1}{x^2 - x + 1}$ (3-5)

24. $f(x) = \dfrac{2x + 8}{x + 2} = \dfrac{n(x)}{d(x)}$

(A) The domain of f is the set of all real numbers x such that $d(x) = x + 2 \neq 0$, that is $(-\infty, -2) \cup (-2, \infty)$ or $x \neq -2$. f has an x intercept where $n(x) = 2x + 8 = 0$, that is, $x = -4$. $f(0) = 4$, hence f has a y intercept at $y = 4$.

(B) *Vertical asymptote:* $x = -2$

Horizontal asymptote: Since $n(x)$ and $d(x)$ have the same degree, the line $y = 2$ is a horizontal asymptote.

(C) *Sign Chart:*

Test numbers	-5	-3	0
Value of f	$\frac{2}{3}$	-2	4
Sign of f	$+$	$-$	$+$

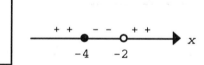

Since $x = -2$ is a vertical asymptote and $f(x) < 0$ for $-4 < x < -2$, and $f(x) > 0$ for $x > -2$, $f(x) \to -\infty$ as $x \to -2^-$ and $f(x) \to \infty$ as $x \to -2^+$.

Complete the sketch:

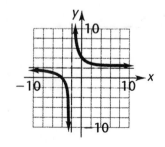

(3-4)

25.
$$2^{x^2} = 4^{x+4}$$
$$2^{x^2} = (2^2)^{x+4}$$
$$2^{x^2} = 2^{2(x+4)}$$
$$x^2 = 2(x + 4)$$
$$x^2 = 2x + 8$$
$$x^2 - 2x - 8 = 0$$
$$(x - 4)(x + 2) = 0$$
$$x - 4 = 0 \qquad x + 2 = 0$$
$$x = 4 \qquad x = -2 \qquad (4\text{-}1)$$

26.
$$2x^2 e^{-x} + x e^{-x} = e^{-x}$$
$$2x^2 e^{-x} + x e^{-x} - e^{-x} = 0$$
$$e^{-x}(2x^2 + x - 1) = 0$$
$$e^{-x}(2x - 1)(x + 1) = 0$$
$$e^{-x} = 0 \qquad 2x - 1 = 0 \qquad x + 1 = 0$$
$$\text{never} \qquad x = \frac{1}{2} \qquad x = -1$$

Solutions: $\frac{1}{2}$, -1 $\qquad (4\text{-}2)$

27. $e^{\ln x} = 2.5$
$\qquad x = 2.5 \qquad (4\text{-}3)$

28. $\log_x 10^4 = 4$
$$x^4 = 10^4$$
$$x = 10 \qquad (4\text{-}3)$$

29. $\log_9 x = -\frac{3}{2}$
$$9^{-3/2} = x$$
$$x = \frac{1}{27} \quad (4\text{-}3)$$

30. $\ln(x + 4) - \ln(x - 4) = 2 \ln 3$
$$\ln \frac{x + 4}{x - 4} = \ln 3^2$$
$$\frac{x + 4}{x - 4} = 3^2$$
$$\frac{x + 4}{x - 4} = 9 \qquad x \neq 4$$
$$x + 4 = 9(x - 4)$$
$$x + 4 = 9x - 36$$
$$-8x = -40$$
$$x = 5$$

Check: $\ln(5 + 4) - \ln(5 - 4) \overset{?}{=} 2 \ln 3$
$$\ln 9 - \ln 1 \overset{?}{=} 2 \ln 3$$
$$\ln 9 - 0 \overset{\surd}{=} \ln 9$$
Solution: $x = 5$ $\qquad (4\text{-}5)$

31. $\ln(2x^2 + 2) = 2 \ln(2x - 4)$
$$\ln(2x^2 + 2) = \ln(2x - 4)^2$$
$$2x^2 + 2 = (2x - 4)^2$$
$$2x^2 + 2 = 4x^2 - 16x + 16$$
$$0 = 2x^2 - 16x + 14$$
$$0 = 2(x - 7)(x - 1)$$
$$x = 7, 1$$

Check: $x = 7$
$$\ln(2 \cdot 7^2 + 2) \overset{?}{=} 2 \ln(2 \cdot 7 - 4)$$
$$\ln 100 \overset{?}{=} 2 \ln 10$$
$$\ln 100 \overset{\surd}{=} \ln 100$$
$x = 1$
$$\ln(2 \cdot 1^2 + 2) \overset{?}{=} 2 \ln(2 \cdot 1 - 4)$$
$$\ln(4) \neq 2 \ln(-2)$$
Solution: $x = 7$ $\qquad (4\text{-}5)$

32. $\log x + \log(x + 15) = 2$
$$\log[x(x + 15)] = 2$$
$$x(x + 15) = 10^2$$
$$x^2 + 15x = 100$$
$$x^2 + 15x - 100 = 0$$
$$(x - 5)(x + 20) = 0$$
$$x = 5, -20$$

Check: $x = 5$
$$\log 5 + \log(5 + 15) \overset{?}{=} 2$$
$$\log 5 + \log 20 \overset{?}{=} 2$$
$$\log 100 \overset{\surd}{=} 2$$
$x = -20$
$$\log(-20) + \log(-20 + 15) \neq 2$$
Solution: $x = 5$ $\qquad (4\text{-}5)$

33.
$$\log(\ln x) = -1$$
$$\ln x = 10^{-1}$$
$$\ln x = 0.1$$
$$x = e^{0.1} \qquad (4\text{-}4)$$

34.
$$4(\ln x)^2 = \ln x^2$$
$$4(\ln x)^2 = 2\ln x$$
$$4(\ln x)^2 - 2\ln x = 0$$
$$2\ln x(2\ln x - 1) = 0$$

$$2\ln x = 0 \qquad 2\ln x - 1 = 0$$
$$\ln x = 0 \qquad \ln x = \frac{1}{2}$$
$$x = 1 \qquad x = e^{0.5}$$

Check: $x = 1$
$$4(\ln 1)^2 \overset{?}{=} \ln 1^2$$
$$0 \overset{\surd}{=} 0$$

$x = e^{0.5}$
$$4(\ln e^{0.5})^2 \overset{?}{=} \ln(e^{0.5})^2$$
$$4(0.5)^2 \overset{?}{=} \ln(e^{2(0.5)})$$
$$4(0.25) \overset{?}{=} \ln e$$
$$1 \overset{\surd}{=} 1$$

Solution: $1,\ e^{0.5} \qquad (4\text{-}5)$

35. $x = \log_3 41$

We use the change of base formula
$$x = \frac{\log 41}{\log 3}$$
$$x = 3.38 \qquad (4\text{-}5)$$

36.
$$\ln x = 1.45$$
$$x = e^{1.45}$$
$$x = 4.26 \qquad (4\text{-}4)$$

37.
$$4(2^x) = 20$$
$$2^x = 5$$
$$x\log 2 = \log 5$$
$$x = \frac{\log 5}{\log 2}$$
$$x = 2.32 \qquad (4\text{-}4)$$

38.
$$10e^{-0.5x} = 1.6$$
$$e^{-0.5x} = 0.16$$
$$-0.5x = \ln 0.16$$
$$x = \frac{\ln 0.16}{-0.5}$$
$$x = 3.67 \qquad (4\text{-}5)$$

39.
$$\frac{e^x - e^{-x}}{e^x + e^{-x}} = \frac{1}{2}$$
$$\frac{e^x - \frac{1}{e^x}}{e^x + \frac{1}{e^x}} = \frac{1}{2}$$
$$\frac{e^x(e^x - \frac{1}{e^x})}{e^x(e^x + \frac{1}{e^x})} = \frac{1}{2}$$
$$\frac{(e^x)^2 - 1}{(e^x)^2 + 1} = \frac{1}{2}$$
$$\frac{e^{2x} - 1}{e^{2x} + 1} = \frac{1}{2}$$
$$2(e^{2x} + 1)\frac{e^{2x} - 1}{e^{2x} + 1} = 2(e^{2x} + 1)\frac{1}{2}$$
$$2(e^{2x} - 1) = e^{2x} + 1$$
$$2e^{2x} - 2 = e^{2x} + 1$$
$$e^{2x} = 3$$
$$2x = \ln 3$$
$$x = \frac{1}{2}\ln 3$$
$$x = 0.549 \qquad (4\text{-}5)$$

40.

x	y
-1	9
0	3
1	1
2	0.33
3	0.11

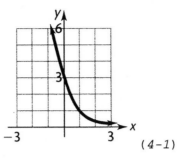

(4-1)

41.

x	f(x)
1	0
0	0.69
-1	1.10
-2	1.39
-3	1.61
-4	1.79
-5	1.95

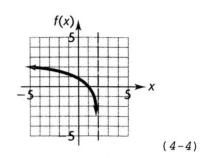

(4-4)

42.

t	A(t)
-2	182
0	100
2	55
4	30
6	17
8	9
10	5

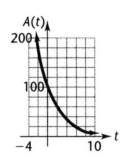

(4-2)

43. The graph is the same as the graph of $y = e^{-x}$ expanded by a factor of 2, reflected with respect to the x axis, and shifted up 3 units. The line $y = 3$ is a horizontal asymptote.

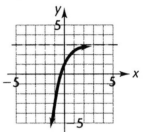

(4-2)

44. If the graph of $y = \ln x$ is reflected in the x axis, y is replaced by $-y$ and the graph becomes the graph of $-y = \ln x$ or $y = -\ln x$.
If the graph of $y = \ln x$ is reflected in the y axis, x is replaced by $-x$ and the graph becomes the graph of $y = \ln(-x)$. (2-5, 4-3)

45. (A) For $x > 0$, $y = e^{-x}$ decreases from 1 to 0 while $\ln x$ increases from $-\infty$ to ∞. Consequently, the graphs can intersect at exactly one point.

(B) Here is a computer-generated graph of e^{-x} and $\ln x$, $0 < x \le 4$.

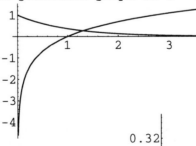

There is a point of intersection between 1.2 and 1.4. To estimate the solution to two decimal places, we zoom in on it. Here is the computer-generated graph of e^{-x} and $\ln x$, $1.3 \le x \le 1.4$.

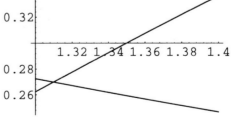

To two decimal places, the solution is 1.31. (4-3)

46. Yes, for example:

$P(x) = (x + i)(x - i)(x + \sqrt{2})(x - \sqrt{2}) = x^4 - x^2 - 2$ has irrational zeros $\sqrt{2}$ and $-\sqrt{2}$. $\hspace{2cm}$ *(3-2)*

47. (A) We form a synthetic division table:

	1	9	-500	0	20,000
0	1	9	-500	0	20,000
10	1	19	-310	-3,100	-11,000
20	1	29	80	1,600	52,000
-10	1	-1	-490	4,900	-29,000
-20	1	-11	-280	5,600	-92,000
-30	1	-21	130	-3,900	137,000
-40	1	-31	740	-29,600	1,204,000

From the table, 20 is an upper bound and -30 is a lower bound.

(B) Here is a computer-generated graph of P, $-30 \leq x \leq 20$.

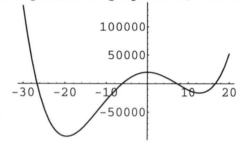

The four real zeros indicated in the table are shown in the graph to lie in $(-30, -20)$, $(-10, 0)$, $(0, 10)$, and $(10, 20)$. To estimate the zero in $(-30, -20)$, we zoom in. Here is a computer-generated graph of P on $(-26.7, -26.6)$

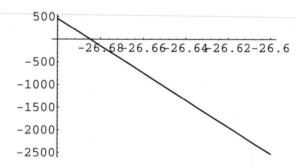

To two decimal places, the zero is -26.68. Similarly, the other zeros can be estimated as -6.22, 7.23, and 16.67 (details omitted). $\hspace{1cm}$ *(3-3)*

48. $f(x) = \dfrac{x^2 + 4x + 8}{x + 2} = \dfrac{n(x)}{d(x)}$

Intercepts. There are no real zeros of $n(x) = x^2 + 4x + 8$. No x intercept

$\qquad$ $f(0) = 4$ $\;y$ intercept

Vertical asymptotes. Real zeros of $d(x) = x + 2$. $\;x = -2$

Sign Chart.

Test numbers	−3	0
Value of f	−5	4
Sign of f	−	+

$$\begin{array}{c}\text{- - -} \quad \text{+ + +} \\ \xrightarrow{\hspace{1cm}\circ\hspace{2cm}} x \\ \qquad -2 \end{array}$$

Since $x = -2$ is a vertical asymptote and $f(x) < 0$ for $x < -2$ and $f(x) > 0$ for $x > -2$,

$\qquad f(x) \to -\infty$ as $x \to -2^-$ and $f(x) \to \infty$ as $x \to -2^+$

Horizontal asymptote. Since the degree of $n(x)$ is greater than the degree of $d(x)$, there is no horizontal asymptote.

Oblique asymptote.

$$\begin{array}{r}
x + 2 \\
x + 2 \overline{)\; x^2 + 4x + 8\;} \\
\underline{x^2 + 2x} \\
2x + 8 \\
\underline{2x + 4} \\
4
\end{array}$$

Thus, $f(x) = x + 2 + \dfrac{4}{x + 2}$

Hence, the line $y = x + 2$ is an oblique asymptote.

Complete the sketch.

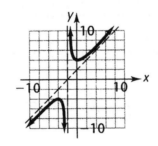

$(3\text{-}4)$

49. $[x - (-1)]^2 (x - 0)^3 [x - (3 + 5i)][x - (3 - 5i)] = (x + 1)^2 x^3 (x - 3 - 5i)(x - 3 + 5i)$

degree 7 $\hspace{6cm}$ $(3\text{-}3)$

50. The possible rational zeros are ±1, ±2, ±3, ±4, ±6, ±12. We form a synthetic division table:

	1	−4	3	10	−10	−12	
1	1	−3	0	10	0	−12	
2	1	−2	−1	8	6	0	2 is a zero

We continue to examine $x^4 - 2x^3 - x^2 + 8x + 6$. The possible rational zeros are −1, ±2, ±3, ±6. We form a synthetic division table:

	1	−2	−1	8	6	
2	1	0	−1	6	18	Not a double zero
3	1	1	2	14	48	an upper bound
−1	1	−3	2	6	0	−1 is a zero

We continue to examine $x^3 - 3x^2 + 2x + 6$. The possible rational zeros are −1, −2, −3, −6. We form a synthetic division table.

	1	-3	2	6	
-1	1	-4	6	0	a double zero

We complete the solution by solving $x^2 - 4x + 6 = 0$ by completing the square.

$$x^2 - 4x = -6$$
$$x^2 - 4x + 4 = -2$$
$$(x - 2)^2 = -2$$
$$x - 2 = \pm i\sqrt{2}$$
$$x = 2 \pm i\sqrt{2}$$

Thus the zeros of $P(x)$ are 2, -1 (double), and $2 \pm i\sqrt{2}$.

$$P(x) = (x - 2)(x + 1)^2[x - (2 + i\sqrt{2})][x - (2 - i\sqrt{2})]$$
$$= (x - 2)(x + 1)^2(x - 2 - i\sqrt{2})(x - 2 + i\sqrt{2}) \qquad (3\text{-}3)$$

51. $P(x) = x^5 + 4x^4 + x^3 - 11x^2 - 8x + 4$. The possible rational zeros are ± 1, ± 2, ± 4. We form a synthetic division table:

	1	4	1	-11	-8	4	
1	1	5	6	-5	-13	-9	
2	1	6	13	15	22	48	an upper bound
-1	1	3	-2	-9	1	3	
-2	1	2	-3	-5	2	0	-2 is a zero

We continue to examine $x^4 + 2x^3 - 3x^2 - 5x + 2$. The only possible rational zero remaining is -2. We form a synthetic division table.

	1	2	-3	-5	2	
-2	1	0	-3	1	0	-2 is a double zero

We contine to examine $x^3 - 3x + 1$.
There are no further rational zeros. The real zeros can be found by the bisection method (details omitted) to be -1.88, 0.35, and 1.53 to two decimal place accuracy. $\qquad (3\text{-}3)$

52. First we must factor $x^4 - x^3 + x^2 - 3x + 2$. The possible rational zeros are ± 1, ± 2. We form a synthetic division table:

	1	-1	1	-3	2
1	1	0	1	-2	0

We examine $x^3 + x - 2$.

	1	0	1	-2
1	1	1	2	0

Thus $x^4 - x^3 + x^2 - 3x + 2 = (x - 1)^2(x^2 + x + 2)$. $x^2 + x + 2$ cannot be factored further in the real numbers, so

$$\frac{x^2 - 4x + 11}{x^4 - x^3 + x^2 - 3x + 2} = \frac{A}{x - 1} + \frac{B}{(x - 1)^2} + \frac{Cx + D}{x^2 + x + 2}$$
$$= \frac{A(x - 1)(x^2 + x + 2) + B(x^2 + x + 2) + (Cx + D)(x - 1)^2}{(x - 1)^2(x^2 + x + 2)}$$

Thus for all x

$$x^2 - 4x + 11 = A(x - 1)(x^2 + x + 2) + B(x^2 + x + 2) + (Cx + D)(x - 1)^2$$

If $x = 1$

$$8 = 4B$$
$$B = 2$$

$$x^2 - 4x + 11 = A(x^3 + x - 2) + B(x^2 + x + 2) + (Cx + D)(x^2 - 2x + 1)$$
$$= (A + C)x^3 + (B - 2C + D)x^2 + (A + B + C - 2D)x - 2A + 2B + D$$

We have already $B = 2$, so equating coefficients of like terms,
$$0 = A + C$$
$$1 = 2 - 2C + D$$
$$-4 = A + 2 + C - 2D$$
$$11 = -2A + 4 + D$$

Since $C = -A$, we can write
$$1 = 2 + 2A + D$$
$$-4 = 2 - 2D$$
$$11 = -2A + 4 + D$$
So $D = 3$, $A = -2$, $C = 2$

Hence $\dfrac{x^2 - 4x + 11}{x^4 - x^3 + x^2 - 3x + 2} = \dfrac{-2}{x - 1} + \dfrac{2}{(x - 1)^2} + \dfrac{2x + 3}{x^2 + x + 2}$ (3-5)

53. $f(x) = 3 \ln(x - 2)$

(A) Since the graph of f is the same as the graph of $y = \ln x$ shifted 2 units to the right and expanded by a factor of 3, it passes the horizontal line test, and f is one-to-one.

Solve $y = f(x)$ for x
$$y = 3 \ln(x - 2)$$
$$\frac{y}{3} = \ln(x - 2)$$
$$x - 2 = e^{y/3}$$
$$x = e^{y/3} + 2 = f^{-1}(y)$$

Interchange x and y:
$$y = f^{-1}(x) = e^{x/3} + 2 \quad \text{Domain: } (-\infty, \infty)$$

Check: $f^{-1}[f(x)] = e^{3\ln(x-2)/3} + 2$
$$= e^{\ln(x-2)} + 2$$
$$= x - 2 + 2$$
$$= x$$
$$f[f^{-1}(x)] = 3 \ln(e^{x/3} + 2 - 2)$$
$$= 3 \ln e^{x/3}$$
$$= 3\left(\frac{x}{3}\right)$$
$$= x$$

(B) Domain of $f = (2, \infty)$ = Range of f^{-1}
    Range of $f = (-\infty, \infty)$ = Domain of f^{-1}

(C)

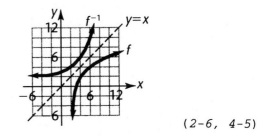

           (2-6, 4-5)

54.
$$A = P\frac{(1 + i)^n - 1}{i}$$
$$Ai = P[(1 + i)^n - 1]$$
$$\frac{Ai}{P} = (1 + i)^n - 1$$
$$1 + \frac{Ai}{P} = (1 + i)^n$$
$$\ln\left(1 + \frac{Ai}{P}\right) = n \ln(1 + i)$$
$$n = \frac{\ln(1 + \frac{Ai}{P})}{\ln(1 + i)} \quad (4-5)$$

55.
$$\ln y = 5x + \ln A$$
$$\ln y - \ln A = 5x$$
$$\ln\left(\frac{y}{A}\right) = 5x$$
$$\frac{y}{A} = e^{5x}$$
$$y = Ae^{5x} \quad (4-5)$$

56.

$$y = \frac{e^x - 2e^{-x}}{2}$$

$$2y = e^x - 2e^{-x}$$

$$2y = e^x - \frac{2}{e^x}$$

$$2ye^x = e^xe^x - e^x\left(\frac{2}{e^x}\right)$$

$$2ye^x = (e^x)^2 - 2$$

$$0 = (e^x)^2 - 2ye^x - 2$$

This equation is quadratic in e^x

$$e^x = \frac{-b \pm \sqrt{b^2 - 4ac}}{2a} \quad a = 1, \ b = -2y, \ c = -2$$

$$e^x = \frac{-(-2y) \pm \sqrt{(-2y)^2 - 4(1)(-2)}}{2(1)}$$

$$e^x = \frac{2y \pm \sqrt{4y^2 + 8}}{2}$$

$$e^x = y \pm \sqrt{y^2 + 2}$$

Note: Since $0 < 2$, $y^2 < y^2 + 2$, $\sqrt{y^2} < \sqrt{y^2 + 2}$ and $y < \sqrt{y^2 + 2}$ for all real y. Hence $y - \sqrt{y^2 + 2}$ is always negative. Also, $y + \sqrt{y^2 + 2}$ is always positive.

$$x = \ln(y + \sqrt{y^2 + 2}) \tag{4-5}$$

57. We are given y = length of container

x = width of one end

Hence $4x$ = girth of container

Length + girth = $y + 4x = 10$

So $y = 10 - 4x$

Since Volume = $8 = x^2y$, we have

$$8 = x^2(10 - 4x)$$

$$8 = 10x^2 - 4x^3$$

$$4x^3 - 10x^2 + 8 = 0$$

$$2x^3 - 5x^2 + 4 = 0$$

The possible rational solutions of this equation are ±1, ±2, ±4, $\pm\frac{1}{2}$.

We form a synthetic division table.

	2	-5	0	4	
1	2	-3	-3	1	
2	2	-1	-2	0	2 is a zero

Thus the equation can be factored

$$(x - 2)(2x^2 - x - 2) = 0$$

To find the remaining zeros, we solve $2x^2 - x - 2 = 0$ by the quadratic formula.

$$2x^2 - x - 2 = 0$$

$$x = \frac{-b \pm \sqrt{b^2 - 4ac}}{2a} \quad a = 2, \ b = -1, \ c = -2$$

$$x = \frac{-(-1) \pm \sqrt{(-1)^2 - 4(2)(-2)}}{2(2)}$$

$$x = \frac{1 + \sqrt{17}}{4} \quad \text{(discarding the negative solution)}$$

Thus, the positive zeros are $x = 2$, 1.3.

Thus, the dimensions of the package are $x = 2$ feet and $y = 2$ feet, or $x = 1.3$ feet and $y = 4.8$ feet.

(3-2)

58.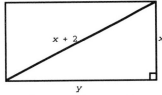

Labelling the rectangle as in the diagram, we have

$$\text{Area} = xy = 6$$

Hence $y = \dfrac{6}{x}$

Applying the Pythagorean theorem, we have

$$(x + 2)^2 = x^2 + y^2$$

$$(x + 2)^2 = x^2 + \left(\frac{6}{x}\right)^2$$

$$x^2 + 4x + 4 = x^2 + \frac{36}{x^2}$$

$$4x + 4 = \frac{36}{x^2}$$

$$4x^3 + 4x^2 = 36$$

$$4x^3 + 4x^2 - 36 = 0$$

$$x^3 + x^2 - 9 = 0$$

There are no rational zeros of $P(x) = x^3 + x^2 - 9$. We form a synthetic division table.

	1	1	0	-9	
0	1	1	0	-9	
1	1	2	2	-7	
2	1	3	6	3	There is a zero between 1 and 2

2 is also an upper bound for the zeros. We apply the bisection method to locate the zero to one decimal place. We organize our calculations in a table.

Sign Change Interval (a, b)	Midpoint m	Sign of P $P(a)$	$P(m)$	$P(b)$
(1, 2)	1.5	−	−	+
(1.5, 2)	1.75	−	−	+
(1.75, 2)	1.875	−	+	+
(1.75, 1.875)	1.8125	−	+	+
(1.75, 1.8125)	We stop here	−		+

Since each point rounds to 1.8, a real zero lies on this interval and is given by 1.8 to one decimal place accuracy. Hence, $x = 1.8$ feet.

Thus, $y = \dfrac{6}{x} = \dfrac{6}{1.8} = 3.3$ feet to one decimal place. (3-3)

59. We use the Doubling Time Growth Model:

$$P = P_0 2^{t/d}$$

Substituting $P_0 = 40$ million and $d = 22$, we have

$$P = 40(2^{t/22}) \text{ million}$$

(A) $t = 5$, hence $P = 40(2^{5/22})$ million
$$= 46.8 \text{ million}$$

(B) $t = 30$, hence $P = 40(2^{30/22})$ million
$$= 103 \text{ million}$$ (4-1)

60. We solve $P = P_0(1.07)^t$ for t, using $P = 2P_0$

$$2P_0 = P_0(1.07)^t$$

$$2 = (1.07)^t$$

$$\ln 2 = t \ln 1.07$$

$$\frac{\ln 2}{\ln 1.07} = t$$

$$t = 10.2 \text{ years}$$ (4-5)

61. We solve $P = P_0 e^{0.07t}$ for t, using $P = 2P_0$

$$2P_0 = P_0 e^{0.07t}$$
$$2 = e^{0.07t}$$
$$\ln 2 = 0.07t$$
$$t = \frac{\ln 2}{0.07}$$
$$t = 9.90 \text{ years}$$

$(4-5)$

62. First, we solve $M = \frac{2}{3} \log \left(\frac{E}{E_0} \right)$ for E.

$$M = \frac{2}{3} \log \left(\frac{E}{E_0} \right)$$
$$\frac{3M}{2} = \log \frac{E}{E_0}$$
$$\frac{E}{E_0} = 10^{3m/2}$$
$$E = E_0 (10^{3m/2})$$

We now compare E_1 for $M = 8.3$ with E_2 for $M = 7.1$.

$$E_1 = E_0 (10^{3 \cdot 8.3/2}) \qquad E_2 = E_0 (10^{3 \cdot 7.1/2})$$
$$E_1 = E_0 (10^{12.45}) \qquad E_2 = E_0 (10^{10.65})$$

Hence $\dfrac{E_1}{E_2} = \dfrac{E_0 (10^{12.45})}{E_0 (10^{10.65})} = 10^{12.45-10.65} = 10^{1.8}$

$$E_1 = 10^{1.8} E_2 \text{ or } 63.1 E_2.$$

The 1906 earthquake was 63.1 times as powerful. $(4-4)$

63. We solve $D = 10 \log \dfrac{I}{I_0}$ for I, with $D = 88$, $I_0 = 10^{-12}$

$$88 = 10 \log \frac{I}{10^{-12}}$$
$$8.8 = \log \frac{I}{10^{-12}}$$
$$10^{8.8} = \frac{I}{10^{-12}}$$
$$I = 10^{8.8} \cdot 10^{-12}$$
$$I = 10^{-3.2}$$
$$I = 6.31 \times 10^{-4} \text{ w/m}^2$$

$(4-4)$

CHAPTER 5 Trigonometric Functions

Exercise 5-1

Key Ideas and Formulas

The wrapping function, W, pairs with each real number x a unique point $W(x) = P(a, b)$, by starting at $A(1, 0)$ and moving $|x|$ along the unit circle ($u^2 + v^2 = 1$), counterclockwise if x is positive and clockwise if x is negative.

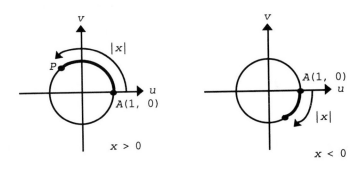

Since the circumference of the unit circle is 2π, we have the following special values of the wrapping function

$$W(0) = (1, 0)$$
$$W\left(\frac{\pi}{2}\right) = (0, 1)$$
$$W(\pi) = (-1, 0)$$
$$W\left(\frac{3\pi}{2}\right) = (0, -1)$$
$$W(2\pi) = (1, 0)$$

Coordinates of Key Circular Points

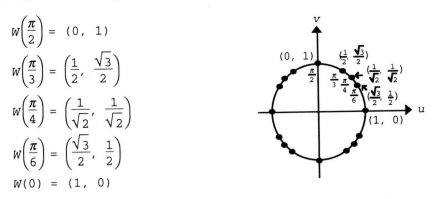

$$W\left(\frac{\pi}{2}\right) = (0, 1)$$
$$W\left(\frac{\pi}{3}\right) = \left(\frac{1}{2}, \frac{\sqrt{3}}{2}\right)$$
$$W\left(\frac{\pi}{4}\right) = \left(\frac{1}{\sqrt{2}}, \frac{1}{\sqrt{2}}\right)$$
$$W\left(\frac{\pi}{6}\right) = \left(\frac{\sqrt{3}}{2}, \frac{1}{2}\right)$$
$$W(0) = (1, 0)$$

For all real numbers x, $W(x) = W(x + 2k\pi)$ k any integer

For Problems 1—12, the following sketches are useful:

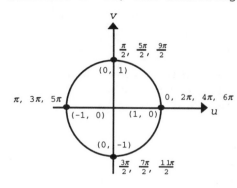 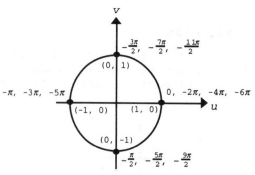

Sketch 1: counterclockwise wrapping   Sketch 2: clockwise wrapping

1. From sketch 1, $W(\pi) = (-1, 0)$

3. From sketch 1, $W(6\pi) = (1, 0)$

5. From sketch 2, $W(-\pi) = (-1, 0)$

7. From sketch 1, $W\left(\dfrac{3\pi}{2}\right) = (0, -1)$

9. From sketch 2, $W\left(-\dfrac{\pi}{2}\right) = (0, -1)$

11. From sketch 1, $W\left(\dfrac{11\pi}{2}\right) = (0, -1)$

13.

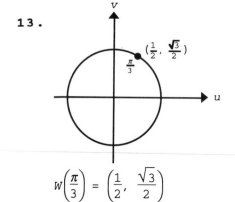

$$W\left(\dfrac{\pi}{3}\right) = \left(\dfrac{1}{2}, \dfrac{\sqrt{3}}{2}\right)$$

15.

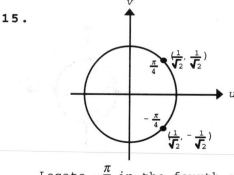

Locate $-\dfrac{\pi}{4}$ in the fourth quadrant and use text Figure 8 and symmetry with respect to the horizontal axis to find

$$W\left(-\dfrac{\pi}{4}\right) = \left(\dfrac{1}{\sqrt{2}}, -\dfrac{1}{\sqrt{2}}\right)$$

17.

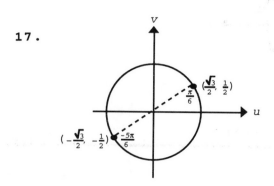

Note that $-\dfrac{5\pi}{6}$ is $\dfrac{\pi}{6}$ more than $-\pi = -\dfrac{6\pi}{6}$. Locate $-\dfrac{5\pi}{6}$ in the third quadrant and use text Figure 8 and symmetry with respect to the origin to find

$$W\left(-\dfrac{5\pi}{6}\right) = \left(-\dfrac{\sqrt{3}}{2}, -\dfrac{1}{2}\right)$$

19.

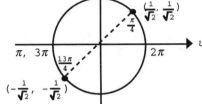

Note that $\frac{13\pi}{4}$ is $\frac{\pi}{4}$ more than $3\pi = \frac{12\pi}{4}$. Locate $\frac{13\pi}{4}$ in the third quadrant and use text Figure 8 and symmetry with respect to the origin to find

$$W\left(\frac{13\pi}{4}\right) = \left(-\frac{1}{\sqrt{2}},\ -\frac{1}{\sqrt{2}}\right)$$

21.

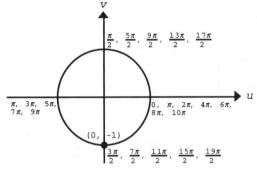

Locate $\frac{19\pi}{2}$ on the negative vertical axis.

$$W\left(\frac{19\pi}{2}\right) = (0,\ -1).$$

23.

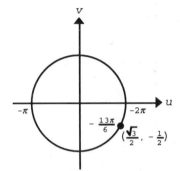

Note that $-\frac{13\pi}{6}$ is $\frac{\pi}{6}$ less (that is, $\frac{\pi}{6}$ farther in the clockwise direction) than $-2\pi = -\frac{12\pi}{6}$. Locate $-\frac{13\pi}{6}$ in the fourth quadrant and use text Figure 8 to find

$$W\left(-\frac{13\pi}{6}\right) = \left(\frac{\sqrt{3}}{2},\ -\frac{1}{2}\right)$$

For Problems 25—34, the following sketches are useful.

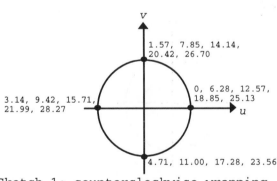

Sketch 1: counterclockwise wrapping

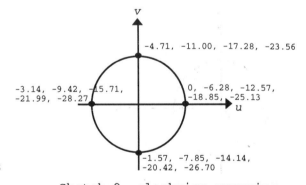

Sketch 2: clockwise wrapping

25. Using sketch 1 and 3.14 < 3.2 < 4.71, we see that $W(3.2)$ is in the third quadrant. Hence, if $W(3.2) = (a,\ b)$, then $a < 0$ and $b < 0$, that is, a and b are negative.

27. Using sketch 2 and -6.28 < -5.7 < -4.71, we see that $W(-5.7)$ is in the first quadrant. Hence, if $W(-5.7) = (a,\ b)$, then $a > 0$ and $b > 0$, that is, a and b are positive.

29. Using sketch 1 and $0 < \sqrt{2} < 1.57$, we see that $W(\sqrt{2})$ is in the first quadrant. Hence, if $W(\sqrt{2}) = (a,\ b)$, then $a > 0$ and $b > 0$, that is, a and b are positive.

31. Using sketch 1 and 11.00 < 12.5 < 12.57, we see that $W(12.5)$ is in the fourth quadrant. Hence, if $W(12.5) = (a, b)$, then $a > 0$ and $b < 0$, that is, a is positive and b is negative.

33. Using sketch 2 and -28.27 < -27 < -26.70, we see that $W(-27)$ is in the third quadrant. Hence, if $W(-27) = (a, b)$, then $a < 0$ and $b < 0$, that is, a and b are negative.

35.

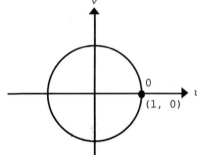

From the sketch, if $W(x) = (1, 0)$, $0 \le x < 2\pi$, then $x = 0$. Since $W(x) = W(x + 2k\pi)$, k any integer, if there are no restrictions on x, then $x = 0 + 2k\pi$ or $x = 2k\pi$, k any integer.

37.

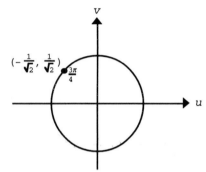

From the sketch, if $W(x) = \left(-\dfrac{1}{\sqrt{2}}, \dfrac{1}{\sqrt{2}}\right)$, $0 \le x < 2\pi$, then $x = \dfrac{3\pi}{4}$. Since $W(x) = W(x + 2k\pi)$, k any integer, if there are no restrictions on x, then $x = \dfrac{3\pi}{4} + 2k\pi$, k any integer.

39. $W(x)$ is the coordinates of a point on a unit circle that is $|x|$ units from $(1, 0)$, in a counterclockwise direction if x is positive and in a clockwise direction if x is negative. $W(x + 4\pi)$ has the same coordinates as $W(x)$, since we return to the same point every time we go around the unit circle any integer multiple of 2π units (the circumference of the circle) in either direction.

41.

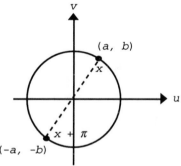

Since (see sketch) $W(x)$ and $W(x + \pi)$ are symmetrically placed with respect to the origin, $W(x) = (a, b)$ requires $W(x + \pi) = (-a, -b)$. T

43.

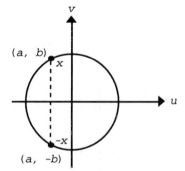

Since (see sketch) $W(x)$ and $W(-x)$ are symmetrically placed with respect to the horizontal axis, $W(x) = (a, b)$ requires $W(-x) = (a, -b)$. F

45. Since for all real numbers x, $W(x) = W(x + 2\pi)$, $W(x) = (a, b)$ requires $W(x + 2\pi) = (a, b)$. T

47.

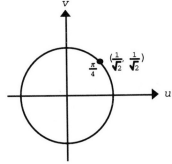

Since (see sketch) $\frac{\pi}{4}$ is a solution,

$\frac{\pi}{4} + 2k\pi$ must be a solution for any integer k. Only $k = -1$ leads to a solution between -2π and 2π, hence, the solutions in the required range are $\frac{\pi}{4} - 2\pi = \frac{-7\pi}{4}$ and $\frac{\pi}{4}$.

49.

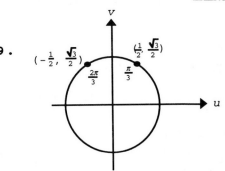

Since (see sketch) $\frac{2\pi}{3}$ is a solution,

$\frac{2\pi}{3} + 2k\pi$ must be a solution for any integer k. Only $k = -1$ leads to a solution between -2π and 2π, hence, the solutions in the required range are $\frac{2\pi}{3} - 2\pi = \frac{-4\pi}{3}$ and $\frac{2\pi}{3}$.

51. Since (see sketch at right) $\frac{7\pi}{6}$ is a solution,

$\frac{7\pi}{6} + 2k\pi$ must be a solution for any integer k.
Only $k = -1$ leads to a solution between -2π and 2π, hence, the solutions in the required range are
$\frac{7\pi}{6} - 2\pi = \frac{-5\pi}{6}$ and $\frac{7\pi}{6}$.

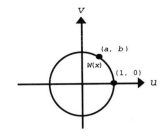

53. Since $\frac{\pi}{4}$ is a solution, $\frac{\pi}{4} + 2k\pi$ must be a solution for any integer k.

Exercise 5-2

Key Ideas and Formulas

Definitions of the Circular Functions:

If x is a real number and (a, b) are the coordinates of the circular point $W(x)$, then

$$\sin x = b \qquad \qquad \csc x = \frac{1}{b}, \quad b \neq 0$$

$$\cos x = a \qquad \qquad \sec x = \frac{1}{a}, \quad a \neq 0$$

$$\tan x = \frac{b}{a}, \quad a \neq 0 \qquad \cot x = \frac{a}{b}, \quad b \neq 0$$

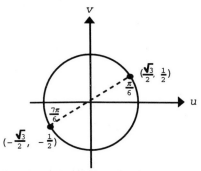

Sign Properties				
Circular	Sign in Quadrant			
Function	I	II	III	IV
$\sin x = b$	+	+	−	−
$\csc x = \frac{1}{b}$	+	+	−	−
$\cos x = a$	+	−	−	+
$\sec x = \frac{1}{a}$	+	−	−	+
$\tan x = \frac{b}{a}$	+	−	+	−
$\cot x = \frac{a}{b}$	+	−	+	−

Basic Identities

Reciprocal Identities

1. $\csc x = \dfrac{1}{\sin x}$ 2. $\sec x = \dfrac{1}{\cos x}$ 3. $\cot x = \dfrac{1}{\tan x}$

Quotient Identities

4. $\tan x = \dfrac{\sin x}{\cos x}$ 5. $\cot x = \dfrac{\cos x}{\sin x}$

Identities for Negatives

6. $\sin(-x) = -\sin x$ 7. $\cos(-x) = \cos x$ 8. $\tan(-x) = -\tan x$

Pythagorean Identity

9. $\sin^2 x + \cos^2 x = 1$

---

1. Directly from the definitions of the circular functions, we have:

(A) $\cos x = a$ (B) $\csc x = \dfrac{1}{b}$, $b \neq 0$ (C) $\cot x = \dfrac{a}{b}$, $b \neq 0$

(D) $\sec x = \dfrac{1}{a}$, $a \neq 0$ (E) $\tan x = \dfrac{b}{a}$, $a \neq 0$ (F) $\sin x = b$

3.

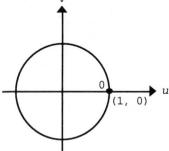

Since (see sketch)
$W(0) = (1, 0) = (\cos 0, \sin 0)$,
$\cos 0 = 1$.

5.

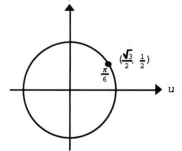

Since (see sketch)

$W\left(\dfrac{\pi}{6}\right) = \left(\dfrac{\sqrt{3}}{2}, \dfrac{1}{2}\right) = \left(\cos \dfrac{\pi}{6}, \sin \dfrac{\pi}{6}\right)$,

$\sin \dfrac{\pi}{6} = \dfrac{1}{2}$.

7.

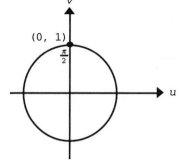

Since (see sketch)

$W\left(\dfrac{\pi}{2}\right) = (0, 1) = \left(\cos \dfrac{\pi}{2}, \sin \dfrac{\pi}{2}\right)$,

$\sin \dfrac{\pi}{2} = 1$.

9.

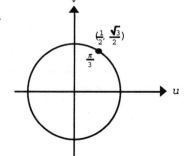

From the sketch,

$W\left(\dfrac{\pi}{3}\right) = \left(\dfrac{1}{2}, \dfrac{\sqrt{3}}{2}\right) = (a, b)$.

So $\tan \dfrac{\pi}{3} = \dfrac{b}{a} = \dfrac{\frac{\sqrt{3}}{2}}{\frac{1}{2}} = \sqrt{3}$.

11. From the sketch in Problem 7, $W\left(\dfrac{\pi}{2}\right) = (0, 1) = (a, b)$. So $\tan\dfrac{\pi}{2} = \dfrac{b}{a} = \dfrac{1}{0}$ is not defined.

13. From the sketch in Problem 3, $W(0) = (1, 0) = (a, b)$. So $\sec 0 = \dfrac{1}{a} = \dfrac{1}{1} = 1$.

15. From the sketch,

$W\left(\dfrac{\pi}{4}\right) = \left(\dfrac{1}{\sqrt{2}}, \dfrac{1}{\sqrt{2}}\right) = (a, b)$.

So $\sec\dfrac{\pi}{4} = \dfrac{1}{a} = \dfrac{1}{\frac{1}{\sqrt{2}}} = \sqrt{2}$.

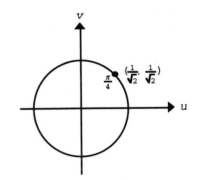

17. From the sketch in Problem 15, $W\left(\dfrac{\pi}{4}\right) = \left(\dfrac{1}{\sqrt{2}}, \dfrac{1}{\sqrt{2}}\right) = (a, b)$.

So $\tan\dfrac{\pi}{4} = \dfrac{b}{a} = \left(\dfrac{1}{\sqrt{2}}\right) \div \left(\dfrac{1}{\sqrt{2}}\right) = 1$.

19. From the sketch in Problem 3, $W(0) = (1, 0) = (a, b)$. So $\csc 0 = \dfrac{1}{b} = \dfrac{1}{0}$ is not defined.

21. Since $\cos x = a$, it is negative in quadrants II and III.

23. Since $\sin x = b$, it is positive in quadrants I and II.

25. Since $\cot x = \dfrac{a}{b}$, it is negative if a and b have opposite signs. This occurs in quadrants II and IV.

27. -64.05 **Common Error:** $\tan(4.728) \neq 0.0827$; calculator must be in *radian* mode.

29. $\sec(-1.489) = \dfrac{1}{\cos(-1.489)} = 12.24$ **31.** 0.4043

33.

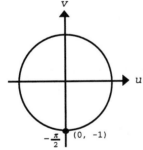

From the sketch, $W\left(-\dfrac{\pi}{2}\right) = (0, -1) = (a, b)$. So $\tan\left(-\dfrac{\pi}{2}\right) = \dfrac{b}{a} = \dfrac{-1}{0}$ is not defined.

35.

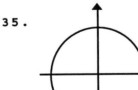

From the sketch, $W\left(\dfrac{3\pi}{2}\right) = (0, -1) = (a, b)$. So $\csc\left(\dfrac{3\pi}{2}\right) = \dfrac{1}{b} = \dfrac{1}{-1} = -1$.

37.

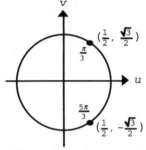

From the sketch,

$$W\left(\frac{5\pi}{3}\right) = \left(\frac{1}{2},\ -\frac{\sqrt{3}}{2}\right) = (a,\ b).$$

So $\cos\left(\dfrac{5\pi}{3}\right) = a = \dfrac{1}{2}$

39.

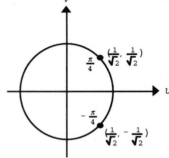

From the sketch,

$$W\left(-\frac{\pi}{4}\right) = \left(\frac{1}{\sqrt{2}},\ -\frac{1}{\sqrt{2}}\right) = (a,\ b).$$

So $\sec\left(-\dfrac{\pi}{4}\right) = \dfrac{1}{a} = \dfrac{1}{1/\sqrt{2}} = \sqrt{2}.$

41.

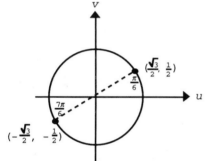

From the sketch,

$$W\left(\frac{7\pi}{6}\right) = \left(-\frac{\sqrt{3}}{2},\ -\frac{1}{2}\right) = (a,\ b).$$

So $\sin\left(\dfrac{7\pi}{6}\right) = b = -\dfrac{1}{2}.$

43.

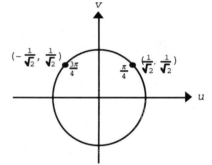

From the sketch,

$$W\left(\frac{3\pi}{4}\right) = \left(-\frac{1}{\sqrt{2}},\ \frac{1}{\sqrt{2}}\right) = (a,\ b).$$

So $\cot\left(\dfrac{3\pi}{4}\right) = \dfrac{a}{b} = \dfrac{-1/\sqrt{2}}{1/\sqrt{2}} = -1.$

45.

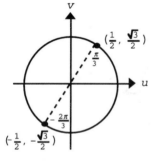

From the sketch,

$$W\left(-\frac{2\pi}{3}\right) = \left(-\frac{1}{2},\ -\frac{\sqrt{3}}{2}\right) = (a,\ b).$$

So $\tan\left(-\dfrac{2\pi}{3}\right) = \dfrac{b}{a} = \dfrac{-\sqrt{3}/2}{-1/2} = \sqrt{3}$

47.

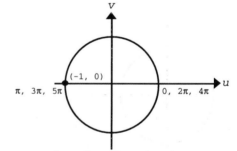

From the sketch,
$W(5\pi) = (-1,\ 0) = (a,\ b).$

So $\csc 5\pi = \dfrac{1}{b} = \dfrac{1}{0}$ is not defined.

49. Locate $W(0.4) = (a,\ b) = (0.9,\ 0.4)$ on the figure.

    (A) $\sin 0.4 = b = 0.4$ (B) $\cos 0.4 = a = 0.9$ (C) $\tan 0.4 = \dfrac{b}{a} = \dfrac{0.4}{0.9} = 0.4$

51. (A) Locate $W(2.2) = (a, b) = (-0.6, 0.8)$ on the figure.

$\sec 2.2 = \dfrac{1}{a} = \dfrac{1}{-0.6} = -2$ (to one significant digit)

(B) Locate $W(5.9) = (a, b) = (0.9, -0.4)$ on the figure.

$\tan 5.9 = \dfrac{b}{a} = \dfrac{-0.4}{0.9} = -0.4$

(C) Locate $W(3.8) = (a, b) = (-0.8, -0.6)$ on the figure.

$\cot 3.8 = \dfrac{a}{b} = \dfrac{-0.8}{-0.6} = 1$ (to one significant digit)

53. $\sin x < 0$ in quadrants III and IV; $\cot x < 0$ in quadrants II and IV; therefore, both are true in quadrant IV.

55. $\cos x < 0$ in quadrants II and III; $\sec x > 0$ in quadrants I and IV; therefore, it is not possible to have both true for the same value of x.

57. $\cos x = a$ is always defined. There are no values for which it is undefined.

59. $\tan x = \dfrac{b}{a}$ is undefined if and only if $a = 0$. This occurs at points on the vertical axis. The only values of x between 0 and 2π for which $W(x)$ is on the vertical axis are $\dfrac{\pi}{2}$ and $\dfrac{3\pi}{2}$.

61. $\sec x = \dfrac{1}{a}$ is undefined if and only if $a = 0$. This occurs at points on the vertical axis. The only values of x between 0 and 2π for which $W(x)$ is on the vertical axis are $\dfrac{\pi}{2}$ and $\dfrac{3\pi}{2}$.

63. Consulting the sketch at the right, and noting that $\sin x$ is the second component of the ordered pair $(a, b) = W(x)$, we see that:

(A) as x varies from 0 to $\dfrac{\pi}{2}$, $\sin x = b$ varies from 0 to 1;

(B) as x varies from $\dfrac{\pi}{2}$ to π, $\sin x = b$ varies from 1 to 0;

(C) as x varies from π to $\dfrac{3\pi}{2}$, $\sin x = b$ varies from 0 to -1;

(D) as x varies from $\dfrac{3\pi}{2}$ to 2π, $\sin x = b$ varies from -1 to 0.

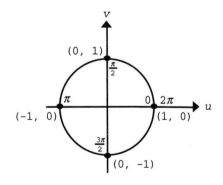

65. 0.8138

67. 0.5290

69. $\sin(-x) = -\sin x = -\left(-\dfrac{1}{3}\right) = \dfrac{1}{3}$

71. $\tan(-x) = -\tan x = -(-\sqrt{5}) = \sqrt{5}$

73. $\cot(-x) = \dfrac{1}{\tan(-x)} = \dfrac{1}{-\tan x} = -\dfrac{1}{\tan x} = -\cot x = -25$

75. We first note that the circular point $W(x)$ is in quadrant IV, since that is the only quadrant in which $\cos x > 0$ and $\tan x < 0$. We next find $\sin x$.
$$\sin^2 x + \cos^2 x = 1$$

$$\sin^2 x + \left(\frac{1}{2}\right)^2 = 1$$

$$\sin^2 x = \frac{3}{4}$$

$$\sin x = -\frac{\sqrt{3}}{2} \text{ since } W(x) \text{ is in quadrant IV.}$$

We can now find values for the other four circular functions.

$$\tan x = \frac{\sin x}{\cos x} = \frac{-\frac{\sqrt{3}}{2}}{\frac{1}{2}} = -\sqrt{3} \qquad \cot x = \frac{\cos x}{\sin x} = \frac{\frac{1}{2}}{-\frac{\sqrt{3}}{2}} = -\frac{1}{\sqrt{3}}$$

$$\csc x = \frac{1}{\sin x} = \frac{1}{-\frac{\sqrt{3}}{2}} = -\frac{2}{\sqrt{3}} \qquad \sec x = \frac{1}{\cos x} = \frac{1}{\frac{1}{2}} = 2$$

77. We first note that $W(x)$ is in quadrant III, since that is the only quadrant in which $\sin x < 0$ and $\cos x < 0$. We next find $\cos x$.
$$\sin^2 x + \cos^2 x = 1$$

$$\left(-\frac{1}{\sqrt{2}}\right)^2 + \cos^2 x = 1$$

$$\cos^2 x = \frac{1}{2}$$

$$\cos x = -\frac{1}{\sqrt{2}} \text{ since } W(x) \text{ is in quadrant III.}$$

We can now find values for the other four circular functions.

$$\tan x = \frac{\sin x}{\cos x} = \frac{-\frac{1}{\sqrt{2}}}{-\frac{1}{\sqrt{2}}} = 1 \qquad \cot x = \frac{\cos x}{\sin x} = \frac{-\frac{1}{\sqrt{2}}}{-\frac{1}{\sqrt{2}}} = 1$$

$$\csc x = \frac{1}{\sin x} = \frac{1}{-\frac{1}{\sqrt{2}}} = -\sqrt{2} \qquad \sec x = \frac{1}{\cos x} = \frac{1}{-\frac{1}{\sqrt{2}}} = -\sqrt{2}$$

79. We first note that $W(x)$ is in quadrant III, since that is the only quadrant in which $\tan x > 0$ and $\sin x < 0$. From the given information, we can write:
$$\tan x = \frac{\sin x}{\cos x} = \sqrt{3} \qquad\qquad \sin x = (\cos x)\sqrt{3}$$
Substituting this into identity (9), we have
$$\sin^2 x + \cos^2 x = 1$$
$$[(\cos x)\sqrt{3}]^2 + \cos^2 x = 1$$
$$3\cos^2 x + \cos^2 x = 1$$
$$\cos^2 x = \frac{1}{4}$$

$$\cos x = -\frac{1}{2} \text{ since } W(x) \text{ is in quadrant III.}$$

We can now find values for the other four circular functions.

$$\sin x = (\cos x)\sqrt{3} = \left(-\frac{1}{2}\right)\sqrt{3} = \frac{-\sqrt{3}}{2} \qquad\qquad \cot x = \frac{1}{\tan x} = \frac{1}{\sqrt{3}}$$

$$\csc x = \frac{1}{\sin x} = \frac{1}{-\frac{\sqrt{3}}{2}} = -\frac{2}{\sqrt{3}} \qquad\qquad \sec x = \frac{1}{\cos x} = \frac{1}{-\frac{1}{2}} = -2$$

81. From the sketch, it is clear that the least positive value of x must be π.

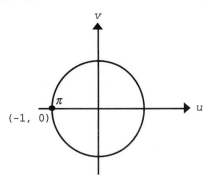

83. Note first that $\cot \frac{\pi}{6} = \sqrt{3}$. The least positive value of x for which $\cot x = -\sqrt{3}$ must have $W(x)$ in quadrant II. Therefore (see sketch), $x = \frac{5\pi}{6}$.

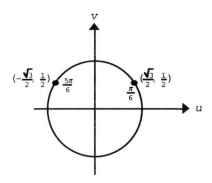

85. See sketch for Problem 83. Note that $\sec \frac{\pi}{6} = \frac{2}{\sqrt{3}}$. The least positive value of x for which $\sec x = -\frac{2}{\sqrt{3}}$ must have $W(x)$ in quadrant II. Therefore, $x = \frac{5\pi}{6}$.

87. (A) Identity (5)   (B) Identity (9)   (C) Identity (1)

89. Given $n = 12$, $r = 5$, we use the given formula $A = \frac{1}{2}nr^2 \sin \frac{2\pi}{n}$ to obtain:

$A = \frac{1}{2}(12)(5)^2 \sin \frac{2\pi}{12}$

$\quad = 150 \sin \frac{\pi}{6}$

$\quad = 150\left(\frac{1}{2}\right)$

$\quad = 75$ square meters

91. Given $n = 3$, $r = 4$, we use the given formula $A = \frac{1}{2}nr^2 \sin \frac{2\pi}{n}$ to obtain:

$A = \frac{1}{2}(3)(4)^2 \sin \frac{2\pi}{3}$

$\quad = 24 \sin \frac{2\pi}{3}$

$\quad = 24\left(\frac{\sqrt{3}}{2}\right)$

$\quad = 12\sqrt{3} \approx 20.78$ square inches

93. $a_1 = 0.5$

$a_2 = a_1 + \cos a_1 = 0.5 + \cos 0.5 = 1.377583$

$a_3 = a_2 + \cos a_2 = 1.377583 + \cos 1.377583 = 1.569596$

$a_4 = a_3 + \cos a_3 = 1.569596 + \cos 1.569596 = 1.570796$

$a_5 = a_4 + \cos a_4 = 1.570796 + \cos 1.570796 = 1.570796$

$\frac{\pi}{2} = 1.570796$ to six decimal places.

Exercise 5-3

Key Ideas and Formulas

Degree Measure: An angle formed by one complete rotation is said to have a measure of 360 degrees (360°). An angle formed by $\frac{1}{360}$ of a complete rotation is said to have a measure of 1 degree (1°).

Radian Measure: If the vertex of an angle θ is placed at the center of a circle with radius $r > 0$, and the length of the arc opposite to θ on the circumference is s, then the radian measure of θ is given by

$\theta = \dfrac{s}{r}$ radians

$s = r\theta$

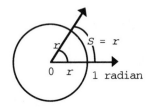

If $s = r$, then

$\theta = \dfrac{r}{r} = 1$ radian

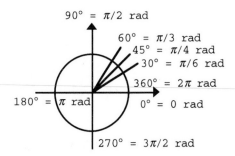

To convert degree measure to radian measure and vice versa: if θ_{deg} represents degree measure and θ_{rad} represents radian measure, then

$\dfrac{\theta_{deg}}{180°} = \dfrac{\theta_{rad}}{\pi \text{ rad}}$ or $\theta_{deg} = \dfrac{180°}{\pi \text{ rad}} \theta_{rad}$

$\theta_{rad} = \dfrac{\pi \text{ rad}}{180°} \theta_{deg}$

Measures of Special Angles:

- $90° = \pi/2$ rad
- $60° = \pi/3$ rad
- $45° = \pi/4$ rad
- $30° = \pi/6$ rad
- $360° = 2\pi$ rad
- $0° = 0$ rad
- $180° = \pi$ rad
- $270° = 3\pi/2$ rad

---

1. Since 1 rotation corresponds to 360°, $\frac{1}{9}$ rotation corresponds to $\frac{1}{9}(360°) = 40°$

3. Since 1 rotation corresponds to 360°, $\frac{3}{4}$ rotation corresonds to $\frac{3}{4}(360°) = 270°$

5. $\theta = \dfrac{s}{r} = \dfrac{24 \text{ centimeters}}{4 \text{ centimeters}} = 6$ radians **7.** $\theta = \dfrac{s}{r} = \dfrac{30 \text{ feet}}{12 \text{ feet}} = 2.5$ radians

9. Since 1 rotation corresponds to 2π radians, $\frac{1}{8}$ rotation corresponds to $\frac{1}{8}(2\pi) = \dfrac{\pi}{4}$ radians

11. As in problem 9, $\frac{3}{4}$ rotation corresponds to $\frac{3}{4}(2\pi) = \dfrac{3\pi}{2}$ radians

13. Using the relation $\theta_{\text{rad}} = \dfrac{\pi \text{ rad}}{180°}\theta_{\text{deg}}$, we have:

$$\frac{\pi \text{ rad}}{180°}30° = \frac{\pi}{6} \text{ rad}$$

$$\frac{\pi \text{ rad}}{180°}60° = \frac{\pi}{3} \text{ rad}$$

$$\frac{\pi \text{ rad}}{180°}90° = \frac{\pi}{2} \text{ rad}$$

$$\frac{\pi \text{ rad}}{180°}120° = \frac{2\pi}{3} \text{ rad}$$

$$\frac{\pi \text{ rad}}{180°}150° = \frac{5\pi}{6} \text{ rad}$$

$$\frac{\pi \text{ rad}}{180°}180° = \pi \text{ rad}$$

15. Using the relation $\theta_{\text{rad}} = \dfrac{\pi \text{ rad}}{180°}\theta_{\text{deg}}$, we have:

$$\frac{\pi \text{ rad}}{180°}(-45°) = -\frac{\pi}{4} \text{ rad}$$

$$\frac{\pi \text{ rad}}{180°}(-90°) = -\frac{\pi}{2} \text{ rad}$$

$$\frac{\pi \text{ rad}}{180°}(-135°) = -\frac{3\pi}{4} \text{ rad}$$

$$\frac{\pi \text{ rad}}{180°}(-180°) = -\pi \text{ rad}$$

17. Using the relation $\theta_{\text{deg}} = \dfrac{180°}{\pi \text{ rad}}\theta_{\text{rad}}$, we have:

$$\frac{180°}{\pi}\frac{\pi}{3} = 60°$$

$$\frac{180°}{\pi}\frac{2\pi}{3} = 120°$$

$$\frac{180°}{\pi}\pi = 180°$$

$$\frac{180°}{\pi}\frac{4\pi}{3} = 240°$$

$$\frac{180°}{\pi}\frac{5\pi}{3} = 300°$$

$$\frac{180°}{\pi}2\pi = 360°$$

19. Using the relation $\theta_{\text{deg}} = \dfrac{180°}{\pi \text{ rad}}\theta_{\text{rad}}$, we have:

$$\frac{180°}{\pi}\left(-\frac{\pi}{2}\right) = -90°$$

$$\frac{180°}{\pi}(-\pi) = -180°$$

$$\frac{180°}{\pi}\left(-\frac{3\pi}{2}\right) = -270°$$

$$\frac{180°}{\pi}(-2\pi) = -360°$$

21. False. For example, two angles of 90° are mutually supplementary.

23. False. For example, an angle of 90° and an angle of 450° are coterminal.

25. True. The terminal side of an angle of 90° in standard position lies along the positive y axis, so this is a quadrantal angle.

27. $5°51'33'' = \left(5 + \dfrac{51}{60} + \dfrac{33}{3600}\right)°$
$ = 5.859°$

29. $354°8'29'' = \left(354 + \dfrac{8}{60} + \dfrac{29}{3600}\right)°$
$ = 354.141°$

31. $3.042° = 3°(0.042 \cdot 60)'$
$ = 3°2.52'$
$ = 3°2'(0.52 \cdot 60)''$
$ = 3°2'31''$

33. $403.223° = 403°(0.223 \cdot 60)'$
$ = 403°13.38'$
$ = 403°13'(0.38 \cdot 60)''$
$ = 403°13'23''$

35. Using the relation $\theta_{\text{rad}} = \dfrac{\pi \text{ rad}}{180°}\theta_{\text{deg}}$, we have

$$\frac{\pi \text{ rad}}{180°}18° = \frac{\pi}{10} = 0.314$$

37. First we convert 23°45'32'' to decimal degrees:

$$23°45'32'' = \left(23 + \frac{45}{60} + \frac{32}{3600}\right)°$$

$$= 23.758\overline{8}°$$

Then, using the relation $\theta_{\text{rad}} = \dfrac{\pi \text{ rad}}{180°}\theta_{\text{deg}}$, we have

$$\frac{\pi \text{ rad}}{180°}23.758\overline{8}° = 0.415$$

39. Using the relation $\theta_{\text{deg}} = \dfrac{180°}{\pi \text{ rad}} \theta_{\text{rad}}$, we have $\dfrac{180°}{\pi} (1.52) = 87.09°$

41. Using the relation $\theta_{\text{deg}} = \dfrac{180°}{\pi \text{ rad}} \theta_{\text{rad}}$, we have $\dfrac{180°}{\pi} (-0.83) = -47.56°$

For Problems 43—66 the following sketch is useful:

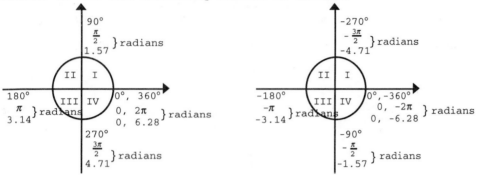

43. From the sketch, we find, since $90° < 130° < 180°$, $130°$ is a II quadrantal angle.

45. Since $-1.57 < -1.34 < 0$, -1.34 is a IV quadrant angle.

47. Since $\dfrac{3\pi}{2} < \dfrac{7\pi}{4} < 2\pi$, $\dfrac{7\pi}{4}$ is a IV quadrant angle.

49. $-\dfrac{5\pi}{2}$ is coterminal with $-\dfrac{5\pi}{2} + 2\pi = -\dfrac{\pi}{2}$. It is therefore a quadrantal angle.

51. $-835°$ is coterminal with $-835° + 2(360°) = -115°$. Since $-180° < -115° < -90°$, $-115°$ and $-835°$ are III quadrant angles.

53. 9.73 is coterminal with $9.73 - 2\pi \approx 3.45$. Since $3.14 < 3.45 < 5.71$, 3.45 and 9.73 are III quadrant angles.

55. $-600° + 2(360°) = 120°$. Coterminal.   **57.** $960° - 2(360°) = 240°$. Not coterminal.

59. since $\dfrac{180°}{\pi}\left(\dfrac{2\pi}{3}\right) = 120°$, $\dfrac{2\pi}{3}$ is coterminal with $120°$.

61. $\dfrac{11\pi}{4} - 2\pi = \dfrac{3\pi}{4}$. Not coterminal.   **63.** $\dfrac{13\pi}{4} - 4\pi = -\dfrac{3\pi}{4}$. Coterminal.

65. Since $\dfrac{\pi}{180°}(-495°) = -\dfrac{11\pi}{4}$, and $-\dfrac{11\pi}{4} + 2\pi = -\dfrac{3\pi}{4}$, $-495°$ is coterminal with $-\dfrac{3\pi}{4}$.

67. We apply $\dfrac{s}{c} = \dfrac{\theta°}{360°}$ with $s = 500$ mi, and $\theta° = 7.5°$

Then $\dfrac{500}{c} = \dfrac{7.5°}{360°}$ or $\dfrac{500}{c} = \dfrac{7.5}{360}$

$360c \cdot \dfrac{500}{c} = 360c \cdot \dfrac{7.5}{360}$

$180,000 = 7.5c$

$c = 24,000$ mi

69. The 7.5° angle and θ have a common side. (An extended vertical pole in Alexandria will pass through the center of the earth.) The sun's rays are essentially parallel when they arrive at the earth. Thus, the other two sides of the angles are parallel, since a sun ray to the bottom of the well, when extended, will pass through the center of the earth. From geometry we know that the alternate interior angles made by a line intersecting two parallel lines are equal. Therefore, $\theta = 7.5°$.

71. From the figure, it should be clear that the minute (larger) hand is displaced π radians from noon, while the hour (smaller) hand is displaced $\theta = \dfrac{4\frac{1}{2}}{12}$ of a full revolution, or $\dfrac{4\frac{1}{2}}{12} \cdot 2\pi$ radians from noon. Hence the larger angle between the hands has measure $\theta + \pi$, or

$$\frac{4\frac{1}{2}}{12} \cdot 2\pi + \pi = \frac{3}{4}\pi + \pi = \frac{7}{4}\pi \text{ radians.}$$

73. We use $\theta = \dfrac{s}{r}$ with $r = \dfrac{1}{2}(10) = 5$ centimeters and $s = 10$ meters $\times\ 100\dfrac{\text{centimeters}}{\text{meters}}$ $= 1000$ centimeters. Then $\theta = \dfrac{1000}{5} = 200$ radians.

75. In one year the line sweeps out one full revolution, or 2π radians. In one week the line sweeps out $\dfrac{1}{52}$ of one full revolution, or $\dfrac{1}{52} \cdot 2\pi = \dfrac{\pi}{26}$ radians.

$\dfrac{\pi}{26} = 0.12$ radian to two decimal places.

77. Following example 4, we reason that points on the circumference of each wheel travel the same distance. Then $s_1 = r_1\theta_1 = r_2\theta_2 = s_2$. The radius of the front wheel is $\dfrac{1}{2}(40)$ or 20 centimeters. The radius of the back wheel is $\dfrac{1}{2}(60)$ or 30 centimeters. so
$$20\theta_1 = 8(30)$$
$$\theta_1 = \frac{240}{20}$$
$$\theta_1 = 12 \text{ radians}$$

79. We use $c \approx s = r\theta$ with $r = 9.3 \times 10^7$ mi and $\theta = 9.3 \times 10^{-3}$ rad.
Then $c \approx (9.3 \times 10^7)(9.3 \times 10^{-3}) = 865,000$ mi

81. We use $c \approx s = r\theta$. $r = 750$ ft. θ must be converted to radians to use the formula, thus, using $\theta_{\text{rad}} = \dfrac{\pi \text{ rad}}{180°}2.5°$ we have $\theta_{\text{rad}} = \dfrac{\pi}{72}$ rad.

Then $c \approx 750 \cdot \dfrac{\pi}{72} = 33$ ft.

83. From the figure, it should be clear that the minute (larger) hand is displaced π radians from noon, while the hour (smaller) hand is displaced $\theta = \dfrac{4\frac{1}{2}}{12}$ of a full revolution, or $\dfrac{4\frac{1}{2}}{12} \cdot 2\pi$ radians from noon. Hence the larger angle between the hands has measure $\theta + \pi$, or

$$\frac{4\frac{1}{2}}{12} \cdot 2\pi + \pi = \frac{3}{4}\pi + \pi = \frac{7}{4}\pi \text{ radians.}$$

85. We use $\theta = \dfrac{s}{r}$ with $r = \dfrac{1}{2}(10) = 5$ centimeters and $s = 10$ meters $\times\ 100\ \dfrac{\text{centimeters}}{\text{meter}}$

$= 1000$ centimeters. Then $\theta = \dfrac{1000}{5} = 200$ radians.

87. In one year the line sweeps out one full revolution, or 2π radians. In one week the line sweeps out $\dfrac{1}{52}$ of one full revolution, or $\dfrac{1}{52} \cdot 2\pi = \dfrac{\pi}{26}$ radians.

$\dfrac{\pi}{26} = 0.12$ radian to two decimal places.

89. Following example 4, we reason that points on the circumference of each wheel travel the same distance. Then $s_1 = r_1\theta_1 = r_2\theta_2 = s_2$. The radius of the front wheel is $\dfrac{1}{2}(40)$ or 20 centimeters. The radius of the back wheel is $\dfrac{1}{2}(60)$ or 30 centimeters. so

$20\theta_1 = 8(30)$

$\theta_1 = \dfrac{240}{20}$

$\theta_1 = 12$ radians

91. We use $c \approx s = r\theta$ with $r = 9.3 \times 10^7$ mi and $\theta = 9.3 \times 10^{-3}$ rad.
Then $c \approx (9.3 \times 10^7)(9.3 \times 10^{-3}) = 865,000$ mi

93. We use $c \approx s = r\theta$. $r = 750$ ft. θ must be converted to radians to use the formula, thus, using $\theta_R = \dfrac{\pi\ \text{rad}}{180°}\,2.5°$ we have $\theta_R = \dfrac{\pi}{72}$ rad.

Then $c \approx 750 \cdot \dfrac{\pi}{72} = 33$ ft.

Exercise 5-4

Key Ideas and Formulas

Trigonometric Functions with Angle Domains

If θ is an angle with radian measure x, then the value of each trigonometric function at θ is given by its value at the real number x.

Trigonometric Function	Circular Function	
$\sin\theta$	$= \sin x$	$= b$
$\cos\theta$	$= \cos x$	$= a$
$\tan\theta$	$= \tan x$	$= \dfrac{b}{a}$
$\cot\theta$	$= \cot x$	$= \dfrac{a}{b}$
$\sec\theta$	$= \sec x$	$= \dfrac{1}{a}$
$\csc\theta$	$= \csc x$	$= \dfrac{1}{b}$

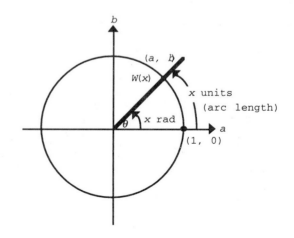

Alternate form of the above:

For an arbitrary angle θ in standard position in a rectangular coordinate system, with $P(a, b)$ an arbitrary point r units from the origin on the terminal side of θ,

$$\sin \theta = \frac{b}{r} \qquad\qquad \csc \theta = \frac{r}{b} \quad b \neq 0 \qquad r = \sqrt{a^2 + b^2} > 0$$

$$\cos \theta = \frac{a}{r} \qquad\qquad \sec \theta = \frac{r}{a} \quad a \neq 0 \quad (a, b) \neq (0, 0)$$

$$\tan \theta = \frac{b}{a} \quad a \neq 0 \quad \cot \theta = \frac{a}{b} \quad b \neq 0$$

For any real number x, we define

$\sin x = \sin (x\ radians)$ $\qquad \csc x = \csc (x\ rad)$
$\cos x = \cos (x\ rad)$ $\qquad \sec x = \sec (x\ rad)$
$\tan x = \tan (x\ rad)$ $\qquad \cot x = \cot (x\ rad)$

Reciprocal Identities

$$\csc x = \frac{1}{\sin x}, \ \sin x \neq 0 \qquad \sec x = \frac{1}{\cos x}, \ \cos x \neq 0 \qquad \cot x = \frac{1}{\tan x}, \ \tan x \neq 0$$

Quadrantal angles are any angles with their terminal side lying along a coordinate axis.

Reference Triangles and Reference Angles

To form a reference triangle for θ, drop a perpendicular from a point $P(a, b)$ on the terminal side of θ to the horizontal axis.

The reference angle α is the acute angle (always taken positive) between the terminal side of θ and the horizontal axis.

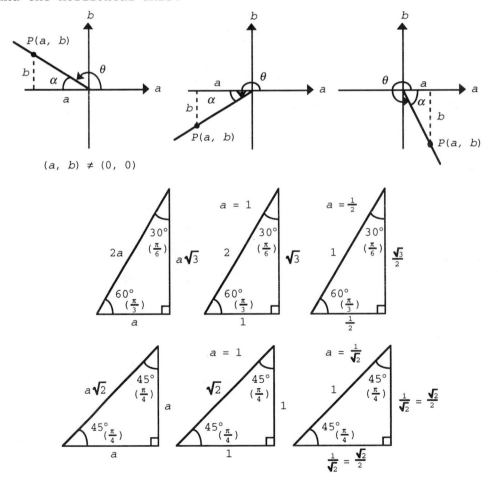

$(a, b) \neq (0, 0)$

$30°-60°$ and $45°$ Special Triangles

Special Angle Values

θ	$\sin \theta$	$\cos \theta$	$\tan \theta$
0°	0	1	0
30°	$\frac{1}{2}$	$\sqrt{3}/2$	$1/\sqrt{3}$ or $\sqrt{3}/3$
45°	$1/\sqrt{2}$ or $\sqrt{2}/2$	$1/\sqrt{2}$ or $\sqrt{2}/2$	1
60°	$\sqrt{3}/2$	$\frac{1}{2}$	$\sqrt{3}$
90°	1	0	Not defined

1. $(a, b) = (6, 8)$

$r = \sqrt{a^2 + b^2} = \sqrt{6^2 + 8^2} = \sqrt{100} = 10$

$\sin \theta = \dfrac{b}{r} = \dfrac{8}{10} = \dfrac{4}{5}$ $\qquad$ $\csc \theta = \dfrac{r}{b} = \dfrac{10}{8} = \dfrac{5}{4}$

$\cos \theta = \dfrac{a}{r} = \dfrac{6}{10} = \dfrac{3}{5}$ $\qquad$ $\sec \theta = \dfrac{r}{a} = \dfrac{10}{6} = \dfrac{5}{3}$

$\tan \theta = \dfrac{b}{a} = \dfrac{8}{6} = \dfrac{4}{3}$ $\qquad$ $\cot \theta = \dfrac{a}{b} = \dfrac{6}{8} = \dfrac{3}{4}$

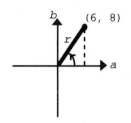

3. $(a, b) = (-1, \sqrt{3})$

$r = \sqrt{a^2 + b^2} = \sqrt{(-1)^2 + (\sqrt{3})^2} = \sqrt{4} = 2$

$\sin \theta = \dfrac{b}{r} = \dfrac{\sqrt{3}}{2}$ $\qquad$ $\csc \theta = \dfrac{r}{b} = \dfrac{2}{\sqrt{3}}$

$\cos \theta = \dfrac{a}{r} = \dfrac{-1}{2} = -\dfrac{1}{2}$ $\qquad$ $\sec \theta = \dfrac{r}{a} = \dfrac{2}{-1} = -2$

$\tan \theta = \dfrac{b}{a} = \dfrac{\sqrt{3}}{-1} = -\sqrt{3}$ $\qquad$ $\cot \theta = \dfrac{a}{b} = \dfrac{-1}{\sqrt{3}} = -\dfrac{1}{\sqrt{3}}$

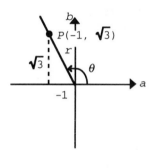

5. 0.9272 (calculator in degree mode) $\qquad$ **7.** −0.2958 (calculator in radian mode)

9. 0.2038 $\qquad$ **11.** $\csc(365°52'48") = \dfrac{1}{\sin(365°52'48")} = \dfrac{1}{\sin\left(365 + \frac{52}{60} + \frac{48}{3600}\right)}$

$= 9.761$

13. 108.6

15.

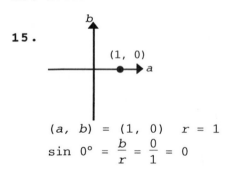

$(a, b) = (1, 0)$ $\quad$ $r = 1$

$\sin 0° = \dfrac{b}{r} = \dfrac{0}{1} = 0$

17.

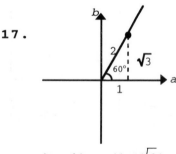

$(a, b) = (1, \sqrt{3})$ $\quad$ $r = 2$

$\tan 60° = \dfrac{b}{a} = \dfrac{\sqrt{3}}{1} = \sqrt{3}$

19.

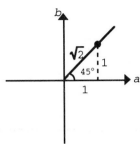

$(a, b) = (1, 1)$ $r = \sqrt{2}$

$\sin 45° = \dfrac{b}{r} = \dfrac{1}{\sqrt{2}}$

21. See sketch for Problem 19.

$\sec 45° = \dfrac{r}{a} = \dfrac{\sqrt{2}}{1} = \sqrt{2}$

23. See sketch for Problem 15. $\cot 0° = \dfrac{a}{b} = \dfrac{1}{0}$ is not defined.

25.

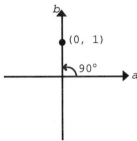

$(a, b) = (0, 1)$ $r = 1$

$\tan 90° = \dfrac{b}{a} = \dfrac{1}{0}$ is not defined

27.

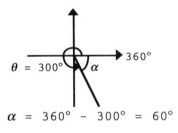

$\alpha = 360° - 300° = 60°$

29.

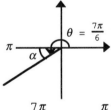

$\alpha = \dfrac{7\pi}{6} - \pi = \dfrac{\pi}{6}$

31.

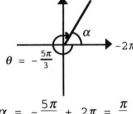

$\alpha = -\dfrac{5\pi}{3} + 2\pi = \dfrac{\pi}{3}$

33.

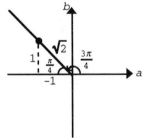

$(a, b) = (-1, 1)$ $r = \sqrt{2}$

$\tan\left(\dfrac{3\pi}{4}\right) = \dfrac{b}{a} = \dfrac{1}{-1} = -1$

35.

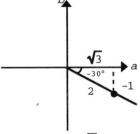

$(a, b) = (\sqrt{3}, -1)$ $r = 2$

$\sin(-30°) = \dfrac{b}{r} = \dfrac{-1}{2} = -\dfrac{1}{2}$

37.

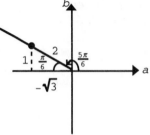

$(a, b) = (-\sqrt{3}, 1) \quad r = 2$

$\sec\left(\dfrac{5\pi}{6}\right) = \dfrac{r}{a} = \dfrac{2}{-\sqrt{3}} = -\dfrac{2}{\sqrt{3}}$

39.

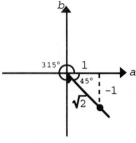

$(a, b) = (1, -1) \quad r = \sqrt{2}$

$\cot 315° = \dfrac{a}{b} = \dfrac{1}{-1} = -1$

41.

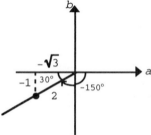

$(a, b) = (-\sqrt{3}, -1) \quad r = 2$

$\csc(-150°) = \dfrac{r}{b} = \dfrac{2}{-1} = -2$

43.

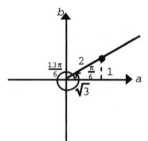

$(a, b) = (\sqrt{3}, 1) \quad r = 2$

$\cos\left(\dfrac{13\pi}{6}\right) = \dfrac{a}{r} = \dfrac{\sqrt{3}}{2}$

45.

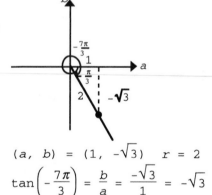

$(a, b) = (1, -\sqrt{3}) \quad r = 2$

$\tan\left(-\dfrac{7\pi}{3}\right) = \dfrac{b}{a} = \dfrac{-\sqrt{3}}{1} = -\sqrt{3}$

47.

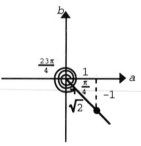

$(a, b) = (1, -1) \quad r = \sqrt{2}$

$\sec\left(\dfrac{23\pi}{4}\right) = \dfrac{r}{a} = \dfrac{\sqrt{2}}{1} = \sqrt{2}$

49. $\cos \theta = \dfrac{a}{r}$. Since r is never zero, $\cos \theta$ is always defined. There are no values that are not defined.

51. $\tan \theta = \dfrac{b}{a}$. This is undefined when $a = 0$. This occurs at 90°, 270°.

53. $\csc \theta = \dfrac{r}{b}$. This is undefined when $b = 0$. This occurs at 0°, 180°.

55. $\cos \theta = \dfrac{a}{r} = \dfrac{-1}{2}$ because $r > 0$. a is negative in the II and III quadrants. The smallest positive θ is associated with a 60° reference triangle in the II quadrant as shown in the figure below.

$\theta = 120°$ or $\dfrac{2\pi}{3}$ radians

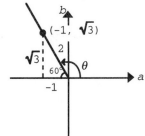

57. $\sin \theta = \dfrac{b}{r} = \dfrac{-1}{2}$ because $r > 0$. b is negative in the III and IV quadrants. The smallest positive θ is associated with a 30° reference triangle in the III quadrant as drawn.

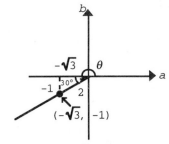

$\theta = 210°$ or $\dfrac{7\pi}{6}$ radians

59. $\csc \theta = \dfrac{r}{b} = \dfrac{2}{-\sqrt{3}}$ because $r > 0$.

b is negative in the III and IV quadrants. The smallest positive θ is associated with a 60° reference triangle in the III quadrant as drawn.

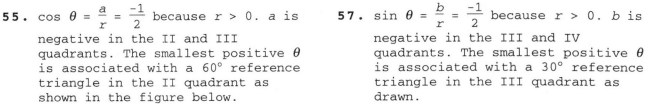

$\theta = 240°$ or $\dfrac{4\pi}{3}$ radians

61. Since $\sin \theta = \dfrac{3}{5} > 0$ and $\cos \theta < 0$, θ is a II quadrant angle. We sketch a reference triangle. Since $\sin \theta = \dfrac{b}{r} = \dfrac{3}{5}$, we know that $b = 3$ and $r = 5$. Use the Pythagorean theorem to find a.

$a^2 + 3^2 = 5^2$
$ a^2 = 16$
$ a = -4$

(a must be negative because θ is a II quadrant angle)

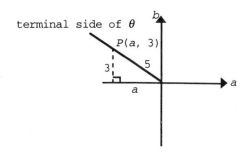

Using $(a, b) = (-4, 3)$ and $r = 5$, we have

$\cos \theta = \dfrac{a}{r} = \dfrac{-4}{5} = -\dfrac{4}{5}$ $\qquad \tan \theta = \dfrac{b}{a} = \dfrac{3}{-4} = -\dfrac{3}{4}$

$\sec \theta = \dfrac{r}{a} = \dfrac{5}{-4} = -\dfrac{5}{4}$ $\qquad \cot \theta = \dfrac{a}{b} = \dfrac{-4}{3} = -\dfrac{4}{3}$

$\csc \theta = \dfrac{r}{b} = \dfrac{5}{3}$

63. Since $\cos\theta = -\dfrac{\sqrt{5}}{3} < 0$ and $\cot\theta > 0$, θ is a III quadrant angle. We sketch a

reference triangle. Since $\cos\theta = \dfrac{a}{r} = -\dfrac{\sqrt{5}}{3}$ we know that $a = -\sqrt{5}$ and $r = 3$. Use

the Pythagorean theorem to find b.

$(-\sqrt{5})^2 + b^2 = 3^2$

$\qquad b^2 = 4$

$\qquad b = -2$

(b must be negative because θ is a III
quadrant angle)

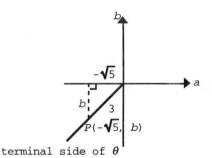

Using $(a, b) = (-\sqrt{5}, -2)$ and $r = 3$, we have

$\sin\theta = \dfrac{b}{r} = -\dfrac{2}{3}$ $\qquad$ $\sec\theta = \dfrac{r}{a} = -\dfrac{3}{\sqrt{5}}$

$\tan\theta = \dfrac{b}{a} = \dfrac{-2}{-\sqrt{5}} = \dfrac{2}{\sqrt{5}}$ $\qquad$ $\cot\theta = \dfrac{a}{b} = \dfrac{-\sqrt{5}}{-2} = \dfrac{\sqrt{5}}{2}$

$\csc\theta = \dfrac{r}{b} = \dfrac{3}{-2} = -\dfrac{3}{2}$

65. In these situations $P(a, b)$ is restricted so that $a = 0$. In this case,
functions for which a is in the denominator are not defined. These functions
are tangent and secant.

67. $\cos\theta = \dfrac{a}{r} = -\dfrac{\sqrt{3}}{2} = \dfrac{-\sqrt{3}}{2}$

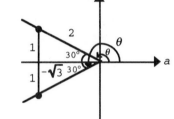

Thus $(a, b) = (-\sqrt{3}, 1)$ or $(-\sqrt{3}, -1)$
θ is associated with a 30° reference triangle in
the II quadrant or the III quadrant as drawn:

$\theta = 150°$ or $210°$

69. $\tan\theta = \dfrac{b}{a} = \dfrac{1}{1} = \dfrac{-1}{-1}$

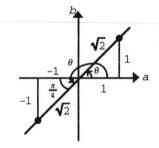

Thus $(a, b) = (1, 1)$ or $(-1, -1)$

θ is associated with a $\dfrac{\pi}{4}$ reference triangle in

the I quadrant or the III quadrant as drawn:

$\theta = \dfrac{\pi}{4}$ or $\dfrac{5\pi}{4}$

71. (A) Since $\theta_R = \dfrac{s}{r}$ and $r =$ radius of circle $= 4$, we have $\theta_R = \dfrac{7}{4}$ or 1.75 radians

(B) Since $\sin\theta = \dfrac{b}{r}$ and $\cos\theta = \dfrac{a}{r}$, we can write

$a = r\cos\theta = 4\cos\dfrac{7}{4} = -0.713$

$b = r\sin\theta = 4\sin\dfrac{7}{4} = 3.936$

$(a, b) = (-0.713, 3.936)$

73. We know that $s = r\theta$. $(a, b) = (6\sqrt{3}, 6)$.
From the Pythagorean theorem,

$r = \sqrt{a^2 + b^2} = \sqrt{(6\sqrt{3})^2 + 6^2} = 12$

Since $\tan \theta = \dfrac{b}{a} = \dfrac{6}{6\sqrt{3}} = \dfrac{1}{\sqrt{3}}$, $\theta = \dfrac{\pi}{6}$.

Hence $s = r\theta = 12\left(\dfrac{\pi}{6}\right) = 2\pi$ units.

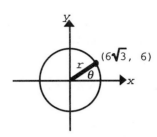

75. $I = k \cos \theta$

For $\theta = 0°$, $I = k \cos 0° = k$

For $\theta = 30°$, $I = k \cos 30° = 0.866k$

For $\theta = 60°$, $I = k \cos 60° = 0.5k$

77. From the figure, the following relations are clear:

$a^2 + b^2 = r^2$ $r = 1$ $\dfrac{a}{1} = \cos \theta$ $\dfrac{b}{1} = \sin \theta$

Using the Pythagorean theorem in the right triangle whose hypotenuse is the rod connecting the piston to the wheel, we have

$(y - b)^2 + a^2 = 4^2$

$\qquad (y - b)^2 = 4^2 - a^2$

$\qquad\quad y - b = \sqrt{4^2 - a^2}$

$\qquad\qquad y = b + \sqrt{4^2 - a^2}$

Since $a = \cos \theta$ and $b = \sin \theta$ and $\theta = 6\pi t$, we have

$y = \sin 6\pi t + \sqrt{16 - (\cos 6\pi t)^2}$

79. $A = n \tan\left(\dfrac{180°}{n}\right)$

(A) For $n = 8$, $A = 8 \tan \dfrac{180°}{8} = 3.31371$

For $n = 100$, $A = 100 \tan \dfrac{180°}{100} = 3.14263$

For $n = 1000$, $A = 1000 \tan \dfrac{180°}{1000} = 3.14160$

For $n = 10,000$, $A = 10,000 \tan \dfrac{180°}{10,000} = 3.14159$

(B) as $n \to \infty$, A seems to approach $\pi \,(= 3.1415926...)$, the area of the circle.

81. (A) Using the formula given:

For $\theta = 88.7°$, $m = \tan \theta = \tan 88.7° = 44.07$

For $\theta = 162.3°$, $m = \tan \theta = \tan 162.3° = -0.32$

(B) Using the formula for inclination, the slope m is given by $m = \tan 137°$
= -0.93. We now use the point-slope form of the equation of a line.

$y - y_0 = m(x - x_0)$

$\quad y - 5 = -0.93[x - (-4)]$

$\quad y - 5 = -0.93x - 3.72$

$\qquad\; y = -0.93x + 1.28$

Exercise 5-5

Key Ideas and Formulas

If two triangles are similar, their corresponding sides are proportional.

Trigonometric Functions of Acute Angles

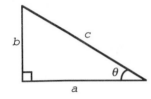

$$\sin \theta = \frac{b}{c} \qquad \csc \theta = \frac{c}{b}$$

$$\cos \theta = \frac{a}{c} \qquad \sec \theta = \frac{c}{a}$$

$$\tan \theta = \frac{b}{a} \qquad \cot \theta = \frac{a}{b}$$

Trigonometric Functions of Acute Angles (Alternate Form)

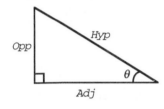

$$\sin \theta = \frac{\text{Opp}}{\text{Hyp}} \qquad \csc \theta = \frac{\text{Hyp}}{\text{Opp}}$$

$$\cos \theta = \frac{\text{Adj}}{\text{Hyp}} \qquad \sec \theta = \frac{\text{Hyp}}{\text{Adj}}$$

$$\tan \theta = \frac{\text{Opp}}{\text{Adj}} \qquad \cot \theta = \frac{\text{Adj}}{\text{Opp}}$$

Complementary Relationships

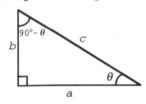

$$\sin \theta = \frac{b}{c} = \cos(90° - \theta)$$

$$\tan \theta = \frac{b}{a} = \cot(90° - \theta)$$

$$\sec \theta = \frac{c}{a} = \csc(90° - \theta)$$

Reciprocal Relationships for $0° < \theta < 90°$

For $0° < \theta < 90°$:

$$\csc \theta = \frac{1}{\sin \theta} \qquad \csc \theta \sin \theta = \frac{\text{Hyp}}{\text{Opp}} \cdot \frac{\text{Opp}}{\text{Hyp}} = 1$$

$$\sec \theta = \frac{1}{\cos \theta} \qquad \sec \theta \cos \theta = \frac{\text{Hyp}}{\text{Adj}} \cdot \frac{\text{Adj}}{\text{Hyp}} = 1$$

$$\cot \theta = \frac{1}{\tan \theta} \qquad \cot \theta \tan \theta = \frac{\text{Adj}}{\text{Opp}} \cdot \frac{\text{Opp}}{\text{Adj}} = 1$$

Pythagorean Theorem

In a right triangle,

$$c^2 = a^2 + b^2$$

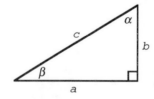

---

1. b/c **3.** c/b **5.** b/a **7.** $\frac{b}{a} = \tan \theta = \cot(90° - \theta)$

9. $\frac{a}{c} = \cos \theta = \sin(90° - \theta)$ **11.** $\frac{b}{c} = \sin \theta = \cos(90° - \theta)$

13. $67.56°$ **15.** $84.01°$ **17.** $42.06°$

19.

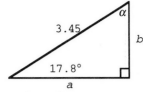

Solve for α:
$\alpha = 90° - 17.8° = 72.2°$

Solve for b: $\sin \beta = \dfrac{b}{c}$

$\sin 17.8° = \dfrac{b}{3.45}$

$\qquad b = 3.45 \sin 17.8°$
$\qquad\ \ = 1.05$

Solve for a: $\cos \beta = \dfrac{a}{c}$

$\cos 17.8° = \dfrac{a}{3.45}$

$\qquad a = 3.45 \cos 17.8°$
$\qquad\ \ = 3.28$

21.

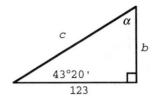

Solve for α:
$\alpha = 90° - 43°20' = 46°40'$

Solve for b: $\tan \beta = \dfrac{b}{a}$

$\tan 43°20' = \dfrac{b}{123}$

$\qquad b = 123 \tan 43°20'$
$\qquad\ \ = 116$

Solve for c: $\sec \beta = \dfrac{c}{a}$

$\sec 43°20' = \dfrac{c}{123}$

$\qquad c = 123 \sec 43°20'$
$\qquad\ \ = 169$

23.

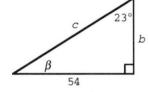

Solve for β:
$\beta = 90° - 23°0' = 67°0'$

Solve for b: $\tan \beta = \dfrac{b}{a}$

$\tan 67°0' = \dfrac{b}{54}$

$\qquad b = 54 \tan 67°0'$
$\qquad\ \ = 127$

Solve for c: $\csc \alpha = \dfrac{c}{a}$

$\csc 23°0' = \dfrac{c}{54}$

$\qquad c = 54 \csc 23°0'$
$\qquad\ \ = 138$

25.

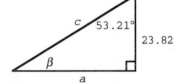

Solve for β:
$\beta = 90° - 53.21° = 36.79°$

Solve for a: $\tan \alpha = \dfrac{a}{b}$

$\tan 53.21° = \dfrac{a}{23.82}$

$\qquad a = 23.82 \tan 53.21°$
$\qquad\ \ = 31.85$

Solve for c: $\sec \alpha = \dfrac{c}{b}$

$\sec 53.21° = \dfrac{c}{23.82}$

$\qquad c = 23.82 \sec 53.21°$
$\qquad\ \ = 39.77$

27.

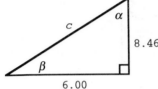

Solve for β: $\tan \beta = \dfrac{b}{a}$

$\tan \beta = \dfrac{8.46}{6.00}$

$\beta = \tan^{-1} \dfrac{8.46}{6.00}$

$ = 54.7°$ or $54°40'$

Solve for α:
$\alpha = 90° - 54°40' = 35°20'$

Solve for c: $\csc \beta = \dfrac{c}{b}$

$\csc 54°40' = \dfrac{c}{8.46}$

$ c = 8.46 \csc 54°40'$

$ = 10.4$

29.

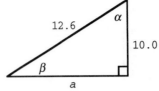

Solve for β: $\sin \beta = \dfrac{b}{c}$

$\sin \beta = \dfrac{10.0}{12.6}$

$\beta = \sin^{-1} \dfrac{10.0}{12.6}$

$ = 52.5°$ or $52°30'$

Solve for α:
$\alpha = 90° - 52°30' = 37°30'$

Solve for a: $\cot \beta = \dfrac{a}{b}$

$\cot 52°30' = \dfrac{a}{10.0}$

$ a = 10.0 \cot 52°30'$

$ = 7.67$

31. (A) In triangle OAD, $\cos \theta = \dfrac{\text{Adj}}{\text{Hyp}} = \dfrac{OA}{1} = {}'OA$

(B) In triangle OED, angle $EOD = 90° - \theta$, angle $OED = 90° - (90° - \theta) = \theta$.
Thus $\cot OED = \dfrac{\text{Adj}}{\text{Opp}} = \dfrac{DE}{1} = DE = \cot \theta$

(C) In triangle ODC, $\sec \theta = \dfrac{\text{Hyp}}{\text{Adj}} = \dfrac{OC}{1} = OC$

33. (A) As θ approaches $90°$, $OA = \cos \theta$ approaches 0.
(B) As θ approaches $90°$, $DE = \cot \theta$ approaches 0.
(C) As θ approaches $90°$, $OC = \sec \theta$ increases without bound.

35. (A) As θ approaches $0°$, $AD = \sin \theta$ approaches 0.
(B) As θ approaches $0°$, $CD = \tan \theta$ approaches 0.
(C) As θ approaches $0°$, $OE = \csc \theta$ increases without bound.

37.

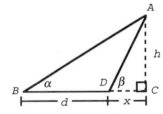

Label as shown at the left.
In right triangle ADC, $\cot \beta = \dfrac{x}{h}$

In right triangle ABC, $\cot \alpha = \dfrac{d + x}{h}$

Hence $x = h \cot \beta$ and $d + x = h \cot \alpha$
$d = h \cot \alpha - x$
$d = h \cot \alpha - h \cot \beta$
$d = h(\cot \alpha - \cot \beta)$
$h = \dfrac{d}{\cot \alpha - \cot \beta}$

39. Sketch a figure:

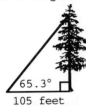

Let h = height of tree
From the figure, it should be clear
that $\tan 65.3° = \dfrac{h}{105}$

$$h = 105 \tan 65.3°$$
$$= 228 \text{ feet}$$

41. Sketch a figure:

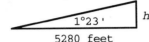

Let h = how far train climbs.
$$\tan 1°23' = \dfrac{h}{5280}$$
$$h = 5280 \tan 1°23'$$
$$= 127.5 \text{ feet}$$

43. Sketch a figure:

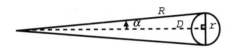

Note $\alpha = \dfrac{1}{2}(32') = 16'$

$$\text{diameter} = d = 2r$$
$$\tan \alpha = \dfrac{r}{D}$$
$$\tan 16' = \dfrac{r}{239,000}$$
$$r = 239,000 \tan 16'$$
$$d = 2(239,000)\tan 16'$$
$$= 2225 \text{ miles}$$

Alternatively, we can write
$$\sin \alpha = \dfrac{r}{R}$$
$$\sin 16' = \dfrac{r}{239,000}$$
$$r = 239,000 \sin 16'$$
$$d = 2(239,000)\sin 16'$$
$$= 2225 \text{ miles}$$

Although it is not clear whether 239,000 miles is to be interpreted as D, R, or $D - r$, at this accuracy it does not matter.

45. Sketch a figure:

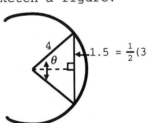

We will find $\dfrac{1}{2}\theta$ and double it.

$$\sin \dfrac{1}{2}\theta = \dfrac{1.5}{4}$$
$$\dfrac{1}{2}\theta = \sin^{-1}\dfrac{1.5}{4}$$
$$\theta = 2\,\sin^{-1}\dfrac{1.5}{4}$$
$$= 44°$$

47. We use $g = \dfrac{v}{t \sin \theta}$ with $v = 4.1$,

$t = 3.0$, $\theta = 8.0°$

$$g = \dfrac{4.1}{3.0 \sin 8.0°}$$

$$g = 9.8 \text{ meters/second}^2$$

49.

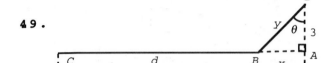

(A) We note that the cable consists of water section IB, and shore section $CB = 20$ mi $- AB$.

Let $y = IB$ and $d = CB = 20 - x$
In right triangle ABI

$$\sec \theta = \frac{y}{3 \text{ mi}} \quad \text{and} \quad \tan \theta = \frac{x}{3 \text{ mi}}$$
$$y = 3 \sec \theta \text{ mi} \qquad x = 3 \tan \theta \text{ mi}$$

Thus the cost of the cable =

$$\left(\begin{array}{c} \text{Cost of Water} \\ \text{Section Per Mile} \end{array} \right) \left(\begin{array}{c} \text{Number of} \\ \text{Water miles} = y \end{array} \right) + \left(\begin{array}{c} \text{Cost of Shore} \\ \text{Section Per Mile} \end{array} \right) \left(\begin{array}{c} \text{Number of} \\ \text{Shore miles} = d \end{array} \right)$$

$$C(\theta) = (25{,}000 \, \frac{\text{dollars}}{\text{mi}})(3 \sec \theta \text{ mi}) + (15{,}000 \, \frac{\text{dollars}}{\text{mi}})(20 - 3 \tan \theta \text{ mi})$$
$$C(\theta) = 75{,}000 \sec \theta + 300{,}000 - 45{,}000 \tan \theta$$

(B)

θ	$C(\theta)$
10°	\$368,222
20°	\$363,435
30°	\$360,622
40°	\$360,146
50°	\$363,050

51. Sketch a figure:

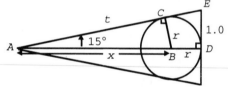

In triangle ABC, we have $\dfrac{r}{x} = \sin 15°$

In triangle ADE, we have $\dfrac{1.0}{r + x} = \tan 15°$

Eliminating x, we see that
$$\frac{r + x}{1.0} = \frac{1}{\tan 15°} = \cot 15° \quad \text{(reciprocal identity)}$$
$$x = \cot 15° - r$$

Substituting, we have
$$\frac{r}{\cot 15° - r} = \sin 15°$$
$$r = \sin 15°(\cot 15° - r)$$
$$r = \sin 15° \cot 15° - \sin 15° r$$
$$r(1 + \sin 15°) = \sin 15° \cot 15°$$
$$r = \frac{\sin 15° \cot 15°}{1 + \sin 15°}$$
$$r = 0.77 \text{ meters}$$

Exercise 5-6

Key Ideas and Formulas

The student should become thoroughly familiar with the graphs of the six functions $y = \sin x$, $y = \cos x$, $y = \tan x$, $y = \cot x$, $y = \sec x$, $y = \csc x$, and their basic characteristics. These are given in Figures 5, 7, 11, 12, 14, 13, respectively of Section 5-6 of the text. They are omitted here for lack of space.

---

1. sine: 2π; cotangent: π; cosecant: 2π

3. (A) Since the range of the cosine function is $[-1, 1]$, the largest and smallest y values on its graph are, respectively, 1 and -1. The largest deviation of the function from the x axis is therefore 1 unit.

   (B) Since the range of the tangent function is all real numbers, the graph deviates indefinitely far from the x axis.

   (C) Since the range of the cosecant function is all real numbers $y \geq 1$ or $y \leq -1$, the graph deviates indefinitely far from the x axis.

5. (A) -2π, $-\pi$, 0, π, 2π (B) $-\dfrac{3\pi}{2}$, $-\dfrac{\pi}{2}$, $\dfrac{\pi}{2}$, $\dfrac{3\pi}{2}$ (C) No x intercepts

7. $y = \cot x$ and $y = \csc x$ are undefined at $x = 0$.

9. (A) There are no vertical asymptotes   (B) $-\dfrac{3\pi}{2}$, $-\dfrac{\pi}{2}$, $\dfrac{\pi}{2}$, $\dfrac{3\pi}{2}$

   (C) -2π, $-\pi$, 0, π, 2π

11. (A) $y = \cos x$ (B) $y = \tan x$ (C) $y = \csc x$

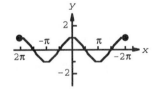   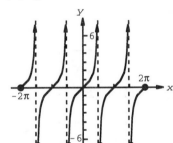   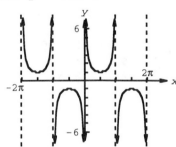

13. (A) A shift of $\pi/2$ to the left will transform the cosecant graph into the secant graph. [The answer is not unique--see part (B).]

    (B) The graph of $y = -\csc(x - \pi/2)$ is a $\pi/2$ shift to the right and a reflection in the x axis of the graph of $y = \csc x$. The result is the graph of $y = \sec x$.

    The graph of $y = -\csc\left(x + \dfrac{\pi}{2}\right)$ is a $\pi/2$ shift to the left and a reflection in the x axis of the graph of $y = \csc x$. The result is not the graph of $y = \sec x$.

15. (A)

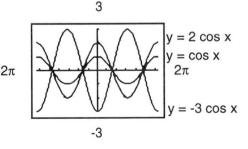

(B) The x intercepts do not change.

(C) The deviation of $y = \cos x$ from the x axis is 1 unit; the deviation of $y = 2 \cos x$ from the x axis is 2 units; the deviation of $y = -3 \cos x$ from the x axis is 3 units.

(D) The deviation of the graph from the x axis is changed by changing A. The deviation appears to be $|A|$.

17. (A)

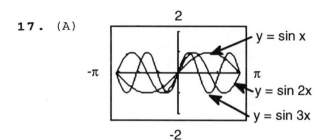

(B) 1 period of $y = \sin x$ appears.
2 periods of $y = \sin 2x$ appear.
3 periods of $y = \sin 3x$ appear.

(C) n periods of $y = \sin nx$ would appear.

19. (A)

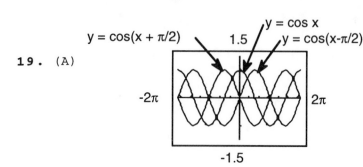

(B) The graph of $y = \cos x$ is shifted $|C|$ units to the right if $C < 0$ and $|C|$ units to the left if $C > 0$.

21. For each case, the number is not in the domain of the function and an error message of some type will appear.

23. If a function is periodic, then $f(x + p) = f(x)$ for some positive p. If the function is increasing on its domain, however, $f(x + p) > f(x)$. These two statements are mutually contradictory.

25. Here are graphs of
$f(x) = \sin x$ and
$g(x) = x$, $-1 \le x \le 1$.

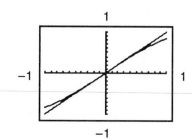

(A) The graphs become more indistinguishable the closer x is to the origin.

(B)

x	-0.3	-0.2	-0.1	0.0	0.1	0.2	0.3
$\sin x$	-0.296	-0.199	-0.100	0.000	0.100	0.199	0.296

27. True. If f and g are periodic with the same period p, then
$f(x + p) = f(x)$ and $g(x + p) = g(x)$.
Then $(f + g)(x + p) = f(x + p) + g(x + p) = f(x) + g(x) = (f + g)(x)$.
Thus, $f + g$ is periodic. Similarly,
  $(f - g)(x + p) = f(x + p) - g(x + p) = f(x) - g(x) = (f - g)(x)$.
Thus, $f - g$ is also periodic.

29. True. If g is periodic with period p, then $g(x + p) = g(x)$.
Then $f \circ g(x + p) = f[g(x + p)] = f[g(x)] = f \circ g(x)$.
Thus, $f \circ g$ is periodic.

31. True. If f is periodic with period p, then $f(x + p) = f(x)$.
Then $g(x + p) = 5f(x + p) = 5f(x) = g(x)$.
Thus, g is periodic.

33. False. The statement is almost true, in the sense that if f is periodic with period p, then $f(x + p) = f(x)$, $f(x + 5p) = f(x)$ (why?) and $j(x + p) = f[5(x + p)] = f(5x + 5p) = f(5x) = j(x)$.
However, since the period of a periodic function is defined to be the *smallest* possible period, and we can write

$$j\left(x + \frac{p}{5}\right) = f\left[5\left(x + \frac{p}{5}\right)\right] = f(5x + p) = f(5x) = j(x)$$

the period of j is $\dfrac{p}{5}$, not p, and j is not periodic with the same period as f.

Thus, $j(x) = \sin 5x$ is a counterexample. $f(x) = \sin x$ has period π, but

$$j\left(x + \frac{2\pi}{5}\right) = \sin\left[5\left(x + \frac{2\pi}{5}\right)\right] = \sin[5x + 2\pi] = \sin(5x) = j(x),$$

thus j has period $\dfrac{2\pi}{5}$.

Exercise 5-7

Key Ideas and Formulas

Graphs of $y = A \sin(Bx + C)$ and $y = A \cos(Bx + C)$

$y = A \sin(Bx + C)$ $y = A \cos(Bx + C)$ $(B > 0)$

Amplitude $= |A|$ Period $= \dfrac{2\pi}{B}$ Phase Shift $= -\dfrac{C}{B}$ (if $C = 0$, there is no phase shift)

The graph completes one full cycle as $Bx + C$ varies from 0 to 2π, that is, as x varies over the interval $\left[-\dfrac{C}{B}, -\dfrac{C}{B} + \dfrac{2\pi}{B}\right]$

Graphs of $y = k + A \sin(Bx + C)$ and $y = k + A \cos(Bx + C)$

These graphs are the same as the graph of $y = A \sin(Bx + C)$ and $y = A \cos(Bx + C)$, translated vertically—up k units if k is positive and down $|k|$ units if k is negative.

1.

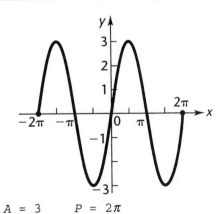

$A = 3$ $P = 2\pi$

3.

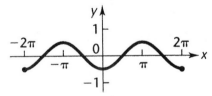

The graph of $-\dfrac{1}{2}\cos x$ is the graph of $\dfrac{1}{2}\cos x$ reflected across the x axis.

$|A| = \dfrac{1}{2}$ $P = 2\pi$

5.

$A = 1$ $P = \dfrac{2\pi}{3}$

7.

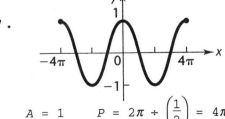

$A = 1$ $P = 2\pi \div \left(\dfrac{1}{2}\right) = 4\pi$

9.

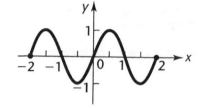

$$A = 1 \qquad P = \frac{2\pi}{\pi} = 2$$

11.

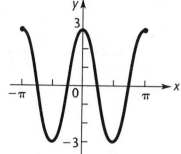

$$A = 3 \qquad P = \frac{2\pi}{2} = \pi$$

13.

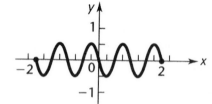

$$A = \frac{1}{2} \qquad P = \frac{2\pi}{2\pi} = 1$$

The graph of $-\frac{1}{2} \sin 2\pi x$ is the graph of $\frac{1}{2} \sin 2\pi x$ reflected across the x axis.

15.

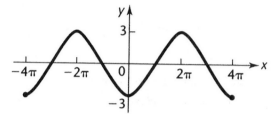

$$A = 3 \qquad P = 2\pi \div \left(\frac{1}{2}\right) = 4\pi$$

The graph of $-3 \cos \frac{x}{2}$ is the graph of $3 \cos \frac{x}{2}$ reflected across the x axis.

17.

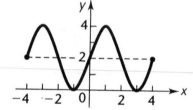

The graph is the same as the graph of $y = 2 \sin \frac{\pi x}{2}$ shifted 2 units up.

$$A = 2 \qquad P = 2\pi \div \left(\frac{\pi}{2}\right) = 4$$

19.

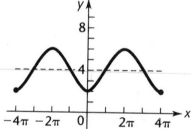

The graph is the same as the graph of $y = 2 \cos \frac{x}{2}$, reflected across the x axis and shifted up 4 units.

$$A = 2 \qquad P = 2\pi \div \frac{1}{2} = 4\pi$$

21. $A = 3 \quad P = \frac{\pi}{2} = \frac{2\pi}{4} = \frac{2\pi}{B}$. Hence $B = 4$, $y = 3 \sin 4x$, $-\frac{\pi}{4} \le x \le \frac{\pi}{2}$.

23. $|A| = 10 \quad P = 2 = \frac{2\pi}{\pi} = \frac{2\pi}{B}$. Hence $B = \pi$, $A = -10$, since the graph has the form of the standard sine curve reflected across the x axis.
$y = -10 \sin \pi x \quad -1 \le x \le 2$.

25. $A = 5 \quad P = 8\pi = 2\pi \cdot 4 = 2\pi \div \frac{1}{4}$. Hence $B = \frac{1}{4}$, $y = 5 \cos \frac{1}{4} x \quad -4\pi \le x \le 8\pi$.

27. $|A| = 0.5$, $P = 8 = 2\pi \cdot \frac{4}{\pi} = 2\pi \div \frac{\pi}{4}$. Hence $B = \frac{\pi}{4}$, $A = -0.5$, since the graph has the form of the standard cosine curve reflected across the x axis.
$y = -0.5 \cos \frac{\pi x}{4} \quad -4 \le x \le 8$

29. Here is a computer-generated graph of $y = \cos^2 x - \sin^2 x$, $-\pi \le x \le \pi$.

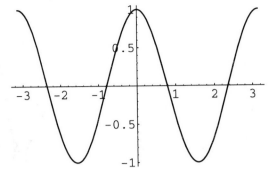

The graph has amplitude 1 and period π. It appears to be the graph of $y = A \cos Bx$ with $A = 1$, and $B = 2\pi \div P = 2\pi \div \pi = 2$, that is, $y = \cos 2x$.

31. Here is a computer-generated graph of $y = 2 \sin^2 x$, $-\pi \le x \le \pi$.

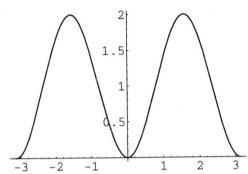

The graph has amplitude $\dfrac{2 - 0}{2} = 1$ and period π. It appears to be the graph of $y = \cos 2x$ reflected across the x axis and shifted up one unit, that is, $y = -\cos 2x + 1$ or $y = 1 - \cos 2x$.

33. $y = \sin(x + \pi)$
$A = 1$
Solve $x + \pi = 0 \qquad x + \pi = 2\pi$
$x = -\pi \qquad\quad x = -\pi + 2\pi = \pi$

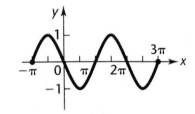

$\underbrace{}$
$\qquad$ Phase shift $\qquad$ Period $P = 2\pi$

The graph completes one full cycle as x varies over the interval $[-\pi,\ \pi]$.

35. $y = \dfrac{1}{2} \cos\left(x - \dfrac{\pi}{4}\right)$

$A = \dfrac{1}{2}$

Solve $x - \dfrac{\pi}{4} = 0 \qquad x - \dfrac{\pi}{4} = 2\pi$

$x = \dfrac{\pi}{4} \qquad\quad x = \dfrac{\pi}{4} + 2\pi = \dfrac{9\pi}{4}$

$\underbrace{}$
$\qquad$ Phase shift $\qquad$ Period $P = 2\pi$

The graph completes one full cycle as x varies over the interval $\left[\dfrac{\pi}{4},\ \dfrac{9\pi}{4}\right]$.

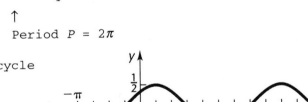

37. $y = \sin[\pi(x - 1)]$
$A = 1$
Solve $\pi(x - 1) = 0 \qquad \pi(x - 1) = 2\pi$
$x = 1 \qquad\qquad x - 1 = 2$
$x = 1 + 2 = 3$

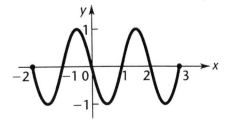

$\underbrace{}$
$\qquad$ Phase shift $\qquad$ Period $P = 2$

The graph completes one full cycle as x varies over the interval $[1, 3]$.

39. $y = 3\cos\left(\pi x + \dfrac{\pi}{2}\right)$

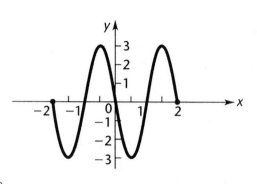

$A = 3$

Solve $\pi x + \dfrac{\pi}{2} = 0 \qquad \pi x + \dfrac{\pi}{2} = 2\pi$

$\qquad\qquad \pi x = -\dfrac{\pi}{2} \qquad\quad \pi x = -\dfrac{\pi}{2} + 2\pi$

$\qquad\qquad x = -\dfrac{1}{2} \qquad\quad x = -\dfrac{1}{2} + 2 = \dfrac{3}{2}$

$\qquad\qquad\quad \uparrow\underline{}\uparrow \quad \uparrow$

$\qquad\qquad\quad$ Phase shift $\qquad$ Period $P = 2$

The graph completes one full cycle as x varies over the interval $\left[-\dfrac{1}{2}, \dfrac{3}{2}\right]$

41. $y = -1 + \sin(x + \pi)$

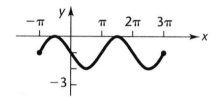

For the graph of $y = \sin(x + \pi)$, we note: $A = 1$,
$P = 2\pi$, Phase shift $= -\pi$. We graph $y = \sin(x + \pi)$,
then vertically translate the graph down 1 unit.

43. $y = 2 - 4\cos(2x - \pi)$

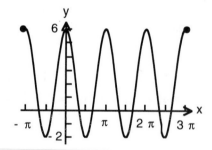

For the graph of $y = 4\cos(2x - \pi)$, we note:
$A = 4$, $P = \pi$, Phase shift $= \dfrac{\pi}{2}$. We graph
$y = 4\cos(2x - \pi)$ turned upside down, then
vertically translate the graph up 2 units.

45. $|A| = 4$. Hence $A = 4$ or -4. The graph completes one full cycle as x varies over the interval $[-1, 3]$. Since $-\dfrac{C}{B}$ is required between 0 and 2, we cannot simply set $-\dfrac{C}{B} = -1$. We must (mentally) extend the curve so that the phase shift is positive. Then the (extended) graph is that of an upside down sine curve that completes one full cycle as x varies over the interval $[1, 5]$. Hence

$A = -4$

$-\dfrac{C}{B} = 1 \qquad -\dfrac{C}{B} + \dfrac{2\pi}{B} = 5$

$C = -B \qquad\qquad \dfrac{2\pi}{B} = 4 \qquad B = \dfrac{2\pi}{4} = \dfrac{\pi}{2} \qquad C = -\dfrac{\pi}{2}$

The equation is then $y = A\sin(Bx + C)$

$$y = -4\sin\left(\dfrac{\pi}{2}x - \dfrac{\pi}{2}\right)$$

47. $|A| = \dfrac{1}{2}$. Hence $A = \dfrac{1}{2}$ or $-\dfrac{1}{2}$. The graph completes one full cycle of the cosine function as x varies over the (mentally extended) intervals $[-\pi,\ 7\pi]$ or $[3\pi,\ 11\pi]$. Since the phase shift is required between 0 and 4π, we must set $-\dfrac{C}{B} = 3\pi$. Then the (extended) graph has the form of a standard cosine curve. Hence

$$A = \frac{1}{2} \qquad\qquad -\frac{C}{B} + \frac{2\pi}{B} = 11\pi$$

$$-\frac{C}{B} = 3\pi \qquad\qquad \frac{2\pi}{B} = 8\pi \quad B = \frac{2\pi}{8\pi} = \frac{1}{4} \quad C = -3\pi B = -\frac{3\pi}{4}$$

$$C = -3\pi B$$

The equation is then
$$y = A \cos(Bx + C)$$

$$y = \frac{1}{2} \cos\left(\frac{1}{4}x - \frac{3\pi}{4}\right)$$

49. If the amplitude is 4, then $|A| = 4$, that is, $A = 4$ or -4. If the period is π, then $\dfrac{2\pi}{B} = \pi$, hence $2\pi = B\pi$ and $B = 2$. If the phase shift is $-\dfrac{\pi}{3}$, then $-\dfrac{C}{B} = -\dfrac{C}{2} = -\dfrac{\pi}{3}$ and $C = \dfrac{2\pi}{3}$. Thus B and C are uniquely determined by the given data, and there are two possibilities for A. Therefore there are two possible functions and their equations are $y = 4 \sin\left(2x + \dfrac{2\pi}{3}\right)$ and $y = -4 \sin\left(2x + \dfrac{2\pi}{3}\right)$.

51. $y = 3.5 \sin\left[\dfrac{\pi}{2}(t + 0.5)\right]$

$A = 3.5$

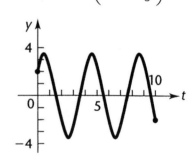

Solve $\dfrac{\pi}{2}(t + 0.5) = 0 \qquad \dfrac{\pi}{2}(t + 0.5) = 2\pi$

$\qquad\qquad t + 0.5 = 0 \qquad\qquad t + 0.5 = 4$

$\qquad\qquad\qquad t = -0.5 \qquad\qquad\quad t = -0.5 + 4 = 3.5$

$\qquad\qquad\qquad\qquad \underbrace{\qquad\qquad\qquad\qquad}\ \ \uparrow$

$\qquad\qquad\qquad$ Phase shift $\quad$ Period $P = 4$

The graph completes one full cycle as t varies over the interval $[-0.5,\ 3.5]$.

53. $y = 50 \cos[2\pi(t - 0.25)]$

$A = 50$

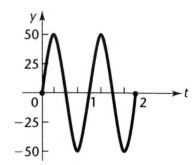

Solve $2\pi(t - 0.25) = 0 \qquad 2\pi(t - 0.25) = 2\pi$

$\qquad\qquad t - 0.25 = 0 \qquad\qquad t - 0.25 = 1$

$\qquad\qquad\qquad t = 0.25 \qquad\qquad\qquad t = 0.25 + 1 = 1.25$

$\qquad\qquad\qquad\qquad \underbrace{\qquad\qquad\qquad\qquad}\ \ \uparrow$

$\qquad\qquad\qquad$ Phase shift $\quad$ Period $P = 1$

The graph completes one full cycle as t varies over the interval $[0.25,\ 1.25]$.

55. Here is a computer-generated graph of $y = \sqrt{2} \sin x + \sqrt{2} \cos x$, $-2\pi \leq x \leq 2\pi$.

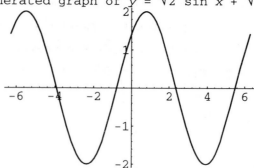

It appears that this is a sine curve shifted to the left, with $A = 2$ and, since $P = \dfrac{2\pi}{B}$ and P appears to be 2π, $B = \dfrac{2\pi}{P} = \dfrac{2\pi}{2\pi} = 1$.

From the graphing utility, we find that the x intercept closest to the origin, to three decimal places, is -0.785. To find C, substitute $B = 1$ and $x = -0.785$ into the phase-shift formula $x = -\dfrac{C}{B}$ and solve for C:

$$x = -\dfrac{C}{B}$$
$$-0.785 = -\dfrac{C}{1}$$
$$C = 0.785$$

The equation is thus $y = 2 \sin(x + 0.785)$.

57. Here is a computer-generated graph of $y = \sqrt{3} \sin x - \cos x$, $-2\pi \leq x \leq 2\pi$.

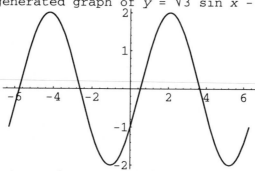

It appears that this is a sine curve shifted to the right, with $A = 2$ and, since $P = \dfrac{2\pi}{B}$ and P appears to be 2π, $B = \dfrac{2\pi}{P} = \dfrac{2\pi}{2\pi} = 1$.

From the graphing utility, we find that the x intercept closest to the origin, to three decimal places, is 0.524. To find C, substitute $B = 1$ and $x = 0.524$ into the phase-shift formula $x = -\dfrac{C}{B}$ and solve for C:

$$x = -\dfrac{C}{B}$$
$$0.524 = -\dfrac{C}{1}$$
$$C = -0.524$$

The equation is thus $y = 2 \sin(x - 0.524)$.

59. Here is a computer-generated graph of $y = 4.8 \sin 2x - 1.4 \cos 2x$, $-\pi \le x \le \pi$.

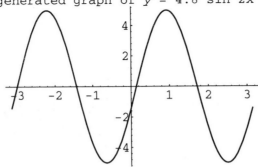

It appears that this is a sine curve shifted to the right, with $A = 5$ and,

since $P = \dfrac{2\pi}{B}$ and P appears to be π, $B = \dfrac{2\pi}{P} = \dfrac{2\pi}{\pi} = 2$.

From the graphing utility, we find that the x intercept closest to the origin, to three decimal places, is 0.142. To find C, substitute $B = 2$ and $C = 0.142$

into the phase-shift formula $x = -\dfrac{C}{B}$ and solve for C:

$$x = -\frac{C}{B}$$
$$0.142 = -\frac{C}{2}$$
$$C = -0.284$$

The equation is thus $y = 5 \sin(2x - 0.284)$.

61. True. If $f(x) = A \sin(Bx + C)$, then $g(x) = f(3x) = A \sin[B(3x) + C]$ $= A \sin(3Bx + C)$. This is a simple harmonic. The same reasoning holds if $f(x) = A \cos(Bx + C)$.

63. True. If $f(x) = A \sin(Bx + C)$, then $j(x) = 3f(x) = 3A \sin(Bx + C)$. This is a simple harmonic. The same reasoning holds if $f(x) = A \cos(Bx + C)$.

65.

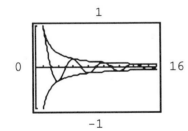

The amplitude is decreasing with time. This is often referred to as a *damped sine wave*. Examples are the vertical motion of a car after going over a bump (which is damped by the suspension system) and the slowing down of a pendulum that is released away from the vertical line of suspension (air resistance and friction).

67.

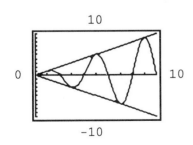

The amplitude is increasing with time. In physical and electrical systems this is referred to as *resonance*. Some examples are the swinging of a bridge during high winds and the movement of tall buildings during an earthquake. Some bridges and buildings are destroyed when the resonance reaches the elastic limits of the structure.

69.

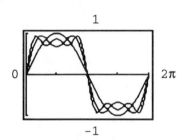

71. $A = \dfrac{1}{3}$

$P = \dfrac{2\pi}{8} = \dfrac{\pi}{4}$

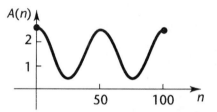

73. When $t = 0$ $y = -8$. Hence $-8 = A \cos B(0)$, that is, $A = -8$. Since the period is 0.5 seconds, $\dfrac{2\pi}{B} = 0.5$, $B = \dfrac{2\pi}{0.5} = 4\pi$. Hence the equation is $y = -8 \cos 4\pi t$.

75. The graph is the same as the graph of $y = \cos \dfrac{n\pi}{26}$, shifted 1.5 units up.

$A = 1$

$P = 2\pi \div \dfrac{\pi}{26} = 52$

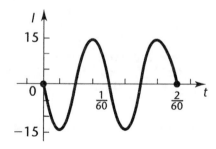

The graph shows the seasonal changes of sulfur dioxide pollutant in the atmosphere; more is produced during winter months because of increased heating.

77. $I = 15 \cos\left(120\pi t + \dfrac{\pi}{2}\right)$

$A = 15$

Solve $120\pi t + \dfrac{\pi}{2} = 0$ $\qquad$ $120\pi t + \dfrac{\pi}{2} = 2\pi$

$\qquad\qquad 120\pi t = -\dfrac{\pi}{2}$ $\qquad\qquad 120\pi t = -\dfrac{\pi}{2} + 2\pi$

$\qquad\qquad t = -\dfrac{\pi}{2} \div 120\pi$ $\qquad t = \left(-\dfrac{\pi}{2} + 2\pi\right) \div 120\pi$

$\qquad\qquad t = -\dfrac{1}{240}$ $\qquad\qquad t = -\dfrac{1}{240} + \dfrac{1}{60} = \dfrac{3}{240}$ or $\dfrac{1}{80}$

$\qquad$ Phase Shift $= -\dfrac{1}{240}$ $\qquad$ Period $P = \dfrac{1}{60}$

The graph completes one full cycle as t varies over the interval $\left[-\dfrac{1}{240}, \dfrac{3}{240}\right]$.

79. If the disk rotates through an angle θ in t seconds, we see

$$\theta = 3\,\frac{\text{revolutions}}{\text{second}} \cdot 2\pi\,\frac{\text{radians}}{\text{revolution}} \cdot t\ \text{seconds}$$
$$= 6\pi t\ \text{radians}$$

Then $\dfrac{y}{R} = \sin\theta = \sin 6\pi t$

$y = R\sin 6\pi t$

Since the disk has radius 3, $y = 3\sin 6\pi t$.

This function has $A = 3$, $P = \dfrac{2\pi}{6\pi} = \dfrac{1}{3}$.

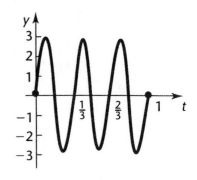

81. (A)

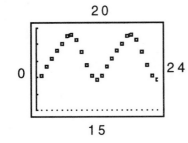

(B) From the table, Max $y = 19{:}35$ and Min $y = 16{:}51$.

Hence $A = \dfrac{\text{Max } y - \text{Min } y}{2} = \dfrac{19\frac{35}{60} - 16\frac{51}{60}}{2} = 1.37$

$B = \dfrac{2\pi}{\text{Period}} = \dfrac{2\pi}{12} = \dfrac{\pi}{6}$

$C = -1.75$ from the graph

$k = \text{Min } y + A = 16\frac{51}{60} + 1.37 = 18.22$

Thus, $y = 18.22 + 1.37\sin\left(\dfrac{\pi x}{6} - 1.75\right)$

(C)

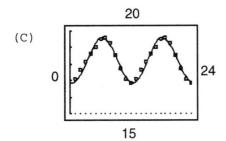

Exercise 5-8

Key Ideas and Formulas

Graphs of $y = A\tan(Bx + C)$ and $y = A\cot(Bx + C)$

$y = A\tan(Bx + C)$ $\qquad$ $y = A\cot(Bx + C)$ $\qquad$ $(B > 0)$

Period $= \dfrac{\pi}{B}$ $\qquad$ Phase Shift $= -\dfrac{C}{B}$ $\quad$ amplitude is not defined

Graphs of $y = A\sec(Bx + C)$ and $y = A\csc(Bx + C)$

$\qquad\quad y = A\sec(Bx + C)$ $\qquad$ $y = A\csc(Bx + C)$ $\quad$ $(B > 0)$

Period $= \dfrac{2\pi}{B}$ $\qquad$ Phase Shift $= -\dfrac{C}{B}$ $\quad$ amplitude is not defined

These graphs are obtained by graphing $y = \dfrac{1}{A}\cos(Bx + C)$ and $y = \dfrac{1}{A}\sin(Bx + C)$, then taking reciprocals.

1. Period = $\frac{\pi}{4}$

One cycle of $y = 2 \cot 4x$ is completed as x varies from 0 to $\frac{\pi}{4}$.

Sketch the graph for one period $\left(0, \frac{\pi}{4}\right)$, then extend over the interval $\left(0, \frac{\pi}{2}\right)$.

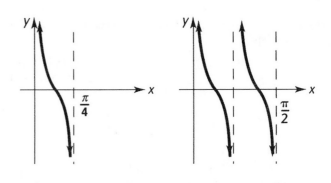

3. Period = $\frac{\pi}{8\pi} = \frac{1}{8}$

One cycle of $y = -\frac{1}{4} \tan 8\pi x$ is completed as x varies from 0 to $\frac{1}{8}$. Sketch the graph (the graph of the basic tangent function turned upside down) for one period $\left[0, \frac{1}{8}\right]$, then extend over the interval $\left[0, \frac{1}{2}\right]$.

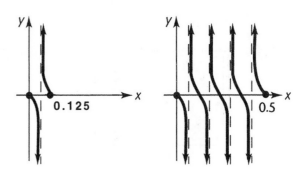

5. Period = $2\pi \div \frac{1}{2} = 4\pi$

Since $\csc \frac{x}{2} = \frac{1}{\sin \frac{x}{2}}$, we graph $y = \sin \frac{x}{2}$ for one cycle from 0 to 4π, and then take reciprocals. We place vertical asymptotes through the x intercepts of the sine graph. Then we extend over the required interval $[-3\pi, 3\pi]$, deleting the part of the graph from 3π to 4π.

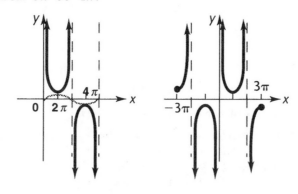

7. Period = $\dfrac{2\pi}{\pi}$ = 2

Since 2 sec $\pi x = \dfrac{1}{\frac{1}{2}\cos \pi x}$,

we graph $y = \dfrac{1}{2}\cos \pi x$ for
one cycle from 0 to 2, and
then take reciprocals. We
place vertical asymptotes
through the x intercepts of
the cosine graph. Then we
extend over the required
interval (-1, 3).

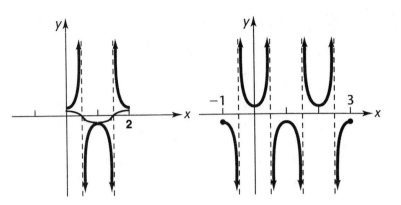

9. One cycle of $y = \cot\left(x + \dfrac{\pi}{2}\right)$ is completed as $x + \dfrac{\pi}{2}$ varies from 0 to π. Solve

each equation for x:

$x + \dfrac{\pi}{2} = 0$ $\qquad\qquad$ $x + \dfrac{\pi}{2} = \pi$

$\qquad x = -\dfrac{\pi}{2}$ $\qquad\qquad$ $x = -\dfrac{\pi}{2} + \pi$

Phase shift = $-\dfrac{\pi}{2}$ $\qquad$ Period = π

Sketch the graph for one period $\left(-\dfrac{\pi}{2}, \dfrac{\pi}{2}\right)$, then extend over the interval $\left(-\dfrac{\pi}{2}, \dfrac{3\pi}{2}\right)$.

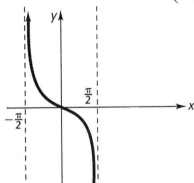
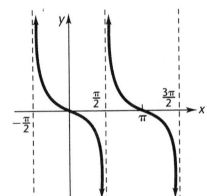

11. One cycle of $y = \tan(2x + \pi)$ is
completed as $2x + \pi$ varies from 0 to
π. Solve each equation for x:

$2x + \pi = 0$ $\qquad$ $2x + \pi = \pi$

$\qquad 2x = -\pi$ $\qquad\quad$ $2x = \pi - \pi$

$\qquad\quad x = -\dfrac{\pi}{2}$ $\qquad\quad$ $x = \dfrac{\pi}{2} - \dfrac{\pi}{2}$

Phase shift = $-\dfrac{\pi}{2}$ $\quad$ Period = $\dfrac{\pi}{2}$

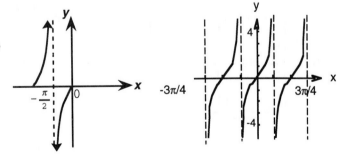

Sketch the graph for one period $\left[-\dfrac{\pi}{2}, 0\right]$, then extend over the interval $\left(-\dfrac{3\pi}{4}, \dfrac{3\pi}{4}\right)$.

13. Find the period and phase shift by solving $\pi x + \frac{\pi}{2} = 0$ and $\pi x + \frac{\pi}{2} = 2\pi$ for x.

$$\pi x + \frac{\pi}{2} = 0 \qquad\qquad \pi x + \frac{\pi}{2} = 2\pi$$

$$\pi x = -\frac{\pi}{2} \qquad\qquad \pi x = 2\pi - \frac{\pi}{2}$$

$$x = -\frac{1}{2} \qquad\qquad x = 2 - \frac{1}{2}$$

$$\text{Phase shift} = -\frac{1}{2} \qquad \text{Period} = 2$$

$y = \sec(\pi x + \frac{\pi}{2}) = \dfrac{1}{\cos(\pi x + \frac{\pi}{2})}$,

we graph $y = \cos(\pi x + \frac{\pi}{2})$ for one cycle from $-\frac{1}{2}$ to $\frac{3}{2}$, and then take reciprocals. We place vertical asymptotes through the x intercepts of the graph.

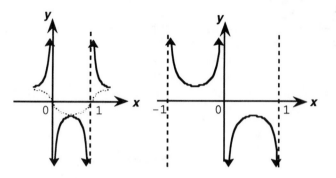

15. False. The graph of $y = \cos(\pi x)$ has all ordinates between -1 and 1. The graph of $y = \csc(\pi x)$ has all ordinates greater than or equal to 1 or less than or equal to -1. The graphs could only intersect if both functions simultaneously took on values of 1 or -1. This occurs for no value of x, since $\cos(\pi x) = 1$ if $x = 2n$ and $\cos(\pi x) = -1$ if $x = 2n + 1$, while $\csc(\pi x) = 1$ if $x = \frac{1}{2} + 2n$ and $\csc(\pi x) = -1$ if $x = \frac{3}{2} + 2n$. (n any integer).

17. False. For example, the x axis never intersects the graph of $y = 0.1 \sec(5x + 1)$.

19. Here is a graph of $y = \cot x - \tan x$ drawn by a graphing utility.

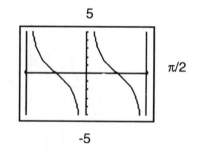

The graph appears to have the form $y = A \cot Bx$. Since the period is $\frac{\pi}{2}$, set $\frac{\pi}{2} = \frac{\pi}{B}$ to obtain $B = 2$. The graph of $y = A \cot 2x$ shown appears to pass through $\left(\frac{\pi}{8},\ 2\right)$, thus

$$2 = A \cot 2\left(\frac{\pi}{8}\right)$$

$$2 = A \cot \frac{\pi}{4}$$

$$2 = A$$

The equation of the graph can be written $y = 2 \cot 2x$.

21. Here is a graph of
$y = \csc x + \cot x$
drawn by a graphing
utility.

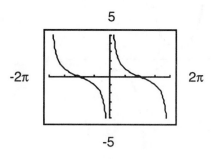

Sketch the graph for
one period $[\pi, 5\pi]$,
then extend over the
interval $[-4\pi, 4\pi]$,
deleting the part of
the graph from 4π to 5π.

The graph appears to have the form
$y = A \cot Bx$. Since the period is 2π, set $2\pi = \dfrac{\pi}{B}$
to obtain $B = \dfrac{1}{2}$. The graph of $y = A \cot \dfrac{1}{2}x$ shown
appears to pass through $\left(\dfrac{\pi}{2}, 1\right)$, thus

$$1 = A \cot \frac{1}{2}\left(\frac{\pi}{2}\right)$$
$$1 = A \cot \frac{\pi}{4}$$
$$1 = A$$

The equation of the graph can be written
$y = \cot \dfrac{1}{2}x$.

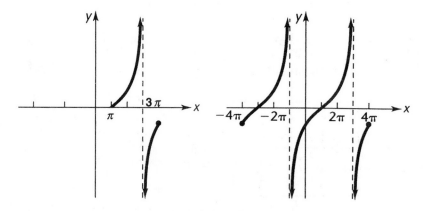

23. One cycle of $y = -2 \tan\left(\dfrac{\pi}{4}x - \dfrac{\pi}{4}\right)$ is completed as $\dfrac{\pi}{4}x - \dfrac{\pi}{4}$ varies from 0 to π.

Solve each equation for x:

$$\frac{\pi}{4}x - \frac{\pi}{4} = 0 \qquad \frac{\pi}{4}x - \frac{\pi}{4} = \pi$$
$$\frac{\pi}{4}x = \frac{\pi}{4} \qquad\quad \frac{\pi}{4}x = \frac{\pi}{4} + \pi$$
$$x = 1 \qquad\qquad\quad x = 1 + 4$$

Phase shift = 1   Period = 4

Sketch the graph (the graph of
the basic tangent function
turned upside down) for one
period
$[1, 5]$, then extend over the
interval $(-1, 7)$.

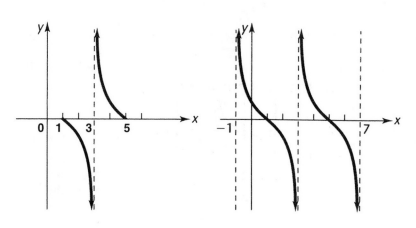

25. Find the period and phase shift by solving $\frac{\pi}{2}x + \frac{\pi}{2} = 0$ and $\frac{\pi}{2}x + \frac{\pi}{2} = 2\pi$ for x:

$$\frac{\pi}{2}x + \frac{\pi}{2} = 0 \qquad\qquad \frac{\pi}{2}x + \frac{\pi}{2} = 2\pi$$

$$\frac{\pi}{2}x = -\frac{\pi}{2} \qquad\qquad \frac{\pi}{2}x = -\frac{\pi}{2} + 2\pi$$

$$x = -1 \qquad\qquad\quad x = -1 + 4$$

Phase shift = -1   Period = 4

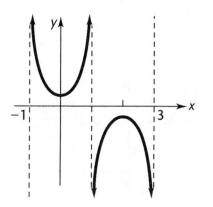

Since $3\csc\left(\dfrac{\pi}{2}x + \dfrac{\pi}{2}\right) = \dfrac{1}{\frac{1}{3}\sin\left(\frac{\pi}{2}x + \frac{\pi}{2}\right)}$ we graph

$y = \dfrac{1}{3}\sin\left(\dfrac{\pi}{2}x + \dfrac{\pi}{2}\right)$ for one cycle from -1 to 3,

and then take reciprocals. We place vertical asymptotes through the x intercepts of the sine graph.

27. Here is a graph of
$y = \sin 3x + \cos 3x \cot 3x$
drawn by a graphing utility.

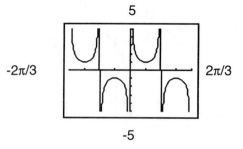

The graph appears to have the form

$y = A \csc Bx$. Since the period is $\dfrac{2\pi}{3}$,

set $\dfrac{2\pi}{3} = \dfrac{2\pi}{B}$ to obtain $B = 3$. The graph of

$y = A \csc 3x$ shown appears to pass through

$\left(\dfrac{\pi}{6},\ 1\right)$, thus

$$1 = A \csc 3\left(\frac{\pi}{6}\right)$$

$$1 = A \csc \frac{\pi}{2}$$

$$1 = A$$

The equation of the graph can be written
$y = \csc 3x$.

29. Here is a graph of
$y = \dfrac{\sin 4x}{1 + \cos 4x}$
drawn by a graphing utility.

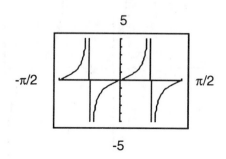

The graph appears to have the form

$y = A \tan Bx$. Since the period is $\dfrac{\pi}{2}$, set $\dfrac{\pi}{B} = \dfrac{\pi}{2}$

to obtain $B = 2$. The graph of $y = A \tan 2x$

shown appears to pass through $\left(\dfrac{\pi}{8},\ 1\right)$, thus

$$1 = A \tan 2\left(\frac{\pi}{8}\right)$$

$$1 = A \tan \frac{\pi}{4}$$

$$1 = A$$

The equation of the graph can be written
$y = \tan 2x$.

31. (A) In triangle MNP, we can write

$$\cos \theta = \frac{20}{c}$$
$$c \cos \theta = 20$$
$$c = \frac{20}{\cos \theta}$$
$$c = 20 \sec \theta$$

Since $\theta = \frac{\pi t}{2}$, the equation for c in terms of t is

$c = 20 \sec \frac{\pi t}{2}$. The equation is valid for $0 \le t < 1$, since $\sec \frac{\pi t}{2}$ is undefined when $t = 1$.

(B) The graph has an asymptote at $t = 1$. Sketch the portion of $y = \sec \frac{\pi t}{2}$ on $[0, 1)$.

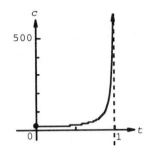

(C) The length of the light beam starts at 20 ft and increases slowly at first, then increases rapidly without end.

Exercise 5-9

Key Ideas and Formulas

The inverse sine function, $\sin^{-1}$ or arcsin, is defined as the inverse of the restricted sine function $y = \sin x \quad -\frac{\pi}{2} \le x \le \frac{\pi}{2}$. Thus,

$$y = \sin^{-1} x \quad \text{and} \quad y = \arcsin x$$

are equivalent to

$$x = \sin y \quad -1 \le x \le 1, \ -\frac{\pi}{2} \le y \le \frac{\pi}{2}$$

$\sin(\sin^{-1} x) = x \quad -1 \le x \le 1$
$\sin^{-1}(\sin y) = y \quad -\frac{\pi}{2} \le y \le \frac{\pi}{2}$

The inverse cosine function, $\cos^{-1}$ or arccos, is defined as the inverse of the restricted cosine function, $y = \cos x \quad 0 \le x \le \pi$. Thus,

$$y = \cos^{-1} x \quad \text{and} \quad y = \arccos x$$

are equivalent to

$$x = \cos y \quad -1 \le x \le 1, \quad 0 \le y \le \pi$$

$\cos(\cos^{-1} x) = x \quad -1 \le x \le 1$
$\cos^{-1}(\cos y) = y \quad 0 \le y \le \pi$

The inverse tangent function, $\tan^{-1}$ or arctan, is defined as the inverse of the restricted tangent function $y = \tan x \quad -\frac{\pi}{2} < x < \frac{\pi}{2}$. Thus,

$$y = \tan^{-1} x \quad \text{and} \quad y = \arctan x$$

are equivalent to

$$x = \tan y \quad -\frac{\pi}{2} < y < \frac{\pi}{2}, \ x \text{ a real number}$$

$\tan(\tan^{-1} x) = x \quad x$ a real number
$\tan^{-1}(\tan y) = y \quad -\frac{\pi}{2} < y < \frac{\pi}{2}$

Common Errors:
$\sin^{-1} x \ne \frac{1}{\sin x}$ or $\csc x$. Also note, although $\sin^{-1} 0 = 0$ and $\sin 0 = 0$, in general $\sin^{-1} a$ and $\sin a$ are different numbers, and $\sin^{-1} a$ is not always defined.

1. $y = \cos^{-1} 0$ is equivalent to
$\cos y = 0 \qquad 0 \le y \le \pi$

$$y = \frac{\pi}{2}$$

3. $y = \arcsin \frac{\sqrt{3}}{2}$ is equivalent to

$$\sin y = \frac{\sqrt{3}}{2} \qquad -\frac{\pi}{2} \le y \le \frac{\pi}{2}$$

$$y = \frac{\pi}{3}$$

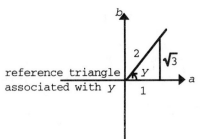

reference triangle associated with y

5. $y = \arctan \sqrt{3}$ is equivalent to
$\tan y = \sqrt{3} \qquad -\frac{\pi}{2} < y < \frac{\pi}{2}$

$$y = \frac{\pi}{3}$$
(see sketch for problem 3)

7. $y = \sin^{-1} \frac{\sqrt{2}}{2}$ is equivalent to

$$\sin y = \frac{\sqrt{2}}{2} \qquad -\frac{\pi}{2} \le y \le \frac{\pi}{2}$$

$$y = \frac{\pi}{4}$$

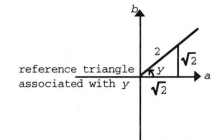

reference triangle associated with y

9. $y = \arccos 1$ is equivalent to
$\cos y = 1 \qquad 0 \le y \le \pi$
$y = 0$

11. $y = \sin^{-1} \frac{1}{2}$ is equivalent to

$$\sin y = \frac{1}{2} \qquad -\frac{\pi}{2} \le y \le \frac{\pi}{2}$$

$$y = \frac{\pi}{6}$$

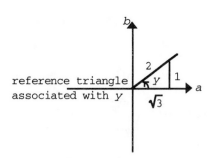

reference triangle associated with y

13. 0.6064   **15.** 1.563   **17.** $\sin^{-1} x$ is not defined if $x > 1$.

19. $y = \arccos\left(-\frac{\sqrt{3}}{2}\right)$ is
equivalent to

$$\cos y = -\frac{\sqrt{3}}{2} \qquad 0 \le y \le \pi$$

$$y = \frac{5\pi}{6}$$

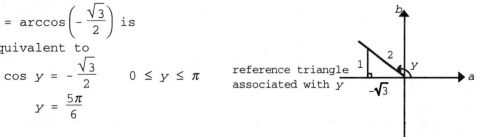

reference triangle associated with y

21. $y = \arctan(-1)$ is equivalent to

$\quad$ $\tan y = -1 \qquad -\dfrac{\pi}{2} < y < \dfrac{\pi}{2}$

$\qquad\quad y = -\dfrac{\pi}{4}$

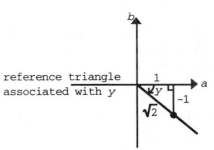

reference triangle
associated with y

23. $y = \sin^{-1}\left(-\dfrac{1}{2}\right)$ is equivalent to

$\quad$ $\sin y = -\dfrac{1}{2} \qquad -\dfrac{\pi}{2} \le y \le \dfrac{\pi}{2}$

$\qquad\quad y = -\dfrac{\pi}{6}$

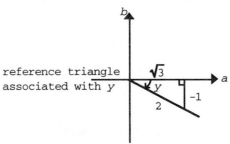

reference triangle
associated with y

25. $\tan(\tan^{-1} 25) = 25$ by the tangent-inverse tangent identity.

27. Let $y = \cos^{-1}\dfrac{\sqrt{3}}{2}$; then $\cos y = \dfrac{\sqrt{3}}{2}$, $0 \le y \le \pi$. Draw the reference triangle

associated with y; then $\sin y = \sin\left[\cos^{-1}\dfrac{\sqrt{3}}{2}\right]$ can be determined directly from
the triangle.

$\quad$ $\cos y = \dfrac{a}{r} = \dfrac{\sqrt{3}}{2} \qquad a = \sqrt{3} \qquad r = 2$

$\qquad (\sqrt{3})^2 + b^2 = 2^2$

$\qquad\qquad\quad b^2 = 1$

$\qquad\qquad\quad\, b = 1$

(positive since y is a quadrant I
or II angle)

Therefore

$\sin y = \sin\left[\cos^{-1}\dfrac{\sqrt{3}}{2}\right] = \dfrac{1}{2}$.

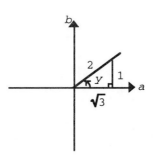

29. $\cot[\cos^{-1}(-0.7003)] = \dfrac{1}{\tan[\cos^{-1}(-0.7003)]} = -0.9810$

31. 2.645

33. $y = \sin^{-1}\left(-\dfrac{\sqrt{2}}{2}\right)$ is equivalent to

$\quad$ $\sin y = -\dfrac{\sqrt{2}}{2} \qquad -\dfrac{\pi}{2} \le y \le \dfrac{\pi}{2}$

$\qquad\quad y = -\dfrac{\pi}{4}$

Using the relation $\theta_{\text{deg}} = \dfrac{180°}{\pi \text{ rad}}\, \theta_{\text{rad}}$, we have

$\sin^{-1}\left(-\dfrac{\sqrt{2}}{2}\right) = \dfrac{180°}{\pi}\left(-\dfrac{\pi}{4}\right) = -45°$.

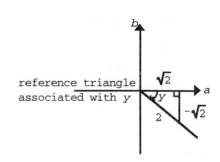

reference triangle
associated with y

35. $y = \arctan(-\sqrt{3}) = \tan^{-1}(-\sqrt{3})$ is equivalent to

$\tan y = -\sqrt{3} \qquad -\dfrac{\pi}{2} < y < \dfrac{\pi}{2}$

$y = -\dfrac{\pi}{3}$

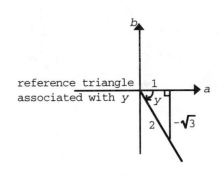

reference triangle associated with y

Using the relation $\theta_{\text{deg}} = \dfrac{180°}{\pi \text{ rad}} \theta_{\text{rad}}$, we have

$\arctan(-\sqrt{3}) = \dfrac{180°}{\pi}\left(-\dfrac{\pi}{3}\right) = -60°.$

37. $y = \cos^{-1}(-1)$ is equivalent to

$\cos y = -1 \qquad 0 \le y \le \pi$

$y = \pi$

Since π rad $= 180°$, we have $\cos^{-1}(-1) = 180°$

Calculator in degree mode for problems 39—42.

39. $43.51°$ **41.** $-21.48°$

43. $\sin^{-1}(\sin 2) = 1.1416 \ne 2$. For the identity $\sin^{-1}(\sin x) = x$ to hold, x must be in the restricted domain of the sine function; that is, $-\dfrac{\pi}{2} \le x \le \dfrac{\pi}{2}$. The number 2 is not in the restricted domain.

45.

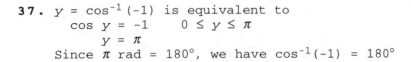

47.

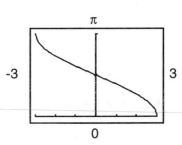

49.

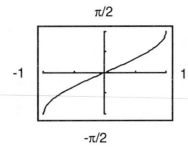

51.

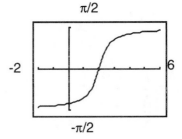

53. (A)

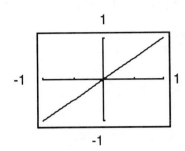

(B) The domain of $\cos^{-1}$ is restricted to $-1 \leq x \leq 1$; hence no graph will appear for other x.

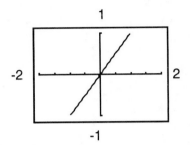

55. $\sin^{-1} x = \cos^{-1} x$
$y = \sin^{-1} x$ is equivalent to
$x = \sin y \qquad -\dfrac{\pi}{2} \leq y \leq \dfrac{\pi}{2}$
$y = \cos^{-1} x$ is equivalent to
$x = \cos y \qquad 0 \leq y \leq \pi$
Hence $\sin^{-1} x = \cos^{-1} x$ is equivalent to
$\sin y = \cos y \qquad 0 \leq y \leq \dfrac{\pi}{2}$
or
$\dfrac{\sin y}{\cos y} = \tan y = 1 \quad 0 \leq y \leq \dfrac{\pi}{2}$
The only possible value of y is $\dfrac{\pi}{4}$.
Then
$\sin y = \cos y = x = \sin \dfrac{\pi}{4} = \dfrac{\sqrt{2}}{2}$.

57. $\sin^{-1}(\sqrt{3}x) = \cos^{-1} x$
$y = \sin^{-1}(\sqrt{3}x)$ is equivalent to
$\sqrt{3}x = \sin y \qquad -\dfrac{\pi}{2} \leq y \leq \dfrac{\pi}{2}$
or
$x = \dfrac{1}{\sqrt{3}} \sin y$
$y = \cos^{-1} x$ is equivalent to
$x = \cos y \qquad 0 \leq y \leq \pi$
Hence $\sin^{-1}(\sqrt{3}x) = \cos^{-1} x$ is equivalent to
$\dfrac{1}{\sqrt{3}} \sin y = \cos y \qquad 0 \leq y \leq \dfrac{\pi}{2}$
or
$\dfrac{\sin y}{\cos y} = \tan y = \sqrt{3} \qquad 0 \leq y \leq \dfrac{\pi}{2}$
The only possible value of y is $\dfrac{\pi}{3}$.
Then
$\dfrac{1}{\sqrt{3}} \sin y = \cos y = x = \cos \dfrac{\pi}{3} = \dfrac{1}{2}$.

59. Let $y = \sin^{-1} x$. Then $x = \sin y$, $-\dfrac{\pi}{2} \leq y \leq \dfrac{\pi}{2}$. If $\sin y = \dfrac{b}{r} = x = \dfrac{x}{1}$, then let
$b = x$, $r = 1$.
$a^2 + b^2 = r^2$
$a^2 + x^2 = 1^2$
$\qquad a^2 = 1 - x^2$
$\qquad a = \sqrt{1 - x^2}$ (a must be positive since y is a I or IV quandrant angle)
$\cos y = \cos(\sin^{-1} x) = \dfrac{a}{r} = \dfrac{\sqrt{1 - x^2}}{1} = \sqrt{1 - x^2}$

61. Let $y = \arctan x$. Then $x = \tan y$, $-\dfrac{\pi}{2} < y < \dfrac{\pi}{2}$. If $\tan y = \dfrac{b}{a} = x = \dfrac{x}{1}$, then let
$b = x$, $a = 1$.
$a^2 + b^2 = r^2$
$1^2 + x^2 = r^2$
$\qquad r = \sqrt{1 + x^2}$ (r is always taken positive)
$\cos y = \cos(\arctan x) = \dfrac{a}{r} = \dfrac{1}{\sqrt{1 + x^2}}$

63. $f(x) = 4 + 2 \cos(x - 3)$. $3 \leq x \leq 3 + \pi$. With this restriction, f is one-to-one (proof omitted). Solve $y = f(x)$ for x:

$$y = 4 + 2 \cos(x - 3) \qquad 3 \leq x \leq 3 + \pi$$
$$y - 4 = 2 \cos(x - 3) \qquad 0 \leq x - 3 \leq \pi$$
$$\frac{y - 4}{2} = \cos(x - 3) \qquad -1 \leq \cos(x - 3) \leq 1$$
$$\cos^{-1} \frac{y - 4}{2} = x - 3 \qquad -2 \leq 2 \cos(x - 3) \leq 2$$
$$x = 3 + \cos^{-1} \frac{y - 4}{2} = f^{-1}(y) \qquad 2 \leq 4 + 2 \cos(x - 3) \leq 6$$

range of f: $2 \leq y \leq 6$
Interchange x and y

$$f^{-1}(x) = 3 + \cos^{-1} \frac{x - 4}{2} \qquad 2 \leq 4 + 2 \cos(y - 3) \leq 6. \text{ Thus range of } f = \text{domain of}$$

f^{-1}: $2 \leq x \leq 6$.

65. (A)

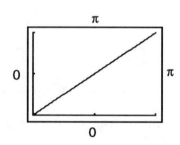

(B) The domain for $\cos x$ is $(-\infty, \infty)$ and the range is $[-1, 1]$, which is the domain for $\cos^{-1} x$. Thus, $y = \cos^{-1}(\cos x)$ has a graph over the interval $(-\infty, \infty)$, but $\cos^{-1}(\cos x) = x$ only on the restricted domain of $\cos x$, $[0, \pi]$.

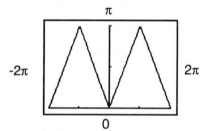

67. For a 28mm lens, $x = 28$, thus $\theta = 2 \tan^{-1} \frac{21.634}{28} = 1.31567$ radians. In decimal degrees, $\theta_D = \frac{180°}{\pi} (1.31567 \text{ rad}) = 75.38°$.

For a 100 mm lens, $x = 100$, thus $\theta = 2 \tan^{-1} \frac{21.634}{100} = 0.42611$ radians. In decimal degrees, $\theta_D = \frac{180°}{\pi} (0.42611 \text{ rad}) = 24.41°$.

69. (A)

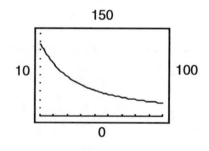

(B) 59.44 mm

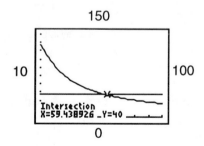

71.

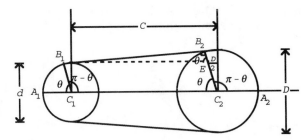

From the above figure, the following should be clear:

Length of belt = $2[arc \overset{\frown}{A_1B_1} + B_1B_2 + arc \overset{\frown}{B_2A_2}]$

$arc\ A_1B_1 = \dfrac{d}{2}(\theta)$

$arc\ B_2A_2 = \dfrac{D}{2}(\pi - \theta)$

To find B_1B_2 we note:

C_1C_2 has length C

B_1E is constructed parallel to C_1C_2. EC_2 is parallel to B_1C_1. Hence $EB_1C_1C_2$ is a parallelogram. EB_1 has length C. EB_1B_2 is a right triangle. Thus

(1) $\cos\theta = \dfrac{B_2E}{EB_1} = \dfrac{\frac{D}{2} - \frac{d}{2}}{C} = \dfrac{D - d}{2C}$

(2) $\sin\theta = \dfrac{B_1B_2}{EB_1}$, so $B_1B_2 = EB_1 \sin\theta = C \sin\theta$

Finally,

Length of belt

$L = 2[arc \overset{\frown}{A_1B_1} + B_1B_2 + arc \overset{\frown}{B_2A_2}]$

$\quad = 2\left[\dfrac{d}{2}\theta + C\sin\theta + \dfrac{D}{2}(\pi - \theta)\right]$

$\quad = d\theta + 2C\sin\theta + D(\pi - \theta)$

$L = \pi D + (d - D)\theta + 2C\sin\theta$

and (from (1) above)

$\theta = \cos^{-1}\dfrac{D - d}{2C}$

Substituting the given values, we have

$D = 4,\ d = 2,\ C = 6$ (calculator in radian mode)

$\theta = \cos^{-1}\dfrac{4 - 2}{2\cdot 6} = \cos^{-1}\dfrac{1}{6}$

$L = 4\pi + (2 - 4)\cos^{-1}\dfrac{1}{6} + 2\cdot 6\ \sin\left(\cos^{-1}\dfrac{1}{6}\right)$

$\quad \approx 21.59$ inches

73. (A)   (B) 7.22 inches

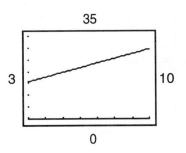

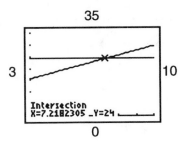

75. (A) Following the hint, we draw AC. Then, since the central angle in a circle subtended by an arc is twice any inscribed angle subtended by the same arc, angle ACB has measure 2θ. Thus, $d = r \cdot 2\theta = 2r\theta$

In triangle ECP, $\tan \theta = \dfrac{x}{r}$, hence $\theta = \tan^{-1} \dfrac{x}{r}$

$d = 2r \tan^{-1} \dfrac{x}{r}$

(B) Substituting the given values, we have
$r = 100 \quad x = 40$

$d = 2 \cdot 100 \tan^{-1} \dfrac{40}{100}$

$\quad = 200 \tan^{-1} \dfrac{40}{100}$

$\quad = 200 \tan^{-1}(0.4)$ (calculator in radian mode)

$\quad = 76.10$ feet

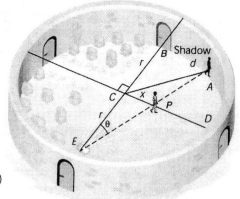

CHAPTER 5 REVIEW

1. $\theta = \dfrac{s}{r} = \dfrac{15 \text{ centimeters}}{6 \text{ centimeters}} = \dfrac{15}{6} = 2.5$ radians ⁤ (5-3)

2. $s = r\theta = (3 \text{ centimeters})(2.5 \text{ radians}) = 7.5$ centimeters ⁤ (5-3)

3. Solve for α:
$\alpha = 90° - 35.2° = 54.8°$

Solve for a:

$\cos \beta = \dfrac{a}{c}$

$\cos 35.2° = \dfrac{a}{20.2}$

$\qquad a = 20.2 \cos 35.2°$

$\qquad = 16.5$ ft

Solve for b:

$\sin \beta = \dfrac{b}{c}$

$\sin 35.2° = \dfrac{b}{20.2}$

$\qquad b = 20.2 \sin 35.2°$

$\qquad = 11.6$ ft ⁤ (5-5)

4. (A)

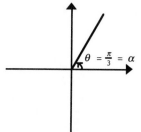

$\alpha = \dfrac{\pi}{3}$

(B)

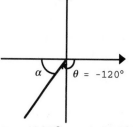

$\alpha = 180° + (-120°) = 60°$

(C)

$\alpha = \left| 2\pi - \dfrac{13\pi}{6} \right| = \dfrac{\pi}{6}$

(D)

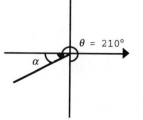

$\alpha = 210° - 180° = 30°$ ⁤ (5-4)

5. (A) $\dfrac{b}{r} = \sin \theta < 0$ if $b < 0$. This occurs in quadrants III, IV.

(B) $\dfrac{a}{r} = \cos \theta < 0$ if $a < 0$. This occurs in quadrants II, III.

(C) $\dfrac{b}{a} = \tan \theta < 0$ if a and b have opposite signs. This occurs in quadrants II, IV.

(5-2, 5-4)

6.
$$a^2 + b^2 = r^2$$
$$4^2 + (-3)^2 = r^2$$
$$25 = r^2$$
$$r = 5$$
$$\sin \theta = \frac{b}{r} = \frac{-3}{5} = -\frac{3}{5}$$
$$\sec \theta = \frac{r}{a} = \frac{5}{4}$$
$$\cot \theta = \frac{a}{b} = \frac{4}{-3} = -\frac{4}{3}$$

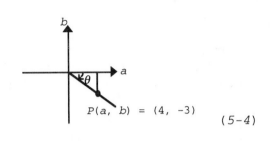

$P(a, b) = (4, -3)$

(5-4)

7.

$\theta°$	θ rad	$\sin \theta$	$\cos \theta$	$\tan \theta$	$\csc \theta$	$\sec \theta$	$\cot \theta$
0°	0	0	1	0	ND*	1	ND
30°	$\pi/6$	1/2	$\sqrt{3}/2$	$1/\sqrt{3}$	2	$2/\sqrt{3}$	$\sqrt{3}$
45°	$\pi/4$	$1/\sqrt{2}$	$1/\sqrt{2}$	1	$\sqrt{2}$	$\sqrt{2}$	1
60°	$\pi/3$	$\sqrt{3}/2$	1/2	$\sqrt{3}$	$2/\sqrt{3}$	2	$1/\sqrt{3}$
90°	$\pi/2$	1	0	ND	1	ND	0
180°	π	0	-1	0	ND	-1	ND
270°	$3\pi/2$	-1	0	ND	-1	ND	0
360°	2π	0	1	0	ND	1	ND

*ND = not defined

(5-2, 5-4)

8. (A) 2π (B) 2π (C) π (5-6)

9. (A) Domain = all real numbers, Range = $[-1, 1]$

(B) Domain is set of all real numbers except $x = \dfrac{2k + 1}{2}\pi$, k an integer, Range = all real numbers

(5-6)

10.

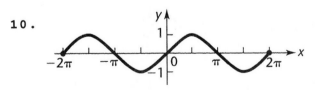

(5-6)

11.

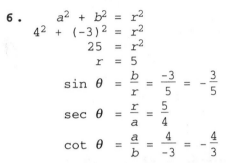

(5-6)

12. The central angle in a circle subtended by an arc of half the length of the radius.

(5-3)

13. If the graph of $y = \sin x$ is shifted $\dfrac{\pi}{2}$ units to the left, the result will be the graph of $y = \cos x$.

(5-6, 5-7)

14. $\theta_D = \dfrac{180°}{\pi}\theta_R = \dfrac{180°}{\pi}(1.37) = 78.50°$ (5-3)

15. Solve for β:

$\tan \beta = \dfrac{b}{a}$

$\tan \beta = \dfrac{13.3}{15.7}$

$\beta = \tan^{-1} \dfrac{13.3}{15.7}$

$= 40.3°$

Solve for α:

$\alpha = 90° - 40.3° = 49.7°$

Solve for c:

$\sec \beta = \dfrac{c}{a}$

$\sec 40.3° = \dfrac{c}{15.7}$

$c = 15.7 \sec 40.3°$

$= 20.6 \text{ cm}$ (5-5)

16. (A) Since $-270° < -210° < -180°$, this is a II quadrant angle.

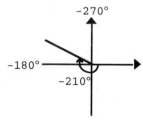

(C) Since $3.14 < 4.2 < 4.71$, this is a III quadrant angle.

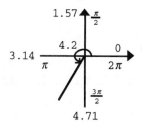

(B) Since $\dfrac{5\pi}{2}$ is coterminal with $\dfrac{\pi}{2}$, this is a quadrantal angle. (5-3)

17. (A) Since $-240° + 360° = 120°$, this is coterminal with $120°$.

(B) Since $-\dfrac{7\pi}{6} + 2\pi = \dfrac{5\pi}{6}$, which is equivalent to $150°$, this is not coterminal with $120°$.

(C) Since $840° - 2(360°) = 120°$, this is coterminal with $120°$. (5-3)

18. (B) and (C), since 3 radians is equivalent to the real number 3, and cosine is periodic with period 2π. (A) is not the same as cos 3, since $3°$ is equivalent to $\dfrac{\pi}{180°}3°$, not 3. (5-4)

19. (A) $\dfrac{b}{a} = \tan x$ is not defined if $a = 0$. This occurs if $\theta = \dfrac{\pi}{2}, \dfrac{3\pi}{2}$.

(B) $\dfrac{a}{b} = \cot x$ is not defined if $b = 0$. This occurs if $\theta = 0, \pi$.

(C) $\dfrac{r}{b} = \csc x$ is not defined if $b = 0$. This occurs if $\theta = 0, \pi$. (5-3, 5-4)

20. Since the coordinates of a point on a unit circle are given by $P(a, b) = P(\cos x, \sin x)$, we evaluate $P(\cos(-8.305), \sin(-8.305))$--using a calculator set in radian mode--to obtain $P(-0.436, -0.900)$. Note that $x = -8.305$, since P is moving clockwise. The quadrant in which $P(a, b)$ lies can be determined by the signs of a and b. In this case P is in the third quadrant, since a is negative and b is negative. (5-1, 5-2)

21.

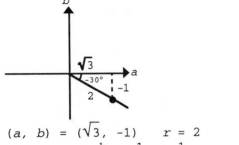

$(a,\ b) = (\sqrt{3},\ -1) \qquad r = 2$

$\sin(-30°) = \dfrac{b}{r} = \dfrac{-1}{2} = -\dfrac{1}{2}$ $\qquad$ (5-4)

22.

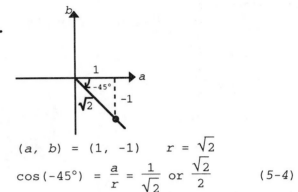

$(a,\ b) = (1,\ -1) \qquad r = \sqrt{2}$

$\cos(-45°) = \dfrac{a}{r} = \dfrac{1}{\sqrt{2}}$ or $\dfrac{\sqrt{2}}{2}$ $\qquad$ (5-4)

23.

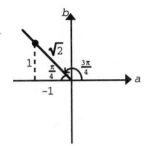

$(a,\ b) = (-1,\ 1) \qquad r = \sqrt{2}$

$\tan\dfrac{3\pi}{4} = \dfrac{b}{a} = \dfrac{1}{-1} = -1$ $\qquad$ (5-4)

24.

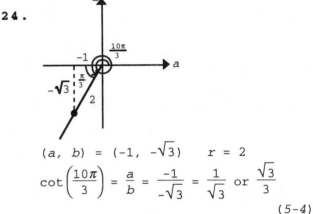

$(a,\ b) = (-1,\ -\sqrt{3}) \qquad r = 2$

$\cot\left(\dfrac{10\pi}{3}\right) = \dfrac{a}{b} = \dfrac{-1}{-\sqrt{3}} = \dfrac{1}{\sqrt{3}}$ or $\dfrac{\sqrt{3}}{3}$

$\qquad\qquad\qquad$ (5-4)

25. $y = \arccos 0$ is equivalent to
$\cos y = 0 \qquad 0 \le y \le \pi$

$\qquad y = \dfrac{\pi}{2}$ $\qquad$ (5-4, 5-9)

26. $y = \sin^{-1}(-1)$ is equivalent to
$\sin y = -1 \qquad -\dfrac{\pi}{2} \le y \le \dfrac{\pi}{2}$

$\qquad y = -\dfrac{\pi}{2}$ $\qquad$ (5-4, 5-9)

27. $y = \arctan 1$ is equivalent to
$\tan y = 1 \qquad -\dfrac{\pi}{2} < y < \dfrac{\pi}{2}$

$\qquad y = \dfrac{\pi}{4}$

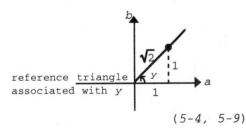

$\qquad\qquad\qquad$ (5-4, 5-9)

28. $y = \tan^{-1}(-\sqrt{3})$ is equivalent to
$\tan y = -\sqrt{3} \qquad -\dfrac{\pi}{2} < y < \dfrac{\pi}{2}$

$\qquad y = -\dfrac{\pi}{3}$

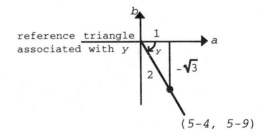

$\qquad\qquad\qquad$ (5-4, 5-9)

29. $\cos^{-1} x$ is not defined if $x = \sqrt{2}$ $\qquad$ (5-4, 5-9)

30. $y = \arccos\left(-\dfrac{1}{2}\right)$ is equivalent to

$\cos y = -\dfrac{1}{2} \quad 0 \le y \le \pi$

$y = \dfrac{2\pi}{3}$

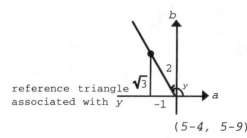

reference triangle associated with y

$(5-4, \ 5-9)$

31. $y = \arcsin\left(\dfrac{\sqrt{3}}{2}\right)$ is equivalent to

$\sin y = \dfrac{\sqrt{3}}{2} \quad -\dfrac{\pi}{2} \le y \le \dfrac{\pi}{2}$

$y = \dfrac{\pi}{3}$

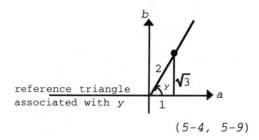

reference triangle associated with y

$(5-4, \ 5-9)$

32. $\sin^{-1} x$ is not defined if $x = -\sqrt{3}$. $(5-4, \ 5-9)$

33. $\sin\left(\sin^{-1}\dfrac{2}{3}\right) = \dfrac{2}{3}$ by the sine-inverse sine identity. $(5-4, \ 5-9)$

34. To find $\cos^{-1}\left(\cos\dfrac{7\pi}{6}\right)$, first

determine $\cos\dfrac{7\pi}{6}$.

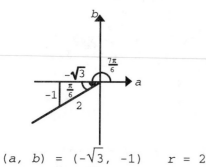

$(a, \ b) = (-\sqrt{3}, \ -1) \quad r = 2$

$\cos\dfrac{7\pi}{6} = -\dfrac{\sqrt{3}}{2} \quad 0 \le y \le \pi$

Then $\cos^{-1}\left(\cos\dfrac{7\pi}{6}\right) = \cos^{-1}\left(-\dfrac{\sqrt{3}}{2}\right) = y$ is

equivalent to

$\cos y = \dfrac{a}{r} = -\dfrac{\sqrt{3}}{2} \quad 0 \le y \le \pi$

$y = \dfrac{5\pi}{6}$

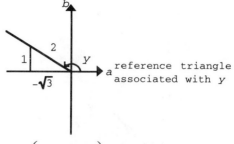

reference triangle associated with y

$\cos\left(\cos^{-1}\dfrac{7\pi}{6}\right) = \dfrac{5\pi}{6}$

> **Common Error:**
> $\cos^{-1}\left(\cos\dfrac{7\pi}{6}\right) \ne \cos\dfrac{7\pi}{6}$
> $\dfrac{7\pi}{6}$ is not in the range of the inverse cosine function.

$(5-4, \ 5-9)$

35. Let $y = \arctan 2$, then $\tan y = 2$, $-\dfrac{\pi}{2} < y < \dfrac{\pi}{2}$. Draw the reference triangle associated with y, then $\sec y = \sec(\arctan 2)$ can be determined directly from the triangle.

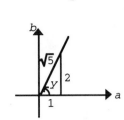

$$\tan y = \frac{b}{a} = \frac{2}{1} \quad b = 2 \quad a = 1$$
$$r^2 = 1^2 + 2^2$$
$$r^2 = 5$$
$$r = \sqrt{5}$$
$$\sec(\arctan 2) = \sec y = \frac{r}{a} = \frac{\sqrt{5}}{1}$$
$$= \sqrt{5} \qquad (5\text{-}4, \ 5\text{-}9)$$

36. Let $y = \arccos \dfrac{3}{4}$, then $\cos y = \dfrac{3}{4}$,

$0 \le y \le \pi$. Draw the reference triangle associated with y, then

$\sin y = \sin\left(\arccos \dfrac{3}{4}\right)$ can be determined directly from the triangle.

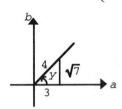

$$\cos y = \frac{3}{4} = \frac{a}{r} \qquad a = 3 \quad r = 4$$
$$4^2 = 3^2 + b^2$$
$$16 = 9 + b^2$$
$$b^2 = 7$$
$$b = \sqrt{7}$$
$$\sin\left(\arccos \frac{3}{4}\right) = \sin y = \frac{b}{r} = \frac{\sqrt{7}}{4} \qquad (5\text{-}4, \ 5\text{-}9)$$

37. 0.4431 $(5\text{-}4)$ **38.** $\tan\left(93 + \dfrac{46}{60} + \dfrac{17}{3600}\right)^\circ = -15.17$ $(5\text{-}4)$

39. $\sec(-2.073) = \dfrac{1}{\cos(-2.073)} = -2.077$ $(5\text{-}2)$ **40.** -0.9750 $(5\text{-}5, \ 5\text{-}9)$

41. $\arccos x$ is not defined if $x < -1$ $(5\text{-}5, \ 5\text{-}9)$ **42.** 1.557 $(5\text{-}5, \ 5\text{-}9)$ **43.** 1.095 $(5\text{-}9)$

44. Since $\tan 1.345 = 4.353 > 1$, $\sin^{-1}(\tan 1.345)$ is not defined. $(5\text{-}9)$

45. (A) $\theta = \arcsin^{-1}\left(-\dfrac{1}{2}\right)$ is equivalent to

$$\sin \theta = -\frac{1}{2} \qquad -\frac{\pi}{2} \le \theta \le \frac{\pi}{2}$$
$$\theta = -\frac{\pi}{6}$$

Using the relation $\theta_D = \dfrac{180^\circ}{\pi \ \text{rad}} \theta_R$, we have

$$\sin^{-1}\left(-\frac{1}{2}\right) = \frac{180^\circ}{\pi} \cdot \left(-\frac{\pi}{6}\right) = -30^\circ$$

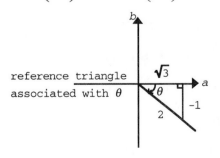

reference triangle associated with θ

(B) $\theta = \arccos\left(-\dfrac{1}{2}\right)$ is equivalent to

$$\cos \theta = -\frac{1}{2} \qquad 0 \le \theta \le \pi$$
$$\theta = \frac{2\pi}{3}$$

Using the relation $\theta_D = \dfrac{180^\circ}{\pi \ \text{rad}} \theta_R$, we have

$$\arccos\left(-\frac{1}{2}\right) = \frac{180^\circ}{\pi} \cdot \frac{2\pi}{3} = 120^\circ.$$

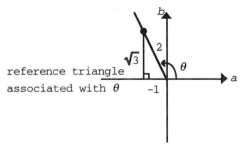

reference triangle associated with θ

$(5\text{-}9)$

46. Calculator in degree mode: (A) Θ = 151.20°   (B) Θ = 82.28°   (5-9)

47. $\cos^{-1}[\cos(-2)] = 2$ For the identity $\cos^{-1}(\cos x) = x$ to hold, x must be in the restricted domain of the cosine function; that is, $0 \le x \le \pi$. The number -2 is not in the restricted domain.   (5-9)

48. Amplitude = $|-2|$ = 2

Period = $\dfrac{2\pi}{\pi}$ = 2

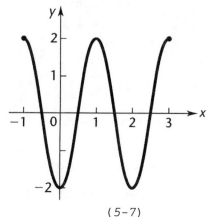

(5-7)

49. $y = -2 + 3\sin\left(\dfrac{x}{2}\right)$

For the graph of $y = 3\sin\dfrac{x}{2}$, we note: $A = 3$, $P = 2\pi \div \dfrac{1}{2} = 4\pi$, phase shift = 0. We graph $y = 3\sin\dfrac{x}{2}$, then vertically translate the graph down 2 units.

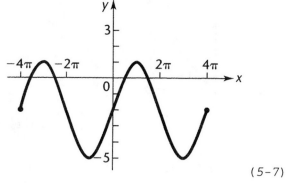

(5-7)

50. $A = 6$ $P = \pi = 2\pi \div 2$.
Hence $B = 2$ $y = 6\cos 2x$;
$-\dfrac{\pi}{2} \le x \le \pi$
Check:

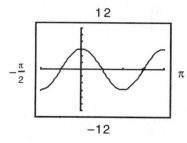

(5-7)

51. $|A| = 0.5$ $P = 2 = 2\pi \div \pi$. Hence $B = \pi$
$A = -0.5$, since the graph has the form of the standard sine curve turned upside down.
$y = -0.5\sin \pi x$; $-1 \le x \le 2$
Check:

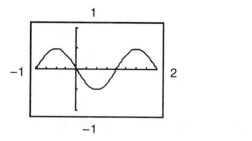

(5-7)

52. If the graph of $y = \tan x$ is shifted $\dfrac{\pi}{2}$ units to the right and reflected in the x axis, the result will be the graph of $y = \cot x$.   (5-6, 5-7)

53. (A) $\sin(-x)\cot(-x) = \sin(-x)\dfrac{\cos(-x)}{\sin(-x)}$ Quotient Identity

$\qquad\qquad\qquad = \cos(-x)$ Algebra
$\qquad\qquad\qquad = \cos x$ Identities for negatives

(B) $\dfrac{\sin^2 x}{1 - \sin^2 x} = \dfrac{\sin^2 x}{\cos^2 x}$ Pythagorean Identity

$\qquad\qquad = \left(\dfrac{\sin x}{\cos x}\right)^2$ Algebra

$\qquad\qquad = \tan^2 x$ Quotient Identity   (5-2)

54. $y = 3 \sin\left(\dfrac{x}{2} + \dfrac{\pi}{2}\right)$

$A = 3$

Solve $\dfrac{x}{2} + \dfrac{\pi}{2} = 0 \qquad \dfrac{x}{2} + \dfrac{\pi}{2} = 2\pi$

$\qquad\qquad \dfrac{x}{2} = -\dfrac{\pi}{2} \qquad\quad \dfrac{x}{2} = -\dfrac{\pi}{2} + 2\pi$

$\qquad\qquad x = -\pi \qquad\qquad x = -\pi + 4\pi$

$\qquad\qquad\quad \uparrow \underline{\hspace{3cm}} \uparrow \quad \uparrow$

$\qquad\qquad\qquad$ Phase shift $\qquad$ Period

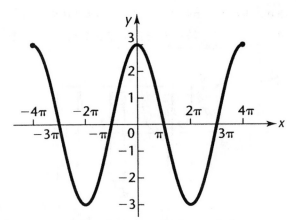

The graph completes one full cycle as x varies over the interval $[-\pi, 3\pi]$.

$(5\text{-}7)$

55. $y = -2 \cos\left(\dfrac{\pi}{2}x - \dfrac{\pi}{4}\right)$

amplitude $= |A| = |-2| = 2$

Solve $\dfrac{\pi}{2}x - \dfrac{\pi}{4} = 0 \qquad \dfrac{\pi}{2}x - \dfrac{\pi}{4} = 2\pi$

$\qquad\quad \dfrac{\pi}{2}x = \dfrac{\pi}{4} \qquad\qquad \dfrac{\pi}{2}x = \dfrac{\pi}{4} + 2\pi$

$\qquad\quad x = \dfrac{\pi}{4} \div \dfrac{\pi}{2} \qquad\quad x = \left(\dfrac{\pi}{4} + 2\pi\right) \div \dfrac{\pi}{2}$

$\qquad\quad x = \dfrac{1}{2} \qquad\qquad\quad x = \dfrac{1}{2} + 4$

$\qquad\quad\ \uparrow \underline{\hspace{3cm}} \uparrow \quad \uparrow$

$\qquad\qquad$ Phase shift $\qquad$ Period $= 4 \qquad\qquad (5\text{-}7)$

56. Domain $= [-1, 1]$

Range $= [0, \pi]$

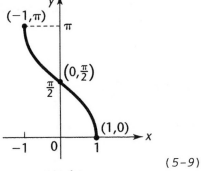

$(5\text{-}9)$

57. Here is the graph of $y = \dfrac{1}{1 + \tan^2 x}$ in a graphing utility.

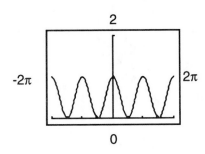

The graph has amplitude $\dfrac{1 - 0}{2} = \dfrac{1}{2}$ and period π. It appears to be the graph of $y = \dfrac{1}{2} \cos 2x$ shifted up $\dfrac{1}{2}$ unit, that is, $y = \dfrac{1}{2} \cos 2x + \dfrac{1}{2}$. $\qquad (5\text{-}7)$

58. (A) Here is the graph of
$y = \dfrac{2\sin^2 x}{\sin 2x}$ in a graphing utility.

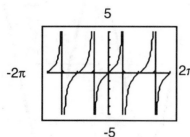

The graph appears to have the form
$y = A \tan Bx$. Since the period is π,
$B = 1$. The graph of $y = A \tan x$
shown appears to pass through
$\left(\dfrac{\pi}{4},\ 1\right)$, thus

$$1 = A \tan \dfrac{\pi}{4}$$
$$1 = A$$

The equation of the graph can be
written $y = \tan x$.

(B) Here is the graph of
$y = \dfrac{2\cos^2 x}{\sin 2x}$ in a graphing utility.

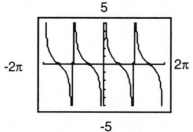

The graph appears to have the form
$y = A \cot Bx$. Since the period is π,
$B = 1$. The graph of $y = A \cot x$
shown appears to pass through
$\left(\dfrac{\pi}{4},\ 1\right)$, thus

$$1 = A \cot \dfrac{\pi}{4}$$
$$1 = A$$

The equation of the graph can be
written $y = \cot x$.   (5-7)

59. (A) Since $\theta_R = \dfrac{s}{r}$ and r = radius of circle = distance of A from center = 8,

we have $\theta_R = \dfrac{20}{8} = 2.5$ radians.

(B) Since $\sin \theta = \dfrac{b}{r}$ and $\cos \theta = \dfrac{a}{r}$, we can write

$$a = r \cos \theta = 8 \cos 2.5 = -6.41 \text{ (calculator in radian mode)}$$
$$b = r \sin \theta = 8 \sin 2.5 = 4.79$$
$$(a,\ b) = (-6.41,\ 4.79)$$

                                                                        (5-1, 5-3)

60. (A) $\cos x = \dfrac{a}{r} = -\dfrac{1}{2} = \dfrac{-1}{2}$. a is

negative in the II and III quadrants.
The least positive x is associated

with a $\dfrac{\pi}{3}$ reference triangle in the II

quadrant as drawn: $x = \dfrac{2\pi}{3}$

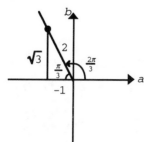

(B) $\csc x = -\sqrt{2} = \dfrac{\sqrt{2}}{-1} = \dfrac{r}{b}$,

b is negative in the III and IV
quadrants. The least positive x is

associated with a $\dfrac{\pi}{4}$ reference

triangle in the III quadrant as

drawn: $x = \dfrac{5\pi}{4}$

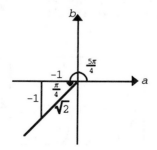

                                                                        (5-2)

61. The dashed curve is the graph of $y = \cos x$. The solid curve is the required graph of $y = \sec x$, $-\dfrac{\pi}{2} < x < \dfrac{3\pi}{2}$

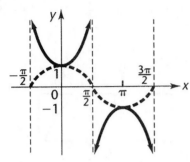

$(5\text{-}6)$

62.

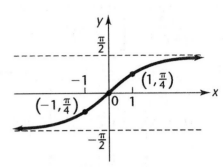

Domain = all real numbers

Range = $\left(-\dfrac{\pi}{2}, \dfrac{\pi}{2}\right)$

$(5\text{-}9)$

63. $y = -5 \tan\left(\pi x + \dfrac{\pi}{2}\right)$

Solve $\pi x + \dfrac{\pi}{2} = 0 \qquad \pi x + \dfrac{\pi}{2} = \pi$

$\qquad\qquad \pi x = -\dfrac{\pi}{2} \qquad\quad \pi x = -\dfrac{\pi}{2} + \pi$

$\qquad\qquad x = -\dfrac{1}{2} \qquad\qquad x = -\dfrac{1}{2} + 1$

Phase Shift = $-\dfrac{1}{2}$ \qquad Period $P = 1$ \quad $(5\text{-}8)$

64. One cycle of $y = 3 \csc\left(\dfrac{x}{2} - \dfrac{\pi}{4}\right)$ is completed as $\dfrac{x}{2} - \dfrac{\pi}{4}$ varies from 0 to 2π. Solve each equation for x:

$\dfrac{x}{2} - \dfrac{\pi}{4} = 0 \qquad \dfrac{x}{2} - \dfrac{\pi}{4} = 2\pi$

$\qquad \dfrac{x}{2} = \dfrac{\pi}{4} \qquad\qquad \dfrac{x}{2} = \dfrac{\pi}{4} + 2\pi$

$\qquad x = \dfrac{\pi}{2} \qquad\qquad x = \dfrac{\pi}{2} + 4\pi$

Phase shift = $\dfrac{\pi}{2}$ \qquad Period = 4π \quad $(5\text{-}8)$

65. From the figure, it should be clear that $\cos(-x) = \cos x$ (P and P_1 have the same x-coordinate)
$\sin(-x) = -\sin x$ (P and P_1 have opposite y-coordinates)
Therefore $\tan(-x) = \dfrac{\sin(-x)}{\cos(-x)} = \dfrac{-\sin x}{\cos x}$
$\qquad\qquad\qquad = -\tan x$

It follows that the graph of
(A) sine has origin symmetry
(B) cosine has y axis symmetry
(C) tangent has origin symmetry

$(5\text{-}6)$

66. Let $y = \sin^{-1} x$. Then $x = \sin y$, $-\dfrac{\pi}{2} \le y \le \dfrac{\pi}{2}$. If $\sin y = \dfrac{b}{r} = \dfrac{x}{1}$, then let

$b = x, \; r = 1$
$a^2 + x^2 = 1^2$
$\qquad a^2 = 1 - x^2$
$\qquad\; a = \sqrt{1 - x^2}$

We choose the positive sign because y is a I or IV quadrant angle.

$\sec y = \sec(\sin^{-1} x) = \dfrac{r}{a} = \dfrac{1}{\sqrt{1 - x^2}}$

$(5\text{-}9)$

67. For each case, the number is not in the domain of the function and an error message of some type will appear.

$(5\text{-}2, \; 5\text{-}9)$

68. $|A| = 2$. Hence $A = 2$ or -2. The graph completes one full cycle as x varies over the interval $\left[-\frac{5}{4}, \frac{3}{4}\right]$. Since $-\frac{C}{B}$ is required between -1 and 0, we cannot simply set $-\frac{C}{B} = -\frac{5}{4}$. We must (mentally) extend the curve so that the (extended) graph is that of a standard sine curve that completes one full cycle as x varies over the interval $\left[-\frac{1}{4}, \frac{7}{4}\right]$. Hence

$A = 2$

$-\frac{C}{B} = -\frac{1}{4}$ $\qquad -\frac{C}{B} + \frac{2\pi}{B} = \frac{7}{4}$

$C = \frac{B}{4}$ $\qquad \frac{2\pi}{B} = 2$ $\qquad B = \frac{2\pi}{2} = \pi$ $\qquad C = \frac{\pi}{4}$

The equation is then
$y = A \sin(Bx + C)$

$y = 2 \sin\left(\pi x + \frac{\pi}{4}\right)$

Check:

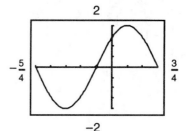

(5-7)

69. True. α and β are complementary if their sum is 90°. If $\alpha + \beta = 90°$, then $2\alpha + 2\beta = 180°$, hence the sum of 2α and 2β is 180°, and they are supplementary. (5-3)

70. True. If α and β are two angles of a triangle with $\alpha < \beta$, then either:
1. $\beta = 90°$ and α is acute. Then $\sin \alpha < 1 = \sin \beta$.

or 2. α and β are both acute. Then $\sin \alpha = \dfrac{\text{side opposite } \alpha}{\text{hypotenuse}}$ and

$\sin \beta = \dfrac{\text{side opposite } \beta}{\text{hypotenuse}}$. But the side opposite a larger angle is always greater than the side opposite a smaller angle, hence $\sin \alpha < \sin \beta$. (5-4)

71. False. For example, let $\alpha = \frac{\pi}{2}$ and $\beta = \frac{3\pi}{4}$. Then $0 < \alpha < \beta < \pi$

but $1 = \sin \alpha > \sin \beta = \dfrac{1}{\sqrt{2}}$. (5-4)

72. False. For example, let $f(x) = \sin x$ and $g(x) = \cos x$. Then f and g are periodic functions with fundamental period 2π. $\dfrac{f}{g}(x) = \dfrac{\sin x}{\cos x} = \tan x$, however, is periodic with fundamental period π. (5-6)

73. True. (5-6)

74. False. None of the inverses is periodic since all are one-to-one functions. Any of the six inverse functions is a counterexample. (5-9)

75. False. For example let $f(x) = \sin(x + 1)$. $A = 1$, $B = 1$, $C = 1$.

The function g defined by $g(x) = f\left(x - \frac{C}{B}\right) = f(x - 1) = \sin x$ has a graph that is shifted 1 unit to the right from the graph of f. The two graphs are different. (5-7)

76. True. f has period $\frac{2\pi}{B}$, that is $f\left(x - \frac{2\pi}{B}\right) = f(x)$ for all x; therefore,

$g(x) = f\left(x - \frac{2\pi}{B}\right) = f(x)$ and f and g have the same graph. (5-7)

77. The wrapping function is a periodic function since $W(x + 2\pi) = W(x)$ for all x, that is, the same point is determined by both values of the variable. *(5-1)*

78.

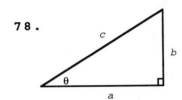

The area of the triangle is given by $A = \frac{1}{2} ab$. If $c = 1$, we can write

$$A = \frac{1}{2} ab$$
$$= \frac{1}{2} \frac{a}{1} \frac{b}{1}$$
$$= \frac{1}{2} \frac{a}{c} \frac{b}{c}$$

But $\frac{a}{c} = \cos \theta$ and $\frac{b}{c} = \sin \theta$. Hence, $A = \frac{1}{2} \cos \theta \sin \theta$. *(5-5)*

79. The area of the triangle is given by $A = \frac{1}{2} ab$. If $a = 1$, we can write

$$A = \frac{1}{2} ab$$
$$= \frac{1}{2} \cdot 1 \cdot \frac{b}{1}$$
$$= \frac{1}{2} \frac{b}{a}$$

But $\frac{b}{a} = \tan \theta$. Hence $A = \frac{1}{2} \tan \theta$. *(5-5)*

80. Draw a figure (the case $n = 8$ is shown, but the reasoning is independent of n).

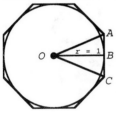

We note: Area of polygon = n (Area of triangle OAC)
$$= 2n \text{ (Area of triangle } OAB)$$
$$= 2n \left(\frac{1}{2} \cdot AB \cdot OB \right)$$
$$= n \cdot AB \cdot OB$$

In triangle OAB, angle $AOB = \frac{1}{2}$ (angle AOC) $= \frac{1}{2} \cdot \frac{1}{n}$ (a full circle)
$$= \frac{1}{2} \cdot \frac{1}{n} (360°) = \frac{180°}{n}$$

$$OB = r = 1 \qquad \frac{AB}{OB} = \frac{AB}{1} = \tan AOB = \tan \frac{180°}{n}$$

Hence,

Area of polygon $= n \cdot \tan \frac{180°}{n} \cdot 1$
$$= n \tan \frac{180°}{n}$$
(5-3, 5-5)

81. Draw a figure (the case $n = 8$ is shown, but the reasoning is independent of n).

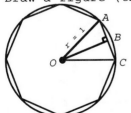

We note: Area of polygon = n (Area of triangle OAC)
$$= 2n \text{ (Area of triangle } OAB)$$
$$= 2n \left(\frac{1}{2} \cdot OB \cdot AB \right)$$
$$= n \cdot OB \cdot AB$$

In triangle OAB, angle $AOB = \frac{1}{2}$ (angle AOC) $= \frac{1}{2}\left[\frac{1}{n} \text{ (full circle)}\right]$

$$= \frac{1}{2}\left[\frac{1}{n} \ (360°)\right] = \frac{180°}{n}$$

$OB = \frac{OB}{1} = \cos AOB = \cos \frac{180°}{n}$ $AB = \frac{AB}{1} = \sin AOB = \sin \frac{180°}{n}$

Hence,

Area of polygon $= n \cos \frac{180°}{n} \sin \frac{180°}{n}$ (5-3, 5-5)

82. Draw a figure (the case $n = 8$ is shown, but the reasoning is independent of n).

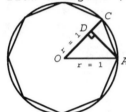

We note: Area of polygon $= n$ (Area of triangle OAC)

$$= n \left(\frac{1}{2} AD \cdot OC\right)$$

$$= \frac{1}{2} n \cdot AD \cdot OC$$

In triangle OAC, angle $AOC = \frac{1}{n}$ (full circle) $= \frac{1}{n}$ (360°) $= \frac{360°}{n}$

$AD = \frac{AD}{1} = \sin AOC = \sin \frac{360°}{n}$ $OC = r = 1$

Hence,

Area of polygon $= \frac{1}{2} n \sin \frac{360°}{n} \cdot 1 = \frac{1}{2} \sin \frac{360°}{n}$ (5-3, 5-5)

83. Here is the graph of $y = 1.2 \sin 2x + 1.6 \cos 2x$ in a graphing utility.

It appears that this is a sine curve shifted to the left, with $A = 2$ and, since $P = \frac{2\pi}{B}$ and P appears to be π, $B = \frac{2\pi}{P} = \frac{2\pi}{\pi} = 2$. From the graphing utility, we find that the x intercept closest to the origin, to three decimal places, is -0.464. To find C, substitute $B = 2$ and $x = -0.464$ into the phase-shift formula $x = -\frac{C}{B}$ and solve for C.

$$x = -\frac{C}{B}$$
$$-0.464 = -\frac{C}{2}$$
$$C = 0.928$$

The equation is thus $y = 2 \sin(2x + 0.928)$.

(5-7)

84. (A)

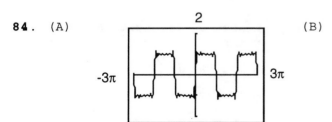

(B)

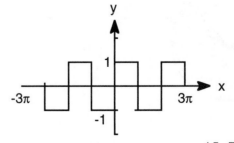

(5-7)

85. In one year the line sweeps out one full revolution, or 2π radians in 365 days. In 73 days the line sweeps out $\frac{73}{365}$ of one full revolution, or $\frac{73}{365} \cdot 2\pi = \frac{2\pi}{5}$ radians.

(5-3)

86. Sketch a figure.
From geometry we know that

$\theta = \frac{1}{8}(360°) = 45°$

$\frac{s}{2} = r \sin 45° = r\frac{1}{\sqrt{2}}$

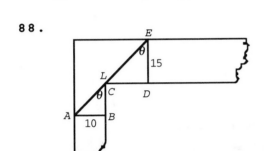

Hence $P = 4s = 8\left(\frac{s}{2}\right) = 8r\left(\frac{1}{\sqrt{2}}\right) = 8(5.00)\left(\frac{1}{\sqrt{2}}\right) = 28.3$ cm

(5-5)

87. When $t = 0$, $I = 30$. Hence $30 = A \cos B(0)$, that is, $A = 30$. Since the period is $\frac{1}{60}$ second, $\frac{2\pi}{B} = \frac{1}{60}$, $B = 2\pi(60) = 120\pi$. Hence the equation is $I = 30 \cos 120\pi t$.

(5-7)

88.

(A) In the figure as labelled, $L = AC + CE$.

In triangle ABC, since $\sin \theta = \frac{10}{AC}$, $AC \sin \theta = 10$,

$AC = \frac{10}{\sin \theta} = 10 \csc \theta$

In triangle CDE, since $\cos \theta = \frac{15}{CE}$, $CE \cos \theta = 15$,

$CE = \frac{15}{\cos \theta} = 15 \sec \theta$

Then $L = AC + CE = 10 \csc \theta + 15 \sec \theta \qquad 0 < \theta < \frac{\pi}{2}$

(B)

θ radians	0.4	0.5	0.6	0.7	0.8	0.9	1.0
L feet	42.0	38.0	35.9	35.1	35.5	36.9	39.6

From the table, the shortest distance L, to the nearest foot, is 35 feet. This is the length of the longest log that can make the corner.

(C) Here is a graph of
$L = 10 \csc \theta + 15 \sec \theta$
from a graphing utility.

The minimum value of L is shown as 35.1 feet, to one decimal place.

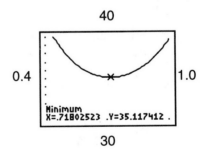

(D) As $\theta \to 0$, $L = 10 \csc \theta + 15 \sec \theta$ approaches an asymptote of $\csc \theta$; L increases without bound.

As $\theta \to \frac{\pi}{2}$, $L = 10 \csc \theta + 15 \sec \theta$ approaches an asymptote of $\sec \theta$; L increases without bound.

(5-5)

89. (A) $|A| = \dfrac{R_{max} - R_{min}}{2} = \dfrac{7 - 1}{2} = 3$. Hence $A = 3$ or -3. The graph appears to be shifted up from a graph of an upside down cosine curve that completes one full cycle as t varies over the interval $[0, 12]$. Thus $A = -3$. Then $P = 12 = 2\pi \cdot \dfrac{6}{\pi} = 2\pi \div \dfrac{\pi}{6}$. Hence $B = \dfrac{\pi}{6}$. $k = \dfrac{R_{max} + R_{min}}{2} = \dfrac{7 + 1}{2} = 4$. Thus $R(t) = 4 - 3 \cos \dfrac{\pi}{6} t$.

(B) The graph shows the seasonal changes in soft drink consumption. Most is consumed in August and the least in February. (5-7)

90. (A)

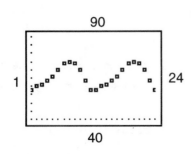

(B) $|A| = \dfrac{Y_{max} - Y_{min}}{2} = \dfrac{75 - 58}{2} = 8.5$. Hence $A = 8.5$ or -8.5.

$k = \dfrac{Y_{max} + Y_{min}}{2} = \dfrac{75 + 58}{2} = 66.5$. $P = 12 = 2\pi \cdot \dfrac{6}{\pi} = 2\pi \div \dfrac{\pi}{6}$. Hence $B = \dfrac{\pi}{6}$.

The x intercept closest to the origin is estimated from the graph as 4.5. To find C, substitute $B = \dfrac{\pi}{6}$ and $x = 4.5$ into the phase shift formula $x = -\dfrac{C}{B}$ and solve for C.

$$x = -\dfrac{C}{B}$$

$$4.5 = -C \div \dfrac{\pi}{6}$$

$$C = -4.5\left(\dfrac{\pi}{6}\right)$$

$$C = -2.4$$

With this value of C, the graph is seen to be shifted up from the graph of a standard sine curve, thus $A = 8.5$. The equation is thus

$$y = 66.5 + 8.5 \sin\left(\dfrac{\pi}{6}x - 2.4\right).$$

(C)

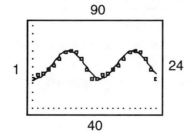

(5-7)

CHAPTER 6

Exercise 6-1

Key Ideas and Formulas

Basic Trigonometric Identities

Reciprocal Identities

$$\csc x = \frac{1}{\sin x} \qquad \sec x = \frac{1}{\cos x} \qquad \cot x = \frac{1}{\tan x}$$

Quotient Identities

$$\tan x = \frac{\sin x}{\cos x} \qquad \cot x = \frac{\cos x}{\sin x}$$

Identities for Negatives

$$\sin(-x) = -\sin x \qquad \cos(-x) = \cos x \qquad \tan(-x) = -\tan x$$

Pythagorean Identities

$$\sin^2 x + \cos^2 x = 1 \qquad \tan^2 x + 1 = \sec^2 x \qquad 1 + \cot^2 x = \csc^2 x$$

$$\sin^2 x = 1 - \cos^2 x$$

$$\cos^2 x = 1 - \sin^2 x$$

Suggested Steps in Verifying Identities

1. Start with the more complicated side of the identity and transform it into the simpler side.

2. Try algebraic operations such as multiplying, factoring, combining fractions, splitting fractions, and so on.

3. If other steps fail, express each function in terms of sine and cosine functions, and then perform appropriate algebraic operations.

4. At each step keep the other side of the identity in mind. This often reveals what you should do in order to get there.

Common Error: Avoid working with identities to be verified as if they were equations to be solved. It is not correct to multiply both members by the same expression, for example, or to perform other steps until the two members are identical. An identity is not verified if it has merely been transformed into a correct statement.

---

1. $\sin \theta \sec \theta = \sin \theta \dfrac{1}{\cos \theta}$ Reciprocal Identity

$$= \frac{\sin \theta}{\cos \theta} \qquad \text{Algebra}$$

$$= \tan \theta \qquad \text{Quotient Identity}$$

3. $\cot u \sec u \sin u = \dfrac{\cos u}{\sin u} \dfrac{1}{\cos u} \sin u$ Reciprocal and Quotient Identities

$$= 1 \qquad \text{Algebra}$$

5. $\dfrac{\sin(-x)}{\cos(-x)} = \dfrac{-\sin x}{\cos x}$ Identities for Negatives

$$= -\tan x \qquad \text{Quotient Identity}$$

7. $\dfrac{\tan\,\alpha\,\cot\,\alpha}{\csc\,\alpha} = \dfrac{\tan\,\alpha\,\frac{1}{\tan\,\alpha}}{\csc\,\alpha}$ Reciprocal Identity

$\qquad\qquad\quad = \dfrac{1}{\csc\,\alpha}$ Algebra

$\qquad\qquad\quad = \dfrac{1}{\frac{1}{\sin\,\alpha}}$ Reciprocal Identity

$\qquad\qquad\quad = \sin\,\alpha$ Algebra

9. $\csc\,u(\cos\,u + \sin\,u) = \dfrac{1}{\sin\,u}\,(\cos\,u + \sin\,u)$ Reciprocal Identity

$\qquad\qquad\qquad\quad = \dfrac{\cos\,u}{\sin\,u} + \dfrac{\sin\,u}{\sin\,u}$ Algebra

$\qquad\qquad\qquad\quad = \dfrac{\cos\,u}{\sin\,u} + 1$ Algebra

$\qquad\qquad\qquad\quad = \cot\,u + 1$ Quotient Identity

Key Algebraic Steps:
$$\dfrac{1}{b}(a + b) = \dfrac{a}{b} + \dfrac{b}{b} = \dfrac{a}{b} + 1$$

11. $\dfrac{\cos\,x - \sin\,x}{\sin\,x\,\cos\,x} = \dfrac{\cos\,x}{\sin\,x\,\cos\,x} - \dfrac{\sin\,x}{\sin\,x\,\cos\,x}$ Algebra

$\qquad\qquad\quad = \dfrac{1}{\sin\,x} - \dfrac{1}{\cos\,x}$ Algebra

$\qquad\qquad\quad = \csc\,x - \sec\,x$ Reciprocal Identities

13. $\dfrac{\sin^2\,t}{\cos\,t} + \cos\,t = \dfrac{\sin^2\,t}{\cos\,t} + \dfrac{\cos\,t}{1}$ Algebra

$\qquad\qquad\qquad = \dfrac{\sin^2\,t}{\cos\,t} + \dfrac{\cos^2\,t}{\cos\,t}$ Algebra

$\qquad\qquad\qquad = \dfrac{\sin^2\,t + \cos^2\,t}{\cos\,t}$ Algebra

$\qquad\qquad\qquad = \dfrac{1}{\cos\,t}$ Pythagorean Identity

$\qquad\qquad\qquad = \sec\,t$ Reciprocal Identity

Key Algebraic Steps:
$$\dfrac{a^2}{b} + b = \dfrac{a^2}{b} + \dfrac{b}{1}$$
$$= \dfrac{a^2}{b} + \dfrac{b^2}{b}$$
$$= \dfrac{a^2 + b^2}{b}$$

15. $\dfrac{\cos\,x}{1 - \sin^2\,x} = \dfrac{\cos\,x}{\sin^2\,x + \cos^2\,x - \sin^2\,x}$ Pythagorean Identity

$\qquad\qquad\quad = \dfrac{\cos\,x}{\cos^2\,x}$ Algebra

$\qquad\qquad\quad = \dfrac{1}{\cos\,x}$ Algebra

$\qquad\qquad\quad = \sec\,x$ Reciprocal Identity

17. $(1 - \cos\,u)(1 + \cos\,u) = 1 - \cos^2\,u$ Algebra

$\qquad\qquad\qquad\quad = \sin^2\,u + \cos^2\,u - \cos^2\,u$ Pythagorean Identity

$\qquad\qquad\qquad\quad = \sin^2\,u$ Algebra

19. $1 - 2 \sin^2 x = \cos^2 x + \sin^2 x - 2 \sin^2 x$ Pythagorean Identity
$= \cos^2 x - \sin^2 x$ Algebra

21. $(\sec t + 1)(\sec t - 1) = \sec^2 t - 1$ Algebra
$= \tan^2 t + 1 - 1$ Pythagorean Identity
$= \tan^2 t$ Algebra

23. $\csc^2 x - \cot^2 x = 1 + \cot^2 x - \cot^2 x$ Pythagorean Identity
$= 1$ Algebra

25.

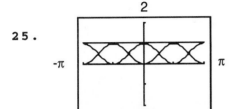

The straight line is the graph of $y = \sin^2 x + \cos^2 x$, illustrating the identity $\sin^2 x + \cos^2 x = 1$.

27.

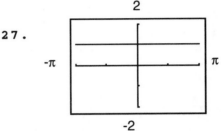

Both graphs are the same horizontal line, illustrating the identity
$$\frac{\cos x}{\cot x \sin x} = 1$$

29. $\dfrac{x}{|x|} = 1$ is not an identity because if $x = -2$, for example, the left side is $\dfrac{-2}{|-2|}$ or -1, and thus is not equal to the right side.

31. The statement is an identity. If $x = 1$ both sides are undefined; otherwise

$\dfrac{1}{\sqrt{x - 1}} = \dfrac{1}{\sqrt{x - 1}} \cdot \dfrac{\sqrt{x - 1}}{\sqrt{x - 1}}$ Fundamental Principle of Fractions

$= \dfrac{\sqrt{x - 1}}{\sqrt{(x - 1)(x - 1)}}$ $\sqrt{a}\sqrt{b} = \sqrt{ab}$ for real a, b

$= \dfrac{\sqrt{x - 1}}{\sqrt{(x - 1)^2}}$ $aa = a^2$

$= \dfrac{\sqrt{x - 1}}{|x - 1|}$ $\sqrt{a^2} = |a|$ for all real a

33. $\sin x + \cos x = 1$ is not an identity because if $x = \dfrac{\pi}{4}$, for example, the left side is $\dfrac{\sqrt{2}}{2} + \dfrac{\sqrt{2}}{2}$ or $\sqrt{2}$, and thus is not equal to the right side.

35. $\sin^4 x + \cos^4 x = 1$ is not an identity because if $x = \dfrac{\pi}{4}$, for example, the left side is $\left(\dfrac{\sqrt{2}}{2}\right)^4 + \left(\dfrac{\sqrt{2}}{2}\right)^4 = \dfrac{4}{16} + \dfrac{4}{16} = \dfrac{1}{2}$, and thus is not equal to the right side.

37. $\dfrac{1 - (\sin x - \cos x)^2}{\sin x} = \dfrac{1 - (\sin^2 x - 2\sin x \cos x + \cos^2 x)}{\sin x}$ Algebra

$\qquad = \dfrac{1 - (-2\sin x \cos x + \sin^2 x + \cos^2 x)}{\sin x}$ Algebra

$\qquad = \dfrac{1 - (-2\sin x \cos x + 1)}{\sin x}$ Pythagorean Identity

$\qquad = \dfrac{2\sin x \cos x}{\sin x}$ Algebra

$\qquad = 2\cos x$ Algebra

Key Algebraic Steps:
$$(a - b)^2 = a^2 - 2ab + b^2 = -2ab + a^2 + b^2$$

39. $\dfrac{\cot \theta + 1}{\csc \theta} = \dfrac{\frac{\cos \theta}{\sin \theta} + 1}{\frac{1}{\sin \theta}}$ Reciprocal and Quotient Identities

$\qquad = \dfrac{\sin \theta \frac{\cos \theta}{\sin \theta} + 1 \cdot \sin \theta}{\sin \theta \frac{1}{\sin \theta}}$ Algebra

$\qquad = \cos \theta + \sin \theta$ Algebra

Key Algebraic Steps:
$$\dfrac{\frac{b}{a} + 1}{\frac{1}{a}} = \dfrac{a \frac{b}{a} + 1a}{a \frac{1}{a}} = b + a$$

41. $\dfrac{1 + \cos y}{1 - \cos y} = \dfrac{(1 + \cos y)}{(1 - \cos y)} \dfrac{(1 - \cos y)}{(1 - \cos y)}$ Algebra

$\qquad = \dfrac{1 - \cos^2 y}{(1 - \cos y)^2}$ Algebra

$\qquad = \dfrac{\sin^2 y + \cos^2 y - \cos^2 y}{(1 - \cos y)^2}$ Pythagorean Identity

$\qquad \dfrac{\sin^2 y}{(1 - \cos y)^2}$ Algebra

43. $\tan^2 x - \sin^2 x = \dfrac{\sin^2 x}{\cos^2 x} - \sin^2 x$ Quotient Identity

$\qquad = \sin^2 x \dfrac{1}{\cos^2 x} - \sin^2 x \cdot 1$ Algebra

$\qquad = \sin^2 x \left(\dfrac{1}{\cos^2 x} - 1 \right)$ Algebra

$\qquad = \sin^2 x (\sec^2 x - 1)$ Reciprocal Identity

$\qquad = \sin^2 x \tan^2 x$ Pythagorean Identity

$\qquad = \tan^2 x \sin^2 x$ Algebra

45. $\dfrac{\csc\,\theta}{\cot\,\theta + \tan\,\theta} = \dfrac{\frac{1}{\sin\,\theta}}{\frac{\cos\,\theta}{\sin\,\theta} + \frac{\sin\,\theta}{\cos\,\theta}}$ Reciprocal and Quotient Identities

$\qquad\qquad = \dfrac{\cos\,\theta\,\sin\,\theta\,\frac{1}{\sin\,\theta}}{\cos\,\theta\,\sin\,\theta\,\frac{\cos\,\theta}{\sin\,\theta} + \cos\,\theta\,\sin\,\theta\,\frac{\sin\,\theta}{\cos\,\theta}}$ Algebra

$\qquad\qquad = \dfrac{\cos\,\theta}{\cos^2\,\theta + \sin^2\,\theta}$ Algebra

$\qquad\qquad = \dfrac{\cos\,\theta}{1}$ Pythagorean Identity

$\qquad\qquad = \cos\,\theta$

Key Algebraic Steps:

$$\dfrac{\frac{1}{a}}{\frac{b}{a} + \frac{a}{b}} = \dfrac{ab\cdot\frac{1}{a}}{ab\cdot\frac{b}{a} + ab\cdot\frac{a}{b}} = \dfrac{b}{b^2 + a^2}$$

47. $\ln\,\tan\,x = \ln\left(\dfrac{\sin\,x}{\cos\,x}\right)$ Quotient Identity

$\qquad\qquad = \ln\,\sin\,x - \ln\,\cos\,x$ Algebra

49. $\ln\,\cot\,x = \ln\left(\dfrac{1}{\tan\,x}\right)$ Reciprocal Identity

$\qquad\qquad = \ln\,1 - \ln\,\tan\,x$ Algebra

$\qquad\qquad = 0 - \ln\,\tan\,x$ Algebra

$\qquad\qquad = -\ln\,\tan\,x$ Algebra

51. $\dfrac{\sec\,A - 1}{\sec\,A + 1} = \dfrac{\frac{1}{\cos\,A} - 1}{\frac{1}{\cos\,A} + 1}$ Reciprocal Identity

$\qquad\qquad = \dfrac{\cos\,A\,\frac{1}{\cos\,A} - 1\cdot\cos\,A}{\cos\,A\,\frac{1}{\cos\,A} + 1\cdot\cos\,A}$ Algebra

$\qquad\qquad = \dfrac{1 - \cos\,A}{1 + \cos\,A}$ Algebra

Key Algebraic Steps:

$$\dfrac{\frac{1}{a} - 1}{\frac{1}{a} + 1} = \dfrac{a\cdot\frac{1}{a} - 1\cdot a}{a\cdot\frac{1}{a} + 1\cdot a} = \dfrac{1 - a}{1 + a}$$

53. $\sin^4\,w - \cos^4\,w = (\sin^2\,w)^2 - (\cos^2\,w)^2$ Algebra

$\qquad\qquad = (\sin^2\,w + \cos^2\,w)(\sin^2\,w - \cos^2\,w)$ Algebra

$\qquad\qquad = 1(\sin^2\,w - \cos^2\,w)$ Pythagorean Identity

$\qquad\qquad = \sin^2\,w - \cos^2\,w$ Algebra

$\qquad\qquad = 1 - \cos^2\,w - \cos^2\,w$ Pythagorean Identity

$\qquad\qquad = 1 - 2\,\cos^2\,w$ Algebra

Key Algebraic Steps:

$$a^4 - b^4 = (a^2)^2 - (b^2)^2 = (a^2 + b^2)(a^2 - b^2)$$

55. $\dfrac{\cos^2 z - 3 \cos z + 2}{\sin^2 z} = \dfrac{\cos^2 z - 3 \cos z + 2}{\sin^2 z + \cos^2 z - \cos^2 z}$ Algebra

$\qquad\qquad\qquad\quad = \dfrac{\cos^2 z - 3 \cos z + 2}{1 - \cos^2 z}$ Pythagorean Identity

$\qquad\qquad\qquad\quad = \dfrac{(\cos z - 1)(\cos z - 2)}{(1 - \cos z)(1 + \cos z)}$ Algebra

$\qquad\qquad\qquad\quad = \dfrac{-(\cos z - 2)}{1 + \cos z}$ Algebra

$\qquad\qquad\qquad\quad = \dfrac{2 - \cos z}{1 + \cos z}$ Algebra

Key Algebraic Steps:

$$\frac{a^2 - 3a + 2}{1 - a^2} = \frac{(a - 1)(a - 2)}{(1 - a)(1 + a)} = \frac{-(a - 2)}{1 + a} = \frac{2 - a}{1 + a}$$

57. $\dfrac{\cos^3 \theta - \sin^3 \theta}{\cos \theta - \sin \theta} = \dfrac{(\cos \theta - \sin \theta)(\cos^2 \theta + \cos \theta \sin \theta + \sin^2 \theta)}{\cos \theta - \sin \theta}$ Algebra

$\qquad\qquad\quad = \cos^2 \theta + \cos \theta \sin \theta + \sin^2 \theta$ Algebra

$\qquad\qquad\quad = 1 + \cos \theta \sin \theta$ Pythagorean Identity

$\qquad\qquad\quad = 1 + \sin \theta \cos \theta$ Algebra

Key Algebraic Steps:

$$\frac{a^3 - b^3}{a - b} = \frac{(a - b)(a^2 + ab + b^2)}{a - b}$$
$$\qquad\quad = a^2 + ab + b^2$$

> **Common Error:**
> $\dfrac{a^3 - b^3}{a - b} \neq a^2 + b^2$

59. $(\sec x - \tan x)^2 = \left(\dfrac{1}{\cos x} - \dfrac{\sin x}{\cos x}\right)^2$ Quotient and Reciprocal Identities

$\qquad\qquad\qquad = \left(\dfrac{1 - \sin x}{\cos x}\right)^2$ Algebra

$\qquad\qquad\qquad = \dfrac{(1 - \sin x)^2}{\cos^2 x}$ Algebra

$\qquad\qquad\qquad = \dfrac{(1 - \sin x)^2}{1 - \sin^2 x}$ Pythagorean Identity

$\qquad\qquad\qquad = \dfrac{(1 - \sin x)(1 - \sin x)}{(1 - \sin x)(1 + \sin x)}$ Algebra

$\qquad\qquad\qquad = \dfrac{1 - \sin x}{1 + \sin x}$ Algebra

61. $\dfrac{\csc^4 x - 1}{\cot^2 x} = \dfrac{(\csc^2 x)^2 - (1)^2}{\cot^2 x}$ Algebra

$\qquad\qquad\quad = \dfrac{(\csc^2 x - 1)(\csc^2 x + 1)}{\cot^2 x}$ Algebra

$\qquad\qquad\quad = \dfrac{(1 + \cot^2 x - 1)(\csc^2 x + 1)}{\cot^2 x}$ Pythagorean Identity

$\qquad\qquad\quad = \dfrac{\cot^2 x(\csc^2 x + 1)}{\cot^2 x}$ Algebra

$\qquad\qquad\quad = \csc^2 x + 1$ Algebra

$\qquad\qquad\quad = 1 + \cot^2 x + 1$ Pythagorean Identity

$\qquad\qquad\quad = 2 + \cot^2 x$ Algebra

63.
$$\frac{1 + \sin v}{\cos v} = \frac{(1 + \sin v)(1 - \sin v)}{\cos v(1 - \sin v)} \qquad \text{Algebra}$$

$$= \frac{1 - \sin^2 v}{\cos v(1 - \sin v)} \qquad \text{Algebra}$$

$$= \frac{\cos^2 v}{\cos v(1 - \sin v)} \qquad \text{Pythagorean Identity}$$

$$= \frac{\cos v}{1 - \sin v} \qquad \text{Algebra}$$

65. Graph both sides of the equation in the same viewing window.

$\dfrac{\sin(-x)}{\cos(-x)\tan(-x)} = -1$ is not an identity, since the graphs do not match.

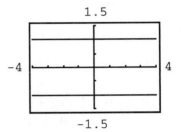

Try $x = -\dfrac{\pi}{4}$

Left side: $\dfrac{\sin[-(-\frac{\pi}{4})]}{\cos[-(-\frac{\pi}{4})]\tan[-(-\frac{\pi}{4})]} = \dfrac{\frac{1}{\sqrt{2}}}{\frac{1}{\sqrt{2}} \cdot 1} = 1$

Right side: $= -1$
This verifies that the equation is not an identity.

67. Graph both sides of the equation in the same viewing window.

$\dfrac{\sin x}{\cos x \tan(-x)} = -1$ appears to be an identity, which we now verify:

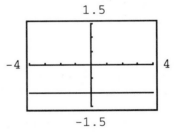

$$\frac{\sin x}{\cos x \tan(-x)} = \frac{\sin x}{\cos x(-\tan x)} \qquad \text{Identities for Negatives}$$

$$= -\frac{\sin x}{\cos x \tan x} \qquad \text{Algebra}$$

$$= -\frac{\sin x}{\cos x \cdot \frac{\sin x}{\cos x}} \qquad \text{Quotient Identity}$$

$$= -\frac{\sin x}{\sin x} \qquad \text{Algebra}$$

$$= -1 \qquad \text{Algebra}$$

69. Graph both sides of the equation in the same viewing window.

$\sin x + \dfrac{\cos^2 x}{\sin x} = \sec x$ is not an identity, since the graphs do not match.

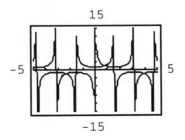

Try $x = -\dfrac{\pi}{4}$

Left side: $\sin\left(-\frac{\pi}{4}\right) + \frac{\cos^2\left(-\frac{\pi}{4}\right)}{\sin\left(-\frac{\pi}{4}\right)} = -\frac{1}{\sqrt{2}} + \frac{\left(\frac{1}{\sqrt{2}}\right)^2}{\left(-\frac{1}{\sqrt{2}}\right)} = -\frac{1}{\sqrt{2}} - \frac{1}{\sqrt{2}} = -\sqrt{2}$

Right side: $\sec\left(-\frac{\pi}{4}\right) = \sqrt{2}$

This verifies that the equation is not an identity.

71. Graph both sides of the equation in the same viewing window.

$\sin x + \dfrac{\cos^2 x}{\sin x} = \csc x$ appears to be an identity, which we now verify:

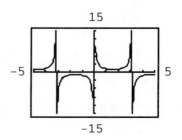

$$\sin x + \frac{\cos^2 x}{\sin x} = \frac{\sin x}{1} + \frac{\cos^2 x}{\sin x} \qquad \text{Algebra}$$

$$= \frac{\sin^2 x}{\sin x} + \frac{\cos^2 x}{\sin x} \qquad \text{Algebra}$$

$$= \frac{\sin^2 x + \cos^2 x}{\sin x} \qquad \text{Algebra}$$

$$= \frac{1}{\sin x} \qquad \text{Pythagorean Identity}$$

$$= \csc x \qquad \text{Reciprocal Identity}$$

73. Graph both sides of the equation in the same viewing window.

$\dfrac{\tan x}{\sin x - 2\tan x} = \dfrac{1}{\cos x - 2}$ appears to be an identity, which we now verify:

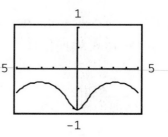

$$\frac{\tan x}{\sin x - 2\tan x} = \frac{\dfrac{\sin x}{\cos x}}{\sin x - 2\dfrac{\sin x}{\cos x}} \qquad \text{Quotient Identity}$$

$$= \frac{\cos x \cdot \dfrac{\sin x}{\cos x}}{\sin x \cos x - 2\cos x \cdot \dfrac{\sin x}{\cos x}} \qquad \text{Algebra}$$

$$= \frac{\sin x}{\sin x \cos x - 2\sin x} \qquad \text{Algebra}$$

$$= \frac{\sin x}{\sin x(\cos x - 2)} \qquad \text{Algebra}$$

$$= \frac{1}{\cos x - 2} \qquad \text{Algebra}$$

75. Graph both sides of the equation in the same viewing window.

$$\frac{\tan x}{\sin x + 2 \tan x} = \frac{1}{\cos x - 2}$$ is not an identity, since the graphs do not match.

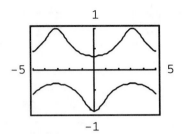

Try $x = \dfrac{\pi}{3}$

Left side: $\dfrac{\tan \frac{\pi}{3}}{\sin \frac{\pi}{3} + 2 \tan \frac{\pi}{3}} = \dfrac{\sqrt{3}}{\frac{\sqrt{3}}{2} + 2\sqrt{3}} = \dfrac{2\sqrt{3}}{\sqrt{3} + 4\sqrt{3}} = \dfrac{2\sqrt{3}}{5\sqrt{3}} = \dfrac{2}{5}$

Right side: $\dfrac{1}{\cos \frac{\pi}{3} - 2} = \dfrac{1}{\frac{1}{2} - 2} = \dfrac{1}{-\frac{3}{2}} = -\dfrac{2}{3}$

This verifies that the equation is not an identity.

77.

$$\frac{2 \sin^2 x + 3 \cos x - 3}{\sin^2 x} = \frac{2(1 - \cos^2 x) + 3 \cos x - 3}{\sin^2 x} \qquad \text{Pythagorean Identity}$$

$$= \frac{2 - 2 \cos^2 x + 3 \cos x - 3}{\sin^2 x} \qquad \text{Algebra}$$

$$= \frac{-2 \cos^2 x + 3 \cos x - 1}{\sin^2 x} \qquad \text{Algebra}$$

$$= \frac{-2 \cos^2 x + 3 \cos x - 1}{1 - \cos^2 x} \qquad \text{Pythagorean Identity}$$

$$= \frac{-(2 \cos^2 x - 3 \cos x + 1)}{-(\cos^2 x - 1)} \qquad \text{Algebra}$$

$$= \frac{-(2 \cos x - 1)(\cos x - 1)}{-(\cos x + 1)(\cos x - 1)} \qquad \text{Algebra}$$

$$= \frac{2 \cos x - 1}{\cos x + 1} \qquad \text{Algebra}$$

$$= \frac{2 \cos x - 1}{1 + \cos x} \qquad \text{Algebra}$$

Key Algebraic Steps:

$$\frac{-2a^2 + 3a - 1}{1 - a^2} = \frac{-(2a^2 - 3a + 1)}{-(a^2 - 1)}$$

$$= \frac{-(2a - 1)(a - 1)}{-(a + 1)(a - 1)}$$

$$= \frac{2a - 1}{a + 1}$$

$$= \frac{2a - 1}{1 + a}$$

79. $\dfrac{\tan u + \sin u}{\tan u - \sin u} - \dfrac{\sec u + 1}{\sec u - 1} = \dfrac{\frac{\sin u}{\cos u} + \sin u}{\frac{\sin u}{\cos u} - \sin u} - \dfrac{\sec u + 1}{\sec u - 1}$ Quotient Identity

$$= \dfrac{\frac{1}{\cos u}\sin u + 1 \cdot \sin u}{\frac{1}{\cos u}\sin u - 1 \cdot \sin u} - \dfrac{\sec u + 1}{\sec u - 1}$$ Algebra

$$= \dfrac{\sin u \left(\frac{1}{\cos u} + 1\right)}{\sin u \left(\frac{1}{\cos u} - 1\right)} - \dfrac{\sec u + 1}{\sec u - 1}$$ Algebra

$$= \dfrac{\frac{1}{\cos u} + 1}{\frac{1}{\cos u} - 1} - \dfrac{\sec u + 1}{\sec u - 1}$$ Algebra

$$= \dfrac{\sec u + 1}{\sec u - 1} - \dfrac{\sec u + 1}{\sec u - 1}$$ Reciprocal Identity

$$= 0$$ Algebra

Key Algebraic Steps:

$$\dfrac{\frac{b}{a} + b}{\frac{b}{a} - b} = \dfrac{b \cdot \frac{1}{a} + b}{b \cdot \frac{1}{a} - b}$$

$$= \dfrac{b\left(\frac{1}{a} + 1\right)}{b\left(\frac{1}{a} - 1\right)}$$

$$= \dfrac{\frac{1}{a} + 1}{\frac{1}{a} - 1}$$

81. $\dfrac{\tan \beta + \cot \alpha}{\tan \beta \cot \alpha} = \dfrac{\tan \beta}{\tan \beta \cot \alpha} + \dfrac{\cot \alpha}{\tan \beta \cot \alpha}$ Algebra

$$= \dfrac{1}{\cot \alpha} + \dfrac{1}{\tan \beta}$$ Algebra

$$= \tan \alpha + \cot \beta$$ Reciprocal Identities

83. Here is a graph of $f(x)$ from a graphing utility.

The graph appears identical with the graph of $g(x) = \cot x$. We attempt to verify that $\dfrac{1 - \sin^2 x}{\tan x} + \sin x \cos x$ = $\cot x$ is an identity.

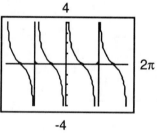

$\dfrac{1 - \sin^2 x}{\tan x} + \sin x \cos x = \dfrac{1}{\tan x} - \dfrac{\sin^2 x}{\tan x} + \sin x \cos x$ Algebra

$$= \cot x - \dfrac{\sin^2 x}{\frac{\sin x}{\cos x}} + \sin x \cos x$$ Quotient Identity

$$= \cot x - \dfrac{\sin^2 x \cos x}{\sin x} + \sin x \cos x$$ Algebra

$$= \cot x - \sin x \cos x + \sin x \cos x$$ Algebra

$$= \cot x$$ Algebra

Thus $f(x) = g(x)$ is an identity for $g(x) = \cot x$.

85. Here is a graph of $f(x)$ from a graphing utility.

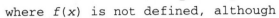

The graph appears similar to the graph of the cosecant function, but shifted down 1 unit, and with

peculiar behavior at $-\dfrac{\pi}{2}$ and $\dfrac{3\pi}{2}$

where $f(x)$ is not defined, although $\csc x$ is defined. We set $g(x) = \csc x - 1$ and attempt to verify that

$\dfrac{\cos^2 x}{1 + \sin x - \cos^2 x} = \csc x - 1$ is an identity.

$$\dfrac{\cos^2 x}{1 + \sin x - \cos^2 x} = \dfrac{1 - \sin^2 x}{1 + \sin x - (1 - \sin^2 x)} \qquad \text{Pythagorean Identity}$$

$$= \dfrac{1 - \sin^2 x}{\sin x + \sin^2 x} \qquad \text{Algebra}$$

$$= \dfrac{(1 - \sin x)(1 + \sin x)}{\sin x(1 + \sin x)} \qquad \text{Algebra}$$

$$= \dfrac{1 - \sin x}{\sin x} \qquad \text{Algebra}$$

$$= \dfrac{1}{\sin x} - \dfrac{\sin x}{\sin x} \qquad \text{Algebra}$$

$$= \csc x - 1 \qquad \text{Reciprocal Identity; Algebra}$$

Thus $f(x) = g(x)$ is an identity for $g(x) = \csc x - 1$.

87. Here is a graph of $f(x)$ from a graphing utility.

The graph appears identical with the graph of $g(x) = 3 \cos x$. We attempt to verify that

$\dfrac{1 + \cos x - 2\cos^2 x}{1 - \cos x} + \dfrac{\sin^2 x}{1 + \cos x}$
$= 3 \cos x$ is an identity.

$$\dfrac{1 + \cos x - 2\cos^2 x}{1 - \cos x} + \dfrac{\sin^2 x}{1 + \cos x}$$

$$= \dfrac{1 + \cos x - 2\cos^2 x}{1 - \cos x} - \dfrac{1 - \cos^2 x}{1 + \cos x}$$

$$= \dfrac{(1 - \cos x)(1 + 2\cos x)}{1 - \cos x} - \dfrac{(1 - \cos x)(1 + \cos x)}{1 + \cos x}$$

$$= 1 + 2\cos x - (1 - \cos x)$$

$$= 1 + 2\cos x - 1 + \cos x$$

$$= 3 \cos x$$

Thus, $f(x) = g(x)$ is an identity for $g(x) = 3 \cos x$.

89. Since $\sqrt{1 - \cos^2 x} = \sqrt{\sin^2 x}$, the equation will be true when $\sqrt{\sin^2 x} = -\sin x$, that is, when $\sin x$ is negative. This occurs in Quadrants III, IV.

91. Since $\sqrt{1 - \cos^2 x} = \sqrt{\sin^2 x}$, the equation will be true when $\sqrt{\sin^2 x} = \sin x$, that is, when $\sin x$ is positive. This occurs in Quadrants I, II.

93. Since $\sqrt{1 - \sin^2 x} = \sqrt{\cos^2 x} = |\cos x|$ is an identity, this will hold in all quadrants.

95. Since $\sqrt{1 - \sin^2 x} = \sqrt{\cos^2 x} = \cos x$ if $\cos x$ is positive, this will hold when $\dfrac{\sin x}{\sqrt{1 - \sin^2 x}} = \dfrac{\sin x}{\sqrt{\cos^2 x}} = \dfrac{\sin x}{\cos x} = \tan x$, that is, when $\cos x$ is positive. This occurs in Quadrants I, IV.

97.
$$\sqrt{a^2 - u^2} = \sqrt{a^2 - (a \sin x)^2}$$
$$= \sqrt{a^2 - a^2 \sin^2 x}$$
$$= \sqrt{a^2(1 - \sin^2 x)}$$
$$= \sqrt{a^2} \sqrt{1 - \sin^2 x}$$
$$= \sqrt{a^2} \sqrt{\cos^2 x}$$

> **Common Error:**
> $\sqrt{1 - \sin^2 x} \neq 1 - \sin x$

Since $a > 0$, $\sqrt{a^2} = a$.
$\cos x$ will be ≥ 0 if x is a quadrant I or IV angle. But the restriction $-\dfrac{\pi}{2} < x < \dfrac{\pi}{2}$ requires that x is such an angle. Therefore $\sqrt{\cos^2 x} = \cos x$
$$\sqrt{a^2 - u^2} = a \cos x$$

99.
$$\sqrt{a^2 + u^2} = \sqrt{a^2 + (a \tan x)^2}$$
$$= \sqrt{a^2 + a^2 \tan^2 x}$$
$$= \sqrt{a^2(1 + \tan^2 x)}$$
$$= \sqrt{a^2} \sqrt{1 + \tan^2 x}$$
$$= \sqrt{a^2} \sqrt{\sec^2 x}$$

Since $a > 0$, $\sqrt{a^2} = a$.
$\sec x$ will be ≥ 0 if x is a quadrant I or IV angle. But the restriction $0 < x < \dfrac{\pi}{2}$ requires that x is a quadrant I angle. Therefore $\sqrt{\sec^2 x} = \sec x$
$$\sqrt{a^2 + u^2} = a \sec x$$

Exercise 6-2

Key Ideas and Formulas

The basic identities listed at the beginning of Exercise 7-1 are used extensively in this section. They are omitted here for lack of space.

Sum Identities:

$$\sin(x + y) = \sin x \cos y + \cos x \sin y$$
$$\cos(x + y) = \cos x \cos y - \sin x \sin y$$
$$\tan(x + y) = \frac{\tan x + \tan y}{1 - \tan x \tan y}$$

Difference Identities:

$$\sin(x - y) = \sin x \cos y - \cos x \sin y$$
$$\cos(x - y) = \cos x \cos y + \sin x \sin y$$
$$\tan(x - y) = \frac{\tan x - \tan y}{1 + \tan x \tan y}$$

Co-function Identities:

(replace $\frac{\pi}{2}$ with 90° if x is in degrees)

$$\sin\left(\frac{\pi}{2} - x\right) = \cos x \qquad \tan\left(\frac{\pi}{2} - x\right) = \cot x \qquad \cos\left(\frac{\pi}{2} - x\right) = \sin x$$

Common Errors:	
$\cos(x + y) \neq \cos x + \cos y$	$\cos(x - y) \neq \cos x - \cos y$
$\sin(x + y) \neq \sin x + \sin y$	$\sin(x - y) \neq \sin x - \sin y$

1. This equation is an identity, since it states the known periodicity of the sine function. Moreover

$$\begin{aligned} \sin(x + 2\pi) &= \sin x \cos 2\pi + \cos x \sin 2\pi & &\text{Sum Identity} \\ &= \sin x(1) + \cos x(0) & &\text{Known Values} \\ &= \sin x & &\text{Algebra} \end{aligned}$$

3. This equation is not an identity, as can be shown:

$$\begin{aligned} \cos(x + \pi) &= \cos x \cos \pi - \sin x \sin \pi & &\text{Sum Identity} \\ &= \cos x(-1) - \sin x(0) & &\text{Known Values} \\ &= -\cos x & &\text{Algebra} \end{aligned}$$

Thus $\cos(x + \pi) = \cos x$ is equivalent to $-\cos x = \cos x$ which is false for many values of x, for example 0. $-\cos 0 = -1 \neq 1 = \cos 0$.

5. This equation is an identity, since it states the know periodicity of the tangent function. Moreover

$$\begin{aligned} \tan(x + \pi) &= \frac{\tan x + \tan \pi}{1 - \tan x \tan \pi} & &\text{Sum Identity} \\ &= \frac{\tan x + 0}{1 - \tan x(0)} & &\text{Known Values} \\ &= \tan x & &\text{Algebra} \end{aligned}$$

7. This equation is not an identity, as can be shown:

$$\begin{aligned} \csc(2\pi - x) &= \frac{1}{\sin(2\pi - x)} & &\text{Reciprocal Identity} \\ &= \frac{1}{\sin 2\pi \cos x - \cos 2\pi \sin x} & &\text{Difference Identity} \\ &= \frac{1}{(0)\cos x - (1)\sin x} & &\text{Known Values} \\ &= -\frac{1}{\sin x} & &\text{Algebra} \\ &= -\csc x & &\text{Reciprocal Identity} \end{aligned}$$

Thus $\csc(2\pi - x) = \csc x$ is equivalent to $-\csc x = \csc x$ which is false for many values of x, for example $\frac{\pi}{2}$. $-\csc \frac{\pi}{2} = -1 \neq 1 = \csc \frac{\pi}{2}$.

9. This equation is an identity, since it states the known periodicity of the sine function. Moreover

$$\begin{aligned} \sin(x + 2k\pi) &= \sin x \cos(2k\pi) + \cos x \sin(2\pi k) & &\text{Sum Identity} \\ &= \sin x(1) + \cos x(0) & &\text{Known Values} \\ &= \sin x & &\text{Algebra} \end{aligned}$$

11. $\cot\left(\dfrac{\pi}{2} - x\right) = \dfrac{\cos\left(\frac{\pi}{2} - x\right)}{\sin\left(\frac{\pi}{2} - x\right)}$ Quotient Identity

$\quad = \dfrac{\cos\frac{\pi}{2}\cos x + \sin\frac{\pi}{2}\sin x}{\sin\frac{\pi}{2}\cos x - \cos\frac{\pi}{2}\sin x}$ Difference Identities

$\quad = \dfrac{0\cos x + 1\sin x}{1\cos x - 0\sin x}$ Known Values

$\quad = \dfrac{\sin x}{\cos x}$ Algebra

$\quad = \tan x$ Quotient Identity

13. $\csc\left(\dfrac{\pi}{2} - x\right) = \dfrac{1}{\sin\left(\frac{\pi}{2} - x\right)}$ Reciprocal Identity

$\quad = \dfrac{1}{\sin\frac{\pi}{2}\cos x - \cos\frac{\pi}{2}\sin x}$ Difference Identity

$\quad = \dfrac{1}{1\cos x - 0\sin x}$ Known Values

$\quad = \dfrac{1}{\cos x}$ Algebra

$\quad = \sec x$ Reciprocal Identity

15. $\cos(x + 45°) = \cos x \cos 45° - \sin x \sin 45°$ Sum Identity

$\quad = \cos x\left(\dfrac{\sqrt{2}}{2}\right) - \sin x\left(\dfrac{\sqrt{2}}{2}\right)$ Known Values

$\quad = \dfrac{\sqrt{2}}{2}(\cos x - \sin x)$ Algebra

17. $\tan\left(\dfrac{\pi}{3} + x\right) = \dfrac{\tan\frac{\pi}{3} + \tan x}{1 - \tan\frac{\pi}{3}\tan x}$ Sum Identity

$\quad = \dfrac{\sqrt{3} + \tan x}{1 - \sqrt{3}\tan x}$ Known Values

19. $\sin(x - 90°) = \sin x \cos 90° - \cos x \sin 90°$ Difference Identity
$\quad = \sin x(0) - \cos x(1)$ Known Values
$\quad = -\cos x$ Algebra

21. $\cos 20° \cos 25° - \sin 20° \sin 25° = \cos(20° + 25°) = \cos 45° = \dfrac{\sqrt{2}}{2}$

23. $\dfrac{\tan 50° - \tan 20°}{1 + \tan 50° \tan 20°} = \tan(50° - 20°) = \tan 30° = \dfrac{1}{\sqrt{3}}$ or $\dfrac{\sqrt{3}}{3}$

25. $\sin 15° = \sin(45° - 30°) = \sin 45° \cos 30° - \cos 45° \sin 30°$

$\quad = \dfrac{\sqrt{2}}{2} \cdot \dfrac{\sqrt{3}}{2} - \dfrac{\sqrt{2}}{2} \cdot \dfrac{1}{2} = \dfrac{\sqrt{2}}{4}(\sqrt{3} - 1)$

27. $\cos \dfrac{11\pi}{12} = \cos\left(\dfrac{2\pi}{3} + \dfrac{\pi}{4}\right) = \cos \dfrac{2\pi}{3} \cos \dfrac{\pi}{4} - \sin \dfrac{2\pi}{3} \sin \dfrac{\pi}{4} = -\dfrac{1}{2} \cdot \dfrac{\sqrt{2}}{2} - \dfrac{\sqrt{3}}{2} \cdot \dfrac{\sqrt{2}}{2}$

$$= -\dfrac{\sqrt{2}}{4}\,(1 + \sqrt{3})$$

29.

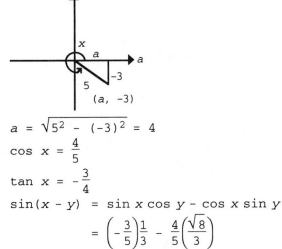

$a = \sqrt{5^2 - (-3)^2} = 4$

$\cos x = \dfrac{4}{5}$

$\tan x = -\dfrac{3}{4}$

$\sin(x - y) = \sin x \cos y - \cos x \sin y$

$$= \left(-\dfrac{3}{5}\right)\dfrac{1}{3} - \dfrac{4}{5}\left(\dfrac{\sqrt{8}}{3}\right)$$

$$= \dfrac{-3 - 4\sqrt{8}}{15}$$

$a = \sqrt{3^2 - (\sqrt{8})^2} = 1$

$\cos y = \dfrac{1}{3}$

$\tan y = \dfrac{\sqrt{8}}{1} = \sqrt{8}$

$\tan(x + y) = \dfrac{\tan x + \tan y}{1 - \tan x \tan y}$

$$= \dfrac{-\frac{3}{4} + \sqrt{8}}{1 - (-\frac{3}{4})\sqrt{8}}$$

$$= \dfrac{-3 + 4\sqrt{8}}{4 + 3\sqrt{8}}$$

$$= \dfrac{4\sqrt{8} - 3}{4 + 3\sqrt{8}}$$

31.

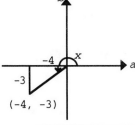

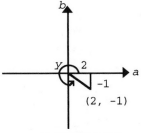

$r = \sqrt{(-4)^2 + (-3)^2} = 5$

$\sin x = -\dfrac{3}{5}$

$\cos x = -\dfrac{4}{5}$

$\sin(x - y) = \sin x \cos y - \cos x \sin y$

$$= \left(-\dfrac{3}{5}\right)\left(\dfrac{2}{\sqrt{5}}\right) - \left(-\dfrac{4}{5}\right)\left(-\dfrac{1}{\sqrt{5}}\right)$$

$$= -\dfrac{6}{5\sqrt{5}} - \dfrac{4}{5\sqrt{5}}$$

$$= \dfrac{-10}{5\sqrt{5}}$$

$$= \dfrac{-2}{\sqrt{5}}$$

$r = \sqrt{2^2 + (-1)^2} = \sqrt{5}$

$\sin y = \dfrac{-1}{\sqrt{5}}$

$\cos y = \dfrac{2}{\sqrt{5}}$

$\tan(x + y) = \dfrac{\tan x + \tan y}{1 - \tan x \tan y}$

$$= \dfrac{\frac{3}{4} + (-\frac{1}{2})}{1 - (\frac{3}{4})(-\frac{1}{2})}$$

$$= \dfrac{6 - 4}{8 + 3}$$

$$= \dfrac{2}{11}$$

33. $\cos 2x = \cos(x + x)$ Algebra
$= \cos x \cos x - \sin x \sin x$ Sum Identity
$= \cos^2 x - \sin^2 x$ Algebra

35. $\cot(x + y) = \dfrac{\cos(x + y)}{\sin(x + y)}$ Quotient Identity

$= \dfrac{\cos x \cos y - \sin x \sin y}{\sin x \cos y + \cos x \sin y}$ Sum Identities

$= \dfrac{\dfrac{\cos x \cos y}{\sin x \sin y} - \dfrac{\sin x \sin y}{\sin x \sin y}}{\dfrac{\sin x \cos y}{\sin x \sin y} + \dfrac{\cos x \sin y}{\sin x \sin y}}$ Algebra

$= \dfrac{\dfrac{\cos x}{\sin x} \cdot \dfrac{\cos y}{\sin y} - 1}{\dfrac{\cos y}{\sin y} + \dfrac{\cos x}{\sin x}}$ Algebra

$= \dfrac{\cot x \cot y - 1}{\cot y + \cot x}$ Quotient Identity

37. $\tan 2x = \tan(x + x)$ Algebra

$= \dfrac{\tan x + \tan x}{1 - \tan x \tan x}$ Sum Identity

$= \dfrac{2 \tan x}{1 - \tan^2 x}$ Algebra

39. $\dfrac{\sin(v + u)}{\sin(v - u)} = \dfrac{\sin v \cos u + \cos v \sin u}{\sin v \cos u - \cos v \sin u}$ Sum and Difference Identities

$= \dfrac{\dfrac{\sin v \cos u}{\sin v \sin u} + \dfrac{\cos v \sin u}{\sin v \sin u}}{\dfrac{\sin v \cos u}{\sin v \sin u} - \dfrac{\cos v \sin u}{\sin v \sin u}}$ Algebra

$= \dfrac{\dfrac{\cos u}{\sin u} + \dfrac{\cos v}{\sin v}}{\dfrac{\cos u}{\sin u} - \dfrac{\cos v}{\sin v}}$ Algebra

$= \dfrac{\cot u + \cot v}{\cot u - \cot v}$ Quotient Identity

41. $\cot x - \tan y = \dfrac{\cos x}{\sin x} - \dfrac{\sin y}{\cos y}$ Quotient Identities

$= \dfrac{\cos x \cos y}{\sin x \cos y} - \dfrac{\sin x \sin y}{\sin x \cos y}$ Algebra

$= \dfrac{\cos x \cos y - \sin x \sin y}{\sin x \cos y}$ Algebra

$= \dfrac{\cos(x + y)}{\sin x \cos y}$ Sum Identity

43. $\tan(x - y) = \dfrac{\tan x - \tan y}{1 + \tan x \tan y}$ Difference Identity

$= \dfrac{\dfrac{1}{\cot x} - \dfrac{1}{\cot y}}{1 + \dfrac{1}{\cot x} \dfrac{1}{\cot y}}$ Reciprocal Identity

$= \dfrac{\cot x \cot y \left(\dfrac{1}{\cot x} - \dfrac{1}{\cot y}\right)}{\cot x \cot y \left(1 + \dfrac{1}{\cot x} \dfrac{1}{\cot y}\right)}$ Algebra

$= \dfrac{\cot y - \cot x}{\cot x \cot y + 1}$ Algebra

45. $\dfrac{\cos(x + h) - \cos x}{h} = \dfrac{\cos x \cos h - \sin x \sin h - \cos x}{h}$ Sum Identity

$= \dfrac{\cos x \cos h - \cos x - \sin x \sin h}{h}$ Algebra

$= \dfrac{\cos x(\cos h - 1) - \sin x \sin h}{h}$ Algebra

$= \cos x\left(\dfrac{\cos h - 1}{h}\right) - \sin x \dfrac{\sin h}{h}$ Algebra

47. $\sin(x - y) = \sin(5.288 - 1.769) = -0.3685$
$\sin x \cos y - \cos x \sin y = \sin 5.288 \cos 1.769 - \cos 5.288 \sin 1.769 = -0.3685$
$\tan(x + y) = \tan(5.288 + 1.769) = 0.9771$
$\dfrac{\tan x + \tan y}{1 - \tan x \tan y} = \dfrac{\tan 5.288 + \tan 1.769}{1 - \tan 5.288 \tan 1.769} = 0.9771$

49. $\sin(x - y) = \sin(42.08° - 68.37°) = -0.4429$
$\sin x \cos y - \cos x \sin y = \sin 42.08° \cos 68.37° - \cos 42.08° \sin 68.37° = -0.4429$
$\tan(x + y) = \tan(42.08° + 68.37°) = -2.682$
$\dfrac{\tan x + \tan y}{1 - \tan x \tan y} = \dfrac{\tan 42.08° + \tan 68.37°}{1 - \tan 42.08° \tan 68.37°} = -2.682$

51. Evaluate each side for a particular set of values of x and y for which each side is defined. If the left side is not equal to the right side, then the equation is not an identity. For example, for $x = 2$ and $y = 1$, both sides are defined, but are not equal.

53. Let $y1 = \sin\left(x + \dfrac{\pi}{6}\right)$

Then $y2 = \sin x \cos \dfrac{\pi}{6} + \cos x \sin \dfrac{\pi}{6}$

$= \sin x \cdot \dfrac{\sqrt{3}}{2} + \cos x \cdot \dfrac{1}{2} = \dfrac{\sqrt{3}}{2} \sin x + \dfrac{1}{2} \cos x.$

The graphs coincide as shown at the right.

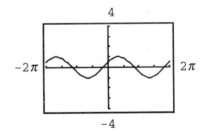

55. Let $y1 = \cos\left(x - \dfrac{3\pi}{4}\right)$

Then $y2 = \cos x \cos \dfrac{3\pi}{4} + \sin x \sin \dfrac{3\pi}{4}$

$= \cos x\left(-\dfrac{\sqrt{2}}{2}\right) + \sin x\left(\dfrac{\sqrt{2}}{2}\right)$

$= -\dfrac{\sqrt{2}}{2} \cos x + \dfrac{\sqrt{2}}{2} \sin x$

The graphs coincide as shown at the right.

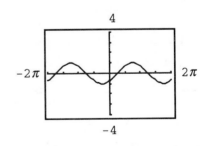

57. Let $y1 = \tan\left(x + \dfrac{2\pi}{3}\right)$

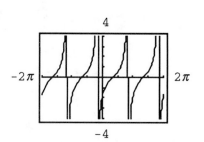

Then $y2 = \dfrac{\tan x + \tan\frac{2\pi}{3}}{1 - \tan x \tan\frac{2\pi}{3}}$

$= \dfrac{\tan x + (-\sqrt{3})}{1 - \tan x(-\sqrt{3})} = \dfrac{\tan x - \sqrt{3}}{1 + \sqrt{3}\,\tan x}$

The graphs coincide as shown at the right.

59. Let $u = \cos^{-1}\left(-\dfrac{4}{5}\right)$, $v = \sin^{-1}\left(-\dfrac{3}{5}\right)$

Then we are asked to evaluate $\sin(u + v)$ which is $\sin u \cos v + \cos u \sin v$

from the sum identity. We know $\sin v = \sin\left[\sin^{-1}\left(-\dfrac{3}{5}\right)\right] = -\dfrac{3}{5}$ and

$\cos u = \cos\left[\cos^{-1}\left(-\dfrac{4}{5}\right)\right] = -\dfrac{4}{5}$ from the function-inverse function identities.

It remains to find $\cos v$ and $\sin u$. Note: $0 \le u \le \pi$ and $-\dfrac{\pi}{2} \le v \le \dfrac{\pi}{2}$

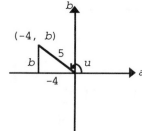

$b = \sqrt{5^2 - (-4)^2} = 3$

$\sin u = \dfrac{3}{5}$

$a = \sqrt{5^2 - (-3)^2} = 4$

$\cos v = \dfrac{4}{5}$

Then $\sin\left[\cos^{-1}\left(-\dfrac{4}{5}\right) + \sin^{-1}\left(-\dfrac{3}{5}\right)\right] = \sin(u + v)$

$= \sin u \cos v + \cos u \sin v$

$= \left(\dfrac{3}{5}\right)\left(\dfrac{4}{5}\right) + \left(-\dfrac{4}{5}\right)\left(-\dfrac{3}{5}\right)$

$= \dfrac{12}{25} + \dfrac{12}{25}$

$= \dfrac{24}{25}$

61. We could proceed as in problem 59. Alternatively, we can shorten the process by recognizing $\arccos \dfrac{1}{2} = \dfrac{\pi}{3}$ and $\arcsin(-1) = -\dfrac{\pi}{2}$. Then $\sin[\arccos \dfrac{1}{2} + \arcsin(-1)] =$

$\sin\left(\dfrac{\pi}{3} + -\dfrac{\pi}{2}\right) = \sin\dfrac{\pi}{3}\cos\left(-\dfrac{\pi}{2}\right) + \cos\dfrac{\pi}{3}\sin\left(-\dfrac{\pi}{2}\right) = \dfrac{\sqrt{3}}{2}(0) + \left(\dfrac{1}{2}\right)(-1) = -\dfrac{1}{2}$.

63. Let $u = \sin^{-1} x$, $v = \cos^{-1} y$. Then $x = \sin u$, $-\dfrac{\pi}{2} \le u \le \dfrac{\pi}{2}$, $y = \cos v$, $0 \le y \le \pi$.

Then $\cos u = \sqrt{1 - \sin^2 u}$ (in Quadrants I, IV) $= \sqrt{1 - x^2}$

$\sin v = \sqrt{1 - \cos^2 v}$ (in Quadrants I, II) $= \sqrt{1 - y^2}$

Hence $\sin(\sin^{-1} x + \cos^{-1} y) = \sin(u + v) = \sin u \cos v + \cos u \sin v$

$= xy + \sqrt{1 - x^2}\sqrt{1 - y^2}$

65. $\cos(x + y + z) = \cos[(x + y) + z]$
$= \cos(x + y)\cos z - \sin(x + y)\sin z$
$= (\cos x \cos y - \sin x \sin y)\cos z$
$\quad - (\sin x \cos y + \cos x \sin y)\sin z$
$= \cos x \cos y \cos z - \sin x \sin y \cos z$
$\quad - \sin x \cos y \sin z - \cos x \sin y \sin z$

67. Let $y1 = \cos 1.2x \cos 0.8x - \sin 1.2x \sin 0.8x$
Then $y2 = \cos(1.2x + 0.8x) = \cos 2x$
The graphs coincide as shown at the right.

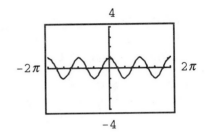

69. $\tan(\theta_2 - \theta_1) = \dfrac{\tan \theta_2 - \tan \theta_1}{1 + \tan \theta_2 \tan \theta_1}$ Difference Identity

$\qquad\qquad = \dfrac{m_2 - m_1}{1 + m_2 m_1}$ Given

$\qquad\qquad = \dfrac{m_2 - m_1}{1 + m_1 m_2}$ Algebra

71.

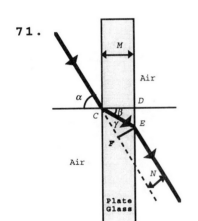

Note: In the text figure we have drawn EF perpendicular to CF, the track of the incident ray. ΔCDE and ΔCEF are right triangles.
Angle $DCF = \beta + \gamma = \alpha$. Hence, $\gamma = \alpha - \beta$.
Denote EC by x. Then,

in ΔCDE $\dfrac{M}{x} = \cos \beta = \sin(90° - \beta)$

in ΔCEF $\dfrac{N}{x} = \sin \gamma = \sin(\alpha - \beta)$

Therefore $x = \dfrac{M}{\sin(90° - \beta)} = \dfrac{N}{\sin(\alpha - \beta)}$

$M \sin(\alpha - \beta) = N \sin(90° - \beta)$ Algebra

$M(\sin \alpha \cos \beta - \cos \alpha \sin \beta) = N(\sin 90° \cos \beta - \cos 90° \sin \beta)$ Difference Identities

$M(\sin \alpha \cos \beta - \cos \alpha \sin \beta) = N(1 \cos \beta - 0 \sin \beta)$ Known Values

$M(\sin \alpha \cos \beta - \cos \alpha \sin \beta) = N \cos \beta$ Algebra

$M \sin \alpha \cos \beta - M \cos \alpha \sin \beta = N \cos \beta$ Algebra

$M \sin \alpha \cos \beta - N \cos \beta = M \cos \alpha \sin \beta$ Algebra

$\cos \beta(M \sin \alpha - N) = M \cos \alpha \sin \beta$ Algebra

$\dfrac{M \sin \alpha - N}{M \cos \alpha} = \dfrac{\sin \beta}{\cos \beta}$ Algebra

$\tan \beta = \dfrac{M \sin \alpha - N}{M \cos \alpha}$ Quotient Identity

$\tan \beta = \dfrac{M \sin \alpha}{M \cos \alpha} - \dfrac{N}{M \cos \alpha}$ Algebra

$\tan \beta = \dfrac{\sin \alpha}{\cos \alpha} - \dfrac{N}{M} \dfrac{1}{\cos \alpha}$ Algebra

$\tan \beta = \tan \alpha - \dfrac{N}{M} \sec \alpha$ Quotient and Reciprocal Identities

73. (A) From the text figure:

In right triangle ABE, we have (1) $\cot \alpha = \dfrac{AB}{AE} = \dfrac{AB}{h}$

In right triangle BCD, we have (2) $\cot \alpha = \dfrac{BC}{CD} = \dfrac{BC}{H}$

In right triangle $EE'D$, we have (3) $\tan \beta = \dfrac{E'D}{EE'} = \dfrac{H - h}{AC} = \dfrac{H - h}{AB + BC}$

From (3), $H - h = (AB + BC)\tan \beta$

From (1) and (2), $AB = h \cot \alpha$ and $BC = H \cot \alpha$

Hence, substituting, we have (4) $H - h = (h \cot \alpha + H \cot \alpha)\tan \beta$, or

$$= (h + H)\cot \alpha \tan \beta$$

Solving for H in terms of h yields:

$$H - h = h \cot \alpha \tan \beta + H \cot \alpha \tan \beta$$

$$H - H \cot \alpha \tan \beta = h \cot \alpha \tan \beta + h$$

$$H(1 - \cot \alpha \tan \beta) = h(\cot \alpha \tan \beta + 1)$$

$$H = h \frac{1 + \cot \alpha \tan \beta}{1 - \cot \alpha \tan \beta}$$

(B) $H = h \dfrac{1 + \frac{\cos \alpha}{\sin \alpha} \frac{\sin \beta}{\cos \beta}}{1 - \frac{\cos \alpha}{\sin \alpha} \frac{\sin \beta}{\cos \beta}}$ Quotient Identities

$H = h \dfrac{\left(1 + \frac{\cos \alpha}{\sin \alpha} \frac{\sin \beta}{\cos \beta}\right) \cdot \sin \alpha \cos \beta}{\left(1 - \frac{\cos \alpha}{\sin \alpha} \frac{\sin \beta}{\cos \beta}\right) \cdot \sin \alpha \cos \beta}$ Algebra

$H = h \dfrac{\sin \alpha \cos \beta + \cos \alpha \sin \beta}{\sin \alpha \cos \beta - \cos \alpha \sin \beta}$ Algebra

$H = h \dfrac{\sin(\alpha + \beta)}{\sin(\alpha - \beta)}$ Sum and Difference Identities

(C) Substitute the given values to obtain

$$H = 4.90 \frac{\sin(46.23° + 46.15°)}{\sin(46.23° - 46.15°)}$$

$$= 4.90 \frac{\sin(92.38°)}{\sin(0.08°)}$$

$$H = 3510 \text{ ft (to three significant digits)}$$

Exercise 6-3

Key Ideas and Formulas

Again, the basic identities listed at the beginning of Exercise 6-1 are used throughout, but not repeated here.

Double-angle Identities

$$\sin 2x = 2 \sin x \cos x$$

$$\cos 2x = \cos^2 x - \sin^2 x = 1 - 2 \sin^2 x = 2 \cos^2 x - 1$$

$$\tan 2x = \frac{2 \tan x}{1 - \tan^2 x} = \frac{2 \cot x}{\cot^2 x - 1} = \frac{2}{\cot x - \tan x}$$

Half-angle Identities

$$\sin \frac{x}{2} = \pm\sqrt{\frac{1 - \cos x}{2}}$$

$$\cos \frac{x}{2} = \pm\sqrt{\frac{1 + \cos x}{2}}$$

$$\tan \frac{x}{2} = \pm\sqrt{\frac{1 - \cos x}{1 + \cos x}} = \frac{\sin x}{1 + \cos x} = \frac{1 - \cos x}{\sin x}$$

where the signs are determined by the quadrant in which $\frac{x}{2}$ lies.

Common Errors:

$\sin 2x \neq 2 \sin x \quad \cos 2x \neq 2 \cos x \qquad \tan 2x \neq 2 \tan x$

$\sin \frac{1}{2}x \neq \frac{1}{2} \sin x \quad \cos \frac{1}{2}x \neq \frac{1}{2} \cos x \qquad \tan \frac{1}{2}x \neq \frac{1}{2} \tan x$

Do not confuse

$\cos^2 x = 1 - \sin^2 x$ with

$\cos 2x = 1 - 2 \sin^2 x$

Both are correct as written, however, the following are common errors:

$\cos^2 x \neq \cos 2x$

$\cos^2 x \neq 1 - 2 \sin^2 x$

$\cos 2x \neq 1 - \sin^2 x$

---

1. $\cos 2(30°) = \cos 60° = \dfrac{1}{2}$

$\cos^2 30° - \sin^2 30° = \left(\dfrac{\sqrt{3}}{2}\right)^2 - \left(\dfrac{1}{2}\right)^2 = \dfrac{3}{4} - \dfrac{1}{4} = \dfrac{2}{4} = \dfrac{1}{2}$

3. $\tan 2\left(\dfrac{\pi}{3}\right) = \tan \dfrac{2\pi}{3} = -\sqrt{3}$

$\dfrac{2}{\cot \frac{\pi}{3} - \tan \frac{\pi}{3}} = \dfrac{2}{\frac{1}{\sqrt{3}} - \sqrt{3}} = \dfrac{2\sqrt{3}}{\sqrt{3}\left(\frac{1}{\sqrt{3}} - \sqrt{3}\right)} = \dfrac{2\sqrt{3}}{1 - 3} = \dfrac{2\sqrt{3}}{-2} = -\sqrt{3}$

5. $\sin \dfrac{\pi}{2} = 1 \qquad \sqrt{\dfrac{1 - \cos \pi}{2}} = \sqrt{\dfrac{1 - (-1)}{2}} = \sqrt{\dfrac{2}{2}} = \sqrt{1} = 1$

7. $\tan 15° = \tan\left(\dfrac{1}{2} \cdot 30°\right)$

$= \dfrac{1 - \cos 30°}{\sin 30°}$

$= \dfrac{1 - \frac{\sqrt{3}}{2}}{\frac{1}{2}}$

$= 2 - \sqrt{3}$

9. $\cos 112.5° = \cos\left(\dfrac{1}{2} \cdot 225°\right)$

$= -\sqrt{\dfrac{1 + \cos 225°}{2}}$

(We use − since 112.5° is a II quadrant angle)

$= -\sqrt{\dfrac{1 + \left(-\frac{\sqrt{2}}{2}\right)}{2}}$

$= -\sqrt{\dfrac{1 - \frac{\sqrt{2}}{2}}{2}}$

$= -\sqrt{\dfrac{2 - \sqrt{2}}{4}}$

$= -\dfrac{\sqrt{2 - \sqrt{2}}}{2}$

11.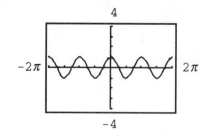

13.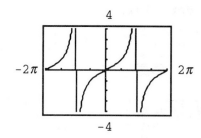

15. $(\sin x + \cos x)^2 = \sin^2 x + 2 \sin x \cos x + \cos^2 x$ Algebra

$\qquad\qquad\qquad\quad\;\; = 1 + 2 \sin x \cos x$ Pythagorean Identity

$\qquad\qquad\qquad\quad\;\; = 1 + \sin 2x$ Double-angle Identity

17. $\dfrac{1}{2}(1 - \cos 2x) = \dfrac{1}{2}[1 - (1 - 2 \sin^2 x)]$ Double-angle Identity

$\qquad\qquad\qquad = \dfrac{1}{2}[1 - 1 + 2 \sin^2 x)]$ Algebra

$\qquad\qquad\qquad = \dfrac{1}{2}(2 \sin^2 x)$ Algebra

$\qquad\qquad\qquad = \sin^2 x$ Algebra

19. $\tan x \sin 2x = \dfrac{\sin x}{\cos x} 2 \sin x \cos x$ Quotient and Double-angle Identities

$\qquad\qquad\quad = 2 \sin^2 x$ Algebra

$\qquad\qquad\quad = 2 \sin^2 x - 1 + 1$ Algebra

$\qquad\qquad\quad = 1 - (1 - 2 \sin^2 x)$ Algebra

$\qquad\qquad\quad = 1 - \cos 2x$ Double-angle Identity

21. $\sin^2 \dfrac{x}{2} = \left(\pm\sqrt{\dfrac{1 - \cos x}{2}}\right)^2$ Half-angle Identity

$\qquad\quad = \left(\sqrt{\dfrac{1 - \cos x}{2}}\right)^2$ Algebra

$\qquad\quad = \dfrac{1 - \cos x}{2}$ Algebra

23. $\cot \dfrac{\theta}{2} = \dfrac{\cos \frac{\theta}{2}}{\sin \frac{\theta}{2}}$ Quotient Identity

$\qquad\quad = \dfrac{2 \sin \frac{\theta}{2} \cos \frac{\theta}{2}}{2 \sin^2 \frac{\theta}{2}}$ Algebra

$\qquad\quad = \dfrac{\sin 2 (\frac{\theta}{2})}{2 \sin^2 \frac{\theta}{2}}$ Dougle-angle Identity

$\qquad\quad = \dfrac{\sin \theta}{2 \sin^2 \frac{\theta}{2}}$ Algebra

$\qquad\quad = \dfrac{\sin \theta}{2\left[\pm\sqrt{\frac{1 - \cos \theta}{2}}\right]^2}$ Half-angle Identity

$\qquad\quad = \dfrac{\sin \theta}{2 \cdot \frac{1 - \cos \theta}{2}}$ Algebra

$\qquad\quad = \dfrac{\sin \theta}{1 - \cos \theta}$ Algebra

25. $\cos 2u = \cos^2 u - \sin^2 u$ Double-angle Identity

$= \dfrac{\cos^2 u - \sin^2 u}{1}$ Algebra

$= \dfrac{\cos^2 u - \sin^2 u}{\cos^2 u + \sin^2 u}$ Pythagorean Identity

$= \dfrac{\frac{\cos^2 u}{\cos^2 u} - \frac{\sin^2 u}{\cos^2 u}}{\frac{\cos^2 u}{\cos^2 u} + \frac{\sin^2 u}{\cos^2 u}}$ Algebra

$= \dfrac{1 - \frac{\sin^2 u}{\cos^2 u}}{1 + \frac{\sin^2 u}{\cos^2 u}}$ Algebra

$= \dfrac{1 - \tan^2 u}{1 + \tan^2 u}$ Quotient Identity

27. $2 \csc 2x = 2 \cdot \dfrac{1}{\sin 2x}$ Reciprocal Identity

$= \dfrac{2}{\sin 2x}$ Algebra

$= \dfrac{2}{2 \sin x \cos x}$ Double-angle Identity

$= \dfrac{1}{\sin x \cos x}$ Algebra

$= \dfrac{\cos^2 x + \sin^2 x}{\sin x \cos x}$ Pythagorean Identity

$= \dfrac{\frac{\cos^2 x}{\cos^2 x} + \frac{\sin^2 x}{\cos^2 x}}{\frac{\sin x \cos x}{\cos^2 x}}$ Algebra

$= \dfrac{1 + \frac{\sin^2 x}{\cos^2 x}}{\frac{\sin x}{\cos x}}$ Algebra

$= \dfrac{1 + \tan^2 x}{\tan x}$ Quotient Identity

29. This equation is not an identity, since the right side $2 \sin x \cos x = \sin 2x$. Thus $\cos 2x = 2 \sin x \cos x$ is equivalent to $\cos 2x = \sin 2x$ which is false for many values of x, for example 0. $\cos(2 \cdot 0) = 1 \neq 0 = \sin(2 \cdot 0)$.

31. This equation is an identity, as can be shown:

$\tan 2x = \dfrac{2 \tan x}{1 - \tan^2 x}$ Dougle-angle Identity

$= \dfrac{-2 \tan x}{-(1 - \tan^2 x)}$ Algebra

$= \dfrac{-2 \tan x}{\tan^2 x - 1}$ Algebra

33. This equation is not an identity. For example, if $x = \dfrac{\pi}{6}$, then

Left side: $\cot 2x = \cot 2\left(\dfrac{\pi}{6}\right) = \cot\left(\dfrac{\pi}{3}\right) = \dfrac{1}{\sqrt{3}}$

Right side: $\dfrac{2 \cot x}{1 - \cot^2 x} = \dfrac{2 \cot\left(\frac{\pi}{6}\right)}{1 - [\cot \frac{\pi}{6}]^2} = \dfrac{2\sqrt{3}}{1 - (\sqrt{3})^2} = \dfrac{2\sqrt{3}}{1 - 3} = \dfrac{2\sqrt{3}}{-2} = -\sqrt{3}$

35.
$$a = -\sqrt{5^2 - 3^2} = -4$$
$$\cos x = -\frac{4}{5} \qquad \tan x = -\frac{3}{4}$$

$$\sin 2x = 2 \sin x \cos x = 2\left(\frac{3}{5}\right)\left(-\frac{4}{5}\right) = -\frac{24}{25}$$

$$\cos 2x = 1 - 2 \sin^2 x = 1 - 2\left(\frac{3}{5}\right)^2 = 1 - \frac{18}{25} = \frac{7}{25}$$

$$\tan 2x = \frac{\sin 2x}{\cos 2x} = \left(-\frac{24}{25}\right) \div \left(\frac{7}{25}\right) = -\frac{24}{25} \cdot \frac{25}{7} = -\frac{24}{7}$$

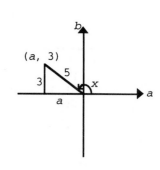

37.
$$r = \sqrt{12^2 + (-5)^2} = 13$$
$$\sin x = -\frac{5}{13} \qquad \cos x = \frac{12}{13}$$

$$\sin 2x = 2 \sin x \cos x = 2\left(-\frac{5}{13}\right)\left(\frac{12}{13}\right) = -\frac{120}{169}$$

$$\cos 2x = 2 \cos^2 x - 1 = 2\left(\frac{12}{13}\right)^2 - 1 = \frac{288}{169} - 1 = \frac{119}{169}$$

$$\tan 2x = \frac{\sin 2x}{\cos 2x} = \left(-\frac{120}{169}\right) \div \left(\frac{119}{169}\right) = -\frac{120}{169} \cdot \frac{169}{119} = -\frac{120}{119}$$

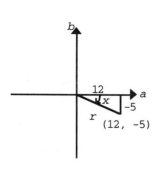

39. Since $\pi < x < \dfrac{3\pi}{2}$, $\dfrac{\pi}{2} < \dfrac{x}{2} < \dfrac{3\pi}{4}$,

$\sin \dfrac{x}{2}$ will be positive, $\cos \dfrac{x}{2}$,

$\tan \dfrac{x}{2}$ will be negative.

$$a = -\sqrt{3^2 - (-1)^2} = -\sqrt{8}$$
$$\cos x = \frac{-\sqrt{8}}{3}$$

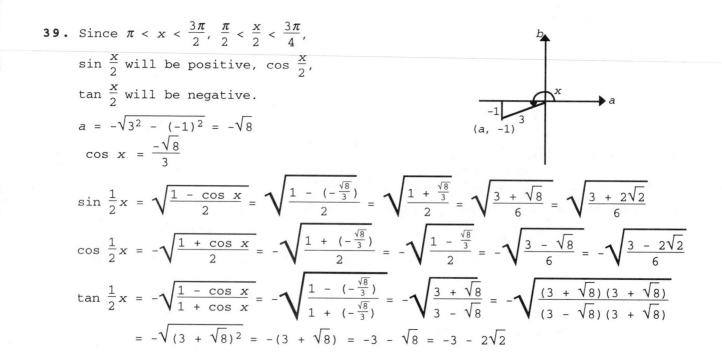

$$\sin \frac{1}{2}x = \sqrt{\frac{1 - \cos x}{2}} = \sqrt{\frac{1 - \left(-\frac{\sqrt{8}}{3}\right)}{2}} = \sqrt{\frac{1 + \frac{\sqrt{8}}{3}}{2}} = \sqrt{\frac{3 + \sqrt{8}}{6}} = \sqrt{\frac{3 + 2\sqrt{2}}{6}}$$

$$\cos \frac{1}{2}x = -\sqrt{\frac{1 + \cos x}{2}} = -\sqrt{\frac{1 + \left(-\frac{\sqrt{8}}{3}\right)}{2}} = -\sqrt{\frac{1 - \frac{\sqrt{8}}{3}}{2}} = -\sqrt{\frac{3 - \sqrt{8}}{6}} = -\sqrt{\frac{3 - 2\sqrt{2}}{6}}$$

$$\tan \frac{1}{2}x = -\sqrt{\frac{1 - \cos x}{1 + \cos x}} = -\sqrt{\frac{1 - \left(-\frac{\sqrt{8}}{3}\right)}{1 + \left(-\frac{\sqrt{8}}{3}\right)}} = -\sqrt{\frac{3 + \sqrt{8}}{3 - \sqrt{8}}} = -\sqrt{\frac{(3 + \sqrt{8})(3 + \sqrt{8})}{(3 - \sqrt{8})(3 + \sqrt{8})}}$$

$$= -\sqrt{(3 + \sqrt{8})^2} = -(3 + \sqrt{8}) = -3 - \sqrt{8} = -3 - 2\sqrt{2}$$

41. Since $-\pi < x < -\frac{\pi}{2}$, $-\frac{\pi}{2} < \frac{x}{2} < -\frac{\pi}{4}$, $\cos \frac{x}{2}$ will be positive, $\sin \frac{x}{2}$, $\tan \frac{x}{2}$ will be negative.

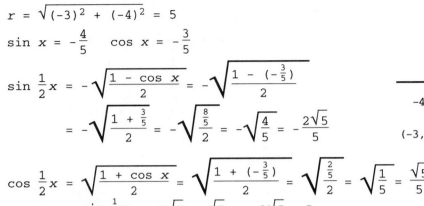

$$r = \sqrt{(-3)^2 + (-4)^2} = 5$$

$$\sin x = -\frac{4}{5} \quad \cos x = -\frac{3}{5}$$

$$\sin \frac{1}{2}x = -\sqrt{\frac{1 - \cos x}{2}} = -\sqrt{\frac{1 - (-\frac{3}{5})}{2}}$$

$$= -\sqrt{\frac{1 + \frac{3}{5}}{2}} = -\sqrt{\frac{\frac{8}{5}}{2}} = -\sqrt{\frac{4}{5}} = -\frac{2\sqrt{5}}{5}$$

$$\cos \frac{1}{2}x = \sqrt{\frac{1 + \cos x}{2}} = \sqrt{\frac{1 + (-\frac{3}{5})}{2}} = \sqrt{\frac{\frac{2}{5}}{2}} = \sqrt{\frac{1}{5}} = \frac{\sqrt{5}}{5}$$

$$\tan \frac{1}{2}x = \frac{\sin \frac{1}{2}x}{\cos \frac{1}{2}x} = \frac{-2\sqrt{5}}{5} \div \frac{\sqrt{5}}{5} = -\frac{2\sqrt{5}}{5} \cdot \frac{5}{\sqrt{5}} = -2$$

43. (A) 2θ is a second quadrant angle, since θ is a first quadrant angle and $\tan 2\theta$ is negative for 2θ in the second quadrant and not for 2θ in the first.

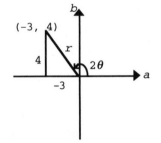

(B) Construct a reference triangle for 2θ in the second quadrant with $(a, b) = (-3, 4)$. Use the Pythagorean theorem to find $r = 5$.

Thus, $\sin 2\theta = \frac{4}{5}$ and $\cos 2\theta = -\frac{3}{5}$.

(C) The double angle identities $\cos 2\theta = 1 - 2 \sin^2 \theta$ and $\cos 2\theta = 2 \cos^2 \theta - 1$.

(D) Use the identities in part (C) in the form

$$\sin \theta = \sqrt{\frac{1 - \cos 2\theta}{2}} \text{ and } \cos \theta = \sqrt{\frac{1 + \cos 2\theta}{2}}$$

The positive radicals are used because θ is in quadrant one.

(E) $\sin \theta = \sqrt{\frac{1 - (-\frac{3}{5})}{2}} = \sqrt{\frac{5 + 3}{10}} = \sqrt{\frac{8}{10}} = \sqrt{\frac{4}{5}} = \frac{2}{\sqrt{5}} = \frac{2\sqrt{5}}{5}$

$\cos \theta = \sqrt{\frac{1 + (-\frac{3}{5})}{2}} = \sqrt{\frac{5 - 3}{10}} = \sqrt{\frac{2}{10}} = \sqrt{\frac{1}{5}} = \frac{1}{\sqrt{5}} = \frac{\sqrt{5}}{5}$

45. (A) $\tan[2(252.06°)] = -0.72335$

$\frac{2 \tan x}{1 - \tan^2 x} = \frac{2 \tan(252.06°)}{1 - \tan^2(252.06°)}$
$= -0.72335$

(B) $\cos \frac{252.06°}{2} = -0.58821$

$-\sqrt{\frac{1 + \cos 252.06°}{2}} = -0.58821$

47. (A) $\tan[2(0.93457)] = -3.2518$

$\frac{2 \tan(0.93457)}{1 - \tan^2(0.93457)} = -3.2518$

(B) $\cos \frac{0.93457}{2} = 0.89279$

$\sqrt{\frac{1 + \cos 0.93457}{2}} = 0.89279$

49.

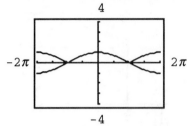

The graphs appear to coincide on the interval $[-\pi, \pi]$.

51.

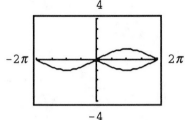

The graphs appear to coincide on the interval $[-2\pi, 0]$.

53.
$$\begin{aligned}
\cos 3x &= \cos(2x + x) && \text{Algebra} \\
&= \cos 2x \cos x - \sin 2x \sin x && \text{Sum Identity} \\
&= (2 \cos^2 x - 1)\cos x - 2 \sin x \cos x \sin x && \text{Double-angle Identities} \\
&= 2 \cos^3 x - \cos x - 2 \sin^2 x \cos x && \text{Algebra} \\
&= 2 \cos^3 x - \cos x - 2(1 - \cos^2 x)\cos x && \text{Pythagorean Identity} \\
&= 2 \cos^3 x - \cos x - 2 \cos x + 2 \cos^3 x && \text{Algebra} \\
&= 4 \cos^3 x - 3 \cos x && \text{Algebra}
\end{aligned}$$

55.
$$\begin{aligned}
\cos 4x &= \cos 2(2x) && \text{Algebra} \\
&= 2 \cos^2 2x - 1 && \text{Double-angle Identity} \\
&= 2(2 \cos^2 x - 1)^2 - 1 && \text{Double-angle Identity} \\
&= 2(4 \cos^4 x - 4 \cos^2 x + 1) - 1 && \text{Algebra} \\
&= 8 \cos^4 x - 8 \cos^2 x + 2 - 1 && \text{Algebra} \\
&= 8 \cos^4 x - 8 \cos^2 x + 1 && \text{Algebra}
\end{aligned}$$

57. Let $u = \cos^{-1} \dfrac{3}{5}$. Then $\cos u = \dfrac{3}{5}$, $0 < u < \pi$. $\cos\left(2 \cos^{-1} \dfrac{3}{5}\right) = \cos 2u$

$= 2 \cos^2 u - 1 = 2\left(\dfrac{3}{5}\right)^2 - 1 = -\dfrac{7}{25}$

59. Let $u = \cos^{-1}\left(-\dfrac{4}{5}\right)$. Then $\cos u = -\dfrac{4}{5}$, $0 < u < \pi$.

$b = \sqrt{5^2 - (-4)^2} = 3$

$\tan u = -\dfrac{3}{4}$

$\tan\left[2 \cos^{-1}\left(-\dfrac{4}{5}\right)\right] \quad = \tan 2u = \dfrac{2 \tan u}{1 - \tan^2 u} = \dfrac{2\left(-\dfrac{3}{4}\right)}{1 - \left(-\dfrac{3}{4}\right)^2}$

$= \dfrac{\frac{-3}{2}}{1 - \frac{9}{16}} = \dfrac{\frac{-3}{2}}{\frac{7}{16}} = \left(-\dfrac{3}{2}\right) \div \dfrac{7}{16}$

$= -\dfrac{3}{2} \cdot \dfrac{16}{7} = -\dfrac{24}{7}$

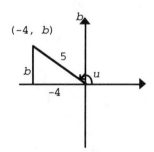

61. Let $u = \cos^{-1}\left(-\dfrac{3}{5}\right)$. Then $\cos u = -\dfrac{3}{5}$, $0 < u < \pi$. Since $0 < \dfrac{1}{2}u < \dfrac{\pi}{2}$,

$\cos \dfrac{1}{2}u$ is positive. So

$\cos\left[\dfrac{1}{2} \cos^{-1}\left(-\dfrac{3}{5}\right)\right] = \cos \dfrac{1}{2}u = \sqrt{\dfrac{1 + \cos u}{2}} = \sqrt{\dfrac{1 + -\frac{3}{5}}{2}} = \sqrt{\dfrac{\frac{2}{5}}{2}} = \sqrt{\dfrac{1}{5}} = \dfrac{\sqrt{5}}{5}$

63. Here is the graph of $f(x) = \csc x - \cot x$ in a graphing utility.

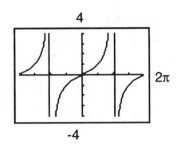

The graph appears to have the form $y = A \tan Bx$. Since the period is 2π, $\frac{\pi}{B} = 2\pi$, thus $B = \frac{1}{2}$. The graph of $A \tan \frac{1}{2} x$ shown appears to pass through $\left(\frac{\pi}{2}, 1 \right)$, thus

$$1 = A \tan \frac{1}{2} \cdot \frac{\pi}{2}$$

$$1 = A \tan \frac{\pi}{4}$$

$$1 = A$$

The equation of the graph appears to be $y = \tan \frac{1}{2} x$. We verify that $\csc x - \cot x = \tan \frac{1}{2} x$ is an identity:

$$\csc x - \cot x = \frac{1}{\sin x} - \frac{\cos x}{\sin x} \qquad \text{Quotient and Reciprocal Identities}$$

$$= \frac{1 - \cos x}{\sin x} \qquad \text{Algebra}$$

$$= \tan \frac{1}{2} x \qquad \text{Half-angle Identity}$$

65. Here is the graph of $f(x) = \frac{1 - 2 \cos 2x}{2 \sin x - 1}$ in a graphing utility.

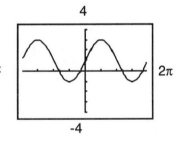

The graph appears to be a standard sine curve shifted up 1 unit. It has amplitude $\frac{3 - (-1)}{2} = 2$ and period 2π, thus, it appears to be the graph of $y = 1 + 2 \sin x$. We verify that $\frac{1 - 2 \cos 2x}{2 \sin x - 1} = 1 + 2 \sin x$ is an identity:

$$\frac{1 - 2 \cos 2x}{2 \sin x + 1} = \frac{1 - 2(1 - 2 \sin^2 x)}{2 \sin x + 1} \qquad \text{Dougle-angle Identity}$$

$$= \frac{1 - 2 + 4 \sin^2 x}{2 \sin x + 1} \qquad \text{Algebra}$$

$$= \frac{4 \sin^2 x - 1}{2 \sin x + 1} \qquad \text{Algebra}$$

$$= \frac{(2 \sin x - 1)(2 \sin x + 1)}{2 \sin x + 1} \qquad \text{Algebra}$$

$$= 2 \sin x - 1 \qquad \text{Algebra}$$

67. Here is the graph of $f(x) = \frac{1}{\cot x \sin 2x - 1}$ in a graphing utility.

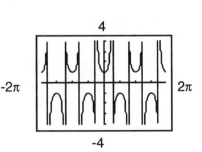

The graph appears to have the form $y = A \sec Bx$. Since the period is π, $\frac{2\pi}{B} = \pi$, thus $B = 2$.

The graph of $y = A \sec 2x$ shown appears to pass through $(0, 1)$, thus

$$1 = A \sec 2 \cdot 0$$
$$1 = A \sec 0$$
$$1 = A$$

The equation of the graph appears to be $y = \sec 2x$. We verify that

$$\frac{1}{\cot x \sin 2x - 1} = \sec 2x \text{ is an identity.}$$

$$\frac{1}{\cot x \sin 2x - 1} = \frac{1}{\frac{\cos x}{\sin x} \sin 2x - 1} \qquad \text{Quotient Identity}$$

$$= \frac{1}{\frac{\cos x}{\sin x} 2 \sin x \cos x - 1} \qquad \text{Double-angle Identity}$$

$$= \frac{1}{2 \cos^2 x - 1} \qquad \text{Algebra}$$

$$= \frac{1}{\cos 2x} \qquad \text{Double-angle Identity}$$

$$= \sec 2x \qquad \text{Reciprocal Identity}$$

69. From the figure we see:

ACD is a right triangle, hence $\cos \theta = \frac{7}{8}$, $\theta = \cos^{-1} \frac{7}{8}$

ABC is a right triangle, with angle $BAC = 2\theta$. Hence $\cos 2\theta = \frac{7}{x}$. Using the hint,

$$\cos 2\theta = 2 \cos^2 \theta - 1$$

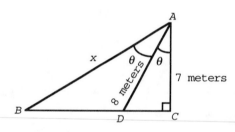

$$= 2 \left(\frac{7}{8}\right)^2 - 1$$

$$= 2 \cdot \frac{49}{64} - 1$$

$$= \frac{34}{64}$$

$$= \frac{17}{32}$$

Thus

$$\frac{17}{32} = \frac{7}{x}$$
$$17x = 32(7)$$
$$x = \frac{32(7)}{17} = \frac{224}{17}$$

To three decimal places

$$x = \frac{224}{17} = 13.176 \text{ meters}$$

$$\theta = \cos^{-1} \frac{7}{8} = 28.955° \quad \text{(calculator in degree mode)}$$

71. (A) Since $2 \sin \theta \cos \theta = \sin 2\theta$ by the double-angle identity, we can write

$$d = \frac{2v_0^2 \sin \theta \cos \theta}{32 \text{ ft/sec}^2}$$

$$d = \frac{v_0^2 (2 \sin \theta \cos \theta)}{32 \text{ ft/sec}^2}$$

$$d = \frac{v_0^2 \sin 2\theta}{32 \text{ ft/sec}^2}$$

(B) For fixed v_0, the quantity $\dfrac{v_0^2 \sin 2\theta}{32}$ will be maximum when $\sin 2\theta$ achieves its maximum value, namely 1. Then

$$\sin 2\theta = 1$$
$$2\theta = 90°$$
$$\theta = 45°$$

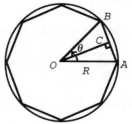

73. (A) Here is a drawing of a typical situation. (The case $n = 8$ is shown, but the reasoning is independent of n):

We note: Area of polygon = n(Area of triangle OAB)
= $2n$(Area of triangle OAC)

$$= 2n\left(\frac{1}{2} \cdot OC \cdot AC\right)$$

$$= n \cdot OC \cdot AC$$

In triangle OAC, angle $AOC = \dfrac{1}{2}\theta = \dfrac{1}{2} \cdot \dfrac{2\pi}{n} = \dfrac{\pi}{n}$

$$\frac{OC}{R} = \cos AOC = \cos \frac{\pi}{n} \qquad \frac{AC}{R} = \sin AOC = \sin \frac{\pi}{n}$$

Thus, $OC = R \cos \dfrac{\pi}{n}$, $AC = R \sin \dfrac{\pi}{n}$

Hence,

Area of polygon = $n\, R \cos \dfrac{\pi}{n}\, R \sin \dfrac{\pi}{n}$

$$= nR^2 \sin \frac{\pi}{n} \cos \frac{\pi}{n}$$

$$= \frac{1}{2} nR^2 \cdot 2 \sin \frac{\pi}{n} \cos \frac{\pi}{n}$$

$$= \frac{1}{2} nR^2 \sin \frac{2\pi}{n} \quad \text{by the double-angle identity for sine}$$

(B) TABLE 1

n	10	100	1,000	10,000
A_n	2.93893	3.13953	3.14157	3.14159

(C) A_n appears to approach π, the area of the circle with radius 1.

(D) A_n will not exactly equal the area of the circumscribing circle for any n no matter how large n is chosen; however, A_n can be made as close to the area of the circumscribing circle as we like by making n sufficiently large.

Exercise 6-4

Key Ideas and Formulas

Product-Sum Identities:

$$\sin x \cos y = \frac{1}{2}[\sin(x + y) + \sin(x - y)]$$

$$\cos x \sin y = \frac{1}{2}[\sin(x + y) - \sin(x - y)]$$

$$\sin x \sin y = \frac{1}{2}[\cos(x - y) - \cos(x + y)]$$

$$\cos x \cos y = \frac{1}{2}[\cos(x + y) + \cos(x - y)]$$

Sum-Product Identities:

$$\sin x + \sin y = 2 \sin \frac{x + y}{2} \cos \frac{x - y}{2}$$

$$\sin x - \sin y = 2 \cos \frac{x + y}{2} \sin \frac{x - y}{2}$$

$$\cos x + \cos y = 2 \cos \frac{x + y}{2} \cos \frac{x - y}{2}$$

$$\cos x - \cos y = -2 \sin \frac{x + y}{2} \sin \frac{x - y}{2}$$

1. $\sin x \cos y = \frac{1}{2}[\sin(x + y) + \sin(x - y)]$

$\sin 3m \cos m = \frac{1}{2}[\sin(3m + m) + \sin(3m - m)]$

$= \frac{1}{2}(\sin 4m + \sin 2m)$

$= \frac{1}{2}\sin 4m + \frac{1}{2}\sin 2m$

3. $\sin x \sin y = \frac{1}{2}[\cos(x - y) - \cos(x + y)]$

$\sin u \sin 3u = \frac{1}{2}[\cos(u - 3u) - \cos(u + 3u)]$

$= \frac{1}{2}(\cos(-2u) - \cos 4u)$

$= \frac{1}{2}(\cos 2u - \cos 4u)$

$= \frac{1}{2}\cos 2u - \frac{1}{2}\cos 4u$

5. $\sin x + \sin y = 2 \sin \dfrac{x + y}{2} \cos \dfrac{x - y}{2}$

$\sin 3t + \sin t = 2 \sin \dfrac{3t + t}{2} \cos \dfrac{3t - t}{2}$

$= 2 \sin 2t \cos t$

7. $\cos x - \cos y = -2 \sin \dfrac{x + y}{2} \sin \dfrac{x - y}{2}$

$\cos 5w - \cos 9w = -2 \sin \dfrac{5w + 9w}{2} \sin \dfrac{5w - 9w}{2}$

$= -2 \sin 7w \sin(-2w)$

$= 2 \sin 7w \sin 2w$

9. $\sin x \sin y = \frac{1}{2}[\cos(x - y) - \cos(x + y)]$

$\sin 75° \sin 15° = \frac{1}{2}[\cos(75° - 15°) - \cos(75° + 15°)]$

$= \frac{1}{2}[\cos 60° - \cos 90°]$

$= \frac{1}{2}\left(\frac{1}{2} - 0\right)$

$= \frac{1}{4}$

11. $\cos x \cos y = \frac{1}{2}[\cos(x + y) + \cos(x - y)]$

$\cos 7.5° \cos 52.5° = \frac{1}{2}[\cos(7.5° + 52.5°) + \cos(7.5° - 52.5°)]$

$= \frac{1}{2}[\cos 60° + \cos(-45°)]$

$= \frac{1}{2}\left[\frac{1}{2} + \frac{\sqrt{2}}{2}\right]$

$= \frac{1 + \sqrt{2}}{4}$

13.
$$\sin x - \sin y = 2 \cos \frac{x + y}{2} \sin \frac{x - y}{2}$$
$$\sin 195° - \sin 105° = 2 \cos \frac{195° + 105°}{2} \sin \frac{195° - 105°}{2}$$
$$= 2 \cos 150° \sin 45°$$
$$= 2\left(-\frac{\sqrt{3}}{2}\right)\left(\frac{\sqrt{2}}{2}\right)$$
$$= -\frac{\sqrt{6}}{2}$$

15.
$$\cos x - \cos y = -2 \sin \frac{x + y}{2} \sin \frac{x - y}{2}$$
$$\cos 165° - \cos 105° = -2 \sin \frac{165° + 105°}{2} \sin \frac{165° - 105°}{2}$$
$$= -2 \sin 135° \sin 30°$$
$$= -2\left(\frac{\sqrt{2}}{2}\right)\left(\frac{1}{2}\right)$$
$$= -\frac{\sqrt{2}}{2}$$

17.
$$\cos(x + y) = \cos x \cos y - \sin x \sin y$$
$$\underline{\cos(x - y) = \cos x \cos y + \sin x \sin y}$$
$$\cos(x + y) + \cos(x - y) = 2 \cos x \cos y \qquad \text{(adding the above)}$$
$$\cos x \cos y = \frac{1}{2}[\cos(x + y) + \cos(x - y)]$$

19. Let $x = u + v$ and $y = u - v$ and solve for u and v in terms of x and y:
$$x = u + v$$
$$\underline{y = u - v}$$
$$x + y = 2u \qquad x - y = 2v$$
$$u = \frac{x + y}{2} \qquad v = \frac{x - y}{2}$$

Substitute these results into $\sin u \sin v = \frac{1}{2}[\cos(u - v) - \cos(u + v)]$ to obtain:
$$\sin \frac{x + y}{2} \sin\left(\frac{x - y}{2}\right) = \frac{1}{2}[\cos y - \cos x]$$
or
$$-\sin \frac{x + y}{2} \sin \frac{x - y}{2} = \frac{1}{2}[\cos x - \cos y]$$
or
$$\cos x - \cos y = -2 \sin \frac{x + y}{2} \sin \frac{x - y}{2}$$

21. $\dfrac{\sin 2t + \sin 4t}{\cos 2t - \cos 4t} = \dfrac{2 \sin \frac{2t + 4t}{2} \cos \frac{2t - 4t}{2}}{-2 \sin \frac{2t + 4t}{2} \sin \frac{2t - 4t}{2}}$ Sum-product Identities

$$= \frac{2 \sin 3t \cos(-t)}{-2 \sin 3t \sin(-t)} \qquad \text{Algebra}$$

$$= \frac{2 \sin 3t \cos t}{2 \sin 3t \sin t} \qquad \text{Identities for Negatives}$$

$$= \frac{\cos t}{\sin t} \qquad \text{Algebra}$$

$$= \cot t \qquad \text{Quotient Identity}$$

23. $$\frac{\sin x - \sin y}{\cos x - \cos y} = \frac{2 \cos \frac{x + y}{2} \sin \frac{x - y}{2}}{-2 \sin \frac{x + y}{2} \sin \frac{x - y}{2}} \qquad \text{Sum-product Identities}$$

$$= -\frac{\cos \frac{x + y}{2}}{\sin \frac{x + y}{2}} \qquad\qquad\qquad \text{Algebra}$$

$$= -\cot \frac{x + y}{2} \qquad\qquad\qquad \text{Quotient Identity}$$

25.   The equation is an identity, as can be shown:

$$2 \sin 2x \sin x = 2\left(\frac{1}{2}\right)[\cos(2x - x) - \cos(2x + x)] \quad \text{Product-sum Identities}$$
$$= \cos(2x - x) - \cos(2x + x) \qquad\qquad \text{Algebra}$$
$$= \cos x - \cos 3x \qquad\qquad\qquad\qquad \text{Algebra}$$

27.   The equation is not an identity. $\sin(x + y) \cos(x - y) - \cos(x + y) \sin(x - y)$ is equivalent to $\sin[(x + y) - (x - y)] = \sin 2y$ by the difference identity for sine. $\cos 2x = \sin 2y$ is not an identity.

29.   The equation is not an identity. For example, let $x = 0$, $y = \frac{\pi}{6}$.

Then Left side $= \sec\left(\frac{\pi}{6}\right) \csc\left(-\frac{\pi}{6}\right)\left[\sin 0 + \sin \frac{2\pi}{6}\right]$

$$= \frac{2}{\sqrt{3}}(-2)\left(\frac{\sqrt{3}}{2}\right) = -2$$

Right side $= 2$

$$-2 \neq 2.$$

31.   (A) $\cos 172.63° \sin 20.177° = -0.34207$

$\frac{1}{2}[\sin(172.63° + 20.177°) - \sin(172.63° - 20.177°)] = -0.34207$

(B) $\cos 172.63° + \cos 20.177° = -0.05311$

$2 \cos \frac{172.63° + 20.177°}{2} \cos \frac{172.63° - 20.177°}{2} = -0.05311$

33.   (A) $\cos 1.1255 \sin 3.6014 = -0.19115$

$\frac{1}{2}[\sin(1.1255 + 3.6014) - \sin(1.1255 - 3.6014)] = -0.19115$

(B) $\cos 1.1255 + \cos 3.6014 = -0.46541$

$2 \cos \frac{1.1255 + 3.6014}{2} \cos \frac{1.1255 - 3.6014}{2} = -0.46541$

35. $\sin 2x + \sin x = 2 \sin \frac{2x + x}{2} \cos \frac{2x - x}{2}$

$$= 2 \sin \frac{3x}{2} \cos \frac{x}{2}$$

Graph $y1 = \sin 2x + \sin x$ and
$$y2 = 2 \sin \frac{3x}{2} \cos \frac{x}{2}$$

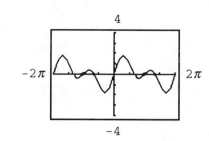

37. $\cos 1.7x - \cos 0.3x$

$$= -2 \sin \frac{1.7x + 0.3x}{2} \sin \frac{1.7x - 0.3x}{2}$$

$$= -2 \sin x \sin 0.7x$$

Graph $y1 = \cos 1.7x - \cos 0.3x$
  $y2 = -2 \sin x \sin 0.7x$

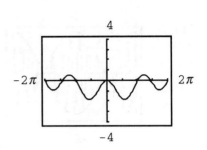

39. $\sin 3x \cos x = \frac{1}{2}[\sin(3x + x) + \sin(3x - x)]$

$$= \frac{1}{2}(\sin 4x + \sin 2x)$$

Graph $y1 = \sin 3x \cos x$ and
  $y2 = \frac{1}{2}(\sin 4x + \sin 2x)$

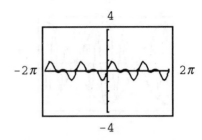

41. $\sin 2.3x \sin 0.7x$

$$= \frac{1}{2}[\cos(2.3x - 0.7x) - \cos(2.3x + 0.7x)]$$

$$= \frac{1}{2}(\cos 1.6x - \cos 3x)$$

Graph $y1 = \sin 2.3x \sin 0.7x$
  $y2 = \frac{1}{2}(\cos 1.6x - \cos 3x)$

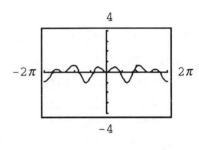

43. $\cos x \cos y \cos z = \cos x \dfrac{1}{2}[\cos(y + z) + \cos(y - z)]$ Product-sum Identity

$$= \frac{1}{2}\cos x \cos(y + z) + \frac{1}{2}\cos x \cos(y - z)$$ Algebra

$$= \frac{1}{2}\left\{\frac{1}{2}[\cos(x + y + z) + \cos(x - \{y + z\})]\right\}$$

$$+ \frac{1}{2}\left\{\frac{1}{2}[\cos(x + y - z) + \cos(x - \{y - z\})]\right\}$$ Product-sum Identity

$$= \frac{1}{4}\cos(x + y + z) + \frac{1}{4}\cos(x - y - z)$$

$$+ \frac{1}{4}\cos(x + y - z) + \frac{1}{4}\cos(x - y + z)$$ Algebra

$$= \frac{1}{4}[\cos(x + y - z) + \cos(x - y - z)$$

$$+ \cos(z + x - y) + \cos(x + y + z)]$$ Algebra

$$= \frac{1}{4}[\cos(x + y - z) + \cos\{-(y + z - x)\}$$

$$+ \cos(z + x - y) + \cos(x + y + z)]$$ Algebra

$$= \frac{1}{4}[\cos(x + y - z) + \cos(y + z - x) + \cos(z + x - y)$$

$$+ \cos(x + y + z)]$$ Identity for Negatives

45. (A)

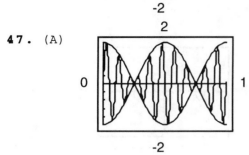

(B) $2 \cos(28\pi x) \cos(2\pi x)$

$= 2 \cdot \dfrac{1}{2} [\cos(28\pi x + 2\pi x) + \cos(28\pi x - 2\pi x)]$

$= \cos 30\pi x + \cos 26\pi x$

Graphing $y1 = \cos 30\pi x + \cos 26\pi x$ will yield the same result as before.

47. (A)

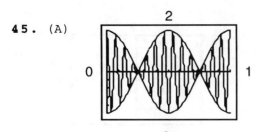

(B) $2 \sin(20\pi x) \cos(2\pi x)$

$= 2 \cdot \dfrac{1}{2} [\sin(20\pi x + 2\pi x) + \sin(20\pi x - 2\pi x)]$

$= \sin 22\pi x + \sin 18\pi x$

Graphing $y1 = \sin 22\pi x + \sin 18\pi x$ will yield the same result as before.

49. (A) $\cos x - \cos y = -2 \sin \dfrac{x + y}{2} \sin \dfrac{x - y}{2}$

In this case

$\cos 128\pi t - \cos 144\pi t = -2 \sin \dfrac{128\pi t + 144\pi t}{2} \sin \dfrac{128\pi t - 144\pi t}{2}$

$\cos 128\pi t - \cos 144\pi t = -2 \sin 136\pi t \sin(-8\pi t)$

$\cos 128\pi t - \cos 144\pi t = 2 \sin 136\pi t \sin 8\pi t$

Multiplying both sides by 0.5, we have
$0.5 \cos 128\pi t - 0.5 \cos 144\pi t = \sin 136\pi t \sin 8\pi t$

(B) $y = 0.5 \cos 128\pi t$ $\qquad\qquad\qquad$ $y = -0.5 \cos 144\pi t$

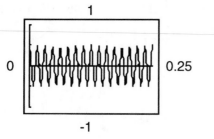

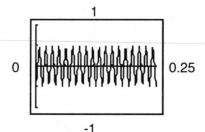

$y = 0.5 \cos 128\pi t - 0.5 \cos 144\pi t$ $\qquad$ $y = \sin 8\pi t \sin 136\pi t$

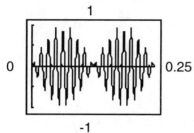

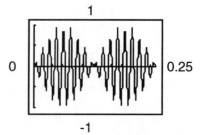

The latter two graphs are the same, illustrating the identity proved in part (A).

Exercise 6-5

Key Ideas and Formulas

Trigonometric Conditional equations can be solved by algebraic methods, which in some cases lead to exact solutions, or by graphing utility methods.

To solve algebraically:
1. Regard one particular trigonometric function as a variable, and solve for it.
   a. Consider using algebraic manipulation such as factoring, combining or separating fractions, and so on.

   b. Consider using identities.

2. After solving for a trigonometric function, solve for the variable.
   a. Solve over one period.
   b. Write an expression for all solutions.

---

1. $2 \sin x + 1 = 0$

$$\sin x = -\frac{1}{2}$$

Sketch a graph of $y = \sin x$ and $y = -\frac{1}{2}$,

$x = [0, 2\pi)$.

$$x = \frac{7\pi}{6}, \frac{11\pi}{6}$$

The checking steps in this and subsequent problems are omitted for lack of space.

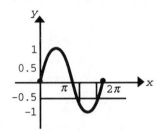

3. We have found all solutions of $\sin x = -\frac{1}{2}$ over one period in the previous problem. Since the sine function is periodic with period 2π, all solutions are given by

$$\left.\begin{array}{l} x = \dfrac{7\pi}{6} + 2k\pi \\[2mm] x = \dfrac{11\pi}{6} + 2k\pi \end{array}\right\} k \text{ any integer}$$

5. $\tan x + \sqrt{3} = 0$

$$\tan x = -\sqrt{3}$$

Sketch a graph of $y = \tan x$ and $y = -\sqrt{3}$
x in $[0, \pi)$.

$$x = \frac{2\pi}{3}$$

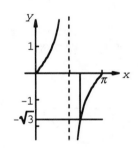

7. We have found all solutions of $\tan x + \sqrt{3} = 0$ over one period in the previous problem. Since the tangent function is periodic with period π, all solutions are given by $x = \frac{2\pi}{3} + k\pi$, k any integer.

9. $2 \cos \theta - \sqrt{3} = 0$

$$\cos \theta = \frac{\sqrt{3}}{2}$$

Sketch a graph of $y = \cos \theta$ and $y = \frac{\sqrt{3}}{2}$,

θ in $[0°, 360°)$.

$\theta = 30°, 330°$

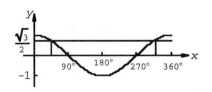

11. $7 \cos x - 3 = 0 \qquad 0 \leq x < 2\pi$

$$\cos x = \frac{3}{7}$$

Sketch a graph of $y = \cos x$ and $y = \frac{3}{7}$,

x in $[0, 2\pi)$.

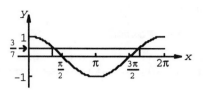

$$x = \begin{cases} \cos^{-1} \dfrac{3}{7} = 1.1279 \quad \text{First quadrant solution} \\ 2\pi - \cos^{-1} \dfrac{3}{7} = 5.1553 \quad \text{Fourth quadrant solution} \end{cases}$$

13. $2 \tan \theta - 7 = 0 \qquad 0° \leq \theta < 180°$

$$\tan \theta = \frac{7}{2}$$

Sketch a graph of $y = \tan \theta$ and $y = \frac{7}{2}$

θ in $[0°, 180°)$

$\theta = \tan^{-1} \dfrac{7}{2} = 74.0546°$

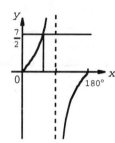

15. $1.3224 \sin x + 0.4732 = 0$

$$\sin x = -\frac{0.4732}{1.3224}$$

Solve over one period $[0, 2\pi)$:

$$\sin x = -0.3578$$

Sketch a graph of $y = \sin x$ and
$y = -0.3578$, x in $[0, 2\pi)$

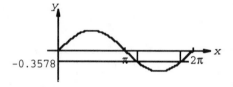

$$x = \begin{cases} \pi - \sin^{-1}(-0.3578) = 3.5075 \quad \text{Third quadrant solution} \\ 2\pi + \sin^{-1}(-0.3578) = 5.9172 \quad \text{Fourth quadrant solution} \end{cases}$$

Since the same function is periodic with period 2π, all solutions are given by

$$x = \begin{cases} 3.5075 + 2k\pi \\ 5.9172 + 2k\pi \end{cases} \quad k \text{ any integer}$$

17. Here is a computer-generated graph
of $y = 1 - x$ and $y = 2 \sin x$,
x in $[0, 2\pi)$.

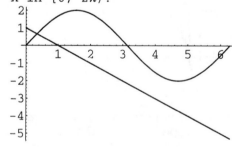

There is a solution in the interval
$[0.2, 0.4]$. Zooming in on this solution,
we obtain the following graph.

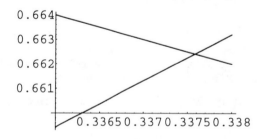

To four decimal places, the solution is
$x = 0.3376$.

19. Here is a computer-generated graph
of $y = \tan \dfrac{x}{2}$ and $y = 8 - x$,
x in $[0, \pi)$.

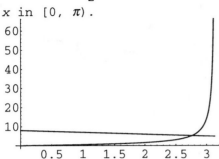

There is a solution in the interval
$[2.7, 2.8]$. Zooming in on this solution,
we obtain the following graph.

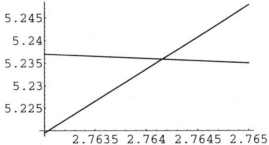

To four decimal places, the solution is
2.7642.

21.
$$2 \sin^2 \theta + \sin 2\theta = 0 \qquad \text{all } \theta$$
$$2 \sin^2 \theta + 2 \sin \theta \cos \theta = 0$$
$$2 \sin \theta(\sin \theta + \cos \theta) = 0$$
Either $\qquad 2 \sin \theta = 0$
$$\sin \theta = 0$$
Solutions over $0° \le \theta < 360°$ are
$\theta = 0°,\ 180°$

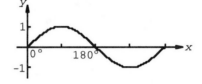

Thus, solutions, if θ is allowed to range over all possible values, are
$\theta = 0° + k180°$ or $\theta = k180°$ k any integer.

Or $\sin \theta + \cos \theta = 0$
$$\sin \theta = -\cos \theta$$
$$\frac{\sin \theta}{\cos \theta} = -1$$
$$\tan \theta = -1$$
Solutions over $0° < \theta < 180°$ are
$\theta = 135°$
Thus, solutions, if θ is allowed to
range over all possible values, are

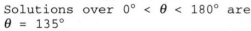

$\theta = 135° + k180°$ k any integer
Solutions: $k180°,\ 135° + k180°$ k any integer

23.
$$\tan x = -2 \sin x \qquad 0 \le x < 2\pi$$
$$\frac{\sin x}{\cos x} = -2 \sin x$$
$$\sin x = -2 \sin x \cos x \quad \cos x \ne 0$$
$$2 \sin x \cos x + \sin x = 0$$
$$\sin x(2 \cos x + 1) = 0$$

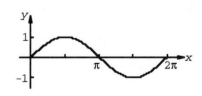

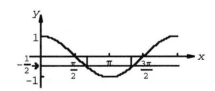

$\sin x = 0 \qquad\qquad 2 \cos x + 1 = 0$
$\quad x = 0,\ \pi \qquad\qquad\quad 2 \cos x = -1$
$$\cos x = -\frac{1}{2}$$
$$x = \frac{2\pi}{3},\ \frac{4\pi}{3}$$

Solutions: $0,\ \dfrac{2\pi}{3},\ \pi,\ \dfrac{4\pi}{3}$

25.
$$2 \cos^2 \theta + 3 \sin \theta = 0 \qquad 0° \le \theta < 360°$$
$$2(1 - \sin^2 \theta) + 3 \sin \theta = 0$$
$$2 - 2 \sin^2 \theta + 3 \sin \theta = 0$$
$$-2 \sin^2 \theta + 3 \sin \theta + 2 = 0$$
$$2 \sin^2 \theta - 3 \sin \theta - 2 = 0$$
$$(2 \sin \theta + 1)(\sin \theta - 2) = 0$$
$$2 \sin \theta + 1 = 0 \qquad \sin \theta - 2 = 0$$
$$2 \sin \theta = -1 \qquad \sin \theta = 2$$
$$\sin \theta = -\frac{1}{2} \qquad \text{No solution}$$
$$\theta = 210°, \ 330°$$

Solutions: $\theta = 210°, \ 330°$

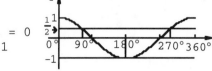

27.
$$\cos 2\theta + \cos \theta = 0 \qquad 0° \le \theta < 360°$$
$$2 \cos^2 \theta - 1 + \cos \theta = 0$$
$$2 \cos^2 \theta + \cos \theta - 1 = 0$$
$$(2 \cos \theta - 1)(\cos \theta + 1) = 0$$
$$2 \cos \theta - 1 = 0 \qquad \cos \theta + 1 = 0$$
$$2 \cos \theta = 1 \qquad \cos \theta = -1$$
$$\qquad\qquad\qquad \theta = 180°$$
$$\cos \theta = \frac{1}{2}$$
$$\theta = 60°, \ 300°$$

Solutions: $\theta = 60°, \ 180°, \ 300°$

29.
$$2 \sin^2 \frac{x}{2} - 3 \sin \frac{x}{2} + 1 = 0 \qquad 0 \le x \le 2\pi \quad \text{is equivalent to}$$
$$2 \sin^2 \frac{x}{2} - 3 \sin \frac{x}{2} + 1 = 0 \qquad 0 \le \frac{x}{2} \le \pi$$
$$\left(2 \sin \frac{x}{2} - 1\right)\left(\sin \frac{x}{2} - 1\right) = 0$$

$2 \sin \frac{x}{2} - 1 = 0$ | **Common Error:** $\sin x \ne 1$ | $\sin \frac{x}{2} - 1 = 0$

$$2 \sin \frac{x}{2} = 1 \qquad\qquad\qquad\qquad\qquad \sin \frac{x}{2} = 1$$
$$\sin \frac{x}{2} = \frac{1}{2} \qquad\qquad\qquad\qquad\qquad \frac{x}{2} = \frac{\pi}{2}$$
$$\frac{x}{2} = \frac{\pi}{6}, \ \frac{5\pi}{6} \qquad\qquad\qquad\qquad x = \pi$$
$$x = \frac{\pi}{3}, \ \frac{5\pi}{3}$$

Solutions: $x = \frac{\pi}{3}, \ \frac{5\pi}{3}, \ \pi$

31.
$$6 \sin^2 \theta + 5 \sin \theta = 6 \qquad 0° \le \theta \le 90°$$
$$6 \sin^2 \theta + 5 \sin \theta - 6 = 0$$
$$(3 \sin \theta - 2)(2 \sin \theta + 3) = 0$$

$$3 \sin \theta - 2 = 0 \quad 0° \le \theta \le 90° \qquad 2 \sin \theta + 3 = 0 \quad 0° \le \theta \le 90°$$
$$3 \sin \theta = 2 \qquad\qquad\qquad\qquad 2 \sin \theta = -3$$
$$\sin \theta = \frac{2}{3} \qquad\qquad\qquad\qquad \sin \theta = -\frac{3}{2}$$
$$\theta = \sin^{-1} \frac{2}{3} \ \text{(calculator in degree mode)} \quad \text{No solution}$$

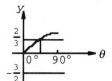

$$\theta = 41.81°$$

Solution: $\theta = 41.81°$

33. $3 \cos^2 x - 8 \cos x = 3$ $0 \leq x \leq \pi$

$3 \cos^2 x - 8 \cos x - 3 = 0$

$(3 \cos x + 1)(\cos x - 3) = 0$

$3 \cos x + 1 = 0$ $0 \leq x \leq \pi$ $\cos x - 3 = 0$ $0 \leq x \leq \pi$

$3 \cos x = -1$ $\cos x = 3$

$\cos x = -\dfrac{1}{3}$ No solution

$x = \cos^{-1}\left(-\dfrac{1}{3}\right)$ (calculator in radian mode)

$x = 1.911$

Solution: $x = 1.911$

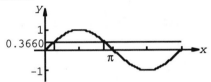

35. $2 \sin x = \cos 2x$ $0 \leq x < 2\pi$

$2 \sin x = 1 - 2 \sin^2 x$

$2 \sin^2 x + 2 \sin x - 1 = 0$ Quadratic in $\sin x$

$\sin x = \dfrac{-b \pm \sqrt{b^2 - 4ac}}{2a}$ $a = 2, \ b = 2, \ c = -1$

$\sin x = \dfrac{-2 \pm \sqrt{2^2 - 4(2)(-1)}}{2(2)} = \dfrac{-2 \pm \sqrt{12}}{4}$

$\sin x = 0.3660$ $\sin x = -1.366$

                              No solution

$x = \begin{cases} \sin^{-1} 0.3660 \\ \pi - \sin^{-1} 0.3660 \end{cases} = \begin{cases} 0.3747 \\ 2.767 \end{cases}$

Solutions: $x = 0.3747, \ 2.767$

37. The equation $\sin^2 x = (1 - \cos x)(1 + \cos x)$ is an identity, since

$(1 - \cos x)(1 + \cos x) = 1 - \cos^2 x$ Algebra

$= \sin^2 x$ Pythagorean Identity

Thus the equation has an infinite number of solutions.

39. The equation $3 \sin x = x - 3$ is not an identity, since it is false if $x = 0$, for example $(0 \neq 0 - 3)$. A graphing utility can be used to produce the following graph:

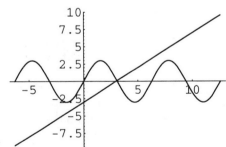

Since $x - 3$ is an increasing function, the graph strongly suggests that there is one and only one solution of the equation.

41. The equation $\sin 2x = 3 \sin x \cos x$ is not an identity, since it is false if $x = \dfrac{\pi}{4}$, for example $\left(\sin \dfrac{2\pi}{4} = 1 \neq 3 \sin \dfrac{\pi}{4} \cos \dfrac{\pi}{4} = \dfrac{3}{2}\right)$. However, since we can write

$\sin 2x = 3 \sin x \cos x$

$2 \sin x \cos x = 3 \sin x \cos x$

$0 = \sin x \cos x$

$\sin x = 0 \ \text{ or } \ \cos x = 0$

the equation has infinitely many solutions (any integer multiple of $\dfrac{\pi}{2}$ is a solution.)

43. Here is a computer-generated graph of $y = 2 \sin x$ and $y = \cos 2x$, $0 \le x < 2\pi$.

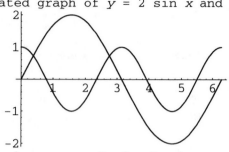

There are solutions of $2 \sin x = \cos 2x$ in the intervals $[0.2, 0.4]$ and $[2.6, 2.8]$. Zooming in on these solutions, we obtain the following graphs.

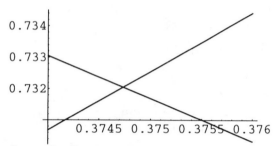
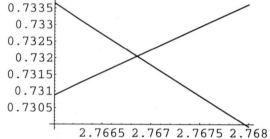

To four decimal places, the solutions are 0.3747 and 2.7669.

45. Here is a computer-generated graph of $y = 2 \sin^2 x$ and $y = 1 - 2 \sin x$, $0 \le x < 2\pi$.

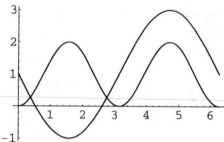

There are solutions of $2 \sin^2 x = 1 - 2 \sin x$ in the intervals $[0.2, 0.4]$ and $[2.6, 2.8]$. Zooming in on these solutions, we obtain the following graphs.

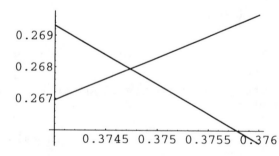
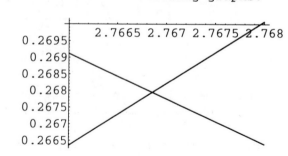

To four decimal places, the solutions in $[0, 2\pi)$ are 0.3747 and 2.7669. Thus, the solutions over all real x are given by

$$x = \begin{cases} 0.3747 + 2k\pi \\ 2.7669 + 2k\pi \end{cases} k \text{ any integer}$$

47. Here is a computer-generated graph
of $y = \cos 2x$ and $y = x^2 - 2$.

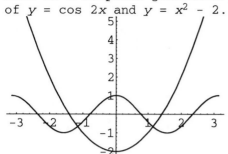

Note that the graphs have y axis symmetry.
There are points of intersection in the
intervals (1, 1.2) and (-1.2, -1).
Zooming in on the right-hand point of
intersection, we obtain the following
graph.

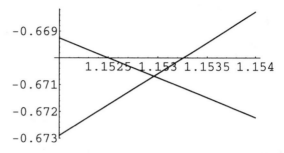

Combining the information from the two
graphs, $\cos 2x > x^2 - 2$ on the interval
(-1.1530, 1.1530), to four decimal places.

49. Here is a computer-generated graph of $y = \cos(2x + 1)$ and $y = 0.5x - 2$.

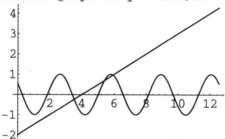

There are points of intersection in the intervals [3.4, 3.8], [5.2, 5.6], and
[5.6, 6]. Zooming in, we obtain the following graphs.

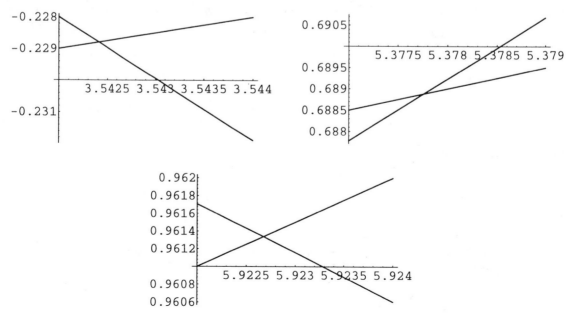

Combining information from all four graphs, $\cos(2x + 1) \le 0.5x - 2$ on the two
intervals [3.5424, 5.3778] and [5.9227, ∞), to four decimal places.

51. Here are computer-generated graphs of $y = e^{\sin x}$ and $y = 2x - 1$.

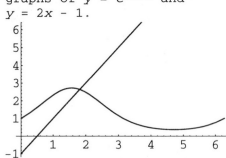

There is a solution of $e^{\sin x} = 2x - 1$ in the interval [1.8, 2]. Zooming in, we obtain the following graph.

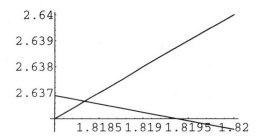

To four decimal places, the solution is 1.8183.

53. $\tan^{-1}(-5.377)$ has exactly one value, -1.387; the equation $\tan x = -5.377$ has infinitely many solutions, which are found by adding $k\pi$, k any integer, to each solution in one period of $\tan x$.

55.

$$\cos x - \sin x = 1 \qquad 0 \le x < 2\pi$$
$$\pm\sqrt{1 - \sin^2 x} - \sin x = 1$$
$$\pm\sqrt{1 - \sin^2 x} = 1 + \sin x$$
$$1 - \sin^2 x = (1 + \sin x)^2 \qquad \text{Squaring both sides}$$
$$1 - \sin^2 x = 1 + 2 \sin x + \sin^2 x \qquad \boxed{\text{Common Error: } (1 + \sin x)^2 \ne 1 + \sin^2 x}$$
$$0 = 2 \sin x + 2 \sin^2 x$$
$$0 = 2 \sin x(1 + \sin x)$$

$$2 \sin x = 0 \qquad\qquad 1 + \sin x = 0$$
$$\sin x = 0 \qquad\qquad\quad \sin x = -1$$
$$x = 0, \pi \qquad\qquad\quad x = \frac{3\pi}{2}$$

In squaring both sides we may have introduced extraneous solutions; hence, it is necessary to check solutions of these equations in the original equation.

$x = 0$	$x = \pi$	$x = \frac{3\pi}{2}$
$\cos x - \sin x = 1$	$\cos x - \sin x = 1$	$\cos x - \sin x = 1$
$\cos 0 - \sin 0 \overset{?}{=} 1$	$\cos \pi - \sin \pi \overset{?}{=} 1$	$\cos \frac{3\pi}{2} - \sin \frac{3\pi}{2} \overset{?}{=} 1$
$1 - 0 \overset{\checkmark}{=} 1$	$-1 - 0 \ne 1$	$0 - (-1) \overset{\checkmark}{=} 1$
A solution	Not a solution	A solution

Solutions: $x = 0, \frac{3\pi}{2}$

57.

$$\tan x - \sec x = 1 \qquad 0 \le x < 2\pi$$
$$\pm\sqrt{\sec^2 x - 1} - \sec x = 1$$
$$\pm\sqrt{\sec^2 x - 1} = 1 + \sec x$$
$$\sec^2 x - 1 = (1 + \sec x)^2 \qquad \text{Squaring both sides}$$
$$\sec^2 x - 1 = 1 + 2 \sec x + \sec^2 x$$
$$0 = 2 + 2 \sec x$$
$$-2 \sec x = 2$$
$$\sec x = -1$$
$$x = \pi$$

In squaring both sides we may have introduced extraneous solutions; hence, it is necessary to check this apparent solution in the original equation.

$\tan x - \sec x = 1$

$\tan \pi - \sec \pi \overset{?}{=} 1$

$\quad\quad 0 - (-1) \overset{\checkmark}{=} 1$

A solution

Solution: $x = \pi$

59. Here are computer-generated graphs of $y = \sin \dfrac{1}{x}$ and $y = 1.5 - 5x$, $0.04 \le x \le 0.2$.

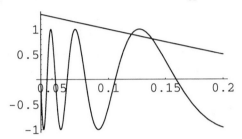

There are solutions of $\sin \dfrac{1}{x} = 1.5 - 5x$ on the intervals $[0.1, 0.125]$ and $[0.125, 0.15]$. Zooming in on these solutions, we obtain the following graphs.

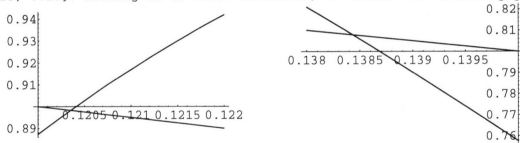

To four decimal places, the solutions are 0.1204 and 0.1384.

61. (A) Here are computer-generated graphs of $f(x) = \sin \dfrac{1}{x}$ on $[0.1, 0.5]$ and $[0.1, 5]$.

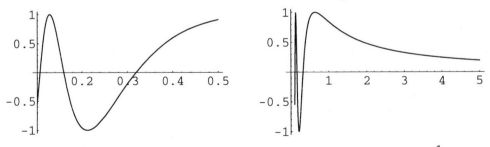

The largest zero for f is 0.3183. As x increases without bound, $\dfrac{1}{x}$ tends to 0 through positive numbers, and $\sin \dfrac{1}{x}$ tends to 0 through positive numbers. $y = 0$ is a horizontal asymptote for the graph of f.

(B) Here is a computer-generated graph of $f(x) = \sin \dfrac{1}{x}$ on $[0.015, 0.1]$.

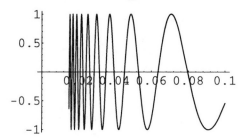

The student may provide other exploratory graphs.

Infinitely many zeros exist between 0 and b, for any b, however small. The exploration graphs suggest this conclusion, which is reinforced by the following reasoning: Note that for each interval $(0, b]$, however small, as x tends to zero through positive numbers, $\dfrac{1}{x}$ increases without bound, and as $\dfrac{1}{x}$ increases without bound, $\sin \dfrac{1}{x}$ will cross the x axis an unlimited number of times. The function f does not have a smallest zero, because, between 0 and b, no matter how small b is, there is always an unlimited number of zeros.

63. $I = 30 \sin 120\pi t \quad I = -10$

$-10 = 30 \sin 120\pi t$

$\sin 120\pi t = -\dfrac{10}{30}$

$120\pi t = \pi + \sin^{-1} \dfrac{10}{30}$ (third quadrant) will yield the least positive solution of the equation.

$t = \dfrac{1}{120\pi} \left(\pi + \sin^{-1} \dfrac{10}{30} \right)$

  $= 0.009235$ sec   (calculator in radian mode)

65. We want the least positive θ so that $I \cos^2 \theta$ is 40% of I, that is

$I \cos^2 \theta = 0.40I \qquad 0° \leq \theta \leq 180°$

  $\cos^2 \theta = 0.40$

   $\cos \theta = \pm\sqrt{0.40}$

     $\theta = \cos^{-1}\sqrt{0.40}$ will yield the least positive solution of the equation
     $\theta = 50.77°$ (calculator in degree mode)

67. We are to solve $3.09 \times 10^7 = \dfrac{3.44 \times 10^7}{1 - 0.206 \cos \theta}$

For convenience, we can divide both sides of this equation by 10^7

$3.09 = \dfrac{3.44}{1 - 0.206 \cos \theta}$

$3.09(1 - 0.206 \cos \theta) = 3.44$

$3.09 - (3.09)(0.206)\cos \theta = 3.44$

$-(3.09)(0.206)\cos \theta = 3.44 - 3.09$

$\cos \theta = \dfrac{3.44 - 3.09}{-(3.09)(0.206)}$

$\cos \theta = -0.5498$

$\theta = 180° - \cos^{-1}(0.5498)$ (second quadrant) will yield the least positive solution of this equation.

or

$\theta = \cos^{-1}(-0.5498)$

$\theta = 123°$

69. Use $A = \frac{1}{2}R^2(\theta - \sin\theta)$ with $R = 8$ and $A = 48$.

$$48 = \frac{1}{2} \cdot 8^2(\theta - \sin\theta)$$
$$48 = 32(\theta - \sin\theta)$$
$$1.5 = \theta - \sin\theta$$

Here is a computer-generated graph of $y = 1.5$ and $y = \theta - \sin\theta$, $0 \le \theta \le \pi$.

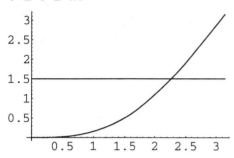

There is a solution of $1.5 = \theta - \sin\theta$ in the interval $[2.2, 2.3]$. Zooming in on this solution, we obtain the following graph.

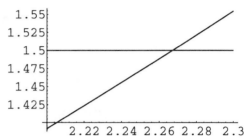

To three decimal places, $\theta = 2.267$ radians.

71. Using the figure in the text, we can write:

$$\sin\theta = \frac{a}{R} \qquad (1)$$
$$a^2 + (R - b)^2 = R^2 \qquad (2) \qquad \text{applying the Pythagorean theorem to the right triangle with sides } R, a, \text{ and } R - b.$$
$$L = R \cdot 2\theta \qquad (3)$$

Thus, given a and b, we can solve equation (2) for R:

$$a^2 + R^2 - 2Rb + b^2 = R^2$$
$$-2Rb = -a^2 - b^2$$
$$R = \frac{a^2 + b^2}{2b}$$

We can solve equation (1) for θ to obtain $\theta = \sin^{-1}\frac{a}{R}$ (since θ is an acute angle)

Hence, from equation (3),

$$L = R \cdot 2\theta = \frac{a^2 + b^2}{2b} \cdot 2\sin^{-1}\frac{a}{\frac{a^2 + b^2}{2b}}$$

$$L = \frac{a^2 + b^2}{b}\sin^{-1}\frac{2ab}{a^2 + b^2} \qquad (4)$$

(A) Substituting the given values $a = 5.5$ and $b = 2.5$,

$$L = \frac{(5.5)^2 + (2.5)^2}{2.5}\sin^{-1}\frac{2(5.5)(2.5)}{(5.5)^2 + (2.5)^2}$$
$$L = 12.4575 \text{ mm.}$$

(B) Substituting $L = 12.4575$ and $a = 5.4$ mm yields

$$12.4575 = \frac{(5.4)^2 + b^2}{b}\sin^{-1}\frac{2(5.4)b}{(5.4)^2 + b^2}$$
$$12.4575 = \frac{29.16 + b^2}{b}\sin^{-1}\frac{10.8b}{29.16 + b^2} \qquad (5)$$

To solve for b, we generate a graph of $y = 12.4575$ and
$$y = \frac{29.16 + b^2}{b} \sin^{-1} \frac{10.86}{29.16 + b^2},$$
$$2 < b \leq 3.$$

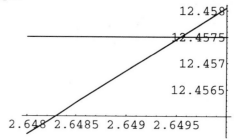

There is a solution of equation (5) in the interval [2.6, 2.7]. Zooming in on this interval, we obtain the following graph.

To four decimal places, $b = 2.6496$ mm.

73. $r = 2 \sin \theta \qquad 0° \leq \theta \leq 360°$

$r = \sin 2\theta$

We solve this system of equations by equating the right sides:

$2 \sin \theta = \sin 2\theta$

$2 \sin \theta = 2 \sin \theta \cos \theta$

$\qquad 0 = 2 \sin \theta \cos \theta - 2 \sin \theta$

$\qquad 0 = 2 \sin \theta (\cos \theta - 1)$

$2 \sin \theta = 0 \qquad\qquad\qquad \cos \theta - 1 = 0$

$\quad \sin \theta = 0 \qquad\qquad\qquad\quad \cos \theta = 1$

$\qquad \theta = 0°,\ 180°,\ 360° \qquad\qquad \theta = 0°,\ 360°$

If we substitute these values of θ in either of the original equations, we obtain

$r = 2 \sin 0° \qquad r = 2 \sin 180° \qquad r = 2 \sin 360°$

$r = 0 \qquad\qquad r = 0 \qquad\qquad r = 0$

Thus, the solutions of the system of equations are

$(r,\ \theta) = (0,\ 0°),\ (r,\ \theta) = (0,\ 180°),$ and $(r,\ \theta) = (0,\ 360°)$

75. $2xy = 1$

$2(u \cos \theta - v \sin \theta)(u \sin \theta + v \cos \theta) = 1$ (substitution)

$2(u \cos \theta u \sin \theta + u \cos \theta v \cos \theta - v \sin \theta u \sin \theta - v \sin \theta v \cos \theta) = 1$

$\qquad\qquad\qquad\qquad\qquad\qquad\qquad\qquad\qquad\qquad$ (multiplication)

$2u^2 \cos \theta \sin \theta + uv(2 \cos^2 \theta - 2 \sin^2 \theta) - 2v^2 \sin \theta \cos \theta = 1$

We are to find the least positive θ so that the coefficient of the uv term will be zero.

$2 \cos^2 \theta - 2 \sin^2 \theta = 0$

$2(\cos^2 \theta - \sin^2 \theta) = 0$

$\qquad\qquad 2 \cos 2\theta = 0$

$\qquad\qquad\quad \cos 2\theta = 0$

$\qquad\qquad\qquad\quad 2\theta = \cos^{-1} 0$ yields the least positive θ

$\qquad\qquad\qquad\quad 2\theta = 90°$

$\qquad\qquad\qquad\quad\ \theta = 45°$

CHAPTER 6 REVIEW

1. $\tan x + \cot x = \dfrac{\sin x}{\cos x} + \dfrac{\cos x}{\sin x}$ Quotient Identities

$\qquad\qquad\qquad = \dfrac{\sin^2 x + \cos^2 x}{\cos x \sin x}$ Algebra

$\qquad\qquad\qquad = \dfrac{1}{\cos x \sin x}$ Pythagorean Identity

$\qquad\qquad\qquad = \dfrac{1}{\cos x}\,\dfrac{1}{\sin x}$ Algebra

$\qquad\qquad\qquad = \sec x \csc x$ Reciprocal Identity   (6-1)

2. $\sec^4 x - 2 \sec^2 x \tan^2 x + \tan^4 x = (\sec^2 x - \tan^2 x)^2$ Algebra

$\qquad\qquad\qquad\qquad\qquad\qquad = (1 + \tan^2 x - \tan^2 x)^2$ Pythagorean Identity

$\qquad\qquad\qquad\qquad\qquad\qquad = 1^2$ Algebra

$\qquad\qquad\qquad\qquad\qquad\qquad = 1$ (6-1)

3. $\dfrac{1}{1 - \sin x} + \dfrac{1}{1 + \sin x} = \dfrac{1 + \sin x + 1 - \sin x}{(1 - \sin x)(1 + \sin x)}$ Algebra

$\qquad\qquad\qquad\qquad\quad = \dfrac{2}{1 - \sin^2 x}$ Algebra

$\qquad\qquad\qquad\qquad\quad = \dfrac{2}{\sin^2 x + \cos^2 x - \sin^2 x}$ Pythagorean Identity

$\qquad\qquad\qquad\qquad\quad = \dfrac{2}{\cos^2 x}$ Algebra

$\qquad\qquad\qquad\qquad\quad = 2 \sec^2 x$ Reciprocal Identity   (6-1)

4. $\cos\left(x - \dfrac{3\pi}{2}\right) = \cos x \cos \dfrac{3\pi}{2} + \sin x \sin \dfrac{3\pi}{2}$ Difference Identity

$\qquad\qquad\qquad = \cos x(0) + \sin x(-1)$ Known Values

$\qquad\qquad\qquad = -\sin x$ Algebra   (6-2)

5. $\cos x \cos y = \dfrac{1}{2}[\cos(x + y) + \cos(x - y)]$

$\cos 2\alpha \cos 3\alpha = \dfrac{1}{2}[\cos(2\alpha + 3\alpha) + \cos(2\alpha - 3\alpha)]$

$\qquad\qquad\quad = \dfrac{1}{2}[\cos(5\alpha) + \cos(-\alpha)]$

$\qquad\qquad\quad = \dfrac{1}{2}(\cos 5\alpha + \cos \alpha)$ by the cosine identity for negatives   (6-4)

6. $\sin x - \sin y = 2 \cos \dfrac{x + y}{2} \sin \dfrac{x - y}{2}$

$\sin 9x - \sin 5x = 2 \cos \dfrac{9x + 5x}{2} \sin \dfrac{9x - 5x}{2}$

$\qquad\qquad\quad = 2 \cos 7x \sin 2x$ (6-4)

7. $\sin\left(x + \dfrac{9\pi}{2}\right) = \sin x \cos \dfrac{9\pi}{2} + \cos x \sin \dfrac{9\pi}{2}$ Sum Identity

$\qquad\qquad = \sin x \cos \dfrac{\pi}{2} + \cos x \sin \dfrac{\pi}{2}$ Periodicity of sine and cosine

$\qquad\qquad = \sin x(0) + \cos x(1)$ Known Values

$\qquad\qquad = \cos x$ *(6-2)*

8. $\sqrt{2} \cos \theta + 1 = 0$ all θ
Solve over one period $[0°, 360°)$:
$\qquad \cos \theta = -\dfrac{1}{\sqrt{2}}$
Sketch a graph of $y = \cos \theta$ and
$y = -\dfrac{1}{\sqrt{2}}$, θ in $[0°, 360°)$.
$\theta = 135°, 225°$

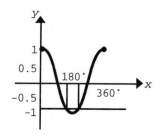

Since the cosine function is periodic with period 360°, all solutions are given
by $\theta = 135° + k360°$, $\theta = 225° + k360°$, k any integer.   *(6-5)*

9. $\sin x \tan x - \sin x = 0$ all x
Solve over one period, $[0, 2\pi)$ for $\sin x$, $[0, \pi)$ for $\tan x$:

$\sin x(\tan x - 1) = 0$
$\sin x = 0 \qquad \tan x - 1 = 0$
$\qquad\qquad\qquad \tan x = 1$
$\qquad\qquad\qquad\quad x = \dfrac{\pi}{4}$
$x = 0, \pi$

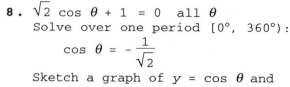

Thus, solutions if x is allowed to range
over all values, are $x = 0 + 2k\pi$, $\pi + 2k\pi$,
k any integer, and $x = \dfrac{\pi}{4} + k\pi$, k any
integer. These solutions can also be
written as $x = k\pi$ or $x = \dfrac{\pi}{4} + k\pi$, k any
integer.

     (6-5)

10. $\cos \theta = 0.8215$ for all θ
Solve over one period, θ in $[0°, 360°)$:
Sketch a graph of $y = \cos \theta$ and $y = 0.8215$,
θ in $[0°, 360°)$.

$\theta = \begin{cases} \cos^{-1} 0.8215 = 34.7648° & \text{First quadrant solution} \\ 360° - \cos^{-1} 0.8215 = 360° - 34.7648° & \text{Fourth quadrant solution} \end{cases}$ (calculator in degree mode)

Since the cosine function is periodic
with period 360°, all solutions are
given by $\theta = 34.7648° + k360°$
and $\theta = 360° - 34.7648° + k360°$,
or $\theta = -34.7648° + (k + 1)360°$, k any
integer.

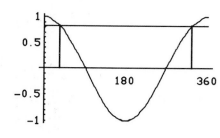

     (6-5)

11. $\cot x = -154.3$ $\quad -\dfrac{\pi}{2} < x < \dfrac{\pi}{2}$

Sketch a graph of $y = \cot x$ and

$y = -154.3$ $\quad x$ in $\left(-\dfrac{\pi}{2}, \dfrac{\pi}{2}\right)$

$x = -\cot^{-1}(154.3)$ fourth quadrant solution
  $= -0.0065$

Note: Only the portion of the graph $-0.5 < x < 0.5$ is shown; however. It is clear from the properties of the cotangent function that there can be no other solution in the entire interval $-\dfrac{\pi}{2} < x < \dfrac{\pi}{2}$.

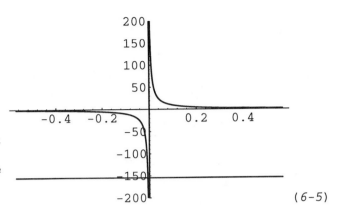

(6-5)

12. $\csc x = 1.786$ all x
Solve over one period $[0, 2\pi)$:
Sketch a graph of $y = \csc x$ and $y = 1.786$, x in $[0, 2\pi)$.

$x = \begin{cases} \csc^{-1} 1.786 = \sin^{-1}\dfrac{1}{1.786} = 0.5943 & \text{First quadrant solution} \\ \pi - \csc^{-1} 1.786 = 2.5473 & \text{Second quadrant solution} \end{cases}$

Since the cosecant function is periodic with period 2π, all solutions are given by

$x = \begin{cases} 0.5943 + 2k\pi & k \text{ any integer} \\ 2.5473 + 2k\pi & k \text{ any integer} \end{cases}$

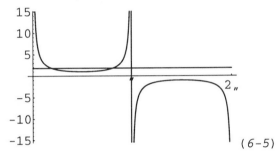

(6-5)

13. $3 \tan(11 - 3x) = 23.46$ $\quad -\dfrac{\pi}{2} < x < \dfrac{\pi}{2}$

$\tan(11 - 3x) = \dfrac{23.46}{3}$

$11 - 3x = \tan^{-1}\left(\dfrac{23.46}{3}\right)$

$-3x = \tan^{-1}\left(\dfrac{23.46}{3}\right) - 11$

$x = \dfrac{\tan^{-1}\left(\frac{23.46}{3}\right) - 11}{-3}$

$x = 3.1855$ $\qquad$ (6-5)

14. (A) Graph both sides of the equation in the same viewing window.
$(\sin x + \cos x)^2 = 1 - 2 \sin x \cos x$ is not an identity, since the graphs do not match.
Try $x = \dfrac{\pi}{4}$.

Left side: $\left(\sin\dfrac{\pi}{4} + \cos\dfrac{\pi}{4}\right)^2 = \left(\dfrac{1}{\sqrt{2}} + \dfrac{1}{\sqrt{2}}\right)^2 = \left(\dfrac{2}{\sqrt{2}}\right)^2 = 2$

Right side: $1 - 2 \sin\dfrac{\pi}{4} \cos\dfrac{\pi}{4} = 1 - 2\left(\dfrac{1}{\sqrt{2}}\right)\left(\dfrac{1}{\sqrt{2}}\right) = 0$

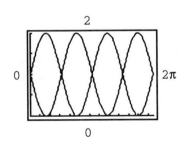

This verifies that the equation is not an identity.

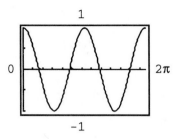

(B) Graph both sides of the equation in the same viewing window.

$\cos^2 x - \sin^2 x = 1 - 2 \sin^2 x$ appears to be an identity, which we now verify:

$$\begin{aligned} \cos^2 x - \sin^2 x &= 1 - \sin^2 x - \sin^2 x \qquad &\text{Pythagorean Identity}\\ &= 1 - 2 \sin^2 x &\text{Algebra} \end{aligned} \qquad (6\text{-}1)$$

15. $\dfrac{1 - 2 \cos x - 3 \cos^2 x}{\sin^2 x} = \dfrac{1 - 2 \cos x - 3 \cos^2 x}{1 - \cos^2 x}$ Pythagorean Identity

$$= \frac{(1 - 3 \cos x)(1 + \cos x)}{(1 + \cos x)(1 - \cos x)} \qquad \text{Algebra}$$

$$= \frac{1 - 3 \cos x}{1 - \cos x} \qquad \text{Algebra} \qquad (6\text{-}1)$$

16. $(1 - \cos x)(\csc x + \cot x) = (1 - \cos x)\left(\dfrac{1}{\sin x} + \dfrac{\cos x}{\sin x}\right)$ Reciprocal and Quotient Identities

$$= (1 - \cos x)\frac{(1 + \cos x)}{\sin x} \qquad \text{Algebra}$$

$$= \frac{(1 - \cos x)(1 + \cos x)}{\sin x} \qquad \text{Algebra}$$

$$= \frac{1 - \cos^2 x}{\sin x} \qquad \text{Algebra}$$

$$= \frac{\sin^2 x}{\sin x} \qquad \text{Pythagorean Identity}$$

$$= \sin x \qquad \text{Algebra}$$

Key Algebraic Steps: $(1 - a)\left(\dfrac{1}{b} + \dfrac{a}{b}\right) = (1 - a)\left(\dfrac{1 + a}{b}\right)$

$$= \frac{(1 - a)(1 + a)}{b}$$

$$= \frac{1 - a^2}{b} \qquad (6\text{-}1)$$

17. $\dfrac{1 + \sin x}{\cos x} = \dfrac{(1 + \sin x)(1 - \sin x)}{\cos x(1 - \sin x)}$ Algebra

$$= \frac{1 - \sin^2 x}{\cos x(1 - \sin x)} \qquad \text{Algebra}$$

$$= \frac{\cos^2 x}{\cos x(1 - \sin x)} \qquad \text{Pythagorean Identity}$$

$$= \frac{\cos x}{1 - \sin x} \qquad \text{Algebra} \qquad (6\text{-}1)$$

18. $\dfrac{1 - \tan^2 x}{1 + \tan^2 x} = \dfrac{1 - \tan^2 x}{\sec^2 x}$ Pythagorean Identity

$$= \dfrac{1 - \dfrac{\sin^2 x}{\cos^2 x}}{\dfrac{1}{\cos^2 x}}$$ Reciprocal and Quotient Identities

$$= \dfrac{\cos^2 x \left[1 - \dfrac{\sin^2 x}{\cos^2 x}\right]}{\cos^2 x \, \dfrac{1}{\cos^2 x}}$$ Algebra

$$= \dfrac{\cos^2 x - \sin^2 x}{1}$$ Algebra

$$= \cos^2 x - \sin^2 x$$ Algebra

$$= \cos 2x$$ Double-angle Identity   *(6-3)*

19. $\cot \dfrac{x}{2} = \dfrac{1}{\tan \dfrac{x}{2}}$ Reciprocal Identity

$$= \dfrac{1}{\dfrac{1 - \cos x}{\sin x}}$$ Half-angle Identity

$$= \dfrac{\sin x}{1 - \cos x}$$ Algebra   *(6-3)*

20. $\cot x - \tan x = \dfrac{\cos x}{\sin x} - \dfrac{\sin x}{\cos x}$ Quotient Identities

$$= \dfrac{\cos^2 x - \sin^2 x}{\sin x \cos x}$$ Algebra

$$= \dfrac{2(\cos^2 x - \sin^2 x)}{2 \sin x \cos x}$$ Algebra

$$= \dfrac{2[\cos^2 x - (1 - \cos^2 x)]}{2 \sin x \cos x}$$ Pythagorean Identity

$$= \dfrac{2[2 \cos^2 x - 1]}{2 \sin x \cos x}$$ Algebra

$$= \dfrac{4 \cos^2 x - 2}{2 \sin x \cos x}$$ Algebra

$$= \dfrac{4 \cos^2 x - 2}{\sin 2x}$$ Double-angle Identity   *(6-3)*

21. $\left(\dfrac{1 - \cot x}{\csc x}\right)^2 = \dfrac{(1 - \cot x)^2}{\csc^2 x}$ Algebra

$$= \dfrac{1 - 2 \cot x + \cot^2 x}{\csc^2 x}$$ Algebra

$$= \dfrac{1 + \cot^2 x - 2 \cot x}{\csc^2 x}$$ Algebra

$$= \dfrac{\csc^2 x - 2 \cot x}{\csc^2 x}$$ Pythagorean Identity

$$= \dfrac{\csc^2 x}{\csc^2 x} - \dfrac{2 \cot x}{\csc^2 x}$$ Algebra

$$= 1 - \dfrac{2 \cot x}{\csc^2 x}$$ Algebra

$$= 1 - \dfrac{2 \dfrac{\cos x}{\sin x}}{\dfrac{1}{\sin^2 x}}$$ Quotient and Reciprocal Identities

$$= 1 - 2 \sin x \cos x$$ Algebra

$$= 1 - \sin 2x$$ Double-angle Identity   *(6-3)*

22. $\tan m + \tan n = \dfrac{\sin m}{\cos m} + \dfrac{\sin n}{\cos n}$ Quotient Identity

$\qquad\qquad = \dfrac{\sin m \cos n + \cos m \sin n}{\cos m \cos n}$ Algebra

$\qquad\qquad = \dfrac{\sin(m + n)}{\cos m \cos n}$ Sum Identity   (6-2)

23. $\tan(x + y) = \dfrac{\tan x + \tan y}{1 - \tan x \tan y}$ Sum Identity

$\qquad\qquad = \dfrac{\dfrac{1}{\cot x} + \dfrac{1}{\cot y}}{1 - \dfrac{1}{\cot x}\dfrac{1}{\cot y}}$ Reciprocal Identity

$\qquad\qquad = \dfrac{\cot y + \cot x}{\cot x \cot y - 1}$ Algebra

$\qquad\qquad = \dfrac{\cot x + \cot y}{\cot x \cot y - 1}$ Algebra   (6-2)

24. $\quad \cos x \sin y = \dfrac{1}{2}[\sin(x + y) - \sin(x - y)]$

$\cos 195° \sin 75° = \dfrac{1}{2}[\sin(195° + 75°) - \sin(195° - 75°)]$

$\qquad\qquad\quad = \dfrac{1}{2}[\sin 270° - \sin 120°]$

$\qquad\qquad\quad = \dfrac{1}{2}\left(-1 - \dfrac{\sqrt{3}}{2}\right)$

$\qquad\qquad\quad = \dfrac{1}{2}\dfrac{-2 - \sqrt{3}}{2}$

$\qquad\qquad\quad = \dfrac{-2 - \sqrt{3}}{4}$ (6-4, 5-4)

25. $\quad \cos x + \cos y = 2 \cos \dfrac{x + y}{2} \cos \dfrac{x - y}{2}$

$\cos 195° + \cos 105° = 2 \cos \dfrac{195° + 105°}{2} \cos \dfrac{195° - 105°}{2}$

$\qquad\qquad\quad = 2 \cos 150° \cos 45°$

$\qquad\qquad\quad = 2\left(-\dfrac{\sqrt{3}}{2}\right)\left(\dfrac{\sqrt{2}}{2}\right)$

$\qquad\qquad\quad = -\dfrac{\sqrt{6}}{2}$ (6-4, 5-4)

26. The equation is not an identity. $\tan^2 x$ is equivalent to $\sec^2 x - 1$ by the Pythagorean identity. $\sec^2 x - 1 = \sec^2 x + 1$ is true for no real x.   (6-1)

27. The equation is an identity, as can be shown:

$\sin 4x = \sin 2(2x)$ Algebra

$\qquad = 2 \sin 2x \cos 2x$ Double-angle Identity

$\qquad = 2 \cdot 2 \sin x \cos x \cos 2x$ Double-angle Identity

$\qquad = 4 \sin x \cos x \cos 2x$ Algebra   (6-3)

28. The equation is an identity, as can be shown:

$\cos\left(x - \dfrac{\pi}{2}\right) = \cos x \cos \dfrac{\pi}{2} + \sin x \sin \dfrac{\pi}{2}$ Sum Identity

$\qquad\qquad = \cos x(0) + \sin x(1)$ Known Values

$\qquad\qquad = \sin x$ Algebra   (6-2)

29. The equation is not an identity. For example, let $x = \dfrac{\pi}{4}$.

Then Left side: $\sin\left(x - \dfrac{\pi}{2}\right) = \sin\left(\dfrac{\pi}{4} - \dfrac{\pi}{2}\right) = \sin\left(-\dfrac{\pi}{4}\right) = -\dfrac{\sqrt{2}}{2}$

Right side: $\cos x = \cos\dfrac{\pi}{4} = \dfrac{\sqrt{2}}{2}$ *(6-2)*

30. $4\sin^2 x - 3 = 0 \quad\quad 0 \le x < 2\pi$

$\quad\quad 4\sin^2 x = 3$

$\quad\quad\quad \sin^2 x = \dfrac{3}{4}$

$\quad\quad\quad\quad \sin x = \pm\dfrac{\sqrt{3}}{2}$

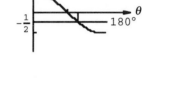

$\sin x = \dfrac{\sqrt{3}}{2} \quad\quad\quad \sin x = -\dfrac{\sqrt{3}}{2}$

$x = \dfrac{\pi}{3}, \dfrac{2\pi}{3} \quad\quad\quad x = \dfrac{4\pi}{3}, \dfrac{5\pi}{3}$

Solutions: $x = \dfrac{\pi}{3}, \dfrac{2\pi}{3}, \dfrac{4\pi}{3}, \dfrac{5\pi}{3}$ *(6-5)*

31. $\quad\quad 2\sin^2\theta + \cos\theta = 1 \quad\quad 0° \le \theta \le 180°$

$\quad\quad 2(1 - \cos^2\theta) + \cos\theta = 1$

$\quad\quad 2 - 2\cos^2\theta + \cos\theta = 1$

$\quad\quad 1 - 2\cos^2\theta + \cos\theta = 0$

$\quad\quad -2\cos^2\theta + \cos\theta + 1 = 0$

$\quad\quad 2\cos^2\theta - \cos\theta - 1 = 0$

$\quad (2\cos\theta + 1)(\cos\theta - 1) = 0$

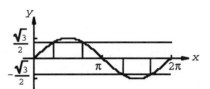

$2\cos\theta + 1 = 0 \quad\quad\quad \cos\theta - 1 = 0$

$\quad \cos\theta = -\dfrac{1}{2} \quad\quad\quad\quad \cos\theta = 1$

$\quad\quad\quad\quad\quad\quad\quad\quad\quad\quad\quad\quad\quad\quad\quad\quad\quad$ *(6-5)*

$\quad\quad\quad \theta = 120° \quad\quad\quad\quad\quad \theta = 0°$

Solutions: $\theta = 0°, 120°$

32. $\quad 2\sin^2 x - \sin x = 0 \quad$ all real x

Solve over one period $[0, 2\pi)$:

$\sin x(2\sin x - 1) = 0$

$\sin x = 0 \quad\quad 2\sin x - 1 = 0$

$\quad x = 0, \pi \quad\quad\quad \sin x = \dfrac{1}{2}$

$\quad\quad\quad\quad\quad\quad\quad\quad x = \dfrac{\pi}{6}, \dfrac{5\pi}{6}$

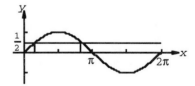

Since the sine function is periodic with period 2π, all solutions are given by

$\quad x = 0 + 2k\pi, \ x = \pi + 2k\pi, \ x = \dfrac{\pi}{6} + 2k\pi, \ x = \dfrac{5\pi}{6} + 2k\pi, \ k$ any integer.

The first two can also be written together as $x = k\pi$, k any integer.   *(6-5)*

33. $\sin 2x = \sqrt{3} \sin x$ all real x
Solve over one period $[0, 2\pi)$:
$$\sin 2x - \sqrt{3} \sin x = 0$$
$$2 \sin x \cos x - \sqrt{3} \sin x = 0$$
$$\sin x(2 \cos x - \sqrt{3}) = 0$$
$\sin x = 0 \qquad\qquad 2 \cos x - \sqrt{3} = 0$
$$x = 0, \pi \qquad\qquad \cos x = \frac{\sqrt{3}}{2}$$
$$x = \frac{\pi}{6}, \frac{11\pi}{6}$$

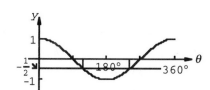

Since the sine and cosine functions are periodic with period 2π, all solutions are given by
$$x = 0 + 2k\pi, \ x = \pi + 2k\pi, \ x = \frac{\pi}{6} + 2k\pi, \ x = \frac{11\pi}{6} + 2k\pi, \ k \text{ any integer.}$$
The first two can also be written together as $x = k\pi$, k any integer.   (6-5)

34. $2 \sin^2 \theta + 5 \cos \theta + 1 = 0$ all θ
Solve over one period $[0°, 360°)$
$$2(1 - \cos^2 \theta) + 5 \cos \theta + 1 = 0$$
$$2 - 2 \cos^2 \theta + 5 \cos \theta + 1 = 0$$
$$-2 \cos^2 \theta + 5 \cos \theta + 3 = 0$$
$$2 \cos^2 \theta - 5 \cos \theta - 3 = 0$$
$$(2 \cos \theta + 1)(\cos \theta - 3) = 0$$

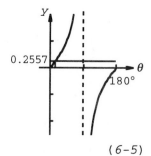

$2 \cos \theta + 1 = 0 \qquad\qquad \cos \theta - 3 = 0$
$$\cos \theta = -\frac{1}{2} \qquad\qquad \cos \theta = 3$$
$\qquad \theta = 120°, 240° \qquad$ No solution
Since the cosine function is periodic with period 360°, all solutions are given
by $\theta = 120° + k360°$, $\theta = 240° + k360°$, k any integer.   (6-5)

35. $\tan \theta = 0.2557$ all θ
Solve over one period $[0°, 180°)$
Sketch a graph of $y = \tan \theta$ and $y = 0.2557$,
θ in $[0°, 180°)$
$$\theta = \tan^{-1} 0.2557 = 14.34°$$

Since the tangent function is periodic with
period 180°, all solutions are given by
$$\theta = 14.34° + k180°$$

(6-5)

36. $\sin^2 x + 2 = 4 \sin x$ all real x
Solve over on period, $[0, 2\pi)$:

$\sin^2 x - 4 \sin x + 2 = 0$ Quadratic in $\sin x$

$\sin x = \dfrac{-b \pm \sqrt{b^2 - 4ac}}{2a}$ $a = 1, \ b = -4, \ c = 2$

$\sin x = \dfrac{-(-4) \pm \sqrt{(-4)^2 - 4(1)(2)}}{2(1)}$

$\sin x = \dfrac{4 \pm \sqrt{8}}{2}$

$\sin x = \dfrac{4 + \sqrt{8}}{2}$ $\sin x = \dfrac{4 - \sqrt{8}}{2}$

$\sin x = 3.414$ $\sin x = 0.5858$

No solution

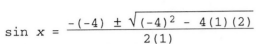

$x = \begin{cases} \sin^{-1} 0.5858 \\ \pi - \sin^{-1} 0.5858 \end{cases} = \begin{cases} 0.6259 \text{ First quadrant solution} \\ 2.516 \text{ Second quadrant solution} \end{cases}$

Since the sine function is periodic with period 2π, all solutions are given by

$x = \begin{cases} 0.6259 + 2k\pi \\ 2.516 + 2k\pi \end{cases}$ k any integer $(6\text{-}5)$

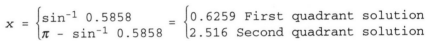

37. $\tan^2 x = 2 \tan x + 1$ $0 \le x < \pi$
$\tan^2 x - 2 \tan x - 1 = 0$ Quadratic in $\tan x$

$\tan x = \dfrac{-b \pm \sqrt{b^2 - 4ac}}{2a}$ $a = 1, \ b = -2, \ c = -1$

$\tan x = \dfrac{-(-2) \pm \sqrt{(-2)^2 - 4(1)(-1)}}{2(1)}$ $\tan x = \dfrac{2 + \sqrt{8}}{2}$ $\tan x = \dfrac{2 - \sqrt{8}}{2}$

$\tan x = \dfrac{(2) \pm \sqrt{8}}{2}$ $\tan x = 2.414$ $\tan x = -0.4142$

$x = \tan^{-1} 2.414$ $x = \pi + \tan^{-1}(-0.4142)$
$x = 1.178$ $x = 2.749$

Solutions: $1.178, \ 2.749$

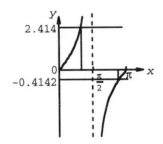

$(6\text{-}5)$

38. Here is a computer-generated graph of
$y = \cos x$ and $y = x^2 - 1$. Since $\cos x$
is periodic, and $x^2 - 1$ is decreasing on
$(-\infty, 0]$ and increasing on $[0, \infty)$, the
graph shows that there are two solutions
of $\cos x = x^2 - 1$.

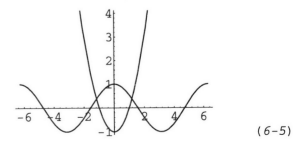

$(6\text{-}5)$

39. Here is a computer-generated graph of $y = e^x$ and $y = 1 + \sin x$. Since $1 + \sin x$ is periodic and e^x has an asymptote along the negative x axis, the graph indicates that the graph of $y = e^x$ will cross the graph of $y = 1 + \sin x$ infinitely many times to the left of the portion shown, although the graph of $y = e^x$ is too close to the x axis to be shown. The equation $e^x = 1 + \sin x$ has infinitely many solutions.

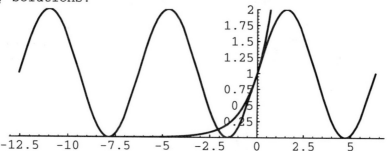

$(6\text{-}5)$

40. We can attempt to solve this equation as follows:

$$\sin x = \sec x$$
$$\sin x = \frac{1}{\cos x}$$
$$\sin x \cos x = 1 \quad \cos x \neq 0$$
$$2 \sin x \cos x = 2$$
$$\sin 2x = 2$$

This is impossible; $\sin 2x$ can never equal 2. The equation has no solutions. $(6\text{-}5)$

41. Here is a computer-generated graph of $y = \tan x$ and $y = 7x - 9$. Since $\tan x$ is periodic with an infinite number of vertical asymptotes, the graph indicates that the graph of $y = 7x - 9$ will cross the graph of $y = \tan x$ infinitely many times to the left and the the right of the portion shown, although the graphs extend too high to be shown. The equation $\tan x = 7x - 9$ has infinitely many solutions.

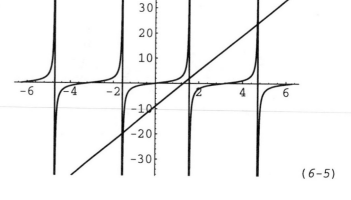

$(6\text{-}5)$

42. Here is a computer-generated graph of $y = 3 \sin 2x$ and $y = 2x - 2.5$, $0 \leq x < 2\pi$.

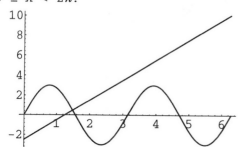

There is a solution of $3 \sin 2x = 2x - 2.5$ in the interval $[1.4, 1.6]$. Zooming in on this solution, we obtain the following graph.

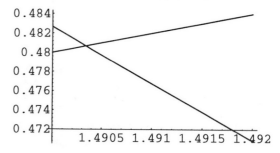

To four decimal places, the solution in $[0, 2\pi)$ is given by 1.4903, and there is no other solution. $(6\text{-}5)$

43. Combining information from the graphs in the previous problem, the graph of
$y = 3 \sin 2x$ is above the graph of $y = 2x - 2.5$, that is, $3 \sin 2x > 2x - 2.5$,
for x in the interval $(-\infty, 1.4903)$. *(6-5)*

44. Here is a computer-generated graph
of $y = 2 \sin^2 x - \cos 2x$ and
$y = 1 - x^2$, $-\pi \le x \le \pi$. Note that
the graphs have y axis symmetry.

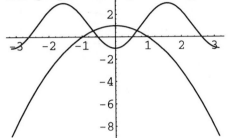

There are points of intersection in
the intervals [0.6, 0.8] and
[-0.8, -0.6]. Zooming in on the
right-hand point of intersection,
we obtain the following graph.

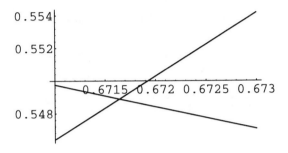

To four decimal places, the solution
is $x = 0.6716$. From the symmetry,
the other solution is given by
$x = -0.6716$. *(6-5)*

45. Combining information from the graphs in the previous problem, the graph of
$y = 2 \sin^2 x - \cos 2x$ is below or intersects the graph of $y = 1 - x^2$, that is,
$2 \sin^2 x - \cos 2x \le 1 - x^2$, for x in the interval $[-0.6716, 0.6716]$. *(6-5)*

46. (A) Testing $x = 0$ and $y = \dfrac{\pi}{4}$ yields

$$\tan\left(0 + \frac{\pi}{4}\right) \overset{?}{=} \tan 0 + \tan \frac{\pi}{4}$$

$$\tan \frac{\pi}{4} \overset{?}{=} 0 + \tan \frac{\pi}{4}$$

$$1 \overset{\checkmark}{=} 0 + 1$$

Yes, $x = 0$ and $y = \dfrac{\pi}{4}$ is a solution.

(B) Consider $x = \dfrac{\pi}{3}$ and $y = \dfrac{\pi}{3}$. Then the left side $= \tan\left(\dfrac{\pi}{3} + \dfrac{\pi}{3}\right) = \tan \dfrac{2\pi}{3} = -\sqrt{3}$.

The right side $= \tan \dfrac{\pi}{3} + \tan \dfrac{\pi}{3} = \sqrt{3} + \sqrt{3} = 2\sqrt{3}$. Since the left side is not
equal to the right side for at least one set of values for which both are
defined, the equation is a conditional equation. *(6-1)*

47. $\sin^{-1} 0.3351$ has exactly one value, while the equation $\sin x = 0.3351$ has
infinitely many solutions. *(5-9, 6-5)*

48. (A) Graph both sides of the equation in the same viewing window.

$$\frac{\tan x}{\sin x + 2 \tan x} = \frac{1}{\cos x - 2}$$ is not an identity, since the graphs do not match.

Try $x = \frac{\pi}{4}$.

Left side: $\dfrac{\tan \frac{\pi}{4}}{\sin \frac{\pi}{4} + 2 \tan \frac{\pi}{4}} = \dfrac{1}{\frac{1}{\sqrt{2}} + 2 \cdot 1} = \dfrac{\sqrt{2}}{1 + 2\sqrt{2}}$

Right side: $\dfrac{1}{\cos \frac{\pi}{4} - 2} = \dfrac{1}{\frac{1}{\sqrt{2}} - 2} = \dfrac{\sqrt{2}}{1 - 2\sqrt{2}}$

This verifies that the equation is not an identity.

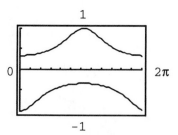

(B) Graph both sides of the equation in the same viewing window.

$$\frac{\tan x}{\sin x - 2 \tan x} = \frac{1}{\cos x - 2}$$ appears to be an identity, which we now verify.

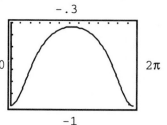

$\dfrac{1}{\cos x - 2} = \dfrac{\tan x}{\tan x(\cos x - 2)}$ Algebra

$\phantom{\dfrac{1}{\cos x - 2}} = \dfrac{\tan x}{\tan x \cos x - 2 \tan x}$ Algebra

$\phantom{\dfrac{1}{\cos x - 2}} = \dfrac{\tan x}{\frac{\sin x}{\cos x}\cos x - 2 \tan x}$ Quotient Identity

$\phantom{\dfrac{1}{\cos x - 2}} = \dfrac{\tan x}{\sin x - 2 \tan x}$ Algebra   *(6-1)*

49. Let $y1 = \cos\left(x - \dfrac{\pi}{3}\right)$

Then $y2 = \cos x \cos \dfrac{\pi}{3} + \sin x \sin \dfrac{\pi}{3}$

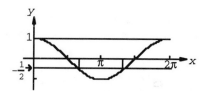

$ = \cos x\left(\dfrac{1}{2}\right) + \sin x\left(\dfrac{\sqrt{3}}{2}\right)$

$ = \dfrac{1}{2} \cos x + \dfrac{\sqrt{3}}{2} \sin x.$

The graphs coincide as shown at the right.   *(6-2)*

50. (A) $\tan \dfrac{x}{2} = 2 \sin x$ $0 \le x < 2\pi$

$\dfrac{1 - \cos x}{\sin x} = 2 \sin x$

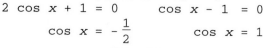

$1 - \cos x = 2 \sin^2 x$ $\sin x \ne 0$

$1 - \cos x = 2(1 - \cos^2 x)$

$1 - \cos x = 2 - 2 \cos^2 x$

$2 \cos^2 x - \cos x - 1 = 0$

$(2 \cos x + 1)(\cos x - 1) = 0$

$2 \cos x + 1 = 0 \qquad \cos x - 1 = 0$

$\cos x = -\dfrac{1}{2} \qquad\qquad \cos x = 1$

$x = \dfrac{2\pi}{3}, \dfrac{4\pi}{3} \qquad\qquad x = 0$

Note that even though we multiplied both sides of the original equation by sin x, which is 0 when $x = 0$, $x = 0$ is still a solution of the original equation, as is shown by checking:

Left side: $\tan \dfrac{0}{2} = \tan 0 = 0$

Right side: $2 \sin 0 = 2 \cdot 0 = 0$

(B) Here is a computer-generated graph of $y = \tan \dfrac{x}{2}$ and $y = 2 \sin x$, $0 \le x < 2\pi$.

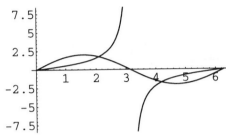

The graph shows that 0 is apparently a solution (checked in the previous part of the problem). There are also solutions in the intervals [2, 2.2] and [4, 4.2]. Zooming in on these solutions, we obtain the following graphs.

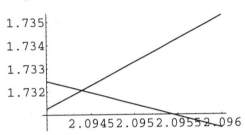 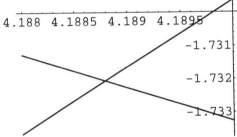

To four decimal places, the solutions are 2.0944 and 4.1888. *(6-5)*

51. Here is a computer-generated graph of $y = 3 \cos(x - 1)$ and $y = 2 - x^2$, $-\pi \le x \le \pi$.

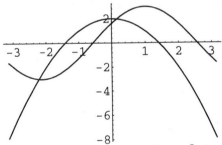

There are solutions of $3 \cos(x - 1) = 2 - x^2$ in the intervals [-2.4, -2.2] and [0, 0.2]. Zooming in on these solutions, we obtain the following graphs.

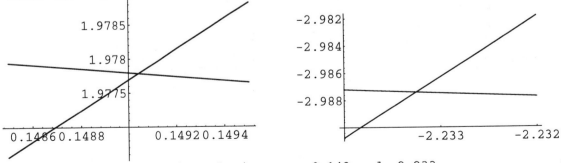

To three decimal places, the solutions are 0.149 and -2.233. *(6-5)*

52. Since $\frac{\pi}{2} \leq x \leq \pi$, $\frac{\pi}{4} \leq \frac{x}{2} \leq \frac{\pi}{2}$, hence $\sin \frac{x}{2}$ is positive.

$\sin x = \frac{3}{5}$ $\cos x = -\frac{4}{5}$

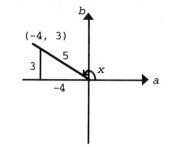

(A) $\sin \frac{x}{2} = \sqrt{\dfrac{1 - \cos x}{2}} = \sqrt{\dfrac{1 - \left(-\frac{4}{5}\right)}{2}} = \sqrt{\dfrac{\frac{9}{5}}{2}} = \sqrt{\dfrac{9}{10}}$

$\qquad\qquad = \dfrac{3}{\sqrt{10}}$ or $\dfrac{3\sqrt{10}}{10}$

(B) $\cos 2x = \cos^2 x - \sin^2 x = \left(-\frac{4}{5}\right)^2 - \left(\frac{3}{5}\right)^2 = \frac{16}{25} - \frac{9}{25} = \frac{7}{25}$ $\qquad$ $(6\text{-}3)$

53. Let $u = \tan^{-1}\left(-\frac{3}{4}\right)$.

Then $-\frac{3}{4} = \tan u$ $-\frac{\pi}{2} < u < \frac{\pi}{2}$.

$\sin u = -\frac{3}{5}$ $\cos u = \frac{4}{5}$.

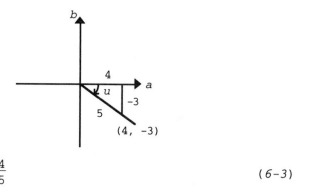

$\sin\left[2 \tan^{-1}\left(-\frac{3}{4}\right)\right] = \sin 2u = 2 \sin u \cos u$

$\qquad\qquad\qquad = 2\left(-\frac{3}{5}\right)\left(\frac{4}{5}\right) = -\frac{24}{25}$ $\qquad$ $(6\text{-}3)$

54. Let $u = \sin^{-1} \frac{3}{5}$

Then $\frac{3}{5} = \sin u$, $-\frac{\pi}{2} \leq u \leq \frac{\pi}{2}$.

Ordinarily we would also set $v = \cos^{-1} \frac{4}{5}$, but a glance at the reference triangle indicates that $\cos^{-1} \frac{4}{5} = \sin^{-1} \frac{3}{5} = u$

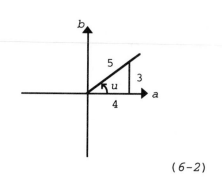

$\sin\left(\sin^{-1} \frac{3}{5} + \cos^{-1} \frac{4}{5}\right) = \sin(u + u) = \sin 2u$

$\qquad\qquad\qquad\qquad = 2 \sin u \cos u$

$\qquad\qquad\qquad\qquad = 2\left(\frac{3}{5}\right)\left(\frac{4}{5}\right) = \frac{24}{25}$ $\qquad$ $(6\text{-}2)$

55. (A) $\cos^2 2x = \cos 2x + \sin^2 2x$ $0 \leq x < \pi$ is equivalent to
$\cos^2 2x = \cos 2x + \sin^2 2x$ $0 \leq 2x < 2\pi$

$\cos^2 2x = \cos 2x + 1 - \cos^2 2x$
$2 \cos^2 2x - \cos 2x - 1 = 0$
$(2 \cos 2x + 1)(\cos 2x - 1) = 0$
$2 \cos 2x + 1 = 0$ or $\cos 2x - 1 = 0$

$\cos 2x = -\frac{1}{2}$ $\qquad\qquad$ $\cos 2x = 1$

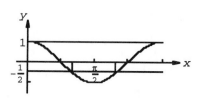

$2x = \frac{2\pi}{3}, \frac{4\pi}{3}$ $\qquad\qquad$ $2x = 0$

$x = \frac{\pi}{3}, \frac{2\pi}{3}$ $\qquad\qquad$ $x = 0$

Solutions: $x = 0, \frac{\pi}{3}, \frac{2\pi}{3}$

(B) Here is a computer-generated
graph of $y = \cos^2 2x$ and
$y = \cos 2x + \sin^2 2x$, $0 \le x < \pi$.

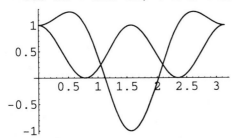

The graph shows that 0 is apparently a
solution. This apparent solution must be
checked:
Left side: $\cos^2 2 \cdot 0 = \cos^2 0 = 1$
Right side: $\cos 2 \cdot 0 + \sin^2 2 \cdot 0$
$= \cos 0 + \sin^2 0 = 1 + 0^2 = 1$

There are also solutions in the intervals [1, 1.1] and [2, 2.1]. Zooming in on
these solutions, we obtain the following graphs.

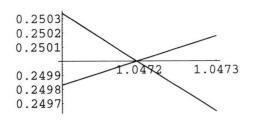

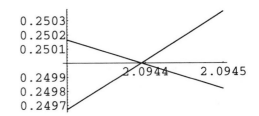

To four decimal places, the solutions are 1.0472 and 2.0944.   (6-5)

56. (A) Here is a computer-generated graph of $f(x) = \sin \dfrac{1}{x-1}$, $0 \le x \le 2$.

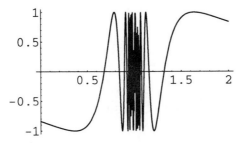

The function appears to have a smallest zero in the interval [0.6, 0.7] and a
largest zero in the interval [1.3, 1.4]. Zooming in on these zeros, we obtain
the following graphs.

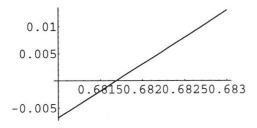

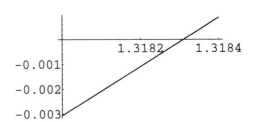

To four decimal places, the smallest zero is 0.6817 and the largest is 1.3183.

(B) As x increases without bound, $\dfrac{1}{x-1}$ tends to 0 through positive numbers and

$\sin \dfrac{1}{x-1}$ tends to 0 through positive numbers. $y = 0$ is a horizontal asymptote
for the graph of f.

(C) The exploratory graphs are left to the student. There are infinitely many zeros in any interval containing $x = 1$. The number $x = 1$ is not a zero because $\sin \dfrac{1}{x - 1}$ is not defined at $x = 1$.

(6-5)

57. We note that $\tan \theta = \dfrac{3}{x}$ and $\tan 2\theta = \dfrac{3 + 6}{x} = \dfrac{9}{x}$ (see figure).

Then, $\tan 2\theta = \dfrac{2 \tan \theta}{1 - \tan^2 \theta}$

$$\dfrac{9}{x} = \dfrac{2\left(\dfrac{3}{x}\right)}{1 - \left(\dfrac{3}{x}\right)^2} = \dfrac{\dfrac{6}{x}}{1 - \dfrac{9}{x^2}} = \dfrac{x^2 \cdot \dfrac{6}{x}}{x^2 \cdot 1 - x^2 \cdot \dfrac{9}{x^2}} = \dfrac{6x}{x^2 - 9}$$

$$x(x^2 - 9)\dfrac{9}{x} = x(x^2 - 9)\dfrac{6x}{x^2 - 9} \quad x \neq 0,\ 3,\ -3$$

$$9(x^2 - 9) = 6x \cdot x$$
$$9x^2 - 81 = 6x^2$$
$$3x^2 - 81 = 0$$
$$x^2 = 27$$

$$x = \sqrt{27} \quad \text{(we discard the negative solution)}$$

To 3 decimal places, $x = 5.196$ cm.

Since $\tan \theta = \dfrac{3}{x} = \dfrac{3}{\sqrt{27}} = \dfrac{1}{\sqrt{3}}$, $\theta = 30.000°$.

(6-3)

58. $I = 50 \sin 120\pi(t - 0.001)$ $\quad I = 40$

$40 = 50 \sin 120\pi(t - 0.001)$

$\sin 120\pi(t - 0.001) = \dfrac{40}{50}$

$120\pi(t - 0.001) = \sin^{-1} \dfrac{40}{50}$ will yield the least positive solution of the equation

$t - 0.001 = \dfrac{1}{120\pi} \sin^{-1} \dfrac{40}{50}$

$t = 0.001 + \dfrac{1}{120\pi} \sin^{-1} \dfrac{40}{50}$

$= 0.00346$ sec (calculator in radian mode)

(6-5)

59. (A) $0.6 \cos 184\pi t - 0.6 \cos 208\pi t = 0.6(\cos 184\pi t - \cos 208\pi t)$ Algebra

$$= 0.6\left[(-2)\sin \dfrac{184\pi t + 208\pi t}{2} \sin \dfrac{184\pi t - 208\pi t}{2}\right] \quad \begin{array}{l}\text{Sum-Product}\\ \text{Identity}\end{array}$$

$= -1.2 \sin 196\pi t \sin(-12\pi t)$ Algebra

$= -1.2 \sin 196\pi t(-\sin 12\pi t)$ Identities for Negatives

$= 1.2 \sin 12\pi t \sin 196\pi t$ Algebra

(B) The required graphs are as shown.

$y = 0.6 \cos 184\pi t$

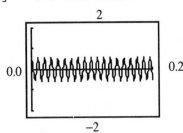

$y = -0.6 \cos 208\pi t$

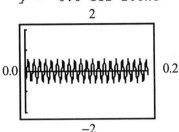

$y = 0.6 \cos 184\pi t - 0.6 \cos 208\pi t$

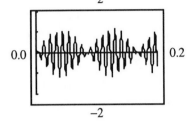

$y = 1.2 \sin 12\pi t \sin 196\pi t$

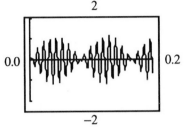

(6-4)

60.

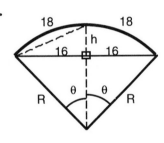

From the figure, $R\theta = 18$ and $\sin \theta = \dfrac{16}{R}$. From these two equations, solving each for R in terms of θ and setting the results equal to each other, we obtain

$$R = \frac{18}{\theta} \qquad R = \frac{16}{\sin \theta}$$

$$\frac{18}{\theta} = \frac{16}{\sin \theta}$$

$$18 \sin \theta = 16\theta$$

$$\sin \theta = \frac{8}{9} \theta \quad \text{as required.}$$

Here is a computer-generated graph of $y = \sin \theta$ and $y = \dfrac{8}{9} \theta$, $0 \leq \theta \leq \dfrac{\pi}{2}$.

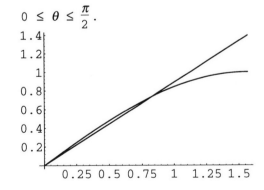

The equation has a solution in the interval [0.75, 1]. Zooming in on this solution, we obtain the following graph.

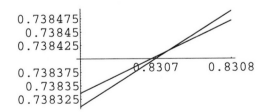

To four decimal places, $\theta = 0.8307$.

Then $R = \dfrac{18}{\theta} = \dfrac{18}{0.8307} = 21.668$ ft.

Since $\dfrac{R - h}{16} = \cot \theta$, $h = R - 16 \cot \theta$

$= 7.057$ ft.

(6-5)

CHAPTER 7

Exercise 7-1

Key Ideas and Formulas

Law of Sines

$$\frac{\sin\ \alpha}{a} = \frac{\sin\ \beta}{b} = \frac{\sin\ \gamma}{c}$$

$(\alpha + \beta = \gamma = 180°)$

The law of sines is generally used to solve the ASA, AAS, and SSA cases for oblique triangles.

SSA VARIATIONS

α	a $(h = b \sin\ \alpha)$	Number of triangles	Figure
Acute	$0 < a < h$	0	
Acute	$a = h$	1	
Acute	$h < a < b$	2	Ambiguous Case
Acute	$a \geq b$	1	
Obtuse	$0 < a \leq b$	0	
Obtuse	$a > b$	1	

1.

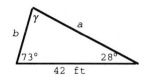

Solve for γ:
$$\alpha + \beta + \gamma = 180°$$
$$\gamma = 180° - (\alpha + \beta)$$
$$= 180° - (73° + 28°)$$
$$= 79°$$

Solve for a:
$$\frac{\sin \alpha}{a} = \frac{\sin \gamma}{c}$$
$$a = \frac{c \sin \alpha}{\sin \gamma}$$
$$= \frac{42 \sin 73°}{\sin 79°}$$
$$= 41 \text{ ft}$$

Solve for b:
$$\frac{\sin \beta}{b} = \frac{\sin \gamma}{c}$$
$$b = \frac{c \sin \beta}{\sin \gamma}$$
$$= \frac{42 \sin 28°}{\sin 79°}$$
$$= 20 \text{ ft}$$

3.

Solve for β:
$$\alpha + \beta + \gamma = 180°$$
$$\beta = 180° - (\alpha + \gamma)$$
$$= 180° - (122° + 18°)$$
$$= 40°$$

Solve for a:
$$\frac{\sin \alpha}{a} = \frac{\sin \beta}{b}$$
$$a = \frac{b \sin \alpha}{\sin \beta}$$
$$= \frac{12 \sin 122°}{\sin 40°}$$
$$= 16 \text{ km}$$

Solve for c:
$$\frac{\sin \beta}{b} = \frac{\sin \gamma}{c}$$
$$c = \frac{b \sin \gamma}{\sin \beta}$$
$$= \frac{12 \sin 18°}{\sin 40°}$$
$$= 5.8 \text{ km}$$

5.

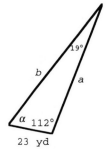

Solve for α:
$$\alpha + \beta + \gamma = 180°$$
$$\alpha = 180° - (\beta + \gamma)$$
$$= 180° - (112° + 19°)$$
$$= 49°$$

Solve for a:
$$\frac{\sin \alpha}{a} = \frac{\sin \gamma}{c}$$
$$a = \frac{c \sin \alpha}{\sin \gamma}$$
$$= \frac{23 \sin 49°}{\sin 19°}$$
$$= 53 \text{ yd}$$

Solve for b:
$$\frac{\sin \beta}{b} = \frac{\sin \gamma}{c}$$
$$b = \frac{c \sin \beta}{\sin \gamma}$$
$$= \frac{23 \sin 112°}{\sin 19°}$$
$$= 66 \text{ yd}$$

7.

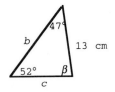

Solve for β:
$$\alpha + \beta + \gamma = 180°$$
$$\beta = 180° - (\alpha + \gamma)$$
$$= 180° - (52° + 47°)$$
$$= 81°$$

Solve for c:
$$\frac{\sin \alpha}{a} = \frac{\sin \gamma}{c}$$
$$c = \frac{a \sin \gamma}{\sin \alpha}$$
$$= \frac{13 \sin 47°}{\sin 52°}$$
$$= 12 \text{ cm}$$

Solve for b:
$$\frac{\sin \alpha}{a} = \frac{\sin \beta}{b}$$
$$b = \frac{a \sin \beta}{\sin \alpha}$$
$$= \frac{13 \sin 81°}{\sin 52°}$$
$$= 16 \text{ cm}$$

9.

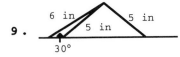

We are given two sides and a non-included angle (SSA). α is acute.
$h = b \sin \alpha = 6 \sin 30° = 3$
$h < a < b$
Two triangles can be constructed; case (c).

11.

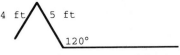

We are given two sides and a non-included angle (SSA). α is acute.
$h = b \sin \alpha = 8 \sin 30° = 4$
$a = 4 = h$
On triangle can be constructed; case (b).

13.

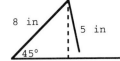

We are given two sides and a non-included angle (SSA). α is acute.
$h = b \sin \alpha = 8 \sin 45° = 4\sqrt{2}$
$0 < a = 5 < 4\sqrt{2} = h$
No triangle can be constructed; case (a).

15.

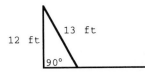

We are given two sides and a non-included angle (SSA). α is obtuse.
$0 < a = 4 < 5 = b$
No triangle can be constructed; case (e).

17.

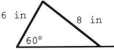

We are given two sides and a non-included angle (SSA). α is acute.
$8 = a \geq b = 6$
One triangle can be constructed; case (d).

19.

We are given two sides and a non-included angle (SSA). α is a right angle, so none of the cases of Table 2 (text) applies. From the given data, we can construct one right triangle, with hypotenuse 13 ft.

21.

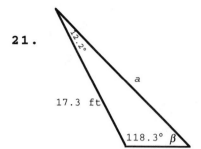

Solve for β:

$\alpha + \beta + \gamma = 180°$
$\beta = 180° - (\alpha + \gamma)$
$= 180° - (118.3° + 12.2°)$
$= 49.5°$

Solve for a:

$$\frac{\sin \alpha}{a} = \frac{\sin \beta}{b}$$

$$a = \frac{b \sin \alpha}{\sin \beta}$$

$$= \frac{17.3 \sin 118.3°}{\sin 49.5°}$$

$$= 20.0 \text{ ft}$$

Solve for c:

$$\frac{\sin \beta}{b} = \frac{\sin \gamma}{c}$$

$$c = \frac{b \sin \gamma}{\sin \beta}$$

$$= \frac{17.3 \sin 12.2°}{\sin 49.5°}$$

$$= 4.81 \text{ ft}$$

23.

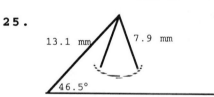

Solve for γ:

$\alpha + \beta + \gamma = 180°$
$\gamma = 180° - (\alpha + \beta)$
$= 180° - (67.7° + 54.2°)$
$= 58.1°$

Solve for c:

$$\frac{\sin \beta}{b} = \frac{\sin \gamma}{c}$$

$$c = \frac{b \sin \gamma}{\sin \beta}$$

$$= \frac{123 \sin 58.1°}{\sin 54.2°}$$

$$= 129 \text{ m}$$

Solve for a:

$$\frac{\sin \alpha}{a} = \frac{\sin \beta}{b}$$

$$a = \frac{b \sin \alpha}{\sin \beta}$$

$$= \frac{123 \sin 67.7°}{\sin 54.2°}$$

$$= 140 \text{ m}$$

25.

α is acute
$h = b \sin \alpha = 13.1 \sin 46.5° = 9.50$
$a = 7.9 < 9.50 = h$
No triangle is possible.

If we try to draw a triangle with these values, we will be unsuccessful. No solution.

27.

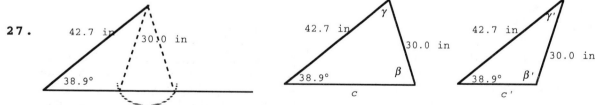

If we try to draw a triangle with these values, we find that two triangles are possible. This is verified by the fact that $h < a < b$, where $h = b \sin \alpha = 26.8$.

We start by calculating angle β and angle β', from the law of sines.

$$\frac{\sin \beta}{b} = \frac{\sin \alpha}{a}$$

$$\sin \beta = \frac{b \sin \alpha}{a} = \frac{42.7 \sin 38.9°}{30.0}$$

$$\beta = \sin^{-1}\left(\frac{42.7 \sin 38.9°}{30.0}\right) \qquad\qquad \beta' = 180° - \sin^{-1}\left(\frac{42.7 \sin 38.9°}{30.0}\right)$$

$$= 63.4° \qquad\qquad\qquad\qquad\qquad\qquad = 116.6°$$

Solve for γ: 　　　　　　　　　　　　　Solve for γ':

$$\alpha + \beta + \gamma = 180° \qquad\qquad\qquad \alpha + \beta' + \gamma' = 180°$$

$$\gamma = 180° - (\alpha + \beta) \qquad\qquad \gamma' = 180° - (\alpha' + \beta')$$

$$= 180° - (38.9° + 63.4°) \qquad\qquad = 180° - (38.9° + 116.6°)$$

$$= 77.7° \qquad\qquad\qquad\qquad\qquad\qquad = 24.5°$$

Solve for c: 　　　　　　　　　　　　　　Solve for c':

$$\frac{\sin \alpha}{a} = \frac{\sin \gamma}{c} \qquad\qquad\qquad\qquad \frac{\sin \alpha}{a} = \frac{\sin \gamma'}{c'}$$

$$c = \frac{a \sin \gamma}{\sin \alpha} \qquad\qquad\qquad\qquad c' = \frac{a \sin \gamma'}{\sin \alpha}$$

$$= \frac{30.0 \sin 77.7°}{\sin 38.9°} \qquad\qquad\qquad = \frac{30.0 \sin 24.5°}{\sin 38.9°}$$

$$= 46.7 \text{ in} \qquad\qquad\qquad\qquad\qquad = 19.8 \text{ in}$$

29.

α is obtuse
$101 = a < b = 152$
No triangle is possible.

If we try to draw a triangle with these values, we will be unsuccessful.
No solution.

31. From a rough sketch, we see that there is only one triangle possible:

Solve for α:

$$\frac{\sin \alpha}{a} = \frac{\sin \beta}{b}$$

$$\sin \alpha = \frac{a \sin \beta}{b} = \frac{43.2 \sin 29°30'}{56.5}$$

$$\alpha = \sin^{-1}\left(\frac{43.2 \sin 29°30'}{56.5}\right)$$

$$= 22°10'$$

Solve for γ: 　　　　　　　　　　　　Solve for c:

$$\alpha + \beta + \gamma = 180° \qquad\qquad\qquad \frac{\sin \beta}{b} = \frac{\sin \gamma}{c}$$

$$\gamma = 180° - (\alpha + \beta) \qquad\qquad\qquad c = \frac{b \sin \gamma}{\sin \beta}$$

$$= 180° - (22°10' + 29°30') \qquad\qquad = \frac{56.5 \sin 128°20'}{\sin 29°30'}$$

$$= 128°20' \qquad\qquad\qquad\qquad\qquad = 89.9 \text{ mm}$$

33.

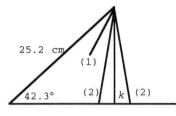

From the diagram, we can see that k = altitude of any possible triangle. Thus,

$$\sin \alpha = \frac{k}{b} \qquad k = b \sin \alpha$$

$$k = (25.2 \text{ cm})\sin 42.3° = 17.0 \text{ cm}$$

If $0 < a < k$, there is no solution: (1) in diagram.
If $a = k$, there is one solution.
If $k < a < b$, there are two solutions: (2) in diagram.

35. Using the given and the calculated data, we have:

$$(a - b)\cos \frac{\gamma}{2} = c \sin \frac{\alpha - \beta}{2}$$

$$(41 - 20)\cos \frac{79°}{2} = 42 \sin \frac{73° - 28°}{2}$$

$$16.204 \approx 16.073$$

37. Sketch a figure:

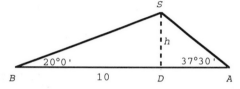

Using the law of sines, we have, in triangle ABS:

$$\frac{AS}{\sin ABS} = \frac{AB}{\sin ASB}$$

We are given angle $ABS = 20°0'$ and $AB = 10$. To find angle ASB, we use $ABS + ASB + BAS = 180°$, hence

$$\begin{aligned}
ASB &= 180° - (ABS + BAS) \\
&= 180° - (20°0' + 37°30') \\
&= 180° - 57°30' \\
&= 122°30'
\end{aligned}$$

Hence

$$\frac{AS}{\sin 20°0'} = \frac{10}{\sin 122°30'}$$

$$\begin{aligned}
AS &= \frac{10 \sin 20°0'}{\sin 122°30'} \\
&= 4.06 \text{ miles} = \text{distance of ship from point } A.
\end{aligned}$$

To find
$h = SD$ = distance of ship from shore, we note in right triangle ADS

$$\frac{h}{AS} = \sin DAS$$

$$\begin{aligned}
h &= AS \sin DAS \\
&= 4.06 \sin 37°30' \\
&= 2.47 \text{ miles}
\end{aligned}$$

39. Redrawing the figure and labeling the sides, we have:

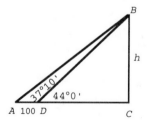

In triangle ABC, we have $\dfrac{AC}{h} = \cot 37°10'$.

Hence $AC = h$ cot 37°10'

In triangle DBC, we have $\dfrac{DC}{h}$ = cot 44°. Hence $DC = h$ cot 44°. Hence

$$100 = AC - DC = h \text{ cot } 37°10' - h \text{ cot } 44°$$

Solving for h, we have

$$100 = h(\text{cot } 37°10' - \text{cot } 44°)$$

So $h = \dfrac{100}{\text{cot } 37°10 - \text{cot } 44°}$

$\qquad$ = 353 feet

Alternatively, we can use the law of sines in triangle ADB to find BD, after noting that angle $ADB = 180° -$ angle $BDC = 180° - 44° = 136°$, and therefore angle $ABD = 180° - (136° + 37°10') = 6°50'$. Hence

$$\dfrac{BD}{\sin DAB} = \dfrac{AD}{\sin ABD}$$

$$BD = \dfrac{AD \sin DAB}{\sin ABD}$$

$$\quad = \dfrac{100 \sin 37°10'}{\sin 6°50'}$$

$$\quad = 508 \text{ feet}$$

Hence, in right triangle BCD, we have

$$\dfrac{h}{BD} = \sin BDC$$

$$h = BD \sin BDC$$

$$\quad = 508 \sin 44°$$

$$\quad = 353 \text{ feet}$$

41.

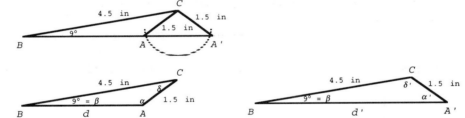

There are two possible triangles, hence two possible values of the required distance. Call them d and d'.

The law of sines gives two possible values for the angle that the crankshaft makes with the center line; we denote them α and α' in the figures.

We start by calculating angle BAC, or α, and angle $BA'C$, or α', from the law of sines.

$$\dfrac{BC}{\sin \alpha} = \dfrac{AC}{\sin CBA}$$

$$\sin \alpha = \dfrac{BC \sin CBA}{AC}$$

$$\quad = \dfrac{4.5 \sin 9°}{1.5}$$

$$\quad = 0.4693$$

Hence the two possibilities are

α obtuse $\qquad\qquad\qquad\qquad \alpha$ acute

$\quad \alpha = 180° - \sin^{-1} 0.4693 \qquad\quad \alpha' = \sin^{-1} 0.4693$

$\qquad = 152° \qquad\qquad\qquad\qquad\qquad = 28°$

Hence the two possibilities for angle BCA, or δ, become

$\quad \delta = 180° - (\alpha + \beta) \qquad\qquad \delta' = 180° - (\alpha' + \beta)$

$\qquad = 180° - (152° + 9°) \qquad\qquad = 180° - (28° + 9°)$

$\qquad = 19° \qquad\qquad\qquad\qquad\qquad = 143°$

Applying the law of sines again to calculate d and d' from these two values of δ, we have

$$\frac{d}{\sin \delta} = \frac{AC}{\sin \beta}$$

$$d = \frac{AC \sin \delta}{\sin \beta}$$

$$= \frac{1.5 \sin 19°}{\sin 9°}$$

$$= 3.1 \text{ in}$$

$$\frac{d'}{\sin \delta'} = \frac{AC}{\sin \beta}$$

$$d' = \frac{AC \sin \delta'}{\sin \beta}$$

$$= \frac{1.5 \sin 143°}{\sin 9°}$$

$$= 5.8 \text{ in}$$

43. Sketch a figure.

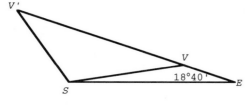

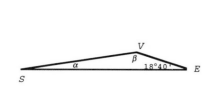 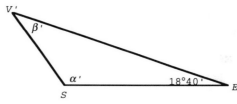

We are given $SE = 1.495 \times 10^8$, $SV = SV' = 1.085 \times 10^8$, angle $SEV = 18°40'$. There are two possible triangles, hence two possible values of the required distance. Call them EV and EV'.

The law of sines gives two possible values for angle VSE; we denote them α and α' in the figures.

We start by calculating angle EVS, or β, and angle $EV'S$, or β', from the law of sines.

$$\frac{SE}{\sin \beta} = \frac{SV}{\sin SEV}$$

$$\sin \beta = \frac{SE \sin SEV}{SV}$$

$$= \frac{1.495 \times 10^8 \sin 18°40'}{1.085 \times 10^8}$$

$$= 0.4410$$

Hence the two possibilities are

β obtuse

$$\beta = 180° - \sin^{-1} 0.4410$$

$$= 153°50'$$

β acute

$$\beta' = \sin^{-1} 0.4410$$

$$= 26.2° = 26°10'$$

Hence the two possibilities for angle VSE, or α, become

$$\alpha = 180° - (\beta + \gamma)$$

$$= 180° - (153°50' + 18°40')$$

$$= 7°30'$$

$$\alpha' = 180° - (\beta' + \gamma')$$

$$= 180° - (26°10' + 18°40')$$

$$= 135°10'$$

Applying the law of sines again to calculate EV and EV' from these two values of α, we have

$$\frac{EV}{\sin \alpha} = \frac{SV}{\sin SEV} \qquad\qquad \frac{EV'}{\sin \alpha'} = \frac{SV'}{\sin SEV}$$

$$EV = \frac{SV \sin \alpha}{\sin SEV} \qquad\qquad EV' = \frac{SV' \sin \alpha'}{\sin SEV}$$

$$= \frac{1.085 \times 10^8 \sin(7°30')}{\sin(18°40')}) \qquad = \frac{1.085 \times 10^8 \sin(135°10')}{\sin(18°40')}$$

$$= 4.42 \times 10^7 \text{ kilometers} \qquad = 2.39 \times 10^8 \text{ kilometers}$$

45. Redrawing the figure and labeling, we have:

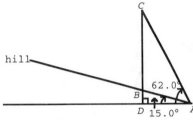

In the figure, note: Triangle ADC is a right triangle, hence $\angle ACB = 90° - 62.0° = 28.0°$. Triangle ABC is not a right triangle; but, from the law of sines,

$$\frac{\sin CAB}{BC} = \frac{\sin ACB}{AB}$$

$$\angle CAB = \angle CAD - \angle BAD = 62.0° - 15.0° = 47.0°$$

$$BC = \frac{AB \sin CAB}{\sin ACB} = \frac{(102 \text{ ft})\sin 47.0°}{\sin 28.0°} = 159 \text{ ft.}$$

47. Labeling the diagram as shown, we note: triangle OAB is isosceles, hence $\alpha = \beta$. Thus, $\alpha + \beta + \theta = \alpha + \alpha + 98.9° = 180°$

$$2\alpha = 81.1°$$

$$\alpha = 40.55°$$

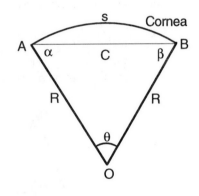

Now apply the law of sines to find R.

$$\frac{\sin \alpha}{R} = \frac{\sin \theta}{AB}$$

$$R = \frac{AB \sin \alpha}{\sin \theta} = \frac{(11.8 \text{ mm})\sin 40.55°}{\sin 98.9°} = 7.76 \text{ mm}$$

To find s, we use the formula $s = R\theta_{deg} = R\frac{\pi_{rad}}{180°}\theta_{deg}$, thus

$$s = (7.76 \text{ mm})\frac{\pi}{180°}(98.9°) = 13.4 \text{ mm}$$

49. Redrawing the figure and labeling, we have:

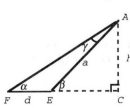

Triangle ACE is a right triangle, hence $\frac{h}{a} = \sin \beta$ and $h = a \sin \beta$. Triangle AEF is not a right triangle, but from the law of sines,

$$\frac{\sin \gamma}{d} = \frac{\sin \alpha}{a}$$

Hence, $a = \dfrac{d \sin \alpha}{\sin \gamma}$

We can find γ since the exterior angle of a triangle has measure equal to the sum of the two nonadjacent interior angles. Hence

$\alpha + \gamma = \beta$

$\qquad \gamma = \beta - \alpha$

Hence

$h = a \sin \beta$

$\quad = \dfrac{d \sin \alpha}{\sin \gamma} \sin \beta$

$\quad = d \dfrac{\sin \alpha \sin \beta}{\sin \gamma}$

$h = d \dfrac{\sin \alpha \sin \beta}{\sin(\beta - \alpha)}$

Exercise 7-2

Key Ideas and Formulas

Law of Cosines (three equivalent formulations)

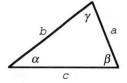

$a^2 = b^2 + c^2 - 2bc \cos \alpha$
$b^2 = a^2 + c^2 - 2ac \cos \beta$
$c^2 = a^2 + b^2 - 2ab \cos \gamma$

The law of cosines is generally used to solve the SAS and SSS cases for oblique triangles.

Law of Sines (restated here for convenience)

$$\frac{\sin \alpha}{a} = \frac{\sin \beta}{b} = \frac{\sin \gamma}{c}$$

---

1. Angle γ is acute. A triangle can have at most one obtuse angle. Since α is acute, then, if the triangle has an obtuse angle it must be the angle opposite the longer of the two sides, b and c. Thus, γ, the angle opposite the shorter of the two sides, c, must be acute.

3.

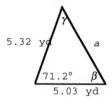

We are given two sides and the included angle (SAS).
Solve for a:
We use the law of cosines
$\quad a^2 = b^2 + c^2 - 2bc \cos \alpha$
$\qquad = (5.32)^2 + (5.03)^2 - 2(5.32)(5.03)\cos 71.2°$
$\qquad = 36.355898...$
$\quad a = 6.03$ yd.

> **Common Error:**
> "$a^2 + b^2 = c^2$"
> The Pythagorean relationship
> $\quad a^2 + b^2 = c^2$
> applies only in right triangles. It is not applicable to oblique triangles.

Solve for β:
Note that since a is the longest side, the angle opposite a, that is, α, must be the largest angle. Hence both β and γ must be less than 71.2°, hence, β and γ must be acute.

We solve for β using the law of sines.

$$\frac{\sin \alpha}{a} = \frac{\sin \beta}{b}$$

$$\sin \beta = \frac{b \sin \alpha}{a} = \frac{5.32 \sin 71.2°}{6.03}$$

$$\beta = \sin^{-1} \frac{5.32 \sin 71.2°}{6.03}$$

$$= 56.6°$$

Solve for γ:

$$\alpha + \beta + \gamma = 180°$$

$$\gamma = 180° - (\alpha + \beta)$$

$$= 180° - (71.2° + 56.6°)$$

$$= 52.2°$$

5.

We are given two sides and the included angle (SAS).

Solve for c:

We use the law of cosines

$$c^2 = a^2 + b^2 - 2ab \cos \gamma$$

$$= (5.73)^2 + (10.2)^2 - 2(5.73)(10.2)\cos 120°20'$$

$$= 195.90685…$$

$$c = 14.0 \text{ mm}$$

Solve for β:

Note that since the sum of β and α is $180° - 120°20' = 59°40'$, each must be less than 90°. That is, both β and α are acute. We use the law of sines.

$$\frac{\sin \beta}{b} = \frac{\sin \gamma}{c}$$

$$\sin \beta = \frac{b \sin \gamma}{c} = \frac{10.2 \sin 120°20'}{14.0}$$

$$\beta = \sin^{-1}\left(\frac{10.2 \sin 120°20'}{14.0}\right)$$

$$= 39°0'$$

Solve for α:

$$\alpha + \beta + \gamma = 180°$$

$$\alpha = 180° - (\beta + \gamma)$$

$$= 180° - (39°0' + 120°20')$$

$$= 20°40'$$

7. If the triangle has an obtuse angle, then it must be the angle opposite the longest side; in this case, β.

9.

We are given three sides of the triangle (SSS). We solve for the largest angle first, using the law of cosines.

Solve for β:

$$b^2 = a^2 + c^2 - 2ac \cos \beta$$

$$\cos \beta = \frac{a^2 + c^2 - b^2}{2ac}$$

$$\beta = \cos^{-1}\left(\frac{a^2 + c^2 - b^2}{2ac}\right)$$

$$= \cos^{-1}\left(\frac{(4.00)^2 + (9.05)^2 - (10.2)^2}{2(4.00)(9.05)}\right)$$

$$= 94.9°$$

Solve for α:
We use the law of sines. Since the sum of α and γ is $180° - 94.9° = 85.1°$, both α and γ must be acute.

$$\frac{\sin \alpha}{a} = \frac{\sin \beta}{b}$$

$$\sin \alpha = \frac{a \sin \beta}{b}$$

$$= \frac{4.00 \sin 94.9°}{10.2}$$

$$\alpha = \sin^{-1}\left(\frac{4.00 \sin 94.9°}{10.2}\right)$$

$$= 23.0°$$

Solve for γ:
$$\alpha + \beta + \gamma = 180°$$
$$\gamma = 180° - (\alpha + \beta)$$
$$= 180° - (23.0° + 94.9°)$$
$$= 62.1°$$

11.

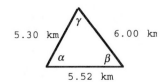

We are given three sides of the triangle (SSS). We solve for the largest angle first, using the law of cosines.

Solve for α:
$$a^2 = b^2 + c^2 - 2bc \cos \alpha$$
$$\cos \alpha = \frac{b^2 + c^2 - a^2}{2bc}$$
$$\alpha = \cos^{-1}\left(\frac{b^2 + c^2 - a^2}{2bc}\right)$$
$$= \cos^{-1}\left(\frac{(5.30)^2 + (5.52)^2 - (6.00)^2}{2(5.30)(5.52)}\right)$$
$$= 67.3°$$

Solve for β:
We use the law of sines. Both β and γ must be less than 67.3°, hence acute.

$$\frac{\sin \alpha}{a} = \frac{\sin \beta}{b}$$

$$\sin \beta = \frac{b \sin \alpha}{a}$$

$$= \frac{5.30 \sin 67.3°}{6.00}$$

$$\beta = \sin^{-1}\left(\frac{5.30 \sin 67.3°}{6.00}\right)$$

$$= 54.6°$$

Solve for γ:
$$\alpha + \beta + \gamma = 180°$$
$$\gamma = 180° - (\alpha + \beta)$$
$$= 180° - (67.3° + 54.6°)$$
$$= 58.1°$$

13.

It is impossible to form a triangle with this data, since angles α and γ together add up to more than 180°. No solution.

15.

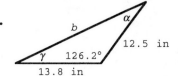

We are given two sides and the included angle (SAS). We use the law of cosines to find the third side, then the law of sines to find a second angle.

Solve for b:
$$b^2 = a^2 + c^2 - 2ac \cos \beta$$
$$= (13.8)^2 + (12.5)^2 - 2(13.8)(12.5)\cos 126.2°$$
$$= 550.44895\ldots$$
$$b = 23.5 \text{ in}$$

Solve for α:
Note that since the sum of α and γ is $180° - 126.2° = 53.8°$, both must be acute.

$$\frac{\sin \alpha}{a} = \frac{\sin \beta}{b}$$
$$\sin \alpha = \frac{a \sin \beta}{b}$$
$$= \frac{13.8 \sin 126.2°}{23.5}$$
$$\alpha = \sin^{-1}\left(\frac{13.8 \sin 126.2°}{23.5}\right)$$
$$= 28.3°$$

Solve for γ:
$$\alpha + \beta + \gamma = 180°$$
$$\gamma = 180° - (\alpha + \beta)$$
$$= 180° - (126.2° + 28.3°)$$
$$= 25.5°$$

17. It is impossible to draw a triangle with this data, since the triangle inequality is violated. (The triangle inequality states that any side of a triangle, for example c, must be less than the sum of the other two sides, thus, $a + b$. Here $c > a + b$.)

19. We are given two sides and a non-included angle (SSA). From a rough sketch, we see that there is only one triangle possible.

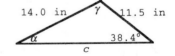

We use the law of sines.

Solve for α:
$$\frac{\sin \alpha}{a} = \frac{\sin \beta}{b}$$
$$\sin \alpha = \frac{a \sin \beta}{b}$$
$$= \frac{11.5 \sin 38.4°}{14.0}$$
$$\alpha = \sin^{-1}\left(\frac{11.5 \sin 38.4°}{14.0}\right) \quad \alpha \text{ is acute}$$
$$= 30.7°$$

Solve for γ:
$$\alpha + \beta + \gamma = 180°$$
$$\gamma = 180° - (\alpha + \beta)$$
$$= 180° - (30.7° + 38.4°)$$
$$= 110.9°$$

Solve for c:
$$\frac{\sin \alpha}{a} = \frac{\sin \gamma}{c}$$
$$c = \frac{a \sin \gamma}{\sin \alpha}$$
$$= \frac{11.5 \sin 110.9°}{\sin 30.7°}$$
$$= 21.0 \text{ in}$$

21.

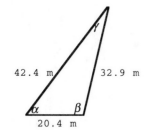

We are given three sides (SSS). We solve for the largest angle, β (largest because it is opposite the largest side, b) using the law of cosines. We then solve for a second angle using the law of sines.

Solve for β:
$$b^2 = a^2 + c^2 - 2ac \cos \beta$$
$$\cos \beta = \frac{a^2 + c^2 - b^2}{2ac}$$
$$\beta = \cos^{-1}\left(\frac{a^2 + c^2 - b^2}{2ac}\right)$$
$$= \cos^{-1}\left(\frac{(32.9)^2 + (20.4)^2 - (42.4)^2}{2(32.9)(20.4)}\right)$$
$$= 102.9°$$

Solve for α:
We use the law of sines. Since the sum of α and γ is $180° - 102.9° = 77.1°$, both α and γ must be acute.
$$\frac{\sin \alpha}{a} = \frac{\sin \beta}{b}$$
$$\sin \alpha = \frac{a \sin \beta}{b}$$
$$= \frac{32.9 \sin 102.9°}{42.4}$$
$$\alpha = \sin^{-1}\left(\frac{32.9 \sin 102.9°}{42.4}\right)$$
$$= 49.1°$$

Solve for γ:
$$\alpha + \beta + \gamma = 180°$$
$$\gamma = 180° - (\alpha + \beta)$$
$$= 180° - (49.1° + 102.9°)$$
$$= 28.0°$$

23.

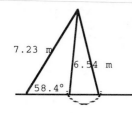   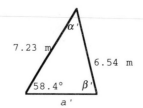

We are given two sides and a non-included angle (SSA). If we try to draw a triangle with these values, we find that two triangles are possible. We use the law of sines to find the two possible values of a second angle.

Solve for β and β':
$$\frac{\sin \beta}{b} = \frac{\sin \gamma}{c}$$
$$\sin \beta = \frac{b \sin \gamma}{c} = \frac{7.23 \sin 58.4°}{6.54} = 0.9416$$

Angle β can be either obtuse or acute:
$$\beta = 180° - \sin^{-1} 0.9416 \qquad \beta' = \sin^{-1} 0.9416$$
$$= 180° - 70.3° \qquad\qquad = 70.3°$$
$$= 109.7°$$

Solve for α and α':
$$\alpha = 180° - (\beta + \gamma) \qquad \alpha' = 180° - (\beta' + \gamma)$$
$$= 180° - (109.7° + 58.4) \qquad = 180° - (70.3° + 58.4°)$$
$$= 11.9° \qquad\qquad\qquad = 51.3°$$

Solve for a and a':

$$\frac{\sin \alpha}{a} = \frac{\sin \gamma}{c}$$

$$a = \frac{c \sin \alpha}{\sin \gamma}$$

$$= \frac{6.54 \sin 11.9°}{\sin 58.4°}$$

$$= 1.58 \text{ m}$$

$$\frac{\sin \alpha'}{a'} = \frac{\sin \gamma}{c}$$

$$a' = \frac{c \sin \alpha'}{\sin \gamma}$$

$$= \frac{6.54 \sin 51.3°}{\sin 58.4°}$$

$$= 5.99 \text{ m}$$

In summary:

Triangle I: $\beta = 109.7°$, $\alpha = 11.9°$, $a = 1.58$ m

Triangle II: $\beta' = 70.3°$, $\alpha' = 51.3°$, $a' = 5.99$ m

25.

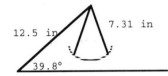

β is acute

$h = a \sin \beta = 12.5 \sin 39.8° = 8.00$

$b = 7.31 < 8.00 = h$

We are given two sides and a non-included angle (SSA). If we try to draw a triangle with these values, we will be unsuccessful. No solution.

27. The law of cosines states that $c^2 = a^2 + b^2 - 2ab \cos \gamma$ for any triangle. If $\gamma = 90°$, $\cos \gamma = 0$; hence,

$c^2 = a^2 + b^2 - 2ac \cos 90°$

$c^2 = a^2 + b^2 - 0$

$c^2 = a^2 + b^2$

29. We write the law of cosines in its three equivalent forms:

$$
\begin{aligned}
c^2 &= a^2 + b^2 && - 2ab \cos \gamma \\
a^2 &= \quad\;\; b^2 + c^2 && - 2bc \cos \alpha \\
b^2 &= a^2 \;\; + c^2 && - 2ac \cos \beta
\end{aligned}
$$

$a^2 + b^2 + c^2 = 2a^2 + 2b^2 + 2c^2 - 2ab \cos \gamma - 2bc \cos \alpha - 2ac \cos \beta$ (Adding)

Subtracting $2a^2 + 2b^2 + 2c^2$ from both sides, we obtain:

$-a^2 - b^2 - c^2 = -2ab \cos \gamma - 2bc \cos \alpha - 2ac \cos \beta$

Dividing both sides by $-2abc$, we obtain

$$\frac{-a^2 - b^2 - c^2}{-2abc} = \frac{-2ab \cos \gamma}{-2abc} + \frac{-2bc \cos \alpha}{-2abc} + \frac{-2ac \cos \beta}{-2abc}$$

Removing common factors and rearranging the terms on the right, we obtain

$$\frac{a^2 + b^2 + c^2}{2abc} = \frac{\cos \alpha}{a} + \frac{\cos \beta}{b} + \frac{\cos \gamma}{c}$$

as required.

31. Redrawing and labeling the drawing in the text, we have:

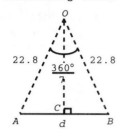

Since triangle OAB is isosceles, triangle OAC is a right triangle, $AC = \dfrac{d}{2}$, and angle $AOC = \dfrac{1}{2}\left(\dfrac{360°}{7}\right) = \dfrac{360°}{14}$.

In right triangle OAC,

$$\sin AOC = \frac{AC}{OA}$$

$$\sin \frac{360°}{14} = \frac{AC}{22.8}$$

$$AC = 22.8 \sin \frac{360°}{14}$$

$$\frac{d}{2} = 22.8 \sin \frac{360°}{14}$$

Then $\qquad d = 45.6 \sin \dfrac{360°}{14}$

$$d = 19.8 \text{ cm}$$

33. From the law of cosines, we have
$$
\begin{aligned}
(AB)^2 &= (AC)^2 + (BC)^2 - 2(AC)(BC)\cos ACB \\
&= 91^2 + 71^2 - 2(91)(71)\cos 96° \\
&= 14{,}672.7168\ldots \\
AB &= 120 \text{ yds}
\end{aligned}
$$

35. Sketch a figure:

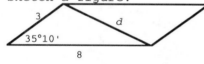

From the law of cosines, we have
$$
\begin{aligned}
d^2 &= 3^2 + 8^2 - 2(3)(8)\cos 35°10' \\
&= 33.760960\ldots \\
d &= 5.81 \text{ feet}
\end{aligned}
$$

37. Sketch a figure:

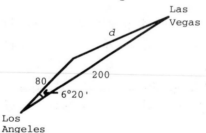

From the law of cosines, we have
$$
\begin{aligned}
d^2 &= 80^2 + 200^2 - 2(80)(200)\cos(6°20') \\
&= 14595.296\ldots \\
d &= 121 \text{ miles}
\end{aligned}
$$

39. Sketch a figure:

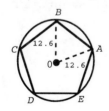

The perimeter $ABCDE$ is 5 times the length of side AB.

Angle $AOB = \dfrac{1}{5}(360°) = 72°$

We use the law of cosines in triangle AOB to determine side AB.
$$
\begin{aligned}
(AB)^2 &= (12.6)^2 + (12.6)^2 - 2(12.6)(12.6)\cos 72° \\
&= 219.40092\ldots \\
AB &= 14.812
\end{aligned}
$$

Perimeter $= 5(AB) = 74.1$ meters

41.

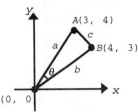

From the given information and the distance formula (Section 2-1), we can determine the lengths of the three sides of the triangle.

$$d(P_1, P_2) = \sqrt{(x_2 - x_1)^2 + (y_2 - y_1)^2}$$

$$c = d(A, B) = \sqrt{(4 - 3)^2 + (3 - 4)^2} = \sqrt{1 + 1} = \sqrt{2}$$

$$b = d(O, B) = \sqrt{(4 - 0)^2 + (3 - 0)^2} = \sqrt{16 + 9} = 5$$

$$a = d(O, A) = \sqrt{(3 - 0)^2 + (4 - 0)^2} = \sqrt{9 + 16} = 5$$

From the law of cosines we can now determine θ.

$$c^2 = a^2 + b^2 - 2ab \cos \theta$$

$$\cos \theta = \frac{a^2 + b^2 - c^2}{2ab}$$

$$\theta = \cos^{-1} \frac{a^2 + b^2 - c^2}{2ab}$$

$$= \cos^{-1} \frac{5^2 + 5^2 - (\sqrt{2})^2}{2(5)(5)}$$

$$= \cos^{-1} 0.96$$

$$= 0.284 \text{ rad}$$

43. From the figure it should be clear that the sides of the triangle are:

$a = 2.03 + 5.00 = 7.03$
$b = 2.03 + 8.20 = 10.23$
$c = 8.20 + 5.00 = 13.20$

Solve for α:
We use the law of cosines.

$$a^2 = b^2 + c^2 - 2bc \cos \alpha$$

$$\cos \alpha = \frac{b^2 + c^2 - a^2}{2bc}$$

$$\alpha = \cos^{-1}\left(\frac{b^2 + c^2 - a^2}{2bc}\right)$$

$$\alpha = \cos^{-1}\left(\frac{(10.23)^2 + (13.20)^2 - (7.03)^2}{2(10.23)(13.20)}\right)$$

$$= \cos^{-1} 0.8497$$

$$= 31°50'$$

Solve for β:
We use the law of sines.

$$\frac{\sin \alpha}{a} = \frac{\sin \beta}{b}$$

$$\sin \beta = \frac{b \sin \alpha}{a}$$

$$\beta = \sin^{-1} \frac{b \sin \alpha}{a} \quad \beta \text{ is acute (see figure.)}$$

$$= \sin^{-1}\left(\frac{(10.23)(\sin 31°50')}{7.03}\right)$$

$$= \sin^{-1} 0.7673$$

$$= 50°10'$$

Solve for γ:
$\gamma = 180° - (\alpha + \beta)$
$\quad = 180° - (31°50' + 50°10')$
$\quad = 98°0'$

45. The three sides of triangle ABC are each in turn the hypotenuse of a right triangle formed with two edges of the solid. Hence, by the Pythagorean theorem,

$AB^2 = 4.3^2 + 8.1^2 = 84.10 \qquad AB = 9.17$ cm

$AC^2 = 2.8^2 + 8.1^2 = 73.45 \qquad AC = 8.57$ cm

$BC^2 = 2.8^2 + 4.3^2 = 26.33 \qquad BC = 5.13$ cm

To find $\angle CAB$, we apply the law of cosines.

$$BC^2 = AC^2 + AB^2 - 2(AC)(AB)\cos CAB$$

$$\cos CAB = \frac{AC^2 + AB^2 - BC^2}{2(AC)(AB)} = \frac{73.45 + 84.10 - 26.33}{2(8.57)(9.17)}$$

$$\angle CAB = \cos^{-1}\left(\frac{73.45 + 84.10 - 26.33}{2(8.57)(9.17)}\right)$$

$$= 33°$$

47. Redrawing the figure and relabeling somewhat:

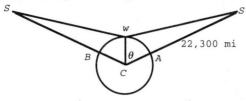

22,300 mi

Angle $\theta = \dfrac{1}{2}$ angle $ACB = \dfrac{1}{2}(130°) = 65°$

CW = radius of earth = 3,964 mi

$CS = CA + AS = 3,964 + 22,300 = 26,264$ mi

From the law of cosines applied to triangle SWC, we have

$$SW^2 = CS^2 + CW^2 - 2(CS)(CW)\cos\theta$$

$$= (26,264)^2 + (3,964)^2 - 2(26,264)(3,964)\cos 65°$$

$$\approx 617,513,000$$

$$SW = 24,800 \text{ mi}$$

Exercise 7-3

Key Ideas and Formulas

A line segment to which a direction has been assigned is called a directed line segment.

A geometric vector is a directed line segment.

The **magnitude** of a vector $\overrightarrow{OP}$ or $\overrightarrow{v}$ is the length of the line segment from O to P and is denoted by $|\overrightarrow{OP}|$ or $|\overrightarrow{v}|$.

The zero vector, denoted by $\overrightarrow{0}$, has a magnitude of zero and an arbitrary direction.

Two vectors are equal if they have the same magnitude and direction.

To find the sum of two vectors, use

Tail-to-tip Rule   *or*   Parallelogram Rule

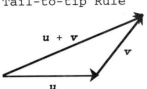 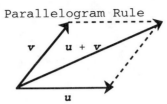

u + **v** is called the sum, or resultant, of the two vectors **u** and **v**.

A vector that represents the direction and speed of an object in motion is called a **velocity vector**.

A vector that represents the direction and magnitude of an applied force is called a **force vector**. If an object is subjected to two forces, then the sum of these two forces, the **resultant force**, is a single force.

Figure for problems 1, 3, 5, 7:

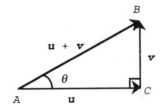

1. To find $|\mathbf{u} + \mathbf{v}|$: Apply the Pythagorean theorem to triangle ABC.

$$|\mathbf{u} + \mathbf{v}|^2 = AB^2 = AC^2 + BC^2$$
$$= |\mathbf{u}|^2 + |\mathbf{v}|^2$$
$$= 30^2 + 72^2$$
$$= 6{,}084$$
$$|\mathbf{u} + \mathbf{v}| = \sqrt{6{,}084}$$
$$= 78 \text{ mi/hr}$$

Solve triangle ABC for θ

$$\tan \theta = \frac{BC}{AC} = \frac{|\mathbf{v}|}{|\mathbf{u}|}$$
$$\theta = \tan^{-1} \frac{|\mathbf{v}|}{|\mathbf{u}|} \quad \theta \text{ is acute}$$
$$= \tan^{-1} \frac{72}{30}$$
$$= 67°$$

3. To find $|\mathbf{u} + \mathbf{v}|$: Apply the Pythagorean theorem to triangle ABC.

$$|\mathbf{u} + \mathbf{v}|^2 = AB^2 = AC^2 + BC^2$$
$$= |\mathbf{u}|^2 + |\mathbf{v}|^2$$
$$= 29^2 + 29^2$$
$$= 1{,}682$$
$$|\mathbf{u} + \mathbf{v}| = \sqrt{1{,}682}$$
$$= 41 \text{ kg}$$

Solve triangle ABC for θ

$$\tan \theta = \frac{BC}{AC} = \frac{|\mathbf{v}|}{|\mathbf{u}|}$$
$$\theta = \tan^{-1} \frac{|\mathbf{v}|}{|\mathbf{u}|} \quad \theta \text{ is acute}$$
$$= \tan^{-1} \frac{29}{29}$$
$$= 45°$$

5. Horizontal component $|\mathbf{u}|$: $\cos \theta = \dfrac{AC}{AB} = \dfrac{|\mathbf{u}|}{|\mathbf{u} + \mathbf{v}|}$

$|\mathbf{u}| = |\mathbf{u} + \mathbf{v}|\cos \theta = 24 \cos 60° = 12 \text{ lb}$

Vertical component $|\mathbf{v}|$: $\sin \theta = \dfrac{BC}{AB} = \dfrac{|\mathbf{v}|}{|\mathbf{u} + \mathbf{v}|}$

$|\mathbf{v}| = |\mathbf{u} + \mathbf{v}|\sin \theta = 24 \sin 60° = 21 \text{ lb}$

7. Horizontal component $|\mathbf{u}|$: $\cos \theta = \dfrac{AC}{AB} = \dfrac{|\mathbf{u}|}{|\mathbf{u} + \mathbf{v}|}$

$|\mathbf{u}| = |\mathbf{u} + \mathbf{v}|\cos \theta = 390 \cos 6° = 388 \text{ mph}$

Vertical component $|\mathbf{v}|$: $\sin \theta = \dfrac{BC}{AB} = \dfrac{|\mathbf{v}|}{|\mathbf{u} + \mathbf{v}|}$

$|\mathbf{v}| = |\mathbf{u} + \mathbf{v}|\sin \theta = 390 \sin 6° = 41 \text{ mph}$

Figures for problems
9, 11, 13, 15:

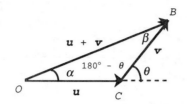 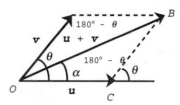

9. $\theta = 68°$. Hence, $\angle OCB = 180° - \theta = 180° - 68° = 112°$

We can find $|\boldsymbol{u} + \boldsymbol{v}|$ using the law of cosines:

$$|\boldsymbol{u} + \boldsymbol{v}|^2 = |\boldsymbol{u}|^2 + |\boldsymbol{v}|^2 - 2|\boldsymbol{u}||\boldsymbol{v}|\cos(OCB)$$
$$= 66^2 + 22^2 - 2(66)(22)\cos 112°$$
$$= 5927.8575\ldots$$
$$|\boldsymbol{u} + \boldsymbol{v}| = \sqrt{5927.8575\ldots}$$
$$= 77g$$

To find α, we use the law of sines:

$$\frac{\sin \alpha}{|\boldsymbol{v}|} = \frac{\sin(OCB)}{|\boldsymbol{u} + \boldsymbol{v}|}$$

$$\frac{\sin \alpha}{22} = \frac{\sin 112°}{77}$$

$$\sin \alpha = \frac{22}{77} \sin 112°$$

$$\alpha = \sin^{-1}\left(\frac{22}{77} \sin 112°\right)$$
$$= 15°$$

11. $\theta = 53°$. Hence, $\angle OCB = 180° - \theta = 180° - 53° = 127°$

We can find $|\boldsymbol{u} + \boldsymbol{v}|$ using the law of cosines:

$$|\boldsymbol{u} + \boldsymbol{v}|^2 = |\boldsymbol{u}|^2 + |\boldsymbol{v}|^2 - 2|\boldsymbol{u}||\boldsymbol{v}|\cos(OCB)$$
$$= 21^2 + 3.2^2 - 2(21)(3.2)\cos 127°$$
$$= 532.12394\ldots$$
$$|\boldsymbol{u} + \boldsymbol{v}| = \sqrt{532.12394\ldots}$$
$$= 23 \text{ knots}$$

To find α, we use the law of sines:

$$\frac{\sin \alpha}{|\boldsymbol{v}|} = \frac{\sin(OCB)}{|\boldsymbol{u} + \boldsymbol{v}|}$$

$$\frac{\sin \alpha}{3.2} = \frac{\sin 127°}{23}$$

$$\sin \alpha = \frac{3.2}{23} \sin 127°$$

$$\alpha = \sin^{-1}\left(\frac{3.2}{23} \sin 127°\right)$$

$$= 6°$$

13. $\theta = 79°$. Hence, $\angle OCB = 180° - \theta = 180° - 79° = 101°$. Also, $\angle OBC = \beta = \theta - \alpha = 79° - 25° = 54°$.

We can now find $|\boldsymbol{u}|$ and $|\boldsymbol{v}|$ from the law of sines.

$$\frac{\sin \beta}{|\boldsymbol{u}|} = \frac{\sin(OCB)}{|\boldsymbol{u} + \boldsymbol{v}|} \qquad\qquad \frac{\sin \alpha}{|\boldsymbol{v}|} = \frac{\sin(OCB)}{|\boldsymbol{u} + \boldsymbol{v}|}$$

$$|\boldsymbol{u}| = \frac{|\boldsymbol{u} + \boldsymbol{v}|\sin \beta}{\sin(OCB)} \qquad\qquad |\boldsymbol{v}| = \frac{|\boldsymbol{u} + \boldsymbol{v}|\sin \alpha}{\sin(OCB)}$$

$$= \frac{14 \sin 54°}{\sin 101°} \qquad\qquad = \frac{14 \sin 25°}{\sin 101°}$$

$$= 12 \text{ kg} \qquad\qquad = 6.0 \text{ kg}$$

15. $\theta = 69.4°$. Hence, $\angle OCB = 180° - \theta = 180° - 69.4° = 110.6°$. Also, $\angle OBC = \beta = \theta - \alpha = 69.4° - 42.3° = 27.1°$.

We can now find $|u|$ and $|v|$ from the law of sines:

$$\frac{\sin \beta}{|u|} = \frac{\sin(OCB)}{|u + v|} \qquad\qquad \frac{\sin \alpha}{|v|} = \frac{\sin(OCB)}{|u + v|}$$

$$|u| = \frac{|u + v|\sin \beta}{\sin(OCB)} \qquad\qquad |v| = \frac{|u + v|\sin \alpha}{\sin(OCB)}$$

$$= \frac{223 \sin 27.1°}{\sin 110.6°} \qquad\qquad = \frac{223 \sin 42.3°}{\sin 110.6°}$$

$$= 109 \text{ mi/hr} \qquad\qquad = 160 \text{ mi/hr}$$

17. True. Since the zero vector has an arbitrary direction, it is perpendicular to every vector.

19. False. Two vectors of magnitude 5 lbs, one pointing east and the other west, have the same magnitude, but are not equal.

21. True. If vectors are equal, they have the same magnitude and direction.

23. False. The single exception is the zero vector, which has magnitude zero.

25.

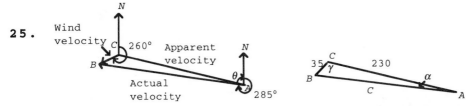

In the (left-hand) vector diagram above, we use the tail-to-tip method of vector addition to obtain the resultant (actual) velocity. Note:

angle $\theta = 360° - 285° = 75°$
angle $ACN = 180° - 75° = 105°$
angle $\gamma = 260° - 105° = 155°$

Then in triangle ABC, we solve for c and α.

Solve for c:
Use the law of cosines.

$$\begin{aligned} c^2 &= a^2 + b^2 - 2ab \cos \gamma \\ &= 35^2 + 230^2 - 2(35)(230) \cos 155° \\ &= 68,716.5… \\ c &= \sqrt{68,716.5…} \\ &= 260 \text{ mph} \end{aligned}$$

Solve for α:
Use the law of sines

$$\frac{\sin \alpha}{a} = \frac{\sin \gamma}{c}$$

$$\sin \alpha = \frac{a \sin \gamma}{c}$$

$$\alpha = \sin^{-1}\left(\frac{a \sin \gamma}{c}\right) \quad \alpha \text{ is acute since } \gamma \text{ is obtuse}$$

$$= \sin^{-1}\left(\frac{35 \sin 155°}{260}\right)$$

$$= 3°$$

Actual heading $= 285° - \alpha = 285° - 3° = 282°$.
Thus, the plane, relative to the ground, is traveling at 260 mph in a direction of 282°.

27.

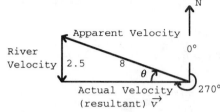

Using the Pythagorean theorem, we find the magnitude of the resultant vector to be

$$|\boldsymbol{v}| = \sqrt{8^2 - 2.5^2} = 7.6 \text{ knots}$$

To find θ, we see that

$$\sin \theta = \frac{2.5}{8}$$

$$\theta = \sin^{-1}\left(\frac{2.5}{8}\right) = 18°$$

Compass heading = $270° + \theta = 270° + 18° = 288°$

Actual speed: 7.6 knots.

29.

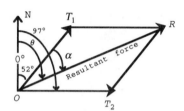

In the figure, we are given $OT_1 = 2{,}300$, $T_1R = OT_2 = 1{,}900$. $\angle T_1OT_2 = 97° - 52° = 45°$, hence $\angle OT_1R = 180° - 45° = 135°$. Since we know two sides and the included angle in triangle OT_1R, we can use the law of cosines to find OR, the magnitude of the resultant force.

$$OR^2 = OT_1{}^2 + T_1R^2 - 2(OT_1)(T_1R)\cos(\angle OT_1R)$$
$$= (2{,}300)^2 + (1{,}900)^2 - 2(2{,}300)(1{,}900)\cos 135°$$
$$= 15{,}080{,}113\ldots$$
$$OR = \sqrt{15{,}080{,}113\ldots}$$
$$= 3{,}900 \text{ pounds}$$

We use the law of sines to find α

$$\frac{\sin \alpha}{T_1R} = \frac{\sin(OT_1R)}{OR}$$

$$\frac{\sin \alpha}{1{,}900} = \frac{\sin 135°}{3{,}900}$$

$$\sin \alpha = \frac{1{,}900}{3{,}900} \sin 135°$$

$$\alpha = \sin^{-1}\left(\frac{1{,}900}{3{,}900} \sin 135°\right)$$
$$= 20°$$

Then the compass direction $\theta = 52° + \alpha = 72°$.

31.

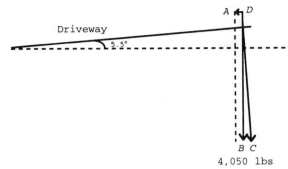

(A) To keep the car at D from rolling down the hill, we need a force with the magnitude of **DA** but oppositely directed. To find $|DA|$, we observe that DAB is a right triangle with $\angle ABD = 5.5°$. Hence

$$\sin 5.5° = \frac{|DA|}{4,050}$$
$$|DA| = 4,050 \sin 5.5°$$
$$= 388 \text{ lb}$$

(B) To find $|DC|$, the magnitude of the force perpendicular to the driveway, we note that DCB is a right triangle with $\angle BDC = 5.5°$. Hence

$$\cos 5.5° = \frac{|DC|}{4,050}$$
$$|DC| = 4,050 \cos 5.5°$$
$$= 4,030 \text{ lb}$$

33.

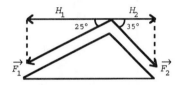

The weights will slide to the left if the horizontal component H_1 pointing left is greater than the horizontal component H_2 pointing right. They will slide to the right if H_2 is greater than H_1.

$$H_1 = |F_1| \cos 25° \qquad\qquad H_2 = |F_2| \cos 35°$$
$$= 110 \cos 25° \qquad\qquad\quad = 85 \cos 35°$$
$$\approx 100 \text{ lb} \qquad\qquad\qquad \approx 70 \text{ lb}$$

Since H_1 is greater than H_2, they will slide to the left.

Exercise 7-4

Key Ideas and Formulas

A geometric vector $\overrightarrow{AB}$ translated so that its initial point is at the origin is said to be in **standard position**. The vector $\overrightarrow{OP}$ such that $OP = AB$ is said to be the **standard vector** for AB.

An **algebraic vector** is an ordered pair of real numbers $\langle a, b \rangle$. It corresponds to the standard geometric vector with terminal point (a, b) and initial point $(0, 0)$. a and b are the scalar components of $\langle a, b \rangle$. $\langle a, b \rangle = \langle c, d \rangle$ if $a = c$ and $b = d$. The zero vector is denoted by $O = \langle 0, 0 \rangle$. The **magnitude**, or **norm**, of a vector $v = \langle a, b \rangle$ is denoted by $|v|$ and is given by $|v| = \sqrt{a^2 + b^2}$

If $\mathbf{u} = \langle a, b \rangle$ and $\mathbf{v} = \langle c, d \rangle$, then

$\mathbf{u} + \mathbf{v} = \langle a + c, b + d \rangle$ (Vector addition)

$k\mathbf{u} = k\langle a, b \rangle = \langle ka, kb \rangle$ (Scalar multiplication)

If $|\mathbf{v}| = 1$ then $\mathbf{v}$ is called a unit vector. If $\mathbf{v}$ is a non-zero vector, then $\mathbf{u} = \dfrac{1}{|\mathbf{v}|}\,\mathbf{v}$ is a unit vector with the same direction as $\mathbf{v}$.

The $\mathbf{i}$ and $\mathbf{j}$ unit vectors:

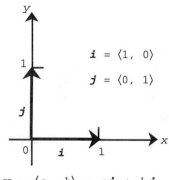

$\mathbf{i} = \langle 1, 0 \rangle$

$\mathbf{j} = \langle 0, 1 \rangle$

$\mathbf{v} = \langle a, b \rangle = a\mathbf{i} + b\mathbf{j}$

Thus, any vector $\mathbf{v}$ can be expressed as a linear combination of the two unit vectors $\mathbf{i}$ and $\mathbf{j}$.

```
┌─────────────────────────────────────────────────────────────┐
│ Algebraic Properties of Vectors                             │
│ I  Addition Properties  For all vectors u, v, and w:        │
│    1. u + v = v + u                    Commutative property  │
│    2. u + (v + w) = (u + v) + w   Associative property      │
│    3. u + 0 = 0 + u = u              Additive identity       │
│    4. u + (-u) = (-u) + u = 0        Additive inverse        │
│                                                             │
│ II Scalar Multiplication Properties  For all vectors u and  │
│    v and all scalars m and n:                               │
│    1. m(nu) = (mn)u                  Associative property    │
│    2. m(u + v) = mu + mv             Distributive property   │
│    3. (m + n)u = mu + nu             Distributive property   │
│    4. 1u = u                         Multiplicative identity │
└─────────────────────────────────────────────────────────────┘
```

Conditions for Static Equilibrium:

1. An object at rest is said to be in *static equilibrium*.

2. For an object located at the origin in a rectangular coordinate system to remain in static equilibrium, at rest, it is necessary that the sum of all the force vectors acting on the object be the zero vector.

---

1. The algebraic vector $\langle a, b \rangle$ corresponds to the standard geometric vector with terminal point $P(a, b)$ and initial point $O(0, 0)$. In this case, $P = B$ has coordinates $(7, 2)$. Hence $\langle a, b \rangle = \langle 7, 2 \rangle$.

3. The algebraic vector $\langle a, b \rangle$ corresponds to the standard geometric vector with terminal point $P(a, b)$ and initial point $O(0, 0)$. The coordinates of the point $P(a, b)$ are given by

$a = x_B - x_A = 0 - 4 = -4$

$b = y_B - y_A = 8 - 0 = 8$

Thus, $\langle a, b \rangle = \langle -4, 8 \rangle$

5. The algebraic vector $\langle a,\ b \rangle$ corresponds to the standard geometric vector with terminal point $P(a,\ b)$ and initial point $O(0,\ 0)$. The coordinates of the point $P(a,\ b)$ are given by

$$a = x_B - x_A = 7 - 9 = -2$$
$$b = y_B - y_A = 5 - (-4) = 9$$

Thus, $\langle a,\ b \rangle = \langle -2,\ 9 \rangle$

7. $|\boldsymbol{v}| = \sqrt{(-15)^2 + 0^2} = \sqrt{225} = 15$ **9.** $|\boldsymbol{v}| = \sqrt{(-21)^2 + (72)^2} = \sqrt{5625} = 75$

11. $|\boldsymbol{v}| = \sqrt{(-155)^2 + 468^2} = \sqrt{243,049} = 493$

13. (A) $\boldsymbol{u} + \boldsymbol{v} = \langle 2,\ 1 \rangle + \langle -1,\ 3 \rangle = \langle 1,\ 4 \rangle$

     (B) $\boldsymbol{u} - \boldsymbol{v} = \langle 2,\ 1 \rangle - \langle -1,\ 3 \rangle = \langle 2,\ 1 \rangle + \langle 1,\ -3 \rangle = \langle 3,\ -2 \rangle$

     (C) $2\boldsymbol{u} - \boldsymbol{v} + 3\boldsymbol{w} = 2\langle 2,\ 1 \rangle - \langle -1,\ 3 \rangle + 3\langle 3,\ 0 \rangle$
$$= \langle 4,\ 2 \rangle + \langle 1,\ -3 \rangle + \langle 9,\ 0 \rangle$$
$$= \langle 14,\ -1 \rangle$$

15. (A) $\boldsymbol{u} + \boldsymbol{v} = \langle -4,\ -1 \rangle + \langle 2,\ 2 \rangle = \langle -2,\ 1 \rangle$

     (B) $\boldsymbol{u} - \boldsymbol{v} = \langle -4,\ -1 \rangle - \langle 2,\ 2 \rangle = \langle -4,\ -1 \rangle + \langle -2,\ -2 \rangle = \langle -6,\ -3 \rangle$

     (C) $2\boldsymbol{u} - \boldsymbol{v} + 3\boldsymbol{w} = 2\langle -4,\ -1 \rangle - \langle 2,\ 2 \rangle + 3\langle 0,\ 1 \rangle$
$$= \langle -8,\ -2 \rangle + \langle -2,\ -2 \rangle + \langle 0,\ 3 \rangle$$
$$= \langle -10,\ -1 \rangle$$

17. $\boldsymbol{v} = \langle -8,\ 0 \rangle = -8\langle 1,\ 0 \rangle = -8\boldsymbol{i}$ **19.** $\boldsymbol{v} = \langle 6,\ -12 \rangle = \langle 6,\ 0 \rangle + \langle 0,\ -12 \rangle$
$$= 6\langle 1,\ 0 \rangle - 12\langle 0,\ 1 \rangle$$
$$= 6\boldsymbol{i} - 12\boldsymbol{j}$$

21. $\boldsymbol{v} = \overrightarrow{AB} = \langle -3 - 2,\ 1 - 3 \rangle$ **23.** $\boldsymbol{u} + \boldsymbol{v} = 3\boldsymbol{i} - 2\boldsymbol{j} + 2\boldsymbol{i} + 4\boldsymbol{j} = 5\boldsymbol{i} + 2\boldsymbol{j}$
$$= \langle -5,\ -2 \rangle$$
$$= \langle -5,\ 0 \rangle + \langle 0,\ -2 \rangle$$
$$= -5\langle 1,\ 0 \rangle - 2\langle 0,\ 1 \rangle$$
$$= -5\boldsymbol{i} - 2\boldsymbol{j}$$

25. $2\boldsymbol{u} - 3\boldsymbol{v} = 2(3\boldsymbol{i} - 2\boldsymbol{j}) - 3(2\boldsymbol{i} + 4\boldsymbol{j})$
$$= 6\boldsymbol{i} - 4\boldsymbol{j} - 6\boldsymbol{i} - 12\boldsymbol{j}$$
$$= -16\boldsymbol{j}$$

27. $2\boldsymbol{u} - \boldsymbol{v} - 2\boldsymbol{w} = 2(3\boldsymbol{i} - 2\boldsymbol{j}) - (2\boldsymbol{i} + 4\boldsymbol{j}) - 2(2\boldsymbol{i})$
$$= 6\boldsymbol{i} - 4\boldsymbol{j} - 2\boldsymbol{i} - 4\boldsymbol{j} - 4\boldsymbol{i}$$
$$= -8\boldsymbol{j}$$

29. $|\boldsymbol{v}| = \sqrt{(-1)^2 + 1^2} = \sqrt{2}$

     Then $\boldsymbol{u} = \dfrac{1}{|\boldsymbol{v}|}\boldsymbol{v} = \dfrac{1}{\sqrt{2}}\langle -1,\ 1 \rangle = \left\langle -\dfrac{1}{\sqrt{2}},\ \dfrac{1}{\sqrt{2}} \right\rangle$

31. $|\boldsymbol{v}| = \sqrt{(-12)^2 + 5^2} = \sqrt{169} = 13$

     Then $\boldsymbol{u} = \dfrac{1}{|\boldsymbol{v}|}\boldsymbol{v} = \dfrac{1}{13}\langle -12,\ 5 \rangle = \left\langle -\dfrac{12}{13},\ \dfrac{5}{13} \right\rangle$

33. False. $-1\boldsymbol{v}$ is a scalar multiple of $\boldsymbol{v}$ that has the opposite direction to $\boldsymbol{v}$.

35. False. For example, $\boldsymbol{i} + -\boldsymbol{i} = \boldsymbol{0}$. The sum of two unit vectors is a zero vector in this case.

37. $\mathbf{u} + (\mathbf{v} + \mathbf{w}) = \langle a, b \rangle + (\langle c, d \rangle + \langle e, f \rangle)$

$\qquad = \langle a, b \rangle + \langle c + e, d + f \rangle$ Definition of vector addition

$\qquad = \langle a + (c + e), b + (d + f) \rangle$ Definition of vector addition

$\qquad = \langle (a + c) + e, (b + d) + f \rangle$ Associative property for addition of real numbers

$\qquad = \langle a + c, b + d \rangle + \langle e, f \rangle$ Definition of vector addition

$\qquad = (\langle a, b \rangle + \langle c, d \rangle) + \langle e, f \rangle$ Definition of vector addition

$\qquad = (\mathbf{u} + \mathbf{v}) + \mathbf{w}$

39. $\mathbf{u} + \mathbf{O} = \langle a, b \rangle + \langle 0, 0 \rangle$

$\qquad = \langle a + 0, b + 0 \rangle$ Definition of vector addition

$\qquad = \langle a, b \rangle$ Additive identity property for real numbers

$\qquad = \mathbf{u}$

41. $(m + n)\mathbf{u} = (m + n)\langle a, b \rangle$

$\qquad = \langle (m + n)a, (m + n)b \rangle$ Definition of scalar multiplication

$\qquad = \langle ma + na, mb + nb \rangle$ Distributive property for real numbers

$\qquad = \langle ma, mb \rangle + \langle na, nb \rangle$ Definition of vector addition

$\qquad = m\langle a, b \rangle + n\langle a, b \rangle$ Definition of scalar multiplication

$\qquad = m\mathbf{u} + n\mathbf{u}$

43. $m(n\mathbf{u}) = m(n\langle a, b \rangle)$

$\qquad = m\langle na, nb \rangle$ Definition of scalar multiplication

$\qquad = \langle m(na), m(nb) \rangle$ Definition of scalar multiplication

$\qquad = \langle (mn)a, (mn)b \rangle$ Associative property for multiplication of real numbers

$\qquad = (mn)\langle a, b \rangle$ Definition of scalar multiplication

$\qquad = (mn)\mathbf{u}$

45. First, form a force diagram with all force vectors in standard position at the origin.

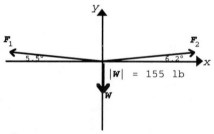

Let $\mathbf{F}_1$ = the tension in left rope
$\mathbf{F}_2$ = the tension in right rope

Write each force vector in terms of $\mathbf{i}$ and $\mathbf{j}$ unit vectors.

$\mathbf{F}_1 = |\mathbf{F}_1|(-\cos 5.5°)\mathbf{i} + |\mathbf{F}_1|(\sin 5.5°)\mathbf{j}$
$\mathbf{F}_2 = |\mathbf{F}_2|(\cos 6.2°)\mathbf{i} + |\mathbf{F}_2|(\sin 6.2°)\mathbf{j}$
$\mathbf{W} = -155\mathbf{j}$

For the system to be in static equilibrium, we must have
$\mathbf{F}_1 + \mathbf{F}_2 + \mathbf{W} = \mathbf{O}$

which becomes, on addition,
$[-|\mathbf{F}_1|(\cos 5.5°) + |\mathbf{F}_2|(\cos 6.2°)]\mathbf{i} + [|\mathbf{F}_1|(\sin 5.5°) + |\mathbf{F}_2|(\sin 6.2°) - 155]\mathbf{j}$
$\qquad = 0\mathbf{i} + 0\mathbf{j}$

Since two vectors are equal if and only if their corresponding components are equal, we are led to the following system of equations in $|F_1|$ and $|F_2|$:

$$-|F_1|\cos 5.5° + |F_2|\cos 6.2° = 0$$
$$|F_1|\sin 5.5° + |F_2|\sin 6.2° - 155 = 0$$

Solving by standard methods, we obtain

$$|F_1| = \frac{155}{\sin 5.5° + \cos 5.5° \tan 6.2°} = 760 \text{ lb to the left}$$

$$|F_2| = 760 \frac{\cos 5.5°}{\cos 6.2°} = 761 \text{ lb to the right.}$$

47. First, form a force diagram with all force vectors in standard position at the origin.

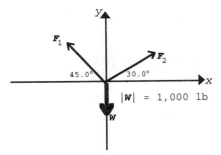

Let F_1 = the tension in left cable
    F_2 = the tension in right cable

Write each force vector in terms of i and j unit vectors.

$$F_1 = |F_1|(-\cos 45.0°)i + |F_1|(\sin 45.0°)j$$
$$F_2 = |F_2|(\cos 30.0°)i + |F_2|(\sin 30.0°)j$$
$$W = -1,000j$$

For the system to be in static equilibrium, we must have
$$F_1 + F_2 + W = O$$

which becomes, on addition,
$$[-|F_1|(\cos 45.0°) + |F_2|(\cos 30.0°)]i + [|F_1|(\sin 45.0°)$$
$$+ |F_2|(\sin 30.0°) - 1,000]j = 0i + 0j$$

Since two vectors are equal if and only if their corresponding components are equal, we are led to the following system of equations in $|F_1|$ and $|F_2|$:

$$-|F_1|\cos 45.0° + |F_2|\cos 30.0° = 0$$
$$|F_1|\sin 45.0° + |F_2|\sin 30.0° - 1,000 = 0$$

Solving by standard methods, we obtain
$$|F_1| = \frac{1,000}{\sin 45.0° + \cos 45.0° \tan 30.0°} = 897 \text{ lb to the left}$$

$$|F_2| = 897 \frac{\cos 45.0°}{\cos 30.0°} = 732 \text{ lb to the right.}$$

49. In the force diagram, figure (b), we are given $|c|$ = 400 lb. In triangle ABC, figure (a), we note: $\cos ABC = \frac{1}{2}$, $ABC = 60°$. Then write each force in terms of i and j unit vectors.

$$1a = |a|i$$

$$b = |b|(-\cos 60°)i + |b|\sin 60° \, j = -\frac{1}{2}|b|i + \frac{\sqrt{3}}{2}|b|j$$

$$c = -400j$$

For the system to be in static equilibrium, we must have
$$a + b + c = 0$$
which becomes, on addition

$$\left[|a| - \frac{1}{2}|b|\right]i + \left[\frac{\sqrt{3}}{2}|b| - 400\right]j = 0i + 0j$$

Since two vectors are equal if and only if their corresponding components are equal, we are led to the following system of equations in $|a|$ and $|b|$:

$$|a| - \frac{1}{2}|b| = 0$$

$$\frac{\sqrt{3}}{2}|b| - 400 = 0$$

Solving, we obtain

$$|b| = \frac{2}{\sqrt{3}}(400) = 462 \text{ lb.}$$ This corresponds to a tension force of 462 lb in member CB.

$$|a| = \frac{1}{2}|b| = 231 \text{ lb}$$ This corresponds to a compression force of 231 lb in member AB.

51. First, form a force diagram with all force vectors in standard position at the origin.

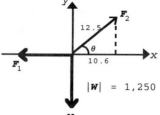

Let F_1 = the force on the horizontal member BC
F_2 = the force on the supporting member AB
W = the downward force (1,250 lb)

We note: $\cos \theta = \dfrac{10.6}{12.5}$ $\theta = \cos^{-1}\dfrac{10.6}{12.5} = 32.0°$

Then write each force vector in terms of i and j unit vectors.

$F_1 = -|F_1|i$
$F_2 = |F_2|(\cos 32.0°)i + |F_2|(\sin 32.0°)j$
$W = -1,250j$

For the system to be in static equilibrium, we must have
$F_1 + F_2 + W = 0$

which becomes, on addition,
$[-|F_1| + |F_2|(\cos 32.0°)]i + [|F_2|(\sin 32.0°) - 1,250]j = 0i + 0j$

Since two vectors are equal if and only if their corresponding components are equal, we are led to the following system of equations in $|F_1|$ and $|F_2|$:

$-|F_1| + |F_2|(\cos 32.0°) = 0$
$|F_2|(\sin 32.0°) - 1,250 = 0$

Solving,
$$|F_2| = \frac{1,250}{\sin 32.0°} = 2,360 \text{ lb}$$
$$|F_1| = |F_2|\cos 32.0° = 2,000 \text{ lb}$$

The force in the member AB is directed oppositely to the diagram—a compression of 2,360 lb. The force in the member BC is also directed oppositely to the diagram—a tension of 2,000 lb.

Exercise 7-5

Key Ideas and Formulas

Polar Coordinate System:

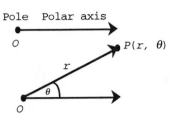

Polar-rectangular Relationships:

$r^2 = x^2 + y^2$

$\sin \theta = \dfrac{y}{r}$ or $y = r \sin \theta$

$\cos \theta = \dfrac{x}{r}$ or $x = r \cos \theta$

$\tan \theta = \dfrac{y}{x}$

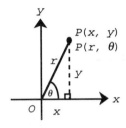

The signs of x and y determine the quadrant for θ. The angle θ is chosen so that $-\pi < \theta \le \pi$ or $-180° < \theta \le 180°$, unless directed otherwise.

To graph a polar coordinate equation, make a table of values that satisfy the equation, plot these, and then join the points with a smooth curve.

If only a rough sketch of a polar equation involving $\sin \theta$ or $\cos \theta$ is desired, set up a table that indicates how r varies as θ varies through each set of quadrant-values. As θ increases from 0 to $\dfrac{\pi}{2}$, $\sin \theta$ increases from 0 to 1, $\cos \theta$ decreases from 1 to 0, and so on. Then sketch these results.

The student should become familiar with the standard polar graphs illustrated in Table 1, Section 7-5, of the text. They are not reproduced here for lack of space.

1.

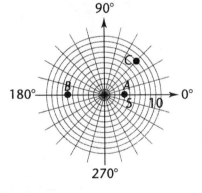

3.

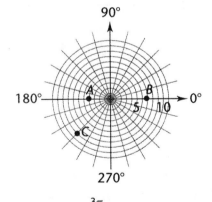

5.

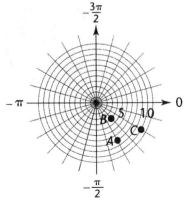

7.

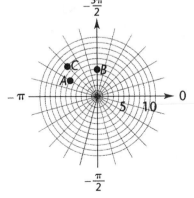

9. See figure.

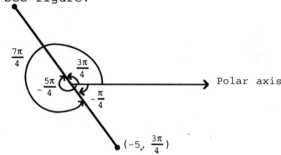

$(-5, \frac{3\pi}{4})$

The point with coordinates $\left(-5, \frac{3\pi}{4}\right)$ can equally well be described as $\left(5, -\frac{\pi}{4}\right)$ or $\left(5, \frac{7\pi}{4}\right)$ or $\left(-5, -\frac{5\pi}{4}\right)$. Thus,

$\left(5, -\frac{\pi}{4}\right)$: The polar axis is rotated $\frac{\pi}{4}$ radians clockwise (negative direction) and the point is located 5 units from the pole along the positive polar axis;

$\left(5, \frac{7\pi}{4}\right)$: The polar axis is rotated $\frac{7\pi}{4}$ radians counterclockwise (positive direction) and the point is located 5 units from the pole along the positive polar axis;

$\left(-5, -\frac{5\pi}{4}\right)$: The polar axis is rotated $\frac{5\pi}{4}$ radians clockwise (negative direction) and the point is located 5 units from the pole along the negative polar axis.

11.

θ	0	$\frac{\pi}{6}$	$\frac{\pi}{4}$	$\frac{\pi}{3}$	$\frac{\pi}{2}$	$\frac{2\pi}{3}$	$\frac{3\pi}{4}$	$\frac{5\pi}{6}$	π
r	0	5	$5\sqrt{2}$ ≈ 7.1	$5\sqrt{3}$ ≈ 8.7	10	$5\sqrt{3}$ ≈ 8.7	$5\sqrt{2}$ ≈ 7.1	5	0

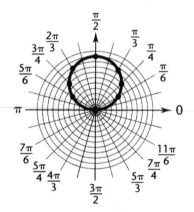

13. The graph consists of all points whose distance from the pole is 8: a circle with center at the pole, and radius 8.

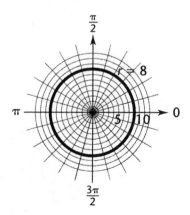

15. The graph consists of all points on a line which forms an angle of $\frac{\pi}{3}$ with the polar axis, and passes through the pole.

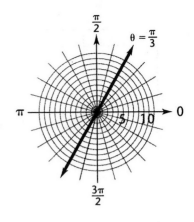

In problems 17—22, use a calculator set in radian mode.

17. $(r, \theta) = \left(6, \frac{\pi}{6}\right)$

$x = r \cos \theta = 6 \cos \frac{\pi}{6} = 5.196$

$y = r \sin \theta = 6 \sin \frac{\pi}{6} = 3.000$

$(x, y) = (5.196, 3.000)$

19. $(r, \theta) = \left(-2, \frac{7\pi}{8}\right)$

$x = r \cos \theta = -2 \cos \frac{7\pi}{8} = 1.848$

$y = r \sin \theta = -2 \sin \frac{7\pi}{8} = -0.765$

$(x, y) = (1.848, -0.765)$

21. $(r, \theta) = (-4.233, -2.084)$

$x = r \cos \theta = -4.233 \cos(-2.084) = 2.078$

$y = r \sin \theta = -4.233 \sin(-2.084) = 3.688$

$(x, y) = (2.078, 3.688)$

In problems 23—28, use a calculator set in degree mode.

23. $(x, y) = (-8, 0)$

$r = \sqrt{x^2 + y^2} = \sqrt{(-8)^2 + 0^2} = \sqrt{64} = 8$

$\tan \theta = \frac{y}{x} = \frac{0}{-8} = 0$

θ is a quadrantal angle and is to be chosen so that $-180° < \theta \leq 180°$. Since $(-8, 0)$ is on the negative x axis, $\theta = 180°$.

$(r, \theta) = (8, 180°)$

25. $(x, y) = (-5, -5)$

$r = \sqrt{x^2 + y^2} = \sqrt{(-5)^2 + (-5)^2} = \sqrt{50} = 5\sqrt{2}$

$\tan \theta = \frac{y}{x} = \frac{-5}{-5} = 1$

θ is a third quadrant angle and is to be chosen so that $-180° < \theta \leq 180°$

$\theta = -135°$

$(r, \theta) = (5\sqrt{2}, -135°)$

27. $(x, y) = (9.79, 5.13)$

$r = \sqrt{x^2 + y^2} = \sqrt{9.79^2 + 5.13^2} = 11.05$

$\tan \theta = \frac{y}{x} = \frac{5.13}{9.79}$

θ is a first quadrant angle and is to be chosen so that $-180° < \theta \leq 180°$.

$\theta = \tan^{-1}\left(\frac{5.13}{9.79}\right) = 27.7°$

$(r, \theta) = (11.05, 27.7°)$

29.

θ	$\sin \theta$	$4 \sin \theta$
0 to $\frac{\pi}{2}$	0 to 1	0 to 4
$\frac{\pi}{2}$ to 0	1 to 0	4 to 0
π to $\frac{3\pi}{2}$	0 to −1	0 to −4
$\frac{3\pi}{2}$ to 2π	−1 to 0	−4 to 0

Curve is traced out a second time in this region although coordinates seem different.

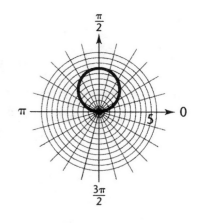

31.

θ	2θ	$\sin 2\theta$	$10 \sin 2\theta$
0 to $\frac{\pi}{4}$	0 to $\frac{\pi}{2}$	0 to 1	0 to 10
$\frac{\pi}{4}$ to $\frac{\pi}{2}$	$\frac{\pi}{2}$ to π	1 to 0	10 to 0
$\frac{\pi}{2}$ to $\frac{3\pi}{4}$	π to $\frac{3\pi}{2}$	0 to −1	0 to −10
$\frac{3\pi}{4}$ to π	$\frac{3\pi}{2}$ to 2π	−1 to 0	−10 to 0
π to $\frac{5\pi}{4}$	2π to $\frac{5\pi}{2}$	0 to 1	0 to 10
$\frac{5\pi}{4}$ to $\frac{3\pi}{2}$	$\frac{5\pi}{2}$ to 3π	1 to 0	10 to 0
$\frac{3\pi}{2}$ to $\frac{7\pi}{4}$	3π to $\frac{7\pi}{2}$	0 to −1	0 to −10
$\frac{7\pi}{4}$ to 2π	$\frac{7\pi}{2}$ to 4π	−1 to 0	−10 to 0

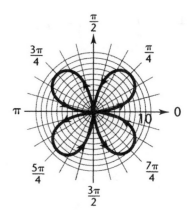

33.

θ	3θ	$\cos 3\theta$	$5 \cos 3\theta$
0 to $\frac{\pi}{6}$	0 to $\frac{\pi}{2}$	1 to 0	5 to 0
$\frac{\pi}{6}$ to $\frac{\pi}{3}$	$\frac{\pi}{2}$ to π	0 to −1	0 to −5
$\frac{\pi}{3}$ to $\frac{\pi}{2}$	π to $\frac{3\pi}{2}$	−1 to 0	−5 to 0
$\frac{\pi}{2}$ to $\frac{2\pi}{3}$	$\frac{3\pi}{2}$ to 2π	0 to 1	0 to 5
$\frac{2\pi}{3}$ to $\frac{5\pi}{6}$	2π to $\frac{5\pi}{2}$	1 to 0	5 to 0
$\frac{5\pi}{6}$ to π	$\frac{5\pi}{2}$ to 3π	0 to −1	0 to −5

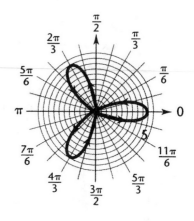

35.

θ	$\sin \theta$	$2 \sin \theta$	$2 + 2 \sin \theta$
0 to $\frac{\pi}{2}$	0 to 1	0 to 2	2 to 4
$\frac{\pi}{2}$ to π	1 to 0	2 to 0	4 to 2
π to $\frac{3\pi}{2}$	0 to −1	0 to −2	2 to 0
$\frac{3\pi}{2}$ to π	−1 to 0	−2 to 0	0 to 2

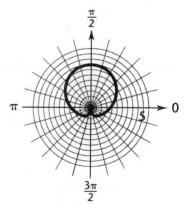

37.

θ	$\sin\theta$	$4\sin\theta$	$2 + 4\sin\theta$
0 to $\frac{\pi}{2}$	0 to 1	0 to 4	2 to 6
$\frac{\pi}{2}$ to π	1 to 0	4 to 0	6 to 2
π to $\frac{3\pi}{2}$	0 to -1	0 to -4	2 to -2
$\frac{3\pi}{2}$ to π	-1 to 0	-4 to 0	-2 to 2

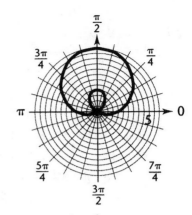

39. $r = 2 + 2\sin\theta$ $r = 4 + 2\sin\theta$ $r = 2 + 4\sin\theta$

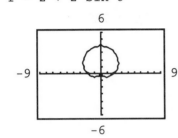   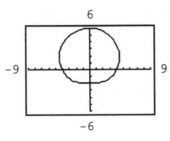   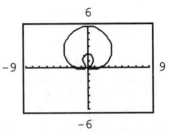

41. (A) $r = 4\sin\theta$ $r = 4\sin 3\theta$ $r = 4\sin 5\theta$

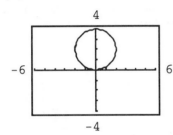   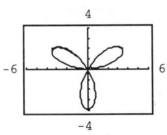   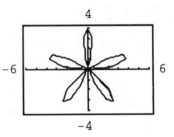

(B) and (C) Since there is one "leaf" for $r = 4\sin\theta$, three leaves for $r = 4\sin 3\theta$, and 5 leaves for $r = 4\sin 5\theta$, reasonable guesses would be seven leaves for $r = 4\sin 7\theta$ and n leaves for $r = a\sin n\theta$, n odd, $a > 0$.

43. (A) $r = 4\sin 2\theta$ $r = 4\sin 4\theta$ $r = 4\sin 6\theta$

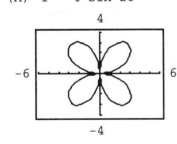   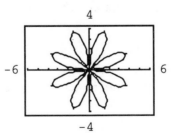   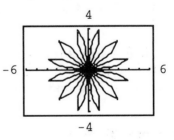

(B) and (C) Since there are four leaves for $r = 4\sin 2\theta$, eight leaves for $r = 4\sin 4\theta$, and twelve leaves for $r = 4\sin 6\theta$, reasonable guesses would be 16 leaves for $r = 4\sin 8\theta$ and $2n$ leaves for $r = a\sin n\theta$, n even, $a > 0$.

45. $x^2 + y^2 = 4$
Use $r^2 = x^2 + y^2$
         $r^2 = 4$
          $r = 2$ or $r = -2$
The graph of $r = -2$ is the same circle as the graph of $r = 2$, so we can discard $r = -2$ and keep only $r = 2$ (circle).

47. $x - \sqrt{3}y = 0$
Use $x = r \cos \theta$ and $y = r \sin \theta$

$$r \cos \theta - \sqrt{3} \, r \sin \theta = 0$$
$$r(\cos \theta - \sqrt{3} \sin \theta) = 0$$
$$r = 0 \quad \text{or} \quad \cos \theta - \sqrt{3} \sin \theta = 0$$

The graph of $r = 0$ is the pole. Since the pole is included in the graph of $\cos \theta - \sqrt{3} \sin \theta = 0$, we can discard $r = 0$ and keep only

$$\cos \theta - \sqrt{3} \sin \theta = 0$$
$$\cos \theta = \sqrt{3} \sin \theta$$
$$1 = \sqrt{3} \tan \theta$$
$$\tan \theta = \frac{1}{\sqrt{3}} \quad \text{or} \quad \theta = \frac{\pi}{6} \qquad \text{This is a line through the pole.}$$

49. $5y = x^2$
$y = r \sin \theta$ and $x = r \cos \theta$, hence
$$5r \sin \theta = (r \cos \theta)^2$$
$$r^2 \cos^2 \theta = 5r \sin \theta$$
$$r^2 \cos^2 \theta - 5r \sin \theta = 0$$
$$r(r \cos^2 \theta - 5 \sin \theta) = 0$$
$$r = 0 \quad \text{or} \quad r \cos^2 \theta - 5 \sin \theta = 0$$

The graph of $r = 0$ is the pole. Since the pole is included in the graph of $r \cos^2 \theta - 5 \sin \theta = 0$, we can discard $r = 0$ and keep only

$$r \cos^2 \theta - 5 \sin \theta = 0$$
$$r \cos^2 \theta = 5 \sin \theta$$
$$r = \frac{5 \sin \theta}{\cos^2 \theta}$$
$$r = 5 \frac{\sin \theta}{\cos \theta} \frac{1}{\cos \theta}$$
$$r = 5 \tan \theta \sec \theta \qquad \text{The graph is a parabola.}$$

51. $r = 3 \cos \theta$
We multiply both sides by r, which adds the pole to the graph. But the pole is already part of the graph, so we have changed nothing.
$$r^2 = 3r \cos \theta$$
But $r^2 = x^2 + y^2$ and $r \cos \theta = x$. Hence $x^2 + y^2 = 3x$. This is the equation of a circle.

53. $r(4 \sin \theta - \cos \theta) = 1$
$$4r \sin \theta - r \cos \theta = 1$$
But $r \sin \theta = y$ and $r \cos \theta = x$. Hence $4y - x = 1$. This is the equation of a line.

55. $r(2 + \cos \theta) = 1$
$$2r + r \cos \theta = 1$$
But $r = \sqrt{x^2 + y^2}$ and $r \cos \theta = x$. Hence $2\sqrt{x^2 + y^2} + x = 1$

$$2\sqrt{x^2 + y^2} = 1 - x$$
$$4(x^2 + y^2) = (1 - x)^2$$
$$4x^2 + 4y^2 = x^2 - 2x + 1$$
$$3x^2 + 4y^2 = 1 - 2x$$

This is the equation of an ellipse.

57. Here are computer-generated graphs of $r = 1 + 2 \sin (n\theta)$ for $n = 1$, $n = 2$, $n = 3$, and $n = 4$.

$n = 1$

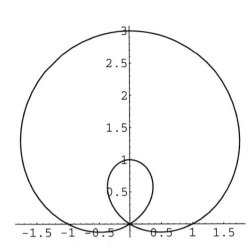

$n = 2$

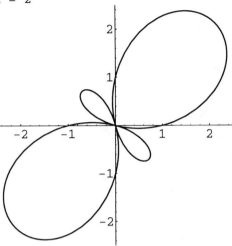

$n = 3$

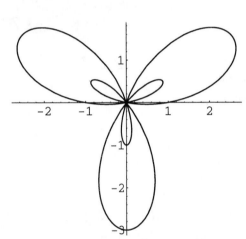

$n = 4$

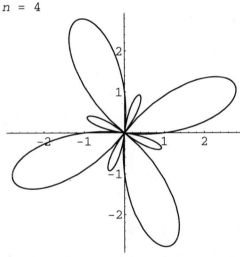

Generalizing from these, we can say that for each value of n, there are n large petals and n small petals. For n odd, the small petals are within the large petals; for n even, the small petals are between the large petals.

59. We solve the system

$r = 4 \cos \theta$

$r = -4 \sin \theta$

by equating the right sides:

$4 \cos \theta = -4 \sin \theta$

$\cos \theta = -\sin \theta$

$-1 = \tan \theta$

The only solution of this equation,

$0 \le \theta \le \pi$, is

$\theta = \dfrac{3\pi}{4}$. If we substitute this in either of

the original equations, we get

$r = 4 \cos \dfrac{3\pi}{4} = -4 \sin \dfrac{3\pi}{4} = 4\left(\dfrac{-\sqrt{2}}{2}\right) = -2\sqrt{2}$

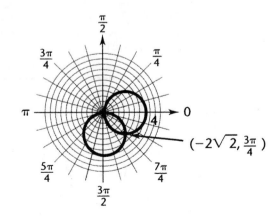

$\left(-2\sqrt{2}, \dfrac{3\pi}{4}\right)$

A sketch of the two graphs shows that the pole is on both graphs; however, the pole has no ordered pairs of coordinates that simultaneously satisfy both equations. As $\left(0, \dfrac{\pi}{2}\right)$ it satisfies the first; as $(0,0)$ it satisfies the second; it is not a solution of the system.

Solution: $\left(-2\sqrt{2}, \dfrac{3\pi}{4}\right)$

61. We solve the system
$r = 6 \cos \theta \qquad 0° \le \theta \le 360°$
$r = 6 \sin 2\theta$
by equating the right sides:
$6 \cos \theta = 6 \sin 2\theta$
$\cos \theta = \sin 2\theta$
$\cos \theta = 2 \sin \theta \cos \theta$
$0 = 2 \sin \theta \cos \theta - \cos \theta$
$0 = \cos \theta (2 \sin \theta - 1)$

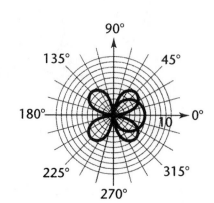

$\cos \theta = 0 \qquad\qquad 2 \sin \theta - 1 = 0$
$\theta = 90°,\ 270° \qquad\qquad \sin \theta = \dfrac{1}{2}$
$\theta = 30°,\ 150°$

If we substitute these values of θ in either of the original equations, we get the corresponding values of r:

$\theta = 90° \qquad r = 6 \cos 90° = 0$
$\theta = 270° \qquad r = 6 \cos 270° = 0$
$\theta = 30° \qquad r = 6 \cos 30° = 3\sqrt{3}$
$\theta = 150° \qquad r = 6 \cos 150° = -3\sqrt{3}$

The four solutions of the system are $(0,\ 90°)$, $(0,\ 270°)$, $(3\sqrt{3},\ 30°)$, and $(-3\sqrt{3},\ 150°)$. Note that two of these $(0,\ 90°)$ and $(0,\ 270°)$ name the same point (the pole). A sketch of the two graphs shows three points of intersection.

63.
$$d = \sqrt{r_1^2 + r_2^2 - 2r_1 r_2 \cos(\theta_2 - \theta_1)}$$

$(r_1,\ \theta_1) = \left(4,\ \dfrac{\pi}{4}\right) \quad (r_2,\ \theta_2) = \left(1,\ \dfrac{\pi}{2}\right)$

$$d = \sqrt{4^2 + 1^2 - 2(4)(1)\cos\left(\dfrac{\pi}{2} - \dfrac{\pi}{4}\right)}$$

$$d = \sqrt{17 - 8 \cos \dfrac{\pi}{4}}$$

$d = 3.368$ units

65. 6 knots at 30°, 13 knots at 75°, 12 knots at 135°, 9 knots at 180°

67. (A) $e = 0.4$ (B) $e = 1$ (C) $e = 1.6$

$$r = \frac{8}{1 - 0.4 \cos \theta}$$

$$r = \frac{8}{1 - \cos \theta}$$

$$r = \frac{8}{1 - 1.6 \cos \theta}$$

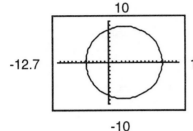

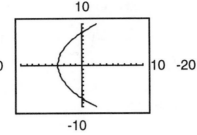

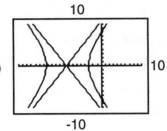

An ellipse   A parabola   A hyperbola

69. (A)

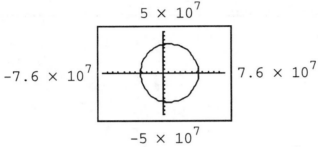

At aphelion, the distance is 4.34×10^7 mi; at perihelion it is 2.85×10^7 mi.

(B) Faster at perihelion. Since the distance from the sun to Mercury is less at perihelion than at aphelion, the planet must move faster near perihelion in order for the line joining Mercury to the sun to sweep out equal areas in equal intervals of time.

Exercise 7-6

Key Ideas and Formulas

Associated with each complex number $a + bi$ is a unique ordered pair of real numbers (a, b) and vice versa.

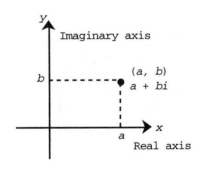

Complex numbers can be written in polar form.

$z = x + iy = r(\cos \theta + i \sin \theta)$

Writing $e^{i\theta} = \cos \theta + i \sin \theta$, $z = re^{i\theta}$

General Polar Form of a complex number:

For k any integer

$z = x + iy = r[\cos(\theta + 2k\pi) + i \sin(\theta + 2k\pi)]$
$\qquad = re^{i(\theta + 2k\pi)}$

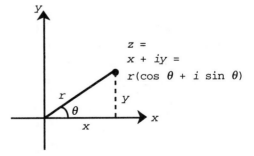

r is called the **modulus**, or **absolute value** of z, and is denoted by mod z or $|z|$. The polar angle θ that the line joining z to the origin makes with the polar axis is called the **argument** of z and is denoted by arg z.

$$\text{mod } z = r = \sqrt{x^2 + y^2} \qquad \text{Never negative}$$
$$\text{arg } z = \theta + 2k\pi \qquad k \text{ any integer}$$

where $\sin \theta = \dfrac{y}{r}$ and $\cos \theta = \dfrac{x}{r}$. The argument θ is usually chosen so that $-180° < \theta \leq 180°$ or $-\pi < \theta \leq \pi$.

If $z_1 = r_1 e^{i\theta_1}$ and $z_2 = r_2 e^{i\theta_2}$, then

1. $z_1 z_2 = r_1 e^{i\theta_1} r_2 e^{i\theta_2} = r_1 r_2\, e^{i(\theta_1 + \theta_2)}$

2. $\dfrac{z_1}{z_2} = \dfrac{r_1 e^{i\theta_1}}{r_2 e^{i\theta_2}} = \dfrac{r_1}{r_2}\, e^{i(\theta_1 - \theta_2)}$

Letting $\theta = \pi$ in $e^{i\theta} = \cos \theta + i \sin \theta$ yields $e^{i\pi} + 1 = 0$.

1.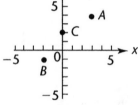

3.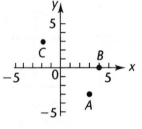

5.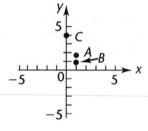

7.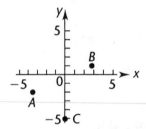

9. (A) A sketch shows that $\sqrt{3} + i$ is associated with a special 30°-60° triangle. Thus, by inspection, $r = 2$, $\theta = 30°$, and

$$\sqrt{3} + i = 2(\cos 30° + i \sin 30°)$$
$$= 2e^{30°i}$$

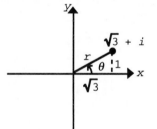

(B) A sketch shows that $-1 - i$ is associated with a special 45° triangle. Thus, by inspection, $r = \sqrt{2}$, $\theta = -135°$, and

$$-1 - i = \sqrt{2}\,[\cos(-135°) + i \sin(-135°)]$$
$$= \sqrt{2}\, e^{-135°i}$$

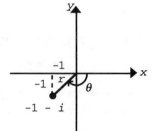

(C) A sketch shows that $5 - 6i$ is not associated with a special triangle.

$r = \sqrt{5^2 + (-6)^2} = 7.81$

$\theta = \tan^{-1} \frac{-6}{5} = -50.19°$

$5 - 6i = 7.81[\cos(-50.19°) + i \sin(-50.19°)]$

$\qquad = 7.81e^{-50.19°i}$

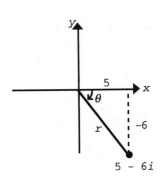

11. (A) A sketch shows that $-i\sqrt{3}$ is associated with the special quadrantal angle $-\frac{\pi}{2}$. Thus, by inspection,

$r = \sqrt{3}, \ \theta = -\frac{\pi}{2}$, and

$-i\sqrt{3} = \sqrt{3}\left[\cos\left(-\frac{\pi}{2}\right) + i \sin\left(-\frac{\pi}{2}\right)\right]$

$\qquad = \sqrt{3}\, e^{(-\pi/2)i}$

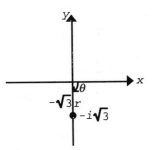

(B) A sketch shows that $-\sqrt{3} - i$ is associated with a special 30°-60° triangle. Thus, by inspection,

$r = 2, \ \theta = -\frac{5\pi}{6}$, and

$-\sqrt{3} - i = 2\left[\cos\left(-\frac{5\pi}{6}\right) + i \sin\left(-\frac{5\pi}{6}\right)\right]$

$\qquad = 2e^{(-5\pi/6)i}$

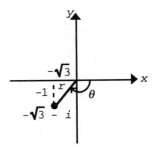

(C) A sketch shows that $-8 + 5i$ is not associated with a special triangle.

$r = \sqrt{(-8)^2 + 5^2} = 9.43$

$\theta = \pi + \tan^{-1}\frac{5}{(-8)} = 2.58$

$-8 + 5i = 9.43(\cos 2.58 + i \sin 2.58)$

$\qquad = 9.43e^{2.58i}$

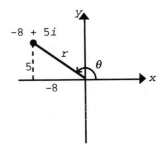

13. (A) $x + iy = 2e^{(\pi/3)i}$

$\qquad = 2\left(\cos\frac{\pi}{3} + i \sin\frac{\pi}{3}\right)$

$\qquad = 2\left(\frac{1}{2} + i\frac{\sqrt{3}}{2}\right)$

$\qquad = 1 + i\sqrt{3}$

(B) $x + iy = \sqrt{2}\, e^{(-45°)i}$

$\qquad = \sqrt{2}\,[\cos(-45°) + i \sin(-45°)]$

$\qquad = \sqrt{2}\left[\frac{1}{\sqrt{2}} + i\left(-\frac{1}{\sqrt{2}}\right)\right]$

$\qquad = 1 - i$

(C) $x + iy = 3.08e^{2.44i}$

$\qquad = 3.08(\cos 2.44 + i \sin 2.44)$

$\qquad = -2.35 + 1.99i$

15. (A) $x + iy = 6e^{(\pi/6)i}$

$$= 6\left(\cos\frac{\pi}{6} + i\sin\frac{\pi}{6}\right)$$

$$= 6\left(\frac{\sqrt{3}}{2} + i\frac{1}{2}\right)$$

$$= 3\sqrt{3} + 3i$$

(B) $x + iy = \sqrt{7}\,e^{(-90°)i}$

$$= \sqrt{7}\,[\cos(-90°) + i\sin(-90°)]$$

$$= \sqrt{7}\,(0 - 1i)$$

$$= -i\sqrt{7}$$

(C) $4.09e^{(-122.88°)i} = 4.09[\cos(-122.88°) + i\sin(-122.88°)]$

$$= -2.22 - 3.43i$$

17. $z_1 z_2 = 7e^{82°i} \cdot 2e^{31°i}$

$$= 7 \cdot 2e^{i(82° + 31°)} = 14e^{113°i}$$

$$\frac{z_1}{z_2} = \frac{7e^{82°i}}{2e^{31°i}}$$

$$= \frac{7}{2}\,e^{i(82° - 31°)} = 3.5e^{51°i}$$

19. $z_1 z_2 = 5e^{52°i} \cdot 2e^{83°i}$

$$= 5 \cdot 2e^{i(52° + 83°)} = 10e^{135°i}$$

$$\frac{z_1}{z_2} = \frac{5e^{52°i}}{2e^{83°i}}$$

$$= \frac{5}{2}\,e^{i(52° - 83°)} = 2.5e^{(-31°)i}$$

21. $z_1 z_2 = 3.05e^{1.76i} \cdot 11.94e^{2.59i}$

$$= 3.05 \cdot 11.94e^{(1.76 + 2.59)i} = 36.42e^{4.35i}$$

$$\frac{z_1}{z_2} = \frac{3.05e^{1.76i}}{11.94e^{2.59i}}$$

$$= \frac{3.05}{11.94}\,e^{(1.76 - 2.59)i} = 0.26e^{-0.83i}$$

23. Directly:

$(1 + i)(2 + 2i) = 2 + 2i + 2i + 2i^2 = 2 + 4i - 2 = 4i$

Using polar forms:

$z_1 = 1 + i \quad r = \sqrt{1^2 + 1^2} = \sqrt{2} \quad \tan\theta = \frac{1}{1} = 1 \quad \theta = 45° \quad z_1 = \sqrt{2}\,e^{45°i}$

$z_2 = 2 + 2i \quad r = \sqrt{2^2 + 2^2} = \sqrt{8} \quad \tan\theta = \frac{2}{2} = 1 \quad \theta = 45° \quad z_2 = \sqrt{8}\,e^{45°i}$

$z_1 z_2 = \sqrt{2}\,e^{45°i}\sqrt{8}\,e^{45°i} = \sqrt{2}\sqrt{8}e^{(45° + 45°)i} = 4e^{90°i}$

25. Directly:

$(-1 + i)^3 = (-1 + i)(-1 + i)^2 = (-1 + i)(1 - 2i + i^2) = (-1 + i)(1 - 2i - 1)$

$$= (-1 + i)(-2i) = 2i - 2i^2 = 2 + 2i$$

Using polar forms:

$z = -1 + i \quad r = \sqrt{(-1)^2 + 1^2} = \sqrt{2} \quad \tan\theta = \frac{1}{-1} = -1 \quad \theta = 135° \quad z = \sqrt{2}\,e^{135°i}$

$z^3 = (\sqrt{2}\,e^{135°i})^3 = \sqrt{2}\,e^{135°i}\sqrt{2}\,e^{135°i}\sqrt{2}\,e^{135°i} = \sqrt{2}\sqrt{2}\sqrt{2}\,e^{(135° + 135° + 135°)i} = 2\sqrt{2}\,e^{405°i}$

$$= 2\sqrt{2}\,e^{(45° + 360°)i} = 2\sqrt{2}\,e^{45°i}$$

27. Directly:

$$\frac{1 + i}{1 - i} = \frac{(1 + i)}{(1 - i)}\frac{(1 + i)}{(1 + i)} = \frac{1 + 2i + i^2}{1 - i^2} = \frac{1 + 2i - 1}{1 + 1} = \frac{2i}{2} = i$$

Using polar forms:

$z_1 = 1 + i = \sqrt{2}\,e^{45°i} \quad$ (see Problem 23)

$z_2 = 1 - i \quad r = \sqrt{1^2 + (-1)^2} = \sqrt{2} \quad \tan\theta = \frac{-1}{1} = -1 \quad \theta = 315° \quad z_2 = \sqrt{2}\,e^{315°i}$

$$\frac{z_1}{z_2} = \frac{\sqrt{2}\,e^{45°i}}{\sqrt{2}\,e^{315°i}} = e^{(45° - 315°)i} = e^{(-270°)i} = e^{(-270° + 360°)i} = e^{90°i}$$

29. Since the conjugate of a complex number can be found by replacing i with $-i$, transforming $a + bi$ into $a - bi$, the conjugate of $re^{i\theta}$ is $re^{(-i)\theta}$ or $re^{-i\theta}$.

31. To show that $r^{1/3}e^{i(\theta/3)}$ is a cube root of $re^{i\theta}$ we need only show that $[r^{1/3}e^{i(\theta/3)}]^3 = re^{i\theta}$. But $[r^{1/3}e^{i(\theta/3)}]^3 = r^{1/3}e^{i(\theta/3)}r^{1/3}e^{i(\theta/3)}r^{1/3}e^{i(\theta/3)} = r^{1/3}r^{1/3}r^{1/3}e^{i(\theta/3 + \theta/3 + \theta/3)} = re^{i\theta}$

Hence
$r^{1/3}e^{i(\theta/3)}$ is a cube root of $re^{i\theta}$.

33. $z = re^{i\theta}$
$z^2 = (re^{i\theta})^2 = re^{i\theta}re^{i\theta} = r^2e^{i(\theta + \theta)} = r^2e^{2\theta i}$
$z^3 = zz^2 = re^{i\theta}r^2e^{2\theta i} = rr^2e^{i(\theta + 2\theta)} = r^3e^{3\theta i}$

Generalizing,
$z^n = r^ne^{n\theta i}$ for n a natural number. (This is a hypothesis at this stage, restated in the following section of the text—DeMoivre's Theorem and provable by the methods of Section 11-2.)

35. $\vec{u}: 20e^{0°i}$
$\vec{v}: 10e^{60°i}$

(A) $\vec{u}: 20e^{0°i} = 20(\cos 0° + i \sin 0°)$
$= 20(1 + 0i) = 20 + 0i$

$\vec{v}: 10e^{60°i} = 10(\cos 60° + i \sin 60°) = 10\left(\frac{1}{2} + i\frac{\sqrt{3}}{2}\right)$
$= 5 + 5i\sqrt{3}$

$(20 + 0i) + (5 + 5i\sqrt{3}) = 25 + 5i\sqrt{3}$

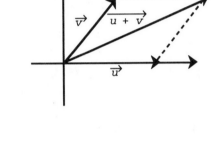

(B) $r = \sqrt{(25)^2 + (5\sqrt{3})^2} = \sqrt{625 + 75} = \sqrt{700} = 26.5$

$\tan \theta = \frac{5\sqrt{3}}{25} = \frac{\sqrt{3}}{5}$ $\theta = \tan^{-1}\frac{y}{x} = \tan^{-1}\frac{\sqrt{3}}{5} = 19.1°$

$25 + 5i\sqrt{3} = 26.5e^{19.1°i}$

(C) $26.5e^{19.1°i}$ represents a force of 26.5 pounds at an angle of 19.1° with the positive x axis.

Exercise 7-7

Key Ideas and Formulas

DeMoivre's Theorem: If $z = x + iy = re^{i\theta}$, and n is a natural number, then
$$z^n = (x + iy)^n = (re^{i\theta})^n = r^ne^{n\theta i}$$

nth-root Theorem: For n a positive integer greater than 1,
$$r^{1/n}e^{[(\theta/n) + k(360°/n)]i} \quad k = 0, 1, …, n - 1$$

are the n distinct nth roots of $re^{i\theta}$, and there are no others.

1. $(3e^{40°i})^4 = 3^4e^{4 \cdot 40°i} = 81e^{160°i}$

3. $(1 - i)^6 = (\sqrt{2}e^{315°i})^6 = (\sqrt{2})^6e^{6 \cdot 315°i} = (2^{1/2})^6e^{1890°i}$
$= 2^3e^{(90° + 5 \cdot 360°)i} = 8e^{90°i}$

5. $\left(\frac{\sqrt{3} - i}{2}\right)^{20} = \left(\frac{\sqrt{3}}{2} - \frac{1}{2}i\right)^{20} = (1e^{330°i})^{20} = 1^{20}e^{20 \cdot 330°i} = e^{6600°i}$
$= e^{(120° + 18 \cdot 360°)i} = e^{120°i}$

7. $(-\sqrt{3} - i)^4 = (2e^{-150°i})^4 = 2^4e^{4(-150°i)} = 16e^{-600°i}$

$$= 16[\cos(-600°) + i\sin(-600°)] = 16\left(-\frac{1}{2} + i\frac{\sqrt{3}}{2}\right) = -8 + 8i\sqrt{3}$$

9. $(1 - i)^8 = (\sqrt{2}e^{-45°i})^8 = (\sqrt{2})^8e^{8(-45°i)} = 16e^{-360°i}$

$$= 16[\cos(-360°) + i\sin(-360°)] = 16(1 + 0i) = 16$$

11. $\left(-\frac{1}{2} + \frac{\sqrt{3}}{2}i\right) = (e^{120°i})^3 = e^{360°i} = \cos 360° + i\sin 360° = 1 + 0i = 1$

13. Using the nth-root theorem, all three cube roots of $8e^{30°i}$ are given by
$$8^{1/3}e^{(30°/3 + k360°/3)i} = 8^{1/3}e^{(10° + k120°)i} \qquad k = 0, 1, 2$$
Thus,
$$w_1 = 8^{1/3}e^{(10° + 0 \cdot 120°)i} = 2e^{10°i}$$
$$w_2 = 8^{1/3}e^{(10° + 1 \cdot 120°)i} = 2e^{130°i}$$
$$w_3 = 8^{1/3}e^{(10° + 2 \cdot 120°)i} = 2e^{250°i}$$

15. Using the nth-root theorem, all four fourth roots of $81e^{60°i}$ are given by
$$81^{1/4}e^{(60°/4 + k360°/4)i} = 81^{1/4}e^{(15° + k90°)i} \qquad k = 0, 1, 2, 3$$
Thus,
$$w_1 = 81^{1/4}e^{(15° + 0 \cdot 90°)i} = 3e^{15°i}$$
$$w_2 = 81^{1/4}e^{(15° + 1 \cdot 90°)i} = 3e^{105°i}$$
$$w_3 = 81^{1/4}e^{(15° + 2 \cdot 90°)i} = 3e^{195°i}$$
$$w_4 = 81^{1/4}e^{(15° + 3 \cdot 90°)i} = 3e^{285°i}$$

17. First write $1 - i$ in polar form.
$$1 - i = \sqrt{2}e^{(-45°)i}$$
Using the nth-root theorem, all five fifth roots of $\sqrt{2}e^{(-45°)i}$ are given by
$$(\sqrt{2})^{1/5}e^{(-45°/5 + k360°/5)i} = (2^{1/2})^{1/5}e^{(-9° + k72°)i} \qquad k = 0, 1, 2, 3, 4$$
Thus,
$$w_1 = 2^{1/10}e^{(-9° + 0 \cdot 72°)i} = 2^{1/10}e^{(-9°)i}$$
$$w_2 = 2^{1/10}e^{(-9° + 1 \cdot 72°)i} = 2^{1/10}e^{63°i}$$
$$w_3 = 2^{1/10}e^{(-9° + 2 \cdot 72°)i} = 2^{1/10}e^{135°i}$$
$$w_4 = 2^{1/10}e^{(-9° + 3 \cdot 72°)i} = 2^{1/10}e^{207°i}$$
$$w_5 = 2^{1/10}e^{(-9° + 4 \cdot 72°)i} = 2^{1/10}e^{279°i}$$

19. First write 8 in polar form.
$$8 = 8 + 0i = 8e^{0°i}$$
Using the nth-root theorem, all three cube roots of
$8e^{0°i}$ are given by
$$8^{1/3}e^{(0°/3 + k360°/3)i} = 8^{1/3}e^{k120°i} \qquad k = 0, 1, 2$$
Thus,
$$w_1 = 2e^{0 \cdot 120°i} = 2e^{0°i}$$
$$w_2 = 2e^{1 \cdot 120°i} = 2e^{120°i}$$
$$w_3 = 2e^{2 \cdot 120°i} = 2e^{240°i}$$

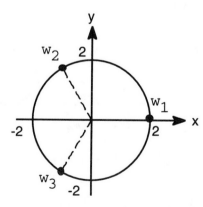

w_1, w_2, w_3 lie on a circle of radius 2, equally spaced so that the angle between successive roots is $120°$.

21. First write -16 in polar form.

$$-16 = -16 + 0i = 16e^{180°i}$$

Using the nth-root theorem, all four fourth roots of $16e^{180°i}$ are given by

$$16^{1/4}e^{(180°/4 + k \cdot 360°/4)i} = 16^{1/4}e^{(45° + k \cdot 90°)i}$$

$$k = 0, 1, 2, 3$$

Thus,

$$w_1 = 2e^{(45° + 0 \cdot 90°)i} = 2e^{45°i}$$
$$w_2 = 2e^{(45° + 1 \cdot 90°)i} = 2e^{135°i}$$
$$w_3 = 2e^{(45° + 2 \cdot 90°)i} = 2e^{225°i}$$
$$w_4 = 2e^{(45° + 3 \cdot 90°)i} = 2e^{315°i}$$

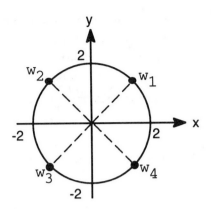

w_1, w_2, w_3, w_4 lie on a circle of radius 2, equally spaced so that the angle between successive roots is 90°.

23. First write i in polar form.

$$i = 0 + 1i = 1e^{90°i}$$

Using the nth-root theorem, all sixth roots of $1e^{90°i}$ are given by

$$1^{1/6}e^{(90°/6 + k \cdot 360°/6)i} = 1e^{(15° + k \cdot 60°)i}$$

$$k = 0, 1, 2, 3, 4, 5$$

Thus,

$$w_1 = 1e^{(15° + 0 \cdot 60°)i} = 1e^{15°i}$$
$$w_2 = 1e^{(15° + 1 \cdot 60°)i} = 1e^{75°i}$$
$$w_3 = 1e^{(15° + 2 \cdot 60°)i} = 1e^{135°i}$$
$$w_4 = 1e^{(15° + 3 \cdot 60°)i} = 1e^{195°i}$$
$$w_5 = 1e^{(15° + 4 \cdot 60°)i} = 1e^{255°i}$$
$$w_6 = 1e^{(15° + 5 \cdot 60°)i} = 1e^{315°i}$$

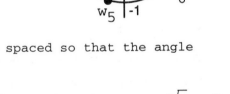

These roots lie on a circle of radius 1, equally spaced so that the angle between successive roots is 60°.

25. True. The square roots of $-p$, where p is a positive real number, are $i\sqrt{p}$ and $-i\sqrt{p}$.

27. False. For example, $e^{30°i}$ is a twelfth root of 1, since $(e^{30°i})^{12} = e^{12 \cdot 30°i} = e^{360°i} = 1$. However, $(e^{30°i})^4 = e^{120°i} = \cos 120° + i \sin 120° = -\frac{1}{2} + i\frac{\sqrt{3}}{2}$, which is not 1, so $e^{30°i}$ is not a fourth root of 1.

29. (A) Substituting $1 + i$ for x in $x^4 + 4 = 0$ yields

$$(1 + i)^4 + 4 \overset{?}{=} 0$$
$$((1 + i)^2)^2 + 4 \overset{?}{=} 0$$
$$(1 + 2i + i^2) + 4 \overset{?}{=} 0$$
$$(2i)^2 + 4 \overset{?}{=} 0$$
$$-4 + 4 \overset{\checkmark}{=} 0$$

Thus, $1 + i$ is a root of $x^4 + 4 = 0$.

(B) The four roots are equally spaced around the circle. Since there are four roots, the angle between successive roots on the circle is $\frac{360°}{4} = 90°$.

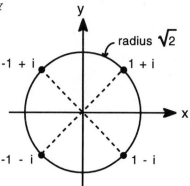

(C) Substituting each root for x in $x^4 + 4 = 0$ yields

$-1 + i$: $(-1 + i)^4 + 4 = ((-1 + i)^2)^2 + 4 = (1 - 2i + i^2)^2 + 4 = (-2i)^2 + 4$
$= -4 + 4 = 0$

$-1 - i$: $(-1 - i)^4 + 4 = ((-1 - i)^2)^2 + 4 = (1 + 2i + i^2)^2 + 4 = (2i)^2 + 4$
$= -4 + 4 = 0$

$1 - i$: $(1 - i)^4 + 4 = ((1 - i)^2)^2 + 4 = (1 - 2i + i^2)^2 + 4 = (-2i)^2 + 4$
$= -4 + 4 = 0$

31. $x^3 + 64 = 0$
$x^3 = -64$

Thus we require the three cube roots of -64.
First write -64 in polar form:
$-64 = 64e^{180°i}$

Using the nth-root theorem, all three cube roots of -64 are given by
$64^{1/3}e^{(180°/3 + k \cdot 360°/3)i} = 4e^{(60° + k \cdot 120°)i}$ $k = 0, 1, 2$

Thus,

$w_1 = 4e^{60°i} = 4(\cos 60° + i \sin 60°) = 2 + 2i\sqrt{3}$
$w_2 = 4e^{180°i} = 4(\cos 180° + i \sin 180°) = -4$
$w_3 = 4e^{300°i} = 4(\cos 300° + i \sin 300°) = 2 - 2i\sqrt{3}$

33. $x^3 - 27 = 0$
$x^3 = 27$

Thus we require the three cube roots of 27.

First write 27 in polar form:
$27 = 27e^{0°i}$

Using the nth-root theorem, all three cube roots of 27 are given by
$27^{1/3}e^{(0°/3 + k \cdot 360°/3)i} = 3e^{k \cdot 120°i}$ $k = 0, 1, 2$

Thus,

$w_1 = 3e^{0°i} = 3(\cos 0° + i \sin 0°) = 3$

$w_2 = 3e^{120°i} = 3(\cos 120° + i \sin 120°) = -\frac{3}{2} + \frac{3\sqrt{3}}{2} i$

$w_3 = 3e^{240°i} = 3(\cos 240° + i \sin 240°) = -\frac{3}{2} - \frac{3\sqrt{3}}{2} i$

35. We require the two square roots of -4. These are clearly given by $2i$ and $-2i$, but we can check this by applying the nth root theorem.

First write -4 in polar form:
$-4 = 4e^{180°i}$

Using the nth root theorem, the two square roots of -4 are given by
$4^{1/2}e^{(180°/2 + k \cdot 360°/2)} = 2e^{(90° + k \cdot 180°)i}$ $k = 0, 1$

Thus
$w_1 = 2e^{90°i} = 2(\cos 90° + i \sin 90°) = 2i$
$w_2 = 2e^{270°i} = 2(\cos 270° + i \sin 270°) = -2i$

37. We require the three cube roots of $8i$.

First write $8i$ in polar form:

$8i = 8e^{90°i}$

Using the nth root theorem, all three cube roots of $8i$ are given by

$8^{1/3}e^{(90°/3 + k \cdot 360°/3)i} = 2e^{(30° + k \cdot 120°)i} \qquad k = 0, 1, 2$

Thus,

$w_1 = 2e^{30°i} = 2(\cos 30° + i \sin 30°) = \sqrt{3} + i$

$w_2 = 2e^{150°i} = 2(\cos 150° + i \sin 150°) = -\sqrt{3} + i$

$w_3 = 2e^{270°i} = 2(\cos 270° + i \sin 270°) = -2i$

39. To show $(r^{1/n}e^{(\theta/n + k \cdot 360°/n)i})^n = re^{i\theta}$ we apply DeMoivre's Theorem:

$$
\begin{aligned}
(r^{1/n}e^{(\theta/n + k \cdot 360°/n)i})^n &= (r^{1/n})^n e^{n(\theta/n + k \cdot 360°/n)i} \\
&= re^{(\theta + k360°)i} \\
&= r[\cos(\theta + k360°) + i \sin(\theta + k \cdot 360°)] \\
&= r(\cos \theta + i \sin \theta) \quad \text{since } k \text{ is an integer and sine and} \\
&\qquad\qquad\qquad\qquad\qquad \text{cosine are periodic with period } 360° \\
&= re^{i\theta}
\end{aligned}
$$

41. The complex zeros of $P(x) = x^5 - 32$ are the values of x for which $x^5 - 32 = 0$, or $x^5 = 32$. Thus we require the five fifth roots of 32. First write 32 in polar form:

$32 = 32e^{0°i}$

Using the nth-root theorem, all five fifth roots of 32 are given by

$32^{1/5}e^{(0°/5 + k \cdot 360°/5)i} = 2e^{k \cdot 72°i} \qquad k = 0, 1, 2, 3, 4$

Thus,

$w_1 = 2e^{0°i}$

$w_2 = 2e^{72°i}$

$w_3 = 2e^{144°i}$

$w_4 = 2e^{216°i}$

$w_5 = 2e^{288°i}$

43. $x^5 + 1 = 0$

$\qquad x^5 = -1$

Thus we require the five fifth roots of -1. First write -1 in polar form:

$\qquad -1 = 1e^{180°i}$

Using the nth-root theorem, all five fifth roots of -1 are given by

$1^{1/5}e^{(180°/5 + k \cdot 360°/5)i} = 1e^{(36° + k \cdot 72°)i} \qquad k = 0, 1, 2, 3, 4$

Thus,

$w_1 = 1e^{(36° + 0 \cdot 72°)i} = e^{36°i}$

$w_2 = 1e^{(36° + 1 \cdot 72°)i} = e^{108°i}$

$w_3 = 1e^{(36° + 2 \cdot 72°)i} = e^{180°i}$

$w_4 = 1e^{(36° + 3 \cdot 72°)i} = e^{252°i}$

$w_5 = 1e^{(36° + 4 \cdot 72°)i} = e^{324°i}$

45. The linear factors of $x^6 + 64$ will, by the factor theorem, be $x - x_i$, where x_i are zeros of $x^6 + 64$, that is $x^6 + 64 = 0$ or $x^6 = -64$. Hence the x_i are the six sixth roots of -64, or $64e^{180°i}$.

Using the nth-root theorem, all six sixth roots of -64 are given by

$64^{1/6}e^{(180°/6 + k \cdot 360°/6)i} = 2e^{(30° + k \cdot 60°)i} \qquad k = 0, 1, 2, 3, 4, 5$

Thus,

$$w_1 = 2e^{30°i} = 2(\cos 30° + i \sin 30°) \quad = 2\left(\frac{\sqrt{3}}{2} + i\frac{1}{2}\right) \quad = \sqrt{3} + i$$

$$w_2 = 2e^{90°i} = 2(\cos 90° + i \sin 90°) \quad = 2(0 + i1) \quad = 2i$$

$$w_3 = 2e^{150°i} = 2(\cos 150° + i \sin 150°) = 2\left(-\frac{\sqrt{3}}{2} + i\frac{1}{2}\right) = -\sqrt{3} + i$$

$$w_4 = 2e^{210°i} = 2(\cos 210° + i \sin 210°) = 2\left(-\frac{\sqrt{3}}{2} - i\frac{1}{2}\right) = -\sqrt{3} - i$$

$$w_5 = 2e^{270°i} = 2(\cos 270° + i \sin 270°) = 2(0 - i1) \quad = -2i$$

$$w_6 = 2e^{330°i} = 2(\cos 330° + i \sin 330°) = 2\left(\frac{\sqrt{3}}{2} - i\frac{1}{2}\right) \quad = \sqrt{3} - i$$

Hence, $x^6 + 64 = (x - 2i)(x + 2i)[x - (-\sqrt{3} + i)][x - (-\sqrt{3} - i)]$
$$[x - (\sqrt{3} + i)][x - (\sqrt{3} - i)]$$

CHAPTER 7 REVIEW

1.

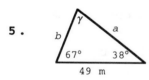

We are given two sides and a non-included angle (SSA). α is acute
$$11 = a \geq b = 3.7$$
One triangle can be constructed.

(7-1)

2. It is impossible to draw or form a triangle with this data, since angles α and β together add up to more than 180°. No triangle can be constructed.

(7-1)

3.

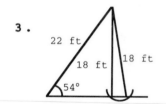

We are given two sides and a non-included angle (SSA). α is acute.
$$h = b \sin \alpha = 22 \sin 54° = 17.8$$
$$h < a < b$$
Two triangles can be constructed.

(7-1)

4. Angle β is acute. A triangle can have at most one obtuse angle. Since α is acute, then, if the triangle has an obtuse angle it must be the angle opposite the longer of the two sides, b and c. Thus, β, the angle opposite the shorter of the two sides, b, must be acute.

(7-2)

5.

We are given two angles and the included side (ASA). We use the law of sines.

Solve for γ:
$$\alpha + \beta + \gamma = 180°$$
$$\gamma = 180° - (\alpha + \beta)$$
$$= 180° - (67° + 38°)$$
$$= 75°$$

Solve for b:
$$\frac{\sin \beta}{b} = \frac{\sin \gamma}{c}$$
$$b = \frac{c \sin \beta}{\sin \gamma}$$
$$= \frac{49 \sin 38°}{\sin 75°}$$
$$= 31 \text{ m}$$

Solve for a:
$$\frac{\sin \alpha}{a} = \frac{\sin \gamma}{c}$$
$$a = \frac{c \sin \alpha}{\sin \gamma}$$
$$= \frac{49 \sin 67°}{\sin 75°}$$
$$= 47 \text{ m}$$

(7-1)

6. 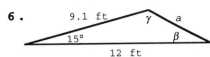 We are given two sides and the included angle (SAS).
We use the law of cosines to find the third side,
then the law of sines to find a second angle.

Solve for a:

$$a^2 = b^2 + c^2 - 2bc \cos \alpha$$
$$= (9.1)^2 + (12)^2 - 2(9.1)(12)\cos 15°$$
$$= 15.8518$$
$$a = 4.00 \text{ ft}$$

Solve for β:

$$\frac{\sin \alpha}{a} = \frac{\sin \beta}{b}$$
$$\sin \beta = \frac{b \sin \alpha}{a}$$
$$= \frac{9.1 \sin 15°}{4.00}$$
$$= 0.5915$$
$$\beta = \sin^{-1}(0.5915)$$
$$= 36°$$

There is another solution of $\sin \beta = 0.5915$ which deserves brief consideration,
that is, $\beta = 180° - \sin^{-1}(0.5915) = 144°$. This would lead to a contradiction,
however, since the largest side of the triangle, c, must be opposite the
largest angle, γ, and this would lead to two obtuse angles in the triangle.

Solve for γ:
$$\alpha + \beta + \gamma = 180°$$
$$\gamma = 180° - (\alpha + \beta)$$
$$= 180° - (15° + 36°)$$
$$= 129°$$

(7-1, 7-2)

7. We are given two sides and a non-included angle
(SSA). From a rough sketch we see that there is
only one triangle possible.

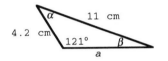

We use the law of sines.
Solve for β:
$$\frac{\sin \beta}{b} = \frac{\sin \gamma}{c}$$
$$\sin \beta = \frac{b \sin \gamma}{c}$$
$$= \frac{4.2 \sin 121°}{11}$$
$$\beta = \sin^{-1} \frac{4.2 \sin 121°}{11} \quad \beta \text{ is acute}$$
$$= 19°$$

Solve for α:
$$\alpha + \beta + \gamma = 180°$$
$$\alpha = 180° - (\beta + \gamma)$$
$$= 180° - (19° + 121°)$$
$$= 40°$$

Solve for a:
$$\frac{\sin \alpha}{a} = \frac{\sin \gamma}{c}$$
$$a = \frac{c \sin \alpha}{\sin \gamma}$$
$$= \frac{11 \sin 40°}{\sin 121°}$$
$$= 8.2 \text{ cm} \qquad (7-1)$$

8. 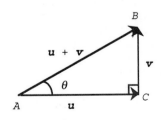 To find $|\mathbf{u} + \mathbf{v}|$: Apply the Pythagorean theorem to
triangle ABC.
$$|\mathbf{u} + \mathbf{v}|^2 = AB^2 = AC^2 + BC^2$$
$$= |\mathbf{u}|^2 + |\mathbf{v}|^2$$
$$= 160^2 + 55^2$$
$$= 28,625$$
$$|\mathbf{u} + \mathbf{v}| = \sqrt{28,625}$$
$$= 170 \text{ mi/hr}$$

Solve triangle ABC for θ: $\tan \theta = \dfrac{BC}{AC} = \dfrac{|\boldsymbol{v}|}{|\boldsymbol{u}|}$

$\qquad \theta = \tan^{-1} \dfrac{|\boldsymbol{v}|}{|\boldsymbol{u}|}$ θ is acute

$\qquad = \tan^{-1} \dfrac{55}{160}$

$\qquad = 19°$ $\hspace{4cm}$ (7-3)

9. The algebraic vector $\langle a,\ b \rangle$ corresponds to the standard geometric vector with terminal point $P(a,\ b)$ and initial point $O(0,\ 0)$. The coordinates of the point $P(a,\ b)$ are given by

$\qquad a = x_B - x_A = 5 - 2 = 3$
$\qquad b = y_B - y_A = -1 - 6 = -7$

Thus, $\langle a,\ b \rangle = \langle 3,\ -7 \rangle$ $\hspace{4cm}$ (7-4)

10. $|\boldsymbol{v}| = \sqrt{(-3)^2 + (-5)^2} = \sqrt{34}$ $\qquad$ (7-4)

11. The graph is the set of all points on a line forming an angle of $\dfrac{\pi}{6}$ with the polar axis.

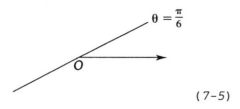

(7-5)

12. The graph is the set of all points at distance 6 from the pole—a circle of radius 6 with center at the pole.

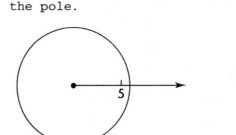

(7-5)

13.

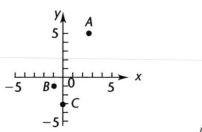

(7-6)

14. See figure.

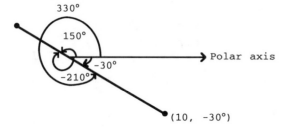

The point with coordinates (10, -30°) can equally well be described as (-10°, -210°) or (-10, 150°) or (10, 330°). Thus, (-10, -210°): The polar axis is rotated 210° clockwise (negative direction) and the point is located 10 units from the pole along the negative polar axis. (-10, 150°): The polar axis is rotated 150° counterclockwise (positive direction) and the point is located 10 units from the pole along the negative polar axis. (10, 330°): The polar axis is rotated 330° counterclockwise and the point is located 10 units from the pole along the positive polar axis. $\hspace{1cm}$ (7-5)

15.

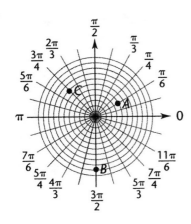

(7-6)

16. (A) A sketch shows that $1 - i\sqrt{3}$ is associated with a special 30°-60° triangle. Thus, by inspection, $r = 2$, $\theta = -60°$, and

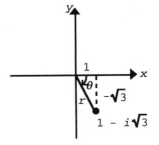

$$1 - i\sqrt{3} = 2[\cos(-60°) + i \sin(-60°)]$$
$$= 2e^{(-60°)i}$$

(B) $4e^{(-30°)i} = 4[\cos(-30°) + i \sin(-30°)]$
$$= 4\left[\frac{\sqrt{3}}{2} + i\left(-\frac{1}{2}\right)\right]$$
$$= 2\sqrt{3} - 2i$$

(7-6)

17. (A) $\left[-\frac{1}{2} - \frac{\sqrt{3}}{2}i\right]^3 = (e^{-120°i})^3 = e^{-360°i} = \cos(-360°) + i \sin(-360°) = 1$

(B) is left to the student

(7-7)

18. $(2e^{15°i})^4 = 2^4 e^{4(15°i)} = 16e^{60°i} = 16(\cos 60° + i \sin 60°) = 16\left(\frac{1}{2} + i\frac{\sqrt{3}}{2}\right)$
$$= 8 + 8i\sqrt{3}$$

(7-7)

19. If the triangle has an obtuse angle, then it must be the angle opposite the longest side; in this case, α.

(7-2)

20.

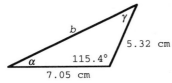

We are given two sides and the included angle (SAS). We use the law of cosines to find the third side, then the law of sines to find a second angle.

Solve for b:
$$b^2 = a^2 + c^2 - 2ac \cos \beta$$
$$= (5.32)^2 + (7.05)^2 - 2(5.32)(7.05)\cos 115.4°$$
$$= 110.18018\ldots$$
$$b = 10.5 \text{ cm}$$

Solve for α:
Note that since the sum of α and γ is 180° - 115.4° = 64.6°, both must be acute.

$$\frac{\sin \alpha}{a} = \frac{\sin \beta}{b}$$

$$\sin \alpha = \frac{a \sin \beta}{b}$$

$$= \frac{5.32 \sin 115.4°}{10.5}$$

$$\alpha = \sin^{-1}\left(\frac{5.32 \sin 115.4°}{10.5}\right)$$

$$= 27.2°$$

Solve for γ:

$$\alpha + \beta + \gamma = 180°$$
$$\gamma = 180° - (\alpha + \beta)$$
$$= 180° - (27.2° + 115.4°)$$
$$= 37.4° \qquad\qquad (7\text{-}2)$$

21.

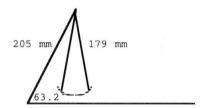

α is acute
$h = b \sin \alpha = 205 \sin 63.2° = 183$
$a = 179 < 183 = h$

We are given two sides and a non-included angle (SSA). If we try to draw a triangle with these values, we will be unsuccessful.
No solution. $\qquad\qquad (7\text{-}1)$

22.

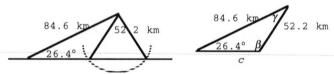

We are given two sides and a non-included angle (SSA). If we try to draw a triangle with these values, we find that two triangles are possible. Following the instructions, we choose the solution for which β is obtuse.

Solve for β:

$$\frac{\sin \beta}{b} = \frac{\sin \alpha}{a}$$

$$\sin \beta = \frac{b \sin \alpha}{a} = \frac{84.6 \sin 26.4°}{52.2} = 0.7206$$

β is chosen obtuse

$$\beta = 180° - \sin^{-1} 0.7206$$
$$= 180° - 46.1°$$
$$= 133.9°$$

Solve for γ:

$$\gamma = 180° - (\alpha + \beta)$$
$$= 180° - (26.4° + 133.9°)$$
$$= 19.7°$$

Solve for c:

$$\frac{\sin \alpha}{a} = \frac{\sin \gamma}{c}$$

$$c = \frac{a \sin \gamma}{\sin \alpha}$$

$$= \frac{52.2 \sin 19.7°}{\sin 26.4°}$$

$$= 39.6 \text{ km} \qquad\qquad (7\text{-}1)$$

23. 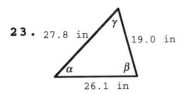 We are given three sides (SSS). We solve for the largest angle, β (largest because it is opposite the largest side, b) using the law of cosines. We then solve for a second angle using the law of sines.

Solve for β:
$$b^2 = a^2 + c^2 - 2ac \cos \beta$$
$$\cos \beta = \frac{a^2 + c^2 - b^2}{2ac}$$
$$\beta = \cos^{-1}\left(\frac{a^2 + c^2 - b^2}{2ac}\right)$$
$$= \cos^{-1}\left(\frac{(19.0)^2 + (26.1)^2 - (27.8)^2}{2(19.0)(26.1)}\right)$$
$$= 74.2°$$

Solve for α:
We use the law of sines. Since α must be smaller than β, α is acute.
$$\frac{\sin \alpha}{a} = \frac{\sin \beta}{b}$$
$$\sin \alpha = \frac{a \sin \beta}{b}$$
$$= \frac{19.0 \sin 74.2°}{27.8}$$
$$\alpha = \sin^{-1}\left(\frac{19.0 \sin 74.2°}{27.8}\right)$$
$$= 41.1°$$

Solve for γ:
$$\alpha + \beta + \gamma = 180°$$
$$\gamma = 180° - (\alpha + \beta)$$
$$= 180° - (41.1° + 74.2°)$$
$$= 64.7°$$

Note: these answers differ slightly from those given in the text. This results from having solved for the angles in a different order. *(7-1, 7-2)*

24. The sum of all of the force vectors must be the zero vector for the object to remain at rest. *(7-4)*

25.

$\theta = 57.2°$.
Hence, $\angle OCB = 180° - \theta = 180° - 57.2° = 122.8°$
We can find $|\mathbf{u} + \mathbf{v}|$ using the law of cosines:
$$|\mathbf{u} + \mathbf{v}|^2 = |\mathbf{u}|^2 + |\mathbf{v}|^2 - 2|\mathbf{u}||\mathbf{v}|\cos(OCB)$$
$$= (75.2)^2 + (34.2)^2 - 2(75.2)(34.2)\cos 122.8°$$
$$= 9611.0537...$$
$$|\mathbf{u} + \mathbf{v}| = \sqrt{9611.0537...}$$
$$= 98.0 \text{ kg}$$

To find α, we use the law of sines:
$$\frac{\sin \alpha}{|\mathbf{v}|} = \frac{\sin(OCB)}{|\mathbf{u} + \mathbf{v}|}$$
$$\frac{\sin \alpha}{34.2} = \frac{\sin 122.8°}{98.0}$$
$$\sin \alpha = \frac{34.2}{98.0} \sin 122.8°$$
$$\alpha = \sin^{-1}\left(\frac{34.2}{98.0} \sin 122.8°\right)$$
$$= 17.1°$$

(7-3)

26. $\mathbf{u} = \langle 11, -7 \rangle = \langle 11, 0 \rangle + \langle 0, -7 \rangle$
$$= 11\langle 1, 0 \rangle - 7\langle 0, 1 \rangle$$
$$= 11\mathbf{i} - 7\mathbf{j} \qquad (7\text{-}4)$$

27. $6\mathbf{u} - 3\mathbf{v} = 6\langle 13, 18 \rangle - 3\langle -9, 21 \rangle$
$$= \langle 78, 108 \rangle + \langle 27, -63 \rangle$$
$$= \langle 105, 45 \rangle \qquad (7\text{-}4)$$

28. $|\mathbf{v}| = \sqrt{8^2 + (-6)^2} = \sqrt{100} = 10$

Then $\mathbf{u} = -\dfrac{1}{|\mathbf{v}|}\mathbf{v} = -\dfrac{1}{10}\langle 8, -6 \rangle = \left\langle -\dfrac{8}{10}, \dfrac{6}{10} \right\rangle = \left\langle -\dfrac{4}{5}, \dfrac{3}{5} \right\rangle \qquad (7\text{-}4)$

29. $2\mathbf{j} + k\langle 1, -3 \rangle = 2\langle 0, 1 \rangle + k\langle 1, -3 \rangle$
$$= \langle 0, 2 \rangle + \langle k, -3k \rangle$$
$$= \langle k, 2 - 3k \rangle$$
$\langle k, 2 - 3k \rangle$ will be a unit vector if and only if its magnitude equals 1. Thus,
$$k^2 + (2 - 3k)^2 = 1$$
$$k^2 + 4 - 12k + 9k^2 = 1$$
$$10k^2 - 12k + 3 = 0$$

$k = \dfrac{-b \pm \sqrt{b^2 - 4ac}}{2a} \qquad \begin{array}{l} a = 10 \\ b = -12 \\ c = 3 \end{array}$

$k = \dfrac{-(-12) \pm \sqrt{(-12)^2 - 4(10)(3)}}{2(10)}$

$k = \dfrac{12 \pm \sqrt{24}}{20}$

$k = \dfrac{6 \pm \sqrt{6}}{10} \qquad (7\text{-}4)$

30. $k_1\langle 1, -2 \rangle + k_2\langle 2, 1 \rangle = \mathbf{i}$ if and only if
$$\langle k_1, -2k_1 \rangle + \langle 2k_2, k_2 \rangle = \langle 1, 0 \rangle$$
$$\langle k_1 + 2k_2, -2k_1 + k_2 \rangle = \langle 1, 0 \rangle$$
Thus, $k_1 + 2k_2 = 1$
$$-2k_1 + k_2 = 0$$
$$k_2 = 2k_1$$
$$k_1 + 2(2k_1) = 1$$
$$5k_1 = 1$$

$$k_1 = \dfrac{1}{5}, \quad k_2 = 2k_1 = 2\left(\dfrac{1}{5}\right) = \dfrac{2}{5} \qquad (7\text{-}4)$$

31. Let $\mathbf{u} = 5\mathbf{i} - 12\mathbf{j}$. Then $|\mathbf{u}| = \sqrt{5^2 + (-12)^2} = \sqrt{169} = 13$. Thus, a *unit* vector having the same direction as $\mathbf{u}$ will be $\mathbf{v} = \dfrac{1}{|\mathbf{u}|}\mathbf{u} = \dfrac{1}{13}(5\mathbf{i} - 12\mathbf{j})$. Therefore, a vector with this direction and magnitude 3 will be

$$3\mathbf{v} = 3 \cdot \dfrac{1}{13}(5\mathbf{i} - 12\mathbf{j})$$
$$= \dfrac{3}{13}(5\mathbf{i} - 12\mathbf{j})$$
$$= \dfrac{15}{13}\mathbf{i} - \dfrac{36}{13}\mathbf{j} \quad \text{or} \quad \left\langle \dfrac{15}{13}, -\dfrac{36}{13} \right\rangle \qquad (7\text{-}4)$$

32. We set up a table that indicates how r varies as we let θ vary through each set of quadrant values.

θ	$\cos\theta$	$4\cos\theta$	$6 + 4\cos\theta$
0 to $\frac{\pi}{2}$	1 to 0	4 to 0	10 to 6
$\frac{\pi}{2}$ to π	0 to -1	0 to -4	6 to 2
π to $\frac{3\pi}{2}$	-1 to 0	-4 to 0	2 to 6
$\frac{3\pi}{2}$ to 2π	0 to 1	0 to 4	6 to 10

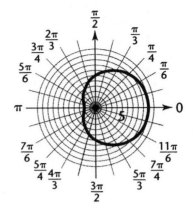

$(7\text{-}5)$

33. We set up a table that indicates how r varies as we let θ vary through each set of quadrant values.

θ	$\sin\theta$	$8\sin\theta$	$8 + 8\sin\theta$
0 to $\frac{\pi}{2}$	0 to 1	0 to 8	8 to 16
$\frac{\pi}{2}$ to π	1 to 0	8 to 0	16 to 8
π to $\frac{3\pi}{2}$	0 to -1	0 to -8	8 to 0
$\frac{3\pi}{2}$ to 2π	-1 to 0	-8 to 0	0 to 8

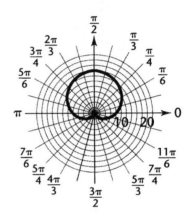

$(7\text{-}5)$

34. We set up a table that indicates how r varies as we let 2θ vary through each set of quadrant values.

θ	2θ	$\cos 2\theta$	$10\cos 2\theta$
0 to $\frac{\pi}{4}$	0 to $\frac{\pi}{2}$	1 to 0	10 to 0
$\frac{\pi}{4}$ to $\frac{\pi}{2}$	$\frac{\pi}{2}$ to π	0 to -1	0 to -10
$\frac{\pi}{2}$ to $\frac{3\pi}{4}$	π to $\frac{3\pi}{2}$	-1 to 0	-10 to 0
$\frac{3\pi}{4}$ to π	$\frac{3\pi}{2}$ to 2π	0 to 1	0 to 10
π to $\frac{5\pi}{4}$	2π to $\frac{5\pi}{2}$	1 to 0	10 to 0
$\frac{5\pi}{4}$ to $\frac{3\pi}{2}$	$\frac{5\pi}{2}$ to 3π	0 to -1	0 to -10
$\frac{3\pi}{2}$ to $\frac{7\pi}{4}$	3π to $\frac{7\pi}{2}$	-1 to 0	-10 to 0
$\frac{7\pi}{4}$ to 2π	$\frac{7\pi}{2}$ to 4π	0 to 1	0 to 10

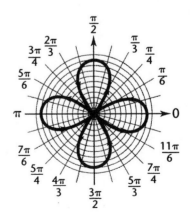

$(7\text{-}5)$

35. We set up a table that indicates how r varies as we let 3θ vary through each set of quadrant values.

θ	3θ	$\sin 3\theta$	$8 \sin 3\theta$
0 to $\frac{\pi}{6}$	0 to $\frac{\pi}{2}$	0 to 1	0 to 8
$\frac{\pi}{6}$ to $\frac{\pi}{3}$	$\frac{\pi}{2}$ to π	1 to 0	8 to 0
$\frac{\pi}{3}$ to $\frac{\pi}{2}$	π to $\frac{3\pi}{2}$	0 to -1	0 to -8
$\frac{\pi}{2}$ to $\frac{2\pi}{3}$	$\frac{3\pi}{2}$ to 2π	-1 to 0	-8 to 0
$\frac{2\pi}{3}$ to $\frac{5\pi}{6}$	2π to $\frac{5\pi}{2}$	0 to 1	0 to 8
$\frac{5\pi}{6}$ to π	$\frac{5\pi}{2}$ to 3π	1 to 0	8 to 0
π to 2π	repeats curve already drawn		

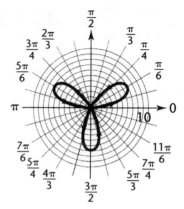

(7-5)

36.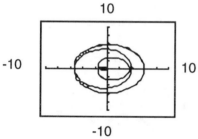

(7-5)

37.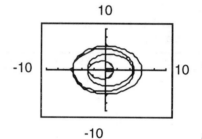

(7-5)

38. $n = 1$ $n = 2$ $n = 3$

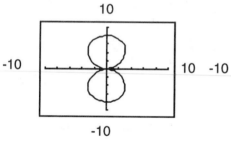   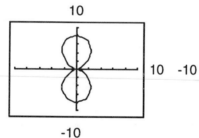   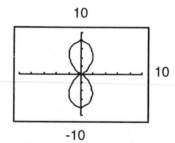

Generalizing from these graphs, we expect 2 leaves for all n.   (7-5)

39. (A) $e = 0.55$
$$r = \frac{3}{1 - 0.55 \cos \theta}$$

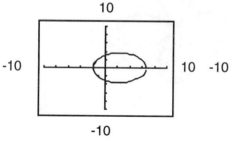

An ellipse

(B) $e = 1$
$$r = \frac{3}{1 - \cos \theta}$$

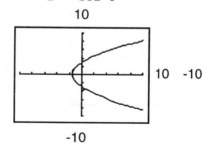

A parabola

(C) $e = 1.7$
$$r = \frac{3}{1 - 1.7 \cos \theta}$$

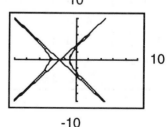

A hyperbola   (7-5)

40. $x^2 + y^2 = 6x$
Use $r^2 = x^2 + y^2$ and $x = r \cos \theta$.
$r^2 = 6r \cos \theta$
$r^2 - 6r \cos \theta = 0$
$r(r - 6 \cos \theta) = 0$
$r = 0$ or $r - 6 \cos \theta = 0$

The graph of $r = 0$ is the pole. Since the pole is included in the graph of
$r - 6 \cos \theta = 0$, we can discard $r = 0$ and keep only
$r - 6 \cos \theta = 0$ or $r = 6 \cos \theta$ $\qquad$ (7-5)

41. $r = 5 \cos \theta$
We multiply both sides by r, which adds the pole to the graph. But the pole is
already part of the graph, so we have changed nothing.
$r^2 = 5r \cos \theta$
But $r^2 = x^2 + y^2$ and $r \cos \theta = x$. Hence, $x^2 + y^2 = 5x$. $\qquad$ (7-5)

42. A sketch shows that $-1 + i$ is associated
with a special 45° triangle. Thus, by
inspection, $r = \sqrt{2}$, $\theta = 135°$, and
$-1 + i = \sqrt{2}(\cos 135° + i \sin 135°)$
$\qquad = \sqrt{2}\,e^{135°i}$

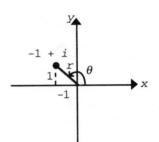

A sketch shows that $-1 - i\sqrt{3}$ is
associated with a special 30°-60°
triangle. Thus, by inspection, $r = 2$,
$\theta = -120°$, and
$-1 - i\sqrt{3} = 2[\cos(-120°) + i \sin(-120°)]$
$\qquad = 2e^{(-120°)i}$

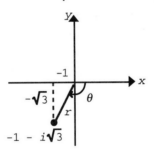

A sketch shows that 5 is associated
with the special quadrantal angle 0°.
Thus, by inspection, $r = 5$, $\theta = 0°$, and
$5 = 5 + 0i = 5(\cos 0° + i \sin 0°)$
$\qquad = 5e^{0°i}$

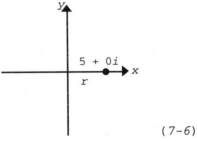

(7-6)

43. $z_1 = x + iy = \sqrt{2}\,e^{(\pi/4)i}$
$\qquad = \sqrt{2}\left(\cos \dfrac{\pi}{4} + i \sin \dfrac{\pi}{4}\right)$
$\qquad = \sqrt{2}\left(\dfrac{1}{\sqrt{2}} + i\,\dfrac{1}{\sqrt{2}}\right)$
$\qquad = 1 + i$

$z_2 = x + iy = 3e^{210°i}$
$\qquad = 3(\cos 210° + i \sin 210°)$
$\qquad = 3\left[-\dfrac{\sqrt{3}}{2} + i\left(-\dfrac{1}{2}\right)\right]$
$\qquad = \dfrac{-3\sqrt{3}}{2} - \dfrac{3}{2}\,i$

$z_3 = x + iy = 2e^{(-2\pi/3)i} = 2\left[\cos\left(-\dfrac{2\pi}{3}\right) + i \sin\left(-\dfrac{2\pi}{3}\right)\right]$
$\qquad\qquad\qquad = 2\left[-\dfrac{1}{2} + i\left(-\dfrac{\sqrt{3}}{2}\right)\right]$
$\qquad\qquad\qquad = -1 - i\sqrt{3}$ $\qquad$ (7-6)

44. (A) $z_1 z_2 = 8e^{25°i} \cdot 4e^{19°i}$

$= 8 \cdot 4e^{i(25° + 19°)} = 32e^{44°i}$

(B) $\dfrac{z_1}{z_2} = \dfrac{8e^{25°i}}{4e^{19°i}}$

$= \dfrac{8}{4} e^{(25° - 19°)i} = 2e^{6°i}$ $\hspace{2cm}$ (7-6)

45. (A) $(1 + i\sqrt{3})^4 = (2e^{60°i})^4 = 2^4 e^{4 \cdot 60°i} = 16e^{240°i}$

$= 16[\cos 240° + i \sin 240°] = 16\left[-\dfrac{1}{2} + i\left(-\dfrac{\sqrt{3}}{2} \right) \right] = -8 - 8i\sqrt{3}$

(B) The calculator steps are not shown here. $-8 - 8i\sqrt{3}$ is approximated on a calculator as $-8 - 13.86i$. $\hspace{3cm}$ (7-7)

46. First write i in polar form.

$i = 0 + 1i = 1e^{90°i}$

Using the nth-root theorem, all three cube roots of $1e^{90°i}$ are given by

$1^{1/3}e^{(90°/3 + k360°/3)i} = 1^{1/3}e^{(30° + k120°)i}$ $\hspace{1cm}$ $k = 0, 1, 2$

Thus,

$w_1 = 1^{1/3}e^{(30° + 0 \cdot 120°)i} = e^{30°i}$

$w_2 = 1^{1/3}e^{(30° + 1 \cdot 120°)i} = e^{150°i}$

$w_3 = 1^{1/3}e^{(30° + 2 \cdot 120°)i} = e^{270°i}$

These roots lie on a circle of radius 1, equally spaced so that the angle between successive roots is 120°.

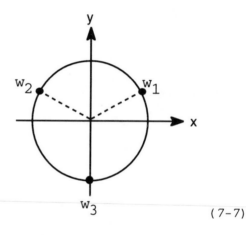

(7-7)

47. First write $-4\sqrt{3} + 4i$ in polar form.

$-4\sqrt{3} + 4i = 8e^{150°i}$

Using the nth-root theorem, all three cube roots of $8e^{150°i}$ are given by

$8^{1/3}e^{(150°/3 + k360°/3)i} = 8^{1/3}e^{(50° + k \cdot 120°)i}$ $\hspace{1cm}$ $k = 0, 1, 2$

Thus,

$w_1 = 8^{1/3}e^{(50° + 0 \cdot 120°)i} = 2e^{50°i}$

$w_2 = 8^{1/3}e^{(50° + 1 \cdot 120°)i} = 2e^{170°i}$

$w_3 = 8^{1/3}e^{(50° + 2 \cdot 120°)i} = 2e^{290°i}$ $\hspace{2cm}$ (7-7)

48. $(4e^{15°i})^2 = 4^2 e^{2 \cdot 15°i} = 16e^{30°i} = 16(\cos 30° + i \sin 30°) = 16\left(\dfrac{\sqrt{3}}{2} + i \cdot \dfrac{1}{2} \right)$

$= 8\sqrt{3} + 8i$

Thus, $4e^{15°i}$ is a square root of $8\sqrt{3} + 8i$. $\hspace{2cm}$ (7-7)

49. $(x, y) = (5.17, -2.53)$

$r = \sqrt{x^2 + y^2} = \sqrt{5.17^2 + (-2.53)^2} = 5.76$

$\tan \theta = \dfrac{y}{x} = \dfrac{-2.53}{5.17}$

θ is a fourth quadrant angle and is to be chosen so that $-180° < \theta \leq 180°$.

$\theta = \tan^{-1} \dfrac{-2.53}{5.17} = -26.08°$

$(r, \theta) = (5.76, -26.08°)$ $\hspace{3cm}$ (7-5)

50. $(r, \theta) = (5.81, -2.72)$

$\qquad x = r \cos \theta = 5.81 \cos(-2.72) = -5.30$

$\qquad y = r \sin \theta = 5.81 \sin(-2.72) = -2.38$

$(x, y) = (-5.30, -2.38)$ (7-5)

51. $r = \sqrt{(-3.18)^2 + (4.19)^2} = 5.26$

$\qquad \theta = 180° + \tan^{-1}\dfrac{4.19}{-3.18} = 127.20°$ since $(-3.18, 4.19)$ is in the second quadrant.

$\qquad -3.18 + 4.19i = 5.26(\cos 127.20° + i \sin 127.20°) = 5.26e^{127.20°i}$ (7-6)

52. $x + iy = 7.63e^{(-162.27°)i}$

$\qquad\qquad = 7.63[\cos(-162.27°) + i \sin(-162.27°)]$

$\qquad\qquad = -7.27 - 2.32i$ (7-6)

53. (A) There are a total of three cube roots and they are spaced equally around a circle of radius 2, so that the angle between successive roots is 120°.

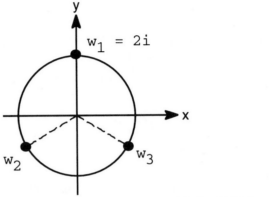

(B) Since $w_1 = 2e^{90°i}$, $w_2 = 2e^{(90° + 120°)i} = 2e^{210°i}$, and $w_3 = 2e^{(210° + 120°)i} = 2e^{330°i}$

$\qquad$ Thus, $w_2 = 2(\cos 210° + i \sin 210°) = 2\left[-\dfrac{\sqrt{3}}{2} + i\left(-\dfrac{1}{2}\right)\right] = -\sqrt{3} - i$

$\qquad\qquad w_3 = 2(\cos 330° + i \sin 330°) = 2\left[\dfrac{\sqrt{3}}{2} + i\left(-\dfrac{1}{2}\right)\right] = \sqrt{3} - i$

(C) $(2i)^3 = 8i^3 = 8(-i) = -8i$

$\qquad (-\sqrt{3} - i)^3 = (-\sqrt{3} - i)(-\sqrt{3} - i)(-\sqrt{3} - i)$

$\qquad\qquad\qquad = (3 + 2i\sqrt{3} + i^2)(-\sqrt{3} - i)$

$\qquad\qquad\qquad = (2 + 2i\sqrt{3})(-\sqrt{3} - i)$

$\qquad\qquad\qquad = -2\sqrt{3} - 2i - 6i + 2\sqrt{3}$

$\qquad\qquad\qquad = -8i$

$\qquad (\sqrt{3} - i)^3 = (\sqrt{3} - i)(\sqrt{3} - i)(\sqrt{3} - i)$

$\qquad\qquad\qquad = (3 - 2i\sqrt{3} + i^2)(\sqrt{3} - i)$

$\qquad\qquad\qquad = (2 - 2i\sqrt{3})(\sqrt{3} - i)$

$\qquad\qquad\qquad = 2\sqrt{3} - 2i - 6i + 2\sqrt{3}i^2$

$\qquad\qquad\qquad = -8i$ (7-7)

54.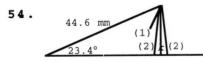

From the diagram, we can see that k = altitude of any possible triangle. Thus,

$$\sin \alpha = \frac{k}{b} \qquad k = b \sin \alpha$$

$$k = (44.6 \text{ mm}) \sin 23.4° = 17.7 \text{ mm}$$

If $0 < a < k$, there is no solution: (1) in above diagram.

If $a = k$, there is one solution.

If $k < a < b$, there are two solutions: (2) in above diagram. (7-1)

55. See Exercise 7-2, Problem 29, for a solution. (7-2)

56. (A) $\mathbf{u} + \mathbf{v} = \langle a, b \rangle + \langle c, d \rangle$

$= \langle a + c, b + d \rangle$ Definition of vector addition

$= \langle c + a, d + b \rangle$ Commutative property for addition of real numbers

$= \langle c, d \rangle + \langle a, b \rangle$ Definition of vector addition

$= \mathbf{v} + \mathbf{u}$

(B) $m(\mathbf{u} + \mathbf{v}) = m(\langle a, b \rangle + \langle c, d \rangle)$

$= m\langle a + c, b + d \rangle$ Definition of vector addition

$= \langle m(a + c), m(b + d) \rangle$ Definition of scalar multiplication

$= \langle ma + mc, mb + md \rangle$ Distributive property for real numbers

$= \langle ma, mb \rangle + \langle mc, md \rangle$ Definition of vector addition

$= m\langle a, b \rangle + m\langle c, d \rangle$ Definition of scalar multiplication

$= m\mathbf{u} + m\mathbf{v}$

                                                      (7-4)

57. False. For example, $\frac{1}{2}$ is a cube root of $\frac{1}{8}$, yet $\left|\frac{1}{2}\right| = \frac{1}{2} > \frac{1}{8} = \left|\frac{1}{8}\right|$.   (7-7)

58. Flase. For example, $e^{36°i}$ is a tenth root of 1, but $(e^{36°i})^5 = e^{180°i} = -1$, hence $e^{36°i}$ is not a fifth root of 1.

                                                       (7-7)

59. True. If w is a cube root of z, then $w^3 = z$. If w is also a fourth root of z, then $w^4 = z$. Then

$$w^4 = w^3$$
$$w^4 - w^3 = 0$$
$$w^3(w - 1) = 0$$
$$w^3 = 0 \quad \text{or} \quad w - 1 = 0$$
$$w = 0 \quad \text{or} \quad w = 1$$

                                                       (7-7)

60. False. For example, i is a non-real square root of the real number -1.   (7-7)

61. We set up a table that indicates how r varies as we let $\frac{\theta}{2}$ vary through each set of quadrant values.

θ	$\frac{\theta}{2}$	$\cos \frac{\theta}{2}$	$4 \cos \frac{\theta}{2}$	$4 + 4 \cos \frac{\theta}{2}$
0 to $\frac{\pi}{2}$	0 to $\frac{\pi}{4}$	1 to $\frac{\sqrt{2}}{2}$	4 to $2\sqrt{2}$	8 to $4 + 2\sqrt{2}$
$\frac{\pi}{2}$ to π	$\frac{\pi}{4}$ to $\frac{\pi}{2}$	$\frac{\sqrt{2}}{2}$ to 0	$2\sqrt{2}$ to 0	$4 + 2\sqrt{2}$ to 4
π to $\frac{3\pi}{2}$	$\frac{\pi}{2}$ to $\frac{3\pi}{4}$	0 to $-\frac{\sqrt{2}}{2}$	0 to $-2\sqrt{2}$	4 to $4 - 2\sqrt{2}$
$\frac{3\pi}{2}$ to 2π	$\frac{3\pi}{4}$ to π	$-\frac{\sqrt{2}}{2}$ to -1	$-2\sqrt{2}$ to -4	$4 - 2\sqrt{2}$ to 0
2π to $\frac{5\pi}{2}$	π to $\frac{5\pi}{4}$	-1 to $-\frac{\sqrt{2}}{2}$	-4 to $-2\sqrt{2}$	0 to $4 - 2\sqrt{2}$
$\frac{5\pi}{2}$ to 3π	$\frac{5\pi}{4}$ to $\frac{3\pi}{2}$	$-\frac{\sqrt{2}}{2}$ to 0	$-2\sqrt{2}$ to 0	$4 - 2\sqrt{2}$ to 4
3π to $\frac{7\pi}{2}$	$\frac{3\pi}{2}$ to $\frac{7\pi}{4}$	0 to $\frac{\sqrt{2}}{2}$	0 to $2\sqrt{2}$	4 to $4 + 2\sqrt{2}$
$\frac{7\pi}{2}$ to 4π	$\frac{7\pi}{4}$ to 2π	$\frac{\sqrt{2}}{2}$ to 1	$2\sqrt{2}$ to 4	$4 + 2\sqrt{2}$ to 8

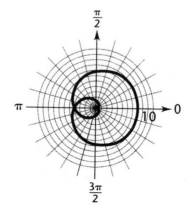

(B)

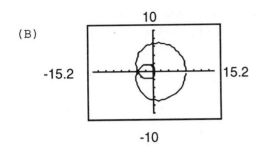

(7-5)

62. (A)

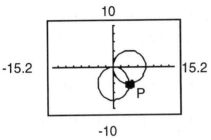

The coordinates of P represent a simultaneous solution.

(B) We solve the system
$$r = -8 \sin \theta \qquad 0 \leq \theta \leq \pi$$
$$r = 8 \cos \theta$$
by equating the right sides:
$$-8 \sin \theta = 8 \cos \theta$$
$$-\sin \theta = \cos \theta$$
$$-1 = \tan \theta$$
The only solution of this equation, $0 \leq \theta \leq \pi$, is $\theta = \dfrac{3\pi}{4}$. If we substitute this in either of the original equations, we get
$$r = -8 \sin \frac{3\pi}{4} = 8 \cos \frac{3\pi}{4} = 8\left(-\frac{\sqrt{2}}{2}\right) = -4\sqrt{2}$$

The coordinates of P are $\left(-4\sqrt{2}, \dfrac{3\pi}{4}\right)$.

(C) The pole is not a simultaneous solution, because the two graphs go through the pole at different values of θ. As $(0, 0)$ the pole satisfies the first equation; as $\left(0, \dfrac{\pi}{2}\right)$ it satisfies the second; it is not a solution of the system.

(7-5)

63. The solution to $x^8 - 1 = 0$ or $x^8 = 1$ will be the eight eighth roots of 1. First write 1 in polar form:
$$1 = 1e^{0°i}$$

Using the nth-root theorem, all eight eighth roots of 1 are given by
$$1^{1/8}e^{(0°/8 + k \cdot 360°/8)i} = e^{k \cdot 45°i} \qquad k = 0, 1, 2, 3, 4, 5, 6, 7$$

Thus,
$$w_1 = e^{0 \cdot 45°i} = e^{0°i} = \cos 0° + i \sin 0° = 1$$
$$w_2 = e^{1 \cdot 45°i} = e^{45°i} = \cos 45° + i \sin 45° = \frac{\sqrt{2}}{2} + i\frac{\sqrt{2}}{2}$$
$$w_3 = e^{2 \cdot 45°i} = e^{90°i} = \cos 90° + i \sin 90° = i$$
$$w_4 = e^{3 \cdot 45°i} = e^{135°i} = \cos 135° + i \sin 135° = -\frac{\sqrt{2}}{2} + i\frac{\sqrt{2}}{2}$$
$$w_5 = e^{4 \cdot 45°i} = e^{180°i} = \cos 180° + i \sin 180° = -1$$
$$w_6 = e^{5 \cdot 45°i} = e^{225°i} = \cos 225° + i \sin 225° = -\frac{\sqrt{2}}{2} - i\frac{\sqrt{2}}{2}$$
$$w_7 = e^{6 \cdot 45°i} = e^{270°i} = \cos 270° + i \sin 270° = -i$$
$$w_8 = e^{7 \cdot 45°i} = e^{315°i} = \cos 315° + i \sin 315° = \frac{\sqrt{2}}{2} - i\frac{\sqrt{2}}{2}$$

(7-7)

64. The linear factors of $x^3 - 8i$ will, by the factor theorem, be $x - x_i$, where x_i are zeros of $x^3 - 8i$, that is, $x^3 - 8i = 0$ or $x^3 = 8i$. Hence the x_i are the three cube roots of $8i$, or $8e^{90°i}$.

Using the nth-root theorem, all three cube roots of $8e^{90°i}$ are given by
$$8^{1/3}e^{(90°/3 + k \cdot 360°/3)i} = 2e^{(30° + k \cdot 120°)i} \qquad k = 0, 1, 2$$
Thus,
$$w_1 = 2e^{(30° + 0 \cdot 120°)i} = 2e^{30°i} = 2(\cos 30° + i \sin 30°) = 2\left(\frac{\sqrt{3}}{2} + i\frac{1}{2}\right) = \sqrt{3} + i$$

$$w_2 = 2e^{(30° + 1 \cdot 120°)i} = 2e^{150°i} = 2(\cos 150° + i \sin 150°) = 2\left(-\frac{\sqrt{3}}{2} + i\frac{1}{2}\right) = -\sqrt{3} + i$$

$$w_3 = 2e^{(30° + 2 \cdot 120°)i} = 2e^{270°i} = 2(\cos 270° + i \sin 270°) = 2(0 - 1i) = -2i$$

Hence, $x^3 - 8i = [x - (\sqrt{3} + i)][x - (-\sqrt{3} + i)](x + 2i)$ $\hspace{2em}$ (7-7)

65. Sketch a figure.

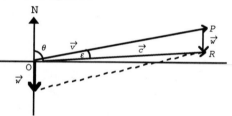

Using $D = rt$, we have

Plane 1: $r = 256$ miles per hour, $t = 2$ hours, $D = 256(2) = 512$ miles $= SA$

Plane 2: $r = 304$ miles per hour, $t = 2$ hours, $D = 304(2) = 608$ miles $= SB$

The angle α between east and south-east is 45°

Given two sides and the included angle, we find d from the law of cosines:
$$d^2 = (SA)^2 + (SB)^2 - 2(SA)(SB)\cos \alpha$$
$$= (512)^2 + (608)^2 - 2(512)(608)\cos 45°.$$
$$= 191,568.97…$$
$$d = 438 \text{ miles}$$
$\hspace{40em}$ (7-3)

66. Sketch a figure.

The actual course $\vec{c}$ will be the resultant of $\vec{v}$, the plane's heading, and $\vec{w}$, the wind velocity. In triangle OPR, we know $OP = |\vec{v}| = 450$, $PR = |\vec{w}| = 65$, and angle OPR, which is congruent to θ, the heading of 75°.

Given two sides and the included angle, we use the law of cosines to find $|\vec{c}| = OR$
$$|\vec{c}|^2 = OP^2 + PR^2 - 2(OP)(PR)\cos OPR$$
$$= (450)^2 + (65)^2 - 2(450)(65)\cos 75°$$
$$= 191,584.09…$$
$$|\vec{c}| = 438 \text{ miles per hour.}$$

To find the actual direction of the plane, we must find angle POR, or ε. Then the heading will be $\theta + \varepsilon$. We use the law of sines.

$$\frac{\sin \varepsilon}{PR} = \frac{\sin OPR}{OR}$$
$$\frac{\sin \varepsilon}{65} = \frac{\sin 75°}{438}$$
$$\sin \varepsilon = \frac{65 \sin 75°}{438}$$
$$\sin \varepsilon = 0.1434$$
$$\varepsilon = \sin^{-1} 0.1434 \text{ since } \varepsilon \text{ is acute}$$
$$\varepsilon = 8°$$

Thus the heading of the plane's actual direction $= \theta + \varepsilon = 75° + 8° = 83°$. (7-3)

67.

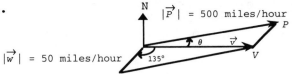

We wish to find $|\vec{v}|$ and θ such that $\vec{v}$, the resultant of $\vec{P}$, the plane's velocity, and $\vec{w}$, the wind's velocity, is in the direction due east. Applying the law of sines to triangle OPV in which we have given two sides, OP and PV, and the angle OVP opposite one of them,

we obtain $\dfrac{\sin \theta}{PV} = \dfrac{\sin \; OVP}{OP}$

$$\sin \theta = \frac{PV \sin \; OVP}{OP}$$

$$= \frac{50 \sin 135°}{500}$$

$$= 0.0707$$

Since θ must be acute, there is only one possibility:
$\theta = \sin^{-1} 0.0707$
  $= 4°$

The heading of the plane will therefore be 90° - 4°, or 86°. We apply the law of sines again to find $|\vec{v}|$, noting that

$$\angle OPV = 180° - (\angle OVP + \theta)$$
$$= 180° - (135° + 4°)$$
$$= 41°$$

$$\frac{|\vec{v}|}{\sin \; OPV} = \frac{OP}{\sin \; OVP}$$

$$|\vec{v}| = \frac{OP \sin \; OPV}{\sin \; OVP}$$

$$|\vec{v}| = \frac{500 \sin 41°}{\sin 135°}$$

$$|\vec{v}| = 464 \text{ miles per hour} \hspace{3cm} (7-3)$$

68.

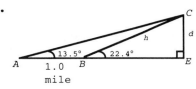

We note: In triangle ABC, angle ABC is the supplement of given angle CBE. Angle ABC has measure 180° - 22.4° = 157.6°. In triangle ABC we know two angles and the included side, hence we can find angle ACB and side BC. Side BC is then the hypotenuse of right triangle BCE, so that we can find d.

Solve for angle ACB: 13.5° + 157.6° + $\angle ACB$ = 180°

$$\angle ACB = 180° - 13.5° - 157.6°$$
$$= 8.9°$$

Solve for CB = h: $\dfrac{\sin A}{h} = \dfrac{\sin \; ACB}{AB}$ 　　　Solve for d: $\sin \; CBE = \dfrac{d}{h}$

$$h = \frac{AB \sin A}{\sin \; ACB}$$ 　　　　　　$$d = h \sin \; CBE$$

$$= \frac{1.0 \sin 13.5°}{\sin 8.9°}$$ 　　　　$$d = 1.51 \sin 22.4°$$

$$= 1.51 \text{ miles}$$ 　　　　　$$d = 0.6 \text{ miles} \quad (7-3)$$

69.

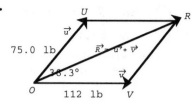

In triangle OVR, angle OVR has measure $180° - 38.3° = 141.7°$. We have $OV = 112$, $VR = 75.0$, hence we can use the law of cosines to find $OR = |\vec{R}|$

$$|\vec{R}|^2 = (OV)^2 + (VR)^2 - 2(OV)(VR)\cos \alpha$$
$$= (112)^2 + (75.0)^2 - 2(112)(75.0)\cos 141.7°$$
$$= 31,353.243...$$
$$|\vec{R}| = 177 \text{ pounds}$$

To find the direction of $\vec{R}$ relative to v, we find angle VOR from the law of sines.

$$\frac{\sin \alpha}{|\vec{R}|} = \frac{\sin VOR}{VR}$$
$$\sin VOR = \frac{VR \sin \alpha}{|\vec{R}|}$$
$$\sin VOR = \frac{75.0 \sin 141.7°}{177}$$
$$\sin VOR = 0.2626$$
$$\angle VOR = \sin^{-1} 0.2626 \text{ since this angle must be acute}$$
$$\angle VOR = 15.2°$$

(7-3)

70.

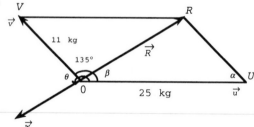

From the figure it should be clear that:

The two forces **u** and **v** act together as their resultant **R**. Needed to balance them, then, is a force **w** in precisely the opposite direction to **R** with the same magnitude as **R**. We calculate $|\mathbf{R}|$, the magnitude of **R**, then its direction β relative to **u** from triangle OUR. Then the direction of **w** relative to **u** is $\theta = 180° + \beta$.

Solve for $|\mathbf{R}|$: We use the law of cosines, with $\alpha = 180° - 135° = 45°$.
$$|\mathbf{R}|^2 = (OU)^2 + (UR)^2 - 2(OU)(UR)\cos \alpha$$
$$= 25^2 + 11^2 - 2(25)(11)\cos 45°$$
$$= 357.09127...$$
$$|\mathbf{R}| = 19 \text{ kg}$$

Solve for β: We use the law of sines.
$$\frac{\sin \beta}{UR} = \frac{\sin \alpha}{|\mathbf{R}|}$$
$$\sin \beta = \frac{UR \sin \alpha}{|\mathbf{R}|}$$
$$= \frac{11 \sin 45°}{19}$$
$$= 0.4094$$
$$\beta = \sin^{-1} 0.4094 \text{ since } \beta \text{ must be acute}$$
$$\beta = 24°$$

Thus the direction of **w** is $\theta = 180° + 24° = 204°$ relative to **u**.

(7-4)

71. (A)

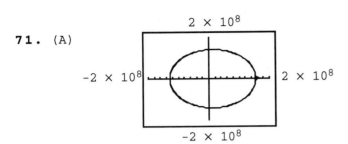

From the trace function of the utility as applied to the graph, the maximum distance (aphelion) is 1.56×10^8 miles, and the minimum distance (perihelion) is 1.29×10^8 miles.

(B) r is maximum when the denominator $1 - 0.0934 \cos \theta$ is minimum. This occurs when $\cos \theta$ is as large as possible, that is, $\cos \theta = 1$.

Then $r = \dfrac{1.41 \times 10^8}{1 - 0.0934} = 1.56 \times 10^8$ miles (aphelion)

r is minimum when the denominator $1 - 0.0934 \cos \theta$ is maximum. This occurs when $\cos \theta = -1$.

Then $r = \dfrac{1.41 \times 10^8}{1 + 0.0934} = 1.29 \times 10^8$ miles (perihelion) $\hfill (7-5)$

72. First, form a force diagram with all forces in standard position at the origin.

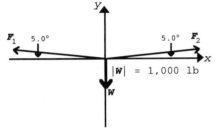

Let $\boldsymbol{F}_1$ = the tension in left cable
$\boldsymbol{F}_2$ = the tension in right cable

Write each force vector in terms of $\boldsymbol{i}$ and $\boldsymbol{j}$ unit vectors.

$\boldsymbol{F}_1 = |\boldsymbol{F}_1|(-\cos 5.0°)\boldsymbol{i} + |\boldsymbol{F}_1|(\sin 5.0°)\boldsymbol{j}$
$\boldsymbol{F}_2 = |\boldsymbol{F}_2|(\cos 5.0°)\boldsymbol{i} + |\boldsymbol{F}_2|(\sin 5.0°)\boldsymbol{j}$
$\boldsymbol{W} = -1,000\boldsymbol{j}$

For the system to be in static equilibrium, we must have
$\boldsymbol{F}_1 + \boldsymbol{F}_2 + \boldsymbol{W} = \boldsymbol{O}$
which becomes, on addition,
$[-|\boldsymbol{F}_1|(\cos 5.0°) + |\boldsymbol{F}_2|(\cos 5.0°)]\boldsymbol{i} + [|\boldsymbol{F}_1|(\sin 5.0°)$
$\qquad\qquad\qquad + |\boldsymbol{F}_2|(\sin 5.0°) - 1,000]\boldsymbol{j} = 0\boldsymbol{i} + 0\boldsymbol{j}$

Since two vectors are equal if and only if their corresponding components are equal, we are led to the following system of equations in $|\boldsymbol{F}_1|$ and $|\boldsymbol{F}_2|$:

$-|\boldsymbol{F}_1|\cos 5.0° + |\boldsymbol{F}_2|\cos 5.0° = 0$
$|\boldsymbol{F}_1|\sin 5.0° + |\boldsymbol{F}_2|\sin 5.0° - 1,000 = 0$

Solving, we obtain
$|\boldsymbol{F}_1| = |\boldsymbol{F}_2|$ (as we expect from the symmetry of the situation)
$|\boldsymbol{F}_1| = \dfrac{1,000}{2 \sin 5.0°} = 5,740$ lb = tension in each half of the cable. $\hfill (7-4)$

CUMULATIVE REVIEW EXERCISE (Chapters 5, 6 and 7)

1. $s = r\theta = $ (6 meters)(0.31 radians) $= 1.86$ meters *(5-3)*

2. Solve for θ:
$$\theta = 90° - 32.7° = 57.3°$$

Solve for a:

$$\tan \theta = \frac{b}{a}$$

$$\tan 57.3° = \frac{12.2}{a}$$

$$a = \frac{12.2}{\tan 57.3°}$$

$$= 7.83 \text{ cm}$$

Solve for c:

$$\sin \theta = \frac{b}{c}$$

$$\sin 57.3° = \frac{12.2}{c}$$

$$c = \frac{12.2}{\sin 57.3°}$$

$$c = 14.5 \text{ cm} \quad (5-5)$$

3. (A) $\dfrac{b}{r} = \sin \theta > 0$ if $b > 0$. This occurs in quadrants I, II.

(B) $\dfrac{a}{r} = \cos \theta > 0$ if $a > 0$. This occurs in quadrants I, IV.

(C) $\dfrac{b}{a} = \tan \theta > 0$ if a and b have the same sign. This occurs in quadrants I, III. *(5-4)*

4.
$$a^2 + b^2 = r^2$$
$$(-3)^2 + 4^2 = r^2$$
$$25 = r^2$$
$$r = 5$$

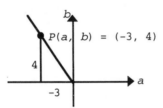
$P(a, b) = (-3, 4)$

(A) $\cos \theta = \dfrac{a}{r} = \dfrac{-3}{5} = -\dfrac{3}{5}$

(B) $\csc \theta = \dfrac{r}{b} = \dfrac{5}{4}$ (C) $\tan \theta = \dfrac{b}{a} = \dfrac{4}{-3} = -\dfrac{4}{3}$ *(5-4)*

5. (A)

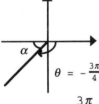

$\theta = -\dfrac{3\pi}{4}$

$$\alpha = \pi + -\dfrac{3\pi}{4} = \dfrac{\pi}{4}$$

(B)

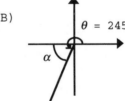

$\theta = 245°$

$$\alpha = 245° - 180° = 65°$$

(C)

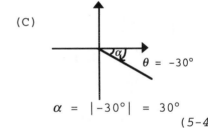

$\theta = -30°$

$$\alpha = |-30°| = 30° \quad (5-4)$$

6. (A) Domain: all real numbers; Range: $-1 \le y \le 1$; Period: 2π

(B) Domain: all real numbers; Range: $-1 \le y \le 1$; Period: 2π

(C) Domain: all real numbers except $x = \dfrac{\pi}{2} + k\pi$, k an integer; Range: all real numbers; Period: π *(5-6)*

7.

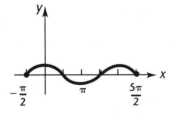

(5-6)

8.

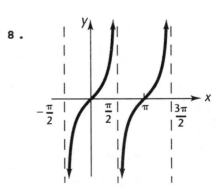

(5-6)

9. See figure.
A central angle in a circle with radian measure 2 is the angle subtended by an arc of length twice the measure of the radius.

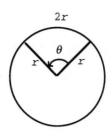

(5-3)

10. See figure for problem 7 above. If the graph of $y = \cos x$ is shifted $\frac{\pi}{2}$ units to the right, the result will be the graph of $y = \sin x$. *(5-7)*

11. $\cot \theta \sec \theta = \dfrac{\cos \theta}{\sin \theta} \dfrac{1}{\cos \theta}$ Quotient and Reciprocal Identities

$\qquad\qquad = \dfrac{1}{\sin \theta}$ Algebra

$\qquad\qquad = \csc \theta$ Reciprocal Identity   *(6-1)*

12. $\sec x - \cos x = \dfrac{1}{\cos x} - \cos x$ Reciprocal Identity

$\qquad\qquad = \dfrac{1}{\cos x} - \dfrac{\cos^2 x}{\cos x}$ Algebra

$\qquad\qquad = \dfrac{1 - \cos^2 x}{\cos x}$ Algebra

$\qquad\qquad = \dfrac{\sin^2 x + \cos^2 x - \cos^2 x}{\cos x}$ Pythagorean Identity

$\qquad\qquad = \dfrac{\sin^2 x}{\cos x}$ Algebra

$\qquad\qquad = \dfrac{\sin x}{\cos x} \sin x$ Algebra

$\qquad\qquad = \tan x \sin x$ Quotient Identity   *(6-1)*

13. $\sin\left(x - \dfrac{\pi}{2}\right) = \sin x \cos \dfrac{\pi}{2} - \cos x \sin \dfrac{\pi}{2}$ Sum Identity

$\qquad\qquad = \sin x(0) - \cos x(1)$ Known Values

$\qquad\qquad = -\cos x$ Algebra   *(6-2)*

14. $\csc 2x = \dfrac{1}{\sin 2x}$ Reciprocal Identity

$\qquad\quad = \dfrac{1}{2 \sin x \cos x}$ Double-angle Identity

$\qquad\quad = \dfrac{1}{2} \dfrac{1}{\sin x} \dfrac{1}{\cos x}$ Algebra

$\qquad\quad = \dfrac{1}{2} \csc x \sec x$ Reciprocal Identities   *(6-3)*

15. (A) Graph both sides of the equation in the same viewing window.

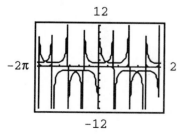

$\dfrac{\sin^2 x}{\cos x} + \cos x = \csc x$ is not an identity, since the graphs do not match. Try $x = \dfrac{3\pi}{4}$.

Left side: $\dfrac{\sin^2 \left(\frac{3\pi}{4}\right)}{\cos \left(\frac{3\pi}{4}\right)} + \cos \left(\dfrac{3\pi}{4}\right) = \dfrac{\frac{1}{2}}{-\frac{1}{\sqrt{2}}} + \left(-\dfrac{1}{\sqrt{2}}\right) = -\sqrt{2}$

Right side: $\csc \left(\dfrac{3\pi}{4}\right) = \sqrt{2}$

This verifies that the equation is not an identity.

(B) Graph both sides of the equation in the same viewing window.

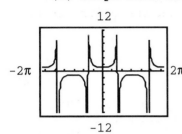

$\dfrac{\sin^2 x}{\cos x} + \cos x = \sec x$ appears to be an identity, which we now verify:

$$\dfrac{\sin^2 x}{\cos x} + \cos x = \dfrac{\sin^2 x}{\cos x} + \dfrac{\cos^2 x}{\cos x} \qquad \text{Algebra}$$

$$= \dfrac{\sin^2 x + \cos^2 x}{\cos x} \qquad \text{Algebra}$$

$$= \dfrac{1}{\cos x} \qquad \text{Pythagorean Theorem}$$

$$= \sec x \qquad \text{Reciprocal Identity}$$

(6-1)

16. Angle α is acute. A triangle can have at most one obtuse angle. Since β is acute, then, if the triangle has an obtuse angle it must be the angle opposite the longer of the two sides, a and c. Thus, α, the angle opposite the shorter of the two sides, a, must be acute.

(7-2)

17. $\sin x = 0.3188 \qquad 0 \le x < 2\pi$
Sketch a graph of $y = \sin x$ and $y = 0.3188$
$x = [0, 2\pi)$.

$$x = \begin{cases} \sin^{-1} 0.3188 = 0.3245 \ \text{First quadrant solution} \\ \pi - \sin^{-1} 0.3188 = 2.8171 \ \text{Second quadrant solution} \end{cases}$$

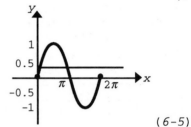

(6-5)

18. $\tan \theta = -4.076 \qquad -90° < \theta < 90°$
Sketch a graph of $y = \tan \theta$ and $y = -4.076$
$\theta = (-90°, 90°)$
$\theta = \tan^{-1}(-4.076) = -76.2154°$

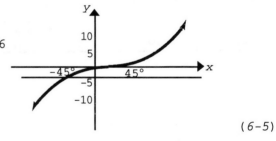

(6-5)

19. We are given two sides and the included angle (SAS). We use the law of cosines to find the third side, then the law of sines to find a second angle.

Solve for b:
$$\begin{aligned} b^2 &= a^2 + c^2 - 2ac \cos \beta \\ &= 12^2 + 13^2 - 2(12)(13)\cos 121° \\ &= 473.69188\ldots \\ b &= 22 \ \text{ft} \end{aligned}$$

Solve for α:
Note that since the sum of α and γ is $180° - 121° = 59°$, both must be acute.

$$\dfrac{\sin \alpha}{a} = \dfrac{\sin \beta}{b}$$

$$\sin \alpha = \dfrac{a \sin \beta}{b}$$

$$= \dfrac{12 \sin 121°}{22}$$

$$\alpha = \sin^{-1}\left(\dfrac{12 \sin 121°}{22}\right)$$

$$= 28°$$

Solve for γ:
$$\begin{aligned} \alpha + \beta + \gamma &= 180° \\ \gamma &= 180° - (\alpha + \beta) \\ &= 180° - (28° + 121°) \\ &= 31° \end{aligned}$$

(7-2, 7-1)

20. The algebraic vector $\langle a, b \rangle$ corresponds to the standard geometric vector with terminal point $P(a, b)$ and initial point $O(0, 0)$. The coordinates of the point $P(a, b)$ are given by

$$a = x_B - x_A = 3 - (-3) = 6$$
$$b = y_B - y_A = -1 - 2 = -3$$

Thus, $\langle a, b \rangle = \langle 6, -3 \rangle$

(7-4)

21. See figure.

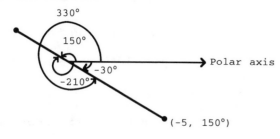

The point with coordinates $(-5, 150°)$ can equally well be described as $(5, -30°)$ or $(-5, -210°)$ or $(5, 330°)$. Thus, $(5, -30°)$: The polar axis is rotated 30° clockwise (negative direction) and the point is located 5 units from the pole along the positive polar axis. $(-5, -210°)$: The polar axis is rotated 210° clockwise (negative direction) and the point is located 5

units from the pole along the negative polar axis. $(5, 330°)$: The polar axis is rotated 330° counterclockwise (positive direction) and the point is located 5 units from the pole along the positive polar axis.

(7-5)

22. We set up a table that indicates how r varies as we let θ vary through each set of quadrant values.

θ	$\cos \theta$	$6 \cos \theta$
0 to $\frac{\pi}{2}$	1 to 0	6 to 0
$\frac{\pi}{2}$ to π	0 to -1	0 to -6
π to $\frac{3\pi}{2}$	-1 to 0	-6 to 0
$\frac{3\pi}{2}$ to 2π	0 to 1	0 to 6

Curve is traced out a second time in this region though coordinates seem different.

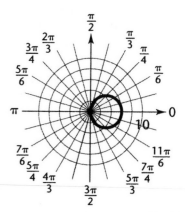

(7-5)

23.

```
      y
 A•   5┊    • B
       ┊
───┼┼┼┼┼┼┼┼┼┼── x
 -5   0┊    5
       ┊
      -5┊
```

(7-7)

24. $(2e^{10°i})^3 = 2^3 e^{3 \cdot 10°i} = 8e^{30°i}$
$$= 8(\cos 30° + i \sin 30°)$$
$$= 8\left(\frac{\sqrt{3}}{2} + i\frac{1}{2}\right)$$
$$= 4\sqrt{3} + 4i$$

(7-7)

25. 30° is not coterminal with 150°; 30° is a I quadrant angle, but 150° is a II quadrant angle.

Since $-\frac{7\pi}{6} + 2\pi = \frac{5\pi}{6}$, which is equivalent to 150°, $-\frac{7\pi}{6}$ is coterminal with 150°.

Since $870° - 2(360°) = 150°$, 870° is coterminal with 150°.

(5-3)

26. $\theta_D = \frac{180°}{\pi}\theta_R = \frac{180°}{\pi}(1.31) = 75.06°$

(5-3)

27. (A) and (C), since 8 radians is equivalent to the real number 8, and cosine is periodic with period 2π. (B) is not the same as cos 8, since 8° is equivalent to $\frac{\pi}{180°}8°$, not 8.

(5-4)

28.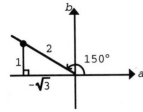

$(a, b) = (-\sqrt{3}, 1) \quad r = 2$

$\cos 150° = \dfrac{a}{r} = \dfrac{-\sqrt{3}}{2} = -\dfrac{\sqrt{3}}{2}$ $(5\text{-}4)$

29.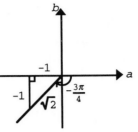

$(a, b) = (-1, -1) \quad r = \sqrt{2}$

$\csc\left(-\dfrac{3\pi}{4}\right) = \dfrac{r}{b} = \dfrac{\sqrt{2}}{-1} = -\sqrt{2}$ $(5\text{-}2)$

30.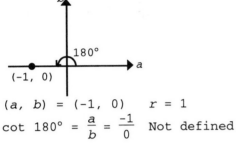

$(a, b) = (-1, 0) \quad r = 1$

$\cot 180° = \dfrac{a}{b} = \dfrac{-1}{0}$ Not defined

$(5\text{-}4)$

31.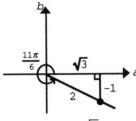

$(a, b) = (\sqrt{3}, -1) \quad r = 2$

$\tan \dfrac{11\pi}{6} = \dfrac{b}{a} = \dfrac{-1}{\sqrt{3}} = -\dfrac{1}{\sqrt{3}}$ or $-\dfrac{\sqrt{3}}{3}$

$(5\text{-}2)$

32. $y = \cos^{-1}\dfrac{1}{2}$ is equivalent to

$\cos y = \dfrac{1}{2} \qquad 0 \le y \le \pi$

$\cos y = \dfrac{a}{r} = \dfrac{1}{2} \quad a = 1 \quad r = 2$

$2^2 = 1^2 + b^2$
$3 = b^2$
$b = \sqrt{3}$
$y = \dfrac{\pi}{3}$

reference triangle associated with y

$(5\text{-}9)$

33. $y = \arctan(-\sqrt{3})$ is equivalent to

$\tan y = -\sqrt{3} \qquad -\dfrac{\pi}{2} < y < \dfrac{\pi}{2}$

$\tan y = \dfrac{b}{a} = \dfrac{-\sqrt{3}}{1} \qquad b = -\sqrt{3} \quad a = 1$

$r^2 = 1^2 + (-\sqrt{3})^2$
$r^2 = 4$
$r = 2$
$y = -\dfrac{\pi}{3}$

reference triangle associated with y

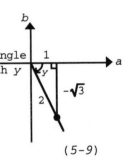

$(5\text{-}9)$

34. arcsin x is not defined if $x > 1$. $(5\text{-}9)$

35. Let $y = \tan^{-1}(-3)$, then $\tan y = -3$, $-\dfrac{\pi}{2} < y < \dfrac{\pi}{2}$.

Draw the reference triangle associated with y, then $\sin y = \sin[\tan^{-1}(-3)]$ can be determined directly from the triangle.

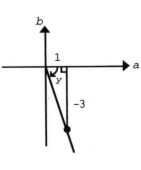

$\tan y = \dfrac{b}{a} = -3 = \dfrac{-3}{1} \quad b = -3 \quad a = 1$

$r^2 = 1^2 + (-3)^2$
$r^2 = 10$
$r = \sqrt{10}$

$\sin[\tan^{-1}(-3)] = \sin y = \dfrac{b}{r} = \dfrac{-3}{\sqrt{10}} = \dfrac{-3\sqrt{10}}{10}$ $(5\text{-}9)$

36. $\cos\left(\cos^{-1}\dfrac{\sqrt{2}}{3}\right) = \dfrac{\sqrt{2}}{3}$ by the cosine-inverse cosine identity.   *(5-9)*

37. To find $\cos^{-1}\left[\cos\left(-\dfrac{\pi}{4}\right)\right]$, first determine $\cos\left(-\dfrac{\pi}{4}\right)$.

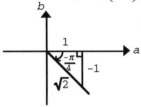

$(a,\ b) = (1,\ -1) \quad r = \sqrt{2}$

$\cos\left(-\dfrac{\pi}{4}\right) = \dfrac{a}{r} = \dfrac{1}{\sqrt{2}}$

Then $\cos^{-1}\left[\cos\left(-\dfrac{\pi}{4}\right)\right] = \cos^{-1}\left(\dfrac{1}{\sqrt{2}}\right) = y$ is equivalent to

$\cos y = \dfrac{1}{\sqrt{2}} \qquad 0 \le y \le \pi$

$\cos y = \dfrac{a}{r} = \dfrac{1}{\sqrt{2}} \quad a = 1 \quad r = \sqrt{2}$

$1^2 + b^2 = (\sqrt{2})^2$

$b^2 = 1$

$b = 1$

$\cos^{-1}\left[\cos\left(-\dfrac{\pi}{4}\right)\right] = y = \dfrac{\pi}{4}$

reference triangle associated with y

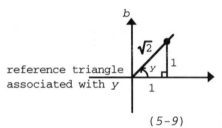

(5-9)

38. (A) 9.871   (B) -3.748   (C) -1.559
(D) Since $\tan 2.314 = -1.088 < -1$, $\cos^{-1}(\tan 2.314)$ is not defined.
    (5-4)

39. $y = 2 - 2\cos\dfrac{\pi x}{2}$

For the graph of $y = 2\cos\dfrac{\pi x}{2}$, we note: $A = 2$

$P = 2\pi \div \dfrac{\pi}{2} = 4$, Phase shift = 0. We graph

$y = 2\cos\dfrac{\pi x}{2}$ turned upside down, then vertically translate the graph up 2 units.

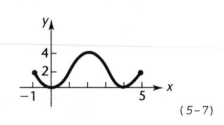

(5-7)

40. (A) $y = \cos^{-1}\left(-\dfrac{\sqrt{3}}{2}\right)$ is equivalent to   (B) -19.755°

$\cos y = -\dfrac{\sqrt{3}}{2} \quad 0 \le y \le \pi$

$y = \dfrac{5\pi}{6}$

Using the relation $\theta_D = \dfrac{180°}{\pi \text{rad}}\theta_R$, we have

reference triangle associated with y

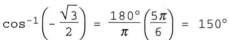

$\cos^{-1}\left(-\dfrac{\sqrt{3}}{2}\right) = \dfrac{180°}{\pi}\left(\dfrac{5\pi}{6}\right) = 150°$

(5-9)

41. $\sin^{-1}(\sin 3) = 0.142$. For the identity $\sin^{-1}(\sin x) = x$ to hold, x must be in the restricted domain of the sine function; that is, $-\dfrac{\pi}{2} \le x \le \dfrac{\pi}{2}$. The number 3 is not in the restricted domain.
    (5-9)

42. Since the coordinates of a point on a unit circle are given by $P(a, b) = P(\cos x, \sin x)$, we evaluate $P(\cos(11.205), \sin(11.205))$--using a calculator set in radian mode--to obtain $P(0.208, -0.978)$. The quadrant in which $P(a, b)$ lies can be determined by the signs of a and b. In this case P is in the fourth quadrant, since a is positive and b is negative. *(5-2)*

43. The equation has infinitely many solutions [$x = \tan^{-1}(-24.5) + k\pi$, k any integer]; $\tan^{-1}(-24.5)$ has a unique value (-1.530 to three decimal places). *(5-9, 6-5)*

44. $k = 3$. $A = \frac{1}{2}(5 - 1) = 2$ $P = 2 = \frac{2\pi}{\pi}$. Hence $B = \pi$, $y = 3 + 2 \sin \pi x$ *(5-7)*

45. $y = 3 \sin(2x - \pi)$
$A = 3$

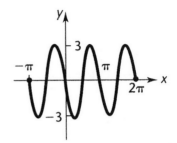

Solve $2x - \pi = 0$ $\qquad$ $2x - \pi = 2\pi$
$\qquad\quad 2x = \pi$ $\qquad\qquad 2x = \pi + 2\pi$
$\qquad\qquad x = \dfrac{\pi}{2}$ $\qquad\qquad\quad x = \dfrac{\pi}{2} + \pi$

$\qquad\qquad\uparrow\underline{\hspace{3cm}}\uparrow\quad\uparrow$
$\qquad\quad$ Phase shift $\qquad$ Period $P = \pi$

The graph completes one full cycle as x varies over the interval $\left[\dfrac{\pi}{2}, \dfrac{3\pi}{2}\right]$ *(5-7)*

46. One cycle of $y = 2 \tan\left(\dfrac{\pi x}{2} - \dfrac{\pi}{2}\right)$ is completed as $\dfrac{\pi x}{2} - \dfrac{\pi}{2}$ varies from 0 to π.

Solve each equation for x:

$\dfrac{\pi x}{2} - \dfrac{\pi}{2} = 0$ $\qquad\qquad \dfrac{\pi x}{2} - \dfrac{\pi}{2} = \pi$

$\qquad \dfrac{\pi x}{2} = \dfrac{\pi}{2}$ $\qquad\qquad\quad \dfrac{\pi x}{2} = \dfrac{\pi}{2} + \pi$

$\qquad\quad x = 1$ $\qquad\qquad\qquad x = 1 + 2$

Phase shift = 1 $\qquad\qquad$ Period = 2

Sketch the graph for one period [1, 3], then extend over the interval (0, 4).

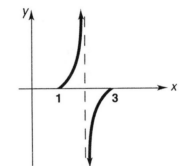

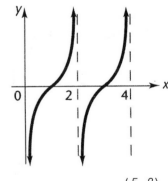

(5-8)

47. The dashed curve is $y = \sin x$. The solid curve is $y = \csc x$.

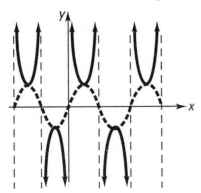

(5-6)

48. See figure.

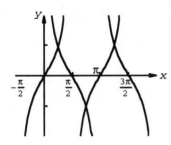

If the graph of $y = \cot x$ is shifted to the left $\frac{\pi}{2}$ units and reflected in the x axis, the result will be the graph of $y = \tan x$.

(5-6, 5-7)

49. Here is the graph of $y = \dfrac{1}{\cot^2 x + 1}$ in a graphing utility.

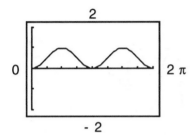

The graph has amplitude $\dfrac{1 - 0}{2} = \dfrac{1}{2}$ and period π. It appears to be the graph of $y = -\dfrac{1}{2} \cos 2x$ shifted up $\dfrac{1}{2}$ unit, that is, $y = \dfrac{1}{2} - \dfrac{1}{2} \cos 2x$.

(5-7)

50. Here is the graph of $y = \dfrac{2 - 2 \sin^2 x}{\sin 2x}$ in a graphing utility.

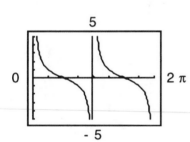

The graph appears to have the form $y = A \cot Bx$. Since the period is π, $B = 1$. The graph of $y = A \cot x$ shown appears to pass through the point $\left(\dfrac{\pi}{4}, 1\right)$, thus

$$1 = A \cot \frac{\pi}{4}$$
$$1 = A$$

The equation of the graph can be written $y = \cot x$.

(5-7, 5-8)

51. (A) Check $x = 0$
Left side:
$\quad \sin 2x = \sin 2 \cdot 0 = \sin 0 = 0$
Right side:
$\quad 2 \sin x = 2 \sin 0 = 2 \cdot 0 = 0$
0 is a solution

Check $x = \pi$
Left side:
$\quad \sin 2x = \sin 2\pi = 0$
Right side:
$\quad 2 \sin x = 2 \sin \pi = 2 \cdot 0 = 0$
π is a solution

(B) The equation is not an identity because both sides are defined at $x = \dfrac{\pi}{2}$; for example, but $\dfrac{\pi}{2}$ is not a solution.

Left side: $\sin 2x = \sin 2\left(\dfrac{\pi}{2}\right) = \sin \pi = 0$

Right side: $2 \sin x = 2 \sin \dfrac{\pi}{2} = 2 \cdot 1 = 2$

The equation is conditional.

(6-1)

52.

$$\frac{\sin u}{1 + \cos u} + \cot u = \frac{\sin u}{1 + \cos u} + \frac{\cos u}{\sin u} \quad \text{Quotient Identity}$$

$$= \frac{\sin^2 u}{\sin u(1 + \cos u)} + \frac{\cos u(1 + \cos u)}{\sin u(1 + \cos u)} \quad \text{Algebra}$$

$$= \frac{\sin^2 u + \cos u(1 + \cos u)}{\sin u(1 + \cos u)} \quad \text{Algebra}$$

$$= \frac{\sin^2 u + \cos u + \cos^2 u}{\sin u(1 + \cos u)} \quad \text{Algebra}$$

$$= \frac{\sin^2 u + \cos^2 u + \cos u}{\sin u(1 + \cos u)} \quad \text{Algebra}$$

$$= \frac{1 + \cos u}{\sin u(1 + \cos u)} \quad \text{Pythagorean Identity}$$

$$= \frac{1}{\sin u} \quad \text{Algebra}$$

$$= \csc u \quad \text{Reciprocal Identity}$$

Key Algebraic Steps:

$$\frac{a}{1 + b} + \frac{b}{a} = \frac{a^2}{a(1 + b)} + \frac{b(1 + b)}{a(1 + b)}$$

$$= \frac{a^2 + b(1 + b)}{a(1 + b)}$$

$$= \frac{a^2 + b + b^2}{a(1 + b)}$$

$$= \frac{a^2 + b^2 + b}{a(1 + b)} \quad (6\text{-}1)$$

53.

$$\sec x + \tan x = \frac{1}{\cos x} + \frac{\sin x}{\cos x} \quad \text{Reciprocal and Quotient Identities}$$

$$= \frac{1 + \sin x}{\cos x} \quad \text{Algebra}$$

$$= \frac{(1 + \sin x)(1 - \sin x)}{\cos x(1 - \sin x)} \quad \text{Algebra}$$

$$= \frac{1 - \sin^2 x}{\cos x(1 - \sin x)} \quad \text{Algebra}$$

$$= \frac{\cos^2 x + \sin^2 x - \sin^2 x}{\cos x(1 - \sin x)} \quad \text{Pythagorean Identity}$$

$$= \frac{\cos^2 x}{\cos x(1 - \sin x)} \quad \text{Algebra}$$

$$= \frac{\cos x}{1 - \sin x} \quad \text{Algebra} \quad (6\text{-}1)$$

54.

$$\tan \frac{x}{2} = \frac{1 - \cos x}{\sin x} \quad \text{Half-angle Identity}$$

$$= \frac{1}{\sin x} - \frac{\cos x}{\sin x} \quad \text{Algebra}$$

$$= \csc x - \cot x \quad \text{Reciprocal and Quotient Identities} \quad (6\text{-}3)$$

55. $\csc^2 \dfrac{x}{2} = \dfrac{1}{\sin^2 \frac{x}{2}}$ Reciprocal Identity

$= \dfrac{1}{\left(\pm\sqrt{\frac{1 - \cos x}{2}}\,\right)^2}$ Half-angle Identity

$= \dfrac{1}{\frac{1 - \cos x}{2}}$ Algebra

$= 1 \div \dfrac{1 - \cos x}{2}$ Algebra

$= 1 \cdot \dfrac{2}{1 - \cos x}$ Algebra

$= \dfrac{2}{1 - \cos x}$ Algebra

$= \dfrac{2}{1 - \cos x} \dfrac{1 + \cos x}{1 + \cos x}$ Algebra

$= \dfrac{2(1 + \cos x)}{1 - \cos^2 x}$ Algebra

$= \dfrac{2(1 + \cos x)}{\sin^2 x + \cos^2 x - \cos^2 x}$ Pythagorean Identity

$= \dfrac{2(1 + \cos x)}{\sin^2 x}$ Algebra

$= 2 \cdot \dfrac{1}{\sin x} \dfrac{1 + \cos x}{\sin x}$ Algebra

$= 2 \cdot \dfrac{1}{\sin x}\left(\dfrac{1}{\sin x} + \dfrac{\cos x}{\sin x}\right)$ Algebra

$= 2 \csc x(\csc x + \cot x)$ Reciprocal and Quotient Identities

Key Algebraic Steps:

$$\dfrac{1}{\frac{1 - a}{2}} = 1 \div \dfrac{1 - a}{2} = 1 \cdot \dfrac{2}{1 - a} = \dfrac{2}{1 - a} = \dfrac{2}{1 - a}\dfrac{1 + a}{1 + a} = \dfrac{2(1 + a)}{1 - a^2} \qquad (6\text{-}3)$$

56. $\dfrac{2}{1 + \cos 2x} = \dfrac{2}{1 + 2\cos^2 x - 1}$ Double-angle Identity

$= \dfrac{2}{2\cos^2 x}$ Algebra

$= \dfrac{1}{\cos^2 x}$ Algebra

$= \sec^2 x$ Reciprocal Identity $(6\text{-}2)$

57. $\dfrac{\cos x + \cos y}{\sin x - \sin y} = \dfrac{2\cos \frac{x + y}{2}\cos \frac{x - y}{2}}{2\cos \frac{x + y}{2}\sin \frac{x - y}{2}}$ Sum-product Identity

$= \dfrac{\cos \frac{x - y}{2}}{\sin \frac{x - y}{2}}$ Algebra

$= \cot \dfrac{x - y}{2}$ Quotient Identity $(6\text{-}4)$

58. (A) Graph both sides of the equation in the same viewing window.

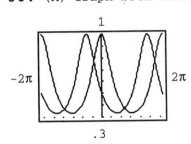

$\dfrac{\tan x}{2 \tan x - \sin x} = \dfrac{1}{2 + \sin x}$ is not an identity, since the graphs do not match.

Try $x = \dfrac{\pi}{4}$

Left side: $\dfrac{\tan \frac{\pi}{4}}{2 \tan \frac{\pi}{4} - \sin \frac{\pi}{4}} = \dfrac{1}{2 \cdot 1 - \frac{1}{\sqrt{2}}} = \dfrac{\sqrt{2}}{2\sqrt{2} - 1}$

Right side: $\dfrac{1}{2 + \sin \frac{\pi}{4}} = \dfrac{1}{2 + \frac{1}{\sqrt{2}}} = \dfrac{\sqrt{2}}{2\sqrt{2} + 1}$

This verifies that the equation is not an identity.

(B) Graph both sides of the equation in the same viewing window.

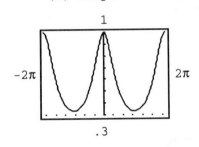

$\dfrac{\tan x}{2 \tan x - \sin x} = \dfrac{1}{2 - \cos x}$ appears to be an identity, which we now verify.

$\dfrac{1}{2 - \cos x} = \dfrac{\tan x}{\tan x(2 - \cos x)}$ Algebra

$= \dfrac{\tan x}{2 \tan x - \cos x \tan x}$ Algebra

$= \dfrac{\tan x}{2 \tan x - \cos x \cdot \frac{\sin x}{\cos x}}$ Quotient Identity

$= \dfrac{\tan x}{2 \tan x - \sin x}$ Algebra   *(6-1)*

59.

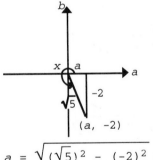

$a = \sqrt{(\sqrt{5})^2 - (-2)^2} = 1$

$\cos x = \dfrac{1}{\sqrt{5}}$

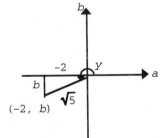

$b = -\sqrt{(\sqrt{5})^2 - (-2)^2} = -1$

$\sin y = \dfrac{-1}{\sqrt{5}}$

$\cos(x - y) = \cos x \cos y + \sin x \sin y$

$= \dfrac{1}{\sqrt{5}}\left(-\dfrac{2}{\sqrt{5}}\right) + \left(\dfrac{-2}{\sqrt{5}}\right)\left(-\dfrac{1}{\sqrt{5}}\right)$

$= \dfrac{-2}{5} + \dfrac{2}{5}$

$= 0$

(6-2)

60.

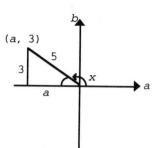

$a = -\sqrt{5^2 - 3^2} = -4$

$\cos x = -\dfrac{4}{5}$

$\sin 2x = 2 \sin x \cos x$

$\qquad = 2\left(\dfrac{3}{5}\right)\left(-\dfrac{4}{5}\right)$

$\qquad = \dfrac{-24}{25}$

Since $\dfrac{\pi}{2} \le x \le \pi$, $\dfrac{\pi}{4} \le \dfrac{x}{2} \le \dfrac{\pi}{2}$,

$\cos \dfrac{x}{2}$ will be positive

$\cos \dfrac{x}{2} = \sqrt{\dfrac{1 + \cos x}{2}}$

$\qquad = \sqrt{\dfrac{1 + \left(-\dfrac{4}{5}\right)}{2}} = \sqrt{\dfrac{\dfrac{1}{5}}{2}} = \sqrt{\dfrac{1}{10}}$ or $\dfrac{\sqrt{10}}{10}$

$(5\text{-}4, \; 6\text{-}3)$

61.

$\qquad 2 \sin^2 \theta + \sin \theta = 1 \qquad\qquad 0° \le \theta < 360°$

$\qquad 2 \sin^2 \theta + \sin \theta - 1 = 0$

$(2 \sin \theta - 1)(\sin \theta + 1) = 0$

$2 \sin \theta - 1 = 0 \qquad\qquad \sin \theta + 1 = 0$

$\qquad \sin \theta = \dfrac{1}{2} \qquad\qquad \sin \theta = -1$

$\qquad \theta = 30°, \; 150° \qquad\qquad \theta = 270°$

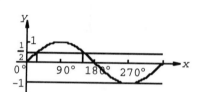

Solutions: $\theta = 30°, \; 150°, \; 270°$

$(6\text{-}5)$

62.

$\qquad\qquad\qquad \sin 2x = \sin x$

$\qquad\qquad \sin 2x - \sin x = 0$

$\quad 2 \sin x \cos x - \sin x = 0$

$\qquad \sin x(2 \cos x - 1) = 0$

$\sin x = 0 \qquad\qquad 2 \cos x - 1 = 0$

$\quad x = k\pi \qquad\qquad\quad \cos x = \dfrac{1}{2}$

k any integer

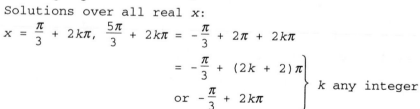

Solutions over one period

$\qquad x = \dfrac{\pi}{3}, \; \dfrac{5\pi}{3}$

Solutions over all real x:

$x = \dfrac{\pi}{3} + 2k\pi, \; \dfrac{5\pi}{3} + 2k\pi = -\dfrac{\pi}{3} + 2\pi + 2k\pi$

$\qquad\qquad\qquad\qquad = -\dfrac{\pi}{3} + (2k + 2)\pi$

$\qquad\qquad\qquad\text{or} \; -\dfrac{\pi}{3} + 2k\pi$ $\bigg\}$ k any integer

Solutions: $x = k\pi, \; \dfrac{\pi}{3} + 2k\pi, \; -\dfrac{\pi}{3} + 2k\pi, \; k$ any integer.

$(6\text{-}5)$

63. (A) $\cot x = -2 \cos x$

$$\frac{\cos x}{\sin x} = -2 \cos x$$

$$\frac{\cos x}{\sin x} + 2 \cos x = 0$$

$$\frac{\cos x}{\sin x} + \frac{2 \sin x \cos x}{\sin x} = 0$$

$$\frac{\cos x + 2 \sin x \cos x}{\sin x} = 0$$

We multiply both sides by sin x, keeping in mind that this could introduce extraneous solutions, hence it is necessary to check the solutions in the original equation.

$\cos x + 2 \sin x \cos x = 0$
  $\cos x (1 + 2 \sin x) = 0$
$\cos x = 0$ $1 + 2 \sin x = 0$

$$x = \frac{\pi}{2}, \frac{3\pi}{2} \qquad \sin x = -\frac{1}{2}$$

$$x = \frac{7\pi}{6}, \frac{11\pi}{6}$$

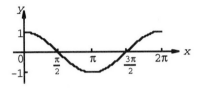

Check:

$$x = \frac{\pi}{2}$$ $$x = \frac{3\pi}{2}$$ $$x = \frac{7\pi}{6}$$ $$x = \frac{11\pi}{6}$$

$$\cot \frac{\pi}{2} \overset{?}{=} -2 \cos \frac{\pi}{2} \qquad \cot \frac{3\pi}{2} \overset{?}{=} -2 \cos \frac{3\pi}{2} \qquad \cot \frac{7\pi}{6} \overset{?}{=} -2 \cos \frac{7\pi}{6} \qquad \cot \frac{11\pi}{6} \overset{?}{=} -2 \cos \frac{11\pi}{6}$$

$$0 \overset{\checkmark}{=} 0 \qquad\qquad 0 \overset{\checkmark}{=} 0 \qquad\qquad \sqrt{3} \overset{\checkmark}{=} -2\left(-\frac{\sqrt{3}}{2}\right) \qquad -\sqrt{3} \overset{\checkmark}{=} -2\left(\frac{\sqrt{3}}{2}\right)$$

   A solution   A solution   A solution   A solution

Solutions: $\dfrac{\pi}{2}, \dfrac{3\pi}{2}, \dfrac{7\pi}{6}, \dfrac{11\pi}{6}$

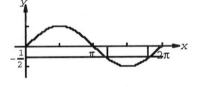

(B) Here is a computer-generated graph of $y = \cot x$ and $y = -2 \cos x$, $0 \le x \le 2\pi$.

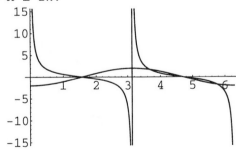

There are points of intersection in the intervals [1.4, 1.6], [3.6, 3.8], [4.6, 4.8], and [5.6, 5.8]. Zooming in on one of these, we obtain the following graph.

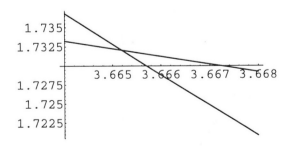

To three decimal places, the solution is 3.665. Similarly, the other solutions can be approximated as 1.571, 4.712, and 5.760 (graphs not shown).   (6-5)

64. Here is a computer-generated graph of $y = 2 \cos x$ and $y = x - \cos 2x$, $-2\pi \le x \le 2\pi$.

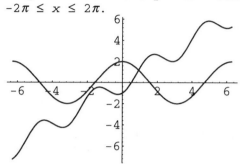

There is a solution in the interval [0.8, 1.2]. Zooming in on this solution, we obtain the following graph.

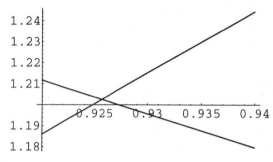

To three decimal places, the solution is 0.926. *(6-5)*

65.

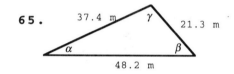

We are given three sides (SSS). We solve for the largest angle, γ (largest because it is opposite the longest side, c) using the law of cosines. We then solve for a second angle using the law of sines.

Solve for γ:
$$c^2 = a^2 + b^2 - 2ab \cos \gamma$$
$$\cos \gamma = \frac{a^2 + b^2 - c^2}{2ab}$$
$$\gamma = \cos^{-1}\left(\frac{a^2 + b^2 - c^2}{2ab}\right)$$
$$= \cos^{-1}\left(\frac{(21.3)^2 + (37.4)^2 - (48.2)^2}{2(21.3)(37.4)}\right)$$
$$= 107.2°$$

Solve for α:
We use the law of sines. Since the sum of α and β is $180° - 107.2° = 72.8°$, both must be acute.
$$\frac{\sin \alpha}{a} = \frac{\sin \gamma}{c}$$
$$\sin \alpha = \frac{a \sin \gamma}{c}$$
$$= \frac{21.3 \sin 107.2°}{48.2}$$
$$\alpha = \sin^{-1}\left(\frac{21.3 \sin 107.2°}{48.2}\right)$$
$$= 25.0°$$

Solve for β:
$$\alpha + \beta + \gamma = 180°$$
$$\beta = 180° - (\alpha + \gamma)$$
$$= 180° - (25.0° + 107.2°)$$
$$= 47.8° \qquad (7\text{-}1,\ 7\text{-}2)$$

66. 20.3 m   25.4 m

125.4°

α is obtuse

$20.3 = a < b = 25.4$

We are given two sides and a non-included angle (SSA). If we try to draw a triangle with these values, we will be unsuccessful.
No solution. *(7-1)*

67.

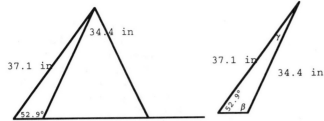

We are given two sides and a non-included angle (SSA). If we try to draw a triangle with these values, we find that two triangles are possible. Following the instructions, we choose the solution for which β is obtuse.

Solve for β:

$$\frac{\sin \beta}{b} = \frac{\sin \alpha}{a}$$

$$\sin \beta = \frac{b \sin \alpha}{a}$$

$$= \frac{37.1 \sin 52.9°}{34.4}$$

$$= 0.8602 \qquad \beta \text{ is chosen obtuse}$$

$$\beta = 180° - \sin^{-1} 0.8602$$

$$= 180° - 59.3°$$

$$= 120.7°$$

Solve for γ:

$$\alpha + \beta + \gamma = 180°$$

$$\gamma = 180° - (\alpha + \beta)$$

$$= 180° - (52.9° + 120.7°)$$

$$= 6.4°$$

Solve for c:

$$\frac{\sin \alpha}{a} = \frac{\sin \gamma}{c}$$

$$c = \frac{a \sin \gamma}{\sin \alpha}$$

$$= \frac{34.4 \sin 6.4°}{\sin 52.9°}$$

$$= 4.81 \text{ in} \qquad (7\text{-}1)$$

68. β must be acute. A triangle can have at most one obtuse angle, and since γ is acute, the obtuse angle, if present, must be opposite the longer of the two sides a and b.

$(7\text{-}2)$

69.

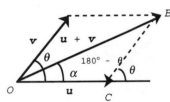

$\theta = 48.3°$. Hence, $\angle OCB = 180° - \theta = 180° - 48.3° = 131.7°$. We can find $|\mathbf{u} + \mathbf{v}|$ using the law of cosines:

$$|\mathbf{u} + \mathbf{v}|^2 = |\mathbf{u}|^2 + |\mathbf{v}|^2 - 2|\mathbf{u}||\mathbf{v}|\cos(OCB)$$

$$= (25.3)^2 + (13.4)^2 - 2(25.3)(13.4)\cos 131.7°$$

$$= 1270.7028\ldots$$

$$|\mathbf{u} + \mathbf{v}| = 35.6 \text{ lb}$$

To find α, we use the law of sines:

$$\frac{\sin \alpha}{|\mathbf{v}|} = \frac{\sin(OCB)}{|\mathbf{u} + \mathbf{v}|}$$

$$\frac{\sin \alpha}{13.4} = \frac{\sin 131.7°}{35.6}$$

$$\sin \alpha = \frac{13.4}{35.6} \sin 131.7°$$

$$\alpha = \sin^{-1}\left(\frac{13.4}{35.6} \sin 131.7°\right)$$

$$= 16.3° \qquad\qquad (7\text{-}1,\ 7\text{-}2,\ 73)$$

70. (A) $2\mathbf{u} - \mathbf{v} + 3\mathbf{w} = 2\langle -1,\ 2 \rangle - \langle 0,\ -2 \rangle + 3\langle 1,\ -1 \rangle$

$$= \langle -2,\ 4 \rangle + \langle 0,\ 2 \rangle + \langle 3,\ -3 \rangle$$

$$= \langle 1,\ 3 \rangle$$

(B) $2\mathbf{u} - \mathbf{v} + 3\mathbf{w} = 2(2\mathbf{i} - \mathbf{j}) - (\mathbf{i} + 3\mathbf{j}) + 3(2\mathbf{j})$

$$= 4\mathbf{i} - 2\mathbf{j} - \mathbf{i} - 3\mathbf{j} + 6\mathbf{j}$$

$$= 3\mathbf{i} + \mathbf{j}$$

$(7\text{-}4)$

71. $x^2 + y^2 = 8y$
Use $r^2 = x^2 + y^2$ and $y = r \sin \theta$
$$r^2 = 8r \sin \theta$$
$$r^2 - 8r \sin \theta = 0$$
$$r(r - 8 \sin \theta) = 0$$
$$r = 0 \text{ or } r - 8 \sin \theta = 0$$

The graph of $r = 0$ is the pole. Since the pole is included in the graph of
$r - 8 \sin \theta = 0$, we can discard $r = 0$ and keep only
$$r - 8 \sin \theta = 0 \text{ or}$$
$$r = 8 \sin \theta \qquad (7\text{-}5)$$

72. $r = -4 \cos \theta$
We multiply both sides by r, which adds the pole to the graph. But the pole is already part of the graph, so we have changed nothing.
$$r^2 = -4r \cos \theta$$
But $r^2 = x^2 + y^2$ and $r \cos \theta = x$. Hence, $x^2 + y^2 = -4x$.
$\qquad (7\text{-}5)$

73. We set up a table that indicates how r varies as we let θ vary through each set of quadrant values.

θ	$\cos \theta$	$4 \cos \theta$	$4 + 4 \cos \theta$
0 to $\frac{\pi}{2}$	1 to 0	4 to 0	8 to 4
$\frac{\pi}{2}$ to π	0 to -1	0 to -4	4 to 0
π to $\frac{3\pi}{2}$	-1 to 0	-4 to 0	0 to 4
$\frac{3\pi}{2}$ to 2π	0 to 1	0 to 4	4 to 8

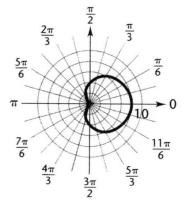

$(7\text{-}5)$

74. We set up a table that indicates how r varies as we let 3θ vary through each set of quadrant values.

θ	3θ	$\sin 3\theta$	$6 \sin 3\theta$
0 to $\frac{\pi}{6}$	0 to $\frac{\pi}{2}$	0 to 1	0 to 6
$\frac{\pi}{6}$ to $\frac{\pi}{3}$	$\frac{\pi}{2}$ to π	1 to 0	6 to 0
$\frac{\pi}{3}$ to $\frac{\pi}{2}$	π to $\frac{3\pi}{2}$	0 to -1	0 to -6
$\frac{\pi}{2}$ to $\frac{2\pi}{3}$	$\frac{3\pi}{2}$ to 2π	-1 to 0	-6 to 0
$\frac{2\pi}{3}$ to $\frac{5\pi}{6}$	2π to $\frac{5\pi}{2}$	0 to 1	0 to 6
$\frac{5\pi}{6}$ to π	$\frac{5\pi}{2}$ to 3π	1 to 0	6 to 0
π to 2π	repeats curve already drawn		

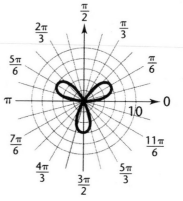

$(7\text{-}5)$

75. $n = 1$ $\qquad\qquad$ $n = 2$ $\qquad\qquad$ $n = 3$

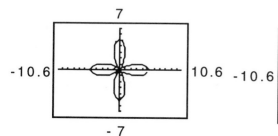

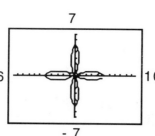

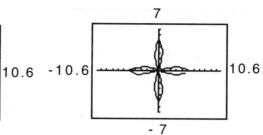

Generalizing from these graphs, we expect 4 leaves for all n. $\qquad (7\text{-}5)$

76.

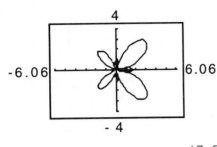

(7-5)

77. $(x, y) = (-2.78, -3.19)$

$r = \sqrt{x^2 + y^2} = \sqrt{(-2.78)^2 + (-3.19)^2} = 4.23$

$\tan \theta = \dfrac{y}{x} = \dfrac{-3.19}{-2.78}$

θ is a third quadrant angle and is to be chosen so that $-180° < \theta \leq 180°$.

$\theta = \tan^{-1} \dfrac{-3.19}{-2.78} - 180° = -131.07°$

$(r, \theta) = (4.23, -131.07°)$ (7-5)

78. $(r, \theta) = (6.22, -4.08)$

$x = r \cos \theta = 6.22 \cos(-4.08) = -3.68$

$y = r \sin \theta = 6.22 \sin(-4.08) = 5.02$

$(x, y) = (-3.68, 5.02)$ (7-5)

79. $2e^{(-\pi/6)i} = 2\left[\cos\left(-\dfrac{\pi}{6}\right) + i \sin\left(-\dfrac{\pi}{6}\right)\right]$

$= 2\left[\dfrac{\sqrt{3}}{2} + i\left(-\dfrac{1}{2}\right)\right]$

$= \sqrt{3} - i$ (7-6)

80. A sketch shows that $-1 + i\sqrt{3}$ is associated with a special 30°-60° triangle. Thus, by inspection, $r = 2$, $\theta = 120°$, and

$-1 + i\sqrt{3} = 2(\cos 120° + i \sin 120°)$

$= 2e^{120°i}$

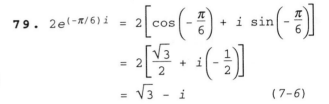

(7-6)

81. $(1 - i\sqrt{3})^6 = (2e^{-60°i})^6 = 2^6 e^{6(-60°i)} = 64e^{-360°i} = 64[\cos(-360°) + i \sin(-360°)]$

$= 64(1) = 64$ or $64 + 0i$ (7-7)

82. First write $-i$ in polar form

$-i = 0 - i = 1e^{-90°i}$

Using the nth-root theorem, all three cube roots of $1e^{-90°i}$ are given by

$1^{1/3}e^{(-90°/3 + k \cdot 360°/3)i} = 1^{1/3}e^{(-30° + k120°)i}$ $\quad k = 0, 1, 2$

Thus,

$w_1 = 1^{1/3}e^{(-30° + 0 \cdot 120°)i} = e^{-30°i} = \cos(-30°) + i \sin(-30°) = \dfrac{\sqrt{3}}{2} - i\dfrac{1}{2}$

$w_2 = 1^{1/3}e^{(-30° + 1 \cdot 120°)i} = e^{90°i} = \cos 90° + i \sin 90° = i$

$w_3 = 1^{1/3}e^{(-30° + 2 \cdot 120°)i} = e^{210°i} = \cos 210° + i \sin 210° = -\dfrac{\sqrt{3}}{2} - i\dfrac{1}{2}$

These roots lie on a circle of radius 1, equally spaced so that the angle between successive roots is 120°.

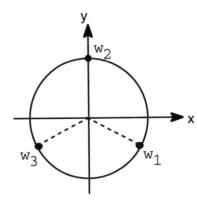

(7-7)

83. $-4.88 - 3.17i$ is not associated with a special triangle.

$$r = \sqrt{(-4.88)^2 + (-3.17)^2} = 5.82$$

$$\theta = \tan^{-1}\frac{-3.17}{-4.88} - 180° = -146.99°$$

$$-4.88 - 3.17i = 5.82[\cos(-146.99°) + i\sin(-146.99°)]$$
$$= 5.82e^{(-146.99°)i}$$

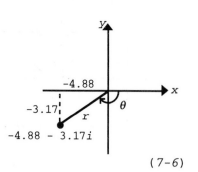

$(7\text{-}6)$

84. $x + iy = 6.97e^{163.87°i}$
$$= 6.97(\cos 163.87° + i\sin 163.87°)$$
$$= -6.70 + 1.94i$$

$(7\text{-}6)$

85. (A) There are a total of four fourth roots and they are spaced equally around a circle of radius $\sqrt{2}$, so that the angle between successive roots is 90°.

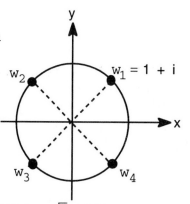

(B) Since $w_1 = \sqrt{2}e^{45°i}$, $w_2 = \sqrt{2}e^{(45° + 90°)i} = \sqrt{2}e^{135°i}$,
$w_3 = \sqrt{2}e^{(135° + 90°)i} = \sqrt{2}e^{225°i}$, and $w_4 = \sqrt{2}e^{(225° + 90°)i} = \sqrt{2}e^{315°i}$

Thus, $w_2 = \sqrt{2}(\cos 135° + i\sin 135°) = -1 + i$

$w_3 = \sqrt{2}(\cos 225° + i\sin 225°) = -1 - i$

$w_4 = \sqrt{2}(\cos 315° + i\sin 315°) = 1 - i$

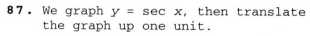

(C) $(1 + i)^4 = [(1 + i)^2]^2 = (1 + 2i + i^2)^2 = (2i)^2 = -4$
$(-1 + i)^4 = [(-1 + i)^2]^2 = (1 - 2i + i^2)^2 = (-2i)^2 = -4$
$(-1 - i)^4 = [(-1 - i)^2]^2 = (1 + 2i + i^2)^2 = (2i)^2 = -4$
$(1 - i)^4 = [(1 - i)^2]^2 = (1 - 2i + i^2)^2 = (-2i)^2 = -4$

$(7\text{-}7)$

86. On this unit circle,

$P(a, b) = P(\cos\theta, \sin\theta)$
$= P(\cos(s \text{ rad}), \sin(s \text{ rad}))$
$= P(\cos 1.2, \sin 1.2)$
$= (0.362, 0.932)$

$(5\text{-}2)$

87. We graph $y = \sec x$, then translate the graph up one unit.

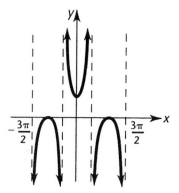

$(5\text{-}8)$

88. $|A| = 3$. The graph completes one full cycle as x varies over the (mentally extended) intervals $\left[-\frac{7}{8}, \frac{1}{8}\right]$ or $\left[\frac{1}{8}, \frac{9}{8}\right]$. Since the phase shift is required between 0 and 1, we must set $-\frac{C}{B} = \frac{1}{8}$. Then the graph has the form of a standard cosine curve. Hence

$$A = 3$$
$$-\frac{C}{B} = \frac{1}{8} \qquad -\frac{C}{B} + \frac{2\pi}{B} = \frac{9}{8}$$
$$\frac{2\pi}{B} = 1 \qquad B = 2\pi \qquad C = -\frac{1}{8}B = -\frac{\pi}{4}$$

The equation is then
$$y = A \cos(Bx + C)$$
$$y = 3 \cos\left(2\pi x - \frac{\pi}{4}\right) \tag{5-7}$$

89. Here is a graph of $y = 1.6 \sin 2x - 1.2 \cos 2x$ in a graphing utility.

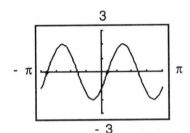

It appears that this is a sine curve shifted to the right, with $A = 2$ and, since $P = \frac{2\pi}{B}$ and P appears to be π, $B = \frac{2\pi}{P} = \frac{2\pi}{\pi} = 2$.

From the graphing utility, we find that the x intercept closest to the origin, to three decimal places, is 0.322. To find C, substitute $B = 2$ and $x = 0.322$ into the phase shift formula $x = -\frac{C}{B}$ and solve for C:

$$x = -\frac{C}{B}$$
$$0.322 = -\frac{C}{2}$$
$$C = -0.644$$

The equation is thus $y = 2 \sin(2x - 0.644)$. $\tag{5-7}$

90. Let $y = \cos^{-1} x$, then $x = \cos y \qquad 0 \le y \le \pi$

Then $\csc y = \dfrac{1}{\sin y} = \dfrac{1}{\sqrt{1 - \cos^2 y}}$ from the reciprocal and Pythagorean identities. (The positive sign is chosen for the square root since $\sin y$ is positive in Quadrants I and II.)

Hence $\csc y = \csc(\cos^{-1} x) = \dfrac{1}{\sqrt{1 - \cos^2 y}} = \dfrac{1}{\sqrt{1 - x^2}}$. $\tag{5-9}$

91. Let $u = \cot^{-1} \dfrac{3}{4}$. Then $\cot u = \dfrac{3}{4}$, $0 < u < \pi$.

$$r = \sqrt{3^2 + 4^2} = 5$$
$$\sin u = \frac{4}{5} \qquad \cos u = \frac{3}{5}$$
$$\sin\left[2 \cot^{-1}\left(\frac{3}{4}\right)\right] = \sin 2u = 2 \sin u \cos u = 2\left(\frac{4}{5}\right)\left(\frac{3}{5}\right) = \frac{24}{25}$$

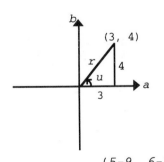

$$(5\text{-}9, \ 6\text{-}3)$$

92. $b = \sqrt{5^2 - (-3)^2} = 4$

$\cos x = -\dfrac{3}{5}$

(A) Since $\dfrac{\pi}{2} \le x \le \pi$, $\dfrac{\pi}{4} \le \dfrac{x}{2} \le \dfrac{\pi}{2}$, $\sin \dfrac{x}{2}$ will be positive.

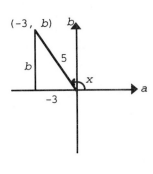

$$\sin \frac{x}{2} = \sqrt{\frac{1 - \cos x}{2}} = \sqrt{\frac{1 - \left(-\frac{3}{5}\right)}{2}}$$

$$= \sqrt{\frac{\frac{8}{5}}{2}}$$

$$= \sqrt{\frac{4}{5}}$$

$$= \frac{2}{\sqrt{5}} \text{ or } \frac{2\sqrt{5}}{5}$$

(B) $\cos 2x = 2\cos^2 x - 1 = 2\left(-\dfrac{3}{5}\right)^2 - 1 = 2\left(\dfrac{9}{25}\right) - 1 = \dfrac{18}{25} - 1 = -\dfrac{7}{25}$ (6-3)

93. (A) $2 \sin^2 x = 3 \cos x$ $0 \le x < 2\pi$

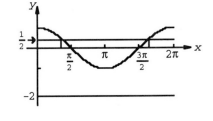

$$2(1 - \cos^2 x) = 3 \cos x$$
$$2 - 2\cos^2 x = 3 \cos x$$
$$0 = 2\cos^2 x + 3 \cos x - 2$$
$$0 = (2 \cos x - 1)(\cos x + 2)$$

$2 \cos x - 1 = 0$ $\cos x + 2 = 0$

$\cos x = \dfrac{1}{2}$ $\cos x = -2$

$x = \dfrac{\pi}{3}, \dfrac{5\pi}{3}$ No solution

Solutions: $\dfrac{\pi}{3}, \dfrac{5\pi}{3}$

(B) Here is a computer-generated graph of $y = 2 \sin^2 x$ and $y = 3 \cos x$, $0 \le x \le 2\pi$.

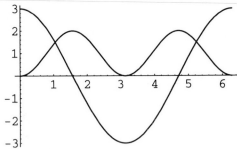

There are solutions of $2 \sin^2 x = 3 \cos x$ in the intervals $[1, 1.2]$ and $[5.2, 5.4]$. Zooming in on these solutions, we obtain the following graphs.

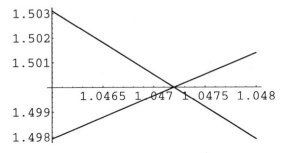

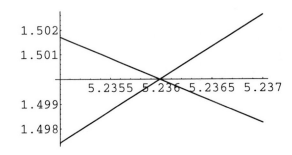

To four decimal places, the solutions are 1.0472 and 5.2360.   (6-5)

94. (A) If $r^2 = 36 \cos 2\theta$, we can write $r = \pm\sqrt{36 \cos 2\theta}$. We then set up a table that indicates how r varies as we let 2θ vary through each set of quadrant values.

θ	2θ	$\cos 2\theta$	$36 \cos 2\theta$	$\pm\sqrt{36 \cos 2\theta}$
0 to $\frac{\pi}{4}$	0 to $\frac{\pi}{2}$	1 to 0	36 to 0	$\left\{\begin{matrix} 6 \text{ to } 0 \\ -6 \text{ to } 0 \end{matrix}\right.$
$\frac{\pi}{4}$ to $\frac{\pi}{2}$	$\frac{\pi}{2}$ to π	0 to -1	0 to -36	Not defined
$\frac{\pi}{2}$ to $\frac{3\pi}{4}$	π to $\frac{3\pi}{2}$	-1 to 0	-36 to 0	Not defined
$\frac{3\pi}{4}$ to π	$\frac{3\pi}{2}$ to 2π	0 to 1	0 to 36	$\left\{\begin{matrix} 0 \text{ to } 6 \\ 0 \text{ to } -6 \end{matrix}\right.$

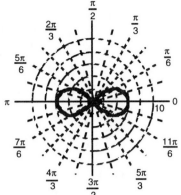

(B)

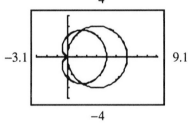

$(7-5)$

95. (A)

(B) 6

(C) We solve the system
$$r = 2 + 2 \cos \theta \qquad 0 \le \theta \le 2\pi$$
$$r = 6 \cos \theta$$
by equating the right sides:
$$6 \cos \theta = 2 + 2 \cos \theta$$
$$4 \cos \theta = 2$$
$$\cos \theta = \frac{1}{2}$$
$$\theta = \frac{\pi}{3}, \frac{5\pi}{3}$$

If we substitute these values of θ in either of the original equations, we get the corresponding values of r.

$$\theta = \frac{\pi}{3} \qquad r = 6 \cos \frac{\pi}{3} = 3$$
$$\theta = \frac{5\pi}{3} \qquad r = 6 \cos \frac{5\pi}{3} = 3$$

The two solutions of the system are $\left(3, \frac{\pi}{3}\right)$ and $\left(3, \frac{5\pi}{3}\right)$.

(D) The points on r2 and r1 arrive at the intersection points for different values of θ, except for the two found in part (C). $(7-5)$

96. The linear factors of $x^3 + i$ will, by the factor theorem, be $x - x_i$, where x_i are zeros of $x^3 + i$, that is, $x^3 + i = 0$ or $x^3 = -i$. Hence the x_i are the three cube roots of $-i$, or $1e^{(-90°)i}$.

Using the nth-root theorem, all three cube roots of $-i$ are given by
$$1^{1/3}e^{(-90°/3 + k \cdot 360°/3)i} = 1e^{(-30° + k \cdot 120°)i} \qquad k = 0, 1, 2$$

Thus,

$$w_1 = 1e^{(-30° + 0 \cdot 120°)i} = e^{-30°i} = \cos(-30°) + i\sin(-30°) = \frac{\sqrt{3}}{2} - \frac{i}{2}$$

$$w_2 = 1e^{(-30° + 1 \cdot 120°)i} = e^{90°i} = \cos 90° + i\sin 90° = i$$

$$w_3 = 1e^{(-30° + 2 \cdot 120°)i} = e^{210°i} = \cos 210° + i\sin 210° = -\frac{\sqrt{3}}{2} - \frac{i}{2}$$

Hence, $x^3 + i = \left[x - \left(\dfrac{\sqrt{3}}{2} - \dfrac{i}{2}\right)\right](x - i)\left[x - \left(-\dfrac{\sqrt{3}}{2} - \dfrac{i}{2}\right)\right]$ *(7-7)*

97. If three angles of a triangle are known, the sides of the triangle cannot be determined by any technique. There are an infinite number of triangles, all similar, that share the same three angles.   *(7-1, 7-2)*

98. False. For example, $\sin^{-1}(\sin \pi) = \sin^{-1} 0 = 0 \neq \pi$.   *(5-9)*

99. True. $\sin^{-1} x$ is defined for all x, $-1 \leq x \leq 1$. For these values of x, $\sin(\sin^{-1} x) = x$ by the sine-inverse sine identity.   *(5-9)*

100. True. If w is a cube root of 1, then $w^3 = 1$.
It follows that $w^6 = (w^3)^2 = 1^2 = 1$, hence w is a sixth root of 1.   *(7-7)*

101. False. For example, -1 is a cube root of -1, since $(-1)^3 = -1$.
However, $(-1)^6 = 1$, hence -1 is not a sixth root of -1.   *(7-7)*

102. False. It is not possible to give a counterexample, since this is not a general statement. However, i and j are perpendicular and do not have opposite directions.   *(7-3, 7-4)*

103. False. For example, $i = \dfrac{1}{2}(2i)$, thus i is a positive scalar multiple of $2i$. However $|i| = 1 < 2 = |2i|$.   *(7-4)*

104. False. For example, the graph of $y = \sin x$ does not intersect the graph of $y = 10$ at all.   *(5-6)*

105. True. Since $f(x + p) = f(x)$ for all x, if $f(x) = c$ for some value of x, then $f(x + p) = c$, $f(x + np) = c$ for any integer n, and the graph of f intersects the graph of $y = c$ infinitely many times.   *(5-6)*

106. In one year the line sweeps out one full revolution, or 2π radians. In 5 days the line sweeps out $\dfrac{5}{365}$ of a full revolution, or $\dfrac{5}{365} \cdot 2\pi = \dfrac{2\pi}{73}$ radians.   *(5-3)*

107. In the diagram, let *E* be the point from which the balloon was released. Then *BDE* and *BCE* are right triangles.

$$\text{In triangle } BDE, \cot 37° = \frac{DE}{h}$$

$$\text{In triangle } BCE, \cot 24° = \frac{1{,}000 + DE}{h}$$

$$\text{Thus, } \cot 24° - \cot 37° = \frac{1{,}000 + DE}{h} - \frac{DE}{h}$$

$$\cot 24° - \cot 37° = \frac{1{,}000}{h}$$

$$h(\cot 24° - \cot 37°) = 1{,}000$$

$$h = \frac{1{,}000}{\cot 24° - \cot 37°}$$

$$h = 1{,}088 \text{ m}$$

(5-5)

108. Sketch a figure:

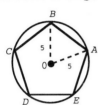

$$\text{Angle } AOB = \frac{1}{5}(360°) = 72°$$

We use the law of cosines in triangle *AOB* to determine side *AB*.

$$(AB)^2 = 5^2 + 5^2 - 2(5)(5)\cos 72°$$
$$= 34.54915…$$
$$AB = 5.88 \text{ in}$$

(5-5, 7-2)

109. The three sides of triangle *ABC* are each in turn the hypotenuse of a right triangle formed with two edges of the solid. Hence, by the Pythagorean theorem,

$$AB^2 = 12^2 + 14^2 = 340 \qquad AB = 18.44 \text{ cm}$$
$$AC^2 = 12^2 + 42^2 = 1{,}908 \qquad AC = 43.68 \text{ cm}$$
$$BC^2 = 14^2 + 42^2 = 1{,}960 \qquad BC = 44.27 \text{ cm}$$

To find ∠*ABC*, we apply the law of cosines.

$$AC^2 = AB^2 + BC^2 - 2(AB)(BC)\cos ABC$$

$$\cos ABC = \frac{AB^2 + BC^2 - AC^2}{2(AB)(BC)} = \frac{340 + 1{,}960 - 1{,}908}{2(18.44)(44.27)}$$

$$\angle ABC = \cos^{-1}\left(\frac{340 + 1{,}960 - 1{,}908}{2(18.44)(44.27)}\right)$$

$$= 76°$$

(7-2)

110. When *t* = 0, *I* = 50. Hence 50 = *A* cos *B*(0), that is, *A* = 50. Since the period is $\frac{1}{110}$ seconds, $\frac{2\pi}{B} = \frac{1}{110}$, *B* = 220π. Hence the equation is

$$I = 50 \cos 220\pi t$$

(5-7)

111. Sketch a
figure.

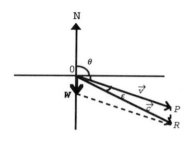

The actual course $\vec{c}$ will be the resultant of $\vec{v}$, the plane's
heading, and $\vec{w}$, the wind velocity. In triangle OPR, we know
$OP = |\vec{v}| = 260$, $PR = |\vec{w}| = 36$, and angle OPR, which is
congruent to θ, the heading of 110°. Given two sides and the
included angle, we use the law of cosines to find

$$|\vec{c}| = OR$$
$$|\vec{c}|^2 = OP^2 + PR^2 - 2(OP)(PR)\cos OPR$$
$$= (260)^2 + (36)^2 - 2(260)(36)\cos 110°$$
$$= 75,298.617...$$
$$|\vec{c}| = 274 \text{ miles per hour.}$$

To find the actual direction of the plane, we must find angle POR, or ε.
Then the heading will be $\theta + \varepsilon$. We use the law of sines.

$$\frac{\sin \varepsilon}{PR} = \frac{\sin OPR}{OR}$$
$$\frac{\sin \varepsilon}{36} = \frac{\sin 110°}{274}$$
$$\sin \varepsilon = \frac{36 \sin 110°}{274}$$
$$\sin \varepsilon = 0.1233$$
$$\varepsilon = \sin^{-1} 0.1233 \text{ since } \varepsilon \text{ is acute}$$
$$\varepsilon = 7°$$

Thus the heading of the plane's actual direction $= \theta + \varepsilon = 110° + 7° = 117°$.

(7-3)

112. First, form a force diagram with all forces in standard position at the origin.

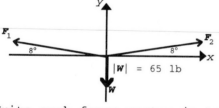

Let F_1 = the tension in left cable
F_2 = the tension in right cable

Write each force vector in terms of i and j unit vectors.

$$F_1 = |F_1|(-\cos 8°)i + |F_1|(\sin 8°)j$$
$$F_2 = |F_2|(\cos 8°)i + |F_2|(\sin 8°)j$$
$$W = -65j$$

For the system to be in static equilibrium, we must have
$$F_1 + F_2 + W = O$$
which becomes, on addition,
$$[-|F_1|\cos 8° + |F_2|\cos 8°]i + [|F_1|\sin 8° + |F_2|\sin 8° - 65]j = 0i + 0j$$

Since two vectors are equal if and only if their corresponding components are
equal, we are led to the following system of equations in $|F_1|$ and $|F_2|$:

$$-|F_1|\cos 8° + |F_2|\cos 8° = 0$$
$$|F_1|\sin 8° + |F_2|\sin 8° - 65 = 0$$

Solving, we obtain
$$|F_1| = |F_2| \text{ (as we expect from the symmetry of the situation)}$$
$$|F_1| = \frac{65}{2 \sin 8°} = 234 \text{ lb = tension in each half of the cable.}$$

(7-4)

113. (A) Add the perpendicular bisector of the chord as shown in the figure. Then, $\sin \theta = \frac{4}{R}$ and $\theta = \frac{5}{R}$.

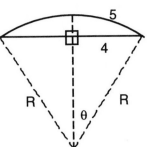

Substituting the second into the first, we obtain
$\sin \frac{5}{R} = \frac{4}{R}$.

(B) R cannot be isolated on one side of the equation.

(C) Plot $y1 = \sin \frac{5}{R}$ and $y2 = \frac{4}{R}$ in the same viewing window and solve for R at the point of intersection using a built-in routine (see figure).
To three decimal places, $R = 4.420$ cm.

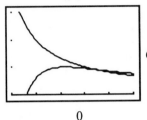

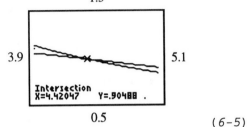

$(6-5)$

114. (A)

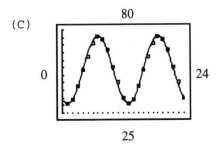

(B) $|A| = \frac{Y_{\max} - Y_{\min}}{2} = \frac{76 - 31}{2} = 22.5$. Hence $A = 22.5$ or -22.5.

$k = \frac{Y_{\max} + Y_{\min}}{2} = \frac{76 + 31}{2} = 53.5$ $P = 12 = 2\pi \cdot \frac{6}{\pi} = 2\pi \div \frac{\pi}{6}$. Hence $B = \frac{\pi}{6}$.

The x intercept closest to the origin is estimated from the graph as 4.0. To find C, substitute $B = \frac{\pi}{6}$ and $x = 4.0$ into the phase shift formula $x = -\frac{C}{B}$ and solve for C.

$$x = -\frac{C}{B}$$

$$4.0 = -C \div \frac{\pi}{6}$$

$$C = -4.0\left(\frac{\pi}{6}\right)$$

$$C = -2.1$$

With this value of C, the graph is seen to be shifted up from the graph of a standard sine curve, thus $A = 22.5$. The equation is thus

$$y = 53.5 + 22.5 \sin\left(\frac{\pi}{6}x - 2.1\right).$$

(C)

$(5-7)$

CHAPTER 8

Exercise 8-1

Key Ideas and Formulas

Systems of Linear Equations: Basic Terms

A system of linear equations is **consistent** if it has one or more solutions and **inconsistent** if no solutions exist. Furthermore, a consistent system is said to be **independent** if it has exactly one solution (often referred to as the **unique solution**) and **dependent** if it has more than one solution.

A linear system of equations in two variables

$$ax + by = h$$
$$cx + dy = k$$

must have

1. exactly one solution (consistent and independent) or
2. no solution (inconsistent) or
3. infinitely many solutions (consistent and dependent).

There are no other possibilities.

A linear system in two variables can be solved by graphing, by substitution, or by elimination using addition. (See text for descriptions of each method.)

Equivalent systems of equations are systems with the same solution set.

A system of linear equations is transformed into an equivalent system if:

1. Two equations are interchanged.
2. An equation is multiplied by a non-zero constant.
3. A constant multiple of another equation is added to a given equation.

A **matrix** is a rectangular array of numbers written within brackets. Each number in a matrix is called an **element** of the matrix.

If a matrix has m rows and n columns, it is called an $m \times n$ **matrix** (read "m by n matrix"). The expression $m \times n$ is called the **size** of the matrix and the numbers m and n are called the **dimensions** of the matrix.

A matrix with n rows and n columns is called a **square matrix of order n**. A matrix with only one column is called a **column matrix**, and a matrix with only one row is called a **row matrix**.

Associated with each linear system of the form:

$$a_{11}x_1 + a_{12}x_2 = k_1$$
$$a_{21}x_1 + a_{22}x_2 = k_2$$

where x_1 and x_2 are variables, is the **augmented matrix** of the system:

Column 1 (C_1)
Column 2 (C_2)
Column 3 (C_3)

$$\begin{bmatrix} a_{11} & a_{12} & k_1 \\ a_{21} & a_{22} & k_2 \end{bmatrix} \begin{matrix} \leftarrow \text{Row 1 } (R_1) \\ \leftarrow \text{Row 2 } (R_2) \end{matrix}$$

We solve a linear system by using row operations to transform it into a row equivalent matrix of one of these types:

$$\begin{bmatrix} 1 & 0 & m \\ 0 & 1 & n \end{bmatrix}$$ a unique solution (consistent and independent)

$$\begin{bmatrix} 1 & m & n \\ 0 & 0 & 0 \end{bmatrix}$$ infinitely many solutions (consistent and dependent)

$$\begin{bmatrix} 1 & m & \bigm| & n \\ 0 & 0 & \bigm| & p \end{bmatrix} \quad \text{no solution (inconsistent)}$$

An augmented matrix is transformed into a row-equivalent matrix if any of the following row operations is performed.

1. Two rows are interchanged. ($R_i \leftrightarrow R_j$ means "interchange row i and row j")

2. A row is multiplied by a non-zero constant ($kR_i \to R_i$ means "multiply row i by the constant k")

3. A constant multiple of another row is added to a given row ($kR_j + R_i \to R_i$ means "multiply row j by the constant k and add to R_i.")

---

1. Both lines in the given system are different, but they have the same slope $\left(\dfrac{1}{2}\right)$ and are therefore parallel. This system corresponds to (B) and has no solution.

3. In slope-intercept form, these equations are $y = 2x - 5$ and $y = -\dfrac{3}{2}x - \dfrac{3}{2}$. Thus, one has slope 2 and y intercept -5; the other has slope $-\dfrac{3}{2}$ and y intercept $-\dfrac{3}{2}$. This system corresponds to (D) and its solution can be read from the graph as $(1, -3)$. Checking, we see that
$$2x - y = 2 \cdot 1 - (-3) = 5$$
$$3x + 2y = 3 \cdot 1 + 2(-3) = -3$$

5.

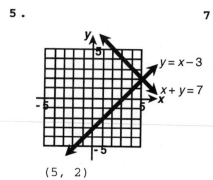

(5, 2)

7.

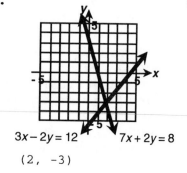

(2, -3)

9.

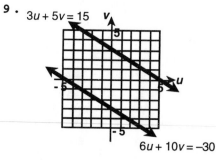

The lines are parallel.
No solution.

11. $2x + 3y = 1$
$3x - y = 7$
If we multiply the bottom equation by 3 and add, we can eliminate y.

$$\begin{aligned} 2x + 3y &= 1 \\ 9x - 3y &= 21 \\ \hline 11x &= 22 \\ x &= 2 \end{aligned}$$

Now substitute $x = 2$ back into the top equation and solve for y.
$$2(2) + 3y = 1$$
$$3y = -3$$
$$y = -1$$
$(2, -1)$

13. $4x - 3y = 15$
$3x + 4y = 5$
If we multiply the top equation by 4, the bottom by 3, and add, we can eliminate y.

$$\begin{aligned} 16x - 12y &= 60 \\ 9x + 12y &= 15 \\ \hline 25x &= 75 \\ x &= 3 \end{aligned}$$

Substituting in the bottom equation, we have
$$3(3) + 4y = 5$$
$$9 + 4y = 5$$
$$4y = -4$$
$$y = -1$$
$(3, -1)$

15. Matrix A has 2 rows and 3 columns, so its size is 2×3.
Matrix C has 1 row and 3 columns, so its size is 1×3.

17. C is the only row matrix.

19. B is the only square matrix.

21. a_{12} in the element is row 1, column 2 of A. This is -2.
a_{23} is the element in row 2, column 3 of A. This is -6.

23. The elements on the principal diagonal are b_{11} = -2, b_{22} = 6, and b_{33} = 0.

25. $R_1 \leftrightarrow R_2$ means interchange
Rows 1 and 2.
$$\begin{bmatrix} 4 & -6 & | & -8 \\ 1 & -3 & | & 2 \end{bmatrix}$$

27. $-4R_1 \rightarrow R_1$ means multiply
Row 1 by -4.
$$\begin{bmatrix} -4 & 12 & | & -8 \\ 4 & -6 & | & -8 \end{bmatrix}$$

29. $2R_2 \rightarrow R_2$ means multiply Row 2 by 2.
$$\begin{bmatrix} 1 & -3 & | & 2 \\ 8 & -12 & | & -16 \end{bmatrix}$$

31. $(-4)R_1 + R_2 \rightarrow R_2$ means replace Row 2
by itself plus -4 times Row 1.
$$\begin{bmatrix} 1 & -3 & | & 2 \\ 4 & -6 & | & -8 \end{bmatrix} \rightarrow \begin{bmatrix} 1 & -3 & | & 2 \\ 0 & 6 & | & -16 \end{bmatrix}$$
$-4 \quad 12 \quad -8$

33. $(-2)R_1 + R_2 \rightarrow R_2$ means replace Row
2 by itself plus -2 times Row 1.
$$\begin{bmatrix} 1 & -3 & | & 2 \\ 4 & -6 & | & -8 \end{bmatrix} \rightarrow \begin{bmatrix} 1 & -3 & | & 2 \\ 2 & 0 & | & -12 \end{bmatrix}$$
$-2 \quad 6 \quad -4$

35. $(-1)R_1 + R_2 \rightarrow R_2$ means replace Row 2
by itself plus -1 times Row 1.
$$\begin{bmatrix} 1 & -3 & | & 2 \\ 4 & -6 & | & -8 \end{bmatrix} \rightarrow \begin{bmatrix} 1 & -3 & | & 2 \\ 3 & -3 & | & -10 \end{bmatrix}$$
$-1 \quad 3 \quad -2$

37. We write the augmented matrix:
$$\begin{bmatrix} 1 & 1 & | & 7 \\ 1 & -1 & | & 1 \end{bmatrix} (-1)R_1 + R_2 \rightarrow R_2$$
$-1 \quad -1 \quad -7$
$\uparrow$
Need a 0 here

$\sim \begin{bmatrix} 1 & 1 & | & 7 \\ 0 & -2 & | & -6 \end{bmatrix} -\frac{1}{2}R_2 \rightarrow R_2$ corresponds to the linear system $\begin{matrix} x_1 + x_2 = 7 \\ -2x_2 = -6 \end{matrix}$
$\uparrow$
Need a 1 here
Need a 0 here
$\downarrow$

$\sim \begin{bmatrix} 1 & 1 & | & 7 \\ 0 & 1 & | & 3 \end{bmatrix} (-1)R_2 + R_1 \rightarrow R_1$ corresponds to the linear system $\begin{matrix} x_1 + x_2 = 7 \\ x_2 = 3 \end{matrix}$
$0 \quad -1 \quad -3$

$\sim \begin{bmatrix} 1 & 0 & | & 4 \\ 0 & 1 & | & 3 \end{bmatrix}$ corresponds to the linear system $\begin{matrix} x_1 = 4 \\ x_2 = 3 \end{matrix}$

The solution is $x_1 = 4$, $x_2 = 3$. Each pair of lines graphed below has the same intersection point, $(4, 3)$.

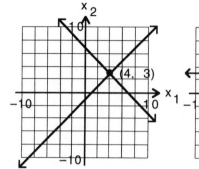 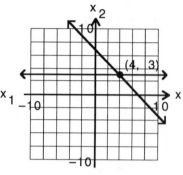 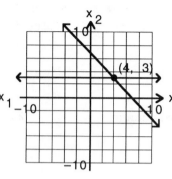 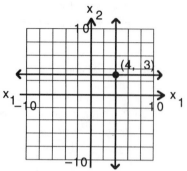

$$x_1 + x_2 = 7 \qquad x_1 + x_2 = 7 \qquad x_1 + x_2 = 7 \qquad x_1 = 4$$
$$x_1 - x_2 = 1 \qquad -2x_2 = -6 \qquad x_2 = 3 \qquad x_2 = 3$$

39. $\begin{bmatrix} 1 & -4 & | & -2 \\ -2 & 1 & | & -3 \end{bmatrix}$ $\quad 2R_1 + R_2 \rightarrow R_2$
↑
Need a 0 here
$\quad$ 2 $\quad$ -8 $\quad$ -4

$\sim \begin{bmatrix} 1 & -4 & | & -2 \\ 0 & -7 & | & -7 \end{bmatrix}$ $\quad -\frac{1}{7}R_2 \rightarrow R_2$
↑
Need a 1 here
Need a 0 here
↓

$\sim \begin{bmatrix} 1 & -4 & | & -2 \\ 0 & 1 & | & 1 \end{bmatrix}$ $\quad 4R_2 + R_1 \rightarrow R_1$

$\quad$ 0 $\quad$ 4 $\quad$ 4

$\sim \begin{bmatrix} 1 & 0 & | & 2 \\ 0 & 1 & | & 1 \end{bmatrix}$

Therefore $x_1 = 2$ and $x_2 = 1$

41. $\begin{bmatrix} 3 & -1 & | & 2 \\ 1 & 2 & | & 10 \end{bmatrix}$ $\quad R_1 \leftrightarrow R_2$

Need a 1 here

$\sim \begin{bmatrix} 1 & 2 & | & 10 \\ 3 & -1 & | & 2 \end{bmatrix}$ $\quad (-3)R_1 + R_2 \rightarrow R_2$
↑
Need a 0 here
$\quad$ -3 $\quad$ -6 $\quad$ -30

$\sim \begin{bmatrix} 1 & 2 & | & 10 \\ 0 & -7 & | & -28 \end{bmatrix}$ $\quad -\frac{1}{7}R_2 \rightarrow R_2$
↑
Need a 1 here
Need a 0 here
↓

$\sim \begin{bmatrix} 1 & 2 & | & 10 \\ 0 & 1 & | & 4 \end{bmatrix}$ $\quad (-2)R_2 + R_1 \rightarrow R_1$

$\quad$ 0 $\quad$ -2 $\quad$ -8

$\sim \begin{bmatrix} 1 & 0 & | & 2 \\ 0 & 1 & | & 4 \end{bmatrix}$

Therefore $x_1 = 2$ and $x_2 = 4$

43. $\begin{bmatrix} 1 & 2 & | & 4 \\ 2 & 4 & | & -8 \end{bmatrix}$ $\quad (-2)R_1 + R_2 \rightarrow R_2$
↑
Need a 0 here
$\quad$ -2 $\quad$ -4 $\quad$ -8

$\sim \begin{bmatrix} 1 & 2 & | & 4 \\ 0 & 0 & | & -16 \end{bmatrix}$

This matrix corresponds to the system
$$x_1 + 2x_2 = 4$$
$$0x_1 + 0x_2 = -16$$
This system has no solution.

45. $\begin{bmatrix} 2 & 1 & | & 6 \\ 1 & -1 & | & -3 \end{bmatrix}$ $R_1 \leftrightarrow R_2$

Need a 1 here

$\sim \begin{bmatrix} 1 & -1 & | & -3 \\ 2 & 1 & | & 6 \end{bmatrix}$ $(-2)R_1 + R_2 \rightarrow R_2$

↑
Need a 0 here
$-2 \quad 2 \quad 6$

$\sim \begin{bmatrix} 1 & -1 & | & -3 \\ 0 & 3 & | & 12 \end{bmatrix}$ $\frac{1}{3}R_2 \rightarrow R_2$

↑
Need a 1 here
Need a 0 here
↓

$\sim \begin{bmatrix} 1 & -1 & | & -3 \\ 0 & 1 & | & 4 \end{bmatrix}$ $R_2 + R_1 \rightarrow R_1$

$\sim \begin{bmatrix} 1 & 0 & | & 1 \\ 0 & 1 & | & 4 \end{bmatrix}$

Therefore $x_1 = 1$ and $x_2 = 4$.

47. $\begin{bmatrix} 3 & -6 & | & -9 \\ -2 & 4 & | & 6 \end{bmatrix}$ $\frac{1}{3}R_1 \rightarrow R_1$

Need a 1 here

$\sim \begin{bmatrix} 1 & -2 & | & -3 \\ -2 & 4 & | & 6 \end{bmatrix}$ $2R_1 + R_2 \rightarrow R_2$

↑
Need a 0 here
$2 \quad -4 \quad -6$

$\sim \begin{bmatrix} 1 & -2 & | & -3 \\ 0 & 0 & | & 0 \end{bmatrix}$

This matrix corresponds to the system
$x_1 - 2x_2 = -3$
$0x_1 + 0x_2 = 0$
Thus $x_1 = 2x_2 - 3$.
Hence there are infinitely many solutions: for any real number s, $x_2 = s$, $x_1 = 2s - 3$ is a solution.

49. $\begin{bmatrix} 4 & -2 & | & 2 \\ -6 & 3 & | & -3 \end{bmatrix}$ $\frac{1}{4}R_1 \rightarrow R_1$

Need a 1 here

$\sim \begin{bmatrix} 1 & -\frac{1}{2} & | & \frac{1}{2} \\ -6 & 3 & | & -3 \end{bmatrix}$ $6R_1 + R_2 \rightarrow R_2$

↑
Need a 0 here
$6 \quad -3 \quad 3$

$\sim \begin{bmatrix} 1 & -\frac{1}{2} & | & \frac{1}{2} \\ 0 & 0 & | & 0 \end{bmatrix}$

This matrix corresponds to the system
$x_1 - \frac{1}{2}x_2 = \frac{1}{2}$
$0x_1 + 0x_2 = 0$
Thus $x_1 = \frac{1}{2}x_2 + \frac{1}{2}$.
Hence there are infinitely many solutions: for any real number s, $x_2 = s$, $x_1 = \frac{1}{2}s + \frac{1}{2}$ is a solution.

51. Here is a computer-generated graph of the system , entered as

$y = \dfrac{2x + 5}{3}$

$y = \dfrac{13 - 3x}{4}$

After zooming in (graph not shown) the intersection point is located at (1.12, 2.41) to two decimal places.

53. Here is a computer-generated graph of the system, entered as

$$y = \frac{3.5x - 0.1}{2.4}$$

$$y = \frac{2.6x + 0.2}{1.7}$$

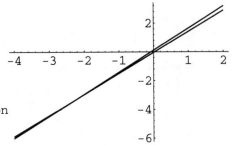

After zooming in (graph not shown) the intersection point is located at (-2.24, -3.31) to two decimal places.

55. Here are computer-generated graphs of the three systems:

(A)

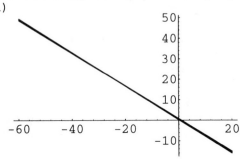

(B)

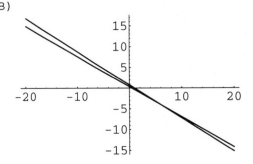

(C)

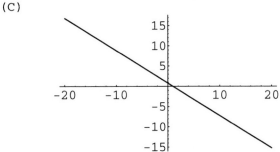

Clearly, the solutions are quite different even though the graph of the second equation changes only slightly from case to case. Solving by elimination yields:

(A)

$4x + 5y = 4$
$9x + 11y = 4$

Multiply the top equation by $-\frac{9}{4}$ and add to eliminate x.

$-9x - \frac{45}{4}y = -9$

$\underline{9x + 11y = 4}$

$-\frac{1}{4}y = -5$

$y = 20$

Now substitute $y = 20$ back into the top equation and solve for x.

$4x + 5(20) = 4$

$4x = -96$

$x = -24$

$(-24, 20)$

(B)

$4x + 5y = 4$
$8x + 11y = 4$

Multiply the top equation by -2 and add to eliminate x.

$-8x - 10y = -8$

$\underline{8x + 11y = 4}$

$y = -4$

Now substitute $y = -4$ back into the top equation and solve for x.

$4x + 5(-4) = 4$

$4x = 24$

$x = 6$

$(6, -4)$

(C)

$4x + 5y = 4$
$8x + 10y = 4$

Multiply the top equation by -2 and add.

$-8x - 10y = -8$

$\underline{8x + 10y = 4}$

$0 = -4$

No solution.

In Problems 57 and 59, steps are shown for the TI-82/3 series calculator. The student should consult the manual for the corresponding steps on the other models.

57. Enter the augmented matrix
Matrix [A]   2 × 3
[0.8 2.88 4]
[1.25 4.34 5]

          Need a 1 here

Perform $\dfrac{1}{0.8} R_1 \rightarrow R_1$:   Enter   *row(1/.8,[A],1)→[A]

[[1   3.6   5
  1.25 4.34   5]]
  ↑
Need a 0 here

Perform $(-1.25) R_1 + R_2 \rightarrow R_2$:   Enter   *row+(-1.25,[A],1,2)→[A]

[[1   3.6   5   ]
 [0   -.16 -1.25]]
        ↑
     Need a 1 here

Perform $\dfrac{1}{-0.16} R_2 \rightarrow R_2$: Enter   *row(-1/0.16,[A],2)→[A]

[[1   3.6   5   ]
 [0   1   7.8125]]

          Need a 0 here

Perform $(-3.6) R_2 + R_1 \rightarrow R_1$:   Enter   *row+(-3.6,[A],2,1)→[A]

[[1   0   -23.125]
 [0   1   7.8125]]

This corresponds to the system
  x_1 = -23.125
      x_2 = 7.8125
The solution is (-23.125, 7.8125).

59. Enter the augmented matrix
Matrix [A]   2 × 3
[4.8   -40.32   295.2]
[-3.75   28.7   -211.2]

          Need a 1 here

Perform $\dfrac{1}{4.8} R_1 \rightarrow R_1$:   Enter   *row(1/4.8,[A],1)→[A]

[[1   -8.4   61.5
 [-3.75   28.7   -211.2]]
  ↑
Need a 0 here

Perform $3.75 R_1 + R_2 \rightarrow R_2$:   Enter   *row+(3.75,[A],1,2)→[A]

[[1   -8.4   61.5
 [-1E-13 -2.8   19.425]]
            ↑
          Need a 1 here

Perform $\dfrac{1}{-2.8} R_2 \rightarrow R_2$:   Enter   *row(1/-2.8,[A],2)→[A]

```
[[1          -8.4      61.5]
 [3.57…E-14    1      -6.9375]]
```

Need a 0 here

Perform $8.4R_2 + R_1 \rightarrow R_1$:   Enter   *row+(8.4,[A],2,1)→[A]

```
[[1          1E-13     3.225]
 [3.57…E-14    1      -6.9375]]
```

This corresponds to the system

$x_1 \quad\ = 3.225$

$\quad x_2 = -6.9375$

The solution is $(3.225, -6.9375)$.

61. Let x = number of 32-cent stamps

$\quad\ y$ = number of 23-cent stamps

She bought 75 stamps, hence

$x + y = 75 \qquad\qquad$ (1)

She spent \$19.50, hence

$0.32x + 0.23y = 19.50 \quad$ (2)

We solve the system of equations (1), (2) using elimination by addition. If we multiply the top equation by -0.23 and add, we can eliminate y.

$$
\begin{array}{r}
-0.23x - 0.23y = -17.25 \\
0.32x + 0.23y = 19.50 \\
\hline
0.09x = 2.25 \\
x = 25
\end{array}
$$

Now substitute $x = 25$ back into equation (1) and solve for y

$25 + y = 75$

$\qquad y = 50$

25 32-cent stamps, 50 23-cent stamps

63. Let x = amount invested at 6%

$\quad\ y$ = amount invested at 9%

Summarize the given information in a table:

	Bond A	Bond B	Total
amount invested	x	y	200,000
interest earned	$0.06x$	$0.09y$	14,775

Form equations from the information:

$$
\left(\begin{array}{c}\text{Amount invested}\\ \text{in Bond } A\end{array}\right) + \left(\begin{array}{c}\text{Amount invested}\\ \text{in Bond } B\end{array}\right) = (\text{Total investment})
$$

$\qquad\ x \qquad\qquad + \qquad\quad y \qquad\quad = \qquad 200,000$

$$
\left(\begin{array}{c}\text{Interest earned}\\ \text{from Bond } A\end{array}\right) + \left(\begin{array}{c}\text{Interest earned}\\ \text{from Bond } B\end{array}\right) = \left(\begin{array}{c}\text{Total}\\ \text{Interest}\end{array}\right)
$$

$\quad\ 0.06x \qquad\ + \qquad\quad 0.09y \qquad = \qquad 14,775$

Solve this system using elimination by addition.

$$
\begin{array}{r}
-0.06x - 0.06y = -12,000 \\
0.06x + 0.09y = 14,775 \\
\hline
0.03y = 2,775 \\
y = 92,500 \\
x + 92,500 = 200,000 \\
x = 107,500
\end{array}
$$

\$107,500 in bond A and \$92,500 in bond B.

65. Let x = amount of 20% solution
y = amount of 80% solution

Summarize the given information in a table:

	20% solution	80% solution	62% solution
amount of solution	x	y	100
amount of acid	$0.2x$	$0.8y$	$0.62(100) = 62$

Form equations from the information:

$$\begin{pmatrix} \text{Amount of} \\ \text{20\% solution} \end{pmatrix} + \begin{pmatrix} \text{Amount of} \\ \text{80\% solution} \end{pmatrix} = \begin{pmatrix} \text{Amount of} \\ \text{62\% solution} \end{pmatrix}$$
$$x \quad + \quad y \quad = \quad 100$$
$$\begin{pmatrix} \text{Amount of acid} \\ \text{in 20\% solution} \end{pmatrix} + \begin{pmatrix} \text{Amount of acid} \\ \text{in 80\% solution} \end{pmatrix} = \begin{pmatrix} \text{Amount of acid} \\ \text{in 62\% solution} \end{pmatrix}$$
$$0.2x \quad + \quad 0.8y \quad = \quad 62$$

Solve this system using elimination by addition.

$$\begin{aligned} -0.2x - 0.2y &= -20 \\ \underline{0.2x + 0.8y} &= \underline{\ 62} \\ 0.6y &= 42 \\ y &= 70 \\ x + 70 &= 100 \\ x &= 30 \end{aligned}$$

30 liters of 20% solution and 70 liters of 80% solution.

67. Let x = number of grams of mix A
y = number of grams of mix B

Summarize the given information in a table:

	mix A	mix B
protein	0.15	0.3
fat	0.1	0.05

Form equations from the information:

$$\begin{pmatrix} \text{protein in} \\ x \text{ gm of mix } A \end{pmatrix} + \begin{pmatrix} \text{protein in} \\ y \text{ gm of mix } B \end{pmatrix} = \begin{pmatrix} \text{total} \\ \text{protein} \end{pmatrix}$$
$$0.15x \quad + \quad 0.3y \quad = \quad 54$$
$$\begin{pmatrix} \text{fat in } x \text{ gm} \\ \text{of mix } A \end{pmatrix} + \begin{pmatrix} \text{fat in } y \text{ gm} \\ \text{of mix } B \end{pmatrix} = \begin{pmatrix} \text{total} \\ \text{fat} \end{pmatrix}$$
$$0.1x \quad + \quad 0.05y \quad = \quad 24$$

Solve using elimination by addition.

$$\begin{aligned} 0.15x + 0.3y &= 54 \\ \underline{-0.6x - 0.3y} &= \underline{-144} \\ -0.45x &= -90 \\ x &= 200 \\ 0.1(200) + 0.05y &= 24 \\ 0.05y &= 4 \\ y &= 80 \end{aligned}$$

200 grams of mix A and 80 grams of mix B.

69. Let x = base price

y = surcharge for each additional pound.

Since a 5-pound package costs the base price plus 4 surcharges,

$x + 4y = 27.75$

Since a 20-pound package costs the base price plus 19 surcharges,

$x + 19y = 64.50$

Solve using elimination by addition.

$$
\begin{array}{r}
-x - 4y = -27.75 \\
\underline{x + 19y = 64.50} \\
15y = 36.75 \\
y = 2.45
\end{array}
$$

$x + 4(2.45) = 27.75$

$x = 17.95$

The base price is \$17.95 and the surcharge per pound is \$2.45.

71. Let x = number of pounds of robust blend

y = number of pounds of mild blend

Summarize the given information in a table:

	Robust blend	Mild blend
ozs. of Columbian beans	12	6
ozs. of Brazilian beans	4	10

Form equations from the information:

$$
\begin{pmatrix} \text{pounds of Columbian} \\ \text{beans needed for} \\ \text{robust blend} \end{pmatrix} + \begin{pmatrix} \text{pounds of Columbian} \\ \text{beans needed for} \\ \text{mild blend} \end{pmatrix} = \begin{pmatrix} \text{Total} \\ \text{Columbian beans} \\ \text{available} \end{pmatrix}
$$

$$
\frac{12}{16}x + \frac{6}{16}Y = 50(132)
$$

$$
\begin{pmatrix} \text{pounds of Brazilian} \\ \text{beans needed for} \\ \text{robust blend} \end{pmatrix} + \begin{pmatrix} \text{pounds of Brazilian} \\ \text{beans needed for} \\ \text{mild blend} \end{pmatrix} = \begin{pmatrix} \text{Total} \\ \text{Brazilian beans} \\ \text{available} \end{pmatrix}
$$

$$
\frac{4}{16}x + \frac{10}{16}Y = 40(132)
$$

Solve using elimination by addition:

$$
\begin{array}{r}
\frac{12}{16}x + \frac{6}{16}y = 6,600 \\
-\frac{12}{16}x - \frac{30}{16}y = -15,840 \\
-\frac{24}{16}y = -9,240 \\
y = 6,160
\end{array}
$$

$\frac{4}{16}x + \frac{10}{16}(6,160) = 40(132)$

$\frac{1}{4}x + 3,850 = 5,280$

$\frac{1}{4}x = 1,430$

$x = 5,720$

5,720 pounds of the robust blend and 6,160 pounds of the mild blend.

Exercise 8-2

Key Ideas and Formulas

Any linear system must have exactly one solution, no solution, or an infinite number of solutions, regardless of the number of equations or the number of variables in the system. The terms **unique solution**, **consistent**, **inconsistent**, **dependent**, and **independent** are used to describe these solutions, just as they are for systems with two variables.

The method used for solving large systems of linear equations is called Gauss-Jordan elimination. We use a step-by-step procedure to transform the augmented matrix into reduced form, from which the solution to the system can be read by inspection. A matrix is in reduced form if:

1. Each row consisting entirely of 0's is below any row having at least one non-zero element.
2. The leftmost non-zero element in each row is 1.
3. The column containing the leftmost 1 of a given row has 0's above and below the 1.
4. The leftmost 1 in any row is to the right of the leftmost 1 in the preceding row.

Gauss-Jordan Elimination:

Write the augmented matrix of the system. Then

Step 1. Choose the leftmost non-zero column and use appropriate row operations to get a 1 at the top.

Step 2. Use multiples of the row containing the 1 from step 1 to get zeros in all remaining places in the column containing this 1.

Step 3. Repeat step 1 with the **submatrix** formed by (mentally) deleting the row used in step 2 and all rows above this row.

Step 4. Repeat step 2 with the **entire matrix**, including the mentally deleted rows. Continue this process until it is impossible to go further.

Note 1: (inconsistent system)
If at any point in this process we obtain a row with all zeros to the left of the vertical line and a non-zero number to the right, we can stop, since we will have a contradiction: $0 = n$, $n \neq 0$. We can then conclude that the system has no solution.

Note 2: (dependent system)
If the number of leftmost 1's in a reduced augmented coefficient matrix is less than the number of variables in the system and there are no contradictions, then the system is dependent and has infinitely many solutions.

1. No. Condition 2 is violated.

3. Yes

5. No. Condition 4 is violated.

7. Yes

9. $x_1 \qquad = -2$
$\qquad x_2 \qquad = 3$
$\qquad \qquad x_3 = 0$

The system is already solved.

11. $x_1 \qquad - 2x_3 = 3$
$\qquad x_2 + x_3 = -5$

Solution:
$x_3 = t$
$x_2 = -5 - x_3 = -5 - t$
$x_1 = 3 + 2x_3 = 3 + 2t$
Thus $x_1 = 2t + 3$, $x_2 = -t - 5$,
$x_3 = t$ is the solution for t any real number.

13.
$$
\begin{aligned}
x_1 \quad\quad &= 0 \\
x_2 \quad &= 0 \\
0 &= 1
\end{aligned}
$$

The system has no solution.

15.
$$
\begin{aligned}
x_1 - 2x_2 \quad\quad - 3x_4 &= -5 \\
x_3 + 3x_4 &= 2
\end{aligned}
$$

Solution:
$x_4 = t$
$x_3 = 2 - 3x_4 = 2 - 3t$
$x_2 = s$
$x_1 = -5 + 2x_2 + 3x_4 = -5 + 2s + 3t$
Thus $x_1 = 2s + 3t - 5$, $x_2 = s$,
$x_3 = -3t + 2$, $x_4 = t$ is the
solution, for s and t any real
numbers.

17. $\begin{bmatrix} 1 & 2 & | & -1 \\ 0 & 1 & | & 3 \end{bmatrix}$ $(-2)R_2 + R_1 \rightarrow R_1$

Need a 0 here

$\sim \begin{bmatrix} 1 & 0 & | & -7 \\ 0 & 1 & | & 3 \end{bmatrix}$

19. $\begin{bmatrix} 1 & 0 & -3 & | & 1 \\ 0 & 1 & 2 & | & 0 \\ 0 & 0 & 3 & | & -6 \end{bmatrix}$ $\frac{1}{3}R_3 \rightarrow R_3$

Need a 1 here
Need 0's here

$\sim \begin{bmatrix} 1 & 0 & -3 & | & 1 \\ 0 & 1 & 2 & | & 0 \\ 0 & 0 & 1 & | & -2 \end{bmatrix}$ $3R_3 + R_1 \rightarrow R_1$
$(-2)R_3 + R_2 \rightarrow R_2$

$\sim \begin{bmatrix} 1 & 0 & 0 & | & -5 \\ 0 & 1 & 0 & | & 4 \\ 0 & 0 & 1 & | & -2 \end{bmatrix}$

21. $\begin{bmatrix} 1 & 2 & -2 & | & -1 \\ 0 & 3 & -6 & | & 1 \\ 0 & -1 & 2 & | & -\frac{1}{3} \end{bmatrix}$ $\frac{1}{3}R_2 \rightarrow R_2$

Need a 1 here

$\sim \begin{bmatrix} 1 & 2 & -2 & | & -1 \\ 0 & 1 & -2 & | & \frac{1}{3} \\ 0 & -1 & 2 & | & -\frac{1}{3} \end{bmatrix}$ $(-2)R_2 + R_1 \rightarrow R_1$
$R_3 + R_2 \rightarrow R_3$

Need 0's here

$\sim \begin{bmatrix} 1 & 0 & 2 & | & -\frac{5}{3} \\ 0 & 1 & -2 & | & \frac{1}{3} \\ 0 & 0 & 0 & | & 0 \end{bmatrix}$

23. $\begin{bmatrix} 2 & 4 & -10 & | & -2 \\ 3 & 9 & -21 & | & 0 \\ 1 & 5 & -12 & | & 1 \end{bmatrix}$ $R_1 \leftrightarrow R_3$

Need a 1 here

$\sim \begin{bmatrix} 1 & 5 & -12 & | & 1 \\ 3 & 9 & -21 & | & 0 \\ 2 & 4 & -10 & | & -2 \end{bmatrix}$ $\begin{array}{l}(-3)R_1 + R_2 \rightarrow R_2 \\ (-2)R_1 + R_3 \rightarrow R_3\end{array}$

Need 0's here

$\sim \begin{bmatrix} 1 & 5 & -12 & | & 1 \\ 0 & -6 & 15 & | & -3 \\ 0 & -6 & 14 & | & -4 \end{bmatrix}$ $-\frac{1}{6}R_2 \rightarrow R_2$

Need a 1 here

$\sim \begin{bmatrix} 1 & 5 & -12 & | & 1 \\ 0 & 1 & -\frac{5}{2} & | & \frac{1}{2} \\ 0 & -6 & 14 & | & -4 \end{bmatrix}$ $\begin{array}{l}(-5)R_2 + R_1 \rightarrow R_1 \\ 6R_2 + R_3 \rightarrow R_3\end{array}$

Need 0's here

$\sim \begin{bmatrix} 1 & 0 & \frac{1}{2} & | & -\frac{3}{2} \\ 0 & 1 & -\frac{5}{2} & | & \frac{1}{2} \\ 0 & 0 & -1 & | & -1 \end{bmatrix}$ $-R_3 \rightarrow R_3$

Need a 1 here

$\sim \begin{bmatrix} 1 & 0 & \frac{1}{2} & | & -\frac{3}{2} \\ 0 & 1 & -\frac{5}{2} & | & \frac{1}{2} \\ 0 & 0 & 1 & | & 1 \end{bmatrix}$ $\begin{array}{l}(-\frac{1}{2})R_3 + R_1 \rightarrow R_1 \\ \frac{5}{2}R_3 + R_2 \rightarrow R_2\end{array}$

Need 0's here

$\sim \begin{bmatrix} 1 & 0 & 0 & | & -2 \\ 0 & 1 & 0 & | & 3 \\ 0 & 0 & 1 & | & 1 \end{bmatrix}$

Therefore $x_1 = -2$, $x_2 = 3$, and $x_3 = 1$.

25. $\begin{bmatrix} 3 & 8 & -1 & | & -18 \\ 2 & 1 & 5 & | & 8 \\ 2 & 4 & 2 & | & -4 \end{bmatrix}$ $\begin{array}{l}\frac{1}{2}R_3 \rightarrow R_3 \\ R_3 \leftrightarrow R_1\end{array}$

Need a 1 here

$\sim \begin{bmatrix} 1 & 2 & 1 & | & -2 \\ 2 & 1 & 5 & | & 8 \\ 3 & 8 & -1 & | & -18 \end{bmatrix}$ $\begin{array}{l}(-2)R_1 + R_2 \rightarrow R_2 \\ (-3)R_1 + R_3 \rightarrow R_3\end{array}$

Need 0's here

$\sim \begin{bmatrix} 1 & 2 & 1 & | & -2 \\ 0 & -3 & 3 & | & 12 \\ 0 & 2 & -4 & | & -12 \end{bmatrix}$ $-\frac{1}{3}R_2 \rightarrow R_2$

Need a 1 here

$\sim \begin{bmatrix} 1 & 2 & 1 & | & -2 \\ 0 & 1 & -1 & | & -4 \\ 0 & 2 & -4 & | & -12 \end{bmatrix}$ $\begin{array}{l}(-2)R_2 + R_1 \rightarrow R_1 \\ (-2)R_2 + R_3 \rightarrow R_3\end{array}$

Need 0's here

$\sim \begin{bmatrix} 1 & 0 & 3 & | & 6 \\ 0 & 1 & -1 & | & -4 \\ 0 & 0 & -2 & | & -4 \end{bmatrix}$ $-\frac{1}{2}R_3 \rightarrow R_3$

Need a 1 here

$\sim \begin{bmatrix} 1 & 0 & 3 & | & 6 \\ 0 & 1 & -1 & | & -4 \\ 0 & 0 & 1 & | & 2 \end{bmatrix}$ $\begin{array}{l}(-3)R_3 + R_1 \rightarrow R_1 \\ R_3 + R_2 \rightarrow R_2\end{array}$

Need 0's here

$\begin{bmatrix} 1 & 0 & 0 & | & 0 \\ 0 & 1 & 0 & | & -2 \\ 0 & 0 & 1 & | & 2 \end{bmatrix}$

Therefore $x_1 = 0$, $x_2 = -2$, and $x_3 = 2$.

27. $\begin{bmatrix} 2 & -1 & -3 & | & 8 \\ 1 & -2 & 0 & | & 7 \end{bmatrix}$ $R_1 \leftrightarrow R_2$

$\sim \begin{bmatrix} 1 & -2 & 0 & | & 7 \\ 2 & -1 & -3 & | & 8 \end{bmatrix}$ $(-2)R_1 + R_2 \rightarrow R_2$

$\sim \begin{bmatrix} 1 & -2 & 0 & | & 7 \\ 0 & 3 & -3 & | & -6 \end{bmatrix}$ $\frac{1}{3}R_2 \rightarrow R_2$

$\sim \begin{bmatrix} 1 & -2 & 0 & | & 7 \\ 0 & 1 & -1 & | & -2 \end{bmatrix}$ $2R_2 + R_1 \rightarrow R_1$

$\sim \begin{bmatrix} 1 & 0 & -2 & | & 3 \\ 0 & 1 & -1 & | & -2 \end{bmatrix}$

Let $x_3 = t$. Then

$x_2 - x_3 \quad\quad = -2$

$\quad x_2 = x_3 - 2 = t - 2$

$x_1 - 2x_3 \quad = 3$

$\quad x_1 = 2x_3 + 3 = 2t + 3$

Solution: $x_1 = 2t + 3$, $x_2 = t - 2$, $x_3 = t$, t any real number.

31. $\begin{bmatrix} 3 & -4 & -1 & | & 1 \\ 2 & -3 & 1 & | & 1 \\ 1 & -2 & 3 & | & 2 \end{bmatrix}$ $R_1 \leftrightarrow R_3$

$\sim \begin{bmatrix} 1 & -2 & 3 & | & 2 \\ 2 & -3 & 1 & | & 1 \\ 3 & -4 & -1 & | & 1 \end{bmatrix}$ $(-2)R_1 + R_2 \rightarrow R_2$ $(-3)R_1 + R_3 \rightarrow R_3$

$\sim \begin{bmatrix} 1 & -2 & 3 & | & 2 \\ 0 & 1 & -5 & | & -3 \\ 0 & 2 & -10 & | & -5 \end{bmatrix}$ $(-2)R_2 + R_3 \rightarrow R_3$

$\sim \begin{bmatrix} 1 & -2 & 3 & | & 2 \\ 0 & 1 & -5 & | & -3 \\ 0 & 0 & 0 & | & 1 \end{bmatrix}$

Since the last row corresponds to the equation $0x_1 + 0x_2 + 0x_3 = 1$, there is no solution.

33. $\begin{bmatrix} -2 & 1 & 3 & | & -7 \\ 1 & -4 & 2 & | & 0 \\ 1 & -3 & 1 & | & 1 \end{bmatrix}$ $R_1 \leftrightarrow R_2$

$\sim \begin{bmatrix} 1 & -4 & 2 & | & 0 \\ -2 & 1 & 3 & | & -7 \\ 1 & -3 & 1 & | & 1 \end{bmatrix}$ $2R_1 + R_2 \rightarrow R_2$ $(-1)R_1 + R_3 \rightarrow R_3$

$\sim \begin{bmatrix} 1 & -4 & 2 & | & 0 \\ 0 & -7 & 7 & | & -7 \\ 0 & 1 & -1 & | & 1 \end{bmatrix}$ $R_2 \leftrightarrow R_3$

$\sim \begin{bmatrix} 1 & -4 & 2 & | & 0 \\ 0 & 1 & -1 & | & 1 \\ 0 & -7 & 7 & | & -7 \end{bmatrix}$ $4R_2 + R_1 \rightarrow R_1$ $7R_2 + R_3 \rightarrow R_3$

$\sim \begin{bmatrix} 1 & 0 & -2 & | & 4 \\ 0 & 1 & -1 & | & 1 \\ 0 & 0 & 0 & | & 0 \end{bmatrix}$

29. $\begin{bmatrix} 2 & -1 & | & 0 \\ 3 & 2 & | & 7 \\ 1 & -1 & | & -1 \end{bmatrix}$ $R_1 \leftrightarrow R_3$

$\sim \begin{bmatrix} 1 & -1 & | & -1 \\ 3 & 2 & | & 7 \\ 2 & -1 & | & 0 \end{bmatrix}$ $(-3)R_1 + R_2 \rightarrow R_2$ $(-2)R_1 + R_3 \rightarrow R_3$

$\sim \begin{bmatrix} 1 & -1 & | & -1 \\ 0 & 5 & | & 10 \\ 0 & 1 & | & 2 \end{bmatrix}$ $R_2 \leftrightarrow R_3$

$\sim \begin{bmatrix} 1 & -1 & | & -1 \\ 0 & 1 & | & 2 \\ 0 & 5 & | & 10 \end{bmatrix}$ $R_2 + R_1 \rightarrow R_1$ $(-5)R_2 + R_3 \rightarrow R_3$

$\sim \begin{bmatrix} 1 & 0 & | & 1 \\ 0 & 1 & | & 2 \\ 0 & 0 & | & 0 \end{bmatrix}$

Therefore, $x_1 = 1$ and $x_2 = 2$.

Let $x_3 = t$. Then

$\quad x_2 - x_3 = 1$

$\quad\quad x_2 = x_3 + 1 = t + 1$

$\quad x_1 - 2x_3 = 4$

$x_1 = 2x_3 + 4 = 2t + 4$

Solution: $x_1 = 2t + 4$, $x_2 = t + 1$, $x_3 = t$, t any real number.

35. $\begin{bmatrix} 2 & -2 & -4 & | & -2 \\ -3 & 3 & 6 & | & 3 \end{bmatrix}$ $\frac{1}{2}R_1 \rightarrow R_1$
$\frac{1}{3}R_2 \rightarrow R_2$

$\sim \begin{bmatrix} 1 & -1 & -2 & | & -1 \\ -1 & 1 & 2 & | & 1 \end{bmatrix}$ $R_1 + R_2 \rightarrow R_2$

$\sim \begin{bmatrix} 1 & -1 & -2 & | & -1 \\ 0 & 0 & 0 & | & 0 \end{bmatrix}$

Let $x_3 = t$, $x_2 = s$. Then
$x_1 - x_2 - 2x_3 = -1$
$$x_1 = x_2 + 2x_3 - 1$$
$$= s + 2t - 1$$
Solution: $x_1 = s + 2t - 1$, $x_2 = s$, $x_3 = t$, s and t any real numbers.

37. $\begin{bmatrix} 4 & -1 & 2 & | & 3 \\ -4 & 1 & -3 & | & -10 \\ 8 & -2 & 9 & | & -1 \end{bmatrix}$ $R_1 + R_2 \rightarrow R_2$
$(-2)R_1 + R_3 \rightarrow R_3$

$\sim \begin{bmatrix} 4 & -1 & 2 & | & 3 \\ 0 & 0 & -1 & | & -7 \\ 0 & 0 & 5 & | & -7 \end{bmatrix}$ $5R_2 + R_3 \rightarrow R_3$

$\sim \begin{bmatrix} 4 & -1 & 2 & | & 3 \\ 0 & 0 & -1 & | & -7 \\ 0 & 0 & 0 & | & -42 \end{bmatrix}$ Since the last row corresponds to the equation $0x_1 + 0x_2 + 0x_3 = -42$, there is no solution.

39. $\begin{bmatrix} 2 & -5 & -3 & | & 7 \\ -4 & 10 & 2 & | & 6 \\ 6 & -15 & -1 & | & -19 \end{bmatrix}$ $2R_1 + R_2 \rightarrow R_2$
$(-3)R_1 + R_3 \rightarrow R_3$

$\sim \begin{bmatrix} 2 & -5 & -3 & | & 7 \\ 0 & 0 & -4 & | & 20 \\ 0 & 0 & 8 & | & -40 \end{bmatrix}$ $\frac{1}{2}R_1 \rightarrow R_1$
$-\frac{1}{4}R_2 \rightarrow R_2$

$\sim \begin{bmatrix} 1 & -2.5 & -1.5 & | & 3.5 \\ 0 & 0 & 1 & | & -5 \\ 0 & 0 & 8 & | & -40 \end{bmatrix}$ $1.5R_2 + R_1 \rightarrow R_1$

$(-8)R_2 + R_3 \rightarrow R_3$

$\sim \begin{bmatrix} 1 & -2.5 & 0 & | & -4 \\ 0 & 0 & 1 & | & -5 \\ 0 & 0 & 0 & | & 0 \end{bmatrix}$

Let $x_2 = t$. Then $x_3 = -5$ and
$x_1 - 2.5x_2 = -4$
$$x_1 = 2.5x_2 - 4$$
$$x_1 = 2.5t - 4$$

Solution: $x_1 = 2.5t - 4$, $x_2 = t$, $x_3 = -5$, t any real number.

41. $\begin{bmatrix} 5 & -3 & 2 & | & 13 \\ 2 & -1 & -3 & | & 1 \\ 4 & -2 & 4 & | & 12 \end{bmatrix}$ $\frac{1}{4}R_3 \rightarrow R_3$

$\sim \begin{bmatrix} 5 & -3 & 2 & | & 13 \\ 2 & -1 & -3 & | & 1 \\ 1 & -\frac{1}{2} & 1 & | & 3 \end{bmatrix}$ $R_1 \leftrightarrow R_3$

$$\sim \begin{bmatrix} 1 & -\frac{1}{2} & 1 & \Big| & 3 \\ 2 & -1 & -3 & \Big| & 1 \\ 5 & -3 & 2 & \Big| & 13 \end{bmatrix} \quad \begin{array}{l} (-2)\,R_1 \,+\, R_2 \to R_2 \\ \\ (-5)\,R_1 \,+\, R_3 \to R_3 \end{array}$$

$$\sim \begin{bmatrix} 1 & -\frac{1}{2} & 1 & \Big| & 3 \\ 0 & 0 & -5 & \Big| & -5 \\ 0 & -\frac{1}{2} & -3 & \Big| & -2 \end{bmatrix} \quad \begin{array}{l} (-1)\,R_3 \,+\, R_1 \to R_1 \\ \\ -\frac{1}{5}R_2 \to R_2 \end{array}$$

$$\sim \begin{bmatrix} 1 & 0 & 4 & \Big| & 5 \\ 0 & 0 & 1 & \Big| & 1 \\ 0 & -\frac{1}{2} & -3 & \Big| & -2 \end{bmatrix} \quad \begin{array}{l} (-4)\,R_2 \,+\, R_1 \to R_1 \\ \\ 3R_2 \,+\, R_3 \to R_3 \end{array}$$

$$\sim \begin{bmatrix} 1 & 0 & 0 & \Big| & 1 \\ 0 & 0 & 1 & \Big| & 1 \\ 0 & -\frac{1}{2} & 0 & \Big| & 1 \end{bmatrix} \quad (-2)\,R_3 \to R_3$$

$$\sim \begin{bmatrix} 1 & 0 & 0 & \Big| & 1 \\ 0 & 0 & 1 & \Big| & 1 \\ 0 & 1 & 0 & \Big| & -2 \end{bmatrix} \quad R_2 \leftrightarrow R_3$$

$$\sim \begin{bmatrix} 1 & 0 & 0 & \Big| & 1 \\ 0 & 1 & 0 & \Big| & -2 \\ 0 & 0 & 1 & \Big| & 1 \end{bmatrix}$$

Solution: $x_1 = 1$, $x_2 = -2$, $x_3 = 1$.

43. (A) The reduced form matrix will have the form

$$\begin{bmatrix} 1 & a & b & \Big| & c \\ 0 & 0 & 0 & \Big| & 0 \\ 0 & 0 & 0 & \Big| & 0 \end{bmatrix}$$

Thus, the system has been shown equivalent to

$$x_1 + ax_2 + bx_3 = c$$
$$0 = 0$$
$$0 = 0$$

The system is dependent, and x_2 and x_3 may assume any real values. Thus, there are two parameters in the solution.

(B) The reduced form matrix will have the form

$$\begin{bmatrix} 1 & 0 & a & \Big| & b \\ 0 & 1 & c & \Big| & d \\ 0 & 0 & 0 & \Big| & 0 \end{bmatrix}$$

Thus, the system has been shown equivalent to

$$x_1 + ax_3 = b$$
$$x_2 + cx_3 = d$$
$$0 = 0$$

The system is dependent, with a solution for any real value of x_3. Thus, there is one parameter in the solution.

(C) The reduced form matrix will have the form

$$\begin{bmatrix} 1 & 0 & 0 & \Big| & a \\ 0 & 1 & 0 & \Big| & b \\ 0 & 0 & 1 & \Big| & c \end{bmatrix}$$

Thus, there is only one solution, $x_1 = a$, $x_2 = b$, $x_3 = c$, and the system is independent.

(D) This is impossible; there are only 3 equations.

45. $\begin{bmatrix} 1 & 2 & -4 & -1 & \bigm| & 7 \\ 2 & 5 & -9 & -4 & \bigm| & 16 \\ 1 & 5 & -7 & -7 & \bigm| & 13 \end{bmatrix}$ $\begin{aligned} (-2)R_1 + R_2 &\to R_2 \\ (-1)R_1 + R_3 &\to R_3 \end{aligned}$

$\sim \begin{bmatrix} 1 & 2 & -4 & -1 & \bigm| & 7 \\ 0 & 1 & -1 & -2 & \bigm| & 2 \\ 0 & 3 & -3 & -6 & \bigm| & 6 \end{bmatrix}$ $\begin{aligned} (-2)R_2 + R_1 &\to R_1 \\[1em] (-3)R_2 + R_3 &\to R_3 \end{aligned}$

$\sim \begin{bmatrix} 1 & 0 & -2 & 3 & \bigm| & 3 \\ 0 & 1 & -1 & -2 & \bigm| & 2 \\ 0 & 0 & 0 & 0 & \bigm| & 0 \end{bmatrix}$

Let $x_4 = t$, $x_3 = s$. Then

$$x_2 - x_3 - 2x_4 = 2$$
$$x_2 = s + 2t + 2$$
$$x_1 - 2x_3 + 3x_4 = 3$$
$$x_1 = 2x_3 - 3x_4 + 3$$
$$= 2s - 3t + 3$$

Solution: $x_1 = 2s - 3t + 3$, $x_2 = s + 2t + 2$, $x_3 = s$, $x_4 = t$, s and t any real numbers.

47. $\begin{bmatrix} 1 & -1 & 3 & -2 & \bigm| & 1 \\ -2 & 4 & -3 & 1 & \bigm| & 0.5 \\ 3 & -1 & 10 & -4 & \bigm| & 2.9 \\ 4 & -3 & 8 & -2 & \bigm| & 0.6 \end{bmatrix}$ $\begin{aligned} 2R_1 + R_2 &\to R_2 \\ (-3)R_1 + R_3 &\to R_3 \\ (-4)R_1 + R_4 &\to R_4 \end{aligned}$

$\sim \begin{bmatrix} 1 & -1 & 3 & -2 & \bigm| & 1 \\ 0 & 2 & 3 & -3 & \bigm| & 2.5 \\ 0 & 2 & 1 & 2 & \bigm| & -0.1 \\ 0 & 1 & -4 & 6 & \bigm| & -3.4 \end{bmatrix}$ $R_4 \leftrightarrow R_2$

$\sim \begin{bmatrix} 1 & -1 & 3 & -2 & \bigm| & 1 \\ 0 & 1 & -4 & 6 & \bigm| & -3.4 \\ 0 & 2 & 1 & 2 & \bigm| & -0.1 \\ 0 & 2 & 3 & -3 & \bigm| & 2.5 \end{bmatrix}$ $\begin{aligned} R_2 + R_1 &\to R_1 \\[1em] (-2)R_2 + R_3 &\to R_3 \\ (-2)R_2 + R_4 &\to R_4 \end{aligned}$

$\sim \begin{bmatrix} 1 & 0 & -1 & 4 & \bigm| & -2.4 \\ 0 & 1 & -4 & 6 & \bigm| & -3.4 \\ 0 & 0 & 9 & -10 & \bigm| & 6.7 \\ 0 & 0 & 11 & -15 & \bigm| & 9.3 \end{bmatrix}$ $(-1)R_4 + R_3 \to R_3$

$\sim \begin{bmatrix} 1 & 0 & -1 & 4 & \bigm| & -2.4 \\ 0 & 1 & -4 & 6 & \bigm| & -3.4 \\ 0 & 0 & -2 & 5 & \bigm| & -2.6 \\ 0 & 0 & 11 & -15 & \bigm| & 9.3 \end{bmatrix}$ $-\tfrac{1}{2}R_3 \leftrightarrow R_3$

$\sim \begin{bmatrix} 1 & 0 & -1 & 4 & \bigm| & -2.4 \\ 0 & 1 & -4 & 6 & \bigm| & -3.4 \\ 0 & 0 & 1 & -2.5 & \bigm| & 1.3 \\ 0 & 0 & 11 & -15 & \bigm| & 9.3 \end{bmatrix}$ $\begin{aligned} R_3 + R_1 &\to R_1 \\ 4R_3 + R_2 &\to R_2 \\[0.5em] (-11)R_3 + R_4 &\to R_4 \end{aligned}$

$\sim \begin{bmatrix} 1 & 0 & 0 & 1.5 & \bigm| & -1.1 \\ 0 & 1 & 0 & -4 & \bigm| & 1.8 \\ 0 & 0 & 1 & -2.5 & \bigm| & 1.3 \\ 0 & 0 & 0 & 12.5 & \bigm| & -5 \end{bmatrix}$ $\tfrac{1}{12.5}R_4 \to R_4$

$$\sim \begin{bmatrix} 1 & 0 & 0 & 1.5 \\ 0 & 1 & 0 & -4 \\ 0 & 0 & 1 & -2.5 \\ 0 & 0 & 0 & 1 \end{bmatrix} \left| \begin{array}{r} -1.1 \\ 1.8 \\ 1.3 \\ -0.4 \end{array} \right. \quad \begin{array}{l} (-1.5)R_4 + R_1 \rightarrow R_1 \\ 4R_4 + R_2 \rightarrow R_2 \\ 2.5R_4 + R_3 \rightarrow R_3 \end{array}$$

$$\sim \begin{bmatrix} 1 & 0 & 0 & 0 \\ 0 & 1 & 0 & 0 \\ 0 & 0 & 1 & 0 \\ 0 & 0 & 0 & 1 \end{bmatrix} \left| \begin{array}{r} -0.5 \\ 0.2 \\ 0.3 \\ -0.4 \end{array} \right.$$

Solution: $x_1 = -0.5$, $x_2 = 0.2$, $x_3 = 0.3$, $x_4 = -0.4$.

49. $\begin{bmatrix} 1 & -2 & 1 & 1 & 2 \\ -2 & 4 & 2 & 2 & -2 \\ 3 & -6 & 1 & 1 & 5 \\ -1 & 2 & 3 & 1 & 1 \end{bmatrix} \left| \begin{array}{r} 2 \\ 0 \\ 4 \\ 3 \end{array} \right. \quad \begin{array}{l} 2R_1 + R_2 \rightarrow R_2 \\ (-3)R_1 + R_3 \rightarrow R_3 \\ R_1 + R_4 \rightarrow R_4 \end{array}$

$$\sim \begin{bmatrix} 1 & -2 & 1 & 1 & 2 \\ 0 & 0 & 4 & 4 & 2 \\ 0 & 0 & -2 & -2 & -1 \\ 0 & 0 & 4 & 2 & 3 \end{bmatrix} \left| \begin{array}{r} 2 \\ 4 \\ -2 \\ 5 \end{array} \right. \quad \tfrac{1}{4}R_2 \rightarrow R_2$$

$$\sim \begin{bmatrix} 1 & -2 & 1 & 1 & 2 \\ 0 & 0 & 1 & 1 & 0.5 \\ 0 & 0 & -2 & -2 & -1 \\ 0 & 0 & 4 & 2 & 3 \end{bmatrix} \left| \begin{array}{r} 2 \\ 1 \\ -2 \\ 5 \end{array} \right. \quad \begin{array}{l} (-1)R_2 + R_1 \rightarrow R_1 \\ \\ 2R_2 + R_3 \rightarrow R_3 \\ (-4)R_2 + R_4 \rightarrow R_4 \end{array}$$

$$\sim \begin{bmatrix} 1 & -2 & 0 & 0 & 1.5 \\ 0 & 0 & 1 & 1 & 0.5 \\ 0 & 0 & 0 & 0 & 0 \\ 0 & 0 & 0 & -2 & 1 \end{bmatrix} \left| \begin{array}{r} 1 \\ 1 \\ 0 \\ 1 \end{array} \right. \quad R_3 \leftrightarrow R_4$$

$$\sim \begin{bmatrix} 1 & -2 & 0 & 0 & 1.5 \\ 0 & 0 & 1 & 1 & 0.5 \\ 0 & 0 & 0 & -2 & 1 \\ 0 & 0 & 0 & 0 & 0 \end{bmatrix} \left| \begin{array}{r} 1 \\ 1 \\ 1 \\ 0 \end{array} \right. \quad (-\tfrac{1}{2})R_3 \rightarrow R_3$$

$$\sim \begin{bmatrix} 1 & -2 & 0 & 0 & 1.5 \\ 0 & 0 & 1 & 1 & 0.5 \\ 0 & 0 & 0 & 1 & -0.5 \\ 0 & 0 & 0 & 0 & 0 \end{bmatrix} \left| \begin{array}{r} 1 \\ 1 \\ -0.5 \\ 0 \end{array} \right. \quad (-1)R_3 + R_2 \rightarrow R_2$$

$$\sim \begin{bmatrix} 1 & -2 & 0 & 0 & 1.5 \\ 0 & 0 & 1 & 0 & 1 \\ 0 & 0 & 0 & 1 & -0.5 \\ 0 & 0 & 0 & 0 & 0 \end{bmatrix} \left| \begin{array}{r} 1 \\ 1.5 \\ -0.5 \\ 0 \end{array} \right.$$

Let $x_5 = t$. Then

$$x_4 - 0.5x_5 = -0.5$$
$$x_4 = 0.5x_5 - 0.5 = 0.5t - 0.5$$
$$x_3 + x_5 = 1.5$$
$$x_3 = -x_5 + 1.5 = -t + 1.5$$

Let $x_2 = s$. Then
$$x_1 - 2x_2 + 1.5x_5 = 1$$
$$x_1 = 2x_2 - 1.5x_5 + 1 = 2s - 1.5t + 1$$

Solution: $x_1 = 2s - 1.5t + 1$, $x_2 = s$, $x_3 = -t + 1.5$, $x_4 = 0.5t - 0.5$, $x_5 = t$, s and t any real numbers.

51. Let x_1 = number of 15-cent stamps
x_2 = number of 20-cent stamps
x_3 = number of 35-cent stamps
Then $x_1 + x_2 + x_3 = 45$ (total number of stamps)
$15x_1 + 20x_2 + 35x_3 = 1400$ (total value of stamps)

We write the augmented matrix and solve by Gauss-Jordan elimination.

$$\begin{bmatrix} 1 & 1 & 1 & | & 45 \\ 15 & 20 & 35 & | & 1400 \end{bmatrix} \quad (-15)R_1 + R_2 \rightarrow R_2$$

$$\sim \begin{bmatrix} 1 & 1 & 1 & | & 45 \\ 0 & 5 & 20 & | & 725 \end{bmatrix} \quad \tfrac{1}{5}R_2 \rightarrow R_2$$

$$\sim \begin{bmatrix} 1 & 1 & 1 & | & 45 \\ 0 & 1 & 4 & | & 145 \end{bmatrix} \quad (-1)R_2 + R_1 \rightarrow R_1$$

$$\sim \begin{bmatrix} 1 & 0 & -3 & | & -100 \\ 0 & 1 & 4 & | & 145 \end{bmatrix}$$

This augmented matrix is in reduced form. It corresponds to the system:
$$x_1 - 3x_3 = -100$$
$$x_2 + 4x_3 = 145$$

Let $x_3 = t$. Then
$$x_2 = -4x_3 + 145$$
$$= -4t + 145$$
$$x_1 = 3x_3 - 100$$
$$= 3t - 100$$

A solution is achieved, not for every real value of t, but for integer values of t that give rise to non-negative x_1, x_2, x_3.
$x_1 \geq 0$ means $3t - 100 \geq 0$ or $t \geq 33\tfrac{1}{3}$
$x_2 \geq 0$ means $-4t + 145 \geq 0$ or $t \leq 36\tfrac{1}{4}$
The only integer values of t that satisfy these conditions are 34, 35, 36. Thus we have the solutions:
$x_1 = (3t - 100)$ 15-cent stamps
$x_2 = (145 - 4t)$ 20-cent stamps
$x_3 = t$ 35-cent stamps
where $t = 34$, 35, or 36.

53. Let x_1 = number of 500-cc containers of 10% solution
x_2 = number of 500-cc containers of 20% solution
x_3 = number of 1,000-cc containers of 50% solution
Then $500x_1 + 500x_2 + 1,000x_3 = 12,000$ (total number of cc)
$0.10(500x_1) + 0.20(500x_2) + 0.50(1,000x_3) = 0.30(12,000)$ (total amount of ingredient in solution.)

After simplification, we have
$$x_1 + x_2 + 2x_3 = 24$$
$$x_1 + 2x_2 + 10x_3 = 72$$

We write the augmented matrix and solve by Gauss-Jordan elimination:

$$\begin{bmatrix} 1 & 1 & 2 & | & 24 \\ 1 & 2 & 10 & | & 72 \end{bmatrix} \quad (-1)R_1 + R_2 \rightarrow R_2$$

$$\sim \begin{bmatrix} 1 & 1 & 2 & | & 24 \\ 0 & 1 & 8 & | & 48 \end{bmatrix} \quad (-1)R_2 + R_1 \rightarrow R_1$$

$$\sim \begin{bmatrix} 1 & 0 & -6 & | & -24 \\ 0 & 1 & 8 & | & 48 \end{bmatrix}$$

This augmented matrix is in reduced form. It corresponds to the system:

$x_1 - 6x_3 = -24$

$x_2 + 8x_3 = 48$

Let $x_3 = t$. Then

$x_2 = -8x_3 + 48$

$\quad = -8t + 48$

$x_1 = 6x_3 - 24$

$\quad = 6t - 24$

A solution is achieved, not for every real value of t, but for integer values of t that give rise to non-negative x_1, x_2, x_3.

$x_1 \geq 0$ means $6t - 24 \geq 0$ or $t \geq 4$

$x_2 \geq 0$ means $-8t + 48 \geq 0$ or $t \leq 6$

Thus we have the solution:

$x_1 = (6t - 24)$ 500-cc containers of 10% solution

$x_2 = (48 - 8t)$ 500-cc containers of 20% solution

$x_3 = t$ 1000-cc containers of 50% solution

where t = 4, 5, or 6.

55. If the curve passes through a point, the coordinates of the point satisfy the equation of the curve. Hence,

$3 = a + b(-2) + c(-2)^2$

$2 = a + b(-1) + c(-1)^2$

$6 = a + b(1) + c(1)^2$

After simplication, we have

$a - 2b + 4c = 3$

$a - b + c = 2$

$a + b + c = 6$

We write the augmented matrix and solve by Gauss-Jordan elimination.

$$\begin{bmatrix} 1 & -2 & 4 & | & 3 \\ 1 & -1 & 1 & | & 2 \\ 1 & 1 & 1 & | & 6 \end{bmatrix} \quad \begin{matrix} \\ (-1)R_1 + R_2 \rightarrow R_2 \\ (-1)R_1 + R_3 \rightarrow R_3 \end{matrix}$$

$$\sim \begin{bmatrix} 1 & -2 & 4 & | & 3 \\ 0 & 1 & -3 & | & -1 \\ 0 & 3 & -3 & | & 3 \end{bmatrix} \quad \begin{matrix} 2R_2 + R_1 \rightarrow R_1 \\ \\ (-3)R_2 + R_3 \rightarrow R_3 \end{matrix}$$

$$\sim \begin{bmatrix} 1 & 0 & -2 & | & 1 \\ 0 & 1 & -3 & | & -1 \\ 0 & 0 & 6 & | & 6 \end{bmatrix} \quad \frac{1}{6}R_3 \rightarrow R_3$$

$$\sim \begin{bmatrix} 1 & 0 & -2 & | & 1 \\ 0 & 1 & -3 & | & -1 \\ 0 & 0 & 1 & | & 1 \end{bmatrix} \quad \begin{matrix} 2R_3 + R_1 \rightarrow R_1 \\ 3R_3 + R_2 \rightarrow R_2 \end{matrix}$$

$$\sim \begin{bmatrix} 1 & 0 & 0 & | & 3 \\ 0 & 1 & 0 & | & 2 \\ 0 & 0 & 1 & | & 1 \end{bmatrix}$$

Thus $a = 3$, $b = 2$, $c = 1$.

57. If the curve passes through a point, the coordinates of the point satisfy the equation of the curve. Hence,

$$6^2 + 2^2 + a(6) + b(2) + c = 0$$
$$4^2 + 6^2 + a(4) + b(6) + c = 0$$
$$(-3)^2 + (-1)^2 + a(-3) + b(-1) + c = 0$$

After simplification, we have

$$6a + 2b + c = -40$$
$$4a + 6b + c = -52$$
$$-3a - b + c = -10$$

We write the augmented matrix and solve by Gauss-Jordan elimination.

$$\begin{bmatrix} 6 & 2 & 1 & | & -40 \\ 4 & 6 & 1 & | & -52 \\ -3 & -1 & 1 & | & -10 \end{bmatrix} \quad \tfrac{1}{6}R_1 \to R_1$$

$$\sim \begin{bmatrix} 1 & \tfrac{1}{3} & \tfrac{1}{6} & | & -\tfrac{20}{3} \\ 4 & 6 & 1 & | & -52 \\ -3 & -1 & 1 & | & -10 \end{bmatrix} \quad \begin{array}{l} (-4)R_1 + R_2 \to R_2 \\ 3R_1 + R_3 \to R_3 \end{array}$$

$$\sim \begin{bmatrix} 1 & \tfrac{1}{3} & \tfrac{1}{6} & | & -\tfrac{20}{3} \\ 0 & \tfrac{14}{3} & \tfrac{1}{3} & | & -\tfrac{76}{3} \\ 0 & 0 & \tfrac{3}{2} & | & -30 \end{bmatrix} \quad \tfrac{2}{3}R_3 \to R_3$$

$$\sim \begin{bmatrix} 1 & \tfrac{1}{3} & \tfrac{1}{6} & | & -\tfrac{20}{3} \\ 0 & \tfrac{14}{3} & \tfrac{1}{3} & | & -\tfrac{76}{3} \\ 0 & 0 & 1 & | & -20 \end{bmatrix} \quad \begin{array}{l} (-\tfrac{1}{6})R_3 + R_1 \to R_1 \\ (-\tfrac{1}{3})R_3 + R_2 \to R_2 \end{array}$$

$$\sim \begin{bmatrix} 1 & \tfrac{1}{3} & 0 & | & -\tfrac{10}{3} \\ 0 & \tfrac{14}{3} & 0 & | & -\tfrac{56}{3} \\ 0 & 0 & 1 & | & -20 \end{bmatrix} \quad \tfrac{3}{14}R_2 \to R_2$$

$$\sim \begin{bmatrix} 1 & \tfrac{1}{3} & 0 & | & -\tfrac{10}{3} \\ 0 & 1 & 0 & | & -4 \\ 0 & 0 & 1 & | & -20 \end{bmatrix} \quad (-\tfrac{1}{3})R_2 + R_1 \to R_1$$

$$\begin{bmatrix} 1 & 0 & 0 & | & -2 \\ 0 & 1 & 0 & | & -4 \\ 0 & 0 & 1 & | & -20 \end{bmatrix}$$

Thus $a = -2$, $b = -4$, and $c = -20$.

59. Let x_1 = number of one-person boats

$\quad\quad x_2$ = number of two-person boats

$\quad\quad x_3$ = number of four-person boats

We have

$$0.5x_1 + 1.0x_2 + 1.5x_3 = 380 \quad \text{cutting department}$$
$$0.6x_1 + 0.9x_2 + 1.2x_3 = 330 \quad \text{assembly department}$$
$$0.2x_1 + 0.3x_2 + 0.5x_3 = 120 \quad \text{packing department}$$

> **Common Error:**
> The facts in this problem do not justify the equation
> $0.5x_1 + 0.6x_2 + 0.2x_3 = 380$

Clearing of decimals for convenience:

$$x_1 + 2x_2 + 3x_3 = 760$$
$$6x_1 + 9x_2 + 12x_3 = 3300$$
$$2x_1 + 3x_2 + 5x_3 = 1200$$

We write the augmented matrix and solve by Gauss-Jordan elimination:

$$\begin{bmatrix} 1 & 2 & 3 & | & 760 \\ 6 & 9 & 12 & | & 3300 \\ 2 & 3 & 5 & | & 1200 \end{bmatrix} \quad \begin{matrix} (-6)R_1 + R_2 \to R_2 \\ (-2)R_1 + R_3 \to R_3 \end{matrix}$$

$$\sim \begin{bmatrix} 1 & 2 & 3 & | & 760 \\ 0 & -3 & -6 & | & -1260 \\ 0 & -1 & -1 & | & -320 \end{bmatrix} \quad -\tfrac{1}{3}R_2 \to R_2$$

$$\sim \begin{bmatrix} 1 & 2 & 3 & | & 760 \\ 0 & 1 & 2 & | & 420 \\ 0 & -1 & -1 & | & -320 \end{bmatrix} \quad \begin{matrix} (-2)R_2 + R_1 \to R_1 \\ \\ R_2 + R_3 \to R_3 \end{matrix}$$

$$\sim \begin{bmatrix} 1 & 0 & -1 & | & -80 \\ 0 & 1 & 2 & | & 420 \\ 0 & 0 & 1 & | & 100 \end{bmatrix} \quad \begin{matrix} R_3 + R_1 \to R_1 \\ (-2)R_3 + R_2 \to R_2 \end{matrix}$$

$$\begin{bmatrix} 1 & 0 & 0 & | & 20 \\ 0 & 1 & 0 & | & 220 \\ 0 & 0 & 1 & | & 100 \end{bmatrix}$$

Therefore
$x_1 = 20$ one-person boats
$x_2 = 220$ two-person boats
$x_3 = 100$ four-person boats

61. This assumption discards the third equation. The system, cleared of decimals, reads

$$x_1 + 2x_2 + 3x_3 = 760$$
$$6x_1 + 9x_2 + 12x_3 = 3300$$

The augmented matrix becomes

$$\begin{bmatrix} 1 & 2 & 3 & | & 760 \\ 6 & 9 & 12 & | & 3300 \end{bmatrix}$$

We solve by Gauss-Jordan elimination. We start by introducing a 0 into the lower left corner using $(-6)R_1 + R_2$ as in the previous problem:

$$\sim \begin{bmatrix} 1 & 2 & 3 & | & 760 \\ 0 & -3 & -6 & | & -1260 \end{bmatrix} \quad -\tfrac{1}{3}R_2 \to R_2$$

$$\sim \begin{bmatrix} 1 & 2 & 3 & | & 760 \\ 0 & 1 & 2 & | & 420 \end{bmatrix} \quad (-2)R_2 + R_1 \to R_1$$

$$\sim \begin{bmatrix} 1 & 0 & -1 & | & -80 \\ 0 & 1 & 2 & | & 420 \end{bmatrix}$$

This augmented matrix is in reduced form. It corresponds to the system:

$$x_1 - x_3 = -80$$
$$x_2 + 2x_3 = 420$$

Let $x_3 = t$. Then
$$x_2 = -2x_3 + 420$$
$$\quad = -2t + 420$$
$$x_1 = x_3 - 80$$
$$\quad = t - 80$$

A solution is achieved, not for every real value of t, but for integer values of t that give rise to non-negative x_1, x_2, x_3.

$x_1 \geq 0$ means $t - 80 \geq 0$ or $t \geq 80$
$x_2 \geq 0$ means $-2t + 420 \geq 0$ or $210 \geq t$

Thus we have the solution
$x_1 = (t - 80)$ one-person boats
$x_2 = (-2t + 420)$ two-person boats
$x_3 = t$ four-person boats
$80 \leq t \leq 210$, t an integer

63. In this case we have $x_3 = 0$ from the beginning. The three equations of problem 59, cleared of decimals, read:

$$x_1 + 2x_2 = 760$$
$$6x_1 + 9x_2 = 3300$$
$$2x_1 + 3x_2 = 1200$$

The augmented matrix becomes:

$$\begin{bmatrix} 1 & 2 & | & 760 \\ 6 & 9 & | & 3300 \\ 2 & 3 & | & 1200 \end{bmatrix}$$

Notice that the row operation

$$(-3)R_3 + R_2 \rightarrow R_2$$

transforms this into the equivalent augmented matrix:

$$\begin{bmatrix} 1 & 2 & | & 760 \\ 0 & 0 & | & -300 \\ 2 & 3 & | & 1200 \end{bmatrix}$$

Therefore, since the second row corresponds to the equation

$$0x_1 + 0x_2 = -300$$

there is no solution.

No production schedule will use all the work-hours in all departments.

65. Let x_1 = number of ounces of food A.

$\quad\quad\ x_2$ = number of ounces of food B.

$\quad\quad\ x_3$ = number of ounces of food C.

> **Common Error:**
> The facts in this problem do not justify the equation
> $30x_1 + 10x_2 + 10x_3 = 340$

Then

$$30x_1 + 10x_2 + 20x_3 = 340 \quad \text{(calcium)}$$
$$10x_1 + 10x_2 + 20x_3 = 180 \quad \text{(iron)}$$
$$10x_1 + 30x_2 + 20x_3 = 220 \quad \text{(vitamin A)}$$

or

$$3x_1 + x_2 + 2x_3 = 34$$
$$x_1 + x_2 + 2x_3 = 18$$
$$x_1 + 3x_2 + 2x_3 = 22$$

is the system to be solved. We form the augmented matrix and solve by Gauss-Jordan elimination.

$$\begin{bmatrix} 3 & 1 & 2 & | & 34 \\ 1 & 1 & 2 & | & 18 \\ 1 & 3 & 2 & | & 22 \end{bmatrix} \quad R_1 \leftrightarrow R_2$$

$$\sim \begin{bmatrix} 1 & 1 & 2 & | & 18 \\ 3 & 1 & 2 & | & 34 \\ 1 & 3 & 2 & | & 22 \end{bmatrix} \quad \begin{array}{l} (-3)R_1 + R_2 \rightarrow R_2 \\ (-1)R_1 + R_3 \rightarrow R_3 \end{array}$$

$$\sim \begin{bmatrix} 1 & 1 & 2 & | & 18 \\ 0 & -2 & -4 & | & -20 \\ 0 & 2 & 0 & | & 4 \end{bmatrix} \quad -\tfrac{1}{2}R_2 \rightarrow R_2$$

$$\sim \begin{bmatrix} 1 & 1 & 2 & | & 18 \\ 0 & 1 & 2 & | & 10 \\ 0 & 2 & 0 & | & 4 \end{bmatrix} \quad \begin{array}{l} (-1)R_2 + R_1 \rightarrow R_1 \\ (-2)R_2 + R_3 \rightarrow R_3 \end{array}$$

$$\sim \begin{bmatrix} 1 & 0 & 0 & | & 8 \\ 0 & 1 & 2 & | & 10 \\ 0 & 0 & -4 & | & -16 \end{bmatrix} \quad -\tfrac{1}{4}R_3 \rightarrow R_3$$

$$\sim \begin{bmatrix} 1 & 0 & 0 & | & 8 \\ 0 & 1 & 2 & | & 10 \\ 0 & 0 & 1 & | & 4 \end{bmatrix} \quad (-2)R_3 + R_2 \to R_2$$

$$\sim \begin{bmatrix} 1 & 0 & 0 & | & 8 \\ 0 & 1 & 0 & | & 2 \\ 0 & 0 & 1 & | & 4 \end{bmatrix}$$

Thus

$x_1 = 8$ ounces food A

$x_2 = 2$ ounces food B

$x_3 = 4$ ounces food C

67. In this case we have $x_3 = 0$ from the beginning. The three equations of problem 65 become

$$30x_1 + 10x_2 = 340$$
$$10x_1 + 10x_2 = 180$$
$$10x_1 + 30x_2 = 220$$

or

$$3x_1 + x_2 = 34$$
$$x_1 + x_2 = 18$$
$$x_1 + 3x_2 = 22$$

The augmented matrix becomes

$$\begin{bmatrix} 3 & 1 & | & 34 \\ 1 & 1 & | & 18 \\ 1 & 3 & | & 22 \end{bmatrix}$$

We solve by Gauss-Jordan elimination, starting by the row operation

$R_1 \leftrightarrow R_2$

$$\begin{bmatrix} 1 & 1 & | & 18 \\ 3 & 1 & | & 34 \\ 1 & 3 & | & 22 \end{bmatrix} \quad \begin{matrix} (-3)R_1 + R_2 \to R_2 \\ (-1)R_1 + R_3 \to R_3 \end{matrix}$$

$$\sim \begin{bmatrix} 1 & 1 & | & 18 \\ 0 & -2 & | & -20 \\ 0 & 2 & | & 4 \end{bmatrix} \quad R_2 + R_3 \to R_3$$

$$\sim \begin{bmatrix} 1 & 1 & | & 18 \\ 0 & -2 & | & -20 \\ 0 & 0 & | & -16 \end{bmatrix}$$

Since the third row corresponds to the equation

$$0x_1 + 0x_2 = -16$$

there is no solution.

69. In this case we discard the third equation. The system becomes

$$30x_1 + 10x_2 + 20x_3 = 340$$
$$10x_1 + 10x_2 + 20x_3 = 180$$

or

$$3x_1 + x_2 + 2x_3 = 34$$
$$x_1 + x_2 + 2x_3 = 18$$

The augmented matrix becomes

$$\begin{bmatrix} 3 & 1 & 2 & | & 34 \\ 1 & 1 & 2 & | & 18 \end{bmatrix}$$

We solve by Gauss-Jordan elimination, starting by the row operation $R_1 \leftrightarrow R_2$.

$$\begin{bmatrix} 1 & 1 & 2 & | & 18 \\ 3 & 1 & 2 & | & 34 \end{bmatrix} \quad (-3)R_1 + R_2 \to R_2$$

$$\sim \begin{bmatrix} 1 & 1 & 2 & | & 18 \\ 0 & -2 & -4 & | & -20 \end{bmatrix} \quad -\tfrac{1}{2}R_2 \rightarrow R_2$$

$$\sim \begin{bmatrix} 1 & 1 & 2 & | & 18 \\ 0 & 1 & 2 & | & 10 \end{bmatrix} \quad (-1)R_2 + R_1 \rightarrow R_1$$

$$\sim \begin{bmatrix} 1 & 0 & 0 & | & 8 \\ 0 & 1 & 2 & | & 10 \end{bmatrix}$$

This augmented matrix is in reduced form. It corresponds to the system

$$x_1 = 8$$
$$x_2 + 2x_3 = 10$$

Let $x_3 = t$

Then $x_2 = -2x_3 + 10$
$$= -2t + 10$$

A solution is achieved, not for every real value t, but for values of t that give rise to non-negative x_2, x_3.

$x_3 \geq 0$ means $t \geq 0$
$x_2 \geq 0$ means $-2t + 10 \geq 0$, $5 \geq t$
Thus we have the solution
$x_1 = 8$ ounces food A
$x_2 = -2t + 10$ ounces food B
$x_3 = t$ ounces food C
$0 \leq t \leq 5$

71. Let x_1 = number of hours company A is to be scheduled
x_2 = number of hours company B is to be scheduled
In x_1 hours, company A can handle $30x_1$ telephone and $10x_1$ house contacts.
In x_2 hours, company B can handle $20x_2$ telephone and $20x_2$ house contacts.
We therefore have:
$$30x_1 + 20x_2 = 600 \text{ telephone contacts}$$
$$10x_1 + 20x_2 = 400 \text{ house contacts}$$

We form the augmented matrix and solve by Gauss-Jordan elimination.

$$\begin{bmatrix} 30 & 20 & | & 600 \\ 10 & 20 & | & 400 \end{bmatrix} \quad \begin{array}{l} \tfrac{1}{10}R_1 \rightarrow R_1 \\ \tfrac{1}{10}R_2 \rightarrow R_2 \end{array}$$

$$\sim \begin{bmatrix} 3 & 2 & | & 60 \\ 1 & 2 & | & 40 \end{bmatrix} \quad R_1 \leftrightarrow R_2$$

$$\sim \begin{bmatrix} 1 & 2 & | & 40 \\ 3 & 2 & | & 60 \end{bmatrix} \quad (-3)R_1 + R_2 \rightarrow R_2$$

$$\sim \begin{bmatrix} 1 & 2 & | & 40 \\ 0 & -4 & | & -60 \end{bmatrix} \quad -\tfrac{1}{4}R_3 \rightarrow R_3$$

$$\sim \begin{bmatrix} 1 & 2 & | & 40 \\ 0 & 1 & | & 15 \end{bmatrix} \quad (-2)R_2 + R_1 \rightarrow R_1$$

$$\sim \begin{bmatrix} 1 & 0 & | & 10 \\ 0 & 1 & | & 15 \end{bmatrix}$$

Therefore
$x_1 = 10$ hours company A
$x_2 = 15$ hours company B

Exercise 8-3

Key Ideas and Formulas

Nonlinear systems are systems that contain at least one nonlinear equation.

Nonlinear systems involving second degree terms can have at most four solutions, some of which may be imaginary.

Nonlinear systems can be solved by the substitution and, in some cases, the elimination methods of Section 8-1.

It is important to check the apparent solutions of any nonlinear system to insure that extraneous roots have not been introduced.

1.
$$x^2 + y^2 = 169$$
$$x = -12$$
$$(-12)^2 + y^2 = 169$$
$$y^2 = 25$$
$$y = \pm 5$$

Solution: $(-12, 5)$, $(-12, -5)$

Check: $-12 \overset{\surd}{=} -12$
$$(-12)^2 + (\pm 5)^2 \overset{\surd}{=} 169$$

3.
$$8x^2 - y^2 = 16$$
$$y = 2x$$

Substitute y from the second equation into the first equation.
$$8x^2 - (2x)^2 = 16$$
$$8x^2 - 4x^2 = 16$$
$$4x^2 = 16$$
$$x^2 = 4$$
$$x = \pm 2$$

For $x = 2$ For $x = -2$
   $y = 2(2)$ $y = 2(-2)$
   $y = 4$ $y = -4$

Solutions: $(2, 4)$, $(-2, -4)$

Check:
For $(2, 4)$ For $(-2, -4)$
$$4 \overset{\surd}{=} 2 \cdot 2 \qquad\qquad -4 \overset{\surd}{=} 2(-2)$$
$$8(2)^2 - 4^2 \overset{\surd}{=} 16 \qquad 8(-2)^2 - (-4)^2 \overset{\surd}{=} 16$$

5. $3x^2 - 2y^2 = 25$
$$x + y = 0$$

Solve for y in the first degree equation
$$y = -x$$
Substitute into the second degree equation.
$$3x^2 - 2(-x)^2 = 25$$
$$x^2 = 25$$
$$x = \pm 5$$

For $x = 5$ For $x = -5$
   $y = -5$ $y = 5$

Solutions: $(5, -5)$, $(-5, 5)$

Check:
For $(5, -5)$ For $(-5, 5)$
$$5 + (-5) \overset{\surd}{=} 0 \qquad\qquad (-5) + 5 \overset{\surd}{=} 0$$
$$3(5)^2 - 2(-5)^2 \overset{\surd}{=} 25 \quad 3(-5)^2 - 2(5)^2 \overset{\surd}{=} 25$$

From this point on we will not show the checking steps for lack of space. The student should perform these checking steps, however.

7. $y^2 = x$
$x - 2y = 2$

Solve for x in the first degree equation.
$x = 2y + 2$
Substitute into the second degree equation.
$$y^2 = 2y + 2$$
$y^2 - 2y - 2 = 0$

$$y = \frac{-b \pm \sqrt{b^2 - 4ac}}{2a} \quad a = 1, \ b = -2, \ c = -2$$

$$y = \frac{-(-2) \pm \sqrt{(-2)^2 - 4(1)(-2)}}{2(1)}$$

$$y = \frac{2 \pm \sqrt{12}}{2}$$

$$y = 1 \pm \sqrt{3}$$

For $y = 1 + \sqrt{3}$ For $y = 1 - \sqrt{3}$

$\quad x = 2(1 + \sqrt{3}) + 2$ $x = 2(1 - \sqrt{3}) + 2$

$\quad x = 4 + 2\sqrt{3}$ $x = 4 - 2\sqrt{3}$

Solutions: $(4 + 2\sqrt{3}, \ 1 + \sqrt{3}), \ (4 - 2\sqrt{3}, \ 1 - \sqrt{3})$

9. $2x^2 + y^2 = 24$
$x^2 - y^2 = -12$

Solve using elimination by addition. Adding, we obtain:
$3x^2 = 12$
$x^2 = 4$
$x = \pm 2$

For $x = 2$ For $x = -2$
$4 - y^2 = -12$ $4 - y^2 = -12$
$\quad -y^2 = -16$ Similarly
$\quad\quad y^2 = 16$ $y = \pm 4$
$\quad\quad y = \pm 4$

Solutions: $(2, \ 4), \ (2, \ -4), \ (-2, \ 4), \ (-2, \ -4)$

11. $x^2 + y^2 = 10$
$16x^2 + y^2 = 25$

Solve using elimination by addition. Multiply the top equation by -1 and add.

$\begin{aligned} -x^2 - y^2 &= -10 \\ 16x^2 + y^2 &= \ \ 25 \\ \hline 15x^2 \quad\quad &= \ \ 15 \\ x^2 \quad\quad &= \ \ \ 1 \\ x \quad &= \ \pm 1 \end{aligned}$

For $x = 1$ For $x = -1$
$1 + y^2 = 10$ $1 + y^2 = 10$
$\quad\ y^2 = 9$ $y = \pm 3$
$\quad\ y = \pm 3$

Solutions: $(1, \ 3), \ (1, \ -3),$
$(-1, \ 3), \ (-1, \ -3)$

13. $xy - 4 = 0$
$x - y = 2$

Solve for x in the first degree equation.
$x = y + 2$
Substitute into the second degree equation
$(y + 2)y - 4 = 0$
$y^2 + 2y - 4 = 0$

$$y = \frac{-b \pm \sqrt{b^2 - 4ac}}{2a} \quad a = 1, \ b = 2, \ c = -4$$

$$y = \frac{-2 \pm \sqrt{(2)^2 - 4(1)(-4)}}{2(1)}$$

$$y = \frac{-2 \pm \sqrt{20}}{2}$$

$$y = -1 \pm \sqrt{5}$$

For $y = -1 + \sqrt{5}$ For $y = -1 - \sqrt{5}$

$\quad x = -1 + \sqrt{5} + 2$ $x = -1 - \sqrt{5} + 2$

$\quad x = 1 + \sqrt{5}$ $x = 1 - \sqrt{5}$

Solutions: $(1 + \sqrt{5}, \ -1 + \sqrt{5}),$
$(1 - \sqrt{5}, \ -1 - \sqrt{5})$

15. $x^2 + 2y^2 = 6$
$\qquad xy = 2$

Solve for y in the second equation

$y = \dfrac{2}{x}$

Substitute into the first equation

$$x^2 + 2\left(\dfrac{2}{x}\right)^2 = 6$$

$$x^2 + \dfrac{8}{x^2} = 6 \qquad x \ne 0$$

$$x^2 \cdot x^2 + x^2 \cdot \dfrac{8}{x^2} = 6x^2$$

$$x^4 + 8 = 6x^2$$
$$x^4 - 6x^2 + 8 = 0$$
$$(x^2 - 2)(x^2 - 4) = 0$$
$$(x - \sqrt{2})(x + \sqrt{2})(x - 2)(x + 2) = 0$$

$$x = \sqrt{2},\ -\sqrt{2},\ 2,\ -2$$

For $x = \sqrt{2}$	For $x = -\sqrt{2}$	For $x = 2$	For $x = -2$
$y = \dfrac{2}{\sqrt{2}}$	$y = -\dfrac{2}{\sqrt{2}}$	$y = \dfrac{2}{2}$	$y = \dfrac{2}{-2}$
$y = \sqrt{2}$	$y = -\sqrt{2}$	$y = 1$	$y = -1$

Solutions: $(\sqrt{2},\ \sqrt{2})$, $(-\sqrt{2},\ -\sqrt{2})$, $(2,\ 1)$, $(-2,\ -1)$

17. $2x^2 + 3y^2 = -4$
$\qquad 4x^2 + 2y^2 = 8$

Solve using elimination by addition. Multiply the second equation by $-\dfrac{1}{2}$ and add.

$$\begin{array}{r} 2x^2 + 3y^2 = -4 \\ -2x^2 - y^2 = -4 \\ \hline 2y^2 = -8 \end{array}$$

$$y^2 = -4$$
$$y = \pm 2i$$

For $y = 2i$ $\qquad$ For $y = -2i$
$2x^2 + 3(2i)^2 = -4$ $\quad$ $2x^2 + 3(-2i)^2 = -4$
$\quad 2x^2 - 12 = -4$ $\qquad$ $2x^2 - 12 = -4$
$\qquad\quad 2x^2 = 8$ $\qquad$ Similarly
$\qquad\quad\ x^2 = 4$ $\qquad\qquad\qquad x = \pm 2$
$\qquad\quad\ \ x = \pm 2$

Solutions: $(2,\ 2i)$, $(-2,\ 2i)$, $(2,\ -2i)$, $(-2,\ -2i)$

19. $x^2 - y^2 = 2$
$\qquad y^2 = x$

Substitute y^2 from the second equation into the first equation.

$$x^2 - x = 2$$
$$x^2 - x - 2 = 0$$
$$(x - 2)(x + 1) = 0$$
$$x = 2,\ -1$$

For $x = 2$ $\qquad$ For $x = -1$
$\quad y^2 = 2$ $\qquad\qquad y^2 = -1$
$\quad\ y = \pm\sqrt{2}$ $\qquad\quad y = \pm i$

Solutions: $(2,\ \sqrt{2})$, $(2,\ -\sqrt{2})$, $(-1,\ i)$, $(-1,\ -i)$

21. $x^2 + y^2 = 9$
$\qquad x^2 = 9 - 2y$

Substitute x^2 from the second equation into the first equation.

$$9 - 2y + y^2 = 9$$
$$y^2 - 2y = 0$$
$$y(y - 2) = 0$$
$$y = 0,\ 2$$

For $y = 0$ $\qquad\qquad$ For $y = 2$
$\quad x^2 = 9 - 2(0)$ $\qquad x^2 = 9 - 2(2)$
$\quad x^2 = 9$ $\qquad\qquad\ x^2 = 5$
$\quad\ x = \pm 3$ $\qquad\qquad\ x = \pm\sqrt{5}$

Solutions: $(3,\ 0)$, $(-3,\ 0)$, $(\sqrt{5},\ 2)$, $(-\sqrt{5},\ 2)$

23. $x^2 - y^2 = 3$
$xy = 2$

Solve for y in the second equation.

$y = \dfrac{2}{x}$

Substitute into the first equation:

$$x^2 - \left(\dfrac{2}{x}\right)^2 = 3$$

$$x^2 - \dfrac{4}{x^2} = 3 \quad x \neq 0$$

$$x^4 - 4 = 3x^2$$

$$x^4 - 3x^2 - 4 = 0$$

$$(x^2 - 4)(x^2 + 1) = 0$$

$x^2 - 4 = 0 \qquad x^2 + 1 = 0$

$x^2 = 4 \qquad x^2 = -1$

$x = \pm 2 \qquad x = \pm i$

For $x = 2$ For $x = -2$ For $x = i$ For $x = -i$

$\quad y = \dfrac{2}{2} \qquad\quad y = \dfrac{2}{-2} \qquad\quad y = \dfrac{2}{i} \qquad\quad y = \dfrac{2}{-i}$

$\quad y = 1 \qquad\qquad y = -1 \qquad\quad y = -2i \qquad\quad y = 2i$

Solutions: $(2, 1)$, $(-2, -1)$, $(i, -2i)$, $(-i, 2i)$

---

25. $y = 5 - x^2$
$y = 2 - 2x$

Substitute y from the first equation into the second equation.

$5 - x^2 = 2 - 2x$

$0 = x^2 - 2x - 3$

$0 = (x - 3)(x + 1)$

$x = 3, -1$

For $x = 3$ For $x = -1$

$\quad y = 2 - 2(3) \qquad y = 2 - 2(-1)$

$\quad y = -4 \qquad\qquad y = 4$

Solutions: $(3, -4)$, $(-1, 4)$

---

27. $y = x^2 - x$
$y = 2x$

Substitute y from the first equation into the second equation.

$\quad x^2 - x = 2x$

$\quad x^2 - 3x = 0$

$x(x - 3) = 0$

$\qquad\quad x = 0, 3$

For $x = 0$ For $x = 3$

$\quad y = 2(0) \qquad\quad y = 2(3)$

$\quad y = 0 \qquad\qquad y = 6$

Solutions: $(0, 0)$, $(3, 6)$

---

29. $y = x^2 - 6x + 9$
$y = 5 - x$

Substitute y from the first equation into the second equation.

$\quad x^2 - 6x + 9 = 5 - x$

$\quad x^2 - 5x + 4 = 0$

$(x - 1)(x - 4) = 0$

$\qquad\qquad\quad x = 1, 4$

For $x = 1$ For $x = 4$

$\quad y = 5 - 1 \qquad\quad y = 5 - 4$

$\quad y = 4 \qquad\qquad y = 1$

Solutions: $(1, 4)$, $(4, 1)$

---

31. $y = 8 + 4x - x^2$
$y = x^2 - 2x$

Substitute y from the first equation into the second equation.

$8 + 4x - x^2 = x^2 - 2x$

$0 = 2x^2 - 6x - 8$

$0 = x^2 - 3x - 4$

$0 = (x - 4)(x + 1)$

$x = 4, -1$

For $x = 4$ For $x = -1$

$\quad y = 4^2 - 2(4) \qquad y = (-1)^2 - 2(-1)$

$\quad y = 8 \qquad\qquad\quad y = 3$

Solutions: $(4, 8)$, $(-1, 3)$

33. (A) The lines are tangent to the circle.

(B) To find values of b such that
$$x^2 + y^2 = 5$$
$$2x - y = b$$
has exactly one solution, we solve the system for arbitrary b. Solve for y in the second equation.
$$y = 2x - b$$
Substitute into the first equation:
$$x^2 + (2x - b)^2 = 5$$
$$x^2 + 4x^2 - 4bx + b^2 = 5$$
$$5x^2 - 4bx + b^2 - 5 = 0$$

This quadratic equation will have one solution if the discriminant $B^2 - 4AC = (-4b)^2 - 4(5)(b^2 - 5)$ is equal to 0.

This will occur when
$$16b^2 - 20b^2 + 100 = 0$$
$$-4b^2 + 100 = 0$$
$$b^2 = 25$$
$$b = \pm 5$$

Consider $b = 5$

Then the solution of the system
$$x^2 + y^2 = 5$$
$$2x - y = 5$$
will be given by solving
$$5x^2 - 4bx + b^2 - 5 = 0$$
for $b = 5$.
$$5x^2 - 4 \cdot 5x + 5^2 - 5 = 0$$
$$5x^2 - 20x + 20 = 0$$
$$5(x - 2)^2 = 0$$
$$x - 2 = 0$$
$$x = 2$$

Since $2x - y = 5$
$$2 \cdot 2 - y = 5$$
$$y = -1$$

The intersection point is $(2, -1)$ for $b = 5$.

Consider $b = -5$

Then the solution of the system
$$x^2 + y^2 = 5$$
$$2x - y = -5$$
will be given by solving
$$5x^2 - 4bx + b^2 - 5 = 0$$
for $b = -5$.
$$5x^2 - 4(-5)x + (-5)^2 - 5 = 0$$
$$5x^2 + 20x + 20 = 0$$
$$5(x + 2)^2 = 0$$
$$x + 2 = 0$$
$$x = -2$$

Since $2x - y = -5$
$$2(-2) - y = -5$$
$$y = 1$$

The intersection point is $(-2, 1)$ for $b = -5$.

(C) The line $x + 2y = 0$ is perpendicular to all the lines in the family and intersects the circle at the intersection points found in part B, since this line passes through the center of the circle and thus includes a diameter of the circle, which is perpendicular to the tangent line at their mutual point of intersection with the circle. Solving the system $x^2 + y^2 = 5$, $x + 2y = 0$ would determine the intersection points.

35. $2x + 5y + 7xy = 8$
$xy - 3 = 0$

Solve for y in the second equation.
$xy = 3$
$y = \dfrac{3}{x}$

Substitute into the first equation.

$2x + 5\left(\dfrac{3}{x}\right) + 7x\left(\dfrac{3}{x}\right) = 8$

$2x + \dfrac{15}{x} + 21 = 8 \qquad x \neq 0$

$2x^2 + 15 + 21x = 8x$
$2x^2 + 13x + 15 = 0$
$(2x + 3)(x + 5) = 0$
$x = -\dfrac{3}{2}, \ -5$

For $x = -\dfrac{3}{2}$ \qquad For $x = -5$

$y = 3 \div \left(-\dfrac{3}{2}\right) \qquad y = \dfrac{3}{-5}$

$y = -2 \qquad\qquad y = -\dfrac{3}{5}$

Solutions: $\left(-\dfrac{3}{2}, \ -2\right), \ \left(-5, \ -\dfrac{3}{5}\right)$

37. $x^2 - 2xy + y^2 = 1$
$x - 2y = 2$

Solve for x in terms of y in the first-degree equation.
$x = 2y + 2$

Substitute into the second-degree equation.
$(2y + 2)^2 - 2(2y + 2)y + y^2 = 1$
$4y^2 + 8y + 4 - 4y^2 - 4y + y^2 = 1$
$y^2 + 4y + 3 = 0$
$(y + 1)(y + 3) = 0$
$y = -1, \ -3$

For $y = -1$ \qquad\qquad For $y = -3$
$x = 2(-1) + 2 \qquad x = 2(-3) + 2$
$= 0 \qquad\qquad\qquad = -4$

Solutions: $(0, \ -1), \ (-4, \ -3)$

39. $2x^2 - xy + y^2 = 8$
$x^2 - y^2 = 0$

Factor the left side of the equation that has a zero constant term.

$(x - y)(x + y) = 0$
$x = y \text{ or } x = -y$

> **Common Error:**
> It is incorrect to replace
> $x^2 - y^2 = 0$ or $x^2 = y^2$ by $x = y$.
> This neglects the possibility $x = -y$.

Thus, the original system is equivalent to the two systems
$2x^2 - xy + y^2 = 8 \qquad 2x^2 - xy + y^2 = 8$
$x = y \qquad\qquad\qquad x = -y$

These systems are solved by substitution.

First system: \qquad\qquad\qquad Second system:
$2x^2 - xy + y^2 = 8 \qquad\qquad\quad 2x^2 - xy + y^2 = 8$
$x = y \qquad\qquad\qquad\qquad\qquad x = -y$
$2y^2 - yy + y^2 = 8 \qquad\quad 2(-y)^2 - (-y)y + y^2 = 8$
$2y^2 = 8 \qquad\qquad\qquad 2y^2 + y^2 + y^2 = 8$
$y^2 = 4 \qquad\qquad\qquad\qquad 4y^2 = 8$
$y = \pm 2 \qquad\qquad\qquad\qquad y^2 = 2$
$y = \pm\sqrt{2}$

For $y = 2$ \qquad For $y = -2$ \qquad For $y = \sqrt{2}$ \qquad For $y = -\sqrt{2}$
$x = 2 \qquad\qquad x = -2 \qquad\qquad x = -\sqrt{2} \qquad\qquad x = \sqrt{2}$

Solutions: $(2, \ 2), \ (-2, \ -2), \ (-\sqrt{2}, \ \sqrt{2}), \ (\sqrt{2}, \ -\sqrt{2})$

41. $x^2 + xy - 3y^2 = 3$
$x^2 + 4xy + 3y^2 = 0$

Factor the left side of the equation that has a zero constant term.
$(x + y)(x + 3y) = 0$
$$x = -y \text{ or } x = -3y$$
Thus the original system is equivalent to the two systems

$x^2 + xy - 3y^2 = 3 \qquad\qquad x^2 + xy - 3y^2 = 3$
$\qquad x = -y \qquad\qquad\qquad\qquad x = -3y$

These systems are solved by substitution.

First system: Second system:

$x^2 + xy - 3y^2 = 3 \qquad\qquad x^2 + xy - 3y^2 = 3$
$\qquad\quad x = -y \qquad\qquad\qquad\qquad\qquad x = -3y$
$(-y)^2 + (-y)y - 3y^2 = 3 \qquad (-3y)^2 + (-3y)y - 3y^2 = 3$
$\quad y^2 - y^2 - 3y^2 = 3 \qquad\qquad 9y^2 - 3y^2 - 3y^2 = 3$
$\qquad\qquad -3y^2 = 3 \qquad\qquad\qquad\qquad 3y^2 = 3$
$\qquad\qquad\quad y^2 = -1 \qquad\qquad\qquad\qquad\quad y^2 = 1$
$\qquad\qquad\quad\; y = \pm i \qquad\qquad\qquad\qquad\quad\; y = \pm 1$

For $y = i$ For $y = -i$ For $y = 1$ For $y = -1$
$\quad x = -i$ $x = i$ $x = -3$ $x = 3$

Solutions: $(-i, i)$, $(i, -i)$, $(-3, 1)$, $(3, -1)$

43. Before we can enter these equations in our graphing utility, we must solve for y:

$-x^2 + 2xy + y^2 = 1 \qquad\qquad 3x^2 - 4xy + y^2 = 2$
$y^2 + 2xy - 1 - x^2 = 0 \qquad\qquad y^2 - 4xy + 3x^2 - 2 = 0$

Applying the quadratic formula to each equation, we have

$y = \dfrac{-2x \pm \sqrt{4x^2 - 4(-1 - x^2)}}{2} \qquad y = \dfrac{4x \pm \sqrt{16x^2 - 4(3x^2 - 2)}}{2}$

$y = \dfrac{-2x \pm \sqrt{8x^2 + 4}}{2} \qquad\qquad y = \dfrac{4x \pm \sqrt{4x^2 + 8}}{2}$

$y = -x \pm \sqrt{2x^2 + 1} \qquad\qquad\quad y = 2x \pm \sqrt{x^2 + 2}$

Entering each of these four equations into a graphing utility produces the graph shown at the right.

Zooming in on the four intersection points, or using a built-in intersection routine (details omitted), yields $(-1.41, -0.82)$, $(-0.13, 1.15)$, $(0.13, -1.15)$, and $(1.41, 0.82)$ to two decimal places.

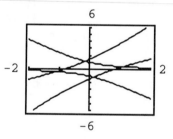

45. Before we can enter these equations in our graphing utility, we must solve for y:

$3x^2 - 4xy - y^2 = 2 \qquad\qquad 2x^2 + 2xy + y^2 = 9$
$y^2 + 4xy + 2 - 3x^2 = 0 \qquad\qquad y^2 + 2xy + 2x^2 - 9 = 0$

Applying the quadratic formula to each equation, we have

$y = \dfrac{-4x \pm \sqrt{16x^2 - 4(2 - 3x^2)}}{2} \qquad y = \dfrac{-2x \pm \sqrt{4x^2 - 4(2x^2 - 9)}}{2}$

$y = \dfrac{-4x \pm \sqrt{28x^2 - 8}}{2} \qquad\qquad y = \dfrac{-2x \pm \sqrt{36 - 4x^2}}{2}$

$y = -2x \pm \sqrt{7x^2 - 2} \qquad\qquad\quad y = -x \pm \sqrt{9 - x^2}$

Entering each of these four equations into a graphing utility produces the graph shown at the right.

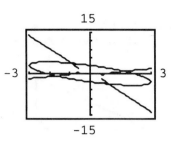

Zooming in on the four intersection points, or using a built-in intersection routine (details omitted), yields (-1.66, -0.84), (-0.91, 3.77), (0.91, -3.77), and (1.66, 0.84) to two decimal places.

47. Before we can enter these equations in our graphing utility, we must solve for y:

$$2x^2 - 2xy + y^2 = 9 \qquad\qquad 4x^2 - 4xy + y^2 + x = 3$$
$$y^2 - 2xy + 2x^2 - 9 = 0 \qquad\qquad y^2 - 4xy + 4x^2 + x - 3 = 0$$

Applying the quadratic formula to each equation, we have

$$y = \frac{2x \pm \sqrt{4x^2 - 4(2x^2 - 9)}}{2} \qquad\qquad y = \frac{4x \pm \sqrt{16x^2 - 4(4x^2 + x - 3)}}{2}$$

$$y = \frac{2x \pm \sqrt{36 - 4x^2}}{2} \qquad\qquad y = \frac{4x \pm \sqrt{12 - 4x}}{2}$$

$$y = x \pm \sqrt{9 - x^2} \qquad\qquad y = 2x \pm \sqrt{3 - x}$$

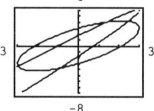

Entering each of these four equations into a graphing utility produces the graph shown at the right.

Zooming in on the four intersection points, or using a built-in intersection routine (details omitted), yields (-2.96, -3.47), (-0.89, -3.76), (1.39, 4.05), and (2.46, 4.18) to two decimal places.

49. Let x and y equal the two numbers. We have the system

$$x + y = 3$$
$$xy = 1$$

Solve the first equation for y in terms of x, then substitute into the second degree equation.

$$y = 3 - x$$
$$x(3 - x) = 1$$
$$3x - x^2 = 1$$
$$-x^2 + 3x - 1 = 0$$
$$x^2 - 3x + 1 = 0$$

$$x = \frac{-b \pm \sqrt{b^2 - 4ac}}{2a} \qquad a = 1, \ b = -3, \ c = 1$$

$$x = \frac{-(-3) \pm \sqrt{(-3)^2 - 4(1)(1)}}{2(1)}$$

$$x = \frac{3 \pm \sqrt{5}}{2}$$

For $x = \dfrac{3 + \sqrt{5}}{2}$ \qquad For $x = \dfrac{3 - \sqrt{5}}{2}$

$$y = 3 - x \qquad\qquad y = 3 - x$$

$$= 3 - \frac{3 + \sqrt{5}}{2} \qquad\qquad = 3 - \frac{3 - \sqrt{5}}{2}$$

$$= \frac{6 - 3 - \sqrt{5}}{2} \qquad\qquad = \frac{6 - 3 + \sqrt{5}}{2}$$

$$= \frac{3 - \sqrt{5}}{2} \qquad\qquad = \frac{3 + \sqrt{5}}{2}$$

Thus the two numbers are $\frac{1}{2}(3 - \sqrt{5})$ and $\frac{1}{2}(3 + \sqrt{5})$.

51. Sketch a figure.
Let x and y represent the lengths of the two legs.

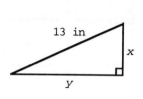

From the Pythagorean Theorem we have
$x^2 + y^2 = 13^2$
From the formula for the area of a triangle we have
$\frac{1}{2}xy = 30$
Thus the system of equations is
$x^2 + y^2 = 169$

$\frac{1}{2}xy = 30$

Solve the second equation for y in terms of x, then substitute into the first equation.

$$xy = 60$$
$$y = \frac{60}{x}$$
$$x^2 + \left(\frac{60}{x}\right)^2 = 169$$
$$x^2 + \frac{3600}{x^2} = 169 \quad x \neq 0$$
$$x^4 + 3600 = 169x^2$$
$$x^4 - 169x^2 + 3600 = 0$$
$$(x^2 - 144)(x^2 - 25) = 0$$
$$(x - 12)(x + 12)(x - 5)(x + 5) = 0$$
$$x = \pm 12, \pm 5$$

Discarding the negative solutions, we have
$x = 12$ or $x = 5$

For $x = 12$ For $x = 5$

$y = \dfrac{60}{x}$ $y = \dfrac{60}{x}$

$y = 5$ $y = 12$

The lengths of the legs are 5 inches and 12 inches.

53. Let x = width of screen.
     y = height of screen.
From the Pythagorean Theorem, we have
$x^2 + y^2 = (7.5)^2$
From the formula for the area of a rectangle we have
$xy = 27$
Thus the system of equations is:
$x^2 + y^2 = 56.25$
    $xy = 27$

Solve the second equation for y in terms of x, then substitute into the first equation.

$$y = \frac{27}{x}$$
$$x^2 + \left(\frac{27}{x}\right)^2 = 56.25 \quad x \neq 0$$
$$x^2 + \frac{729}{x^2} = 56.25 \quad x \neq 0$$
$$x^4 + 729 = 56.25x^2$$
$$x^4 - 56.25x^2 + 729 = 0 \quad \text{quadratic in } x^2$$

$$x^2 = \frac{-b \pm \sqrt{b^2 - 4ac}}{2a} \quad a = 1, \ b = -56.25, \ c = 729$$

$$x^2 = \frac{-(-56.25) \pm \sqrt{(-56.25)^2 - 4(1)(729)}}{2(1)}$$

$$x^2 = \frac{56.25 \pm 15.75}{2}$$

$$x^2 = 36, \ 20.25$$

$$x = 6, \ 4.5 \ \text{(discarding the negative solutions)}$$

For $x = 6$ For $x = 4.5$

$$y = \frac{27}{6} = 4.5 \qquad\qquad y = \frac{27}{4.5} = 6$$

The dimensions of the screen must be 6 inches by 4.5 inches.

55.

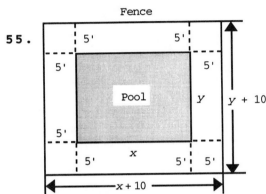

Redrawing and labeling the figure as shown, we have

Area of pool = 572
$$xy = 572$$
Area enclosed by fence = 1,152
$$(x + 10)(y + 10) = 1,152$$

We solve this system by solving for y in terms of x in the first equation, then substituting into the second equation.

$$y = \frac{572}{x}$$

$$(x + 10)\left(\frac{572}{x} + 10\right) = 1,152$$

$$572 + 10x + \frac{5,720}{x} + 100 = 1,152$$

$$10x + \frac{5,720}{x} - 480 = 0 \quad x \neq 0$$

$$10x^2 + 5,720 - 480x = 0$$

$$x^2 - 48x + 572 = 0$$

$$(x - 26)(x - 22) = 0$$

$$x = 26, \ 22$$

For $x = 26$ For $x = 22$

$$y = \frac{572}{26} \qquad\qquad y = \frac{572}{22}$$

$$y = 22 \qquad\qquad\qquad y = 26$$

The dimensions of the pool are 22 feet by 26 feet.

57.   Let x = average speed of Boat B

Then $x + 5$ = average speed of Boat A

Let y = time of Boat B, then $y - \frac{1}{2}$ = time of Boat A

Using Distance = rate × time, we have

$$75 = xy$$

$$75 = (x + 5)\left(y - \frac{1}{2}\right)$$

Note: The *faster* boat, A, has the *shorter* time. It is a common error to confuse the signs here. Another common error: if rates are expressed in miles per hour, then $y - 30$ is not the correct time for boat A. Times must be expressed in hours.

Solve the first equation for y in terms of x, then substitute into the second equation.

$$y = \frac{75}{x}$$

$$75 = (x + 5)\left(\frac{75}{x} - \frac{1}{2}\right)$$

$$75 = 75 - \frac{1}{2}x + \frac{375}{x} - \frac{5}{2}$$

$$0 = -\frac{1}{2}x + \frac{375}{x} - \frac{5}{2} \qquad x \neq 0$$

$$2x(0) = 2x\left(-\frac{1}{2}x\right) + 2x\left(\frac{375}{x}\right) - 2x\left(\frac{5}{2}\right)$$

$$0 = -x^2 + 750 - 5x$$

$$x^2 + 5x - 750 = 0$$

$$(x - 25)(x + 30) = 0$$

$$x = 25, \; -30$$

Discarding the negative solution, we have

$$x = 25 \text{ mph} = \text{average speed of Boat } B$$

$$x + 5 = 30 \text{ mph} = \text{average speed of Boat } A$$

Exercise 8-4

Key Ideas and Formulas

A line divides a plane into two halves called **half-planes**. A vertical line divides a plane into **left** and **right half-planes**; a nonvertical line divides a plane into **upper** and **lower half-planes**.

The graph of a linear inequality $Ax + By < C$ or $Ax + By > C$ (with $B \neq 0$) is either the upper half-plane or the lower half-plane (but not both) determined by the line $Ax + By = C$. If $B = 0$, then the graph of $Ax < C$ or $Ax > C$ is either the left half-plane or the right half-plane (but not both) determined by the line $Ax = C$. The line $Ax + By = C$ is called the **boundary line** for the half-plane.

Linear inequalities are graphed by the procedure given in the text (not repeated here for reasons of space.)

To graph systems of linear inequalities:

1. Graph each inequality, shading the solution half-plane lightly.

2. The multiply shaded region is the graph of the system, called the solution region or the feasible region.

3. Determine the corner points from the graph, and, as necessary, solving the systems of equations formed by pairs of equations of the appropriate lines.

(A **corner point** is a point in the solution region that is the intersection of two boundary lines.)

A solution region of a system of linear inequalities is bounded if it can be enclosed in a circle; if it cannot be enclosed within a circle, then it is unbounded.

1. Graph $2x - 3y = 6$ as a dashed line, since equality is not included in the original statement. The origin is a suitable test point.
$2x - 3y < 6$
$2(0) - 3(0) = 0 < 6$
Hence $(0, 0)$ is in the solution set. The graph is the half-plane containing $(0, 0)$.

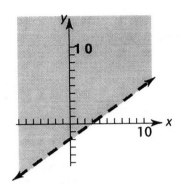

3. Graph $3x + 2y = 18$ as a solid line, since equality is included in the original statement. The origin is a suitable test point.
$3(0) + 2(0) = 0 \ngeqslant 18$
Hence $(0, 0)$ is not in the solution set. The graph is the line $3x + 2y = 18$ and the half-plane not containing the origin.

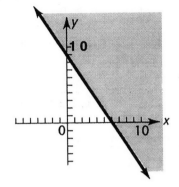

5. Graph $y = \frac{2}{3}x + 5$ as a solid line, since equality is included in the original statement. The origin is a suitable test point.
$0 \overset{?}{\leq} \frac{2}{3}(0) + 5$
$0 \leq 5$
Hence $(0, 0)$ is in the solution set. The graph is the line $y = \frac{2}{3}x + 5$ and the half-plane containing the origin.

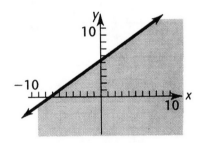

7. Graph $y = 8$ as a dashed line, since equality is not included in the original statement. Clearly the graph consists of all points whose y-coordinates are less than 8, that is, the lower half-plane.

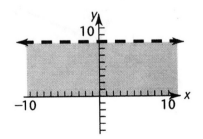

9. This system is equivalent to the system
$y \geq -3$
$y < 2$
and its graph is the intersection of the graphs of these inequalities.

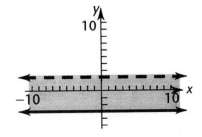

11. $x + 2y \leq 8$
$3x - 2y \geq 0$
Choose a suitable test point that lies on neither line, for example, (2, 0).
$2 + 2(0) = 2 \leq 8$ Hence, the solution region is *below* the graph
of $x + 2y = 8$.
$3(2) - 2(0) = 6 \geq 0$ Hence, the solution region is *below* the graph
of $3x - 2y = 0$.
Thus the solution region is region IV in the diagram.

13. $x + 2y \geq 8$
$3x - 2y \geq 0$
Choose a suitable test point that lies on neither line, for example, (2, 0).
$2 + 2(0) = 2 \ngeq 8$ Hence, the solution region is *above* the graph
of $x + 2y = 8$.
$3(2) - 2(0) = 6 \geq 0$ Hence, the solution region is *below* the graph
of $3x - 2y = 0$.
Thus the solution region is region I in the diagram.

15.

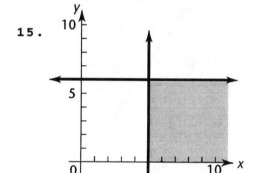

17.

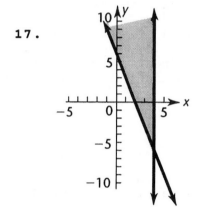

19.

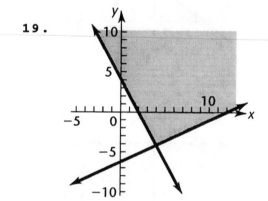

21. Choose a suitable test point that lies on none of the lines, say (5, 1).
$5 + 3(1) = 8 \leq 18$
Hence, the solution region is *below* the graph of $x + 3y = 18$.

$2(5) + 1 = 11 \ngeq 16$
Hence, the solution region is *above* the graph of $2x + y = 16$.

$5 \geq 0$
$1 \geq 0$
Thus the solution region is region IV in the diagram. The corner points are the labelled points (6, 4), (8, 0), and (18, 0).

23. Choose a suitable test point that lies on none of the lines, say (5, 1).
$5 + 3(1) = 8 \ngeq 18$ Hence, the solution region is *above* the graph
of $x + 3y = 18$.
$2(5) + 1 = 11 \ngeq 16$ Hence, the solution region is *above* the graph
of $2x + y = 16$.
$5 \geq 0$
$1 \geq 0$
Thus the solution region is region I in the diagram. The corner points are the labelled points (0, 16), (6, 4), and (18, 0).

25. The solution region is bounded (contained in, for example, the circle $x^2 + y^2 = 16$). The corner points are obvious from the graph: (0, 0), (0, 2), (3, 0).

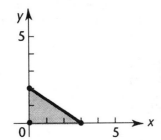

27. The solution region is unbounded. The corner points are obvious from the graph: (0, 4) and (5, 0).

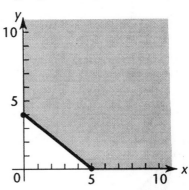

29. The solution region is bounded. Three corner points are obvious from the graph: (0, 4), (0, 0), (4, 0). The fourth corner point is obtained by solving the system
$x + 3y = 12$

$2x + y = 8$ to obtain $\left(\dfrac{12}{5},\ \dfrac{16}{5}\right)$.

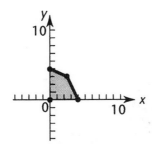

31. The solution region is unbounded. The corner points are obvious from the graph: (9, 0) and (0, 8). The third corner point is obtained by solving the system
$4x + 3y = 24$
$2x + 3y = 18$ to obtain (3, 4).

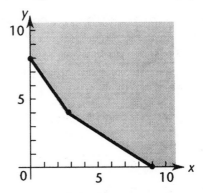

33. The solution region is bounded. Three corner points are obvious from the graph: (6, 0), (0, 0), and (0, 5). The other corner points are obtained by solving:
$2x + y = 12$ and $x + y = 7$
 $x + y = 7$ $x + 2y = 10$
to obtain   to obtain
  (5, 2)   (4, 3)

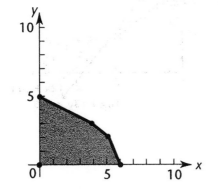

35. The solution region is unbounded. Two of the corner points are obvious from the graph: (16, 0) and (0, 14). The other corner points are obtained by solving:
$x + 2y = 16$ and $x + y = 12$
 $x + y = 12$ $2x + y = 14$
to obtain   to obtain
  (8, 4)   (2, 10)

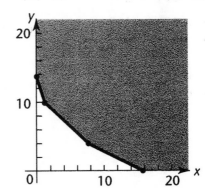

37. The solution region is bounded. The corner points are obtained by solving:

$\begin{cases} x + y = 11 \\ 5x + y = 15 \end{cases}$ to obtain (1, 10)

$\begin{cases} 5x + y = 15 \\ x + 2y = 12 \end{cases}$ to obtain (2, 5), and

$\begin{cases} x + y = 11 \\ x + 2y = 12 \end{cases}$ to obtain (10, 1)

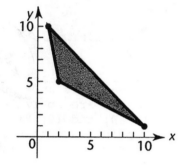

39. From the graph it should be clear that there is no point with x coordinate greater than 4 which satisfies both $3x + 2y \leq 24$ (arrows pointing, roughly, northeast) and $3x + y \leq 15$ (arrows pointing, roughly, southwest). The feasible region is empty.

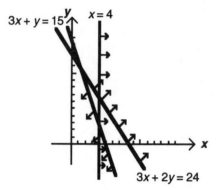

41. The feasible region is bounded. The corner points are obtained by solving:

$\begin{cases} x + y = 10 \\ 3x - 2y = 15 \end{cases}$ to obtain (7, 3)

$\begin{cases} 3x - 2y = 15 \\ 3x + 5y = 15 \end{cases}$ to obtain (5, 0),

$\begin{cases} 3x + 5y = 15 \\ -5x + 2y = 6 \end{cases}$ to obtain (0, 3), and

$\begin{cases} -5x + 2y = 6 \\ x + y = 10 \end{cases}$ to obtain (2, 8)

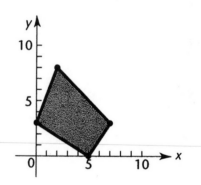

43. The feasible region is bounded. The corner points are obtained by solving:

$\begin{cases} 16x + 13y = 119 \\ 12x + 16y = 101 \end{cases}$ to obtain (5.91, 1.88)

$\begin{cases} 16x + 13y = 119 \\ -4x + 3y = 11 \end{cases}$ to obtain (2.14, 6.53)

and

$\begin{cases} 12x + 16y = 101 \\ -4x + 3y = 11 \end{cases}$ to obtain (1.27, 5.36)

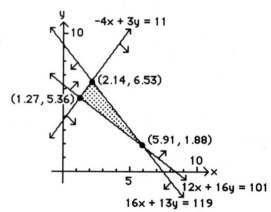

45. Let x = number of trick skis produced per day.
$\qquad y$ = number of slalom skis produced per day

Clearly x and y must be non-negative.
Hence $x \geq 0 \qquad$ (1)
$\qquad y \geq 0 \qquad$ (2)
To fabricate x trick skis requires $6x$ hours.
To fabricate y slalom skis requires $4y$ hours.
108 hours are available for fabricating; hence
$6x + 4y \leq 108 \qquad$ (3)
To finish x trick skis requires $1x$ hours.
To finish y slalom skis requires $1y$ hours.
24 hours are available for finishing, hence
$x + y \leq 24 \qquad$ (4)

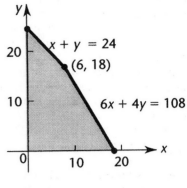

Graphing the inequality system (1), (2), (3), (4), we have the diagram.

47. (A) All production schedules in the feasible region that are on the graph
of $50x + 60y = 1,100$ will result in a profit of \$1,100.

(B) There are many possible choices. For example, producing 5 trick and 15
slalom skis will produce a profit of \$1,150. The graph of the line
$50x + 60y = 1,150$ includes all the production schedules in the feasible
region that result in a profit of \$1,150.

(C) A graphical approach would involve drawing other lines of the type
$50x + 60y = A$. The graphs of these lines include all production schedules
that will result in a profit of A. Increase A until the line either
intersects the feasible region only in 1 corner point or contains an edge
of the feasible region. This value of A will be the maximum profit
possible. For more details, see Section 8-5 of the text.

49. Clearly x and y must be non-negative.
Hence $x \geq 0 \qquad$ (1)
$\qquad y \geq 0 \qquad$ (2)
x cubic yards of mix A contains $20x$ pounds of phosphoric acid.
y cubic yards of mix B contains $10y$ pounds of phosphoric acid.

At least 460 pounds of phosphoric acid are required, hence
$20x + 10y \geq 460 \qquad$ (3)
x cubic yards of mix A contains $30x$ pounds of nitrogen.
y cubic yards of mix B contains $30y$ pounds of nitrogen.

At least 960 pounds of nitrogen are required, hence
$30x + 30y \geq 960 \qquad$ (4)
x cubic yards of mix A contains $5x$ pounds of potash.
y cubic yards of mix B contains $10y$ pounds of potash.

At least 220 pounds of potash are required,
hence
$5x + 10y \geq 220 \qquad$ (5)

Graphing the inequality system (1), (2),
(3), (4), (5), we have the diagram:

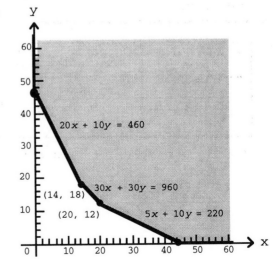

51. Clearly x and y must be non-negative.
Hence $x \geq 0$ (1)
          $y \geq 0$ (2)
Each sociologist will spend 10 hours collecting
data: $10x$ hours.
Each research assistant will spend 30 hours
collecting data: $30y$ hours.
At least 280 hours must be spent collecting data;
hence
$10x + 30y \geq 280$ (3)
Each sociologist will spend 30 hours analyzing data:
$30x$ hours.
Each research assistant will spend 10 hours
analyzing data: $10y$ hours.
At least 360 hours must be spent analyzing data;
hence
$30x + 10y \geq 360$ (4)
Graphing the inequality system (1), (2), (3), (4), we have the diagram.

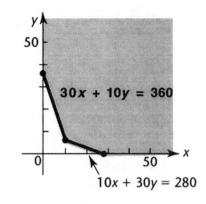

Exercise 8-5

Key Ideas and Formulas

A linear programming problem involves maximizing or minimizing an objective function
subject to linear inequalities known as problem constraints.

Fundamental Theorem of Linear Programming:

Let S be the feasible region for a linear programming problem and let $z = ax + by$ be
the objective function. If S is bounded, then z has both a maximum and a minimum
value on S and each of these occurs at a corner point of S. If S is unbounded, then a
maximum or minimum value of z on S may not exist. However, if either does exist, then
it must occur at a corner point of S.

Solution of Linear Programming Problems:

1. Form a mathematical model for the problem:
   (A) Introduce decision variables and write a linear objective function.
   (B) Write problem constraints in the form of linear inequalities.
   (C) Write non-negative constraints.

2. Graph the feasible region and find the corner points.

3. Evaluate the objective function at each corner point to determine the optimal
   solution.

1.

Corner Point (x, y)	Objective Function $z = x + y$	
(0, 12)	12	
(7, 9)	16	Maximum value
(10, 0)	10	
(0, 0)	0	

The maximum value of z on S is 16 at (7, 9).

3.

Corner Point (x, y)	Objective Function z = 3x + 7y	
(0, 12)	84	Maximum value
(7, 9)	84	Maximum value
(10, 0)	30	
(0, 0)	0	

Maximum value ⎫ Multiple optimal solutions
Maximum value ⎭

The maximum value of z on S is 84 at both (0, 12) and (7, 9).

5.

Corner Point (x, y)	Objective Function z = 7x + 4y	
(0, 12)	48	
(12, 0)	84	
(4, 3)	40	
(0, 8)	32	Minimum value

The minimum value of z on S is 32 at (0, 8).

7.

Corner Point (x, y)	Objective Function 3x + 8y	
(0, 12)	96	
(12, 0)	36	Minimum value
(4, 3)	36	Minimum value
(0, 8)	64	

Minimum value ⎫ Multiple optimal solutions
Minimum value ⎭

The minimum value of z on S is 36 at both (12, 0) and (4, 3).

9. The feasible region is graphed as follows:
The corner points (0, 5), (5, 0) and (0, 0) are
obvious from the graph. The corner point (4, 3) is
obtained by solving the system
$x + 2y = 10$
$3x + y = 15$
We now evaluate the objective function at each
corner point.

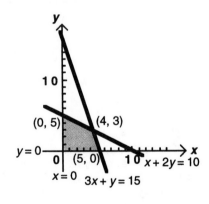

Corner Point (x, y)	Objective Function z = 3x + 2y	
(0, 5)	10	
(0, 0)	0	
(5, 0)	15	
(4, 3)	18	Maximum value

The maximum value of z on S is 18 at (4, 3).

11. The feasible region is graphed as follows:
The corner points $(4, 0)$ and $(10, 0)$ are obvious
from the graph. The corner point $(2, 4)$ is obtained
by solving the system
$x + 2y = 10$
$2x + y = 8$
We now evaluate the objective function at each
corner point.

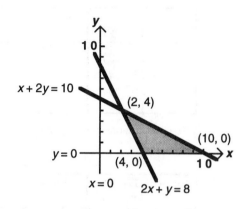

Corner Point (x, y)	Objective Function $z = 3x + 4y$	
$(4, 0)$	12	Minimum value
$(10, 0)$	30	
$(2, 4)$	22	

The minimum value of z on S is 12 at $(4, 0)$.

13. The feasible region is graphed below. The corner points $(0, 12)$, $(0, 0)$, and
$(12, 0)$ are obvious from the graph. The other corner points are obtained by
solving:
$x + 2y = 24$ and $x + y = 14$
$x + y = 14$ to obtain $(4, 10)$ $2x + y = 24$ to obtain $(10, 4)$
We now evaluate the objective function at each corner point.

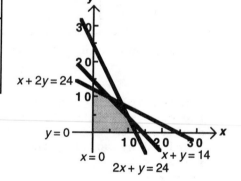

Corner Point (x, y)	Objective Function $z = 3x + 4y$	
$(0, 12)$	48	
$(0, 0)$	0	
$(12, 0)$	36	
$(10, 4)$	46	
$(4, 10)$	52	Maximum value

The maximum value of z on S is 52 at $(4, 10)$.

15. The feasible region is graphed as follows:
The corner points $(0, 20)$ and $(20, 0)$ are obvious
from the graph. The third corner point is obtained
by solving:
$x + 4y = 20$
$4x + y = 20$ to obtain $(4, 4)$
We now evaluate the objective function at each
corner point.

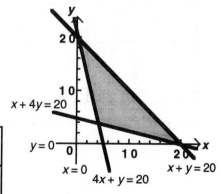

Corner Point (x, y)	Objective Function $z = 5x + 6y$	
$(0, 20)$	120	
$(20, 0)$	100	
$(4, 4)$	44	Minimum value

The minimum value of z on S is 44 at $(4, 4)$.

17. The feasible region is graphed as follows:
The corner points (60, 0) and (120, 0) are
obvious from the graph. The other corner
points are obtained by solving:

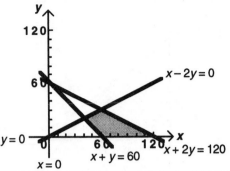

$x + y = 60$ and $x + 2y = 120$
$x - 2y = 0$ $x - 2y = 0$
to obtain (40, 20)   to obtain (60, 30)

We now evaluate the objective function at
each corner point.

Corner Point (x, y)	Objective Function $25x + 50y$	
(60, 0)	1,500	Minimum value
(40, 20)	2,000	
(60, 30)	3,000	Maximum value ⎫ Multiple optimal solutions
(120, 0)	3,000	Maximum value ⎭

The minimum value of z on S is 1,500 at (60, 0). The maximum value of z on S is
3,000 at (60, 30) and (120, 0).

19. The feasible region is graphed as follows:
The corner points (0, 45), (0, 20), (25, 0), and (60, 0) are obvious from the
graph shown below. The other corner points are obtained by solving:

$3x + 4y = 240$ and $3x + 4y = 240$
        $y = 45$ to obtain (20, 45) $x = 60$ to obtain (60, 15)

We now evaluate the objective function at each corner point.

Corner Point (x, y)	Objective Function $25x + 15y$	
(0, 45)	675	
(0, 20)	300	Minimum value
(25, 0)	625	
(60, 0)	1,500	
(60, 15)	1,725	Maximum value
(20, 45)	1,175	

The minimum value of z on S is 300 at (0, 20). The maximum value of z on S is
1,725 at (60, 15).

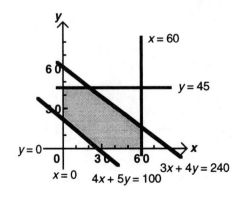

21. The feasible region is graphed as follows:

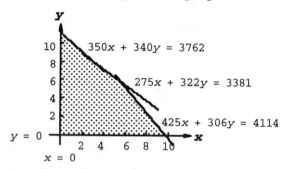

The corner point $(0, 0)$ is obvious from the graph. The other corner points are obtained by solving:

$x = 0$
$275x + 322y = 3381$
to obtain $(0, 10.5)$

$350x + 340y = 3762$
$275x + 322y = 3381$
to obtain $(3.22, 7.75)$

$350x + 340y = 3762$
$425x + 306y = 4114$
to obtain $(6.62, 4.25)$

$425x + 306y = 4114$
$y = 0$
to obtain $(9.68, 0)$

We now evaluate the objective function at each corner point.

Corner Point (x, y)	Objective Function $525x + 478y$	
$(0, 0)$	0	Minimum value
$(0, 10.5)$	5019	
$(3.22, 7.75)$	5395	
$(6.62, 4.25)$	5507	Maximum value
$(9.68, 0)$	5082	

The maximum value of P is 5507 at the corner point $(6.62, 4.25)$.

23. The feasible region is graphed as follows: (heavily outlined for clarity)

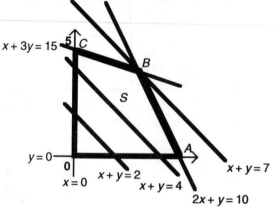

Consider the objective function $x + y$. It should be clear that it takes on the value 2 along $x + y = 2$, the value 4 along $x + y = 4$, the value 7 along $x + y = 7$, and so on. The maximum value of the objective function, then, on S, is 7, which occurs at B. Graphically this occurs when the line $x + y = c$ coincides with the boundary of S. Thus to answer questions (A)-(E) we must determine values of a and b such that the appropriate line $ax + by = c$ coincides with the boundary of S only at the specified points.

(A) The line $ax + by = c$ must have slope negative, but greater in absolute value that that of line segment AB, $2x + y = 10$. Therefore $a > 2b$.

(B) The line $ax + by = c$ must have slope negative but between that of $x + 3y = 15$ and $2x + y = 10$. Therefore $\frac{1}{3}b < a < 2b$.

(C) The line $ax + by = c$ must have slope greater than that of line segment BC, $x + 3y = 15$. Therefore $a < \frac{1}{3}b$ or $b > 3a$

(D) The line $ax + by = c$ must be parallel to line segment AB, therefore $a = 2b$.

(E) The line $ax + by = c$ must be parallel to line segment BC, therefore $b = 3a$.

25. Much of the mathematical model for this
problem was formed in Section 8-4 problem 45.
We let x = the number of trick skis
$\quad\quad\quad y$ = the number of slalom skis
The problem constraints were
$6x + 4y \leq 108$
$\quad x + y \leq 24$
The non-negative constraints were
$x \geq 0$
$y \geq 0$
The feasible region was graphed there.

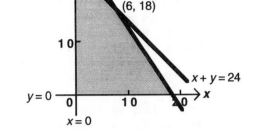

(A) We note now: the linear objective function
$\quad\quad P = 40x + 30y$ represents the profit.

Three of the corner points are obvious from the graph: (0, 24), (0, 0), and
(18, 0). The fourth corner point is obtained by solving:
$\quad\quad 6x + 4y = 108$
$\quad\quad\: x + y = 24$ to obtain (6, 18).

Summarizing: the mathematical model for this problem is:
Maximize $P = 40x + 30y$
subject to: $6x + 4y \leq 108$
$\quad\quad\quad\quad\quad\; x + y \leq 24$
$\quad\quad\quad\quad\quad\;\; x, y \geq 0$

We now evaluate the objective function $40x + 30y$ at each corner point.

Corner Point (x, y)	Objective Function $40x + 30y$	
(0, 0)	0	
(18, 0)	720	
(6, 18)	780	Maximum value
(0, 24)	720	

The optimal value is 780 at the corner point (6, 18). Thus, 6 trick skis
and 18 slalom skis should be manufactured to obtain the maximum profit of
$780.

(B) The objective function now becomes $40x + 25y$. We evaluate this at each
corner point.

Corner Point (x, y)	Objective Function $40x + 25y$	
(0, 0)	0	
(18, 0)	720	Maximum value
(6, 18)	690	
(0, 24)	600	

The optimal value is now 720 at the corner point (18, 0). Thus, 18 trick
skis and no slalom skis should be produced to obtain a maximum profit of
$720.

(C) The objective function now becomes $40x + 45y$. We evaluate this at each
corner point.

Corner Point (x, y)	Objective Function $40x + 45y$	
(0, 0)	0	
(18, 0)	720	
(6, 18)	1,050	
(0, 24)	1,080	Maximum value

The optimal value is now 1,080 at the corner point (0, 24). Thus, no trick
skis and 24 slalom skis should be produced to obtain a maximum profit of
$1,080.

27. Let x = number of model A trucks
    y = number of model B trucks
We form the linear objective function
$C = 15,000x + 24,000y$
We wish to minimize C, the cost of buying
x trucks @ \$15,000 and y trucks @ \$24,000,
subject to the constraints.
    $x + y \leq 15$ maximum number of trucks constraint
$2x + 3y \geq 36$ capacity constraint
    $x, y \geq 0$ non-negative constraints.

Solving the system of constraint inequalities
graphically, we obtain the feasible region S
shown in the diagram.

Next we evaluate the objective function at each corner point.

Corner Point (x, y)	Objective Function $C = 15,000x + 24,000y$	
(0, 12)	288,000	
(0, 15)	360,000	
(9, 6)	279,000	Minimum value

The optimal value is \$279,000 at the corner point (9, 6). Thus, the company
should purchase 9 model A trucks and 6 model B trucks to realize the minimum
cost of \$279,000.

29. (A) Let x = number of tables
        y = number of chairs
We form the linear objective function
$P = 90x + 25y$
We wish to maximize P, the profit from x
tables @ \$90 and y chairs @ \$25, subject to
the constraints
$8x + 2y \leq 400$ assembly department constraint
 $2x + y \leq 120$ finishing department constraint
    $x, y \geq 0$ non-negative constraints

Solving the system of constraint inequalities
graphically, we obtain the feasible region S
shown in the diagram.

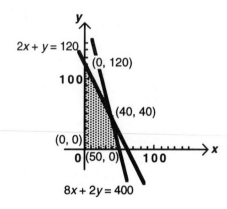

Next we evaluate the objective function at each corner point.

Corner Point (x, y)	Objective Function $P = 90x + 25y$	
(0, 0)	0	
(50, 0)	4,500	
(40, 40)	4,600	Maximum value
(0, 120)	3,000	

The optimal value is 4,600 at the corner point (40, 40). Thus, the company
should manufacture 40 tables and 40 chairs for a maximum profit of \$4,600.

(B) We are faced with the further condition
that $y \geq 4x$. We wish, then, to maximize
$P = 90x + 25y$ under the constraints
$$8x + 2y \leq 400$$
$$2x + y \leq 120$$
$$y \geq 4x$$
$$x,\ y \geq 0$$
The feasible region is now S' as graphed.

Note that the new condition has the effect
of excluding (40, 40) from the feasible
region.

We now evaluate the objective function at
the new corner points.

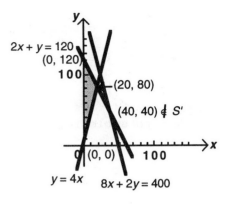

Corner Point (x, y)	Objective Function $P = 90x + 25y$	
(0, 120)	3,000	
(0, 0)	0	
(20, 80)	3,800	Maximum value

The optimal value is now 3,800 at the corner point (20, 80). Thus the company
should manufacture 20 tables and 80 chairs for a maximum profit of $3,800.

31. Let x = number of gallons produced using the old process
y = number of gallons produced using the new process

We form the linear objective function
$$P = 0.6x + 0.2y$$

(A) We wish to maximize P, the profit from x gallons using the old process and
y gallons using the new process, subject to the contraints
$$20x + 5y \leq 16,000 \qquad \text{sulfur dioxide constraint}$$
$$40x + 20y \leq 30,000 \qquad \text{particulate matter contraint}$$
$$x,\ y \geq 0 \qquad\qquad\ \text{non-negative contraints}$$

Solving the system of constraint
inequalities graphically, we obtain the
feasible region S shown in the diagram.
Note that no corner points are
determined by this (very weak) sulfur
dioxide constraint.

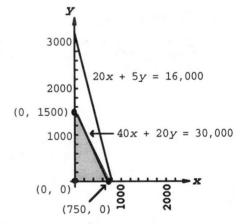

We evaluate the objective function at each corner point.

Corner Point (x, y)	Objective Function $P = 0.6x + 0.2y$	
(0, 0)	0	
(0, 1500)	300	
(750, 0)	450	Maximum value

The optimal value is 450 at the corner point (750, 0). Thus, the company
should manufacture 750 gallons by the old process exclusively, for a profit
of $450.

(B) The sulfur dioxide constraint is now
$20x + 5y \leq 11{,}500$. We now wish to
maximize P subject to the constraints
$$20x + 5y \leq 11{,}500$$
$$40x + 20y \leq 30{,}000$$
$$x,\ y \geq 0$$

The feasible region is now S_1 as shown.

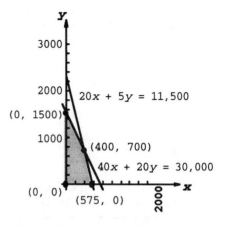

We evaluate the objective function at the new corner points.

Corner Point $(x,\ y)$	Objective Function $P = 0.6x + 0.2y$	
(0, 0)	0	
(575, 0)	345	
(400, 700)	380	Maximum value
(0, 1500)	300	

The optimal value is now 380 at the corner point (400, 700). Thus, the company should manufacture 400 gallons by the old process and 700 gallons by the new process, for a profit of $380.

(C) The sulfur dioxide constraint is now
$20x + 5y \leq 7{,}200$. We now wish to
maximize P subject to the constraints
$$20x + 5y \leq 7{,}200$$
$$40x + 20y \leq 30{,}000$$
$$x,\ y \geq 0$$

The feasible region is now S_2 as shown.

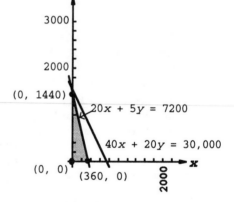

Note that now no corner points are
determined by the particulate matter
constraint.

We evaluate the objective function at the new corner points.

Corner Point $(x,\ y)$	Objective Function $P = 0.6x + 0.2y$	
(0, 0)	0	
(360, 0)	216	
(0, 1440)	288	Maximum value

The optimal value is now 288 at the corner point (0, 1440). Thus, the company should manufacture 1,440 gallons by the new process exclusively, for a profit of $288.

33. (A) Let x = number of bags of Brand A
y = number of bags of Brand B
We form the objective function
$N = 6x + 7y$
N represents the amount of nitrogen in x bags @ 6 pounds per bag and y bags @ 7 pounds per bag.

We wish to optimize N subject to the constraints
$2x + 4y \geq 480$ phosphoric acid constraint
$6x + 3y \geq 540$ potash constraint
$3x + 4y \leq 620$ chlorine constraint
$x, y \geq 0$ non-negative contraints
Solving the system of constraint inequalities graphically, we obtain the feasible region S shown in the diagram.

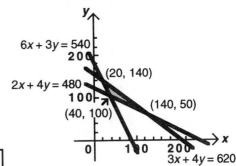

Next we evaluate the objective function at the corner points.

Corner Point (x, y)	Objective Function $N = 6x + 7y$	
(20, 140)	1,100	
(40, 100)	940	Minimum value
(140, 50)	1,190	Maximum value

Hence, the nitrogen will range from a minimum of 940 pounds when 40 bags of brand A and 100 bags of Brand B are used to a maximum of 1,190 pounds when 140 bags of brand A and 50 bags of brand B are used.

CHAPTER 8 REVIEW

1. $2x + y = 7$
$3x - 2y = 0$
We multiply the top equation by 2 and add.

$4x + 2y = 14$
$\underline{3x - 2y = 0}$
$7x = 14$
$x = 2$

Substituting $x = 2$ in the top equation, we have
$2(2) + y = 7$
$y = 3$
Solution: (2, 3)   *(8-1)*

2. $3x - 6y = 5$
$-2x + 4y = 1$
We multiply the top equation by 2, the bottom by 3, and add.

$6x - 12y = 10$
$\underline{-6x + 12y = 3}$
$0 = 13$

No solution   *(8-1)*

3. $4x - 3y = -8$
$-2x + \dfrac{3}{2}y = 4$

We multiply the bottom equation by 2 and add.

$4x - 3y = -8$
$\underline{-4x + 3y = 8}$
$0 = 0$

There are infinitely many solutions. For any real number t, $4t - 3y = -8$, hence,
$-3y = -4t - 8$
$y = \dfrac{4t + 8}{3}$

Thus, $\left(t, \dfrac{4t + 8}{3}\right)$ is a solution for any real number t.   *(8-1)*

4. $y = x^2 - 5x - 3$

$y = -x + 2$

We multiply the bottom equation by -1 and add.

$$y = x^2 - 5x - 3$$
$$\underline{-y = x - 2}$$
$$0 = x^2 - 4x - 5$$
$$0 = (x - 5)(x + 1)$$
$$x = 5, -1$$

For $x = 5$ For $x = -1$

    $y = -5 + 2$ $y = -(-1) + 2$

    $y = -3$ $y = 3$

Solutions: $(5, -3)$, $(-1, 3)$

Check: For $(5, -3)$ For $(-1, 3)$

   $y = x^2 - 5x - 3$ $y = x^2 - 5x - 3$

   $-3 \overset{?}{=} 5^2 - 5 \cdot 5 - 3$ $3 \overset{?}{=} (-1)^2 - 5(-1) - 3$

   $-3 \overset{\surd}{=} -3$ $3 \overset{\surd}{=} 1 + 5 - 3$

                                                 (8-3)

5. $x^2 + y^2 = 2$

$2x - y = 3$

We solve the first-degree equation for y in terms of x, then substitute into the second-degree equation.

$$2x - y = 3$$
$$-y = 3 - 2x$$
$$y = 2x - 3$$
$$x^2 + (2x - 3)^2 = 2$$
$$x^2 + 4x^2 - 12x + 9 = 2$$
$$5x^2 - 12x + 7 = 0$$
$$(5x - 7)(x - 1) = 0$$
$$x = \frac{7}{5}, 1$$

For $x = \dfrac{7}{5}$ For $x = 1$

$y = 2\left(\dfrac{7}{5}\right) - 3$ $y = 2(1) - 3$

$= -\dfrac{1}{5}$ $= -1$

Solutions: $(1, -1)$, $\left(\dfrac{7}{5}, -\dfrac{1}{5}\right)$ *(8-3)*

The checking steps are omitted for lack of space.

6. $3x^2 - y^2 = -6$

$2x^2 + 3y^2 = 29$

We use elimination by addition. We multiply the top equation by 3 and add.

$$9x^2 - 3y^2 = -18$$
$$\underline{2x^2 + 3y^2 = 29}$$
$$11x^2 = 11$$
$$x^2 = 1$$
$$x = \pm 1$$

For $x = 1$ For $x = -1$

$3(1)^2 - y^2 = -6$ $3(-1)^2 - y^2 = -6$

     $-y^2 = -9$ $-y^2 = -9$

      $y^2 = 9$ $y = \pm 3$

      $y = \pm 3$

Solutions: $(1, 3)$, $(1, -3)$, $(-1, 3)$, $(-1, -3)$

                                (8-3)

7.

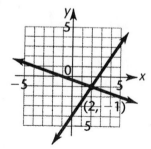

                (8-1)

8. Graph $3x - 4y = 24$ as a solid line since
equality is included in the original statement.
$(0, 0)$ is a suitable test point.
$3(0) - 4(0) = 0 \not\geq 24$
Therefore $(0, 0)$ is not in the solution set.
The line $3x - 4y = 24$ and the half-plane not
containing $(0, 0)$ form the graph.

9.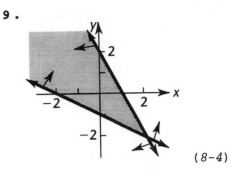

$(8-4)$

$(8-4)$

10. $R_1 \leftrightarrow R_2$ means interchange Rows 1 and
2.
$$\begin{bmatrix} 3 & -6 & | & 12 \\ 1 & -4 & | & 5 \end{bmatrix}$$ $(8-1)$

11. $\frac{1}{3} R_2 \rightarrow R_2$ means multiply Row 2 by $\frac{1}{3}$.
$$\begin{bmatrix} 1 & -4 & | & 5 \\ 1 & -2 & | & 4 \end{bmatrix}$$ $(8-1)$

12. $(-3)R_1 + R_2 \rightarrow R_2$ means replace Row 2
by itself plus -3 times Row 1.
$$\begin{bmatrix} 1 & -4 & | & 5 \\ 0 & 6 & | & -3 \end{bmatrix}$$ $(8-1)$

13. $x_1 = 4$
$x_2 = -7$
The solution is $(4, -7)$ $(8-2)$

14. $x_1 - x_2 = 4$
$\qquad 0 = 1$

No solution $(8-2)$

15. $x_1 - x_2 = 4$
$\qquad 0 = 0$
Solution:
$x_2 = t$
$x_1 = x_2 + 4 = t + 4$
Thus $x_1 = t + 4$, $x_2 = t$ is the
solution, for t any real number.
$(8-2)$

16.

Corner Point (x, y)	Objective Function $z = 5x + 3y$	
$(0, 10)$	30	
$(0, 6)$	18	Minimum value
$(4, 2)$	26	
$(6, 4)$	42	Maximum value

The maximum value of z on S is 42 at $(6, 4)$. The minimum value of z on S is 18
at $(0, 6)$.
$(8-5)$

17. We write the augmented matrix:

$$\begin{bmatrix} 1 & -1 & | & 4 \\ 2 & 1 & | & 2 \end{bmatrix} \quad (-2)R_1 + R_2 \to R_2$$

-2   2   -8

Need a 0 here

$$\sim \begin{bmatrix} 1 & -1 & | & 4 \\ 0 & 3 & | & -6 \end{bmatrix} \tfrac{1}{3} R_2 \to R_2 \qquad \text{corresponds to the linear system} \quad \begin{array}{r} x_1 - x_2 = 4 \\ 3x_2 = -6 \end{array}$$

↑
Need a 1 here
Need a 0 here
↓

$$\sim \begin{bmatrix} 1 & -1 & | & 4 \\ 0 & 1 & | & -2 \end{bmatrix} R_2 + R_1 \to R_1 \qquad \text{corresponds to the linear system} \quad \begin{array}{r} x_1 - x_2 = 4 \\ x_2 = -2 \end{array}$$

$$\sim \begin{bmatrix} 1 & 0 & | & 2 \\ 0 & 1 & | & -2 \end{bmatrix} \qquad \text{corresponds to the linear system} \quad \begin{array}{r} x_1 = 2 \\ x_2 = -2 \end{array}$$

The solution is $x_1 = 2$, $x_2 = -2$. Each pair of lines graphed below has the same intersection point, $(2, -2)$.

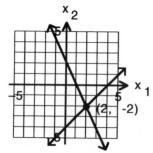

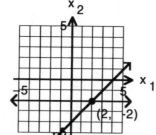

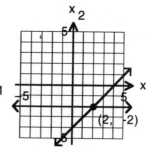

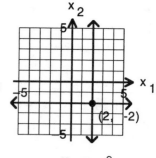

$$\begin{array}{l} x_1 - x_2 = 4 \\ 2x_1 + x_2 = 2 \end{array} \qquad \begin{array}{l} x_1 - x_2 = 4 \\ 3x_2 = -6 \end{array} \qquad \begin{array}{l} x_1 - x_2 = 4 \\ x_2 = -2 \end{array} \qquad \begin{array}{l} x_1 = 2 \\ x_2 = -2 \quad (8\text{-}1) \end{array}$$

18. Here is a computer-generated graph of the system , entered as

$$y = \frac{9 - x}{3}$$

$$y = \frac{10 + 2x}{7}$$

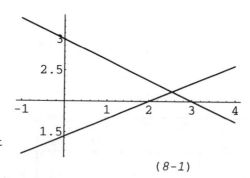

After zooming in (graph not shown) the intersection point is located at $(2.54, 2.15)$ to two decimal places.

(8-1)

19. $\begin{bmatrix} 3 & 2 & | & 3 \\ 1 & 3 & | & 8 \end{bmatrix}$ $R_1 \leftrightarrow R_2$

$\sim \begin{bmatrix} 1 & 3 & | & 8 \\ 3 & 2 & | & 3 \end{bmatrix}$ $(-3)R_1 + R_2 \rightarrow R_2$

$-3 \quad -9 \quad -24$

$\sim \begin{bmatrix} 1 & 3 & | & 8 \\ 0 & -7 & | & -21 \end{bmatrix}$ $-\frac{1}{7}R_2 \rightarrow R_2$

$\sim \begin{bmatrix} 1 & 3 & | & 8 \\ 0 & 1 & | & 3 \end{bmatrix}$ $(-3)R_2 + R_1 \rightarrow R_1$

$0 \quad -3 \quad -9$

$\sim \begin{bmatrix} 1 & 0 & | & -1 \\ 0 & 1 & | & 3 \end{bmatrix}$

Solution: $x_1 = -1$, $x_2 = 3$ (8-2)

20. $\begin{bmatrix} 1 & 1 & 0 & | & 1 \\ 1 & 0 & -1 & | & -2 \\ 0 & 1 & 2 & | & 4 \end{bmatrix}$ $(-1)R_1 + R_2 \rightarrow R_2$

$\sim \begin{bmatrix} 1 & 1 & 0 & | & 1 \\ 0 & -1 & -1 & | & -3 \\ 0 & 1 & 2 & | & 4 \end{bmatrix}$ $(-1)R_2 \rightarrow R_2$

$\sim \begin{bmatrix} 1 & 1 & 0 & | & 1 \\ 0 & 1 & 1 & | & 3 \\ 0 & 1 & 2 & | & 4 \end{bmatrix}$ $\begin{matrix}(-1)R_2 + R_1 \rightarrow R_1 \\ (-1)R_2 + R_3 \rightarrow R_3\end{matrix}$

$\sim \begin{bmatrix} 1 & 0 & -1 & | & -2 \\ 0 & 1 & 1 & | & 3 \\ 0 & 0 & 1 & | & 1 \end{bmatrix}$ $\begin{matrix}R_3 + R_1 \rightarrow R_1 \\ (-1)R_3 + R_2 \rightarrow R_2\end{matrix}$

$\sim \begin{bmatrix} 1 & 0 & 0 & | & -1 \\ 0 & 1 & 0 & | & 2 \\ 0 & 0 & 1 & | & 1 \end{bmatrix}$

Solution: $x_1 = -1$, $x_2 = 2$, $x_3 = 1$
(8-2)

21. $\begin{bmatrix} 1 & 2 & 3 & | & 1 \\ 2 & 3 & 4 & | & 3 \\ 1 & 2 & 1 & | & 3 \end{bmatrix}$ $\begin{matrix}(-2)R_1 + R_2 \rightarrow R_2 \\ (-1)R_1 + R_3 \rightarrow R_3\end{matrix}$

$\sim \begin{bmatrix} 1 & 2 & 3 & | & 1 \\ 0 & -1 & -2 & | & 1 \\ 0 & 0 & -2 & | & 2 \end{bmatrix}$ $\begin{matrix}(-1)R_2 \rightarrow R_2 \\ -\frac{1}{2}R_3 \rightarrow R_3\end{matrix}$

$\sim \begin{bmatrix} 1 & 2 & 3 & | & 1 \\ 0 & 1 & 2 & | & -1 \\ 0 & 0 & 1 & | & -1 \end{bmatrix}$ $(-2)R_2 + R_1 \rightarrow R_1$

$\sim \begin{bmatrix} 1 & 0 & -1 & | & 3 \\ 0 & 1 & 2 & | & -1 \\ 0 & 0 & 1 & | & -1 \end{bmatrix}$ $\begin{matrix}R_3 + R_1 \rightarrow R_1 \\ (-2)R_3 + R_2 \rightarrow R_2\end{matrix}$

$\sim \begin{bmatrix} 1 & 0 & 0 & | & 2 \\ 0 & 1 & 0 & | & 1 \\ 0 & 0 & 1 & | & -1 \end{bmatrix}$

Solution: $x_1 = 2$, $x_2 = 1$, $x_3 = -1$
(8-2)

22. $\begin{bmatrix} 1 & 2 & -1 & | & 2 \\ 2 & 3 & 1 & | & -3 \\ 3 & 5 & 0 & | & -1 \end{bmatrix}$ $\begin{matrix}(-2)R_1 + R_2 \rightarrow R_2 \\ (-3)R_1 + R_3 \rightarrow R_3\end{matrix}$

$\sim \begin{bmatrix} 1 & 2 & -1 & | & 2 \\ 0 & -1 & 3 & | & -7 \\ 0 & -1 & 3 & | & -7 \end{bmatrix}$ $(-1)R_2 + R_3 \rightarrow R_3$

$\sim \begin{bmatrix} 1 & 2 & -1 & | & 2 \\ 0 & -1 & 3 & | & -7 \\ 0 & 0 & 0 & | & 0 \end{bmatrix}$ $2R_2 + R_1 \rightarrow R_1$

$\sim \begin{bmatrix} 1 & 0 & 5 & | & -12 \\ 0 & -1 & 3 & | & -7 \\ 0 & 0 & 0 & | & 0 \end{bmatrix}$ $(-1)R_2 \rightarrow R_2$

$\sim \begin{bmatrix} 1 & 0 & 5 & | & -12 \\ 0 & 1 & -3 & | & 7 \\ 0 & 0 & 0 & | & 0 \end{bmatrix}$

This corresponds to the system
$x_1 + \quad\quad 5x_3 = -12$
$\quad\quad x_2 - 3x_3 = 7$

Let $x_3 = t$
Then $x_2 = 3x_3 + 7$
$\quad\quad = 3t + 7$
$\quad x_1 = -5x_3 - 12$
$\quad\quad = -5t - 12$
Hence $x_1 = -5t - 12$, $x_2 = 3t + 7$,
$x_3 = t$ is a solution for every real
number t. There are infinitely many
solutions.   (8-2)

23. $\begin{bmatrix} 1 & -2 & | & 1 \\ 2 & -1 & | & 0 \\ 1 & -3 & | & -2 \end{bmatrix}$ $(-2)R_1 + R_2 \to R_2$
$(-1)R_1 + R_3 \to R_3$

$\sim \begin{bmatrix} 1 & -2 & | & 1 \\ 0 & 3 & | & -2 \\ 0 & -1 & | & -3 \end{bmatrix}$ $3R_3 + R_2 \to R_2$

$\sim \begin{bmatrix} 1 & -2 & | & 1 \\ 0 & 0 & | & -11 \\ 0 & -1 & | & -3 \end{bmatrix}$

The second row corresponds to the equation
$0x_1 + 0x_2 = -11$,
hence there is no solution. *(8-2)*

24. $\begin{bmatrix} 1 & 2 & -1 & | & 2 \\ 3 & -1 & 2 & | & -3 \end{bmatrix}$ $(-3)R_1 + R_2 \to R_2$

$\sim \begin{bmatrix} 1 & 2 & -1 & | & 2 \\ 0 & -7 & 5 & | & -9 \end{bmatrix}$ $-\frac{1}{7}R_2 \to R_2$

$\sim \begin{bmatrix} 1 & 2 & -1 & | & 2 \\ 0 & 1 & -\frac{5}{7} & | & \frac{9}{7} \end{bmatrix}$ $(-2)R_2 + R_1 \to R_1$

$\sim \begin{bmatrix} 1 & 0 & \frac{3}{7} & | & -\frac{4}{7} \\ 0 & 1 & -\frac{5}{7} & | & \frac{9}{7} \end{bmatrix}$

This corresponds to the system
$$x_1 + \tfrac{3}{7}x_3 = -\tfrac{4}{7}$$
$$x_2 - \tfrac{5}{7}x_3 = \tfrac{9}{7}$$

Let $x_3 = t$
Then $x_2 = \tfrac{5}{7}x_3 + \tfrac{9}{7}$
$\qquad = \tfrac{5}{7}t + \tfrac{9}{7}$
$\qquad x_1 = -\tfrac{3}{7}x_3 - \tfrac{4}{7}$
$\qquad\quad = -\tfrac{3}{7}t - \tfrac{4}{7}$
Hence $x_1 = -\tfrac{3}{7}t - \tfrac{4}{7}$, $x_2 = \tfrac{5}{7}t + \tfrac{9}{7}$,
$x_3 = t$ is a solution for every real
number t. There are infinitely many
solutions. *(8-2)*

The checking steps are omitted in problems 25—27 for lack of space.

25. $x^2 - y^2 = 2$
$\qquad y^2 = x$

Substitute y^2 in the second equation into the first equation.
$$x^2 - x = 2$$
$$x^2 - x - 2 = 0$$
$$(x - 2)(x + 1) = 0$$
$$x = 2, -1$$

For $x = 2$ $\qquad\qquad$ For $x = -1$
$\quad y^2 = 2$ $\qquad\qquad\quad y^2 = -1$
$\quad y = \pm\sqrt{2}$ $\qquad\qquad\quad y = \pm i$
Solutions: $(2, \sqrt{2})$, $(2, -\sqrt{2})$, $(-1, i)$, $(-1, -i)$ *(8-3)*

26. $x^2 + 2xy + y^2 = 1$
$\qquad\qquad xy = -2$

Solve for y in the second equation.
$$y = \frac{-2}{x}$$
Substitute into the first equation
$$x^2 + 2x\left(-\frac{2}{x}\right) + \left(-\frac{2}{x}\right)^2 = 1$$
$$x^2 - 4 + \frac{4}{x^2} = 1 \quad x \neq 0$$
$$x^4 - 4x^2 + 4 = x^2$$
$$x^4 - 5x^2 + 4 = 0$$
$$(x^2 - 4)(x^2 - 1) = 0$$
$$(x - 2)(x + 2)(x - 1)(x + 1) = 0$$
$$x = 2, -2, 1, -1$$

For $x = 2$ For $x = -2$ For $x = 1$ For $x = -1$

$y = \dfrac{-2}{2}$ $y = \dfrac{-2}{-2}$ $y = \dfrac{-2}{1}$ $y = \dfrac{-2}{-1}$

$y = -1$ $y = 1$ $y = -2$ $y = 2$

Solutions: $(2, -1)$, $(-2, 1)$, $(1, -2)$, $(-1, 2)$ (8-3)

27. $2x^2 + xy + y^2 = 8$
$ x^2 - y^2 = 0$

We factor the left side of the equation that has a zero constant term.
$(x - y)(x + y) = 0$
$x = y$ or $x = -y$
Thus, the original system is equivalent to the two systems
$2x^2 + xy + y^2 = 8 \qquad 2x^2 + xy + y^2 = 8$
$ x = y x = -y$
These systems are solved by substitution.

First System: Second System:
$2x^2 + xy + y^2 = 8 \qquad\qquad 2x^2 + xy + y^2 = 8$
$ x = y \qquad\qquad\qquad\qquad x = -y$
$2y^2 + yy + y^2 = 8 \qquad 2(-y)^2 + (-y)y + y^2 = 8$
$ 4y^2 = 8 \qquad\qquad\qquad 2y^2 = 8$
$ y^2 = 2 \qquad\qquad\qquad\quad y^2 = 4$
$ y = \pm\sqrt{2} \qquad\qquad\qquad y = \pm 2$

For $y = \sqrt{2}$ For $y = -\sqrt{2}$ For $y = 2$ For $y = -2$

$x = \sqrt{2}$ $x = -\sqrt{2}$ $x = -2$ $x = 2$

Solutions: $(\sqrt{2}, \sqrt{2})$, $(-\sqrt{2}, -\sqrt{2})$, $(2, -2)$, $(-2, 2)$ (8-3)

28. The solution region is bounded (contained in, for example, the circle $x^2 + y^2 = 36$). Three corner points are obvious from the graph: $(0, 4)$, $(0, 0)$, and $(4, 0)$. The fourth corner point is obtained by solving the system
$2x + y = 8$
$2x + 3y = 12$ to obtain $(3, 2)$.

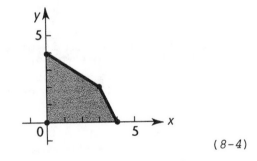

(8-4)

29. The solution region is unbounded. Two corner points are obvious from the graph: $(0, 8)$ and $(12, 0)$. The third corner point is obtained by solving the system
$2x + y = 8$
$x + 3y = 12$

to obtain $\left(\dfrac{12}{5}, \dfrac{16}{5}\right)$.

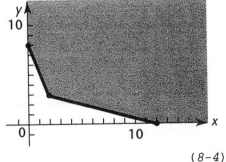

(8-4)

30. The solution region is bounded. The corner point (20, 0) is obvious from the graph. The other corner points are found by solving the system

$$\begin{cases} x + y = 20 \\ x - y = 0 \end{cases} \text{ to obtain (10, 10)}$$

and

$$\begin{cases} x - y = 0 \\ x + 4y = 20 \end{cases} \text{ to obtain (4, 4)}$$

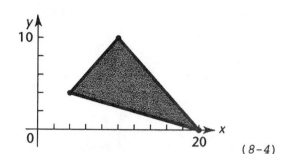

(8-4)

31. The feasible region is graphed as follows: The corner points (0, 4), (0, 0) and (5, 0) are obvious from the graph. The corner point (4, 2) is obtained by solving the system

$x + 2y = 8$
$2x + y = 10$

We now evaluate the objective function at each corner point.

Corner Point (x, y)	Objective Function $z = 7x + 9y$	
(0, 4)	36	
(0, 0)	0	
(5, 0)	35	
(4, 2)	46	Maximum value

The maximum value of z on S is 46 at (4, 2).

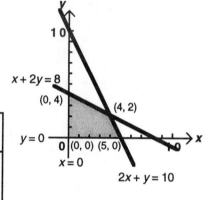

(8-5)

32. The feasible region is graphed as follows: The corner points (0, 20), (0, 15), (15, 0) and (20, 0) are obvious from the graph. The corner point (3, 6) is obtained by solving the system

$3x + y = 15$
$x + 2y = 15$

We now evaluate the objective function at each corner point.

Corner Point (x, y)	Objective Function $z = 5x + 10y$	
(0, 20)	200	
(0, 15)	150	
(3, 6)	75	Minimum value ⎫ Multiple optimal solutions
(15, 0)	75	Minimum value ⎭
(20, 0)	100	

The minimum value of z on S is 75 at (3, 6) and (15, 0).

(8-5)

33. The feasible region is graphed below. The corner points (0, 10) and (0, 7) are obvious from the graph. The other corner points are obtained by solving the systems:

$x + 2y = 20$ and $3x + y = 15$
$3x + y = 15$ to obtain (2, 9) $x + y = 7$ to obtain (4, 3)

We now evaluate the objective function at each corner point.

Corner Point (x, y)	Objective Function $z = 5x + 8y$	
(0, 10)	80	
(0, 7)	56	
(4, 3)	44	Minimum value
(2, 9)	82	Maximum value

The minimum value of z on S is 44 at (4, 3). The maximum value of z on S is 82 at (2, 9).

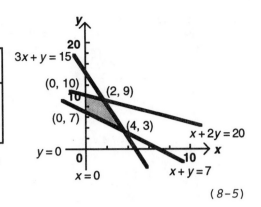

$(8-5)$

34. $\begin{bmatrix} 1 & 1 & 1 & | & 7000 \\ 0.04 & 0.05 & 0.06 & | & 360 \\ 0.04 & 0.05 & -0.06 & | & 120 \end{bmatrix}$ $\begin{array}{l}(-0.04)R_1 + R_2 \rightarrow R_2 \\ (-0.04)R_1 + R_3 \rightarrow R_3\end{array}$

$\sim \begin{bmatrix} 1 & 1 & 1 & | & 7000 \\ 0 & 0.01 & 0.02 & | & 80 \\ 0 & 0.01 & -0.1 & | & -160 \end{bmatrix}$ $100R_2 \rightarrow R_2$

$\sim \begin{bmatrix} 1 & 1 & 1 & | & 7000 \\ 0 & 1 & 2 & | & 8000 \\ 0 & 0.01 & -0.1 & | & -160 \end{bmatrix}$ $\begin{array}{l}(-1)R_2 + R_1 \rightarrow R_1 \\ (-0.01)R_2 + R_3 \rightarrow R_3\end{array}$

$\sim \begin{bmatrix} 1 & 0 & -1 & | & -1000 \\ 0 & 1 & 2 & | & 8000 \\ 0 & 0 & -0.12 & | & -240 \end{bmatrix}$ $-\frac{25}{3}R_3 \rightarrow R_3$

$\sim \begin{bmatrix} 1 & 0 & -1 & | & -1000 \\ 0 & 1 & 2 & | & 8000 \\ 0 & 0 & 1 & | & 2000 \end{bmatrix}$ $\begin{array}{l}R_3 + R_1 \rightarrow R_1 \\ (-2)R_3 + R_2 \rightarrow R_2\end{array}$

$\sim \begin{bmatrix} 1 & 0 & 0 & | & 1000 \\ 0 & 1 & 0 & | & 4000 \\ 0 & 0 & 1 & | & 2000 \end{bmatrix}$

Solution: $x_1 = 1000$, $x_2 = 4000$, $x_3 = 2000$

$(8-2)$

35. $x^2 - xy + y^2 = 4$
$x^2 + xy - 2y^2 = 0$

Factor the left side of the equation that has a zero constant term.
$(x + 2y)(x - y) = 0$
$x = -2y$ or $x = y$
Thus, the original system is equivalent to the two systems
$x^2 - xy + y^2 = 4$ $x^2 - xy + y^2 = 4$
$x = -2y$ $x = y$
These systems are solved by substitution

First System:
$$x^2 - xy + y^2 = 4$$
$$x = -2y$$
$$(-2y)^2 - (-2y)y + y^2 = 4$$
$$4y^2 + 2y^2 + y^2 = 4$$
$$7y^2 = 4$$
$$y = \pm\sqrt{\frac{4}{7}}$$
$$y = \pm\frac{2\sqrt{7}}{7}$$

Second System:
$$x^2 - xy + y^2 = 4$$
$$x = y$$
$$y^2 - yy + y^2 = 4$$
$$y^2 = 4$$
$$y = \pm 2$$

For $y = \dfrac{2\sqrt{7}}{7}$ | For $y = -\dfrac{2\sqrt{7}}{7}$ | For $y = -2$ | For $y = 2$

$$x = -2\left(\dfrac{2\sqrt{7}}{7}\right) \qquad\qquad x = -2\left(-\dfrac{2\sqrt{7}}{7}\right) \qquad\qquad x = -2 \qquad\qquad x = 2$$

$$= -\dfrac{4\sqrt{7}}{7} \qquad\qquad\qquad = \dfrac{4\sqrt{7}}{7}$$

Solutions: $\left(\dfrac{4\sqrt{7}}{7},\ -\dfrac{2\sqrt{7}}{7}\right),\ \left(-\dfrac{4\sqrt{7}}{7},\ \dfrac{2\sqrt{7}}{7}\right),\ (2,\ 2),\ (-2,\ -2)$. The checking steps are omitted for lack of space.

(8-3)

36. For convenience we rewrite the constraint conditions:

$1.2x + 0.6y \le 960$	becomes	$2x + y \le 1{,}600$
$0.04x + 0.03y \le 36$	becomes	$4x + 3y \le 3{,}600$
$0.2x + 0.3y \le 270$	becomes	$2x + 3y \le 2{,}700$
$x,\ y \ge 0$	are unaltered.	

The feasible region is graphed in the diagram. The corner points $(0, 900)$, $(0, 0)$, and $(800, 0)$ are obvious from the graph. The other corner points are obtained by solving the system:

$2x + y = 1{,}600$ and $4x + 3y = 3{,}600$
$4x + 3y = 3{,}600$ to obtain $(600, 400)$ $2x + 3y = 2{,}700$ to obtain $(450, 600)$

We now evaluate the objective function at each corner point.

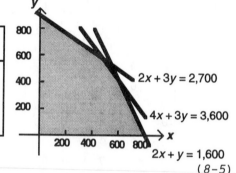

Corner Point $(x,\ y)$	Objective Function $z = 30x + 20y$	
$(0,\ 900)$	$18{,}000$	
$(0,\ 0)$	0	
$(800,\ 0)$	$24{,}000$	
$(600,\ 400)$	$26{,}000$	Maximum value
$(450,\ 600)$	$25{,}500$	

The maximum value of z on S is $26{,}000$ at $(600, 400)$.

(8-5)

37. Before we can enter these equations in our graphing utility, we must solve for y:

$$x^2 + 4xy + y^2 = 8 \qquad\qquad 5x^2 + 2xy + y^2 = 25$$
$$y^2 + 4xy + x^2 - 8 = 0 \qquad\qquad y^2 + 2xy + 5x^2 - 25 = 0$$

Applying the quadratic formula to each equation, we have

$$y = \dfrac{-4x \pm \sqrt{16x^2 - 4(x^2 - 8)}}{2} \qquad\qquad y = \dfrac{-2x \pm \sqrt{4x^2 - 4(5x^2 - 25)}}{2}$$

$$y = \dfrac{-4x \pm \sqrt{12x^2 + 32}}{2} \qquad\qquad\qquad y = \dfrac{-2x \pm \sqrt{100 - 16x^2}}{2}$$

$$y = -2x \pm \sqrt{3x^2 + 8} \qquad\qquad\qquad\qquad y = -x \pm \sqrt{25 - 4x^2}$$

Entering each of these four equations into a graphing utility produces the graph shown at the right.

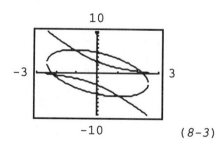

Zooming in on the four intersection points, or using a built-in intersection routine (details omitted), yields $(-2.16, -0.37)$, $(-1.09, 5.59)$, $(1.09, -5.59)$, and $(2.16, 0.37)$ to two decimal places.

(8-3)

38. (A) The system is independent. There is one solution.

(B) The matrix is in fact

$$\begin{bmatrix} 1 & 0 & -3 & | & 4 \\ 0 & 1 & 2 & | & 5 \\ 0 & 0 & 0 & | & n \end{bmatrix}$$

The third row corresponds to the equation $0x_1 + 0x_2 + 0x_3 = n$. This is impossible. The system has no solution.

(C) The matrix is in fact

$$\begin{bmatrix} 1 & 0 & -3 & | & 4 \\ 0 & 1 & 2 & | & 5 \\ 0 & 0 & 0 & | & 0 \end{bmatrix}$$

Thus, there are an infinite number of solutions ($x_3 = t$, $x_2 = 5 - 2t$, $x_1 = 4 + 3t$, for t any real number.) *(8-2)*

39. Let x = number of $\frac{1}{2}$ pound packages

y = number of $\frac{1}{3}$ pound packages

There are 120 packages. Hence

$x + y = 120$ (1)

Since x $\frac{1}{2}$-pound packages weigh $\frac{1}{2}x$ pounds and y $\frac{1}{3}$-pound packages weigh $\frac{1}{3}y$ pounds, we have

$\frac{1}{2}x + \frac{1}{3}y = 48$ (2)

We solve the system (1), (2) using elimination by addition. We multiply the second equation by -3 and add.

$x + y = 120$

$-\frac{3}{2}x - y = -144$

$-\frac{1}{2}x = -24$

$x = 48$

Substituting into equation (1), we have

$48 + y = 120$

$y = 72$

48 $\frac{1}{2}$-pound packages and 72 $\frac{1}{3}$-pound packages. *(8-1)*

40. Using the formulas for perimeter and area of a rectangle, we have

$2a + 2b = 28$

$ab = 48$

Solving the first-degree equation for b and substituting into the second-degree equation, we have

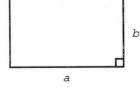

$2b = 28 - 2a$

$b = 14 - a$

$a(14 - a) = 48$

$-a^2 + 14a = 48$

$a^2 - 14a + 48 = 0$

$(a - 6)(a - 8) = 0$

$a = 6, 8$

For $a = 6$ For $a = 8$

$b = 14 - 6$ $b = 14 - 8$

$= 8$ $= 6$

Dimensions: 6 meters by 8 meters. *(8-3)*

41. Let x_1 = number of grams of mix A
x_2 = number of grams of mix B
x_3 = number of grams of mix C

We have

$0.30x_1 + 0.20x_2 + 0.10x_3 = 27$ (protein)
$0.03x_1 + 0.05x_2 + 0.04x_3 = 5.4$ (fat)
$0.10x_1 + 0.20x_2 + 0.10x_3 = 19$ (moisture)

Clearing of decimals for convenience, we have

$3x_1 + 2x_2 + x_3 = 270$
$3x_1 + 5x_2 + 4x_3 = 540$
$x_1 + 2x_2 + x_3 = 190$

Form the augmented matrix and solve by Gauss-Jordan elimination.

$$\begin{bmatrix} 3 & 2 & 1 & | & 270 \\ 3 & 5 & 4 & | & 540 \\ 1 & 2 & 1 & | & 190 \end{bmatrix} \quad R_3 \leftrightarrow R_1$$

$$\sim \begin{bmatrix} 1 & 2 & 1 & | & 190 \\ 3 & 5 & 4 & | & 540 \\ 3 & 2 & 1 & | & 270 \end{bmatrix} \quad \begin{matrix} (-3)R_1 + R_2 \rightarrow R_2 \\ (-3)R_1 + R_3 \rightarrow R_3 \end{matrix}$$

$$\sim \begin{bmatrix} 1 & 2 & 1 & | & 190 \\ 0 & -1 & 1 & | & -30 \\ 0 & -4 & -2 & | & -300 \end{bmatrix} \quad (-1)R_2 \rightarrow R_2$$

$$\sim \begin{bmatrix} 1 & 2 & 1 & | & 190 \\ 0 & 1 & -1 & | & 30 \\ 0 & -4 & -2 & | & -300 \end{bmatrix} \quad \begin{matrix} (-2)R_2 + R_1 \rightarrow R_1 \\ \\ 4R_2 + R_3 \rightarrow R_3 \end{matrix}$$

$$\sim \begin{bmatrix} 1 & 0 & 3 & | & 130 \\ 0 & 1 & -1 & | & 30 \\ 0 & 0 & -6 & | & -180 \end{bmatrix} \quad -\tfrac{1}{6}R_3 \rightarrow R_3$$

$$\sim \begin{bmatrix} 1 & 0 & 3 & | & 130 \\ 0 & 1 & -1 & | & 30 \\ 0 & 0 & 1 & | & 30 \end{bmatrix} \quad \begin{matrix} (-3)R_3 + R_1 \rightarrow R_1 \\ R_3 + R_2 \rightarrow R_2 \end{matrix}$$

$$\sim \begin{bmatrix} 1 & 0 & 0 & | & 40 \\ 0 & 1 & 0 & | & 60 \\ 0 & 0 & 1 & | & 30 \end{bmatrix}$$

Therefore

$x_1 = 40$ grams Mix A
$x_2 = 60$ grams Mix B
$x_3 = 30$ grams Mix C

$(8-2)$

42. (A) Let x_1 = number of nickels
x_2 = number of dimes

Then $x_1 + x_2 = 30$ (total number of coins)
$5x_1 + 10x_2 = 190$ (total value of coins)

We form the augmented matrix and solve by Gauss-Jordan elimination.

$$\begin{bmatrix} 1 & 1 & | & 30 \\ 5 & 10 & | & 190 \end{bmatrix} \quad (-5)R_1 + R_2 \rightarrow R_2$$

$$\sim \begin{bmatrix} 1 & 1 & | & 30 \\ 0 & 5 & | & 40 \end{bmatrix} \quad \tfrac{1}{5}R_2 \rightarrow R_2$$

$$\sim \begin{bmatrix} 1 & 1 & | & 30 \\ 0 & 1 & | & 8 \end{bmatrix} \quad (-1)R_2 + R_1 \rightarrow R_1$$

$$\sim \begin{bmatrix} 1 & 0 & | & 22 \\ 0 & 1 & | & 8 \end{bmatrix}$$

The augmented matrix is in reduced form. It corresponds to the system
$x_1 = 22$ nickels
$x_2 = 8$ dimes

(B) Let x_1 = number of nickels
           x_2 = number of dimes
           x_3 = number of quarters
       Then $x_1 + x_2 + x_3 = 30$ (total number of coins)
           $5x_1 + 10x_2 + 25x_3 = 190$ (total value cf coins)
We form the augmented matrix and solve by Gauss-Jordan elimination

$$\begin{bmatrix} 1 & 1 & 1 & | & 30 \\ 5 & 10 & 25 & | & 190 \end{bmatrix} \quad (-5)R_1 + R_2 \rightarrow R_2$$

$$\sim \begin{bmatrix} 1 & 1 & 1 & | & 30 \\ 0 & 5 & 20 & | & 40 \end{bmatrix} \quad \tfrac{1}{5}R_2 \rightarrow R_2$$

$$\sim \begin{bmatrix} 1 & 1 & 1 & | & 30 \\ 0 & 1 & 4 & | & 8 \end{bmatrix} \quad (-1)R_2 + R_1 \rightarrow R_1$$

$$\sim \begin{bmatrix} 1 & 0 & -3 & | & 22 \\ 0 & 1 & 4 & | & 8 \end{bmatrix}$$

The augmented matrix is in reduced form. It corresponds to the system:
$x_1 - \quad 3x_3 = 22$
$\quad x_2 + 4x_3 = 8$
Let $x_3 = t$. Then $x_2 = -4x_3 + 8$
                         $= -4t + 8$
                    $x_1 = 3x_3 + 22$
                         $= 3t + 22$

A solution is achieved, not for every real value of t, but for integer values
of t that give rise to non-negative x_1, x_2, x_3.

$x_1 \geq 0$ means $3t + 22 \geq 0$ or $t \geq -7\frac{1}{3}$
$x_2 \geq 0$ means $-4t + 8 \geq 0$ or $t \leq 2$
$x_3 \geq 0$ means $t \geq 0$
The only integer values of t that satisfy these conditions are 0, 1, 2. Thus we
have the solutions

$x_1 = 3t + 22$ nickels
$x_2 = 8 - 4t$ dimes
$x_3 = t$ quarters
where $t = 0$, 1, or 2 $(8-1, 8-2)$

43. Let x = number of regular sails
y = number of competition sails

(A) We form the linear objective function
$P = 60x + 100y$
We wish to maximize P, the profit from x regular sails @ $60 and y competition sails @ $100, subject to the constraints

$x + 2y \leq 140$ cutting department constraint
$3x + 4y \leq 360$ sewing department constraint
$x, y \geq 0$ non-negative constraints

Solving the system of constraint inequalities graphically, we obtain the feasible region S shown.

Next we evaluate the objective function at each corner point.

Corner Point (x, y)	Objective Function $P = 60x + 100y$	
(0, 0)	0	
(0, 70)	7,000	
(80, 30)	7,800	Maximum value
(120, 0)	7,200	

The optimal value is 7,800 at the corner point (80, 30). Thus, the company should manufacture 80 regular sails and 30 competition sails for a maximum profit of $7,800.

(B) The objective function now becomes $60x + 125y$. We evaluate this at each corner point.

Corner Point (x, y)	Objective Function $P = 60x + 125y$	
(0, 0)	0	
(0, 70)	8,750	Maximum value
(80, 30)	8,550	
(120, 0)	7,200	

The optimal value is now 8,750 at the corner point (0, 70). Thus, the company should manufacture 70 competition sails and no regular sails for a maximum profit of $8,750.

(C) The objective function now becomes $60x + 75y$. We evaluate this at each corner point.

Corner Point (x, y)	Objective Function $P = 60x + 75y$	
(0, 0)	0	
(0, 70)	5,250	
(80, 30)	7,050	
(120, 0)	7,200	Maximum value

The optimal value is now 7,200 at the corner point (120, 0). Thus, the company should manufacture 120 regular sails and no competition sails for a maximum profit of $7,200.

$(8-5)$

44. Let x = number of grams of mix A
y = number of grams of mix B

(A) We form the objective function
$$C = 0.07x + 0.04y$$

C represents the cost of x grams at \$0.07 per gram and y grams at \$0.04 per gram. We wish to minimize C subject to

$5x + 2y \geq 800$	vitamin constraint
$2x + 4y \geq 800$	mineral contraint
$4x + 4y \leq 1300$	calorie constraint
$x,\ y \geq 0$	non-negative contraints

Solving the system of constraint inequalities graphically, we obtain the feasible region S shown in the diagram.

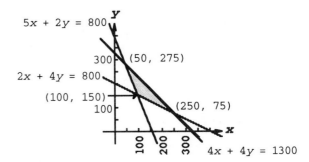

Next we evaluate the objective function at each corner point.

Corner Point $(x,\ y)$	Objective Function $C = 0.07x + 0.04y$	
(100, 150)	13	Minimum value
(250, 75)	20.5	
(50, 275)	14.5	

The optimal value is 13 at the corner point (100, 150). Thus, 100 grams of mix A and 150 grams of mix B should be used for a cost of \$13.

(B) The objective function now becomes $0.07x + 0.02y$. We evaluate this at each corner point.

Corner Point $(x,\ y)$	Objective Function $C = 0.07x + 0.02y$	
(100, 150)	10	
(250, 75)	19	
(50, 275)	9	Minimum value

The optimal value is now 9 at the corner point (50, 275). Thus, 50 grams of mix A and 275 grams of mix B should be used for a cost of \$9.

(C) The objective function now becomes $0.07x + 0.15y$. We evaluate this at each corner point.

Corner Point $(x,\ y)$	Objective Function $C = 0.07x + 0.15y$	
(100, 150)	29.5	
(250, 75)	28.75	Minimum value
(50, 275)	44.75	

The optimal value is now 28.75 at the corner point (250, 75). Thus, 250 grams of mix A and 75 grams of mix B should be used for a cost of \$28.75.

$(8-5)$

CHAPTER 9

Exercise 9-1

Key Ideas and Formulas

Two matrices are equal if they have the same size and their corresponding elements are equal.

The sum of two matrices of the same size, A and B, denoted $A + B$, is a matrix with elements that are the sums of the corresponding elements of A and B.

$A + B = B + A$ (Commutative property of addition).

$(A + B) + C = A + (B + C)$ (Associative property of addition).

The negative of a matrix M, denoted $-M$, is a matrix with elements that are the negatives of the elements of M.

A matrix with elements that are all 0's is called a zero matrix, denoted 0.

$M + (-M) = 0$.

If A and B are matrices of the same size, we define subtraction as follows:

$A - B = A + (-B)$.

The product of a number k and a matrix M, denoted kM, is a matrix formed by multiplying each element of M by k.

The product of a $1 \times n$ row matrix and an $n \times 1$ column matrix is a 1×1 matrix given by

$$[a_1 \quad a_2 \; \cdots \; a_n] \begin{bmatrix} b_1 \\ b_2 \\ \vdots \\ b_n \end{bmatrix} = [a_1 b_1 + a_2 b_2 + \cdots + a_n b_n]$$

(A 1×1 matrix is often referred to as a real number.)

The product of two matrices A and B is defined only on the assumption that the number of columns in A is equal to the number of rows in B.

$$\begin{array}{ccc} A & B & = AB \\ m \times p & p \times n & m \times n \end{array}$$

number of columns in $A = p =$ number of rows in B

If A is an $m \times p$ matrix and B is a $p \times n$ matrix, then the **matrix product** of A and B, denoted AB, is an $m \times n$ matrix whose element in the ith row and jth column is the real number obtained from the product of the ith row of A and the jth column of B. If the number of columns in A does not equal the number of rows in B, then the matrix product AB is **not defined**.

Matrix multiplication is not commutative. Depending on A and B, either AB or BA could be not defined, or defined but of different dimensions, or of the same dimension but AB still unequal to BA.

Also, AB could be zero without $A = 0$ or $B = 0$.

---

1. $\begin{bmatrix} 5 & -2 \\ 3 & 0 \end{bmatrix} + \begin{bmatrix} -3 & 7 \\ 1 & -6 \end{bmatrix} = \begin{bmatrix} 5 + (-3) & (-2) + 7 \\ 3 + 1 & 0 + (-6) \end{bmatrix} = \begin{bmatrix} 2 & 5 \\ 4 & -6 \end{bmatrix}$

3. $\begin{bmatrix} 4 & 0 \\ -2 & 3 \\ 8 & 1 \end{bmatrix} + \begin{bmatrix} -1 & 2 \\ 0 & 5 \\ 4 & -6 \end{bmatrix} = \begin{bmatrix} 4 + (-1) & 0 + 2 \\ (-2) + 0 & 3 + 5 \\ 8 + 4 & 1 + (-6) \end{bmatrix} = \begin{bmatrix} 3 & 2 \\ -2 & 8 \\ 12 & -5 \end{bmatrix}$

5. These matrices have different sizes, hence the sum is not defined.

7. $\begin{bmatrix} 5 & -1 & 0 \\ 4 & 6 & 3 \end{bmatrix} - \begin{bmatrix} 2 & 4 & -6 \\ 3 & 5 & -5 \end{bmatrix} = \begin{bmatrix} 5-2 & (-1)-4 & 0-(-6) \\ 4-3 & 6-5 & 3-(-5) \end{bmatrix} = \begin{bmatrix} 3 & -5 & 6 \\ 1 & 1 & 8 \end{bmatrix}$

9. $\begin{bmatrix} 12 & -16 & 28 \\ -8 & 36 & 20 \end{bmatrix}$ **11.** $[5 \quad 3] \begin{bmatrix} 4 \\ 7 \end{bmatrix} = [5 \cdot 4 + 3 \cdot 7] = [41]$

13. $\begin{bmatrix} -6 & 3 \\ 2 & -5 \end{bmatrix} \begin{bmatrix} 1 \\ 3 \end{bmatrix} = \begin{bmatrix} (-6)1 + 3 \cdot 3 \\ 2 \cdot 1 + (-5)3 \end{bmatrix} = \begin{bmatrix} 3 \\ -13 \end{bmatrix}$

15. $\begin{bmatrix} 5 & 1 \\ 4 & 6 \end{bmatrix} \begin{bmatrix} 2 & 0 \\ 3 & 8 \end{bmatrix} = \begin{bmatrix} 5 \cdot 2 + 1 \cdot 3 & 5 \cdot 0 + 1 \cdot 8 \\ 4 \cdot 2 + 6 \cdot 3 & 4 \cdot 0 + 6 \cdot 8 \end{bmatrix} = \begin{bmatrix} 13 & 8 \\ 26 & 48 \end{bmatrix}$

17. $\begin{bmatrix} 8 & -3 \\ -5 & 3 \end{bmatrix} \begin{bmatrix} 2 & 0 \\ 0 & 6 \end{bmatrix} = \begin{bmatrix} 8 \cdot 2 + (-3)0 & 8 \cdot 0 + (-3)6 \\ (-5)2 + 3 \cdot 0 & (-5)0 + 3 \cdot 6 \end{bmatrix} = \begin{bmatrix} 16 & -18 \\ -10 & 18 \end{bmatrix}$

19. $[3 \quad -6] \begin{bmatrix} -2 \\ 5 \end{bmatrix} = [3(-2) + (-6)5] = [-36]$

21. $\begin{bmatrix} -2 \\ 5 \end{bmatrix} [3 \quad -6] = \begin{bmatrix} (-2)3 & (-2)(-6) \\ 5 \cdot 3 & 5(-6) \end{bmatrix} = \begin{bmatrix} -6 & 12 \\ 15 & -30 \end{bmatrix}$

> **Common Error:** $[(-2)3 + 5(-6)]$ The product of a 2 × 1 matrix with a 1 × 2 matrix is a 2 × 2 matrix, not a 1 × 1 matrix.

23. $[5 \quad 0 \quad -3] \begin{bmatrix} 1 \\ -2 \\ 6 \end{bmatrix} = [5 \cdot 1 + 0(-2) + (-3)6] = [-13]$

25. $\begin{bmatrix} 1 \\ -2 \\ 6 \end{bmatrix} [5 \quad 0 \quad -3] = \begin{bmatrix} 1 \cdot 5 & 1 \cdot 0 & 1(-3) \\ (-2)5 & (-2)0 & (-2)(-3) \\ 6 \cdot 5 & 6 \cdot 0 & 6(-3) \end{bmatrix} = \begin{bmatrix} 5 & 0 & -3 \\ -10 & 0 & 6 \\ 30 & 0 & -18 \end{bmatrix}$

27. *A* has 3 columns. *B* has 2 rows. Therefore, *AB* is not defined.

29. $AC = \begin{bmatrix} 3 & 2 & 0 \\ -1 & 4 & -6 \end{bmatrix} \begin{bmatrix} 1 & 2 & -4 \\ 0 & -2 & 3 \\ 5 & 0 & 4 \end{bmatrix}$

$= \begin{bmatrix} 3 \cdot 1 + 2 \cdot 0 + 0 \cdot 5 & 3 \cdot 2 + 2(-2) + 0 \cdot 0 & 3(-4) + 2 \cdot 3 + 0 \cdot 4 \\ (-1)1 + 4 \cdot 0 + (-6)5 & (-1)2 + 4(-2) + (-6)0 & (-1)(-4) + 4 \cdot 3 + (-6)4 \end{bmatrix}$

$= \begin{bmatrix} 3 & 2 & -6 \\ -31 & -10 & -8 \end{bmatrix}$

31. $A^2 = AA$ is not defined, because *A* has 3 columns and 2 rows.

33. $C^2 = CC = \begin{bmatrix} 1 & 2 & -4 \\ 0 & -2 & 3 \\ 5 & 0 & 4 \end{bmatrix} \begin{bmatrix} 1 & 2 & -4 \\ 0 & -2 & 3 \\ 5 & 0 & 4 \end{bmatrix}$

$= \begin{bmatrix} 1 \cdot 1 + 2 \cdot 0 + (-4)5 & 1 \cdot 2 + 2(-2) + (-4)0 & 1(-4) + 2 \cdot 3 + (-4)4 \\ 0 \cdot 1 + (-2)0 + 3 \cdot 5 & 0 \cdot 2 + (-2)(-2) + 3 \cdot 0 & 0(-4) + (-2)3 + 3 \cdot 4 \\ 5 \cdot 1 + 0 \cdot 0 + 4 \cdot 5 & 5 \cdot 2 + 0(-2) + 4 \cdot 0 & 5(-4) + 0 \cdot 3 + 4 \cdot 4 \end{bmatrix}$

$= \begin{bmatrix} -19 & -2 & -14 \\ 15 & 4 & 6 \\ 25 & 10 & -4 \end{bmatrix}$

35. $2CD = (2C)D = \left(2\begin{bmatrix} 1 & 2 & -4 \\ 0 & -2 & 3 \\ 5 & 0 & 4 \end{bmatrix}\right)\begin{bmatrix} 2 & 0 \\ -1 & 6 \\ -3 & 7 \end{bmatrix} = \begin{bmatrix} 2 & 4 & -8 \\ 0 & -4 & 6 \\ 10 & 0 & 8 \end{bmatrix}\begin{bmatrix} 2 & 0 \\ -1 & 6 \\ -3 & 7 \end{bmatrix}$

$= \begin{bmatrix} 2\cdot 2 + 4(-1) + (-8)(-3) & 2\cdot 0 + 4\cdot 6 + (-8)7 \\ 0\cdot 2 + (-4)(-1) + 6(-3) & 0\cdot 0 + (-4)6 + 6\cdot 7 \\ 10\cdot 2 + 0(-1) + 8(-3) & 10\cdot 0 + 0\cdot 6 + 8\cdot 7 \end{bmatrix} = \begin{bmatrix} 24 & -32 \\ -14 & 18 \\ -4 & 56 \end{bmatrix}$

37. $3AC - 4BD$ is not defined, because BD is not defined. (B has 2 columns and D has 3 rows.)

39. $DA = \begin{bmatrix} 2 & 0 \\ -1 & 6 \\ -3 & 7 \end{bmatrix}\begin{bmatrix} 3 & 2 & 0 \\ -1 & 4 & -6 \end{bmatrix} = \begin{bmatrix} 2\cdot 3 + 0(-1) & 2\cdot 2 + 0\cdot 4 & 2\cdot 0 + 0(-6) \\ (-1)3 + 6(-1) & (-1)2 + 6\cdot 4 & (-1)0 + 6(-6) \\ (-3)3 + 7(-1) & (-3)2 + 7\cdot 4 & (-3)0 + 7(-6) \end{bmatrix}$

$= \begin{bmatrix} 6 & 4 & 0 \\ -9 & 22 & -36 \\ -16 & 22 & -42 \end{bmatrix}$

$5DA - 6C = 5\begin{bmatrix} 6 & 4 & 0 \\ -9 & 22 & -36 \\ -16 & 22 & -42 \end{bmatrix} - 6\begin{bmatrix} 1 & 2 & -4 \\ 0 & -2 & 3 \\ 5 & 0 & 4 \end{bmatrix}$

$= \begin{bmatrix} 30 & 20 & 0 \\ -45 & 110 & -180 \\ -80 & 110 & -210 \end{bmatrix} - \begin{bmatrix} 6 & 12 & -24 \\ 0 & -12 & 18 \\ 30 & 0 & 24 \end{bmatrix} = \begin{bmatrix} 24 & 8 & 24 \\ -45 & 122 & -198 \\ -110 & 110 & -234 \end{bmatrix}$

41. DA was calculated in the previous problem.

$DAC = (DA)C = \begin{bmatrix} 6 & 4 & 0 \\ -9 & 22 & -36 \\ -16 & 22 & -42 \end{bmatrix}\begin{bmatrix} 1 & 2 & -4 \\ 0 & -2 & 3 \\ 5 & 0 & 4 \end{bmatrix}$

$= \begin{bmatrix} 6\cdot 1 + 4\cdot 0 + 0\cdot 5 & 6\cdot 2 + 4(-2) + 0\cdot 0 & 6(-4) + 4\cdot 3 + 0\cdot 4 \\ (-9)1 + 22\cdot 0 + (-36)5 & (-9)2 + 22(-2) + (-36)0 & (-9)(-4) + 22\cdot 3 + (-36)4 \\ (-16)1 + 22\cdot 0 + (-42)5 & (-16)2 + 22(-2) + (-42)0 & (-16)(-4) + 22\cdot 3 + (-42)4 \end{bmatrix}$

$= \begin{bmatrix} 6 & 4 & -12 \\ -189 & -62 & -42 \\ -226 & -76 & -38 \end{bmatrix}$

43. $ADB = (AD)B = \begin{bmatrix} 3 & 2 & 0 \\ -1 & 4 & -6 \end{bmatrix}\begin{bmatrix} 2 & 0 \\ -1 & 6 \\ -3 & 7 \end{bmatrix}B$

$= \begin{bmatrix} 3\cdot 2 + 2(-1) + 0(-3) & 3\cdot 0 + 2\cdot 6 + 0\cdot 7 \\ (-1)2 + 4(-1) + (-6)(-3) & (-1)0 + 4\cdot 6 + (-6)7 \end{bmatrix}B$

$= \begin{bmatrix} 4 & 12 \\ 12 & -18 \end{bmatrix}B = \begin{bmatrix} 4 & 12 \\ 12 & -18 \end{bmatrix}\begin{bmatrix} 5 & -2 \\ 1 & 3 \end{bmatrix} = \begin{bmatrix} 4\cdot 5 + 12\cdot 1 & 4(-2) + 12\cdot 3 \\ 12\cdot 5 + (-18)1 & 12(-2) + (-18)3 \end{bmatrix}$

$= \begin{bmatrix} 32 & 28 \\ 42 & -78 \end{bmatrix}$

45. $\begin{bmatrix} 5 & -2 \\ 8 & 4 \end{bmatrix} + \begin{bmatrix} a & b \\ c & d \end{bmatrix} = \begin{bmatrix} 5 + a & -2 + b \\ 8 + c & 4 + d \end{bmatrix} = \begin{bmatrix} 9 & 3 \\ -1 & 0 \end{bmatrix}$

if and only if corresponding elements are equal.

$\begin{matrix} 5 + a = 9 & -2 + b = 3 & 8 + c = -1 & 4 + d = 0 \\ a = 4 & b = 5 & c = -9 & d = -4 \end{matrix}$

47. $\begin{bmatrix} 1 & 2x \\ -3x & -1 \end{bmatrix} + \begin{bmatrix} 3 & -3y \\ 6y & 4 \end{bmatrix} = \begin{bmatrix} 4 & 2x - 3y \\ -3x + 6y & 3 \end{bmatrix} = \begin{bmatrix} 4 & 4 \\ -3 & 3 \end{bmatrix}$

if and only if corresponding elements are equal.

$\left. \begin{array}{cc} 4 = 4 & 2x - 3y = 4 \\ -3x + 6y = -3 & 3 = 3 \end{array} \right\}$ Two conditions are already met.

To find x and y, we solve the system
$$2x - 3y = 4$$
$$-3x + 6y = -3$$
to obtain $x = 5$, $y = 2$.

49. $\begin{bmatrix} 1 & 4 \\ 0 & 1 \end{bmatrix} \begin{bmatrix} a & b \\ c & d \end{bmatrix} = \begin{bmatrix} a + 4c & b + 4d \\ c & d \end{bmatrix} = \begin{bmatrix} 1 & 0 \\ 0 & 1 \end{bmatrix}$

if and only if

$a + 4c = 1 \qquad b + 4d = 0$
$\qquad c = 0 \qquad \qquad d = 1$

Solving, we obtain $a = 1$, $b = -4$, $c = 0$, $d = 1$.

51. $\begin{bmatrix} 2 & 5 \\ 1 & 3 \end{bmatrix} \begin{bmatrix} a & b \\ c & d \end{bmatrix} = \begin{bmatrix} 2a + 5c & 2b + 5d \\ a + 3c & b + 3d \end{bmatrix} = \begin{bmatrix} 4 & 7 \\ 3 & 3 \end{bmatrix}$

if and only if

$2a + 5c = 4 \qquad 2b + 5d = 7$
$\ a + 3c = 3 \qquad \ b + 3d = 3$

Solving these systems, we obtain $a = -3$, $b = 6$, $c = 2$, $d = -1$.

53. True. $\begin{bmatrix} a & 0 \\ 0 & d \end{bmatrix} + \begin{bmatrix} a' & 0 \\ 0 & d' \end{bmatrix} = \begin{bmatrix} a + a' & 0 \\ 0 & d + d' \end{bmatrix}$.

55. True. If $A = \begin{bmatrix} a & 0 \\ 0 & d \end{bmatrix}$ and $B = \begin{bmatrix} a' & 0 \\ 0 & d' \end{bmatrix}$, then $AB = \begin{bmatrix} a & 0 \\ 0 & d \end{bmatrix} \begin{bmatrix} a' & 0 \\ 0 & d' \end{bmatrix} = \begin{bmatrix} aa' & 0 \\ 0 & dd' \end{bmatrix}$

$BA = \begin{bmatrix} a' & 0 \\ 0 & d' \end{bmatrix} \begin{bmatrix} a & 0 \\ 0 & d \end{bmatrix} = \begin{bmatrix} a'a & 0 \\ 0 & d'd \end{bmatrix} = \begin{bmatrix} aa' & 0 \\ 0 & dd' \end{bmatrix} = AB$

57. True. Take A and B as in the previous problem,

then $A + B = \begin{bmatrix} a & 0 \\ 0 & d \end{bmatrix} + \begin{bmatrix} a' & 0 \\ 0 & d' \end{bmatrix} = \begin{bmatrix} a + a' & 0 \\ 0 & d + d' \end{bmatrix}$.

$B + A = \begin{bmatrix} a' & 0 \\ 0 & d' \end{bmatrix} + \begin{bmatrix} a & 0 \\ 0 & d \end{bmatrix} = \begin{bmatrix} a' + a & 0 \\ 0 & d' + d \end{bmatrix} = \begin{bmatrix} a + a' & 0 \\ 0 & d + d' \end{bmatrix} = A + B$

59. True. $\begin{bmatrix} 0 & 0 \\ 0 & 0 \end{bmatrix}$ is a special case of $\begin{bmatrix} a & 0 \\ 0 & d \end{bmatrix}$.

61. True. $\begin{bmatrix} 1 & 0 \\ 0 & 1 \end{bmatrix} \begin{bmatrix} a & b \\ c & d \end{bmatrix} = \begin{bmatrix} a & b \\ c & d \end{bmatrix} \begin{bmatrix} 1 & 0 \\ 0 & 1 \end{bmatrix} = \begin{bmatrix} a & b \\ c & d \end{bmatrix}$ is always true.

63. False. $\begin{bmatrix} 0 & 0 \\ 0 & 0 \end{bmatrix} \begin{bmatrix} 0 & 0 \\ 0 & 0 \end{bmatrix} = \begin{bmatrix} 0 & 0 \\ 0 & 0 \end{bmatrix}$, hence $\begin{bmatrix} 0 & 0 \\ 0 & 0 \end{bmatrix}$ is a counter example.

65. $\frac{1}{2}(A + B) = \frac{1}{2}\left(\begin{bmatrix} 30 & 25 \\ 60 & 80 \end{bmatrix} + \begin{bmatrix} 36 & 27 \\ 54 & 74 \end{bmatrix} \right) = \frac{1}{2} \begin{bmatrix} 66 & 52 \\ 114 & 154 \end{bmatrix} = \begin{bmatrix} 33 & 26 \\ 57 & 77 \end{bmatrix} \begin{array}{l} \text{Materials} \\ \text{Labor} \end{array}$

67. If a quantity is increased by 15%, the result is a multiplication by 1.15.
If a quantity is increased by 10%, the result is a multiplication by 1.1.
Thus we must calculate $1.1N - 1.15M$. The mark-up matrix is:

$$1.1N - 1.15M = 1.1\begin{bmatrix} 13,900 & 783 & 263 & 215 \\ 15,000 & 838 & 395 & 236 \\ 18,300 & 967 & 573 & 248 \end{bmatrix} - 1.15\begin{bmatrix} 10,400 & 682 & 215 & 182 \\ 12,500 & 721 & 295 & 182 \\ 16,400 & 827 & 443 & 192 \end{bmatrix}$$

$$= \begin{bmatrix} 15,290 & 861.3 & 289.3 & 236.5 \\ 16,500 & 921.8 & 434.5 & 259.6 \\ 20,130 & 1,063.7 & 630.3 & 272.8 \end{bmatrix} - \begin{bmatrix} 11,960 & 784.3 & 247.25 & 209.3 \\ 14,375 & 829.15 & 339.25 & 209.3 \\ 18,860 & 951.05 & 509.45 & 220.8 \end{bmatrix}$$

	Basic Car	Air	AM/FM Radio	Cruise Control
= Model A	$3,330	$ 77	$ 42	$27
Model B	$2,125	$ 93	$ 95	$50
Model C	$1,270	$113	$121	$52

= Mark up

69. (A) $[0.6 \quad 0.6 \quad 0.2]\begin{bmatrix} 8 \\ 10 \\ 5 \end{bmatrix} = (0.6)8 + (0.6)10 + (0.2)5 = 11.80$ dollars per boat

(B) $[1.5 \quad 1.2 \quad 0.4]\begin{bmatrix} 9 \\ 12 \\ 6 \end{bmatrix} = (1.5)9 + (1.2)12 + (0.4)6 = 30.30$ dollars per boat

(C) The matrix NM has no obvious meaning, but the matrix MN gives the labor costs per boat at each plant.

(D) $MN = \begin{bmatrix} 0.6 & 0.6 & 0.2 \\ 1.0 & 0.9 & 0.3 \\ 1.5 & 1.2 & 0.4 \end{bmatrix}\begin{bmatrix} 8 & 9 \\ 10 & 12 \\ 5 & 6 \end{bmatrix}$

$$= \begin{bmatrix} (0.6)8 + (0.6)10 + (0.2)5 & (0.6)9 + (0.6)12 + (0.2)6 \\ (1.0)8 + (0.9)10 + (0.3)5 & (1.0)9 + (0.9)12 + (0.3)6 \\ (1.5)8 + (1.2)10 + (0.4)5 & (1.5)9 + (1.2)12 + (0.4)6 \end{bmatrix}$$

	Plant I	Plant II	
=	$11.80	$13.80	One-person boat
	$18.50	$21.60	Two-person boat
	$26.00	$30.30	Four-person boat

This matrix gives the labor costs for each type of boat at each plant.

71. (A) $A^2 = AA = \begin{bmatrix} 0&1&0&1&0 \\ 0&0&1&0&0 \\ 1&0&0&0&1 \\ 0&0&1&0&0 \\ 0&0&0&1&0 \end{bmatrix}\begin{bmatrix} 0&1&0&1&0 \\ 0&0&1&0&0 \\ 1&0&0&0&1 \\ 0&0&1&0&0 \\ 0&0&0&1&0 \end{bmatrix} = \begin{bmatrix} 0&0&2&0&0 \\ 1&0&0&0&1 \\ 0&1&0&2&0 \\ 1&0&0&0&1 \\ 0&0&1&0&0 \end{bmatrix}$

The 1 in row 2 and column 1 of A^2 indicates that there is one way to travel from Baltimore to Atlanta with one intermediate connection. The 2 in row 1 and column 3 indicates that there are two ways to travel from Atlanta to Chicago with one intermediate connection. In general, the elements in A^2 indicate the number of different ways to travel from the ith city to the jth city with one intermediate connection.

(B) $A^3 = A^2 A = \begin{bmatrix} 0 & 0 & 2 & 0 & 0 \\ 1 & 0 & 0 & 0 & 1 \\ 0 & 1 & 0 & 2 & 0 \\ 1 & 0 & 0 & 0 & 1 \\ 0 & 0 & 1 & 0 & 0 \end{bmatrix} \begin{bmatrix} 0 & 1 & 0 & 1 & 0 \\ 0 & 0 & 1 & 0 & 0 \\ 1 & 0 & 0 & 0 & 1 \\ 0 & 0 & 1 & 0 & 0 \\ 0 & 0 & 0 & 1 & 0 \end{bmatrix} = \begin{bmatrix} 2 & 0 & 0 & 0 & 2 \\ 0 & 1 & 0 & 2 & 0 \\ 0 & 0 & 3 & 0 & 0 \\ 0 & 1 & 0 & 2 & 0 \\ 1 & 0 & 0 & 0 & 1 \end{bmatrix}$

The 1 in row 4 and column 2 of A^3 indicates that there is one way to travel from Denver to Baltimore with two intermediate connections. The 2 in row 1 and column 5 indicates that there are two ways to travel from Atlanta to El Paso with two intermediate connections. In general, the elements in A^3 indicate the number of different ways to travel from the ith city to the the jth city with two intermediate connections.

(C) A is given above.

$A + A^2 = \begin{bmatrix} 0 & 1 & 0 & 1 & 0 \\ 0 & 0 & 1 & 0 & 0 \\ 1 & 0 & 0 & 0 & 1 \\ 0 & 0 & 1 & 0 & 0 \\ 0 & 0 & 0 & 1 & 0 \end{bmatrix} + \begin{bmatrix} 0 & 0 & 2 & 0 & 0 \\ 1 & 0 & 0 & 0 & 1 \\ 0 & 1 & 0 & 2 & 0 \\ 1 & 0 & 0 & 0 & 1 \\ 0 & 0 & 1 & 0 & 0 \end{bmatrix} = \begin{bmatrix} 0 & 1 & 2 & 1 & 0 \\ 1 & 0 & 1 & 0 & 1 \\ 1 & 1 & 0 & 2 & 1 \\ 1 & 0 & 1 & 0 & 1 \\ 0 & 0 & 1 & 1 & 0 \end{bmatrix}$

$A + A^2 + A^3 = \begin{bmatrix} 0 & 1 & 2 & 1 & 0 \\ 1 & 0 & 1 & 0 & 1 \\ 1 & 1 & 0 & 2 & 1 \\ 1 & 0 & 1 & 0 & 1 \\ 0 & 0 & 1 & 1 & 0 \end{bmatrix} + \begin{bmatrix} 2 & 0 & 0 & 0 & 2 \\ 0 & 1 & 0 & 2 & 0 \\ 0 & 0 & 3 & 0 & 0 \\ 0 & 1 & 0 & 2 & 0 \\ 1 & 0 & 0 & 0 & 1 \end{bmatrix} = \begin{bmatrix} 2 & 1 & 2 & 1 & 2 \\ 1 & 1 & 1 & 2 & 1 \\ 1 & 1 & 3 & 2 & 1 \\ 1 & 1 & 1 & 2 & 1 \\ 1 & 0 & 1 & 1 & 1 \end{bmatrix}$

A zero element remains, so we must compute A^4.

$A^4 = A^3 A = \begin{bmatrix} 2 & 0 & 0 & 0 & 2 \\ 0 & 1 & 0 & 2 & 0 \\ 0 & 0 & 3 & 0 & 0 \\ 0 & 1 & 0 & 2 & 0 \\ 1 & 0 & 0 & 0 & 1 \end{bmatrix} \begin{bmatrix} 0 & 1 & 0 & 1 & 0 \\ 0 & 0 & 1 & 0 & 0 \\ 1 & 0 & 0 & 0 & 1 \\ 0 & 0 & 1 & 0 & 0 \\ 0 & 0 & 0 & 1 & 0 \end{bmatrix} = \begin{bmatrix} 0 & 2 & 0 & 4 & 0 \\ 0 & 0 & 3 & 0 & 0 \\ 3 & 0 & 0 & 0 & 3 \\ 0 & 0 & 3 & 0 & 0 \\ 0 & 1 & 0 & 2 & 0 \end{bmatrix}$

Then $A + A^2 + A^3 + A^4 = \begin{bmatrix} 2 & 1 & 2 & 1 & 2 \\ 1 & 1 & 1 & 2 & 1 \\ 1 & 1 & 3 & 2 & 1 \\ 1 & 1 & 1 & 2 & 1 \\ 1 & 0 & 1 & 1 & 1 \end{bmatrix} + \begin{bmatrix} 0 & 2 & 0 & 4 & 0 \\ 0 & 0 & 3 & 0 & 0 \\ 3 & 0 & 0 & 0 & 3 \\ 0 & 0 & 3 & 0 & 0 \\ 0 & 1 & 0 & 2 & 0 \end{bmatrix} = \begin{bmatrix} 2 & 3 & 2 & 5 & 2 \\ 1 & 1 & 4 & 2 & 1 \\ 4 & 1 & 3 & 2 & 4 \\ 1 & 1 & 4 & 2 & 1 \\ 1 & 1 & 1 & 3 & 1 \end{bmatrix}$

This matrix indicates that it is possible to travel from any origin to any destination with at most 3 intermediate connections.

73. (A) $[1{,}000 \quad 500 \quad 5{,}000] \begin{bmatrix} \$0.80 \\ \$1.50 \\ \$0.40 \end{bmatrix} = 1{,}000(\$0.80) + 500(\$1.50) + 5{,}000(\$0.40)$
$= \$3{,}550$

(B) $[2{,}000 \quad 800 \quad 8{,}000] \begin{bmatrix} \$0.80 \\ \$1.50 \\ \$0.40 \end{bmatrix} = 2{,}000(\$0.80) + 800(\$1.50) + 8{,}000(\$0.40)$
$= \$6{,}000$

(C) The matrix MN has no obvious interpretations, but the matrix NM represents the total cost of all contacts in each town.

(D) $NM = \begin{bmatrix} 1{,}000 & 500 & 5{,}000 \\ 2{,}000 & 800 & 8{,}000 \end{bmatrix} \begin{bmatrix} \$0.80 \\ \$1.50 \\ \$0.40 \end{bmatrix} = \begin{bmatrix} 1{,}000(0.80) + 500(1.50) + 5{,}000(0.40) \\ 2{,}000(0.80) + 800(1.50) + 8{,}000(0.40) \end{bmatrix}$

$= \begin{bmatrix} \$3{,}550 \\ \$6{,}000 \end{bmatrix} \begin{matrix} \text{Berkeley} \\ \text{Oakland} \end{matrix} = $ Cost of all contacts in each town.

(E) The matrix [1   1]N can be used to find the total number of each of the three types of contact:

$$[1 \quad 1]\begin{bmatrix} 1,000 & 500 & 5,000 \\ 2,000 & 800 & 8,000 \end{bmatrix} = [1,000 + 2,000 \quad 500 + 800 \quad 5,000 + 8,000]$$

[Telephone House Letter] = [3,000 1,300 13,000]

(F) The matrix $N\begin{bmatrix} 1 \\ 1 \\ 1 \end{bmatrix}$ can be used to find the total number of contacts in each town:

$$\begin{bmatrix} 1,000 & 500 & 5,000 \\ 2,000 & 800 & 8,000 \end{bmatrix}\begin{bmatrix} 1 \\ 1 \\ 1 \end{bmatrix}$$

$$= \begin{bmatrix} 1,000 + 500 + 5,000 \\ 2,000 + 800 + 8,000 \end{bmatrix} = \begin{bmatrix} 6,500 \\ 10,800 \end{bmatrix} = \begin{bmatrix} \text{Berkeley contacts} \\ \text{Oakland contacts} \end{bmatrix}$$

Exercise 9-2

Key Ideas and Formulas

The identity element for multiplication for square matrices of order n (size $n \times n$), denoted I, is the square matrix of order n with 1's along the principal diagonal and 0's elsewhere. For A a square matrix of order n,

$$IA = AI = A$$

If M is a square matrix of order n and if there exists a square matrix M^{-1} such that $MM^{-1} = M^{-1}M = I$, M^{-1} is called the (multiplicative) inverse of M.

To find M^{-1} if it exists, form the augmented matrix $[M \mid I]$ and use row operations to transform into $[I \mid B]$ if possible. Then $B = M^{-1}$. If, however, all 0's are obtained at some stage in one or more rows to the left of the vertical line, then M^{-1} will not exist.

---

1. $\begin{bmatrix} 6 & -3 \\ 5 & 8 \end{bmatrix}$ **3.** $\begin{bmatrix} 2 & -5 \\ 8 & -4 \end{bmatrix}$ **5.** $\begin{bmatrix} 4 & -2 & 5 \\ 6 & 0 & -3 \\ -1 & 2 & 7 \end{bmatrix}$ **7.** $\begin{bmatrix} -3 & 2 & -5 \\ 4 & -1 & 6 \\ -3 & 4 & -2 \end{bmatrix}$

9. $\begin{bmatrix} 1 & 6 \\ 0 & 1 \end{bmatrix}\begin{bmatrix} 1 & -6 \\ 0 & 1 \end{bmatrix} = \begin{bmatrix} 1 \cdot 1 + 6 \cdot 0 & 1(-6) + 6 \cdot 1 \\ 0 \cdot 1 + 1 \cdot 0 & 0(-6) + 1 \cdot 1 \end{bmatrix} = \begin{bmatrix} 1 & 0 \\ 0 & 1 \end{bmatrix}$

11. $\begin{bmatrix} 3 & 4 \\ 5 & 7 \end{bmatrix}\begin{bmatrix} 7 & -4 \\ -5 & 3 \end{bmatrix} = \begin{bmatrix} 3 \cdot 7 + 4(-5) & 3(-4) + 4 \cdot 3 \\ 5 \cdot 7 + 7(-5) & 5(-4) + 7 \cdot 3 \end{bmatrix} = \begin{bmatrix} 1 & 0 \\ 0 & 1 \end{bmatrix}$

13. $\begin{bmatrix} 1 & -3 & 5 \\ 0 & 1 & 4 \\ 0 & 0 & 1 \end{bmatrix}\begin{bmatrix} 1 & 3 & -17 \\ 0 & 1 & -4 \\ 0 & 0 & 1 \end{bmatrix}$

$$= \begin{bmatrix} 1 \cdot 1 + (-3)0 + 5 \cdot 0 & 1 \cdot 3 + (-3)1 + 5 \cdot 0 & 1(-17) + (-3)(-4) + 5 \cdot 1 \\ 0 \cdot 1 + 1 \cdot 0 + 4 \cdot 0 & 0 \cdot 3 + 1 \cdot 1 + 4 \cdot 0 & 0(-17) + 1(-4) + 4 \cdot 1 \\ 0 \cdot 1 + 0 \cdot 0 + 1 \cdot 0 & 0 \cdot 3 + 0 \cdot 1 + 1 \cdot 0 & 0(-17) + 0(-4) + 1 \cdot 1 \end{bmatrix}$$

$$= \begin{bmatrix} 1 & 0 & 0 \\ 0 & 1 & 0 \\ 0 & 0 & 1 \end{bmatrix}$$

15. $\begin{bmatrix} 1 & 9 & | & 1 & 0 \\ 0 & 1 & | & 0 & 1 \end{bmatrix}$ $(-9)R_2 + R_1 \rightarrow R_1$

$\sim \begin{bmatrix} 1 & 0 & | & 1 & -9 \\ 0 & 1 & | & 0 & 1 \end{bmatrix}$

Hence, $M^{-1} = \begin{bmatrix} 1 & -9 \\ 0 & 1 \end{bmatrix}$

Check: $M^{-1}M = \begin{bmatrix} 1 & -9 \\ 0 & 1 \end{bmatrix}\begin{bmatrix} 1 & 9 \\ 0 & 1 \end{bmatrix} = \begin{bmatrix} 1 \cdot 1 + (-9)0 & 1 \cdot 9 + (-9)1 \\ 0 \cdot 1 + 1 \cdot 0 & 0 \cdot 9 + 1 \cdot 1 \end{bmatrix} = \begin{bmatrix} 1 & 0 \\ 0 & 1 \end{bmatrix}$

17. $\begin{bmatrix} -1 & -2 & | & 1 & 0 \\ 2 & 5 & | & 0 & 1 \end{bmatrix}$ $2R_1 + R_2 \rightarrow R_2$

$\sim \begin{bmatrix} -1 & -2 & | & 1 & 0 \\ 0 & 1 & | & 2 & 1 \end{bmatrix}$ $2R_2 + R_1 \rightarrow R_1$

$\sim \begin{bmatrix} -1 & 0 & | & 5 & 2 \\ 0 & 1 & | & 2 & 1 \end{bmatrix}$ $(-1)R_1 \rightarrow R_1$

$\sim \begin{bmatrix} 1 & 0 & | & -5 & -2 \\ 0 & 1 & | & 2 & 1 \end{bmatrix}$

Hence, $M^{-1} = \begin{bmatrix} -5 & -2 \\ 2 & 1 \end{bmatrix}$

Check: $M^{-1}M = \begin{bmatrix} -5 & -2 \\ 2 & 1 \end{bmatrix} \begin{bmatrix} -1 & -2 \\ 2 & 5 \end{bmatrix} = \begin{bmatrix} (-5)(-1) + (-2)2 & (-5)(-2) + (-2)5 \\ 2(-1) + 1 \cdot 2 & 2(-2) + 1 \cdot 5 \end{bmatrix}$

$= \begin{bmatrix} 1 & 0 \\ 0 & 1 \end{bmatrix}$

19. $\begin{bmatrix} -5 & 7 & | & 1 & 0 \\ 2 & -3 & | & 0 & 1 \end{bmatrix}$ $3R_2 + R_1 \rightarrow R_1$

$\sim \begin{bmatrix} 1 & -2 & | & 1 & 3 \\ 2 & -3 & | & 0 & 1 \end{bmatrix}$ $(-2)R_1 + R_2 \rightarrow R_2$

$\sim \begin{bmatrix} 1 & -2 & | & 1 & 3 \\ 0 & 1 & | & -2 & -5 \end{bmatrix}$ $2R_2 + R_1 \rightarrow R_1$

$\sim \begin{bmatrix} 1 & 0 & | & -3 & -7 \\ 0 & 1 & | & -2 & -5 \end{bmatrix}$

Hence, $M^{-1} = \begin{bmatrix} -3 & -7 \\ -2 & -5 \end{bmatrix}$

Check: $M^{-1}M = \begin{bmatrix} -3 & -7 \\ -2 & -5 \end{bmatrix}\begin{bmatrix} -5 & 7 \\ 2 & -3 \end{bmatrix} = \begin{bmatrix} (-3)(-5) + (-7)2 & (-3)7 + (-7)(-3) \\ (-2)(-5) + (-5)2 & (-2)7 + (-5)(-3) \end{bmatrix} = \begin{bmatrix} 1 & 0 \\ 0 & 1 \end{bmatrix}$

21.
$$\begin{bmatrix} 1 & -1 & 0 & | & 1 & 0 & 0 \\ -1 & 1 & -1 & | & 0 & 1 & 0 \\ 0 & -1 & 1 & | & 0 & 0 & 1 \end{bmatrix} \quad R_1 + R_2 \to R_2$$

$$\sim \begin{bmatrix} 1 & -1 & 0 & | & 1 & 0 & 0 \\ 0 & 0 & -1 & | & 1 & 1 & 0 \\ 0 & -1 & 1 & | & 0 & 0 & 1 \end{bmatrix} \quad R_2 \leftrightarrow R_3$$

$$\sim \begin{bmatrix} 1 & -1 & 0 & | & 1 & 0 & 0 \\ 0 & -1 & 1 & | & 0 & 0 & 1 \\ 0 & 0 & -1 & | & 1 & 1 & 0 \end{bmatrix} \quad (-1)R_2 + R_1 \to R_1$$

$$\sim \begin{bmatrix} 1 & 0 & -1 & | & 1 & 0 & -1 \\ 0 & -1 & 1 & | & 0 & 0 & 1 \\ 0 & 0 & -1 & | & 1 & 1 & 0 \end{bmatrix} \quad \begin{matrix} (-1)R_3 + R_1 \to R_1 \\ R_3 + R_2 \to R_2 \end{matrix}$$

$$\sim \begin{bmatrix} 1 & 0 & 0 & | & 0 & -1 & -1 \\ 0 & -1 & 0 & | & 1 & 1 & 1 \\ 0 & 0 & -1 & | & 1 & 1 & 0 \end{bmatrix} \quad \begin{matrix} (-1)R_2 \to R_2 \\ (-1)R_3 \to R_3 \end{matrix}$$

$$\sim \begin{bmatrix} 1 & 0 & 0 & | & 0 & -1 & -1 \\ 0 & 1 & 0 & | & -1 & -1 & -1 \\ 0 & 0 & 1 & | & -1 & -1 & 0 \end{bmatrix}$$

Hence, $M^{-1} = \begin{bmatrix} 0 & -1 & -1 \\ -1 & -1 & -1 \\ -1 & -1 & 0 \end{bmatrix}$

Check: $M^{-1}M = \begin{bmatrix} 0 & -1 & -1 \\ -1 & -1 & -1 \\ -1 & -1 & 0 \end{bmatrix}\begin{bmatrix} 1 & -1 & 0 \\ -1 & 1 & -1 \\ 0 & -1 & 1 \end{bmatrix}$

$$= \begin{bmatrix} 0 \cdot 1 + (-1)(-1) + (-1)0 & 0(-1) + (-1)1 + (-1)(-1) & 0 \cdot 0 + (-1)(-1) + (-1)1 \\ (-1)1 + (-1)(-1) + (-1)0 & (-1)(-1) + (-1)1 + (-1)(-1) & (-1)0 + (-1)(-1) + (-1)1 \\ (-1)1 + (-1)(-1) + 0 \cdot 0 & (-1)(-1) + (-1)1 + 0(-1) & (-1)0 + (-1)(-1) + 0 \cdot 1 \end{bmatrix}$$

$$= \begin{bmatrix} 1 & 0 & 0 \\ 0 & 1 & 0 \\ 0 & 0 & 1 \end{bmatrix}$$

23.
$$\begin{bmatrix} 1 & 2 & 5 & | & 1 & 0 & 0 \\ 3 & 5 & 9 & | & 0 & 1 & 0 \\ 1 & 1 & -2 & | & 0 & 0 & 1 \end{bmatrix} \quad \begin{matrix} (-3)R_1 + R_2 \to R_2 \\ (-1)R_1 + R_3 \to R_3 \end{matrix}$$

$$\sim \begin{bmatrix} 1 & 2 & 5 & | & 1 & 0 & 0 \\ 0 & -1 & -6 & | & -3 & 1 & 0 \\ 0 & -1 & -7 & | & -1 & 0 & 1 \end{bmatrix} \quad \begin{matrix} 2R_2 + R_1 \to R_1 \\ \\ (-1)R_2 + R_3 \to R_3 \end{matrix}$$

$$\sim \begin{bmatrix} 1 & 0 & -7 & | & -5 & 2 & 0 \\ 0 & -1 & -6 & | & -3 & 1 & 0 \\ 0 & 0 & -1 & | & 2 & -1 & 1 \end{bmatrix} \quad \begin{matrix} (-7)R_3 + R_1 \to R_1 \\ (-6)R_3 + R_2 \to R_2 \end{matrix}$$

$$\sim \begin{bmatrix} 1 & 0 & 0 & | & -19 & 9 & -7 \\ 0 & -1 & 0 & | & -15 & 7 & -6 \\ 0 & 0 & -1 & | & 2 & -1 & 1 \end{bmatrix} \quad \begin{matrix} (-1)R_2 \to R_2 \\ (-1)R_3 \to R_3 \end{matrix}$$

$$\sim \begin{bmatrix} 1 & 0 & 0 & \bigm| & -19 & 9 & -7 \\ 0 & 1 & 0 & \bigm| & 15 & -7 & 6 \\ 0 & 0 & 1 & \bigm| & -2 & 1 & -1 \end{bmatrix}$$

Hence, $M^{-1} = \begin{bmatrix} -19 & 9 & -7 \\ 15 & -7 & 6 \\ -2 & 1 & -1 \end{bmatrix}$

Check: $M^{-1}M = \begin{bmatrix} -19 & 9 & -7 \\ 15 & -7 & 6 \\ -2 & 1 & -1 \end{bmatrix}\begin{bmatrix} 1 & 2 & 5 \\ 3 & 5 & 9 \\ 1 & 1 & -2 \end{bmatrix}$

$$= \begin{bmatrix} (-19)1 + 9\cdot 3 + (-7)1 & (-19)2 + 9\cdot 5 + (-7)1 & (-19)5 + 9\cdot 9 + (-7)(-2) \\ 15\cdot 1 + (-7)3 + 6\cdot 1 & 15\cdot 2 + (-7)5 + 6\cdot 1 & 15\cdot 5 + (-7)9 + 6(-2) \\ (-2)1 + 1\cdot 3 + (-1)1 & (-2)2 + 1\cdot 5 + (-1)1 & (-2)5 + 1\cdot 9 + (-1)(-2) \end{bmatrix}$$

$$= \begin{bmatrix} 1 & 0 & 0 \\ 0 & 1 & 0 \\ 0 & 0 & 1 \end{bmatrix}$$

25. If the inverse existed we would find it by row operations on the following matrix:

$$= \begin{bmatrix} 3 & 9 & \bigm| & 1 & 0 \\ 2 & 6 & \bigm| & 0 & 1 \end{bmatrix}$$

But consider what happens if we perform $\left(-\frac{2}{3}\right) R_1 + R_2 \to R_2$

$$= \begin{bmatrix} 3 & 9 & \bigm| & 1 & 0 \\ 0 & 0 & \bigm| & -\frac{2}{3} & 1 \end{bmatrix}$$

Since a row of zeros results to the left of the vertical line, no inverse exists.

27. $\begin{bmatrix} 2 & 3 & \bigm| & 1 & 0 \\ 3 & 5 & \bigm| & 0 & 1 \end{bmatrix} \frac{1}{2} R_1 \to R_1$

$$\sim \begin{bmatrix} 1 & 1.5 & \bigm| & 0.5 & 0 \\ 3 & 5 & \bigm| & 0 & 1 \end{bmatrix} (-3) R_1 + R_2 \to R_2$$

$$\sim \begin{bmatrix} 1 & 1.5 & \bigm| & 0.5 & 0 \\ 0 & 0.5 & \bigm| & -1.5 & 1 \end{bmatrix} 2 R_2 \to R_2$$

$$\sim \begin{bmatrix} 1 & 1.5 & \bigm| & 0.5 & 0 \\ 0 & 1 & \bigm| & -3 & 2 \end{bmatrix} (-1.5) R_2 + R_1 \to R_1$$

$$\sim \begin{bmatrix} 1 & 0 & \bigm| & 5 & -3 \\ 0 & 1 & \bigm| & -3 & 2 \end{bmatrix}$$

The inverse is $\begin{bmatrix} 5 & -3 \\ -3 & 2 \end{bmatrix}$

The checking steps are omitted for lack of space in this and subsequent problems.

29. $\begin{bmatrix} 2 & 2 & -1 & | & 1 & 0 & 0 \\ 0 & 4 & -1 & | & 0 & 1 & 0 \\ -1 & -2 & 1 & | & 0 & 0 & 1 \end{bmatrix} R_1 \leftrightarrow R_3$

$\sim \begin{bmatrix} -1 & -2 & 1 & | & 0 & 0 & 1 \\ 0 & 4 & -1 & | & 0 & 1 & 0 \\ 2 & 2 & -1 & | & 1 & 0 & 0 \end{bmatrix} 2R_1 + R_3 \rightarrow R_3$

$\sim \begin{bmatrix} -1 & -2 & 1 & | & 0 & 0 & 1 \\ 0 & 4 & -1 & | & 0 & 1 & 0 \\ 0 & -2 & 1 & | & 1 & 0 & 2 \end{bmatrix} \begin{matrix} (-1)R_3 + R_1 \rightarrow R_1 \\ 2R_3 + R_2 \rightarrow R_2 \end{matrix}$

$\sim \begin{bmatrix} -1 & 0 & 0 & | & -1 & 0 & -1 \\ 0 & 0 & 1 & | & 2 & 1 & 4 \\ 0 & -2 & 1 & | & 1 & 0 & 2 \end{bmatrix} (-1)R_2 + R_3 \rightarrow R_3$

$\sim \begin{bmatrix} -1 & 0 & 0 & | & -1 & 0 & -1 \\ 0 & 0 & 1 & | & 2 & 1 & 4 \\ 0 & -2 & 0 & | & -1 & -1 & -2 \end{bmatrix} R_2 \leftrightarrow R_3$

$\sim \begin{bmatrix} -1 & 0 & 0 & | & -1 & 0 & -1 \\ 0 & -2 & 0 & | & -1 & -1 & -2 \\ 0 & 0 & 1 & | & 2 & 1 & 4 \end{bmatrix} \begin{matrix} (-1)R_1 \rightarrow R_1 \\ (-\frac{1}{2})R_2 \rightarrow R_2 \end{matrix}$

$\sim \begin{bmatrix} 1 & 0 & 0 & | & 1 & 0 & 1 \\ 0 & 1 & 0 & | & \frac{1}{2} & \frac{1}{2} & 1 \\ 0 & 0 & 1 & | & 2 & 1 & 4 \end{bmatrix}$

The inverse is $\begin{bmatrix} 1 & 0 & 1 \\ \frac{1}{2} & \frac{1}{2} & 1 \\ 2 & 1 & 4 \end{bmatrix}$

31. If the inverse existed we would find it by row operations on the following matrix:

$\begin{bmatrix} 2 & 1 & 1 & | & 1 & 0 & 0 \\ 1 & 1 & 0 & | & 0 & 1 & 0 \\ -1 & -1 & 0 & | & 0 & 0 & 1 \end{bmatrix}$

But consider what happens if we perform $R_2 + R_3 \rightarrow R_2$

$\begin{bmatrix} 2 & 1 & 1 & | & 1 & 0 & 0 \\ 0 & 0 & 0 & | & 0 & 1 & 1 \\ -1 & -1 & 0 & | & 0 & 0 & 1 \end{bmatrix}$

Since a row of zeros results to the left of the vertical line, no inverse exists.

33. $\begin{bmatrix} 1 & 5 & 10 & | & 1 & 0 & 0 \\ 0 & 1 & 4 & | & 0 & 1 & 0 \\ 1 & 6 & 15 & | & 0 & 0 & 1 \end{bmatrix} (-1)R_1 + R_3 \rightarrow R_3$

$\sim \begin{bmatrix} 1 & 5 & 10 & | & 1 & 0 & 0 \\ 0 & 1 & 4 & | & 0 & 1 & 0 \\ 0 & 1 & 5 & | & -1 & 0 & 1 \end{bmatrix} \begin{matrix} (-5)R_2 + R_1 \rightarrow R_1 \\ \\ (-1)R_2 + R_3 \rightarrow R_3 \end{matrix}$

$\sim \begin{bmatrix} 1 & 0 & -10 & | & 1 & -5 & 0 \\ 0 & 1 & 4 & | & 0 & 1 & 0 \\ 0 & 0 & 1 & | & -1 & -1 & 1 \end{bmatrix} \begin{matrix} 10R_3 + R_1 \rightarrow R_1 \\ (-4)R_3 + R_2 \rightarrow R_2 \end{matrix}$

$$\sim \begin{bmatrix} 1 & 0 & 0 & \vline & -9 & -15 & 10 \\ 0 & 1 & 0 & \vline & 4 & 5 & -4 \\ 0 & 0 & 1 & \vline & -1 & -1 & 1 \end{bmatrix}$$

The inverse is $\begin{bmatrix} -9 & -15 & 10 \\ 4 & 5 & -4 \\ -1 & -1 & 1 \end{bmatrix}$

35. We calculate A^{-1} by row operations on

$$\begin{bmatrix} 3 & 4 & \vline & 1 & 0 \\ 2 & 3 & \vline & 0 & 1 \end{bmatrix} \quad (-1)R_2 + R_1 \rightarrow R_1$$

$$\sim \begin{bmatrix} 1 & 1 & \vline & 1 & -1 \\ 2 & 3 & \vline & 0 & 1 \end{bmatrix} \quad (-2)R_1 + R_2 \rightarrow R_2$$

$$\sim \begin{bmatrix} 1 & 1 & \vline & 1 & -1 \\ 0 & 1 & \vline & -2 & 3 \end{bmatrix} \quad (-1)R_2 + R_1 \rightarrow R_1$$

$$\sim \begin{bmatrix} 1 & 0 & \vline & 3 & -4 \\ 0 & 1 & \vline & -2 & 3 \end{bmatrix}$$

Hence $A^{-1} = \begin{bmatrix} 3 & -4 \\ -2 & 3 \end{bmatrix}$

We calculate $(A^{-1})^{-1}$ by row operations on

$$\begin{bmatrix} 3 & -4 & \vline & 1 & 0 \\ -2 & 3 & \vline & 0 & 1 \end{bmatrix} \quad R_2 + R_1 \rightarrow R_1$$

$$\sim \begin{bmatrix} 1 & -1 & \vline & 1 & 1 \\ -2 & 3 & \vline & 0 & 1 \end{bmatrix} \quad 2R_1 + R_2 \rightarrow R_2$$

$$\sim \begin{bmatrix} 1 & -1 & \vline & 1 & 1 \\ 0 & 1 & \vline & 2 & 3 \end{bmatrix} \quad R_2 + R_1 \rightarrow R_1$$

$$\sim \begin{bmatrix} 1 & 0 & \vline & 3 & 4 \\ 0 & 1 & \vline & 2 & 3 \end{bmatrix}$$

Hence $(A^{-1})^{-1} = \begin{bmatrix} 3 & 4 \\ 2 & 3 \end{bmatrix} = A$

37. Consider the matrix $\begin{bmatrix} a^{-1} & 0 \\ 0 & d^{-1} \end{bmatrix}$

Since $\begin{bmatrix} a^{-1} & 0 \\ 0 & d^{-1} \end{bmatrix}\begin{bmatrix} a & 0 \\ 0 & d \end{bmatrix} = \begin{bmatrix} 1 & 0 \\ 0 & 1 \end{bmatrix}$ $\begin{bmatrix} a^{-1} & 0 \\ 0 & d^{-1} \end{bmatrix}$ is an inverse for M whenever it exists. This will hold if and only if a and d are non-zero.

Steps are not shown for Problems 39 - 41.

39. $\begin{bmatrix} -0.4 & 0.8 & 1.8 & -1.4 \\ 0.1 & -0.2 & -0.7 & 0.6 \\ -0.85 & 1.7 & 3.95 & -3.6 \\ -0.1 & -0.8 & -1.3 & 1.4 \end{bmatrix}$ **41.** $\begin{bmatrix} -0.75 & 3.75 & -4 & 0.75 & 3.5 \\ -0.5 & -0.5 & 1 & -0.5 & -1 \\ 2.5 & -10.5 & 11.5 & -1.5 & -9.5 \\ 0.125 & 1.375 & -1.75 & 0.375 & 1.5 \\ -0.75 & 0.75 & -0.5 & -0.25 & 0 \end{bmatrix}$

43. Using the assignation numbers 1 to 27 with the letters of the alphabet and a blank as in the text, write

C A T   I N   T H E   H A T
3 1 20 27 9 14 27 20 8 5 27 8 1 20

and calculate

$$\begin{bmatrix} 3 & 5 \\ 1 & 2 \end{bmatrix}\begin{bmatrix} 3 & 20 & 9 & 27 & 8 & 27 & 1 \\ 1 & 27 & 14 & 20 & 5 & 8 & 20 \end{bmatrix} = \begin{bmatrix} 14 & 195 & 97 & 181 & 49 & 121 & 103 \\ 5 & 74 & 37 & 67 & 18 & 43 & 41 \end{bmatrix}$$

The encoded message is thus
14 5 195 74 97 37 181 67 49 18 121 43 103 41

45. The inverse of matrix A is easily calculated to be

$$A^{-1} = \begin{bmatrix} 2 & -5 \\ -1 & 3 \end{bmatrix}$$

Putting the coded message into matrix form and multiplying by A^{-1} yields:

$$\begin{bmatrix} 2 & -5 \\ -1 & 3 \end{bmatrix}\begin{bmatrix} 111 & 40 & 177 & 50 & 116 & 86 & 62 & 121 & 68 \\ 43 & 15 & 68 & 19 & 45 & 29 & 22 & 43 & 27 \end{bmatrix}$$

$$= \begin{bmatrix} 7 & 5 & 14 & 5 & 7 & 27 & 14 & 27 & 1 \\ 18 & 5 & 27 & 7 & 19 & 1 & 4 & 8 & 13 \end{bmatrix}$$

This decodes to   7 18 5 5 14 27 5 7 7 19 27 1 14 4 27 8 1 13
G R E E N   E G G S   A N D   H A M

47. Using the assignation of numbers 1 to 27 with the letters of the alphabet and a blank as in the text, write

D W I G H T   D A V I D   E I S E N H O W E R
4 23 9 7 8 20 27 4 1 22 9 4 27 5 9 19 5 14 8 15 23 5 18 27 27

and calculate

$$\begin{bmatrix} 1 & 0 & 1 & 0 & 1 \\ 0 & 1 & 1 & 0 & 3 \\ 2 & 1 & 1 & 1 & 1 \\ 0 & 0 & 1 & 0 & 2 \\ 1 & 1 & 1 & 2 & 1 \end{bmatrix}\begin{bmatrix} 4 & 20 & 9 & 19 & 23 \\ 23 & 27 & 4 & 5 & 5 \\ 9 & 4 & 27 & 14 & 18 \\ 7 & 1 & 5 & 8 & 27 \\ 8 & 22 & 9 & 15 & 27 \end{bmatrix} = \begin{bmatrix} 21 & 46 & 45 & 48 & 68 \\ 56 & 97 & 58 & 64 & 104 \\ 55 & 94 & 63 & 80 & 123 \\ 25 & 48 & 45 & 44 & 72 \\ 58 & 75 & 59 & 69 & 127 \end{bmatrix}$$

The encoded message is thus
21 56 55 25 58   46   97 94 48 75 45 58 63 45 59 48 64
80 44 69 68 104 123 72 127

49. The inverse of B is calculated to be

$$B^{-1} = \begin{bmatrix} -2 & -1 & 2 & 2 & -1 \\ 3 & 2 & -2 & -4 & 1 \\ 6 & 2 & -4 & -5 & 2 \\ -2 & -1 & 1 & 2 & 0 \\ -3 & -1 & 2 & 3 & -1 \end{bmatrix}$$

Putting the coded message into matrix form and multiplying by B^{-1} yields

$$\begin{bmatrix} -2 & -1 & 2 & 2 & -1 \\ 3 & 2 & -2 & -4 & 1 \\ 6 & 2 & -4 & -5 & 2 \\ -2 & -1 & 1 & 2 & 0 \\ -3 & -1 & 2 & 3 & -1 \end{bmatrix}\begin{bmatrix} 41 & 25 & 43 & 44 & 68 \\ 84 & 56 & 54 & 67 & 135 \\ 82 & 67 & 89 & 86 & 136 \\ 44 & 20 & 39 & 44 & 81 \\ 74 & 54 & 102 & 90 & 149 \end{bmatrix} = \begin{bmatrix} 12 & 14 & 14 & 15 & 14 \\ 25 & 27 & 5 & 8 & 27 \\ 14 & 2 & 19 & 14 & 27 \\ 4 & 1 & 27 & 19 & 27 \\ 15 & 9 & 10 & 15 & 27 \end{bmatrix}$$

This decodes to
12 25 14 4 15 14 27 2 1 9 14 5 19 27 10 15 8 14 19 15 14 27 27 27
L Y N D O N   B A I N E S   J O H N S O N

Exercise 9-3

Key Ideas and Formulas

Basic Properties of Matrices

Assuming all products and sums are defined for the indicated matrices A, B, C, I, and 0, then

ADDITION PROPERTIES

ASSOCIATIVE:	$(A + B) + C = A + (B + C)$
COMMUTATIVE:	$A + B = B + A$
ADDITIVE IDENTITY:	$A + 0 = 0 + A = A$
ADDITIVE INVERSE:	$A + (-A) = (-A) + A = 0$

MULTIPLICATION PROPERTIES

ASSOCIATIVE PROPERTY:	$A(BC) = (AB)C$
MULTIPLICATIVE IDENTITY:	$AI = IA = A$
MULTIPLICATIVE INVERSE:	If A is a square matrix and A^{-1} exists, then $AA^{-1} = A^{-1}A = I$.

COMBINED PROPERTIES

LEFT DISTRIBUTIVE:	$A(B + C) = AB + AC$
RIGHT DISTRIBUTIVE:	$(B + C)A = BA + CA$

EQUALITY

ADDITION:	If $A = B$, then $A + C = B_i + C$
LEFT MULTIPLICATION:	If $A = B$, then $CA = CB$.
RIGHT MULTIPLICATION:	If $A = B$, then $AC = BC$.

A system of n equations in n variables can be written as

$$AX = B$$

where A is the coefficient matrix of the system,

$$X = \begin{bmatrix} x_1 \\ x_2 \\ \vdots \\ x_n \end{bmatrix}$$

and B is the column matrix of constant terms. Then if A^{-1} exists, the solution of the system can be written as $X = A^{-1}B$.

---

1. $2x_1 - x_2 = 3$
$\quad\ x_1 + 3x_2 = -2$

3. $-2x_1 \quad\quad + x_3 = 3$
$\quad\ x_1 + 2x_2 + x_3 = -4$
$\quad\quad\quad x_2 - x_3 = 2$

5. $\begin{bmatrix} 4 & -3 \\ 1 & 2 \end{bmatrix}\begin{bmatrix} x_1 \\ x_2 \end{bmatrix} = \begin{bmatrix} 2 \\ 1 \end{bmatrix}$

7. $\begin{bmatrix} 1 & -2 & 1 \\ -1 & 1 & 0 \\ 2 & 3 & 1 \end{bmatrix}\begin{bmatrix} x_1 \\ x_2 \\ x_3 \end{bmatrix} = \begin{bmatrix} -1 \\ 2 \\ -3 \end{bmatrix}$

9. Since $\begin{bmatrix} 3 & -2 \\ 1 & 4 \end{bmatrix}\begin{bmatrix} -2 \\ 1 \end{bmatrix} = \begin{bmatrix} 3(-2) + (-2)1 \\ 1(-2) + 4\cdot1 \end{bmatrix}$

$= \begin{bmatrix} -8 \\ 2 \end{bmatrix}, \begin{bmatrix} x_1 \\ x_2 \end{bmatrix} = \begin{bmatrix} -8 \\ 2 \end{bmatrix}$ if and only if $x_1 = -8$ and $x_2 = 2$.

11. Since $\begin{bmatrix} -2 & 3 \\ 2 & -1 \end{bmatrix}\begin{bmatrix} 3 \\ 2 \end{bmatrix} = \begin{bmatrix} (-2)3 + 3\cdot2 \\ 2\cdot3 + (-1)2 \end{bmatrix} = \begin{bmatrix} 0 \\ 4 \end{bmatrix}$

$\begin{bmatrix} x_1 \\ x_2 \end{bmatrix} = \begin{bmatrix} 0 \\ 4 \end{bmatrix}$ if and only if $x_1 = 0$ and $x_2 = 4$.

13. $\begin{bmatrix} -1 & -2 \\ 2 & 5 \end{bmatrix}\begin{bmatrix} x_1 \\ x_2 \end{bmatrix} = \begin{bmatrix} k_1 \\ k_2 \end{bmatrix}$

$AX = K$ has solution $X = A^{-1}K$.

We find $A^{-1}K$ for each given K. From problem 17, Exercise 9-2, $A^{-1} = \begin{bmatrix} -5 & -2 \\ 2 & 1 \end{bmatrix}$

(A) $K = \begin{bmatrix} 2 \\ 5 \end{bmatrix}$ $A^{-1}K = \begin{bmatrix} -5 & -2 \\ 2 & 1 \end{bmatrix}\begin{bmatrix} 2 \\ 5 \end{bmatrix} = \begin{bmatrix} -20 \\ 9 \end{bmatrix}$ $x_1 = -20$, $x_2 = 9$

(B) $K = \begin{bmatrix} -4 \\ 1 \end{bmatrix}$ $A^{-1}K = \begin{bmatrix} -5 & -2 \\ 2 & 1 \end{bmatrix}\begin{bmatrix} -4 \\ 1 \end{bmatrix} = \begin{bmatrix} 18 \\ -7 \end{bmatrix}$ $x_1 = 18$, $x_2 = -7$

(C) $K = \begin{bmatrix} -3 \\ -2 \end{bmatrix}$ $A^{-1}K = \begin{bmatrix} -5 & -2 \\ 2 & 1 \end{bmatrix}\begin{bmatrix} -3 \\ -2 \end{bmatrix} = \begin{bmatrix} 19 \\ -8 \end{bmatrix}$ $x_1 = 19$, $x_2 = -8$

15. $\begin{bmatrix} -5 & 7 \\ 2 & -3 \end{bmatrix}\begin{bmatrix} x_1 \\ x_2 \end{bmatrix} = \begin{bmatrix} k_1 \\ k_2 \end{bmatrix}$

$AX = K$ has solution $X = A^{-1}K$.

We find $A^{-1}K$ for each given K. From problem 19, Exercise 9-2, $A^{-1} = \begin{bmatrix} -3 & -7 \\ -2 & -5 \end{bmatrix}$

(A) $K = \begin{bmatrix} -5 \\ 1 \end{bmatrix}$ $A^{-1}K = \begin{bmatrix} -3 & -7 \\ -2 & -5 \end{bmatrix}\begin{bmatrix} -5 \\ 1 \end{bmatrix} = \begin{bmatrix} 8 \\ 5 \end{bmatrix}$ $x_1 = 8$, $x_2 = 5$

(B) $K = \begin{bmatrix} 8 \\ -4 \end{bmatrix}$ $A^{-1}K = \begin{bmatrix} -3 & -7 \\ -2 & -5 \end{bmatrix}\begin{bmatrix} 8 \\ -4 \end{bmatrix} = \begin{bmatrix} 4 \\ 4 \end{bmatrix}$ $x_1 = 4$, $x_2 = 4$

(C) $K = \begin{bmatrix} 6 \\ 0 \end{bmatrix}$ $A^{-1}K = \begin{bmatrix} -3 & -7 \\ -2 & -5 \end{bmatrix}\begin{bmatrix} 6 \\ 0 \end{bmatrix} = \begin{bmatrix} -18 \\ -12 \end{bmatrix}$ $x_1 = -18$, $x_2 = -12$

17. $\begin{bmatrix} 1 & -1 & 0 \\ -1 & 1 & -1 \\ 0 & -1 & 1 \end{bmatrix}\begin{bmatrix} x_1 \\ x_2 \\ x_3 \end{bmatrix} = \begin{bmatrix} k_1 \\ k_2 \\ k_3 \end{bmatrix}$

$AX = K$ has solution $X = A^{-1}K$.

We find $A^{-1}K$ for each given K. From problem 21, Exercise 9-2, $A^{-1} = \begin{bmatrix} 0 & -1 & -1 \\ -1 & -1 & -1 \\ -1 & -1 & 0 \end{bmatrix}$

(A) $K = \begin{bmatrix} 1 \\ 1 \\ 2 \end{bmatrix}$ $A^{-1}K = \begin{bmatrix} 0 & -1 & -1 \\ -1 & -1 & -1 \\ -1 & -1 & 0 \end{bmatrix}\begin{bmatrix} 1 \\ 1 \\ 2 \end{bmatrix} = \begin{bmatrix} -3 \\ -4 \\ -2 \end{bmatrix}$ $x_1 = -3$, $x_2 = -4$, $x_3 = -2$

(B) $K = \begin{bmatrix} -1 \\ 0 \\ -4 \end{bmatrix}$ $A^{-1}K = \begin{bmatrix} 0 & -1 & -1 \\ -1 & -1 & -1 \\ -1 & -1 & 0 \end{bmatrix}\begin{bmatrix} -1 \\ 0 \\ -4 \end{bmatrix} = \begin{bmatrix} 4 \\ 5 \\ 1 \end{bmatrix}$ $x_1 = 4$, $x_2 = 5$, $x_3 = 1$

(C) $K = \begin{bmatrix} 3 \\ -2 \\ 0 \end{bmatrix}$ $A^{-1}K = \begin{bmatrix} 0 & -1 & -1 \\ -1 & -1 & -1 \\ -1 & -1 & 0 \end{bmatrix}\begin{bmatrix} 3 \\ -2 \\ 0 \end{bmatrix} = \begin{bmatrix} 2 \\ -1 \\ -1 \end{bmatrix}$ $x_1 = 2$, $x_2 = -1$, $x_3 = -1$

19. $\begin{bmatrix} 1 & 2 & 5 \\ 3 & 5 & 9 \\ 1 & 1 & -2 \end{bmatrix} \begin{bmatrix} x_1 \\ x_2 \\ x_3 \end{bmatrix} = \begin{bmatrix} k_1 \\ k_2 \\ k_3 \end{bmatrix}$

$AX = K$ has solution $X = A^{-1}K$.

We find $A^{-1}K$ for each given K. From problem 23, Exercise 9-2, $A^{-1} = \begin{bmatrix} -19 & 9 & -7 \\ 15 & -7 & 6 \\ -2 & 1 & -1 \end{bmatrix}$

(A) $K = \begin{bmatrix} 0 \\ 1 \\ 4 \end{bmatrix}$ $A^{-1}K = \begin{bmatrix} -19 & 9 & -7 \\ 15 & -7 & 6 \\ -2 & 1 & -1 \end{bmatrix} \begin{bmatrix} 0 \\ 1 \\ 4 \end{bmatrix} = \begin{bmatrix} -19 \\ 17 \\ -3 \end{bmatrix}$ $x_1 = -19,\ x_2 = 17,\ x_3 = -3$

(B) $K = \begin{bmatrix} 5 \\ -1 \\ 0 \end{bmatrix}$ $A^{-1}K = \begin{bmatrix} -19 & 9 & -7 \\ 15 & -7 & 6 \\ -2 & 1 & -1 \end{bmatrix} \begin{bmatrix} 5 \\ -1 \\ 0 \end{bmatrix} = \begin{bmatrix} -104 \\ 82 \\ -11 \end{bmatrix}$ $x_1 = -104,\ x_2 = 82,\ x_3 = -11$

(C) $K = \begin{bmatrix} -6 \\ 0 \\ 2 \end{bmatrix}$ $A^{-1}K = \begin{bmatrix} -19 & 9 & -7 \\ 15 & -7 & 6 \\ -2 & 1 & -1 \end{bmatrix} \begin{bmatrix} -6 \\ 0 \\ 2 \end{bmatrix} = \begin{bmatrix} 100 \\ -78 \\ 10 \end{bmatrix}$ $x_1 = 100,\ x_2 = -78,\ x_3 = 10$

21. $AX = BX + C$

$AX + (-BX) = (-BX) + BX + C$ Addition property of equality

$AX - BX = 0 + C$ Additive inverse property; definition of subtraction

$AX - BX = C$

$(A - B)X = C$ Right distributive property

[To be more careful, we should write
$AX - BX = AX + -BX$ by the definition of subtraction
           $= [A + (-B)]X$ by the right distributive property
           $= (A - B)X$ by the definition of subtraction
but the distributive properties are generally
understood as applying to subtraction also.]

$(A - B)^{-1}[(A - B)X] = (A - B)^{-1}C$ Left multiplication property of equality

$[(A - B)^{-1}(A - B)]X = (A - B)^{-1}C$ Associative property

$IX = (A - B)^{-1}C$ Multiplicative inverse property

$X = (A - B)^{-1}C$ Multiplicative identity property

23.

$X = AX + C$

$X + (-AX) = (-AX) + AX + C$ Addition property of equality

$X - AX = 0 + C$ Additive inverse property; definition of subtraction

$X - AX = C$ Additive identity property

$IX - AX = C$ Multiplicative identity property

$(I - A)X = C$ Right distributive property

$(I - A)^{-1}[(I - A)X] = (I - A)^{-1}C$ Left multiplication property of equality

$[(I - A)^{-1}(I - A)]X = (I - A)^{-1}C$ Associative property

$IX = (I - A)^{-1}C$ Multiplicative inverse property

$X = (I - A)^{-1}C$ Multiplicative identity property

25.
$$AX + C = 3X$$
$$AX + (-AX) + C = 3X + (-AX) \qquad \text{Addition property of equality}$$
$$0 + C = 3X - AX \qquad \text{Additive inverse property;}$$
definition of subtraction
$$C = 3X - AX \qquad \text{Additive identity property}$$
$$C = 3IX - AX \qquad \text{Multiplicative identity property}$$
$$C = (3I - A)X \qquad \text{Right distributive property}$$
$$(3I - A)^{-1}C = (3I - A)^{-1}[(3I - A)X] \qquad \text{Left multiplication property of}$$
equality
$$(3I - A)^{-1}C = [(3I - A)^{-1}(3I - A)]X \qquad \text{Associative property}$$
$$(3I - A)^{-1}C = IX \qquad \text{Multiplicative inverse property}$$
$$(3I - A)^{-1}C = X \qquad \text{Multiplicative inverse property}$$

27. $\begin{bmatrix} 1 & 5.001 \\ 1 & 5 \end{bmatrix} \begin{bmatrix} x_1 \\ x_2 \end{bmatrix} = \begin{bmatrix} k_1 \\ k_2 \end{bmatrix}$

$AX = K$ has solution $X = A^{-1}K$

$A^{-1} = \begin{bmatrix} -5000 & 5001 \\ 1000 & -1000 \end{bmatrix}$

We find $A^{-1}K$ for each given K.

(A) $K = \begin{bmatrix} 1 \\ 1 \end{bmatrix}$ $A^{-1}K = \begin{bmatrix} -5000 & 5001 \\ 1000 & -1000 \end{bmatrix} \begin{bmatrix} 1 \\ 1 \end{bmatrix} = \begin{bmatrix} 1 \\ 0 \end{bmatrix} = \begin{bmatrix} x_1 \\ x_2 \end{bmatrix}$ $x_1 = 1, \ x_2 = 0$

(B) $K = \begin{bmatrix} 1 \\ 0 \end{bmatrix}$ $A^{-1}K = \begin{bmatrix} -5000 & 5001 \\ 1000 & -1000 \end{bmatrix} \begin{bmatrix} 1 \\ 0 \end{bmatrix} = \begin{bmatrix} -5000 \\ 1000 \end{bmatrix} = \begin{bmatrix} x_1 \\ x_2 \end{bmatrix}$ $x_1 = -5000, \ x_2 = 1000$

(C) $K = \begin{bmatrix} 0 \\ 1 \end{bmatrix}$ $A^{-1}K = \begin{bmatrix} -5000 & 5001 \\ 1000 & -1000 \end{bmatrix} \begin{bmatrix} 0 \\ 1 \end{bmatrix} = \begin{bmatrix} 5001 \\ -1000 \end{bmatrix} = \begin{bmatrix} x_1 \\ x_2 \end{bmatrix}$ $x_1 = 5001, \ x_2 = -1000$

Because of the size of the entries of A^{-1}, a small change in K has a magnified effect on the solutions of $AX = K$. Geometrically, the lines are very close to being parallel, so a small change in the position (y intercept) of one line displaces their point of intersection a great distance.

29. $\begin{bmatrix} 2 & 4 & 7 \\ 5 & -6 & 9 \\ 3 & 6 & -5 \end{bmatrix} \begin{bmatrix} x_1 \\ x_2 \\ x_3 \end{bmatrix} = \begin{bmatrix} 26 \\ -169 \\ 225 \end{bmatrix}$

$AX = B$ has solution $X = A^{-1}B$.

From a computer mathematics system or a graphing calculator, A^{-1} is found to be

$\begin{bmatrix} -0.048\ldots & 0.125 & 0.157\ldots \\ 0.1048\ldots & -0.0625 & 0.034\ldots \\ 0.0967\ldots & 0 & -0.064\ldots \end{bmatrix}$

and

$X = A^{-1}B = A^{-1}\begin{bmatrix} 26 \\ -169 \\ 225 \end{bmatrix} = \begin{bmatrix} 13 \\ 21 \\ -12 \end{bmatrix}$ $x_1 = 13, \ x_2 = 21, \ x_3 = -12$

31.
$$\begin{bmatrix} 1 & 1 & -1 & 3 \\ 2 & -1 & 3 & -1 \\ 5 & 6 & 1 & 1 \\ 5 & 1 & 7 & 2 \end{bmatrix} \begin{bmatrix} x_1 \\ x_2 \\ x_3 \\ x_4 \end{bmatrix} = \begin{bmatrix} 75 \\ 100 \\ 180 \\ 315 \end{bmatrix}$$

$AX = B$ has solution $X = A^{-1}B$.

From a computer mathematics system or a graphing calculator, A^{-1} is found to be

$$\begin{bmatrix} 0.3898\ldots & 0.7288\ldots & 0.1016\ldots & -0.2711\ldots \\ -0.3107\ldots & -0.5084\ldots & 0.1073\ldots & 0.1581\ldots \\ -0.2937\ldots & -0.3898\ldots & -0.0621\ldots & 0.2768\ldots \\ 0.2090\ldots & -0.2033\ldots & -0.0903\ldots & 0.1299\ldots \end{bmatrix}$$

and

$$X = A^{-1}B = A^{-1} \begin{bmatrix} 75 \\ 100 \\ 180 \\ 315 \end{bmatrix} = \begin{bmatrix} 35 \\ -5 \\ 15 \\ 20 \end{bmatrix} \qquad x_1 = 35, \ x_2 = -5, \ x_3 = 15, \ x_4 = 20$$

33. The system to be solved, for an arbitrary return, is derived as follows:

Let x_1 = number of \$4 tickets sold

$\quad x_2$ = number of \$8 tickets sold

Then $x_1 + x_2 = 10,000$ number of seats

$\quad 4x_1 + 8x_2 = k_2$ return required

We solve the system by writing it as a matrix equation.

$$\begin{matrix} A & X & B \end{matrix}$$
$$\begin{bmatrix} 1 & 1 \\ 4 & 8 \end{bmatrix} \begin{bmatrix} x_1 \\ x_2 \end{bmatrix} = \begin{bmatrix} 10,000 \\ k_2 \end{bmatrix}$$

If A^{-1} exists, then $X = A^{-1}B$. To find A^{-1}, we perform row operations on

$$\begin{bmatrix} 1 & 1 & | & 1 & 0 \\ 4 & 8 & | & 0 & 1 \end{bmatrix} \quad (-4)R_1 + R_2 \to R_2$$

$$\sim \begin{bmatrix} 1 & 1 & | & 1 & 0 \\ 0 & 4 & | & -4 & 1 \end{bmatrix} \quad 0.25R_2 \to R_2$$

$$\sim \begin{bmatrix} 1 & 1 & | & 1 & 0 \\ 0 & 1 & | & -1 & 0.25 \end{bmatrix} \quad (-1)R_2 + R_1 \to R_1$$

$$\sim \begin{bmatrix} 1 & 0 & | & 2 & -0.25 \\ 0 & 1 & | & -1 & 0.25 \end{bmatrix}$$

Hence $A^{-1} = \begin{bmatrix} 2 & -0.25 \\ -1 & 0.25 \end{bmatrix}$

Check: $A^{-1}A = \begin{bmatrix} 2 & -0.25 \\ -1 & 0.25 \end{bmatrix} \begin{bmatrix} 1 & 1 \\ 4 & 8 \end{bmatrix} = \begin{bmatrix} 1 & 0 \\ 0 & 1 \end{bmatrix}$

We can now solve the system as

$$\begin{matrix} X & A^{-1} & B \end{matrix}$$
$$\begin{bmatrix} x_1 \\ x_2 \end{bmatrix} = \begin{bmatrix} 2 & -0.25 \\ -1 & 0.25 \end{bmatrix} \begin{bmatrix} 10,000 \\ k_2 \end{bmatrix}$$

If $k_2 = 56,000$ (Concert 1),

$$\begin{bmatrix} x_1 \\ x_2 \end{bmatrix} = \begin{bmatrix} 2 & -0.25 \\ -1 & 0.25 \end{bmatrix} \begin{bmatrix} 10,000 \\ 56,000 \end{bmatrix} = \begin{bmatrix} 6,000 \\ 4,000 \end{bmatrix}$$

Concert 1: 6,000 \$4 tickets and 4,000 \$8 tickets

If $k_2 = 60,000$ (Concert 2),

$$\begin{bmatrix} x_1 \\ x_2 \end{bmatrix} = \begin{bmatrix} 2 & -0.25 \\ -1 & 0.25 \end{bmatrix} \begin{bmatrix} 10,000 \\ 60,000 \end{bmatrix} = \begin{bmatrix} 5,000 \\ 5,000 \end{bmatrix}$$

Concert 2: 5,000 \$4 tickets and 5,000 \$8 tickets

If $k_2 = 68,000$ (Concert 3),

$$\begin{bmatrix} x_1 \\ x_2 \end{bmatrix} = \begin{bmatrix} 2 & -0.25 \\ -1 & 0.25 \end{bmatrix} \begin{bmatrix} 10,000 \\ 68,000 \end{bmatrix} = \begin{bmatrix} 3,000 \\ 7,000 \end{bmatrix}$$

Concert 3: 3,000 \$4 tickets and 7,000 \$8 tickets

35. We solve the system, for arbitrary V_1 and V_2, by writing it as a matrix equation.

$$\overset{A}{\begin{bmatrix} 1 & -1 & 1 \\ 1 & 1 & 0 \\ 0 & 1 & 2 \end{bmatrix}} \overset{J}{\begin{bmatrix} I_1 \\ I_2 \\ I_3 \end{bmatrix}} = \overset{B}{\begin{bmatrix} 0 \\ V_1 \\ V_2 \end{bmatrix}}$$

If A^{-1} exists, then $J = A^{-1}B$. To find A^{-1}, we perform row operations on

$$\begin{bmatrix} 1 & -1 & 1 & | & 1 & 0 & 0 \\ 1 & 1 & 0 & | & 0 & 1 & 0 \\ 0 & 1 & 2 & | & 0 & 0 & 1 \end{bmatrix} \quad (-1)R_1 + R_2 \rightarrow R_2$$

$$\sim \begin{bmatrix} 1 & -1 & 1 & | & 1 & 0 & 0 \\ 0 & 2 & -1 & | & -1 & 1 & 0 \\ 0 & 1 & 2 & | & 0 & 0 & 1 \end{bmatrix} \quad R_2 \leftrightarrow R_3$$

$$\sim \begin{bmatrix} 1 & -1 & 1 & | & 1 & 0 & 0 \\ 0 & 1 & 2 & | & 0 & 0 & 1 \\ 0 & 2 & -1 & | & -1 & 1 & 0 \end{bmatrix} \quad \begin{matrix} R_2 + R_1 \rightarrow R_1 \\ \\ (-2)R_2 + R_3 \rightarrow R_3 \end{matrix}$$

$$\sim \begin{bmatrix} 1 & 0 & 3 & | & 1 & 0 & 1 \\ 0 & 1 & 2 & | & 0 & 0 & 1 \\ 0 & 0 & -5 & | & -1 & 1 & -2 \end{bmatrix} \quad -\frac{1}{5}R_3 \rightarrow R_3$$

$$\sim \begin{bmatrix} 1 & 0 & 3 & | & 1 & 0 & 1 \\ 0 & 1 & 2 & | & 0 & 0 & 1 \\ 0 & 0 & 1 & | & \frac{1}{5} & -\frac{1}{5} & \frac{2}{5} \end{bmatrix} \quad \begin{matrix} (-3)R_3 + R_1 \rightarrow R_1 \\ (-2)R_3 + R_2 \rightarrow R_2 \end{matrix}$$

$$\sim \begin{bmatrix} 1 & 0 & 0 & | & \frac{2}{5} & \frac{3}{5} & -\frac{1}{5} \\ 0 & 1 & 0 & | & -\frac{2}{5} & \frac{2}{5} & \frac{1}{5} \\ 0 & 0 & 1 & | & \frac{1}{5} & -\frac{1}{5} & \frac{2}{5} \end{bmatrix}$$

Hence $A^{-1} = \dfrac{1}{5}\begin{bmatrix} 2 & 3 & -1 \\ -2 & 2 & 1 \\ 1 & -1 & 2 \end{bmatrix}$

Check: $A^{-1}A = \dfrac{1}{5}\begin{bmatrix} 2 & 3 & -1 \\ -2 & 2 & 1 \\ 1 & -1 & 2 \end{bmatrix} \begin{bmatrix} 1 & -1 & 1 \\ 1 & 1 & 0 \\ 0 & 1 & 2 \end{bmatrix} = \begin{bmatrix} 1 & 0 & 0 \\ 0 & 1 & 0 \\ 0 & 0 & 1 \end{bmatrix}$

We can now solve the system as

$$\overset{J}{\begin{bmatrix} I_1 \\ I_2 \\ I_3 \end{bmatrix}} = \frac{1}{5} \overset{A^{-1}}{\begin{bmatrix} 2 & 3 & -1 \\ -2 & 2 & 1 \\ 1 & -1 & 2 \end{bmatrix}} \overset{B}{\begin{bmatrix} 0 \\ V_1 \\ V_2 \end{bmatrix}}$$

(A) $V_1 = 10$ $V_2 = 10$

$$\begin{bmatrix} I_1 \\ I_2 \\ I_3 \end{bmatrix} = \frac{1}{5} \begin{bmatrix} 2 & 3 & -1 \\ -2 & 2 & 1 \\ 1 & -1 & 2 \end{bmatrix} \begin{bmatrix} 0 \\ 10 \\ 10 \end{bmatrix} = \begin{bmatrix} 4 \\ 6 \\ 2 \end{bmatrix} \qquad I_1 = 4, \; I_2 = 6, \; I_3 = 2 \text{ (amperes)}$$

(B) $V_1 = 10$ $V_2 = 15$

$$\begin{bmatrix} I_1 \\ I_2 \\ I_3 \end{bmatrix} = \frac{1}{5} \begin{bmatrix} 2 & 3 & -1 \\ -2 & 2 & 1 \\ 1 & -1 & 2 \end{bmatrix} \begin{bmatrix} 0 \\ 10 \\ 15 \end{bmatrix} = \begin{bmatrix} 3 \\ 7 \\ 4 \end{bmatrix} \qquad I_1 = 3, \; I_2 = 7, \; I_3 = 4 \text{ (amperes)}$$

(C) $V_1 = 15$ $V_2 = 10$

$$\begin{bmatrix} I_1 \\ I_2 \\ I_3 \end{bmatrix} = \frac{1}{5} \begin{bmatrix} 2 & 3 & -1 \\ -2 & 2 & 1 \\ 1 & -1 & 2 \end{bmatrix} \begin{bmatrix} 0 \\ 15 \\ 10 \end{bmatrix} = \begin{bmatrix} 7 \\ 8 \\ 1 \end{bmatrix} \qquad I_1 = 7, \; I_2 = 8, \; I_3 = 1 \text{ (amperes)}$$

37. If the graph of $f(x) = ax^2 + bx + c$ passes through a point, the coordinates of the point must satisfy the equation of the graph. Hence
$k_1 = a(1)^2 + b(1) + c$
$k_2 = a(2)^2 + b(2) + c$
$k_3 = a(3)^2 + b(3) + c$
After simplification, we obtain:
  $a + b + c = k_1$
  $4a + 2b + c = k_2$
  $9a + 3b + c = k_3$

We solve this system, for arbitrary k_1, k_2, k_3, by writing it as a matrix equation.

$$\overset{A}{\begin{bmatrix} 1 & 1 & 1 \\ 4 & 2 & 1 \\ 9 & 3 & 1 \end{bmatrix}} \overset{X}{\begin{bmatrix} a \\ b \\ c \end{bmatrix}} = \overset{B}{\begin{bmatrix} k_1 \\ k_2 \\ k_3 \end{bmatrix}}$$

If A^{-1} exists, then $X = A^{-1}B$. To find A^{-1} we perform row operations on

$$\begin{bmatrix} 1 & 1 & 1 & | & 1 & 0 & 0 \\ 4 & 2 & 1 & | & 0 & 1 & 0 \\ 9 & 3 & 1 & | & 0 & 0 & 1 \end{bmatrix} \begin{matrix} \\ (-4)R_1 + R_2 \to R_2 \\ (-9)R_1 + R_3 \to R_3 \end{matrix}$$

$$\sim \begin{bmatrix} 1 & 1 & 1 & | & 1 & 0 & 0 \\ 0 & -2 & -3 & | & -4 & 1 & 0 \\ 0 & -6 & -8 & | & -9 & 0 & 1 \end{bmatrix} \quad -\tfrac{1}{2}R_2 \to R_2$$

$$\sim \begin{bmatrix} 1 & 1 & 1 & | & 1 & 0 & 0 \\ 0 & 1 & \tfrac{3}{2} & | & 2 & -\tfrac{1}{2} & 0 \\ 0 & -6 & -8 & | & -9 & 0 & 1 \end{bmatrix} \begin{matrix} (-1)R_2 + R_1 \to R_1 \\ \\ 6R_2 + R_3 \to R_3 \end{matrix}$$

$$\sim \begin{bmatrix} 1 & 0 & -\tfrac{1}{2} & | & -1 & \tfrac{1}{2} & 0 \\ 0 & 1 & \tfrac{3}{2} & | & 2 & -\tfrac{1}{2} & 0 \\ 0 & 0 & 1 & | & 3 & -3 & 1 \end{bmatrix} \begin{matrix} \tfrac{1}{2}R_3 + R_1 \to R_1 \\ (-\tfrac{3}{2})R_3 + R_2 \to R_2 \end{matrix}$$

$$\sim \begin{bmatrix} 1 & 0 & 0 & | & \tfrac{1}{2} & -1 & \tfrac{1}{2} \\ 0 & 1 & 0 & | & -\tfrac{5}{2} & 4 & -\tfrac{3}{2} \\ 0 & 0 & 1 & | & 3 & -3 & 1 \end{bmatrix}$$

Hence $A^{-1} = \frac{1}{2} \begin{bmatrix} 1 & -2 & 1 \\ -5 & 8 & -3 \\ 6 & -6 & 2 \end{bmatrix}$

Check: $A^{-1}A = \frac{1}{2} \begin{bmatrix} 1 & -2 & 1 \\ -5 & 8 & -3 \\ 6 & -6 & 2 \end{bmatrix} \begin{bmatrix} 1 & 1 & 1 \\ 4 & 2 & 1 \\ 9 & 3 & 1 \end{bmatrix} = \begin{bmatrix} 1 & 0 & 0 \\ 0 & 1 & 0 \\ 0 & 0 & 1 \end{bmatrix}$

We can now solve the system as

$$\underset{X}{\begin{bmatrix} a \\ b \\ c \end{bmatrix}} = \frac{1}{2} \underset{A^{-1}}{\begin{bmatrix} 1 & -2 & 1 \\ -5 & 8 & -3 \\ 6 & -6 & 2 \end{bmatrix}} \underset{B}{\begin{bmatrix} k_1 \\ k_2 \\ k_3 \end{bmatrix}}$$

(A) $\begin{bmatrix} a \\ b \\ c \end{bmatrix} = \frac{1}{2} \begin{bmatrix} 1 & -2 & 1 \\ -5 & 8 & -3 \\ 6 & -6 & 2 \end{bmatrix} \begin{bmatrix} -2 \\ 1 \\ 6 \end{bmatrix} = \begin{bmatrix} 1 \\ 0 \\ -3 \end{bmatrix}$ $a = 1$, $b = 0$, $c = -3$

(B) $\begin{bmatrix} a \\ b \\ c \end{bmatrix} = \frac{1}{2} \begin{bmatrix} 1 & -2 & 1 \\ -5 & 8 & -3 \\ 6 & -6 & 2 \end{bmatrix} \begin{bmatrix} 4 \\ 3 \\ -2 \end{bmatrix} = \begin{bmatrix} -2 \\ 5 \\ 1 \end{bmatrix}$ $a = -2$, $b = 5$, $c = 1$

(C) $\begin{bmatrix} a \\ b \\ c \end{bmatrix} = \frac{1}{2} \begin{bmatrix} 1 & -2 & 1 \\ -5 & 8 & -3 \\ 6 & -6 & 2 \end{bmatrix} \begin{bmatrix} 8 \\ -5 \\ 4 \end{bmatrix} = \begin{bmatrix} 11 \\ -46 \\ 43 \end{bmatrix}$ $a = 11$, $b = -46$, $c = 43$

39. The system to be solved, for an arbitrary diet, is derived as follows:
Let x_1 = amount of mix A
  x_2 = amount of mix B
Then $0.20x_1 + 0.10x_2 = k_1$ (k_1 = amount of protein)
  $0.02x_1 + 0.06x_2 = k_2$ (k_2 = amount of fat)
We solve the system by writing it as a matrix equation.

$$\underset{A}{\begin{bmatrix} 0.20 & 0.10 \\ 0.02 & 0.06 \end{bmatrix}} \underset{X}{\begin{bmatrix} x_1 \\ x_2 \end{bmatrix}} = \underset{B}{\begin{bmatrix} k_1 \\ k_2 \end{bmatrix}}$$

If A^{-1} exists, then $X = A^{-1}B$. To find A^{-1}, we perform row operations on

$\begin{bmatrix} 0.20 & 0.10 & | & 1 & 0 \\ 0.02 & 0.06 & | & 0 & 1 \end{bmatrix}$ $\begin{matrix} 5R_1 \to R_1 \\ 50R_2 \to R_2 \end{matrix}$

$\sim \begin{bmatrix} 1 & 0.5 & | & 5 & 0 \\ 1 & 3 & | & 0 & 50 \end{bmatrix}$ $(-1)R_1 + R_2 \to R_2$

$\sim \begin{bmatrix} 1 & 0.5 & | & 5 & 0 \\ 0 & 2.5 & | & -5 & 50 \end{bmatrix}$ $0.4R_2 \to R_2$

$\sim \begin{bmatrix} 1 & 0.5 & | & 5 & 0 \\ 0 & 1 & | & -2 & 20 \end{bmatrix}$ $(-0.5)R_2 + R_1 \to R_1$

$\sim \begin{bmatrix} 1 & 0 & | & 6 & -10 \\ 0 & 1 & | & -2 & 20 \end{bmatrix}$

Hence $A^{-1} = \begin{bmatrix} 6 & -10 \\ -2 & 20 \end{bmatrix}$

Check: $A^{-1}A = \begin{bmatrix} 6 & -10 \\ -2 & 20 \end{bmatrix} \begin{bmatrix} 0.20 & 0.10 \\ 0.02 & 0.06 \end{bmatrix} = \begin{bmatrix} 1 & 0 \\ 0 & 1 \end{bmatrix}$

We can now solve the system as

$$\begin{matrix} X & A^{-1} & B \end{matrix}$$

$$\begin{bmatrix} x_1 \\ x_2 \end{bmatrix} = \begin{bmatrix} 6 & -10 \\ -2 & 20 \end{bmatrix}\begin{bmatrix} k_1 \\ k_2 \end{bmatrix}$$

For Diet 1, $k_1 = 20$ and $k_2 = 6$

$$\begin{bmatrix} x_1 \\ x_2 \end{bmatrix} = \begin{bmatrix} 6 & -10 \\ -2 & 20 \end{bmatrix}\begin{bmatrix} 20 \\ 6 \end{bmatrix} = \begin{bmatrix} 60 \\ 80 \end{bmatrix}$$ Diet 1: 60 ounces Mix A and 80 ounces Mix B

For Diet 2, $k_1 = 10$ and $k_2 = 4$

$$\begin{bmatrix} x_1 \\ x_2 \end{bmatrix} = \begin{bmatrix} 6 & -10 \\ -2 & 20 \end{bmatrix}\begin{bmatrix} 10 \\ 4 \end{bmatrix} = \begin{bmatrix} 20 \\ 60 \end{bmatrix}$$ Diet 2: 20 ounces Mix A and 60 ounces Mix B

For Diet 3, $k_1 = 10$ and $k_2 = 6$

$$\begin{bmatrix} x_1 \\ x_2 \end{bmatrix} = \begin{bmatrix} 6 & -10 \\ -2 & 20 \end{bmatrix}\begin{bmatrix} 10 \\ 6 \end{bmatrix} = \begin{bmatrix} 0 \\ 100 \end{bmatrix}$$ Diet 3: 0 ounces Mix A and 100 ounces Mix B

Exercise 9-4

Key Ideas and Formulas

The determinant of a square matrix A is a number, denoted det A, or by writing the array of elements in A using vertical lines instead of square brackets.

Second-order determinant:

$$\begin{vmatrix} a_{11} & a_{12} \\ a_{21} & a_{22} \end{vmatrix} = a_{11}a_{22} - a_{21}a_{12}$$

Third-order determinant:

$$\begin{vmatrix} a_{11} & a_{12} & a_{13} \\ a_{21} & a_{22} & a_{23} \\ a_{31} & a_{32} & a_{33} \end{vmatrix}$$

The minor of an element a_{ij} is the determinant array obtained by deleting the row and column that contain a_{ij}, that is, row i and column j.

The cofactor of $a_{ij} = (-1)^{i+j}$ (Minor of a_{ij}).

The value of a determinant of order 3 (or $n > 3$) is the sum of the three (or n) products obtained by multiplying each element of any one row (or each element of any one column) by its cofactor.

---

1. $\begin{vmatrix} 5 & 4 \\ 2 & 3 \end{vmatrix} = 5 \cdot 3 - 2 \cdot 4 = 7$ **3.** $\begin{vmatrix} 3 & -7 \\ -5 & 6 \end{vmatrix} = 3 \cdot 6 - (-5)(-7) = -17$

5. $\begin{vmatrix} 4.3 & -1.2 \\ -5.1 & 3.7 \end{vmatrix} = (4.3)(3.7) - (-5.1)(-1.2) = 9.79$

7. $\begin{vmatrix} 5 & -1 & -3 \\ 3 & 4 & 6 \\ 0 & -2 & 8 \end{vmatrix} = \begin{vmatrix} 4 & 6 \\ -2 & 8 \end{vmatrix}$ **9.** $\begin{vmatrix} 5 & -1 & -3 \\ 3 & 4 & 6 \\ 0 & -2 & 8 \end{vmatrix} = \begin{vmatrix} 5 & -1 \\ 0 & -2 \end{vmatrix}$

11. $(-1)^{1+1}\begin{vmatrix} 4 & 6 \\ -2 & 8 \end{vmatrix} = (-1)^2[4 \cdot 8 - (-2)6] = 44$

13. $(-1)^{2+3} \begin{vmatrix} 5 & -1 \\ 0 & -2 \end{vmatrix} = (-1)^5[5(-2) - 0(-1)] = 10$

15. We expand by row 1

$\begin{vmatrix} 1 & 0 & 0 \\ -2 & 4 & 3 \\ 5 & -2 & 1 \end{vmatrix} =$

a_{11} (cofactor of a_{11}) + a_{12} (cofactor of a_{12}) + a_{13} (cofactor of a_{13})

$= 1(-1)^{1+1} \begin{vmatrix} 4 & 3 \\ -2 & 1 \end{vmatrix} + 0(\ \diagup\ ) + (\ \diagdown\ )$

It is unnecessary to evaluate these since they are multiplied by 0.

$= (-1)^2[4 \cdot 1 - (-2)3]$
$= 10$

17. We expand by column 1

$\begin{vmatrix} 0 & 1 & 5 \\ 3 & -7 & 6 \\ 0 & -2 & -3 \end{vmatrix} = a_{11}$ (cofactor of a_{11}) + a_{21} (cofactor of a_{21}) + a_{31} (cofactor of a_{31})

$= 0(\ \diagup\ ) + 3(-1)^{2+1} \begin{vmatrix} 1 & 5 \\ -2 & -3 \end{vmatrix} + 0(\ \diagdown\ )$

It is unnecessary to evaluate these since they are multiplied by 0.

$= 3(-1)^3[1(-3) - (-2)5]$
$= -21$

> **Common Error:** Neglecting the sign of the cofactor. The cofactor is often called the "signed" minor.

19. We expand by column 2

$\begin{vmatrix} -1 & 2 & -3 \\ -2 & 0 & -6 \\ 4 & -3 & 2 \end{vmatrix} =$

a_{12} (cofactor of a_{12}) + a_{22} (cofactor of a_{22}) + a_{32} (cofactor of a_{32})

$= 2(-1)^{1+2} \begin{vmatrix} -2 & -6 \\ 4 & 2 \end{vmatrix} + 0(\ \ ) + (-3)(-1)^{3+2} \begin{vmatrix} -1 & -3 \\ -2 & -6 \end{vmatrix}$

$= 2(-1)^3[(-2)2 - 4(-6)] + (-3)(-1)^5[(-1)(-6) - (-2)(-3)]$

$= (-2)(20) + 3(0) = -40$

21. $(-1)^{1+1} \begin{vmatrix} a_{11} & a_{12} & a_{13} & a_{14} \\ a_{21} & a_{22} & a_{23} & a_{24} \\ a_{31} & a_{32} & a_{33} & a_{34} \\ a_{41} & a_{42} & a_{43} & a_{44} \end{vmatrix} = (-1)^{1+1} \begin{vmatrix} a_{22} & a_{23} & a_{24} \\ a_{32} & a_{33} & a_{34} \\ a_{42} & a_{43} & a_{44} \end{vmatrix}$

23. $(-1)^{4+3} \begin{vmatrix} a_{11} & a_{12} & a_{13} & a_{14} \\ a_{21} & a_{22} & a_{23} & a_{24} \\ a_{31} & a_{32} & a_{33} & a_{34} \\ a_{41} & a_{42} & a_{43} & a_{44} \end{vmatrix} = (-1)^{4+3} \begin{vmatrix} a_{11} & a_{12} & a_{14} \\ a_{21} & a_{22} & a_{24} \\ a_{31} & a_{32} & a_{34} \end{vmatrix}$

25. We expand by the second column

$$\begin{vmatrix} 3 & -2 & -8 \\ -2 & 0 & -3 \\ 1 & 0 & -4 \end{vmatrix} =$$

a_{12} (cofactor of a_{12}) + a_{22} (cofactor of a_{22}) + a_{32} (cofactor of a_{32})

$$= (-2)(-1)^{1+2} \begin{vmatrix} -2 & -3 \\ 1 & -4 \end{vmatrix} + 0 + 0$$
$$= (-2)(-1)^3[(-2)(-4) - 1(-3)]$$
$$= 2(11)$$
$$= 22$$

27. We expand by the first row

$$\begin{vmatrix} 1 & 4 & 1 \\ 1 & 1 & -2 \\ 2 & 1 & -1 \end{vmatrix} = a_{11} \text{ (cofactor of } a_{11}) + a_{12} \text{ (cofactor of } a_{12}) + a_{13} \text{ (cofactor of } a_{13})$$

$$= 1(-1)^{1+1} \begin{vmatrix} 1 & -2 \\ 1 & -1 \end{vmatrix} + 4(-1)^{1+2} \begin{vmatrix} 1 & -2 \\ 2 & -1 \end{vmatrix} + 1(-1)^{1+3} \begin{vmatrix} 1 & 1 \\ 2 & 1 \end{vmatrix}$$
$$= (-1)^2[1(-1) - 1(-2)] + 4(-1)^3[1(-1) - 2(-2)] + (-1)^4[1 \cdot 1 - 2 \cdot 1]$$
$$= 1 + (-12) + (-1)$$
$$= -12$$

29. We expand by the first row

$$\begin{vmatrix} 1 & 4 & 3 \\ 2 & 1 & 6 \\ 3 & -2 & 9 \end{vmatrix} = a_{11} \text{ (cofactor of } a_{11}) + a_{12} \text{ (cofactor of } a_{12}) + a_{13} \text{ (cofactor of } a_{13})$$

$$= 1(-1)^{1+1} \begin{vmatrix} 1 & 6 \\ -2 & 9 \end{vmatrix} + 4(-1)^{1+2} \begin{vmatrix} 2 & 6 \\ 3 & 9 \end{vmatrix} + 3(-1)^{1+3} \begin{vmatrix} 2 & 1 \\ 3 & -2 \end{vmatrix}$$
$$= (-1)^2[1 \cdot 9 - (-2)6] + 4(-1)^3[2 \cdot 9 - 3 \cdot 6] + 3(-1)^4[2(-2) - 1 \cdot 3]$$
$$= 21 + 0 - 21$$
$$= 0$$

31. We expand by the second row. Clearly the only non-zero term will be a_{22} (cofactor of a_{22}), which is

$$3(-1)^{2+2} \begin{vmatrix} 2 & 1 & 7 \\ 3 & 2 & 5 \\ 0 & 0 & 2 \end{vmatrix}$$

The order 3 determinant is expanded by the third row. Again there is only one non-zero term, a_{33} (cofactor of a_{33}). So the original determinant is reduced to

$$3(-1)^{2+2} 2(-1)^{3+3} \begin{vmatrix} 2 & 1 \\ 3 & 2 \end{vmatrix} = 6(-1)^{10}(2 \cdot 2 - 3 \cdot 1) = 6$$

33.

$$\begin{vmatrix} -2 & 0 & 0 & 0 & 0 \\ 9 & -1 & 0 & 0 & 0 \\ 2 & 1 & 3 & 0 & 0 \\ -1 & 4 & 2 & 2 & 0 \\ 7 & -2 & 3 & 5 & 5 \end{vmatrix} = (-2)(-1)^{1+1} \begin{vmatrix} -1 & 0 & 0 & 0 \\ 1 & 3 & 0 & 0 \\ 4 & 2 & 2 & 0 \\ -2 & 3 & 5 & 5 \end{vmatrix} + 0 \text{ terms}$$

$$= -2 \begin{vmatrix} -1 & 0 & 0 & 0 \\ 1 & 3 & 0 & 0 \\ 4 & 2 & 2 & 0 \\ -2 & 3 & 5 & 5 \end{vmatrix} = (-2)\left[(-1)(-1)^{1+1} \begin{vmatrix} 3 & 0 & 0 \\ 2 & 2 & 0 \\ 3 & 5 & 5 \end{vmatrix} + 0 \text{ terms} \right]$$

$$= (-2)(-1) \begin{vmatrix} 3 & 0 & 0 \\ 2 & 2 & 0 \\ 3 & 5 & 5 \end{vmatrix} = (-2)(-1) \left[3(-1)^{1+1} \begin{vmatrix} 2 & 0 \\ 5 & 5 \end{vmatrix} + 0 \text{ terms} \right]$$

$$= (-2)(-1)3 \begin{vmatrix} 2 & 0 \\ 5 & 5 \end{vmatrix}$$

$$= (-2)(-1)(3)[2 \cdot 5 - 5 \cdot 0]$$

$$= (-2)(-1)(3)(2)(5)$$

$$= 60$$

35. $\begin{vmatrix} 2 & 6 & -1 & 2 & 6 \\ 5 & 3 & -7 & 5 & 3 \\ -4 & 2 & 1 & -4 & -2 \end{vmatrix}$

$$2 \cdot 3 \cdot 1 + 6(-7)(-4) + (-1)(5)(-2) - (-4)(3)(-1) - (-2)(-7)2 - 6 \cdot 5 \cdot 1$$

$$= 6 + 168 + 10 - 12 - 28 - 30$$

$$= 114$$

37. False. $\begin{vmatrix} 10 & 10 \\ 0 & 0 \end{vmatrix}$ is a counterexample.

39. True. Expanding $\begin{vmatrix} a_{11} & a_{12} & a_{13} & a_{14} \\ 0 & a_{22} & a_{23} & a_{24} \\ 0 & 0 & a_{33} & a_{34} \\ 0 & 0 & 0 & a_{44} \end{vmatrix}$ by the first column, we obtain

successively

$$a_{11} \begin{vmatrix} a_{22} & a_{23} & a_{24} \\ 0 & a_{33} & a_{34} \\ 0 & 0 & a_{44} \end{vmatrix} = a_{11}a_{22} \begin{vmatrix} a_{33} & a_{34} \\ 0 & a_{44} \end{vmatrix} = a_{11}a_{22}a_{33}a_{44} .$$

Similarly for the determinant of an $n \times n$ upper triangular matrix, we would obtain $a_{11}a_{22}a_{33} \cdot \ldots \cdot a_{nn}$ as proposed.

41. $\begin{vmatrix} a & b \\ c & d \end{vmatrix} = ad - bc$ $\begin{vmatrix} c & d \\ a & b \end{vmatrix} = cb - ad = -(ad - bc)$

Hence, $\begin{vmatrix} a & b \\ c & d \end{vmatrix} = - \begin{vmatrix} c & d \\ a & b \end{vmatrix}$; interchanging the rows of this determinant changes its sign.

43. $\begin{vmatrix} a & b \\ c & d \end{vmatrix} = ad - bc$ $\begin{vmatrix} ka & b \\ kc & d \end{vmatrix} = kad - kcb = k(ad - bc)$

Hence, $\begin{vmatrix} ka & b \\ kc & d \end{vmatrix} = k \begin{vmatrix} a & b \\ c & d \end{vmatrix}$; multiplying a column of this determinant by a number k multiplies the value of the determinant by k.

45. $\begin{vmatrix} a & b \\ c & d \end{vmatrix} = ad - bc$ $\begin{vmatrix} kc + a & kd + b \\ c & d \end{vmatrix} = (kc + a)d - (kd + b)c$

$$= kcd + ad - kdc - bc$$

$$= ad - bc$$

Hence, $\begin{vmatrix} kc + a & kd + b \\ c & d \end{vmatrix} = \begin{vmatrix} a & b \\ c & d \end{vmatrix}$; adding a multiple of one row to the other row does not change the value of this determinant.

47. Expanding by the first column

$$\begin{vmatrix} a_{11} & a_{12} & a_{13} \\ a_{21} & a_{22} & a_{23} \\ a_{31} & a_{32} & a_{33} \end{vmatrix}$$

$$= a_{11}(-1)^{1+1}\begin{vmatrix} a_{22} & a_{23} \\ a_{32} & a_{33} \end{vmatrix} + a_{21}(-1)^{2+1}\begin{vmatrix} a_{12} & a_{13} \\ a_{32} & a_{33} \end{vmatrix} + a_{31}(-1)^{3+1}\begin{vmatrix} a_{12} & a_{13} \\ a_{22} & a_{23} \end{vmatrix}$$

$$= a_{11}\begin{vmatrix} a_{22} & a_{23} \\ a_{32} & a_{33} \end{vmatrix} - a_{21}\begin{vmatrix} a_{12} & a_{13} \\ a_{32} & a_{33} \end{vmatrix} + a_{31}\begin{vmatrix} a_{12} & a_{13} \\ a_{22} & a_{23} \end{vmatrix}$$

$$= a_{11}(a_{22}a_{33} - a_{32}a_{23}) - a_{21}(a_{12}a_{33} - a_{32}a_{13}) + a_{31}(a_{12}a_{23} - a_{22}a_{13})$$

$$\quad\quad ① \quad\quad\quad\quad ② \quad\quad\quad\quad ③ \quad\quad\quad\quad ④ \quad\quad\quad\quad ⑤ \quad\quad\quad\quad ⑥$$

$$= a_{11}a_{22}a_{33} - a_{11}a_{32}a_{23} - a_{21}a_{12}a_{33} + a_{21}a_{32}a_{13} + a_{31}a_{12}a_{23} - a_{31}a_{22}a_{13}$$

Expanding by the third row

$$\begin{vmatrix} a_{11} & a_{12} & a_{13} \\ a_{21} & a_{22} & a_{23} \\ a_{31} & a_{32} & a_{33} \end{vmatrix}$$

$$= a_{31}(-1)^{3+1}\begin{vmatrix} a_{12} & a_{13} \\ a_{22} & a_{23} \end{vmatrix} + a_{32}(-1)^{3+2}\begin{vmatrix} a_{11} & a_{13} \\ a_{21} & a_{23} \end{vmatrix} + a_{33}(-1)^{3+3}\begin{vmatrix} a_{11} & a_{12} \\ a_{21} & a_{22} \end{vmatrix}$$

$$= a_{31}\begin{vmatrix} a_{12} & a_{13} \\ a_{22} & a_{23} \end{vmatrix} - a_{32}\begin{vmatrix} a_{11} & a_{13} \\ a_{21} & a_{23} \end{vmatrix} + a_{33}\begin{vmatrix} a_{11} & a_{12} \\ a_{21} & a_{22} \end{vmatrix}$$

$$= a_{31}(a_{12}a_{23} - a_{13}a_{22}) - a_{32}(a_{11}a_{23} - a_{13}a_{21}) + a_{33}(a_{11}a_{22} - a_{12}a_{21})$$

$$\quad\quad ⑤ \quad\quad\quad\quad ⑥ \quad\quad\quad\quad ② \quad\quad\quad\quad ④ \quad\quad\quad\quad ① \quad\quad\quad\quad ③$$

$$= a_{31}a_{12}a_{23} - a_{31}a_{13}a_{22} - a_{32}a_{11}a_{23} + a_{32}a_{13}a_{21} + a_{33}a_{11}a_{22} - a_{33}a_{12}a_{21}$$

Comparing the two expressions, with the aid of the numbers over the terms, shows that the expressions are the same.

49. $A = \begin{bmatrix} 2 & 3 \\ 1 & -2 \end{bmatrix}$ $B = \begin{bmatrix} -1 & 3 \\ 2 & 1 \end{bmatrix}$

We calculate $AB = \begin{bmatrix} 2 & 3 \\ 1 & -2 \end{bmatrix}\begin{bmatrix} -1 & 3 \\ 2 & 1 \end{bmatrix} = \begin{bmatrix} 2(-1) + 3\cdot2 & 2\cdot3 + 3\cdot1 \\ 1(-1) + (-2)2 & 1\cdot3 + (-2)\cdot1 \end{bmatrix} = \begin{bmatrix} 4 & 9 \\ -5 & 1 \end{bmatrix}$

$$\det\ (AB)\ =\ \begin{vmatrix} 4 & 9 \\ -5 & 1 \end{vmatrix} = 4\cdot1 - (-5)9 = 49$$

$$\det A\ =\ \begin{vmatrix} 2 & 3 \\ 1 & -2 \end{vmatrix} = 2(-2) - 1\cdot3 = -7$$

$$\det B\ =\ \begin{vmatrix} -1 & 3 \\ 2 & 1 \end{vmatrix} = (-1)1 - 2\cdot3 = -7$$

Therefore $\det\ (AB)\ =\ 49\ =\ (-7)(-7)\ =\ \det A\cdot\det B$

51. The matrix $xI - A$ is calculated as

$$x\begin{bmatrix} 1 & 0 \\ 0 & 1 \end{bmatrix} - \begin{bmatrix} 5 & -4 \\ 2 & -1 \end{bmatrix} = \begin{bmatrix} x & 0 \\ 0 & x \end{bmatrix} - \begin{bmatrix} 5 & -4 \\ 2 & -1 \end{bmatrix} = \begin{bmatrix} x - 5 & 4 \\ -2 & x + 1 \end{bmatrix}$$

The characteristic polynomial is the determinant of this matrix:

$$\begin{vmatrix} x - 5 & 4 \\ -2 & x + 1 \end{vmatrix} = (x - 5)(x + 1) - (4)(-2) = x^2 - 4x - 5 + 8 = x^2 - 4x + 3$$

The zeros of this polynomial are the solutions of $x^2 - 4x + 3 = 0$

$$x^2 - 4x + 3 = 0$$
$$(x - 1)(x - 3) = 0$$
$$x = 1, \ 3$$

The eigenvalues of this matrix are 1 and 3.

53. The matrix $xI - A$ is calculated as

$$x\begin{bmatrix} 1 & 0 & 0 \\ 0 & 1 & 0 \\ 0 & 0 & 1 \end{bmatrix} - \begin{bmatrix} 4 & -4 & 0 \\ 2 & -2 & 0 \\ 4 & -8 & -4 \end{bmatrix} = \begin{bmatrix} x & 0 & 0 \\ 0 & x & 0 \\ 0 & 0 & x \end{bmatrix} - \begin{bmatrix} 4 & -4 & 0 \\ 2 & -2 & 0 \\ 4 & -8 & -4 \end{bmatrix} = \begin{bmatrix} x - 4 & 4 & 0 \\ -2 & x + 2 & 0 \\ -4 & 8 & x + 4 \end{bmatrix}$$

The characteristic polynomial is the determinant of this matrix.

$$\begin{vmatrix} x - 4 & 4 & 0 \\ -2 & x + 2 & 0 \\ -4 & 8 & x + 4 \end{vmatrix} = (x + 4)(-1)^{3+3}\begin{vmatrix} x - 4 & 4 \\ -2 & x + 2 \end{vmatrix} \text{ expanding by the third column}$$

$$= (x + 4)(1)[(x - 4)(x + 2) - 4(-2)]$$
$$= (x + 4)(x^2 - 2x - 8 + 8)$$
$$= (x + 4)(x^2 - 2x)$$
$$= x^3 + 2x^2 - 8x$$

The zeros of this polynomial are the solutions of

$$x^3 + 2x^2 - 8x = 0$$
$$x(x^2 + 2x - 8) = 0$$
$$x(x + 4)(x - 2) = 0$$
$$x = 0, \ -4, \ 2$$

The eigenvalues of this matrix are 0, -4, and 2.

Exercise 9-5

Key Ideas and Formulas

If each element of any row (or column) of a determinant is multiplied by a constant k, the new determinant is k times the original. (Theorem 1) Caution: This operation and its result is very different from the operation of multiplying a matrix by a number k.

If every element in a row (or column) is 0, the value of the determinant is 0. (Theorem 2)

If two rows (or two columns) of a determinant are interchanged, the new determinant is the negative of the original. (Theorem 3)

If the corresponding elements are equal in two rows (or columns) the value of the determinant is 0. (Theorem 4)

If a multiple of any row (or column) of a determinant is added to any other row (or column) the value of the determinant is unchanged. (Theorem 5)

---

1. Theorem 1   **3.** Theorem 1   **5.** Theorem 2   **7.** Theorem 3   **9.** Theorem 5

11. $3C_1 + C_2 \to C_2$ has been used. Hence $x = 3(-1) + 3 = 0$

13. $3C_1 + C_3 \rightarrow C_3$ has been used. Hence $x = 3(1) + 2 = 5$

15. Interchanging two rows of a determinant changes the sign of the determinant (Theorem 3).
Hence,
$$\begin{vmatrix} c & d \\ a & b \end{vmatrix} = -10$$

17. Adding a multiple (in this case a multiple by 1) of a row to another row does not change the value of the determinant (Theorem 5).
Hence,
$$\begin{vmatrix} a + c & b + d \\ c & d \end{vmatrix} = 10$$

19. Adding a multiple of a column to another column does not change the value of the determinant (Theorem 5). Hence,
$$\begin{vmatrix} a & a - b \\ c & c - d \end{vmatrix} = \begin{vmatrix} a & -b \\ c & -d \end{vmatrix}$$
If every element of a column of a determinant is multiplied by -1, the value of the determinant is -1 times the original. Hence,
$$\begin{vmatrix} a & a - b \\ c & c - d \end{vmatrix} = \begin{vmatrix} a & -b \\ c & -d \end{vmatrix} = (-1)\begin{vmatrix} a & b \\ c & d \end{vmatrix} = -1(10) = -10$$

21.
$$\begin{vmatrix} -1 & 0 & 3 \\ 2 & 5 & 4 \\ 1 & 5 & 2 \end{vmatrix} = \begin{vmatrix} -1 & 0 & 3 \\ 1 & 0 & 2 \\ 1 & 5 & 2 \end{vmatrix} \quad (-1)R_3 + R_2 \rightarrow R_2$$
$$= 5(-1)^{3+2}\begin{vmatrix} -1 & 3 \\ 1 & 2 \end{vmatrix} = -5[(-1)2 - 1(3)] = 25$$

23.
$$\begin{vmatrix} 3 & 5 & 0 \\ 1 & 1 & -2 \\ 2 & 1 & -1 \end{vmatrix} = \begin{vmatrix} 3 & 5 & 0 \\ -3 & -1 & 0 \\ 2 & 1 & -1 \end{vmatrix} \quad (-2)R_3 + R_2 \rightarrow R_2$$
$$= (-1)(-1)^{3+3}\begin{vmatrix} 3 & 5 \\ -3 & -1 \end{vmatrix} = (-1)[3(-1) - (-3)5] = -12$$

25. Theorem 1   **27.** Theorem 2   **29.** Theorem 5

31. $2C_3 + C_1 \rightarrow C_1$ has been used. Hence $x = 2 \cdot 1 + 3 = 5$
$C_3 + C_2 \rightarrow C_2$ has been used. Hence $y = (-2) + 2 = 0$

33. $(-4)R_2 + R_1 \rightarrow R_1$ has been used. Hence $x = (-4)3 + 9 = -3$
$2R_2 + R_3 \rightarrow R_3$ has been used. Hence $y = 2 \cdot 3 + 4 = 10$

35. We will generate zeros in the first column by row operations.
$$\begin{vmatrix} 1 & 5 & 3 \\ 4 & 2 & 1 \\ 3 & 1 & 2 \end{vmatrix} = \begin{vmatrix} 1 & 5 & 3 \\ 0 & -18 & -11 \\ 0 & -14 & -7 \end{vmatrix} \quad \begin{array}{l} (-4)R_1 + R_2 \rightarrow R_2 \\ (-3)R_1 + R_3 \rightarrow R_3 \end{array}$$
$$= 1(-1)^{1+1}\begin{vmatrix} -18 & -11 \\ -14 & -7 \end{vmatrix} + 0 + 0$$
$$= (-1)^2[(-18)(-7) - (-14)(-11)] = -28$$

37. We will generate zeros in the second row by column operations.
$$\begin{vmatrix} 5 & 2 & -3 \\ -2 & 4 & 4 \\ 1 & -1 & 3 \end{vmatrix} = \begin{vmatrix} 5 & 12 & 7 \\ -2 & 0 & 0 \\ 1 & 1 & 5 \end{vmatrix} \quad \begin{array}{l} 2C_1 + C_2 \rightarrow C_2 \\ 2C_1 + C_3 \rightarrow C_3 \end{array}$$
$$= (-2)(-1)^{2+1}\begin{vmatrix} 12 & 7 \\ 1 & 5 \end{vmatrix} + 0 + 0$$
$$= (-2)(-1)^3[12 \cdot 5 - 1 \cdot 7] = 106$$

39. The column operation $(-3)C_3 + C_1 \rightarrow C_1$ transforms this determinant into

$$\begin{vmatrix} 0 & -4 & 1 \\ 0 & -1 & 2 \\ 0 & 2 & 3 \end{vmatrix}$$

By Theorem 2, the value of this determinant is 0.

41. We start by generating one more zero in the first row.

$$\begin{vmatrix} 0 & 1 & 0 & 1 \\ 1 & -2 & 4 & 3 \\ 2 & 1 & 5 & 4 \\ 1 & 2 & 1 & 2 \end{vmatrix} = \begin{vmatrix} 0 & 0 & 0 & 1 \\ 1 & -5 & 4 & 3 \\ 2 & -3 & 5 & 4 \\ 1 & 0 & 1 & 2 \end{vmatrix} \qquad (-1)C_4 + C_2 \rightarrow C_2$$

$$= 1(-1)^{1+4} \begin{vmatrix} 1 & -5 & 4 \\ 2 & -3 & 5 \\ 1 & 0 & 1 \end{vmatrix} + 0 + 0 + 0$$

$$= (-1) \begin{vmatrix} 1 & -5 & 4 \\ 2 & -3 & 5 \\ 1 & 0 & 1 \end{vmatrix}$$

$$= \begin{vmatrix} 1 & 5 & 4 \\ 2 \cdot & 3 & 5 \\ 1 & 0 & 1 \end{vmatrix} \quad \text{by Theorem 1}$$

We now generate one more zero in the third row.

$$\begin{vmatrix} 1 & 5 & 4 \\ 2 & 3 & 5 \\ 1 & 0 & 1 \end{vmatrix} = \begin{vmatrix} 1 & 5 & 3 \\ 2 & 3 & 3 \\ 1 & 0 & 0 \end{vmatrix} \qquad (-1)C_1 + C_3 \rightarrow C_3$$

$$= 1(-1)^{3+1} \begin{vmatrix} 5 & 3 \\ 3 & 3 \end{vmatrix} = (-1)^4[5 \cdot 3 - 3 \cdot 3] = 6$$

43. We start by generating zeros in the third row.

$$\begin{vmatrix} 3 & 2 & 3 & 1 \\ 3 & -2 & 8 & 5 \\ 2 & 1 & 3 & 1 \\ 4 & 5 & 4 & -3 \end{vmatrix} = \begin{vmatrix} 1 & 1 & 0 & 1 \\ -7 & -7 & -7 & 5 \\ 0 & 0 & 0 & 1 \\ 10 & 8 & 13 & -3 \end{vmatrix} \qquad \begin{array}{l} (-2)C_4 + C_1 \rightarrow C_1 \\ (-1)C_4 + C_2 \rightarrow C_2 \\ (-3)C_4 + C_3 \rightarrow C_3 \end{array}$$

$$= 1(-1)^{3+4} \begin{vmatrix} 1 & 1 & 0 \\ -7 & -7 & -7 \\ 10 & 8 & 13 \end{vmatrix} + 0 + 0 + 0$$

$$= (-1) \begin{vmatrix} 1 & 1 & 0 \\ -7 & -7 & -7 \\ 10 & 8 & 13 \end{vmatrix}$$

$$= \begin{vmatrix} 1 & 1 & 0 \\ 7 & 7 & 7 \\ 10 & 8 & 13 \end{vmatrix} \quad \text{by Theorem 1}$$

We now generate one more zero in the first row.

$$\begin{vmatrix} 1 & 1 & 0 \\ 7 & 7 & 7 \\ 10 & 8 & 13 \end{vmatrix} = \begin{vmatrix} 1 & 0 & 0 \\ 7 & 0 & 7 \\ 10 & -2 & 13 \end{vmatrix} \qquad (-1)C_1 + C_2 \rightarrow C_2$$

$$= (-2)(-1)^{3+2} \begin{vmatrix} 1 & 0 \\ 7 & 7 \end{vmatrix} + 0 + 0$$

$$= (-2)(-1)^5[1 \cdot 7 - 0 \cdot 7] = 14$$

45. Expand, for example, by the first column.

$$\begin{vmatrix} a & b & a \\ d & e & d \\ g & h & g \end{vmatrix} = a(-1)^{1+1}\begin{vmatrix} e & d \\ h & g \end{vmatrix} + d(-1)^{2+1}\begin{vmatrix} b & a \\ h & g \end{vmatrix} + g(-1)^{3+1}\begin{vmatrix} b & a \\ e & d \end{vmatrix}$$

$$= a\begin{vmatrix} e & d \\ h & g \end{vmatrix} - d\begin{vmatrix} b & a \\ h & g \end{vmatrix} + g\begin{vmatrix} b & a \\ e & d \end{vmatrix}$$

$$= a(eg - hd) - d(bg - ha) + g(bd - ea)$$

$$= aeg - adh - bdg + adh + bdg - aeg$$

$$= 0$$

47. We expand the left side by the first column, the right side by the second column.

$$\begin{vmatrix} a_1 & b_1 & c_1 \\ a_2 & b_2 & c_2 \\ a_3 & b_3 & c_3 \end{vmatrix} = a_1(-1)^{1+1}\begin{vmatrix} b_2 & c_2 \\ b_3 & c_3 \end{vmatrix} + a_2(-1)^{2+1}\begin{vmatrix} b_1 & c_1 \\ b_3 & c_3 \end{vmatrix} + a_3(-1)^{3+1}\begin{vmatrix} b_1 & c_1 \\ b_2 & c_2 \end{vmatrix}$$

$$= a_1\begin{vmatrix} b_2 & c_2 \\ b_3 & c_3 \end{vmatrix} - a_2\begin{vmatrix} b_1 & c_1 \\ b_3 & c_3 \end{vmatrix} + a_3\begin{vmatrix} b_1 & c_1 \\ b_2 & c_2 \end{vmatrix}$$

$$-\begin{vmatrix} b_1 & a_1 & c_1 \\ b_2 & a_2 & c_2 \\ b_3 & a_3 & c_3 \end{vmatrix} = -\left[a_1(-1)^{1+2}\begin{vmatrix} b_2 & c_2 \\ b_3 & c_3 \end{vmatrix} + a_2(-1)^{2+2}\begin{vmatrix} b_1 & c_1 \\ b_3 & c_3 \end{vmatrix} + a_3(-1)^{3+2}\begin{vmatrix} b_1 & c_1 \\ b_2 & c_2 \end{vmatrix} \right]$$

$$= -\left[-a_1\begin{vmatrix} b_2 & c_2 \\ b_3 & c_3 \end{vmatrix} + a_2\begin{vmatrix} b_1 & c_1 \\ b_3 & c_3 \end{vmatrix} - a_3\begin{vmatrix} b_1 & c_1 \\ b_2 & c_2 \end{vmatrix} \right]$$

$$= a_1\begin{vmatrix} b_2 & c_2 \\ b_3 & c_3 \end{vmatrix} - a_2\begin{vmatrix} b_1 & c_1 \\ b_3 & c_3 \end{vmatrix} + a_3\begin{vmatrix} b_1 & c_1 \\ b_2 & c_2 \end{vmatrix}$$

Hence the two original expressions are equal.

49. The statements: $(2, 5)$ satisfies the equation $\begin{vmatrix} x & y & 1 \\ 2 & 5 & 1 \\ -3 & 4 & 1 \end{vmatrix} = 0$

and $(-3, 4)$ satisfies the equation $\begin{vmatrix} x & y & 1 \\ 2 & 5 & 1 \\ -3 & 4 & 1 \end{vmatrix} = 0$

are equivalent to the statements $\begin{vmatrix} 2 & 5 & 1 \\ 2 & 5 & 1 \\ -3 & 4 & 1 \end{vmatrix} = 0$ and $\begin{vmatrix} -3 & 4 & 1 \\ 2 & 5 & 1 \\ -3 & 4 & 1 \end{vmatrix} = 0$.

The latter statements are true by Theorem 4.

51. The statement $\begin{vmatrix} x & y & 1 \\ x_1 & y_1 & 1 \\ x_2 & y_2 & 1 \end{vmatrix} = 0$ is the equation of a line because, expanding by the

first row, we have $x(-1)^{1+1}\begin{vmatrix} y_1 & 1 \\ y_2 & 1 \end{vmatrix} + y(-1)^{1+2}\begin{vmatrix} x_1 & 1 \\ x_2 & 1 \end{vmatrix} + 1(-1)^{1+3}\begin{vmatrix} x_1 & y_1 \\ x_2 & y_2 \end{vmatrix} = 0$

This is in the standard form for the equation of a line
$Ax + By + C = 0$

To show that the line passes through (x_1, y_1) and (x_2, y_2), we note merely that (x_1, y_1) and (x_2, y_2) satisfy the equation, because

$$\begin{vmatrix} x_1 & y_1 & 1 \\ x_1 & y_1 & 1 \\ x_2 & y_2 & 1 \end{vmatrix} = 0 \text{ and } \begin{vmatrix} x_2 & y_2 & 1 \\ x_1 & y_1 & 1 \\ x_2 & y_2 & 1 \end{vmatrix} = 0 \text{ are true by Theorem 4.}$$

53. Using the result stated in problem 52, we have

$$\begin{vmatrix} x_1 & y_1 & 1 \\ x_2 & y_2 & 1 \\ x_3 & y_3 & 1 \end{vmatrix} = 2 \times \text{(area of triangle formed by the three points)}.$$

If the determinant is 0, then the area of the triangle formed by the three points is zero. The only way this can happen is if the three points are on the same line; that is, the points are collinear.

Exercise 9-6

Key Ideas and Formulas

Cramer's Rule for Two Equations and Two Variables

Given the system

$$a_{11}x + a_{12}y = k_1$$
$$a_{21}x + a_{22}y = k_2$$

with

$$D = \begin{vmatrix} a_{11} & a_{12} \\ a_{21} & a_{22} \end{vmatrix} \neq 0$$

then

$$x = \frac{\begin{vmatrix} k_1 & a_{12} \\ k_2 & a_{22} \end{vmatrix}}{D} \text{ and } y = \frac{\begin{vmatrix} a_{11} & k_1 \\ a_{21} & k_2 \end{vmatrix}}{D}$$

Cramer's Rule for Three Equations and Three Variables

Given the system

$$a_{11}x + a_{12}y + a_{13}z = k_1$$
$$a_{21}x + a_{22}y + a_{23}z = k_2$$
$$a_{31}x + a_{32}y + a_{33}z = k_3$$

with

$$D = \begin{vmatrix} a_{11} & a_{12} & a_{13} \\ a_{21} & a_{22} & a_{23} \\ a_{31} & a_{32} & a_{33} \end{vmatrix} \neq 0$$

then

$$x = \frac{\begin{vmatrix} k_1 & a_{12} & a_{13} \\ k_2 & a_{22} & a_{23} \\ k_3 & a_{32} & a_{33} \end{vmatrix}}{D} \quad y = \frac{\begin{vmatrix} a_{11} & k_1 & a_{13} \\ a_{21} & k_2 & a_{23} \\ a_{31} & k_3 & a_{33} \end{vmatrix}}{D} \quad z = \frac{\begin{vmatrix} a_{11} & a_{12} & k_1 \\ a_{21} & a_{22} & k_2 \\ a_{31} & a_{32} & k_3 \end{vmatrix}}{D}$$

The determinant D is called the coefficient determinant. If $D \neq 0$, the system has exactly one solution, given by Cramer's rule. If $D = 0$, the system is either inconsistent or dependent. Cramer's rule does not apply. We use methods of Chapter 8 to find solutions (if any) of the system.

1. $D = \begin{vmatrix} 1 & 2 \\ 1 & 3 \end{vmatrix} = 1$ $x = \dfrac{\begin{vmatrix} 1 & 2 \\ -1 & 3 \end{vmatrix}}{D} = \dfrac{5}{1} = 5$ $y = \dfrac{\begin{vmatrix} 1 & 1 \\ 1 & -1 \end{vmatrix}}{D} = \dfrac{-2}{1} = -2$ $x = 5, \; y = -2$

3. $D = \begin{vmatrix} 2 & 1 \\ 5 & 3 \end{vmatrix} = 1$ $x = \dfrac{\begin{vmatrix} 1 & 1 \\ 2 & 3 \end{vmatrix}}{D} = \dfrac{1}{1} = 1$ $y = \dfrac{\begin{vmatrix} 2 & 1 \\ 5 & 2 \end{vmatrix}}{D} = \dfrac{-1}{1} = -1$ $x = 1, \; y = -1$

5. $D = \begin{vmatrix} 2 & -1 \\ -1 & 3 \end{vmatrix} = 5$ $x = \dfrac{\begin{vmatrix} -3 & -1 \\ 3 & 3 \end{vmatrix}}{D} = \dfrac{-6}{5} = -\dfrac{6}{5}$ $y = \dfrac{\begin{vmatrix} 2 & -3 \\ -1 & 3 \end{vmatrix}}{D} = \dfrac{3}{5}$

7. $D = \begin{vmatrix} 4 & -3 \\ 3 & 2 \end{vmatrix} = 17$ $x = \dfrac{\begin{vmatrix} 4 & -3 \\ -2 & 2 \end{vmatrix}}{D} = \dfrac{2}{17}$ $y = \dfrac{\begin{vmatrix} 4 & 4 \\ 3 & -2 \end{vmatrix}}{D} = \dfrac{-20}{17} = -\dfrac{20}{17}$

9. $D = \begin{vmatrix} 0.9925 & -0.9659 \\ 0.1219 & 0.2588 \end{vmatrix} = 0.37460$

$x = \dfrac{\begin{vmatrix} 0 & -0.9659 \\ 2,500 & 0.2588 \end{vmatrix}}{D} = \dfrac{2,414.75}{0.37460} = 6,400$ to two significant digits

$y = \dfrac{\begin{vmatrix} 0.9925 & 0 \\ 0.1219 & 2,500 \end{vmatrix}}{D} = \dfrac{2,481.25}{0.37460} = 6,600$ to two significant digits

11. $D = \begin{vmatrix} 0.9954 & -0.9942 \\ 0.0958 & 0.1080 \end{vmatrix} = 0.20275$

$x = \dfrac{\begin{vmatrix} 0 & -0.9942 \\ 155 & 0.1080 \end{vmatrix}}{D} = \dfrac{154.10}{0.20275} = 760$ to two significant digits

$y = \dfrac{\begin{vmatrix} 0.9954 & 0 \\ 0.0958 & 155 \end{vmatrix}}{D} = \dfrac{154.29}{0.20275} = 760$ to two significant digits

13. $D = \begin{vmatrix} 1 & 1 & 0 \\ 0 & 2 & 1 \\ -1 & 0 & 1 \end{vmatrix} = 1$ $x = \dfrac{\begin{vmatrix} 0 & 1 & 0 \\ -5 & 2 & 1 \\ -3 & 0 & 1 \end{vmatrix}}{D} = \dfrac{2}{1} = 2$ $y = \dfrac{\begin{vmatrix} 1 & 0 & 0 \\ 0 & -5 & 1 \\ -1 & -3 & 1 \end{vmatrix}}{D} = \dfrac{-2}{1} = -2$

$z = \dfrac{\begin{vmatrix} 1 & 1 & 0 \\ 0 & 2 & -5 \\ -1 & 0 & -3 \end{vmatrix}}{D} = \dfrac{-1}{1} = -1$

15. $D = \begin{vmatrix} 1 & 1 & 0 \\ 0 & 2 & 1 \\ 0 & -1 & 1 \end{vmatrix} = 3$ $x = \dfrac{\begin{vmatrix} 1 & 1 & 0 \\ 0 & 2 & 1 \\ 1 & -1 & 1 \end{vmatrix}}{D} = \dfrac{4}{3}$ $y = \dfrac{\begin{vmatrix} 1 & 1 & 0 \\ 0 & 0 & 1 \\ 0 & 1 & 1 \end{vmatrix}}{D} = \dfrac{-1}{3} = -\dfrac{1}{3}$

$$z = \dfrac{\begin{vmatrix} 1 & 1 & 1 \\ 0 & 2 & 0 \\ 0 & -1 & 1 \end{vmatrix}}{D} = \dfrac{2}{3}$$

17. $D = \begin{vmatrix} 0 & 3 & 1 \\ 1 & 0 & 2 \\ 1 & -3 & 0 \end{vmatrix} = 3$ $x = \dfrac{\begin{vmatrix} -1 & 3 & 1 \\ 3 & 0 & 2 \\ -2 & -3 & 0 \end{vmatrix}}{D} = \dfrac{-27}{3} = -9$ $y = \dfrac{\begin{vmatrix} 0 & -1 & 1 \\ 1 & 3 & 2 \\ 1 & -2 & 0 \end{vmatrix}}{D} = \dfrac{-7}{3} = -\dfrac{7}{3}$

$$z = \dfrac{\begin{vmatrix} 0 & 3 & -1 \\ 1 & 0 & 3 \\ 1 & -3 & -2 \end{vmatrix}}{D} = \dfrac{18}{3} = 6$$

19. $D = \begin{vmatrix} 0 & 2 & -1 \\ 1 & -1 & -1 \\ 1 & -1 & 2 \end{vmatrix} = -6$ $x = \dfrac{\begin{vmatrix} -3 & 2 & -1 \\ 2 & -1 & -1 \\ 4 & -1 & 2 \end{vmatrix}}{D} = \dfrac{-9}{-6} = \dfrac{3}{2}$ $y = \dfrac{\begin{vmatrix} 0 & -3 & -1 \\ 1 & 2 & -1 \\ 1 & 4 & 2 \end{vmatrix}}{D} = \dfrac{7}{-6} = -\dfrac{7}{6}$

$$z = \dfrac{\begin{vmatrix} 0 & 2 & -3 \\ 1 & -1 & 2 \\ 1 & -1 & 4 \end{vmatrix}}{-6} = \dfrac{-4}{-6} = \dfrac{2}{3}$$

21. $x = \dfrac{\begin{vmatrix} -3 & -3 & 1 \\ -11 & 3 & 2 \\ 3 & -1 & -1 \end{vmatrix}}{\begin{vmatrix} 2 & -3 & 1 \\ -4 & 3 & 2 \\ 1 & -1 & -1 \end{vmatrix}} = \dfrac{20}{5} = 4$

23. $y = \dfrac{\begin{vmatrix} 12 & 5 & 11 \\ 15 & -13 & -9 \\ 5 & 0 & 2 \end{vmatrix}}{\begin{vmatrix} 12 & -14 & 11 \\ 15 & 7 & -9 \\ 5 & -3 & 2 \end{vmatrix}} = \dfrac{28}{14} = 2$

25. $z = \dfrac{\begin{vmatrix} 3 & -4 & 18 \\ -9 & 8 & -13 \\ 5 & -7 & 33 \end{vmatrix}}{\begin{vmatrix} 3 & -4 & 5 \\ -9 & 8 & 7 \\ 5 & -7 & 10 \end{vmatrix}} = \dfrac{5}{2}$

27. $D = \begin{vmatrix} 1 & -4 & 9 \\ 4 & -1 & 6 \\ 1 & -1 & 3 \end{vmatrix} = \begin{vmatrix} -2 & -1 & 0 \\ 2 & 1 & 0 \\ 1 & -1 & 3 \end{vmatrix} \begin{matrix} (-3)R_3 + R_1 \to R_1 \\ (-2)R_3 + R_2 \to R_2 \end{matrix} = -\begin{vmatrix} 2 & 1 & 0 \\ 2 & 1 & 0 \\ 1 & -1 & 3 \end{vmatrix}$ by Theorem 1

$= 0$ by Theorem 4

Since $D = 0$, the system either has no solution or infinitely many. Since $x = 0$, $y = 0$, $z = 0$ is a solution, the second case must hold.

29. We start with
$$a_{11}x + a_{12}y = k_1$$
$$a_{21}x + a_{22}y = k_2$$
We wish to eliminate x. We multiply the top equation by $-a_{21}$ and the bottom equation by a_{11}, then add.

$$-a_{11}a_{21}x - a_{21}a_{12}y = -k_1 a_{21}$$
$$\underline{a_{11}a_{21}x + a_{11}a_{22}y = k_2 a_{11}}$$
$$0x + a_{11}a_{22}y - a_{21}a_{12}y = k_2 a_{11} - k_1 a_{21}$$
$$(a_{11}a_{22} - a_{21}a_{12})y = a_{11}k_2 - a_{21}k_1$$

$$y = \frac{a_{11}k_2 - a_{21}k_1}{a_{11}a_{22} - a_{21}a_{12}} = \frac{\begin{vmatrix} a_{11} & k_1 \\ a_{21} & k_2 \end{vmatrix}}{\begin{vmatrix} a_{11} & a_{12} \\ a_{21} & a_{22} \end{vmatrix}}$$

31. (A) $R = xp + yq = (200 - 6p + 4q)p + (300 + 2p - 3q)q$
$$= 200p - 6p^2 + 4pq + 300q + 2pq - 3q^2 = 200p + 300q - 6p^2 + 6pq - 3q^2$$

(B) Rewrite the demand equations as
$$6p - 4q = 200 - x$$
$$-2p + 3q = 300 - y$$

Apply Cramer's rule: $D = \begin{vmatrix} 6 & -4 \\ -2 & 3 \end{vmatrix} = 10$

$$p = \frac{\begin{vmatrix} 200 - x & -4 \\ 300 - y & 3 \end{vmatrix}}{D} = \frac{1800 - 3x - 4y}{10} = -0.3x - 0.4y + 180$$

$$q = \frac{\begin{vmatrix} 6 & 200 - x \\ -2 & 300 - y \end{vmatrix}}{D} = \frac{2200 - 2x - 6y}{10} = -0.2x - 0.6y + 220$$

Then
$$R = xp + yq = x(-0.3x - 0.4y + 180) + y(-0.2x - 0.6y + 220)$$
$$= -0.3x^2 - 0.4xy + 180x - 0.2xy - 0.6y^2 + 220y$$
$$= 180x + 220y - 0.3x^2 - 0.6xy - 0.6y^2$$

CHAPTER 9 REVIEW

1. $AB = \begin{bmatrix} 4 & -2 \\ 0 & 3 \end{bmatrix}\begin{bmatrix} -1 & 5 \\ -4 & 6 \end{bmatrix} = \begin{bmatrix} 4(-1) + (-2)(-4) & 4 \cdot 5 + (-2)6 \\ 0(-1) + 3(-4) & 0 \cdot 5 + 3 \cdot 6 \end{bmatrix} = \begin{bmatrix} 4 & 8 \\ -12 & 18 \end{bmatrix}$ (9-1)

2. $CD = \begin{bmatrix} -1 & 4 \end{bmatrix}\begin{bmatrix} 3 \\ -2 \end{bmatrix} = [(-1)3 + 4(-2)] = [-11]$ (9-1)

3. $CB = \begin{bmatrix} -1 & 4 \end{bmatrix}\begin{bmatrix} -1 & 5 \\ -4 & 6 \end{bmatrix} = [(-1)(-1) + 4(-4) \quad (-1)5 + 4 \cdot 6] = [-15 \quad 19]$ (9-1)

4. $AD = \begin{bmatrix} 4 & -2 \\ 0 & 3 \end{bmatrix}\begin{bmatrix} 3 \\ -2 \end{bmatrix} = \begin{bmatrix} 4 \cdot 3 + (-2)(-2) \\ 0 \cdot 3 + 3(-2) \end{bmatrix} = \begin{bmatrix} 16 \\ -6 \end{bmatrix}$ (9-1)

5. $A + B = \begin{bmatrix} 4 & -2 \\ 0 & 3 \end{bmatrix} + \begin{bmatrix} -1 & 5 \\ -4 & 6 \end{bmatrix} = \begin{bmatrix} 4 + (-1) & (-2) + 5 \\ 0 + (-4) & 3 + 6 \end{bmatrix} = \begin{bmatrix} 3 & 3 \\ -4 & 9 \end{bmatrix}$ (9-1)

6. $C + D$ is not defined (9-1)   **7.** $A + C$ is not defined (9-1)

8. $2A - 5B = 2\begin{bmatrix} 4 & -2 \\ 0 & 3 \end{bmatrix} - 5\begin{bmatrix} -1 & 5 \\ -4 & 6 \end{bmatrix} = \begin{bmatrix} 8 & -4 \\ 0 & 6 \end{bmatrix} - \begin{bmatrix} -5 & 25 \\ -20 & 30 \end{bmatrix} = \begin{bmatrix} 13 & -29 \\ 20 & -24 \end{bmatrix}$ (9-1)

9. $CA + C = [-1 \quad 4]\begin{bmatrix} 4 & -2 \\ 0 & 3 \end{bmatrix} + [-1 \quad 4] = [(-1)4 + 4 \cdot 0 \quad (-1)(-2) + 4 \cdot 3] + [-1 \quad 4]$

$= [-4 \quad 14] + [-1 \quad 4] = [-5 \quad 18]$ (9-1)

10. $\begin{bmatrix} 4 & 7 & | & 1 & 0 \\ -1 & -2 & | & 0 & 1 \end{bmatrix}$ $R_1 \leftrightarrow R_2$

$\sim \begin{bmatrix} -1 & -2 & | & 0 & 1 \\ 4 & 7 & | & 1 & 0 \end{bmatrix}$ $4R_1 + R_2 \to R_2$

$\sim \begin{bmatrix} -1 & -2 & | & 0 & 1 \\ 0 & -1 & | & 1 & 4 \end{bmatrix}$ $(-2)R_2 + R_1 \to R_1$

$\sim \begin{bmatrix} -1 & 0 & | & -2 & -7 \\ 0 & -1 & | & 1 & 4 \end{bmatrix}$ $\begin{matrix} (-1)R_1 \to R_1 \\ (-1)R_2 \to R_2 \end{matrix}$

$\sim \begin{bmatrix} 1 & 0 & | & 2 & 7 \\ 0 & 1 & | & -1 & -4 \end{bmatrix}$

Hence, $A^{-1} = \begin{bmatrix} 2 & 7 \\ -1 & -4 \end{bmatrix}$

$A^{-1}A = \begin{bmatrix} 2 & 7 \\ -1 & -4 \end{bmatrix}\begin{bmatrix} 4 & 7 \\ -1 & -2 \end{bmatrix} = \begin{bmatrix} 2 \cdot 4 + 7(-1) & 2 \cdot 7 + 7(-2) \\ (-1)4 + (-4)(-1) & (-1)7 + (-4)(-2) \end{bmatrix} = \begin{bmatrix} 1 & 0 \\ 0 & 1 \end{bmatrix} = I$

   (9-2)

11. As a matrix equation the system becomes

$\begin{matrix} A & X & B \end{matrix}$
$\begin{bmatrix} 3 & 2 \\ 4 & 3 \end{bmatrix}\begin{bmatrix} x_1 \\ x_2 \end{bmatrix} = \begin{bmatrix} k_1 \\ k_2 \end{bmatrix}$

The solution of $AX = B$ is $X = A^{-1}B$.

Applying matrix methods, we obtain $A^{-1} = \begin{bmatrix} 3 & -2 \\ -4 & 3 \end{bmatrix}$ (details omitted). Applying this inverse, we have

$X = \begin{bmatrix} 3 & -2 \\ -4 & 3 \end{bmatrix}\begin{bmatrix} k_1 \\ k_2 \end{bmatrix}$

(A) $\begin{bmatrix} x_1 \\ x_2 \end{bmatrix} = \begin{bmatrix} 3 & -2 \\ -4 & 3 \end{bmatrix}\begin{bmatrix} 3 \\ 5 \end{bmatrix} = \begin{bmatrix} 3 \cdot 3 + (-2)5 \\ (-4)3 + 3 \cdot 5 \end{bmatrix} = \begin{bmatrix} -1 \\ 3 \end{bmatrix}$ $x_1 = -1, \; x_2 = 3$

(B) $\begin{bmatrix} x_1 \\ x_2 \end{bmatrix} = \begin{bmatrix} 3 & -2 \\ -4 & 3 \end{bmatrix}\begin{bmatrix} 7 \\ 10 \end{bmatrix} = \begin{bmatrix} 3 \cdot 7 + (-2)10 \\ (-4)7 + 3 \cdot 10 \end{bmatrix} = \begin{bmatrix} 1 \\ 2 \end{bmatrix}$ $x_1 = 1, \; x_2 = 2$

(C) $\begin{bmatrix} x_1 \\ x_2 \end{bmatrix} = \begin{bmatrix} 3 & -2 \\ -4 & 3 \end{bmatrix}\begin{bmatrix} 4 \\ 2 \end{bmatrix} = \begin{bmatrix} 3 \cdot 4 + (-2)2 \\ (-4)4 + 3 \cdot 2 \end{bmatrix} = \begin{bmatrix} 8 \\ -10 \end{bmatrix}$ $x_1 = 8, \; x_2 = -10$ (9-3)

12. $\begin{vmatrix} 2 & -3 \\ -5 & -1 \end{vmatrix} = 2(-1) - (-5)(-3) = -17$ (9-4)

13. $\begin{vmatrix} 2 & 3 & -4 \\ 0 & 5 & 0 \\ 1 & -4 & -2 \end{vmatrix} = 0 + 5(-1)^{2+2}\begin{vmatrix} 2 & -4 \\ 1 & -2 \end{vmatrix} + 0 = 5(-1)^4[2(-2) - 1(-4)] = 0$ (9-4, 9-5)

14. $D = \begin{vmatrix} 3 & -2 \\ 1 & 3 \end{vmatrix} = 11$

$x = \dfrac{\begin{vmatrix} 8 & -2 \\ -1 & 3 \end{vmatrix}}{D} = \dfrac{22}{11} = 2 \qquad y = \dfrac{\begin{vmatrix} 3 & 8 \\ 1 & -1 \end{vmatrix}}{D} = \dfrac{-11}{11} = -1 \qquad (9\text{-}6)$

15. (A) Interchanging two rows of a determinant changes the sign of the determinant (Theorem 3, Section 9-5). Hence,

$\begin{vmatrix} g & h & i \\ d & e & f \\ a & b & c \end{vmatrix} = -2$

(B) If every element of a column of a determinant is multiplied by 3, (Theorem 5, Section 9-5), the value of the determinant is 3 times the original. Hence,

$\begin{vmatrix} a & 3b & c \\ d & 3e & f \\ g & 3h & i \end{vmatrix} = 3 \cdot 2 = 6$

(C) Adding a multiple of a column to another column does not change the value of the determinant (Theorem 5, Section 9-5). Hence,

$\begin{vmatrix} a & b & a+b+c \\ d & e & d+e+f \\ g & h & g+h+i \end{vmatrix} = \begin{vmatrix} a & b & b+c \\ d & e & e+f \\ g & h & h+i \end{vmatrix} = \begin{vmatrix} a & b & c \\ d & e & f \\ g & h & i \end{vmatrix} = 2 \qquad (9\text{-}5)$

16. $AD = \begin{bmatrix} 1 & 2 \\ 4 & 5 \\ -3 & -1 \end{bmatrix}\begin{bmatrix} 7 & 0 & -5 \\ 0 & 8 & -2 \end{bmatrix} = \begin{bmatrix} 1\cdot 7 + 2\cdot 0 & 1\cdot 0 + 2\cdot 8 & 1(-5) + 2(-2) \\ 4\cdot 7 + 5\cdot 0 & 4\cdot 0 + 5\cdot 8 & 4(-5) + 5(-2) \\ (-3)7 + (-1)0 & (-3)0 + (-1)8 & (-3)(-5) + (-1)(-2) \end{bmatrix}$

$= \begin{bmatrix} 7 & 16 & -9 \\ 28 & 40 & -30 \\ -21 & -8 & 17 \end{bmatrix} \qquad (9\text{-}1)$

17. $DA = \begin{bmatrix} 7 & 0 & -5 \\ 0 & 8 & -2 \end{bmatrix}\begin{bmatrix} 1 & 2 \\ 4 & 5 \\ -3 & -1 \end{bmatrix} = \begin{bmatrix} 7\cdot 1 + 0\cdot 4 + (-5)(-3) & 7\cdot 2 + 0\cdot 5 + (-5)(-1) \\ 0\cdot 1 + 8\cdot 4 + (-2)(-3) & 0\cdot 2 + 8\cdot 5 + (-2)(-1) \end{bmatrix}$

$= \begin{bmatrix} 22 & 19 \\ 38 & 42 \end{bmatrix} \qquad (9\text{-}1)$

18. $BC = \begin{bmatrix} 6 \\ 0 \\ -4 \end{bmatrix}[2 \quad 4 \quad -1] = \begin{bmatrix} 6\cdot 2 & 6\cdot 4 & 6(-1) \\ 0\cdot 2 & 0\cdot 4 & 0(-1) \\ (-4)2 & (-4)4 & (-4)(-1) \end{bmatrix} = \begin{bmatrix} 12 & 24 & -6 \\ 0 & 0 & 0 \\ -8 & -16 & 4 \end{bmatrix} \qquad (9\text{-}1)$

19. $CB = [2 \quad 4 \quad -1]\begin{bmatrix} 6 \\ 0 \\ -4 \end{bmatrix} = [2\cdot 6 + 4\cdot 0 + (-1)(-4)] = [16] \qquad (9\text{-}1)$

20. Since D has 3 columns and E has 2 rows, DE is not defined. $\qquad (9\text{-}1)$

21. $ED = \begin{bmatrix} 9 & -3 \\ -6 & 2 \end{bmatrix}\begin{bmatrix} 7 & 0 & -5 \\ 0 & 8 & -2 \end{bmatrix}$

$= \begin{bmatrix} 9\cdot 7 + (-3)0 & 9\cdot 0 + (-3)8 & 9(-5) + (-3)(-2) \\ (-6)7 + 2\cdot 0 & (-6)0 + 2\cdot 8 & (-6)(-5) + 2(-2) \end{bmatrix} = \begin{bmatrix} 63 & -24 & -39 \\ -42 & 16 & 26 \end{bmatrix} \qquad (9\text{-}1)$

22. $\begin{bmatrix} 1 & 0 & 4 & | & 1 & 0 & 0 \\ -2 & 1 & 0 & | & 0 & 1 & 0 \\ 4 & -1 & 4 & | & 0 & 0 & 1 \end{bmatrix}$ $\quad 2R_1 + R_1 \rightarrow R_2$
$\qquad\qquad\qquad\qquad\qquad\qquad (-4)R_1 + R_3 \rightarrow R_3$

$\sim \begin{bmatrix} 1 & 0 & 4 & | & 1 & 0 & 0 \\ 0 & 1 & 8 & | & 2 & 1 & 0 \\ 0 & -1 & -12 & | & -4 & 0 & 1 \end{bmatrix}$ $\quad R_2 + R_3 \rightarrow R_3$

$\sim \begin{bmatrix} 1 & 0 & 4 & | & 1 & 0 & 0 \\ 0 & 1 & 8 & | & 2 & 1 & 0 \\ 0 & 0 & -4 & | & -2 & 1 & 1 \end{bmatrix}$ $\quad \begin{matrix} R_3 + R_1 \rightarrow R_1 \\ 2R_3 + R_2 \rightarrow R_2 \end{matrix}$

$\sim \begin{bmatrix} 1 & 0 & 0 & | & -1 & 1 & 1 \\ 0 & 1 & 0 & | & -2 & 3 & 2 \\ 0 & 0 & -4 & | & -2 & 1 & 1 \end{bmatrix}$ $\quad \left(-\frac{1}{4}\right) R_3 \rightarrow R_3$

$\sim \begin{bmatrix} 1 & 0 & 0 & | & -1 & 1 & 1 \\ 0 & 1 & 0 & | & -2 & 3 & 2 \\ 0 & 0 & 1 & | & \frac{1}{2} & -\frac{1}{4} & -\frac{1}{4} \end{bmatrix}$

Hence

$A^{-1} = \begin{bmatrix} -1 & 1 & 1 \\ -2 & 3 & 2 \\ \frac{1}{2} & -\frac{1}{4} & -\frac{1}{4} \end{bmatrix}$

$A^{-1}A = \begin{bmatrix} -1 & 1 & 1 \\ -2 & 3 & 2 \\ \frac{1}{2} & -\frac{1}{4} & -\frac{1}{4} \end{bmatrix} \begin{bmatrix} 1 & 0 & 4 \\ -2 & 1 & 0 \\ 4 & -1 & 4 \end{bmatrix}$

$= \begin{bmatrix} (-1)1 + 1(-2) + 1\cdot 4 & (-1)0 + 1\cdot 1 + 1(-1) & (-1)4 + 1\cdot 0 + 1\cdot 4 \\ (-2)1 + 3(-2) + 2\cdot 4 & (-2)0 + 3\cdot 1 + 2(-1) & (-2)4 + 3\cdot 0 + 2\cdot 4 \\ (\frac{1}{2})1 + (-\frac{1}{4})(-2) + (-\frac{1}{4})4 & (\frac{1}{2})0 + (-\frac{1}{4})1 + (-\frac{1}{4})(-1) & (\frac{1}{2})4 + (-\frac{1}{4})0 + (-\frac{1}{4})4 \end{bmatrix}$

$= \begin{bmatrix} 1 & 0 & 0 \\ 0 & 1 & 0 \\ 0 & 0 & 1 \end{bmatrix}$ $\qquad\qquad\qquad\qquad\qquad\qquad\qquad\qquad (9\text{-}2)$

$\qquad\qquad A \qquad\qquad X \quad\;\; B$

23. $\begin{bmatrix} 1 & 2 & 3 \\ 2 & 3 & 4 \\ 1 & 2 & 1 \end{bmatrix} \begin{bmatrix} x_1 \\ x_2 \\ x_3 \end{bmatrix} = \begin{bmatrix} k_1 \\ k_2 \\ k_3 \end{bmatrix}$

The solution to $AX = B$ is $X = A^{-1}B$.

Applying matrix methods, we obtain $A^{-1} = \begin{bmatrix} -\frac{5}{2} & 2 & -\frac{1}{2} \\ 1 & -1 & 1 \\ \frac{1}{2} & 0 & -\frac{1}{2} \end{bmatrix}$ (details omitted).
Applying the inverse, we have

(A) $B = \begin{bmatrix} 1 \\ 3 \\ 3 \end{bmatrix}$ $X = \begin{bmatrix} x_1 \\ x_2 \\ x_3 \end{bmatrix} = \begin{bmatrix} -\frac{5}{2} & 2 & -\frac{1}{2} \\ 1 & -1 & 1 \\ \frac{1}{2} & 0 & -\frac{1}{2} \end{bmatrix} \begin{bmatrix} 1 \\ 3 \\ 3 \end{bmatrix} = \begin{bmatrix} 2 \\ 1 \\ -1 \end{bmatrix}$ $x_1 = 2, \; x_2 = 1, \; x_3 = -1$

(B) $B = \begin{bmatrix} 0 \\ 0 \\ -2 \end{bmatrix}$ $X = \begin{bmatrix} x_1 \\ x_2 \\ x_3 \end{bmatrix} = \begin{bmatrix} -\frac{5}{2} & 2 & -\frac{1}{2} \\ 1 & -1 & 1 \\ \frac{1}{2} & 0 & -\frac{1}{2} \end{bmatrix} \begin{bmatrix} 0 \\ 0 \\ -2 \end{bmatrix} = \begin{bmatrix} 1 \\ -2 \\ 1 \end{bmatrix}$ $x_1 = 1, \; x_2 = -2, \; x_3 = 1$

(C) $B = \begin{bmatrix} -3 \\ -4 \\ 1 \end{bmatrix}$ $X = \begin{bmatrix} x_1 \\ x_2 \\ x_3 \end{bmatrix} = \begin{bmatrix} -\frac{5}{2} & 2 & -\frac{1}{2} \\ 1 & -1 & 1 \\ \frac{1}{2} & 0 & -\frac{1}{2} \end{bmatrix} \begin{bmatrix} -3 \\ -4 \\ 1 \end{bmatrix} = \begin{bmatrix} -1 \\ 2 \\ -2 \end{bmatrix}$ $x_1 = -1, \; x_2 = 2, \; x_3 = -2$ $\qquad (9\text{-}3)$

24. $\begin{vmatrix} -\frac{1}{4} & \frac{3}{2} \\ \frac{1}{2} & \frac{2}{3} \end{vmatrix} = \left(-\frac{1}{4}\right)\left(\frac{2}{3}\right) - \left(\frac{1}{2}\right)\left(\frac{3}{2}\right) = -\frac{1}{6} - \frac{3}{4} = -\frac{11}{12}$ (9-4)

25. $\begin{vmatrix} 2 & -1 & 1 \\ -3 & 5 & 2 \\ 1 & -2 & 4 \end{vmatrix} = \begin{vmatrix} 0 & 0 & 1 \\ -7 & 7 & 2 \\ -7 & 2 & 4 \end{vmatrix}$ $\begin{matrix} (-2)C_3 + C_1 \rightarrow C_1 \\ C_3 + C_2 \rightarrow C_2 \end{matrix}$

$= 0 + 0 + 1(-1)^{1+3}\begin{vmatrix} -7 & 7 \\ -7 & 2 \end{vmatrix} = (-1)^4[(-7)2 - (-7)7] = 35$ (9-4, 9-5)

26. $y = \dfrac{\begin{vmatrix} 1 & -6 & 1 \\ 0 & 4 & -1 \\ 2 & 2 & 1 \end{vmatrix}}{\begin{vmatrix} 1 & -2 & 1 \\ 0 & 1 & -1 \\ 2 & 2 & 1 \end{vmatrix}} = \dfrac{\begin{vmatrix} 1 & -6 & 1 \\ 0 & 4 & -1 \\ 0 & 14 & -1 \end{vmatrix}}{\begin{vmatrix} 1 & -2 & -1 \\ 0 & 1 & 0 \\ 2 & 2 & 3 \end{vmatrix}} = \dfrac{1(-1)^{1+1}\begin{vmatrix} 4 & -1 \\ 14 & -1 \end{vmatrix}}{1(-1)^{2+2}\begin{vmatrix} 1 & -1 \\ 2 & 3 \end{vmatrix}} = \dfrac{(-1)^2[4(-1) - 14(-1)]}{(-1)^4[1\cdot 3 - 2(-1)]}$

$= \dfrac{10}{5} = 2$ (9-6)

27. (A) If the coefficient matrix has an inverse, then the system can be written as $AX = B$ and its solution can be written $X = A^{-1}B$. Thus the system has one solution.

(B) If the coefficient matrix does not have an inverse, then the system can be solved by Gauss-Jordan elimination, but it will not have exactly one solution. The other possibilities are that the system has no solution or an infinite number of solutions, and either possibility may occur. (9-3)

28. If we assume that A is a non-zero matrix with an inverse A^{-1}, then if $A^2 = 0$ we can write $A^{-1}A^2 = A^{-1}0$ or $A^{-1}AA = 0$ or $IA = 0$ or $A = 0$. But A was assumed non-zero, so there is a contradiction. Hence A^{-1} cannot exist for such a matrix. (9-3)

29.

$AX - B = CX$

$AX - B + B - CX = CX - CX + B$ Addition property

$AX + 0 - CX = 0 + B$ $M + (-M) = 0$

$AX - CX = B$ $M + 0 = M$

$(A - C)X = B$ Right distributive property
(applied to subtraction of matrices; see problem 21, Exercise 9-3)

$(A - C)^{-1}[(A - C)X] = (A - C)^{-1}B$ Left multiplication property

$[(A - C)^{-1}(A - C)]X = (A - C)^{-1}B$ Associative property

$IX = (A - C)^{-1}B$ $A^{-1}A = I$

$X = (A - C)^{-1}B$ $IX = X$

Common Errors: $(A - C)^{-1}B \neq B(A - C)^{-1}$
$(A - C)^{-1} \neq A^{-1} - C^{-1}$

(9-3)

30. $\begin{bmatrix} 4 & 5 & 6 & | & 1 & 0 & 0 \\ 4 & 5 & -6 & | & 0 & 1 & 0 \\ 1 & 1 & 1 & | & 0 & 0 & 1 \end{bmatrix}$ $(-1)R_1 + R_2 \rightarrow R_2$

$\sim \begin{bmatrix} 4 & 5 & 6 & | & 1 & 0 & 0 \\ 0 & 0 & -12 & | & -1 & 1 & 0 \\ 1 & 1 & 1 & | & 0 & 0 & 1 \end{bmatrix}$ $R_1 \leftrightarrow R_3$

$\sim \begin{bmatrix} 1 & 1 & 1 & | & 0 & 0 & 1 \\ 0 & 0 & -12 & | & -1 & 1 & 0 \\ 4 & 5 & 6 & | & 1 & 0 & 0 \end{bmatrix}$ $(-4)R_1 + R_3 \rightarrow R_3$

$\sim \begin{bmatrix} 1 & 1 & 1 & | & 0 & 0 & 1 \\ 0 & 0 & -12 & | & -1 & 1 & 0 \\ 0 & 1 & 2 & | & 1 & 0 & -4 \end{bmatrix}$ $R_2 \leftrightarrow R_3$

$\sim \begin{bmatrix} 1 & 1 & 1 & | & 0 & 0 & 1 \\ 0 & 1 & 2 & | & 1 & 0 & -4 \\ 0 & 0 & -12 & | & -1 & 1 & 0 \end{bmatrix}$ $(-1)R_2 + R_1 \rightarrow R_1$

$\sim \begin{bmatrix} 1 & 0 & -1 & | & -1 & 0 & 5 \\ 0 & 1 & 2 & | & 1 & 0 & -4 \\ 0 & 0 & -12 & | & -1 & 1 & 0 \end{bmatrix}$ $-\frac{1}{12}R_3 \rightarrow R_3$

$\sim \begin{bmatrix} 1 & 0 & -1 & | & -1 & 0 & 5 \\ 0 & 1 & 2 & | & 1 & 0 & -4 \\ 0 & 0 & 1 & | & \frac{1}{12} & -\frac{1}{12} & 0 \end{bmatrix}$ $\begin{array}{l} R_3 + R_1 \rightarrow R_1 \\ (-2)R_3 + R_2 \rightarrow R_2 \end{array}$

$\sim \begin{bmatrix} 1 & 0 & 0 & | & -\frac{11}{12} & -\frac{1}{12} & 5 \\ 0 & 1 & 0 & | & \frac{10}{12} & \frac{2}{12} & -4 \\ 0 & 0 & 1 & | & \frac{1}{12} & -\frac{1}{12} & 0 \end{bmatrix}$

Hence

$A^{-1} = \begin{bmatrix} -\frac{11}{12} & -\frac{1}{12} & 5 \\ \frac{10}{12} & \frac{2}{12} & -4 \\ \frac{1}{12} & -\frac{1}{12} & 0 \end{bmatrix}$ or $\frac{1}{12}\begin{bmatrix} -11 & -1 & 60 \\ 10 & 2 & -48 \\ 1 & -1 & 0 \end{bmatrix}$

$A^{-1}A = \frac{1}{12}\begin{bmatrix} -11 & -1 & 60 \\ 10 & 2 & -48 \\ 1 & -1 & 0 \end{bmatrix}\begin{bmatrix} 4 & 5 & 6 \\ 4 & 5 & -6 \\ 1 & 1 & 1 \end{bmatrix}$

$= \frac{1}{12}\begin{bmatrix} (-11)4 + (-1)4 + 60\cdot1 & (-11)5 + (-1)5 + 60\cdot1 & (-11)6 + (-1)(-6) + 60\cdot1 \\ 10\cdot4 + 2\cdot4 + (-48)1 & 10\cdot5 + 2\cdot5 + (-48)1 & 10\cdot6 + 2(-6) + (-48)1 \\ 1\cdot4 + (-1)4 + 0\cdot1 & 1\cdot5 + (-1)5 + 0\cdot1 & 1\cdot6 + (-1)(-6) + 0\cdot1 \end{bmatrix}$

$= \frac{1}{12}\begin{bmatrix} 12 & 0 & 0 \\ 0 & 12 & 0 \\ 0 & 0 & 12 \end{bmatrix} = \begin{bmatrix} 1 & 0 & 0 \\ 0 & 1 & 0 \\ 0 & 0 & 1 \end{bmatrix} = I$ $(9\text{-}2)$

31. Multiplying the first two equations by 100, the system becomes

$4x_1 + 5x_2 + 6x_3 = 36,000$

$4x_1 + 5x_2 - 6x_3 = 12,000$

$x_1 + x_2 + x_3 = 7,000$

As a matrix equation, we have

$$\overset{A}{\begin{bmatrix} 4 & 5 & 6 \\ 4 & 5 & -6 \\ 1 & 1 & 1 \end{bmatrix}} \overset{X}{\begin{bmatrix} x_1 \\ x_2 \\ x_3 \end{bmatrix}} = \overset{B}{\begin{bmatrix} 36,000 \\ 12,000 \\ 7,000 \end{bmatrix}}$$

The solution to $AX = B$ is $X = A^{-1}B$. Using A^{-1} from problem 30, we have

$$X = \begin{bmatrix} x_1 \\ x_2 \\ x_3 \end{bmatrix} = \frac{1}{12} \begin{bmatrix} -11 & -1 & 60 \\ 10 & 2 & -48 \\ 1 & -1 & 0 \end{bmatrix} \begin{bmatrix} 36,000 \\ 12,000 \\ 7,000 \end{bmatrix}$$

$$= \frac{1}{12} \begin{bmatrix} (-11)(36,000) + (-1)(12,000) + (60)(7,000) \\ (10)(36,000) + (2)(12,000) + (-48)(7,000) \\ 1(36,000) + (-1)(12,000) + (0)(7,000) \end{bmatrix} = \frac{1}{12} \begin{bmatrix} 12,000 \\ 48,000 \\ 24,000 \end{bmatrix} = \begin{bmatrix} 1,000 \\ 4,000 \\ 2,000 \end{bmatrix}$$

Hence, $x_1 = 1,000$, $x_2 = 4,000$, $x_3 = 2,000$ $\hspace{2cm}$ (9-3)

32.
$$\begin{vmatrix} -1 & 4 & 1 & 1 \\ 5 & -1 & 2 & -1 \\ 2 & -1 & 0 & 3 \\ -3 & 3 & 0 & 3 \end{vmatrix} = \begin{vmatrix} -1 & 4 & 1 & 1 \\ 7 & -9 & 0 & -3 \\ 2 & -1 & 0 & 3 \\ -3 & 3 & 0 & 3 \end{vmatrix} \quad (-2)R_1 + R_2 \rightarrow R_2$$

$$= 1(-1)^{1+3} \begin{vmatrix} 7 & -9 & -3 \\ 2 & -1 & 3 \\ -3 & 3 & 3 \end{vmatrix} + 0 + 0 + 0$$

$$= \begin{vmatrix} 7 & -9 & -3 \\ 2 & -1 & 3 \\ -3 & 3 & 3 \end{vmatrix} \quad \begin{array}{l} R_1 + R_2 \rightarrow R_2 \\ R_1 + R_3 \rightarrow R_3 \end{array}$$

$$= \begin{vmatrix} 7 & -9 & -3 \\ 9 & -10 & 0 \\ 4 & -6 & 0 \end{vmatrix} = (-3)(-1)^{1+3} \begin{vmatrix} 9 & -10 \\ 4 & -6 \end{vmatrix} + 0 + 0$$

$$= (-3)(-1)^4[9(-6) - 4(-10)]$$

$$= (-3)(-14) = 42 \hspace{2cm} (9-5)$$

33.
$$\begin{vmatrix} u + kv & v \\ w + kx & x \end{vmatrix} = (u + kv)x - (w + kx)v = ux + kvx - wv - kvx = ux - wv$$

$$= \begin{vmatrix} u & v \\ w & x \end{vmatrix} \hspace{2cm} (9-5)$$

34. The statements: (1, 2) satisfies the equation $\begin{vmatrix} x & y & 1 \\ 1 & 2 & 1 \\ -1 & 5 & 1 \end{vmatrix} = 0$ and (-1, 5)

satisfies the equation $\begin{vmatrix} x & y & 1 \\ 1 & 2 & 1 \\ -1 & 5 & 1 \end{vmatrix} = 0$ are equivalent to the statements

$\begin{vmatrix} 1 & 2 & 1 \\ 1 & 2 & 1 \\ -1 & 5 & 1 \end{vmatrix} = 0$ and $\begin{vmatrix} -1 & 5 & 1 \\ 1 & 2 & 1 \\ -1 & 5 & 1 \end{vmatrix} = 0$. The latter statements are true by

Theorem 4, Section 9-5. All other points on the line through the given points will also satisfy the equation. $\hspace{2cm}$ (9-5)

35. Let x_1 = number of tons at Big Bend

x_2 = number of tons at Saw Pit

Then

$0.05x_1 + 0.03x_2$ = number of tons of nickel at both mines = k_1

$0.07x_1 + 0.04x_2$ = number of tons of copper at both mines = k_2

We solve

$0.05x_1 + 0.03x_2 = k_1$

$0.07x_1 + 0.04x_4 = k_2$,

For arbitrary k_1 and k_2, by writing the system as a matrix equation.

$$\underset{A}{\begin{bmatrix} 0.05 & 0.03 \\ 0.07 & 0.04 \end{bmatrix}} \underset{X}{\begin{bmatrix} x_1 \\ x_2 \end{bmatrix}} = \underset{B}{\begin{bmatrix} k_1 \\ k_2 \end{bmatrix}}$$

If A^{-1} exists, then $X = A^{-1}B$. To find A^{-1}, we perform row operations on

$$\begin{bmatrix} 0.05 & 0.03 & | & 1 & 0 \\ 0.07 & 0.04 & | & 0 & 1 \end{bmatrix} \quad 20R_1 \rightarrow R_1$$

$$\sim \begin{bmatrix} 1 & 0.6 & | & 20 & 0 \\ 0.07 & 0.04 & | & 0 & 1 \end{bmatrix} \quad (-0.07)R_1 + R_2 \rightarrow R_2$$

$$\sim \begin{bmatrix} 1 & 0.6 & | & 20 & 0 \\ 0 & -0.002 & | & -1.4 & 1 \end{bmatrix} \quad -500R_2 \rightarrow R_2$$

$$\sim \begin{bmatrix} 1 & 0.6 & | & 20 & 0 \\ 0 & 1 & | & 700 & -500 \end{bmatrix} \quad (-0.6)R_2 + R_1 \rightarrow R_1$$

$$\sim \begin{bmatrix} 1 & 0 & | & -400 & 300 \\ 0 & 1 & | & 700 & -500 \end{bmatrix}$$

Hence $A^{-1} = \begin{bmatrix} -400 & 300 \\ 700 & -500 \end{bmatrix}$

Check: $A^{-1}A = \begin{bmatrix} -400 & 300 \\ 700 & -500 \end{bmatrix} \begin{bmatrix} 0.05 & 0.03 \\ 0.07 & 0.04 \end{bmatrix} = \begin{bmatrix} 1 & 0 \\ 0 & 1 \end{bmatrix}$

We can now solve the system as:

$$\underset{X}{\begin{bmatrix} x_1 \\ x_2 \end{bmatrix}} = \underset{A^{-1}}{\begin{bmatrix} -400 & 300 \\ 700 & -500 \end{bmatrix}} \underset{B}{\begin{bmatrix} k_1 \\ k_2 \end{bmatrix}}$$

(A) If $k_1 = 3.6$, $k_2 = 5$,

$$\begin{bmatrix} x_1 \\ x_2 \end{bmatrix} = \begin{bmatrix} -400 & 300 \\ 700 & -500 \end{bmatrix} \begin{bmatrix} 3.6 \\ 5 \end{bmatrix} = \begin{bmatrix} 60 \\ 20 \end{bmatrix}$$

60 tons of ore must be produced at Big Bend, 20 tons of ore at Saw Pit.

(B) If $k_1 = 3$, $k_2 = 4.1$,

$$\begin{bmatrix} x_1 \\ x_2 \end{bmatrix} = \begin{bmatrix} -400 & 300 \\ 700 & -500 \end{bmatrix} \begin{bmatrix} 3 \\ 4.1 \end{bmatrix} = \begin{bmatrix} 30 \\ 50 \end{bmatrix}$$

30 tons of ore must be produced at Big Bend, 50 tons of ore at Saw Pit.

(C) If $k_1 = 3.2$, $k_2 = 4.4$,

$$\begin{bmatrix} x_1 \\ x_2 \end{bmatrix} = \begin{bmatrix} -400 & 300 \\ 700 & -500 \end{bmatrix} \begin{bmatrix} 3.2 \\ 4.4 \end{bmatrix} = \begin{bmatrix} 40 \\ 40 \end{bmatrix}$$

40 tons of ore must be produced at Big Bend, 40 tons of ore at Saw Pit. *(9-3)*

36. (A) The labor cost of producing one printer stand at the South Carolina plant is the product of the stand row of L with South Carolina column of H.

$$[0.9 \quad 1.8 \quad 0.6] \begin{bmatrix} 10.00 \\ 8.50 \\ 4.50 \end{bmatrix} = 27 \text{ dollars}$$

(B) The matrix HL has no obvious meaning, but the matrix LH represents the total labor costs for each item at each plant.

(C) $LH = \begin{bmatrix} 1.7 & 2.4 & 0.8 \\ 0.9 & 1.8 & 0.6 \end{bmatrix} \begin{bmatrix} 11.50 & 10.00 \\ 9.50 & 8.50 \\ 5.00 & 4.50 \end{bmatrix}$

$$= \begin{bmatrix} (1.7)(11.50) + (2.4)(9.50) + (0.8)(5.00) & (1.7)(10.00) + (2.4)(8.50) + (0.8)(4.50) \\ (0.9)(11.50) + (1.8)(9.50) + (0.6)(5.00) & (0.9)(10.00) + (1.8)(8.50) + (0.6)(4.50) \end{bmatrix}$$

$$\begin{array}{cc} \text{N.C.} & \text{S.C.} \end{array}$$
$$\begin{bmatrix} \$46.35 & \$41.00 \\ \$30.45 & \$27.00 \end{bmatrix} \begin{array}{l} \text{Desk} \\ \text{Stands} \end{array}$$

(9-1)

37. (A) The average monthly production for the months of January and February is represented by the matrix $\frac{1}{2}(J + F)$

$$\frac{1}{2}(J + F) = \frac{1}{2}\left(\begin{bmatrix} 1,500 & 1,650 \\ 850 & 700 \end{bmatrix} + \begin{bmatrix} 1,700 & 1,810 \\ 930 & 740 \end{bmatrix}\right) = \frac{1}{2}\begin{bmatrix} 3,200 & 3,460 \\ 1,780 & 1,440 \end{bmatrix}$$

$$\begin{array}{cc} \text{N.C.} & \text{S.C.} \end{array}$$
$$= \begin{bmatrix} 1,600 & 1,730 \\ 890 & 720 \end{bmatrix} \begin{array}{l} \text{Desks} \\ \text{Stands} \end{array}$$

(B) The increase in production from January to February is represented by the matrix $F - J$.

$$F - J = \begin{bmatrix} 1,700 & 1,810 \\ 930 & 740 \end{bmatrix} - \begin{bmatrix} 1,500 & 1,650 \\ 850 & 700 \end{bmatrix}$$

$$\begin{array}{cc} \text{N.C.} & \text{S.C.} \end{array}$$
$$= \begin{bmatrix} 200 & 160 \\ 80 & 40 \end{bmatrix} \begin{array}{l} \text{Desks} \\ \text{Stands} \end{array}$$

(C) $J\begin{bmatrix} 1 \\ 1 \end{bmatrix} = \begin{bmatrix} 1,500 & 1,650 \\ 850 & 700 \end{bmatrix}\begin{bmatrix} 1 \\ 1 \end{bmatrix} = \begin{bmatrix} 3,150 \\ 1,550 \end{bmatrix} \begin{array}{l} \text{Desks} \\ \text{Stands} \end{array}$

This matrix represents the total production of each item in January.

(9-1)

38. The inverse of matrix B is calculated to be

$$B^{-1} = \begin{bmatrix} 1 & 1 & -1 \\ 0 & -1 & 1 \\ -1 & 0 & 1 \end{bmatrix}$$

Putting the coded message into matrix form and multiplying by B^{-1} yields

$$\begin{bmatrix} 1 & 1 & -1 \\ 0 & -1 & 1 \\ -1 & 0 & 1 \end{bmatrix}\begin{bmatrix} 25 & 24 & 21 & 41 & 21 & 52 \\ 8 & 25 & 41 & 30 & 32 & 52 \\ 26 & 33 & 48 & 50 & 41 & 79 \end{bmatrix} = \begin{bmatrix} 7 & 16 & 14 & 21 & 12 & 25 \\ 18 & 8 & 7 & 20 & 9 & 27 \\ 1 & 9 & 27 & 9 & 20 & 27 \end{bmatrix}$$

This decodes to

7 18 1 16 8 9 14 7 27 21 20 9 12 9 20 25 27 27

G R A P H I N G   U T I L I T Y

(9-2)

CUMULATIVE REVIEW EXERCISE (Chapters 8 and 9)

1. We choose elimination by addition. We multiply the top equation by 3, the bottom by 5, and add.

$$9x - 15y = 33$$
$$\underline{10x + 15y = \;\;5}$$
$$19x \qquad\;\; = 38$$
$$\qquad x \;= \;\; 2$$

Substituting $x = 2$ in the bottom equation, we have

$$2(2) + 3y = 1$$
$$4 + 3y = 1$$
$$3y = -3$$
$$y = -1$$

Solution: $(2, -1)$ *(8-1)*

2.

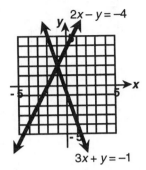

$(-1, 2)$ *(8-1)*

3. $x^2 + y^2 = 2$
$2x - y = 1$

We choose substitution, solving the first-degree equation for y in terms of x, then substituting into the second-degree equation.

$$2x - y = 1$$
$$-y = 1 - 2x$$
$$y = 2x - 1$$
$$x^2 + (2x - 1)^2 = 2$$
$$x^2 + 4x^2 - 4x + 1 = 2$$
$$5x^2 - 4x - 1 = 0$$
$$(5x + 1)(x - 1) = 0$$
$$x = -\frac{1}{5}, \; 1$$

For $x = -\dfrac{1}{5}$ For $x = 1$

$$y = 2\left(-\frac{1}{5}\right) - 1 \qquad y = 2(1) - 1$$
$$= -\frac{7}{5} \qquad\qquad\qquad = 1$$

Solutions: $\left(-\dfrac{1}{5}, \; -\dfrac{7}{5}\right)$, $(1, 1)$ *(8-3)*

The checking steps are omitted for lack of space.

4.

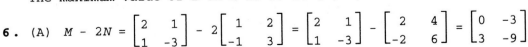

(8-4)

5.

Corner Point (x, y)	Objective Function $z = 2x + 3y$	
$(0, 4)$	12	
$(5, 0)$	10	Minimum value
$(6, 7)$	33	Maximum value
$(0, 10)$	30	

The minimum value of z on S is 10 at $(5, 0)$.
The maximum value of z on S is 33 at $(6, 7)$.   *(8-5)*

6. (A) $M - 2N = \begin{bmatrix} 2 & 1 \\ 1 & -3 \end{bmatrix} - 2\begin{bmatrix} 1 & 2 \\ -1 & 3 \end{bmatrix} = \begin{bmatrix} 2 & 1 \\ 1 & -3 \end{bmatrix} - \begin{bmatrix} 2 & 4 \\ -2 & 6 \end{bmatrix} = \begin{bmatrix} 0 & -3 \\ 3 & -9 \end{bmatrix}$

     (B) $P + Q$ is not defined

(C) $PQ = \begin{bmatrix} 1 & 2 \end{bmatrix} \begin{bmatrix} -1 \\ 2 \end{bmatrix} = [1(-1) + 2 \cdot 2] = [3]$

(D) $MN = \begin{bmatrix} 2 & 1 \\ 1 & -3 \end{bmatrix} \begin{bmatrix} 1 & 2 \\ -1 & 3 \end{bmatrix} = \begin{bmatrix} 2 \cdot 1 + 1(-1) & 2 \cdot 2 + 1 \cdot 3 \\ 1 \cdot 1 + (-3)(-1) & 1 \cdot 2 + (-3)3 \end{bmatrix} = \begin{bmatrix} 1 & 7 \\ 4 & -7 \end{bmatrix}$

(E) $PN = \begin{bmatrix} 1 & 2 \end{bmatrix} \begin{bmatrix} 1 & 2 \\ -1 & 3 \end{bmatrix} = [1 \cdot 1 + 2(-1) \quad 1 \cdot 2 + 2 \cdot 3] = \begin{bmatrix} -1 & 8 \end{bmatrix}$

(F) QM is not defined

$(9-1)$

7. $\begin{vmatrix} 0 & 2 & 0 \\ 1 & 3 & 2 \\ -1 & 4 & 3 \end{vmatrix} = 0 + 2(-1)^{1+2} \begin{vmatrix} 1 & 2 \\ -1 & 3 \end{vmatrix} + 0 = 2(-1)^3 [1 \cdot 3 - (-1)2] = -10$

$(9-4)$

8. (A) $x_1 = 3$
$x_2 = -4$

(B) $x_1 - 2x_2 = 3$
Let $x_2 = t$
Then $x_1 = 2x_2 + 3$
$= 2t + 3$
Hence, $x_1 = 2t + 3$, $x_2 = t$ is a
solution for every real number t.

(C) $x_1 - 2x_2 = 3$
$0x_1 + 0x_2 = 1$
No solution.

$(8-1)$

9. (A) $\begin{bmatrix} 1 & 1 & | & 3 \\ -1 & 1 & | & 5 \end{bmatrix}$

(B) $\begin{bmatrix} 1 & 1 & | & 3 \\ -1 & 1 & | & 5 \end{bmatrix}$ $R_1 + R_2 \rightarrow R_2$

$\sim \begin{bmatrix} 1 & 1 & | & 3 \\ 0 & 2 & | & 8 \end{bmatrix}$ $\frac{1}{2} R_2 \rightarrow R_2$

$\sim \begin{bmatrix} 1 & 1 & | & 3 \\ 0 & 1 & | & 4 \end{bmatrix}$ $(-1)R_2 + R_1 \rightarrow R_1$

$\sim \begin{bmatrix} 1 & 0 & | & -1 \\ 0 & 1 & | & 4 \end{bmatrix}$

(C) Solution: $x_1 = -1$, $x_2 = 4$

$(8-1, 8-2)$

10. (A) As a matrix equation the system becomes

$\begin{array}{ccc} A & X & B \end{array}$

$\begin{bmatrix} 1 & -3 \\ 2 & -5 \end{bmatrix} \begin{bmatrix} x_1 \\ x_2 \end{bmatrix} = \begin{bmatrix} k_1 \\ k_2 \end{bmatrix}$

(B) To find A^{-1}, we perform row operations on

$\begin{bmatrix} 1 & -3 & | & 1 & 0 \\ 2 & -5 & | & 0 & 1 \end{bmatrix}$ $(-2)R_1 + R_2 \rightarrow R_2$

$\sim \begin{bmatrix} 1 & -3 & | & 1 & 0 \\ 0 & 1 & | & -2 & 1 \end{bmatrix}$ $3R_2 + R_1 \rightarrow R_1$

$\sim \begin{bmatrix} 1 & 0 & | & -5 & 3 \\ 0 & 1 & | & -2 & 1 \end{bmatrix}$

Hence $A^{-1} = \begin{bmatrix} -5 & 3 \\ -2 & 1 \end{bmatrix}$

Check: $A^{-1}A = \begin{bmatrix} -5 & 3 \\ -2 & 1 \end{bmatrix} \begin{bmatrix} 1 & -3 \\ 2 & -5 \end{bmatrix} = \begin{bmatrix} (-5)1 + 3 \cdot 2 & (-5)(-3) + 3(-5) \\ (-2)1 + 1 \cdot 2 & (-2)(-3) + 1(-5) \end{bmatrix} = \begin{bmatrix} 1 & 0 \\ 0 & 1 \end{bmatrix}$

(C) The solution of $AX = B$ is $X = A^{-1}B$. Using the result of (B), we have

$$X = \begin{bmatrix} -5 & 3 \\ -2 & 1 \end{bmatrix}\begin{bmatrix} k_1 \\ k_2 \end{bmatrix}$$

If $k_1 = -2$, $k_2 = 1$, we have

$$\begin{bmatrix} x_1 \\ x_2 \end{bmatrix} = \begin{bmatrix} -5 & 3 \\ -2 & 1 \end{bmatrix}\begin{bmatrix} -2 \\ 1 \end{bmatrix} = \begin{bmatrix} (-5)(-2) + 3\cdot 1 \\ (-2)(-2) + 1\cdot 1 \end{bmatrix} = \begin{bmatrix} 13 \\ 5 \end{bmatrix} \quad x_1 = 13,\ x_2 = 5$$

(D) If $k_1 = 1$, $k_2 = -2$, reasoning as in (C) we have

$$\begin{bmatrix} x_1 \\ x_2 \end{bmatrix} = \begin{bmatrix} -5 & 3 \\ -2 & 1 \end{bmatrix}\begin{bmatrix} 1 \\ -2 \end{bmatrix} = \begin{bmatrix} (-5)1 + 3(-2) \\ (-2)1 + 1(-2) \end{bmatrix} = \begin{bmatrix} -11 \\ -4 \end{bmatrix} \quad x_1 = -11,\ x_2 = -4 \qquad (9\text{-}3)$$

11. (A) $D = \begin{vmatrix} 2 & -3 \\ 4 & -5 \end{vmatrix} = 2(-5) - 4(-3) = 2$

(B) $x = \dfrac{\begin{vmatrix} 1 & -3 \\ 2 & -5 \end{vmatrix}}{D} = \dfrac{1}{2} \qquad y = \dfrac{\begin{vmatrix} 2 & 1 \\ 4 & 2 \end{vmatrix}}{D} = \dfrac{0}{2} = 0$ \qquad (9\text{-}6)

12. We write the augmented matrix:

$$\begin{bmatrix} 1 & 3 & | & 10 \\ 2 & -1 & | & -1 \end{bmatrix} \ (-2)R_1 + R_2 \to R_2$$

$$-2 \quad -6 \quad -20$$

$\uparrow$
Need a 0 here

$\sim \begin{bmatrix} 1 & 3 & | & 10 \\ 0 & -7 & | & -21 \end{bmatrix} -\tfrac{1}{7}R_2 \to R_2$ \quad corrresponds to the linear system $\begin{aligned} x_1 + 3x_2 &= 10 \\ -7x_2 &= -21 \end{aligned}$

$\uparrow$
Need a 1 here

Need a 0 here
$\downarrow$

$\begin{bmatrix} 1 & 3 & | & 10 \\ 0 & 1 & | & 3 \end{bmatrix} \ (-3)R_2 + R_1 \to R_1$ \quad corresponds to the linear system $\begin{aligned} x_1 + 3x_2 &= 10 \\ x_2 &= 3 \end{aligned}$

$$0 \quad -3 \quad -9$$

$\sim \begin{bmatrix} 1 & 0 & | & 1 \\ 0 & 1 & | & 3 \end{bmatrix}$ \quad corresponds to the linear system $\begin{aligned} x_1 &= 1 \\ x_2 &= 3 \end{aligned}$

The solution is $x_1 = 1$, $x_2 = 3$. Each pair of lines graphed below has the same intersection point, $(1, 3)$.

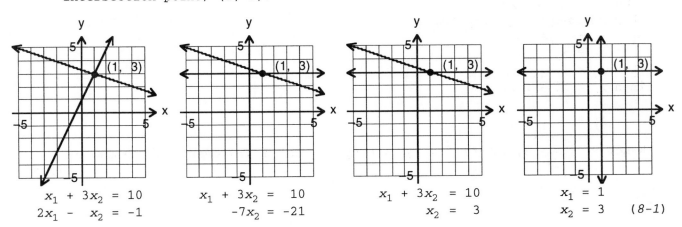

$\begin{aligned} x_1 + 3x_2 &= 10 \\ 2x_1 - x_2 &= -1 \end{aligned}$ \qquad $\begin{aligned} x_1 + 3x_2 &= 10 \\ -7x_2 &= -21 \end{aligned}$ \qquad $\begin{aligned} x_1 + 3x_2 &= 10 \\ x_2 &= 3 \end{aligned}$ \qquad $\begin{aligned} x_1 &= 1 \\ x_2 &= 3 \end{aligned}$ \quad (8\text{-}1)

13. Here is a computer-generated graph of the system, entered as

$$y = \frac{7 + 2x}{3}$$

$$y = \frac{18 - 3x}{4}$$

After zooming in (graph not shown) the intersection point is located at (1.53, 3.35) to two decimal places.

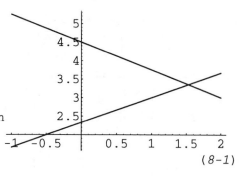

(8-1)

14.
$$\begin{bmatrix} 1 & 2 & -1 & | & 3 \\ 0 & 1 & 1 & | & -2 \\ 2 & 3 & 1 & | & 0 \end{bmatrix} \quad (-2)R_1 + R_3 \rightarrow R_3$$

$$\sim \begin{bmatrix} 1 & 2 & -1 & | & 3 \\ 0 & 1 & 1 & | & -2 \\ 0 & -1 & 3 & | & -6 \end{bmatrix} \begin{array}{l} (-2)R_2 + R_1 \rightarrow R_1 \\ \\ R_2 + R_3 \rightarrow R_3 \end{array}$$

$$\sim \begin{bmatrix} 1 & 0 & -3 & | & 7 \\ 0 & 1 & 1 & | & -2 \\ 0 & 0 & 4 & | & -8 \end{bmatrix} \quad \frac{1}{4}R_3 \rightarrow R_3$$

$$\sim \begin{bmatrix} 1 & 0 & -3 & | & 7 \\ 0 & 1 & 1 & | & -2 \\ 0 & 0 & 1 & | & -2 \end{bmatrix} \begin{array}{l} 3R_3 + R_1 \rightarrow R_1 \\ \\ (-1)R_3 + R_2 \rightarrow R_2 \end{array}$$

$$\sim \begin{bmatrix} 1 & 0 & 0 & | & 1 \\ 0 & 1 & 0 & | & 0 \\ 0 & 0 & 1 & | & -2 \end{bmatrix}$$

Solution: $x_1 = 1$, $x_2 = 0$, $x_3 = -2$

(8-2)

15.
$$\begin{bmatrix} 1 & 1 & -2 & | & 2 \\ 0 & 4 & 6 & | & -1 \\ 0 & 6 & 9 & | & 0 \end{bmatrix} \quad (-\frac{2}{3})R_2 + R_3 \rightarrow R_3$$

$$\sim \begin{bmatrix} 1 & 1 & -2 & | & 2 \\ 0 & 4 & 6 & | & -1 \\ 0 & 0 & 0 & | & \frac{2}{3} \end{bmatrix}$$

The last row corresponds to the equation

$$0x_1 + 0x_2 + 0x_3 = \frac{2}{3},$$

hence there is no solution.   *(8-2)*

16.
$$\begin{bmatrix} 1 & -2 & 1 & | & 1 \\ 3 & -2 & -1 & | & -5 \end{bmatrix} \quad (-3)R_1 + R_2 \rightarrow R_2$$

$$\sim \begin{bmatrix} 1 & -2 & 1 & | & 1 \\ 0 & 4 & -4 & | & -8 \end{bmatrix} \quad \frac{1}{4}R_2 \rightarrow R_2$$

$$\sim \begin{bmatrix} 1 & -2 & 1 & | & 1 \\ 0 & 1 & -1 & | & -2 \end{bmatrix} \quad 2R_2 + R_1 \rightarrow R_1$$

$$\sim \begin{bmatrix} 1 & 0 & -1 & | & -3 \\ 0 & 1 & -1 & | & -2 \end{bmatrix}$$

This corresponds to the system

$$x_1 \quad\quad - x_3 = -3$$
$$x_2 - x_3 = -2$$

Let $x_3 = t$

Then $x_2 = x_3 - 2$
$$\quad\quad = t - 2$$
$$x_1 = x_3 - 3$$
$$\quad\quad = t - 3$$

Hence $x_1 = t - 3$, $x_2 = t - 2$, $x_3 = t$ is a solution for every real number t.

(8-2)

17. $x^2 - 3xy + 3y^2 = 1$
$xy = 1$

Solve for y in the second equation.

$$y = \frac{1}{x}$$

$$x^2 - 3x\left(\frac{1}{x}\right) + 3\left(\frac{1}{x}\right)^2 = 1$$

$$x^2 - 3 + \frac{3}{x^2} = 1 \quad x \neq 0$$

$$x^4 - 3x^2 + 3 = x^2$$

$$x^4 - 4x^2 + 3 = 0$$

$$(x^2 - 1)(x^2 - 3) = 0$$

$x^2 - 1 = 0 \qquad x^2 - 3 = 0$

$x^2 = 1 \qquad\qquad x^2 = 3$

$x = \pm 1 \qquad\qquad x = \pm\sqrt{3}$

For $x = 1$ For $x = -1$ For $x = \sqrt{3}$ For $x = -\sqrt{3}$

$y = \dfrac{1}{1}$ $y = \dfrac{1}{-1}$ $y = \dfrac{1}{\sqrt{3}}$ $y = \dfrac{1}{-\sqrt{3}}$

$y = 1$ $y = -1$ $y = \dfrac{\sqrt{3}}{3}$ $y = -\dfrac{\sqrt{3}}{3}$

Solutions: $(1, 1)$, $(-1, -1)$, $\left(\sqrt{3}, \dfrac{\sqrt{3}}{3}\right)$, $\left(-\sqrt{3}, -\dfrac{\sqrt{3}}{3}\right)$ *(8-3)*

18. $x^2 - 3xy + y^2 = -1$
$x^2 - xy = 0$

Factor the left side of the equation that has a zero constant term.

$x(x - y) = 0$

$x = 0$ or $x = y$

Thus the original system is equivalent to the two systems

$x^2 - 3xy + y^2 = -1$ $x^2 - 3xy + y^2 = -1$
$x = 0$ $x = y$

These systems are solved by substitution.

First system: Second system:

$x^2 - 3xy + y^2 = -1$ $x^2 - 3xy + y^2 = -1$
$x = 0$ $x = y$

$0^2 - 3(0)y + y^2 = -1$ $y^2 - 3yy + y^2 = -1$

$y^2 = -1$ $-y^2 = -1$

$y = \pm i$ $y^2 = 1$

$x = 0$ $y = \pm 1$

For $y = 1$ For $y = -1$
$x = 1$ $x = -1$

Solutions: $(1, 1)$, $(-1, -1)$, $(0, i)$, $(0, -i)$ *(8-3)*

19. (A) $MN = \begin{bmatrix} 1 & 2 & -1 \end{bmatrix} \begin{bmatrix} 1 \\ -1 \\ 2 \end{bmatrix} = [1\cdot 1 + 2(-1) + (-1)2] = [-3]$

(B) $NM = \begin{bmatrix} 1 \\ -1 \\ 2 \end{bmatrix} \begin{bmatrix} 1 & 2 & -1 \end{bmatrix} = \begin{bmatrix} 1\cdot 1 & 1\cdot 2 & 1(-1) \\ (-1)1 & (-1)2 & (-1)(-1) \\ 2\cdot 1 & 2\cdot 2 & 2(-1) \end{bmatrix} = \begin{bmatrix} 1 & 2 & -1 \\ -1 & -2 & 1 \\ 2 & 4 & -2 \end{bmatrix}$ *(9-1)*

20. (A) $LM = \begin{bmatrix} 2 & -1 & 0 \\ 1 & 2 & 1 \end{bmatrix} \begin{bmatrix} 1 & 2 \\ -1 & 0 \\ 1 & 1 \end{bmatrix} = \begin{bmatrix} 2 \cdot 1 + (-1)(-1) + 0 \cdot 1 & 2 \cdot 2 + (-1)0 + 0 \cdot 1 \\ 1 \cdot 1 + 2(-1) + 1 \cdot 1 & 1 \cdot 2 + 2 \cdot 0 + 1 \cdot 1 \end{bmatrix}$

$= \begin{bmatrix} 3 & 4 \\ 0 & 3 \end{bmatrix}$

$LM - 2N = \begin{bmatrix} 3 & 4 \\ 0 & 3 \end{bmatrix} - 2\begin{bmatrix} 2 & 1 \\ -1 & 0 \end{bmatrix} = \begin{bmatrix} 3 & 4 \\ 0 & 3 \end{bmatrix} - \begin{bmatrix} 4 & 2 \\ -2 & 0 \end{bmatrix} = \begin{bmatrix} -1 & 2 \\ 2 & 3 \end{bmatrix}$

(B) Since ML is a 3 × 3 matrix and N is a 2 × 2 matrix, $ML + N$ is not defined.

(9-1)

21. The solution region is unbounded. Two corner points are obvious from the graph: (0, 6) and (8, 0). The third corner point is obtained by solving the system
$3x + 2y = 12$
$x + 2y = 8$ to obtain (2, 3).

(8-4)

22. The feasible region is graphed as follows: The corner points (0, 7), (0, 0) and (8, 0) are obvious from the graph. The corner point (6, 4) is obtained by solving the system
$x + 2y = 14$
$2x + y = 16$
We now evaluate the objective function at each corner point.

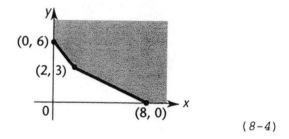

Corner Point (x, y)	Objective Function $z = 4x + 9y$	
(0, 7)	63	Maximum value
(0, 0)	0	
(8, 0)	32	
(6, 4)	60	

(8-5)

23. (A) As a matrix equation the system becomes

$A X B$
$\begin{bmatrix} 1 & 4 & 2 \\ 2 & 6 & 3 \\ 2 & 5 & 2 \end{bmatrix} \begin{bmatrix} x_1 \\ x_2 \\ x_3 \end{bmatrix} = \begin{bmatrix} k_1 \\ k_2 \\ k_3 \end{bmatrix}$

(B) To find A^{-1}, we perform row operations on

$\begin{bmatrix} 1 & 4 & 2 & | & 1 & 0 & 0 \\ 2 & 6 & 3 & | & 0 & 1 & 0 \\ 2 & 5 & 2 & | & 0 & 0 & 1 \end{bmatrix} \begin{matrix} \\ (-2)R_1 + R_2 \to R_2 \\ (-2)R_1 + R_3 \to R_3 \end{matrix}$

$$\sim \begin{bmatrix} 1 & 4 & 2 & | & 1 & 0 & 0 \\ 0 & -2 & -1 & | & -2 & 1 & 0 \\ 0 & -3 & -2 & | & -2 & 0 & 1 \end{bmatrix} \quad -\tfrac{1}{2} R_2 \to R_2$$

$$\sim \begin{bmatrix} 1 & 4 & 2 & | & 1 & 0 & 0 \\ 0 & 1 & \tfrac{1}{2} & | & 1 & -\tfrac{1}{2} & 0 \\ 0 & -3 & -2 & | & -2 & 0 & 1 \end{bmatrix} \quad \begin{array}{l} (-4) R_2 + R_1 \to R_1 \\ \\ 3R_2 + R_3 \to R_3 \end{array}$$

$$\sim \begin{bmatrix} 1 & 0 & 0 & | & -3 & 2 & 0 \\ 0 & 1 & \tfrac{1}{2} & | & 1 & -\tfrac{1}{2} & 0 \\ 0 & 0 & -\tfrac{1}{2} & | & 1 & -\tfrac{3}{2} & 1 \end{bmatrix} \quad R_3 + R_2 \to R_2$$

$$\sim \begin{bmatrix} 1 & 0 & 0 & | & -3 & 2 & 0 \\ 0 & 1 & 0 & | & 2 & -2 & 1 \\ 0 & 0 & -\tfrac{1}{2} & | & 1 & -\tfrac{3}{2} & 1 \end{bmatrix} \quad -2R_3 \to R_3$$

$$\sim \begin{bmatrix} 1 & 0 & 0 & | & -3 & 2 & 0 \\ 0 & 1 & 0 & | & 2 & -2 & 1 \\ 0 & 0 & 1 & | & -2 & 3 & -2 \end{bmatrix}$$

Hence $A^{-1} = \begin{bmatrix} -3 & 2 & 0 \\ 2 & -2 & 1 \\ -2 & 3 & -2 \end{bmatrix}$

Check: $A^{-1}A = \begin{bmatrix} -3 & 2 & 0 \\ 2 & -2 & 1 \\ -2 & 3 & -2 \end{bmatrix}\begin{bmatrix} 1 & 4 & 2 \\ 2 & 6 & 3 \\ 2 & 5 & 2 \end{bmatrix}$

$$= \begin{bmatrix} (-3)1 + 2\cdot2 + 0\cdot2 & (-3)4 + 2\cdot6 + 0\cdot5 & -3\cdot2 + 2\cdot3 + 0\cdot2 \\ 2\cdot1 + (-2)2 + 1\cdot2 & 2\cdot4 + (-2)6 + 1\cdot5 & 2\cdot2 + (-2)3 + 1\cdot2 \\ (-2)1 + 3\cdot2 + (-2)2 & (-2)4 + 3\cdot6 + (-2)5 & (-2)2 + 3\cdot3 + (-2)2 \end{bmatrix}$$

$$= \begin{bmatrix} 1 & 0 & 0 \\ 0 & 1 & 0 \\ 0 & 0 & 1 \end{bmatrix}$$

(C) The solution of $AX = B$ is $X = A^{-1}B$. Using the result of (B), we have

$$X = \begin{bmatrix} -3 & 2 & 0 \\ 2 & -2 & 1 \\ -2 & 3 & -2 \end{bmatrix}\begin{bmatrix} k_1 \\ k_2 \\ k_3 \end{bmatrix}$$

If $k_1 = -1$, $k_2 = 2$, $k_3 = 1$, we have

$$\begin{bmatrix} x_1 \\ x_2 \\ x_3 \end{bmatrix} = \begin{bmatrix} -3 & 2 & 0 \\ 2 & -2 & 1 \\ -2 & 3 & -2 \end{bmatrix}\begin{bmatrix} -1 \\ 2 \\ 1 \end{bmatrix} = \begin{bmatrix} 7 \\ -5 \\ 6 \end{bmatrix} \quad x_1 = 7,\ x_2 = -5,\ x_3 = 6$$

(D) If $k_1 = 2$, $k_2 = 0$, $k_3 = -1$, reasoning as in (C) we have

$$\begin{bmatrix} x_1 \\ x_2 \\ x_3 \end{bmatrix} = \begin{bmatrix} -3 & 2 & 0 \\ 2 & -2 & 1 \\ -2 & 3 & -2 \end{bmatrix}\begin{bmatrix} 2 \\ 0 \\ -1 \end{bmatrix} = \begin{bmatrix} -6 \\ 3 \\ -2 \end{bmatrix} \quad x_1 = -6,\ x_2 = 3,\ x_3 = -2 \qquad (9\text{-}3)$$

24. (A) $D = \begin{vmatrix} 1 & 2 & -1 \\ 2 & 8 & 1 \\ -1 & 3 & 5 \end{vmatrix} = 1(-1)^{1+1}\begin{vmatrix} 8 & 1 \\ 3 & 5 \end{vmatrix} + 2(-1)^{1+2}\begin{vmatrix} 2 & 1 \\ -1 & 5 \end{vmatrix} + (-1)(-1)^{1+3}\begin{vmatrix} 2 & 8 \\ -1 & 3 \end{vmatrix}$

$$= (-1)^2(8\cdot5 - 3\cdot1) + 2(-1)^3[2\cdot5 - (-1)1] + (-1)^5[2\cdot3 - (-1)8]$$

$$= 37 - 22 - 14 = 1$$

(B) $\quad z = \dfrac{\begin{vmatrix} 1 & 2 & 1 \\ 2 & 8 & -2 \\ -1 & 3 & 2 \end{vmatrix}}{D} = \dfrac{1(-1)^{1+1}\begin{vmatrix} 8 & -2 \\ 3 & 2 \end{vmatrix} + 2(-1)^{1+2}\begin{vmatrix} 2 & -2 \\ -1 & 2 \end{vmatrix} + 1(-1)^{1+3}\begin{vmatrix} 2 & 8 \\ -1 & 3 \end{vmatrix}}{1}$

$\quad\quad = 32$ $\hfill$ *(9-5, 9-6)*

25. Before we can enter these equations in our graphing utility, we must solve
for y:

$$x^2 + 2xy - y^2 = 1 \hspace{3cm} 9x^2 + 4xy + y^2 = 15$$
$$y^2 - 2xy - x^2 + 1 = 0 \hspace{2.5cm} y^2 + 4xy + 9x^2 - 15 = 0$$

Applying the quadratic formula to each equation, we have

$$y = \frac{2x \pm \sqrt{4x^2 - 4(-x^2 + 1)}}{2} \hspace{2cm} y = \frac{-4x \pm \sqrt{16x^2 - 4(9x^2 - 15)}}{2}$$

$$y = \frac{2x \pm \sqrt{8x^2 - 4}}{2} \hspace{3cm} y = \frac{-4x \pm \sqrt{60 - 20x^2}}{2}$$

$$y = x \pm \sqrt{2x^2 - 1} \hspace{3.5cm} y = -2x \pm \sqrt{15 - 5x^2}$$

Entering each of these four equations into
a graphing utility produces the graph
shown at the right.

Zooming in on the four intersection
points, or using a built-in intersection
routine (details omitted), yields (-1.35,
0.28), (-0.87, -1.60), (0.87, 1.60), and
(1.35, -0.28) to two decimal places.

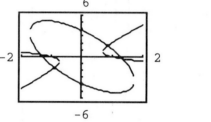

$\hfill$ *(8-3)*

26. (A) The matrix is in fact

$$\begin{bmatrix} 1 & 0 & -5 & | & 2 \\ 0 & 1 & 3 & | & 6 \\ 0 & 0 & 0 & | & 0 \end{bmatrix}$$

Thus there are an infinite number of solutions ($x_3 = t$, $x_2 = 6 - 3t$,
$x_1 = 2 + 5t$, for t any real number.)

(B) The matrix is in fact

$$\begin{bmatrix} 1 & 0 & -5 & | & 2 \\ 0 & 1 & 3 & | & 6 \\ 0 & 0 & 0 & | & n \end{bmatrix}$$

The third row corresponds to the equation $0x_1 + 0x_2 + 0x_3 = n$. This is
impossible. The system has no solution.

(C) The system is independent. There is one solution. $\hfill$ *(8-2)*

27. If $A^2 = A$ and A^{-1} exists, then $A^{-1}A^2 = A^{-1}A$ or $A^{-1}AA = I$ or $IA = I$. Thus, $A = I$,
the $n \times n$ identity matrix. $\hfill$ *(9-3)*

28. L, M, and P are in reduced form.
N is not in reduced form; it violates condition 1 of the definition of reduced
matrix, Section 8-2: there is a row of 0's above rows having non-zero
elements. $\hfill$ *(8-2)*

29. $k\begin{vmatrix} a & b \\ c & d \end{vmatrix} = k(ad - bc) = kad - kbc = kad - kcb = \begin{vmatrix} ka & b \\ kc & d \end{vmatrix}$ $\hfill$ *(9-5)*

30. $\begin{vmatrix} a & b \\ c & d \end{vmatrix} = ad - bc = ad + akb - akb - bc = ad + akb - (bc + bka)$

$= a(d + kb) - b(c + ka) = \begin{vmatrix} a & b \\ c + ka & d + kb \end{vmatrix}$ $\qquad (9\text{-}5)$

31. We assume that $\det M = ad - bc \neq 0$ and find M^{-1} by applying row operations to

$\begin{bmatrix} a & b & | & 1 & 0 \\ c & d & | & 0 & 1 \end{bmatrix}$ Case I: $a \neq 0$

$\qquad\qquad\qquad\qquad (-\frac{c}{a})R_1 + R_2 \rightarrow R_2$

$\sim \begin{bmatrix} a & b & | & 1 & 0 \\ 0 & \frac{ad - bc}{a} & | & -\frac{c}{a} & 1 \end{bmatrix}$ $\frac{a}{ad - bc}R_2 \rightarrow R_2$

$\sim \begin{bmatrix} a & b & | & 1 & 0 \\ 0 & 1 & | & \frac{-c}{ad - bc} & \frac{a}{ad - bc} \end{bmatrix}$ $(-b)R_2 + R_1 \rightarrow R_1$

$\sim \begin{bmatrix} a & 0 & | & 1 + \frac{bc}{ad - bc} & \frac{-ab}{ad - bc} \\ 0 & 1 & | & \frac{-c}{ad - bc} & \frac{a}{ad - bc} \end{bmatrix}$

$\sim \begin{bmatrix} a & 0 & | & \frac{ad}{ad - bc} & \frac{-ab}{ad - bc} \\ 0 & 1 & | & \frac{-c}{ad - bc} & \frac{a}{ad - bc} \end{bmatrix}$ $\frac{1}{a}R_1 \rightarrow R_1$

$\sim \begin{bmatrix} 1 & 0 & | & \frac{d}{ad - bc} & \frac{-b}{ad - bc} \\ 0 & 1 & | & \frac{-c}{ad - bc} & \frac{a}{ad - bc} \end{bmatrix}$

Case II: $a = 0$ (b, $c \neq 0$, since $b = 0$ or $c = 0$ would imply $\det M = ad - bc = 0$)

$\begin{bmatrix} 0 & b & | & 1 & 0 \\ c & d & | & 0 & 1 \end{bmatrix}$ $R_1 \leftrightarrow R_2$

$\sim \begin{bmatrix} c & d & | & 0 & 1 \\ 0 & b & | & 1 & 0 \end{bmatrix}$ $\frac{1}{c}R_1 \rightarrow R_1$

$\sim \begin{bmatrix} 1 & \frac{d}{c} & | & 0 & \frac{1}{c} \\ 0 & b & | & 1 & 0 \end{bmatrix}$ $(-\frac{d}{bc})R_2 + R_1 \rightarrow R_1$

$\sim \begin{bmatrix} 1 & 0 & | & -\frac{d}{bc} & \frac{1}{c} \\ 0 & b & | & 1 & 0 \end{bmatrix}$ $\frac{1}{b}R_2 \rightarrow R_2$

$\sim \begin{bmatrix} 1 & 0 & | & -\frac{d}{bc} & \frac{1}{c} \\ 0 & 1 & | & \frac{1}{b} & 0 \end{bmatrix}$

$\sim \begin{bmatrix} 1 & 0 & | & \frac{d}{ad - bc} & \frac{-b}{ad - bc} \\ 0 & 1 & | & \frac{-c}{ad - bc} & \frac{a}{ad - bc} \end{bmatrix}$ since $a = 0$.

In either case,

$M^{-1} = \begin{bmatrix} \frac{d}{ad - bc} & \frac{-b}{ad - bc} \\ \frac{-c}{ad - bc} & \frac{a}{ad - bc} \end{bmatrix} = \frac{1}{ad - bc}\begin{bmatrix} d & -b \\ -c & a \end{bmatrix} = \frac{1}{\det M}\begin{bmatrix} d & -b \\ -c & a \end{bmatrix}$ $\qquad (9\text{-}2,\ 9\text{-}4)$

32. True. For example,

$$\begin{bmatrix} a_{11} & a_{12} & a_{13} \\ 0 & a_{22} & a_{23} \\ 0 & 0 & a_{33} \end{bmatrix} + \begin{bmatrix} b_{11} & b_{12} & b_{13} \\ 0 & b_{22} & b_{23} \\ 0 & 0 & b_{33} \end{bmatrix} = \begin{bmatrix} a_{11} + b_{11} & a_{12} + b_{12} & a_{13} + b_{13} \\ 0 & a_{22} + b_{22} & a_{23} + b_{23} \\ 0 & 0 & a_{33} + b_{33} \end{bmatrix}$$

proves the statement for 3 × 3 matrices. A similar proof can be given for any order matrix. *(9-1)*

33. True. For example, $\begin{bmatrix} a_{11} & 0 \\ a_{21} & a_{22} \end{bmatrix}\begin{bmatrix} b_{11} & 0 \\ b_{21} & b_{22} \end{bmatrix} = \begin{bmatrix} a_{11}b_{11} & 0 \\ a_{21}b_{11} + a_{22}b_{21} & a_{22}b_{22} \end{bmatrix}$ proves the statement for 2 × 2 matrices. A similar proof can be given for any order matrix. *(9-1)*

34. False. For example, $\begin{bmatrix} 1 & 1 \\ 0 & 1 \end{bmatrix} + \begin{bmatrix} 1 & 0 \\ 1 & 1 \end{bmatrix} = \begin{bmatrix} 2 & 1 \\ 1 & 2 \end{bmatrix}$ which is not a diagonal matrix. *(9-1)*

35. False. For example, $\begin{bmatrix} 1 & 1 \\ 0 & 1 \end{bmatrix}\begin{bmatrix} 1 & 0 \\ 1 & 1 \end{bmatrix} = \begin{bmatrix} 2 & 1 \\ 1 & 1 \end{bmatrix}$ which is not a diagonal matrix. *(9-1)*

36. True. If all elements above and all elements below the principal diagonal are zero, then all elements not on the principal diagonal are zero. *(9-2)*

37. True. For example, the inverse of $\begin{bmatrix} a_{11} & 0 & 0 \\ 0 & a_{22} & 0 \\ 0 & 0 & a_{33} \end{bmatrix}$ is $\begin{bmatrix} \dfrac{1}{a_{11}} & 0 & 0 \\ 0 & \dfrac{1}{a_{22}} & 0 \\ 0 & 0 & \dfrac{1}{a_{33}} \end{bmatrix}$.

This is defined if none of elements on the principal diagonal are zero. A similar situation holds for any order matrix. *(9-2)*

38. True. For example, $\begin{vmatrix} a_{11} & 0 & 0 & 0 \\ 0 & a_{22} & 0 & 0 \\ 0 & 0 & a_{33} & 0 \\ 0 & 0 & 0 & a_{44} \end{vmatrix}$ can be expanded by the first column to

yield, successively, $a_{11}\begin{vmatrix} a_{22} & 0 & 0 \\ 0 & a_{33} & 0 \\ 0 & 0 & a_{44} \end{vmatrix} = a_{11}a_{22}\begin{vmatrix} a_{33} & 0 \\ 0 & a_{44} \end{vmatrix} = a_{11}a_{12}a_{33}a_{44}$.

A similar result holds for any order matrix. *(9-4)*

39. True. For example, $\begin{vmatrix} a_{11} & 0 & 0 & 0 \\ a_{22} & a_{22} & 0 & 0 \\ a_{31} & a_{32} & a_{33} & 0 \\ a_{41} & a_{42} & a_{43} & a_{44} \end{vmatrix}$ can be expanded by the first row to

yield, successively, $a_{11}\begin{vmatrix} a_{22} & 0 & 0 \\ a_{32} & a_{33} & 0 \\ a_{42} & a_{43} & a_{44} \end{vmatrix} = a_{11}a_{22}\begin{vmatrix} a_{33} & 0 \\ a_{43} & a_{44} \end{vmatrix} = a_{11}a_{22}a_{33}a_{44}$.

A similar result holds for any order matrix. *(9-4)*

40. Let x = amount invested at 8%
y = amount invested at 14%
Then $x + y = 12{,}000$ (total amount invested) $\qquad$ (1)
$0.08x + 0.14y = 0.10(12{,}000)$ (total yield on investment) $\qquad$ (2)

We solve the system of equations (1), (2) using elimination by addition.

$\begin{array}{ll} -0.08x - 0.08y = -0.08(12{,}000) & -0.08[\text{equation (1)}] \\ \underline{0.08x + 0.14y = 0.10(12{,}000)} & \text{equation (2)} \end{array}$

$\qquad\qquad 0.06y = -0.08(12{,}000) + 0.10(12{,}000)$
$\qquad\qquad 0.06y = 240$
$\qquad\qquady = 4{,}000$
$\qquad\quad x + y = 12{,}000$
$\qquad\qquad x = 8{,}000$

$8,000 at 8% and $4,000 at 14%. $\hfill$ (8-1, 8-2)

41. Let x_1 = number of ounces of mix A.
x_2 = number of ounces of mix B.
x_3 = number of ounces of mix C.

Then
$\begin{aligned} 0.2x_1 + 0.1x_2 + 0.15x_3 &= 23 &&\text{(protein)} \\ 0.02x_1 + 0.06x_2 + 0.05x_3 &= 6.2 &&\text{(fat)} \\ 0.15x_1 + 0.1x_2 + 0.05x_3 &= 16 &&\text{(moisture)} \end{aligned}$

or
$\begin{aligned} 4x_1 + 2x_2 + 3x_3 &= 460 \\ 2x_1 + 6x_2 + 5x_3 &= 620 \\ 3x_1 + 2x_2 + x_3 &= 320 \end{aligned}$

is the system to be solved. We form the augmented matrix and solve by Gauss-Jordan elimination.

$$\begin{bmatrix} 4 & 2 & 3 & 460 \\ 2 & 6 & 5 & 620 \\ 3 & 2 & 1 & 320 \end{bmatrix} \quad \tfrac{1}{4}R_1 \to R_1$$

$$\sim \begin{bmatrix} 1 & \tfrac{1}{2} & \tfrac{3}{4} & 115 \\ 2 & 6 & 5 & 620 \\ 3 & 2 & 1 & 320 \end{bmatrix} \quad \begin{array}{l}(-2)R_1 + R_2 \to R_2 \\ (-3)R_1 + R_3 \to R_3\end{array}$$

$$\sim \begin{bmatrix} 1 & \tfrac{1}{2} & \tfrac{3}{4} & 115 \\ 0 & 5 & \tfrac{7}{2} & 390 \\ 0 & \tfrac{1}{2} & -\tfrac{5}{4} & -25 \end{bmatrix} \quad R_2 \leftrightarrow R_3$$

$$\sim$$

$$\begin{bmatrix} 1 & \tfrac{1}{2} & \tfrac{3}{4} & 115 \\ 0 & \tfrac{1}{2} & -\tfrac{5}{4} & -25 \\ 0 & 5 & \tfrac{7}{2} & 390 \end{bmatrix} \quad \begin{array}{l}(-1)R_2 + R_1 \to R_1 \\ \\ (-10)R_2 + R_3 \to R_3\end{array}$$

$$\sim \begin{bmatrix} 1 & 0 & 2 & 140 \\ 0 & \tfrac{1}{2} & -\tfrac{5}{4} & -25 \\ 0 & 0 & 16 & 640 \end{bmatrix} \quad \begin{array}{l}2R_2 \to R_2 \\ \tfrac{1}{16}R_3 \to R_3\end{array}$$

$$\sim \begin{bmatrix} 1 & 0 & 2 & 140 \\ 0 & 1 & -\tfrac{5}{2} & -50 \\ 0 & 0 & 1 & 40 \end{bmatrix} \quad \begin{array}{l}(-2)R_3 + R_1 \to R_1 \\ (\tfrac{5}{2})R_3 + R_2 \to R_2\end{array}$$

$$\sim \begin{bmatrix} 1 & 0 & 0 & 60 \\ 0 & 1 & 0 & 50 \\ 0 & 0 & 1 & 40 \end{bmatrix}$$

Thus, x_1 = 60g Mix A, x_2 = 50g Mix B, x_3 = 40g Mix C $\hfill$ (8-2)

42. Let x_1 = number of model A trucks
x_2 = number of model B trucks
x_3 = number of model C trucks

Then $x_1 + x_2 + x_3 = 12$ (total number of trucks)
$18,000x_1 + 22,000x_2 + 30,000x_3 = 300,000$ (total funds needed)

We form the augmented matrix and solve by Gauss-Jordan elimination.

$$\begin{bmatrix} 1 & 1 & 1 & \bigm| & 12 \\ 18,000 & 22,000 & 30,000 & \bigm| & 300,000 \end{bmatrix} \quad \tfrac{1}{2,000} R_2 \to R_2$$

$$\sim \begin{bmatrix} 1 & 1 & 1 & \bigm| & 12 \\ 9 & 11 & 15 & \bigm| & 150 \end{bmatrix} \quad (-9)R_1 + R_2 \to R_2$$

$$\sim \begin{bmatrix} 1 & 1 & 1 & \bigm| & 12 \\ 0 & 2 & 6 & \bigm| & 42 \end{bmatrix} \quad -\tfrac{1}{2} R_2 + R_1 \to R_1$$

$$\sim \begin{bmatrix} 1 & 0 & -2 & \bigm| & -9 \\ 0 & 2 & 6 & \bigm| & 42 \end{bmatrix} \quad \tfrac{1}{2} R_2 \to R_2$$

$$\sim \begin{bmatrix} 1 & 0 & -2 & \bigm| & -9 \\ 0 & 1 & 3 & \bigm| & 21 \end{bmatrix}$$

Let $x_3 = t$. Then $x_2 = -3x_3 + 21$
$\qquad\qquad\qquad = -3t + 21$
$\qquad\qquad x_1 = 2x_3 - 9$
$\qquad\qquad\qquad = 2t - 9$

A solution is achieved, not for every value of t, but for integer values of t that give rise to non-negative x_1, x_2, x_3.

$x_1 \geq 0$ means $\quad 2t - 9 \geq 0 \quad$ or $\quad t \geq 4\tfrac{1}{2}$
$x_2 \geq 0$ means $\quad -3t + 21 \geq 0 \quad$ or $\quad t \leq 7$

The only integer values of t that satisfy these conditions are 5, 6, 7. Thus we have the solutions

$x_1 = 2t - 9 \qquad$ model A trucks
$x_2 = -3t + 21 \quad$ model B trucks
$x_3 = t \qquad\qquad$ model C trucks

where t = 5, 6, or 7. Thus the distributor can purchase
1 model A truck, 6 model B trucks, and 5 model C trucks, or
3 model A trucks, 3 model B trucks, and 6 model C trucks, or
5 model A trucks and 7 model C trucks. $\qquad\qquad\qquad\qquad (8-2)$

43. Let a = length of rectangle
b = width of rectangle
Then $ab = 32$ (area)
$2a + 2b = 24$ (perimeter)

We solve by substitution, solving the first-degree equation for b and substituting into the second-degree equation.
$\qquad 2b = 24 - 2a$
$\qquad\quad b = 12 - a$
$a(12 - a) = 32$
$\quad 12a - a^2 = 32$
$\qquad\quad 0 = a^2 - 12a + 32$
$\qquad\quad 0 = (a - 4)(a - 8)$
$a = 4 \qquad\qquad$ or $\qquad a = 8$
$b = 12 - a = 8 \qquad\qquad b = 12 - a = 4$

Dimensions: 8 meters by 4 meters $\qquad\qquad\qquad\qquad\qquad\qquad (8-3)$

44. Let x = number of standard day packs
y = number of deluxe day packs

(A) We form the linear objective function
$P = 8x + 12y$
We wish to maximize P, the profit from x standard packs @ \$8 and y deluxe packs @ \$12, subject to the constraints
$0.5x + 0.5y \leq 300$ fabricating constraint
$0.3x + 0.6y \leq 240$ sewing constraint
$x, y \geq 0$ non-negative constraints

Solving the system of constraint inequalities graphically, we obtain the feasible region S shown in the diagram.

The three corner points $(0, 400)$, $(0, 0)$, and $(600, 0)$ are obvious from the diagram. The corner point $(400, 200)$ is found by solving the system
$0.5x + 0.5y = 300$
$0.3x + 0.6y = 240$

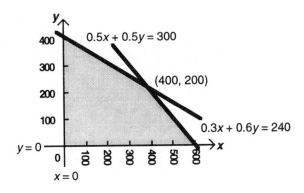

Next we evaluate the objective function at each corner point.

Corner Point (x, y)	Objective Function $P = 8x + 12y$	
(0, 0)	0	
(600, 0)	4,800	
(400, 200)	5,600	Maximum value
(0, 400)	4,800	

The optimal value is 5,600 at the corner point $(400, 200)$. Thus, the company should manufacture 400 standard packs and 200 deluxe packs for a maximum profit of \$5,600.

(B) The objective function now becomes $5x + 15y$. We evaluate this at each corner point.

Corner Point (x, y)	Objective Function $P = 5x + 15y$	
(0, 0)	0	
(600, 0)	3,000	
(400, 200)	5,000	
(0, 400)	6,000	Maximum value

The optimal value is now 6,000 at the corner point $(0, 400)$. Thus, the company should manufacture no standard packs and 400 deluxe packs for a maximum profit of \$6,000.

(C) The objective function now becomes $11x + 9y$. We evaluate this at each corner point.

Corner Point (x, y)	Objective Function $P = 11x + 9y$	
(0, 0)	0	
(600, 0)	6,600	Maximum value
(400, 200)	6,200	
(0, 400)	3,600	

The optimal value is now 6,600 at the corner point $(600, 0)$. Thus, the company should manufacture 600 standard packs and no deluxe packs for a maximum profit of \$6,600.

$(8-5)$

45. (A) $M \begin{bmatrix} 0.25 \\ 0.25 \\ 0.25 \\ 0.25 \end{bmatrix} = \begin{bmatrix} 78 & 84 & 81 & 86 \\ 91 & 65 & 84 & 92 \\ 95 & 90 & 92 & 91 \\ 75 & 82 & 87 & 91 \\ 83 & 88 & 81 & 76 \end{bmatrix} \begin{bmatrix} 0.25 \\ 0.25 \\ 0.25 \\ 0.25 \end{bmatrix}$

$$= \begin{bmatrix} 0.25(78 + 84 + 81 + 86) \\ 0.25(91 + 65 + 84 + 92) \\ 0.25(95 + 90 + 92 + 91) \\ 0.25(75 + 82 + 87 + 91) \\ 0.25(83 + 88 + 81 + 76) \end{bmatrix} = \begin{bmatrix} 82.25 \\ 83 \\ 92 \\ 83.75 \\ 82 \end{bmatrix} \begin{matrix} \text{Ann} \\ \text{Bob} \\ \text{Carol} \\ \text{Dan} \\ \text{Eric} \end{matrix}$$

(B) $M \begin{bmatrix} 0.2 \\ 0.2 \\ 0.2 \\ 0.4 \end{bmatrix} = \begin{bmatrix} 78 & 84 & 81 & 86 \\ 91 & 65 & 84 & 92 \\ 95 & 90 & 92 & 91 \\ 75 & 82 & 87 & 91 \\ 83 & 88 & 81 & 76 \end{bmatrix} \begin{bmatrix} 0.2 \\ 0.2 \\ 0.2 \\ 0.4 \end{bmatrix}$

$$= \begin{bmatrix} 0.2(78 + 84 + 81) + 0.4 \cdot 86 \\ 0.2(91 + 65 + 84) + 0.4 \cdot 92 \\ 0.2(95 + 90 + 92) + 0.4 \cdot 91 \\ 0.2(75 + 82 + 87) + 0.4 \cdot 91 \\ 0.2(83 + 88 + 81) + 0.4 \cdot 76 \end{bmatrix} = \begin{bmatrix} 83 \\ 84.8 \\ 91.8 \\ 85.2 \\ 80.8 \end{bmatrix} \begin{matrix} \text{Ann} \\ \text{Bob} \\ \text{Carol} \\ \text{Dan} \\ \text{Eric} \end{matrix}$$

(C) $[0.2 \quad 0.2 \quad 0.2 \quad 0.2 \quad 0.2]M = [0.2 \quad 0.2 \quad 0.2 \quad 0.2 \quad 0.2] \begin{bmatrix} 78 & 84 & 81 & 86 \\ 91 & 65 & 84 & 92 \\ 95 & 90 & 92 & 91 \\ 75 & 82 & 87 & 91 \\ 83 & 88 & 81 & 76 \end{bmatrix}$

$= [0.2(78 + 91 + 95 + 75 + 83) \quad 0.2(84 + 65 + 90 + 82 + 88) \quad 0.2(81 + 84 + 92 + 87 + 81) \quad 0.2(86 + 92 + 91 + 91 + 76)]$

$\begin{matrix} \quad \text{Test 1} & \text{Test 2} & \text{Test 3} & \text{Test 4} \\ = [\quad 84.4 & 81.8 & 85 & 87.2] \end{matrix}$

$(9\text{-}1)$

CHAPTER 10

Exercise 10-1

Key Ideas and Formulas

A sequence is a function with domain a set of successive integers.

Notation: for $f(n)$ we write a_n. The elements of the range, $f(n)$, are called terms of the sequence, a_n. a_1 is the first term, a_2 is the second term, a_n is the nth term or the general term.

Finite sequence: domain is finite set of successive integers.

Infinite sequence: domain is infinite set of successive integers.

Some sequences are specified by a recursion formula, a formula that defines each term in terms of one or more preceding terms.

If a_1, a_2, a_3, ..., a_n, ... is a sequence, then the expression $a_1 + a_2 + a_3 + \cdots + a_n$ is called a series. If a_1, a_2, ..., a_n are the terms of the sequence, the series is written

$$\sum_{k=1}^{n} a_k = a_1 + a_2 + \cdots + a_n \qquad k \text{ is called the summing index.}$$

---

1. $a_1 = 2 \cdot 1 + 5 = 7 \qquad a_2 = 2 \cdot 2 + 5 = 9 \qquad a_3 = 2 \cdot 3 + 5 = 11 \qquad a_4 = 2 \cdot 4 + 5 = 13$

7, 9, 11, 13, ...

3. $a_1 = \dfrac{3 \cdot 1 + 1}{1 + 1} = 2 \qquad a_2 = \dfrac{3 \cdot 2 + 1}{2 + 1} = \dfrac{7}{3} \qquad a_3 = \dfrac{3 \cdot 3 + 1}{3 + 1} = \dfrac{10}{4} = \dfrac{5}{2} \qquad a_4 = \dfrac{3 \cdot 4 + 1}{4 + 1} = \dfrac{13}{5}$

$2, \dfrac{7}{3}, \dfrac{5}{2}, \dfrac{13}{5}, \ldots$

5. $a_1 = 3^1 + (-2)^1 = 1 \qquad a_2 = 3^2 + (-2)^2 = 13 \qquad a_3 = 3^3 + (-2)^3 = 19 \qquad a_4 = 3^4 + (-2)^4 = 97$

1, 13, 19, 97, ...

7. $a_{100} = 2 \cdot 100 + 5 = 205$ 　　　　　　**9.** $a_{200} = \dfrac{3 \cdot 200 + 1}{200 + 1} = \dfrac{601}{201}$

11. $S_5 = 1 + 2 + 3 + 4 + 5$ 　　　**13.** $S_3 = \dfrac{1}{10^1} + \dfrac{1}{10^2} + \dfrac{1}{10^3} = \dfrac{1}{10} + \dfrac{1}{100} + \dfrac{1}{1000}$

15. $S_4 = (-1)^1 + (-1)^2 + (-1)^3 + (-1)^4 = (-1) + 1 + (-1) + 1 = -1 + 1 - 1 + 1$

17. $a_1 = (-1)^{1+1} 1^2 = 1 \qquad a_2 = (-1)^{2+1} 2^2 = -4 \qquad a_3 = (-1)^{3+1} 3^2 = 9$

$a_4 = (-1)^{4+1} 4^2 = -16 \qquad a_5 = (-1)^{5+1} 5^2 = 25$

19.
$$a_1 = \frac{1}{3}\left[1 - \frac{1}{10^1}\right] = \frac{1}{3} \cdot \frac{9}{10} = \frac{3}{10} = 0.3$$

$$a_2 = \frac{1}{3}\left[1 - \frac{1}{10^2}\right] = \frac{1}{3} \cdot \frac{99}{100} = \frac{33}{100} = 0.33$$

$$a_3 = \frac{1}{3}\left[1 - \frac{1}{10^3}\right] = \frac{1}{3} \cdot \frac{999}{1,000} = \frac{333}{1,000} = 0.333$$

$$a_4 = \frac{1}{3}\left[1 - \frac{1}{10^4}\right] = \frac{1}{3} \cdot \frac{9,999}{10,000} = \frac{3,333}{10,000} = 0.3333$$

$$a_5 = \frac{1}{3}\left[1 - \frac{1}{10^5}\right] = \frac{1}{3} \cdot \frac{99,999}{100,000} = \frac{33,333}{100,000} = 0.33333$$

21. $a_1 = \left(-\dfrac{1}{2}\right)^{1-1} = \left(-\dfrac{1}{2}\right)^0 = 1$

$a_2 = \left(-\dfrac{1}{2}\right)^{2-1} = -\dfrac{1}{2}$

$a_3 = \left(-\dfrac{1}{2}\right)^{3-1} = \dfrac{1}{4}$

$a_4 = \left(-\dfrac{1}{2}\right)^{4-1} = -\dfrac{1}{8}$

$a_5 = \left(-\dfrac{1}{2}\right)^{5-1} = \dfrac{1}{16}$

23. $a_1 = 7$

$a_2 = a_{2-1} - 4 = a_1 - 4 = 7 - 4 = 3$

$a_3 = a_{3-1} - 4 = a_2 - 4 = 3 - 4 = -1$

$a_4 = a_{4-1} - 4 = a_3 - 4 = -1 - 4 = -5$

$a_5 = a_{5-1} - 4 = a_4 - 4 = -5 - 4 = -9$

> **Common Error:**
> $a_2 \neq a_2 - 1 - 4$. The 1 is in
> the subscript: a_{2-1}; that is, a_1

25. $a_1 = 4$

$a_2 = \dfrac{1}{4} a_{2-1} = \dfrac{1}{4} a_1 = \dfrac{1}{4} \cdot 4 = 1$

$a_3 = \dfrac{1}{4} a_{3-1} = \dfrac{1}{4} a_2 = \dfrac{1}{4} \cdot 1 = \dfrac{1}{4}$

$a_4 = \dfrac{1}{4} a_{4-1} = \dfrac{1}{4} a_3 = \dfrac{1}{4} \cdot \dfrac{1}{4} = \dfrac{1}{16}$

$a_5 = \dfrac{1}{4} a_{5-1} = \dfrac{1}{4} a_4 = \dfrac{1}{4} \cdot \dfrac{1}{16} = \dfrac{1}{64}$

27. a_n: 3, 5, 7, 9, $\cdots$

$a_n - 1$: 2, 4, 6, 8, $\cdots$

$n = 1, 2, 3, 4, \cdots$

Comparing $a_n - 1$ with n, we see that
$a_n - 1 = 2n$; thus, $a_n = 1 + 2n$

29. a_n: 4, 9, 16, 25, $\cdots$

$n = 1, 2, 3, 4, \cdots$

$n + 1 = 2, 3, 4, 5, \cdots$

Comparing a_n with $n + 1$, we see
that $a_n = (n + 1)^2$

31. a_n: 4, 8, 16, 32, $\cdots$

$n = 2^2, 2^3, 2^4, 2^5, \cdots$

Comparing a_n with n, we see that
$a_n = 2^{n+1}$

33. a_n: 1, $\dfrac{3}{2}$, $\dfrac{5}{3}$, $\dfrac{7}{4}$, $\cdots$

$n = 1, 2, 3, 4, \cdots$

na_n: = 1, 3, 5, 7, $\cdots$

Comparing na_n with n, we see that $na_n = 2n - 1$; thus, $a_n = \dfrac{2n - 1}{n}$

35. a_n: 7, -7, 7, -7, $\cdots$

a_n: $7 \cdot 1$, $7(-1)$, $7 \cdot 1$, $7(-1)$, $\cdots$

$n = 1, 2, 3, 4, \cdots$

Comparing a_n with n, we see that a_n involves 7 times -1 to successively even
and odd powers, that is, to powers that depend on n. We could write
$a_n = 7(-1)^{n+1}$ or $a_n = 7(-1)^{n-1}$
or other choices: $7(-1)^{n+1}$ is one of many correct answers.

37. a_n: x, $\dfrac{x^2}{2}$, $\dfrac{x^3}{4}$, $\dfrac{x^4}{8}$, $\cdots$

a_n: $\dfrac{x^1}{2^0}$, $\dfrac{x^2}{2^1}$, $\dfrac{x^3}{2^2}$, $\dfrac{x^4}{2^3}$, $\cdots$

$n = 1, 2, 3, 4, \cdots$

Comparing a_n with n,

we see that $a_n = \dfrac{x^n}{2^{n-1}}$

39.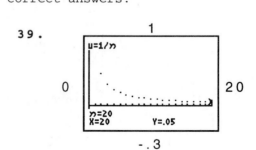

41.

```
        1.5
   ┌─────────────────────┐
   │u=(-0.9)^n           │
   │ .                   │
 0 │   . . . . . . . . . │ 20
   │ .  . .  . . . . .   │
   │  .                  │
   │n=20                 │
   │X=20      Y=.12157665│
   └─────────────────────┘
       -1.5
```

43. $S_4 = \dfrac{(-2)^{1+1}}{1} + \dfrac{(-2)^{2+1}}{2} + \dfrac{(-2)^{3+1}}{3} + \dfrac{(-2)^{4+1}}{4}$

$= \dfrac{4}{1} - \dfrac{8}{2} + \dfrac{16}{3} - \dfrac{32}{4}$

45. $S_3 = \dfrac{1}{1}x^{1+1} + \dfrac{1}{2}x^{2+1} + \dfrac{1}{3}x^{3+1}$

$= x^2 + \dfrac{x^3}{2} + \dfrac{x^4}{3}$

47. $S_5 = \dfrac{(-1)^{1+1}}{1}x^1 + \dfrac{(-1)^{2+1}}{2}x^2 + \dfrac{(-1)^{3+1}}{3}x^3 + \dfrac{(-1)^{4+1}}{4}x^4 + \dfrac{(-1)^{5+1}}{5}x^5$

$= x - \dfrac{x^2}{2} + \dfrac{x^3}{3} - \dfrac{x^4}{4} + \dfrac{x^5}{5}$

49. $S_4 = 1^2 + 2^2 + 3^2 + 4^2$

$k = 1, 2, 3, 4, \cdots$

Clearly, $a_k = k^2$, $k = 1, 2, 3, 4$

$S_4 = \displaystyle\sum_{k=1}^{4} k^2$

51. $S_5 = \dfrac{1}{2^1} + \dfrac{1}{2^2} + \dfrac{1}{2^3} + \dfrac{1}{2^4} + \dfrac{1}{2^5}$

$k = 1, 2, 3, 4, \cdots$

Clearly, $a_k = \dfrac{1}{2^k}$, $k = 1, 2, 3, 4, 5$

$S_5 = \displaystyle\sum_{k=1}^{5} \dfrac{1}{2^k}$

53. $S_n = 1 + \dfrac{1}{2^2} + \dfrac{1}{3^2} + \ldots + \dfrac{1}{n^2}$

$k = 1, 2, 3, \cdots, n$

Clearly, $a_k = \dfrac{1}{k^2}$, $k = 1, 2, 3, \cdots, n$

$S_n = \displaystyle\sum_{k=1}^{n} \dfrac{1}{k^2}$

55. $S_n = 1 - 4 + 9 + \cdots + (-1)^{n+1}n^2$

$k = 1, 2, 3, \cdots, n$

Clearly, $a_k = (-1)^{k+1}k^2$, $k = 1, 2, 3, \cdots, n$

$S_n = \displaystyle\sum_{k=1}^{n} (-1)^{k+1}k^2$

57. (A) $a_1 = 3$

$a_2 = \dfrac{a_{2-1}^2 + 2}{2a_{2-1}} = \dfrac{a_1^2 + 2}{2a_1} = \dfrac{3^2 + 2}{2 \cdot 3} \approx 1.83$

$a_3 = \dfrac{a_{3-1}^2 + 2}{2a_{3-1}} = \dfrac{a_2^2 + 2}{2a_2} = \dfrac{(1.83)^2 + 2}{2(1.83)} \approx 1.46$

$a_4 = \dfrac{a_{4-1}^2 + 2}{2a_{4-1}} = \dfrac{a_3^2 + 2}{2a_3} = \dfrac{(1.46)^2 + 2}{2(1.46)} \approx 1.415$

(B) Calculator $\sqrt{2} = 1.4142135\ldots$

(C) $a_1 = 1$

$a_2 = \dfrac{a_1^2 + 2}{2a_1} = \dfrac{1^2 + 2}{2 \cdot 1} = 1.5$

$a_3 = \dfrac{a_2^2 + 2}{2a_2} = \dfrac{(1.5)^2 + 2}{2(1.5)} \approx 1.417$

$a_4 = \dfrac{a_3^2 + 2}{2a_3} = \dfrac{(1.417)^2 + 2}{2(1.417)} \approx 1.414$

59. The first ten terms of the Fibonacci sequence a_n are

1, 1, 2, 3, 5, 8, 13, 21, 34, 55

The first ten terms of b_n are

1, 3, 4, 7, 11, 18, 29, 47, 76, 123

The first ten terms of $c_n = \dfrac{b_n}{a_n}$ are

$$\frac{1}{1}, \; \frac{3}{1}, \; \frac{4}{2}, \; \frac{7}{3}, \; \frac{11}{5}, \; \frac{18}{8}, \; \frac{29}{13}, \; \frac{47}{21}, \; \frac{76}{34}, \; \frac{123}{55}$$

In decimal notation this becomes

1, 3, 2, 2.33..., 2.2, 2.25, 2.23..., 2.238..., 2.235..., 2.236...

61.

$$e^{0.2} = 1 + \frac{0.2}{1!} + \frac{(0.2)^2}{2!} + \frac{(0.2)^3}{3!} + \frac{(0.2)^4}{4!}$$

$$= 1 + 0.2 + \frac{(0.2)^2}{1 \cdot 2} + \frac{(0.2)^3}{1 \cdot 2 \cdot 3} + \frac{(0.2)^4}{1 \cdot 2 \cdot 3 \cdot 4}$$

$$= 1.2 + 0.2 \left[\frac{0.2}{2} + \frac{(0.2)^2}{6} + \frac{(0.2)^3}{24} \right]$$

$$= 1.2 + 0.2 \left\{ 0.2 \left[\frac{1}{2} + \frac{0.2}{6} + \frac{(0.2)^2}{24} \right] \right\}$$

$$= 1.2 + 0.2 \left\{ 0.2 \left[\frac{1}{2} + 0.2 \left(\frac{1}{6} + \frac{0.2}{24} \right) \right] \right\} \quad \text{(In this form this is more easily entered into a scientific calculator.)}$$

$$= 1.2214000$$

$e^{0.2} = 1.2214028$ (calculator—direct evaluation)

63. $\displaystyle\sum_{k=1}^{n} ca_k = ca_1 + ca_2 + ca_3 + \cdots + ca_n = c(a_1 + a_2 + a_3 + \cdots a_n) = c\sum_{k=1}^{n} a_k$

Exercise 10-2

Key Ideas and Formulas

Principle of Mathematical Induction

Let P_n be a statement associated with each positive integer n and suppose the following conditions are satisfied:

1. P_1 is true.

2. For any positive integer k, if P_k is true, then P_{k+1} is also true.

Then the statement P_n is true for all positive integers n.

Extended Principle of Mathematical Induction

Let m be a positive integer, let P_n be a statement associated with each integer $n \geq m$, and suppose the following conditions are satisfied:

1. P_m is true.

2. For any integer $k \geq m$, if P_k is true, then P_{k+1} is also true.

Then the statement P_n is true for all integers $n \geq m$.

Proof by Mathematical Induction: To prove that a statement P_n holds for all integers $n \geq p$: (p is often, but not always, equal to 1.) State the conjecture, P_n.

Part 1. Show that P_p is true.

Part 2. Write out P_k and P_{k+1}. Show that if P_k is true, then P_{k+1} is true.

Then P_n is true for all integers $n \geq p$.

1. $3^1 + 4^1 \geq 5^1$ (or $7 \geq 5$) True
$\quad 3^2 + 4^2 \geq 5^2$ (or $25 \geq 25$) True
$\quad 3^3 + 4^3 \geq 5^3$ (or $91 \geq 125$) False
Fails at $n = 3$

3. $17^1 - 1 = 16$ is divisible by 2^1 True $(16 \div 2 = 8)$
$\quad 17^2 - 1 = 288$ is divisible by 2^2 True $(288 \div 4 = 72)$
$\quad 17^3 - 1 = 4{,}912$ is divisible by 2^3 True $(4912 \div 8 = 614)$
$\quad 17^4 - 1 = 83{,}520$ is divisible by 2^4 True $(83{,}520 \div 16 = 5220)$
$\quad 17^5 - 1 = 1{,}419{,}856$ is divisible by 2^5 False $(1{,}419{,}856 \div 32 = 44{,}370.5)$
Fails at $n = 5$

5. P_1: $2 = 2 \cdot 1^2$ $2 = 2$
$\quad\quad\quad P_2$: $2 + 6 = 2 \cdot 2^2$ $8 = 8$
$\quad P_3$: $2 + 6 + 10 = 2 \cdot 3^2$ $18 = 18$

7. P_1: $a^5 a^1 = a^{5+1}$ $a^6 = a^6$
$\quad P_2$: $a^5 a^2 = a^{5+2}$ $a^5 a^2 = a^5(a^1 a) = (a^5 a)a = a^6 a = a^7 = a^{5+2}$
$\quad P_3$: $a^5 a^3 = a^{5+3}$ $a^5 a^3 = a^5(a^2 a) = a^5(a^1 a)a = [(a^5 a)a]a = a^8 = a^{5+3}$

9. P_1: $9^1 - 1 = 8$ is divisible by 4
$\quad P_2$: $9^2 - 1 = 80$ is divisible by 4
$\quad P_3$: $9^3 - 1 = 728$ is divisible by 4

11. P_k: $2 + 6 + 10 + \cdots + (4k - 2) = 2k^2$
$\quad P_{k+1}$: $2 + 6 + 10 + \cdots + (4k - 2) + [4(k + 1) - 2] = 2(k + 1)^2$
or $2 + 6 + 10 + \cdots + (4k - 2) + (4k + 2) = 2(k + 1)^2$

13. P_k: $a^5 a^k = a^{5+k}$
$\quad P_{k+1}$: $a^5 a^{k+1} = a^{5+k+1}$

15. P_k: $9^k - 1 = 4r$ for some integer r
$\quad P_{k+1}$: $9^{k+1} - 1 = 4s$ for some integer s

17. Prove: $2 + 6 + 10 + \cdots + (4n - 2) = 2n^2$, $n \in N$
Write: P_n: $2 + 6 + 10 + \cdots + (4n - 2) = 2n^2$

Part 1: Show that P_1 is true:
P_1: $2 = 2 \cdot 1^2$
$\quad\quad = 2$ P_1 is true

Part 2: Show that if P_k is true, then P_{k+1} is true:
Write out P_k and P_{k+1}.
$\quad P_k$: $2 + 6 + 10 + \cdots + (4k - 2) = 2k^2$
$\quad P_{k+1}$: $2 + 6 + 10 + \cdots + (4k - 2) + (4k + 2) = 2(k + 1)^2$

We start with P_k:
$\quad 2 + 6 + 10 + \cdots + (4k - 2) = 2k^2$
Adding $(4k + 2)$ to both sides, we get
$\quad 2 + 6 + 10 + \cdots + (4k - 2) + (4k + 2) = 2k^2 + 4k + 2$
$\quad\quad\quad\quad\quad\quad\quad\quad\quad\quad\quad\quad\quad\quad\quad\quad\quad = 2(k^2 + 2k + 1)$
$\quad\quad\quad\quad\quad\quad\quad\quad\quad\quad\quad\quad\quad\quad\quad\quad\quad = 2(k + 1)^2$

We have shown that if P_k is true, then P_{k+1} is true.

Conclusion: P_n is true for all positive integers n.

> **Common Errors:**
> P_1 is not the nonsensical statement
> $2 + 6 + 10 + \cdots + (4 \cdot 1 - 2) = 2 \cdot 1^2$
> achieved by "substitution" of 1 for n. The left side of P_n has n terms; the left side of P_1 must
> have 1 term. Also, $P_{k+1} \neq P_k + 1$
> Nor is P_{k+1} simply its $k + 1$st term. $4k + 2 \neq 2(k + 1)^2$

19. Prove: $a^5 a^n = a^{5+n}$, $n \in N$

Write: P_n: $a^5 a^n = a^{5+n}$

Part 1: Show that P_1 is true:

P_1: $a^5 a^1 = a^{5+1}$

$\qquad a^5 a = a^{5+1}$. True by the recursive definition of a^n.

Part 2: Show that if P_k is true, then P_{k+1} is true:

Write out P_k and P_{k+1}.

$\quad P_k$: $a^5 a^k = a^{5+k}$

$\quad P_{k+1}$: $a^5 a^{k+1} = a^{5+k+1}$

We start with P_k:

$\quad a^5 a^k = a^{5+k}$

Multiply both sides by a:

$\quad a^5 a^k a = a^{5+k} a$

$\quad a^5 a^{k+1} = a^{5+k+1}$ by the recursive definition of a^n.

We have shown that if P_k is true, then P_{k+1} is true.

Conclusion: P_n is true for all positive integers n.

21. Prove: $9^n - 1$ is divisible by 4.

Write: P_n: $9^n - 1$ is divisible by 4

Part 1: Show that P_1 is true:

P_1: $9^1 - 1 = 8 = 4 \cdot 2$ is divisible by 4. Clearly true.

Part 2: Show that if P_k is true, then P_{k+1} is true:

Write out P_k and P_{k+1}.

$\quad P_k$: $9^k - 1 = 4r$ for some integer r

$\quad P_{k+1}$: $9^{k+1} - 1 = 4s$ for some integer s

We start with P_k:

$\quad 9^k - 1 = 4r$ for some integer r

Now, $9^{k+1} - 1 = 9^{k+1} - 9^k + 9^k - 1$

$\qquad\qquad\qquad = 9^k (9 - 1) + 9^k - 1$

$\qquad\qquad\qquad = 9^k \cdot 8 + 9^k - 1$

$\qquad\qquad\qquad = 4 \cdot 9^k \cdot 2 + 4r$

$\qquad\qquad\qquad = 4(9^k \cdot 2 + r)$

Therefore,

$9^{k+1} - 1 = 4s$ for some integer s ($= 2 \cdot 9^k + r$)

$9^{k+1} - 1$ is divisible by 4.

We have shown that if P_k is true, then P_{k+1} is true.

Conclusion: P_n is true for all positive integers n.

23. The polynomial $x^4 + 1$ is a counterexample, since it has degree 4, and no real zeros.

25. 23 is a counterexample, since there are no prime numbers p such that $23 < p < 29$.

27. P_n: $2 + 2^2 + 2^3 + \cdots + 2^n = 2^{n+1} - 2$

Part 1: Show that P_1 is true:

P_1: $2 = 2^{1+1} - 2$

$\qquad = 2^2 - 2$

$\qquad = 4 - 2$

$\qquad = 2 \qquad P_1$ is true

Part 2: Show that if P_k is true, then P_{k+1} is true:

Write out P_k and P_{k+1}.

P_k: $2 + 2^2 + 2^3 + \cdots + 2^k = 2^{k+1} - 2$

P_{k+1}: $2 + 2^2 + 2^3 + \cdots + 2^k + 2^{k+1} = 2^{k+2} - 2$

We start with P_k:

$2 + 2^2 + 2^3 + \cdots + 2^k = 2^{k+1} - 2$

Adding 2^{k+1} to both sides,

$$2 + 2^2 + 2^3 + \cdots + 2^k + 2^{k+1} = 2^{k+1} - 2 + 2^{k+1}$$
$$= 2^{k+1} + 2^{k+1} - 2$$
$$= 2 \cdot 2^{k+1} - 2$$
$$= 2^1 \cdot 2^{k+1} - 2$$
$$= 2^{k+2} - 2$$

We have shown that if P_k is true, then P_{k+1} is true.

Conclusion: P_n is true for all positive integers n.

29. P_n: $1^2 + 3^2 + 5^2 + \cdots + (2n - 1)^2 = \frac{1}{3}(4n^3 - n)$

Part 1: Show that P_1 is true:

P_1: $1^2 = \frac{1}{3}(4 \cdot 1^3 - 1)$

$\qquad = \frac{1}{3}(4 - 1)$

$\qquad = 1 \qquad$ True.

Part 2: Show that if P_k is true, then P_{k+1} is true:

Write out P_k and P_{k+1}.

P_k: $1^2 + 3^2 + 5^2 + \cdots + (2k - 1)^2 = \frac{1}{3}(4k^3 - k)$

P_{k+1}: $1^2 + 3^2 + 5^2 + \cdots + (2k - 1)^2 + (2k + 1)^2 = \frac{1}{3}[4(k + 1)^3 - (k + 1)]$

We start with P_k:

$1^2 + 3^2 + 5^2 + \cdots + (2k - 1)^2 = \frac{1}{3}(4k^3 - k)$

Adding $(2k + 1)^2$ to both sides,

$$1^2 + 3^2 + 5^2 + \cdots + (2k - 1)^2 + (2k + 1)^2 = \frac{1}{3}(4k^3 - k) + (2k + 1)^2$$
$$= \frac{1}{3}[4k^3 - k + 3(2k + 1)^2]$$
$$= \frac{1}{3}[4k^3 - k + 3(4k^2 + 4k + 1)]$$
$$= \frac{1}{3}[4k^3 - k + 12k^2 + 12k + 3]$$
$$= \frac{1}{3}[4k^3 + 12k^2 + 12k - k + 3]$$
$$= \frac{1}{3}[4k^3 + 12k^2 + 12k + 4 - k - 1]$$
$$= \frac{1}{3}[4(k^3 + 3k^2 + 3k + 1) - (k + 1)]$$
$$= \frac{1}{3}[4(k + 1)^3 - (k + 1)]$$

We have shown that if P_k is true, then P_{k+1} is true.

Conclusion: P_n is true for all positive integers n.

31. P_n: $1^2 + 2^2 + 3^2 + \cdots + n^2 = \dfrac{n(n + 1)(2n + 1)}{6}$

Part 1: Show that P_1 is true:

P_1: $1^2 = \dfrac{1(1 + 1)(2 \cdot 1 + 1)}{6}$

$ = \dfrac{1 \cdot 2 \cdot 3}{6}$

$ = 1 \qquad P_1$ is true.

Part 2: Show that if P_k is true, then P_{k+1} is true:

Write out P_k and P_{k+1}.

$\quad P_k$: $1^2 + 2^2 + 3^2 + \cdots + k^2 = \dfrac{k(k + 1)(2k + 1)}{6}$

$\quad P_{k+1}$: $1^2 + 2^2 + 3^2 + \cdots + k^2 + (k + 1)^2 = \dfrac{(k + 1)(k + 2)(2k + 3)}{6}$

We start with P_k:

$\quad 1^2 + 2^2 + 3^2 + \cdots + k^2 = \dfrac{k(k + 1)(2k + 1)}{6}$

Adding $(k + 1)^2$ to both sides,

$1^2 + 2^2 + 3^2 + \cdots + k^2 + (k + 1)^2 = \dfrac{k(k + 1)(2k + 1)}{6} + (k + 1)^2$

$ = \dfrac{k(k + 1)(2k + 1)}{6} + \dfrac{6(k + 1)^2}{6}$

$ = \dfrac{k(k + 1)(2k + 1) + 6(k + 1)^2}{6}$

$ = \dfrac{(k + 1)[k(2k + 1) + 6(k + 1)]}{6}$

$ = \dfrac{(k + 1)[2k^2 + 7k + 6]}{6}$

$ = \dfrac{(k + 1)(k + 2)(2k + 3)}{6}$

We have shown that if P_k is true, then P_{k+1} is true.

Conclusion: P_n is true for all positive integers n.

33. P_n: $\dfrac{a^n}{a^3} = a^{n-3} \qquad n > 3$

Part 1: Show that P_4 is true:

P_4: $\dfrac{a^4}{a^3} = a^{4-3}$

But $\dfrac{a^4}{a^3} = \dfrac{a^3 a}{a^3} = a = a^1 = a^{4-3}$ $\qquad$ So, P_4 is true.

Part 2: Show that if P_k is true, then P_{k+1} is true:

Write out P_k and P_{k+1}.

$\quad P_k$: $\dfrac{a^k}{a^3} = a^{k-3}$

$\quad P_{k+1}$: $\dfrac{a^{k+1}}{a^3} = a^{k-2}$

We start with P_k:

$\dfrac{a^k}{a^3} = a^{k-3}$

Multiply both sides by a:

$$\frac{a^k}{a^3}a = a^{k-3}a$$

$$\frac{a^{k+1}}{a^3} = a^{k-3+1} \text{ by the recursive definition of } a^n$$

$$\frac{a^{k+1}}{a^3} = a^{k-2}$$

We have shown that if P_k is true, then P_{k+1} is true.

Conclusion: P_n is true for all integers $n > 3$.

35. Write: P_n: $a^m a^n = a^{m+n}$ m an arbitrary element of N.

Part 1: Show that P_1 is true:

P_1: $a^m a^1 = a^{m+1}$ This is true by the recursive definition of a^n.

Part 2: Show that if P_k is true, then P_{k+1} is true:

Write out P_k and P_{k+1}.

$\quad P_k$: $a^m a^k = a^{m+k}$

$\quad P_{k+1}$: $a^m a^{k+1} = a^{m+k+1}$

We start with P_k:

$\quad a^m a^k = a^{m+k}$

Multiplying both sides by a:

$\quad a^m a^k a = a^{m+k} a$

$\quad a^m a^{k+1} = a^{m+k+1}$ by the recursive definition of a^n.

We have shown that if P_k is true, then P_{k+1} is true.

Conclusion: $a^m a^n = a^{m+n}$ is true for arbitrary m, n positive integers.

37. Write: P_n: $x^n - 1 = (x - 1)Q_n(x)$ for some polynomial $Q_n(x)$.

Part 1: Show that P_1 is true:

P_1: $x^1 - 1 = (x - 1)Q_1(x)$ for some $Q_1(x)$. This is true since we can choose $Q_1(x) = 1$.

Part 2: Show that if P_k is true, then P_{k+1} is true:

Write out P_k and P_{k+1}.

$\quad P_k$: $x^k - 1 = (x - 1)Q_k(x)$ for some $Q_k(x)$

$\quad P_{k+1}$: $x^{k+1} - 1 = (x - 1)Q_{k+1}(x)$ for some $Q_{k+1}(x)$

We start with P_k:

$x^k - 1 = (x - 1)Q_k(x)$ for some $Q_k(x)$

Now, $x^{k+1} - 1 = x^{k+1} - x^k + x^k - 1$

$$\begin{aligned} &= x^k(x - 1) + x^k - 1 \\ &= x^k(x - 1) + (x - 1)Q_k(x) \\ &= (x - 1)(x^k + Q_k(x)) \\ &= (x - 1)Q_{k+1}(x), \text{ where } Q_{k+1}(x) = x^k + Q_k(x) \end{aligned}$$

Conclusion: $x^n - 1 = (x - 1)Q_n(x)$ for all positive integers n. Thus $x^n - 1$ is divisible by $x - 1$ for all positive integers n.

39. Write: P_n: $x^{2n} - 1 = (x - 1)Q_n(x)$ for some polynomial $Q_n(x)$.

Part 1: Show that P_1 is true:

P_1: $x^2 - 1 = (x - 1)Q_1(x)$ for some polynomial $Q_1(x)$. This is true since we can choose $Q_1(x) = x + 1$.

Part 2: Show that if P_k is true, then P_{k+1} is true:

Write out P_k and P_{k+1}.

$\quad P_k$: $x^{2k} - 1 = (x - 1)Q_k(x)$ for some $Q_k(x)$

$\quad P_{k+1}$: $x^{2k+2} - 1 = (x - 1)Q_{k+1}(x)$ for some $Q_{k+1}(x)$

Now, $x^{2k+2} - 1 = x^{2k+2} - x^{2k} + x^{2k} - 1$

$$\begin{aligned}
&= x^{2k}(x^2 - 1) + x^{2k} - 1 \\
&= x^{2k}(x + 1)(x - 1) + (x - 1)Q_k(x) \\
&= (x - 1)[x^{2k}(x + 1) + Q_k(x)] \\
&= (x - 1)Q_{k+1}(x), \text{ where } Q_{k+1}(x) = x^{2k}(x + 1) + Q_k(x)
\end{aligned}$$

Conclusion: $x^{2n} - 1 = (x - 1)Q_n(x)$ for all positive integers n. Thus $x^{2n} - 1$ is divisible by $x - 1$ for all positive integers n.

41. P_n: $1^3 + 2^3 + 3^3 + \cdots + n^3 = (1 + 2 + 3 + \cdots + n)^2$

Part 1: Show that P_1 is true:

P_1: $1^3 = 1^2$ True

Part 2: Show that if P_k is true, then P_{k+1} is true:

Write out P_k and P_{k+1}.

$\quad P_k$: $1^3 + 2^3 + 3^3 + \cdots + k^3 = (1 + 2 + 3 + \cdots + k)^2$

$\quad P_{k+1}$: $1^3 + 2^3 + 3^3 + \cdots + k^3 + (k + 1)^3 = (1 + 2 + 3 + \cdots + k + k + 1)^2$

We take it as proved in text Matched Problem 1 that

$$1 + 2 + 3 + \cdots + n = \frac{n(n + 1)}{2} \text{ for } n \in N$$

Thus, $1 + 2 + 3 + \cdots + k = \dfrac{k(k + 1)}{2}$

$\quad 1 + 2 + 3 + \cdots + k + (k + 1) = \dfrac{(k + 1)(k + 2)}{2}$ are known

We start with P_k:

$\quad 1^3 + 2^3 + 3^3 + \cdots + k^3 = (1 + 2 + 3 + \cdots + k)^2$

Adding $(k + 1)^3$ to both sides:

$$\begin{aligned}
1^3 + 2^3 + 3^3 + \cdots + k^3 + (k + 1)^3 &= (1 + 2 + 3 + \cdots + k)^2 + (k + 1)^3 \\
&= \left[\frac{k(k + 1)}{2}\right]^2 + (k + 1)^3 \\
&= (k + 1)^2\left[\left(\frac{k}{2}\right)^2 + (k + 1)\right] \\
&= (k + 1)^2\left[\frac{k^2}{4} + \frac{4(k + 1)}{4}\right] \\
&= (k + 1)^2\left[\frac{k^2 + 4k + 4}{4}\right] \\
&= \frac{(k + 1)^2(k + 2)^2}{4} \\
&= \left[\frac{(k + 1)(k + 2)}{2}\right]^2 \\
&= [1 + 2 + 3 + \cdots + k + (k + 1)]^2
\end{aligned}$$

Conclusion: P_n is true for all positive integers n.

43. We note:

$$2 = 2 = 1 \cdot 2 \quad n = 1$$
$$2 + 4 = 6 = 2 \cdot 3 \quad n = 2$$
$$2 + 4 + 6 = 12 = 3 \cdot 4 \quad n = 3$$
$$2 + 4 + 6 + 8 = 20 = 4 \cdot 5 \quad n = 4$$

Hypothesis: P_n: $2 + 4 + 6 + \cdots + 2n = n(n + 1)$

Proof: Part 1: Show P_1 is true

P_1: $2 = 1 \cdot 2$ True

Part 2: Show that if P_k is true, then P_{k+1} is true:

Write out P_k and P_{k+1}.

  P_k: $2 + 4 + 6 + \cdots + 2k = k(k + 1)$

  P_{k+1}: $2 + 4 + 6 + \cdots + 2k + (2k + 2) = (k + 1)(k + 2)$

We start with P_k:

  $2 + 4 + 6 + \cdots + 2k = k(k + 1)$

Adding $2k + 2$ to both sides:

$$\begin{aligned}
2 + 4 + 6 + \cdots + 2k + (2k + 2) &= k(k + 1) + 2k + 2 \\
&= (k + 1)k + (k + 1)2 \\
&= (k + 1)(k + 2)
\end{aligned}$$

Conclusion: The hypothesis P_n is true for all positive integers n.

45. $n = 1$: no line is determined.

$n = 2$: one line is determined.

$n = 3$: three lines are determined.

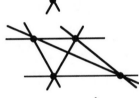

$n = 4$: six lines are determined.

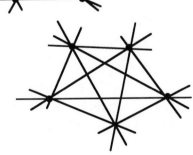

$n = 5$: ten lines are determined.

Hypothesis: P_n: n points (no three collinear) determine

$1 + 2 + 3 + \cdots + (n - 1) = \dfrac{n(n - 1)}{2}$ lines, $n \geq 2$.

Proof: Part 1: Show that P_2 is true.

P_2: 2 points determine one line:

$$\begin{aligned}
1 &= \frac{2(2 - 1)}{2} \\
&= \frac{2 \cdot 1}{2} \\
&= 1 \text{ is true}
\end{aligned}$$

Part 2: Show that if P_k is true, then P_{k+1} is true:
Write out P_k and P_{k+1}.

P_k: k points determine $1 + 2 + 3 + \cdots + (k - 1) = \dfrac{k(k - 1)}{2}$ lines

P_{k+1}: $k + 1$ points determine $1 + 2 + 3 + \cdots + (k - 1) + k = \dfrac{(k + 1)k}{2}$ lines

We start with P_k:

k points determine $1 + 2 + 3 + \cdots + (k - 1) = \dfrac{k(k - 1)}{2}$ lines

Now, the $k + 1$st point will determine a total of k new lines, one with each of the previously existing k points. These k new lines will be added to the previously existing lines. Hence, $k + 1$ points determine

$$1 + 2 + 3 + \cdots + (k - 1) + k = \frac{k(k - 1)}{2} + k \text{ lines}$$

$$= k\left[\frac{k - 1}{2} + 1\right] \text{ lines}$$

$$= k\left[\frac{k - 1 + 2}{2}\right] \text{lines}$$

$$= \frac{k(k + 1)}{2} \text{ lines}$$

Conclusion: The hypothesis P_n is true for $n \geq 2$.

47. P_n: $a > 1 \Rightarrow a^n > 1$, $n \in N$.

Part 1: Show that P_1 is true:

P_1: $a > 1 \Rightarrow a^1 > 1$. This is automatically true.

Part 2: Show that if P_k is true, then P_{k+1} is true:

Write out P_k and P_{k+1}.

P_k: $a > 1 \Rightarrow a^k > 1$

P_{k+1}: $a > 1 \Rightarrow a^{k+1} > 1$

We start with P_k. Further, assume $a > 1$ and try to derive $a^{k+1} > 1$. If this succeeds, we have proved P_{k+1}.

Assume $a > 1$.

From P_k we know that $a^k > 1$, also $1 > 0$, hence $a > 0$. We may therefore multiply both sides of the inequality $a^k > 1$ by a without changing the sense of the inequality. Hence,

$a^k a > 1a$

$a^{k+1} > a$

But, $a > 1$ by assumption. Hence, $a^{k+1} > 1$. We have derived this from P_k and $a > 1$. Thus, if P_k is true, then P_{k+1} is true.

Conclusion: P_n is true for all $n \in N$.

49. P_n: $n^2 > 2n$, $n \geq 3$.

Part 1: Show that P_3 is true:

P_3: $3^2 > 2 \cdot 3$

$\quad$ 9 > 6 $\quad$ True

Part 2: Show that if P_k is true, then P_{k+1} is true:

Write out P_k and P_{k+1}.

$\quad P_k$: $k^2 > 2k$

$\quad P_{k+1}$: $(k + 1)^2 > 2(k + 1)$

We start with P_k:

$\quad k^2 > 2k$

Adding $2k + 1$ to both sides: $k^2 + 2k + 1 > 2k + 2k + 1$

$\quad (k + 1)^2 > 2k + 2 + 2k - 1$

Now, $2k - 1 > 0$, $k \in N$, hence $2k + 2 + 2k - 1 > 2k + 2$.

Therefore, $(k + 1)^2 > 2k + 2$

$\qquad\qquad (k + 1)^2 > 2(k + 1)$

Thus, if P_k is true, then P_{k+1} is true.

Conclusion: P_n is true for all integers $n \geq 3$.

51. $3^4 + 4^4 + 5^4 + 6^4 = 2{,}258$

$\qquad\qquad\qquad 7^4 = 2{,}401$

$3^4 + 4^4 + 5^4 + 6^4 \neq 7^4$

Thus, there is no true obvious generalization of the given facts.

53. To prove $a_n = b_n$, $n \in N$, write:

P_n: $a_n = b_n$.

Proof: Part 1: Show P_1 is true.

P_1: $a_1 = b_1$ $\quad a_1 = 1$ $\quad b_1 = 2 \cdot 1 - 1 = 1$

Thus, $a_1 = b_1$

Part 2: Show that if P_k is true, then P_{k+1} is true:

Write out P_k and P_{k+1}.

$\quad P_k$: $a_k = b_k$

$\quad P_{k+1}$: $a_{k+1} = b_{k+1}$

We start with P_k:

$\quad a_k = b_k$

Now, $a_{k+1} = a_{k+1-1} + 2 = a_k + 2 = b_k + 2 = 2k - 1 + 2 = 2k + 1 = 2(k + 1) - 1 = b_{k+1}$

Therefore, $a_{k+1} = b_{k+1}$

Thus, if P_k is true, then P_{k+1} is true.

Conclusion: P_n is true for all $n \in N$. Hence, $\{a_n\} = \{b_n\}$.

55. To prove: $a_n = b_n$, $n \in N$, write:

P_n: $a_n = b_n$.

Proof: Part 1: Show P_1 is true.

P_1: $a_1 = b_1$ $a_1 = 2$ $b_1 = 2^{2 \cdot 1 - 1} = 2^1 = 2$

Thus, $a_1 = b_1$

Part 2: Show that if P_k is true, then P_{k+1} is true:

Write out P_k and P_{k+1}.

$\quad P_k$: $a_k = b_k$

$\quad\ P_{k+1}$: $a_{k+1} = b_{k+1}$

We start with P_k:

$\quad a_k = b_k$

Now, $a_{k+1} = 2^2 a_{k+1-1} = 2^2 a_k = 2^2 b_k = 2^2 2^{2k-1} = 2^{2+2k-1} = 2^{2(k+1)-1} = b_{k+1}$

Therefore, $a_{k+1} = b_{k+1}$

Thus, if P_k is true, then P_{k+1} is true.

Conclusion: P_n is true for all $n \in N$. Hence, $\{a_n\} = \{b_n\}$.

Exercise 10-3

Key Ideas and Formulas

A sequence a_1, a_2, a_3 $\cdots$, a_n, $\cdots$ is called an arithmetic sequence or arithmetic progression if there exists a constant d, called the *common difference*, such that $a_n - a_{n-1} = d$, that is, $a_n = a_{n-1} + d$ for every $n > 1$. Then $a_n = a_1 + (n - 1)d$ for every $n > 1$.

A sequence a_1, a_2, a_3, $\cdots$, a_n is called a geometric sequence or geometric progression if there exists a constant r, called the common ratio, such that

$$\frac{a_n}{a_{n-1}} = r, \text{ that is, } a_n = ra_{n-1}$$

Then $a_n = a_1 r^{n-1}$ for every $n > 1$

Sum of an arithmetic series: Let $S_n = \sum\limits_{k=1}^{n} a_n$. Then

$$S_n = \frac{n}{2}[2a_1 + (n - 1)d] = \frac{n}{2}(a_1 + a_n)$$

Sum of a geometric series: Let $S_n = \sum\limits_{k=1}^{n} a_k$, then

$$S_n = \frac{a_1 - a_1 r^n}{1 - r} = \frac{a_1 - ra_n}{1 - r} \quad r \neq 1$$

Sum of an infinite geometric series

$$S_\infty = \frac{a_1}{1 - r} \quad |r| < 1 \text{ only}. \text{ If } r > 1 \text{ or } r < -1 \text{ the infinite geometric series has no sum.}$$

---

1. (A) Since $(-16) - (-11) = -5$ and $(-21) - (-16) = -5$, the given terms can start an arithmetic sequence with $d = -5$. Then next terms are then $-21 + (-5) = -26$, and $(-26) + (-5) = -31$.

(B) Since $(-4) - 2 = -6$ and $8 - (-4) = 12$, there is no common difference. Since $(-4) \div 2 = -2$ and $8 \div (-4) = -2$, the given terms can start a geometric sequence with $r = -2$. The next terms are then $8 \cdot (-2) = -16$, and $(-16) \cdot (-2) = 32$.

(C) Since $4 - 1 \neq 9 - 4$, there is no common difference, so the sequence is not an arithmetic sequence. Since $4 \div 1 \neq 9 \div 4$, there is no common ratio, so the sequence is not geometric either.

(D) Since $\frac{1}{6} - \frac{1}{2} = -\frac{1}{3}$ and $\frac{1}{18} - \frac{1}{6} = -\frac{1}{9}$, there is no common difference. Since $\frac{1}{6} \div \frac{1}{2} = \frac{1}{3}$ and $\frac{1}{18} \div \frac{1}{6} = \frac{1}{3}$, the given terms can start a geometric sequence with $r = \frac{1}{3}$. The next terms are then $\frac{1}{18} \cdot \frac{1}{3} = \frac{1}{54}$ and $\frac{1}{54} \cdot \frac{1}{3} = \frac{1}{162}$.

3. $a_2 = a_1 + d = 6 + 5 = 11$
$a_3 = a_2 + d = 11 + 5 = 16$
$a_4 = a_3 + d = 16 + 5 = 21$

5. $a_n = a_1 + (n - 1)d$
$a_{21} = a_1 + 20d = (-13) + 20(-7) = -153$
$S_n = \frac{n}{2}[2a_1 + (n - 1)d]$
$S_{21} = \frac{21}{2}[2(-13) + (21 - 1)(-7)]$
$= \frac{21}{2}(-166)$
$= -1,743$

7. $a_2 - a_1 = 5 - (-1) = 6 = d$
$S_n = \frac{n}{2}[2a_1 + (n - 1)d]$
$S_{25} = \frac{25}{2}[2(-1) + (25 - 1)6]$
$= \frac{25}{2}(142)$
$= 1,775$

9. $a_2 - a_1 = 95 - 40 = 55 = d$
$S_n = \frac{n}{2}[2a_1 + (n - 1)d]$
$S_{15} = \frac{15}{2}[2 \cdot 40 + (15 - 1)55]$
$= \frac{15}{2}(850)$
$= 6,375$

11. $a_2 = a_1 r = (-6)\left(-\frac{1}{2}\right) = 3$
$a_3 = a_2 r = 3\left(-\frac{1}{2}\right) = -\frac{3}{2}$
$a_4 = a_3 r = \left(-\frac{3}{2}\right)\left(-\frac{1}{2}\right) = \frac{3}{4}$

13. $a_n = a_1 r^{n-1}$
$a_{10} = 81\left(\frac{1}{3}\right)^{10-1}$
$= \frac{1}{243}$

15. $S_n = \frac{a_1 - ra_n}{1 - r}$
$S_7 = \frac{3 - 3(2,187)}{1 - 3}$
$= 3,279$

17. $a_n = a_1 + (n - 1)d$
$a_5 = a_1 + 4d$
$17 = 5 + 4d$
So, $d = 3$. Therefore
$S_{50} = \frac{50}{2}[2 \cdot 5 + (50 - 1)3]$
$= 3,925$

19. $a_n = a_1 + (n - 1)d$
$a_{15} = a_1 + 14d$
$32 = 25 + 14d$
So, $d = 0.5$. Therefore
$a_{64} = a_1 + 63d$
$a_{64} = 25 + 63(0.5)$
$= 56.5$

21. $a_n = a_1 + (n - 1)d$
$a_9 = a_1 + 8d$
$a_4 = a_1 + 3d$

Eliminating d between these two statements by addition, we have
$$3a_9 = 3a_1 + 24d$$
$$\underline{-8a_4 = -8a_1 - 24d}$$
$$3a_9 - 8a_4 = -5a_1$$
$$a_1 = \frac{3a_9 - 8a_4}{-5} = \frac{3 \cdot 62 - 8 \cdot 27}{-5}$$
$$= 6$$

23. $S_n = \frac{n}{2}(a_1 + a_n)$

$S_{10} = \frac{10}{2}(a_1 + a_{10})$

$35 = \frac{10}{2}(4 + a_{10})$

$35 = 5(4 + a_{10})$

So, $a_{10} = 3$

$a_n = a_1 + (n - 1)d$

$a_{10} = a_1 + 9d$

$3 = 4 + 9d$

$-1 = 9d$

$d = -\frac{1}{9}$

25. $a_n = a_1 r^{n-1}$
$a_5 = a_1 r^4$
$100 = 1r^4$
$r^4 = 100$
$r^2 = 10$
$r = \pm\sqrt{10}$

27. $S_n = \frac{a_1 - a_1 r^n}{1 - r}$

$S_{10} = \frac{64,000 - 64,000\left(-\frac{1}{2}\right)^{10}}{1 - \left(-\frac{1}{2}\right)}$

$= \frac{64,000 - 64,000\left(\frac{1}{1024}\right)}{\frac{3}{2}}$

$= \frac{64,000 - 62.5}{1.5}$

$= 42,625$

29. First find r:
$a_n = a_1 r^{n-1}$
$a_4 = a_1 r^3$
$135 = 40r^3$
$\frac{27}{8} = r^3$
$r = \frac{3}{2}$

$a_2 = ra_1 = \frac{3}{2}(40) = 60$

$a_3 = ra_2 = \frac{3}{2}(60) = 90$

31. $a_n = a_1 + (n - 1)d$
$d = a_2 - a_1$
$= (3 \cdot 2 + 3) - (3 \cdot 1 + 3) = 3$
$a_{51} = a_1 + (51 - 1)d$
$= (3 \cdot 1 + 3) + 50 \cdot 3$
$= 156$

$S_n = \frac{n}{2}(a_1 + a_n)$

$S_{51} = \frac{51}{2}(3 \cdot 1 + 3 + 156)$

$= \frac{51}{2} \cdot 162$

$= 4,131$

33. $S_n = \frac{a_1 - a_1 r^n}{1 - r}$

First, note $a_1 = (-3)^{1-1} = 1$

$r = \frac{a_2}{a_1} = \frac{(-3)^{2-1}}{(-3)^{1-1}} = -3$

$S_7 = \frac{1 - 1(-3)^7}{1 - (-3)}$

$= \frac{2,188}{4}$

$= 547$

35. $g(t) = 5 - t$
$g(1) = 5 - 1 = 4$
$g(51) = 5 - 51 = -46$
$g(1) + g(2) + g(3) + \cdots + g(51) = S_{51}$

$S_n = \frac{n}{2}(a_1 + a_n)$

$S_{51} = \frac{51}{2}(g(1) + g(51))$

$= \frac{51}{2}[4 + (-46)]$

$= -1,071$

37. $g(1) + g(2) + \cdots + g(10)$ is a geometric series, with

$$g(1) = a_1 = \left(\frac{1}{2}\right)^1 = \frac{1}{2}$$

$$r = \frac{g(2)}{g(1)} = \frac{\left(\frac{1}{2}\right)^2}{\left(\frac{1}{2}\right)^1} = \frac{1}{2}$$

$$S_n = \frac{a_1 - a_1 r^n}{1 - r}$$

$$S_{10} = \frac{\frac{1}{2} - \frac{1}{2}\left(\frac{1}{2}\right)^{10}}{1 - \frac{1}{2}}$$

$$= \frac{1 - \left(\frac{1}{2}\right)^{10}}{2 - 1}$$

$$= \frac{1,023}{1,024}$$

39. First, find n:

$$a_n = a_1 + (n - 1)d$$
$$134 = 22 + (n - 1)2$$
$$n = 57$$

Now, find S_{57}

$$S_n = \frac{n}{2}(a_1 + a_n)$$

$$S_{57} = \frac{57}{2}(22 + 134)$$

$$= 4,446$$

41. To prove:

$$1 + 3 + 5 + \cdots + (2n - 1) = n^2$$

The sequence 1, 3, 5, $\cdots$ is an arithmetic sequence, with $d = 2$. We are to find S_n.

But, $S_n = \frac{n}{2}(a_1 + a_n)$

$$= \frac{n}{2}[1 + (2n - 1)]$$

$$= \frac{n}{2} \cdot 2n$$

So, $S_n = n^2$

43. $\dfrac{a_2}{a_1} = \dfrac{a_3}{a_2} = r$ for $a_1 + a_2 + a_3$ to be a geometric series. Hence,

$$\frac{x}{-2} = \frac{-6}{x}$$
$$x^2 = 12$$
$$x = 2\sqrt{3},$$

since x is specified positive.

45. A sequence can be both arithmetic and geometric if $d = 0$ and $r = 1$. Then the sequence is $a_1, a_1, a_1, a_1, \cdots$.

47. Here are computer-generated graphs of $\{a_n\}$ (the dots suggest a straight line) and $\{b_n\}$.

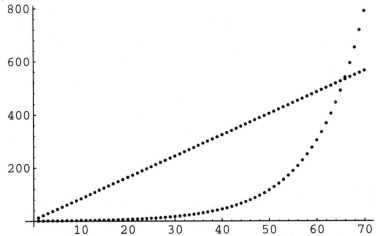

The graph indicates that 66 is the least positive integer n such that $a_n < b_n$. From the table display we note

$$n = 65 \qquad a_n = 525 \qquad b_n = 490.371$$
$$n = 66 \qquad a_n = 533 \qquad b_n = 539.408$$

confirming the graph.

49. Here are computer-generated graphs of $\{a_n\}$ (the dots suggest a straight line) and $\{b_n\}$.

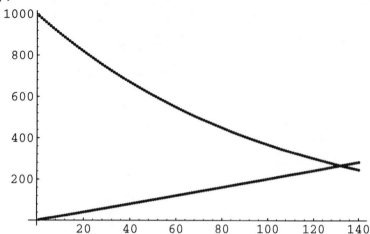

The graph indicates that the least positive integer n such that $a_n < b_n$ is betwen 132 and 136. From the table display we note

$\quad n = 132 \qquad a_n = 265.366 \qquad b_n = 265$

$\quad n = 133 \qquad a_n = 262.713 \qquad b_n = 267$

confirming that 133 is the least positive integer n required.

51. $\dfrac{a_2}{a_1} = 1 \div 3 = \dfrac{1}{3} = r. \quad |r| < 1$

Therefore, this infinite geometric series has a sum.

$S_\infty = \dfrac{a_1}{1 - r}$

$\quad = \dfrac{3}{1 - \frac{1}{3}} = \dfrac{9}{2}$

53. $\dfrac{a_2}{a_1} = \dfrac{4}{2} = 2 = r \geq 1 \quad$ Therefore, this infinite geometric series has no sum.

55. $\dfrac{a_2}{a_1} = \left(-\dfrac{1}{2}\right) \div 2 = -\dfrac{1}{4} = r. \quad |r| < 1$

Therefore, this infinite geometric series has a sum.

$S_\infty = \dfrac{a_1}{1 - r}$

$\quad = \dfrac{2}{1 - (-\frac{1}{4})}$

$\quad = \dfrac{8}{5}$

57. $0.\overline{7} = 0.777... = 0.7 + 0.07 + 0.007 + 0.0007 + ...$

This is an infinite geometric series with

$a_1 = 0.7$ and $r = 0.1$.

Thus,

$S_\infty = \dfrac{a_1}{1 - r} = \dfrac{0.7}{1 - 0.1} = \dfrac{0.7}{0.9} = \dfrac{7}{9}$

59. $0.\overline{54}... = 0.54 + 0.0054 + 0.000054 + ...$

This is an infinite geometric series with $a_1 = 0.54$ and $r = 0.01$. Thus,

$S_\infty = \dfrac{a_1}{1 - r} = \dfrac{0.54}{1 - 0.01} = \dfrac{0.54}{0.99} = \dfrac{6}{11}$

61. $3.\overline{216} = 3.216216216... = 3 + 0.216 + 0.000216 + 0.000000216 + ...$

Therefore, we note: $0.216 + 0.000216 + 0.000000216 + ...$ is an infinite geometric series with $a_1 = 0.216$ and $r = 0.001$. Thus,

$3.\overline{216} = 3 + S_\infty = 3 + \dfrac{a_1}{1 - r} = 3 + \dfrac{0.216}{1 - 0.001} = 3 + \dfrac{0.216}{0.999} = 3\dfrac{8}{37}$ or $\dfrac{119}{37}$

63. Write: P_n: $a_n = a_1 + (n - 1)d$

Proof: Part 1: Show that P_1 is true:
P_1: $a_1 = a_1 + (1 - 1)d = a_1$ P_1 is true.

Part 2: Show that if P_k is true, then P_{k+1} is true:
Write out P_k and P_{k+1}.

    P_k: $a_k = a_1 + (k - 1)d$
    P_{k+1}: $a_{k+1} = a_1 + kd$
We start with P_k:
    $a_k = a_1 + (k - 1)d$
Adding d to both sides:
    $a_k + d = a_1 + (k - 1)d + d$
      $a_{k+1} = a_1 + d[(k - 1) + 1]$
      $a_{k+1} = a_1 + kd$
Thus, if P_k is true, then P_{k+1} is true.

Conclusion: P_n is true for all positive integers n.

65. This is a geometric sequence with ratio $\dfrac{a_n}{a_{n-1}} = -3$. Hence,

    $a_n = a_1 r^{n-1}$
      $= (-2)(-3)^{n-1}$

67. Assume x, y, z are consecutive terms of an arithmetic progression.
Then, $y - x = d$, $z - y = d$
To show $a = x^2 + xy + y^2$, $b = z^2 + xz + x^2$, $c = y^2 + yz + z^2$, are consecutive terms of an arithmetic progression, we need only show that
$b - a = c - b$
$b - a = z^2 + xz + x^2 - (x^2 + xy + y^2)$
      $= z^2 + xz - xy - y^2$
      $= z^2 + x(z - y) - y^2$
      $= z^2 - y^2 + x(z - y)$
      $= (z - y)(z + y + x)$
      $= d(x + y + z)$
$c - b = y^2 + yz + z^2 - (z^2 + xz + x^2)$
      $= y^2 + yz - xz - x^2$
      $= y^2 - x^2 + yz - xz$
      $= y^2 - x^2 + z(y - x)$
      $= (y - x)[y + x + z]$
      $= d(x + y + z)$
Hence, a, b, c, that is, $x^2 + xy + y^2$, $z^2 + xz + x^2$, $y^2 + yz + z^2$ are consecutive terms of an arithmetic progression.

69. Write: P_n: $a_n = a_1 r^{n-1}$, $n \in N$

Proof: Part 1: Show that P_1 is true:
P_1: $a_1 = a_1 r^{1-1}$
      $= a_1 r^0$
      $= a_1$ True.

Part 2: Show that if P_k is true, then P_{k+1} is true:
Write out P_k and P_{k+1}.

    P_k: $a_k = a_1 r^{k-1}$
    P_{k+1}: $a_{k+1} = a_1 r^k$
We start with P_k:
    $a_k = a_1 r^{k-1}$

Multiply both sides by r:
$$a_k r = a_1 r^{k-1} r$$
$$a_k r = a_1 r^k$$

But, we are given that $a_{k+1} = ra_k = a_k r$.

Hence, $a_{k+1} = a_1 r^k$

Thus, if P_k is true, then P_{k+1} is true.

Conclusion: P_n is true for all $n \in N$.

71. Given a, b, c, d, e, f is an arithmetic progression, let D be the common difference. Then, assuming $D \neq 0$:

$b = a + D$, $c = a + 2D$, $d = a + 3D$, $e = a + 4D$, $f = a + 5D$

We know from Cramer's Rule that there will be a unique solution if

$$\begin{vmatrix} a & b \\ d & e \end{vmatrix} \neq 0$$

$$\begin{vmatrix} a & b \\ d & e \end{vmatrix} = \begin{vmatrix} a & a + D \\ a + 3D & a + 4D \end{vmatrix} = a(a + 4D) - (a + 3D)(a + D)$$

$$= a^2 + 4aD - (a^2 + 4aD + 3D^2)$$

$$= -3D^2 \neq 0$$

We can now calculate x and y, since

$$x = \frac{\begin{vmatrix} c & b \\ f & e \end{vmatrix}}{-3D^2} = \frac{ce - fb}{-3D^2} = \frac{(a + 2D)(a + 4D) - (a + 5D)(a + D)}{-3D^2}$$

$$= \frac{a^2 + 6aD + 8D^2 - (a^2 + 6aD + 5D^2)}{-3D^2}$$

$$= \frac{3D^2}{-3D^2}$$

$$= -1$$

$$y = \frac{\begin{vmatrix} a & c \\ d & f \end{vmatrix}}{-3D^2} = \frac{af - dc}{-3D^2} = \frac{a(a + 5D) - (a + 3D)(a + 2D)}{-3D^2}$$

$$= \frac{a^2 + 5aD - (a^2 + 5aD + 6D^2)}{-3D^2}$$

$$= \frac{-6D^2}{-3D^2}$$

$$= 2$$

73. With each firm, the salaries form an arithmetic sequence. We are asked for the sum of fifteen terms, or S_{15}, given a_1 and d.

$$S_n = \frac{n}{2}[2a_1 + (n - 1)d]$$

$$S_{15} = \frac{15}{2}[2a_1 + 14d] = 15(a_1 + 7d)$$

Firm A: $a_1 = 25{,}000$ $d = 1{,}200$ $S_{15} = 15(25{,}000 + 7 \cdot 1{,}200) = \$501{,}000$

Firm B: $a_1 = 28{,}000$ $d = 800$ $S_{15} = 15(28{,}000 + 7 \cdot 800) = \$504{,}000$

75. We are asked for the sum of an infinite geometric series.

$a_1 = \$800,000$

$r = 0.8 \quad |r| \le 1,$

so the series has a sum,

$S_\infty = \dfrac{a_1}{1 - r}$

$\quad = \dfrac{\$800,000}{1 - 0.8}$

$\quad = \dfrac{\$800,000}{0.2}$

$\quad = \$4,000,000$

77. After one year, $P(1 + r)$ is present. Hence, the geometric sequence has $a_1 = P(1 + r)$. The ratio is given as $(1 + r)$, hence

$a_n = a_1(1 + r)^{n-1}$

$\quad = P(1 + r)(1 + r)^{n-1}$

$\quad = P(1 + r)^n$ is the amount present after n years.

The time taken for P to double is represented by n years. We set $A = 2P$, $r = 0.06$, then solve

$2P = P(1 + 0.06)^n$

$2 = (1.06)^n$

$\log 2 = n \log 1.06$

$n = \dfrac{\log 2}{\log 1.06}$

$\quad \approx 12$ years

79. This involves an arithmetic sequence. Let d = increase in earnings each year.

$a_1 = 7,000$

$a_{11} = 14,000$

$a_n = a_1 + (n - 1)d$

$14,000 = 7,000 + (11 - 1)d$

$d = \$700$

The amount of money received over the 11 years is

$S_{11} = a_1 + a_2 + \cdots + a_{11}$

$S_n = \dfrac{n}{2}(a_1 + a_n)$

$S_{11} = \dfrac{11}{2}(7,000 + 14,000) = \$115,500$

81. We are asked for the sum of an infinite geometric series.

a_1 = number of revolutions in the first minute = 300

$r = \dfrac{2}{3} \quad |r| < 1$, so the series has a sum,

$S_\infty = \dfrac{a_1}{1 - r} = \dfrac{300}{1 - \frac{2}{3}} = 900$ revolutions.

83. This involves a geometric sequence. Let $a_n = 2,000$ calories. There are five stages: $n = 5$. We require a_1 on the assumption that $r = 20\% = \dfrac{1}{5}$.

$a_n = a_1 r^{n-1}$

$2000 = a_1 \left(\dfrac{1}{5}\right)^{5-1}$

$a_1 = 2,000 \cdot 5^4$

$\quad = 1,250,000$ calories

85. We have an arithmetic sequence, 16, 48, 80, $\cdots$, with $a_1 = 16$, $d = 32$.

(A) This requires a_{11}: $a_n = a_1 + (n - 1)d$

$a_{11} = 16 + (11 - 1)32$

$\quad = 336$ feet

(B) This requires s_{11}: $s_n = \dfrac{n}{2}(a_1 + a_n)$

$s_{11} = \dfrac{11}{2}(16 + 336)$

$\quad = 1,936$ feet

(C) This requires s_t: $s_n = \dfrac{n}{2}[2a_1 + (n - 1)d]$

$$s_t = \dfrac{t}{2}[2a_1 + (t - 1)32]$$

$$= \dfrac{t}{2}[32 + (t - 1)32]$$

$$= \dfrac{t}{2} \cdot 32t$$

$$= 16t^2 \text{ feet}$$

87. This involves a geometric sequence. Let $a_1 = 2A_0$ = number present after 1 half-hour period. In t hours, $2t$ half-hours will have elapsed, hence, $n = 2t$; $r = 2$, since the number of bacteria doubles in each period.

$$a_n = a_1 r^{n-1}$$
$$a_{2t} = 2A_0 2^{2t-1}$$
$$= A_0 2^1 2^{2t-1}$$
$$= A_0 2^{2t}$$

89. If b_n is the brightness of an nth-magnitude star, we find r for the geometric progression $b_1, b_2, b_3, \dots,$ given $b_1 = 100b_6$.

$$b_n = b_1 r^{n-1}$$
$$b_6 = b_1 r^{6-1}$$
$$b_6 = b_1 r^5$$
$$b_6 = 100b_6 r^5$$
$$\frac{1}{100} = r^5$$
$$r = \sqrt[5]{0.01} = 10^{-0.4} = 0.398$$

91. This involves a geometric sequence. a_1 = amount of money on first square; $a_1 = \$0.01$, $r = 2$; a_n = amount of money on nth square.

$$a_n = a_1 r^{n-1}$$
$$a_{64} = 0.01 \cdot 2^{64-1}$$
$$= 9.22 \times 10^{16} \text{ dollars}$$

The amount of money on the whole board = $a_1 + a_2 + a_3 + \dots + a_{64} = S_{64}$.

$$S_n = \frac{a_1 - ra_n}{1 - r}$$
$$S_{64} = \frac{0.01 - 2(9.223 \times 10^{16})}{1 - 2}$$
$$= 1.845 \times 10^{17} \text{ dollars}$$

93. This involves a geometric sequence. Let $a_1 = 15$ = pressure at sea level, $r = \dfrac{1}{10}$ = factor of decrease. We require a_5 = pressure after four 10-mile increases in altitude.

$$a_5 = 15\left(\frac{1}{10}\right)^{5-1}$$

$$= 0.0015 \text{ pounds per square inch.}$$

95. This involves an infinite geometric series. a_1 = perimeter of first triangle = 1, $r = \dfrac{1}{2}$, $|r| < 1$, so the series has a sum.

$$S_\infty = \frac{a_1}{1 - r} = \frac{1}{1 - \frac{1}{2}} = 2$$

97.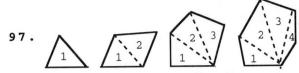

From the figure, and ordinary induction, it should be clear that the interior angles of an $n + 2$-sided polygon, $n = 1, 2, 3, \cdots$, are those of n triangles, that is, $180°n$. Thus, for the sequence of interior angles $\{a_n\}$,

$a_n - a_{n-1} = 180°n - 180°(n - 1) = 180° = d$.

A detailed proof by mathematical induction is omitted.

For a 21-sided polygon, we use $a_n = a_1 + (n - 1)d$ with $a_1 = 180°$, $d = 180°$, and $n + 2 = 21$, hence, $n = 19$.

$a_{19} = 180° + (19 - 1)180°$

$\quad\ = 3,420°$

Exercise 10-4

Key Ideas and Formulas

For $n \in N$, $n! = n(n - 1) \cdots 2 \cdot 1$

$\qquad\qquad\quad 1! = 0! = 1$

$\qquad\qquad\quad n! = n \cdot (n - 1)!$

$(a + b)^0 = 1$

$(a + b)^1 = a + b$

$(a + b)^2 = a^2 + 2ab + b^2$

$(a + b)^3 = a^3 + 3a^2b + 3ab^2 + b^3$

$(a + b)^4 = a^4 + 4a^3b + 6a^2b^2 + 4ab^3 + b^4$

$(a + b)^5 = a^5 + 5a^4b + 10a^3b^2 + 10a^2b^3 + 5ab^4 + b^5$

$$\binom{n}{r} = \frac{n!}{r!(n - r)!} = \frac{n(n - 1)(n - 2)\cdots(n - r + 1)}{r(r - 1)\cdots 2 \cdot 1} = {}_nC_r$$

Binomial Formula

$$(a + b)^n = \sum_{k=0}^{n} \binom{n}{k}a^{n-k}b^k \quad n \in N, \ n \geq 1$$

---

1. $7! = 7 \cdot 6 \cdot 5 \cdot 4 \cdot 3 \cdot 2 \cdot 1 = 5,040$

3. $\dfrac{15!}{13!} = \dfrac{15 \cdot 14 \cdot 13!}{13!} = 210$

5. $4! + 5! = 4 \cdot 3 \cdot 2 \cdot 1 + 5 \cdot 4 \cdot 3 \cdot 2 \cdot 1 = 24 + 120 = 144$

7. $\dfrac{8!}{3!5!} = \dfrac{8 \cdot 7 \cdot 6 \cdot 5 \cdot 4 \cdot 3 \cdot 2 \cdot 1}{3 \cdot 2 \cdot 1 \cdot 5 \cdot 4 \cdot 3 \cdot 2 \cdot 1} = 56$

9. $\dfrac{7!}{0!(7 - 0)!} = \dfrac{7!}{1 \cdot 7!} = 1$

11. $\dfrac{10!}{7!} = \dfrac{10 \cdot 9 \cdot 8 \cdot 7!}{7!} = 10 \cdot 9 \cdot 8 = 720$

13. Since $n! = n(n - 1)!$, $\dfrac{n!}{(n - 1)!} = n$. Hence, $9 = \dfrac{9!}{(9 - 1)!} = \dfrac{9!}{8!}$.

15. $6 \cdot 7 \cdot 8 = \dfrac{1 \cdot 2 \cdot 3 \cdot 4 \cdot 5 \cdot 6 \cdot 7 \cdot 8}{1 \cdot 2 \cdot 3 \cdot 4 \cdot 5} = \dfrac{8!}{5!}$

17. $\dbinom{13}{9} = \dfrac{13!}{9!(13 - 9)!} = \dfrac{13!}{9!4!} = \dfrac{13 \cdot 12 \cdot 11 \cdot 10 \cdot 9!}{9!4 \cdot 3 \cdot 2 \cdot 1} = 715$

19. $\dbinom{14}{7} = \dfrac{14!}{7!(14-7)!} = \dfrac{14!}{7!7!} = \dfrac{14\cdot13\cdot12\cdot11\cdot10\cdot9\cdot8\cdot7!}{7\cdot6\cdot5\cdot4\cdot3\cdot2\cdot1\cdot7!} = 3,432$

21. $\dbinom{100}{97} = \dfrac{100!}{97!(100-97)!} = \dfrac{100!}{97!3!} = \dfrac{100\cdot99\cdot98\cdot97!}{97!\cdot3\cdot2\cdot1} = 161,700$

23. The answer to this question depends on the model of calculator. An example: on a TI-82 or a TI-83 model calculator, $69! = 1.711 \times 10^{98}$, but $70!$ produces an overflow error.

25. $(2x - 3y)^3 = [2x + (-3y)]^3 = \displaystyle\sum_{k=0}^{3}\dbinom{3}{k}(2x)^{3-k}(-3y)^k$

$= \dbinom{3}{0}(2x)^3 + \dbinom{3}{1}(2x)^2(-3y)^1 + \dbinom{3}{2}(2x)(-3y)^2 + \dbinom{3}{3}(-3y)^3$

$= 8x^3 + 3(4x^2)(-3y) + 3(2x)(9y^2) + (-27y^3)$

$= 8x^3 - 36x^2y + 54xy^2 - 27y^3$

27. $(x - 2)^4 = [x + (-2)]^4 = \displaystyle\sum_{k=0}^{4}\dbinom{4}{k}x^{4-k}(-2)^k$

$= \dbinom{4}{0}x^4 + \dbinom{4}{1}x^3(-2)^1 + \dbinom{4}{2}x^2(-2)^2 + \dbinom{4}{3}x(-2)^3 + \dbinom{4}{4}(-2)^4$

$= x^4 - 8x^3 + 24x^2 - 32x + 16$

29. $(2x - y)^5 = [2x + (-y)]^5 = \displaystyle\sum_{k=0}^{5}\dbinom{5}{k}(2x)^{5-k}(-y)^k$

$= \dbinom{5}{0}(2x)^5 + \dbinom{5}{1}(2x)^4(-y)^1 + \dbinom{5}{2}(2x)^3(-y)^2 + \dbinom{5}{3}(2x)^2(-y)^3$

$\qquad\qquad\qquad + \dbinom{5}{4}(2x)(-y)^4 + \dbinom{5}{5}(-y)^5$

$= 32x^5 - 80x^4y + 80x^3y^2 - 40x^2y^3 + 10xy^4 - y^5$

31. In the expansion of $(a + b)^n$, the exponent of b in the rth term is $r - 1$ and the exponent of a is $n - (r - 1)$. Here, $r = 7$, $n = 15$.
$r - 1 = 6$ and $n - (r - 1) = 9$.

Seventh term $= \dbinom{15}{6}u^9v^6$

$= \dfrac{15!}{9!6!}u^9v^6$

$= \dfrac{15\cdot14\cdot13\cdot12\cdot11\cdot10}{6\cdot5\cdot4\cdot3\cdot2\cdot1}u^9v^6$

$= 5,005u^9v^6$

33. In the expansion of $(a + b)^n$, the exponent of b in the rth term is $r - 1$ and the exponent of a is $n - (r - 1)$. Here, $r = 9$, $n = 10$.
$r - 1 = 8$ and $n - (r - 1) = 2$.

Ninth term $= \dbinom{10}{8}(2m)^2n^8$

$= \dfrac{10}{8!2!}4m^2n^8$

$= \dfrac{10\cdot9}{2\cdot1}4m^2n^8$

$= 180m^2n^8$

35. In the expansion of $(a + b)^n$, the exponent of b in the rth term is $r - 1$ and the exponent of a is $n - (r - 1)$. Here, $r = 5$, $n = 20$.
$r - 1 = 4$ and $n - (r - 1) = 16$.

$$\text{Fifth term} = \binom{20}{4} w^{16} (-3)^4$$

$$= \frac{20!}{4!16!} \, 81 w^{16}$$

$$= \frac{20 \cdot 19 \cdot 18 \cdot 17}{4 \cdot 3 \cdot 2 \cdot 1} \, 81 w^{16}$$

$$= 392,445 w^{16}$$

37. In the expansion of $(a + b)^n$, the exponent of b in the rth term is $r - 1$ and the exponent of a is $n - (r - 1)$. Here, $r = 6$, $n = 8$.
$r - 1 = 5$ and $n - (r - 1) = 3$.

$$\text{Sixth term} = \binom{8}{5} (3x)^3 (-2y)^5$$

$$= \frac{8!}{5!3!} (3x)^3 (-2y)^5$$

$$= \frac{8 \cdot 7 \cdot 6 \cdot 5!}{5! \cdot 3 \cdot 2 \cdot 1} (27x^3)(-32y^5)$$

$$= -(56)(27)(32) x^3 y^5$$

$$= -48,384 x^3 y^5$$

39. Here is a computer-generated graph of the sequence

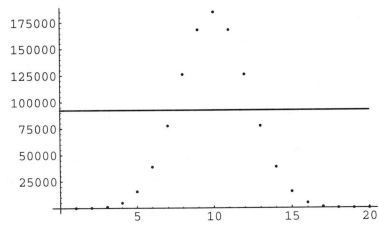

The solid line indicates $\frac{1}{2}(184,756) = 92,378$, that is, half the largest term.

Thus there are 5 terms greater than half the largest term. From the table display we note

$$\binom{20}{8} = 125,970 \qquad \binom{20}{9} = 167,960 \qquad \binom{20}{10} = 184,756$$

$$\binom{20}{11} = 167,960 \qquad \binom{20}{12} = 125,970$$

are all greater than 92,378, but the other terms are less than 92,378.

41. Here is a computer-generated graph of the sequence

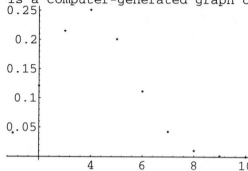

From the table display we find the largest term to be $\binom{10}{4}(.6)^6(.4)^4 = 0.2508$ as displayed in the graph.

(B) According to the binomial formula,

$$\sum_{k=0}^{n}\binom{n}{k}a^{n-k}b^k = (a + b)^n$$

Thus, $a_0 + a_1 + a_2 + \cdots + a_{10} = \sum_{k=0}^{10}\binom{10}{k}(0.6)^{10-k}(0.4)^k = (0.6 + 0.4)^{10} = 1^{10} = 1$

43. $(1.01)^{10} = (1 + 0.01)^{10} = \sum_{k=0}^{10}\binom{10}{k}1^{10-k}(0.01)^k,\ 1^{10-k} = 1$

$$= \sum_{k=0}^{10}\binom{10}{k}(0.01)^k$$

$$= \binom{10}{0} + \binom{10}{1}(0.01)^1 + \binom{10}{2}(0.01)^2 + \binom{10}{3}(0.01)^3$$

$$+ \binom{10}{4}(0.01)^4 + \binom{10}{5}(0.01)^5 + \binom{10}{6}(0.01)^6 + \binom{10}{7}(0.01)^7$$

$$+ \binom{10}{8}(0.01)^8 + \binom{10}{9}(0.01)^9 + \binom{10}{10}(0.01)^{10}$$

$$= 1 + 0.1 + 0.0045 + 0.00012 + (0.0000021 + \text{other terms}$$
$$\text{with no effect in fourth decimal place})$$

$$= 1.1046$$

45. True. $\binom{n}{r} = \dfrac{n!}{r!(n-r)!} = \dfrac{n!}{(n-r)!r!} = \dfrac{n!}{(n-r)![n-(n-r)]!} = \binom{n}{n-r}$

47. True. $\binom{p}{0} = \binom{p}{p} = 1$; otherwise, for $1 \le r \le p - 1$, $\binom{p}{r}$ is an integer,

but $\binom{p}{r} = \dfrac{p(p-1)\cdots 1}{r!(p-r)!}$ and therefore is divisible by p.

49. $\binom{k}{r-1} + \binom{k}{r} = \dfrac{k!}{(r-1)!(k-r+1)!} + \dfrac{k!}{r!(k-r)!}$

$\qquad\qquad\qquad = \dfrac{rk! + (k-r+1)k!}{r!(k-r+1)!} = \dfrac{(r+k-r+1)k!}{r!(k-r+1)!}$

$\qquad\qquad\qquad = \dfrac{(k+1)k!}{r!(k-r+1)!} = \dfrac{(k+1)!}{r!(k-r+1)!}$

$\qquad\qquad\qquad = \binom{k+1}{r}$

51. $\binom{k}{k} = \dfrac{k!}{k!(k-k)!} = 1 = \dfrac{(k+1)!}{(k+1)![(k+1)-(k+1)]!} = \binom{k+1}{k+1}$

53. $2^n = (1+1)^n = \displaystyle\sum_{k=0}^{n}\binom{n}{k}1^{n-k}1^k, \quad 1^{n-k} = 1^k = 1$

$\qquad\qquad\quad = \displaystyle\sum_{k=0}^{n}\binom{n}{k}$

$\qquad\qquad\quad = \binom{n}{0} + \binom{n}{1} + \binom{n}{2} + \cdots + \binom{n}{n}$

Exercise 10-5

Key Ideas and Formulas

Multiplication Principle (Fundamental Counting Principle)

1. If two operations O_1 and O_2 are performed in order, with N_1 possible outcomes for the first operation and N_2 possible outcomes for the second operation, then there are

$$N_1 \cdot N_2$$

possible combined operations of the first operation followed by the second.

2. In general, if n operations O_1, O_2, …, O_n are performed in order, with possible number of outcomes N_1, N_2, …, N_n, respectively, then there are

$$N_1 \cdot N_2 \cdot \ldots \cdot N_n$$

possible combined outcomes of the operations performed in the given order.

Number of Permutations of n objects: $P_{n,n} = n!$

Number of Permutations of n objects, taken r at a time:

$$P_{n,r} = n(n-1)\ldots(n-r+1) = \frac{n!}{(n-r)!}$$

Number of Combinations of n objects, taken r at a time:

$$C_{n,r} = \binom{n}{r} = \frac{P_{n,r}}{r!} = \frac{n!}{r!(n-r)!} \quad 0 \le r \le n$$

In a permutation, the order of the objects counts. In a combination, the order of the objects does not count.

---

1. $\dfrac{15!}{12!} = \dfrac{15 \cdot 14 \cdot 13 \cdot 12!}{12!} = 2{,}730$

3. $\dfrac{32!}{0!32!} = \dfrac{32!}{1 \cdot 32!} = 1$

5. $\dfrac{9!}{6!3!} = \dfrac{9 \cdot 8 \cdot 7 \cdot 6!}{6! \cdot 3 \cdot 2 \cdot 1} = 84$

7. $\dfrac{16!}{4!(16-4)!} = \dfrac{16!}{4!12!} = \dfrac{16 \cdot 15 \cdot 14 \cdot 13 \cdot 12!}{4 \cdot 3 \cdot 2 \cdot 1 \cdot 12!} = 1{,}820$

9. $P_{8,5} = \dfrac{8!}{(8-5)!} = \dfrac{8!}{3!} = \dfrac{8 \cdot 7 \cdot 6 \cdot 5 \cdot 4 \cdot 3!}{3!} = 6{,}720$

11. $P_{52,3} = \dfrac{52!}{(52-3)!} = \dfrac{52!}{49!} = \dfrac{52 \cdot 51 \cdot 50 \cdot 49!}{49!} = 132{,}600$

13. $C_{13,5} = \dfrac{13!}{5!(13-5)!} = \dfrac{13!}{5!8!} = \dfrac{13 \cdot 12 \cdot 11 \cdot 10 \cdot 9 \cdot 8!}{5 \cdot 4 \cdot 3 \cdot 2 \cdot 1 \cdot 8!} = 1{,}287$

15. $C_{52,5} = \dfrac{52!}{5!(52-5)!} = \dfrac{52!}{5!47!} = \dfrac{52!}{5!47!} = \dfrac{52 \cdot 51 \cdot 50 \cdot 49 \cdot 48 \cdot 47!}{5 \cdot 4 \cdot 3 \cdot 2 \cdot 1 \cdot 47!} = 2{,}598{,}960$

17.
O_1: Selecting the color	N_1: 5 ways
O_2: Selecting the transmission	N_2: 3 ways
O_3: Selecting the interior	N_3: 4 ways
O_4: Selecting the engine	N_4: 2 ways

Applying the multiplication principle, there are $5 \cdot 3 \cdot 4 \cdot 2 = 120$ variations of the car.

19. Order is important here. We use permutations, selecting, in order, three horses out of ten:
$P_{10,3} = 10 \cdot 9 \cdot 8 = 720$ different finishes

21. For the subcommittee, order is not important. We use combinations, selecting three persons out of seven:
$C_{7,3} = \dfrac{7!}{3!(7-3)!} = \dfrac{7!}{3!4!} = \dfrac{7 \cdot 6 \cdot 5 \cdot 4!}{3 \cdot 2 \cdot 1 \cdot 4!} = 35$ subcommittees
In choosing a president, a vice-president, and a secretary, we can use permutations, or apply the multiplication principle.
O_1: Selecting the president N_1: 7 ways
O_2: Selecting the vice-president N_2: 6 ways (the president is not considered)
O_3: Selecting the secretary N_3: 5 ways (the president and vice-president
                                                are not considered)
Thus, there are $N_1 \cdot N_2 \cdot N_3 = 7 \cdot 6 \cdot 5\ (= P_{7,3}) = 210$ ways.

23. For each game, we are selecting two teams out of ten to be opponents. Since the order of the opponents does not matter (this has nothing to do with the order in which the games might be played, which is not under discussion here), we use combinations.
$C_{10,2} = \dfrac{10!}{2!(10-2)!} = \dfrac{10!}{2!8!} = \dfrac{10 \cdot 9 \cdot 8!}{2 \cdot 1 \cdot 8!} = 45$ games

25.

	No letter can be repeated	Allowing letters to repeat
O_1: Selecting first letter	N_1: 6 ways	N_1: 6 ways
O_2: Selecting second letter	N_2: 5 ways	N_2: 6 ways
O_3: Selecting third letter	N_3: 4 ways	N_3: 6 ways
O_4: Selecting fourth letter	N_4: 3 ways	N_4: 6 ways
	$P_{6,4} = 6 \cdot 5 \cdot 4 \cdot 3 = 360$ possible code words	$6 \cdot 6 \cdot 6 \cdot 6 = 1{,}296$ possible code words

27. We are selecting five cards out of the 13 hearts in the deck. The order is not important, so we use combinations.
$C_{13,5} = \dfrac{13!}{5!(13-5)!} = \dfrac{13!}{5!8!} = \dfrac{13 \cdot 12 \cdot 11 \cdot 10 \cdot 9 \cdot 8!}{5 \cdot 4 \cdot 3 \cdot 2 \cdot 1 \cdot 8!} = 1{,}287$

29.

	Repeats allowed	No repeats allowed
O_1: Selecting first letter	N_1: 26 ways	N_1: 26 ways
O_2: Selecting second letter	N_2: 26 ways	N_2: 25 ways
O_3: Selecting third letter	N_3: 26 ways	N_3: 24 ways
O_4: Selecting first digit	N_4: 10 ways	N_4: 10 ways
O_5: Selecting second digit	N_5: 10 ways	N_5: 9 ways
O_6: Selecting third digit	N_6: 10 ways	N_6: 8 ways

$$26 \cdot 26 \cdot 26 \cdot 10 \cdot 10 \cdot 10 = \qquad 26 \cdot 25 \cdot 24 \cdot 10 \cdot 9 \cdot 8 =$$
$$17,576,000 \text{ license plates} \qquad 11,232,000 \text{ license plates}$$

31. O_1: Choosing 5 spades out of 13 possible (order is not important)

N_1: $C_{13,5}$

O_2: Choosing 2 hearts out of 13 possible (order is not important)

N_2: $C_{13,2}$

Using the multiplication principle, we have:

$$\text{Number of hands} = C_{13,5} \cdot C_{13,2} = \frac{13!}{5!(13-5)!} \cdot \frac{13!}{2!(13-2)!}$$
$$= 1,287 \cdot 78$$
$$= 100,386$$

33. O_1: Choosing 3 appetizers out of 8 possible (order is not important)

N_1: $C_{8,3}$

O_2: Choosing 4 main courses out of 10 possible (order is not important)

N_2: $C_{10,4}$

O_3: Choosing 2 desserts out of 7 possible (order is not important)

N_3: $C_{7,2}$

Using the multiplication principle, we have:

$$\text{Number of banquets} = C_{8,3} \cdot C_{10,4} \cdot C_{7,2} = \frac{8!}{3!(8-3)!} \cdot \frac{10!}{4!(10-4)!} \cdot \frac{7!}{2!(7-2)!}$$
$$= 56 \cdot 210 \cdot 21$$
$$= 246,960$$

35. (A)

$P_{10,0} = 1$	$0! = 1$
$P_{10,1} = 10$	$1! = 1$
$P_{10,2} = 90$	$2! = 2$
$P_{10,3} = 720$	$3! = 6$
$P_{10,4} = 5,040$	$4! = 24$
$P_{10,5} = 30,240$	$5! = 120$
$P_{10,6} = 151,200$	$6! = 720$
$P_{10,7} = 604,800$	$7! = 5,040$
$P_{10,8} = 1,814,400$	$8! = 40,320$
$P_{10,9} = 3,628,800$	$9! = 362,880$
$P_{10,10} = 3,628,800$	$10! = 3,628,800$

Thus, $P_{10,r} \geq r!$ for $r = 0, 1, \ldots, 10$

(B) If $r = 0$, $P_{10,0} = 0! = 1$. If $r = 10$, $P_{10,10} = 10! = 3,628,800$

(C) $P_{n,r}$ and $r!$ are each the product of r consecutive integers, the largest of which is n for $P_{n,r}$ and r for $r!$. Thus if $r \leq n$, $P_{n,r} \geq r!$

37. O_1: Choosing a left glove out of 12 possible
N_1: 12
O_2: Choosing a right glove out of all the right gloves that do *not* match the left glove already chosen
N_2: 11
Using the multiplication principle, we have:
Number of ways to mismatch gloves = $12 \cdot 11 = 132$.

39. (A) We are choosing 2 points out of the 8 to join by a chord. Order is not important.
$$C_{8,2} = \frac{8!}{2!(8-2)!} = 28 \text{ chords}$$

(B) No three of the points can be collinear, since no line intersects a circle at more than two points. Thus, we can select any three of the 8 to use as vertices of the triangle. Order is not important.
$$C_{8,3} = \frac{8!}{3!(8-3)!} = 56 \text{ triangles}$$

(C) We can select any four of the eight points to use as vertices of a quadrilateral. Order is not important.
$$C_{8,4} = \frac{8!}{4!(8-4)!} = 70 \text{ quadrilaterals}$$

41. To seat two people, we can seat one person, then the second person.
O_1: Seat the first person in any chair
N_1: 5 ways
O_2: Seat the second person in any remaining chair
N_2: 4 ways

Thus, applying the multiplication principle, we can seat two persons in $N_1 \cdot N_2 = 5 \cdot 4 = 20$ ways.

We can continue this reasoning for a third person.
O_3: Seat the third person in any of the three remaining chairs
N_3: 3 ways

Thus we can seat 3 persons in $5 \cdot 4 \cdot 3 = 60$ ways.

For a fourth person:
O_4: Seat the fourth person in any of the two remaining chairs
N_4: 2 ways

Thus we can seat 4 persons in $5 \cdot 4 \cdot 3 \cdot 2 = 120$ ways.
For a fifth person:
O_5: Seat the fifth person. There will be only one chair remaining.
N_5: 1 way

Thus we can seat 5 persons in $5 \cdot 4 \cdot 3 \cdot 2 \cdot 1 = 120$ ways

43. (A) Order is important, so we use permutations, selecting, in order, 5 persons out of 8:
$$P_{8,5} = 8 \cdot 7 \cdot 6 \cdot 5 \cdot 4 = 6,720 \text{ teams.}$$

(B) Order is not important, so we use combinations, selecting 5 persons out of 8
$$C_{8,5} = \frac{8!}{5!(8-5)!} = \frac{8!}{5!3!} = \frac{8 \cdot 7 \cdot 6 \cdot 5!}{5! \cdot 3 \cdot 2 \cdot 1} = 56 \text{ teams}$$

(C) O_1: Selecting either Mike or Ken out of {Mike, Ken}

N_1: $C_{2,1}$

O_2: Selecting the 4 remaining players out of the 6 possibilities that do not include either Mike or Ken.

N_2: $C_{6,4}$

Using the multiplication principle, we have

$$N_1 \cdot N_2 = C_{2,1} \cdot C_{6,4} = \frac{2!}{1!(2-1)!} \cdot \frac{6!}{4!(6-4)!}$$
$$= \frac{2!}{1!1!} \cdot \frac{6!}{4!2!}$$
$$= 2 \cdot 15$$
$$= 30 \text{ teams}$$

45. The number of ways to deal a hand containing exactly 1 king is computed as follows:

O_1: Choosing 1 king out of 4 possible (order is not important)

N_1: $C_{4,1}$

O_2: Choosing 4 cards out of 48 possible (order is not important)

N_2: $C_{48,4}$

Using the multiplication principle, we have

$$N_1 \cdot N_2 = C_{4,1} \cdot C_{48,4} = \frac{4!}{1!(4-1)!} \cdot \frac{48!}{4!(48-4)!}$$
$$= \frac{4!}{1!3!} \cdot \frac{48!}{4!44!}$$
$$= 4 \cdot 194,580$$
$$= 778,320 \text{ ways}$$

The number of ways to deal a hand containing no hearts is

$$C_{39,5} = \frac{39!}{5!(39-5)!} = \frac{39!}{5!34!} = 575,757 \text{ ways}$$

Thus the hand that contains exactly one king is more likely.

47. Here is a possible analysis. V = vanilla, S = strawberry, C = chocolate.

a. It is possible to have all five cones alike.
   This can happen in 3 ways (all V, all D, all C).

b. It is possible to have four cones alike, but not five. Then
   O_1: choose the four-flavor. N_1: 3
   O_2: choose the fifth-flavor. N_2: 2
   $N_1 \cdot N_2 = 3 \cdot 2 = 6$ ways.

c. It is possible to have three cones alike, but not four. Then
   O_1: choose the three-flavor N_1: 3
   O_2: choose the other two cones N_2: 3   (both alike in two ways, or
   both different in one way)
   $N_1 \cdot N_2 = 3 \cdot 3 = 9$ ways

d. If no three cones are alike, then this can happen in 3 ways (each of the three flavors single, the other two double).

There are no other possibilities, thus there are 3 + 6 + 9 + 3 = 21 possible orders.

CHAPTER 10 REVIEW

1. (A) Since $\dfrac{-8}{16} = \dfrac{4}{-8} = -\dfrac{1}{2}$, this could start a geometric sequence.

(B) Since $7 - 5 = 9 - 7 = 2$, this could start an arithmetic sequence.

(C) Since $-5 - (-8) = -2 - (-5) = 3$, this could start an arithmetic sequence.

(D) Since $\dfrac{3}{2} \neq \dfrac{5}{3}$ and $3 - 2 \neq 5 - 3$, this could start neither an arithmetic nor a geometric sequence.

(E) Since $\dfrac{2}{-1} = \dfrac{-4}{2} = -2$, this could start a geometric sequence.   *(10-1, 10-3)*

2. $a_n = 2n + 3$

(A) $\begin{aligned} a_1 &= 2 \cdot 1 + 3 = 5 \\ a_2 &= 2 \cdot 2 + 3 = 7 \\ a_3 &= 2 \cdot 3 + 3 = 9 \\ a_4 &= 2 \cdot 4 + 3 = 11 \end{aligned}$

(B) This is an arithmetic sequence with $d = 2$. Hence
$$\begin{aligned} a_n &= a_1 + (n - 1)d \\ a_{10} &= 5 + (10 - 1)d \\ &= 23 \end{aligned}$$

(C) $S_n = \dfrac{n}{2}(a_1 + a_n)$

$\begin{aligned} S_{10} &= \dfrac{10}{2}(5 + 23) \\ &= 140 \end{aligned}$

(10-1, 10-3)

3. $a_n = 32\left(\dfrac{1}{2}\right)^n$

(A) $a_1 = 32\left(\dfrac{1}{2}\right)^1 = 16$

$a_2 = 32\left(\dfrac{1}{2}\right)^2 = 8$

$a_3 = 32\left(\dfrac{1}{2}\right)^3 = 4$

$a_4 = 32\left(\dfrac{1}{2}\right)^4 = 2$

(B) This is a geometric sequence with $r = \dfrac{1}{2}$. Hence
$$a_n = a_1 r^{n-1}$$
$$a_{10} = 16\left(\dfrac{1}{2}\right)^{10-1}$$
$$= \dfrac{1}{32}$$

(C) $S_n = \dfrac{a_1 - r a_n}{1 - r}$

$S_{10} = \dfrac{16 - \frac{1}{2}\left(\frac{1}{32}\right)}{\frac{1}{2}} = \dfrac{16 - \frac{1}{64}}{\frac{1}{2}} = 31\tfrac{31}{32}$

(10-1, 10-3)

4. $a_1 = -8$, $a_n = a_{n-1} + 3$, $n \geq 2$

(A) $\begin{aligned} a_1 &= -8 \\ a_2 &= a_1 + 3 = -8 + 3 = -5 \\ a_3 &= a_2 + 3 = -5 + 3 = -2 \\ a_4 &= a_3 + 3 = -2 + 3 = 1 \end{aligned}$

(B) This is an arithmetic sequence with $d = 3$. Hence
$$\begin{aligned} a_n &= a_1 + (n - 1)d \\ a_{10} &= -8 + (10 - 1)3 \\ &= 19 \end{aligned}$$

(C) $S_n = \dfrac{n}{2}(a_1 + a_n)$

$\begin{aligned} S_{10} &= \dfrac{10}{2}(-8 + 19) \\ &= 55 \end{aligned}$

(10-1, 10-3)

5. $a_1 = -1, \ a_n = (-2)a_{n-1}, \ n \geq 2$

(A) $a_1 = -1$
$a_2 = (-2)a_1 = (-2)(-1) = 2$
$a_3 = (-2)a_2 = (-2)2 = -4$
$a_4 = (-2)a_3 = (-2)(-4) = 8$

(B) This is a geometric sequence with $r = -2$. Hence
$$a_n = a_1 r^{n-1}$$
$$a_{10} = (-1)(-2)^{10-1}$$
$$= 512$$

(C) $S_n = \dfrac{a_1 - ra_n}{1 - r}$

$S_{10} = \dfrac{-1 - (-2)(512)}{1 - (-2)}$

$\quad = 341$

$\hfill (10\text{-}1, \ 10\text{-}3)$

6. This is a geometric sequence with $a_1 = 16$ and $r = \dfrac{1}{2} < 1$, so the sum exists:

$S_\infty = \dfrac{a_1}{1 - r}$

$\quad = \dfrac{16}{1 - \frac{1}{2}}$

$\quad = 32$

$\hfill (10\text{-}3)$

7. $10! = 10 \cdot 9 \cdot 8 \cdot 7 \cdot 6 \cdot 5 \cdot 4 \cdot 3 \cdot 2 \cdot 1 = 3{,}628{,}800 \qquad (10\text{-}4)$

8. $\dfrac{30!}{25!} = \dfrac{30 \cdot 29 \cdot 28 \cdot 27 \cdot 26 \cdot 25!}{25!} = 17{,}100{,}720 \qquad (10\text{-}4)$

9. $\dfrac{13!}{5!(13 - 5)!} = \dfrac{13!}{5!8!} = \dfrac{13 \cdot 12 \cdot 11 \cdot 10 \cdot 9 \cdot 8!}{5 \cdot 4 \cdot 3 \cdot 2 \cdot 1 \cdot 8!} = 1{,}287 \qquad (10\text{-}4)$

10. $P_{8,4} = \dfrac{8!}{(8 - 4)!} = \dfrac{8!}{4!} = \dfrac{8 \cdot 7 \cdot 6 \cdot 5 \cdot 4!}{4!} = 1{,}680$

$C_{8,4} = \dfrac{8!}{4!(8 - 4)!} = \dfrac{8!}{4!4!} = \dfrac{8 \cdot 7 \cdot 6 \cdot 5 \cdot 4!}{4 \cdot 3 \cdot 2 \cdot 1 \cdot 4!} = 70 \qquad (10\text{-}5)$

11. (A) The outcomes can be displayed in a tree diagram as follows:

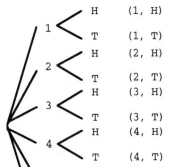

H	(1, H)
T	(1, T)
H	(2, H)
T	(2, T)
H	(3, H)
T	(3, T)
H	(4, H)
T	(4, T)
H	(5, H)
T	(5, T)
H	(6, H)
T	(6, T)

(B) O_1: Rolling the die
N_1: 6 outcomes
O_2: Flipping the coin
N_2: 2 outcomes

Applying the multiplication principle, there are $6 \cdot 2 = 12$ combined outcomes.

$\hfill (10\text{-}5)$

12. O_1: Seating the first person N_1: 6 ways
O_2: Seating the second person N_2: 5 ways
O_3: Seating the third person N_3: 4 ways
O_4: Seating the fourth person N_4: 3 ways
O_5: Seating the fifth person N_5: 2 ways
O_6: Seating the sixth person N_6: 1 way

Applying the multiplication principle, there are $6 \cdot 5 \cdot 4 \cdot 3 \cdot 2 \cdot 1 = 720$ arrangements. *(10-5)*

13. Order is important here. We use permutations to determine the number of arrangements of 6 objects. $P_{6,6} = 6! = 720$ *(10-5)*

14. P_1: $5 = 1^2 + 4 \cdot 1 = 5$
P_2: $5 + 7 = 2^2 + 4 \cdot 2$
      $12 = 12$
P_3: $5 + 7 + 9 = 3^2 + 4 \cdot 3$
        $21 = 21$ *(10-2)*

15. P_1: $2 = 2^{1+1} - 2 = 4 - 2 = 2$
P_2: $2 + 4 = 2^{2+1} - 2$
      $6 = 6$
P_3: $2 + 4 + 8 = 2^{3+1} - 2$
        $14 = 14$ *(10-2)*

16. P_1: $49^1 - 1$ is divisible by 6
$48 = 6 \cdot 8$ true
P_2: $49^2 - 1$ is divisible by 6
$2,400 = 6 \cdot 400$ true
P_3: $49^3 - 1$ is divisible by 6
$117,648 = 19,608 \cdot 6$ true   *(10-2)*

17. P_k: $5 + 7 + 9 + \cdots + (2k + 3) = k^2 + 4k$
P_{k+1}: $5 + 7 + 9 + \cdots + (2k + 3) + (2k + 5) = (k + 1)^2 + 4(k + 1)$ *(10-2)*

18. P_k: $2 + 4 + 8 + \cdots + 2^k = 2^{k+1} - 2$
P_{k+1}: $2 + 4 + 8 + \cdots + 2^k + 2^{k+1} = 2^{k+2} - 2$ *(10-2)*

19. P_k: $49^k - 1 = 6r$ for some integer r
P_{k+1}: $49^{k+1} - 1 = 6s$ for some integer s *(10-2)*

20. Although 1 is less than 4, $1 + \dfrac{1}{2}$ is less than 4, $1 + \dfrac{1}{2} + \dfrac{1}{3}$ is less than 4, and soon, the statement is false. In fact,
$$1 + \frac{1}{2} + \frac{1}{3} + \cdots + \frac{1}{31} \approx 4.027245$$
hence $n = 31$ is a counterexample.   *(10-2)*

21. $S_{10} = (2 \cdot 1 - 8) + (2 \cdot 2 - 8) + (2 \cdot 3 - 8) + (2 \cdot 4 - 8) + (2 \cdot 5 - 8) + (2 \cdot 6 - 8)$
                           $+ (2 \cdot 7 - 8) + (2 \cdot 8 - 8) + (2 \cdot 9 - 8) + (2 \cdot 10 - 8)$
      $= (-6) + (-4) + (-2) + 0 + 2 + 4 + 6 + 8 + 10 + 12$
      $= 30$ *(10-3)*

22. $S_7 = \dfrac{16}{2^1} + \dfrac{16}{2^2} + \dfrac{16}{2^3} + \dfrac{16}{2^4} + \dfrac{16}{2^5} + \dfrac{16}{2^6} + \dfrac{16}{2^7}$
     $= 8 + 4 + 2 + 1 + \dfrac{1}{2} + \dfrac{1}{4} + \dfrac{1}{8}$
     $= 15\dfrac{7}{8}$ *(10-3)*

23. This is an infinite geometric sequence with $a_1 = 27$.

$r = \dfrac{-18}{27} = -\dfrac{2}{3}$ $\quad \left| -\dfrac{2}{3} \right| < 1$, hence

the sum exists

$S_\infty = \dfrac{a_1}{1 - r}$

$\quad = \dfrac{27}{1 - (-\frac{2}{3})}$

$\quad = \dfrac{81}{5}$ $\hspace{3cm}$ *(10-3)*

24. $S_n = \displaystyle\sum_{k=1}^{n} \dfrac{(-1)^{k+1}}{3^k}$

This geometric sequence has $a_1 = \dfrac{1}{3}$.

$r = \left(-\dfrac{1}{9}\right) \div \dfrac{1}{3} = -\dfrac{1}{3}$ $\quad \left| -\dfrac{1}{3} \right| < 1$, hence

the sum exists.

$S_\infty = \dfrac{a_1}{1 - r}$

$\quad = \dfrac{\frac{1}{3}}{1 - (-\frac{1}{3})}$

$\quad = \dfrac{1}{4}$ $\hspace{3cm}$ *(10-3)*

25. We can select any three of the six points to use as vertices of the triangle. Order is not important. $C_{6,3} = \dfrac{6!}{3!(6-3)!} = 20$ triangles $\hspace{1cm}$ *(10-5)*

26. First, find d:

$a_n = a_1 + (n - 1)d$

$31 = 13 + (7 - 1)d$

$31 = 13 + 6d$

$ d = 3$

Hence, $a_5 = 13 + (5 - 1)3$

$ = 25$ $\hspace{3cm}$ *(10-3)*

27. $a_n = a_1 + (n - 1)d$ $\hspace{3cm}$ $S_n = \dfrac{n}{2}[2a_1 + (n - 1)d]$

$S_{10} = \dfrac{10}{2}[2a_1 + (10 - 1)d]$ $\hspace{2cm}$ $a_6 = a_1 + (6 - 1)d$

$S_{10} = 81 = 5(2a_1 + 9d)$ $\hspace{2.5cm}$ $a_6 = 10 = a_1 + 5d$

Solving the system:

$\hspace{2cm} 81 = 5(2a_1 + 9d)$ $\hspace{2cm}$ (1)

$\hspace{2cm} 10 = a_1 + 5d$ $\hspace{3cm}$ (2)

by substitution, we obtain, in turn

$\hspace{2cm} a_1 = 10 - 5d$ $\hspace{2cm}$ from equation (2)

$\hspace{2cm} 81 = 5[2(10 - 5d) + 9d]$ $\hspace{0.5cm}$ substituting into equation (1)

$\hspace{2cm} 81 = 5(20 - d)$

$\hspace{2cm} 81 = 100 - 5d$

$\hspace{2.3cm} d = \dfrac{19}{5}$ and

$\hspace{2cm} a_1 = 10 - 5d = 10 - 5\left(\dfrac{19}{5}\right) = -9$

28.

		Case 1	Case 2	Case 3
O_1:	select the first letter	N_1: 8 ways	8 ways	8 ways
O_2:	select the second letter	N_2: 7 ways	8 ways	7 ways
O_3:	select the third letter	N_3: 6 ways	8 ways	7 ways
		(exclude first and second letter)		(exclude second letter.)
		$8 \cdot 7 \cdot 6 = 336$ words	$8 \cdot 8 \cdot 8 = 512$ words	$8 \cdot 7 \cdot 7 = 392$ words

(10-5)

29. $0.\overline{72} = 0.72 + 0.0072 + 0.000072 + \cdots$
This is an infinite geometric sequence with $a_1 = 0.72$ and $r = 0.01$.

$$0.\overline{72} = S_\infty = \frac{a_1}{1 - r}$$
$$= \frac{0.72}{1 - 0.01}$$
$$= \frac{0.72}{0.99}$$
$$= \frac{8}{11} \qquad\qquad (10\text{-}3)$$

30. (A) Order is important here. We are selecting 3 digits out of 6 possible.
$P_{6,3} = 6 \cdot 5 \cdot 4 = 120$ lock combinations

(B) Order is not important here. We are selecting 2 players out of 5.
$C_{5,2} = \dfrac{5!}{2!(5 - 2)!} = 10$ games $\qquad\qquad (10\text{-}5)$

31. $\dfrac{20!}{18!(20 - 18)!} = \dfrac{20!}{18!2!}$
$$= \frac{20 \cdot 19 \cdot 18!}{18!2 \cdot 1} = 190$$
$$(10\text{-}4)$$

32. $\dbinom{16}{12} = \dfrac{16!}{12!(16 - 12)!} = \dfrac{16!}{12!4!}$
$$= \frac{16 \cdot 15 \cdot 14 \cdot 13 \cdot 12!}{12! \cdot 4 \cdot 3 \cdot 2 \cdot 1} = 1{,}820$$
$$(10\text{-}4)$$

33. $\dbinom{11}{11} = \dfrac{11!}{11!(11 - 11)!} = \dfrac{11!}{11!0!} = 1 \qquad\qquad (10\text{-}4)$

34. $\dfrac{987!}{(987 - 493)!} = \dfrac{987!}{494!}$ Since $494! = 494 \cdot 493! > 493!$, it follows
that $\dfrac{987!}{493!} > \dfrac{987!}{494!}$, that is, $\dfrac{987!}{493!}$ is larger than $\dfrac{987!}{(987 - 493)!}$ $\qquad (10\text{-}4)$

35. $\dbinom{1000}{500} = \dfrac{1000!}{500!500!}$ $\qquad \dbinom{1000}{501} = \dfrac{1000!}{501!499!}$

Since $501!499! = 501 \cdot 500!499! = 501 \cdot 500! \cdot \dfrac{500!}{500} = \dfrac{501}{500} \cdot 500!500!$, it follows

that $501!499! > 500!500!$ and thus $\dfrac{1000!}{500!500!} > \dfrac{1000!}{501!499!} \cdot \dbinom{1000}{501}$ is smaller. $(10\text{-}5)$

36. $(x - y)^5 = [x + (-y)]^5 = \displaystyle\sum_{k=0}^{5} \binom{5}{k}(x)^{5-k}(-y)^k$

$$= \binom{5}{0}x^5 + \binom{5}{1}x^4(-y)^1 + \binom{5}{2}x^3(-y)^2 + \binom{5}{3}x^2(-y)^3$$

$$+ \binom{5}{4}x(-y)^4 + \binom{5}{5}(-y)^5$$

$$= x^5 - 5x^4y + 10x^3y^2 - 10x^2y^3 + 5xy^4 - y^5 \qquad\qquad (10\text{-}4)$$

37. In the expansion of $(a + b)^n$, the exponent of b in the rth term is $r - 1$ and the exponent of a is $n - (r - 1)$. Here, $r = 10$, $n = 12$.

$$\text{Tenth term} = \binom{12}{9}(2x)^3(-y)^9$$

$$= \frac{12!}{9!\,3!}(8x^3)(-y^9)$$

$$= \frac{12 \cdot 11 \cdot 10 \cdot 9!}{9! \cdot 3 \cdot 2 \cdot 1}(-8x^3y^9)$$

$$= -1760x^3y^9$$

$(10\text{-}4)$

38. Write: P_n: $5 + 7 + 9 + \cdots + (2n + 3) = n^2 + 4n$

Proof: Part 1: Show that P_1 is true:

P_1: $5 = 1^2 + 4 \cdot 1$

$\qquad = 1 + 4$ Clearly true.

Part 2: Show that if P_k is true, then P_{k+1} is true:

Write out P_k and P_{k+1}.

$\quad P_k$: $5 + 7 + 9 + \cdots + (2k + 3) = k^2 + 4k$

$\quad P_{k+1}$: $5 + 7 + 9 + \cdots + (2k + 3) + (2k + 5) = (k + 1)^2 + 4(k + 1)$

We start with P_k:

$\quad 5 + 7 + 9 + \cdots + (2k + 3) = k^2 + 4k$

Adding $2k + 5$ to both sides:

$\quad 5 + 7 + 9 + \cdots + (2k + 3) + (2k + 5) = k^2 + 4k + 2k + 5$

$$= k^2 + 6k + 5$$

$$= k^2 + 2k + 1 + 4k + 4$$

$$= (k + 1)^2 + 4(k + 1)$$

We have shown that if P_k is true, then P_{k+1} is true.

Conclusion: P_n is true for all positive integers n.

$(10\text{-}2)$

39. Write: P_n: $2 + 4 + 8 + \cdots + 2^n = 2^{n+1} - 2$

Proof: Part 1: Show that P_1 is true:

P_1: $2 = 2^{1+1} - 2$

$\qquad = 4 - 2$ Clearly true.

Part 2: Show that if P_k is true, then P_{k+1} is true:

Write out P_k and P_{k+1}.

$\quad P_k$: $2 + 4 + 8 + \cdots + 2^k = 2^{k+1} - 2$

$\quad P_{k+1}$: $2 + 4 + 8 + \cdots + 2^k + 2^{k+1} = 2^{k+2} - 2$

We start with P_k:

$\quad 2 + 4 + 8 + \cdots + 2^k = 2^{k+1} - 2$

Adding 2^{k+1} to both sides:

$\quad 2 + 4 + 8 + \cdots + 2^k + 2^{k+1} = 2^{k+1} - 2 + 2^{k+1}$

$$= 2^{k+1} + 2^{k+1} - 2$$

$$= 2 \cdot 2^{k+1} - 2$$

$$= 2^{k+2} - 2$$

We have shown that if P_k is true, then P_{k+1} is true.

Conclusion: P_n is true for all positive integers n.

$(10\text{-}2)$

40. Write: P_n: $49^n - 1 = 6r$ for some r in N.
Part 1: Show that P_1 is true:
P_1: $49^1 - 1 = 6r$ is true ($r = 8$).

Part 2: Show that if P_k is true, then P_{k+1} is true:
Write out P_k and P_{k+1}.
P_k: $49^k - 1 = 6r$, for some integer r
P_{k+1}: $49^{k+1} - 1 = 6s$ for some integer s
We start with P_k:
$49^k - 1 = 6r$ for some integer r
Now, $49^{k+1} - 1 = 49^{k+1} - 49^k + 49^k - 1$
$= 49^k(49 - 1) + 49^k - 1$
$= 49^k \cdot 8 \cdot 6 + 6r$
$= 6(49^k \cdot 8 + r)$
$= 6s$ with $s = 49^k \cdot 8 + r$
We have shown that if P_k is true, then P_{k+1} is true.

Conclusion: P_n is true for all positive integers n. *(10-2)*

41. Here are computer-generated graphs of $\{a_n\}$ and $\{b_n\}$.

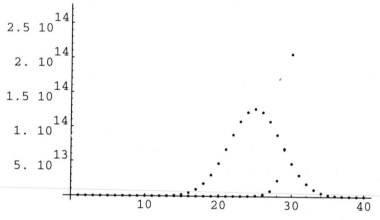

The graph indicates that the least positive integer n such that $a_n < b_n$ is between 28 and 30. From the table display we note
$n = 28$ $a_n = 8.875 \times 10^{13}$ $b_n = 2.287 \times 10^{13}$
$n = 29$ $a_n = 6.733 \times 10^{13}$ $b_n = 6.863 \times 10^{13}$
confirming that 29 is the least positive integer n required. *(10-4)*

42. Here are computer-generated graphs of $\{a_n\}$ and $\{b_n\}$ (the dots suggest a straight line).

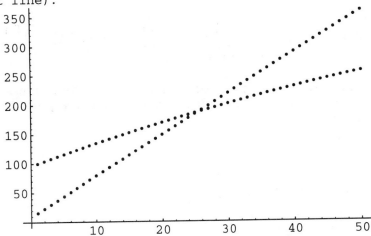

The graph indicates that the least positive integer n such that $a_n < b_n$ is between 25 and 27. From the table display we note

$n = 25$ $a_n = 188.87$ $b_n = 184$

$n = 26$ $a_n = 191.98$ $b_n = 193$

confirming that 26 is the least positive integer n required.   *(10-1)*

43. In the first case, the order matters. We have five successive events, each of which can happen two ways (girl or boy). Applying the multiplication principle, there are $2 \cdot 2 \cdot 2 \cdot 2 \cdot 2 = 32$ possible families. In the second case, the possible families can be listed as {0 girls, 1 girl, 2 girls, 3 girls, 4 girls, 5 girls}, thus there are 6 possibilities.   *(10-5)*

44. An arithmetic sequence is involved, with $a_1 = \dfrac{g}{2}$, $d = \dfrac{3g}{2} - \dfrac{g}{2} = g$. Distance fallen during the twenty-fifth second $= a_{25}$.

$$a_n = a_1 + (n - 1)d$$

$$a_{25} = \frac{g}{2} + (25 - 1)g$$

$$= \frac{49g}{2} \text{ feet}$$

Total distance fallen after twenty-five seconds $= a_1 + a_2 + a_3 + \cdots + a_{25} = S_{25}$

$$S_n = \frac{n}{2}(a_1 + a_n)$$

$$S_{25} = \frac{25}{2}\left(\frac{g}{2} + \frac{49g}{2}\right)$$

$$= \frac{625g}{2} \text{ feet}$$ *(10-3)*

45. To seat two people, we can seat one person, then the second person.

O_1: Seat the first person in any chair.

N_1: 4 ways

O_2: Seat the second person in any remaining chair.

N_2: 3 ways

Thus, applying the multiplication principle, there are $4 \cdot 3 = 12$ ways to seat two persons.   *(10-5)*

46. $(x + i)^6 = \sum_{k=0}^{6} \binom{6}{k} x^{6-k} i^k$

$$= \binom{6}{0} x^6 + \binom{6}{1} x^5 i^1 + \binom{6}{2} x^4 i^2 + \binom{6}{3} x^3 i^3 + \binom{6}{4} x^2 i^4 + \binom{6}{5} x i^5 + \binom{6}{6} i^6$$
$$= x^6 + 6ix^5 - 15x^4 - 20ix^3 + 15x^2 + 6ix - 1 \qquad (10\text{-}4)$$

47. A route plan can be regarded as a series of choices of stores, thus an arrangement of the 5 stores. Since the order matters, we use permutations: $P_{5,5} = 5! = 120$ route plans. $\qquad (10\text{-}5)$

48. Write: P_n: $\displaystyle\sum_{k=1}^{n} k^3 = \left(\sum_{k=1}^{n} k\right)^2$

Proof: Part 1: Show that P_1 is true.

$$P_1: \sum_{k=1}^{1} k^3 = 1^3 = 1^2 = \left(\sum_{k=1}^{1} k\right)^2$$

Part 2: Show that if P_j is true, then P_{j+1} is true.

Write out P_j and P_{j+1}.

$$P_j: \sum_{k=1}^{j} k^3 = \left(\sum_{k=1}^{j} k\right)^2$$

$$P_{j+1}: \sum_{k=1}^{j+1} k^3 = \left(\sum_{k=1}^{j+1} k\right)^2$$

We start with P_j:

$$\sum_{k=1}^{j} k^3 = \left(\sum_{k=1}^{j} k\right)^2$$

Adding $(j + 1)^3$ to both sides:

$$\sum_{k=1}^{j} k^3 + (j + 1)^3 = \left(\sum_{k=1}^{j} k\right)^2 + (j + 1)^3$$

$$\sum_{k=1}^{j+1} k^3 = (1 + 2 + 3 + \dots + j)^2 + (j + 1)^3$$

$$= \left[\frac{j(j + 1)}{2}\right]^2 + (j + 1)^3 \text{ using Matched Problem 1, Section 10-2}$$

$$= (j + 1)^2 \frac{j^2}{4} + (j + 1)^2 (j + 1)$$

$$= (j + 1)^2 \left[\frac{j^2}{4} + j + 1\right]$$

$$= (j + 1)^2 \left[\frac{j^2 + 4j + 4}{4}\right]$$

$$= (j + 1)^2 \frac{(j + 2)^2}{2^2}$$

$$= \left[\frac{(j + 1)(j + 2)}{2}\right]^2$$

$$= [1 + 2 + 3 + \dots + (j + 1)]^2 \text{ using Matched Problem 1, Section 10-2}$$

$$= \left(\sum_{k=1}^{j+1} k\right)^2$$

We have shown that if P_j is true, then P_{j+1} is true.

Conclusion: P_n is true for all positive integers n. $\qquad (10\text{-}2)$

49. Write: P_n: $x^{2n} - y^{2n} = (x - y)Q_n(x, y)$, where $Q_n(x, y)$ denotes some polynomial in x and y.

Proof: Part 1: Show that P_1 is true.
P_1: $x^{2 \cdot 1} - y^{2 \cdot 1} = (x - y)(x + y) = (x - y)Q_1(x, y)$ P_1 is true.

Part 2: Show that if P_k is true, then P_{k+1} is true.
Write out P_k and P_{k+1}.
P_k: $x^{2k} - y^{2k} = (x - y)Q_k(x, y)$
P_{k+1}: $x^{2k+2} - y^{2k+2} = (x - y)Q_{k+1}(x, y)$

We start with P_k:
$x^{2k} - y^{2k} = (x - y)Q_k(x, y)$

Now, $x^{2k+2} - y^{2k+2} = x^{2k+2} - x^{2k}y^2 + x^{2k}y^2 - y^{2k+2}$
$= x^{2k}(x^2 - y^2) + y^2(x^{2k} - y^{2k})$
$= (x - y)x^{2k}(x + y) + y^2(x - y)Q_k(x, y)$ by P_k
$= (x - y)[x^{2k}(x + y) + y^2Q_k(x, y)]$
$= (x - y)Q_{k+1}(x, y)$

We have shown that if P_k is true, then P_{k+1} is true.
Conclusion: P_n is true for all positive integers n.   (10-2)

50. Write: P_n: $\dfrac{a^n}{a^m} = a^{n-m}$, m an arbitrary positive integer, $n > m$.

Proof: Part 1: Show P_{m+1} is true.

P_{m+1}: $\dfrac{a^{m+1}}{a^m} = a^{m+1-m}$

$\dfrac{a^m a}{a^m} = a^1$ by the recursive definition of a^n

$a = a$ P_{m+1} is true.

Part 2: Show that if P_k is true, then P_{k+1} is true.
Write out P_k and P_{k+1}.

P_k: $\dfrac{a^k}{a^m} = a^{k-m}$

P_{k+1}: $\dfrac{a^{k+1}}{a^m} = a^{k-m+1}$

We start with P_k:

$\dfrac{a^k}{a^m} = a^{k-m}$

Multiplying both sides by a:

$\dfrac{a^k}{a^m}a = a^{k-m}a$

$\dfrac{a^k a}{a^m} = a^{k-m+1}$

$\dfrac{a^{k+1}}{a^m} = a^{k-m+1}$

We have show that if P_k is true, then P_{k+1} is true, for m an arbitrary positive integer.
Conclusion: P_n is true for all positive integers m, n.   (10-2)

51. To prove $a_n = b_n$, n a positive integer, write:

P_n: $a_n = b_n$

Proof: Part 1: Show P_1 is true.

$\quad a_1 = -3 \qquad b_1 = -5 + 2 \cdot 1 = -3$.

Thus, $a_1 = b_1$

Part 2: Show that if P_k is true, then P_{k+1} is true.

Write out P_k and P_{k+1}.

$\quad P_k$: $a_k = b_k$

$\quad P_{k+1}$: $a_{k+1} = b_{k+1}$

We start with P_k:

$\quad a_k = b_k$

Now, $a_{k+1} = a_k + 2 = b_k + 2$

$$\begin{aligned} &= -5 + 2k + 2 \\ &= -5 + 2(k + 1) \\ &= b_{k+1} \end{aligned}$$

Therefore, $a_{k+1} = b_{k+1}$.

Thus, if P_k is true, then P_{k+1} is true.

Conclusion: P_n is true for all $n \in N$. Hence, $\{a_n\} = \{b_n\}$ $\qquad$ (10-2)

52. Write: P_n: $(1!)1 + (2!)2 + (3!)3 + \cdots + (n!)n = (n + 1)! - 1$.

Proof: Part 1: Show that P_1 is true.

$\quad P_1$: $(1!)1 = 1 = 2 - 1 = 2! - 1$ is true.

Part 2: Show that if P_k is true, then P_{k+1} is true.

Write out P_k and P_{k+1}.

$\quad P_k$: $(1!)1 + (2!)2 + (3!)3 + \cdots + (k!)k = (k + 1)! - 1$

$\quad P_{k+1}$: $(1!)1 + (2!)2 + (3!)3 + \cdots + (k!)k + (k + 1)!(k + 1) = (k + 2)! - 1$

We start with P_k:

$\quad (1!)1 + (2!)2 + (3!)3 + \cdots + (k!)k = (k + 1)! - 1$

Adding $(k + 1)!(k + 1)$ to both sides:

$\quad (1!)1 + (2!)2 + (3!)3 + \cdots + (k!)k + (k + 1)!(k + 1)$

$$\begin{aligned} &= (k + 1)! - 1 + (k + 1)!(k + 1) \\ &= (k + 1)!(1 + k + 1) - 1 \\ &= (k + 2)(k + 1)! - 1 \\ &= (k + 2)! - 1 \end{aligned}$$

Thus, if P_k is true, then P_{k+1} is true.

Conclusion: P_n is true for all positive integers n. $\qquad$ (10-2)

53. $a_n = \begin{pmatrix} 2n \\ n \end{pmatrix}$ $\qquad\qquad$ $b_n = 4^{n-1}$

$a_1 = \begin{pmatrix} 2 \\ 1 \end{pmatrix} = \dfrac{2!}{1!\,1!} = 2$ $\qquad\qquad$ $b_1 = 4^{1-1} = 4^0 = 1$

$a_2 = \begin{pmatrix} 4 \\ 2 \end{pmatrix} = \dfrac{4!}{2!\,2!} = 6$ $\qquad\qquad$ $b_2 = 4^{2-1} = 4^1 = 4$

$a_3 = \begin{pmatrix} 6 \\ 3 \end{pmatrix} = \dfrac{6!}{3!\,3!} = 20$ $\qquad\qquad$ $b_3 = 4^{3-1} = 4^2 = 16$

$a_4 = \begin{pmatrix} 8 \\ 4 \end{pmatrix} = \dfrac{8!}{4!\,4!} = 70$ $\qquad\qquad$ $b_4 = 4^{4-1} = 4^3 = 64$

$a = \begin{pmatrix} 10 \\ 5 \end{pmatrix} = \dfrac{10!}{5!\,5!} = 252$ $\qquad\qquad$ $b_5 = 4^{5-1} = 4^4 = 256$

Thus $a_4 > b_4$ but $a_5 \le b_5$. The smallest positive integer such that $a_n \le b_n$ is 5. *(10-1)*

54. $\{a_n\}$ is neither arithmetic nor geometric, since $a_2 - a_1 = 6 - 2 = 4$ but

$a_3 - a_2 = 20 - 6 = 14$, and $\dfrac{a_2}{a_1} = \dfrac{6}{2} = 3$ but $\dfrac{a_3}{a_2} = \dfrac{20}{6} = \dfrac{10}{3}$.

$\{b_n\}$ is geometric, since $\dfrac{b_{n+1}}{b_n} = \dfrac{4^{(n+1)-1}}{4^{n-1}} = \dfrac{4^n}{4^{n-1}} = 4 = r$ $\qquad\qquad$ *(10-3)*

55. $\dfrac{a_{n+1}}{a_n} = \begin{pmatrix} 2(n+1) \\ n+1 \end{pmatrix} \div \begin{pmatrix} 2n \\ n \end{pmatrix} = \dfrac{[2(n+1)]!}{(n+1)!\,[2(n+1)-(n+1)]!} \div \dfrac{(2n)!}{n!\,(2n-n)!}$

$= \dfrac{(2n+2)!}{(n+1)!\,(n+1)!} \div \dfrac{(2n)!}{n!\,n!}$

$= \dfrac{(2n+2)!}{(n+1)!\,(n+1)!} \cdot \dfrac{n!\,n!}{(2n)!}$

$= \dfrac{(2n+2)(2n+1)(2n)!}{(n+1)n!\,(n+1)n!} \cdot \dfrac{n!\,n!}{(2n)!}$

$= \dfrac{(2n+2)(2n+1)}{(n+1)(n+1)}$

$= \dfrac{2(n+1)(2n+1)}{(n+1)(n+1)}$

$= \dfrac{4n+2}{n+1}$

$= \dfrac{4n+4-2}{n+1}$

$= \dfrac{4(n+1)-2}{n+1}$

$= 4 - \dfrac{2}{n+1}$

This quantity is clearly less than 4 for all positive integers n. $\qquad\qquad$ *(10-1)*

56. Note first that if c, d, x, y are all positive, then if $c \leq d$ and $x < y$, it follows that $cx \leq dy$, since we can write:

$$c \leq d$$
$$cx \leq dx$$
$$dx < dy$$
$$cx \leq dy$$

Then write: P_n: $a_n \leq b_n$ for $n \geq 5$

Proof: Part 1: Show that P_5 is true

$$P_5: a_5 \leq b_5 \qquad a_5 = \binom{10}{5} = 252 < 256 = 4^{5-1} = b_5. \ P_5 \text{ is true.}$$

Part 2: Show that if P_k is true then P_{k+1} is true.
Write out P_k and P_{k+1}.

$$P_k: a_k \leq b_k \qquad \text{for} \quad k \geq 5$$
$$P_{k+1}: a_{k+1} \leq b_{k+1} \quad \text{for} \quad k + 1 \geq 5$$

We start with P_k:

$$a_k \leq b_k$$

We know from problem 55 that $\dfrac{a_{k+1}}{a_k} < 4$. But $\dfrac{b_{k+1}}{b_k} = 4$ since $\{b_n\}$ is a geometric sequence with ratio $r = 4$.

Therefore, $\dfrac{a_{k+1}}{a_k} < \dfrac{b_{k+1}}{b_k}$. By the inequality noted at the beginning, since $a_k \leq b_k$ it follows that

$$a_k \cdot \frac{a_{k+1}}{a_k} \leq b_k \cdot \frac{b_{k+1}}{b_k}$$

that is, $a_{k+1} \leq b_{k+1}$

Thus, if P_k is true, then P_{k+1} is true.

Conclusion: P_n is true for all integers $n \geq 5$.

(10-2)

CHAPTER 11

Exercise 11-1

Key Ideas and Formulas

The graphs of a second degree equation in two variables $Ax^2 + Bxy + Cy^2 + Dx + Ey + F = 0$ (for different values of the coefficients: A, B, C not all zero) are plane curves called conic sections. They include the circle, parabola, ellipse, and hyperbola.

A parabola is the set of all points in a plane equidistant from a fixed point F (the focus) and a fixed line L (the directrix) in the plane. A line through the focus perpendicular to the plane is called the **axis** and the point on the axis halfway between the focus and directrix is called the **vertex**.

Standard Equations of a Parabola with Vertex at (0, 0)

1. $y^2 = 4ax$
   Vertex: (0, 0)
   Focus: $(a, 0)$
   Directrix: $x = -a$
   Axis: the x axis

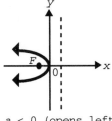

a < 0 (opens left)   a > 0 (opens right)

Symmetric with respect to the x axis.

2. $x^2 = 4ay$
   Vertex: (0, 0)
   Focus: $(0, a)$
   Directrix: $y = -a$
   Axis: the y axis

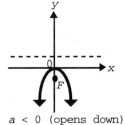

a < 0 (opens down)   a > 0 (opens up)

Symmetric with respect to the y axis.

---

1. To graph $y^2 = 4x$, assign x values that make the right side a perfect square (x must be non-negative for y to be real) and solve for y. Since the coefficient of x is positive, a must be positive, and the parabola opens right.

x	0	1	4
y	0	±2	±4

To find the focus and directrix, solve
$4a = 4$
$a = 1$
Focus: (1, 0) Directrix: $x = -1$

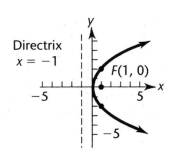

3. To graph $x^2 = 8y$, assign y values that make the right side a perfect square (y must be non-negative for x to be real) and solve for x. Since the coefficient of y is positive, a must be positive, and the parabola opens up.

x	0	±4	±2
y	0	2	$\frac{1}{2}$

To find the focus and directrix, solve
$4a = 8$
$a = 2$
Focus: (0, 2) Directrix: $y = -2$

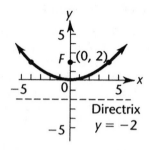

5. To graph $y^2 = -12x$, assign x values that make the right side a perfect square (x must be zero or negative for y to be real) and solve for y. Since the coefficient of x is negative, a must be negative, and the parabola opens left.

x	0	-3	$-\frac{1}{3}$
y	0	±6	±2

To find the focus and directrix, solve
$4a = -12$
$a = -3$
Focus: (-3, 0) Directrix: $x = -(-3) = 3$

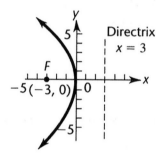

7. To graph $x^2 = -4y$, assign y values that make the right side a perfect square (y must be zero or negative for x to be real) and solve for x. Since the coefficient of y is negative, a must be negative, and the parabola opens down.

x	0	±2	±4
y	0	-1	-4

To find the focus and directrix, solve
$4a = -4$
$a = -1$
Focus: (0, -1) Directrix: $y = -(-1) = 1$

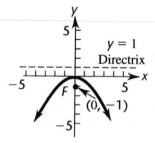

9. To graph $y^2 = -20x$, we may proceed as in problem 5. Alternatively, after noting that x must be zero or negative for y to be real, we may pick convenient values for x and solve for y using a calculator. Since the coefficient of x is negative, a must be negative, and the parabola opens left.

x	0	-1	-2
y	0	$\pm\sqrt{20} \approx \pm 4.5$	$\pm\sqrt{40} \approx \pm 6.3$

To find the focus and directrix, solve
$4a = -20$
$a = -5$
Focus: (-5, 0) Directrix: $x = -(-5) = 5$

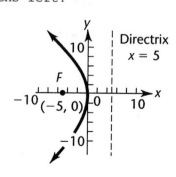

11. To graph $x^2 = 10y$, we may proceed as in Problem 3. Alternatively, after noting that y must be non-negative for x to be real, we may pick convenient values for y and solve for x using a calculator. Since the coefficient of y is positive, a must be positive, and the parabola opens up.

x	0	$\pm\sqrt{10} \approx \pm 3.2$	$\pm\sqrt{20} \approx \pm 4.5$
y	0	1	2

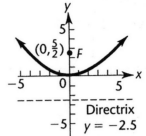

To find the focus and directrix, solve
$4a = 10$
$\;a = 2.5$
Focus: $(0, 2.5)$ Directrix: $y = -2.5$

13. Comparing $y^2 = 39x$ with $y^2 = 4ax$, the standard equation of a parabola symmetric with respect to the x axis, we have
$4a = 39$ Focus on x axis
$\;a = 9.75$
Focus: $(9.75, 0)$

15. Comparing $x^2 = -105y$ with $x^2 = 4ay$, the standard equation of a parabola symmetric with respect to the y axis, we have
$4a = -105$ Focus on y axis
$\;a = -26.25$
Focus: $(0, -26.25)$

17. Comparing $y^2 = -77x$ with $y^2 = 4ax$, the standard equation of a parabola symmetric with respect to the x axis, we have
$4a = -77$ Focus on x axis
$\;a = -19.25$
Focus: $(-19.25, 0)$

19. Comparing focus $(-6, 0)$ with the information in the chart of standard equations for the parabola, we see:
$a = -6$. Axis: the x axis.
Equation: $y^2 = 4ax$.
Thus the equation of the parabola must be $y^2 = 4(-6)x$, or $y^2 = -24x$.

21. Comparing directrix $y = -5$ with the information in the chart of standard equations for the parabola, we see:
$a = 5$. Axis: the y axis.
Equation: $x^2 = 4ay$.
Thus the equation of the parabola must be $x^2 = 4 \cdot 5y$, or $x^2 = 20y$.

23. Comparing focus $\left(0, -\dfrac{1}{3}\right)$ with the information in the chart of standard equations for the parabola, we see:
$a = -\dfrac{1}{3}$. Axis: the y axis.
Equation: $x^2 = 4ay$.
Thus the equation of the parabola must be $x^2 = 4\left(-\dfrac{1}{3}\right)y$, or $x^2 = -\dfrac{4}{3}y$

25. The parabola is opening left and has an equation of the form $y^2 = 4ax$.
Since $(-4, -20)$ is on the graph, we have:
$$y^2 = 4ax$$
$$(-20)^2 = 4a(-4)$$
$$-25 = a$$
Thus, the equation of the parabola is
$$y^2 = 4(-25)x$$
$$y^2 = -100x$$

27. The parabola is opening down and has an equation of the form $x^2 = 4ay$. Since $(9, -27)$ is on the graph, we have
$$x^2 = 4ay$$
$$9^2 = 4a(-27)$$
$$a = -\frac{3}{4}$$
Thus, the equation of the parabola is
$$x^2 = 4\left(-\frac{3}{4}\right)y$$
$$x^2 = -3y$$

29. The parabola is opening left and has an equation of the form $y^2 = 4ax$. Since $(-8, -2)$ is on the graph, we have
$$y^2 = 4ax$$
$$(-2)^2 = 4a(-8)$$
$$a = -\frac{1}{8}$$
Thus, the equation of the parabola is
$$y^2 = 4\left(-\frac{1}{8}\right)x$$
$$y^2 = -\frac{1}{2}x$$

31. $x^2 = 4y$
$y^2 = 4x$

Solve for y in the first equation, then substitute into the second equation.

$$y = \frac{x^2}{4}$$
$$\left(\frac{x^2}{4}\right)^2 = 4x$$
$$\frac{x^4}{16} = 4x$$
$$x^4 = 64x$$
$$x^4 - 64x = 0$$
$$x(x^3 - 64) = 0$$
$$x(x - 4)(x^2 + 4x + 8) = 0$$
$$x = 0 \quad x - 4 = 0 \quad x^2 + 4x + 8 = 0$$
$$x = 4 \quad \text{No real solutions}$$

For $x = 0$ For $x = 4$
$$y = 0 \qquad y = \frac{4^2}{4}$$
$$y = 4$$
Solutions: $(0, 0)$, $(4, 4)$

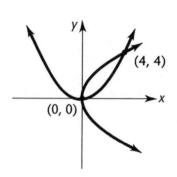

The exact solutions yield more than 3-digit accuracy.

33. $y^2 = 6x$
$x^2 = 5y$

Solve for x in the first equation, then substitute into the second equation.

$$x = \frac{y^2}{6}$$
$$\left(\frac{y^2}{6}\right)^2 = 5y$$
$$\frac{y^4}{36} = 5y$$
$$y^4 = 180y$$
$$y^4 - 180y = 0$$
$$y(y^3 - 180) = 0$$
$$y(y - \sqrt[3]{180})(y^2 + y\sqrt[3]{180} + (\sqrt[3]{180})^2) = 0$$

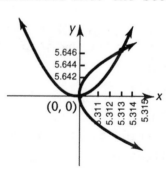

$$y = 0 \quad y - \sqrt[3]{180} = 0 \quad\quad y^2 + y\sqrt[3]{180} + (\sqrt[3]{180})^2 = 0$$

$$y = \sqrt[3]{180} \quad \text{No real solutions}$$

$$\approx 5.646$$

For $y = 0$ For $y = \sqrt[3]{180}$

$$x = 0 \quad\quad x = \frac{(\sqrt[3]{180})^2}{6}$$

$$\approx 5.313$$

Solutions: $(0, 0)$, $(5.313, 5.646)$

35. (A) The line $x = 0$ intersects the parabola $x^2 = 4ay$ only at $(0, 0)$.
The line $y = 0$ intersects the parabola $x^2 = 4ay$ only at $(0, 0)$.
Only these 2 lines intersect the parabola at exactly one point (see part (B)).

(B) A line through $(0, 0)$ with slope $m \neq 0$ has equation $y = mx$.
Solve the system:

$$y = mx$$
$$x^2 = 4ay$$

by substituting y from the first equation into the second equation.

$$x^2 = 4amx$$
$$x^2 - 4amx = 0$$
$$x(x - 4am) = 0$$
$$x = 0 \quad x = 4am$$

For $x = 0$ For $x = 4am$
$$y = 0 \quad\quad\quad y = m(4am)$$
$$= 4am^2$$

Solutions: $(0, 0)$, $(4am, 4am^2)$ are the required coordinates.

37. Since A and B lie on the curve $x^2 = 4ay$, their coordinates must satisfy the equation of the curve. Clearly, the y coordinate of A, F, and B is a.
Substituting a for y, we have

$$x^2 = 4aa$$
$$x^2 = 4a^2$$
$$x = \pm 2a$$

Therefore, A has coordinates $(-2a, a)$ and B has coordinates $(2a, a)$.

39. False. For example, if $a = 0$, then the graph of $y^2 = 4ax$, or $y^2 = 0$, is the x axis.

41. True. A vertical line with equation $x = h$ will intersect the graph of $x^2 = 4y$ at $\left(h, \dfrac{h^2}{4}\right)$.

43. Let $P(x, y)$ be a point on the parabola.

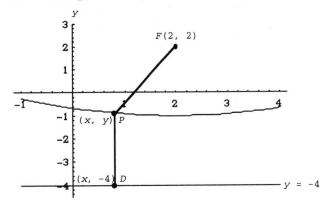

Then, by the definition of the parabola, the distance from $P(x, y)$ to the focus $F(2, 2)$ must equal the perpendicular distance from P to the directrix at $D(x, -4)$. Applying the distance formula, we have

$$d(P, F) = d(P, D)$$
$$\sqrt{(x - 2)^2 + (y - 2)^2} = \sqrt{(x - x)^2 + [y - (-4)]^2}$$
$$(x - 2)^2 + (y - 2)^2 = (x - x)^2 + (y + 4)^2$$
$$x^2 - 4x + 4 + y^2 - 4y + 4 = 0 + y^2 + 8y + 16$$
$$x^2 - 4x - 12y - 8 = 0$$

45. Let $P(x, y)$ be a point on the parabola.

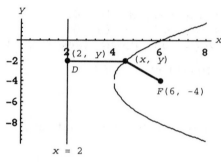

Then, by the definition of the parabola, the distance from $P(x, y)$ to the focus $F(6, -4)$ must equal the perpendicular distance from P to the directrix at $D(2, y)$. Applying the distance formula, we have

$$d(P, F) = d(P, D)$$
$$\sqrt{(x - 6)^2 + [y - (-4)]^2} = \sqrt{(x - 2)^2 + (y - y)^2}$$
$$(x - 6)^2 + (y + 4)^2 = (x - 2)^2 + (y - y)^2$$
$$x^2 - 12x + 36 + y^2 + 8y + 16 = x^2 - 4x + 4 + 0$$
$$y^2 + 8y - 8x + 48 = 0$$

47. Here are computer-generated graphs of $x^2 = 8y$ and $y = 5x + 4$, $-15 \le x \le 45$.

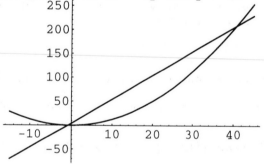

The curves intersect for x between -2 and 0, and between 40 and 42. Zooming in on these intervals, we obtain the following graphs.

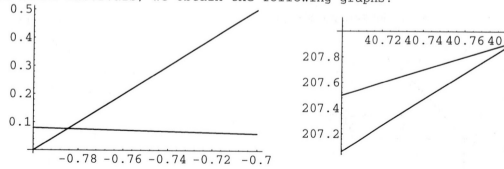

To two decimal places, the coordinates of the points of intersection are $(-0.78, 0.08)$ and $(40.78, 207.92)$.

49. Here are computer-generated
graphs of $x^2 = -8y$ and $y^2 = -5x$
(entered as $y = \pm\sqrt{-5x}$).

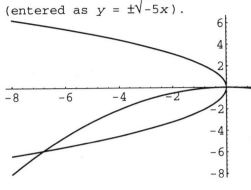

The curves intersect at (0, 0), as can be
easily checked by substitution. The other
point of intersection occurs for x between
-7.2 and -6.8. Zooming in on this interval,
we obtain the following graph.

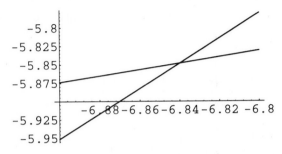

To two decimal places, the coordinates of the
other point of intersection are (-6.84, -5.85).

51. From the figure, we see that the coordinates
of P must be (-100, -50). The parabola is
opening down with axis the y axis, hence it
has an equation of the form $x^2 = 4ay$. Since
(-100, -50) is on the graph, we have
$$(-100)^2 = 4a(-50)$$
$$10,000 = -200a$$
$$a = -50$$
Thus, the equation of the parabola is
$$x^2 = 4(-50)y$$
$$x^2 = -200y$$

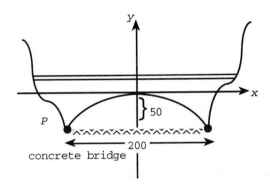

53. (A) From the figure, we take the parabola to be
opening up with axis the y axis, it has an equation
of the form $x^2 = 4ay$. Since the focus is at
$(0, a) = (0, 100)$, $a = 100$ and the equation of the
parabola is $x^2 = 400y$ or $y = 0.0025x^2$

(B) Since the depth represents the y coordinate y_1
of a point on the parabola with $x = 100$, we have

$$y_1 = 0.0025(100)^2$$
$$\text{depth} = 25 \text{ feet}$$

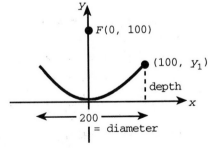

Exercise 11-2

Key Ideas and Formulas

An ellipse is the set of all points P in a plane such that the sum of the distances of P from two fixed points (the foci) in the plane is constant.

The line through the foci intersects the ellipse at two points called vertices; the line segment connecting the vertices is called the major axis; its perpendicular bisector is called the minor axis. The axes intersect at a point called the center of the ellipse. Each end of the major axis is called a vertex.

Standard Equations of an Ellipse — Center at (0, 0)

1. $\dfrac{x^2}{a^2} + \dfrac{y^2}{b^2} = 1 \quad a > b > 0$

   x intercepts: $\pm a$ (vertices)
   y intercepts: $\pm b$
   Foci: $F'(-c, 0)$, $F(c, 0)$
   $c^2 = a^2 - b^2$
   Major axis length = $2a$
   Minor axis length = $2b$

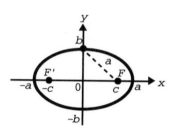

2. $\dfrac{x^2}{b^2} + \dfrac{y^2}{a^2} = 1 \quad a > b > 0$

   x intercepts: $\pm b$
   y intercepts: $\pm a$ (vertices)
   Foci: $F'(0, -c)$, $F(0, c)$
   $c^2 = a^2 - b^2$
   Major axis length = $2a$
   Minor axis length = $2b$

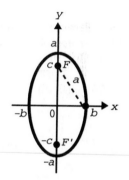

---

1. When $y = 0$, $\dfrac{x^2}{25} = 1$. x intercepts: ± 5

When $x = 0$, $\dfrac{y^2}{4} = 1$. y intercepts: ± 2

Thus, $a = 5$, $b = 2$, and the major axis is on the x axis.

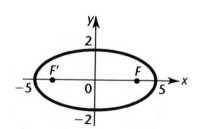

Foci: $c^2 = a^2 - b^2$
$\qquad c^2 = 25 - 4$
$\qquad c^2 = 21$
$\qquad\quad c = \sqrt{21}$ Foci: $F'(-\sqrt{21}, 0)$, $F(\sqrt{21}, 0)$

Major axis length = $2(5) = 10$
Minor axis length = $2(2) = 4$

> **Common Error:**
> The relationship $c^2 = a^2 + b^2$ does not apply to a, b, c as defined for ellipses.

3. When $y = 0$, $\frac{x^2}{4} = 1$. x intercepts: ± 2

When $x = 0$, $\frac{y^2}{25} = 1$. y intercepts: ± 5

Thus, $a = 5$, $b = 2$, and the major axis is on the y axis.
Foci: $c^2 = a^2 - b^2$
$\quad\quad c^2 = 25 - 4$
$\quad\quad c^2 = 21$
$\quad\quad\quad c = \sqrt{21}$ Foci: $F'(0, -\sqrt{21})$, $F(0, \sqrt{21})$
Major axis length = $2(5) = 10$ Minor axis length = $2(2) = 4$

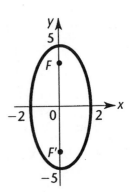

5. First, write the equation in standard form by dividing both sides by 9.
$x^2 + 9y^2 = 9$
$\frac{x^2}{9} + \frac{y^2}{1} = 1$
Locate the intercepts.
When $y = 0$, $\frac{x^2}{9} = 1$. x intercepts: ± 3

When $x = 0$, $\frac{y^2}{1} = 1$. y intercepts: ± 1

Thus $a = 3$, $b = 1$, and the major axis is on the
x axis.
Foci: $c^2 = a^2 - b^2$
$\quad\quad c^2 = 9 - 1$
$\quad\quad c^2 = 8$
$\quad\quad\quad c = \sqrt{8}$ Foci: $F'(-\sqrt{8}, 0)$, $F(\sqrt{8}, 0)$
Major axis length = $2(3) = 6$
Minor axis length = $2(1) = 2$

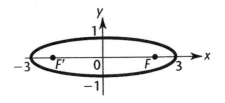

7. First, write the equation in standard form by dividing both sides by 225.
$25x^2 + 9y^2 = 225$
$\frac{x^2}{9} + \frac{y^2}{25} = 1$
Locate the intercepts.
When $y = 0$, $\frac{x^2}{9} = 1$. x intercepts: ± 3

When $x = 0$, $\frac{y^2}{25} = 1$. y intercepts: ± 5

Thus $a = 5$, $b = 3$, and the major axis is on the y axis.
Foci: $c^2 = a^2 - b^2$
$\quad\quad c^2 = 25 - 9$
$\quad\quad c^2 = 16$
$\quad\quad\quad c = 4$ Foci: $F'(0, -4)$, $F(0, 4)$
Major axis length = $2(5) = 10$
Minor axis length = $2(3) = 6$

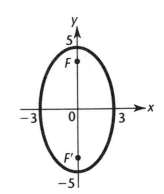

9. First, write the equation in standard form by dividing both sides by 12.

$2x^2 + y^2 = 12$

$\dfrac{x^2}{6} + \dfrac{y^2}{12} = 1$

Locate the intercepts.

When $y = 0$, $\dfrac{x^2}{6} = 1$. x intercepts: $\pm\sqrt{6}$

When $x = 0$, $\dfrac{y^2}{12} = 1$. y intercepts: $\pm\sqrt{12}$

Thus $a = \sqrt{12}$, $b = \sqrt{6}$, and the major axis is on the y axis.

Foci: $c^2 = a^2 - b^2$

$\quad c^2 = 12 - 6$

$\quad c^2 = 6$

$\quad\quad c = \sqrt{6}$ Foci: $F'(0, -\sqrt{6})$, $F(0, \sqrt{6})$

Major axis length = $2\sqrt{12} \approx 6.93$

Minor axis length = $2\sqrt{6} \approx 4.90$

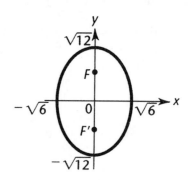

11. First, write the equation in standard form by dividing both sides by 28.

$4x^2 + 7y^2 = 28$

$\dfrac{x^2}{7} + \dfrac{y^2}{4} = 1$

Locate the intercepts.

When $y = 0$, $\dfrac{x^2}{7} = 1$. x intercepts: $\pm\sqrt{7}$

When $x = 0$, $\dfrac{y^2}{4} = 1$. y intercepts: ± 2

Thus $a = \sqrt{7}$, $b = 2$, and the major axis is on the x axis.

Foci: $c^2 = a^2 - b^2$

$\quad c^2 = 7 - 4$

$\quad c^2 = 3$

$\quad\quad c = \sqrt{3}$ Foci: $F'(-\sqrt{3}, 0)$, $F(\sqrt{3}, 0)$

Major axis length = $2\sqrt{7} \approx 5.29$

Minor axis length = $2(2) = 4$

13. Make a rough sketch of the ellipse and compute x and y intercepts.

$\dfrac{x^2}{b^2} + \dfrac{y^2}{a^2} = 1$

$a = \dfrac{6}{2} = 3$ $b = \dfrac{2}{2} = 1$

$\dfrac{x^2}{1} + \dfrac{y^2}{9} = 1$ or $x^2 + \dfrac{y^2}{9} = 1$

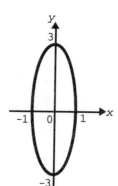

15. Make a rough sketch of the ellipse, locate focus and y intercepts, then determine x intercepts from the special triangle relationship.

$$\frac{x^2}{a^2} + \frac{y^2}{b^2} = 1$$

$$b = \frac{10}{2} = 5 \quad c = 4$$

$$a^2 = b^2 + c^2$$
$$a^2 = 5^2 + 4^2 = 25 + 16 = 41$$

$$a = \sqrt{41}$$

$$\frac{x^2}{41} + \frac{y^2}{25} = 1$$

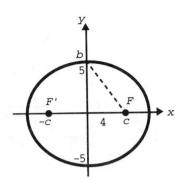

17. Make a rough sketch of the ellipse, locate focus and y intercepts, then determine x intercepts using the special triangle relationship.

$$\frac{x^2}{b^2} + \frac{y^2}{a^2} = 1$$

$$a = \frac{24}{2} = 12 \quad c = \frac{2}{2} = 1$$

$$b^2 = a^2 - c^2$$
$$b^2 = 12^2 - 1^2 = 143$$

$$b = \sqrt{143}$$

$$\frac{x^2}{143} + \frac{y^2}{144} = 1$$

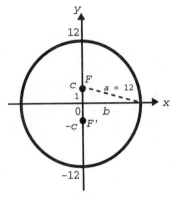

19. The graph does not pass the vertical line test; most vertical lines that intersect an ellipse do so in two places; hence, the equation does not define a function.

21. $16x^2 + 25y^2 = 400$
$\quad\;\; 2x - 5y = 10$

We solve for x in terms of y in the second equation, then substitute the expression for x into the first equation.

$$2x = 5y + 10$$
$$x = \frac{5y + 10}{2}$$

$$16\left(\frac{5y + 10}{2}\right)^2 + 25y^2 = 400$$

$$\frac{16(25y^2 + 100y + 100)}{4} + 25y^2 = 400$$

$$100y^2 + 400y + 400 + 25y^2 = 400$$
$$125y^2 + 400y = 0$$
$$25y(5y + 16) = 0$$

$$y = 0 \qquad\qquad y = -\frac{16}{5} \text{ or } -3.2$$

For $y = 0$ For $y = -\frac{16}{5}$

$$x = \frac{5 \cdot 0 + 10}{2} \qquad\qquad x = \frac{5(-\frac{16}{5}) + 10}{2}$$

$$x = 5 \qquad\qquad\qquad\;\; x = \frac{-16 + 10}{2}$$

$$x = -3$$

Solutions: $(5, 0)$, $(-3, -3.2)$.

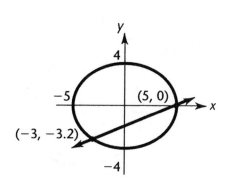

Graphs:
$$16x^2 + 25y^2 = 400$$
$$\frac{x^2}{25} + \frac{y^2}{16} = 1 \quad \text{Ellipse}$$
$a = 5$, $b = 4$, major axis on the x axis.
x intercepts: ± 5
y intercepts: ± 4
$$2x - 5y = 10$$
Straight line, intercepts $(0, -2)$ and $(5, 0)$

23. $25x^2 + 16y^2 = 400$
$25x^2 - 36y = 0$

Subtract the second equation from the first.
$16y^2 + 36y = 400$
Solve this equation for y.
$16y^2 + 36y - 400 = 0$
$4y^2 + 9y - 100 = 0$
$(4y + 25)(y - 4) = 0$
$4y + 25 = 0 \qquad y - 4 = 0$
$$y = -\frac{25}{4} \qquad y = 4$$

For $y = -\dfrac{25}{4}$ | For $y = 4$

$25x^2 - 36\left(-\dfrac{25}{4}\right) = 0$ | $25x^2 - 36(4) = 0$

$25x^2 + 225 = 0$ | $25x^2 - 144 = 0$

$x^2 = -9$ | $x^2 = \dfrac{144}{25}$

$x = \pm 3i$ | $x = \pm\dfrac{12}{5}$ or ± 2.4

No real solution | Solutions: $(2.4, 4)$ and $(-2.4, 4)$

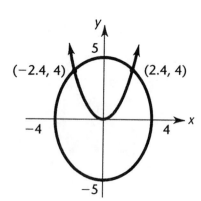

Graphs:
$$25x^2 + 16y^2 = 400$$
$$\frac{x^2}{16} + \frac{y^2}{25} = 1$$
$a = 5$, $b = 4$, major axis on the y axis.
x intercepts: ± 4
y intercepts: ± 5
$$25x^2 - 36y = 0$$
$$x^2 = \frac{36}{25}y$$
Parabola, opens up.

25. $5x^2 + 2y^2 = 63$
$2x - y = 0$

Solve for y in the second equation, then substitute into the first equation.

$$y = 2x$$
$$5x^2 + 2(2x)^2 = 63$$
$$5x^2 + 8x^2 = 63$$
$$13x^2 = 63$$
$$x^2 = \frac{63}{13}$$
$$x = \pm\sqrt{\frac{63}{13}}$$
$$\approx \pm 2.201$$

We discard the negative solution, since we are asked for first quadrant solutions only.

For $x = 2.201$
$$y = 2(2.201)$$
$$= 4.402$$
Solution: $(2.201, 4.402)$

27. $2x^2 + 3y^2 = 33$
$x^2 - 8y = 0$

Solve for y in the second equation, then substitute into the first equation.

$$x^2 = 8y$$
$$y = \frac{x^2}{8}$$
$$2x^2 + 3\left(\frac{x^2}{8}\right)^2 = 33$$
$$2x^2 + \frac{3x^4}{64} = 33$$
$$3x^4 + 128x^2 = 2112$$
$$3x^4 + 128x^2 - 2112 = 0 \quad \text{Quadratic in } x^2.$$
$$x^2 = \frac{-b \pm \sqrt{b^2 - 4ac}}{2a} \quad a = 3, \ b = 128, \ c = -2112$$
$$x^2 = \frac{-128 \pm \sqrt{128^2 - 4(3)(-2112)}}{2(3)}$$
$$x^2 = \frac{-128 + \sqrt{41728}}{6} \quad \text{discarding the negative solution}$$
$$x = \sqrt{\frac{-128 + \sqrt{41728}}{6}} \quad \text{discarding the negative solution}$$
$$x \approx 3.565$$
$$y = \frac{x^2}{8}$$
$$= \frac{(3.565)^2}{8}$$
$$\approx 1.589$$
Solution: $(3.565, 1.589)$

29. False. For example, in the ellipse $\frac{x^2}{5} + \frac{y^2}{4} = 1$, the minor axis has length 2.
Since $a^2 = 5$ and $b^2 = 4$, the line segment joining the foci has length $2c$, where $c^2 = a^2 - b^2 = 5 - 4 = 1$. Thus $2c = 2$ and the two quantities are equal.

31. True. Since both the ellipse and any line through its center (the origin) have origin symmetry, they intersect at a point and at the reflection of this point through the origin, thus, at two points.

33. From the figure, we see that the point $P(x, y)$ is a point on the curve if and only if

$$d_1 = \frac{1}{2}d_2$$

$$d(P, F) = \frac{1}{2}d(P, M)$$

$$\sqrt{(x - 2)^2 + (y - 0)^2} = \frac{1}{2}\sqrt{(x - 8)^2 + (y - y)^2}$$

$$(x - 2)^2 + y^2 = \frac{1}{4}(x - 8)^2$$

$$x^2 - 4x + 4 + y^2 = \frac{1}{4}(x^2 - 16x + 64)$$

$$4x^2 - 16x + 16 + 4y^2 = x^2 - 16x + 64$$

$$3x^2 + 4y^2 = 48$$

$$\frac{x^2}{16} + \frac{y^2}{12} = 1$$

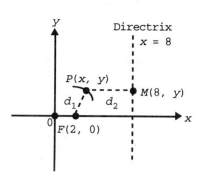

The curve must be an ellipse since its equation can be written in standard form for an ellipse.

35. Here are computer-generated graphs of $x^2 + 3y^2 = 20$ $\left(\text{entered as } y = \pm\sqrt{\dfrac{20 - x^2}{3}}\right)$

and $4x + 5y = 11$ $\left(\text{entered as } y = \dfrac{11 - 4x}{5}\right)$.

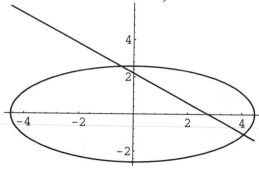

The curves intersect for x between -0.8 and -0.4 and between 4 and 4.4. Zooming in on these intervals, we obtain the following graphs:

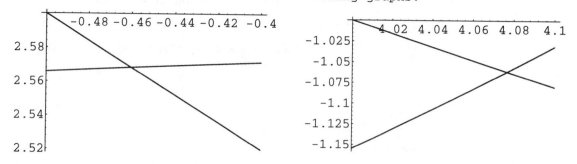

To two decimal places, the coordinates of the points of intersection are $(-0.46, 2.57)$ and $(4.08, -1.06)$.

37. Here are computer-generated graphs of
$50x^2 + 4y^2 = 1025$

$\left(\text{entered as } y = \pm\sqrt{\dfrac{1025 - 50x^2}{4}}\right)$

and $9x^2 + 2y^2 = 300$

$\left(\text{entered as } y = \pm\sqrt{\dfrac{300 - 9x^2}{2}}\right).$

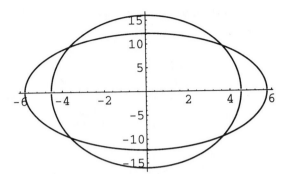

Note that the graphs are symmetric with respect to both axes and the origin. The curves intersect in the first quadrant for x between 3.6 and 4. Zooming in on this interval, we obtain the graph shown at the right.

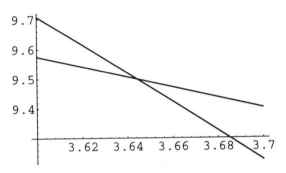

To two decimal places, the coordinates of the point of intersection are (3.64, 9.50). From symmetry, we obtain the other three coordinate pairs: (-3.64, 9.50), (-3.64, -9.50), and (3.64, -9.50).

39. From the figure we see that the x and y intercepts of the ellipse must be 20 and 12 respectively. Hence the equation of the ellipse must be

$$\frac{x^2}{(20)^2} + \frac{y^2}{(12)^2} = 1$$

or

$$\frac{x^2}{400} + \frac{y^2}{144} = 1$$

To find the clearance above the water 5 feet from the bank, we need the y coordinate y_1 of the point P whose x coordinate is $a - 5 = 15$. Since P is on the ellipse, we have

$$\frac{15^2}{400} + \frac{y_1^2}{144} = 1$$

$$\frac{225}{400} + \frac{y_1^2}{144} = 1$$

$$0.5625 + \frac{y_1^2}{144} = 1$$

$$\frac{y_1^2}{144} = 0.4375$$

$$y_1^2 = 144(0.4375)$$

$$y_1^2 = 63$$

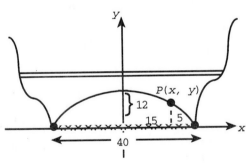

$y_1 \approx \pm 7.94$

Therefore, the clearance is 7.94 feet, approximately.

41. (A) From the figure we see that the x intercept of the ellipse must be 24.0. Hence the equation of the ellipse must have the form

$$\frac{x^2}{(24.0)^2} + \frac{y^2}{b^2} = 1$$

Since the point (23.0, 1.14) is on the ellipse, its coordinates must satisfy

the equation of the ellipse. Hence

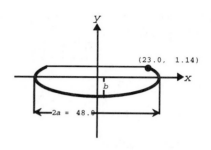

$$\frac{(23.0)^2}{(24.0)^2} + \frac{(1.14)^2}{b^2} = 1$$

$$\frac{(1.14)^2}{b^2} = 1 - \frac{(23.0)^2}{(24.0)^2}$$

$$\frac{(1.14)^2}{b^2} = \frac{47}{576}$$

$$b^2 = \frac{576(1.14)^2}{47}$$

$$b^2 = 15.9$$

Thus, the equation of the ellipse must be $\dfrac{x^2}{576} + \dfrac{y^2}{15.9} = 1$

(B) From the figure, we can see that the width of the wing must equal
1.14 + b = 1.14 + $\sqrt{15.9}$ = 5.13 feet.

Exercise 11-3

Key Ideas and Formulas

A hyperbola is the set of all points P in a plane such that the absolute value of the difference of the distances of P to two-fixed points (the foci) is a positive constant. The line through the foci intersects the hyperbola at two points called vertices; the line segment connecting the vertices is called the transverse axis. The midpoint of the transverse axis is called the center of the hyperbola.

Standard Equations of a Hyperbola — Center at (0, 0)

1. $\dfrac{x^2}{a^2} - \dfrac{y^2}{b^2} = 1$

   x intercepts: $\pm a$ (vertices)
   y intercepts: none
   Foci: $F'(-c,\ 0)$, $F(c,\ 0)$
   $c^2 = a^2 + b^2$
   Transverse axis length = $2a$
   Conjugate axis length = $2b$

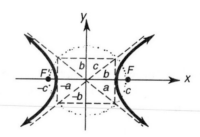

Asymptotes: $y = \pm\dfrac{b}{a}x$

2. $\dfrac{y^2}{a^2} - \dfrac{x^2}{b^2} = 1$

   x intercepts: none
   y intercepts: $\pm a$ (vertices)
   Foci: $F'(0,\ -c)$, $F(0,\ c)$
   $c^2 = a^2 + b^2$
   Transverse axis length = $2a$
   Conjugate axis length = $2b$

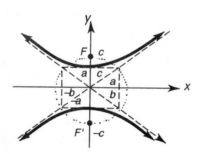

Asymptotes: $y = \pm\dfrac{a}{b}x$

Note: Both graphs are symmetric with respect to the x axis, y axis, and origin.

1. When $y = 0$, $\frac{x^2}{9} = 1$. x intercepts: ±3 $a = 3$

When $x = 0$, $-\frac{y^2}{4} = 1$. There are no y intercepts, but $b = 2$.

Sketch the asymptotes using the
asymptote rectangle, then sketch in
the hyperbola.
Foci: $c^2 = 3^2 + 2^2$
        $c^2 = 13$
        $c = \sqrt{13}$
$F'(-\sqrt{13},\ 0)$, $F(\sqrt{13},\ 0)$
Transverse axis length $= 2(3) = 6$
Conjugate axis length $= 2(2) = 4$

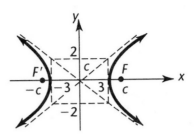

3. When $y = 0$, $-\frac{x^2}{9} = 1$. There are no x intercepts, but $b = 3$.

When $x = 0$, $\frac{y^2}{4} = 1$. y intercepts: ±2, $a = 2$

Sketch the asymptotes using the
asymptote rectangle, then sketch in
the hyperbola.
Foci: $c^2 = 2^2 + 3^2$
        $c^2 = 13$
        $c = \sqrt{13}$
$F'(0,\ -\sqrt{13})$, $F(0,\ \sqrt{13})$
Transverse axis length $= 2(2) = 4$
Conjugate axis length $= 2(3) = 6$

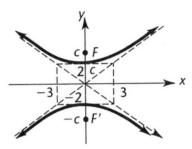

5. First, write the equation in standard form by dividing both sides by 16.
$4x^2 - y^2 = 16$
$$\frac{x^2}{4} - \frac{y^2}{16} = 1$$

Locate intercepts:
When $y = 0$, $x = \pm2$. x intercepts: ±2 $a = 2$

When $x = 0$, $-\frac{y^2}{16} = 1$. There are no y intercepts,

but $b = 4$.

Sketch the asymptotes using the
asymptote rectangle, then sketch in
the hyperbola.
Foci: $c^2 = 2^2 + 4^2$
        $c^2 = 20$
        $c = \sqrt{20}$
$F'(-\sqrt{20},\ 0)$, $F(\sqrt{20},\ 0)$
Transverse axis length $= 2(2) = 4$
Conjugate axis length $= 2(4) = 8$

7. First, write the equation in standard form by dividing both sides by 144.
$9y^2 - 16x^2 = 144$
$$\frac{y^2}{16} - \frac{x^2}{9} = 1$$

Locate intercepts:
When $y = 0$, $-\frac{x^2}{9} = 1$. There are no x intercepts, but $b = 3$.

When $x = 0$, $\frac{y^2}{16} = 1$, $y = \pm 4$.

y intercepts: ± 4 $a = 4$

Sketch the asymptotes using the asymptote rectangle, then sketch in the hyperbola.
Foci: $c^2 = 4^2 + 3^2$
$\qquad c^2 = 25$
$\qquad\quad c = 5$
$F'(0, -5)$, $F(0, 5)$
Transverse axis length $= 2(4) = 8$
Conjugate axis length $= 2(3) = 6$

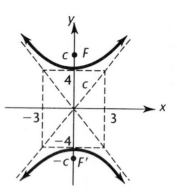

9. First, write the equation in standard form by dividing both sides by 12.
$3x^2 - 2y^2 = 12$
$$\frac{x^2}{4} - \frac{y^2}{6} = 1$$

Locate intercepts:

When $y = 0$, $\frac{x^2}{4} = 1$. $x = \pm 2$.

x intercepts: ± 2 $a = 2$

When $x = 0$, $-\frac{y^2}{6} = 1$. There are no y intercepts, but $b = \sqrt{6}$.

Sketch the asymptotes using the asymptote rectangle, then sketch in the hyperbola.
Foci: $c^2 = 2^2 + (\sqrt{6})^2$
$\qquad c^2 = 10$
$\qquad\quad c = \sqrt{10}$
$F'(-\sqrt{10}, 0)$, $F(\sqrt{10}, 0)$
Transverse axis length $= 2(2) = 4$
Conjugate axis length $= 2\sqrt{6} \approx 4.90$

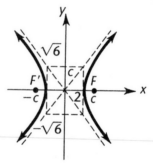

11. First, write the equation in standard form by dividing both sides by 28.
$7y^2 - 4x^2 \qquad = 28$
$$\frac{y^2}{4} - \frac{x^2}{7} \qquad = 1$$

Locate intercepts:

When $y = 0$, $-\frac{x^2}{7} = 1$. There are no x intercepts, but $b = \sqrt{7}$.

When $x = 0$, $\frac{y^2}{4} = 1$, $y = \pm 2$. y intercepts: ± 2 $a = 2$

Sketch the asymptotes using the asymptote rectangle, then sketch in the hyperbola.
Foci: $c^2 = 2^2 + (\sqrt{7})^2$
$\qquad c^2 = 11$
$\qquad\quad c = \sqrt{11}$
$F'(0, -\sqrt{11})$, $F(0, \sqrt{11})$
Transverse axis length $= 2(2) = 4$
Conjugate axis length $= 2\sqrt{7} \approx 5.29$

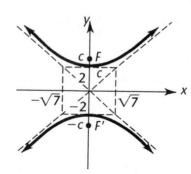

13. Since the transverse axis is on the y axis, start with
$$\frac{y^2}{a^2} - \frac{x^2}{b^2} = 1$$
and find a and b.
$$a = \frac{10}{2} = 5 \text{ and } b = \frac{18}{2} = 9$$
Thus, the equation is
$$\frac{y^2}{25} - \frac{x^2}{81} = 1$$

15. Since the transverse axis is on the x axis, start with
$$\frac{x^2}{a^2} - \frac{y^2}{b^2} = 1$$
and find a and b.
$$a = \frac{14}{2} = 7 \qquad c = 9$$

To find b, sketch the asymptote rectangle, label known parts, and use the Pythagorean Theorem.

$$b^2 = 9^2 - 7^2$$
$$b^2 = 32$$
$$b = \sqrt{32}$$
Thus, the equation is
$$\frac{x^2}{49} - \frac{y^2}{32} = 1$$

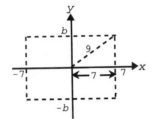

17. Since the conjugate axis is on the x axis, start with
$$\frac{y^2}{a^2} - \frac{x^2}{b^2} = 1$$
and find a and b.
$$b = \frac{12}{2} = 6 \qquad c = \frac{12\sqrt{2}}{2} = 6\sqrt{2}$$

To find a, sketch the asymptote rectangle, label known parts, and use the Pythagorean Theorem.

$$a^2 = (6\sqrt{2})^2 - 6^2$$
$$a^2 = 36$$
$$a = 6$$
Thus, the equation is
$$\frac{y^2}{36} - \frac{x^2}{36} = 1$$

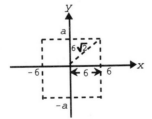

19. (A) If a hyperbola has center at $(0, 0)$ and a focus at $(1, 0)$, its equation must be of form
$$\frac{x^2}{a^2} - \frac{y^2}{b^2} = 1$$
with $c = 1$, thus $a^2 + b^2 = 1$ or $b^2 = 1 - a^2$

Therefore there are an infinite number of such hyperbolas.
Each has an equation of form
$$\frac{x^2}{a^2} - \frac{y^2}{1 - a^2} = 1$$
Note that since $0 < a < c$, we require $0 < a < 1$.

(B) If an ellipse has center at (0, 0) and a focus at (1, 0), its equation must be of form

$$\frac{x^2}{a^2} + \frac{y^2}{b^2} = 1$$

with $c = 1$, thus $a^2 - 1 = b^2$

Therefore there are an infinite number of such ellipses. Each has an equation of form

$$\frac{x^2}{a^2} + \frac{y^2}{a^2 - 1} = 1$$

Note that since $a > c > 0$, we require $a > 1$.

(C) If a parabola has vertex at (0, 0) and focus at (1, 0), its equation must be of form

$$y^2 = 4ax$$

with $a = 1$.

Therefore there is one such parabola; its equation is $y^2 = 4x$.

21. $3y^2 - 4x^2 = 12$
$\quad y^2 + x^2 = 25$

We solve using elimination by addition, multiplying the second equation by 4 and adding to the first equation.

$$
\begin{array}{r}
3y^2 - 4x^2 = 12 \\
4y^2 + 4x^2 = 100 \\
\hline
7y^2 \quad\quad = 112 \\
\end{array}
$$

$$y^2 = 16$$
$$y = \pm 4$$

For $y = 4$ or $y = -4$
$$y^2 + x^2 = 25$$
$$16 + x^2 = 25$$
$$x^2 = 9$$
$$x = \pm 3$$

Solutions: (3, 4), (-3, 4), (-3, -4), (3, -4)

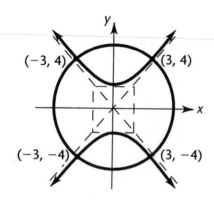

Graphs:
$3y^2 - 4x^2 = 12$ Hyperbola

$$\frac{y^2}{4} - \frac{x^2}{3} = 1$$

y intercepts: ± 2 $a = 2$

x intercepts: none, but $b = \sqrt{3}$
Transverse axis on y axis.
$y^2 + x^2 = 25$ Circle
radius 5, center (0, 0)

23. $2x^2 + y^2 = 24$
$\quad x^2 - y^2 = -12$

We solve by adding the two equations to eliminate y.
$$3x^2 = 12$$
$$x^2 = 4$$
$$x = \pm 2$$

For $x = \pm 2$
$$2x^2 + y^2 = 24$$
$$2(4) + y^2 = 24$$
$$8 + y^2 = 24$$
$$y^2 = 16$$
$$y = \pm 4$$

Solutions: (2, 4), (2, -4), (-2, -4), (-2, 4)

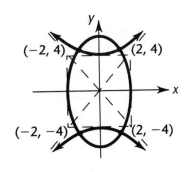

Graphs:

$2x^2 + y^2 = 24$

$\dfrac{x^2}{12} + \dfrac{y^2}{24} = 1$ Ellipse

$a = \sqrt{24}$, $b = \sqrt{12}$, major axis on the y axis

x intercepts: $\pm\sqrt{12}$

y intercepts: $\pm\sqrt{24}$

$x^2 - y^2 = -12$

$\dfrac{y^2}{12} - \dfrac{x^2}{12} = 1$ Hyperbola

y intercepts: $\pm\sqrt{12}$ $a = \sqrt{12}$

x intercepts: none, but $b = \sqrt{12}$

Transverse axis on y axis.

25. $y^2 - x^2 = 9$
$2y - x = 8$

Solve for x in the second equation, then substitute into the first equation.

$$2y = 8 + x$$
$$x = 2y - 8$$
$$y^2 - (2y - 8)^2 = 9$$
$$y^2 - (4y^2 - 32y + 64) = 9$$
$$-3y^2 + 32y - 64 = 9$$
$$3y^2 - 32y + 73 = 0$$

$$y = \frac{-b \pm \sqrt{b^2 - 4ac}}{2a} \quad a = 3,\ b = -32,\ c = 73$$

$$y = \frac{-(-32) \pm \sqrt{(-32)^2 - 4(3)(73)}}{2(3)}$$

$$= \frac{32 \pm \sqrt{148}}{6}$$

$$y \approx 3.306,\ 7.361$$

For $y = 3.306$ For $y = 7.361$
   $x = 2(3.306) - 8$ $x = 2(7.361) - 8$
      ≈ -1.389 ≈ 6.722

Solutions: $(-1.389,\ 3.306)$, $(6.722,\ 7.361)$

27. $y^2 - x^2 = 4$
$y^2 + 2x^2 = 36$

We solve using elimination by addition, multiplying the first equation by 2 and adding to the second equation.

$$2y^2 - 2x^2 = 8$$
$$\underline{y^2 + 2x^2 = 36}$$
$$3y^2 \qquad\quad = 44$$

$$y^2 = \frac{44}{3}$$

$$y = \pm\sqrt{\frac{44}{3}}$$

$$y \approx \pm3.830$$
$$y^2 - x^2 = 4$$
$$x^2 = y^2 - 4$$
$$= \frac{44}{3} - 4$$

$$x^2 = \frac{32}{3}$$

$$x = \pm3.266$$

Solutions: $(\pm3.266,\ \pm3.830)$

29. True. Since $b^2 = c^2 - a^2$, and a is never 0, $b^2 < c^2$, $b < c$, and $2b < 2c$. The line segment joining the foci of a hyperbola has length $2c$, and this is greater than $2b$, the length of the conjugate axis.

31. False. The center of $4x^2 - y^2 = 16$ is the origin. The y axis, $x = 0$, through the origin, does not intersect the hyperbola at all.

33. From the figure, we see that the point $P(x, y)$ is a point on the curve if and only if

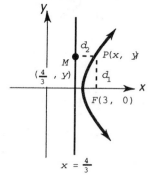

$$d_1 = \tfrac{3}{2} d_2$$

$$d(P, F) = \tfrac{3}{2} d(P, M)$$

$$\sqrt{(x - 3)^2 + (y - 0)^2} = \tfrac{3}{2}\sqrt{\left(x - \tfrac{4}{3}\right)^2 + (y - y)^2}$$

$$(x - 3)^2 + y^2 = \tfrac{9}{4}\left(x - \tfrac{4}{3}\right)^2$$

$$x^2 - 6x + 9 + y^2 = \tfrac{9}{4}\left(x^2 - \tfrac{8}{3}x + \tfrac{16}{9}\right)$$

$$4x^2 - 24x + 36 + 4y^2 = 9x^2 - 24x + 16$$

$$-5x^2 + 4y^2 = -20$$

$$\frac{x^2}{4} - \frac{y^2}{5} = 1$$

The curve must be a hyperbola since its equation can be written in standard form for a hyperbola.

35. Here are computer-generated graphs of $2x^2 - 3y^2 = 20$ $\left(\text{entered as } y = \pm\sqrt{\dfrac{2x^2 - 20}{3}}\right)$ and $7x + 15y = 10$ $\left(\text{entered as } y = \dfrac{10 - 7x}{15}\right)$.

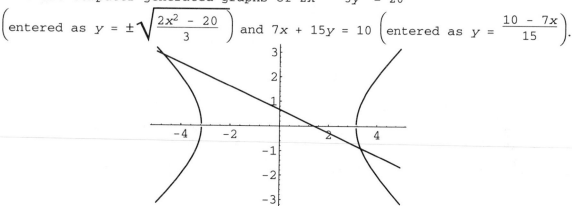

The curves intersect for x between -5.2 and -4.8 and between 3.2 and 3.6. Zooming in on these intervals, we obtain the following graphs.

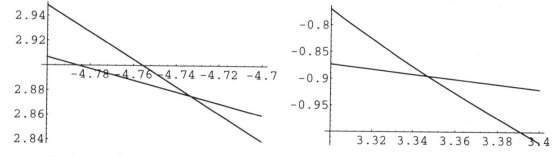

To two decimal places, the coordinates of the points of intersection are $(-4.73, 2.88)$ and $(3.35, -0.90)$.

37. Here are computer-generated graphs of
$24y^2 - 18x^2 = 175$

$\left(\text{entered as } y = \pm\sqrt{\dfrac{175 + 18x^2}{24}} \right)$ and

$90x^2 + 3y^2 = 200$

$\left(\text{entered as } y = \pm\sqrt{\dfrac{200 - 90x^2}{3}} \right)$.

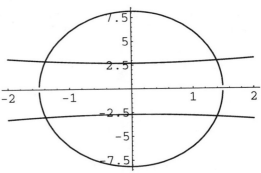

Note that the graphs are symmetric with respect to both axes and the origin. The curves intersect in the first quadrant for x between 1.2 and 1.4. Zooming in on this interval, we obtain the second graph.

To two decimal places, the coordinates of the points of intersection are (1.39, 2.96). From symmetry, we obtain the other three coordinate pairs: (-1.39, 2.96), (-1.39, -2.96), and (1.39, -2.96).

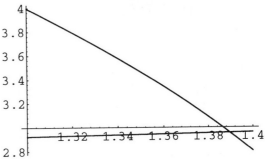

39. From the figure below, we see that the transverse axis of the hyperbola must be on the y axis and $a = 4$. Hence, the equation of the hyperbola must have the form

$$\frac{y^2}{4^2} - \frac{x^2}{b^2} = 1$$

To find b, we note that the point (8, 12) is on the hyperbola, hence its coordinates satisfy the equation. Substituting, we have

$$\frac{12^2}{4^2} - \frac{8^2}{b^2} = 1$$

$$9 - \frac{64}{b^2} = 1$$

$$-\frac{64}{b^2} = -8$$

$$-64 = -8b^2$$

$$b^2 = 8$$

The equation required is

$$\frac{y^2}{16} - \frac{x^2}{8} = 1$$

Using this equation, we can compute y when $x = 6$ to answer the question asked (see figure).

$$\frac{y^2}{16} - \frac{6^2}{8} = 1$$

$$\frac{y^2}{16} - \frac{36}{8} = 1$$

$$y^2 - 72 = 16$$

$$y^2 = 88$$

$$y = 9.38 \text{ to two decimal places}$$

The height above the vertex

$$= y - \text{height of vertex}$$

$$= 9.38 - 4$$

$$= 5.38 \text{ feet}$$

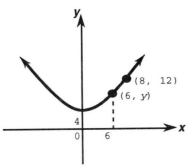

Hyperbola part of dome

41. From the figure below, we can see:

$FF' = 2c = 120 - 20 = 100$

Thus $c = 50$

$FV = c - a = 120 - 110 = 10$

Thus $a = c - 10 = 50 - 10 = 40$

Since $c^2 = a^2 + b^2$

$$50^2 = 40^2 + b^2$$

$$b = 30$$

Thus the equation of the hyperbola, in standard form, is

$$\frac{y^2}{40^2} - \frac{x^2}{30^2} = 1$$

Expressing y in terms of x, we have

$$\frac{y^2}{40^2} = 1 + \frac{x^2}{30^2}$$

$$\frac{y^2}{40^2} = \frac{1}{30^2}(30^2 + x^2)$$

$$y^2 = \frac{40^2}{30^2}(30^2 + x^2)$$

$$y = \frac{4}{3}\sqrt{x^2 + 30^2}$$

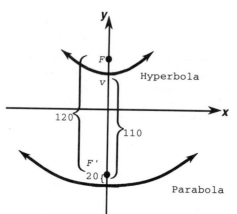

discarding the negative solution, since the reflecting hyperbola is above the x axis.

Exercise 11-4
Key Ideas and Formulas

A translation of coordinates occurs when the new coordinate axes have the same direction as and are parallel to the old coordinate axes.

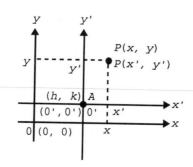

If the coordinates of the origin of the translated system are (h, k) relative to the old system, then (x, y) and (x', y') are related by:

1. $x = x' + h$
   $y = y' + k$

2. $x' = x - h$
   $y' = y - k$

Common Error:
The signs are very easily confused here, especially if h or k is negative in a problem.

Standard Equations for Translated Conics

PARABOLAS

$(x - h)^2 = 4a(y - k)$

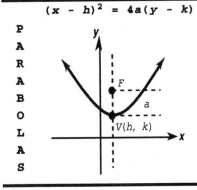

Vertex (h, k)
Focus $(h, k + a)$
$a > 0$ opens up
$a < 0$ opens down

$(y - k)^2 = 4a(x - h)$

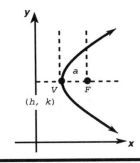

Vertex (h, k)
Focus $(h + a, k)$
$a < 0$ opens left
$a > 0$ opens right

CIRCLES

$(x - h)^2 + (y - k)^2 = r^2$

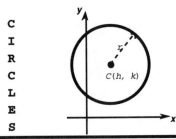

Center (h, k)
Radius r

ELLIPSES

$$\frac{(x - h)^2}{a^2} + \frac{(y - k)^2}{b^2} = 1$$
$$a > b > 0$$

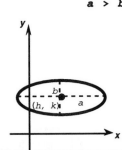

Center (h, k)
Major axis $2a$
Minor axis $2b$

$$\frac{(x - h)^2}{b^2} + \frac{(y - k)^2}{a^2} = 1$$

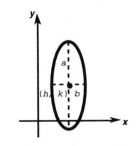

Center (h, k)
Major axis $2a$
Minor axis $2b$

HYPERBOLAS

$$\frac{(x - h)^2}{a^2} - \frac{(y - k)^2}{b^2} = 1$$

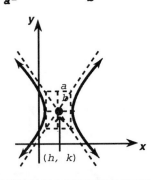

Center (h, k)
Transverse axis $2a$
Conjugate axis $2b$

$$\frac{(y - k)^2}{a^2} - \frac{(x - h)^2}{b^2} = 1$$

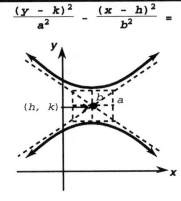

Center (h, k)
Transverse axis $2a$
Conjugate axis $2b$

1. (A) Since $(h, k) = (3, 5)$, use translation formulas
$$x' = x - h = x - 3$$
$$y' = y - k = y - 5$$
(B) $x'^2 + y'^2 = 81$
(C) Circle

3. (A) Since $(h, k) = (-7, 4)$, use translation formulas
$$x' = x - h = x + 7$$
$$y' = y - k = y - 4$$
(B) $\dfrac{x'^2}{9} + \dfrac{y'^2}{16} = 1$
(C) Ellipse

5. (A) Since $(h, k) = (4, -9)$, use translation formulas
$$x' = x - h = x - 4$$
$$y' = y - k = y + 9$$
(B) $y'^2 = 16x'$
(C) Parabola

7. (A) Since $(h, k) = (-8, -3)$, use translation formulas
$$x' = x - h = x + 8$$
$$y' = y - k = y + 3$$
(B) $\dfrac{x'^2}{12} + \dfrac{y'^2}{8} = 1$
(C) Ellipse

9. (A) Divide both sides by 144
$$\frac{(x - 3)^2}{9} - \frac{(y + 2)^2}{16} = 1$$
(B) This is the equation of a hyperbola.

11. (A) Divide both sides by 30.
$$\frac{(x + 5)^2}{5} + \frac{(y + 7)^2}{6} = 1$$
(B) This is the equation of an ellipse.

13. (A) Subtract $24(y - 4)$ from both sides.
$$(x + 6)^2 = -24(y - 4)$$
(B) This is the equation of a parabola.

15.
$$4x^2 + 9y^2 - 16x - 36y + 16 = 0$$
$$4x^2 - 16x + 9y^2 - 36y = -16$$
$$4(x^2 - 4x + ?) + 9(y^2 - 4y + ?) = -16$$
$$4(x^2 - 4x + 4) + 9(y^2 - 4y + 4) = -16 + 16 + 36$$
$$4(x - 2)^2 + 9(y - 2)^2 = 36$$
$$\frac{(x - 2)^2}{9} + \frac{(y - 2)^2}{4} = 1$$

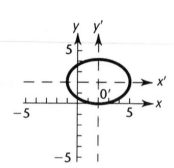

This is the equation of an ellipse with center at $(2, 2)$. The equations of translation are $x' = x - 2$, $y' = y - 2$. Making these substitutions, we obtain
$$\frac{x'^2}{9} + \frac{y'^2}{4} = 1$$
We graph this in the $x'y'$ system, following the process discussed in Section 11-2.

17. $x^2 + 8x + 8y = 0$
$$x^2 + 8x = -8y$$
$$x^2 + 8x + 16 = -8y + 16$$
$$(x + 4)^2 = -8(y - 2)$$

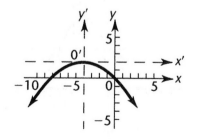

This is the equation of a parabola opening down with vertex at $(h, k) = (-4, 2)$. The equations of translation are $x' = x + 4$, $y' = y - 2$. Making these substitutions, we obtain
$$x'^2 = -8y'$$
We graph this in the $x'y'$ system, following the process discussed in Section 11-1.

19.
$$x^2 + y^2 + 12x + 10y + 45 = 0$$
$$x^2 + 12x + y^2 + 10y = -45$$
$$x^2 + 12x + 36 + y^2 + 10y + 25 = -45 + 36 + 25$$
$$(x + 6)^2 + (y + 5)^2 = 16$$

This is the equation of a circle with center at (-6, -5) and radius 4. The equations of translation are $x' = x + 6$, $y' = y + 5$. Making these substitutions, we obtain
$$x'^2 + y'^2 = 16$$
We graph this in the $x'y'$ system.

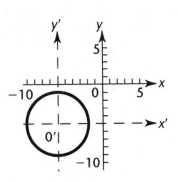

21.
$$-9x^2 + 16y^2 - 72x - 96y - 144 = 0$$
$$-9x^2 - 72x + 16y^2 - 96y = 144$$
$$-9(x^2 + 8x + ?) + 16(y^2 - 6y + ?) = 144$$
$$-9(x^2 + 8x + 16) + 16(y^2 - 6y + 9) = 144 - 144 + 144$$
$$-9(x + 4)^2 + 16(y - 3)^2 = 144$$
$$\frac{(y - 3)^2}{9} - \frac{(x + 4)^2}{16} = 1$$

This is the equation of a hyperbola with center at (-4, 3). The equations of translation are $x' = x + 4$, $y' = y - 3$. Making these substitutions, we obtain
$$\frac{y'^2}{9} - \frac{x'^2}{16} = 1$$
We graph this in the $x'y'$ system, following the process discussed in Section 11-3.

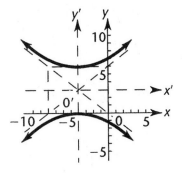

23. If $A \neq 0$, $C = 0$, and $E \neq 0$, write
$$Ax^2 + Dx + Ey + F = 0$$

Complete the square relative to x:

$$A\left(x^2 + \frac{D}{A}x + ?\right) = -Ey - F$$

$$A\left(x^2 + \frac{D}{A}x + \frac{D^2}{4A^2}\right) = -Ey - F + \frac{D^2}{4A}$$

$$\left(x + \frac{D}{2A}\right)^2 = -\frac{E}{A}y - \frac{F}{A} + \frac{D^2}{4A^2}$$

$$\left(x + \frac{D}{2A}\right)^2 = 4 \cdot \left(-\frac{E}{4A}\right)\left[y + \frac{F}{E} - \frac{D^2}{4AE}\right]$$

$$\left[x - \left(-\frac{D}{2A}\right)\right]^2 = 4\left(-\frac{E}{4A}\right)\left[y - \frac{D^2 - 4AF}{4AE}\right]$$

The equations of translation are
$$x' = x - \left(-\frac{D}{2A}\right) \qquad y' = y - \frac{D^2 - 4AF}{4AE}$$

Thus, $h = -\frac{D}{2A}$, $k = \frac{D^2 - 4AF}{4AE}$

The equation would become $x'^2 = 4\left(-\frac{E}{4A}\right)y'$, that is, the equation of a parabola.

25. Locate the vertex and axis in the original coordinate system, then sketch the parabola and translate the origin to the vertex of the parabola. Next write the equation of the parabola in the translated system:

$$x'^2 = 4ay'$$

The origin in the translated system is at $(h, k) = (5, 3)$ and the translation formulas are

$$x' = x - h = x - 5$$
$$y' = y - k = y - 3$$

Thus, the equation of the parabola in the original system is

$$(x - 5)^2 = 4a(y - 3)$$

Since the distance from focus to vertex, a, is given by the distance from $(5, 3)$ to $(5, 11)$, or 8, the equation of the parabola is

$$(x - 5)^2 = 4 \cdot 8(y - 3)$$
$$x^2 - 10x + 25 = 32y - 96$$
$$x^2 - 10x - 32y + 121 = 0$$

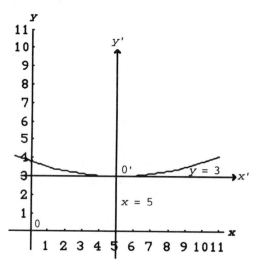

27. Locate the vertices in the original coordinate system, then sketch the ellipse and translate the origin to the center of the ellipse. Next write the equation of the ellipse in the translated system. $2a$ = distance between vertices = 12, hence $a = 6$. $2b$ = length of minor axis = 10, hence $b = 5$. Hence the equation is

$$\frac{x'^2}{5^2} + \frac{y'^2}{6^2} = 1$$

The origin in the translated system is at the midpoint of the major axis, that is, $(h, k) = (-3, 4)$ and the translation formulas are

$$x' = x - h = x - (-3) = x + 3$$
$$y' = y - k = y - 4$$

Thus, the equation of the ellipse in the original system is

$$\frac{(x + 3)^2}{25} + \frac{(y - 4)^2}{36} = 1$$
$$36(x + 3)^2 + 25(y - 4)^2 = 900$$

$$36x^2 + 216x + 324 + 25y^2 - 200y + 400 = 900$$
$$36x^2 + 25y^2 + 216x - 200y - 176 = 0$$

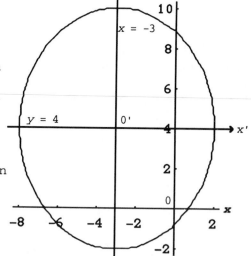

29. Locate the vertices, asymptote rectangle, and asymptotes in the original coordinate system, then sketch the hyperbola and translate the origin to the center of the hyperbola. Since $2a = 5 - 3 = 2$, $a = 1$. Since $2c = 6 - 2 = 4$, $c = 2$. Hence, $b = \sqrt{c^2 - a^2} = \sqrt{2^2 - 1^2} = \sqrt{3}$.

Next write the equation of the hyperbola in the translated system.

$$\frac{x'^2}{1^2} - \frac{y'^2}{(\sqrt{3})^2} = 1$$

The origin in the translated system is at the midpoint of the transverse axis $(4, 1) = (h, k)$, and the translation formulas are

$$x' = x - h = x - 4$$
$$y' = y - k = y - 1$$

Thus, the equation of the hyperbola in the original system is

$$\frac{(x - 4)^2}{1} - \frac{(y - 1)^2}{3} = 1$$
$$3(x - 4)^2 - (y - 1)^2 = 3$$
$$3x^2 - 24x + 48 - y^2 + 2y - 1 = 3$$
$$3x^2 - y^2 - 24x + 2y + 44 = 0$$

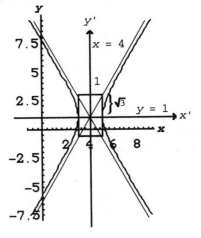

31. Locate the axis in the original coordinate system and position the vertex approximately, on the y axis at $(0, k)$. Next write an equation for the parabola in the translated system.

$$x'^2 = 4ay'$$

The origin in the translated system is at $(0, k)$ and the translation formulas are

$$x' = x - 0 = x$$
$$y' = y - k$$

The equation of the parabola in the original system is

$$x^2 = 4a(y - k)$$

Since the points $(1, 0)$ and $(2, 4)$ are on the parabola, their coordinates must satisfy the equation of the parabola, hence

$$1^2 = 4a(0 - k)$$
$$2^2 = 4a(4 - k)$$

Solving this system of equations for a and k, we obtain

$$1 = -4ak$$
$$4 = 16a - 4ak$$
$$4 = 16a + 1$$
$$a = \frac{3}{16}$$
$$k = -\frac{1}{4a} = -\frac{4}{3}$$

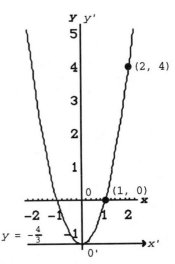

Thus, the equation of the parabola is

$$x^2 = 4 \cdot \frac{3}{16}\left[y - \left(-\frac{4}{3}\right)\right]$$
$$x^2 = \frac{3}{4}\left(y + \frac{4}{3}\right)$$
$$4x^2 = 3y + 4$$
$$4x^2 - 3y - 4 = 0$$

33. Locate the vertices in the original coordinate system, then sketch the ellipse and translate the origin to the center of the ellipse. Since $2a = 1 - (-5) = 6$, $a = 3$. Next, write the equation of the ellipse in the translated system

$$\frac{x'^2}{3^2} - \frac{y'^2}{b^2} = 1$$

The origin in the translated system is at the midpoint of the major axis, at $(h, k) = (-2, -1)$ and the translation formulas are

$$x' = x - h = x - (-2) = x + 2$$
$$y' = y - k = y - (-1) = y + 1$$

Thus, the equation of the ellipse in the original system is

$$\frac{(x + 2)^2}{9} + \frac{(y + 1)^2}{b^2} = 1$$

Since the point $(0, 0)$ is on the ellipse, its coordinates must satisfy the equation of the ellipse, hence

$$\frac{(0 + 2)^2}{9} + \frac{(0 + 1)^2}{b^2} = 1$$

$$\frac{4}{9} + \frac{1}{b^2} = 1$$

$$\frac{1}{b^2} = \frac{5}{9}$$

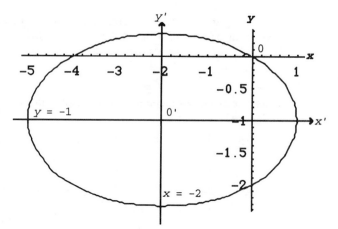

Thus, the equation of the ellipse is

$$\frac{(x + 2)^2}{9} + \frac{5(y + 1)^2}{9} = 1$$
$$x^2 + 4x + 4 + 5y^2 + 10y + 5 = 9$$
$$x^2 + 5y^2 + 4x + 10y = 0$$

35. First find the coordinates of the foci in the translated system.
$$c'^2 = 3^2 - 2^2 = 5$$
$$c' = \sqrt{5}$$
$$-c' = -\sqrt{5}$$
Thus the coordinates in the translated system are
$F'(-\sqrt{5}, 0)$ and $F(\sqrt{5}, 0)$
Now use $x = x' + h = x' + 2$
$y = y' + k = y' + 2$
to obtain $F'(-\sqrt{5} + 2, 2)$ and $F(\sqrt{5} + 2, 2)$ as the coordinates of the foci in the original system.

37. First find the coordinates of the focus in the translated system. Since $a = -2$, and the parabola opens down, the coordinates are $(0, -2)$. Now use
$x = x' + h = x' - 4 = 0 - 4$
$y = y' + k = y' + 2 = -2 + 2$
to obtain $(-4, 0)$ as the coordinates of the focus in the original system.

39. First find the coordinates of the foci in the translated system.
$$c'^2 = 3^2 + 4^2 = 25$$
$$c' = 5$$
$$-c' = -5$$
Thus the coordinates in the translated system are
$F'(0, -5)$ and $F(0, 5)$
Now use $x = x' + h = x' - 4$
$y = y' + k = y' + 3$
to obtain
$F'(0 - 4, -5 + 3) = F'(-4, -2)$ and $F(0 - 4, 5 + 3) = F(-4, 8)$ as the coordinates of the foci in the original system.

41. Before we can enter these equations in our graphing utility, we must solve for y:

$$3x^2 - 5y^2 + 7x - 2y + 11 = 0 \qquad 6x + 4y = 15$$
$$5y^2 + 2y - (3x^2 + 7x + 11) = 0 \qquad 4y = 15 - 6x$$
$$y = \frac{15 - 6x}{4}$$

Applying the quadratic formula yields

$$y = \frac{-2 \pm \sqrt{4 + 4(5)(3x^2 + 7x + 11)}}{10}$$

$$y = \frac{-1 \pm \sqrt{1 + 5(3x^2 + 7x + 11)}}{5}$$

$$y = \frac{-1 \pm \sqrt{15x^2 + 35x + 56}}{5}$$

Entering these three equations into a graphing utility produces the following graph:

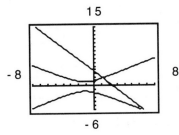

The curves intersect for x between 1 and 1.5 and between 6.5 and 7. Zooming in on these intervals, we obtain the following graphs:

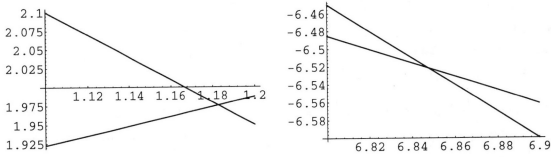

To two decimal places, the coordinates of the points of intersection are $(1.18, 1.98)$ and $(6.85, -6.52)$.

43. Before we can enter these equations in our graphing utility, we must solve for y:

$$7x^2 - 8x + 5y - 25 = 0 \qquad\qquad x^2 + 4y^2 + 4x - y - 12 = 0$$
$$5y = 25 + 8x - 7x^2 \qquad\qquad 4y^2 - y - (12 - 4x - x^2) = 0$$
$$y = \frac{25 + 8x - 7x^2}{5}$$

Applying the quadratic formula yields

$$y = \frac{1 \pm \sqrt{1 + 4(4)(12 - 4x - x^2)}}{8}$$

$$y = \frac{1 \pm \sqrt{193 - 64x - 16x^2}}{8}$$

Entering these three equations into a graphing utility produces the following graph:

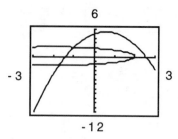

The curves intersect for x between -1.8 and -1.6 and between -1 and -0.8. Zooming in on these intervals, we obtain the following graphs:

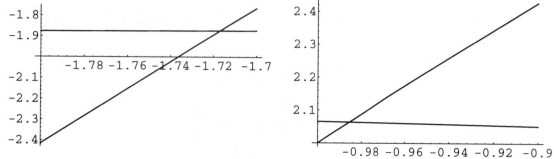

To two decimal places, the coordinates of the points of intersection are $(-1.72, -1.87)$ and $(-0.99, 2.06)$.

Exercise 11-5

Key Ideas and Formulas

A plane curve is the set of all points (x, y) such that

$$x = f(t) \quad y = g(t) \quad t \in I$$

where functions f and g are both defined on the interval I. t is called a parameter.

To graph a plane curve, one can set up a table of values for t, x, and y, then plot the ordered pairs (x, y).

It may be possible to eliminate the parameter t by algebraic substitution methods. The curve may then be identified as a curve that has been studied.

Parametric Equations for a Cycloid:

For a circle of radius a rolled along the x axis, the resulting Cycloid generated by a point on the rim starting at the origin is given by

$$x = a\theta - a \sin \theta$$
$$y = a - a \cos \theta \quad \theta \in (-\infty, \infty)$$

1. Construct a table and graph.

t	0	1	2	3	-1	-2	-3
x	0	-1	-2	-3	1	2	3
y	-2	0	2	4	-4	-6	-8

To eliminate the parameter t we solve
$x = -t$ for t to obtain
  $t = -x$
Then we substitute the expression for t into
$y = 2t - 2$ to obtain
  $y = 2(-x) - 2$
  $y = -2x - 2$
This is the equation of a straight line.

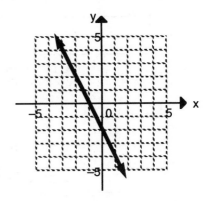

3. Construct a table and graph.

t	0	1	2	3	-1	-2	-3	
x	0	-1	-4	-9	-1	-4	-9	Note: $x \leq 0$
y	-2	0	6	16	0	6	16	

To eliminate the parameter t we solve
$x = -t^2$ for t^2 (we do not need t) to obtain
  $t^2 = -x$
Then we substitute the expression for t^2 into
$y = 2t^2 - 2$ to obtain
  $y = 2(-x) - 2$
  $y = -2x - 2$

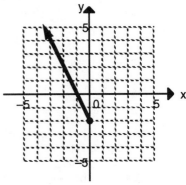

This is the equation of a straight line. However, since $x = -t^2$, x is
restricted so that $x \leq 0$. Therefore the equation of the curve is actually
  $y = -2x - 2, \quad x \leq 0$
This is the equation of a ray (part of a straight line).

5. Construct a table and graph.

t	0	1	2	3	-1	-2	-3
x	0	3	6	9	-3	-6	-9
y	0	-2	-4	-6	2	4	6

To eliminate the parameter t we solve
$x = 3t$ for t to obtain

  $t = \dfrac{x}{3}$

Then we substitute the expression for t into
$y = -2t$ to obtain

  $y = -2\left(\dfrac{x}{3}\right)$

  $y = -\dfrac{2}{3}x$

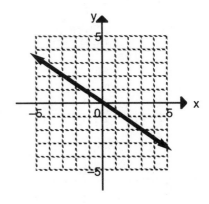

This is the equation of a straight line.

7. Construct a table of values and graph.

t	0	1	2	3	-1	-2	-3
x	0	$\frac{1}{4}$	1	$\frac{9}{4}$	$\frac{1}{4}$	1	$\frac{9}{4}$
y	0	1	2	3	-1	-2	-3

To eliminate the parameter t we substitute $t = y$
(from $y = t$) into the expression for x to obtain
$$x = \frac{1}{4}y^2$$
$$y^2 = 4x$$
This is the equation of a parabola.

9. Construct a table of values and graph.

t	0	1	2	3	-1	-2	-3
x	0	$\frac{1}{4}$	4	$\frac{81}{4}$	$\frac{1}{4}$	4	$\frac{81}{4}$
y	0	1	4	9	1	4	9

To eliminate the parameter we substitute $t^2 = y$
(from $y = t^2$) into the expression for x to obtain
$$x = \frac{1}{4}t^4$$
$$x = \frac{1}{4}(t^2)^2$$
$$x = \frac{1}{4}y^2$$
$$y^2 = 4x$$
This is the equation of a parabola. However, since $y = t^2$, y is restricted to
nonnegative values only. Hence the equation of the curve is actually
$$y^2 = 4x, \ y \geq 0$$
This is the equation of the upper half of a parabola.

11. We eliminate the parameter θ as follows:
$$\frac{x}{3} = \sin \theta \qquad \frac{y}{4} = \cos \theta$$
$$\left(\frac{x}{3}\right)^2 + \left(\frac{y}{4}\right)^2 = \sin^2 \theta + \cos^2 \theta$$
$$\frac{x^2}{9} + \frac{y^2}{16} = 1$$
This is the equation of an ellipse in standard form.
Graph the intercepts $x = \pm 3$, $y = \pm 4$, and sketch in
the ellipse.

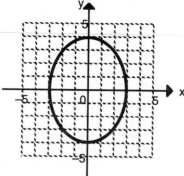

13. We eliminate the parameter θ as follows:
$$x - 2 = 2 \sin \theta \qquad y - 3 = 2 \cos \theta$$
$$\frac{x - 2}{2} = \sin \theta \qquad \frac{y - 3}{2} = \cos \theta$$
$$\left(\frac{x - 2}{2}\right)^2 + \left(\frac{y - 3}{2}\right)^2 = \sin^2 \theta + \cos^2 \theta$$
$$\frac{(x - 2)^2}{4} + \frac{(y - 3)^2}{4} = 1$$
$$(x - 2)^2 + (y - 3)^2 = 4$$
This is the equation of a circle with center
(2, 3) and radius 2.

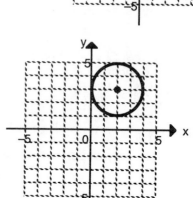

15. We eliminate the parameter t as follows:
$$x = t - 2$$
$$x + 2 = t$$

Substituting this expression for t into the expression for y, we obtain

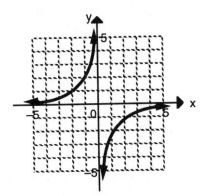

$$y = \frac{2}{2 - t}$$
$$y = \frac{2}{2 - (x + 2)}$$
$$y = \frac{2}{-x}$$
$$y = -\frac{2}{x}$$

This is the equation of a hyperbola with asymptotes x and y axes.

Construct a table of values and graph.

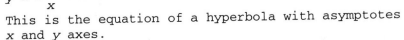

x	1	2	4	-1	-2	-4
y	-2	-1	$-\frac{1}{2}$	2	1	$\frac{1}{2}$

17. We eliminate the parameter t as follows:
Solve $x = t - 1$ for t to obtain $t = x + 1$.

Substitute the expression for t into $y = \sqrt{t}$ to obtain

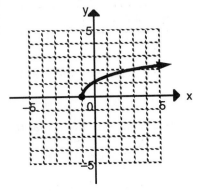

$$y = \sqrt{x + 1} \text{ or}$$
$$y^2 = x + 1, \; y \geq 0.$$

Since t is restricted so that $t \geq 0$, we also have $x + 1 \geq 0$ or $x \geq -1$. This graph is a parabola with vertex at $(-1, 0)$, opening right. Since $y \geq 0$ only the upper half of the parabola is traced out.

19. If $A \neq 0$, $C = 0$, and $E \neq 0$, write
$$Ax^2 + Dx + Ey + F = 0$$
There are many possible parametric equations for this curve. A simple approach is to set $x = t$ and solve for y in terms of t.
$$At^2 + Dt + Ey + F = 0$$
$$At^2 + Dt + F = -Ey$$
$$\frac{At^2 + Dt + F}{-E} = y$$

Then $x = t$, $y = \dfrac{At^2 + Dt + F}{-E}$, $-\infty < t < \infty$, are parametric equations for this curve. The curve is a parabola.

21. We eliminate the parameter t as follows:

Since $y = t^{-2} = \dfrac{1}{t^2}$ and $x = t^2$, we observe that

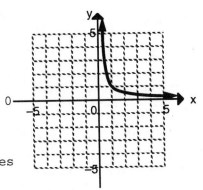

$y = \dfrac{1}{x}$. Since both x and y involve t^2, we see that x and y are restricted to nonnegative values. $t \neq 0$ restricts $x \neq 0$, hence only the part of the curve $y = \dfrac{1}{x}$ for which $x > 0$ is traced out. The graph is one branch of the hyperbola $xy = 1$, with asymptotes the positive x and y axes.

23. We eliminate the parameter θ as follows:
$y = 4 \sin \theta$

$\dfrac{y}{4} = \sin \theta$

From the double angle identity, we obtain
$$x = \cos 2\theta$$
$$x = 1 - 2 \sin^2 \theta$$
$$x = 1 - 2\left(\dfrac{y}{4}\right)^2$$
$$x = 1 - \dfrac{y^2}{8}$$
$$x - 1 = -\dfrac{y^2}{8}$$
$$y^2 = -8(x - 1)$$

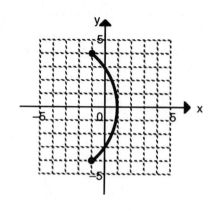

This is the equation of a parabola with vertex at (1, 0), opening left. Since $x = \cos 2\theta$, only values of x between -1 and 1 are taken on, that is $-1 \le x \le 1$.

x	1	0	-1
y	0	$\pm\sqrt{8}$	± 4

25. We eliminate the parameter t as follows: First note that since $x = \dfrac{8}{t^2 + 4}$, $(t^2 + 4)x = 8$ or $t^2 + 4 = \dfrac{8}{x}$. Therefore x cannot be 0. Now square the expressions for x and y.

$$x^2 = \dfrac{64}{(t^2 + 4)^2}$$

$$y^2 = \dfrac{16t^2}{(t^2 + 4)^2}$$

$$x^2 + y^2 = \dfrac{64}{(t^2 + 4)^2} + \dfrac{16t^2}{(t^2 + 4)^2}$$

$$x^2 + y^2 = \dfrac{64 + 16t^2}{(t^2 + 4)^2}$$

$$x^2 + y^2 = \dfrac{16(4 + t^2)}{(t^2 + 4)^2}$$

$$x^2 + y^2 = \dfrac{16}{t^2 + 4}$$

Since $x = \dfrac{8}{t^2 + 4}$, $2x = \dfrac{16}{t^2 + 4}$, so

$$x^2 + y^2 = 2x \quad x \ne 0$$

This is the equation of a circle. Completing the square we see
$$x^2 - 2x + y^2 = 0$$
$$x^2 - 2x + 1 + y^2 = 1$$
$$(x - 1)^2 + y^2 = 1, \quad x \ne 0$$

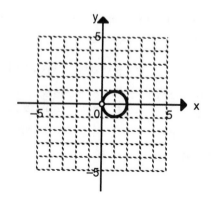

Hence the circle has center (1, 0) and radius 1. The origin is not part of the curve, hence there is a hole in the graph there.

27. Construct a table of values and graph.

θ	0	$\frac{\pi}{6}$	$\frac{\pi}{3}$	$\frac{\pi}{2}$	$\frac{2\pi}{3}$	$\frac{5\pi}{6}$	π	$\frac{7\pi}{6}$	$\frac{4\pi}{3}$	$\frac{3\pi}{2}$	$\frac{5\pi}{3}$	$\frac{11\pi}{6}$	2π
x	0	.02	.18	.57	1.23	2.12	3.14	4.17	5.05	5.71	6.10	6.26	6.28
y	0	.13	.5	1	1.5	1.87	2	1.87	1.5	1	.5	.13	0

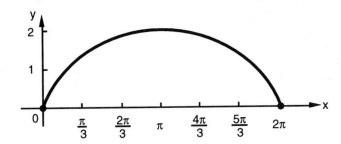

29. Here is a graph of $x = 3 + 6 \cos t$, $y = 2 + 4 \sin t$, $0 \le t \le 2\pi$ in a graphing utility.

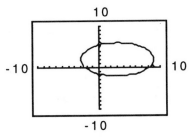

We eliminate the parameter t as follows:
$$x - 3 = 6 \cos t \qquad y - 2 = 4 \sin t$$
$$\frac{x - 3}{6} = \cos t \qquad \frac{y - 2}{4} = \sin t$$
$$\left(\frac{x - 3}{6}\right)^2 + \left(\frac{y - 2}{4}\right)^2 = \cos^2 t + \sin^2 t$$
$$\frac{(x - 3)^2}{36} + \frac{(y - 2)^2}{16} = 1$$
The curve is an ellipse with center (3, 2).

31. Here is a graph of $x = -3 + 2 \tan t$, $y = -1 + 5 \sec t$, $-\frac{\pi}{2} < t < \frac{3\pi}{2}$, $t \ne \frac{\pi}{2}$ in a graphing utility.

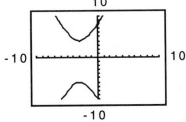

We eliminate the parameter t as follows:
$$x + 3 = 2 \tan t \qquad y + 1 = 5 \sec t$$
$$\frac{x + 3}{2} = \tan t \qquad \frac{y + 1}{5} = \sec t$$
$$\left(\frac{y + 1}{5}\right)^2 - \left(\frac{x + 3}{2}\right)^2 = \sec^2 t - \tan^2 t$$
$$\frac{(y + 1)^2}{25} - \frac{(x + 3)^2}{4} = 1$$
The curve is a hyperbola with center (-3, -1).

33. Since $x = 2 \tan t$, we can write $\tan t = \frac{x}{2}$. Since $y = 5 \tan t$, we can write $\tan t = \frac{y}{5}$. Setting the right sides of these equations equal, we can eliminate $\tan t$ and write
$$\frac{x}{2} = \frac{y}{5}$$
or $\quad 5x - 2y = 0$

35. $25x^2 - 200x - 9y^2 - 18y + 616 = 0$

Complete the square relative to both x and y.

$25x^2 - 200x \quad - 9y^2 - 18y \quad = -616$

$25(x^2 - 8x \quad) - 9(y^2 + 2y \quad) = -616$

$25(x^2 - 8x + 16) - 9(y^2 + 2y + 1) = -616 + 400 - 9$

$25(x - 4)^2 - 9(y + 1)^2 = -225$

$$\frac{(y + 1)^2}{25} - \frac{(x - 4)^2}{9} = 1$$

The curve is a hyperbola with center $(4, -1)$.

We introduce parameter t in analogy with problem 31:

$\dfrac{y + 1}{5} = \sec t \qquad \dfrac{x - 4}{3} = \tan t$

$y + 1 = 5 \sec t \qquad x - 4 = 3 \tan t$

$x = 4 + 3 \tan t \qquad y = -1 + 5 \sec t, \ -\dfrac{\pi}{2} < t < \dfrac{3\pi}{2},$

$$t \neq \frac{\pi}{2}$$

At the right is a graph in a graphing utility.

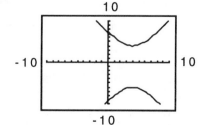

37. $4x^2 - 24x + 49y^2 + 392y + 624 = 0$

Complete the square relative to both x and y.

$4x^2 - 24x \quad + 49y^2 + 392y \quad = -624$

$4(x^2 - 6x \quad) + 49(y^2 + 8y \quad) = -624$

$4(x^2 - 6x + 9) + 49(y^2 + 8y + 16) = -624 + 36 + 784$

$4(x - 3)^2 + 49(y + 4)^2 = 196$

$$\frac{(x - 3)^2}{49} + \frac{(y + 4)^2}{4} = 1$$

The curve is an ellipse with center $(3, -4)$.

We introduce parameter t in analogy with problem 29:

$\dfrac{x - 3}{7} = \cos t \qquad \dfrac{y + 4}{2} = \sin t$

$x - 3 = 7 \cos t \qquad y + 4 = 2 \sin t$

$x = 3 + 7 \cos t \qquad y = -4 + 2 \sin t, \ 0 \leq t \leq 2\pi$

At the right is a graph in a graphing utility.

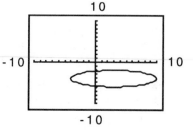

39. (A) When $t = 0.1$

$x = 5 \sin[6\pi(0.1)] = 5 \sin(0.6\pi) = 4.8$

$y = 5 \cos[6\pi(0.1)] = 5 \cos(0.6\pi) = -1.5$

(B) We eliminate the parameter t as follows:

$\dfrac{x}{5} = \sin(6\pi t)$

$\dfrac{y}{5} = \cos(6\pi t)$

$\left(\dfrac{x}{5}\right)^2 + \left(\dfrac{y}{5}\right)^2 = \sin^2(6\pi t) + \cos^2(6\pi t)$

$\dfrac{x^2}{25} + \dfrac{y^2}{25} = 1$

$x^2 + y^2 = 25$

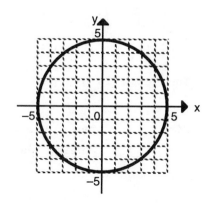

This is the equation of a circle with center at the origin and radius 5.

41. Using equations (5) in the text, we have $v_0 = 300$, $\alpha = 45°$.

$x = (300 \cos 45°)t$
$y = (300 \sin 45°)t - 4.9t^2$

(A) The time of impact is the positive value of t for which $y = 0$.
Solving $0 = (300 \sin 45°)t - 4.9t^2$
we have $t = 0$ (which we discard) or
$0 = 300 \sin 45° - 4.9t$

$t = \dfrac{300 \sin 45°}{4.9} = 43.292$ seconds

(B) The range is the value of x which corresponds to this value of t. Using the

exact value for t we obtain $x = 300 \cos 45° \left(\dfrac{300 \sin 45°}{4.9} \right) = 9{,}183.674$ meters

or, using the rounded value for t, we have $x = 300 \cos 45°(43.292) = 9{,}183.619$
meters. To convert meters to kilometers we divide either by $1{,}000$ to obtain
9.184 kilometers.

(C) To find the maximum height attained by the projectile, we compute the
height of the vertex of the parabola above the ground.
Since $y = (v_0 \sin \alpha)t - 4.9t^2$ or $-4.9t^2 + v_0 \sin \alpha t$ is the equation of the
parabola, the height of the vertex of the parabola is the value of y found by
completing the square relative to t.

$$y = -4.9\left(t^2 - \frac{v_0 \sin \alpha}{4.9} t \right)$$

$$= -4.9\left[t^2 - \frac{v_0 \sin \alpha}{4.9} t + \left(\frac{v_0 \sin \alpha}{9.8} \right)^2 \right] + 4.9\left(\frac{v_0 \sin \alpha}{9.8} \right)^2$$

$$= -4.9\left(t - \frac{v_0 \sin \alpha}{9.8} \right)^2 + 4.9\left(\frac{v_0 \sin \alpha}{9.8} \right)^2$$

Thus the maximum height of the projectile occurs when $t = \dfrac{v_0 \sin \alpha}{9.8}$ and is given by

$$y = 4.9\left(\frac{v_0 \sin \alpha}{9.8} \right)^2$$

$$= 4.9\left(\frac{300 \sin 45°}{9.8} \right)^2$$

$$= 2{,}295.918 \text{ meters}$$

CHAPTER 11 REVIEW

1. First write the equation in standard form by dividing both sides by 225.
$9x^2 + 25y^2 = 225$

$\dfrac{x^2}{25} + \dfrac{y^2}{9} = 1$

In this form the equation is identifiable as that of an ellipse. Locate the
intercepts.

When $y = 0$, $\dfrac{x^2}{25} = 1$. x intercepts: ± 5

When $x = 0$, $\dfrac{x^2}{9} = 1$. y intercepts: ± 3

Thus, $a = 5$, $b = 3$, and the major axis is
on the x axis.

Foci: $c^2 = a^2 - b^2$
$c^2 = 25 - 9$
$c^2 = 16$
$c = 4$
Foci: $F'(-4, 0)$, $F(4, 0)$
Major axis length $= 2(5) = 10$
Minor axis length $= 2(3) = 6$

> **Common Error:**
> The relationship $c^2 = a^2 + b^2$ applies to
> a, b, c as defined for hyperbolas but
> not for ellipses.

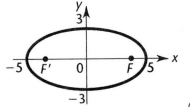

$(11\text{-}2)$

2. $x^2 = -12y$ is the equation of a parabola. To graph, assign y values that make the right side a perfect square (y must be zero or negative for x to be real) and solve for x. Since the coefficient of y is negative, a must be negative, and the parabola opens down.

x	0	±6	±2
y	0	-3	$-\frac{1}{3}$

To find the focus and directrix, solve
$4a = -12$
$\ a = -3$
Focus: $(0, -3)$ Directrix: $y = -(-3) = 3$

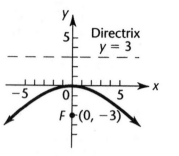

(11-1)

3. First, write the equation in standard form by dividing both sides by 225.
$25y^2 - 9x^2 = 225$
$$\frac{y^2}{9} - \frac{x^2}{25} = 1$$

In this form the equation is identifiable as that of a hyperbola.

When $y = 0$, $-\dfrac{x^2}{25} = 1$. There are no

x intercepts, but $b = 5$.

When $x = 0$, $\dfrac{y^2}{9} = 1$. y intercepts: ±3

Sketch the asymptotes using the asymptote rectangle, then sketch in the hyperbola.
Foci: $c^2 = 3^2 + 5^2$
$\qquad c^2 = 34$
$\qquad\ c = \sqrt{34}$
Foci: $F'(0, -\sqrt{34})$, $F(0, \sqrt{34})$
Transverse axis length $= 2(3) = 6$
Conjugate axis length $= 2(5) = 10$

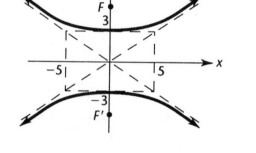

(11-3)

4. (A) Divide both sides by 100
$$\frac{(y + 2)^2}{25} - \frac{(x - 4)^2}{4} = 1$$

(B) This is the equation of a hyperbola. *(11-4)*

5. (A) Subtract $12(y + 4)$ from both sides.
$$(x + 5)^2 = -12(y + 4)$$

(B) This is the equation of a parabola. *(11-4)*

6. (A) Divide both sides by 144
$$\frac{(x - 6)^2}{9} + \frac{(y - 4)^2}{16} = 1$$

(B) This is the equation of an ellipse. *(11-4)*

7. The parabola is opening left and has an equation of the form $y^2 = 4ax$.
Since $(-4, 2\sqrt{7})$ is on the graph, we have:
$\qquad\quad y^2\ = 4ax$
$(2\sqrt{7})^2\ = 4a(-4)$
$\qquad\quad a\ = -\dfrac{7}{4} = a$

Thus, the equation of the parabola is

$\qquad y^2\ = 4\left(-\dfrac{7}{4}\right)x$
$\qquad y^2\ = -7x$

(11-1)

8. Make a rough sketch of the ellipse, locate the focus and y intercepts, then determine x intercepts from the special triangle relationship.

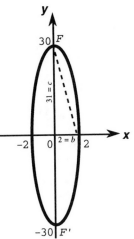

$$\frac{x^2}{b^2} + \frac{y^2}{a^2} = 1$$

$$b = \frac{4}{2} = 2 \qquad c = \frac{62}{2} = 31$$

$$a^2 = b^2 + c^2$$

$$a^2 = 2^2 + 31^2 = 4 + 961 = 965$$

$$a = \sqrt{965}$$

$$\frac{x^2}{4} + \frac{y^2}{965} = 1$$

(11-2)

9. Since the equation is given as $\frac{x^2}{M} - \frac{y^2}{N} = 1$ with $M, N > 0$, start with

$$\frac{x^2}{a^2} - \frac{y^2}{b^2} = 1$$

and find a and b.

$$a = \frac{10}{2} = 5 \qquad c = 6$$

To find b, sketch the asymptote rectangle, label known parts, and use the Pythagorean Theorem.

$$b^2 = c^2 - a^2$$

$$b^2 = 6^2 - 5^2$$

$$b^2 = 11$$

$$b = \sqrt{11}$$

Thus, the equation is

$$\frac{x^2}{25} - \frac{y^2}{11} = 1$$

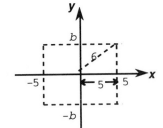

(11-3)

10. Construct a table of values and graph.

t	0	1	2	3	-1	-2	-3
x	0	-1	-4	-9	-1	-4	-9
y	1	$\frac{1}{2}$	-1	$-\frac{7}{2}$	$\frac{1}{2}$	-1	$-\frac{7}{2}$

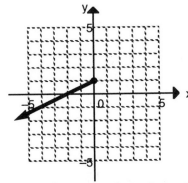

To eliminate the parameter t, we substitute $-t^2 = x$ into the expression

$$y = -\frac{1}{2}t^2 + 1$$

$$y = \frac{1}{2}(-t^2) + 1$$

$$y = \frac{1}{2}x + 1$$

This is the equation of a line. Since $x = -t^2$, however, x is restricted so that $x \le 0$, hence the graph is a ray, part of a straight line:

$$y = \frac{1}{2}x + 1, \ x \le 0$$

(11-5)

11. $x^2 + 4y^2 = 32$
$\quad\ \ x + 2y = 0$

We solve for x in terms of y in the second equation, then substitute the expression for x into the first equation.

$$x = -2y$$
$$(-2y)^2 + 4y^2 = 32$$
$$4y^2 + 4y^2 = 32$$
$$8y^2 = 32$$
$$y^2 = 4$$
$$y = \pm 2$$

For $y = 2$ For $y = -2$
$\quad x = -2(2)$ $x = -2(-2)$
$\qquad = -4$ $= 4$

Solutions: $(-4, 2)$, $(4, -2)$

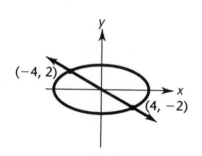

Graphs:
$x^2 + 4y^2 = 32$

$$\frac{x^2}{32} + \frac{y^2}{8} = 1 \quad \text{Ellipse}$$

$a = \sqrt{32}$, $b = \sqrt{8}$, major axis on the x axis

x intercepts: $\pm\sqrt{32}$

y intercepts: $\pm\sqrt{8}$

$x + 2y = 0$

Straight line through origin, slope $-\dfrac{1}{2}$.

$(11\text{-}2,\ 8\text{-}3)$

12. $16x^2 + 25y^2 = 400$
$\quad 16x^2 - 45y = 0$

We eliminate x by multiplying the second equation by -1, then adding the two equations.

$$\begin{aligned} 16x^2 + 25y^2 &= 400 \\ \underline{-16x^2 + 45y\ } &= \underline{0} \\ 25y^2 + 45y &= 400 \end{aligned}$$

$$5y^2 + 9y - 80 = 0$$
$$(5y - 16)(y + 5) = 0$$
$$y = \frac{16}{5} \quad \text{or} \quad y = -5$$

For $y = \dfrac{16}{5}$ For $y = -5$

$16x^2 = 45y$ $16x^2 = 45(-5)$

$16x^2 = 45\left(\dfrac{16}{5}\right)$ $x^2 = -\dfrac{225}{16}$

$\quad x^2 = 9$ No real solution

$\quad\ x = \pm 3$

Solutions: $\left(3,\ \dfrac{16}{5}\right)$, $\left(-3,\ \dfrac{16}{5}\right)$

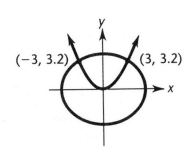

Graphs:
$16x^2 + 25y^2 = 400$
$\dfrac{x^2}{25} + \dfrac{y^2}{16} = 1$ Ellipse
$a = 5$, $b = 4$, major axis on the x axis.
x intercepts: ± 5
y intercepts: ± 4
$16x^2 - 45y = 0$ Parabola, opens up.
$(11\text{-}1,\ 11\text{-}2,\ 8\text{-}3)$

13. $x^2 + y^2 = 10$
$16x^2 + y^2 = 25$

We eliminate y by multiplying the first equation by -1, then adding the two equations.

$$\begin{aligned}
-x^2 - y^2 &= -10 \\
16x^2 + y^2 &= 25 \\
\hline
15x^2 &= 15 \\
x^2 &= 1 \\
x &= \pm 1
\end{aligned}$$

For $x = \pm 1$
$x^2 + y^2 = 10$
$1 + y^2 = 10$
$y^2 = 9$
$y = \pm 3$

Solutions: $(1, 3)$, $(1, -3)$, $(-1, 3)$, $(-1, -3)$

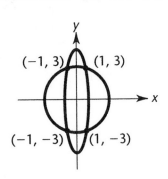

Graphs:
$16x^2 + y^2 = 25$
$\dfrac{x^2}{\frac{25}{16}} + \dfrac{y^2}{25} = 1$ Ellipse
$a = 5$, $b = \dfrac{5}{4}$, major axis on the y axis.
x intercepts: $\pm\dfrac{5}{4}$
y intercepts: ± 5
$x^2 + y^2 = 10$ Circle, radius $\sqrt{10}$
center $(0, 0)$. $(11\text{-}2,\ 8\text{-}3)$

14.
$$\begin{aligned}
16x^2 + 4y^2 + 96x - 16y + 96 &= 0 \\
16x^2 + 96x + 4y^2 - 16y &= -96 \\
16(x^2 + 6x + ?) + 4(y^2 - 4y + ?) &= -96 \\
16(x^2 + 6x + 9) + 4(y^2 - 4y + 4) &= -96 + 144 + 16 \\
16(x + 3)^2 + 4(y - 2)^2 &= 64 \\
\dfrac{(x + 3)^2}{4} + \dfrac{(y - 2)^2}{16} &= 1
\end{aligned}$$

This is the equation of an ellipse with center at $(-3, 2)$. The equations of translation are $x' = x + 3$, $y' = y - 2$. Making these substitutions, we obtain

$$\dfrac{x'^2}{4} + \dfrac{y'^2}{16} = 1$$

We graph this in the $x'y'$ system, following the process discussed in Section 11-2.

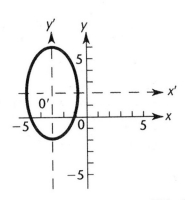

$(11\text{-}4)$

15. $x^2 - 4x - 8y - 20 = 0$
$$x^2 - 4x = 8y + 20$$
$$x^2 - 4x + 4 = 8y + 24$$
$$(x - 2)^2 = 4(2)(y + 3)$$

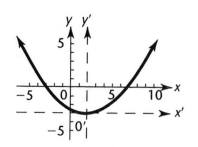

This is the equation of a parabola opening up with vertex at $(h, k) = (2, -3)$. The equations of translation are $x' = x - 2$, $y' = y + 3$. Making these substitutions, we obtain
$$x'^2 = 8y'$$
We graph this in the $x'y'$ system, following the process discussed in Section 11-1.

(11-4)

16. $4x^2 - 9y^2 + 24x - 36y - 36 = 0$
$$4x^2 + 24x - 9y^2 - 36y = 36$$
$$4(x^2 + 6x + ?) - 9(y^2 + 4y + ?) = 36$$
$$4(x^2 + 6x + 9) - 9(y^2 + 4y + 4) = 36 + 36 - 36$$
$$4(x + 3)^2 - 9(y + 2)^2 = 36$$
$$\frac{(x + 3)^2}{9} - \frac{(y + 2)^2}{4} = 1$$

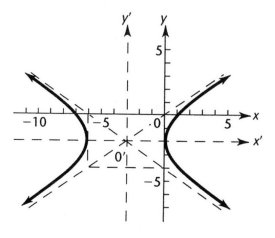

This is the equation of a hyperbola with center at $(-3, -2)$. The equations of translation are $x' = x + 3$, $y' = y + 2$. Making these substitutions, we obtain
$$\frac{x'^2}{9} - \frac{y'^2}{4} = 1$$
We graph this in the $x'y'$ system, following the process discussed in Section 11-3.

(11-4)

17. We eliminate the parameter θ as follows:
$$x = -2 + 2 \sin \theta \qquad y = 3 + 4 \cos \theta$$
$$x + 2 = 2 \sin \theta \qquad y - 3 = 4 \cos \theta$$
$$\frac{x + 2}{2} = \sin \theta \qquad \frac{y - 3}{4} = \cos \theta$$

$$\left(\frac{x + 2}{2}\right)^2 + \left(\frac{y - 3}{4}\right)^2 = \sin^2 \theta + \cos^2 \theta$$

$$\frac{(x + 2)^2}{4} + \frac{(y - 3)^2}{16} = 1$$

To graph, we note that the translation $x' = x + 2$, $y' = y - 3$ transforms this into
$$\frac{x'^2}{4} + \frac{y'^2}{16} = 1$$
In this form, this is identifiable as the equation of an ellipse. The origin O' of the new system is $(h, k) = (-2, 3)$. Sketch the intercepts in the translated system ($x' = \pm 2$, $y' = \pm 4$), then sketch in the graph.

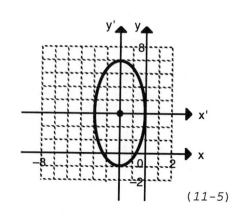

(11-5)

18. Here is a computer-generated graph of
$x^2 = y$ and $x^2 = 50y$, $-10 \leq x \leq 10$,
$-10 \leq y \leq 10$.

In order to make the graph of $x^2 = y$, in
the window $-m \leq x \leq m$, $-m \leq y \leq m$, look
like the graph at the right of $x^2 = 50y$,
we need a magnification by a factor of 50.

Then $\left(\dfrac{x'}{50}\right)^2 = \dfrac{y'}{50}$ becomes $x'^2 = 50y$. A

magnification by a factor of 50 requires

$\dfrac{10}{50} = 0.2 = m$. Here is the required graph

of $x^2 = y$ in the window $-0.2 \leq x \leq 0.2$,
$-0.2 \leq y \leq 0.2$.

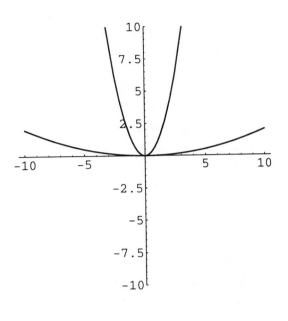

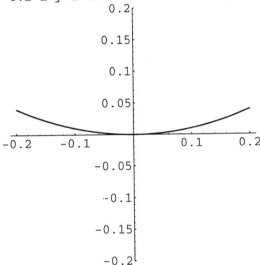

(11-1)

19. From the figure, we see that the point $P(x, y)$ is a point on the curve if and
only if

$$d_1 = d_2$$
$$d(F, P) = d(M, P)$$
$$\sqrt{(x - 2)^2 + (y - 4)^2} = \sqrt{(x - 6)^2 + (y - y)^2}$$
$$(x - 2)^2 + (y - 4)^2 = (x - 6)^2$$
$$x^2 - 4x + 4 + (y - 4)^2 = x^2 - 12x + 36$$
$$(y - 4)^2 = -8x + 32$$
$$(y - 4)^2 = -8(x - 4) \text{ or}$$
$$y^2 - 8y + 16 = -8x + 32$$
$$y^2 - 8y + 8x - 16 = 0$$

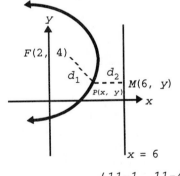

(11-1, 11-4)

20. From the figure, we see that the point $P(x, y)$ is a point on the curve if and only if

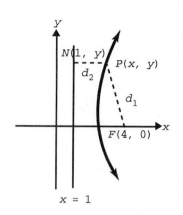

$$d_1 = 2d_2$$

$$d(F, P) = 2d(N, P)$$

$$\sqrt{(x - 4)^2 + (y - 0)^2} = 2\sqrt{(x - 1)^2 + (y - y)^2}$$

$$(x - 4)^2 + y^2 = 4(x - 1)^2$$

$$x^2 - 8x + 16 + y^2 = 4(x^2 - 2x + 1)$$

$$x^2 - 8x + 16 + y^2 = 4x^2 - 8x + 4$$

$$-3x^2 + y^2 = -12$$

$$\frac{x^2}{4} - \frac{y^2}{12} = 1$$

This is the equation of a hyperbola. $\qquad$ *(11-3, 11-4)*

21. From the figure, we see that the point $P(x, y)$ is a point on the curve if and only if

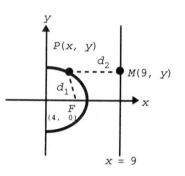

$$d_1 = \frac{2}{3}d_2$$

$$d(F, P) = \frac{2}{3}d(M, P)$$

$$\sqrt{(x - 4)^2 + (y - 0)^2} = \frac{2}{3}\sqrt{(x - 9)^2 + (y - y)^2}$$

$$(x - 4)^2 + y^2 = \frac{4}{9}(x - 9)^2$$

$$9[(x - 4)^2 + y^2] = 4(x - 9)^2$$

$$9(x^2 - 8x + 16 + y^2) = 4(x^2 - 18x + 81)$$

$$9x^2 - 72x + 144 + 9y^2 = 4x^2 - 72x + 324$$

$$5x^2 + 9y^2 = 180$$

$$\frac{x^2}{36} + \frac{y^2}{20} = 1$$

This is the equation of an ellipse. $\qquad$ *(11-2, 11-4)*

22. First find the coordinates of the foci in the translated system.

$$c'^2 = 4^2 - 2^2 = 12$$

$$c' = \sqrt{12}$$

$$-c' = -\sqrt{12}$$

Thus the coordinates in the translated system are

$F'(0, -\sqrt{12})$ and $F(0, \sqrt{12})$

Now use

$x = x' + h = x' - 3$

$y = y' + k = y' + 2$

to obtain

$F'(-3, -\sqrt{12} + 2)$ and $F(-3, \sqrt{12} + 2)$

as the coordinates of the foci in the original system. $\qquad$ *(11-4)*

23. First find the coordinates of the focus in the translated system. Since $a = 2$ and the parabola opens up, they are $(0, 2)$. Now use

$x = x' + h = x' + 2 = 0 + 2 = 2$

$y = y' + k = y' - 3 = 2 - 3 = -1$

to obtain $(2, -1)$ as the coordinates of the focus in the original system. $\qquad$ *(11-4)*

24. First find the coordinates of the foci in the original system.

$$c'^2 = 3^2 + 2^2 = 13$$
$$c' = \sqrt{13}$$
$$-c' = -\sqrt{13}$$

Thus the coordinates in the translated system are

$F'(-\sqrt{13}, 0)$ and $F(\sqrt{13}, 0)$

Now use

$$x = x' + h = x' - 3$$
$$y = y' + k = y' - 2$$

to obtain

$F'(-\sqrt{13} - 3, -2)$ and $F(\sqrt{13} - 3, -2)$ as the coordinates of the foci in the original system. (*11-4*)

25. We eliminate the parameter t as follows: Since

$y = 2^{-t} = \dfrac{1}{2^t}$ and $x = 2^t$, we observe that $y = \dfrac{1}{x}$.

Since both x and y involve 2^t, we see that they are restricted to values in the range of the exponential function, that is, positive values.

Hence only the part of the curve $y = \dfrac{1}{x}$ for which

$x > 0$ is traced out. The graph is one branch on the hyperbola $xy = 1$, with asymptotes the positive x and y axes.

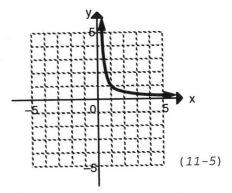

(*11-5*)

26. Before we can enter these equations in our graphing utility, we must solve for y:

$$x^2 - 3y^2 + 9x + 7y - 22 = 0 \qquad 4x^2 + 5x + 10y - 53 = 0$$
$$3y^2 - 7y - (x^2 + 9x - 22) = 0 \qquad \qquad 10y = 53 - 5x - 4x^2$$
$$y = \frac{53 - 5x - 4x^2}{10}$$

Applying the quadratic formula yields

$$y = \frac{7 \pm \sqrt{49 + 4(3)(x^2 + 9x - 22)}}{6}$$

$$y = \frac{7 \pm \sqrt{12x^2 + 108x - 215}}{6}$$

Entering these three equations into a graphing utility produces the graph at the right.

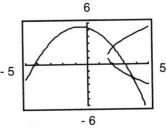

The curves intersect for x between 2 and 2.4 and between 3.6 and 4. Zooming in on these intervals, we obtain the following graphs:

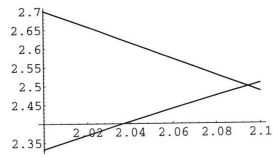 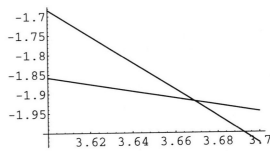

To two decimal places, the coordinates of the points of intersection are (2.09, 2.50) and (3.67, -1.92). (*11-4*)

27. From the figure, we see that the parabola opens up, hence its equation must be of the form $x^2 = 4ay$. Since $(4, 1)$ is on the graph we have

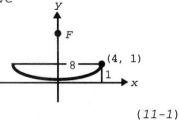

$$4^2 = 4a \cdot 1$$
$$16 = 4a$$
$$a = 4$$

Thus a, the distance of the focus from the vertex, is 4 feet.

$(11\text{-}1)$

28. From the figure, we see that the x intercepts must be at $(-5, 0)$ and $(5, 0)$, the foci at $(-4, 0)$ and $(4, 0)$. Hence $a = 5$ and $c = 4$. We can determine the y intercepts using the special triangle relationship

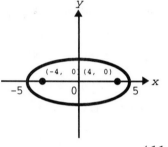

$$5^2 = 4^2 + b^2$$
$$25 = 16 + b^2$$
$$b = 3$$

Hence, the equation of the ellipse is

$$\frac{x^2}{5^2} + \frac{y^2}{3^2} = 1$$

$(11\text{-}2)$

29. From the figure, we can see:

$d + a = y_1 =$ the y coordinate of the point on the hyperbola with x coordinate 15. From the equation of the hyperbola $\dfrac{y^2}{40^2} - \dfrac{x^2}{30^2} = 1$, we have $a = 40$, hence $d + 40 = y_1$, or $d = y_1 - 40$. Since the point $(15, y_1)$ is on the hyperbola, its coordinates must satisfy the equation of the hyperbola. Thus,

$$\frac{y_1^2}{40^2} - \frac{15^2}{30^2} = 1$$
$$\frac{y_1^2}{40^2} = 1 + \frac{15^2}{30^2}$$
$$\frac{y_1^2}{40^2} = \frac{5}{4}$$
$$y_1^2 = 2000$$
$$y_1 = 44.72$$

$$\text{depth} = y_1 - 40 = 4.72 \text{ feet.}$$

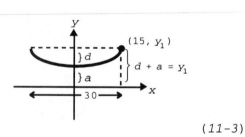

$(11\text{-}3)$

CUMULATIVE REVIEW EXERCISE (Chapters 10 and 11)

1. (A) Since $\frac{25}{5} \neq \frac{100}{25}$ and $25 - 5 \neq 100 - 25$, this could start neither an arithmetic nor a geometric sequence.

   (B) Since $3 - 15 = (-9) - 3 = -12$, this could start an arithmetic sequence.

   (C) Since $1 - 1 = 1 - 1 = 0$ and $\frac{1}{1} = \frac{1}{1} = 1$, this could start both an arithmetic and a geometric sequence.

   (D) Since $\frac{16}{-64} = \frac{-4}{16} = -\frac{1}{4}$, this could start a geometric sequence.

   (E) Since $\frac{119}{12} = \frac{833}{119} = 7$, this could start a geometric sequence.

   (F) Since $\frac{3}{1} \neq \frac{6}{3}$ and $3 - 1 \neq 6 - 3$, this could start neither an arithmetic nor a geometric sequence. *(10-3)*

2. $a_n = (-2)^n$

   (A) $a_1 = (-2)^1 = -2$
   $a_2 = (-2)^2 = 4$
   $a_3 = (-2)^3 = -8$
   $a_4 = (-2)^4 = 16$

   (B) This is a geometric sequence with $r = -2$. Hence,
   $a_n = a_1 r^{n-1}$
   $a_8 = (-2)(-2)^{8-1}$
   $\quad = 256$

   (C) $S_n = \dfrac{a_1 - r a_n}{1 - r}$
   $S_8 = \dfrac{(-2) - (-2)(256)}{1 - (-2)}$
   $\quad = 170$ *(10-3)*

3. $a_n = 6n - 5$

   (A) $a_1 = 6 \cdot 1 - 5 = 1$
   $a_2 = 6 \cdot 2 - 5 = 7$
   $a_3 = 6 \cdot 3 - 5 = 13$
   $a_4 = 6 \cdot 4 - 5 = 19$

   (B) This is an arithmetic sequence with $d = 6$. Hence,
   $a_n = a_1 + (n - 1)d$
   $a_8 = 1 + (8 - 1)6$
   $\quad = 43$

   (C) $S_n = \dfrac{n}{2}(a_1 + a_n)$
   $S_8 = \dfrac{8}{2}(1 + 43)$
   $\quad = 176$ *(10-3)*

4. $a_1 = -20 \quad a_n = a_{n-1} + 4, \quad n \geq 2$

   (A) $a_1 = -20$
   $a_2 = a_1 + 4 = -16$
   $a_3 = a_2 + 4 = -12$
   $a_4 = a_3 + 4 = -8$

   (B) This is an arithmetic sequence with $d = 4$. Hence,
   $a_n = a_1 + (n - 1)d$
   $a_8 = -20 + (8 - 1)4$
   $\quad = 8$

   (C) $S_n = \dfrac{n}{2}(a_1 + a_n)$
   $S_8 = \dfrac{8}{2}[(-20) + 8]$
   $\quad = -48$ *(10-3)*

5. (A) $7! = 7 \cdot 6 \cdot 5 \cdot 4 \cdot 3 \cdot 2 \cdot 1 = 5,040$

   (B) $\dfrac{25!}{22!} = \dfrac{25 \cdot 24 \cdot 23 \cdot 22!}{22!} = 13,800$

   (C) $\dfrac{10!}{(10 - 4)!4!} = \dfrac{10!}{6!4!}$
   $\qquad = \dfrac{10 \cdot 9 \cdot 8 \cdot 7 \cdot 6!}{6! \cdot 4 \cdot 3 \cdot 2 \cdot 1}$
   $\qquad = 210$ *(104)*

6. (A) $\dbinom{12}{6} = \dfrac{12!}{6!\,(12-6)!} = \dfrac{12!}{6!6!} = \dfrac{12\cdot11\cdot10\cdot9\cdot8\cdot7\cdot6!}{6\cdot5\cdot4\cdot3\cdot2\cdot1\cdot6!} = 924$

(B) $C_{9,4} = \dfrac{9!}{4!\,(9-4)!} = \dfrac{9!}{4!5!} = \dfrac{9\cdot8\cdot7\cdot6\cdot5!}{4\cdot3\cdot2\cdot1\cdot5!} = 126$

(C) $P_{8,5} = \dfrac{8!}{(8-5)!} = \dfrac{8!}{3!} = \dfrac{8\cdot7\cdot6\cdot5\cdot4\cdot3!}{3!} = 6{,}720$ 　　　　　　　(10-4, 10-5)

7. First, write the equation in standard form by dividing both sides by 900.

$25x^2 - 36y^2 = 900$

$\dfrac{x^2}{36} - \dfrac{y^2}{25} = 1$

In this form the equation is identifiable as that of a hyperbola.

When $x = 0$, $-\dfrac{y^2}{25} = 1$. There are no y intercepts, but $b = 5$.

When $y = 0$, $\dfrac{x^2}{36} = 1$. x intercepts: ±6.

Sketch the asymptotes using the asymptote rectangle, then sketch in the hyperbola.

Foci: $c^2 = 5^2 + 6^2$

$c^2 = 61$

$c = \sqrt{61}$

Foci: $F'(-\sqrt{61},\ 0)$, $F(\sqrt{61},\ 0)$
Transverse axis length = 2(6) = 12
Conjugate axis length = 2(5) = 10

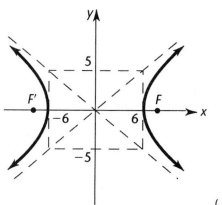

(11-3)

8. First, write the equation in standard form by dividing both sides by 900.

$25x^2 + 36y^2 = 900$

$\dfrac{x^2}{36} + \dfrac{y^2}{25} = 1$

In this form the equation is identifiable as that of an ellipse. Locate the intercepts:

When $y = 0$, $\dfrac{x^2}{36} = 1$. x intercepts: ±6.

When $x = 0$, $\dfrac{y^2}{25} = 1$. y intercepts: ±5

Thus, $a = 6$, $b = 5$, and the major axis is on the x axis.

Foci: $c^2 = a^2 - b^2$

$c^2 = 36 - 25$

$c^2 = 11$

$c = \sqrt{11}$

Foci: $F'(-\sqrt{11},\ 0)$, $F(\sqrt{11},\ 0)$
Major axis length = 2(6) = 12
Minor axis length = 2(5) = 10

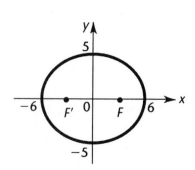

(11-2)

9. $25x^2 - 36y = 0$ is the equation of a parabola. For convenience, we rewrite this as $25x^2 = 36y$. To graph, assign y values that make $25x^2$ a perfect square (y must be positive or zero for x to be real) and solve for x. Since the coefficient of x is positive, a must be positive, and the parabola opens up.

x	0	$\pm\frac{6}{5}$	$\pm\frac{12}{5}$
y	0	1	4

To find the focus and directrix, solve

$$4a = \frac{36}{25}$$

$$a = \frac{9}{25}$$

Focus: $\left(0, \dfrac{9}{25}\right)$. Directrix: $y = -\dfrac{9}{25}$

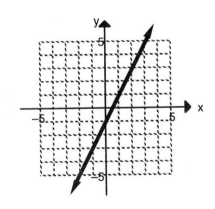

$(11-1)$

10. (A) The outcomes can be displayed in a tree diagram as follows:

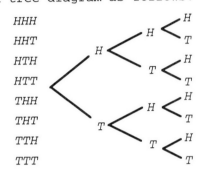

HHH
HHT
HTH
HTT
THH
THT
TTH
TTT

(B) O_1: First flip of the coin
N_1: 2 outcomes
O_2: Second flip of the coin
N_2: 2 outcomes
O_3: Third flip of the coin
N_3: 2 outcomes

Applying the multiplication principle, there are $2 \cdot 2 \cdot 2 = 8$ possible outcomes. $(10-5)$

11. (A) O_1: Place first book
O_2: Place second book
O_3: Place third book
O_4: Place fourth book

N_1: 4 ways
N_2: 3 ways
N_3: 2 ways
N_4: 1 way

Applying the multiplication principle, there are $4 \cdot 3 \cdot 2 \cdot 1$ arrangements.

(B) Order is important here. We use permutations to determine the number of arrangements of 4 objects. $P_{4,4} = 4! = 24$. $(10-5)$

12. Construct a table and graph.

t	0	1	2	3	-1	-2	-3
x	3	5	7	9	1	-1	-3
y	5	9	13	17	1	-3	-7

To eliminate the parameter t we solve $x = 2t + 3$ for t to obtain

$$2t = x - 3$$

$$t = \frac{x - 3}{2}$$

Then we substitute the expression for t into $y = 4t + 5$ to obtain

$$y = 4\left(\frac{x - 3}{2}\right) + 5$$

$$y = 2(x - 3) + 5$$

$$y = 2x - 1$$

This is the equation of a straight line.

$(11-5)$

13. P_1: $1 = 1(2 \cdot 1 - 1) = 1 \cdot 1 = 1$
P_2: $1 + 5 = 2(2 \cdot 2 - 1)$
$\quad\quad 6 = 6$
P_3: $1 + 5 + 9 = 3(2 \cdot 3 - 1)$
$\quad\quad\quad 15 = 15$ *(10-2)*

14. P_1: $1^2 + 1 + 2$ is divisible by 2
$\quad\quad 4 = 2 \cdot 2$ true
P_2: $2^2 + 2 + 2$ is divisible by 2
$\quad\quad 8 = 2 \cdot 4$ true
P_3: $3^2 + 3 + 2$ is divisible by 2
$\quad\quad 14 = 2 \cdot 7$ true *(10-2)*

15. P_k: $1 + 5 + 9 + \cdots + (4k - 3) = k(2k - 1)$
P_{k+1}: $1 + 5 + 9 + \cdots + (4k - 3) + (4k + 1) = (k + 1)(2k + 1)$ *(10-2)*

16. P_k: $k^2 + k + 2 = 2r$ for some integer r
P_{k+1}: $(k + 1)^2 + (k + 1) + 2 = 2s$ for some integer s *(10-2)*

17. The parabola is opening either up or down and has an equation of the form $x^2 = 4ay$. Since $(2, -8)$ is on the graph, we have:
$2^2 = 4a(-8)$
$4 = -32a$
$a = -\dfrac{1}{8}$

Thus, the equation of the parabola is
$x^2 = 4\left(-\dfrac{1}{8}\right)y$
$x^2 = -\dfrac{1}{2}y$
$y = -2x^2$ *(11-1)*

18. Make a rough sketch of the ellipse, locate focus and x intercepts, then determine y intercepts using the special triangle relationship.
$\dfrac{x^2}{a^2} + \dfrac{y^2}{b^2} = 1$
$\quad\quad a = \dfrac{10}{2} = 5$
$\quad\quad b^2 = a^2 - c^2 = 5^2 - 3^2 = 25 - 9 = 16$
$\quad\quad b = 4$
$\dfrac{x^2}{25} + \dfrac{y^2}{16} = 1$

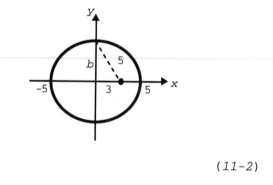

(11-2)

19. Start with $\dfrac{x^2}{a^2} - \dfrac{y^2}{b^2} = 1$ and find a and b.

$a = \dfrac{16}{2} = 8$

To find b, sketch the asymptote rectangle, label known parts, and use the Pythagorean Theorem.
$b^2 = (\sqrt{89})^2 - 8^2$
$\quad = 89 - 64$
$b^2 = 25$
$\quad b = 5$
Thus, the equation is
$\dfrac{x^2}{64} - \dfrac{y^2}{25} = 1$

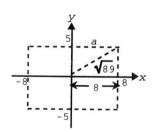

(11-3)

20. $\displaystyle\sum_{k=1}^{5} k^k = 1^1 + 2^2 + 3^3 + 4^4 + 5^5 = 1 + 4 + 27 + 256 + 3{,}125 = 3{,}413$ *(10-1)*

21. $S_6 = \dfrac{2}{2!} - \dfrac{2^2}{3!} + \dfrac{2^3}{4!} - \dfrac{2^4}{5!} + \dfrac{2^5}{6!} - \dfrac{2^6}{7!}$

$k = 1, \ 2, \ 3, \ 4, \ 5, \ 6$

Noting that the terms alternate in sign, we can rewrite as follows:

$S_6 = (-1)^2 \dfrac{2}{2!} + (-1)^3 \dfrac{2^2}{3!} + (-1)^4 \dfrac{2^3}{4!} + (-1)^5 \dfrac{2^4}{5!} + (-1)^6 \dfrac{2^5}{6!} + (-1)^7 \dfrac{2^6}{7!}$

Clearly, $a_k = (-1)^{k+1} \dfrac{2^k}{(k + 1)!}$

$S_6 = \displaystyle\sum_{k=1}^{6} (-1)^{k+1} \dfrac{2^k}{(k + 1)!}$ *(10-1)*

22. $\dfrac{a_2}{a_1} = \dfrac{-36}{108} = -\dfrac{1}{3} = r. \quad |r| < 1.$

Therefore, this infinite geometric series has a sum.

$S_\infty = \dfrac{a_1}{1 - r}$

$\quad = \dfrac{108}{1 - \left(-\frac{1}{3}\right)}$

$\quad = 81$ *(10-3)*

23.

	Case 1	Case 2	Case 3
O_1: Select the first letter	N_1: 6 ways	6 ways	6 ways
O_2: Select the second letter	N_2: 5 ways	6 ways	5 ways (exclude first letter)
O_3: Select the third letter	N_3: 4 ways	6 ways	5 ways (exclude second letter)
O_4: Select the fourth letter	N_4: 3 ways	6 ways	5 ways (exclude third letter)
	$6 \cdot 5 \cdot 4 \cdot 3 =$ 360 words	$6 \cdot 6 \cdot 6 \cdot 6 =$ 1,296 words	$6 \cdot 5 \cdot 5 \cdot 5 = 750$ words

(10-5)

24. Here are computer-generated graphs of $\{a_n\}$ and $\{b_n\}$ (the dots suggest a straight line).

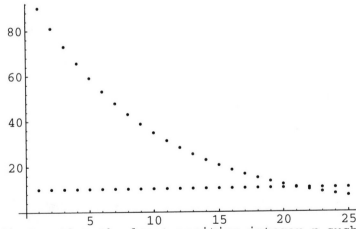

The graph indicates that the least positive integer n such that $a_n < b_n$ is between 21 and 23. From the table display we note

$\quad n = 21 \qquad a_n = 10.9419 \qquad b_n = 10.63$

$\quad n = 22 \qquad a_n = 9.84771 \qquad b_n = 10.66$

confirming that 22 is the least positive integer n required. *(10-3)*

25. We eliminate the parameter θ as follows:

$$x = 2 + 7 \cos \theta \qquad\qquad y = -3 + 5 \sin \theta$$
$$x - 2 = 7 \cos \theta \qquad\qquad y + 3 = 5 \sin \theta$$
$$\frac{x - 2}{7} = \cos \theta \qquad\qquad \frac{y + 3}{5} = \sin \theta$$

$$\left(\frac{x - 2}{7}\right)^2 + \left(\frac{y + 3}{5}\right)^2 = \cos^2 \theta + \sin^2 \theta$$

$$\frac{(x - 2)^2}{49} + \frac{(y + 3)^2}{25} = 1$$

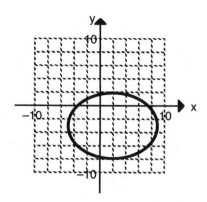

This is the equation of an ellipse with center $(2, -3)$, $a = 7$, $b = 5$. $(11-5)$

26. (A) $P_{25,5} = 25 \cdot 24 \cdot 23 \cdot 22 \cdot 21 = 6{,}375{,}600$

(B) $C_{25,5} = \dfrac{25!}{5!(25 - 5)!} = \dfrac{25!}{5!20!} = \dfrac{25 \cdot 24 \cdot 23 \cdot 22 \cdot 21 \cdot 20!}{5 \cdot 4 \cdot 3 \cdot 2 \cdot 1 \cdot 20!} = 53{,}130$

(C) $\dbinom{25}{5} = \dfrac{25!}{5!(25 - 5)!} = 53{,}130$ $(10\text{-}4,\ 10\text{-}5)$

27. $\left(a + \dfrac{1}{2}b\right)^6 = \displaystyle\sum_{k=0}^{6} \binom{6}{k} a^{6-k}\left(\frac{1}{2}b\right)^k$

$$= \binom{6}{0}a^6 + \binom{6}{1}a^5\left(\frac{1}{2}b\right)^1 + \binom{6}{2}a^4\left(\frac{1}{2}b\right)^2 + \binom{6}{3}a^3\left(\frac{1}{2}b\right)^3 + \binom{6}{4}a^2\left(\frac{1}{2}b\right)^4$$

$$+ \binom{6}{5}a\left(\frac{1}{2}b\right)^5 + \binom{6}{6}\left(\frac{1}{2}b\right)^6$$

$$= a^6 + 3a^5 b + \frac{15}{4}a^4 b^2 + \frac{5}{2}a^3 b^3 + \frac{15}{16}a^2 b^4 + \frac{3}{16}ab^5 + \frac{1}{64}b^6 \qquad (10\text{-}4)$$

28. In the expansion of $(a + b)^n$, the exponent of b in the rth term is $r - 1$ and the exponent of a is $n - (r - 1)$. Here, in the first case, $r = 5$, $n = 10$.

Fifth term $= \dbinom{10}{4}(3x)^6(-y)^4 = 153{,}090x^6 y^4$

In the second case, $r = 8$, n is still 10.

Eighth term $= \dbinom{10}{7}(3x)^3(-y)^7 = -3{,}240x^3 y^7$ $(10\text{-}4)$

29. P_n: $1 + 5 + 9 + \cdots + (4n - 3) = n(2n - 1)$

Part 1: Show that P_1 is true.
P_1: $1 = 1(2 \cdot 1 - 1) = 1$ True

Part 2: Show that if P_k is true, then P_{k+1} is true.
Write out P_k and P_{k+1}.
$\quad P_k$: $1 + 5 + 9 + \cdots + (4k - 3) = k(2k - 1)$
$\quad P_{k+1}$: $1 + 5 + 9 + \cdots + (4k - 3) + (4k + 1) = (k + 1)(2k + 1)$
We start with P_k:
$\quad 1 + 5 + 9 + \cdots + (4k - 3) = k(2k - 1)$
Adding $4k + 1$ to both sides:
$\quad 1 + 5 + 9 + \cdots + (4k - 3) + (4k + 1) = k(2k - 1) + 4k + 1$
$$= 2k^2 - k + 4k + 1$$
$$= 2k^2 + 3k + 1$$
$$= (k + 1)(2k + 1)$$

We have shown that if P_k is true, then P_{k+1} is true.
Conclusion: P_n is true for all positive integers n. $(10\text{-}2)$

30. P_n: $n^2 + n + 2 = 2p$ for some integer p.

Part 1: Show that P_1 is true.

P_1: $1^2 + 1 + 2 = 4 = 2 \cdot 2$ is true.

Part 2: Show that if P_k is true, then P_{k+1} is true.

Write out P_k and P_{k+1}.

P_k: $k^2 + k + 2 = 2r$ for some integer r

P_{k+1}: $(k + 1)^2 + (k + 1) + 2 = 2s$ for some integer s

We start with P_k:

$k^2 + k + 2 = 2r$ for some integer r

$$\begin{aligned}
\text{Now, } (k + 1)^2 + (k + 1) + 2 &= k^2 + 2k + 1 + k + 1 + 2 \\
&= k^2 + k + 2 + 2k + 2 \\
&= 2r + 2k + 2 \\
&= 2(r + k + 2)
\end{aligned}$$

Therefore,

$(k + 1)^2 + (k + 1) + 2 = 2s$ for some integer s $(= r + k + 2)$

$(k + 1)^2 + (k + 1) + 2$ is divisible by 2.

We have shown that if P_k is true, then P_{k+1} is true.

Conclusion: P_n is true for all positive integers n.   (10-2)

31. We are to find $51 + 53 + \cdots + 499$. This is the sum of an arithmetic sequence, S_n, with $d = 2$.

First, find n:

$$a_n = a_1 + (n - 1)d$$
$$499 = 51 + (n - 1)2$$
$$n = 225$$

Now, find S_{225}

$$S_n = \frac{n}{2}(a_1 + a_n)$$

$$S_{225} = \frac{225}{2}(51 + 499)$$

$$= 61{,}875 \qquad (10\text{-}3)$$

32. $2.\overline{45} = 2.454545\ldots$

$= 2 + 0.454545\ldots$

$0.454545\ldots = 0.45 + 0.0045 + 0.000045 + \cdots$

This is an infinite geometric series with $a_1 = 0.45$ and $r = 0.01$. Thus

$$S_\infty = \frac{a_1}{1 - r} = \frac{0.45}{1 - 0.01} = \frac{0.45}{0.99} = \frac{5}{11}$$

Hence, $2.\overline{45} = 2 + \dfrac{5}{11} = \dfrac{27}{11}$

    (10-3)

33. Here is a computer-generated graph of the sequence

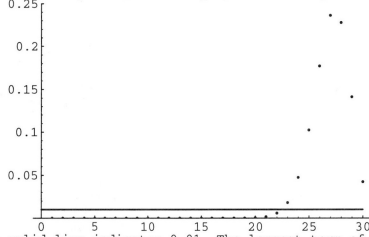

The solid line indicates 0.01. The largest term of the sequence is a_{27}. From the table display,

$$a_{27} = \binom{30}{27}(0.1)^{30-27}(0.9)^{27} = 0.236088$$

There are 8 terms larger than 0.01, as can be seen from the graph or the table display (details omitted).   (10-4)

34.
$$4x + 4y - y^2 + 8 = 0$$
$$4x + 8 = y^2 - 4y$$
$$4x + 8 + 4 = y^2 - 4y + 4$$
$$4x + 12 = (y - 2)^2$$
$$4(x + 3) = (y - 2)^2$$

This is the equation of a parabola opening right with vertex at $(h, k) = (-3, 2)$. The equations of translation are $x' = x + 3$, $y' = y - 2$. Making these substitutions, we obtain
$$4x' = y'^2$$
We graph this in the $x'y'$ system.

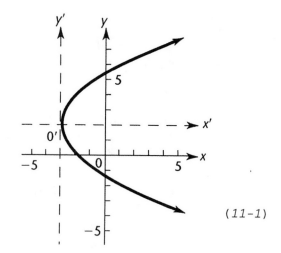

$(11-1)$

35.
$$x^2 + 2x - 4y^2 - 16y + 1 = 0$$
$$x^2 + 2x - 4y^2 - 16y = -1$$
$$x^2 + 2x + ? - 4(y^2 + 4y + ?) = -1$$
$$x^2 + 2x + 1 - 4(y^2 + 4y + 4) = -1 + 1 - 16$$
$$(x + 1)^2 - 4(y + 2)^2 = -16$$
$$\frac{(y + 2)^2}{4} - \frac{(x + 1)^2}{16} = 1$$

This is the equation of a hyperbola with center at $(-1, -2)$. The equations of translation are $x' = x + 1$, $y' = y + 2$. Making these substitutions, we obtain
$$\frac{y'^2}{4} - \frac{x'^2}{16} = 1$$
We graph this in the $x'y'$ system.

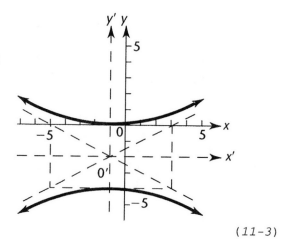

$(11-3)$

36.
$$4x^2 - 16x + 9y^2 + 54y + 61 = 0$$
$$4x^2 - 16x + 9y^2 + 54y = -61$$
$$4(x^2 - 4x + ?) + 9(y^2 + 6y + ?) = -61$$
$$4(x^2 - 4x + 4) + 9(y^2 + 6y + 9) = -61 + 16 + 81$$
$$4(x - 2)^2 + 9(y + 3)^2 = 36$$
$$\frac{(x - 2)^2}{9} + \frac{(y + 3)^2}{4} = 1$$

This is the equation of an ellipse with center at $(2, -3)$. The equations of translation are $x' = x - 2$, $y' = y + 3$. Making these substitutions, we obtain
$$\frac{x'^2}{9} + \frac{y'^2}{4} = 1$$
We graph this in the $x'y'$ system.

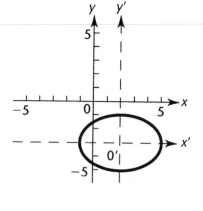

$(11-2)$

37.

	Allowing digits to repeat	No digit can be repeated
O_1: Selecting first digit	N_1: 10 ways	N_1: 10 ways
O_2: Selecting second digit	N_2: 10 ways	N_2: 9 ways
$\vdots$	$\vdots$	$\vdots$
O_9: Selecting ninth digit	N_9: 10 ways	N_9: 2 ways

$10 \cdot 10 \cdot 10 \cdot 10 \cdot 10 \cdot 10 \cdot 10 \cdot 10 \cdot 10$
$= 10^9$ zip codes

$P_{10,9} = 10 \cdot 9 \cdot 8 \cdot 7 \cdot 6 \cdot 5 \cdot 4 \cdot 3 \cdot 2$
$= 3,628,800$ zip codes

$(10-5)$

38. Before we can enter these equations in our graphing utility, we must solve for y:

$$5x^2 + 2y^2 - 7x + 8y - 48 = 0 \qquad e^x - e^{-x} - 2y = 0$$
$$2y^2 + 8y + 5x^2 - 7x - 48 = 0 \qquad e^x - e^{-x} = 2y$$
$$y = \frac{e^x - e^{-x}}{2}$$

Applying the quadratic formula yields

$$y = \frac{-8 \pm \sqrt{64 - 4(2)(5x^2 - 7x - 48)}}{4}$$

$$y = \frac{-8 \pm \sqrt{448 + 56x - 40x^2}}{4}$$

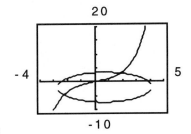

Entering these three equations into a graphing utility produces the graph shown at the right.

The curves intersect for x between -2.4 and -2.2 and between 1.8 and 2. Zooming in on these intervals, we obtain the following graphs:

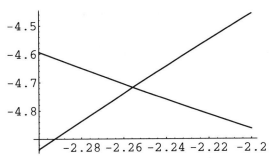
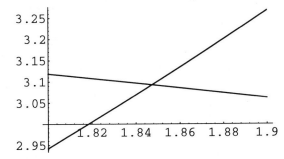

To two decimal places, the coordinates of the points of intersection are $(-2.26, -4.72)$ and $(1.85, 3.09)$.

(11-4)

39. Prove P_n: $\dfrac{1}{1 \cdot 3} + \dfrac{1}{3 \cdot 5} + \dfrac{1}{5 \cdot 7} + \cdots + \dfrac{1}{(2n - 1)(2n + 1)} = \dfrac{n}{2n + 1}$

Part 1: Show that P_1 is true.

$$P_1: \frac{1}{1 \cdot 3} = \frac{1}{2 \cdot 1 + 1}$$
$$\frac{1}{3} = \frac{1}{3} \quad P_1 \text{ is true}$$

Part 2: Show that if P_k is true, then P_{k+1} is true.
Write out P_k and P_{k+1}:

$$P_k: \frac{1}{1 \cdot 3} + \frac{1}{3 \cdot 5} + \frac{1}{5 \cdot 7} + \cdots + \frac{1}{(2k - 1)(2k + 1)} = \frac{k}{2k + 1}$$

$$P_{k+1}: \frac{1}{1 \cdot 3} + \frac{1}{3 \cdot 5} + \frac{1}{5 \cdot 7} + \cdots + \frac{1}{(2k - 1)(2k + 1)} + \frac{1}{(2k + 1)(2k + 3)} = \frac{k + 1}{2k + 3}$$

We start with P_k:

$$\frac{1}{1 \cdot 3} + \frac{1}{3 \cdot 5} + \frac{1}{5 \cdot 7} + \cdots + \frac{1}{(2k - 1)(2k + 1)} = \frac{k}{2k + 1}$$

Adding $\dfrac{1}{(2k + 1)(2k + 3)}$ to both sides, we get

$$\dfrac{1}{1 \cdot 3} + \dfrac{1}{3 \cdot 5} + \dfrac{1}{5 \cdot 7} + \cdots + \dfrac{1}{(2k - 1)(2k + 1)} + \dfrac{1}{(2k + 1)(2k + 3)}$$

$$= \dfrac{k}{2k + 1} + \dfrac{1}{(2k + 1)(2k + 3)}$$

$$= \dfrac{k(2k + 3)}{(2k + 1)(2k + 3)} + \dfrac{1}{(2k + 1)(2k + 3)}$$

$$= \dfrac{k(2k + 3) + 1}{(2k + 1)(2k + 3)}$$

$$= \dfrac{2k^2 + 3k + 1}{(2k + 1)(2k + 3)}$$

$$= \dfrac{(2k + 1)(k + 1)}{(2k + 1)(2k + 3)}$$

$$= \dfrac{k + 1}{2k + 3}$$

We have shown that if P_k is true, then P_{k+1} is true.

Conclusion: P_n is true for positive integers n.

$(10\text{-}2)$

40. $(x - 2i)^6 = [x + (-2i)]^6 = \displaystyle\sum_{k=0}^{6} \binom{6}{k} x^{6-k} (-2i)^k$

$$= \binom{6}{0} x^6 + \binom{6}{1} x^5 (-2i)^1 + \binom{6}{2} x^4 (-2i)^2 + \binom{6}{3} x^3 (-2i)^3$$

$$+ \binom{6}{4} x^2 (-2i)^4 + \binom{6}{5} x (-2i)^5 + \binom{6}{6} (-2i)^6$$

$$= x^6 - 12ix^5 - 60x^4 + 160ix^3 + 240x^2 - 192ix - 64$$

$(10\text{-}4)$

41. Let $P(x, y)$ be a point on the parabola.

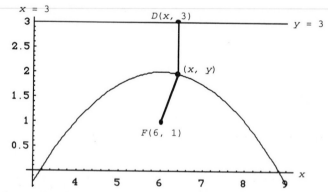

Then by the definition of the parabola, the distance from $P(x, y)$ to the focus $F(6, 1)$ must equal the perpendicular distance from P to the directrix at $D(x, 3)$. Applying the distance formula, we have

$$d(P, F) = d(P, D)$$

$$\sqrt{(x - 6)^2 + (y - 1)^2} = \sqrt{(x - x)^2 + (y - 3)^2}$$

$$(x - 6)^2 + (y - 1)^2 = (x - x)^2 + (y - 3)^2$$

$$x^2 - 12x + 36 + y^2 - 2y + 1 = 0 + y^2 - 6y + 9$$

$$x^2 - 12x + 4y + 28 = 0$$

$(11\text{-}1)$

42. Make a rough sketch of the ellipse, locate focus and x intercepts, then determine y intercepts using the special triangle relationship.

$a = 4$
$b^2 = a^2 - c^2$
$\quad = 4^2 - 2^2$
$\quad = 12$
$b = \sqrt{12} = 2\sqrt{3}$
y intercepts: $\pm 2\sqrt{3}$

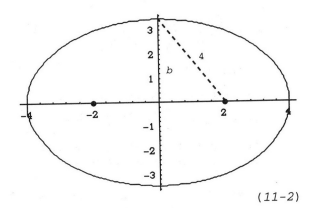

$(11\text{-}2)$

43. Sketch the asymptote rectangle, label known parts, and use the Pythagorean Theorem.

$a = 3 \qquad c = 5$
$b^2 = c^2 - a^2$
$\quad = 5^2 - 3^2$
$\quad = 16$
$b = 4$
Thus, the length of the conjugate axis is $2b = 8$.

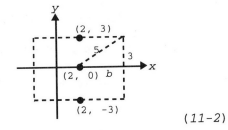

$(11\text{-}2)$

44. $a_n = \dfrac{2^n}{5^{n+1}}$

Arithmetic?: $a_1 = \dfrac{2^1}{5^{1+1}} = \dfrac{2}{25} \qquad a_2 = \dfrac{2^2}{5^{2+1}} = \dfrac{4}{125} \qquad a_3 = \dfrac{2^3}{5^{3+1}} = \dfrac{8}{625}$

$\dfrac{4}{125} - \dfrac{2}{25} \neq \dfrac{8}{625} - \dfrac{4}{125}$

The sequence is not arithmetic.

Geometric?: $a_{n+1} \div a_n = \dfrac{2^{n+1}}{5^{n+2}} \div \dfrac{2^n}{5^{n+1}} = \dfrac{2}{5} = r$

The sequence is geometric.

$(10\text{-}3,\ 10\text{-}4)$

45. $a_1 = 3 \qquad a_{n+1} = a_n + \dfrac{1}{n} \qquad n \geq 1$

Arithmetic?: $a_1 = 3 \qquad a_2 = a_1 + \dfrac{1}{1} = 4 \qquad a_3 = a_2 + \dfrac{1}{2} = 4 + \dfrac{1}{2} = \dfrac{9}{2}$

$4 - 3 \neq \dfrac{9}{2} - 4$

The sequence is not arithmetic.

Geometric?: $4 \div 3 \neq \dfrac{9}{2} \div 4$

The sequence is not geometric.
The sequence is neither arithmetic nor geometric.

$(10\text{-}3)$

46. $a_1 = 1 \qquad a_2 = 1 \qquad a_{n+2} = a_n \cdot a_{n+1},\ n \geq 1$
$a_1 = 1,\ a_2 = 1,\ a_3 = 1 \cdot 1 = 1,\ a_4 = 1 \cdot 1 = 1,\ a_5 = 1 \cdot 1 = 1,\ \dots$

Since $\dfrac{a_3}{a_2} = \dfrac{a_2}{a_1} = 1 = \dfrac{a_{n+1}}{a_n}$ and $a_3 - a_2 = a_2 - a_1 = 0 = a_{n+1} - a_n$, the sequence is both arithmetic and geometric, with $r = 1$ and $d = 0$.

$(10\text{-}3)$

47. $a_n = \binom{n}{2} - \dfrac{n^2}{2} = \dfrac{n!}{2!(n-2)!} - \dfrac{n^2}{2} = \dfrac{n(n-1)}{2} - \dfrac{n^2}{2} = \dfrac{n^2 - n - n^2}{2} = -\dfrac{n}{2}$

Arithmetic?: $a_{n+1} - a_n = -\dfrac{(n+1)}{2} - \left(-\dfrac{n}{2}\right) = -\dfrac{1}{2} = d$

The sequence is arithmetic.

Geometric?: $\dfrac{a_2}{a_1} = \dfrac{-2/2}{-1/2} \neq \dfrac{-3/2}{-2/2} = \dfrac{a_3}{a_2}$

The sequence is not geometric.

The sequence is arithmetic.
$\hfill$ *(10-3)*

48. False. If $n = 15$, $n^{10} = 15^{10} \approx 5.8 \times 10^{11}$. $n! = 15! \approx 1.3 \times 10^{12}$.
$n^{10} \geq n!$ is false for $n = 15$.
$\hfill$ *(10-2, 10-4)*

49. True. n^n and $n!$ both involve n factors. Since for n^n the factors are all equal to n, while for $n!$ all but one are less than n, $n^n \geq n!$. (For example, $1^1 = 1!$, $2^2 = 4 \geq 2 = 2!$, and so on. The statement can be proved by mathematical induction, but a proof is not required.)
$\hfill$ *(10-2, 10-4)*

50. False. For example, 6 is divisible by 2. However, $\binom{6}{2} = 15$ is not divisible by 2.
$\hfill$ *(10-2, 10-4)*

51. True. $\binom{p}{2} = \dfrac{p(p-1)}{2}$. If p is a prime greater than 2, then p is odd, $p - 1$ is even, hence $p - 1$ is divisible by 2, and $p\left(\dfrac{p-1}{2}\right) = \binom{p}{2}$ is divisible by p.
$\hfill$ *(10-2, 10-4)*

52. True. Both quantities are represented by $2a$.
$\hfill$ *(11-2)*

53. False. For example, the parabola $y^2 = 4x$ has vertex $(0, 0)$ and directrix $x = -1$. $(0, 0)$ is a point on the parabola. The distance from $(0, 0)$ to the directrix is 1. The distance from $(0, 0)$ to the vertex is 0.
$\hfill$ *(11-1)*

54. False. For example, choose the point $(2, 0)$ on the hyperbola $\dfrac{x^2}{4} - \dfrac{y^2}{5} = 1$, $a = 2$, $b = \sqrt{5}$, $c = 3$. The foci are $(3, 0)$ and $(-3, 0)$. Thus the absolute value of the differences of the distances from P to the foci $= 2a = 4$. The conjugate axis has length $2b = 2\sqrt{5}$.
$\hfill$ *(11-3)*

55. True. Since b^2 is defined as $a^2 - c^2$, $a > b$, $2a > 2b$ and the length of the major axis is greater than the length of the minor axis.
$\hfill$ *(11-2)*

56. False. For example, in the hyperbola $x^2 - \dfrac{y^2}{4} = 1$, the length of the transverse axis $= 2a = 2$, while the length of the conjugate axis $= 2b = 4$.
$\hfill$ *(11-3)*

57. If the parabola has the x axis as axis, its equation must be
$$y^2 = 4a(x - h)$$
If it passes through the points $(0, -3)$ and $(2, 5)$, the coordinates of these points must satisfy the equation of the parabola. Thus
$$(-3)^2 = 4a(0 - h)$$
$$5^2 = 4a(2 - h)$$
Solving this system, we obtain
$$9 = -4ah$$
$$25 = 8a - 4ah$$
$$25 = 8a + 9$$
$$a = 2$$
$$h = -\frac{9}{4a} = -\frac{9}{8}$$
Thus, the equation of the parabola is uniquely determined by the given conditions. There is one such parabola, and its equation is
$$y^2 = 4 \cdot 2\left[x - \left(-\frac{9}{8}\right)\right]$$
$$y^2 = 8x + 9$$
$(11\text{-}4)$

58. The center of the hyperbola is at the intersection of the transverse and conjugate axes, at $(5, -4)$. The equation of the hyperbola is of the form
$$\frac{(x - 5)^2}{a^2} - \frac{(y + 4)^2}{b^2} = 1$$
The equations of the asymptotes are
$$y + 4 = \pm\frac{b}{a}(x - 5)$$
Since the transverse axis has length twice the length of the conjugate axis, $2a = 2(2b)$, $\frac{b}{a} = \frac{1}{2}$, and the equations of the asymptotes are $y + 4 = \pm\frac{1}{2}(x - 5)$.
$(11\text{-}3, 11\text{-}4)$

59. Since the distance between the foci is $12 = 2c$, $c = 6$. Since the length of the minor axis is $4 = 2b$, $b = 2$. Using the relationship
$$c^2 = a^2 - b^2$$
we have
$$6^2 = a^2 - 2^2$$
$$36 = a^2 - 4$$
$$a^2 = 40$$
$$a = 2\sqrt{10}$$
Thus the sum of the distances from P to the foci $= 2a = 4\sqrt{10}$.
$(11\text{-}2)$

60. We can select any three of the seven points to use as vertices of the triangle. Order is not important.
$$C_{7,3} = \frac{7!}{3!(7 - 3)!} = 35 \text{ triangles}$$
$(10\text{-}5)$

61. We eliminate the parameter t as follows:
Solve $y = 1 - e^t$ for e^t (we do not need t) to obtain
$$e^t = 1 - y$$
Then we substitute the expression for e^t into
$$x = e^{2t} - 4 \text{ or } (e^t)^2 - 4$$
to obtain
$$x = (1 - y)^2 - 4 \quad \text{or}$$
$$x + 4 = (y - 1)^2$$
This is the equation of a parabola with vertex $(-4, 1)$, opening right. However, since $y = 1 - e^t$, y is restricted so that $e^t = 1 - y$ is positive, that is, $1 - y > 0$. Hence $1 > y$. Therefore the equation of the curve is actually $x + 4 = (y - 1)^2$, $y < 1$. This is the equation of the lower half of a parabola, excluding the vertex.
$(11\text{-}5)$

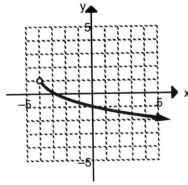

62. P_n: $2^n < n!$ $n \geq 4$ (for n an integer, this is identical with the condition $n > 3$)

Part 1: Show that P_4 is true

P_4: $2^4 < 4!$

$16 < 24$ True.

Part 2: Show that if P_k is true, then P_{k+1} is true.

Write out P_k and P_{k+1}:

P_k: $2^k < k!$

P_{k+1}: $2^{k+1} < (k + 1)!$

We start with P_k:

$2^k < k!$

Multiplying both sides by 2: $2 \cdot 2^k < 2 \cdot k!$

Now, $k > 3$, thus $1 < k$, $2 < k + 1$, hence $2 \cdot k! < (k + 1)k!$

Therefore, $2^{k+1} = 2 \cdot 2^k < 2 \cdot k! < (k + 1)k! = (k + 1)!$

$2^{k+1} < (k + 1)!$

Thus, if P_k is true, then P_{k+1} is true.

Conclusion: P_n is true for all $n \geq 4$.

$(10\text{-}2)$

63. To prove $a_n = b_n$ for all positive integers n, write:

P_n: $a_n = b_n$

Proof: Part 1: Show P_1 is true.

P_1: $a_1 = b_1$. $a_1 = 3$. $b_1 = 2^1 + 1 = 3$.

Thus, $a_1 = b_1$

Part 2: Show that if P_k is true, then P_{k+1} is true.

Write out P_k and P_{k+1}.

P_k: $a_k = b_k$

P_{k+1}: $a_{k+1} = b_{k+1}$

We start with P_k:

$a_k = b_k$

Now, $a_{k+1} = 2a_{k+1-1} - 1 = 2a_k - 1 = 2b_k - 1 = 2(2^k + 1) - 1$

$= 2 \cdot 2^k + 2 - 1 = 2^1 \cdot 2^k + 1 = 2^{k+1} + 1 = b_{k+1}$

Therefore, $a_{k+1} = b_{k+1}$

Thus, if P_k is true, then P_{k+1} is true.

Conclusion: P_n is true for all positive integers n. Hence, $\{a_n\} = \{b_n\}$. $(10\text{-}2)$

64. Make a rough sketch of the situation:

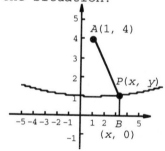

We are given $d(P, A) = 3d(P, B)$. Applying the distance formula, we have

$$\sqrt{(x - 1)^2 + (y - 4)^2} = 3\sqrt{(x - x)^2 + (y - 0)^2}$$

$$(x - 1)^2 + (y - 4)^2 = 9[(x - x)^2 + (y - 0)^2]$$

$$x^2 - 2x + 1 + y^2 - 8y + 16 = 9y^2$$

$$x^2 - 2x - 8y^2 - 8y + 17 = 0$$

is the equation of the curve. Completing the square relative to x and y, we have

$$x^2 - 2x + 1 - 8\left(y^2 + y + \frac{1}{4}\right) - 1 + 2 + 17 = 0$$

$$(x - 1)^2 - 8\left(y + \frac{1}{2}\right)^2 = -18$$

$$\frac{(y + \frac{1}{2})^2}{\frac{18}{8}} - \frac{(x - 1)^2}{18} = 1$$

This is the equation of a hyperbola.

$(11\text{-}3)$

65. We are asked for the sum of an infinite geometric series.

$a = \$2,000,000(0.75)$

$r = 0.75 \quad |r| \le 1$, so the series has a sum,

$$S_\infty = \frac{a_1}{1 - r}$$

$$= \frac{\$2,000,000(0.75)}{1 - 0.75}$$

$$= \frac{\$2,000,000(0.75)}{0.25}$$

$$= \$6,000,000$$

$(10\text{-}3)$

66.

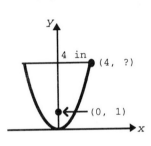

From the figure, we see that the parabola can be positioned so that it is opening up with axis the y axis, hence it has an equation of the form $x^2 = 4ay$. Since the focus is at $(0, a) = (0, 1)$, $a = 1$ and the equation of the parabola is $x^2 = 4y$.

Since the depth represents the y coordinate y_1 of a point on the parabola with $x = 4$, we have

$$4^2 = 4y_1$$

$$y_1 = \text{depth} = 4 \text{ in.}$$

$(11\text{-}1)$

67.

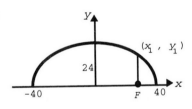

From the figure, we see that the x and y intercepts of the ellipse must be 40 and 24 respectively. Hence the equation of the ellipse must be

$$\frac{x^2}{(40)^2} + \frac{y^2}{(24)^2} = 1$$

or $\dfrac{x^2}{1,600} + \dfrac{y^2}{576} = 1$

To find the distance c of each focus from the center of the arch, we use the special triangle relationship. $a = 40$, $b = 24$

$$c^2 = a^2 - b^2$$

$$= 40^2 - 24^2$$

$$c^2 = 1024$$

$$c = 32 \text{ ft.}$$

To find the height of the arch above each focus, we need the y coordinate y_1 of the point P whose x coordinate is 32. Since P is on the ellipse, we have

$$\frac{32^2}{1,600} + \frac{y_1^2}{576} = 1$$

$$\frac{1,024}{1,600} + \frac{y_1^2}{576} = 1$$

$$0.64 + \frac{y_1^2}{576} = 1$$

$$\frac{y_1^2}{576} = 0.36$$

$$y_1^2 = 576(0.36)$$

$$y_1^2 = 207.36$$

$$y_1 = 14.4 \text{ ft.}$$

$(11\text{-}2)$

APPENDIX A

Exercise A-1

Key Ideas and Formulas

Basic Properties of the Set of Real Numbers

Let R be the set of real numbers and x, y and z arbitrary elements of R.

ADDITION PROPERTIES

CLOSURE:	$x + y$ is a unique element in R.
ASSOCIATIVE:	$(x + y) + z = x + (y + z)$
COMMUTATIVE:	$x + y = y + x$
IDENTITY:	0 is the additive identity; that is, $0 + x = x + 0 = x$ for all x in R, and 0 is the only element in R with this property.
INVERSE:	For each x in R, $-x$ is its unique additive inverse; that is, $x + (-x) = (-x) + x = 0$, and $-x$ is the only element in R relative to x with this property.

MULTIPLICATION PROPERTIES

CLOSURE:	xy is a unique element in R.
ASSOCIATIVE:	$(xy)z = x(yz)$
COMMUTATIVE:	$xy = yx$
IDENTITY:	1 is the multiplicative identity; that is $1x = x1 = x$, and 1 is the only element in R with this property.
INVERSE:	For each x in R, $x \neq 0$, $\frac{1}{x}$ is its unique multiplicative inverse; that is, $x(1/x) = (1/x)x = 1$, and $\frac{1}{x}$ is the only element in R relative to x with this property.

COMBINED PROPERTY

DISTRIBUTIVE:

$x(y + z) = xy + xz$ $(x + y)z = xz + yz$

Properties of Negatives:

For all real numbers a and b

1. $-(-a) = a$
2. $(-a)b = -(ab) = a(-b) = -ab$
3. $(-a)(-b) = ab$
4. $(-1)a = -a$
5. $\dfrac{-a}{b} = -\dfrac{a}{b} = \dfrac{a}{-b}$ $b \neq 0$
6. $\dfrac{-a}{-b} = -\dfrac{-a}{b} = -\dfrac{a}{-b} = \dfrac{a}{b}$ $b \neq 0$

Zero Properties:

For all real numbers a and b

1. $a \cdot 0 = 0$
2. $ab = 0$ if and only if $a = 0$ or $b = 0$ or both.

---

1. Since 4 is an element of $\{3,4,5\}$, the statement is true. T

3. Since 3 is an element of $\{3,4,5\}$, the statement is false. F

5. The statement is false, since not every element of $\{1,2\}$ is an element of $\{1,3,5\}$. 2 is an element of $\{1,2\}$, but not an element of $\{1,3,5\}$. F

7. Since each element of the set $\{7,3,5\}$ is also an element of the set $\{3,5,7\}$, the statement is true. T

9. The commutative property (+) states that, in general, $x + y = y + x$. Comparing this with $x + 7 = ?$, we see that 7 takes the place of y. Hence, $x + 7 = 7 + x$.
$7 + x$

11. The associative property $(\cdot)$ states that, in general, $(xy)z = x(yz)$. Comparing this with $x(yz) = ?$, we see that $x(yz) = (xy)z$. $(xy)z$

13. The identity property $(+)$ states that, in general, $0 + x = x + 0 = x$. Comparing this with $0 + 9m = ?$ we see that $9m$ takes the place of x. Hence, $0 + 9m = 9m$. $9m$

15. Commutative $(\cdot)$ $ym = my$ is a special case of $xy = yx$.

17. Distributive. $7u + 9u = (7 + 9)u$ is a special case of $xz + yz = (x + y)z$

19. Inverse$(\cdot)$. $(-2)(\frac{1}{-2}) = 1$ is a special case of $x(\frac{1}{x}) = 1$.

21. Inverse $(+)$. $w + (-w) = 0$ is a special case of $x + (-x) = 0$.

23. Identity $(+)$. $3(xy + z) + 0 = 3(xy + z)$ is a special case of $x + 0 = x$.

25. Negatives. $\frac{-x}{-y} = \frac{x}{y}$ is a special case of $\frac{-a}{-b} = \frac{a}{b}$. (Theorem 1, Part 6)

27. The even integers between -3 and 5 are -2, 0, 2, and 4. Hence, the set is written $\{-2, 0, 2, 4\}$.

29. The letters in "status" are: s, t, a, u. Hence, the set is written $\{s, t, a, u\}$, or, equivalently, $\{a, s, t, u\}$.

31. Since there are no months starting with B, the set is empty. $\varnothing$

33. (A) The empty set, $\varnothing$, is a subset of every set.
$\{a\}$ and $\{b\}$ are subsets of S_2.
S_2 is a subset of itself.
Thus, there are four subsets of S_2.

(B) The empty set, $\varnothing$, is a subset of every set.
$\{a\}$, $\{b\}$, and $\{c\}$ are one-member subsets of S_3.
$\{b, c\}$, $\{a, c\}$, and $\{a, b\}$ are two-member subsets of S_3.
S_3 is a subset of itself.
Thus, there are $1 + 3 + 3 + 1 = 8$ subsets of S_3.

(C) The empty set, $\varnothing$, is a subset of every set.
$\{a\}$, $\{b\}$, $\{c\}$, and $\{d\}$ are one-member subsets of S_4.
$\{a, b\}$, $\{a, c\}$, $\{a, d\}$, $\{b, c\}$, $\{b, d\}$, and $\{c, d\}$ are two-member subsets of S_4.
$\{b, c, d\}$, $\{a, c, d\}$, $\{a, b, d\}$, and $\{b, c, d\}$ are three-member subsets of S_4.
S_4 is a subset of itself.
Thus, there are $1 + 4 + 6 + 4 + 1 = 16$ subsets of S_4.

35. Negatives. $-(-y) = y$ is a special case of $-(-a) = a$.

37. Associative $(\cdot)$. $(wz)(zw) = w[z(zw)]$ is a special case of $(xy)z = x(yz)$.

39. Distributive. $(n + 2)(m + 3) = n(m + 3) + 2(m + 3)$ is a special case of $(x + y)z = xz + yz$.

41. Zero Property 2. This is a special case of $ab = 0$ if and only if $a = 0$ or $b = 0$.

43. Yes. This restates Zero Property (2). (Theorem 2, Part 2)

45. (A) True.
(B) False, $\frac{2}{3}$ is an example of a real number that is not irrational.
(C) True.

47. $\frac{3}{5}$ and -1.43 are two examples of infinitely many.

49. (A) $S = \{1, 12\}$ (B) $S = \{-3, 0, 1, 12\}$ (C) $S = \{-3, -\frac{2}{3}, 0, 1, \frac{9}{5}, 12\}$

51. (A) 0.888 888…; repeating; repeated digit: 8

    (B) 0.272 727…; repeating; repeated digits: 27

    (C) 2.236 067 977…; nonrepeating and nonterminating

    (D) 1.375; terminating

53. (A) True; commutative property for addition.
    (B) False; for example $3 - 5 \neq 5 - 3$.
    (C) True; commutative property for multiplication.
    (D) False; for example $9 \div 3 \neq 3 \div 9$.

55. (A) List each element of A. Follow these with each element of B that is not yet
       listed. $\{1,2,3,4,6\}$
    (B) List each element of A that is also an element of B $\{2,4\}$.

57.   Let $c = 0.090909\ldots$ **59.**

$$\text{Then } 100c = 9.0909\ldots$$
$$100c - c = (9.0909\ldots) - (0.090909\ldots)$$
$$99c = 9$$
$$c = \frac{9}{99} = \frac{1}{11}$$

23		$23 \cdot 12 = 23(2 + 10)$
12		$= 23 \cdot 2 + 23 \cdot 10$
46	$23 \cdot 2$	
230	$23 \cdot 10$	$= 46 + 230$
276		$= 276$

Exercise A-2

Key Ideas and Formulas

For n a natural number and a any real number, $a^n = \underbrace{a \cdot a \cdots a}_{n \text{ factors}}$

First Property of Exponents: For any natural numbers m and n, and a any real number:
$a^m a^n = a^{m+n}$.

A polynomial in x is an algebraic expression of the form

$$a_n x^n + a_{n-1} x^{n-1} + \cdots + a_1 x + a_0$$

where the coefficients $a_0, a_1, \ldots, a_n$ are real numbers and n is a non-negative
integer.

Two terms in a polynomial are called like terms if they have exactly the same
variable factors to the same powers. Like terms in a polynomial are combined by
adding their numerical coefficients.

$$a - b = a + (-b)$$

Distributive Properties:

$$a(b + c) = (b + c)a = ab + ac$$
$$a(b - c) = (b - c)a = ab - ac$$
$$a(b + c + \cdots + f) = ab + ac + \cdots + af$$

FOIL method:

	F	O	I	L
	First Product	Outer Product	Inner Product	Last Product
$(a + b)(c + d) =$	ac +	ad +	bc +	bd

Special Products:

$$(a - b)(a + b) = a^2 - b^2$$
$$(a + b)^2 = a^2 + 2ab + b^2$$
$$(a - b)^2 = a^2 - 2ab + b^2$$

Order of Operations:

1. Simplify inside the innermost grouping first, then the next innermost, and so on.

2. Unless grouping symbols indicate otherwise, apply exponents before multiplication or division is performed.

3. Unless grouping symbols indicate otherwise, perform multiplication and division before addition and subtraction. In either case, proceed from left to right.

---

1. 3

3. $(2x^3 - 3x^2 + x + 5) + (2x^2 + x - 1)$
$$= 2x^3 - 3x^2 + x + 5 + 2x^2 + x - 1$$
$$= 2x^3 - x^2 + 2x + 4$$

5. $(2x^3 - 3x^2 + x + 5) - (2x^2 + x - 1)$
$$= 2x^3 - 3x^2 + x + 5 - 2x^2 - x + 1$$
$$= 2x^3 - 5x^2 + 6$$

> **Common Error:**
> Reversing terms: $(2x^2 + x - 1) - (2x^3 - 3x^2 + x + 5)$
> Should write: (first) - (second). In this case, $(a) - (b)$.

> **Common Error:**
> $2x^3 - 3x^2 + x + 5 - 2x^2 + x - 1$
> Should change the sign of each term in the second parentheses: $-(2x^2 + x - 1) = -2x^2 - x + 1$

7. $2x^3 - 3x^2 + x + 5$
$\underline{3x - 2}$
$6x^4 - 9x^3 + 3x^2 + 15x$
$\underline{ - 4x^3 + 6x^2 - 2x - 10}$
$6x^4 - 13x^3 + 9x^2 + 13x - 10$

9. $2(x - 1) + 3(2x - 3) - (4x - 5)$
$$= 2x - 2 + 6x - 9 - 4x + 5$$
$$= 4x - 6$$

11. $2y - 3y[4 - 2(y - 1)]$
$$= 2y - 3y[4 - 2y + 2]$$
$$= 2y - 3y[6 - 2y]$$
$$= 2y - 18y + 6y^2$$
$$= -16y + 6y^2$$
$$= 6y^2 - 16y$$

13. $(m - n)(m + n) = m^2 - n^2$

15.

	First Product	Outer Product	Inner Product	Last Product
$(4t - 3)(t - 2) = 4t^2$		$-8t$	$-3t$	$+6$
$= 4t^2 - 11t + 6$				

17.

	First Product	Outer Product	Inner Product	Last Product
$(3x + 2y)(x - 3y) = 3x^2$		$-9xy$	$+2xy$	$-6y^2$
$= 3x^2 - 7xy - 6y^2$				

19. $(2m - 7)(2m + 7) = (2m)^2 - (7)^2 = 4m^2 - 49.$

21.

	First Product	Outer Product	Inner Product	Last Product
$(6x - 4y)(5x + 3y) = 30x^2$		$+18xy$	$-20xy$	$-12y^2$
$= 30x^2 - 2xy - 12y^2$				

23. $(3x - 2y)(3x + 2y) = (3x)^2 - (2y)^2$
$$= 9x^2 - 4y^2$$

25. $(4x - y)^2 = (4x - y)(4x - y)$
$$= (4x)^2 - 2(4x)(y) + (y)^2$$
$$= 16x^2 - 8xy + y^2$$

27. $a^2 - ab + b^2$

$\underline{a + b}$

$a^3 - a^2b + ab^2$

$\underline{\quad\quad a^2b - ab^2 + b^3}$

$a^3 \quad\quad\quad\quad + b^3$

29. $2x - 3\{x + 2[x - (x + 5)] + 1\}$

$\quad = 2x - 3\{x + 2[x - x - 5] + 1\}$

$\quad = 2x - 3\{x + 2[-5] + 1\}$

$\quad = 2x - 3\{x - 10 + 1\}$

$\quad = 2x - 3\{x - 9\}$

$\quad = 2x - 3x + 27$

$\quad = -x + 27$

31. $2\{3[a - 4(1 - a)] - (5 - a)\}$

$\quad = 2\{3[a - 4 + 4a] - 5 + a\}$

$\quad = 2\{3[5a - 4] - 5 + a\}$

$\quad = 2\{15a - 12 - 5 + a\}$

$\quad = 2\{16a - 17\}$

$\quad = 32a - 34$

33. $2x^2 - 3x + 1$

$\underline{x^2 + x - 2}$

$2x^4 - 3x^3 + x^2$

$\quad\quad 2x^3 - 3x^2 + x$

$\quad\quad\quad\quad\underline{- 4x^2 + 6x - 2}$

$2x^4 - x^3 - 6x^2 + 7x - 2$

35. $(x - 2y)^2 = x^2 - 2(x)(2y) + (2y)^2 = x^2 - 4xy + 4y^2$

$\quad (x + 2y)^2 = x^2 + 2(x)(2y) + (2y)^2 = x^2 + 4xy + 4y^2$

The multiplication $(x - 2y)^2(x + 2y)^2 = (x^2 - 4xy + 4y^2)(x^2 + 4xy + 4y^2)$ can now be performed vertically, so we write:

$x^2 - 4xy + 4y^2$

$\underline{x^2 + 4xy + 4y^2}$

$x^4 - 4x^3y + 4x^2y^2$

$\quad\quad 4x^3y - 16x^2y^2 + 16xy^3$

$\quad\quad\quad\quad\underline{4x^2y^2 - 16xy^3 + 16y^4}$

$x^4 \quad\quad - 8x^2y^2 \quad\quad + 16y^4$

Alternatively, we can write:

$(x - 2y)^2(x + 2y)^2 = [(x - 2y)(x + 2y)]^2$

$\quad\quad\quad\quad\quad\quad\quad = [(x)^2 - (2y)^2]^2$

$\quad\quad\quad\quad\quad\quad\quad = (x^2 - 4y^2)^2$

$\quad\quad\quad\quad\quad\quad\quad = (x^2)^2 - 2(x^2)(4y^2) + (4y^2)^2$

$\quad\quad\quad\quad\quad\quad\quad = x^4 - 8x^2y^2 + 16y^4$

37. $(3u - 2v)^2 - (2u - 3v)(2u + 3v) = (3u)^2 - 2(3u)(2v) + (2v)^2 - [(2u)^2 - (3v)^2]$

$\quad\quad\quad\quad\quad\quad\quad\quad\quad\quad\quad = 9u^2 - 12uv + 4v^2 - (4u^2 - 9v^2)$

$\quad\quad\quad\quad\quad\quad\quad\quad\quad\quad\quad = 9u^2 - 12uv + 4v^2 - 4u^2 + 9v^2$

$\quad\quad\quad\quad\quad\quad\quad\quad\quad\quad\quad = 5u^2 - 12uv + 13v^2$

> **Commom Error:** $9u^2 - 12uv + 4v^2 - 4u^2 - 9v^2$ The entire result of multiplying $(2u - 3v)(2u + 3v)$ must be subtracted

39. The multiplication $(z + 2)(z^2 - 2z + 3)$ is best performed vertically, so we write:

$z^2 - 2z + 3$

$\underline{\quad\quad z + 2}$

$z^3 - 2z^2 + 3z$

$\quad\quad\underline{2z^2 - 4z + 6}$

$z^3 \quad\quad - z + 6$

Therefore $(z + 2)(z^2 - 2z + 3) + z - 7 = z^3 - z + 6 + z - 7 = z^3 - 1$

41. $(2m - n)^3 = (2m - n)(2m - n)(2m - n)$

$\quad\quad\quad\quad\quad = [(2m - n)(2m - n)](2m - n)$

$\quad\quad\quad\quad\quad = [(2m)^2 - 2(2m)(n) + n^2](2m - n)$

$\quad\quad\quad\quad\quad = [4m^2 - 4mn + n^2](2m - n)$

The last multiplication is best performed vertically, so we write:

$$
\begin{array}{r}
4m^2 - 4mn + n^2 \\
2m - n \\
\hline
8m^3 - 8m^2n + 2mn^2 \\
- 4m^2n + 4mn^2 - n^3 \\
\hline
\end{array}
$$

$(2m - n)^3 = 8m^3 - 12m^2n + 6mn^2 - n^3$

43. $3(x + h) - 7 - (3x - 7) = 3x + 3h - 7 - 3x + 7 = 3h$

45. $2(x + h)^2 - 3(x + h) - (2x^2 - 3x)$
  $= 2(x^2 + 2xh + h^2) - 3x - 3h - 2x^2 + 3x$
  $= 2x^2 + 4xh + 2h^2 - 3x - 3h - 2x^2 + 3x$
  $= 4xh + 2h^2 - 3h$

> **Commom Error:**
> $2(x + h)^2 \neq (2x + 2h)^2$
> Exponentiation must be performed before multiplication.

47. $2(x + h)^2 - 4(x + h) - 9 - (2x^2 - 4x - 9)$
  $= 2(x^2 + 2xh + h^2) - 4x - 4h - 9 - 2x^2 + 4x + 9$
  $= 2x^2 + 4xh + 2h^2 - 4x - 4h - 9 - 2x^2 + 4x + 9$
  $= 4xh + 2h^2 - 4h$

49. $(x + h)^3 = (x + h)(x + h)(x + h)$
  $= (x + h)^2(x + h)$
  $= [x^2 + 2xh + h^2](x + h)$

The last multiplication is best performed vertically, so we write:

$$
\begin{array}{r}
x^2 + 2xh + h^2 \\
x + h \\
\hline
x^3 + 2x^2h + xh^2 \\
x^2h + 2xh^2 + h^3 \\
\hline
x^3 + 3x^2h + 3xh^2 + h^3
\end{array}
$$

Therefore $(x + h)^3 - 2(x + h)^2 - (x^3 - 2x^2)$
  $= x^3 + 3x^2h + 3xh^2 + h^3 - 2(x^2 + 2xh + h^2) - (x^3 - 2x^2)$
  $= x^3 + 3x^2h + 3xh^2 + h^3 - 2x^2 - 4xh - 2h^2 - x^3 + 2x^2$
  $= 3x^2h + 3xh^2 + h^3 - 4xh - 2h^2$

51. The sum of the first two polynomials:
  $(3m^2 - 2m + 5) + (4m^2 - m) = 3m^2 - 2m + 5 + 4m^2 - m = 7m^2 - 3m + 5$

The sum of the last two polynomials:
  $(3m^2 - 3m - 2) + (m^3 + m^2 + 2) = 3m^2 - 3m - 2 + m^3 + m^2 + 2 = m^3 + 4m^2 - 3m$

Subtract from the sum of the last two polynomials the sum of the first two.
  $(m^3 + 4m^2 - 3m) - (7m^2 - 3m + 5) = m^3 + 4m^2 - 3m - 7m^2 + 3m - 5$
  $= m^3 - 3m^2 - 5$

53. Since $(x - 2)^2 = (x)^2 - 2(x)(2) + (2)^2 = x^2 - 4x + 4$,
  $(x - 2)^3 = (x - 2)^2(x - 2) = (x^2 - 4x + 4)(x - 2)$

$$
\begin{array}{r}
x^2 - 4x + 4 \\
x - 2 \\
\hline
x^3 - 4x^2 + 4x \\
- 2x^2 + 8x - 8 \\
\hline
x^3 - 6x^2 + 12x - 8
\end{array}
$$

Therefore $2(x - 2)^3 - (x - 2)^2 - 3(x - 2) - 4$
  $= 2(x^3 - 6x^2 + 12x - 8) - (x^2 - 4x + 4) - 3(x - 2) - 4$
  $= 2x^3 - 12x^2 + 24x - 16 - x^2 + 4x - 4 - 3x + 6 - 4$
  $= 2x^3 - 13x^2 + 25x - 18$

55. We will multiply $(x + 2)(x^2 - 3)$ using the FOIL method.

	First Product	Outer Product	Inner Product	Last Product
$(x + 2)(x^2 - 3) = x^3$ or $x^3 + 2x^2 - 3x - 6$		$-3x$	$+2x^2$	-6

Then $-3x\{x[x - x(2 - x)] - (x + 2)(x^2 - 3)\}$

$$= -3x\{x[x - 2x + x^2] - [x^3 + 2x^2 - 3x - 6]\}$$
$$= -3x\{x[-x + x^2] - [x^3 + 2x^2 - 3x - 6]\}$$
$$= -3x\{-x^2 + x^3 - x^3 - 2x^2 + 3x + 6\}$$
$$= -3x\{-3x^2 + 3x + 6\}$$
$$= 9x^3 - 9x^2 - 18x$$

57. One example is given by choosing $a = 1$ and $b = 1$. Then $(a + b)^2 = (1 + 1)^2 = 2^2 = 4$, but $a^2 + b^2 = 1^2 + 1^2 = 1 + 1 = 2$. Thus, in general, $(a + b)^2 \neq a^2 + b^2$. In fact, since $(a + b)^2 = a^2 + 2ab + b^2$, this quantity can only equal $a^2 + b^2$ if $2ab = 0$. By the properties of 0, either $a = 0$ or $b = 0$. In these cases $(a + b)^2$ would equal $a^2 + b^2$, but only in these.

59. The non-zero term with the highest degree in the polynomial of degree m has degree m. This term is not changed during the addition, since there is no term in the polynomial of degree n with the degree $m(m > n)$. It will still be the highest degree term in the sum, so the sum will be a polynomial of degree m.

61. Now the non-zero term with the highest degree in each polynomial has degree m. It is possible for the coefficients of these terms to be equal in absolute value but opposite in sign, in which case they would add to 0, leaving a term of degree less than m as the term with the highest degree in the sum. Otherwise the terms of degree m will combine to give another term of degree m, which will be the highest degree term in the sum. Thus, $(2x^4 + x^2 + 1) + (-2x^4 + x^2 + 1) = 2x^2 + 2$, but $(2x^4 + x^2 + 1) + (2x^4 + 2x^2 + 1) = 4x^4 + 3x^2 + 2$.
Summarizing, the degree of the sum may be less than or equal to m.

63. There are three quantities in this problem, perimeter, length, and width. They are related by the perimeter formula $P = 2\ell + 2w$. Since $x = $ length of the rectangle, and the width is 5 meters less than the length, $x - 5 = $ width of the rectangle. So $P = 2x + 2(x - 5)$ represents the perimeter of the rectangle. Simplifying: $P = 2x + 2x - 10 = 4x - 10$ (meters)

65. There are several quantities involved in this problem. It is important to keep them distinct by using enough words. We write:

$$x = \text{number of nickels}$$
$$x - 5 = \text{number of dimes}$$
$$(x - 5) + 2 = \text{number of quarters}$$

This follows because there are five fewer dimes than nickels $(x - 5)$ and 2 more quarters than dimes (2 more than $x - 5$). Each nickel is worth 5 cents, each dime worth 10 cents, and each quarter worth 25 cents. Hence the value of the nickels is 5 times the number of nickels, the value of the dimes is 10 times the number of dimes, and the value of the quarters is 25 times the number of quarters.

$$\text{value of nickels} = 5x$$
$$\text{value of dimes} = 10(x - 5)$$
$$\text{value of quarters} = 25[(x - 5) + 2]$$

The value of the pile = (value of nickels) + (value of dimes)
                                                    + (value of quarters)
$$= 5x + 10(x - 5) + 25[(x - 5) + 2]$$

Simplifying this expression, we get:
The value of the pile $= 5x + 10x - 50 + 25[x - 5 + 2]$
$$= 5x + 10x - 50 + 25[x - 3]$$
$$= 5x + 10x - 50 + 25x - 75$$
$$= 40x - 125 \text{ (cents)}$$

67. The volume of the plastic shell is equal to the volume of the larger sphere $(V = \frac{4}{3}\pi r^3)$ minus the volume of the hole. Since the radius of the hole is x cm and the plastic is 0.3 cm thick, the radius of the larger sphere is $x + 0.3$ cm. Thus, we have

$$\begin{pmatrix} \text{Volume of} \\ \text{shell} \end{pmatrix} = \begin{pmatrix} \text{Volume of} \\ \text{larger sphere} \end{pmatrix} - \begin{pmatrix} \text{Volume of} \\ \text{hole} \end{pmatrix}$$

$$\begin{aligned} \text{Volume} &= \tfrac{4}{3}\pi (x + 0.3)^3 - \tfrac{4}{3}\pi x^3 \\ &= \tfrac{4}{3}\pi (x + 0.3)(x + 0.3)^2 - \tfrac{4}{3}\pi x^3 \\ &= \tfrac{4}{3}\pi (x + 0.3)(x^2 + 0.6x + 0.09) - \tfrac{4}{3}\pi x^3 \\ &= \tfrac{4}{3}\pi (x^3 + 0.6x^2 + 0.09x + 0.3x^2 + 0.18x + 0.027) - \tfrac{4}{3}\pi x^3 \\ &= \tfrac{4}{3}\pi (x^3 + 0.9x^2 + 0.27x + 0.027) - \tfrac{4}{3}\pi x^3 \\ &= \tfrac{4}{3}\pi x^3 + 1.2\pi x^2 + 0.36\pi x + 0.036\pi - \tfrac{4}{3}\pi x^3 \\ &= 1.2\pi x^2 + 0.36\pi x + 0.036\pi \ (\text{cm}^3) \end{aligned}$$

Exercise A-3

Key Ideas and Formulas

$$\begin{aligned} ab + ac &= a(b + c) & \\ u^2 + 2uv + v^2 &= (u + v)^2 & \text{Perfect square} \\ u^2 - 2uv + v^2 &= (u - v)^2 & \text{Perfect square} \\ u^2 - v^2 &= (u - v)(u + v) & \text{Difference of squares} \\ u^3 - v^3 &= (u - v)(u^2 + uv + v^2) & \text{Difference of cubes} \\ u^3 + v^3 &= (u + v)(u^2 - uv + v^2) & \text{Sum of cubes} \end{aligned}$$

Common Errors:

Confusing $u^2 + v^2$ (a prime polynomial) with
$$(u + v)^2 = (u + v)(u + v) = u^2 + 2uv + v^2 \text{ (a perfect square).}$$
Confusing $u^2 - v^2 = (u - v)(u + v)$ (difference of squares) with
$$(u - v)^2 = (u - v)(u - v) = u^2 - 2uv + v^2 \text{ (a perfect square).}$$

---

1. $2x^2(3x^2 - 4x - 1)$ **3.** $5xy(2x^2 + 4xy - 3y^2)$ **5.** $(x + 1)(5x - 3)$ **7.** $(y - 2z)(2w - x)$

9. $\begin{aligned} x^2 - 2x + 3x - 6 &= (x^2 - 2x) + (3x - 6) \\ &= x(x - 2) + 3(x - 2) \\ &= (x - 2)(x + 3) \end{aligned}$

11. $\begin{aligned} 6m^2 + 10m - 3m - 5 &= (6m^2 + 10m) - (3m + 5) = 2m(3m + 5) - 1(3m + 5) \\ &= (3m + 5)(2m - 1) \end{aligned}$

13. $\begin{aligned} 2x^2 - 4xy - 3xy + 6y^2 &= (2x^2 - 4xy) - (3xy - 6y^2) = 2x(x - 2y) - 3y(x - 2y) \\ &= (x - 2y)(2x - 3y) \end{aligned}$

15. $\begin{aligned} 8ac + 3bd - 6bc - 4ad &= 8ac - 4ad - 6bc + 3bd \\ &= (8ac - 4ad) - (6bc - 3bd) \\ &= 4a(2c - d) - 3b(2c - d) \\ &= (2c - d)(4a - 3b) \end{aligned}$

17. $(2x + 3)(x - 1)$ **19.** Prime **21.** $(m - 2)(m + 3)$ **23.** $(2a + 3b)(2a - 3b)$

25. $4x^2 - 20x + 25 = (2x)^2 - 2(2x)(5) + 5^2 = (2x - 5)^2$ **27.** Prime

29. $6x^2 + 48x + 72 = 6(x^2 + 8x + 12) = 6(x + 2)(x + 6)$

31. $2y^3 - 22y^2 + 48y = 2y(y^2 - 11y + 24) = 2y(y - 3)(y - 8)$

33. $16x^2y - 8xy + y = y(16x^2 - 8x + 1) = y[(4x)^2 - 2(4x) + 1] = y(4x - 1)^2$

35. Prime

37. $x^3y - 9xy^3 = xy(x^2 - 9y^2) = xy(x - 3y)(x + 3y)$

39. $3m(m^2 - 2m + 5)$

41. $(m + n)(m^2 - mn + n^2)$

43. $2x(x + 1)^4 + 4x^2(x + 1)^3 = 2x(x + 1)^3[(x + 1) + 2x] = 2x(x + 1)^3(3x + 1)$

45. $6(3x - 5)(2x - 3)^2 + 4(3x - 5)^2(2x - 3) = 2(3x - 5)(2x - 3)[3(2x - 3) + 2(3x - 5)]$
$$= 2(3x - 5)(2x - 3)[6x - 9 + 6x - 10]$$
$$= 2(3x - 5)(2x - 3)(12x - 19)$$

47. $5x^4(9 - x)^4 - 4x^5(9 - x)^3 = x^4(9 - x)^3[5(9 - x) - 4x]$
$$= x^4(9 - x)^3[45 - 5x - 4x]$$
$$= x^4(9 - x)^3(45 - 9x) \quad \text{or} \quad 9x^4(9 - x)^3(5 - x)$$

49. $2(x + 1)(x^2 - 5)^2 + 4x(x + 1)^2(x^2 - 5) = 2(x + 1)(x^2 - 5)[x^2 - 5 + 2x(x + 1)]$
$$= 2(x + 1)(x^2 - 5)[x^2 - 5 + 2x^2 + 2x]$$
$$= 2(x + 1)(x^2 - 5)(3x^2 + 2x - 5)$$
$$= 2(x + 1)(x^2 - 5)(3x + 5)(x - 1)$$

51. $(a - b)^2 - 4(c - d)^2 = (a - b)^2 - [2(c - d)]^2$
$$= [(a - b) - 2(c - d)][(a - b) + 2(c - d)]$$

53. $2am - 3an + 2bm - 3bn = (2am - 3an) + (2bm - 3bn) = a(2m - 3n) + b(2m - 3n)$
$$= (2m - 3n)(a + b)$$

55. Prime

57. $x^3 - 3x^2 - 9x + 27 = (x^3 - 3x^2) - (9x - 27) = x^2(x - 3) - 9(x - 3) = (x - 3)(x^2 - 9)$
$$= (x - 3)(x - 3)(x + 3) = (x - 3)^2(x + 3)$$

59. $a^3 - 2a^2 - a + 2 = (a^3 - 2a^2) - (a - 2) = a^2(a - 2) - 1(a - 2) = (a - 2)(a^2 - 1)$
$$= (a - 2)(a + 1)(a - 1)$$

61. Prime

63. $m^4 - n^4 = (m^2)^2 - (n^2)^2 = (m^2 - n^2)(m^2 + n^2) = (m - n)(m + n)(m^2 + n^2)$

65. Prime

67. $m^2 + 2mn + n^2 - m - n = (m^2 + 2mn + n^2) - (m + n) = (m + n)^2 - 1(m + n)$
$$= (m + n)(m + n - 1)$$

69. $18a^3 - 8a(x^2 + 8x + 16) = 2a[9a^2 - 4(x^2 + 8x + 16)] = 2a\{(3a)^2 - [2(x + 4)]^2\}$
$$= 2a[3a - 2(x + 4)][3a + 2(x + 4)]$$

71. $x^4 + 2x^2 + 1 - x^2 = (x^4 + 2x^2 + 1) - x^2 = (x^2 + 1)^2 - x^2$
$$= (x^2 + 1 - x)(x^2 + 1 + x) = (x^2 - x + 1)(x^2 + x + 1)$$

73. (A) The area of the cardboard can be written as (Original area) - (Removed area), where the original area = $20^2 = 400$ and the removed area consists of 4 squares of area x^2 each; thus,
(Original area) - (Removed area) = $400 - 4x^2$ in expanded form.
In factored form, $400 - 4x^2 = 4(100 - x^2) = 4(10 - x)(10 + x)$.

(B) See figure.

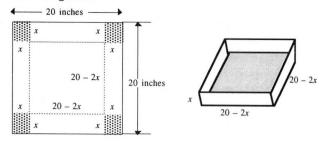

The volume of the box = ℓwh = $x(20 - 2x)(20 - 2x)$
= $4x(10 - x)(10 - x)$ in factored form

In expanded form, $4x(10 - x)(10 - x)$ = $4x[10^2 - 2(10)x + x^2]$
= $4x[100 - 20x + x^2]$
= $400x - 80x^2 + 4x^3$

Exercise A-4

Key Ideas and Formulas

Fundamental Property of Fractions

If a, b, and k are real numbers with b, $k \neq 0$, then

$$\frac{ak}{bk} = \frac{a}{b} \qquad \text{Also,} \quad \frac{a}{b} = \frac{ak}{bk}$$

Multiplication and Division

For a, b, c, and d, real numbers,

$$\frac{a}{b} \cdot \frac{c}{d} = \frac{ac}{bd} \qquad\qquad b, \ d \neq 0$$

$$\frac{a}{b} \div \frac{c}{d} = \frac{a}{b} \cdot \frac{d}{c} \qquad\qquad b, \ c, \ d \neq 0$$

Addition and Subtraction

For a, b, and c real numbers

$$\frac{a}{b} + \frac{c}{b} = \frac{a + c}{b} \qquad\qquad b \neq 0$$

$$\frac{a}{b} - \frac{c}{b} = \frac{a - c}{b} \qquad\qquad b \neq 0$$

Common Errors:

There are many possible errors in applying the fundamental property of fractions. Always be sure that each quantity removed is a factor of the entire numerator and a factor of the entire denominator. Examples of errors abound, but here are a few classic types to avoid:

$$\left. \begin{aligned} \frac{m + 2}{m - 3} &\neq \frac{2}{-3} \\[2mm] \frac{x^2 + 2x - 15}{x^2 + 4x - 5} &\neq \frac{2x - 15}{4x - 5} \end{aligned} \right\} \text{cannot "cancel" common terms}$$

$$\left. \frac{x + 7}{ax - bx} \neq \frac{7}{a - bx} \right\} \text{cannot "cancel" a term against a factor of a term}$$

$$\left. \begin{aligned} \frac{x - (x - 4)}{(x - 4)(x + 4)} &\neq \frac{x}{x + 4} \\[2mm] \frac{x^2 + 8}{8x^3} &\neq \frac{x^2}{x^3} \end{aligned} \right\} \text{cannot "cancel" a term against a factor}$$

1. $\left(\dfrac{d^5}{3a} \div \dfrac{d^2}{6a^2}\right) \cdot \dfrac{a}{4d^3} = \left(\dfrac{d^5}{3a} \cdot \dfrac{6a^2}{d^2}\right) \cdot \dfrac{a}{4d^3} = \left(\dfrac{\cancel{d^5}^{d^3}}{\cancel{3a}_1} \cdot \dfrac{\cancel{6a^2}^{2a}}{\cancel{d^2}_1}\right) \cdot \dfrac{a}{4d^3} = \dfrac{2ad^3}{1} \cdot \dfrac{a}{4d^3} = \dfrac{\cancel{2ad^3}^1}{1} \cdot \dfrac{a}{\cancel{4d^3}_2} = \dfrac{a^2}{2}$

3. $\dfrac{2y}{18} - \dfrac{-1}{28} - \dfrac{y}{42} = \dfrac{28y}{252} - \dfrac{-9}{252} - \dfrac{6y}{252} = \dfrac{22y + 9}{252}$

Note: $\left.\begin{array}{l} 18 = 2\cdot3\cdot3 \\ 28 = 2\cdot2\cdot7 \\ 42 = 2\cdot3\cdot7 \end{array}\right\}$ so LCD $= 2\cdot2\cdot3\cdot3\cdot7 = 252$

5. $\dfrac{3x + 8}{4x^2} - \dfrac{2x - 1}{x^3} - \dfrac{5}{8x} = \dfrac{2x(3x + 8)}{8x^3} - \dfrac{8(2x - 1)}{8x^3} - \dfrac{x^2(5)}{8x^3}$

$\qquad = \dfrac{6x^2 + 16x - 16x + 8 - 5x^2}{8x^3} = \dfrac{x^2 + 8}{8x^3}$

7. $\dfrac{2x^2 - 3x - 2}{x^2 - 4} \div (2x + 1) = \dfrac{2x^2 - 3x - 2}{x^2 - 4} \div \dfrac{(2x + 1)}{1} = \dfrac{(2x + 1)(x - 2)}{(x + 2)(x - 2)} \cdot \dfrac{1}{(2x + 1)}$

$\qquad = \dfrac{\cancel{(2x + 1)}^1 \cancel{(x - 2)}^1}{(x + 2)\cancel{(x - 2)}_1} \cdot \dfrac{1}{\cancel{(2x + 1)}_1} = \dfrac{1}{x + 2}$

9. $\dfrac{2m + 3n}{2m^2 - 5mn - 12n^2} \div \dfrac{2m^2 - 7mn - 4n^2}{m^2 - 8mn + 16n^2} = \dfrac{2m + 3n}{2m^2 - 5mn - 12n^2} \cdot \dfrac{m^2 - 8mn + 16n^2}{2m^2 - 7mn - 4n^2}$

$\qquad = \dfrac{2m + 3n}{(2m + 3n)(m - 4n)} \cdot \dfrac{(m - 4n)(m - 4n)}{(2m + n)(m - 4n)}$

$\qquad = \dfrac{\cancel{(2m + 3n)}^1}{\cancel{(2m + 3n)}_1 \cancel{(m - 4n)}_1} \cdot \dfrac{\cancel{(m - 4n)}^1 \cancel{(m - 4n)}^1}{(2m + n)\cancel{(m - 4n)}_1}$

$\qquad = \dfrac{1}{2m + n}$

11. $\dfrac{2a - b}{a^2 - b^2} - \dfrac{2a + 3b}{a^2 + 2ab + b^2} = \dfrac{2a - b}{(a + b)(a - b)} - \dfrac{2a + 3b}{(a + b)(a + b)}$

$\qquad = \dfrac{(2a - b)(a + b)}{(a + b)(a + b)(a - b)} - \dfrac{(2a + 3b)(a - b)}{(a + b)(a + b)(a - b)}$

$\qquad = \dfrac{(2a - b)(a + b) - (2a + 3b)(a - b)}{(a + b)^2 (a - b)}$

$\qquad = \dfrac{2a^2 + ab - b^2 - (2a^2 + ab - 3b^2)}{(a + b)^2 (a - b)}$

$\qquad = \dfrac{2a^2 + ab - b^2 - 2a^2 - ab + 3b^2}{(a + b)^2 (a - b)}$

$\qquad = \dfrac{2b^2}{(a + b)^2 (a - b)}$

13. $m + 2 - \dfrac{m - 2}{m - 1} = \dfrac{m + 2}{1} - \dfrac{m - 2}{m - 1} = \dfrac{(m + 2)(m - 1)}{m - 1} - \dfrac{(m - 2)}{m - 1}$

$\qquad = \dfrac{(m^2 + m - 2) - (m - 2)}{m - 1} = \dfrac{m^2 + m - 2 - m + 2}{m - 1}$

$\qquad = \dfrac{m^2}{m - 1}$

15. $\dfrac{3}{x - 2} - \dfrac{2}{2 - x} = \dfrac{3}{x - 2} - \dfrac{-2}{x - 2} = \dfrac{3 + 2}{x - 2} = \dfrac{5}{x - 2}$

17. $\dfrac{3}{y+2} + \dfrac{2}{y-2} - \dfrac{4y}{y^2-4} = \dfrac{3(y-2)}{(y+2)(y-2)} + \dfrac{2(y+2)}{(y+2)(y-2)} - \dfrac{4y}{(y+2)(y-2)}$

$$= \dfrac{3(y-2) + 2(y+2) - 4y}{(y+2)(y-2)} = \dfrac{3y - 6 + 2y + 4 - 4y}{(y+2)(y-2)}$$

$$= \dfrac{y-2}{(y+2)(y-2)} = \dfrac{\overset{1}{\cancel{(y-2)}}}{(y+2)\underset{1}{\cancel{(y-2)}}} = \dfrac{1}{y+2}$$

19. $\dfrac{\frac{x^2}{y^2} - 1}{\frac{x}{y} + 1} = \dfrac{y^2\left(\frac{x^2}{y^2} - 1\right)}{y^2\left(\frac{x}{y} + 1\right)} = \dfrac{x^2 - y^2}{xy + y^2} = \dfrac{(x-y)\overset{1}{\cancel{(x+y)}}}{\underset{1}{y\cancel{(x+y)}}} = \dfrac{x-y}{y}$

21. $\dfrac{6x^3(x^2+2)^2 - 2x(x^2+2)^3}{x^4} = \dfrac{2x(x^2+2)^2[3x^2 - (x^2+2)]}{x^4}$

$$= \dfrac{2x(x^2+2)^2[3x^2 - x^2 - 2]}{x^4}$$

$$= \dfrac{2\overset{1}{\cancel{x}}(x^2+2)^2(2x^2 - 2)}{\underset{x^3}{\cancel{x^4}}}$$

$$= \dfrac{2(x^2+2)^2(2x^2 - 2)}{x^3}$$

$$= \dfrac{2(x^2+2)^2 2(x^2 - 1)}{x^3}$$

$$= \dfrac{4(x^2+2)^2(x+1)(x-1)}{x^3}$$

23. $\dfrac{2x(1-3x)^3 + 9x^2(1-3x)^2}{(1-3x)^6} = \dfrac{x(1-3x)^2[2(1-3x) + 9x]}{(1-3x)^6}$

$$= \dfrac{x(1-3x)^2[2 - 6x + 9x]}{(1-3x)^6}$$

$$= \dfrac{x\overset{1}{\cancel{(1-3x)^2}}(2+3x)}{\underset{(1-3x)^4}{\cancel{(1-3x)^6}}}$$

$$= \dfrac{x(2+3x)}{(1-3x)^4}$$

25. $\dfrac{-2x(x+4)^3 - 3(3-x^2)(x+4)^2}{(x+4)^6} = \dfrac{(x+4)^2[-2x(x+4) - 3(3-x^2)]}{(x+4)^6}$

$$= \dfrac{(x+4)^2[-2x^2 - 8x - 9 + 3x^2]}{(x+4)^6}$$

$$= \dfrac{\overset{1}{\cancel{(x+4)^2}}(x^2 - 8x - 9)}{\underset{(x+4)^4}{\cancel{(x+4)^6}}}$$

$$= \dfrac{x^2 - 8x - 9}{(x+4)^4}$$

$$= \dfrac{(x+1)(x-9)}{(x+4)^4}$$

27. $\dfrac{y}{y^2 - 2y - 8} - \dfrac{2}{y^2 - 5y + 4} + \dfrac{1}{y^2 + y - 2}$

$\qquad = \dfrac{y}{(y - 4)(y + 2)} - \dfrac{2}{(y - 4)(y - 1)} + \dfrac{1}{(y + 2)(y - 1)}$

$\qquad = \dfrac{y(y - 1)}{(y - 1)(y - 4)(y + 2)} - \dfrac{2(y + 2)}{(y - 4)(y - 1)(y + 2)} + \dfrac{1(y - 4)}{(y - 4)(y - 1)(y + 2)}$

$\qquad = \dfrac{y^2 - y - 2y - 4 + y - 4}{(y - 4)(y - 1)(y + 2)} = \dfrac{y^2 - 2y - 8}{(y - 4)(y - 1)(y + 2)} = \dfrac{\overset{1}{\cancel{(y - 4)}}\,\overset{1}{\cancel{(y + 2)}}}{\underset{1}{\cancel{(y - 4)}}\,(y - 1)\,\underset{1}{\cancel{(y + 2)}}}$

$\qquad = \dfrac{1}{y - 1}$

29. $\dfrac{16 - m^2}{m^2 + 3m - 4} \cdot \dfrac{m - 1}{m - 4} = \dfrac{\overset{-1}{\cancel{(4 - m)}}\,\overset{1}{\cancel{(4 + m)}}}{\underset{1}{\cancel{(m + 4)}}\,\underset{1}{\cancel{(m - 1)}}} \cdot \dfrac{\overset{1}{\cancel{m - 1}}}{\underset{1}{\cancel{m - 4}}} = \dfrac{-1}{1} = -1$

31. $\dfrac{x + 7}{ax - bx} + \dfrac{y + 9}{by - ay} = \dfrac{x + 7}{x(a - b)} + \dfrac{y + 9}{y(b - a)} = \dfrac{y(x + 7)}{xy(a - b)} + \dfrac{-x(y + 9)}{xy(a - b)}$

$\qquad = \dfrac{xy + 7y - xy - 9x}{xy(a - b)} = \dfrac{7y - 9x}{xy(a - b)}$

33. $\dfrac{x^2 - 16}{2x^2 + 10x + 8} \div \dfrac{x^2 - 13x + 36}{x^3 + 1} = \dfrac{x^2 - 16}{2x^2 + 10x + 8} \cdot \dfrac{x^3 + 1}{x^2 - 13x + 36}$

$\qquad = \dfrac{\overset{1}{\cancel{(x - 4)}}\,\overset{1}{\cancel{(x + 4)}}}{2\,\underset{1}{\cancel{(x + 1)}}\,\underset{1}{\cancel{(x + 4)}}} \cdot \dfrac{\overset{1}{\cancel{(x + 1)}}(x^2 - x + 1)}{\underset{1}{\cancel{(x - 4)}}(x - 9)} = \dfrac{x^2 - x + 1}{2(x - 9)}$

35. $\dfrac{x^2 - xy}{xy + y^2} \div \left(\dfrac{x^2 - y^2}{x^2 + 2xy + y^2} \div \dfrac{x^2 - 2xy + y^2}{x^2y + xy^2} \right)$

$\qquad = \dfrac{x^2 - xy}{xy + y^2} \div \left(\dfrac{x^2 - y^2}{x^2 + 2xy + y^2} \cdot \dfrac{x^2y + xy^2}{x^2 - 2xy + y^2} \right)$

$\qquad = \dfrac{x^2 - xy}{xy + y^2} \div \left(\dfrac{\overset{1}{\cancel{(x - y)}}\,\overset{1}{\cancel{(x + y)}}}{\underset{1}{\cancel{(x + y)}}\,\underset{1}{\cancel{(x + y)}}} \cdot \dfrac{xy\,\overset{1}{\cancel{(x + y)}}}{\underset{1}{\cancel{(x - y)}}(x - y)} \right)$

$\qquad = \dfrac{x^2 - xy}{xy + y^2} \div \dfrac{xy}{x - y} = \dfrac{x^2 - xy}{xy + y^2} \cdot \dfrac{x - y}{xy}$

$\qquad = \dfrac{\overset{1}{\cancel{x}}(x - y)}{y(x + y)} \cdot \dfrac{x - y}{\underset{1}{\cancel{xy}}} = \dfrac{(x - y)^2}{y^2(x + y)}$

37. $\left(\dfrac{x}{x^2 - 16} - \dfrac{1}{x + 4}\right) \div \dfrac{4}{x + 4} = \left(\dfrac{x}{(x - 4)(x + 4)} - \dfrac{1}{x + 4}\right) \div \dfrac{4}{x + 4}$

$$= \left(\dfrac{x}{(x - 4)(x + 4)} - \dfrac{(x - 4)}{(x - 4)(x + 4)}\right) \div \dfrac{4}{x + 4}$$

$$= \dfrac{x - x + 4}{(x - 4)(x + 4)} \div \dfrac{4}{x + 4} = \dfrac{4}{(x - 4)(x + 4)} \div \dfrac{4}{x + 4}$$

$$= \dfrac{\overset{1}{\cancel{4}}}{(x - 4)\underset{1}{\cancel{(x + 4)}}} \cdot \dfrac{\overset{1}{\cancel{x + 4}}}{\underset{1}{\cancel{4}}} = \dfrac{1}{x - 4}$$

39. $\dfrac{1 + \frac{2}{x} - \frac{15}{x^2}}{1 + \frac{4}{x} - \frac{5}{x^2}} = \dfrac{x^2\left(1 + \frac{2}{x} - \frac{15}{x^2}\right)}{x^2\left(1 + \frac{4}{x} - \frac{5}{x^2}\right)} = \dfrac{x^2 + 2x - 15}{x^2 + 4x - 5} = \dfrac{\overset{1}{\cancel{(x + 5)}}(x - 3)}{\underset{1}{\cancel{(x + 5)}}(x - 1)} = \dfrac{x - 3}{x - 1}$

41. $\dfrac{\frac{1}{x + h} - \frac{1}{x}}{h} = \dfrac{\frac{x}{x(x + h)} - \frac{x + h}{x(x + h)}}{h} = \dfrac{\frac{x - x - h}{x(x + h)}}{h} = \dfrac{\frac{-h}{x(x + h)}}{h} = \dfrac{-h}{x(x + h)} \div h$

$$= \dfrac{-h}{x(x + h)} \div \dfrac{h}{1} = \dfrac{\overset{-1}{-\cancel{h}}}{x(x + h)} \cdot \dfrac{1}{\underset{1}{\cancel{h}}} = \dfrac{-1}{x(x + h)}$$

43. $\dfrac{\frac{(x + h)^2}{x + h + 2} - \frac{x^2}{x + 2}}{h} = \dfrac{\frac{(x + h)^2(x + 2)}{(x + h + 2)(x + 2)} - \frac{x^2(x + h + 2)}{(x + h + 2)(x + 2)}}{h} = \dfrac{\frac{(x + h)^2(x + 2) - x^2(x + h + 2)}{(x + h + 2)(x + 2)}}{h}$

$$= \dfrac{\frac{(x^2 + 2xh + h^2)(x + 2) - x^3 - x^2h - 2x^2}{(x + h + 2)(x + 2)}}{h}$$

$$= \dfrac{\frac{x^3 + 2x^2h + h^2x + 2x^2 + 4xh + 2h^2 - x^3 - x^2h - 2x^2}{(x + h + 2)(x + 2)}}{h}$$

$$= \dfrac{\frac{x^2h + h^2x + 4xh + 2h^2}{(x + h + 2)(x + 2)}}{h} = \dfrac{x^2h + h^2x + 4xh + 2h^2}{(x + h + 2)(x + 2)} \div h$$

$$= \dfrac{x^2h + h^2x + 4xh + 2h^2}{(x + h + 2)(x + 2)} \div \dfrac{h}{1} = \dfrac{\overset{1}{\cancel{h}}(x^2 + hx + 4x + 2h)}{(x + h + 2)(x + 2)} \cdot \dfrac{1}{\underset{1}{\cancel{h}}}$$

$$= \dfrac{x^2 + hx + 4x + 2h}{(x + h + 2)(x + 2)}$$

45. (A) The solution is incorrect, because in the first step the quantity 4 has been removed from numerator and denominator. But 4 is a term, and only factors can be cancelled. A correct solution would factor numerator and denominator, then cancel common factors, if any.

(B) $\dfrac{x^2 + 5x + 4}{x + 4} = \dfrac{(x + 1)\overset{1}{\cancel{(x + 4)}}}{\underset{1}{\cancel{x + 4}}} = x + 1$

47. (A) The solution is incorrect, because in the first step the quantity h has been removed from numerator and denominator. Although h is a factor of the denominator it is not a factor of the numerator as shown and only common factors of numerator and denominator can be cancelled. A correct solution would factor numerator and denominator, revealing that h is actually a factor of the numerator in factored form and allowing a correct cancellation of h.

(B) $\dfrac{(x+h)^2 - x^2}{h} = \dfrac{[(x+h) - x][(x+h) + x]}{h}$ using the difference of two squares

$= \dfrac{[x+h-x][x+h+x]}{h}$

$= \dfrac{\overset{1}{\cancel{h}}[2x+h]}{\underset{1}{\cancel{h}}}$

$= 2x + h$

49. (A) The solution is incorrect. In adding a quantity to a fractional expression, the quantity cannot simply be placed in the numerator. A correct solution would rewrite $x - 2$ as a fractional expression, first with denominator 1, and then with denominator equal to the LCD of all denominators in the addition.

(B) $\dfrac{x^2 - 2x}{x^2 - x - 2} + x - 2 = \dfrac{x\overset{1}{\cancel{(x-2)}}}{(x+1)\underset{1}{\cancel{(x-2)}}} + \dfrac{x-2}{1}$

$= \dfrac{x}{x+1} + \dfrac{x-2}{1}$

$= \dfrac{x}{x+1} + \dfrac{(x-2)(x+1)}{x+1}$

$= \dfrac{x + x^2 - x - 2}{x+1}$

$= \dfrac{x^2 - 2}{x-1}$

51. (A)(B) The solution is correct.

53. $\dfrac{Y - \dfrac{y^2}{y-x}}{1 + \dfrac{x^2}{y^2 - x^2}} = \dfrac{(y^2 - x^2)[y - \frac{y^2}{y-x}]}{(y^2 - x^2)[1 + \frac{x^2}{y^2 - x^2}]} = \dfrac{(y^2 - x^2)y - \overset{(y+x)}{\cancel{(y^2 - x^2)}}\frac{y^2}{\cancel{y-x}}\frac{}{1}}{y^2 - x^2 + \underset{1}{\cancel{(y^2 - x^2)}}\frac{x^2}{\cancel{y^2 - x^2}}\frac{}{1}}$

$= \dfrac{y^3 - x^2y - y^3 - xy^2}{y^2 - x^2 + x^2} = \dfrac{-x^2y - xy^2}{y^2} = \dfrac{-x\overset{1}{\cancel{y}}(x+y)}{\underset{y}{\cancel{y^2}}} = \dfrac{-x(x+y)}{y}$

55. $2 - \dfrac{1}{1 - \frac{2}{a+2}} = 2 - \dfrac{(a+2)\cdot 1}{(a+2)[1 - \frac{2}{a+2}]} = 2 - \dfrac{a+2}{a+2 - \cancel{(a+2)}\frac{2}{\cancel{a+2}}\frac{}{1}} = 2 - \dfrac{a+2}{a+2-2}$

$= 2 - \dfrac{a+2}{a} = \dfrac{2}{1} - \dfrac{a+2}{a} = \dfrac{2a}{a} - \dfrac{(a+2)}{a} = \dfrac{2a - a - 2}{a} = \dfrac{a-2}{a}$

57. (A) The multiplicative inverse of x is the unique real number $\frac{1}{x}$ such that $x(\frac{1}{x}) = 1$. Since $\frac{c}{d} \cdot \frac{d}{c} = \frac{cd}{dc} = \frac{cd}{cd} = 1$, $\frac{d}{c}$ is the multiplicative inverse of $\frac{c}{d}$.

(B) By Definition 1 of Section A-1 of the text $a \div b = a(\frac{1}{b}) = a \cdot$ (multiplicative inverse of b). Hence $\frac{a}{b} \div \frac{c}{d} = \frac{a}{b} \cdot$ (multiplicative inverse of $\frac{c}{d}$)

$= \frac{a}{b} \cdot \frac{d}{c}$ by part A.

Exercise A-5

Key Ideas and Formulas

a^n, n an integer and a real

1. For n a positive integer:
$$a^n = \underbrace{a \cdot a \cdot \cdots \cdot a}_{n \text{ factors of } a}$$

2. For $n = 0$: $a^0 = 1$, $a \neq 0$
0^0 is not defined

3. For n a negative integer:
$$a^n = \frac{1}{a^{-n}} \qquad a \neq 0$$

Properties of Exponents
For m, n and p integers and a
and b real numbers, (division by 0
excluded) then:

1. $a^m a^n = a^{m+n}$

2. $(a^n)^m = a^{mn}$

3. $(ab)^m = a^m b^m$

4. $\left(\dfrac{a}{b}\right)^m = \dfrac{a^m}{b^m} \quad b \neq 0$

5. $\dfrac{a^m}{a^n} = \begin{cases} a^{m-n} \\ \dfrac{1}{a^{n-m}} \end{cases} \quad a \neq 0$

6. $a^{-n} = \dfrac{1}{a^n} \quad a \neq 0$

7. $(a^m b^n)^p = a^{pm} b^{pn}$

8. $\left(\dfrac{a^m}{b^n}\right)^p = \dfrac{a^{pm}}{b^{pn}} \quad b \neq 0$

9. $\dfrac{a^{-n}}{b^{-m}} = \dfrac{b^m}{a^n} \quad a, \, b \neq 0$

10. $\left(\dfrac{a}{b}\right)^{-n} = \left(\dfrac{b}{a}\right)^n \quad a, \, b \neq 0$

---

1. $x^{-6} x^6 = x^{-6+6} = x^0 = 1$

3. $(3x^2)(5x^4)(4x^5) = (3 \cdot 5 \cdot 4)(x^2 x^4 x^5) = 60x^{11}$

5. $(2x^4 z^{-3})^2 = 2^2 x^8 z^{-6} = \dfrac{4x^8}{z^6}$

7. $\left(\dfrac{r^2 s^3}{pq^4}\right) = \dfrac{(r^2)^5 (s^3)^5}{p^5 (q^4)^5} = \dfrac{r^{10} s^{15}}{p^5 q^{20}}$

9. $\dfrac{10^{19} \cdot 10^{-14}}{10^{-5} \cdot 10^3} = \dfrac{10^5}{10^{-2}} = 10^{5-(-2)} = 10^7$

11. $\dfrac{6a^{-3} b^{-5}}{2a^{-7} b^{-2}} = 3a^{(-3)-(-7)} b^{(-5)-(-2)} = 3a^4 b^{-3} = \dfrac{3a^4}{b^3}$ $\boxed{\text{Common Error: } \dfrac{a^{-3}}{a^{-7}} \neq a^{-3-7}}$

13. $\left(\dfrac{y^{-2}}{y^{-3}}\right)^{-1} = \dfrac{y^2}{y^3} = \dfrac{1}{y}$

15. $\dfrac{12 \times 10^4}{3 \times 10^{-7}} = 4 \times 10^{4-(-7)} = 4 \times 10^{11}$

17. $45{,}320{,}000 = 4.5320000. \times 10^7$
7 places left
$= 4.532 \times 10^7$

19. $0.066 = 0.06.6 \times 10^{-2} = 6.6 \times 10^{-2}$
2 places right
negative exponent

21. $0.000\ 000\ 084 = 8.4 \times 10^{-8}$
8 places right

23. $9 \times 10^{-5} = 0.00009. = 0.000\ 09$
5 places left

25. $3.48 \times 10^6 = 3.480\ 000. = 3{,}480{,}000$
6 places right

27. $4.2 \times 10^{-9} = 0.000\ 000\ 004.2 = 0.000\ 000\ 004\ 2$
9 places left

29. $\dfrac{27x^{-5}x^5}{18y^{-6}y^2} = \dfrac{\overset{3}{\cancel{27}}x^{-5+5}}{\underset{2}{\cancel{18}}y^{-6+2}} = \dfrac{3x^0}{2y^{-4}} = \dfrac{3y^4}{2}$ **31.** $\left(\dfrac{x^4y^{-1}}{x^{-2}y^3}\right)^2 = \dfrac{x^8y^{-2}}{x^{-4}y^6} = x^{8-(-4)}y^{-2-6} = x^{12}y^{-8} = \dfrac{x^{12}}{y^8}$

33. $\left(\dfrac{2x^{-3}y^2}{4xy^{-1}}\right)^{-2} = \left(\dfrac{x^{-3}y^2}{2xy^{-1}}\right)^{-2} = \dfrac{x^6y^{-4}}{2^{-2}x^{-2}y^2} = 2^2x^{6-(-2)}y^{-4-2} = 4x^8y^{-6} = \dfrac{4x^8}{y^6}$

35. $\left[\left(\dfrac{u^3v^{-1}w^{-2}}{u^{-2}v^{-2}w}\right)^{-2}\right]^2 = \left(\dfrac{u^3v^{-1}w^{-2}}{u^{-2}v^{-2}w^1}\right)^{-4} = \dfrac{u^{-12}v^4w^8}{u^8v^8w^{-4}} = u^{-20}v^{-4}w^{12} = \dfrac{w^{12}}{u^{20}v^4}$

37. $(x + y)^{-2} = \dfrac{1}{(x + y)^2}$

39. $\dfrac{1 + x^{-1}}{1 - x^{-2}} = \dfrac{1 + \dfrac{1}{x}}{1 - \dfrac{1}{x^2}} = \dfrac{x^2\left(1 + \dfrac{1}{x}\right)}{x^2\left(1 - \dfrac{1}{x^2}\right)} = \dfrac{x^2 + x}{x^2 - 1} = \dfrac{x(\cancel{x + 1})}{(x - 1)\underset{1}{(\cancel{x + 1})}} = \dfrac{x}{x - 1}$

41. $\dfrac{x^{-1} - y^{-1}}{x - y} = \dfrac{\dfrac{1}{x} - \dfrac{1}{y}}{x - y} = \dfrac{xy\left(\dfrac{1}{x} - \dfrac{1}{y}\right)}{xy(x - y)} = \dfrac{\overset{-1}{\cancel{y - x}}}{xy\underset{1}{(\cancel{x - y})}} = \dfrac{-1}{xy}$

43. $-3(x^3 + 3)^{-4}(3x^2) = -3 \cdot \dfrac{1}{(x^3 + 3)^4} \cdot 3x^2 = \dfrac{-9x^2}{(x^3 + 3)^4}$ $\boxed{\text{Common Error: } (x^3 + 3)^{-4} \neq x^{-12} + 3^{-4}}$

45. $2^{3^2} = 64$ on the assumption that 2^3^2 is entered, without parentheses.

47. The identity property for multiplication states that for x in R, $x(1) = x$, and 1 is the only element in R with that property. Therefore, if $a^ma^0 = a^m$, since $a^m(1) = a^m$, a^0 should equal 1.

49. $\dfrac{4x^2 - 12}{2x} = \dfrac{4x^2}{2x} - \dfrac{12}{2x} = 2x - \dfrac{6}{x} = 2x - 6x^{-1}$

51. $\dfrac{5x^3 - 2}{3x^2} = \dfrac{5x^3}{3x^2} - \dfrac{2}{3x^2} = \dfrac{5x}{3} - \dfrac{2}{3x^2} = \dfrac{5}{3}x - \dfrac{2}{3}x^{-2}$

53. $\dfrac{2x^3 - 3x^2 + x}{2x^2} = \dfrac{2x^3}{2x^2} - \dfrac{3x^2}{2x^2} + \dfrac{x}{2x^2} = x - \dfrac{3}{2} + \dfrac{1}{2x} = x - \dfrac{3}{2} + \dfrac{1}{2}x^{-1}$

55. $\dfrac{(32.7)(0.000\ 000\ 008\ 42)}{(0.0513)(80,700,000,000)} = \dfrac{(32.7)(8.42 \times 10^{-9})}{(0.0513)(8.07 \times 10^{10})} = 6.65 \times 10^{-17}$

57. $\dfrac{(5,760,000,000)}{(527)(0.000\ 007\ 09)} = \dfrac{5.76 \times 10^9}{(527)(7.09 \times 10^{-6})} = 1.54 \times 10^{12}$

59. 1.0295×10^{11} **61.** -4.3647×10^{-18}

63. $(9,820,000,000)^3 = (9.82 \times 10^9)^3 = 9.4697 \times 10^{29}$

65. $\dfrac{12(a + 2b)^{-3}}{6(a + 2b)^{-8}} = \dfrac{\overset{2}{\cancel{12}}(a + 2b)^{-3-(-8)}}{\underset{1}{\cancel{6}}} = 2(a + 2b)^5$

67. $\dfrac{xy^{-2} - yx^{-2}}{y^{-1} - x^{-1}} = \dfrac{\dfrac{x}{y^2} - \dfrac{y}{x^2}}{\dfrac{1}{y} - \dfrac{1}{x}} = \dfrac{x^2y^2\left(\dfrac{x}{y^2} - \dfrac{y}{x^2}\right)}{x^2y^2\left(\dfrac{1}{y} - \dfrac{1}{x}\right)} = \dfrac{x^3 - y^3}{x^2y - xy^2}$

$\qquad = \dfrac{\overset{1}{(\cancel{x - y})}(x^2 + xy + y^2)}{xy\underset{1}{(\cancel{x - y})}} = \dfrac{x^2 + xy + y^2}{xy}$

69. $\left(\dfrac{x^{-1}}{x^{-1} - y^{-1}}\right)^{-1} = \dfrac{x^{-1} - y^{-1}}{x^{-1}} = \dfrac{\frac{1}{x} - \frac{1}{y}}{\frac{1}{x}} = \dfrac{xy(\frac{1}{x} - \frac{1}{y})}{xy(\frac{1}{x})} = \dfrac{y - x}{y}$

71. mass of earth in pounds

= mass of earth in grams × number of pounds/gram

= $6.1 \times 10^{27} \times 2.2 \times 10^{-3}$

= 13.42×10^{24}

= 1.342×10^{25}

= 1.3×10^{25} pounds to two significant digits

73. 1 operation in 10^{-8} seconds means $1 \div 10^{-8}$ operations/second, that is 10^8 operations in 1 second, that is, 100,000,000 or 100 million. Similarly, 1 operation in 10^{-10} seconds means $1 \div 10^{-10}$, that is 10,000,000,000 or 10 billion operations in 1 second.

Since 1 minute is 60 seconds, multiply each number by 60 to get $60 \times 10^8 = 6 \times 10^9$ or 6,000,000,000 or 6 billion operations in 1 minute and $60 \times 10^{10} = 6 \times 10^{11}$ or 600,000,000,000 or 600 billion operations in 1 minute.

75. one person's share of national debt = $\dfrac{\text{amount of national debt}}{\text{number of persons}}$

= $\dfrac{5{,}680{,}000{,}000{,}000}{274{,}000{,}000} = \dfrac{5.68 \times 10^{12}}{2.74 \times 10^8}$

= 2.07×10^4 dollars per person to three significant digits

= \$20,700 per person

Exercise A-6

Key Ideas and Formulas

Note: All properties of exponents treated in Section A-5 are valid for rational number exponents defined in this section.

For n a natural number and b a real number, $b^{1/n}$ is the principal nth root of b.
1. If n is even and b is positive, then $b^{1/n}$ represents the positive nth root of b.
2. If n is even and b is negative, then $b^{1/n}$ does not represent a real number.
3. If n is odd, then $b^{1/n}$ represents the real nth root of b (there is only one).
4. $0^{1/n} = 0$

For m and n natural numbers and b any real number (except b cannot be negative when n is even):

$b^{m/n} = (b^{1/n})^m \qquad b^{m/n} = (b^m)^{1/n}$

$b^{-m/n} = \dfrac{1}{b^{m/n}} \qquad b \neq 0$

---

1. $25^{1/2} = 5$ **3.** $9^{3/2} = (9^{1/2})^3 = 3^3 = 27$ **5.** $-16^{1/2} = -(16^{1/2}) = -4$

7. $(-16)^{1/2}$ is not a real number   **9.** $\left(\dfrac{27}{125}\right)^{2/3} = \dfrac{27^{2/3}}{125^{2/3}} = \dfrac{(27^{1/3})^2}{(125^{1/3})^2} = \dfrac{3^2}{5^2} = \dfrac{9}{25}$

11. $4^{-5/2} = \dfrac{1}{4^{5/2}} = \dfrac{1}{32}$

13. $a^{1/3}a^{4/3} = a^{1/3 + 4/3} = a^{5/3}$ **15.** $c^{3/5}c^{-1/5} = c^{3/5 + (-1/5)} = c^{2/5}$

17. $(u^{-5})^{1/10} = u^{(1/10)(-5)} = u^{-5/10} = u^{-1/2} = \dfrac{1}{u^{1/2}}$

19. $(16x^8y^{-4})^{1/4} = 16^{1/4}x^{(1/4)(8)}y^{(1/4)(-4)} = 2x^2y^{-1} = \dfrac{2x^2}{y}$

21. $\left(\dfrac{a^{-3}}{b^4}\right)^{1/12} = \dfrac{a^{-3/12}}{b^{4/12}} = \dfrac{a^{-1/4}}{b^{1/3}} = \dfrac{1}{a^{1/4}b^{1/3}}$

23. $\left(\dfrac{4x^{-2}}{y^4}\right)^{-1/2} = \dfrac{4^{-1/2}x^1}{y^{-2}} = \dfrac{xy^2}{4^{1/2}} = \dfrac{xy^2}{2}$ **Common Error:** $\dfrac{4x^1}{y^{-2}}$. This is wrong; the exponent $\left(-\dfrac{1}{2}\right)$ applies to constant as well as variable factors.

25. $\left(\dfrac{8a^{-4}b^3}{27a^2b^{-3}}\right)^{1/3} = \dfrac{8^{1/3}a^{-4/3}b^1}{27^{1/3}a^{2/3}b^{-1}} = \dfrac{2b^{1-(-1)}}{3a^{2/3-(-4/3)}} = \dfrac{2b^2}{3a^2}$

27. $\dfrac{8x^{-1/3}}{12x^{1/4}} = \dfrac{2}{3x^{1/4-(-1/3)}} = \dfrac{2}{3x^{7/12}}$

29. $\left(\dfrac{a^{2/3}b^{-1/2}}{a^{1/2}b^{1/2}}\right)^2 = \dfrac{a^{4/3}b^{-1}}{a^1b^1} = \dfrac{a^{4/3-1}}{b^{1-(-1)}} = \dfrac{a^{1/3}}{b^2}$

31. $2m^{1/3}(3m^{2/3} - m^6) = 6m^{1/3+2/3} - 2m^{(1/3)+6} = 6m - 2m^{19/3}$

33. $(a^{1/2} + 2b^{1/2})(a^{1/2} - 3b^{1/2}) = a^{1/2}a^{1/2} - 3a^{1/2}b^{1/2} + 2a^{1/2}b^{1/2} - 6b^{1/2}b^{1/2}$
$$= a - a^{1/2}b^{1/2} - 6b$$

35. $(2x^{1/2} - 3y^{1/2})(2x^{1/2} + 3y^{1/2}) = (2x^{1/2})^2 - (3y^{1/2})^2 = 4x - 9y$

37. $(x^{1/2} + 2y^{1/2})^2 = (x^{1/2})^2 + 2(x^{1/2})(2y^{1/2}) + (2y^{1/2})^2 = x + 4x^{1/2}y^{1/2} + 4y$

39. $15^{5/4} = 15^{1.25} = 29.52$

41. $103^{-3/4} = 103^{-0.75} = 0.03093$

43. $2.876^{8/5} = 2.876^{1.6} = 5.421$

45. $(0.000\,000\,077\,35)^{-2/7} = (7.735 \times 10^{-8})^{(-2\div7)} = 107.6$

47. There are many examples; one simple choice is $x = y = 1$. Then $(x + y)^{1/2} = (1 + 1)^{1/2} = 2^{1/2} = \sqrt{2}$, and $x^{1/2} + y^{1/2} = 1^{1/2} + 1^{1/2} = 1 + 1 = 2$; thus, the left side is not equal to the right side.

49. There are many examples; one simple choice is $x = y = 1$. Then $(x + y)^{1/3} = (1 + 1)^{1/3} = 2^{1/3} = \sqrt[3]{2}$ and $\dfrac{1}{(x + y)^3} = \dfrac{1}{(1 + 1)^3} = \dfrac{1}{2^3} = \dfrac{1}{8}$; thus, the left side is not equal to the right side.

51. $\dfrac{12x^{1/2} - 3}{4x^{1/2}} = \dfrac{12x^{1/2}}{4x^{1/2}} - \dfrac{3}{4x^{1/2}} = 3 - \dfrac{3}{4}x^{-1/2}$

53. $\dfrac{3x^{2/3} + x^{1/2}}{5x} = \dfrac{3x^{2/3}}{5x^1} + \dfrac{x^{1/2}}{5x^1} = \dfrac{3}{5}x^{-1/3} + \dfrac{1}{5}x^{-1/2}$

55. $\dfrac{x^2 - 4x^{1/2}}{2x^{1/3}} = \dfrac{x^2}{2x^{1/3}} - \dfrac{4x^{1/2}}{2x^{1/3}} = \dfrac{1}{2}x^{5/3} - 2x^{1/6}$

57. $(a^{3/n}b^{3/m})^{1/3} = a^{(1/3)(3/n)}b^{(1/3)(3/m)} = a^{1/n}b^{1/m}$

59. $(x^{m/4}y^{n/3})^{-12} = x^{-12(m/4)}y^{-12(n/3)} = x^{-3m}y^{-4n} = \dfrac{1}{x^{3m}y^{4n}}$

61. (A) Since $(x^2)^{1/2}$ represents the positive real square root of x^2, it will not equal x if x is negative. So any negative value of x, for example $x = -2$, will make the left side $\neq$ right side. $[(-2)^2]^{1/2} = 4^{1/2} = 2 \neq -2$.

    (B) Similarly any positive (or zero) value of x, for example $x = 2$, will make the left side = right side. $[2^2]^{1/2} = 4^{1/2} = 2 = 2$.

(C) Any value of x will make the left side = right side.
$[2^3]^{1/3} = 8^{1/3} = 2$ $[(-2)^3]^{1/3} = [-8]^{1/3} = -2$ $[0^3]^{1/3} = 0^{1/3} = 0$
There is no real value of x such that $(x^3)^{1/3} \neq x$.

63. No. Any negative value of b, for example -4, will make $(b^m)^{1/n}$ not a real number if n is even and m is odd. Thus, $[(-4)^3]^{1/2} = [-64]^{1/2}$ is not a real number.

65.
$$\dfrac{(2x - 1)^{1/2} - (x + 2)(\frac{1}{2})(2x - 1)^{-1/2}(2)}{(2x - 1)} = \dfrac{(2x - 1)^{1/2} - (x + 2)(2x - 1)^{-1/2}}{(2x - 1)^1}$$

> **Common Error:** Do not "cancel" the $(2x - 1)^{1/2}$. It is not a common factor of numerator and denominator.

$$= \dfrac{(2x - 1)^{1/2} - \frac{x + 2}{(2x - 1)^{1/2}}}{(2x - 1)^1}$$

$$= \dfrac{(2x - 1)^{1/2}}{(2x - 1)^{1/2}} \dfrac{(2x - 1)^{1/2} - \frac{x + 2}{(2x - 1)^{1/2}}}{(2x - 1)^1}$$

$$= \dfrac{(2x - 1)^1 - (x + 2)}{(2x - 1)^{3/2}}$$

$$= \dfrac{2x - 1 - x - 2}{(2x - 1)^{3/2}} = \dfrac{x - 3}{(2x - 1)^{3/2}}$$

67.
$$\dfrac{2(3x - 1)^{1/3} - (2x + 1)(\frac{1}{3})(3x - 1)^{-2/3}(3)}{(3x - 1)^{2/3}} = \dfrac{2(3x - 1)^{1/3} - (2x + 1)(3x - 1)^{-2/3}}{(3x - 1)^{2/3}}$$

$$= \dfrac{2(3x - 1)^{1/3} - \frac{(2x + 1)}{(3x - 1)^{2/3}}}{(3x - 1)^{2/3}}$$

$$= \dfrac{(3x - 1)^{2/3}}{(3x - 1)^{2/3}} \dfrac{2(3x - 1)^{1/3} - \frac{(2x + 1)}{(3x - 1)^{2/3}}}{(3x - 1)^{2/3}}$$

$$= \dfrac{2(3x - 1)^1 - (2x + 1)}{(3x - 1)^{4/3}}$$

$$= \dfrac{6x - 2 - 2x - 1}{(3x - 1)^{4/3}}$$

$$= \dfrac{4x - 3}{(3x - 1)^{4/3}}$$

69. We are to calculate $N = 10x^{3/4}y^{1/4}$ given $x = 256$ units of labor and $y = 81$ units of capital.
$N = 10(256)^{3/4}(81)^{1/4}$
$\quad = 10(256^{1/4})^3(81)^{1/4}$
$\quad = 10(4^3)(3)$
$\quad = 1,920$ units of finished product.

71. We are to calculate $d = 0.0212v^{7/3}$ given $v = 70$ miles/hour.
$d = 0.0212(70)^{7/3}$
$\quad = 0.0212(70)^{(7 \div 3)}$
$\quad = 428$ feet (to the nearest foot)

Exercise A-7

Key Ideas and Formulas

$\sqrt[n]{b}$ = the principal nth root of b
$= b^{1/n}$ (b is called the radicand, n the index, in $\sqrt[n]{b}$)

$b^{m/n} = \begin{cases} \sqrt[n]{b^m} \\[2mm] (\sqrt[n]{b})^m \end{cases}$

For n a natural number > 1, and x and y positive real numbers,

$\sqrt[n]{x^n} = x$

$\sqrt[n]{xy} = \sqrt[n]{x}\,\sqrt[n]{y}$

$\sqrt[n]{\dfrac{x}{y}} = \dfrac{\sqrt[n]{x}}{\sqrt[n]{y}}$

Simplified (Radical) form conditions:

1. No radicand (the expression within the radical sign) contains a factor to a power greater than or equal to the index of the radical.

2. No power of the radicand and the index of the radical have a common factor other than 1.

3. No radical appears in a denominator.

4. No fraction appears within a radical.

---

1. $m^{2/3} = \sqrt[3]{m^2}$ or $(\sqrt[3]{m})^2$. The first is usually preferred, and we will use this form.

3. $6x^{3/5} = 6(x)^{3/5} = 6\sqrt[5]{x^3}$ **5.** $(4xy^3)^{2/5} = \sqrt[5]{(4xy^3)^2}$ **7.** $(x+y)^{1/2} = \sqrt{x+y}$

9. $\sqrt[5]{b} = b^{1/5}$ **11.** $5\sqrt[4]{x^3} = 5x^{3/4}$ **13.** $\sqrt[5]{(2x^2y)^3} = (2x^2y)^{3/5}$

15. $\sqrt[3]{x} + \sqrt[3]{y} = x^{1/3} + y^{1/3}$ $\boxed{\textbf{Common Error:}\ \text{not}\ (x+y)^{1/3}}$ **17.** $\sqrt[3]{-8} = -2$

19. $\sqrt{9x^8y^4} = \sqrt{9}\sqrt{x^8}\sqrt{y^4} = 3x^4y^2$ **21.** $\sqrt[4]{16m^4n^8} = \sqrt[4]{16}\sqrt[4]{m^4}\sqrt[4]{n^8} = 2mn^2$

23. $\sqrt{8a^3b^5} = \sqrt{4a^2b^4 \cdot 2ab} = \sqrt{4a^2b^4}\sqrt{2ab} = 2ab^2\sqrt{2ab}$

25. $\sqrt[3]{2^4x^4y^7} = \sqrt[3]{2^3x^3y^6 \cdot 2xy} = \sqrt[3]{2^3x^3y^6}\sqrt[3]{2xy} = 2xy^2\sqrt[3]{2xy}$

27. $\sqrt[4]{m^2} = \sqrt[2 \cdot 2]{m^{2 \cdot 1}} = \sqrt[2]{m^1} = \sqrt{m}$ **29.** $\sqrt[5]{\sqrt[3]{xy}} = \sqrt[15]{xy}$

31. $\sqrt[3]{9x^2}\sqrt[3]{9x} = \sqrt[3]{(9x^2)(9x)} = \sqrt[3]{81x^3} = \sqrt[3]{27x^3 \cdot 3} = \sqrt[3]{27x^3}\sqrt[3]{3} = 3x\sqrt[3]{3}$

33. $\dfrac{1}{\sqrt{5}} = \dfrac{1}{\sqrt{5}}\dfrac{\sqrt{5}}{\sqrt{5}} = \dfrac{\sqrt{5}}{5}$ **35.** $\dfrac{6x}{\sqrt{3x}} = \dfrac{6x}{\sqrt{3x}}\dfrac{\sqrt{3x}}{\sqrt{3x}} = \dfrac{6x\sqrt{3x}}{3x} = 2\sqrt{3x}$

37. $\dfrac{2}{\sqrt{2}-1} = \dfrac{2}{(\sqrt{2}-1)}\dfrac{(\sqrt{2}+1)}{(\sqrt{2}+1)} = \dfrac{2(\sqrt{2}+1)}{2-1} = \dfrac{2(\sqrt{2}+1)}{1} = 2(\sqrt{2}+1) = 2\sqrt{2}+2$

39. $\dfrac{\sqrt{2}}{\sqrt{6}+2} = \dfrac{\sqrt{2}}{(\sqrt{6}+2)}\dfrac{(\sqrt{6}-2)}{(\sqrt{6}-2)} = \dfrac{\sqrt{2}(\sqrt{6}-2)}{6-4} = \dfrac{\sqrt{2}\sqrt{6}-2\sqrt{2}}{2} = \dfrac{\sqrt{12}-2\sqrt{2}}{2} = \dfrac{2\sqrt{3}-2\sqrt{2}}{2}$

$= \dfrac{2(\sqrt{3}-\sqrt{2})}{2} = \sqrt{3}-\sqrt{2}$

41. $x\sqrt[5]{3^6x^7y^{11}} = x\sqrt[5]{(3^5x^5y^{10})(3x^2y)} = x\sqrt[5]{3^5x^5y^{10}}\sqrt[5]{3x^2y} = x(3xy^2)\sqrt[5]{3x^2y} = 3x^2y^2\sqrt[5]{3x^2y}$

43. $\dfrac{\sqrt[4]{32m^7n^9}}{2mn} = \dfrac{\sqrt[4]{16m^4n^8 \cdot 2m^3n}}{2mn} = \dfrac{2mn^2\sqrt[4]{2m^3n}}{2mn} = n\sqrt[4]{2m^3n}$

45. $\sqrt[6]{a^4(b-a)^2} = \sqrt[2\cdot3]{a^{2\cdot2}(b-a)^{2\cdot1}} = \sqrt[3]{a^2(b-a)}$ **47.** $\sqrt[3]{\sqrt[4]{a^9b^3}} = \sqrt[3\cdot4]{a^{3\cdot3}b^{3\cdot1}} = \sqrt[4]{a^3b}$

49. $\sqrt[3]{2x^2y^4}\,\sqrt[3]{3x^5y} = \sqrt[3]{(2x^2y^4)(3x^5y)} = \sqrt[3]{6x^7y^5} = \sqrt[3]{(x^6y^3)(6xy^2)} = \sqrt[3]{x^6y^3}\,\sqrt[3]{6xy^2} = x^2y\,\sqrt[3]{6xy^2}$

51. $\sqrt[3]{a^3+b^3}$ is in simplified form. There is no correct way to simplify this expression further.

53. $\dfrac{\sqrt{2m}\sqrt{5}}{\sqrt{20m}} = \dfrac{\sqrt{10m}}{\sqrt{20m}} = \sqrt{\dfrac{10m}{20m}} = \sqrt{\dfrac{1}{2}} = \sqrt{\dfrac{1}{2}\cdot\dfrac{2}{2}} = \sqrt{\dfrac{2}{4}} = \dfrac{\sqrt{2}}{2}$ or $\dfrac{1}{2}\sqrt{2}$

55. $\dfrac{4a^3b^2}{\sqrt[3]{2ab^2}} = \dfrac{4a^3b^2}{\sqrt[3]{2ab^2}}\cdot\dfrac{\sqrt[3]{2^2a^2b}}{\sqrt[3]{2^2a^2b}} = \dfrac{4a^3b^2\,\sqrt[3]{4a^2b}}{\sqrt[3]{2^3a^3b^3}} = \dfrac{4a^3b^2\,\sqrt[3]{4a^2b}}{2ab} = 2a^2b\,\sqrt[3]{4a^2b}$

Notice that we multiplied numerator and denominator by $\sqrt[3]{2^2a^2b}$ instead of $\sqrt[3]{(2ab^2)^2}$, which would also have rationalized the denominator. We choose the *smallest* exponents that will make the radicand a perfect cube.

57. $\sqrt[4]{\dfrac{3y^3}{4x^3}} = \sqrt[4]{\dfrac{3y^3}{4x^3}\cdot\dfrac{4x}{4x}} = \sqrt[4]{\dfrac{12xy^3}{16x^4}} = \dfrac{\sqrt[4]{12xy^3}}{\sqrt[4]{16x^4}} = \dfrac{\sqrt[4]{12xy^3}}{2x}$ or $\dfrac{1}{2x}\sqrt[4]{12xy^3}$

Notice again that we could have multiplied numerator and denominator by $(4x^3)^3$, but the simpler quantity $4x$ is preferred.

59. $\dfrac{3\sqrt{y}}{2\sqrt{y}-3} = \dfrac{3\sqrt{y}}{(2\sqrt{y}-3)}\dfrac{(2\sqrt{y}+3)}{(2\sqrt{y}+3)} = \dfrac{3\sqrt{y}(2\sqrt{y}+3)}{(2\sqrt{y})^2-(3)^2} = \dfrac{6y+9\sqrt{y}}{4y-9}$

61. $\dfrac{2\sqrt{5}+3\sqrt{2}}{5\sqrt{5}+2\sqrt{2}} = \dfrac{(2\sqrt{5}+3\sqrt{2})}{(5\sqrt{5}+2\sqrt{2})}\dfrac{(5\sqrt{5}-2\sqrt{2})}{(5\sqrt{5}-2\sqrt{2})}$

$= \dfrac{2\sqrt{5}\cdot5\sqrt{5}-2\sqrt{5}\cdot2\sqrt{2}+3\sqrt{2}\cdot5\sqrt{5}-3\sqrt{2}\cdot2\sqrt{2}}{(5\sqrt{5})^2-(2\sqrt{2})^2}$

$= \dfrac{(10)(5)-4\sqrt{10}+15\sqrt{10}-(6)(2)}{(25)(5)-(4)(2)} = \dfrac{50+11\sqrt{10}-12}{125-8} = \dfrac{38+11\sqrt{10}}{117}$

63. $\dfrac{x^2}{\sqrt{x^2+9}-3} = \dfrac{x^2}{(\sqrt{x^2+9}-3)}\cdot\dfrac{(\sqrt{x^2+9}+3)}{(\sqrt{x^2+9}+3)}$

$= \dfrac{x^2(\sqrt{x^2+9}+3)}{(\sqrt{x^2+9})^2-(3)^2} = \dfrac{x^2(\sqrt{x^2+9}+3)}{x^2+9-9}$

$= \dfrac{\overset{1}{\cancel{x^2}}(\sqrt{x^2+9}+3)}{\underset{1}{\cancel{x^2}}} = \sqrt{x^2+9}+3$

65. Here we want to remove the radical terms from the *numerator*.

$\dfrac{\sqrt{t}-\sqrt{x}}{t-x} = \dfrac{(\sqrt{t}-\sqrt{x})}{t-x}\cdot\dfrac{(\sqrt{t}+\sqrt{x})}{\sqrt{t}+\sqrt{x}} = \dfrac{(\sqrt{t})^2-(\sqrt{x})^2}{(t-x)(\sqrt{t}+\sqrt{x})}$

$= \dfrac{\overset{1}{\cancel{t-x}}}{\underset{1}{\cancel{(t-x)}}(\sqrt{t}+\sqrt{x})} = \dfrac{1}{\sqrt{t}+\sqrt{x}}$

67. Here we want to remove the radical terms from the *numerator*.

$$\frac{\sqrt{x+h}-\sqrt{x}}{h} = \frac{(\sqrt{x+h}-\sqrt{x})}{h} \cdot \frac{(\sqrt{x+h}+\sqrt{x})}{(\sqrt{x+h}+\sqrt{x})} = \frac{(\sqrt{x+h})^2 - (\sqrt{x})^2}{h(\sqrt{x+h}+\sqrt{x})} = \frac{x+h-x}{h(\sqrt{x+h}+\sqrt{x})}$$

$$= \frac{\overset{1}{\cancel{h}}}{\underset{1}{\cancel{h}}(\sqrt{x+h}+\sqrt{x})} = \frac{1}{\sqrt{x+h}+\sqrt{x}}$$

69. $\sqrt{0.032\,965} = 0.1816$ **71.** $\sqrt[3]{45.0218} = (45.0218)^{1/3} = 45.0218^{(1\div3)} = 3.557$

73. $\sqrt[8]{5.477 \times 10^{-9}} = (5.477 \times 10^{-9})^{(1\div8)} = 0.09275$

75. $\sqrt[5]{9 + \sqrt[5]{9}} = (9 + 9^{(1\div5)})^{(1\div5)} = 1.602$

77. $\sqrt[4]{\sqrt[5]{100}} = (100^{(1\div5)})^{(1\div4)} = 1.259$

$\sqrt[20]{100} = 100^{(1\div20)} = 1.259$

79. $\dfrac{1}{\sqrt[3]{9}} = 9^{-1/3} = 9^{(-1\div3)} = 0.4807$

$\dfrac{\sqrt[3]{3}}{3} = \dfrac{3^{1/3}}{3} = 3^{(1\div3)} \div 3 = 0.4807$

81. $\sqrt{x^2}$ is the principal square root of x^2, that is, the positive (or zero) square root of x^2. The two square roots of x^2 are x and $-x$. $\sqrt{x^2} = -x$ if and only if $-x$ is positive or zero. Hence $\sqrt{x^2} = -x$ if and only if x is negative or zero. $x \leq 0$.

83. $\sqrt[3]{x^3}$ is the real third root of x^3. This is always equal to x if x is real. All real numbers.

85. (A) $\sqrt{3} + \sqrt{5} \approx 3.968\,118\,785$ (B) $\sqrt{2+\sqrt{3}} + \sqrt{2-\sqrt{3}} \approx 2.449\,489\,743$

(C) $1 + \sqrt{3} \approx 2.732\,050\,808$ (D) $\sqrt[3]{10 + 6\sqrt{3}} \approx 2.732\,050\,808$

(E) $\sqrt{8 + \sqrt{60}} \approx 3.968\,118\,785$ (F) $\sqrt{6} \approx 2.449\,489\,743$

Thus, (A) and (E), (B) and (F), and (C) and (D) have the same value to 9 decimal places. To show that they are actually equal:

(A) and (E): $(\sqrt{3} + \sqrt{5})^2 = (\sqrt{3})^2 + 2\sqrt{3}\sqrt{5} + (\sqrt{5})^2 = 3 + \sqrt{4}\sqrt{3}\sqrt{5} + 5 = 8 + \sqrt{60}$
Therefore, since $\sqrt{3} + \sqrt{5}$ is positive, $\sqrt{3} + \sqrt{5}$ is the positive square root of $8 + \sqrt{60}$. Thus, $\sqrt{8 + \sqrt{60}} = \sqrt{3} + \sqrt{5}$.

(B) and (F):
$$\left(\sqrt{2+\sqrt{3}} + \sqrt{2-\sqrt{3}}\right)^2 = \left(\sqrt{2+\sqrt{3}}\right)^2 + 2\sqrt{2+\sqrt{3}}\sqrt{2-\sqrt{3}} + \left(\sqrt{2-\sqrt{3}}\right)^2$$
$$= 2 + \sqrt{3} + 2\sqrt{(2+\sqrt{3})(2-\sqrt{3})} + 2 - \sqrt{3}$$
$$= 4 + 2\sqrt{2^2 - (\sqrt{3})^2}$$
$$= 4 + 2\sqrt{4-3}$$
$$= 4 + 2 \cdot 1$$
$$= 6$$

Therefore, since $\sqrt{2+\sqrt{3}} + \sqrt{2-\sqrt{3}}$ is positive, $\sqrt{2+\sqrt{3}} + \sqrt{2-\sqrt{3}}$ is the positive square root of 6. Thus, $\sqrt{2+\sqrt{3}} + \sqrt{2-\sqrt{3}} = \sqrt{6}$.

(C) and (D): $(1 + \sqrt{3})^3 = (1 + \sqrt{3})(1 + \sqrt{3})(1 + \sqrt{3})$

$$= (1^2 + 2\sqrt{3} + \sqrt{3}^2)(1 + \sqrt{3})$$

$$= (4 + 2\sqrt{3})(1 + \sqrt{3})$$

$$= 4 + 4\sqrt{3} + 2\sqrt{3} + 2\sqrt{3}\sqrt{3}$$

$$= 10 + 6\sqrt{3}$$

Therefore, $1 + \sqrt{3}$ is the real cube root of $10 + 6\sqrt{3}$. Thus,

$$1 + \sqrt{3} = \sqrt[3]{10 + 6\sqrt{3}}.$$

87. $\dfrac{1}{\sqrt[3]{a} - \sqrt[3]{b}} = \dfrac{1}{(\sqrt[3]{a} - \sqrt[3]{b})} \dfrac{[(\sqrt[3]{a})^2 + \sqrt[3]{a}\sqrt[3]{b} + (\sqrt[3]{b})^2]}{[(\sqrt[3]{a})^2 + \sqrt[3]{a}\sqrt[3]{b} + (\sqrt[3]{b})^2]}$

using $(x - y)(x^2 + xy + y^2) = x^3 - y^3$

$$= \dfrac{1[\sqrt[3]{a^2} + \sqrt[3]{ab} + \sqrt[3]{b^2}]}{(\sqrt[3]{a})^3 - (\sqrt[3]{b})^3} = \dfrac{\sqrt[3]{a^2} + \sqrt[3]{ab} + \sqrt[3]{b^2}}{a - b}$$

89. $\dfrac{1}{\sqrt{x} - \sqrt{y} + \sqrt{z}} = \dfrac{1}{[(\sqrt{x} - \sqrt{y}) + \sqrt{z}]} \dfrac{[(\sqrt{x} - \sqrt{y}) - \sqrt{z}]}{[(\sqrt{x} - \sqrt{y}) - \sqrt{z}]} = \dfrac{\sqrt{x} - \sqrt{y} - \sqrt{z}}{(\sqrt{x} - \sqrt{y})^2 - (\sqrt{z})^2}$

$$= \dfrac{\sqrt{x} - \sqrt{y} - \sqrt{z}}{(\sqrt{x})^2 - 2\sqrt{x}\sqrt{y} + (\sqrt{y})^2 - (\sqrt{z})^2} = \dfrac{\sqrt{x} - \sqrt{y} - \sqrt{z}}{x - 2\sqrt{xy} + y - z}$$

$$= \dfrac{\sqrt{x} - \sqrt{y} - \sqrt{z}}{x - 2\sqrt{xy} + y - z} = \dfrac{\sqrt{x} - \sqrt{y} - \sqrt{z}}{x + y - z - 2\sqrt{xy}}$$

$$= \dfrac{(\sqrt{x} - \sqrt{y} - \sqrt{z})}{[(x + y - z) - 2\sqrt{xy}]} \dfrac{[(x + y - z) + 2\sqrt{xy}]}{[(x + y - z) + 2\sqrt{xy}]}$$

$$= \dfrac{(\sqrt{x} - \sqrt{y} - \sqrt{z})[(x + y - z) + 2\sqrt{xy}]}{(x + y - z)^2 - (2\sqrt{xy})^2}$$

$$= \dfrac{(\sqrt{x} - \sqrt{y} - \sqrt{z})[(x + y - z) + 2\sqrt{xy}]}{(x + y - z)^2 - 4xy}$$

91. $\dfrac{\sqrt[3]{x + h} - \sqrt[3]{x}}{h} = \dfrac{(\sqrt[3]{x + h} - \sqrt[3]{x})}{h} \dfrac{[(\sqrt[3]{x + h})^2 + \sqrt[3]{x + h}\sqrt[3]{x} + (\sqrt[3]{x})^2]}{[(\sqrt[3]{x + h})^2 + \sqrt[3]{x + h}\sqrt[3]{x} + (\sqrt[3]{x})^2]}$

$$= \dfrac{(\sqrt[3]{x + h})^3 - (\sqrt[3]{x})^3}{h[\sqrt[3]{(x + h)^2} + \sqrt[3]{x(x + h)} + \sqrt[3]{x^2}]}$$

Using $(a - b)(a^2 + ab + b^2) = a^3 - b^3$

$$= \dfrac{x + h - x}{h[\sqrt[3]{(x + h)^2} + \sqrt[3]{x(x + h)} + \sqrt[3]{x^2}]}$$

$$= \dfrac{h}{h[\sqrt[3]{(x + h)^2} + \sqrt[3]{x(x + h)} + \sqrt[3]{x^2}]}$$

$$= \dfrac{1}{\sqrt[3]{(x + h)^2} + \sqrt[3]{x(x + h)} + \sqrt[3]{x^2}}$$

93. $\sqrt[kn]{x^{km}} = (x^{km})^{1/kn} = x^{km/kn} = x^{m/n} = \sqrt[n]{x^m}$

95. $\dfrac{M_0}{\sqrt{1 - \dfrac{v^2}{c^2}}} = M_0 \div \sqrt{1 - \dfrac{v^2}{c^2}}$

$\qquad\qquad = M_0 \div \sqrt{\dfrac{c^2 - v^2}{c^2}} = M_0 \div \dfrac{\sqrt{c^2 - v^2}}{c}$

$\qquad\qquad = M_0 \cdot \dfrac{c}{\sqrt{c^2 - v^2}}$

Now rationalize the denominator

$\qquad\qquad = M_0 \cdot \dfrac{c}{\sqrt{c^2 - v^2}} \dfrac{\sqrt{c^2 - v^2}}{\sqrt{c^2 - v^2}}$

$\qquad\qquad = \dfrac{M_0 c \sqrt{c^2 - v^2}}{c^2 - v^2}$

APPENDIX A REVIEW

1. (A) Since 3 is an element of $\{1,2,3,4,5\}$, the statement is true. *T*.

(B) Since 5 is not an element of $\{4,1,2\}$, the statement is true. *T*.

(C) Since B is not an element of $\{1,2,3,4,5\}$, the statement is false. *F*.

(D) Since each element of the set $\{1,2,4\}$ is also an element of the set $\{1,2,3,4,5\}$, the statement is true. *T*.

(E) Since sets B and C have exactly the same elements, the sets are equal. The statement is false. *F*.

(F) The statement is false, since not every element of the set $\{1,2,3,4,5\}$ is an element of the set $\{1,2,4\}$. For example, 3 is an element of the first, but not the second. *F*. *(A-1)*

2. (A) The commutative property ($\cdot$) states that, in general, $xy = yx$.
Comparing this with
$x(y + z) = ?$
we see that $x(y + z) = (y + z)x$
$(y + z)x$

(B) The associative property (+) states that, in general,
$(x + y) + z = x + (y + z)$
Comparing this with
$? = 2 + (x + y)$
we see that
$(2 + x) + y = 2 + (x + y)$
$(2 + x) + y$

(C) The distributive property states that, in general,
$(x + y)z = xz + yz$
Comparing this with
$(2 + 3)x = ?$
we see that
$(2 + 3)x = 2x + 3x$
$2x + 3x$ *(A-1)*

3. $(3x - 4) + (x + 2) + (3x^2 + x - 8) + (x^3 + 8)$
$\qquad = 3x - 4 + x + 2 + 3x^2 + x - 8 + x^3 + 8$
$\qquad = x^3 + 3x^2 + 5x - 2$ *(A-2)*

4. sum of (a) and (c) = $(3x - 4) + (3x^2 + x - 8) = 3x^2 + 4x - 12$
sum of (b) and (d) = $(x + 2) + (x^3 + 8) = x^3 + x + 10$
subtract the sum of (a) and (c) from the sum of (b) and (d)
$= (x^3 + x + 10) - (3x^2 + 4x - 12) = x^3 + x + 10 - 3x^2 - 4x + 12$
$= x^3 - 3x^2 - 3x + 22$ *(A-2)*

5. $3x^2 + x - 8$
$\underline{x^3 \qquad + 8}$
$3x^5 + x^4 - 8x^3$
$\underline{\qquad\qquad\qquad 24x^2 + 8x - 64}$
$3x^5 + x^4 - 8x^3 + 24x^2 + 8x - 64$ *(A-2)*

6. 3 *(A-2)*

7. 1 *(A-2)*

8. $5x^2 - 3x[4 - 3(x - 2)] = 5x^2 - 3x[4 - 3x + 6]$
$= 5x^2 - 3x[10 - 3x]$
$= 5x^2 - 30x + 9x^2$
$= 14x^2 - 30x$

Common Errors:
$-3(x - 2) \neq -3x - 6$
$4 - 3(x - 2) \neq 1(x - 2)$
The first is incorrect use of the distributive property, the second is incorrect order of operations.

(A-2)

9. $(3m - 5n)(3m + 5n) = (3m)^2 - (5n)^2 = 9m^2 - 25n^2$ *(A-2)*

10.

	First Product	Outer Product	Inner Product	Last Product
$(2x + y)(3x - 4y) = 6x^2$		$-8xy$	$+3xy$	$-4y^2$
$\qquad\qquad\qquad = 6x^2 - 5xy - 4y^2$				

(A-2)

11. $(2a - 3b)^2 = (2a)^2 - 2(2a)(3b) + (3b)^2 = 4a^2 - 12ab + 9b^2$ *(A-2)*

12. $(3x - 2)^2$ *(A-3)*

13. Prime *(A-3)*

14. $6n^3 - 9n^2 - 15n = 3n(2n^2 - 3n - 5) = 3n(2n - 5)(n + 1)$ *(A-3)*

15. $\dfrac{2}{5b} - \dfrac{4}{3a^3} - \dfrac{1}{6a^2b^2} = \dfrac{12a^3b}{30a^3b^2} - \dfrac{40b^2}{30a^3b^2} - \dfrac{5a}{30a^3b^2} = \dfrac{12a^3b - 40b^2 - 5a}{30a^3b^2}$ *(A-4)*

16. $\dfrac{3x}{3x^2 - 12x} + \dfrac{1}{6x} = \dfrac{3x}{3x(x - 4)} + \dfrac{1}{6x} = \dfrac{2(3x)}{6x(x - 4)} + \dfrac{1(x - 4)}{6x(x - 4)} = \dfrac{6x + x - 4}{6x(x - 4)} = \dfrac{7x - 4}{6x(x - 4)}$ *(A-4)*

17. $\dfrac{y - 2}{y^2 - 4y + 4} \div \dfrac{y^2 + 2y}{y^2 + 4y + 4} = \dfrac{y - 2}{y^2 - 4y + 4} \cdot \dfrac{y^2 + 4y + 4}{y^2 + 2y}$

$= \dfrac{\overset{1}{\cancel{(y - 2)}}}{\cancel{(y - 2)}(y - 2)} \cdot \dfrac{\overset{1}{\cancel{(y + 2)}}(y + 2)}{y\cancel{(y + 2)}}$

$= \dfrac{y + 2}{y(y - 2)}$ *(A-4)*

18. $\dfrac{u - \dfrac{1}{u}}{1 - \dfrac{1}{u^2}} = \dfrac{u^2\left(u - \dfrac{1}{u}\right)}{u^2\left(1 - \dfrac{1}{u^2}\right)} = \dfrac{u^3 - u}{u^2 - 1} = \dfrac{u\overset{1}{\cancel{(u^2 - 1)}}}{\underset{1}{\cancel{u^2 - 1}}} = u$ *(A-4)*

19. $6(xy^3)^5 = 6x^5(y^3)^5 = 6x^5y^{15}$ *(A-5)*

20. $\dfrac{9u^8v^6}{3u^4v^8} = \dfrac{3u^4}{v^2}$ *(A-5)*

21. $(2 \times 10^5)(3 \times 10^{-3}) = (2 \cdot 3) \times 10^{5+(-3)} = 6 \times 10^2$ *(A-5)*

22. $(x^{-3}y^2)^{-2} = (x^{-3})^{-2}(y^2)^{-2} = x^6 y^{-4} = \dfrac{x^6}{y^4}$ *(A-6)*

23. $u^{5/3}u^{2/3} = u^{5/3+2/3} = u^{7/3}$ *(A-6)*

24. $(9a^4 b^{-2})^{1/2} = 9^{1/2}(a^4)^{1/2}(b^{-2})^{1/2} = 3a^2 b^{-1} = \dfrac{3a^2}{b}$ *(A-7)*

25. $3\sqrt[5]{x^2}$ *(A-7)*

26. $-3(xy)^{2/3}$ *(A-7)*

27. $3x\sqrt[3]{x^5 y^4} = 3x\sqrt[3]{x^3 y^3 \cdot x^2 y} = 3x\sqrt[3]{x^3 y^3}\sqrt[3]{x^2 y} = 3x(xy)\sqrt[3]{x^2 y} = 3x^2 y\sqrt[3]{x^2 y}$ *(A-7)*

28. $\sqrt{2x^2 y^5}\sqrt{18x^3 y^2} = \sqrt{(2x^2 y^5)(18x^3 y^2)} = \sqrt{36x^5 y^7} = \sqrt{36x^4 y^6 \cdot xy} = \sqrt{36x^4 y^6}\sqrt{xy}$
$$= 6x^2 y^3\sqrt{xy}$$
     (A-7)

29. $\dfrac{6ab}{\sqrt{3a}} = \dfrac{6ab}{\sqrt{3a}}\dfrac{\sqrt{3a}}{\sqrt{3a}} = \dfrac{\overset{2}{\cancel{6ab}}\sqrt{3a}}{\underset{1}{\cancel{3a}}} = 2b\sqrt{3a}$ *(A-7)*

30. $\dfrac{\sqrt{5}}{3-\sqrt{5}} = \dfrac{\sqrt{5}}{(3-\sqrt{5})}\dfrac{(3+\sqrt{5})}{(3+\sqrt{5})} = \dfrac{\sqrt{5}(3+\sqrt{5})}{3^2 - (\sqrt{5})^2} = \dfrac{\sqrt{5}(3+\sqrt{5})}{9-5} = \dfrac{\sqrt{5}(3+\sqrt{5})}{4}$
$$= \dfrac{3\sqrt{5}+5}{4}$$
     (A-7)

31. $\sqrt[8]{y^6} = \sqrt[2 \cdot 4]{y^{2 \cdot 3}} = \sqrt[4]{y^3}$ *(A-7)*

32. The odd integers between -4 and 2 are -3, -1, and 1. Hence the set is written $\{-3, -1, 1\}$ *(A-1)*

33. Subtraction.
$(-3) - (-2) = (-3) + [-(-2)]$ is a special case of $a - b = a + (-b)$.   *(A-1)*

34. Commutative $(+)$
$3y + (2x + 5) = (2x + 5) + 3y$ is a special case of $x + y = y + x$.   *(A-1)*

35. Distributive
$(2x + 3)(3x + 5) = (2x + 3)3x + (2x + 3)5$ is a special case of
$x(y + z) = xy + xz$.   *(A-1)*

36. Associative $(\cdot)$
$3 \cdot (5x) = (3 \cdot 5)x$ is a special case of $x(yz) = (xy)z$.   *(A-1)*

37. Negatives
$\dfrac{a}{-(b-c)} = -\dfrac{a}{b-c}$ is a special case of $\dfrac{a}{-b} = -\dfrac{a}{b}$. (Theorem 1, Part 5)   *(A-1)*

38. Identity $(+)$
$3xy + 0 = 3xy$ is a special case of $x + 0 = x$.   *(A-1)*

39. (A) T   (B) F   *(A-1)*

40. 0 and -3 are two examples of infinitely many.   *(A-1)*

41. (A) a and d   (B) None   *(A-2)*

42. $(2x - y)(2x + y) - (2x - y)^2 = (2x)^2 - y^2 - [(2x)^2 - 2(2x)y + y^2]$
$$= 4x^2 - y^2 - [4x^2 - 4xy + y^2]$$
$$= 4x^2 - y^2 - 4x^2 + 4xy - y^2$$
$$= 4xy - 2y^2$$

> **Common Errors:**
> $(2x - y)^2 \neq 4x^2 - y^2$ (exponents do not distribute over subtraction)
> $-(2x - y)^2 \neq (-2x + y)^2$ (square first, then subtract)
> $-(2x - y)^2 \neq -4x^2 - 4xy + y^2$ (must change all signs)

(A-2)

43. A straightforward way to perform this multiplication is vertically:

$m^2 + 2mn - n^2$
$m^2 - 2mn - n^2$
$\overline{}$
$m^4 + 2m^3n - m^2n^2$
$ - 2m^3n - 4m^2n^2 + 2mn^3$
$ - m^2n^2 - 2mn^3 + n^4$
$\overline{}$
$m^4 - 6m^2n^2 + n^4$

Alternatively, we can notice

$(m^2 + 2mn - n^2)(m^2 - 2mn - n^2) = [(m^2 - n^2) + 2mn][(m^2 - n^2) - 2mn]$
$$= (m^2 - n^2)^2 - (2mn)^2 \text{ (difference of squares)}$$
$$= (m^2)^2 - 2m^2n^2 + (n^2)^2 - 4m^2n^2$$
$$= m^4 - 2m^2n^2 + n^4 - 4m^2n^2$$
$$= m^4 - 6m^2n^2 + n^4$$

(A-2)

44. $5(x + h)^2 - 7(x + h) - (5x^2 - 7x) = 5(x^2 + 2xh + h^2) - 7(x + h) - (5x^2 - 7x)$
$$= 5x^2 + 10xh + 5h^2 - 7x - 7h - 5x^2 + 7x$$
$$= 10xh + 5h^2 - 7h$$

(A-2)

45. $-2x\{(x^2 + 2)(x - 3) - x[x - x(3 - x)]\} = -2x\{(x^2 + 2)(x - 3) - x[x - 3x + x^2]\}$
$$= -2x\{(x^2 + 2)(x - 3) - x[-2x + x^2]\}$$
$$= -2x\{x^3 - 3x^2 + 2x - 6 + 2x^2 - x^3\}$$
$$= -2x\{-x^2 + 2x - 6\}$$
$$= 2x^3 - 4x^2 + 12x$$

(A-2)

46. $(x - 2y)^3 = (x - 2y)(x - 2y)(x - 2y)$
$$= [(x - 2y)(x - 2y)](x - 2y)$$
$$= [x^2 - 2(x)(2y) + (2y)^2](x - 2y)$$
$$= [x^2 - 4xy + 4y^2](x - 2y)$$

The last multiplication is best performed vertically, so we write

$x^2 - 4xy + 4y^2$
$x - 2y$
$\overline{}$
$x^3 - 4x^2y + 4xy^2$
$ - 2x^2y + 8xy^2 - 8y^3$
$\overline{}$
$x^3 - 6x^2y + 12xy^2 - 8y^3$

(A-2)

47. $(4x - y)^2 - 9x^2 = (4x - y)^2 - (3x)^2$
$$= [(4x - y) - 3x][(4x - y) + 3x]$$
$$= (x - y)(7x - y)$$

(A-3)

48. Prime   (A-3)

49. $3xy(2x^2 + 4xy - 5y^2)$ (A-3)

50. $(y - b)^2 - y + b = (y - b)(y - b) - 1(y - b) = (y - b)(y - b - 1)$ (A-3)

51. $3x^3 + 24y^3 = 3(x^3 + 8y^3) = 3[x^3 + (2y)^3] = 3(x + 2y)[x^2 - x(2y) + (2y)^2]$
$$= 3(x + 2y)(x^2 - 2xy + 4y^2)$$

(A-3)

52. $y^3 + 2y^2 - 4y - 8 = y^2(y + 2) - 4(y + 2)$
$$= (y + 2)(y^2 - 4) = (y + 2)(y + 2)(y - 2)$$
$$= (y - 2)(y + 2)^2$$

(A-3)